HANDBOOK OF
RADIOACTIVITY ANALYSIS

SECOND EDITION

With forewords by
Dr. Mohamed M. El Baradei, *Director General*
Prof. Dr. Werner Burkart, *Deputy Director General*
International Atomic Energy Agency, Vienna

Edited by

Michael F. L'Annunziata

ACADEMIC PRESS

Amsterdam Boston Heidelberg London New York Oxford
Paris San Diego San Francisco Singapore Sydney Tokyo

Academic Press
An imprint of Elsevier Science
525 B Street, Suite 1900, San Diego, California 92101-4495, USA
http://www.academicpress.com

First impression of second edition 2003

Library of Congress Cataloging in Publication Data
A catalog record from the Liabrary of Congress has been applied for.

British Library Cataloguing in Publication Data
A catalogue record from the British Library has been applied for.

ISBN: 0-12-436603-1

♾ The paper used in this publlication meets the requirements of ANSI/NISO Z39 48-1992 (Permanence of paper)

Printed in Great Britain

CONTENTS

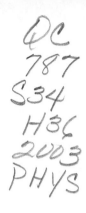

I Nuclear Radiation, Its Interaction with Matter and Radioisotope Decay

MICHAEL F. L'ANNUNZIATA

2   Gas Ionization Detectors

KARL BUCHTELA

3   Solid State Nuclear Track Detectors

RADOMIR ILIĆ AND SAEED A. DURRANI

4   Semiconductor Detectors

PAUL F. FETTWEIS, JAN VERPLANCKE,
RAMKUMAR VENKATARAMAN, BRIAN M. YOUNG AND
HAROLD SCHWENN

5   Liquid Scintillation Analysis: Principles and Practice

MICHAEL F. L'ANNUNZIATA

6 Environmental Liquid Scintillation Analysis

GORDON T. COOK, CHARLES J. PASSO, JR. AND BRIAN CARTER

7   Radioactivity Counting Statistics

AGUSTÍN GRAU MALONDA AND AGUSTÍN GRAU CARLES

8   Sample Preparation Techniques for Liquid Scintillation Analysis

JAMES THOMSON

9 Cherenkov Counting

MICHAEL F. L'ANNUNZIATA

10   Radioisotope Mass Spectrometry

GERHARD HUBER, GERD PASSLER, KLAUS WENDT,
JENS VOLKER KRATZ AND NORBERT TRAUTMANN

II Solid Scintillation Analysis

MICHAEL F. L'ANNUNZIATA

12 Flow Scintillation Analysis

MICHAEL F. L'ANNUNZIATA

13 Radionuclide Imaging
LORAINE V. UPHAM AND DAVID F. ENGLERT

14 Automated Radiochemical Separation, Analysis, and Sensing

JAY W. GRATE AND OLEG B. EGOROV

15 Radiation Dosimetry

DAVID A. SCHAUER, ALLEN BRODSKY, AND JOSEPH A. SAYEG

CONTRIBUTORS

Numbers in parenthesis indicate the page on which the authors' contributions begin.

Allen Brodsky (1165) Science Applications International Corporation, McLean, Virginia, USA

Karl Buchtela (123) Atominstitute of the Austrian Universities, A-1020 Vienna, Austria

Brian Carter (537) Ontario Power Generation Inc., Whitby, Ontario, L1N 1E4, Canada

Gordon T. Cook (537) Scottish Universities Research and Reactor Centre, East Kilbride, Glasgow G75 0QF, Scotland

Saeed A. Durrani (179) School of Physics and Astronomy, University of Birmingham, Birmingham B15 2TT, UK

Oleg B. Egorov (1129) Pacific Northwest National Laboratory, Richland, Washington 99352, USA

David F. Englert (1063) BioConsulting, West Hartford, Connecticut 06107, USA

Paul F. Fettweis (239) CANBERRA Semiconductor N.V., B-2250 Olen, Belgium

Jay W. Grate (1129) Pacific Northwest National Laboratory, Richland, Washington 99352, USA

Agustín Grau Malonda (609) Instituto de Estudios de la Energía, CIEMAT, Avda. Complutense 22, 28040 Madrid, Spain

Agustín Grau Carles (609) Departamento de Fusión y Física de Partículas, CIEMAT, Avda. Complutense 22, 28040 Madrid, Spain

Gerhard Huber (799) Institut für Physik, Universität Mainz, 55099 Mainz, Germany

Radomir Ilić (179) Faculty of Civil Engineering, University of Maribor, Smetanova 17, 2000 Maribor, Slovenia; and Jožef Stefan Institute, Jamova 39, 1000 Ljubljana, Slovenia

Michael J. Kessler (347) (deceased), Packard Instrument Company, Meriden, Connecticut 06450, USA

Jens Volker Kratz (799) Institut für Kernchemie, Universität Mainz, 55099 Mainz, Germany

Michael F. L'Annunziata (1, 347, 719, 845, 989) The Montague Group, P.O. Box 5033, Oceanside, California 92052-5033, USA

Gerd Passler (799) Institut für Physik, Universität Mainz, 55099 Mainz, Germany

Charles J. Passo, Jr. (537) PerkinElmer Life and Analytical Sciences, Downers Grove, Illinois 60515, USA

Joseph A. Sayeg (1165) (Emeritus), Department of Radiation Medicine, University of Kentucky, Lexington, Kentucky, USA

David A. Schauer (1165) Department of Radiology and Radiological Sciences, Uniformed Services University of the Health Sciences, Bethesda, Maryland 20814, USA

Harold Schwenn (239) Canberra Industries, Inc. Meriden, Connecticut 06450, USA

James Thomson (655) PerkinElmer Life and Analytical Sciences, Groningen, The Netherlands

Norbert Trautmann (799) Institut für Kernchemie, Universität Mainz, 55099 Mainz, Germany

Loraine V. Upham (1063) Myriad Proteomics, Salt Lake City, Utah 84108, USA

Ramkumar Venkataraman (239) Canberra Industries, Inc. Meriden, Connecticut 06450, USA

Jan Verplancke (239) CANBERRA Semiconductor N.V., B-2250 Olen, Belgium

Klaus Wendt (799) Institut für Physik, Universität Mainz, 55099 Mainz, Germany

Brian M. Young (239) Canberra Industries, Inc. Meriden, Connecticut 06450, USA

ACRONYMS, ABBREVIATIONS AND SYMBOLS

A	mass number, amplifier
a	years (anni)
Å	angstrom (10^{-10} meters)
AAPM	American Association of Physicists in Medicine
AC	alternating current
ADC	analog to digital converter
ADME	absorption, distribution, metabolism and elimination
AEC	automatic efficiency control
AES	atomic emission spectrometry
AFS	atomic fluorescence spectrometry
AM	β-artemether
AMP	adenosine monophosphates, amplifier
AMS	accelerator mass spectrometry
amu	atomic mass units
ANDA	7-amino-1,3-naphthalenedisulphonic acid
ANSI	American National Standards Institute
α	alpha particle, internal-conversion coefficient
$\propto$	proportional to
APCI	atmospheric pressure chemical ionization
APD	avalanche photodiode
$\sim$	approximately
AQC	automatic quench compensation
AQP(I)	asymmetric quench parameter of the isotope
ATP	adenosine triphosphate
β	particle relative phase velocity
β^-	negatron, negative beta particle
β^+	positron, positive beta particle
BAC	N,N'-bisacrylylcystamine

BBD	2,5-di-(4-biphenylyl)-1,3,4-oxadiazole
BBO	2,5-di(4-biphenylyl)oxazole
BBOT	2,5-*bis*-2-(5-t-butyl-benzoxazoyl) thiophene
BCC	burst counting circuitry
BEGe	broad-energy germanium detector
BGO	bismuth germanate ($Bi_4Ge_3O_{12}$)
bis-MSB	*p-bis*-(*o*-methylstyryl)benzene
bkg, BKG	background
Bq	Becquerel = 1 disintegration per second
BSA	bovine serum albumin
BSF	backscatter factor
BSO	bismuth silicate ($Bi_4Si_3O_{12}$)
BT	bound tritium
butyl-PBD	2-(4-t-butylphenyl)-5-(4-biphenylyl)1,3,4-oxadiazole
c	speed of light in vacuum (2.9979×10^8 m/s)
C	Coulomb
$°C$	degrees Celsius
$CaF_2(Eu)$	europium-activated calcium fluoride
CAI	calcium-aluminum-rich inclusions
CAM	continuous air monitoring
CANDU	Canadian deuterium uranium reactor
CCD	charged coupled device
CD ROM	compact disc read-only memory
CE	chemical etching, capillary electrophoresis
CERN	European Organization for Nuclear Research, Geneva
C_F	feedback capacitor
CFN	cross-flow nebulizer
CGE	Chamber Gram Estimator
Ci	Curie = 2.22×10^{12} dpm = 3.7×10^{10} dps
CICM	conventional integral counting method
CID	collision induced dissociation
CIEMAT	Centro de Investigaciones Energéticas, Medioambientales y Technológicas, Madrid
cm	centimeter
CMPO	octyl(phenyl)-N,N-di-isobutylcarbamoylmethylphosphine oxide
cph, CPH	counts per hour
CPE	charged particle equilibrium
cpm, CPM	counts per minute
cps, CPS	counts per second
CR-39	polyallyldiglycol carbonate plastic SSNTD
CsI(Na)	sodium-activated cesium iodide
CsI(Tl)	thallium-activated cesium iodide
CT	computed tomography
CTF	contrast transfer function
CTFE	chlorotrifluoroethylene
CTR	controlled thermonuclear reactor
cts	counts

CV	core valence, coefficient of variation
CWOSL	continuous wave optically stimulated luminescence
d	days, deuteron
2D	two-dimensional
DAC	derived air concentration
DATDA	diallyltartardiamide
DC	direct current
dc-GDMS	direct current–glow discharge mass spectrometry
DE	double escape
δ	delta rays
DESR	double external standard relation
Det.	detector
DF-ICP-MS	double focusing ICP-MS
DIHEN	direct injection high-efficiency nebulizer
DIM	data interpretation module
dimethyl POPOP	1,4-*bis*-2-(4-methyl-5-phenyloxazolyl)benzene
DIN	di-isopropylnaphthalene
DJD	diffused junction detectors
DLU	digital light units
DMG	dimethylglyoxime
DMSO	dimethyl sulfoxide
DNA	deoxyribonucleic acid
D_2O	heavy water
DOE	United States Department of Energy
DOELAP	Department of Energy Laboratory Accreditation Program
DOT	digital overlay technique
dpm, DPM	disintegrations per minute
dps, DPS	disintegrations per second
dpy, DPY	disintegrations per year
DQP	double quench parameter
DRAM	dynamic random access memory
DSP	digital signal processor
DTPA	diethylenetriamine pentaacetic acid
DU	depleted uranium
DWPF	Defense Waste Processing Facility
E	counting efficiency, energy
e^-	electron
e^-h^+	electron–hole pair
EC	electron capture
ECDL	extended cavity diode laser
ECE	electrochemical etching
EDTA	ethylenediamine tetraacetic acid
E_F	Fermi level
EF	enrichment factor
EIA	enzyme immunoassay
EMA	extra mural absorber
EO	ethylene oxide

EPA	United States Environmental Protection Agency
EPR	electron paramagnetic resonance
ES	external standard
ESCR	external standard channels ratio
ESI	electrospray ionization
ESP	external standard pulse
ET	efficiency tracing
ET-DPM	efficiency tracing disintegrations per minute (method)
eV	electron volt
E_{av}	average energy (beta particle)
E_{max}	maximum energy (beta particle)
E_α	alpha-particle energy
E_p	proton energy
EURADOS	European Radiation Dosimetry Group
EXAFS	x-ray absorption fine structure
°F	degrees Fahrenheit
FDA	United States Food & Drug Administration
FEP	full energy peak
FET	field effect transistor
fmol	femtomoles (10^{-15} moles)
FI	flow injection
FT	fission track
FTD	fission track dating
FOM	figure of merit
fov	field of view
fp	fission products
FSA	flow scintillation analysis
FS-DPM	full-spectrum disintegrations per minute (method)
FWHM	full width at half maximum
FWT	free water tritium
FWTM	full width at tenth maximum
g	gram
G#	G-number (quench indicating parameter)
γ	gamma radiation
GBq	gigabecquerels (10^9 Bq)
GDMS	glow discharge mass spectrometry
Ge(Li)	lithium-compensated germanium
GEM	gas electron multiplier
GeV	giga electron volts (10^9 eV)
GHz	gigahertz
GLP	good laboratory practice
GM	Geiger-Mueller
GS-20	glass scintillator
GSO:Ce	cerium-activated gadolinium orthosilicate (Gd_2SiO_5:Ce)
Gy	Gray
h	Plank's constant (6.626×10^{-34} J s), hours
H#	Horrock's number (quench indicating parapeter)
HBT	2-(2-hydroxyphenyl)-benzothiazole

HDEHP	*bis*(2-ethylhexyl)phosphoric acid
HEN	high efficiency nebulizer
HEP	high energy particle
HEPES	N-2-hydroxyethylpiperazine-N′-2-ethanesulfonic acid
HEX-ICP-MS	hexapole collision cell ICP-MS mass spectrometry
3HF	3-hydroxy flavone
HPGE	high purity germanium
HPIC	high performance ionic chromatography
HKG	housekeeping gene
HPLC	high performance liquid chromatography
HT	high tension
HV	high voltage
HWHM	half width at half maximum
Hz	Hertz
i_{in}	current pulse
IAEA	International Atomic Energy Agency, Vienna
IC	ion chromatography
IC#	Isotope Center Number
ICPs	inductively coupled plasmas
ICP-MS	inductively coupled plasma mass spectrometry
ICP-QMS	inductively coupled plasma quadrupole mass spectrometry
ICRP	International Commission on Radiological Protection
ICRU	International Commission on Radiation Units and Measurements
ID	inner diameter
IEEE	Institute of Electrical and Electronics Engineers
IL-5	interleukin-5
I/O	input/output
IPA	instrument performance assessment
IPRI	Laboratoire Primaire des Rayonnements Ionisants, France
IPT	intramolecular proton transfer
IR	infrared
IS	internal standard
ISOCS	in-situ object calibration software
IT	isomeric or internal transition
ITER	International Thermonuclear Experimental Reactor
J	joule
JET	Joint European Torus reactor
JFET	junction field effect transistor
K	particle kinetic energy
K	degrees Kelvin, Kerma
kcps	kilocounts per second
kBq	kilobecquerels (10^3 Bq)
keV	kiloelectron volts
kGy	kilogray
kHz	kilohertz
kV	kilovolts

L, l	liters
LAB	dodecylbenzene, linear alkyl benzene
LA-ICP-MS	laser ablation inductively coupled plasma mass spectrometry
λ	wavelength, decay constant, microliter (10^{-6} L), free parameter
λ_{nr}	nonrelativistic wavelength
λ_r	relativistic wavelength
LAN	local area network
LAr	liquid argon
LAW	low activity waste
LC	liquid chromatography
LED	light emitting diode
LEGE	low-energy gemanium detector
LET	linear energy transfer
LiI(Eu)	europium-activated lithium iodide
LIST	laser ion source trap
LL	lower level
LLCM	low-level count mode
LLD	lower limit of detection, lower level discriminator
LM-OSL	linear modulation optically stimulated luminescence
LN_2	liquid nitrogen
LOD	limit of detection
LPRI	Laboratoire Primaire des Ionizants, Paris
LPS	lipopolysaccharide
LS	liquid scintillation, liquid scintillator
LSA	liquid scintillation analysis (analyzer)
LSC	liquid scintillation counting (counter)
LSO	cerium-activated lutetium oxyorthosilicate ($Ce:Lu_2SiO_5$)
LSS	liquid scintillation spectrometer
LuAP	cerium-activated lutetium aluminum perovskite ($Ce:LuAlO_3$)
LXe	liquid xenon
m	particle mass
m_0	particle rest mass
m_r	speed-dependent particle mass
m	mass, meters
mA	milliampere (10^{-3} ampere)
MAPMT	multi-anode photomultiplier tube
mCi	millicurie (10^{-3} Ci)
mL, ml	milliliter (10^{-3} L)
MBq	megabecquerels (10^6 Bq)
MCA	multichannel analyzer
MCF	moving curve fitting
MC-ICP-MS	multiple ion collector-ICP-MS
MCN	microconcentric nebulizer
MCP	microchannel plate
MCP-PM	microchannel plate photomultiplier
MD	Molecular Dynamics
MDA	minimal detectable activity

MeV	megaelectron volts
MeVee	electron equivalent energy
MHz	megahertz
MIBK	methyl isobutyl ketone
MICAD	Microchannel Array Detector®
MICM	modified integral counting method
mBq	millibequerels (10^{-3} Bq)
mg	milligram (10^{-3} g)
mGy	milligray
min.	minutes
MLR	multiple linear regression
mm	millimeter (10^{-3} m)
MCNP	Monte Carlo N-particle
MP	MultiPurpose
mRNA	messenger RNA
MS	mass spectrometry
ms	milliseconds (10^{-3} s)
MSB	methylstyrylbenzene
μ	attenuation coefficient
μA	microampere (10^{-6} ampere)
μCi	microcurie (10^{-6} Ci)
μg	microgram (10^{-6} g)
μL	microliter (10^{-6} L)
μm	micrometer (10^{-6} m)
μs	microseconds (10^{-6} s)
MWPC	multiwire proportional chamber
MV	megavolts (10^{6} volts)
MVC	multivariate calibration
n	neutron
n	index of refraction
NAA	neutron activation analysis
NAC	N-acetylcysteine
NaI(Tl)	thallium-activated sodium iodide
nCi	nanocurie (10^{-9} Ci)
NCM	normal count mode
NCRP	National Council on Radiation Protection and Measurements
NIST	National Institute of Standards and Technology, Gaithersburg
NPD	2-(1-naphthyl)-5-phenyl-1,3,4-oxadiazole
NPO	2-(1-naphthyl)-5-phenyloxazole
NRC	United States Nuclear Regulatory Commission
NVLAP	National Voluntary Accreditation Program
ν	neutrino, photon frequency, particle velocity
$\bar{\nu}$	antineutrino
nM	nanomolar (10^{-9} M)
nm	nanometer (10^{-9} m)
NMR	nuclear magnetic resonance

ns, nsec	nanosecond (10^{-9} s)
N-TIMS	negative ion thermal ionization mass spectrometry
NTS	Nevada Test Site
OLLSC	on-line liquid scintillation counting
OSL	optically stimulated luminescence
p	particle momentum
p, p$^+$	proton
PAC	pulse amplitude comparison (comparator)
PAGE	polyacrylamide gel electrophoresis
PBBO	2-(4′-biphenylyl)-6-phenylbenzoxazole
PBD	2-phenyl-5-(4-biphenylyl)-1,3,4-oxadiazole
PBO	2-(4-biphenylyl)-5-phenyloxazole
PBS	phosphate buffered saline
PC	proportional counter(ing), personal computer
PCB	polychlorinated biphenyl
pCi	picocurie (10^{-12} Ci)
PCR	principle component regression
PD	photodiodes
PDA	pulse decay analysis
PDD	pulse decay discriminator
PE	phosphate ester
PEC	power and event controller
PERALS®	Photon Electron Rejecting Alpha Liquid Scintillation
PET	positron emission tomography, polyethylene terephthalate
pF	picofarad (10^{-12} farad)
pg	picogram (10^{-12} gram)
PFA	perfluoroalkoxy
PHA	pulse height analysis
PHOSWICH	PHOSphor sandwich (detector)
PID	particle identification
PIPS	passivated implanted planar silicon
PKC	protein kinase C
PLS	partial least squares
PLSR	partial least squares regression
PM	photomultiplier
PMMA	polymethylmethacrylate
PMP	1-phenyl-3-mesityl-2-pyrazoline
PMT	photomultiplier tube
PN	pneumatic nebulizers
POPOP	1,4-*bis*-2-(5-phenyloxazolyl)benzene
POSL	pulsed optically stimulated luminescence
ppb	parts per billion
PPD	2,5-diphenyl-1,3,4-oxadiazole
PPO	2,5-diphenyloxazole
PS	polystyrene
PSA	pulse shape analysis
PSD	pulse shape discrimination
psi	pounds per square inch

PSL	photostimulable light
P/T	peak-to-total ratio
PTB	Physikalisch-Technische Bundesanstalt, Braunschweig
PTFE	polytetrafluoroethylene
P-TIMS	positive ion thermal ionization mass spectrometry
PTP	p-terphenyl
PUR	pile up rejector
PVC	polyvinyl chloride
PVT	polyvinyl toluene
PWR	pressurized water reactor
PXE	phenyl-ortho-xylylethane
QC	quality control
QC-CPM	quench corrected count rate
QDC	charge-to-digital converter
QIP	quench indicating parameter
RAST	radioallergosorbent test
RBE	relative biological effectiveness
RDC	remote detector chamber
RE	recovery efficiency
REGe	reverse-electrode coaxial Ge detector
RF	radiofrequency
R_F	feedback resister
ρ	density $(g\,cm^{-3})$, neutron absorption cross section, resistivity
RIA	radioimmunoassay
RICH	Ring Imaging Cherenkov (counters)
RIMS	resonance ionization mass spectrometry
RIS	resonant ionization
RNA	ribonucleic acid
RPH	relative pulse height
RSC	renewable separation column
RSD	relative standard deviation
RSF	relative sensitivity factor
RST	reverse spectral transform
s	seconds
SAM	standard analysis method
SCA	single channel analyzer
SCC	squamous cell carcinoma
SCR	sample channels ratio
SD	standard deviation
SDD	silicon drift detector
SDP	silicon drift photodiode
SE	single escape
SF	spontaneous fission
SFD	scintillation fiber detector
SHE	superheavy elements
SI	International System of Units, sequential injection
SIA	sequential injection analysis
SIE	spectral index of the external standard

σ	thermal neutron cross section
Si(Li)	lithium-compensated silicon
SIMS	secondary ion mass spectrometry
SI-RSC	sequential injection renewable separation column
SIS	spectral index of the sample
SLM	standard laboratory module
SLSD	scintillator-Lucite sandwich detector
SMDA	specific minimum detectable activity
S/N	signal-to-noise
SNM	special nuclear materials
SOI	silicon-on-insulator
SPA	scintillation proximity assay
SPC	single photon counting
SPE	single photon event
SPECT	single photon emission computed tomography
SQP(I)	spectral endpoint energy
SQP(E)	spectral quench parameter of the external standard
SQS	self-quenched streamer
SR	super resolution
SRS	Savannah River Site
SSB	silicon surface barrier detector
SSM	selective scintillating microsphere, standard service module
ST	super sensitive
STE	self-trapped excitation
STNTD	solid state nuclear track detection (detectors)
STP	standard temperature and pressure
Sv	sievert
$t_{1/2}$, $T_{1/2}$	half-life
T	particle kinetic energy
TAR	tissue-air ratio
TBP	tributyl phosphate
TCA	trichloroacetic acid
TD	time discriminator
TDCR	triple-to-double coincidence ratio
TEA	triethylamine
TEM	transmission electron microcroscopy
TFTR	Tokamak Fusion Test Reactor
TIMS	thermal ionization mass spectrometry
TL	thermoluminescence
TLC	thin-layer chromatography (chromatogram)
TLD	thermoluminescent dosimeter (dosimetry)
TMOS	tetramethoxysilane
TMS	tetramethylsilane
TNOA	tri-n-octylamine sulfate
TOF	time-of-flight
TOP	time-of-propagation
TOPO	trioctylphosphine
TP	p-terphenyl

TR	Tritium Sensitive
TRACOS	automatic system for nuclear track evaluations
TRE	12-O-tetradecanoyl phorbol-13-acetate responsive element
TR-LSC®	time-resolved liquid scintillation counting
TR-PDA®	time-resolved pulse decay analysis
TRPO	trialkyl phosphine oxide
TSC	task sequence controller
TSEE	thermally stimulated exoelectron emission
tSIE	transformed spectral index of the external standard
tSIS	transformed spectral index of the sample
TTA	thenoyltrifluoroacetone
TU	Tritium Unit (0.118 Bq or 7.19 DPM of ^{3}H L^{-1} H$_2$O)
u	atomic mass unit (1/12 m of ^{12}C = 1.6605402 × 10^{-27} kg)
u	particle speed
u_{nr}	nonrelativistic particle speed
u_r	relativistic particle speed
UL	upper level
ULB	ultra low background
ULD	upper level discriminator
ULEGE	ultra low-energy Ge
U.S.A.E.C.	United States Atomic Energy Commission (now NRC)
USEPA	United States Environmental Protection Agency
USN	ultrasonic nebulizers
UV	ultraviolet
V	volts
V_0	step voltage
VAX	Digital Equipment Corporation tradename
WIMP	weakly interacting massive particle
y	years
YAG:Yb	Yb-doped Y$_3$Al$_5$O$_{12}$
YAP:Ce	cerium activated yttrium aluminum perovskite (Ce:YAlO$_3$)
YSi(Ce)	cerium-activated yttrium silicate
XRF	x-ray fluorescence
XtRA	extended range
Z	atomic number
ZCH	Central Analytical Laboratory, Jülich
ZnS(Ag)	silver-activated zinc sulfide

FOREWORD TO THE FIRST EDITION

One hundred years after the discovery of radioactivity by Becquerel, the analysis of radioactivity has become of great significance to many disciplines and persons working in fields as diverse as nuclear medicine, radiopharmacy, clinical diagnosis, health physics, biological sciences, food preservation, industry, environmental monitoring, nuclear power, and nuclear safety and safeguards. The accurate measurement of the activity of radionuclides is today a *sine qua non* condition for better knowledge of the environment we live in and for progress and advancement in various scientific and technological disciplines.

Since the International Atomic Energy Agency was founded in 1957, global cooperation in the peaceful use of nuclear energy through nuclear power production and the use of radionuclides and radiation sources has played a significant role in world development. The advances being made in the peaceful application of nuclear technology depend to a great extent on the ease and accuracy of radioactivity measurements. The use of radioactive materials, their production, and the safe disposal of radioactive waste rely greatly on these precise measurements.

Several international experts in various aspects of radionuclide analysis have contributed to this valuable book. As a handbook, it integrates the modern principles of radiation detection and measurement with the practical guidelines and procedures needed by scientists, physicians, engineers, and technicians from many diverse disciplines. It provides the information needed to measure all types of radioactivity, from low levels naturally present in the environment to high levels found in the production, applications, and disposal of radionuclides. This book will facilitate further refinements in the measurement and analysis of radioactivity needed either for scientific investigations or for the safe and peaceful applications of radioactive and radiation sources.

Dr. Mohamed M. ElBaradei
Director General
International Atomic Energy Agency

FOREWORD TO THE SECOND EDITION

The use of radioactivity plays an important role in human development. Both natural and man-made radionuclides have been harnessed to provide us with the tools to live longer and lead more productive lives. For example, the use of radioactive sources in medical diagnosis is only one of many ways radioactivity helps to provide a healthier way of life. With this in mind the International Atomic Energy Agency promotes research and development projects that are focused on the peaceful applications of radioactive sources in over a hundred countries of the world. Among these are projects for development in agriculture, food preservation, medicine, hydrology for water resources management, industry, and electric power generation, just to mention a few, which require the safe application of radioactive sources. While radioactive materials have provided us with the means to improve our lives, they have also forced us to assure their safe handling and peaceful use on a global scale. Consequently the IAEA is charged by its Member States to also promote projects in nuclear safety, environmental protection, and nuclear safeguards in order to provide assurances, as best we can through accurate analysis, that radioactive materials are used safely and exclusively for peaceful purposes.

The accurate measurement of radioactivity remains vital to the correct use of nuclear materials and the protection that must be provided against their potential harmful effects. Consequently, I am pleased to see that a new 2nd Edition of the Handbook of Radioactivity Analysis has been completed. Numerous international experts in radioactivity analysis and dosimetry have contributed to the preparation of this book. Many advances have been made since the first edition was published in 1998, and the scientific developments provided in the new edition can help refine the analysis of radioactivity from the very low levels found in the environment to the high levels needed for peaceful applications for development. The new edition should serve as a tool for further advancements in scientific research and peaceful applications of radioactive materials to serve basic human needs around the globe.

Werner Burkart, Prof. Dr.
Deputy Director General
Department of Nuclear Sciences and Applications
International Atomic Energy Agency, Vienna

PREFACE TO THE FIRST EDITION

This book focuses on the techniques and principles used to measure the disintegration rates of radioactive nuclides (radionuclides) and the types and energies of radiation emanating from radionuclides. The determination of the disintegration rate of a radionuclide provides, of course, a quantitative measure of the amount of that radionuclide in a sample. Therefore, activity analysis techniques presented in this book are aimed at determining the activity of radionuclides in units of the curie or becquerel.

The measurement of radionuclide activities is a science of interest to persons working in a wide spectrum of disciplines. These include scientists, engineers, physicians, and technicians whose work entails the preparation, utilization, or disposal of radioactive materials and the measurement of radioactivity in the environment. Among these are persons working in the fields of radiopharmacy, nuclear medicine, clinical analysis, scientific research, industrial applications, health physics, nuclear power, nuclear fuel cycle facilities, nuclear waste management, and nuclear safeguards, to mention only a few. During almost fifteen years with the International Atomic Energy Agency (IAEA) in Vienna, I had the opportunity to meet and work with persons from all of these disciplines and from all corners of the globe. They all shared the common challenge of measuring, as accurately as possible, the activities of radionuclides in many types of samples. The activities ranged from the very low levels of natural or man-made radionuclides encountered in the environment to higher levels used in research, medicine, and the nuclear power-related fields.

While serving as Head of Fellowships and Training of the IAEA in 1987, I was fortunate to publish a book in this field titled *Radionuclide Tracers, Their Detection and Measurement,* which was aimed at providing a reference work for users of radioactive materials. I believe the book achieved its goal as, according to a review by Testuo Sumi, *Isotope News,* **11**(410), 46, November 1987. "This book is a *vade mecum* for the user of radionuclide tracers as well as a reference book for radiation measurement." since then, of course, many advances have been made, and the need emerged to produce yet a more practical text that included not only the modern principles of radiation detection and measurement, but also guidelines and procedures for measuring radionuclides in samples of many types. An authoritative handbook of this kind requires contributions from scientists with expertise in various aspects of radioactivity measurement. With that objective in mind, notable scientists from various parts of the globe have been united in this work, each person an expert in his or her field of radionuclide activity analysis. The outcome of this effort is a handbook containing sample preparation procedures, required calculations, and guidelines on the use of computer-controlled high-sample-throughput activity analysis techniques.

The editor does not claim that this book is exhaustive in its coverage of analytical techniques available in this field. It was decided to limit the scope of the book to the most popular direct methods of radioactivity analysis, which include the detection and counting of the radiation emissions from radionuclides. Direct methods of radioactivity analysis remain today the most commonly utilized by far in laboratories throughout the world. Indirect methods of radionuclide measurement that remain of limited use, such as accelerator mass spectrometry and inductively coupled plasma mass spectrometry, are not described in this book. These methods are not yet widely in use because of the need for an accelerator facility and/or very expensive equipment that is still out of reach of most laboratories.

The importance of semiconductor detectors in radiation spectroscopy warranted a very detailed chapter on the principles and practice of semiconductor detector applications, including sample preparation procedures. A chapter on principles and current applications of gas ionization detectors, a method that has evolved since the very early days of radiation detection and measurement, has also been included.

Liquid scintillation analysis techniques are separated into two chapters, namely, "Radiotracer Liquid Scintillation Analysis," which focuses on the measurement of relatively high levels of radioactivity normally encountered in radionuclide applications, and "Environmental Liquid Scintillation Analysis," which requires certain low-level activity analysis techniques for the measurement of natural and man-made radionuclides in the environment. Glass and plastic scintillators, which by definition may not be solids due to their lack of crystalline structure, are included in the chapter on solid scintillation analysis because these scintillators are used in the state of mechanical rigidity when employed as radiation detectors. A separate chapter on sample preparation techniques for liquid scintillation analysis was needed because of the large number of radionuclides analyzed by this method, as well as to provide guidelines to help the reader optimize counting efficiency and reduce interferences from chemiluminescence and quenching. Because of the random nature of radionuclide decay, a chapter on statistical computations used in radiation counting is included.

There is an ever-increasing need for high-sample-throughput radionuclide analysis at clinical and drug-screening laboratories, among others, which use techniques such as scintillation proximity assay (SPA) in receptor-binding assays, immunoassays, and enzyme assays. With this in mind, multidetector systems for liquid and solid scintillation analysis are included in this handbook with considerable information on high-sample-throughput microplate scintillation analysis techniques.

Advances and guidelines in radionuclide activity analysis by Cherenkov counting techniques are included in this book, as they provide a very practical and inexpensive method of radioactivity analysis whenever radiation energies and activity levels are not limiting factors. The reader will encounter the words Cherenkov and Cerenkov as two variations of the spelling for the characteristic radiation produced by charged particles. The first is the phonetic spelling originating from the Russian pronunciation and the latter is the anglicized version of the word. Both spellings are used currently in the scientific literature. This is explained in more detail in Chapter 9. Because of the widespread interest in flow scintillation analysis, a chapter is included with guidelines and procedures for the real-time, on-line activity analysis of radionuclides in flowing streams such as effluents from high-performance liquid chromatography, high-performance ionic chromatography, and effluents associated with nuclear power and fuel processing plants.

Electronic radionuclide imaging methods, which provide relatively rapid quantitative imaging of radionuclide activities in whole-body sections, sequencing gels, polyacrylamide gel electrophoresis and thin-layer chromatography, among other media, are described in this handbook. Electronic radionuclide imaging methods are replacing in many cases the older, less quantitative and slower method of film autoradiography.

In line with current technology, computer-controlled automation and data processing are described throughout the book. Nevertheless, it was considered necessary to include a separate chapter on robotics and automation in radionuclide analysis to help the working scientist apply the full potential of modern technology to radioactivity analysis.

The fundamental properties of radioactivity, radionuclide decay, and methods of detection are described in this handbook to provide the neophyte scientist with the basis for a thorough explanation of the analytical procedures. The volume can be used, therefore, not only as a handbook but also a teaching text.

For complementary reading on the significance of monitoring radionuclide activity in the environment, the reader is invited to peruse the new fourth edition of *Environmental Radioactivity, From Natural, Industrial, and Military Sources* by Merril Eisenbud and Thomas Gesell published by Academic Press in 1997.

Mention of commercial products in this book does not imply recommendation or endorsement by the authors or editor. Other and more suitable products may be available. The names of these products are included for convenience or information purposes only.

This book project had a very sad beginning with the unexpected passing of Dr. Michael J. Kessler on April 21, 1997, after a heart attack. Mike Kessler was the first person I spoke to about the idea for this book. He was overwhelmingly in favor of the handbook idea, and he planned to contribute to several chapters of this book. Those who knew Mike personally will miss a dear friend and respected scientist of international renown in this field.

I am very grateful to the authors for their contribution and unwavering commitment to this project. Their writings were submitted in a timely fashion, and they have covered their fields of expertise meritoriously. I believe that with their contributions to this effort, we have fulfilled the objectives of this handbook. I gratefully acknowledge the support of Gene Della Vecchia, George Serrano, Michael J. Kessler and Charles J. Passo, Jr. I also thank

David J. Packer, Senior Editor at Academic Press, for encouraging the preparation of a practical handbook for a wide spectrum of users. The assistance of Jock Thomson and Charles J. Passo, Jr. in the review of some of the material in this book is appreciated. Above all, I thank my wife, Reyna, for her support, understanding, and unflagging patience.

Michael F. L'Annunziata

PREFACE TO THE SECOND EDITION

Many advances have been made since the publication of the First Edition of the *Handbook of Radioactivity Analysis* in 1998. This is reflected in the numerous citations found in this new edition. Also it is pleasing to note that the First Edition was well received by many persons from a broad spectrum of disciplines in the academic, research, and applied fields of science where radioactive nuclides are measured. The advances made since the First Edition, together with the demand for the book, sparked interest to produce this new Second Edition with additional chapters and subject matter. It is hoped that broadening the scope of the book and increasing the practical content of the material presented could satisfy more fully the needs of persons from many fields. Radionuclides and the precise measurement of their activity is a subject of concern to persons in many fields including physics, chemistry, hydrology, agricultural research, industry, nuclear medicine, radiopharmacy, biological sciences, electric power production, waste management, environmental conservation, and nuclear safeguards, just to mention a few. Although scientists working in the fields cited are very diverse in their objectives and techniques of study, they have one common need: to measure as accurately as possible the activity or disintegration rate of radionuclides. The radionuclides to be measured and sample types to be analyzed can differ greatly depending on the field of science—a radiopharmaceutical about to be administered to a cancer patient, nuclides in air, water or soil samples taken from the environment, or radioactive waste from a nuclear power plant serve as examples. The objective of this Second Edition is to provide the academic, research, and applied scientists in all fields of endeavor with up-to-date information on the principles and practice of radioactivity analysis that can be applied by persons concerned with peaceful applications of radioactive sources for development and conservation of the environment.

With the Second Edition the scope of the book was expanded with new chapters on Solid State Nuclear Track Detectors, Radioisotope Mass Spectrometry, and Radiation Dosimetry. Solid State Nuclear Track Detectors, as described by the authors, can be applied to a wide range of fields, and it can be one of the least expensive methods available to scientists in planetary, physical, biological, and medical sciences. The number of laboratories in the world that have the capability of using mass spectrometry to measure radioactive nuclides is increasing. It was decided therefore, that the addition of a chapter on Radioisotope Mass Spectrometry is needed. This chapter provides details on the various types and applications of mass spectrometry and the advantages and disadvantages of counting the radionuclide atoms in a sample, provided by mass spectrometry, versus the counting of the radiation emissions from the radionuclides in a sample, provided by the radioactivity counting methods described in other chapters. It was considered appropriate to include also an additional chapter on radiation dosimetry, as it is a field of concern to anyone who must use radioactive materials or who is concerned with their measurement.

All chapters have been updated and expanded. Advances, new topics and concepts have been added to each chapter. The chapter on Automated Radiochemical Separation, Analysis, and Sensing is altogether new with practical methodology for the automated measurement of radionuclide mixtures in nuclear waste and the environment. The new chapter on Radioactivity Counting Statistics addresses issues related to statistical fluctuations observed in radiation measurements, caused by the inherently random nature of the radionuclide decay process. The chapter is relevant to anyone who measures and counts radionuclide emissions.

Mention of commercial products in this book does not imply recommendation or endorsement by the authors or editor. Other and more suitable products may be available. The names of products are included for convenience or information purposes only.

Among the authors of the various chapters are 27 persons from 10 countries of the world with expertise in various disciplines of radioactivity analysis. Their unwavering commitment to this project and the efforts they have made to cover their field of expertise in each chapter were vital to meeting the objectives of this book. I gratefully acknowledge the support and encouragement of Dr. Markku Koskelo, Dr. Egbert M. van Wezenbeek, Carla Kinney and Christine Kloiber, as well as to Derek Coleman and Imran Mirza for their editorial assistance. The assistance of Dr. Ramkumar Venkataraman, Dr. Agustín Grau Malonda, and Dr. Romard Barthel, C.S.C. in the review of some of the material in this book is highly appreciated. Above all, I thank my wife, Reyna, for her support, understanding, and unflagging patience throughout this demanding project.

Michael F. L'Annunziata
February 2003

1

NUCLEAR RADIATION, ITS INTERACTION WITH MATTER AND RADIOISOTOPE DECAY

MICHAEL F. L'ANNUNZIATA

The Montague Group, P.O. Box 5033, Oceanside, CA 92052-5033, USA

I. INTRODUCTION

The analysis of radioactivity is a challenging field. Both the sources of radioactivity (e.g., radionuclides) and the media within which the radionuclides may be found can present themselves in a wide range of complexities. For example, nuclear radiation can occur in various types, percent abundances, and energies. Also, a given radionuclide may have more than one mode of decay. The presence of appreciable activities of more than one radionuclide in a sample can further complicate analysis. In addition, the different parent–daughter nuclide decay schemes, equilibria between parent and daughter radionuclides, and the rates of decay that radioactive nuclides undergo may facilitate or complicate the analysis for a given radionuclide. The problem of radioactivity analysis may be confounded further by the wide range of chemical and/or physical media (i.e., sample matrices) from which the nuclear radiation may emanate.

As we will find in this book, there are many modern methods of radioactivity analysis. The types of detectors available for the measurement of radioactivity are numerous, and they may be designed in the gaseous, liquid, or solid state. They will differ not only in their physical state but also in chemistry. The instrumentation and electronic circuitry associated with radiation detectors will also vary. As a result, radiation detectors and the instrumentation associated with detectors will perform with varying efficiencies of radiation detection depending on many factors, including the characteristics of the instrumentation, the types and energies of the radiation, as well as sample properties.

The proper selection of a particular radiation detector or method of radioactivity analysis requires a good understanding of the properties of nuclear radiation, the mechanisms of interaction of radiation with matter, half-life, decay schemes, decay abundances, and energies of decay. This chapter will cover these concepts as a prelude to the various chapters that follow on radioactivity analysis. Throughout the book reference will be made to the concepts covered in this introductory chapter. For the experienced radioanalytical chemist, this chapter may serve only as a review. However, the newcomer in this field should find this introductory chapter essential to the understanding of the concepts of radiation detection and measurement. He or she will find that the concepts covered in this introductory chapter will facilitate the selection of the most suitable radiation detector and instrumentation required for any particular case.

The properties of nuclear radiation and the mechanisms whereby nuclear radiation dissipates its energy in matter, dealt with in this chapter, form the basis for the methods of detection and measurement of radionuclides.

II. PARTICULATE RADIATION

A. Alpha Particles

The alpha particle, structurally equivalent to the nucleus of a helium atom and denoted by the Greek letter α, consists of two protons and two neutrons. It is emitted as a decay product of many radionuclides predominantly of atomic number greater than 82 (See Appendix A, Table of Radioactive Isotopes). For example, the radionuclide americium-241 decays by alpha particle emission to yield the daughter nuclide ^{237}Np according to the following equation:

$$^{241}_{95}\text{Am} \rightarrow {}^{237}_{93}\text{Np} + {}^{4}_{2}\text{He} + 5.63\,\text{MeV} \tag{1.1}$$

The loss of two protons and two neutrons from the americium nucleus results in a mass reduction of four and a charge reduction of two on the nucleus. In nuclear equations such as the preceding one, the subscript denotes the charge on the nucleus (i.e., the number of protons or atomic number, also referred to as the Z number) and the superscript denotes the mass number (i.e., the number of protons plus neutrons, also referred to as the A number). The energy liberated during nuclear decay is referred to as decay energy. Many reference books report the precise decay energies of radioisotopes. The value reported by Holden (1997a) in the Table of Isotopes for the decay energy of ^{241}Am illustrated in Eq. (1.1) is 5.63 megaelectron volts (MeV). Energy and mass are conserved in the process; that is, the energy liberated in radioactive decay is equivalent to the loss of mass by the parent radionuclide (e.g., ^{241}Am) or, in other words, the difference in masses between the parent radionuclide and the product nuclide and particle.

We can calculate the energy liberated in the decay of ^{241}Am, as well as for any radioisotope decay, by accounting for the mass loss in the decay equation. Using Einstein's equation for equivalence of mass and energy

$$E = mc^2 \tag{1.2}$$

we can write the expression for the energy equivalence to mass loss in the decay of ^{241}Am as

$$Q = (M_{^{241}\text{Am}} - M_{^{237}\text{Np}} - M_\alpha)c^2 \tag{1.3}$$

where Q is the disintegration energy released in joules, $M_{^{241}\text{Am}}$, $M_{^{237}\text{Np}}$ and M_α are the masses of ^{241}Am, ^{237}Np and the alpha particle in kilograms and c is the speed of light in a vacuum, 3.00×10^8 m/s). When the nuclide masses are expressed in the more convenient atomic mass units (u) the energy liberated in decay equations can be calculated in units of megaelectron volts according to the equation

$$Q = (M_{^{241}\text{Am}} - M_{^{237}\text{Np}} - M_\alpha)(931.494\,\text{MeV/u}) \tag{1.4}$$

The precise atomic mass units obtained from reference tables (Holden, 1997a) can be inserted into Eq. (1.4) to obtain

$$Q = (241.056822u - 237.048166u - 4.00260325u)(931.494 \, \text{MeV/u})$$
$$= (0.00605275u)(931.494 \, \text{MeV/u})$$
$$= 5.63 \, \text{MeV}$$

The energy liberated is shared between the daughter nucleus and the alpha particle. If the parent nuclide (e.g., ^{241}Am) is at rest when it decays, most of the decay energy will appear as kinetic energy of the liberated less-massive alpha particle and only a small fraction of the kinetic energy remains with the recoiling massive daughter nuclide (e.g., ^{237}Np). The kinetic energy of the recoiling daughter nuclide is comparable to that of a recoiling canon after a shell is fired; the shell being analogous to that of the alpha particle shooting out of the nucleus. Figure 1.1 illustrates the transitions involved in the decay of ^{241}Am. The interpretation of this figure is given in the following paragraph. There are four possible alpha particle transitions in the decay of ^{241}Am each involving an α-particle emission at different energies and relative abundances. These are illustrated in Fig. 1.1. The decay energy of 5.63 MeV for ^{241}Am calculated above and reported in the literature is slightly higher than any of the α-particle energies provided in Fig. 1.1. This is because there

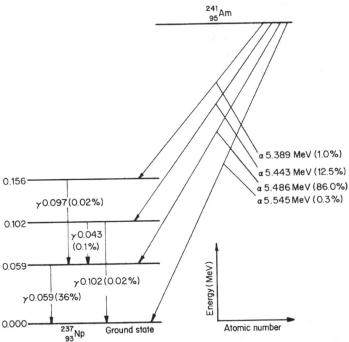

FIGURE 1.1 Decay scheme of ^{241}Am. The relative abundances (intensities) of alpha particle and gamma-ray emissions are expressed in percent beside the radiation energy values in MeV.

remains also the recoil energy of the daughter nucleus and any gamma-ray energy that may be emitted by the daughter, when its nucleus remains at an excited state. The emission of gamma radiation often accompanies radionuclide decay processes that occur by alpha particle emission. Gamma radiation is described in Section III.B of this chapter. The recoil energy, E_{recoil}, of the daughter nucleus can be calculated by the equation

$$E_{recoil} = (M_\alpha/M_{recoil})E_\alpha \qquad (1.5)$$

derived by Ehman and Vance (1991) where M_α is the mass of the alpha particle as defined in Eq. 1.3, M_{recoil} is the mass of the recoil nucleus and E_α is the alpha particle energy. For example, the recoil energy of the ^{237}Np daughter nucleus for the transition of the 5.545 MeV alpha particle (Fig. 1.1) can be calculated according to Eq. 1.5 as

$$
\begin{aligned}
E_{recoil} &= (4.00260325u/237.0481u)5.545\,\text{MeV} \\
&= (0.0168851)(5.545\,\text{MeV}) \\
&= 0.0936\,\text{MeV}
\end{aligned}
$$

The transition energy, E_{trans}, for the above alpha particle emission is the sum of the alpha particle and recoil nuclear energies or

$$
\begin{aligned}
E_{trans} &= E_\alpha + E_{recoil} \\
&= 5.545\,\text{MeV} + 0.0936\,\text{MeV} \\
&= 5.63\,\text{MeV} \qquad (1.6)
\end{aligned}
$$

In the above case the transition energy turns out to be equal to the reported and calculated decay energy, because as illustrated in Fig. 1.1 the ^{241}Am radionuclides decay directly to the ground state whenever 5.545 MeV alpha particles are emitted. This is not the case when alpha particles of other energies are emitted from ^{241}Am. If we take, for example, the 5.486 MeV α-particle transition of Fig. 1.1, the decay energy, E_{decay}, would be the sum of the transition energy plus gamma-ray energy, E_γ, emitted from the daughter nucleus or

$$
\begin{aligned}
E_{decay} &= E_{trans} + E_\gamma \\
&= E_\alpha + E_{recoil} + E_\gamma \\
&= E_\alpha + (M_\alpha/M_{recoil})E_\alpha + E_\gamma \\
&= 5.486\,\text{MeV} + (0.0168851)(5.486\,\text{MeV}) + 0.059\,\text{MeV} \\
&= 5.486\,\text{MeV} + 0.0926\,\text{MeV} + 0.059\,\text{MeV} \\
&= 5.63\,\text{MeV} \qquad (1.7)
\end{aligned}
$$

The gamma-ray energy emitted from the daughter nucleus for the 5.486 MeV α-particle transition in ^{241}Am decay is found in Fig 1.1. Gamma-ray energy values of other radionuclides are available from Appendix A and reference tables (Michael Lederer *et al.*, 1978; Browne *et al.*, 1986, Firestone *et al.*, 1996).

As described in the previous paragraphs alpha particles are emitted with a certain quantum of energy as the parent nuclide decays to a lower energy state. The energy emitted from radionuclides as nuclear radiation can be described by a decay scheme such as that given in Fig. 1.1. Decay schemes are written such that the energy levels of the nuclides are plateaus along the ordinate, and these energy plateaus are distributed along the abscissa according to atomic number. The alpha particles, as the example shows (Fig. 1.1), are emitted with certain magnitudes of kinetic energy, which is most often expressed in units of megaelectron volts (MeV). The definition of MeV is given in Section IV.C of this chapter. The energies of alpha particles from most nuclear decay reactions fall within the range 1 to 10.5 MeV.

Alpha particles are emitted from unstable nuclei with discrete quanta of energy, often leaving the daughter nuclide at an excited energy state. In such cases, when the daughter nuclide occurs at an elevated energy state, it may reach the ground state via the emission of energy in the form of electromagnetic gamma radiation as illustrated in Fig. 1.1.

The nuclei of daughter atoms of alpha particle-emitting nuclides are often unstable themselves and may also decay by further alpha particle emission. Thus, alpha particle-emitting nuclides may consist of a mixture of radionuclides, all part of a decay chain, as illustrated in Fig. 1.38 further on in this chapter. Additional reading on radionuclide alpha decay is available from Das and Ferbel (1994).

Now consider what happens to an alpha particle that dissipates its kinetic energy by interaction with matter. Alpha particles possess a double positive charge due to the two protons present. This permits ionization to occur within a given substance (solid, liquid or gas) by the formation of ion pairs due to coulombic attraction between a traversing alpha particle and atomic electrons of the atoms within the material the alpha particle travels. The two neutrons of the alpha particle give it additional mass, which further facilitates ionization by coulombic interaction or even direct collision of the alpha particle with atomic electrons. The much greater mass of the alpha particle, 4 atomic mass units (u), in comparison with the electron (5×10^{-4} u) facilitates the ejection of atomic electrons of atoms through which it passes, either by direct collision with the electron or by passing close enough to it to cause its ejection by coulombic attraction. The ion pairs formed consist of the positively charged atoms and the negatively charged ejected electrons. The alpha particle continues along its path suffering, for the most part, negligible deflection by these collisions or coulombic interactions because of the large difference in mass between the particle and the electron. Thus, an alpha particle travels through matter producing thousands of ion pairs (see the following calculation) in such a fashion until its kinetic energy has been completely dissipated within the substance it traverses.

In air, an alpha particle dissipates an average of 35 eV (electron volts) of energy per ion pair formed. Before it stops, having lost its energy, an alpha particle produces many ion pairs. For example, as a rough estimate, a 5-MeV alpha particle will produce 1.4×10^5 ion pairs in air before coming to a stop:

$$\frac{5{,}000{,}000 \, \text{eV}}{35 \text{eV/ion pair}} = 1.4 \times 10^5 \text{ ion pairs in air}$$

The thousands of interactions between a traveling alpha particle and atomic electrons can be abstractly compared with a traveling bowling ball colliding with stationary ping-pong balls. Because of the large mass difference of the two, it will take thousands of ping-pong balls to stop a bowling ball. The additional stopping force of electrons is the binding energy of the atomic electrons.

The amount of energy required to produce ion pairs is a function of the absorbing medium. For example, argon gas absorbs approximately 25 eV per ion pair formed and a semiconductor material requires only 2–3 eV to produce an ion pair. Ionization is one of the principal phenomena utilized to detect and measure radionuclides and is treated in more detail in subsequent chapters. The energy threshold for ion pair formation in semiconductor materials is approximately 10 times lower than in gases, which gives semiconductor materials an important advantage as radiation detectors (see Chapter 4) when energy resolution in radioactivity analysis is an important factor.

In addition to ionization, another principal mechanism by which alpha particles and charged particles, in general, may impart their energy in matter is via electron excitation. This occurs when the alpha particle fails to impart sufficient energy to an atomic electron to cause it to be ejected from the atom. Rather, the atoms or molecules of a given material may absorb a portion of the alpha particle energy and become elevated to a higher energy state. Depending on the absorbing material, the excited atoms or molecules of the material may immediately fall back to a lower energy state or ground state by dissipating the absorbed energy as photons of visible light. This process, referred to as fluorescence, was first observed by Sir William Crookes in London in 1903 and soon confirmed by Julius Elster and Hans Geitel the same year in Wolfenbüttel, Germany. They observed fluorescence when alpha particles emitted from radium bombarded a zinc sulfide screen. In darkness, individual flashes of light were observed and counted on the screen with a magnifying glass with the screen positioned a few millimeters from the radium source. The phenomenon of fluorescence and its significance in the measurement of radionuclide tracers are discussed in subsequent chapters. Thus, as described in the previous paragraphs, alpha particles as well as other types of charged particles, dissipate their energy in matter mainly by two mechanisms, ionization and electron excitation.

Because the atomic "radius" is so very much bigger ($\approx 10^{-10}$ m) than the "radius" of the nucleus ($\approx 10^{-14}$ m), the interactions of alpha particles with matter via direct collision with an atomic nucleus are few and far between. In this case, though, the large mass of the nucleus causes deflection or ricocheting of the alpha particle via coulombic repulsion without generating any change within the atom. Such deflection was discovered in the early part of this century by Ernest Rutherford and his students Hans Geiger and Ernest Marsden, who bombarded very thin gold foil (only 6×10^{-5} cm thick) with alpha particles and observed the occasional deflection of an alpha particle by more than 90°, even directly backwards toward the alpha particle source. Lord Rutherford took advantage of this discovery to provide evidence that the greater mass of an atom existed in a minute nucleus. In his own words, Rutherford (1940) related in an essay

"It was quite the most incredible event that ever happened to me in my life. It was almost as incredible as if you fired a 15-inch shell at a piece of tissue paper and it came back and hit you. On consideration, I realized that the scattering backwards must be the result of a single collision, and when I made calculations I saw that it was impossible to get anything of that order of magnitude unless you took a system in which the greater part of the mass of the atom was concentrated in a minute nucleus."

Rutherford went even further to make use of this interaction to determine the nuclear radius of aluminum. By selecting a metal foil of low Z (aluminum, $Z = 13$) and thus low Coulomb barrier to alpha penetration, and applying alpha particles of high energy (7.7 MeV) whereby defined alpha particle-scattering at acute angle due to coulombic repulsion would begin to fail, Rutherford (1919, 1920) was able to demonstrate that the distance of closest approach of these alpha particles according to Coulomb's law was equivalent to the nuclear radius of aluminum, $\sim 5 \times 10^{-15}$ m.

Scattering of alpha particles at angles of less than 90° may occur by coulombic repulsion between a nucleus and a particle that passes in close proximity to the nucleus. These deflected particles continue traveling until sufficient energy is lost via the formation of ion pairs. The formation of ion pairs remains, therefore, the principal interaction between alpha particles and matter.

The high mass and charge of the alpha particle in relation to other forms of nuclear radiation give it greater ionization power but a poorer ability to penetrate matter. In air, alpha particles may travel only a few centimeters. This short range of travel varies depending on the initial energy of the particle. For example, a 5.5-MeV alpha particle, such as that emitted by the radionuclide ^{241}Am previously described, has a range of approximately 4 cm in dry air at standard temperature and pressure, as estimated by empirical formulae, such as Eqs. 1.8 and 1.9 provided below

$$R_{air} = (0.005E + 0.285)E^{3/2} \tag{1.8}$$

where R is the average linear range in cm of the alpha particle in air and E is the energy of the particle in MeV. The empirical formula is applied for alpha particles in the energy range 4–15 MeV. According to calculations of Fenyves and Haiman (1969), the ranges of alpha particles

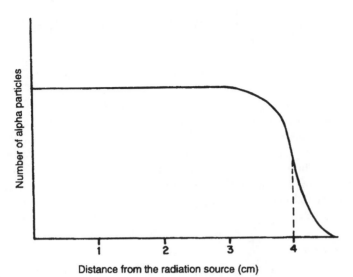

FIGURE I.2   Range of 5.5-MeV alpha particles in air.

with energies between 4 and 7 MeV can be estimated by using a simplified version of Eq. 1.8 as follows:

$$R_{air} = 0.3E^{3/2} \tag{1.9}$$

Ranges of alpha particles in air over a wider range of alpha particle energy can be obtained from Figures B.1 and B.2 of Appendix B. A thorough treatment of range calculations for charged particles is available from Fenyves and Haiman (1969). The approximate 4-cm range of 5.5-MeV alpha particles in air is illustrated in Fig. 1.2. There is no abrupt drop in the number of alpha particles detected at the calculated range of 4 cm owing to statistical variations in the number of collisions that the particles may have with air molecules and to variations in the amount of energy loss by the particles for each ion pair formed. After being halted, an alpha particle acquires two free electrons through coulombic attraction and is converted to helium gas.

In materials other than air, such as liquids and solids, the range of alpha particles is obviously much shorter owing to their higher densities, which enhance the number of collisions a particle may undergo per path length of travel. The range of alpha particles in liquids and solids may be approximated by comparison with ranges in air according to the formula

$$R_{cm} = 0.00032(A^{1/2}/\rho)R_{air} \tag{1.10}$$

described in a previous text (L'Annunziata, 1987), where R_{cm} is the average range in cm of the alpha particle in an absorber other than air, A is the atomic weight of the absorber, ρ is the absorber density in g cm^{-3}, and R_{air} is the calculated average linear range of the alpha particle in air (from Eq. 1.8 or 1.9). For example, the 5.5-MeV alpha particles emitted by ^{241}Am have a calculated linear range of only 2.4×10^{-3} cm or $24 \, \mu m$ in aluminum ($A = 27$ and $\rho = 2.69 \, g \, cm^{-3}$).

The linear ranges of alpha particles in liquids and solid materials are too short to measure with conventional laboratory instrumentation. The alternative is to express range in units of weight of absorber material per unit area, such as $mg \, cm^{-2}$, which is a measure of milligrams of absorber per square centimeter in the absorption path, or in other words, a measure of absorber thickness. If we multiply the linear range of the alpha particle measured in cm of absorber material by the density of the absorber in units of $mg \, cm^{-3}$, the range of the alpha particle in an absorber will be expressed in terms of the weight of absorber per unit area ($mg \, cm^{-2}$) as described by Eq. 1.11, as follows

$$R_{mg \, cm^{-2}} = (R_{cm})(\rho) \tag{1.11}$$

Where $R_{mg \, cm^2}$ is the range of alpha particles of a given energy in units of $mg \, cm^{-2}$, also referred to as mass thickness units or material surface density, R_{cm} is the linear range of the alpha particles, and ρ is the absorber density. For example, the linear range of the 5.5 MeV alpha particles in aluminum calculated above with Eq. 1.10 is converted to range in mass thickness units according to Eq. 1.11 as follows

$$R_{mg \, cm^{-2}} = (2.4 \times 10^{-3} \, cm)(2690 \, mg \, cm^{-3}) = 6.4 \, mg \, cm^{-2}$$

Therefore, the mass thickness of $6.4 \, mg \, cm^{-2}$ of aluminum absorber is sufficient to absorb alpha particles of 5.5 MeV energy.

Ranges of alpha particles as well as other charged particles such as protons and deuterons of a given energy in absorber elements of atomic number $Z > 10$ in units of absorber mass thickness can be calculated directly by comparison to the calculated range of the same charged particles of the same energy in air according to the following formula described by Friedlander et al. (1964)

$$\frac{R_Z}{R_{air}} = 0.90 + 0.0275Z + (0.06 - 0.0086Z) \log \frac{E}{M} \tag{1.12}$$

where R_Z is the range of the charged particle in mass thickness units, $mg \, cm^{-2}$, R_{air} is the range of the charged particle in air in the same mass thickness units, Z is the atomic number of the absorber element, E is the particle energy in MeV, and M is the mass number of the particle (i.e., 1 for protons, 2 for deuterons, and 4 for alpha particles). For example, if we use the empirical formula provided above (Eq. 1.12) to calculate the range of 5.5 MeV alpha particles ($M = 4$) in aluminum ($Z = 13$), we obtain the value of $R_Z = 6.1 \, mg \, cm^{-2}$, which is in close agreement to the mass thickness range calculated previously. In this example, Eq. 1.12 requires the value of R_{air} for 5.5 MeV alpha particles, which is determined according to Eq. 1.11 as the product of the 5.5 MeV alpha particle linear range in air (previously calculated) and the density of air at STP ($\rho = 1.226\,\mathrm{mg\,cm^{-3}}$), that is, $R_{\mathrm{air}} = (4\,\mathrm{cm})(1.226\,\mathrm{mg\,cm^{-3}}) = 4.90\,\mathrm{mg\,cm^{-2}}$. The formula provided by Eq. 1.12 is applicable to charged particles over a wide range of energies (approximately over the range 0.1–1000 MeV) and for absorber elements of $Z > 10$. For lighter absorber elements the term $0.90 + 0.0275Z$ is replaced by the value 1.00 with the exception of hydrogen and helium, where the value of 0.30 and 0.82 are used, respectively (Friedlander *et al.*, 1964).

Where alpha particles alone are concerned, the range in mass thickness units can be calculated according to Eq. 1.13 described by Ehman and Vance (1991), as follows

$$R_{\mathrm{mg\,cm^{-2}}} = 0.173E^{3/2}A^{1/3} \tag{1.13}$$

where E is the energy of the alpha particle in MeV and A is the atomic weight of the absorber. If we continue to use the 5.5 MeV alpha particles emitted from $^{241}\mathrm{Am}$ as an example, we can calculate their range in mass thickness units in aluminum according to Eq. 1.13 as follows

$$R_{\mathrm{mg\,cm^2}} = 0.173(5.5)^{3/2}(27)^{1/3} = 6.6\,\mathrm{mg\,cm^{-2}}.$$

Ranges reported in mass thickness units ($\mathrm{mg\,cm^{-2}}$) of absorber can be converted to linear range (cm) in that same absorber material from the absorber density (ρ) from the relationship described in Eq. 1.11 or

$$R_{\mathrm{cm}} = R_{\mathrm{mg\,cm^{-2}}}/\rho \tag{1.14}$$

For example, the linear range of the 5.5 MeV alpha particles in aluminum ($\rho = 2.69\,\mathrm{g\,cm^{-3}}$) is calculated as

$$R_{\mathrm{cm}} = 6.6\,\mathrm{mg\,cm^{-2}}/2690\,\mathrm{mg\,cm^{-3}} = 0.0024\,\mathrm{cm} = 24\,\mu\mathrm{m}.$$

When the absorber material is not a pure element, but a molecular compound (e.g., water, paper, polyethylene, etc.) or mixture of elements, such as an alloy, the ranges of alpha particles in the absorber are calculated according to Eq. 1.15 on the basis of the atomic weights of the elements and their percent composition in the absorber material or, in other words, the weight fraction of each element in the complex material. Thus, the range in mass-thickness units for alpha particles in absorbers consisting of compounds or mixtures of elements is calculated according to the equation

$$\frac{1}{R_{\mathrm{mg\,cm^{-2}}}} = \frac{w_1}{R_1} + \frac{w_2}{R_2} + \frac{w_3}{R_3} + \cdots + \frac{w_n}{R_n}. \tag{1.15}$$

where $R_{\mathrm{mg\,cm^2}}$ is the range of the alpha particles in mass-thickness of the complex absorber material, and $w_1, w_2, w_3, \ldots, w_n$ are the weight fractions of each element in the absorber, and $R_1, R_2, R_3, \ldots, R_n$ are the ranges in $mg\,cm^{-2}$ of the alpha particle of defined energy in each element of the absorber. For example, the range of 5.5 MeV alpha particles in Mylar (polyethylene terephthalate) in units of mass thickness are calculated as follows

$$\frac{1}{R_{mg\,cm^{-2}}} = \frac{w_C}{R_C} + \frac{w_H}{R_H} + \frac{w_O}{R_O}$$

where w_C, w_H, and w_O are the weight fractions of carbon, hydrogen, and oxygen, respectively, in Mylar and R_C, R_H, and R_O are the mass-thickness ranges of the alpha particles in pure carbon, hydrogen, and oxygen, respectively. The ranges of 5.5 MeV alpha particles in carbon, hydrogen and oxygen are calculated according to Eq. 1.13 as

$$R_C = 0.173(5.5)^{3/2}(12)^{1/3} = 5.10\,mg\,cm^{-2}$$
$$R_H = 0.173(5.5)^{3/2}(1)^{1/3} = 2.23\,mg\,cm^{-2}$$
$$R_O = 0.173(5.5)^{3/2}(16)^{1/3} = 5.62\,mg\,cm^{-2}$$

The weight fractions of the carbon, hydrogen, and oxygen in Mylar $[-(C_{10}H_8O_4)_n-]$ are calculated as

$$w_C = (12 \times 10)/192 = 0.625$$
$$w_H = (1 \times 8)/192 = 0.042$$
$$w_O = (16 \times 4)/192 = 0.333$$

The calculated ranges of the 5.5 MeV alpha particles in each element and the values of the weight fractions of each element in Mylar can now be used to calculate the alpha particle range in Mylar in mass-thickness units according to Eq. 1.15 as

$$\frac{1}{R_{Mylar}} = \frac{0.625}{5.10} + \frac{0.042}{2.23} + \frac{0.333}{5.62} = 0.200$$
$$R_{Mylar} = 1/0.200 = 5.0\,mg\,cm^{-2}$$

The linear range of these alpha particles in Mylar are obtained from range in mass thickness units and the density of Mylar ($\rho = 1.38\,g\,cm^{-3}$) as

$$R_{cm} = 5.0\,mg\,cm^{-2}/1380\,mg\,cm^{-3} = 0.0036\,cm = 36\,\mu m.$$

To provide illustrative examples the values of the ranges of 5.5 MeV alpha particles in units of mass thickness of various absorber materials are provided in Table 1.1. These values represent the milligrams of absorber per square centimeter in the alpha particle absorption path. It can be difficult to envisage alpha particle distance of travel from the values of range when express in units of mass thickness. However, it is intuitively obvious that, the

TABLE I.I   **Ranges of 5.5-MeV Alpha Particles in Various Absorbers in Units of Surface Density or Mass Thickness**

Water[a]	Paper[a,b]	Aluminum[c]	Copper[c]	Gold[c]
$4.8\,mg\,cm^{-2}$	$4.9\,mg\,cm^{-2}$	$6.6\,mg\,cm^{-2}$	$8.9\,mg\,cm^{-2}$	$12.9\,mg\,cm^{-2}$

[a]Calculated with empirical formula provided by Eq. 1.15 on the basis of the weight fraction of each element in the absorber.
[b]Cellulose $(C_6H_{10}O_5)_n$ calculated on the basis of the weight fraction of each element in the monomer.
[c]Calculated with empirical formula provided by Eq. 1.13.

TABLE I.2   **Linear Ranges of 5.5-MeV Alpha Particles in Various Absorbers in Units of cm and μm or 10^{-6} m**

Air[a]	Water[b]	Mylar[b,c]	Paper[b,d]	Aluminum[b]	Copper[b]	Gold[b]
4 cm	0.0048 cm	0.0036 cm	0.0034 cm	0.0024 cm	0.001 cm	0.00075
40,000 μm	48 μm	36 μm	34 μm	24 μm	10 μm	7.5 μm

[a]Calculated with empirical formula provided by Eqs. 1.8 and 1.9.
[b]Calculated by dividing the range in mass thickness by the absorber density according to Eq. 1.14.
[c]Polyethylene terephthalate, $\rho = 1.38\,g\,cm^{-3}$.
[d]Cellulose $(C_6H_{10}O_5)_n$ $\rho = 1.45\,g\,cm^{-3}$.

greater the charge on the nucleus of the absorber (i.e., absorber atomic number, Z), the greater the atomic weight of the absorber (A); and the greater the absorber density (ρ), the shorter will be the path length of travel of the alpha particle through the absorber. This is more evident from the calculated values of linear range of 5.5 MeV alpha particles in various gaseous, liquid and solid absorbers provided in Table 1.2. From the linear ranges we can see that 5.5-MeV alpha particles could not pass through fine commercial aluminum or copper foils 0.0025 cm thick. Although commercial paper varies in thickness and density, the linear range in paper calculated in Table 1.2 illustrates that 5.5 MeV alpha particles would not pass through 0.0034 cm thick paper, which has an average density value of $1.45\,g\,cm^{-3}$. Also, the alpha particles of the same energy would not pass through a layer of Mylar only 0.0036 cm thick. Mylar is a polymer sometimes used as a window for gas ionization detectors. From our previous calculations in this chapter we can see that a Mylar window of mass thickness $5\,mg\,cm^{-2}$ would not allow 5.5-MeV alpha particles to pass into the gas ionization chamber. A sample emitting such alpha particles would have to be placed directly into the chamber in a windowless fashion to be detected and counted.

From the above treatment it is clear that the range of alpha particle-travel depends on several variables including (i) the energy of the alpha particle, (ii) the atomic number and atomic weight of the absorber, and (iii) the density of the absorber. The higher the alpha particle energy, the greater will be its penetration power into or through a given substance as more coulombic interactions of the alpha particle with the electrons of the absorber will be

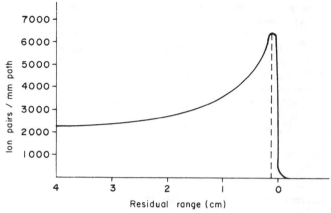

FIGURE 1.3 Specific ionization of an alpha particle in air along its range of travel.

required to dissipate its energy before coming to rest. Also, if we consider an alpha particle of given energy, their ranges will be shorter in absorbers of higher atomic number or atomic weight, as the absorber atoms will contain a higher number of atomic electrons, and consequently increase the number of coulombic interactions of the alpha particle per path length of travel.

As the alpha particle travels through air and undergoes energy loss via numerous collisions, the velocity of the particle obviously diminishes. At reduced velocity and consequently reduced momentum, an alpha particle is more affected by coulombic attraction within the vicinity of a given atom. Progressive reduction in the velocity of travel of the alpha particle therefore results in an increase in the number of ion pairs produced per millimeter of path length of travel. The increase in ionization per path length of travel of an alpha particle is illustrated in Fig. 1.3. The highest specific ionization (number of ion pairs formed per millimeter of path) occurs shortly before termination of the alpha particle's travel, some 2 or 3 mm before the end of its range.

B. Negatrons

A negatron or negative beta particle (β^-) is an electron emitted from the nucleus of a decaying radionuclide that possesses an excess of neutrons or, in other words, a neutron/proton (n/p) imbalance. (See Section II.C.1 for a brief discussion of n/p ratios and nuclear stability.) The nuclear instability caused by the n/p imbalance results in the conversion of a neutron to a proton within the nucleus, where the balance of charge is conserved by the simultaneous formation of an electron (negatron) according to the equation

$$n \rightarrow p^+ + \beta^- + \bar{\nu}. \tag{1.16}$$

A neutrino (ν), which is a particle of zero charge, accompanies beta-particle emission. The neutrino can be identified further as two types with opposite spin, namely, the antineutrino ($\bar{\nu}$), which accompanies negative beta-particle (negatron) emission and the neutrino (ν), which accompanies positive beta-particle (positron) emission (See Section II.C of this chapter). Because the neutrino and antineutrino have similar properties with the exception of spin, it is common to use the word "neutrino" to simplify references to both particles. The explanation for the neutrino and its properties, also emitted from the decaying nucleus, is given further on in this section. The electron formed cannot remain within the nucleus and is thus ejected as a negatron or negative beta particle, β^-, with a maximum energy equivalent to the slight mass difference between the parent and daughter atoms less the mass of the beta particle, antineutrino or neutrino in the case of positron emission, and any gamma-ray energy that may be emitted by the daughter nucleus if it is left in an excited energy state (see Section III.B of this chapter). Tritium (^{3}H), for example, decays with β^- emission according to the following:

$$^3_1\text{H} \rightarrow {}^3_2\text{He} + \beta^- + \bar{\nu} + 0.0186 \, \text{MeV} \tag{1.17}$$

The value of 0.0186 MeV (megaelectron volts) is the maximum energy the beta particle may possess. The unstable tritium nucleus contains two neutrons and one proton. The transformation of a neutron to a proton within the tritium nucleus results in a charge transfer on the nucleus from +1 to +2 without any change in the mass number. Although there is no change in the mass number, the mass of the stable helium isotope produced is slightly less than that of its parent tritium atom. Equations 1.18–1.23 illustrate other examples of β^- decay, and many beta particle-emitting nuclides are listed in the Appendix.

$$^{14}_6\text{C} \rightarrow {}^{14}_7\text{N} + \beta^- + \bar{\nu} + 0.156 \, \text{MeV} \tag{1.18}$$

$$^{32}_{15}\text{P} \rightarrow {}^{32}_{16}\text{S} + \beta^- + \bar{\nu} + 1.710 \, \text{MeV} \tag{1.19}$$

$$^{35}_{16}\text{S} \rightarrow {}^{35}_{17}\text{Cl} + \beta^- + \bar{\nu} + 0.167 \, \text{MeV} \tag{1.20}$$

$$^{36}_{17}\text{Cl} \rightarrow {}^{36}_{18}\text{Ar} + \beta^- + \bar{\nu} + 0.714 \, \text{MeV} \tag{1.21}$$

$$^{45}_{20}\text{Ca} \rightarrow {}^{45}_{21}\text{Sc} + \beta^- + \bar{\nu} + 0.258 \, \text{MeV} \tag{1.22}$$

$$^{89}_{38}\text{Sr} \rightarrow {}^{89}_{39}\text{Y} + \beta^- + \bar{\nu} + 1.490 \, \text{MeV} \tag{1.23}$$

The energies of beta particle-decay processes are usually reported as the maximum energy, E_{max}, that the emitted beta particle or antineutrino may possess. The maximum energy is reported because beta particles are emitted from radionuclides with a broad spectrum of energies. A typical spectrum is illustrated in Fig. 1.4. Unlike alpha particles, which have a discrete energy, beta particles have a wide spectrum of energies ranging from zero to E_{max}.

The majority of beta particles emitted have energies of approximately $1/3$ (E_{max}). Only a very small portion of the beta particles are emitted with the maximum possible energy from any radionuclide sample. In 1930

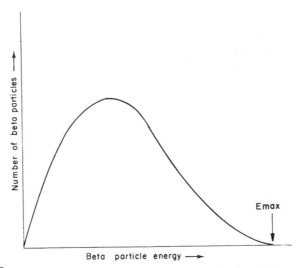

FIGURE 1.4 General energy spectrum of beta particles.

Wolfgang Pauli was the first to postulate why beta particles were not emitted with fixed quanta of energy, quite the contrary to what is observed in alpha particle emission. He proposed the existence of an elusive, neutral, and almost massless particle in a letter to Lise Meitner and Hans Geiger. The neutrino was considered elusive, because if it existed, its zero charge and near zero rest mass would make the neutrino undetectable by conventional means and allow a neutrino to penetrate matter easily and even pass through the entire earth without causing a single interaction. It is the neutrino that would be emitted simultaneously with the beta particle from the decaying nucleus and share the energy of decay with the beta particle. For example, if a beta particle was emitted from tritium (decay energy $= 0.0186$ MeV) with an energy of 0.0086 MeV, the accompanying neutrino would possess the remaining energy of 0.01 MeV, that is, the decay energy less the beta-particle energy (0.0186–0.0086 MeV). Consequently, if we observe any number of beta particles emitted from a tritium sample or other beta-emitting nuclide sample (e.g., ^{14}C, ^{32}P, ^{90}Sr), they would possess different energies and display an energy spectrum from zero to E_{max} as illustrated in Fig. 1.4.

With Pauli's postulation of the neutral particle, Enrico Fermi elaborated the beta-decay theory in 1934 and coined the term "neutrino" from Italian language meaning "little neutral one." The particle remained elusive until the observation of the neutrino was demonstrated by Reines and Cowan in 1956 (see Reines and Cowan 1956, 1957 Cowan *et al.*, 1956; Reines, 1960, 1979, 1994). They confirmed the existence of the neutrino by demonstrating inverse beta decay where an antineutrino interacts with a proton to yield a neutron and positron

$$\bar{\nu} + p^+ \rightarrow n + \beta^+ \qquad (1.24)$$

They used a tank of water containing a solution of ^{113}CdCl$_2$. Neutrinos interacted with the protons of the water to produce neutrons and positrons.

Some of the neutrons produced would be absorbed by the ^{113}Cd with the concomitant emission of characteristic gamma radiation. In coincidence, they observed two 511 keV gamma rays, which originate when a positron comes to rest in the vicinity of an electron, its antiparticle, which result in the annihilation of two electrons into two gamma-ray photons of energy equivalent to the electron masses, 0.511 MeV. In the same year Lee and Yang (1956) proposed that neutrinos and antineutrinos possessed left-handed and right-handed spins, respectively. Reverse beta decay remains an important nuclear process utilized in the measurement of solar neutrinos today (Gratta and Wang, 1999).

Since its inception by Pauli in 1930 up to recent years, the neutrino or antineutrino had been thought to be almost massless or to possess a near-zero rest mass. It was not until June 5, 1998 was it announced by the Super-Kamiokande Collaboration, including scientists from 23 institutions in Japan and the United States, at the "Neutrino 98" International Physics Conference in Takayama, Japan, that neutrinos possessed a definite mass (Gibbs, 1998; Kearns *et al.* 1999; Kesterbaum, 1998; Nakahata, 2000). The mass was not reported, but evidence was provided that the neutrino did possess mass although it was considered to be "very small," at least 0.07 eV, which would be less than a millionth of the electron mass. Evidence for the neutrino mass was provided by demonstrating that neutrinos can "oscillate" from one type into another (i.e., electron,- muon-, and tau-neutrinos) as they travel through space and matter. Oscillation is the changing of neutrino types back and forth from one type to another, and this could occur only if the neutrino possessed mass.

More recently at the "Neutrino 2000" Conference held at Sudbury, Canada June 16–21, 2000, groups from the University of Mainz, Germany (Bonn *et al.*, 2000) and Institute for Nuclear Research, Moscow (Lobashev *et al.*, 2000) reported the mass of the neutrino to be 2.2 and 2.5 eV/c^2, respectively at 95% confidence levels. It is common to express subatomic particle mass in units of energy on the basis of equivalence of mass and energy ($E = mc^2$), so that the particle mass m is measured in units of E/c^2 or eV/c^2. To put the mass of the neutrino in perspective, we can take the experimental value of the neutrino rest mass, $m_\nu = 2.2$ eV/c^2, from the University of Mainz Group and convert this to kilograms as follows:

By definition 1 eV $= 1.60 \times 10^{-19}$ J, and from the equation $E = mc^2$

$$m_\nu = E/c^2 = (2.2 \text{ eV})(1.60 \times 10^{-19} \text{ J/eV})/(3.00 \times 10^8 \text{ m/s})^2 = 3.9 \times 10^{-36} \text{ kg}.$$

If we compare the rest mass of the neutrino, m_ν, to that of the miniscule electron, m_e, we see that the neutrino mass is approximately 4 millionths that of the electron or

$$m_\nu/m_e = 3.9 \times 10^{-36} \text{ kg}/9.1 \times 10^{-31} \text{ kg} = 4.2 \times 10^{-6}$$

Owing to the very low mass of the β^- electron compared with the alpha particle, it travels at a much higher velocity than an alpha particle of equivalent energy. Because of its greater velocity, lower mass and lower charge, the specific ionization produced in air by the traveling beta particle is much lower (by a factor of a thousand) than that of an alpha particle of equivalent energy.

Like the alpha particle, the beta particle interacts with matter via (i) ionization and (ii) electron orbital excitation as it dissipates its kinetic energy. A third mechanism of interaction with matter, which distinguishes the beta particle, is radiative energy dissipation via Bremsstrahlung production (see Section III.F). Thus as described by Turner (1995) the stopping power for beta particles (β^- or β^+) is the sum of the collisional and radiative contributions or

$$\left(-\frac{dE}{dx}\right)_{tot}^{\pm} = \left(-\frac{dE}{dx}\right)_{col}^{\pm} + \left(-\frac{dE}{dx}\right)_{rad}^{\pm} \tag{1.25}$$

where the superscript $\pm$ refers to positively or negatively charged electrons. The radiative contribution, that is, the absorption of beta-particle energy with the concomitant emission of Bremsstrahlung radiation is significant with high-energy beta particles (e.g., ^{32}P or ^{80}Y beta-particle emissions) in absorbers of high atomic number (e.g., Pb-glass). Bremsstrahlung radiation is discussed in Section III.F of this chapter.

Collisional interactions of beta particles are somewhat different than those that occur with alpha particles. A beta particle may collide with an orbital electron or come into close proximity to it and cause the electron to be ejected, resulting in the formation of an ion pair. Considerable scattering of beta particles occurs in such collisions because the mass of the beta particle is equivalent to that of an atomic electron. This is in direct contrast to the alpha particle, which, for the most part, retains a relatively undeviating path while passing through matter and interacting with atomic electrons. The mass equivalence of beta particles and electrons is an important factor that gives bombarding beta particles the power to impart a major portion of their kinetic energy to atomic electrons in a single collision. The atomic electrons ejected upon beta particle collisions themselves cause ionization in a similar fashion. This is referred to as secondary ionization, and the ionization caused by the first beta particle–electron collisions is classified as primary ionization. Because the major portion of beta particle energy may be imparted to an atomic electron upon collision, secondary ionization may account for as much as 80% of the total ionization produced in a given material bombarded by beta particles.

The probability of beta-particle interactions with atomic electrons increases with the density of the absorbing material. Beta particle-absorption is consequently proportional to the density and thickness of an absorber. When we compare substances of similar atomic composition, we find that the range of beta particles (β^- or β^+) expressed in mass thickness units ($mg\,cm^{-2}$) are approximately the same. For example, Fig. B.3 of Appendix B provides a curve where the range in units of $g\,cm^{-2}$ in substances of low atomic number can be estimated for beta particles of energies from 0.01 to 10 MeV. The range of beta particles expressed in terms of surface density or mass thickness ($g\,cm^{-2}$) of absorber can be converted to absorber thickness (cm) when the absorber density ($g\,cm^{-3}$) is known. Several empirical formulas exist for calculating beta particle ranges and are solved on the basis of the E_{max} of the beta particle. The formulas reported by Glendenin (1948) are

$$R = 0.542E - 0.133 \quad \text{for } E > 0.8\,\text{MeV} \tag{1.26}$$

and

$$R = 0.407E^{1.38} \quad \text{for } 0.15\,\text{MeV} < E < 0.8\,\text{MeV} \tag{1.27}$$

where R is the beta particle-range in $g\,cm^{-2}$ and E is the energy of the beta particle (i.e., E_{max}) in MeV. Also, the following empirical formula of Flammersfield (1946) described by Paul and Steinwedel (1955) can be used:

$$R = 0.11\left(\sqrt{1 + 22.4E^2} - 1\right) \quad \text{for } 0 < E < 3\,\text{MeV} \tag{1.28}$$

This formula provides calculated ranges in units of $g\,cm^{-2}$ in close agreement to those obtained from Eqs. 1.26 and 1.27 or those found from Fig. B.3 in Appendix B.

According to Eq. 1.26, a 1.0-MeV beta particle has a calculated range of $0.409\,g\,cm^{-2}$. This value may be divided by the density, ρ, of the absorber material to provide the range in centimeters of absorber thickness. Thus, it can be estimated that a 1.0-MeV beta particle travels approximately 334 cm in dry air ($\rho = 0.001226\,g\,cm^{-3}$ at STP), 0.40 cm in water ($\rho = 1.00\,g\,cm^{-3}$) and 0.15 cm in aluminum ($\rho = 2.7\,g\,cm^{-3}$). The effect of absorber density on beta particle range is obvious from the foregoing examples, which demonstrate that 1 cm of dry air has about the same stopping power as 0.004 mm of aluminum.

The range of beta particles in matter is considerably greater than that of alpha particles of the same energy. Again, this is due to the lower mass, lower charge, and higher velocity of travel of the beta particle in comparison with an alpha particle of equivalent energy. The significance of this difference may be appreciated by reference to Table 1.3, in which the alpha particle and beta particle and/or electron ranges in air as a function of particle energy are compared. To put this data into historical perspective, it is interesting to recall the origin of the names "alpha- and beta-radiation." Before alpha and beta particles were characterized fully, Ernest Rutherford carried out experiments in 1898 that demonstrated two types of radiation existed; one radiation that was most easily absorbed by matter and another that possessed a greater penetrating power. Out of convenience, he named these radiations as "alpha" and "beta." Not much later P. V. Villard in France discovered in 1900 a yet more penetrating radiation, that was named "gamma" in harmony with the nomenclature coined by Rutherford.

It is important to emphasize that, although all beta particles can be completely absorbed by matter, the shields we select can be of great consequence. Hazardous bremsstrahlung radiation can be significant when high-energy beta particles interact with shields of high atomic number.

TABLE I.3 Ranges of Alpha and Beta Particles (or Electrons) of Various Energies in Air

	Range (mg cm^{-2} Air)		Range (cm Air)^c	
Energy (Mev)	Alpha particle^a	Beta particle^b	Alpha particle	Beta particle
0.1	0.013^d	13	0.010	11
0.5	0.4	163	0.3	133
1.0	0.6	412	0.5	336
1.5	0.9	678	0.7	553
2.0	1.2	946	1.0	772
2.5	1.6	1217	1.3	993
3.0	2.1	1484	1.7	1210
4.0	3.1	2014	2.5	1643
5.0	4.2	2544	3.5	2075
6.0	5.6	3074	4.6	2507
7.0	7.2	3604	5.9	2940
8.0	8.7	4134	7.1	3372

^{a}From curve provided in Fig. B.1 Appendix B with the exception of 0.1 MeV particle energy.
^{b}Calculated from the formulas for range (R) in units of g cm^{-2}, $R = 0.412E^{1.27-0.0945\ln E}$ for $0.01 \leq E \leq 2.5$ MeV and $R = 0.530E - 0.106$ for $E > 2.5$ MeV. (See Fig. B.3 of Appendix B.)
^{c}Calculated from the range in mass thickness units (mg cm^{-2}) and the density of dry air at STP, $\rho_{air} = 1.226$ mg cm^{-3} according to Eq. 1.14.
^{d}Calculated from Eqs. 1.13 and 1.15 using weight averages of elements in air according to the following: 78.06% N, 21% O, 0.93% Ar and 0.011% C.

The phenomenon of bremsstrahlung production is discussed further in Section III.F of this chapter.

C. Positrons

In contrast to negatron emission from nuclei having neutron/proton (n/p) ratios too large for stability, positrons, which consist of positively charged electrons (positive beta particles), are emitted from nuclei having n/p ratios too small for stability, that is, those which have an excess of protons. (See Section II.C.1 for a brief discussion of n/p ratios and nuclear stability.)

To attain nuclear stability, the n/p ratio is increased. This is realized by a transformation of a proton to a neutron within the nucleus. The previously discussed alteration of a neutron to a proton in a negatron-emitting nuclide (Eq. 1.16) may now be considered in reverse for the emission of positrons. Equation 1.29 illustrates such a transformation

$$p^+ \rightarrow n + \beta^+ + \nu \qquad (1.29)$$

^{58}Co may be cited as an example of a nuclide that decays by positron emission:

$$^{58}_{27}\text{Co} \rightarrow \, ^{58}_{26}\text{Fe} + \beta^+ + \nu \qquad (1.30)$$

Note that the mass number does not change but the charge on the nucleus (Z number) decreases by 1. As in negatron emission, a neutrino, v, is emitted simultaneously with the positron (beta particle) and shares the decay energy with the positron. Thus, positrons, like negatrons emitted from a given radionuclide sample, may possess a broad spectrum of energies from near zero to E_{max} as illustrated in Fig. 1.4.

Decay by positron emission can occur only when the decay energy is significantly above 1.02 MeV. This is because two electrons of opposite charge are produced (β^+, β^-) within the nucleus, and the energy equivalence of the electron mass is 0.51 MeV (see Section IV.C of this chapter). The positive electron, β^+, is ejected from the nucleus and the negative electron, β^-, combines with a proton to form a neutron:

$$\beta^- + p^+ \rightarrow n \tag{1.31}$$

Thus, the E_{max} of a positron emitted from a nucleus is equivalent to the mass difference of the parent and daughter nuclides, less the mass of the positron and neutrino (albeit, the neutrino mass is very small compared to the mass of the positron) emitted from the nucleus (see equivalence of mass and energy, Section IV.C) and less any gamma-ray energy of the daughter nuclide if left in an excited state (see Section III.B of this chapter).

From the table of nuclides in Appendix A it is possible to cite specific examples of the n/p imbalance in relation to negatron and positron emission. Figure 1.5 illustrates the relative positions of the stable nuclides ^{12}C, ^{13}C,

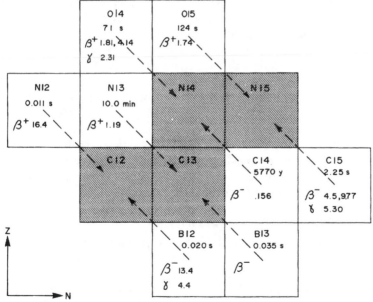

FIGURE I.5 A segment of the chart of the nuclides showing the relative positions of some stable (shaded) and unstable nuclides. The ordinate Z and abscissa N represent the number of protons (atomic number) and the number of neutrons within the nucleus, respectively.

^{14}N, and ^{15}N and of their neighboring radionuclides. The nuclides are positioned as a function of the number of protons, Z, and the number of neutrons, N, in their respective nuclei. Dashed arrows are placed through the blocks that segregate radionuclides interrelated with common daughter nuclides resulting from β^- or β^+ decay processes. For example, the stable nuclide ^{12}C of atomic number 6 has a nucleus with an n/p ratio of 6/6. However, the nuclide ^{12}N of atomic number 7 has an unstable n/p ratio of 5/7, an excess of protons. Thus, this nuclide decays via positron emission according to the equation

$$^{12}_{7}\text{N} \rightarrow ^{12}_{6}\text{C} + \beta^+ + \nu \tag{1.32}$$

to ^{12}C by positron emission as indicated by a dashed arrow of Fig. 1.5.

The nuclide ^{12}B of atomic number 5 has the unstable n/p ratio of 7/5, an excess of neutrons. This nuclide thus decays to ^{12}C by negatron emission according to the equation

$$^{12}_{5}\text{B} \rightarrow ^{12}_{6}\text{C} + \beta^- + \bar{\nu} \tag{1.33}$$

Similar reasoning may be used to explain positron and negatron decay of the unstable nuclides shown in Fig. 1.5 to the stable products ^{13}C, ^{14}N, and ^{15}N. The interrelationship between β^- and β^+ decay leading to the formation of stable nuclides is to be found throughout the chart of the nuclides; however, as the atomic number increases ($Z > 20$) the n/p ratio of the stable nuclides exceeds 1.0 (see the following Section II.C.1).

Positrons dissipate their energy in matter via the same mechanisms as previously described for negatrons, which is understandable, as both are electrons. The stopping powers and ranges of positrons are virtually identical to negatrons and electrons over the broad energy range of $0.03–10^3$ MeV (Turner, 1995). Although two equations (Eqs. 1.131 and 1.132) are cited in Section V.A for calculating the ionization-excitation stopping powers for negatrons and positrons due to collision interactions with absorbers, their difference as noted by Tsoulfanidis (1995) is due only to the second term in the brackets of these two equations, which is much smaller than the logarithmic term, and consequently the differences between negatron and positron stopping powers do not exceed 10%. However, positrons are unique in that these particles produce annihilation gamma radiation in matter discussed in Section III.C of this chapter.

I. N/Z Ratios and Nuclear Stability

In Sections II.B and II.C of this chapter we discussed negatron and positron decay as processes whereby unstable nuclei may achieve stability via neutron or proton transformations, respectively. These processes in the nucleus of the radionuclide result in a change in the neutron/proton or N/Z ratio of the nucleus.

If we look throughout The Chart of The Nuclides we will notice that the stable nuclides of low atomic number will have a N/Z ratio of approximately 1. However, as the atomic number increases ($Z > 20$), the N/Z ratio of the stable nuclides increases gradually and reaches as high as approximately 1.5 (e.g., $^{209}_{83}$Bi, $Z = 83$, $N/Z = 1.518$). Furthermore, there are no stable nuclides of atomic number greater than 83.

The nature of nuclear forces and the relationship of N/Z ratio to nuclear stability are discussed in detail by Serway *et al.* (1997) and Sundaresan (2001). In brief, the importance of N/Z ratio to nuclear stability is explained by the fact that there exists a short-range attractive nuclear force, which extends to a distance of ≈ 2 fm (2 fermi or 2×10^{-15} m). This attractive force has charge independence and is a consequence of the relative spins of the protons and neutrons and their relative positions in the nucleus. These binding exchange forces exist therefore, regardless of charge on the particles, between two protons, two neutrons, and a proton and neutron. While the attractive nuclear forces will tend to hold the nucleus together there exists, at the same time, repelling coulombic forces between the positively charged protons that act to force them apart. For nuclides of low Z, the attractive nuclear forces exceed the repelling coulombic forces when $N \approx Z$. However, increasing the number of protons (e.g., $Z > 20$) further increases the strength of the repelling coulombic forces over a larger nucleus, which will tend to force the nucleus apart. Therefore, additional neutrons, $N > Z$, provide additional attractive nuclear forces needed to overcome the repelling forces of the larger proton population. As the atomic number increases further, $Z > 83$, all nuclides are unstable. Even though N/Z ratios reach 1.5, nuclear stability is not achieved when the number of protons in the nucleus exceeds 83.

2. Positron Emission versus Electron Capture

Another mechanism by which an unstable nucleus can increase its n/p ratio is via the capture by the nucleus of a proximate atomic electron (e.g., K- or L-shell electron). The absorbed electron combines with a proton to yield a neutron within the nucleus as follows:

$$e^- + p^+ \rightarrow n + \nu + \text{inner bremsstrahlung}$$
$$+ \text{ x-rays} + \text{Auger electrons} + (\gamma) \qquad (1.34)$$

The decay process is known as electron capture (EC), or sometimes referred to as K capture, because most of the electrons are captured from the K shell, which is closest to the nucleus. A neutrino, ν, is emitted and this is accompanied by the emission of internal bremsstrahlung, which is a continuous spectrum of electromagnetic radiation that originates from the atomic electron as it undergoes acceleration toward the nucleus. Unlike the beta-decay process, which results in the emission of a neutrino from the nucleus with a broad spectrum of energies, the neutrino emitted from the EC decay process does not share the transition energy with another particle and, therefore, it is emitted with a single quantum of energy equal to the transition energy less the atomic electron binding energy. The capture of an atomic electron by the nucleus leaves a vacancy in an electron shell, and this is usually filled by an electron from an outer shell, resulting in the production of x-radiation (see Sections III.E and F). The electron that fills the vacancy leaves yet another vacancy at a more distant shell. A cascade of electron vacancies and subsequent filling of vacancies from outer electron shells occurs with the production of x-rays characteristic of the daughter atom. The x-rays will either travel out of the atom or interact with orbital electrons to eject these as Auger electrons. Gamma radiation is illustrated in the above Eq. 1.34, because it is emitted only when the daughter nuclide is left at an unstable elevated energy state (see Fig. 1.19 and Section III.B).

The electron capture decay process may compete with β^+ emission. That is, some radionuclides may decay by either electron capture or, β^+ emission. As discussed previously, positron emission requires a transition energy of at least 1.02 MeV, the minimum energy required for pair production in the nucleus (i.e., two electron rest mass energies or 2×0.511 MeV). Positron emission, therefore, will not compete with electron capture for decay transitions less than 1.02 MeV. In general, positron emission will predominate when the transition energy is high (well above 1.02 MeV) and for nuclides of low atomic number, while the EC decay process will predominate for low transition energies and nuclides of higher atomic number. The decay transitions of ^{22}Na and ^{65}Zn serve as examples. In the case of ^{22}Na, decay by β^+ emission predominates (90%) as compared with decay via electron capture (10%),

$$^{22}_{11}\text{Na} \rightarrow {}^{22}_{10}\text{Ne} + \beta^+ + \nu \ (90\%) \tag{1.35}$$

and

$$^{22}_{11}\text{Na} \xrightarrow{EC} {}^{22}_{10}\text{Ne} + \nu \ (10\%) \tag{1.36}$$

The transition energy of ^{22}Na is 2.842 MeV (Holden, 1997a), well above the 1.02 MeV minimum required for positron emission. On the other hand, taking the example of the nuclide ^{65}Zn, we see that electron capture predominates over β^+ emission

$$^{65}_{30}\text{Zn} \rightarrow {}^{65}_{29}\text{Cu} + \beta^+ + \nu \ (1.5\%) \tag{1.37}$$

and

$$^{65}_{30}\text{Zn} \xrightarrow{EC} {}^{65}_{29}\text{Cu} + \nu \ (98\%) \tag{1.38}$$

In the case of ^{65}Zn, the transition energy is only 1.35 MeV (Holden, 1997a), which is not much above the minimum energy of 1.02 MeV required for positron emission. Consequently, EC decay predominates.

It is generally known that, chemical factors do not control nuclear decay processes. However, because the electron capture decay process involves the capture of an orbital electron by the nucleus, atomic or molecular binding effects which vary with chemical structure can influence the electron capture decay process. Ehman and Vance (1991) cite the interesting examples of ^{7}Be and ^{90m}Nb, which display different electron-capture decay rates depending on the chemical state of the nuclides. ^{7}Be as a free metal and in the form ^{7}BeF$_2$ salt display a 0.08% difference in EC decay rates, while ^{90m}Nb as a free metal and the salt form ^{90m}NbF$_3$ exhibit an even greater 3.6% difference in EC decay rates.

D. Beta Particle Absorption and Transmission

Early research work on measuring the range of beta particles involved placing absorbers of increasing thickness between the radioactive source and the detector. The detector would measure the beta particles transmitted through the absorber. Increasing the absorber thickness would increasingly diminish the number of beta particles transmitted on to the detector. The transmission of beta particles was then plotted against absorber thickness as illustrated in Fig. 1.6 in an attempt to determine the thickness of absorber required to fully stop the beta particles. Unfortunately the plots could not be used directly to accurately determine beta particle-ranges; rather they had to be compared to an absorption curve of a beta-emitter of known range by what became known as Feather analysis (Feather, 1938; Glendenin, 1948). An auspicious outcome of this work was the observation that the plots of beta particle-absorption had more or less an exponential character. When plotted semi-logarithmically against distance the beta-particle absorption and/or transmission through the absorber was linear or near linear when plotted against absorber thickness as illustrated in Fig. 1.6. This was a fortuitous outcome of the continuous energy spectrum of beta particles emitted from any given source. These findings are quite the contrast to the absorption curve of alpha particles discussed previously (Fig. 1.2), where the alpha particle intensity remains constant and then comes to an almost abrupt stop.

The curve illustrated in Fig. 1.6 is characteristic of beta particles. The somewhat linear segment of the semilogarithmic plot of activity transmitted versus absorber thickness levels off horizontally due to a background of

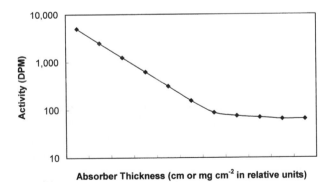

FIGURE I.6 The transmission of beta particles through absorber material of increasing thickness. The semi-logarithmic plot is linear over a specific range of absorber thickness and then levels off horizontally due to a background of bremsstrahlung radiation.

bremsstrahlung radiation. Negatrons and positrons both display a somewhat linear semilogarithmic plot with the exception that, in the case of positrons, the horizontal portion of the plot has an added background due to annihilation radiation (Glendenin, 1948). Because beta particles have a definite range in matter, beta particle-transmission is not a purely logarithmic one as we shall see is the case for gamma radiation (see Section IV.D of this chapter). The curves may not display a purely exponential character and the plots may have a degree of concavity to them depending on the distance of the source and detector to the absorber and on the shape of the beta-particle continuous energy spectrum. The greater the atomic number of the beta particle-emitter, and the more the beta spectrum is displaced toward the lower energies, the more nearly exponential (linear) will be the absorption curve (Glendenin, 1948). It is common to express the amount of absorber in mass thickness units, that is, mass per unit area (e.g., $g\,cm^{-2}$), which is the product of absorber thickness and density, as it is easier to measure accurately very thin absorbers simply from their weight.

On the basis of the exponential character of beta-particle absorption we can describe the transmission of beta particles through the absorber as

$$I = I_0\,e^{-\mu x} \tag{1.39}$$

where I is intensity of the beta particles (DPM) transmitted through the absorber, I_0 is the initial intensity of beta particles (DPM) incident on the absorber, μ is the linear absorption coefficient in units of cm^{-1} and x is the absorber thickness in cm. If we express absorber thickness in mass thickness units (e.g., $mg\,cm^{-2}$ or $g\,cm^{-2}$) we can rewrite Eq. 1.39 as

$$I = I_0 e^{-(\mu/\rho)(\rho x)} \tag{1.40}$$

or

$$\frac{I}{I_0} = e^{-(\mu/\rho)(\rho x)} \tag{1.41}$$

and

$$\ln\frac{I}{I_0} = -(\mu/\rho)(\rho x) \tag{1.42}$$

where μ/ρ is the mass absorption coefficient (also referred to as mass attenuation coefficient) in units of $cm^2\,g^{-1}$, that is, the linear absorption coefficient divided by the absorber density, and ρx is the absorber thickness in mass thickness units $g\,cm^{-2}$, that is, the product of the absorber density and absorber thickness.

Equation 1.42 can be used to determine experimentally the unknown thickness of absorber materials. A standard curve is plotted with the ratio I/I_0 on a logarithmic scale versus mass thickness (ρx) of the absorber on a linear scale. A value for I in units of DPM are determined with a detector by measuring the beta particle-intensity transmitted through a given absorber thickness. This is repeated with absorbers of different thickness. The magnitude of the incident beta-particle intensity, I_0, is a constant value and determined with the detector in the absence of absorber. The linear portion of the plot has a negative slope, such as that illustrated in Fig. 1.6, and from least squares analysis the mass attenuation coefficient μ/ρ is determined (Yi et al., 1999). Consequently, the thickness of an unknown similar material can be determined from measured intensity, I, of the transmitted beta particle-radiation after placing the material between the beta particle-source and detector without altering the counting geometry. The sample thickness is calculated or determined directly from the aforementioned curve (Tumul'kan, 1991 and Clapp et al., 1995).

Beta particle-transmission has many practical applications today in industrial manufacturing. Beta particle-sources and detectors are placed on the production line to test for thickness, uniformity and defects in the manufacture of paper, metal and plastic films as well as on-line inspection of sewn seams in the textile industry (Ogando, 1993; Clapp et al., 1995; Mapleston, 1997; Titus et al., 1997) and in agronomic research to measure leaf water content (Mederski, 1961, 1968; Nakayama, 1964; Obregewitsch, 1975) or to measure the biomass of a prairie (Knapp et al., 1985). These are commonly referred to as beta transmission thickness gauges. The beta particle-sources used depend on the absorber thickness to be measured and the E_{max} of the beta particles. Three sources commonly used are ^{147}Pm ($E_{max}=0.224\,\text{MeV}$), ^{85}Kr ($E_{max}=0.672\,\text{MeV}$) and ^{90}Sr(^{90}Y) in secular equilibrium (E_{max} of ^{90}Sr and ^{90}Y $=0.546$ and $2.280\,\text{MeV}$, respectively). The source with the lowest beta particle E_{max} (e.g., ^{147}Pr) is used to measure the finest thickness of material (Balasubramanian, 1997, 1998), and the sources are changed according to beta-particle energy, penetration power and thickness of material to be tested.

E. Internal Conversion Electrons

Decay by internal conversion (IC) results in the emission of an atomic electron. This electron, called the internal conversion electron, is emitted from an atom after absorbing the excited energy of a nucleus. This mode of decay accompanies and even competes with gamma-ray emission as a deexcitation process of unstable nuclei.

The kinetic energy of the electron emitted is equivalent to the energy lost by the nucleus (energy of transition of the excited nucleus to its ground or lower energy state) less the binding energy of the electron. This is illustrated by the following equation:

$$E_e = (E_i - E_f) - E_b \tag{1.43}$$

where E_e is the kinetic energy of the internal conversion electron, $(E_i - E_f)$ is the energy of transition between the initial, E_i, and the final, E_f, nuclear energies normally associated with gamma ray emission, and E_b is the binding energy of the atomic electron.

An example of radionuclide decay by internal conversion is found in Fig. 1.7, which illustrates the decay of the parent–daughter nuclides $^{109}Cd(^{109m}Ag)$. Note that the ^{109m}Ag daughter decays by internal conversion with a 96% probability (i.e., 45% for IC from the K shell + 48% from the L shell + 3% from higher electron shells) and decay occurs via gamma emission with the remaining 4% probability (Rachinhas *et al.*, 2000).

Because the emission of internal-conversion electrons competes with gamma-ray emission as an alternative mode of nuclear deexcitation, many radioactive nuclei that emit gamma radiation will also emit internal-conversion electrons. The degree to which this competition occurs is expressed as the internal-conversion coefficient, which is the ratio of the rate

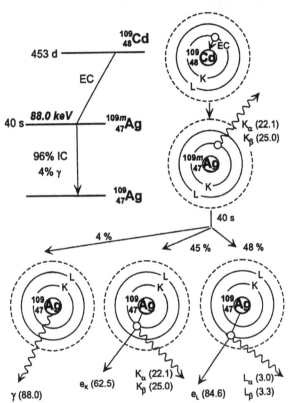

FIGURE 1.7 Decay scheme of $^{109}Cd(^{109m}Ag)$. The numbers in parenthesis indicate energy values in keV. The electron capture (EC) process occurs from K, L and outer shells with probabilities of 79, 17 and 4%, respectively, but only K-capture is represented above. The ^{109m}Ag daughter decays by emission of 88.0 keV gamma rays with a 4% probability or by internal conversion (IC) with the probabilities of 45 and 48% for K and L shells. Internal conversion from shells higher than L contribute the remaining 3%. The K and L IC decay illustrated involve the ejection of a conversion electron with energy $e_K = 62.5$ keV or $e_L = 84.6$ keV, accompanied by the emission of a Ag K- or L-fluorescence x-ray photon ($K_\alpha = 22.1$, $K_\beta = 25.0$ keV, or $L_\alpha = 3.0$, $L_\beta = 3.3$ keV) or by the emission of Auger electrons (not represented) and x-ray photons following Auger electron emissions. (From Rachinhas *et al.*, 2000, reprinted with permission from Elsevier Science.)

of emission of internal conversion electrons to the rate of emission of gamma rays of equivalent energy. In other words, the internal-conversion coefficient is a quantitative measure of the number of internal-conversion electrons divided by the number of gamma rays emitted from a radionuclide sample. The internal-conversion coefficient is denoted by α or e/γ.

Internal-conversion electrons may be emitted from specific electron shells of atoms and may be expressed in terms of internal-conversion electrons and gamma rays of the same energy less the energy difference resulting from the binding energy of the electron. When expressed in terms of electrons emitted from specific shells, the internal-conversion coefficient is written with a subscript denoting the electron shell of origin, for example, α_K or e_K/γ, α_L or e_L/γ, and α_i or e_i/γ, where $i = K, L, M$, and so on electron shells.

Values of internal-conversion coefficients are provided in many reference tables on isotope decay. In general, internal-conversion coefficients are small for gamma ray-emitting nuclides of low Z and high-energy transitions and larger for nuclides of high Z and low-energy transitions.

This relationship is illustrated in Table 1.4, which lists a few radionuclides selected at random as examples in order of increasing Z number. As can be seen, large internal-conversion coefficients occur when internal-conversion electrons are emitted with low-energy nuclear transitions as indicated by the large values of α associated with low gamma-ray energies.

TABLE I.4 Relationship between Gamma Radiation and internal-conversion Electron Radiation, e⁻, Associated with Several Nuclides Listed in Order of Increasing Z-Number

Nuclide $^A_Z X$	Gamma radiation (MeV)^a	e⁻ (MeV)	$\alpha = e/\gamma$	X-rays^a
$^{7}_{4}$Be	0.477 (10%)		7.0×10^{-7}	
$^{22}_{11}$Na	1.275 (100%)		6.7×10^{-6}	
$^{44}_{22}$Ti	0.068 (90%)	0.065	0.12	Sc K
	0.078 (98%)	0.073	0.03	
$^{57}_{27}$Co	0.014 (9%)	0.013	8.2	Fe K (55%)
	0.122 (87%)	0.115	0.02	
	0.136 (11%)	0.129	0.15	
$^{64}_{29}$Cu	1.34 (0.6%)	1.33	1.3×10^{-4}	Ni K (14%)
$^{87m}_{38}$Sr^b	0.388 (80%)	0.386	0.21	Sr K (9.4%)
$^{119m}_{50}$Sn^b	0.024 (16%)	0.020	5.13	Sn K (28%)
$^{125}_{53}$I	0.035 (7%)	0.030	13.6	Te K (138%)
$^{129}_{53}$I	0.040 (9%)	0.034	22	Xe K (69%)
$^{169}_{68}$Er	0.008 (0.3%)	0.006	220	Tm M
$^{181}_{74}$W	0.006 (1%)	0.004	46	Ta K (65%)
$^{203}_{80}$Hg	0.279 (82%)	0.275	0.23	Tl K (13%)
$^{239}_{94}$Pu	0.039 (0.01%)	0.033	461	U. K (0.012%)
	0.052 (0.02%)	0.047	269	

^{a}Values in per cent are radiation intensities or abundances.
^{b}m denotes a metastable state.

It should also be pointed out that the internal-conversion electron (e^-) energies are slightly lower than the gamma-ray energies. This is because the energy of the internal-conversion electron is equal to the energy absorbed from the decaying nucleus (transition energy) less the binding energy of the atomic electron described previously in Eq. 1.43. On the other hand, gamma-ray energies serve as a measure of the exact quanta of energies lost by a nucleus.

The loss of atomic electrons through the emission of internal-conversion electrons leaves vacancies in atomic electron shells. The vacancies are filled by electrons from outer higher-energy shells, whereby there is a concomitant loss of electron energy as internal bremsstrahlung or x-radiation. Emission of x-radiation resulting from electron filling of vacancies in electron shells $(K, L, M \ldots)$ is also listed in Table 1.4. This is a process that occurs in the daughter atoms; the x-rays are a characteristic of the daughter rather than of the parent.

Internal-conversion electrons are identical in their properties to beta particles. They differ, however, in their origin. Beta particles originate from the nucleus of an atom, whereas internal-conversion electrons originate from atomic electron shells. A characteristic difference between these two types of electron is their energy spectra. Beta particles, as discussed previously, are emitted from nuclei with a broad spectrum or smear of energies ranging from near zero to E_{max}. However, internal-conversion electrons are emitted from the atoms of decaying nuclei with discrete lines of energy of a magnitude equivalent to that of the energy lost by the nuclei less the electron binding energy. The energy of an internal-conversion electron can be used to estimate the energy lost by a nucleus.

Like beta particles, internal-conversion electrons dissipate their energy by ionization they cause in matter. The abundance of internal-conversion electrons emitted from some nuclide samples can be significant and should not be ignored. In certain cases it can play a significant role in radionuclide detection and measurement. Internal-conversion electron energies are slightly lower than the true gamma decay energy because of the energy consumed in the ejection of the bound atomic electron (E_b in Eq. 1.43).

An internal-conversion coefficient of large magnitude does not, however, necessarily signify the emission of a high abundance of internal-conversion electrons. For example, ^{239}Pu with a high internal-conversion coefficient $(\alpha = 461)$ corresponding to a 0.039-MeV gamma decay process emits only a trace of internal-conversion electrons because of the low abundance of gamma decay (0.01%, see Table 1.4).

F. Auger Electrons

An Auger (pronounced OH-ZHAY) electron can be considered as the atomic analogue of the internal conversion electron. In the electron-capture (EC) decay processes, vacancies are left in electron shells $(K, L, M \ldots)$ that can be filled by atomic electrons from higher energy levels. In the process of falling to a lower energy shell to fill a vacancy, electron energy is lost as a photon of x-radiation (see Section III.E of this chapter). This x-radiation may either travel on to be emitted from the atom or it may collide with an atomic electron, resulting in the emission of the electron referred to as an Auger electron.

Whenever an x-ray photon causes the ejection of an atomic electron another electron falls from an outer shell to a lower one to fill the vacancy, and there is a cascading effect of electrons falling from yet more distant shells to fill vacancies left behind until the atom reaches the ground or stable state. The downward transitions of electrons in this fashion produce additional x-ray photons of lower energy than the initial x-ray photon. The production of x-ray photons in this fashion is referred to as x-ray fluorescence.

The energy of an Auger electron is low, because it is equivalent to the energy of the x-ray photon less the electron binding energy. For example, an x-ray photon resulting from an electron transition from the L shell to the K shell can produce an Auger electron of energy

$$\mathcal{E}_{\text{Auger}} = (E_L - E_K) - E_b \tag{1.44}$$

where E_L and E_K are the electron energies in the L and K shells, respectively, and E_b is the binding energy of the electron prior to its ejection as an Auger electron. Other transitions may be also described such as E_M–E_L for M and L electron shells. Equation 1.44 may be also written as

$$\mathcal{E}_{\text{Auger}} = h\nu - E_b \tag{1.45}$$

where $h\nu$ is the x-ray photon energy expressed as a product of Planck's constant, h ($h = 6.62 \times 10^{-27}$ erg s $= 4.14 \times 10^{-15}$ eV s $= 6.62 \times 10^{-34}$ J s), and the photon frequency, ν, in units of s^{-1}.

Auger electron emission competes with x-ray emission, and it can accompany any decay process that results in the production of x-rays. Like internal-conversion electron emission described previously, the electron-capture decay process (see Section II.C.2) also results in the emission of appreciable quantities of x-radiation. Thus, Auger electron emissions also accompany electron-capture decay. Because the energies of Auger electrons are low (approximately equivalent to x-ray photon energies), Auger electrons may not play a significant direct role in the detection and measurement of radionuclides. However, Auger electron emission can reduce appreciably the abundance of x-ray emission normally expected to accompany radionuclide decay processes. The two competing processes of Auger electron emission and x-ray emission are important to consider in the detection and measurement of nuclides that decay by electron capture. This is measured by both the fluorescence yield and Auger yield. The fluorescence yield is the fraction of vacancies in a given electron shell that is filled with accompanying x-ray emission, and Auger yield is the fraction of vacancies that are filled resulting in the emission of Auger electrons (Friedlander et al., 1964). The fluorescence yield is important in the measurement of nuclides that decay by electron capture, as it is the x-ray fluorescence photons that are usually detected (Mann, 1978). Figure 1.8 illustrates the K-shell fluorescence yield as a

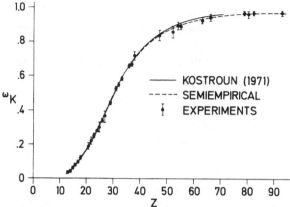

FIGURE I.8   Fluoresence K-shell yields, ω_K, as a function of atomic number, Z: (a) according to Kostroun *et al.*, (1971); (b) a best fit to selected experimental data; and (c) critically evaluated experimental results. **(From Bambynek *et al.*, 1972, reprinted with permission Copyright The American Physical Society.)**

function of nuclide atomic number. The *L*-shell fluorescence yield also varies similarly with atomic number as the *K*-shell fluorescence yield, but is several times lower in magnitude (Friedlander *et al.*, 1964).

G. Neutron Radiation

The neutron is a neutral particle, which is stable only in the confines of the nucleus of the atom. Its mass, like that of the proton, is equivalent to 1 u (atomic mass unit). Unlike the particulate alpha and beta nuclear radiation previously discussed, neutron radiation is not emitted in any significant quantities from radionuclides that undergo the traditional nuclear decay processes with the exception of a few radionuclides such as ^{252}Cf and ^{248}Cm, which decay to a significant extent by spontaneous fission (see Section II.G.2.b). Significant quantities of neutron radiation occur when neutrons are ejected from the nuclei of atoms following reactions between the nuclei and particulate radiation. The lack of charge of the neutron also makes it unable to cause directly any ionization in matter, again unlike alpha and beta radiation. The various sources, properties, and mechanisms of interaction of neutrons with matter are described subsequently.

I. Neutron Classification

Neutrons are generally classified according to their kinetic energies. There is no sharp division or energy line of demarcation between the various classes of neutrons; however, the following is an approximate categorization according to neutron energy:

- Cold neutrons < 0.003 eV
- Slow (thermal) neutrons 0.003–0.4 eV
- Slow (epithermal) neutrons 0.4–100 eV
- Intermediate neutrons 100 eV–200 keV

- Fast neutrons 200 keV–10 MeV
- High energy (relativistic) neutrons > 10 MeV

The energies of neutrons are also expressed in terms of velocity (meters per second) as depicted in the terminology used to classify neutrons. A neutron of specific energy and velocity is also described in terms of wavelength, because particles in motion also have wave properties. It is the wavelength of the neutron that becomes important in studies of neutron diffraction. The values of energy, velocity, and wavelength of the neutron, as with all particles in motion, are interrelated. The velocity of neutrons increases according to the square root of the energy, and the wavelength of the neutron is inversely proportional to its velocity. Knowing only one of the properties, either the energy, velocity, or wavelength of a neutron, we can calculate the other two. We can relate the neutron energy and velocity using the kinetic energy equation

$$E = \tfrac{1}{2}mv^2 \quad \text{or} \quad v = \sqrt{2E/m} \tag{1.46}$$

where E is the particle energy in joules ($1\,\text{eV} = 1.6 \times 10^{-19}\,\text{J}$), m is the mass of the neutron ($1.67 \times 10^{-27}\,\text{kg}$), and v is the particle velocity in meters per second. The wavelength is obtained from the particle mass and velocity according to

$$\lambda = \frac{h}{p} = \frac{h}{mv}, \tag{1.47}$$

where λ is the particle wavelength in meters, h is Planck's constant ($6.63 \times 10^{-34}\,\text{J s}$), p is the particle momentum, and m and v are the particle mass and velocity as previously defined. The correlation between neutron energy, velocity, and wavelength is provided in Fig. 1.9, which is constructed from the classical Eqs. 1.46 and 1.47 relating particle mass, energy, velocity and wavelength. However, calculations involving high-energy particles that approach the speed of light will contain a certain degree of error unless relativistic calculations are used, as the mass of the particle will increase according to the particle speed. In Section IV.C of this chapter we used the Einstein equation $E = mc^2$ to convert the rest mass of the positron or negatron to its rest energy (0.51 MeV). When gauging particles in motion the *total energy* of the particle is the sum of its kinetic (K) and rest energies (mc^2) or

$$E = K + mc^2 = \gamma mc^2 \tag{1.48}$$

where

$$\gamma = \frac{1}{\sqrt{1 - (u^2/c^2)}} \tag{1.49}$$

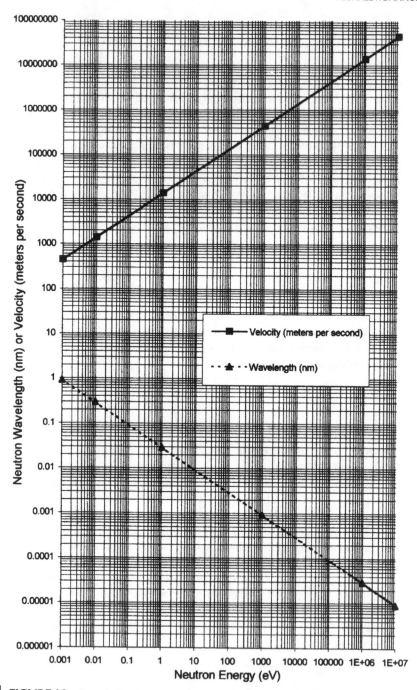

FIGURE 1.9 Correlation between neutron energy in electron volts (eV), velocity (m s^{-1}), and wavelength (nm). From the energy of the neutron in eV on the abscissa a line is drawn vertically to cross the wavelength and velocity curves. The values of neutron wavelength and velocity are obtained directly from the ordinate. For example, to determine the wavelength and velocity of 0.025 eV thermal neutrons, the value of 0.025 eV is found on the abscissa. A line is then drawn vertically from the point of 0.025 eV to cross the values of 0.18 nm wavelength and 2200 m s^{-1} velocity.

u is the particle speed, and $u < c$. If we call the particle rest mass m_0, then the relativistic mass, m_r, which is the speed-dependent mass of the particle is calculated as

$$m_r = \frac{m_0}{\sqrt{1 - (u^2/c^2)}} \tag{1.50}$$

The relativistic speed is defined as

$$u_r = c\sqrt{1 - (K/mc^2 + 1)^{-2}} \tag{1.51}$$

where K is the kinetic energy, and the particle speed u is always less than c (Serway et al., 1997). The nonrelativistic speed is that described by Eq. 1.46 or $u_{nr} = \sqrt{2E/m}$.

To confirm the validity of the use of nonrelativistic calculations of particle speed for the construction of Fig 1.9 let us use Eqs. 1.46 and 1.51 to compare the differences between the nonrelativistic and relativistic speeds of a neutron of 10 MeV kinetic energy. This energy was selected, because it is the highest neutron energy included in Fig. 1.9, and differences between nonrelativistic and relativistic calculations increase with particle energy. The difference between the two calculated speeds is defined by the ratio of the two or

$$\frac{u_{nr}}{u_r} = \frac{\sqrt{2E/m}}{c\sqrt{1 - (K/mc^2 + 1)^{-2}}} \tag{1.52}$$

The rest energy of the neutron, mc^2, is first calculated as

$$mc^2 = (1.6749 \times 10^{-27}\,\text{kg})(2.9979 \times 10^8\,\text{m s}^{-1})^2 = 1.505 \times 10^{-10}\,\text{J}$$

and

$$1.505 \times 10^{-10}\,\text{J}/1.602 \times 10^{-19}\,\text{J eV}^{-1} = 939.5\,\text{MeV}$$

since by definition, $1\,\text{eV} = 1.602 \times 10^{-19}\,\text{J}$. From Eq. 1.52 the ratio of the nonrelativistic and relativistic speeds are calculated as

$$\frac{u_{nr}}{u_r} = \frac{\sqrt{2(10\,\text{MeV})(1.602 \times 10^{-13}\,\text{J MeV}^{-1})/1.6749 \times 10^{-27}\,\text{kg}}}{c\sqrt{1 - ((10\,\text{MeV}/939.5\,\text{MeV}) + 1)^{-2}}}$$

$$= \frac{4.3737 \times 10^7\,\text{m s}^{-1}}{0.144775c} = \frac{4.3737 \times 10^7\,\text{m s}^{-1}}{(0.144775)(2.9979 \times 10^8\,\text{m s}^{-1})}$$

$$= \frac{4.3737 \times 10^7\,\text{m s}^{-1}}{4.340 \times 10^7\,\text{m s}^{-1}} = 1.0077 = 0.77\%\,\text{error}.$$

The error between the nonrelativistic and relativistic calculations is small at this high neutron energy. However, if we consider higher neutron energies in excess of 10 MeV the error of making nonrelativistic calculations increases.

As we observed above in the case of particle speed, we will also see that particle wavelength will also differ for nonrelativistic and relativistic calculations. In 1923 Louis Victor de Broglie first postulated that all particles or matter in motion should have wave characteristics just as photons display both a wave and particle character. We therefore attribute the wavelength of particles in motion as de Broglie wavelengths. Let us then compare calculated nonrelativistic and relativistic wavelengths. From Eq. 1.47, we can describe the nonrelativistic wavelength, λ_{nr}, as

$$\lambda_{nr} = \frac{h}{p} = \frac{hc}{pc} = \frac{hc}{cmv} = \frac{hc}{cm\sqrt{2E/m}} = \frac{hc}{\sqrt{2mc^2E}} \tag{1.53}$$

where $p = mv = m\sqrt{2E/m}$. For relativistic calculations the value of pc is calculated according to the following equation derived by Halpern (1988):

$$pc = \left[2m_0c^2K\left(1 + \frac{K}{2m_0c^2}\right)\right]^{1/2} \tag{1.54}$$

and the calculation for the relativistic de Broglie wavelength, λ_r, then becomes

$$\lambda_r = \frac{hc}{pc} = \frac{hc}{[2m_0c^2K(1 + (K/2m_0c^2))]^{1/2}} \tag{1.55}$$

We can then compare the difference between the nonrelativistic and relativistic wavelengths for the 10 MeV neutron as follows:

$$
\begin{aligned}
\frac{\lambda_{nr}}{\lambda_r} &= \frac{hc/\sqrt{2mc^2E}}{hc/[2m_0c^2K(1 + K/2m_0c^2)]^{1/2}} \\[2mm]
&= \frac{\dfrac{[(6.626 \times 10^{-34}\,\text{J s})(2.9979 \times 10^8\,\text{m s}^{-1})/(1.602 \times 10^{-13}\,\text{J MeV}^{-1})]}{/\sqrt{2(939.5\,\text{MeV})(10\,\text{MeV})}}}{\dfrac{[(6.626 \times 10^{-34}\,\text{J s})(2.9979 \times 10^8\,\text{m s}^{-1})/1.602 \times 10^{-13}\,\text{J MeV}^{-1}]}{/\sqrt{2(939.5\,\text{MeV})(10\,\text{MeV})[1 + (10\,\text{MeV}/2(939.5\,\text{MeV}))]}}} \\[2mm]
&= \frac{12.3995 \times 10^{-4}\,\text{MeV nm}/\sqrt{18790\,\text{MeV}^2}}{12.3995 \times 10^{-4}\,\text{MeV nm}/\sqrt{18889.96\,\text{MeV}^2}} \\[2mm]
&= \frac{9.0457 \times 10^{-6}\,\text{nm}}{9.0217 \times 10^{-6}\,\text{nm}} = 1.0026 = 0.26\%\ \text{error.}
\end{aligned} \tag{1.56}
$$

From the above comparison of nonrelativistic and relativistic calculations of neutron wavelength and velocity, we see that the data provided in Fig. 1.9 based on nonrelativistic calculations are valid with less than 1% error for the highest energy neutron included in that figure. However, if we consider higher energies beyond 10 MeV, where we classify the neutron as relativistic, the errors in making nonrelativistic calculations will increase with neutron energy. It will be clearly obvious to the reader that factors in Eq. 1.56 can be cancelled out readily and the equation simplified to the following, which provides a quick evaluation of the effect of particle energy on the error in nonrelativistic calculation of the de Broglie wavelength:

$$\frac{\lambda_{\mathrm{nr}}}{\lambda_{\mathrm{r}}} = \sqrt{1 + \frac{K}{2m_0 c^2}} \qquad (1.57)$$

where K is the particle kinetic energy in MeV and $m_0 c^2$ is the particle rest energy (e.g., 939.5 MeV for the neutron and 0.511 MeV for the beta particle). For example, a nonrelativistc calculation of the wavelength of a 50-MeV neutron would have the following error:

$$\frac{\lambda_{\mathrm{nr}}}{\lambda_{\mathrm{r}}} = \sqrt{1 + \frac{50\,\mathrm{MeV}}{2(939.5\,\mathrm{MeV})}} = 1.0131 = 1.31\%\ \text{error.}$$

Note that the above-computed errors in nonrelativistic calculations of the de Broglie wavelength increased from 0.26% for a 10-MeV neutron to 1.31% for a 50-MeV neutron, and the error will increase with particle energy. Errors in nonrelativistic calculations are yet greater for particles of smaller mass (e.g., beta particles) of a given energy compared to neutrons of the same energy. This is due obviously to the fact that particles of lower mass and a given energy will travel at higher speeds than particles of the same energy but higher mass. This is illustrated in Fig. 1.10 where the particle speed, u, is a function of the particle kinetic energy, K, and its mass or rest energy, mc^2. The particle energy in Fig. 1.10 is expressed as K/mc^2 to permit the reader to apply the curves for nonrelativistic and relativistic calculations to particles of different mass. For example, from the abscissa of Fig. 1.10, the values of K/mc^2 for a 2-MeV beta particle is 2 MeV/0.51 MeV = 3.9 and that for a 2-Mev neutron is 2 MeV/939.5 MeV = 0.0021. From Fig. 1.10 we see that the nonrelativistic calculation of the speed of a 2-MeV beta particle would be erroneously extreme (well beyond the speed of light), while there would be only a small error in the relativistic calculation of the speed of the massive neutron of the same energy.

2. Sources of Neutrons

The discovery of the neutron had eluded humanity until as late as 1932, because of the particle's neutral charge and high penetrating power when traveling through matter. In 1932 J. Chadwick provided evidence for the existence of the neutron. He placed a source of alpha particle-radiation in close proximity to beryllium. It was known that bombarding beryllium with alpha radiation would produce another source of radiation, which had a penetration power through matter even greater than that of gamma radiation.

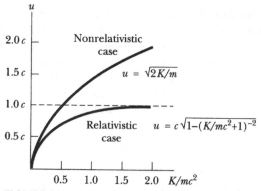

FIGURE 1.10 A graph comparing nonrelativistic and relativistic kinetic energy. The speeds are plotted versus energy. In the relativistic case, u is always less than c. (From *Modern Physics*, 2nd Edition by Serway *et al.*, © 1997, reprinted with permission by Brooks/Cole, an imprint of the Wadsworth Group, a division of Thomas Learning.)

Chadwick observed that, when a sheet of paraffin (wax) was placed in the path of travel of this unknown radiation, he could detect a high degree of ionization in a gas ionization chamber caused by protons emitted from the paraffin. This phenomenon would not occur when other materials such as metals and even lead were placed in the path of this unknown radiation. On the basis of further measurements of the proton velocities and scattering intensities, it was concluded that the unknown radiation had a mass similar to that of the proton, but with a neutral charge. Only a particle with neutral charge would have a high penetration power through matter. As noted in the previous discussion of beta particle decay, the neutron is of mass similar to that of the proton and, within the nucleus of an atom, the particle is a close union between a proton and an electron.

a. Alpha Particle-Induced Nuclear Reactions

It is interesting to note that the method used by Chadwick to produce neutrons by alpha particle-induced reactions, described in the previous paragraph, remains an important method of producing a neutron source, particularly when a relatively small or easily transportable neutron source is required. The source may be prepared by compressing an alpha particle-emitting radioisotope substance with beryllium metal. The nuclear reaction, which occurs between the alpha particle and the beryllium nucleus, terminates with the emission of a neutron and the production of stable carbon as follows

$$^{9}_{4}\text{Be} + ^{4}_{2}\text{He} \rightarrow ^{1}_{0}\text{n} + ^{12}_{6}\text{C} + 5.5\,\text{MeV}_{\text{(average)}} \tag{1.58}$$

Several alpha particle sources are used to produce neutrons via the preceding (α, n) reaction. Among these are the alpha emitters ^{241}Am, ^{242}Cm, ^{210}Po, ^{239}Pu, and ^{226}Ra. The alpha radiation source selected may depend on its half-life as well as its gamma-ray emissions. As noted previously in this chapter, gamma radiation often accompanies alpha decay. The use of an alpha source, which also emits abundant gamma radiation, requires additional protection for the user against penetrating gamma rays. For example, Am–Be sources are preferred over the Ra–Be sources of neutrons used in soil moisture probes (Nielsen and Cassel, 1984; O'Leary and Incerti, 1993), because the latter have a higher output of gamma radiation and require more shielding for operator protection.

The energies of the neutrons emitted from these sources will vary over the broad spectrum of 0 to 10 MeV. The average neutron energy of 5.5 MeV is shown in Eq. 1.58. The neutrons produced by these sources vary in energy as a consequence of several factors, including the sharing of the liberated energy between the neutron and ^{12}C nucleus, the varying directions of emission of neutrons from the nucleus with consequent varying energies and velocities, and the variations in kinetic energies of the bombarding alpha particles.

The neutron activities available from these sources increase up to a maximum as a function of the amounts of alpha emitter and beryllium target material used. For example, as explained by Bacon (1969), the Ra–Be source, prepared by mixing and compressing radium bromide with beryllium powder, increases steadily in neutron activity (neutrons per second) for each gram of radium used as the amount of beryllium is increased to about 10 g; but no significant increase in neutron output is achieved if more beryllium is used. The maximum neutron output achieved is approximately 2×10^7 neutrons per second per gram of radium. Because alpha decay from any alpha particle-emitting source occurs by means of random events, the production of neutrons by (α, n) reactions is also a random event. Therefore, these reactions can be referred to as "not time correlated." This is contrary to the case of neutron sources provided through fission, discussed subsequently.

b. Spontaneous Fission

About 100 radionuclides are known to decay by spontaneous fission (SF) with the emission of neutrons (Karelin *et al.*, 1997) as an alternative to another decay mode, such as alpha decay. Spontaneous fission involves the spontaneous noninduced splitting of the nucleus into two nuclides or fission fragments and the simultaneous emission of more than one neutron on the average. This phenomenon occurs with radionuclides of high mass number, $A \geq 230$. The radionuclide ^{252}Cf is a good example of a commercially available spontaneous fission neutron source. It decays with a half-life of 2.65 years primarily by alpha emission (96.91% probability); the remaining of the ^{252}Cf decay processes occur by spontaneous fission with a probability of 3.09% (Martin *et al.*, 2000, see also Appendix A). Decay of ^{252}Cf by spontaneous fission produces an average number of 3.7 neutrons per fission. Because the sizes of the two fragments resulting from fission are not predictable, average sizes of the two fragments are determined. Consequently, the numbers of neutrons emitted from individual fissions are not the same; and an average number of neutrons produced per fission is determined. The fission rate of ^{252}Cf is 6.2×10^5 SF s^{-1} µg^{-1} (Isotope Products Laboratories, 1995). The neutron emission from ^{252}Cf in units of neutrons per second per unit mass is reported to be 2.314×10^6 s^{-1} µg^{-1} with a specific activity of 0.536 mCi µg^{-1} (Martin *et al.*,

2000). If we know the radionuclide specific activity and the % probability of decay by spontaneous fission, we can calculate the fission rate. For example, taking the specific activity and % probability of spontaneous fission reported above for ^{252}Cf, we can calculate the fission rate as the product of decay rate and probability of SF per decay or

$$(0.536 \, \text{mCi}/\mu\text{g})(3.7 \times 10^7 \, \text{dps/mCi})(0.0309) = 6.13 \times 10^5 \, \text{SF s}^{-1} \, \mu\text{g}^{-1}$$

which is in close agreement with the value cited above. See Section VII.A for a discussion of radioactivity units and calculations.

The variations in fission fragment sizes and number of neutrons emitted per fission provide variable neutron energies over the range 0–5.5 MeV with an average neutron energy from ^{252}Cf of approximately 2.3 MeV. Small sources of ^{252}Cf are commercially available for a wide range of applications such as prompt-gamma neutron activation analysis of coal, cement, minerals, detection of explosives and land mines, neutron radiography and cancer therapy. These sources are described by Martin *et al.* (1997, 2000) among which include 50-mg sources of ^{252}Cf providing a neutron intensity $> 10^{11} \, \text{s}^{-1}$ and measuring only 5 cm in length $\times$ 1 cm diameter. They report also larger sources of mass > 100 mg of ^{252}Cf that approach reactor capabilities for neutrons.

Another standard nuclide source of neutrons is ^{248}Cm, which provides spontaneous fission intensity of only $4.12 \times 10^4 \, \text{s}^{-1} \, \text{mg}^{-1}$ and decays with a half-life of 3.6×10^5 years (Radchenko *et al.*, 2000). The lower neutron flux intensity of this source limits its application, although it has the advantage of a very long half-life providing invariability of sample intensity with time.

Some radionuclides of interest in nuclear energy and safeguards also decay by spontaneous fission. The isotopes of plutonium of even mass number, namely ^{238}Pu, ^{240}Pu, and ^{242}Pu, decay principally by alpha particle-emission but can also undergo spontaneous fission to a lesser extent at rates of 1100, 471, and 800 SF s^{-1} g^{-1}, respectively. The average number of neutrons emitted per fission is between 2.16 and 2.26 of broad energy spectrum (Canberra Nuclear, 1996). Because the neutrons produced with each fission occurrence are emitted simultaneously, we can refer to these emissions as "time correlated." Other isotopes of uranium and plutonium also undergo spontaneous fission but at a much lower rate.

c. Neutron-Induced Fission

When the naturally occurring isotope of uranium, ^{235}U, is exposed to slow neutrons, it can absorb the neutron to form the unstable nuclide ^{236}U (Eq. 1.71 in Section II.G.3.c). The newly formed nucleus may decay by alpha particle and gamma ray emission with the long half-life of 2.4×10^7 years. This occurs in approximately 14% of the cases when ^{235}U absorbs a slow neutron. However, in the remaining 86% of the cases, the absorption of a slow neutron by ^{235}U results in the production of the unstable ^{236}U nuclide, which takes on the characteristics of an unstable oscillating droplet. This oscillating nuclear droplet with the opposing forces of two positively charged nuclides splits into two fragments, not necessarily of equal size, with the liberation of an average energy of 193.6 MeV. The general ^{235}U fission reaction may be illustrated by

$$^{235}\text{U} + \text{n} \rightarrow \text{fp} + \nu\text{n} + E \qquad (1.59)$$

which represents the fission of one atom of ^{235}U by one thermal neutron n to yield the release of fission products fp of varying masses plus an average yield of $\nu = 2.42$ neutrons and an overall average release of energy $E = 193.6$ MeV (Koch, 1995). Most of this energy (over 160 MeV) appears in the form of kinetic energy of the two fission fragments. The remaining energy is shared among the neutrons emitted, with prompt gamma radiation accompanying fission and beta particles and gamma radiation from decaying fission fragments and neutrinos accompanying beta decay. When a sample of ^{235}U is bombarded with slow neutrons, the fission fragments produced are rarely of equal mass. The ^{236}U intermediate nuclide breaks into fragments in as many as 30 different possible ways, producing, therefore, 60 different nuclide fission fragments. In a review Koch (1995) provides a list of the fission fragments and their relative abundances as produced in a typical pressurized water reactor (PWR). The most common fission fragments have a mass difference in the ratio 3:2 (Bacon, 1969). On the average, 2.42 neutrons are emitted per ^{236}U fission (Koch, 1995). Neutrons emitted from this fission process vary in energy over the range 0–10 MeV with an average neutron energy of 2 MeV and are classified as fast neutrons. Because more than one neutron is released per fission, a self-sustaining chain reaction is possible with the liberation of considerable energy, forming the basis for the nuclear reactor as a principal source of neutrons and energy. In the case of ^{235}U, slow neutrons are required for neutron absorption and fission to occur. The nuclear reactor, therefore, will be equipped with a moderator such as heavy water (D_2O) or graphite, which can reduce the energies of the fast neutrons via elastic scattering of the neutrons with atoms of low atomic weight. The protons of water also serve as a good moderator of fast neutrons, provided the neutrons lost via the capture process $^{1}\text{H}(\text{n}, \gamma)^{2}\text{H}$ can be compensated by the use of a suitable enrichment of the ^{235}U in the nuclear reactor fuel (Byrne, 1994). The notation $^{1}\text{H}(\text{n}, \gamma)^{2}\text{H}$ is a form of abbreviating a nuclear reaction according to the format

Target Nucleus(Projectile, Detected Particle)Product Nucleus.

It can be read as follows: The target nucleus of the isotope ^{1}H absorbs a neutron to form the product isotope ^{2}H with the release of gamma radiation.

The previously described fission of ^{235}U represents the one and only fission of a naturally occurring radionuclide that can be induced by slow neutrons. The radionuclides ^{239}Pu and ^{233}U also undergo slow neutron-induced fission; however, these nuclides are man-made via the neutron irradiation and neutron absorption of ^{238}U and ^{232}Th as illustrated in the following (Murray, 1993). The preparation of ^{239}Pu occurs by means of neutron absorption by ^{238}U followed by beta decay as follows:

$$^{238}_{92}\text{U} + ^{1}_{0}\text{n} \rightarrow ^{239}_{92}\text{U} + \gamma \qquad (1.60)$$

$$^{239}_{92}U \xrightarrow{t_{1/2}=23.5\,min} {}^{239}_{93}Np + \beta^- \qquad (1.61)$$

$$^{239}_{93}Np \xrightarrow{t_{1/2}=2.35\,days} {}^{239}_{94}Pu + \beta^- \qquad (1.62)$$

The preparation of ^{233}U is carried out via neutron absorption of ^{232}Th followed by beta decay according to the following:

$$^{232}_{90}Th + {}^{1}_{0}n \rightarrow {}^{233}_{90}Th + \gamma \qquad (1.63)$$

$$^{233}_{90}Th \xrightarrow{t_{1/2}=22.4\,min} {}^{233}_{91}Pa + \beta^- \qquad (1.64)$$

$$^{233}_{91}Pa \xrightarrow{t_{1/2}=27.0\,days} {}^{233}_{92}U + \beta^- \qquad (1.65)$$

Nuclides that undergo slow neutron-induced fission are referred to as fissile materials. Although ^{235}U is the only naturally occurring fissile radionuclide, it stands to reason that if an excess of neutrons is produced in a thermal reactor, it would be possible to produce fissile ^{239}Pu or ^{233}U fuel in a reactor in excess of the fuel actually consumed in the reactor. This is referred to as "breeding" fissile material, and it forms the basis for the new generation of breeder reactors (Murray, 1993).

Other heavy isotopes, such as ^{232}Th, ^{238}U, and ^{237}Np, undergo fission but require bombardment by fast neutrons of at least 1 MeV energy to provide sufficient energy to the nucleus for fission to occur. These radionuclides are referred to as fissionable isotopes.

d. Photoneutron (γ, n) Sources

Many nuclides emit neutrons upon irradiation with gamma or x-radiation; however, most elements require high-energy electromagnetic radiation in the range 10–19 MeV. The gamma or x-ray energy threshold for the production of neutrons varies with target element. Deuterium and beryllium metal are two exceptions, as they can yield appreciable levels of neutron radiation when bombarded by gamma radiation in the energy range of only 1.7–2.7 MeV. The target material of D_2O or beryllium metal is used to enclose a β^--emitting radionuclide, which also emits gamma rays. The gamma radiation bombards the targets deuterium and beryllium to produce neutrons according to the photonuclear reactions $^2H(\gamma, n)^1H$ and $^9Be(\gamma, n)^8Be$, respectively. The photoneutron source $^{124}Sb + Be$ serves as a good example of a relatively high-yielding combination of gamma emitter with beryllium target. The ^{124}Sb gamma radiation of relevance in photoneutron production is emitted with an energy of 1.69 MeV at 50% abundance (i.e., one-half of the ^{124}Sb radionuclides emit the 1.69-MeV gamma radiation with beta decay). A yield of 5.1 neutrons per 106 beta disintegrations per gram of target material has been reported (Byrne, 1994). The half-life ($t_{1/2}$) of ^{124}Sb is only 60.2 days, which limits the lifetime of the photoneutron generator; nevertheless, this isotope of antimony is easily prepared in the nuclear reactor by neutron irradiation of natural stable ^{123}Sb.

e. Accelerator Sources

The accelerator utilizes electric and magnetic fields to accelerate beams of charged particles such as protons, electrons, and deuterons into target materials. Nuclear reactions are made possible when the charged particles have sufficient kinetic energy to react with target nuclei. Some of the reactions between the accelerated charged particles and target material can be used to generate neutrons.

When electrons are accelerated, they gain kinetic energy as a function of the particle velocity. This kinetic energy is lost as bremsstrahlung electromagnetic radiation when the accelerated electrons strike the target material. Bremsstrahlung radiation is described in Section III.F of this chapter. It is the bremsstrahlung photons that interact with nuclei to produce neutrons according to the mechanisms described in the previous section under photoneutron (γ, n) sources. The accelerated electron-generated neutrons have been reported to yield in a uranium target as many as 10^{-2} neutrons per accelerated electron at an electron energy of 30 MeV with a total yield of 2×10^{13} neutrons per second (Byrne, 1994). The accelerator is a good neutron source for the potential generation of nuclear fuels.

Accelerated deuterons can be used to produce high neutron yields when deuterium and tritium are used as target materials according to the reactions ^{2}H(d, n)^{3}He and ^{3}H(d, n)^{4}He, respectively. In the deuterium energy range 100–300 eV it is possible to obtain neutron yields of the order of 10^{10} neutrons per second from these (d, n) reactions (Byrne 1994) with relatively small electrostatic laboratory accelerators. Large accelerators can provide charged particle energies > 300 MeV capable of inducing neutron sources, such as accelerated proton-induced charge exchange reactions in ^{3}H and ^{7}Li target nuclei according to the reactions ^{3}H(p, n)^{3}He and ^{7}Li(p, n)^{7}Be as described by Byrne (1994). Practical implications of these neutron sources for the generation of nuclear fuels were noted in the previous paragraph. Murray (1993) pointed out that a yield of as many as 50 neutrons per single 500-MeV deuteron has been predicted and that this source of neutron could be used to produce new nuclear fuels via neutron capture by ^{238}U and ^{232}Th according to reactions 1.60–1.65 described previously.

f. Nuclear Fusion

The fusion of two atomic nuclei into one nucleus is not possible under standard temperature and pressure. This is because the repulsing coulombic forces between the positive charges of atomic nuclei prevent them from coming into the required close proximity of 10^{-15} m before they can coalesce into one. However, as described by Kudo (1995) in a review on nuclear fusion, if temperatures are raised to 100 million degrees, nuclei can become plasmas in which nuclei and electrons move independently at a speed of 1000 km s^{-1}, thereby overcoming the repulsing forces between nuclei. Nuclear fusion reactors or controlled thermonuclear reactors (CTRs) are under development to achieve nuclear fusion as a practical energy source.

The reactors are based on maintaining plasmas through magnetic or inertia confinement as described by Dolan (1982) and Kudo (1995). Some fusion reactions also produce neutrons.

The energy liberated during nuclear fusion is derived from the fact that the mass of any nucleus is less than the sum of its component protons and neutrons. This is because protons and neutrons in a nucleus are bound together by strong attractive nuclear forces discussed previously in Section II.C.1. As described by Serway *et al.* (1997) this energy is referred to as the binding energy (BE), that is, the energy of work required to pull a bound system apart leaving its component parts free of attractive forces described by the equation

$$Mc^2 + BE = \sum_{i=1}^{n} m_i c^2 \tag{1.66}$$

where M is mass of the bound nucleus, the m_i's are the free component particle masses (e.g., protons and neutrons), and n is the number of component particles of the nucleus. From Eq. 1.66 we can see that if it is possible to overcome the repulsive forces of protons in nuclei and fuse these into a new nucleus or element of lower mass, energy will be liberated.

Nuclear fusion reactions of two types emit neutrons, and these are of prime interest in man-made controlled thermonuclear reactors. The first type is fusion between deuterium and tritium nuclei according to

$$_{1}^{2}\text{H} + _{1}^{3}\text{H} \rightarrow _{2}^{4}\text{He} + _{0}^{1}\text{n} + 17.58\,\text{MeV} \tag{1.67}$$

and the other type involves fusion between two deuterium nuclei according to either of the following equations, which have approximately equal probabilities of occurring (Kudo, 1995):

$$_{1}^{2}\text{H} + _{1}^{2}\text{H} \rightarrow _{2}^{3}\text{He} + _{0}^{1}\text{n} + 3.27\,\text{MeV} \tag{1.68}$$

and

$$_{1}^{2}\text{H} + _{1}^{2}\text{H} \rightarrow _{1}^{3}\text{H} + _{1}^{1}\text{H} + 4.04\,\text{MeV} \tag{1.69}$$

The fusion reaction between deuterium and tritium or D–T reaction (Eq. 1.67) gives rise to a 14.06-MeV neutron and a 3.52-MeV alpha particle. A D–T plasma burning experiment was performed with 0.2 g of tritium fuel with the Joint European Torus (JET) reactor in November 1991; and in December 1993 a higher power D–T experimental program with 20–30 g of tritium was continued on the Tokamak Fusion Test Reactor (TFTR). These are described by JET Team (1994), Strachan *et al.* (1994), Hawryluk *et al.* (1994), and Kudo (1995). The International Thermonuclear Experimental Reactor (ITER) project was set up under the auspices of the International Atomic Energy Agency (IAEA) to develop a prototype fusion reactor by the year 2030.

Fusion energy production via a commercial reactor is assumed to start around the year 2050 (Sheffield, 2001).

Under development are compact neutron sources, which utilize either D–D or D–T fusion reactions. One instrument described by Miley and Sved (1997) is the inertial electrostatic confinement (IEC) device, which accelerates deuteron ions producing fusion reactions as the ions react with a pure deuterium or deuterium–tritium plasma target. The device is compact measuring 12 cm in diameter and 1 m in length and provides a neutron flux of 10^6–10^7 2.5-MeV D–D n s^{-1} or 10^8–10^9 14-MeV D–T n s^{-1}. Another similar device described by Tsybin (1997) utilizes laser irradiation to create a plasma in an ion source. Compact neutron sources of these types can become competitive with other neutron sources previously described such as ^{252}Cf and accelerator solid-target sources, because of advantages including (i) on–off capability, (ii) longer lifetime without diminished neutron flux strength, and (iii) minimum handling of radioactivity.

3. Interactions of Neutrons with Matter

If a neutron possesses kinetic energy it will travel through matter much more easily than other nuclear particles of similar energy, such as alpha particles, negatrons, positrons, protons, or electrons. In great contrast to other nuclear particles, which carry charge, the neutron, because it lacks charge, can pass through the otherwise impenetrable barrier of the atomic electrons and actually collide with nuclei of atoms and be scattered in the process or be captured by the nucleus of an atom. Collision of neutrons with nuclei can result in scattering of the neutrons and recoil nuclei with conservation of momentum (elastic scattering) or loss of kinetic energy of the neutron as gamma radiation (inelastic scattering). The capture of a neutron by a nucleus of an atom may result in the emission of other nuclear particles from the nucleus (nonelastic reactions) or the fragmentation of the nucleus into two (nuclear fission). A brief treatment of the various types of neutron interactions, which are based on their scattering or capture of neutrons by atomic nuclei, is provided next.

a. Elastic Scattering

The elastic scattering of a neutron by collision with an atomic nucleus is similar to that of a billiard ball colliding with another billiard ball. A portion of the kinetic energy of one particle is transferred to the other without loss of kinetic energy in the process. In other words, part of the kinetic energy of the neutron can be transferred to a nucleus via collision with the nucleus, and the sum of the kinetic energies of the scattered neutron and recoil nucleus will be equal to the original energy of the colliding neutron. This process of interaction of neutrons with matter results only in scattering of the neutron and recoil nucleus. It does not leave the recoil nucleus in an excited energy state. Elastic scattering is a common mechanism by which fast neutrons lose their energy when they interact with atomic nuclei of low atomic number, such as hydrogen (^{1}H) in light water or paraffin, deuterium (^{2}H) in heavy water, and ^{12}C in graphite, which may be encountered in nuclear reactor moderators. It is easy to conceptualize what would occur when particles of

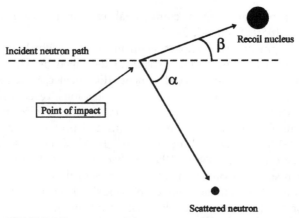

FIGURE I.II Elastic scattering of a neutron by collision of the neutron with an atomic nucleus. The neutron is scattered at an angle α and the nucleus recoils at an angle β to the direction of travel of the incident neutron.

equal or similar mass collide; the event would result in energy transfer and scattering without any other secondary effects, similar to what occurs in billiard ball collisions.

Neutron scattering is the principal mechanism for the slowing of fast neutrons, particularly in media with low atomic number. Let us consider what occurs when a neutron collides with a nucleus and undergoes elastic scattering. Figure 1.11 illustrates the direction of travel of an incident neutron with given kinetic energy (dashed line). The neutron collides with the nucleus. The nucleus is illustrated as undergoing recoil at an angle β while the neutron is scattered at an angle α to the direction of travel of the incident neutron. The kinetic energy (E_k) lost by the neutron in this collision is defined by the equation

$$E_k = \frac{4M\,m_n}{(M + m_n)^2} \cos^2 \beta \qquad (1.70)$$

where M is the mass of the nucleus, m_n is the mass of the neutron, and β is the recoil angle of the nucleus. A derivation of Eq. 1.70 is provided by Bacon (1969). Let us look at two extreme examples of elastic collisions between a neutron and a nucleus. In the first example, it is intuitively obvious from Eq. 1.70 that for a recoil angle $\beta = 90°$, $\cos^2\beta = 0$ and consequently $E_k = 0$. Under such a circumstance, the neutron is undeflected by the nucleus and there is no energy transfer to the nucleus. The neutron continues along its path undeflected until it encounters another nucleus. For the second case, however, let us consider the other extreme in which the recoil angle, $\beta = 0°$ where we have a head-on collision of the neutron with the nucleus of an atom. In this case the maximum possible energy of the neutron is imparted to the nucleus, where $\cos^2\beta = 1$. For example, Table 1.5 provides the maximum fraction of the kinetic energy calculated according to Eq. 1.70 that a neutron can lose upon collision with various atomic nuclei. As illustrated in Table 1.5,

TABLE I.5 The Maximum Fraction of the Kinetic Energy (E_k) that a Neutron Can Lose Upon Collision with the Nucleus of Various Atoms Listed in Increasing Mass in Atomic Mass Units (u)

Nuclide	Nuclide Mass, M	Neutro Mass, m_n	$E_k = 4M\,m_n\,/(M + m_n)^2\cos^2\beta$
^{1}H	1.007825	1.008665	4.065566/4.066232 = 0.999 or 100%
^{2}H	2.014102	1.008665	8.126217/9.137120 = 0.89 or 89%
^{9}Be	9.012182	1.008665	36.36109/100.41737 = 0.362 or 36.2%
^{12}C	12.000000	1.008665	48.41592/169.22536 = 0.286 or 28.6%
^{16}O	15.994915	1.008665	64.53404/289.12173 = 0.223 or 22.3%
^{28}Si	27.976927	1.008665	112.87570/840.16454 = 0.134 or 13.4%
^{55}Mn	54.938047	1.008665	221.65633/3130.0329 = 0.071 or 7.1%
^{197}Au	196.96654	1.008665	787.86616/39194.175 = 0.020 or 2.0%

the neutron can transfer more energy to the nuclei of atoms, which have a low mass; and the highest fraction of its energy can be transferred to the nucleus of the proton, which is almost equal in mass to the neutron. Nuclides of low mass number are, therefore, good moderators for the slowing down of fast neutrons. The substances often used are light water (H_2O), heavy water (D_2O), paraffin (C_nH_{2n+2}), and graphite (C).

b. Inelastic Scattering

We may picture a fast neutron colliding with a nucleus. The neutron is scattered in another direction as described in the previous paragraph; however, part of the neutron's kinetic energy is lost to the recoil nucleus, leaving it in an excited metastable state. Inelastic scattering can occur when fast neutrons collide with nuclei of large atomic number. The recoil nucleus may lose this energy immediately as gamma radiation or remain for a period of time in the excited metastable state. In inelastic scattering, therefore, there is no conservation of momentum between the scattered neutron and recoil nucleus. Inelastic scattering occurs mainly with fast neutron collisions with nuclei of large atomic number.

Neutron scattering is a common mechanism by which fast and intermediate neutrons are slowed down to the thermal neutron energy levels. Thermal neutrons have an energy level at which they are in thermal equilibrium with the surrounding atoms at room temperature. There is an energy range for thermal neutrons as described earlier in this chapter; however, the properties of thermal neutrons are often cited at an energy calculated to be the most probable thermal neutron energy of 0.0253 eV at 20°C corresponding to a velocity of 2200 m s^{-1} (Gibson and Piesch, 1985). Figure 1.9 may be used to find the velocity of the neutron at energy levels over the range 0.001–10 MeV. For example, if we select the position 0.025 eV on the X axis and follow up the graph with a straight line to the upper curve, we find the value 2200 m s^{-1}. At the thermal energy state, the mechanisms of interaction of neutrons with matter change drastically as discussed in the following.

c. Neutron Capture

Because of the neutral charge on the neutron, it is relatively easy for slow neutrons in spite of their low kinetic energy to "find themselves" in the vicinity of the nucleus without having to hurdle the coulombic forces of atomic electrons. Once in close proximity to nuclei, it is easy for slow neutrons to enter into and be captured by nuclei to cause nuclear reactions. The capture of thermal neutrons, therefore, is possible with most radionuclides, and neutron capture is the main reaction of slow neutrons with matter. The power of a nucleus to capture a neutron depends on the type of nucleus as well as the neutron energy. The neutron absorption cross section, σ, with units of $10^{-24}\,cm^2$ or "barns," is used to measure the power of nuclides to absorb neutrons. A more detailed treatment of the absorption cross section and its units and application are given in Section II.G.4 of this chapter. However, because capture of thermal neutrons is possible with most radionuclides, references will cite the neutron cross sections of the nuclides for comparative purposes at the thermal neutron energy of $0.0253\,eV$ equivalent to a neutron velocity of $2200\,m\,s^{-1}$. This is also the energy of the neutron, which is in thermal equilibrium with the surrounding atoms at room temperature. For comparative purposes, therefore, Table 1.6 lists the thermal neutron cross sections for neutron capture reactions in barns ($10^{-24}\,cm^2$)

TABLE I.6 Cross Sections σ in Barns for Thermal Neutron Capture Reactions of Selected Nuclides in Order of Increasing Magnitude

Nuclide	σ (barns)
$^{3}_{1}H$	< 0.000006
$^{2}_{1}H$	0.00052
$^{16}_{8}O$	0.00019
$^{12}_{6}C$	0.0035
$^{1}_{1}H$	0.332
$^{14}_{7}N$	1.8
$^{238}_{92}U$	2.7
$^{232}_{90}Th$	7.4
$^{55}_{25}Mn$	13.3
$^{233}_{92}U$	530
$^{235}_{92}U$	586
$^{239}_{94}Pu$	752
$^{6}_{3}Li$	940
$^{10}_{5}B$	3840
$^{3}_{2}He$	5330
$^{7}_{4}Be$	39,000
$^{155}_{64}Gd$	61,000
$^{157}_{64}Gd$	254,000

Data from Holden (1997).

for several nuclides. The nuclides selected for Table 1.6 show a broad range of power for thermal neutron capture. Some of the nuclides listed have practical applications, which are referred to in various sections of this book.

The capture of a slow neutron by a nucleus results in a compound nucleus, which finds itself in an excited energy state corresponding to an energy slightly higher than the binding energy of the neutron in the new compound nucleus. This energy of excitation is generally emitted as gamma radiation. Neutron capture reactions of this type are denoted as (n, γ) reactions. Two practical examples of (n, γ) neutron capture reactions were provided earlier in this chapter in the neutron irradiation of ^{238}U and ^{232}Th for the preparation of fissile ^{239}Pu and ^{233}U (Eqs. 1.60 and 1.63), respectively. Another interesting example of a (n, γ) reaction is neutron capture by ^{235}U according to

$$^{235}_{92}U + ^{1}_{0}n \rightarrow ^{236}_{92}U + \gamma \qquad (1.71)$$

This neutron capture reaction is interesting, because the ^{236}U product nuclide decays by alpha emission in approximately 14% of the cases and decays by nuclear fission with emission of neutrons in the remaining 86% of the cases as discussed previously in Section II.G.2.c. The subject of neutron capture is treated in more detail in Section II.G.4, which concerns the neutron cross section and neutron attenuation in matter.

d. Nonelastic Reactions

Neutron capture can occur in nuclei resulting in nuclear reactions that entail the emission of nuclear particles such as protons (n, p), deuterons (n, d), alpha particles $(n, \alpha$ and even neutrons $(n, 2n)$. These reactions may not occur in any specific energy range but may be prevalent at specific resonances, which are energy states of the excited compound nuclei that are specific to relatively narrow energies of the incident neutron. The effect of resonance in neutron capture by nuclei is discussed in more detail subsequently in Section II.G.4. The $(n, 2n)$ reactions occur at very high incident neutron energies, $> 10\,\mathrm{MeV}$ (Gibson and Piesch, 1985). The (n, p) and (n, α) reactions can occur in the slow neutron capture and reaction with nuclides of low atomic number (low Z), where the Coulomb forces of the electron shells are limited and present less a hurdle for the escape of charged particles from the confines of the atom. Some practical examples of these reactions are the (n, p) reaction used in the synthesis of ^{14}C by slow (thermal) neutron capture by ^{14}N

$$^{14}_{7}N + ^{1}_{0}n \rightarrow ^{14}_{6}C + ^{1}_{1}H \qquad (1.72)$$

and the (n, p) and (n, α) reactions used to detect neutrons by the interaction of slow neutrons with ^{3}He and ^{10}B, respectively, according to Eqs. 1.73 and 1.74.

$$^{3}_{2}He + ^{1}_{0}n \rightarrow ^{1}_{1}H + ^{3}_{1}H + 0.76\,\mathrm{MeV} \qquad (1.73)$$

$$^{10}_{5}B + ^{1}_{0}n \rightarrow ^{7}_{3}Li + ^{4}_{2}He + 2.8\,\mathrm{MeV} \qquad (1.74)$$

Either of these reactions is used to detect neutrons by using gas proportional detectors containing helium or a gaseous form of boron (e.g., boron trifluoride). Slow neutrons that penetrate these detectors produce either radioactive tritium (Eq. 1.73) or alpha particles (Eq. 1.74), which produce ionization in the gas. The ionization events or ion pairs formed can be collected and counted as described in Chapter 2 to determine a neutron count rate.

e. Nuclear Fission

The reaction of neutron-induced fission occurs when a neutron interacts with a fissile or fissionable nucleus and the nucleus becomes unstable, taking on the characteristics of an oscillating droplet, which then fragments into two nuclides (fission fragments). At the same time there is the release of more than one neutron (2.4 neutrons on the average for ^{235}U fission) and a relatively high amount of energy ($\sim 194\,$MeV). Fission in natural ^{235}U and man-made ^{233}U and ^{239}Pu is optimal at thermal incident neutron energies; whereas fission in ^{238}U and ^{232}Th requires neutron energies of at least 1 MeV. A more detailed treatment of nuclear fission was provided previously in Section II.G.2.c.

4. Neutron Attenuation and Cross Sections

As we have seen in our previous treatment of the neutron, there are several possible interactions of neutrons with nuclei. Among these are elastic scattering, inelastic scattering, neutron capture, nonelastic reactions, and nuclear fission. As we have seen in several examples, probabilities exist for any of these interactions to occur depending on the energy of the incident neutron and the type of nuclide with which the neutron interacts. We can define this probability of interaction by the term cross section, which is a measure of the capturing power of a particular material for neutrons of a particular energy.

The range of neutrons in matter is a function of the neutron energy and the cross section or capturing power of the matter or medium through which the neutrons travel. To define cross section, let us consider an incident beam of neutrons of given intensity or number (I_0), which impinges on a material of unit area (e.g., cm^2) and thickness dx as illustrated in Fig. 1.12.

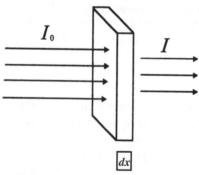

FIGURE 1.12 Attenuation of an incident neutron beam of intensity I_0 by an absorber material of unit area (cm^2) and thickness dx.

The intensity (I) of the neutron beam traveling beyond the thickness dx will be reduced according to the number of nuclei (n) per unit volume in the material and the "area of obstruction" (e.g., cm^2) that the nuclei present to the oncoming beam. This area of obstruction is referred to as the cross section of the material. On the basis of the description previously given, we can write the equation

$$dI/dx = -n\sigma I \qquad (1.75)$$

which defines the change in beam intensity (dI) with respect to absorber thickness (dx) as proportional to the beam intensity (I) times a proportionality factor, which we may call the absorption coefficient or "obstruction coefficient" that the nuclei pose to the oncoming beam. The coefficient is a function of the number of nuclei (n) in the path of the neutron beam and the stopping power of the nuclei to interact with the neutron beam or, in other words, the neutron cross section (σ) of the material through which the neutron beam travels. Equation 1.75 may be written as

$$dI/I = -n\sigma \, dx \qquad (1.76)$$

Equation 1.76 is very similar to Eq. 1.117 defining the attenuation of gamma radiation in matter with the exception that the absorption coefficients and attenuation coefficients involved for neutron and gamma radiation, respectively, are very different. The negative sign of Eqs. 1.75 and 1.76 denotes the diminishing intensity of the neutron beam as a function of absorption coefficient and absorber thickness. The absorption coefficient $n\sigma$ is the combined effect of the number of nuclei (n) in the neutron beam path that might impede the continued travel of neutrons and the power of the nuclei to react with the neutrons. Equation 1.76 can be integrated over the limits of beam intensity from I_0 to I and absorber thickness from 0 to x as follows:

$$\int_{I_0}^{I} dI/I = -n\sigma \int_{0}^{x} dx \qquad (1.77)$$

to give the equation

$$\ln I_0/I = n\sigma x \qquad (1.78)$$

or

$$I = I_0 e^{-n\sigma x} \qquad (1.79)$$

which is the most simplified expression for the calculated beam intensity (I) after passing through an absorber of thickness (x) when the absorber material consists of only one pure nuclide and only one type of reaction between the neutron beam and nuclei is possible. If, however, several types of nuclei and reactions between the neutron beam and nuclei of the absorber material are possible, we must utilize the sum of the neutron cross sections for all reactions that could take place.

We can use Eq. 1.78 to calculate the half-value thickness ($x_{1/2}$) or the thickness of absorber material needed to reduce the incident neutron beam intensity by one-half. If we give the initial beam intensity (I_0) a value of 1 and the transmitted intensity (I) a value of 1/2, we can write

$$\ln 1/0.5 = n\sigma\, x_{1/2} \tag{1.80}$$

and

$$\ln 2 = n\sigma\, x_{1/2} \tag{1.81}$$

or

$$0.693 = n\sigma\, x_{1/2} \tag{1.82}$$

The half-value thickness for neutron beam attenuation may be written

$$x_{1/2} = 0.693/n\sigma \tag{1.83}$$

where $n\sigma$ is the number of nuclei per unit volume (cm^{-3}) and σ the neutron cross section in cm^2. The neutron cross section σ can be defined as the area in cm^2 for which the number of nuclei–neutron reactions taking place is equal to the product of the number of incident neutrons that would pass through the area and the number of target nuclei. The cross section is defined in units of $10^{-24}\,cm^2$ on the basis of the radius of atomic nuclei being about $10^{-12}\,cm$. It provides a measure of the chances for the nuclei of a material being hit by a neutron of a certain energy. The unit of $10^{-24}\,cm^2$ for nuclear cross sections is called the barn. Tables in reference sources of nuclear data provide the neutron cross sections in units of barns for various nuclides and nuclide energies. An example is the reference directory produced by McLane et al. (1988), which provides neutron cross section values in barns and neutron cross section curves for most nuclides over the neutron energy range 0.01eV to 200 MeV.

Let us take an example of 10-eV neutrons incident on a water barrier (i.e., neutrons traveling in water). We may use Eq. 1.83 to estimate the half-value thickness, if we ignore the less significant interactions with oxygen atoms. This is because the neutron cross section for hydrogen at 10 eV is about 20 barns (Fig. 1.13) and that of oxygen is only 3.7 barns (McLane et al., 1988), and there are twice as many hydrogen atoms as oxygen atoms per given volume of water. The half-value thickness may be calculated as follows:

The value of n for the number of hydrogen nuclei per cm^3 of water may be calculated on the basis of Avogadro's number of molecules per mole.

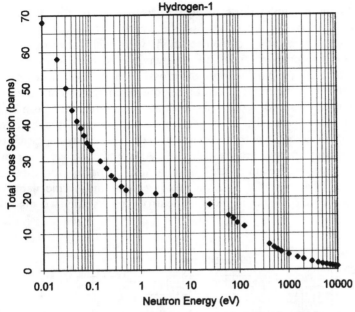

FIGURE 1.13 **Total cross section curve for hydrogen-1 over the neutron energy range 0.01–10 keV.**

If 1 mole of water is equivalent to 18.0 g and the density of water is 1.0 g cm^{-3}, we can calculate the number of hydrogen nuclei per cm^3 as

$$6.22 \times 10^{23} \text{molecules H}_2\text{O}/18 \, \text{cm}^3 = 0.0334 \times 10^{24} \text{molecules H}_2\text{O/cm}^3$$

$$n = (0.0334 \times 10^{24} \text{molecules H}_2\text{O/cm}^3)(2 \text{ proton atoms or } {}_1^1\text{H/molecule})$$

$$= 0.0668 \times 10^{24} \, {}_1^1\text{H nuclei/cm}^3.$$

By definition, 20 barns is equal to $20 \times 10^{-24} \, \text{cm}^2$ and the half-value thickness may then be calculated as

$$x_{1/2} = 0.693/(0.0668 \times 10^{24} \, \text{cm}^{-3})(20 \times 10^{-24} \, \text{cm}^{-2})$$

$$= 0.693/1.34 \, \text{cm}^{-1}$$

$$= 0.51 \, \text{cm}$$

If we make the calculation for 1-MeV neutrons traversing water and use the value 4.1 barns for the neutron cross section of hydrogen nuclei at this neutron energy (McLane *et al.*, 1988), we calculate a half-value thickness of

$$x_{1/2} = 0.693/(0.0668 \times 10^{24} \, \text{cm}^{-3})(4.1 \times 10^{-24} \, \text{cm}^2)$$

$$= 0.693/0.274 \, \text{cm}^{-1}$$

$$= 2.53 \, \text{cm}$$

As the examples illustrate in the case of the proton, the neutron cross section (or barns) decreases as the energy or velocity of the neutron increases. That is, the neutron reactions with nuclei obey the general rule of having some proportionality to $1/v$, where v is the velocity of the neutron. This inverse proportionality of cross section and neutron velocity is particularly pronounced in certain regions of energy as illustrated in the total neutron cross section curves for protons and elemental boron in Figs. 1.13 and 1.14, respectively. However, this is not always the case with many nuclides at certain neutron energies where there exists a resonance between the neutron energy and the nucleus. At specific or very narrow neutron energy ranges, certain nuclei have a high capacity for interaction with neutrons. The elevated neutron cross sections at specific neutron energies appear as sharp peaks in plots of neutron cross section versus energy, such

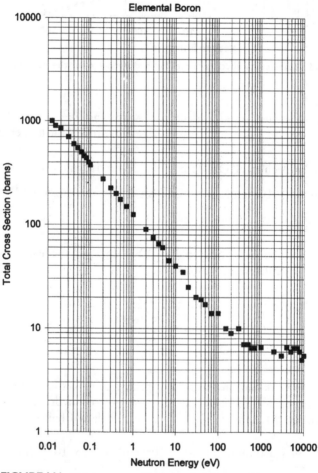

FIGURE 1.14 Total cross section curve for elemental boron over the neutron energy range 0.01–10 keV.

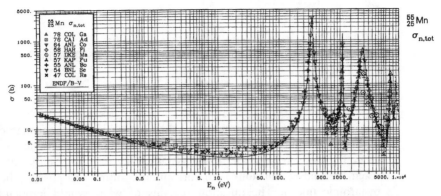

FIGURE 1.15 Total cross section curve for manganese-55 over the neutron energy range 0.01–10 keV. The columns in the upper left-hand corner provide the number of data points and an abbreviation of the laboratory that provided the data. (From McLane et al., 1988, reprinted with permission from Elsevier Science.)

as the cross section curve illustrated in Fig. 1.15 for $^{55}_{25}$Mn. These peaks are called resonances and often occur with (n, γ) reactions. The high cross sections occur when the energy of the incident neutron corresponds exactly to the quantum state of the excited compound nucleus, which is the newly formed nucleus consisting of a compound between the incident neutron and the nucleus. Most nuclides display both the $1/v$ dependence on neutron cross section and the resonance effects over the entire possible neutron energy spectrum. We should keep in mind that neutron cross sections can be specific and differ in value for certain reactions, such as proton (σ_p)- and alpha particle (σ_α)-producing reactions, fission reactions (σ_f), or neutron capture cross sections (σ_c). The total neutron cross section (σ_{tot} or σ_a) would be the cross section representing the sum of all possible neutron reactions at that specific neutron energy. For example, the thermal neutron cross section for ^{235}U, which is the neutron cross section at 0.0253 eV neutron energy corresponding to a neutron velocity of 2200 m s^{-1} at room temperature, can be given as $\sigma_c = 95$ barns for the neutron capture cross section, $\sigma_f = 586$ barns for the fission cross section, and $\sigma_\alpha = 0.0001$ barns for the neutron cross section for the alpha particle-producing reaction. These neutron cross section values indicate that neutron fission would predominate at the thermal neutron energy of 0.0253 eV, although some neutron absorption would also occur. The total neutron cross section, σ_{tot}, would be the total of the three possible reactions or $\sigma_{tot} = 95$ barns $+ 586$ barns $+ 0.0001$ barns $= 681$ barns. In our treatment of slow neutron capture by ^{235}U in Section II.G.3, illustrated by Eq. 1.71, we noted that about 14% of the slow neutron captures by ^{235}U nuclei result in the formation of ^{236}U and gamma radiation and the remaining 86% of the slow neutron captures result in nuclear fission. This is exactly what is predicted by the thermal neutron cross section values just provided; that is, for ^{235}U

$$\sigma_c/\sigma_{tot} = 95 \text{ barns}/681 \text{ barns} = 14\% \text{ neutron capture}$$

and

$$\sigma_f / \sigma_{\text{tot}} = 586 \text{ barns}/681 \text{ barns} = 86\% \text{ fission}$$

5. Neutron Decay

We have seen that fast neutrons may lose their energy through elastic and inelastic collisions with other nuclei, and if these neutrons do not undergo other reactions with nuclei (e.g., fission), they may lose sufficient energy to reach thermal equilibrium with surrounding atoms and possibly be captured by atomic nuclei. The question remains of what would happen to a free neutron that is not absorbed by any atomic nucleus.

Earlier in this chapter (Section II.B) we discussed the transformation of the neutron within nuclei of radioactive atoms, which have a neutron/proton ratio too high for stability. In these unstable nuclides the neutron breaks up into a proton, negatron (negative electron), and antineutrino. However, within the confines of a stable nucleus, that is, one that does not have an n/p imbalance, there is no transformation of the neutron. If the neutron can transform itself in unstable nuclei, it stands to reason that the neutron might be unstable outside the protective boundaries of the stable nucleus. This is just the case, as A. H. Snell and L. C. Miller demonstrated in 1948 followed by further studies by Robson (1950a,b) and Snell et al. (1950) that when neutrons were in free flight in a vacuum, they would indeed decay with a lifetime in the range of 9–25 minutes with a release of 0.782 MeV of energy. More recent and accurate measurements of neutron decay demonstrate the lifetime to be 885.4 ± 0.9 s (Abele, 2000; Arzumanov et al., 2000; Pichlmaier et al., 2000; Snow et al., 2000). The decay of elementary particles is characterized in terms of lifetime. The lifetime, usually symbolized as τ, is related to the term half-life, $t_{1/2}$, the mean time it takes for one-half of the particles to decay (Sundaresan, 2001) according to the relationship

$$t_{1/2} = (\ln 2)\tau = 0.693\tau \tag{1.84}$$

The free neutron decays according to the scheme

$$n \rightarrow p^+ + e^- + \bar{\nu} + 0.782 \text{ MeV} \tag{1.85}$$

The 0.782 MeV of energy released in the neutron decay corresponds to the difference in mass of the neutron (1.0086649 u) and the sum of the masses of the products of the neutron decay, the proton (1.0072765 u) plus the electron (0.0005485 u), or 1.0078250 u. Using Einstein's equation of equivalence of mass and energy (Section IV.C of this chapter), this mass difference of 0.0008399 u can be converted to the equivalent of 0.782 MeV of energy. This calculation provides additional evidence for the decay of the neutron into a proton and an electron. The neutron, therefore, outside the protective confines of a stable nucleus, has a very short lifetime.

III. ELECTROMAGNETIC RADIATION — PHOTONS

A. Dual Nature: Wave and Particle

In the latter part of the 19th century Heinrich Hertz carried out a series of experiments demonstrating that an oscillating electric current sends out electromagnetic waves similar to light waves, but of different wavelength. Hertz proved, thereby, the earlier theory of James Clerk Maxwell, that electric current oscillations would create alternating electric and magnetic fields, and radiated electromagnetic waves would have the same physical properties of light. A subsequent discovery by Pieter Zeeman in 1896 further linked the properties of light with electricity and magnetism when he discovered that a magnetic field would alter the frequency of light emitted by a glowing gas, known as the Zeeman effect (Serway *et al.*, 1997).

Not long after the discoveries of Hertz and Zeeman came the work of Max Planck, who in 1900 proposed a formula to explain that the vibrating particles in the heated walls of a kiln could radiate light only at certain energies. These energies would be defined by the product of a constant having the units of energy × time and the radiation frequency. The constant, which he calculated became known as the universal Planck constant, $h = 6.626 \times 10^{-34}$ J s. Therefore, radiation would be emitted at discrete energies, which were multiples of Planck's constant and the radiation frequency, ν. Planck named the discrete radiation energy as the quantum from the Latin *quantus* meaning "how great."

In 1905 Einstein grasped the calculations of Planck to explain and provide evidence that light not only traveled as waves but also existed as discrete packets of energy or particles, which he named "energy quanta." Today we refer to these energy quanta as photons. Einstein demonstrated the existence of the photon in his explanation of the photoelectric effect (see Section IV.A of this chapter). He demonstrated that the energy of an electron (photoelectron) ejected from its atomic orbital after being struck by light was not dependent on the light intensity, but rather on the wavelength or frequency of the light. In other words, increasing the light intensity would increase the number of photoelectrons, but not their energy. Whereas, altering the frequency, thus energy, of the light would alter the energy of the photoelectron. In summary, Einstein demonstrated that the energy of the photoelectron depended on the energy of the photon that collided with the electron or, the product of Planck's constant times the light frequency according to the formula

$$E = h\nu = \frac{hc}{\lambda} \qquad (1.86)$$

Equation 1.86 is referred to as the Planck–Einstein relation (Woan, 2000). Notice from Eq. 1.86 that the product of the photon frequency, ν, and wavelength, λ, always yields the velocity, c, the speed of light. The photon always travels at the constant speed in a vacuum, $c = 2.9979 \times 10^8$ m s^{-1}; it cannot travel at a speed less than c in a vacuum.

From our previous treatment we see that the photon behaves as a particle, which could knock out an electron from its atomic orbit provided it possessed sufficient energy to do so, that is, an energy in excess of the electron binding energy. Therefore, the photon can be considered also as another elementary particle. In his explanation of the photoelectric effect Einstein was the first to demonstrate the particulate nature of light, and it is for this work he won the Nobel Prize. Since these findings of Einstein, electromagnetic radiation is known to have a dual nature as energy that travels as a wave and particle.

Electromagnetic radiation may be classified according to its wavelength or origin. For example, we will see in this section of the chapter that gamma rays and x-rays are similar, but have different origins. Gamma rays arise from the nucleus of an atom while x-rays come from extranuclear electrons. The classification of electromagnetic radiation according to wavelength and frequency is illustrated in Fig. 1.16.

Since electromagnetic radiations or photons have properties of particles, they should also possess momentum. We calculate momentum as the product of mass and velocity. For relativistic conditions, the mass of a particle is

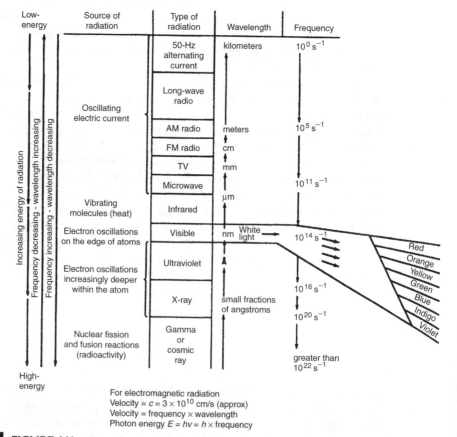

FIGURE 1.16 Electromagnetic radiation spectrum. (From Dean, 1995, reproduced with permission of The McGraw-Hill Companies.)

a function of its speed according to Eq. 1.50 previously described or

$$m = \frac{m_0}{\sqrt{1-(u^2/c^2)}} \tag{1.87}$$

where m and m_0 are the particle relativistic and rest masses, u is the particle speed and c, the speed of light. Grateau and Savin (1999) transform Eq. 1.87 by squaring both sides and then multiplying each side by $c^2[1-(u^2/c^2)]$ to yield the equation

$$m^2 c^4 - m^2 u^2 c^2 = m_0^2 c^4 \tag{1.88}$$

Using $E=mc^2$ and $E_0=m_0c^2$ to define the relativistic and rest energies and $p=mu$ to define the particle momentum together with the fact that the rest energy of the photon is always zero, i.e., $m_0=0$, Eq. 1.88 becomes

$$E^2 - p^2 c^2 = 0 \tag{1.89}$$

and

$$p = \frac{E}{c} \tag{1.90}$$

From Eqs. 1.86 and 1.90 we can further describe the photon momentum as

$$p = \frac{E}{c} = \frac{h\nu}{c} = \frac{h}{\lambda} \tag{1.91}$$

To illustrate the use of the above equations defining the relationships of photon properties, let us calculate the wavelength, frequency and momentum of a 2-MeV gamma-ray photon. From Eq. 1.86 we can write the equation for calculating the wavelength as

$$\lambda = \frac{hc}{E} \tag{1.92}$$

Planck's constant, h, can be converted from units of J s to eV s as

$$h = 6.626 \times 10^{-34} \, \text{J s} / 1.602 \times 10^{-19} \text{J} \, \text{eV}^{-1}$$
$$= 4.136 \times 10^{-15} \, \text{eV s} \tag{1.93}$$

and hc is calculated as

$$hc = (4.136 \times 10^{-15}\,\text{eV}\,\text{s})(2.9979 \times 10^8\,\text{m}\,\text{s}^{-1})$$
$$= 12.399 \times 10^{-7}\,\text{eV}\,\text{m} \tag{1.94}$$
$$= 12.4\,\text{keV}\,\text{Å}$$

The wavelength according to Eq. 1.92 becomes

$$\lambda = \frac{12.4\,\text{keV} \cdot \text{Å}}{2 \times 10^3\,\text{keV}} = 0.0062\,\text{Å}$$

The frequency is calculated according to Eq. 1.86 as

$$\nu = \frac{c}{\lambda} = \frac{2.9979 \times 10^8\,\text{m}\,\text{s}^{-1}}{0.0062 \times 10^{-10}\,\text{m}} = 484 \times 10^{18}\,\text{s}^{-1} = 4.84 \times 10^{20}\,\text{Hz}$$

The momentum is expressed according to Eq. 1.90 as

$$p = \frac{E}{c} = 2.0\,\text{MeV}/c$$

Notice that relativistic calculations of momentum have units of MeV/c, while conventional units of momentum are derived from mass times velocity or kg $\cdot$ m s^{-1}. Units of MeV/c can be converted to the conventional units with the conversion factor $1\,\text{MeV}/c = 0.534 \times 10^{-21}$ kg m s^{-1} (Gautreau and Savin, 1999).

B. Gamma Radiation

Radionuclide decay processes often leave the product nuclide in an excited energy state. The product nuclide in such an excited state either falls directly to the ground state or descends in steps to lower energy states through the dissipation of energy as gamma radiation.

A nuclide in an excited energy state is referred to as a nuclear isomer, and the transition (or decay) from a higher to a lower energy state is referred to as isomeric transition. Gamma rays are emitted in discrete energies corresponding to the energy state transitions a nuclide may undergo when in an excited state. The energy, E_γ, of a gamma ray may be described as the difference in energy states of the nuclear isomers:

$$E_\gamma = h\nu = E_1 - E_2 \tag{1.95}$$

where $h\nu$ is the energy of the electromagnetic radiation described previously in Section III.A, and E_1 and E_2 represent the energy levels of the nuclear isomers.

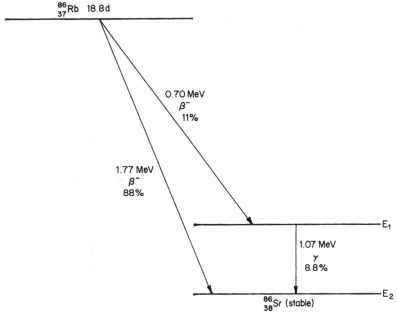

FIGURE 1.17   Decay scheme of $^{86}_{37}$Rb.

Let us consider the decay schemes of some radionuclides to illustrate the process in more detail.

Figure 1.17 shows the decay scheme of $^{86}_{37}$Rb with a half-life of 18.8 days. This nuclide decays by β^- emission with an increase in atomic number to $^{86}_{38}$Sr. Eighty-eight percent of the beta particles emitted have a maximum energy of 1.77 MeV; the remaining 11% have a maximum energy of 0.70 MeV. The percentages cited and illustrated in the figure are referred to as transition probabilities or intensities. Obviously, a greater quantum of energy is released by the 1.77-MeV, β decay process. As a consequence, the ^{86}Sr product nuclides that result from β^- emission of 0.70 MeV (11%) are at a higher energy state than those that result from β^- emission of 1.77 MeV. The energy difference of the two ^{86}Sr product nuclide isomers, $E_1 - E_2$, is equivalent to the difference of the two β^- energies, $1.77 \text{MeV} - 0.70 \text{MeV} = 1.07 \text{MeV}$. Consequently, the ^{86}Sr nuclide isomers, which are products of the 0.70-MeV, β-decay process, can emit the remaining energy as 1.07-MeV gamma-ray photons.

As illustrated in Fig. 1.17, 11% of the parent ^{86}Rb nuclides decay to an ^{86}Sr nuclear isomer at an elevated energy state. Not all of these isomers immediately decay to the ground state. Only 8.8% of the ^{86}Rb $\rightarrow$ ^{86}Sr disintegrations result in the emission of a gamma-ray photon of 1.07 MeV. For example, a 37-kBq sample of ^{86}Rb by definition would emit 2.22×10^6 beta particles in 1 minute $(37,000 \text{ dps} \times 60 \text{ s m}^{-1})$. However, only $(2.22 \times 10^6)(0.088) = 1.95 \times 10^5$ gamma-ray photons of 1.07 MeV can be expected to be emitted in 1 minute from this sample.

Figure 1.18 shows the somewhat more complicated decay scheme of $^{144}_{58}$Ce, which has a half-life of 284.5 days. This nuclide decays by β^- emission

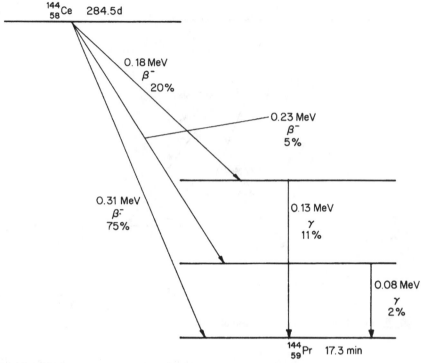

FIGURE I.18 Decay scheme of $^{144}_{58}$Ce.

with an increase in atomic number to $^{144}_{59}$Pr. In this case, three distinct
β-decay processes produce three nuclear isomers of the daughter ^{144}Pr.
Seventy-five percent of the beta particles emitted have a maximum energy
of 0.31 MeV, 20% have a maximum energy of 0.18 MeV, and the remaining
5% have a maximum energy of 0.23 MeV. Obviously, a greater amount
of energy is released by the 0.31-MeV β-decay process. As a consequence,
^{144}Pr nuclides that result from β^{-} emission of 0.23 MeV can decay to the
ground state with the emission of gamma-ray photons with an energy
equivalent to 0.08 MeV (0.31 MeV−0.23 MeV). Likewise, ^{144}Pr isomers at an
even higher energy state are products of the 0.18-MeV β-decay process.
These can decay to the ground state with the emission of gamma-ray photons
of energy 0.13 MeV (0.31 MeV−0.18 MeV). Not all of the product isomers
decay with the immediate emission of gamma radiation, and the abundance
of these transitions is given in Fig. 1.18. The per cent abundances of gamma-
ray emissions that occur in the decay of radionuclides are given in the
Appendix.

     It is also possible that essentially all of the product nuclides of a decay
reaction will be at an excited or elevated energy state and subsequently fall
to a lower energy state by the emission of gamma radiation.

     The decay scheme of the nuclide $^{22}_{11}$Na with a 2.6-year half-life serves
as an example (see Fig. 1.19). The $^{22}_{11}$Na nuclides decay by both electron
capture and β^{+} emission, at relative proportions of 10 and 90%, respectively,
to yield immediate $^{22}_{10}$Ne product nuclides in an elevated energy state.

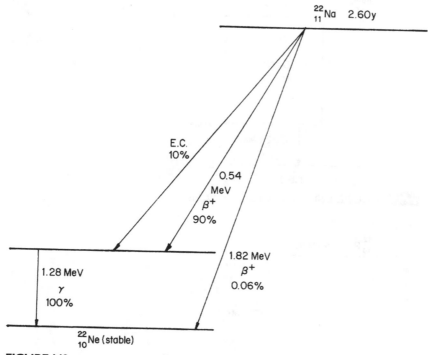

FIGURE 1.19 Decay scheme of $_{11}^{22}$Na.

Only a trace of the ^{22}Na nuclides (0.06%) decay directly to the ground state. All of the $_{10}^{22}$Ne isomers in the excited energy state decay immediately with the emission of gamma-ray photons of 1.28 MeV energy, which is equivalent to the difference of the energy levels of the two $_{10}^{22}$Ne isomers and also equivalent to the difference in energies released by the two β^+-decay processes (1.82–0.54 MeV).

Isomeric transition, as described earlier, is a decay process in which γ emission is the sole process of eliminating energy from an excited nucleus. This mode of decay is referred to as isomeric transition because neither the mass number, A, nor the atomic number, Z, of a nuclide ($_Z^A$X) changes in the decay process, and the nuclides are considered to be in isomeric energy states.

In the previous examples (Figs. 1.17, 1.18, and 1.19) the isomeric energy state transitions are short-lived; that is, they occur virtually immediately after the other decay processes (e.g., β^-, β^+, and EC) and the half-life of the parent nuclide is dependent on these initial processes. If, however, the isomeric transitions are long-lived, the nuclide is considered to be in a metastable state. These nuclides are denoted by a superscript m beside the mass number of the nuclide. The radionuclide $^{119m}_{50}$Sn with a 250-day half-life is an example. Its decay scheme, shown in Fig. 1.20, illustrates the emission of two γ photons of 0.065 and 0.024 MeV energy falling from the 0.089-MeV excited state to the ground (stable) state.

Gamma radiation is not produced in all radionuclide decay processes. Instead, some radionuclides decay by emitting only particulate radiation to

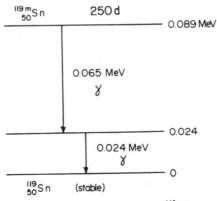

FIGURE 1.20   Decay scheme of $^{119m}_{50}$Sn.

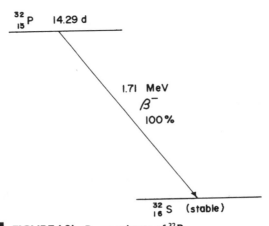

FIGURE 1.21   Decay scheme of $^{32}_{15}$P.

yield a product nuclide at an unexcited ground state. An example is the commonly used radionuclide ^{32}P, whose decay scheme is shown in Fig. 1.21.

C. Annihilation Radiation

The negatron or negative beta particle, produced by β decay or by pair production (see Section IV.C), will travel through matter until it has completely dissipated its kinetic energy via ionization, electron excitation or bremsstrahlung. The negatron then at rest acts as an atomic or free electron in matter.

A positron or positive beta particle, however, may be considered an "antiparticle" of an electron and consequently, in the electron environment of atoms, has a definite instability. A given positron emitted by pair production or by β^+ decay will also dissipate its kinetic energy in matter via interactions described previously for the case of the negatron. However, as the positron loses its kinetic energy and comes to a near stop, it comes into contact with an electron (Fig. 1.22) with nearly simultaneous annihilation of the positron

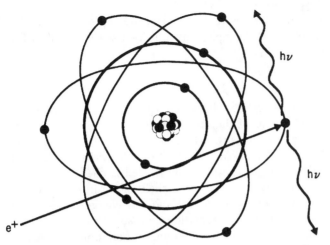

FIGURE I.22   Annihilation. The interaction between a positron and electron and the conversion of their mass into two photons of 0.5I MeV energy.

and the electron masses and their conversion into energy. The annihilation involves the formation of positronium, which is a short-lived association of the positron and electron. Its lifetime is only approximately 10^{-10} or 10^{-7} seconds, depending on whether the spin states of the associated particles are parallel (ortho-positronium) or opposed (para-positronium). The para-positronium is the shorter-lived spin state. The energy released in this annihilation appears as two photons emitted in opposite directions. This transformation of mass into energy, considered as the reverse of pair production, is described as

$$e^{+} + e^{-} = 2h\nu = 2E_{\gamma} \qquad (1.96)$$

where a positron, e^{+}, and electron, e^{-}, combine to form two gamma-ray photons of energy E_{γ}. To maintain the equivalence of mass and energy (see Eq. 1.112), the equivalent of two electron rest masses (0.51 MeV) must appear as photon energies (see Section IV.C). In agreement with Eq. 1.113, the annihilation results in the emission of two 0.51-MeV photons in opposite directions.

D. Cherenkov Radiation

Charged particles, when they possess sufficient energy, may travel through matter at a speed greater than the speed of light in that material. This occurrence causes emission of photons of light. These photons extend over a spectrum of wavelengths from the ultraviolet into the visible portion of the electromagnetic radiation spectrum.

The photon emission is a result of a coherent disturbance of adjacent molecules in matter caused by the traveling charged particle, which must possess a certain threshold energy. This phenomenon has practical applications in the measurement and detection of radionuclides that emit relatively high-energy beta particles (L'Annunziata and Passo, 2002). The theory and applications of Cherenkov photons are discussed in detail in Chapter 9.

E. X-Radiation

Mention has been made of the electron capture decay process whereby an electron from one of the atomic shells (generally the innermost K shell) is absorbed by the nucleus, where it combines with a proton to form a neutron. No particle emission results from this decay process. However, the vacancy left by the electron from the K shell is filled by an electron from an outer shell (generally the adjacent L shell). Transitions produced in electron shell energy levels result in the emission of energy as x-radiation (see also Sections II.E and II.F). This radiation consists of photons of electromagnetic radiation similar to gamma radiation. X-radiation and gamma radiation differ in their origin. X-rays arise from atomic electron energy transitions and gamma rays from transitions between nuclei of different energy states. The production of x-radiation from atomic electron transitions is illustrated in Figs. 1.7 and 1.23.

When an electron transition occurs from the outer L shell to an inner K shell, the energy emitted is equivalent to the difference between the K and L electron binding energies. The electron transitions that ensue in the filling of vacancies are a deexcitation process, and the energy lost by the atom as x-radiation is equivalent to the difference of the electron energies of the outer or excited state, E_{outer}, and its new inner ground state, E_{inner},

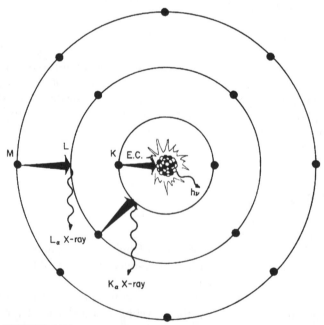

FIGURE 1.23 Electron capture (EC) decay and the accompanying gamma ($h\nu$) and x-radiation.

as described by

$$hv = E_{outer} - E_{inner} \qquad (1.97)$$

The radiation emitted consists of a discrete line of energy characteristic of the electron shell and, consequently, of the atom from which it arises.

The production of x-rays in radionuclide decay is, however, more complex. The filling of one electron vacancy in an inner shell is followed by a series of electron transitions in an overall adjustment of electrons in outer shells. This gives rise to further x-rays with lines characteristic of outer shells. Such electron transitions, each resulting in the emission of discrete lines of characteristic x-rays, are illustrated in Fig. 1.24. The transitions are identified by a letter corresponding to the shell (K, L, M, etc.) with vacancy giving rise to the x-ray photon and a subscript (α, β, γ, etc.) to identify, from among a series of outer electron shells of the atom, the shell from which the electron vacancy is filled. For example, an x-ray arising from an electron transition from the L to the K shell is denoted as K_α and that arising from a transition from the M to the K shell as K_β. Transitions involving the filling of electron vacancies in the L shell from outer M, N and O shells are denoted by L_α, L_β, and L_γ, etc.

Because x-radiation is characteristic of the atom from which it arises, it is customary to identify the element along with the x-ray photon (e.g., Cr K x-rays, Hg L x-rays, and many others as listed in Appendix A). In these examples, the fine structure of the x-ray emissions is not given and the lines are grouped together as K and L x-rays.

The complexity of x-ray lines emitted and their abundances of emission are compounded by the existence of other mechanisms of x-ray production in unstable atoms. One of these mechanisms is the production of Auger electrons. An x-ray emitted from an atom may produce an Auger electron via

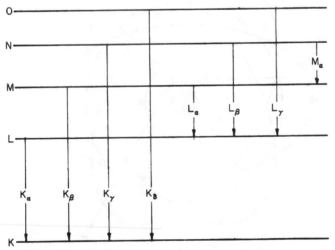

FIGURE I.24 Atomic electron energy levels or shells (K, L, M, etc.) and lines of transition corresponding to characteristic x-rays (K_α K_β K_γ, etc.).

an internal photoelectric effect (see Section II.F), which results in the emission of an atomic electron from a shell farther away from the nucleus. The vacancy left by the Auger electron gives rise to additional x-rays characteristic of outer shells following the electron readjustments that ensue. Auger electrons can be emitted from a variety of electron shells, followed by an equal variety of characteristic x-rays from subsequent electron adjustments in outer shells.

Any process that would cause the ejection of an atomic electron of an inner shell can result in the production of x-radiation. Other processes not yet mentioned in this section that involve the ejection of atomic electrons are the emission of internal-conversion electrons (see Section II.E) and radiation-induced ionization (see Sections II and IV).

F. Bremsstrahlung

Bremsstrahlung is electromagnetic radiation similar to x-radiation. It is emitted by a charged particle as it decelerates in a series of collisions with atomic particles. This mechanism is illustrated in Fig. 1.25, where a beta particle traveling through matter approaches a nucleus and is deflected by it. This deflection causes a deceleration of the beta particle and consequently a reduction in its kinetic energy with the emission of energy as a photon of bremsstrahlung or "braking radiation." The phenomenon is described by

$$h\nu = E_i - E_f \tag{1.98}$$

where $h\nu$ is the energy of the photon of bremsstrahlung, E_i is the initial kinetic energy of the beta particle prior to collision or deflection, producing

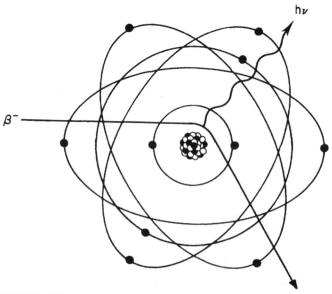

FIGURE 1.25 **Bremsstrahlung production. A beta particle is deflected by an atomic nucleus and loses kinetic energy with the emission of a photon of x-radiation.**

a final kinetic energy E_f of the electron. When beta particles from a particular radionuclide source strike an absorber material a wide spectrum of bremsstrahlung photon wavelengths (or energies) will be produced. The broad spectrum of bremsstrahlung is due to the broad possibilities of different interactions, i.e., deflections or collisions, that the beta particles can have with atomic nuclei of the absorber and the broad spectrum of beta-particle energies emitted from any given radionuclide. In a given spectrum of bremsstrahlung the shortest wavelength, λ_{min}, is observed when a beta particle or electron undergoes a direct collision with the nucleus of an atom and loses all of its kinetic energy, $h\nu_{max}$, as bremsstrahlung or x-radiation according to the relation

$$h\nu_{max} = \frac{hc}{\lambda_{min}}, \tag{1.99}$$

which follows the energy-wavelength relation previously described by Eq. 1.86.

Let us consider an example of a 1710 keV beta particle from ^{32}P ($E_{max} = 1.71$ MeV) striking a nucleus of Pb in a lead–glass shield. If the beta particle loses all of its energy in the collision, the wavelength of the bremsstrahlung emitted from this interaction according to Eq. 1.99 would be

$$\lambda = \frac{hc}{h\nu_{max}} = \frac{12.4 \, \text{keV} \, \text{Å}}{1710 \, \text{keV}} = 0.00725 \, \text{Å}$$

See Eq. 1.94 for the conversion of the constant hc to convenient units of eV m or eV Å. Bremsstrahlung production by high-energy beta particles in absorber material of high atomic number is significant (see Section V). Consequently to avoid the production of bremsstrahlung in radiation shielding against the harmful effects of high-energy beta particles, an absorber of low atomic number (e.g., plastic) may be preferred over one of high atomic number (e.g., Pb-glass).

An apparatus used to artificially produce x-rays such as those employed in medical diagnosis or x-ray diffraction functions on a similar principle of bremsstrahlung described previously. The x-ray tube consists of an evacuated tube containing a cathode filament and a metal anode target such as tungsten ($A = 74$). A voltage potential is applied to the tube so that electrons emitted from the cathode accelerate towards the anode. Upon colliding with the tungsten anode the accelerated electrons lose energy as bremsstrahlung x radiation. For example, an electron accelerated in an x-ray tube to an energy of 40 keV, which loses all of its energy upon impact with a tungsten nucleus would produce a single x-ray photon of wavelength calculated as

$$\lambda = \frac{hc}{h\nu} = \frac{12.4 \, \text{keV} \, \text{Å}}{40 \, \text{keV}} = 0.31 \, \text{Å} = 0.031 \, \text{nm}$$

Ionization and electron excitation were previously described as predominant mechanisms by which a traveling beta particle may lose its kinetic energy in matter (see Sections II.B and V of this chapter). However, the production of bremsstrahlung may also be another significant mechanism for the dissipation of beta-particle energy, particularly as the beta-particle energy and the atomic number of the absorber increase (Kudo, 1995). A more thorough treatment is found in Section V of this chapter, which includes examples of calculations involved to determine the degree of bremsstrahlung production as a function of beta-particle energy and absorber atomic number. In general terms we can state that for a high-energy beta particle such as the "strongest" beta particle emitted from ^{32}P ($E_{\max} = 1.7$ MeV) in a high-atomic-number material such as lead (Pb = 82), bremsstrahlung production is significant. In a substance of low atomic number such as aluminum (Al = 13) bremsstrahlung occurs at a low and often insignificant level.

In view of the wide spectrum of beta-particle energies emitted from radionuclides and the wide variations of degree of beta-particle interactions with atomic particles, the production of a broad spectrum, or smear, of photon energies of bremsstrahlung is characteristic. This contrasts with x-radiation, which is emitted in atomic electron deexcitation processes as discrete lines of energy. We have excluded bremsstrahlung production by charged particles other than beta particles or electrons, because other charged particles are of much greater mass than the beta particle or electron, and consequently they do not undergo such a rapid deceleration and energy loss as they travel through absorber material.

Bremsstrahlung of very low intensity also results from the transforming nucleus in electron capture decay processes (see Section II.C.2). This is referred to as internal or inner bremsstrahlung. Because a neutrino is emitted in these decay processes, the quantum of energy not carried away by the neutrino is emitted as internal bremsstrahlung. Thus, in electron capture decay, internal bremsstrahlung may possess energies between zero and the maximum, or transition energy of a radionuclide. When gamma radiation is also emitted, the internal bremsstrahlung may be masked by the more intense gamma rays and go undetected. In such cases, internal bremsstrahlung may be of insufficient intensity to lend itself to radionuclide detection. However, in the absence of gamma radiation, the upper limit of the internal bremsstrahlung can be used to determine the transition energy of a nuclide in electron capture decay. Some examples of radionuclides that decay by electron capture without the emission of gamma radiation are as follows:

$$^{55}_{26}\text{Fe} \rightarrow {}^{55}_{25}\text{Mn} + \nu + h\nu \quad (0.23 \text{ MeV}) \tag{1.100}$$

$$^{37}_{18}\text{Ar} \rightarrow {}^{37}_{17}\text{Cl} + \nu + h\nu \quad (0.81 \text{ MeV}) \tag{1.101}$$

and

$$^{49}_{23}\text{V} \rightarrow {}^{49}_{22}\text{Ti} + \nu + h\nu \quad (0.60 \text{ MeV}) \tag{1.102}$$

where $h\nu$ *is* the internal bremsstrahlung, the upper energy limits of which are expressed in MeV.

IV. INTERACTION OF ELECTROMAGNETIC RADIATION WITH MATTER

The lack of charge or mass of electromagnetic gamma and x-radiation hinder their interaction with, and dissipation of their energy in, matter. Consequently, gamma radiation and x-rays have greater penetration power and longer ranges in matter than the massive and charged alpha and beta particles of the same energy. Nevertheless, gamma and x-radiation are absorbed by matter, and the principal mechanisms by which this type of radiation interacts with matter are discussed in this section.

A. Photoelectric Effect

The energy of a photon may be completely absorbed by an atom. Under such circumstances, the entire absorbed photon energy is transferred to an electron of the atom and the electron is released, resulting in the formation of an ion pair (see also Section III.A). Consequently, the energy of the emitted electron is equal to the energy of the impinging photon less the binding energy of the electron. This is described by the photoelectric equation of Einstein:

$$E_e = h\nu - \phi \tag{1.103}$$

where E_e is the energy of the ejected electron, $h\nu$ is the energy of the incident photon, and ϕ is the binding energy of the electron or the energy required to remove the electron from the atom. The ejected electron is identical to a beta particle and produces ionization (secondary ionization in this case) as it travels through matter as previously described for beta particles.

When an electron from an inner atomic K or L shell is ejected, electrons from outer shells fall from their higher energy states to fill the resulting gap. These transitions in electron energy states require a release of energy by the atomic electrons, which appears as soft (low-energy) x-rays. X-radiation is identical in properties to gamma radiation. The essential difference lies in its origin. As previously described, gamma radiation originates from energy state transformations of the nucleus of an atom, whereas x-radiation originates from energy state transformations of atomic electrons.

B. Compton Effect

There is a second mechanism by which a photon (e.g., x-ray or gamma ray, etc.) transfers its energy to an atomic orbital electron. In this interaction, illustrated in Fig. 1.26, the photon, E_γ, imparts only a fraction of its energy to the electron and in so doing is deflected with energy E_γ' at an angle Θ, while the bombarded electron is ejected at an angle θ to the trajectory of the primary photon. This interaction is known as the Compton effect and also as Compton scattering. The result of this interaction is the formation of an ion pair as in the case of the photoelectric effect. However, the deflected photon continues traveling through matter until it dissipates its entire kinetic

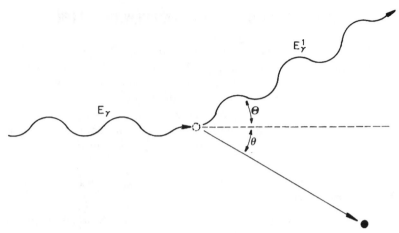

FIGURE 1.26 **The Compton effect. An incident photon collides with an atomic electron and imparts energy to it, the photon and electron being deflected at angles Θ and θ, respectively, to the trajectory of the incident photon.**

energy by interacting with other electrons in a similar fashion or via other mechanisms of interaction with matter discussed in this section. The ejected electron, being identical in properties to a beta particle, loses its energy through the secondary ionization it causes according to mechanisms previously described.

Our understanding of the Compton effect comes from the original work of Arthur H. Compton (1923), who discovered that x-ray photons scattered by thin foils underwent a wavelength shift. The shift in wavelength of the scattered photon with respect to that of the incident photon was a function of the angle of scatter Θ. To interpret this effect he treated the x-radiation as photon particles or quanta according to the Einstein–Planck relation $E = h\nu$ (see Eq. 1.86) and the scattering to occur as photon–electron collisions somewhat like billiard-ball collisions as illustrated in Fig. 1.26. Compton derived the equation, which describes the wavelength shift between the incident and scattered photons and angle of scatter as

$$\lambda' - \lambda = \frac{h}{m_0 c}(1 - \cos \Theta) \qquad (1.104)$$

where λ' and λ are the wavelengths of the incident and deflected photons, h is Planck's constant, m_0 is the rest mass of the electron, c is the speed of light, and Θ is the angle of scatter of the photon relative to its original direction of travel.

The Compton-scatter photon will always be of longer wavelength (lower energy) than the incident photon, because of energy lost in the collision with the electron. For example, let us calculate the wavelength shift and energy loss by an incident photon of wavelength 0.300 nm that collides with a free electron, and where the photon is scattered at an angle of 70°. The wavelength of the scattered photon is calculated according to Eq. 1.104 as

$$\lambda' = \lambda + \frac{h}{m_0 c}(1 - \cos \Theta)$$

$$= 3.0 \times 10^{-10} \text{ m} + \frac{6.626 \times 10^{-34} \text{ J s}}{(9.109 \times 10^{-31} \text{ kg})(2.997 \times 10^8 \text{ m s}^{-1})}(1 - \cos 70°)$$

$$= 3.0 \times 10^{-10} \text{ m} + 2.43 \times 10^{-12} \text{ m}(1 - 0.342)$$

$$= 0.3016 \text{ nm}$$

The energy lost by the incident photon according to the Einstein–Planck relation (Eq. 1.86) is given by

$$\Delta E = E_\gamma - E'_\gamma$$

$$= \frac{hc}{\lambda} - \frac{hc}{\lambda'}$$

$$= \frac{12.4 \text{ keV Å}}{3.00 \text{ Å}} - \frac{12.40 \text{ keV Å}}{3.016 \text{ Å}}$$

$$= 4.133 \text{ keV} - 4.111 \text{ keV} = 0.022 \text{ keV}$$

and the fraction of photon energy lost becomes

$$\frac{\Delta E}{E} = \frac{0.022 \text{ keV}}{4.133 \text{ keV}} = 0.0053 = 0.53\%$$

We can calculate directly the energy of the Compton scatter photon, λ', if we know the incident x-ray or gamma-ray photon energy and angle of scatter of the photon according to the equation

$$E'_\gamma = \frac{E_\gamma}{1 + (E_\gamma/m c^2)(1 - \cos \Theta)} \qquad (1.105)$$

where E'_γ is the energy of the Compton scatter photon, E_γ is the incident photon energy, mc^2 is the rest energy of the electron (511 keV or 0.511 MeV, see Section IV.C), and Θ is the Compton photon angle of scatter (Tait, 1980). If we take the data from the previous example where the incident photon energy was 4.133 keV (3.00 Å) and the angle of scatter was 70°, we can calculate the energy of the Compton photon according to Eq. 1.105 to be

$$E'_\gamma = \frac{4.133 \text{ keV}}{1 + (4.133 \text{ keV}/511 \text{ keV})(1 - \cos 70°)} = 4.111 \text{ keV}$$

The result is in agreement with the calculations above using Eq. 1.104 derived by Compton.

It has been shown by Compton that the angle of deflection of the photon is a function of the energy imparted to the electron. This angle may vary from just above $\Theta = 0°$ for low Compton electron energies to a maximum $\Theta = 180°$ for the highest Compton electron energy. Compton electrons are thus emitted with energies ranging between zero and a maximum energy referred to as the Compton edge. The Compton edge is the Compton electron energy corresponding to complete backscattering of the gamma ray photon. With $\Theta = 180°$ or $\cos\Theta = -1$, Eq. 1.105 is reduced to the following equation describing the energy, E'_γ, of the gamma ray photon at the Compton edge in MeV units:

$$E'_\gamma = \frac{E_\gamma}{1 + (E_\gamma / 0.511 \, \text{MeV})(1 - \cos 180°)} \qquad (1.106)$$

or

$$E'_\gamma = \frac{E_\gamma}{1 + 2E_\gamma / 0.511} \qquad (1.107)$$

or

$$E'_\gamma = \frac{E_\gamma}{1 + 3.914 \, E_\gamma} \qquad (1.108)$$

As an example, the energy of the gamma-ray photon in MeV at the Compton edge for an incident gamma ray from ^{137}Cs ($E_\gamma = 0.662 \, \text{MeV}$) is calculated according to Eq. 1.108 to be

$$E_\gamma = \frac{0.662}{1 + 3.914(0.662)} = 0.184 \, \text{MeV}$$

A Compton scatter photon is of longer wavelength and lower energy than the incident photon. Deflected Compton photons occur with a broad spectrum of energies. Spectra of Compton-scattered photon energies contain a peak known as the backscatter peak (see Fig. 11.18, Chapter 11). The backscatter peak arises from Compton scattering into a gamma photon detector [e.g., NaI(Tl) crystal] from the surrounding detector shielding and housing materials. The backscatter peak occurs at increasing values of energy (MeV) in proportion to the incident photon energy and approaches a constant value of 0.25 MeV, according to Eq. 1.108, for incident photon energies greater than 1 MeV (Tait, 1980). The energy of the Compton electron, E_e, may be described by

$$E_e = E_\gamma - E'_\gamma - \phi \qquad (1.109)$$

where E_γ and E'_γ are the energies of the incident and deflected photons, respectively, and ϕ is the binding energy of the electron. As the binding energy of the atomic electron is relatively small, the energy of the ejected electron is essentially the difference between the incident and deflected photon energies. Substituting the value of E'_γ from Eq. 1.105 and ignoring the electron binding energy, the Compton electron energy can be expressed as

$$E_e = E_\gamma - \frac{E_\gamma}{1 + (E_\gamma/mc^2)(1 - \cos \Theta)} \tag{1.110}$$

$$= E_\gamma - \frac{E_\gamma}{1 + (E_\gamma/0.511\,\text{MeV})(1 - \cos \Theta)} \tag{1.111}$$

where the electron energies are given in MeV. For example, the energy of a Compton electron, E_e, scattered at 180° (Compton edge: $\cos \Theta = -1$) and originating from an incident gamma ray photon from ^{137}Cs ($E_\gamma = 0.662\,\text{MeV}$) is calculated according to Eq. 1.111 as

$$E_e = 0.662 - \frac{0.662}{1 + (0.662/0.511)(1 - \cos 180°)} = 0.478\,\text{MeV}$$

Alternatively, if we ignore the negligible electron binding energy and know the incident photon energy and Compton scatter photon energy, we can calculate the Compton electron energy by difference according to Eq. 1.109

$$E_e = 0.662\,\text{MeV} - 0.184\,\text{MeV} = 0.478\,\text{MeV}$$

which is in agreement with the electron energy calculated above. The Compton edge and backscatter peak due to interactions of Compton electrons and Compton backscatter photons, respectively, in a scintillation crystal detector are illustrated in Fig. 11.18, Chapter 11.

C. Pair Production

The interactions of gamma radiation with matter considered earlier involve the transfer of γ-energy, in whole or in part, to atomic electrons of the irradiated material. Pair production, as another mechanism of γ-energy dissipation in matter, results in the creation of nuclear particles from the γ-energy. The nuclear particles produced are a negatron and a positron from an individual gamma-ray photon that interacts with the coulombic field of a nucleus (see Fig. 1.27). Consequently, this phenomenon involves the creation of mass from energy. The creation of an electron requires a certain quantum of energy of a gamma-ray photon, which may be calculated according to Einstein's equation for the equivalence of mass and energy

$$E = m_e c^2 \tag{1.112}$$

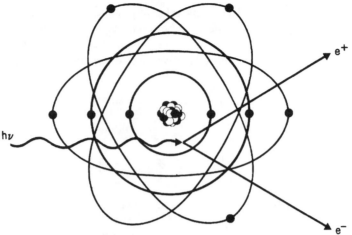

FIGURE 1.27   **Pair production. The conversion of a gamma ray photon into a negatron and positron pair.**

where E is energy, m_e is the electron rest mass, and c is the speed of light in a vacuum. According to Eq. 1.112 the rest energy of the electron (negatron or positron) is calculated as

$$E = (9.109 \times 10^{-31} \text{ kg})(2.997 \times 10^8 \text{ m s}^{-1})^2 = 8.182 \times 10^{-14} \text{ J}$$

Since by definition, $1 \text{ eV} = 1.602 \times 10^{-19}$ J, the electron rest energy in joules is converted to electron volts as

$$8.182 \times 10^{-14} \text{ J}/1.602 \times 10^{-19} \text{ J eV}^{-1} = 0.511 \text{ MeV}$$

Thus, the creation of an electron (negatron) requires a minimum energy of 0.511 MeV. However, a gamma ray of 0.511 MeV energy cannot alone create a negatron, as there must also be simultaneous creation of its antiparticle, the positron of equal mass and opposite charge. The minimum gamma ray photon energy required for the creation of the negatron–positron pair is

$$E_{\text{pair}} = m_{e^-} c^2 + m_{e^+} c^2 = 2mc^2 = 2(0.511 \text{ MeV}) = 1.022 \text{ MeV} \qquad (1.113)$$

where m_{e^-} and m_{e^+} are the rest masses of the negatron and positron, respectively. Thus, the absorption by matter of gamma radiation greater than 1.02 MeV may result in pair production. The probability of pair production increases in proportion to the magnitude of gamma-ray photon energy above 1.02 MeV, and pair production is the predominant mechanism of absorption of photons of energies of 5 MeV and above (see Figs. 1.29 and 1.30). In pair production, gamma-ray energy in excess of 1.02 MeV appears as kinetic

TABLE I.7 Examples of Nuclides That Exhibit Internal Pair Production, Their Gamma Radiations and Relative Intensities of the Positron–Negatron Pairs

	Gamma radiations		Pair/gamma ratio ($e^{\pm}/\gamma$)
	Energy (MeV)	Abundance (%)	
^{24}Na	1.369	100	6×10^{-5}
	2.754	100	7×10^{-4}
^{56}Mn	1.81	29	5.6×10^{-4}
	2.11	15	4.6×10^{-4}
^{59}Fe	1.099	57	1.4×10^{-4}
	1.292	43	1.1×10^{-4}
^{60}Co	1.17	100	3.7×10^{-5}
	1.33	100	(combined)
^{142}Pr	1.576	4	1.1×10^{-4}
^{144}Pr	1.489	0.3	1.9×10^{-4}
	2.186	0.7	6.7×10^{-4}
^{154}Eu	1.274	37	8.0×10^{-5}

energy of the negatron and positron produced, or

$$hv = 2mc^2 + E_{e^-} + E_{e^+} \tag{1.114}$$

where hv is the energy of the gamma ray photon, $2mc^2$ is the 1.02 MeV required for pair production, and E_{e^-} and E_{e^+} are the kinetic energies of the negatron and positron produced. As discussed previously in Section III.C, positrons will produce annihilation radiation when they come to rest in the proximity of a negative electron, i.e., their antiparticle, resulting in the simultaneous conversion of two electron masses into two gamma-ray photons of 0.511 MeV energy.

Pair production does not only occur in the vicinity of atomic nuclei bombarded by gamma radiation. It may also originate from nuclei that emit gamma radiation with transition energies greater than 1.02 MeV. This is referred to as internal pair production, and the mechanism competes to a small extent with the emission of gamma radiation. The degree to which this competition occurs is measured by the ratio of intensities of positron-negatron pairs to gamma radiation or ($e^{\pm}/\gamma$). Some examples of nuclides that emit such positron–negatron pairs and the intensities of these pairs relative to gamma radiation are given in Table 1.7.

D. Combined Photon Interactions

Because of its zero rest mass and zero charge, gamma radiation has an extremely high penetration power in matter in comparison with alpha and beta particles.

Materials of high density and atomic number (such as lead) are used most often as absorbers to reduce x- or gamma-radiation intensity. Radiation intensity, I, is defined here as the number of photons of a radiation beam that traverse a given area per second, the units of which can be photons $cm^{-2} s^{-1}$. Suppose a given absorber material of thickness x attenuates or reduces the intensity of incident gamma radiation by one-half. Placing a similar barrier of the same thickness along the path of the transmitted gamma radiation would reduce the intensity again by one-half. With three barriers each of thickness x and an initial gamma-ray intensity I_0, there is a progressive drop in the transmitted gamma-ray intensities: $I_1 = (1/2)I_0$, $I_2 = (1/2)I_1$, $I_3 = (1/2)I_2$, and $I_n = (1/2)I_{n-1}$. Obviously, incident x- or gamma radiation may be reduced from I_0 to I_3 by using a $3x$ thickness of the same material as an absorber. Consequently, the intensity of the transmitted electromagnetic radiation is proportional to the thickness of the absorber material and to the initial intensity of the radiation. An increasing absorber thickness increases the probability of photon removal because there is a corresponding increase of absorber atoms that may attenuate the incident photons via the photoelectric effect, the Compton effect, and pair production mechanisms.

If gamma-ray attenuation with respect to absorber thickness is considered, the change in gamma-ray intensity, ΔI, with respect to the absorber thickness, Δx, is proportional to the initial gamma-ray photon intensity, I. This may be written as

$$\Delta I / \Delta x = -\mu I \qquad (1.115)$$

where μ is the proportionality constant, referred to as the linear attenuation coefficient or linear absorption coefficient. Its value is dependent on the atomic composition and density of the absorber material. The change in intensity over an infinitely thin section of a given absorber material may be expressed as

$$dI/dx = -\mu I \qquad (1.116)$$

or

$$dI/I = -\mu \, dx \qquad (1.117)$$

Integrating Eq. 1.117 over the limits defined by the initial intensity, I_0, to the transmitted intensity, I, and over the limits of absorber thickness from zero to a finite value x, such as

$$\int_{I_0}^{I} dI/I = -\mu \int_{0}^{x} dx \qquad (1.118)$$

gives

$$\ln I - \ln I_0 = -\mu x \qquad (1.119)$$

or

$$\ln I_0/I = \mu x \qquad (1.120)$$

Equation 1.120 may be written in exponential form as

$$I = I_0\, e^{-\mu x} \qquad (1.121)$$

which is somewhat similar to the exponential attenuation of neutrons discussed earlier in this chapter.

Because gamma-ray absorption is exponential, the term half-value thickness, $x_{1/2}$, is used to define the attenuation of gamma radiation by matter. Half-value thickness is the thickness of a given material of defined density that can reduce the intensity of incident gamma radiation by one-half. The half-value thickness may also be defined according to Eq. 1.120, in which the initial gamma ray intensity, I_0, is given an arbitrary value of 1 and the transmitted intensity must, by definition, have a value of 1/2, or

$$\ln 1/0.5 = \mu\, x_{1/2} \qquad (1.122)$$

or

$$\ln 2 = \mu\, x_{1/2} \qquad (1.123)$$

and

$$x_{1/2} = 0.693/\mu \qquad (1.124)$$

From the linear attenuation coefficient, μ, of a given material and gamma-ray photon energy, it is possible to calculate the half-value thickness, $x_{1/2}$. The linear attenuation coefficient has units of cm^{-1}, so that calculated half-value thickness is provided in units of material thickness (cm). Linear attenuation coefficients for some materials as a function of photon energy are provided in Table 1.8. The table refers to these as total linear attenuation coefficients, because they constitute the sum of coefficients due to Compton, photoelectric and pair production interactions. Calculated half-value thicknesses of various absorber materials as a function of gamma-ray energy are illustrated in Figure 1.28 to illustrate some examples of the varying amounts of absorber material required to attenuate gamma-ray photons. The linear attenuation coefficient is a constant for a given absorber material and gamma-ray photon energy and has units of reciprocal length such as cm^{-1}. It is, however, dependent on the state of the absorber or the number of atoms per unit volume of absorber. A more popular coefficient is the mass attenuation coefficient, μ_m, which is independent of the physical state of the absorber material and is defined as

$$\mu_m = \mu/\rho \qquad (1.125)$$

TABLE I.8   Total Linear Attenuation Coefficients (cm^{-1}) for Gamma-Ray Photons in Various Materials[a]

Photon Energy (MeV)	Water	Aluminum	Iron	Lead
0.1	0.167	0.435	2.704	59.99
0.2	0.136	0.324	1.085	10.16
0.4	0.106	0.2489	0.7223	2.359
0.8	0.0786	0.1844	0.5219	0.9480
1.0	0.0706	0.1658	0.4677	0.7757
1.5	0.0575	0.1350	0.3812	0.5806
2.0	0.0493	0.1166	0.3333	0.5182
4.0	0.0339	0.0837	0.2594	0.4763
8.0	0.0240	0.0651	0.2319	0.5205
10.0	0.0219	0.0618	0.2311	0.5545

[a]Data obtained from Argonne National Laboratory, ANL-5800 (1963), Hubbell (1969), and Serway *et al.* (1997).

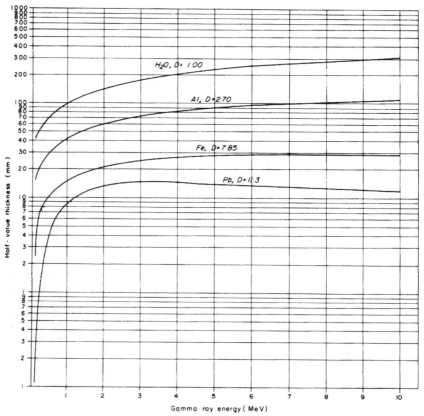

FIGURE I.28   Half-value thicknesses of various materials as a function of gamma-ray energy. *D* is the density of each material.

where ρ is the density of the absorber in units of $g\,cm^{-3}$, and μ_m has units of $cm^2\,g^{-1}$. Some examples of mass attenuation coefficients according to x- and gamma-ray photon energy are provided in Table 1.9. Using the mass attenuation coefficient, Eq. 1.121 changes to

$$I = I_0\,e^{-\mu_m\,\rho x} \tag{1.126}$$

and the half-value thickness is calculated according to Eq. 1.124 as

$$x_{1/2} = 0.693/\mu_m\,\rho \tag{1.127}$$

Mass attenuation coefficients for x- or gamma-ray photons over a wide range of energies from 1 keV to 1000 MeV in 100 elements are available from Berger and Hubbell (1997). A sample of mass attenuation coefficients over the range of 5 keV to 10 MeV in a few materials are listed in Table 1.9. The following calculation illustrates the use of the data from Tables 1.8 and 1.9 to calculate half-value thickness and radiation attenuation:

Let us calculate the half-value thickness of lead ($\rho = 11.3\,g\,cm^{-3}$) for 2.0 MeV gamma radiation, and further calculate what reduction in radiation intensity would result if we positioned four times the half-value thickness of lead in the path of the radiation beam. Firstly, the linear attenuation coefficient, μ, or mass attenuation coefficient, μ_m, for 2.0 MeV photons in

TABLE 1.9 Total Mass Attenuation Coefficients ($cm^2\,g^{-1}$) for X- or Gamma-Ray Photons in Various Materials[a]

Photon Energy (MeV)	Air	Water	Aluminum	Iron	Lead
0.005			193	140	730
0.01			26.2	171	131
0.05			0.368	1.96	8.04
0.1	0.151	0.167	0.170	0.372	5.55
0.2	0.123	0.136	0.122	0.146	0.999
0.4	0.0953	0.106	0.0922	0.0919	0.208
0.8	0.0706	0.0786	0.0683	0.0664	0.0836
1.0	0.0655	0.0706	0.0614	0.0595	0.0684
1.5	0.0517	0.0575	0.0500	0.0485	0.0512
2.0	0.0445	0.0493	0.0432	0.0424	0.0457
4.0	0.0307	0.0339	0.0310	0.0330	0.0420
8.0	0.0220	0.0240	0.0241	0.0295	0.0459
10.0	0.0202	0.0219	0.0229	0.0294	0.0489

[a]Data from Argonne National Laboratory, ANL-5800 (1963), Hubbell (1969), and Berger and Hubbell (1997).

lead are obtained from either Table 1.8 or 1.9 and the half-value thickness of lead for 2.0 MeV photons is calculated as

$$x_{1/2} = \frac{0.693}{\mu} \quad \text{or} \quad \frac{0.693}{\mu_m \rho}$$

or

$$x_{1/2} = \frac{0.693}{0.5182 \, \text{cm}^{-1}} \quad \text{or} \quad \frac{0.693}{(0.0457 \, \text{cm}^2 \, \text{g}^{-1})(11.3 \, \text{g cm}^{-3})}$$
$$x_{1/2} = 1.34 \, \text{cm}$$

Thus, a barrier of 1.34 cm thickness of lead is sufficient to reduce the radiation intensity of 2.0 MeV photons by 1/2 or 50%. According to Eq. 1.121 the relation between the initial radiation intensity, I_0, and the transmitted intensity, I is

$$I/I_0 = e^{-\mu x}$$

and for $x = 1.34$, if the initial radiation intensity is given an arbitrary value of 2, the transmitted intensity would be 50% of the initial intensity or equal to 1. We then can write

$$I/I_0 = 1/2 = e^{-1.34\mu}$$

If we employ four times the half-value thickness of lead or 4×1.34 cm $= 5.36$ cm, we can calculate that the transmitted radiation would be reduced to the following:

$$I/I_0 = (e^{-1.34\mu})^4 = (1/2)^4$$

or

$$e^{-5.36\mu} = 1/16 = 0.0625 = 6.25 \, \% \text{ transmitted}$$

The remaining 15/16 or 93.75% of the initial radiation is attenuated by the 5.36 cm lead barrier. In general, we need not know the half-value thickness of the material or shield, but simply obtain the linear or mass attenuation coefficient for a given energy of x- or gamma radiation from reference tables and use Eqs. 1.121 or 1.126 to calculate the degree of radiation attenuation for any thickness of the absorber material. For example, if we used only 2.5 cm of lead barrier, the attenuation of 2.0 MeV gamma rays could be calculated as

$$I/I_0 = e^{-\mu x} = e^{-\mu_m \rho x}$$

and

$$I/I_0 = e^{-(0.5182\,\text{cm}^{-1})(2.5\,\text{cm})} = e^{-(0.0457\,\text{cm}^2\,\text{g}^{-1})(11.3\,\text{g cm}^{-3})(2.5\,\text{cm})}$$

$$= e^{-1.29} = 0.275 = 27.5\%$$

Thus the 2.0 MeV radiation transmitted through a shield of 2.5 cm of lead would be 27.5% of the initial radiation intensity.

As previously discussed, the absorption of gamma radiation is a process that principally involves three mechanisms of gamma-ray attenuation: the Compton effect, the photoelectric effect, and pair production. The attenuation coefficients just discussed are also referred to as total attenuation coefficients because they consist of the sum of three independent coefficients or

$$\mu = \mu_c + \mu_e + \mu_p \tag{1.128}$$

where μ_c, μ_e, and μ_p are attenuation coefficients for Compton, photoelectric, and pair production processes. The attenuation coefficients are proportional to the probabilities of occurrence of these radiation attenuation processes and can be used as a measure of the relative roles these processes play in the absorption of gamma-ray photons.

Accordingly, the total and partial mass attenuation coefficients can be written as

$$\mu_m/\rho = \mu_c/\rho + \mu_e/\rho + \mu_p/\rho \tag{1.129}$$

Figures 1.29 and 1.30 provide a graphic representation of the relative frequency of occurrence of the Compton, photoelectric, and pair production processes in aluminum and sodium iodide absorbers as a function of photon energy. From these curves, it is seen that the photoelectric effect plays an increasing role in total gamma-ray attenuation at lower gamma-ray energies and with absorber materials of higher atomic number.

As illustrated in Figs. 1.29 and 1.30, the pair production process does not occur at gamma-ray energies below the threshold value of 1.02 MeV as expected in accord with the combined positron and negatron rest energies (2×0.511 MeV) required for pair production. In some absorber materials of relatively high density, absorption edges can be measured for low photon energies such as the K edge illustrated in Fig. 1.30. The absorption edge is a discontinuity in the attenuation coefficient curve for the photoelectric effect that is caused when photon energies are less than the binding energies of electrons of a certain shell (e.g., K shell) and that reduces the number of electrons which may be ejected by the photoelectric effect. When photons possess the threshold binding energy of electrons of that shell, there is a sudden surge in attenuation owing to the ejection of electrons from that shell via photoelectric interactions. A thorough treatment of the attenuation and absorption of gamma radiation in matter is available from Hubbell (1969) and Turner (1995).

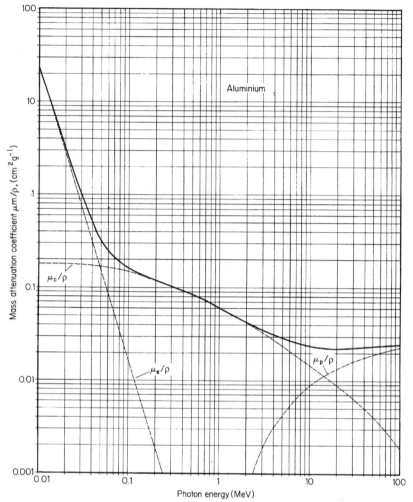

FIGURE I.29 Mass attenuation coefficients for photons in aluminum. The total attenuation is given by the solid line, which is the sum of the partial attenuations due to the Compton effect, μ_c/ρ, the photoelectric effect, μ_e/ρ, and pair production, μ_p/ρ. Linear attenuation coefficients are obtained from these values by multiplying by the density of aluminum, $\rho = 2.70 \, \text{g cm}^{-3}$ (From Evans, 1955, reproduced with permission of The McGraw-Hill Companies.)

For more information on nuclear radiation and its mechanisms of interaction with matter the reader may refer to books by Krane (1988) and Serway *et al.* (1997).

V. STOPPING POWER AND LINEAR ENERGY TRANSFER

The previous paragraphs provide information on the mechanisms of interaction of radiation with matter. In summary, we can state that the principal mechanisms of interaction of charged particles (e.g., alpha particles,

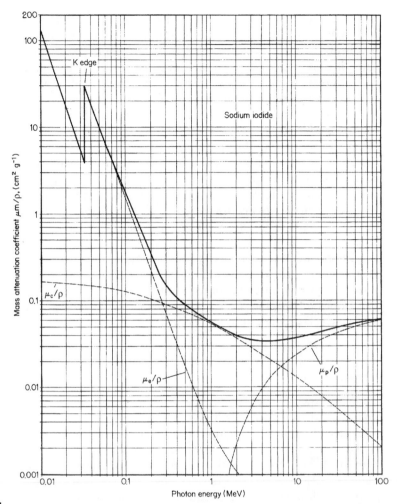

FIGURE I.30 Mass attenuation coefficients for photons in sodium iodide. The total attenuation is given by the solid line, which is the sum of the partial attenuations due to the Compton effect, μ_c/ρ, the photoelectric effect, μ_e/ρ, and pair production, μ_p/ρ. Linear attenuation coefficients are obtained from these values by multiplying by the density of sodium iodide, $\rho = 3.67\,\mathrm{g\,cm^{-3}}$ (From Evans, 1955, reproduced with permission of The McGraw-Hill Companies.)

protons, deuterons and electrons or beta particles) with matter, which result in significant charged-particle energy loss are (i) ionization via coulombic interactions of the charged particles with atomic electrons of the absorbing medium, (ii) electron orbital excitation of the medium, which occurs when the energy transfer through coulombic interaction is not sufficient to actually eject an electron from an atom, and (iii) the radial emission of energy as bremsstrahlung (x-radiation) when an electron or beta particle decelerates as it approaches an atomic nucleus. Release of particle energy by bremsstrahlung radiation becomes increasingly significant as the beta-particle energy and absorber atomic number increase. On the other hand, electromagnetic radiation dissipates its energy in matter via three mechanisms, namely, (i) the photoelectric effect, (ii) Compton scattering, and (iii) pair production. The photoelectric effect and Compton scattering generate ion pairs directly within the absorbing medium, whereas, pair production results in the creation of charged particles (positrons and negatrons) that will subsequently dissipate their energy via ionization, electron excitation, and, in the case of positrons, annihilation. Also, we have seen that energetic neutrons, will dissipate their energy in matter through elastic collisions with atomic nuclei of the absorbing medium. When hydrogen is present in the absorbing material, the bulk of the fast neutron energy is passed on to the hydrogen nuclei. In turn, the kinetic energy of these protons is absorbed in the medium via ionization and excitation processes. We have seen also that low- and high-energy neutrons are absorbed principally via inelastic neutron reactions, which can result in the production of charged particles and gamma radiation.

The radiation properties (e.g., charge, mass, and energy) and mechanisms of interaction previously described govern the rate of dissipation of energy and consequently the range of travel of the nuclear radiation in the absorber. This brings to bare the concepts of stopping power and linear energy transfer (LET), which are described subsequently.

A. Stopping Power

Stopping Power is defined by The International Commission on Radiation Units and Measurements or ICRU (Taylor *et al.*, 1970) as the average energy dissipated by ionizing radiation in a medium per unit path length of travel of the radiation in the medium. It is, of course, impossible to predict how a given charged-particle will interact with any given atom of the absorber medium. Also, when we consider that the coulombic forces of charged particles will interact simultaneously with many atoms as it travels through the absorbed medium, we can only predict an average effect of energy loss per particle distance of travel. Taking into account the charge, mass and speed (energy) of the particle, and the density and atomic number of the absorbing medium, Bethe (1933, 1953) derived the formula for calculating the stopping power resulting from coulombic interactions of heavy charged particles (e.g., alpha particles, protons, and deuterons) traveling through absorber media. Rohrlich and Carlson (1954) have refined the calculations to include energy losses via bremsstrahlung radiation, significant when high-energy electrons and beta particles interact with absorbers of high atomic number. Also, refinements to the stopping power formulae in the low energy ranges of heavy particles have been made by several researchers including Bohr and Lindhard (1954), Lindhard and Scharff (1960, 1961), Northcliffe (1963) and Mozumder *et al.* (1968). Derivations of stopping power formulas can be obtained from texts by Friedlander *et al.* (1964), Roy and Reed (1968), Segrè (1968), and Evans (1972). The formulas for the stopping power of charged particles due to coulombic interactions (i.e., ionization and electron orbital excitation) are most clearly defined by Tsoulfanidis

(1995) as the following:

(i) for heavy charged particles (e.g., protons, deuterons, and alpha particles),

$$\frac{dE}{dx} = 4\pi r_0^2 z^2 \frac{mc^2}{\beta^2} NZ \left[\ln\left(\frac{2mc^2}{I} \beta^2 \gamma^2\right) - \beta^2 \right] \tag{1.130}$$

(ii) for electrons or negatrons (negative beta particles),

$$\frac{dE}{dx} = 4\pi r_0^2 \frac{mc^2}{\beta^2} NZ$$

$$\times \left\{ \ln\left(\frac{\beta\gamma\sqrt{\gamma-1}}{I} mc^2\right) + \frac{1}{2\gamma^2}\left[\frac{(\gamma-1)^2}{8} + 1 - (\gamma^2 + 2\gamma - 1)\ln 2\right] \right\} \tag{1.131}$$

and (iii) positrons (positive beta particles),

$$\frac{dE}{dx} = 4\pi r_0^2 \frac{mc^2}{\beta^2} NZ$$

$$\times \left\{ \ln\left(\frac{\beta\gamma\sqrt{\gamma-1}}{I} mc^2\right) - \frac{\beta^2}{24}\left[23 + \frac{14}{\gamma+1} + \frac{10}{(\gamma+1)^2} + \frac{4}{(\gamma+1)^3}\right] + \frac{\ln 2}{2} \right\} \tag{1.132}$$

where dE/dx is the particle stopping power in units of MeV/m, r_0 is the classical electron radius $= 2.818 \times 10^{-15}$ m, z is the charge on the particle ($z = 1$ for p, d, β^-, β^+ and $z = 2$ for α), mc^2 is the rest energy of the electron $= 0.511$ MeV (see Section IV.C of this chapter), N is the number of atoms per m^3 in the absorber material through which the charged particle travels ($N = \rho(N_A/A)$ where ρ is the absorber density (e.g., for NaI, $\rho = 3.67\,\mathrm{g\,cm^{-3}}$), N_A is Avogadro's number $= 6.022 \times 10^{23}$ atoms per mol, A and Z are the atomic weight and atomic number, respectively, of the absorber, $\gamma = (T + Mc^2)/Mc^2 = 1/\sqrt{1 - \beta^2}$ where T is the particle kinetic energy in MeV and M is the particle rest mass (e.g., proton $= 931.5$ MeV/c^2, deuteron $= 2(931.5)$ MeV/c^2, alpha particle $= 4(931.5)$ MeV/c^2, and β^- or $\beta^+ = 0.511$ MeV/c^2, and β the relative phase velocity of the particle $= v/c$, the velocity of the particle in the medium divided by the speed of light in a vacuum $= \sqrt{1 - (1/\gamma^2)}$ (See also Chapter 9 for a treatment on β), and I is the mean excitation potential of the absorber in units of eV approximated by the equation

$$I = (9.76 + 58.8Z^{-1.19})Z, \quad \text{when } Z > 12 \tag{1.133}$$

where pure elements are involved as described by Tsoulfanidis (1995). However, when a compound or mixture of elements is concerned, a mean excitation energy, $\langle I \rangle$, must be calculated according to Bethe theory as follows

$$\langle I \rangle = \exp\left\{\left[\sum_j w_j(Z_j/A_j)\ln I_j\right]\Big/\sum_j w_j Z_j/A_j\right\} \quad (1.134)$$

where w_j, Z_j, A_j and I_j are the weight fraction, atomic number, atomic weight, and mean excitation energy, respectively, of the jth element (Seltzer and Berger, 1982a). See Anderson *et al.* (1969), Sorensen and Anderson (1973), Janni (1982), Seltzer and Berger (1982a,b, 1984), Berger and Seltzer (1983) and Tsoulfanidis (1995) for experimentally determined values of I for various elements and thorough treatments of stopping power calculations. Values of mean excitation potentials, I, for 100 elements and many inorganic and organic compounds are provided by Seltzer and Berger (1982a, 1984).

An example of the application of one of the above equations would be the following calculation of the stopping power for a 2.280 MeV beta particle (E_{max}) emitted from ^{90}Y traveling through a NaI solid scintillation crystal detector. This would be a practical example, as the NaI detector is used commonly for the measurement of ^{90}Y. The solution is as follows:

Firstly, the calculation of relevant variables are

$$\gamma = \frac{2.280\,\text{MeV} + 0.511\,\text{MeV}}{0.511\,\text{MeV}} = 5.462$$

$$\beta = \sqrt{1 - \frac{1}{\gamma^2}} = 0.9665 \quad \text{and} \quad \beta^2 = 0.9341$$

The atomic weight A for NaI would be the average atomic weight (A_{av}) based on the weight-fraction w_{Na} for Na (15.3%) and w_I for I (84.7%) in NaI or

$$A_{av} = (0.153)(A_{Na}) + (0.847)(A_I) = (0.153)(23) + (0.847)(127) = 111$$

Also, on the basis of the weight averages for Na and I, the atomic number Z would be the effective atomic number Z_{ef} calculated according to the following equation described by Tsoulfanidis (1995):

$$Z_{ef} = \frac{\sum_{i=1}^{L}(w_i/A_i)Z_i^2}{\sum_{i=1}^{L}(w_i/A_i)Z_i} \quad (1.135)$$

where L is the number of elements in the absorber, w_i is the weight fraction of the ith element, A_i is the atomic weight of the ith element, Z_i is the atomic number of the ith element, and $w_i = N_iA_i/M$ where N_i is the number of atoms of the ith element and M is the molecular weight of the absorber. If we apply Eq. 1.135 to the absorber NaI we find

$$Z_{ef} = \frac{(0.153/22.989)(11)^2 + (0.847/126.893)(53)^2}{(0.153/22.989)(11) + (0.847/126.893)(53)} = 45.798$$

For pure elements the value of the mean excitation potential, I, can be calculated according to the empirical formula provided by Eq. 1.133.

However, for the compound NaI, the mean excitation energy, $\langle I \rangle$, will be calculated according to Eq. 1.134 as follows

$$\langle I \rangle = \exp\left\{\frac{[(0.153)(11/22.989)\ln 149 + (0.847)(53/126.893)\ln 491]}{[(0.153 \cdot 11/22.989) + (0.847 \cdot 53/126.893)]}\right\}$$

$$= 400\,\text{eV}$$

From Eq. (1.131) the stopping power for the 2.280 beta particle traveling through a NaI crystal is calculated as

$$\frac{dE}{dx} = 4(3.14)(2.818 \times 10^{-15}\,\text{m})^2 \left(\frac{0.511\,\text{MeV}}{0.9341}\right)(3.67\,\text{g cm}^{-3})$$

$$\times \left(\frac{6.022 \times 10^{23}\,\text{atoms mol}^{-1}}{111\,\text{g mol}^{-1}}\right)\left(\frac{10^6\,\text{cm}^3}{\text{m}^3}\right)$$

$$\times (45.798)\left\{\ln\left(\frac{(0.9665)(5.462)\sqrt{4.462}}{400\,\text{eV}}(0.511\,\text{MeV})(10^6\,\text{eV MeV}^{-1})\right)\right.$$

$$\left. + \frac{1}{2(5.462)}\left[\frac{(4.462)^2}{8} + 1 - (5.462^2 + 2(5.462) - 1)\right]\ln 2\right\}$$

$$= 473.6\,\text{MeV m}^{-1}$$

In SI units the stopping power can be expressed in units of J m^{-1} or

$$(473.6\,\text{MeV m}^{-1})(1.602 \times 10^{-13}\,\text{J MeV}^{-1}) = 7.58 \times 10^{-11}\,\text{J m}^{-1}$$

The stopping power is often expressed in units of MeV/g cm^{-2} or J/kg m^{-2}, which provides values for stopping power without defining the density of the absorber medium (Taylor *et al.*, 1970 and Tsoulfanidis, 1995). In these units the above calculation can also be expressed as

$$\frac{1}{\rho}\left(\frac{dE}{dx}\right) = \frac{4.736\,\text{MeV cm}^{-1}}{3.67\,\text{g cm}^{-3}} = 1.29\,\text{MeV/g cm}^{-2}$$

Equation 1.131 used above to calculate the stopping power for the 2.280 MeV beta particle from ^{90}Y in NaI accounts only for energy of the beta particle lost via collision interactions resulting in ionization and electron-orbital excitations. The equation does not account for radial energy loss via the production of bremsstrahlung radiation, which can be very significant with beta particles of high energy and absorber materials of high atomic number. Thus, a complete calculation of the stopping power must include also the radial energy loss via bremsstrahlung. The ratio of beta-particle energy loss via bremsstrahlung emission to energy loss via collision interactions causing ionization and excitation is described by the relation

$$\frac{E_{\text{Brems.}}}{E_{\text{ioniz.}}} = \frac{EZ}{750} \tag{1.136}$$

where E is the beta-particle energy in MeV and Z is the atomic number of the absorber material (Friedlander *et al.*, 1964 and Evans, 1972). From Eqs. 1.131 and 1.136, we can write

$$
\begin{aligned}
\left(\frac{dE}{dx}\right)_{rad.} &= \frac{ZE}{750}\left(\frac{dE}{dx}\right)_{ion.}\\
&= \frac{(45.798)(2.280)}{750}(4.74\,\mathrm{MeV\,cm^{-1}}) = 0.660\,\mathrm{MeV\,cm^{-1}}
\end{aligned} \tag{1.137}
$$

The total stopping power of the 2.280 MeV beta particle in NaI according to Eq. 1.25 is calculated as

$$
\begin{aligned}
\left(\frac{dE}{dx}\right)_{total} &= \left(\frac{dE}{dx}\right)_{ion.} + \left(\frac{dE}{dx}\right)_{rad.}\\
&= 4.74\,\mathrm{MeV\,cm^{-1}} + 0.660\,\mathrm{MeV\,cm^{-1}} = 5.4\,\mathrm{MeV\,cm^{-1}}
\end{aligned} \tag{1.138}
$$

Beta-particle loss via bremsstrahlung radiation of the 2.280 MeV beta particles from ^{90}Y is significant in NaI, namely, 0.66/5.4 or 12.2% of the total energy loss. Consequently, NaI solid scintillation detectors are at times used for the analysis of ^{90}Y (Coursey *et al.*, 1993). The actual detection efficiencies reported by Coursey *et al.* (1993) for the solid scintillation analysis of ^{90}Y fall in the range of 9.9–18% depending on sample and detector counting geometries. The detection efficiencies exceed the above-calculated 12.2% energy loss via bremsstrahlung production, because the NaI detector will also respond to collision-excitation energy of the beta-particle in addition to bremsstrahlung radiation excitation (See Chapter 11 on Solid Scintillation Analysis). Caution is warranted in making correlations between detector response to beta-particle radiation and stopping-power calculations, because we must keep in mind that each stopping-power calculation, such as the above example, provides values for only one beta-particle energy. Beta particles, on the other hand, are emitted with a broad spectrum of energies from zero to E_{max}, the majority of which may possess an average energy, E_{av}, of approximately one-third of E_{max}.

B. Linear Energy Transfer

The International Commission on Radiation Units and Measurements or ICRU (Taylor *et al.*, 1970) defines linear energy transfer (L) of charged particles in a medium as

$$
L = \frac{dE_L}{dl} \tag{1.139}
$$

where dE_L is the average energy locally imparted to the medium by a charged particle of specified energy in traversing a distance dl. The term "locally imparted" refers either to a maximum distance from the particle track or to a maximum value of discrete energy loss by the particle beyond which losses are no longer considered as local. Linear energy transfer or LET is generally measured in units of $\text{keV}\,\mu\text{m}^{-1}$. The ICRU recommends when a restricted form of LET is desired, that the energy cut-off form of LET be applied because this can be evaluated using restricted stopping-power formulae (Taylor et al., 1970). The energy-restricted form of LET or L_Δ is therefore defined as that part of the total energy loss of a charged particle which is due to energy transfers up to a specified energy cut-off value

$$L_\Delta = \left(\frac{\mathrm{d}E}{\mathrm{d}l}\right)_\Delta \tag{1.140}$$

where the cut-off energy (Δ) in eV units must be defined or stated. If no cut-off energy is applied then the subscript ∞ is used in place of Δ, where L_∞ would signify the value of LET, which includes all energy losses and would therefore be equal to the total mass stopping power.

Fig. 1.31 illustrates charged particle interactions within an absorber involved in the measurement of LET. The possible types of energy loss, ΔE, of a charged particle of specified energy, E, traversing an absorber over a track length Δl is illustrated, where O represents a particle traversing the observer without any energy loss, U is the energy transferred to a localized interaction site, q is the energy transferred to a short-range secondary particle when $q \le \Delta$, and Δ is a selected cut-off energy level (e.g., 100 eV), Q' is the energy transferred to a long-range secondary particle (e.g., formation of delta rays) for which $Q' > \Delta$, γ is the energy transferred to photons (e.g., excitation fluorescence, Cherenkov photons, etc.), r is a selected cut-off distance from the particle's initial trajectory or path of travel, and θ is the angle of particle scatter. The interactions q, Q, and γ are subdivided in Fig. 1.31 when these fall into different compartments of the absorber medium. See Taylor et al. (1970)

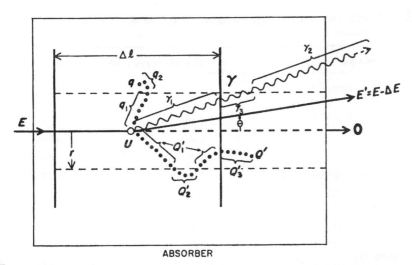

ABSORBER

FIGURE 1.31 Diagram of the passage of particle of energy E through a thickness Δl of material illustrating the several types of energy loss that may occur. (From Taylor et al., 1970.)

for methods used for the precise calculations of LET. Some examples of LET in water for various radiation types are given in Table 1.10. The table clearly illustrates that radiation of a given energy with shorter range in a medium will yield higher values of LET than radiations of the same energy with longer ranges in the same medium. This may be intuitively obvious, because the shorter the range of the radiation the greater is the energy dissipated per unit path length of travel. We can take this further and generalize that the following radiation types will yield LET values of decreasing orders of magnitude (the heavier charged particles are considered here to be of the same energy for purposes of comparison) according to the sequence:

$$
\left[
\begin{array}{l}
\text{Decreasing LET:} \\
\text{FissionProducts} > \text{Alpha Particles} > \text{Deuterons} > \text{Protons} > \\
\quad \text{Low-energy x-Rays and Beta Particles} > \text{High-energy} \\
\quad \text{x-Rays and Beta Particles} > \\
\text{Gamma Radiation and High-energy Beta Particles}
\end{array}
\right] \quad (1.141)
$$

Although the electromagnetic x- and gamma radiations are not charged particles, these radiations do have the characteristics of particles (photons), that produce ionization in matter. They are, therefore, included in the above sequence (1.141) and among the radiations listed in Table 1.10.

The term delta rays, referred to in the previous paragraph, is used to identify energetic electrons that produce secondary ionization. When a charged particle, such as an alpha particle, travels through matter ionization occurs principally through coulombic attraction of orbital electrons to the positive

TABLE 1.10 Track-average Values of LET ($\bar{L}_\Delta$) in Water Irradiated with Various Radiations[a]

Radiation	Cut-off Energy, Δ (eV)	$\bar{L}_\Delta$ (keV μm^{-1})
^{60}Co gamma rays	Unrestricted	0.239
	10,000	0.232
	1,000	0.230
	100	0.229
22-MeV x-rays	100	0.19
2-MeV electrons (whole track)	100	0.20
200-kV x-rays	100	1.7
^{3}H beta particles	100	4.7
50-kV x-rays	100	6.3
5.3 MeV alpha particles (whole track)	100	43

[a]From Taylor *et al.* (1970).

charge on the alpha particle with the ejection of electrons of such low energy that these electrons do not produce further ionization. However, direct head-on collisions of the primary ionizing particle with an electron does occur occasionally whereby a large amount of energy is transferred to the electron. The energetic electron will then travel on in the absorbing matter to produce secondary ionization. These energetic electrons are referred to as delta rays. Delta rays form ionization tracks away from the track produced by the primary ionizing particle. The occurrence and effects of delta rays in radiation absorption are applied to studies of radiation dosimetry (Casnati *et al.*, 1998 and Cucinotta *et al.*, 1998).

When we compare particles of similar energy, we can state that, the ranges of particles of greater mass and charge will obviously be shorter and the magnitude of their LET values would be consequently higher in any given medium. The relationship between mass, charge, energy, range of particles, and their corresponding LET values can be appreciated from Table 1.11. The LET values in Table 1.11 are estimated by dividing the radiation energy by its range or path length in the medium. Such a calculation provides only an estimate of the LET, because the energy dissipated by the radiation will vary along its path of travel, particularly in the case of charged particles, more energy is released when the particle slows down before it comes to a stop as illustrated in Fig. 1.3, when energy liberated in ion-pair formation is the highest. Nevertheless, the LET values provided in Table 1.11 give good orders of magnitude for comparative purposes.

The concept of LET and the calculated values of LET for different radiation types and energies can help us interpret and sometimes even predict the effects of ionizing radiation on matter. For example, we can predict that heavy charged particles, such as alpha radiation, will dissipate their energy at shorter distances within a given absorber body than the more penetrating beta- or gamma radiations. Also, low-energy x-radiation can produce a similar effect as certain beta radiations. The order of magnitude of the LET will help us predict the penetration power and degree of energy dissipation in an absorber body, which is critical information in studies of radiation chemistry, radiation therapy, and dosimetry, among others. For additional information, the reader is referred to works by Ehman and Vance (1991), Farhataziz and Rodgers (1987), and Spinks and Woods (1990).

VI. RADIOISOTOPE DECAY

The activity of a radioactive source or radionuclide sample is, by definition, its strength or intensity or, in other words, the number of nuclei decaying per unit time. The activity decreases with time. A time in which there is an observable change in the rate of radioactivity for a given quantity of radionuclide may be very short, of the order of seconds, or very long, of the order of years. The decay of some nuclides is so slow that it is impossible to observe any change in radioactivity.

TABLE I.II Range and LET Values for Various Charged-Particle Radiations in Water in Order of Decreasing Mass[a]

Nuclide	Radiation Energy (MeV)	Range in Water (mm)	Average LET in Water (keV μm^{-1})
Thorium-232	α, 4.0	0.029^b	138
Americium-241	α, 5.5	0.048^b	114
Thorium-227	α, 6.0	0.055^b	109
Polonium-211	α, 7.4	0.075^b	98
—	d, 4.0	0.219^c	18.3
—	d, 5.5	0.377^c	14.6
—	d, 6.0	0.440^c	13.6
—	d, 7.4	0.611^c	12.1
—	p, 4.0	0.355^d	11.3
—	p, 5.5	0.613^e	9.0
—	p, 6.0	0.699^f	8.6
—	p, 7.4	1.009^g	7.3
Tritium	β^-, 0.0186 (E_{max})	0.00575^h	3.2^h
Carbon-14	β^-, 0.156 (E_{max})	0.280^h	0.56^h
Phosphorus-32	β^-, 1.710 (E_{max})	7.92^h	0.22^h
Yttrium-90	β^-, 2.280 (E_{max})	10.99^h	0.21^h

[a]The deuteron (d) and proton (p) energies were arbitrarily selected to correspond to the alpha particle (α) energies to facilitate the comparison of the effects of particle mass and charge on range and LET.
[b]Calculated according to Eqs. 1.14 and 1.15.
[c]The deuteron range is calculated from the equation $R_{Z,M,E} = M/Z^2 R_{p,E/M}$. The equation provides the range of a particle of charge Z, mass M, and energy E, where $R_{p,E/M}$ is the range in the same absorber of a proton of energy E/M (Friedlander et al., 1964).
[d]Calculated according to Eqs. 1.12, 1.14 and 1.15, $R_{air} = 28.5$ mg cm^{-2} (Fig. B.1, Appendix B).
[e]Calculated according to Eqs. 1.12, 1.14 and 1.15, $R_{air} = 49.5$ mg cm^{-2} (Fig. B.3).
[f]Calculated according to Eqs. 1.12, 1.14 and 1.15, $R_{air} = 56.5$ mg cm^{-2} (Fig. B.1).
[g]Calculated according to Eqs. 1.12, 1.14 and 1.15, $R_{air} = 82.0$ mg cm^{-2} (Fig. B.1).
[h]Calculations are based on the maximum energy (E_{max}) of the beta particles. When the lower value of average beta particle-energy (E_{av}) is used, the calculated value of range would be shorter and LET higher. The range was calculated according to the empirical formula $R = 0.412E^{1.27-0.0954\ln E}$ available from the curve provided in Fig. B.3, Appendix B.

A. Half-Life

Rates of radionuclide decay are usually expressed in terms of half-life. This is the time, t, required for a given amount of radionuclide to lose 50% of its activity. In other words, it is the time required for one-half of a certain number of nuclei to decay. The decay curve of ^{32}P (Fig. 1.32) illustrates the concept of half-life. In Fig. 1.32, the activity of the ^{32}P is plotted against time in days. It can be seen that, after every interval of 14.3 days, the radioactivity of the ^{32}P is reduced by half. Thus, the half-life, $t_{1/2}$, of ^{32}P is 14.3 days. It is not possible to predict when one particular atom of ^{32}P will decay; however,

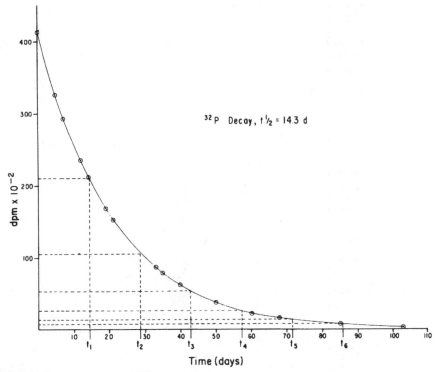

FIGURE I.32 Decay of ^{32}P represented as linear plot. Horizontal and vertical lines between the ordinate and abscissa delineate ^{32}P activities (dpm) for six half-lives identified by the symbols $t_1, t_2, t_3, \ldots, t_6$. (From L'annunziata, 1965, unpublished work.)

it is possible to predict statistically for a large number of ^{32}P radionuclides that one-half of the atoms would decay in 14.3 days.

In cases in which decay can be recorded within a reasonable period of time, the half-life of a nuclide can be determined by means of a semi-logarithmic plot of activity versus time, as shown in Fig. 1.33. Radionuclide decay is a logarithmic relation, and the straight line obtained on the semilogarithmic plot permits a more accurate determination of the half-life.

Radionuclide decay may best be defined in mathematical terms. The number, ΔN, of atoms disintegrating in a given time, Δt, is proportional to the number, N, of radioactive atoms present. This relationship may be written as

$$\Delta N / \Delta t = \lambda N \qquad (1.142)$$

or

$$dN/dt = -\lambda N \qquad (1.143)$$

where λ is a proportionality constant, commonly referred to as the decay constant, and the negative sign signifies a decreasing number of radionuclides with time.

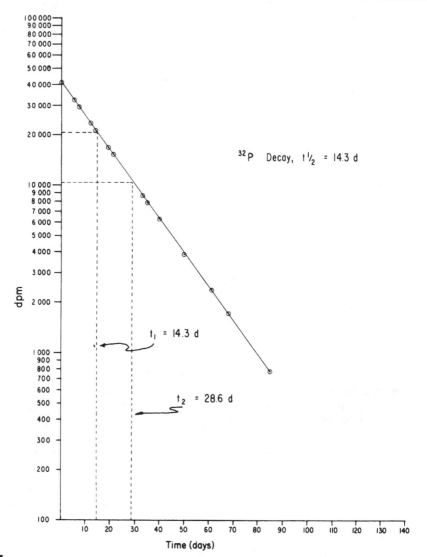

FIGURE 1.33 Semilogarithmic plot of the decay of ^{32}P. Two half-lives (t_1 and t_2) are delineated by horizontal and vertical lines between the ordinate and abscissa. (From L'Annunziata, 1965, unpublished work.)

One condition must be fulfilled for Eq. 1.143 to be rigorously applicable: the total number of radioactive atoms Δt being considered must be large enough to make statistical methods valid. For example, in the case of a single isolated atom of ^{32}P there is no way to predict when the atom will decay. In fact, the atom might decay in the first second after $t = 0$ (the moment observations are initiated) or it might decay days later. The concept of half-life is a statistical one, which, when applied to a large number of atoms, as is usually the case, allows an accurate calculation of the activity of radionuclides after a given time interval.

For radionuclide decay calculations, Eq. 1.143 must be transformed into a more suitable form and may be expressed as

$$dN/N = -\lambda dt \tag{1.114}$$

which can be integrated between the limits N_0 and N and between t_0 and t, where t_0 is 0 (the moment observations are initiated), N_0 is the number of atoms originally present at time t_0, and N is the number of atoms remaining after time t:

$$\int_{N_0}^{N} dN/N = -\lambda \int_{t_0}^{t} dt \tag{1.145}$$

to give

$$\ln N/N_0 = -\lambda t \tag{1.146}$$

Equation 1.146 may be written in exponential form as

$$N = N_0 \, e^{-\lambda t} \tag{1.147}$$

where e is the base of the natural logarithm, λ is the decay constant, and t is the interval of time. Equation 1.147 is the form used to determine the decay of a radionuclide sample after a given time interval. To use Eq. 1.147, the value of the decay constant λ, must be known, and this is different for each radionuclide. To determine λ for a particular radionuclide, a relationship between the decay constant and the half-life may be derived from the decay Eq. 1.146, which may be transposed to

$$\ln N_0/N = \lambda t \tag{1.148}$$

By definition, we know that, after an interval of time corresponding to the half-life, half of the original activity remains. Therefore, we may assign the original activity N_0 as unity whereby after one half-life the remaining activity N would be one-half of unity, and Eq. 1.148 would become

$$\ln 1/(1/2) = \lambda \, t_{1/2} \tag{1.149}$$

or

$$\ln 2 = \lambda t_{1/2} \tag{1.150}$$

and

$$0.693 = \lambda t_{1/2} \tag{1.151}$$

The decay constant can then be defined as

$$\lambda = 0.693/t_{1/2} \tag{1.152}$$

The value of λ can be calculated easily from the half-life of an isotope with Eq. 1.152. The units used for λ are expressed in reciprocal time, s^{-1}, m^{-1}, h^{-1}, d^{-1}, or y^{-1}, depending on the half-life of the radionuclide and also on the time interval t used in Eq. 1.147. For example, if ^{32}P, which has a half-life of 14.3-days, is used in an experiment, λ may be expressed in d^{-1}. The unit of the decay constant must agree with the time interval t of Eq. 1.147.

The following example illustrates the use of Eq. 1.147 to calculate the decay of a radionuclide sample within any time interval.

If a sample contained 3.7 MBq of ^{32}P on a given date and an investigator wished to determine the amount remaining after a 30-day period, he or she would first determine the decay constant for ^{32}P according to Eq. 1.152 and then calculate the activity after the specified time period using the decay equation 1.147 as follows. The decay constant in units of d^{-1} is determined by

$$\lambda = 0.693/14.3\,d = 4.85 \times 10^{-2}\,d^{-1}$$

With the calculated value of λ and the time interval t equal to 30 days, the activity of the remaining ^{32}P is determined according to Eq. 1.147 as

$$
\begin{aligned}
N &= 3.7 \times 10^6 \text{ dps} \cdot e^{-(4.85 \times 10^{-2}\,d^{-1}(30\,d))} \\
&= 3.7 \times 10^6 \text{ dps} \cdot e^{-1.455} \\
&= 3.7 \times 10^6 \text{ dps} \cdot 0.2334 \\
&= 8.64 \times 10^5 \text{ dps} = 0.864 \text{ MBq}
\end{aligned}
$$

where $N_0 = 3.7 \times 10^6$ dps by definition ($1\,MBq = 1 \times 10^6$ dps). This gives the value of the activity of ^{32}P after the 30-day period as $N = 8.64 \times 10^5$ dps $= 0.864$ MBq.

The decay equation has many practical applications, as it can also be used as well to calculate the time required for a given radionuclide sample to decay to a certain level of activity. Let us consider the following example:

A patient was administered intravenously 600 MBq of ^{99m}Tc methylene diphosphate, which is a common agent administered for the purposes of carrying out a diagnostic bone scan. The doctor then wanted to know how much time would be required for the ^{99m}Tc radioactivity in the patient's body to be reduced to 0.6 MBq (0.1% of the original activity) from radionuclide decay alone ignoring any losses from bodily excretion. The half-life $t_{1/2}$ of ^{99m}Tc is 6.00 hours. To calculate the time required we can write Eq. 1.147 as

$$A/A_0 = e^{-\lambda t} \tag{1.153}$$

where A is the activity in dps (disintegrations per second) after time t and A_0 is the initial activity at time t_0. Equation 1.153 can be transposed to

$$\ln A_0/A = \lambda t \tag{1.154}$$

or

$$t = \frac{1}{\lambda} \ln \frac{A_0}{A} \tag{1.155}$$

By definition (Eq. 1.152) the decay constant λ of ^{99m}Tc is $0.693/t_{1/2}$ or $0.693/6.00$ h. Solving Eq. 1.155 after inserting the value of λ and the relevant activities of ^{99m}Tc gives

$$t = (6.00\,\text{h}/0.693)\ln(600\,\text{MBq}/0.6\,\text{MBq}) = 59.8\,\text{hours} = 2.5\,\text{days}$$

In the case of a mixture of independently decaying radionuclides, the rate of decay of each nuclide species does not change. However, the rate of decay of the overall sample is equal to the sum of the decay rates of the individual nuclide species. The cumulative decay of a mixture of independently decaying nuclides from the most simple case of a mixture of two nuclides to a more complex case of n number of nuclides is described by

$$N = N_1^0\,e^{-\lambda t} + N_2^0\,e^{-\lambda t} + \cdots + N_n^0\,e^{-\lambda t} \tag{1.156}$$

where N is the number of atoms remaining after time t, and N_1^0, N_2^0, and N_n^0 are the numbers of atoms originally present at time t_0 of 1, 2, and n number of nuclide species, respectively.

The semilogarithmic decay plot of a mixture of two independently decaying nuclides is not a straight line, contrary to pure radionuclide samples, but is a composite plot, as in the case of a mixture of ^{32}P and ^{45}Ca (see Fig. 1.34). If the half-lives of the two nuclides are significantly different, the composite curve may be analyzed so that these may be determined. If the decay of the composite mixture can be observed over a reasonable period of time, the composite curve will eventually yield a straight line representing the decay of the longer-lived nuclide after the disappearance of the shorter-lived nuclide (depicted in Fig. 1.34). This straight line may be extrapolated to time $t=0$ so that the activity (dpm) of this nuclide at $t=0$ can be found. The difference between the activity at $t=0$ of the longer-lived nuclide and the total activity of the sample at $t=0$ gives the activity at $t=0$ of the shorter-lived nuclide. Likewise, further subtraction of points of the extrapolated decay curve from the composite curve yields the decay curve of the shorter-lived nuclide.

The half-lives of the two radionuclides are determined from the slopes of the two decay curves isolated from the composite curve. Equation 1.148, which is expressed in natural logarithms, may be transformed to logarithms

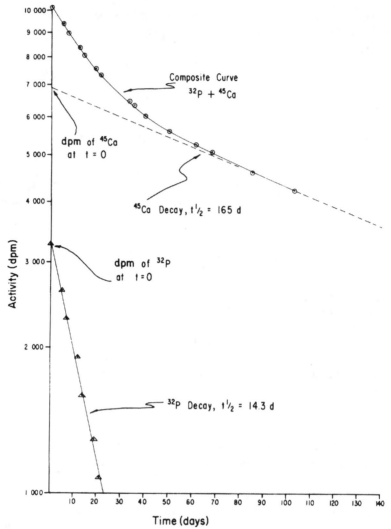

FIGURE 1.34 Semi logarithmic decay curves of ^{32}P and ^{45}Ca isolated from a composite decay curve of a mixture of $^{32}P + ^{45}Ca$. (From L'Annunziata, 1965, unpublished work.)

to the base 10 by

$$2.30 \log(N_1/N_2) = \lambda(t_2 - t_1) \qquad (1.157)$$

or

$$\log(N_1/N_2) = \frac{\lambda}{2.30}(t_2 - t_1) \qquad (1.158)$$

where N_1 and N_2 are the numbers of atoms or activity of the sample at times t_1 and t_2, respectively. Because semilogarithmic paper is used to plot the straight-line decay curves and because $\lambda/2.30$ of Eq. 1.158 is equal to the slope, the decay constant, λ, may be calculated from a graphical determination of the slope. With a calculated value of λ, the half-life of the nuclide is then calculated from Eq. 1.152.

Many radionuclides have very long half-lives, which make the graphic representation of their decay impossible. Some examples are ^{3}H ($t_{1/2} = 12.3$ y), ^{14}C ($t_{1/2} = 5.73 \times 10^{3}$ y), ^{40}K ($t_{1/2} = 1.3 \times 10^{9}$ y), and ^{174}Hf ($t_{1/2} = 2 \times 10^{15}$ y) (see the Appendix). In such cases the half-lives can be calculated from Eqs. 1.143 and 1.152. The decay rate or activity, A, in disintegrations per year (DPY) of a given nuclide sample, defined by dN/dt of Eq. 1.143, is measured experimentally. The number of atoms of the radioassayed sample, defined by N of Eq. 1.143, must be known or determined. This is simple for pure samples. For example, the number of atoms of ^{40}K in a pure sample of KCl is easily calculated from Avogadro's number (6.022×10^{23} molecules mol^{-1}) and the percentage natural abundance of ^{40}K (0.012%). Samples of unknown purity and isotopic abundance require a quantitative analysis of the element such as that provided by a mass spectral analysis of the isotopic abundance. The value of λ in y^{-1} is calculated as

$$\lambda = \frac{dN/dt}{N} = \frac{A}{N} = \frac{CPM/E}{N}(5.25 \times 10^{5} \text{ m y}^{-1}) \qquad (1.159)$$

where A is the sample nuclide activity in DPY, N is the number of atoms of the nuclide in the sample, CPM is the sample count rate provided by the instrument radioactivity detector, E is the instrument counting efficiency, and 5.25×10^{5} m y^{-1} is the factor used to convert counts per minute (CPM) to counts per year (CPY). The half-life can then be calculated according to Eqs. 1.152 and 1.159 both of which define the value of λ.

Let us look at a practical example of the use of the above equations to determine the half-life of ^{40}K taken from the recent work of Grau Malonda and Grau Carles (2002). The accurate determination of the half-life of ^{40}K has very practical implications, as it is currently used by geologists to determine the date of a rock's formation based on the measurement of the quantity of the stable daughter nuclide ^{40}Ar. Grau Malonda and Grau Carles (2002) report the accurate determination of the half-life of ^{40}K by measuring accurately the activity of ^{40}K in a sample of pure KNO_3 and applying the relationships of half-life to λ according to Eqs. 1.152 and 1.159. They measured the ^{40}K specific activity in KNO_3 by the very accurate CIEMAT/NIST efficiency tracing liquid scintillation standardization method (see Chapter 5) to be 12.24 ± 0.014 Bq/g. Also, applying the known isotopic concentration of ^{40}K in KNO_3 of 0.01167% and the value of Avogadro's number 6.022×10^{23} atoms per mole, they could calculate the number of atoms of ^{40}K in 1 g of KNO_3 as follows:

$(6.022 \times 10^{23}$ molecules/101.103 g $KNO_3)(0.0001167) = 6.951 \times 10^{17}$ atoms ^{40}K per gram of KNO_3. From Eqs. 1.152 and 1.159 we can write

$$\frac{1}{\lambda} = \frac{t_{1/2}}{0.693} = \frac{N}{A} \qquad (1.160)$$

or

$$t_{1/2} = 0.693\left(\frac{N}{A}\right) \tag{1.161}$$

From the determined specific radioactivity of ^{40}K in KNO$_3$ and the number of atoms of ^{40}K per gram of KNO$_3$, Grau Malonda and Grau Carles (2002) calculated the half-life of ^{40}K as

$$t_{1/2} = 0.693\left(\frac{6.951 \times 10^{17}\,\text{atoms}\,^{40}\text{K/g KNO}_3}{(12.24\,\text{dps}\,^{40}\text{K/g KNO}_3)(60\,\text{s/m})(5.25 \times 10^5\,\text{m/y})}\right)$$

and

$$t_{1/2} = 1.248 \times 10^9\,\text{y}$$

From the mean of nine determinations, Grau Malonda and Grau Carles (2002) were able to assign the value of the half-life $(t_{1/2})$ of ^{40}K to be $(1.248 \pm 0.004) \times 10^9$ y at a 95% confidence level.

Other radionuclides have very short half-lives such as ^{209}Ra $(t_{1/2} = 4.6\,\text{s})$, ^{215}At $(t_{1/2} = 1.0 \times 10^{-4}\,\text{s})$ and ^{212}Po $(t_{1/2} = 2.98 \times 10^{-7}\,\text{s})$. The methods of determination of half-lives of such short duration can be determined by delayed coincidence methods (Schwarzschild, 1963; Ohm et al., 1990; Morozov et al., 1998), which involve the use of scintillation detectors with detector response times as short as 10^{-11} s. These methods are applicable when a parent nuclide of normally perceptible or long half-life produces a daughter of very short half-life. Radiation detectors with resolving times of fractions of a microsecond are set electronically so that a delay circuit will detect a radiation-induced pulse from the parent in coincidence with a radiation pulse produced from the daughter. Varying the delay time of the coincidence circuit results in a delay of the coincidence pulse rate from which a decay curve of the very short-lived daughter nuclide can be plotted and the half-life determined.

B. General Decay Equations

The simplest decay relationship between parent and daughter nuclides that can be considered is that of a parent nuclide which decays to form a stable daughter nuclide.

The decay of the radionuclide ^{33}P serves as an example. The parent nuclide ^{33}P decays with a half-life of 25 days with the production of the stable daughter ^{33}S, as indicated by

$$^{33}_{15}\text{P} \rightarrow\, ^{33}_{16}\text{S (stable)} + \beta^- + \bar{\nu} \tag{1.162}$$

Numerous radionuclides, such as ^{3}H, ^{14}C, ^{32}P, ^{35}S, ^{36}Cl, ^{45}Ca, and ^{131}I (see Appendix A), decay by this simple parent–daughter relationship.

However, numerous other radionuclides produce unstable daughter nuclides. The simplest case would be that in which the parent nuclide A decays to a daughter nuclide B, which in turn decays to a stable nuclide C:

$$A \rightarrow B \rightarrow C \, (\text{stable}) \tag{1.163}$$

In such decay chains, the rate of decay and production of the daughter must be considered as well as the rate of decay of the parent. The decay of the parent is described by the simple rate equation

$$-dN_A/dt = \lambda_A \, N_A \tag{1.164}$$

which is integrated to the form

$$N_A = N_A^0 \, e^{-\lambda_A t} \tag{1.165}$$

where N_A^0 is the number of atoms of the parent at the time $t = 0$ and N_A is the number of atoms after a given period of time $t = t_1$.

The decay rate of the daughter is dependent on its own decay rate as well as the rate at which it is formed by the parent. It is written as

$$-dN_B/dt = \lambda_B \, N_B - \lambda_A \, N_A \tag{1.166}$$

where $\lambda_B \, N_B$ is the rate of decay of the daughter alone and $\lambda_A \, N_A$ is the rate of decay of the parent or rate of formation of the daughter. Equations 1.165 and 1.166 may be transposed into the linear differential equation

$$d \, N_B/dt + \lambda_B \, N_B - \lambda_A \, N_A^0 \, e^{-\lambda_A t} = 0 \tag{1.167}$$

which is solved for the number of atoms of daughter, N_B, as a function of time to give

$$N_B = \frac{\lambda_A}{\lambda_B - \lambda_A} N_A^0 (e^{-\lambda_A t} - e^{-\lambda_B t}) + N_B^0 \, e^{-\lambda_B t} \tag{1.168}$$

Although unnecessary in this treatment, the solution to Eq. 1.167 is given by Friedlander *et al.* (1964).

In decay schemes of this type, the following three conditions may predominate: (1) secular equilibrium, (2) transient equilibrium, and (3) the state of no equilibrium. Each of these cases will now be considered in detail.

C. Secular Equilibrium

Secular equilibrium is a steady-state condition of equal activities between a long-lived parent radionuclide and its short-lived daughter. The important criteria upon which secular equilibrium depends are:

1. The parent must be long-lived; that is, negligible decay of the parent occurs during the period of observation, and
2. The daughter must have a relatively short half-life. The relative difference in half-life in this latter criterion is further clarified by

$$\lambda_A/\lambda_B \leq\sim 10^4 \tag{1.169}$$

that is,

$$\lambda_A \ll \lambda_B \tag{1.170}$$

where λ_A and λ_B are the respective decay constants of the parent and daughter nuclides. The importance of these two requirements can be clearly seen if the $^{90}\text{Sr}(^{90}\text{Y})$ equilibrium is taken as an example.

The infamous fallout nuclide ^{90}Sr is the parent in the decay scheme

$$^{90}_{38}\text{Sr} \xrightarrow{t_{1/2}=28.8\,\text{y}} {}^{90}_{39}\text{Y} \xrightarrow{t_{1/2}=2.7\,\text{d}} {}^{90}_{40}\text{Zr} \quad (\text{stable}) \tag{1.171}$$

The long half-life of ^{90}Sr definitely satisfies the first requirement for secular equilibrium, because over a quarter of a century is needed for it to lose 50% of its original activity. As will be seen, less than 3 weeks are required for secular equilibrium to be attained and, in this interim period, negligible decay of ^{90}Sr occurs.

To satisfy the second requirement the decay constants for ^{90}Sr and ^{90}Y, λ_A and λ_B, respectively, must be compared. The decay constants for ^{90}Sr and ^{90}Y are easily calculated from their half-lives and Eq. 1.152, and the values are $6.60 \times 10^{-5}\,\text{d}^{-1}$ and $2.57 \times 10^{-1}\,\text{d}^{-1}$, respectively. Consequently, in the comparison $\lambda_A/\lambda_B = 2.57 \times 10^{-4}$, and this is in agreement with the order of magnitude required for secular equilibrium.

An equation for the growth of daughter atoms from the parent can be obtained from Eq. 1.168 by consideration of the limiting requirements for secular equilibrium. Since $\lambda_A \approx 0$ and $\lambda_A \ll \lambda_B$, $e^{-\lambda_A t} = 1$ and λ_A falls out of the denominator in the first term. If the daughter nuclide is separated physically from the parent (L'Annunziata, 1971), $N_B^0 = 0$ at time $t = 0$ (time of parent–daughter separation) and the last term would fall out of Eq. 1.168. Thus, in the case of secular equilibrium, the expression of the ingrowth of daughter atoms with parent can be written as

$$N_B = \frac{\lambda_A N_A^0}{\lambda_B}(1 - e^{-\lambda_B t}) \tag{1.172}$$

If the observation of the ingrowth of the daughter is made over many half-lives of the daughter, it is seen that the number of atoms of daughter approaches a maximum value $\lambda_A N_A^0/\lambda_B$, which is the rate of production of daughter divided by its decay constant. The final form of Eq. 1.172 to be used for the calculation of the ingrowth of daughter can be expressed as

$$N_B = (N_B)_{max}(1 - e^{-\lambda_B t})$$ (1.173)

Since the activity of the daughter atoms, A_B, is proportional to the number of daughter atoms, or $A_B = k \lambda_B N_B$, where k is the coefficient of detection of the daughter atoms, Eq. 1.173 may also be written as

$$A_B = (A_B)_{max}(1 - e^{-\lambda_B t})$$ (1.174)

Arbitrarily selecting activities of 100 dpm of parent ^{90}Sr and 100 dpm of daughter ^{90}Y, it is possible to calculate and graphically represent the ingrowth of ^{90}Y with its parent and also the decay of ^{90}Y subsequent to the separation of parent and daughter nuclides (L'Annunziata, 1971). Identical activities of ^{90}Sr and ^{90}Y are arbitrarily chosen, because their activities are equal while in secular equilibrium prior to their separation. Figure 1.35 illustrates the calculated growth of ^{90}Y as produced by ^{90}Sr (curve B) using Eq. 1.174 with $(A_B)_{max} = 100$. The decay of separated ^{90}Y (curve A) is plotted by simple half-life decay ($t_{1/2} = 2.7$ d). The dashed line (line C) represents the decay of ^{90}Sr, which is negligible during the period of observation ($t_{1/2} = 28.8$ y). The total activity (curve D) is the result of both ^{90}Sr decay and the ingrowth of ^{90}Y after the separation of the latter and is obtained by the addition of curve B to line C. It may be noted from Fig. 1.35 that after approximately six half-lives of ^{90}Y (~ 18 d) the growth of ^{90}Y has reached the activity of ^{90}Sr, after which both nuclides decay with the same half-life, that of the parent ^{90}Sr (28.8 y).

As an example of the practical utility of this phenomenon, the application of secular equilibrium theory to the analysis of ^{90}Sr in biological systems is discussed.

One method reported by the Los Alamos National Laboratory (see Gautier, 1995) entails the initial chelation (complex formation) of the sample strontium with the sodium salt of ethylenediaminetetraacetic acid (EDTA). The complexed strontium is then isolated by elution on an ion exchange column. The eluted strontium is then precipitated as a carbonate. The activity of radioactive strontium, which will include ^{89}Sr + ^{90}Sr in the sample, is determined by low-background counting. Low-background liquid scintillation counting is most often used for the total ^{89}Sr + ^{90}Sr analysis as described by Passo and Cook (1994). The isolated radiostrontium is then allowed to remain in the sample without further treatment for a period of about 2 weeks to allow ingrowth of ^{90}Y. About 2 weeks are needed to ensure the parent and daughter radionuclides are in secular equilibrium before the chemical separation of yttrium from strontium. From Eq. 1.173 it is calculated that after 2 weeks the activity of ^{90}Y grows to 97.4% of its original level. Carrier yttrium is then added to the dissolved radiostrontium, and the yttrium is precipitated as the hydroxide, redissolved, and reprecipitated as an oxalate (see Section VII.C of this chapter for a discussion of the concepts of carrier

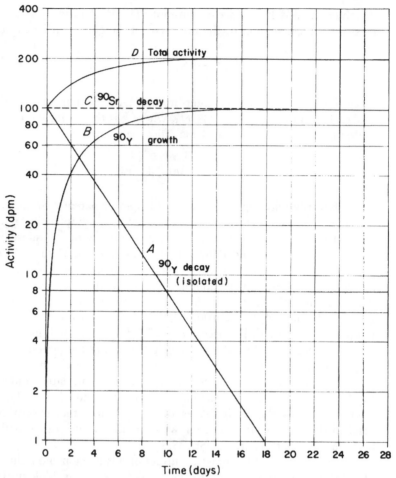

FIGURE 1.35 Growth and decay curves following the separation of ^{90}Sr(^{90}Y) in secular equilibrium. **(A)** Decay of isolated ^{90}Y. **(B)** Ingrowth of ^{90}Y with ^{90}Sr. **(C)** Decay of isolated ^{90}Sr. **(D)** Total activity from isolated ^{90}Sr, representing both ^{90}Sr decay and ^{90}Y growth until secular equilibrium is attained. (From L'Annunziata, 1971, reprinted with permission Copytight American Chemical Society.)

and carrier-free radionuclides). The step involving the precipitation of yttrium from the sample results in the separation of ^{90}Y from the radiostrontium. The separated ^{90}Y can then be assayed by suitable low-background counting using liquid scintillation or Cherenkov counting (Passo and Cook, 1994; L'Annunziata and Passo, 2002). The ^{90}Sr activity in the sample is determined from the activity of ^{90}Y by calculating the ^{90}Y decay from the time of separation (precipitation) of yttrium from strontium. This is possible because the parent and daughter radionuclides were at secular equilibrium (i.e., ^{90}Sr dpm $= {}^{90}$Y dpm) at time $t = t_0$ when the precipitation and separation of yttrium from strontium were carried out. The ^{89}Sr activity in the sample is determined from the difference between the total radiostrontium activity (^{89}Sr $+ {}^{90}$Sr) and the measured activity of ^{90}Sr.

Certain chemical processes in natural and biological systems can preferentially select either the parent or daughter nuclide and, in this manner, separate the two. For example, a research investigator could administer nuclides in secular equilibrium to a soil and plant system. At the time of administration, the nuclides are in secular equilibrium; that is, both the parent and daughter activities are equal. However, if in the course of the experiment the investigator obtains a plant sample for radioassay, which had preferably absorbed either the parent or daughter, problems ensue if the equilibrium phenomenon is not considered. Radioassay of plant tissue that had selectively concentrated the parent could show an initial progressive rise in radioactivity due to ingrowth of daughter, whereas a selective concentration of daughter would result in a sample showing an initial decrease in radioactivity. In cases such as these, it is necessary to isolate the parent radionuclide chemically and wait for a period of time sufficient to permit secular equilibrium to be reached [~ 2 weeks for the ^{90}Sr(^{90}Y) example] before counting a sample.

D. Transient Equilibrium

Like secular equilibrium, transient equilibrium is a steady-state condition between the parent and daughter nuclides. However, in transient equilibrium the parent–daughter nuclides do not possess the same activities, but rather they decay at the same half-life, that of the parent nuclide.

The criterion upon which transient equilibrium rests is that the parent nuclide must be longer lived than its daughter, but not of the order of magnitude described by Eq. 1.169; that is, it is necessary that $\lambda_A < \lambda_B$. However, the ratio λ_A/λ_B should fall within the limits $10^{-4} < \lambda_A/\lambda_B < 1$.

The decay chain of ^{100}Pd serves as an example of parent–daughter nuclides that may attain transient equilibrium. ^{100}Pd decays by electron capture to ^{100}Rh with a half-life of 96 h. The daughter nuclide ^{100}Rh decays by electron capture and positron emission to the stable nuclide ^{100}Ru. The half-life of the daughter nuclide is 21 h. The decay scheme may be represented as

$$^{100}_{46}\text{Pd} \xrightarrow{t_{1/2} = 96\,\text{h}} {}^{100}_{45}\text{Rh} \xrightarrow{t_{1/2} = 21\,\text{h}} {}^{100}_{44}\text{Ru} \quad (\text{stable}) \qquad (1.175)$$

The first criterion for transient equilibrium is satisfied in this case; the half-life of the parent nuclide is greater than that of the daughter. If the decay constants λ_A and λ_B are now calculated, we can determine whether or not the second criterion ($10^{-4} < \lambda_A/\lambda_B < 1$) is satisfied.

The value of λ_A, given by 0.693/96 h, is 7.2×10^{-3} h^{-1}, and that of λ_B, given by 0.693/21 h, is 3.3×10^{-2} h^{-1}. Consequently, the ratio $\lambda_A/\lambda_B = 2.2 \times 10^{-1}$ and lies within the limits of the second criterion.

If the general decay Eq. 1.168 of the daughter nuclide is considered, the term $e^{-\lambda_B t}$ is negligible compared with $e^{-\lambda_A t}$ for sufficiently large values of t. Thus the terms $e^{-\lambda_B t}$ and $N_B^0 e^{-\lambda_B t}$ may be dropped from Eq. 1.168 to give

$$N_B = \frac{\lambda_A}{\lambda_B - \lambda_A}(N_A^0 e^{-\lambda_A t}) \qquad (1.176)$$

for the decay of the daughter nuclide as a function of time. Because $N_A = N_A^0 e^{-\lambda_A t}$, Eq. 1.176 may be written as

$$N_B/N_A = \frac{\lambda_A}{\lambda_B - \lambda_A} \qquad (1.177)$$

From Eq. 1.177, it can be seen that the ratio of the number of atoms or the ratio of the activities of the parent and daughter nuclides is a constant in the case of transient equilibrium.

Since $A_A = k_A \lambda_A N_A$ and $A_B = k_B \lambda_B N_B$, where A_A and A_B are the activities of the parent and daughter nuclides, respectively, and k_A and k_B are the detection coefficients of these nuclides, Eq. 1.177 may be written in terms of activities as

$$\frac{A_B}{k_B \lambda_B}(\lambda_B - \lambda_A) = \frac{A_A}{k_A \lambda_A}\lambda_A \qquad (1.178)$$

or

$$A_B/A_A = \frac{k_B \lambda_B}{k_A(\lambda_B - \lambda_A)} \qquad (1.179)$$

If equal detection coefficients are assumed for the parent and daughter nuclides, Eq. 1.179 may be written as

$$A_B/A_A = \frac{\lambda_B}{(\lambda_B - \lambda_A)} \qquad (1.180)$$

Thus, for transient equilibrium Eq. 1.180 indicates that the activity of the daughter is always greater than that of the parent by the factor $\lambda_B/(\lambda_B - \lambda_A)$. Equation 1.180 may likewise be written as

$$A_A/A_B = 1 - \lambda_A/\lambda_B \qquad (1.181)$$

whereby the ratio A_A/A_B falls within the limits $0 < A_A/A_B < 1$ in transient equilibrium.

If an activity of 100 dpm is arbitrarily chosen for the daughter nuclide ^{100}Rh in transient equilibrium with its parent ^{100}Pd, the activity of ^{100}Pd can be found using either Eq. 1.180 or 1.181. Equation 1.180 gives

$$100\,\text{dpm}/A_A = \frac{3.3 \times 10^{-2}\ \text{h}^{-1}}{3.3 \times 10^{-2}\ \text{h}^{-1} - 7.2 \times 10^{-3}\ \text{h}^{-1}}$$

or

$$A_A = 78\text{dpm}$$

Using Eq. 1.180 or 1.181, the decay of the daughter nuclide may be calculated as a function of parent decay in transient equilibrium. The ^{100}Pd–^{100}Rh parent–daughter decay in transient equilibrium is illustrated by curves A and B, respectively, of Fig. 1.36. The parent and daughter nuclides are shown to have respective activities of 78 dpm and 100 dpm at time $t = 0$. As curves A and B show, the parent and daughter nuclides in transient equilibrium decay with the same half-life, that corresponds to the half-life of the parent.

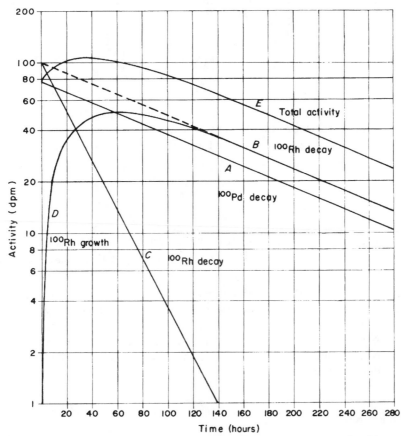

FIGURE 1.36 Growth and decay curves following the separation of ^{100}Pd(^{100}Rh) in transient equilibrium. **(A)** Decay of isolated parent nuclide ^{100}Pd. **(B)** Decay of ^{100}Rh daughter nuclide in transient equilibrium. The dashed portion of this curve represents ^{100}Rh decay if parent and daughter nuclides were not separated. **(C)** Decay of ^{100}Rh after separation from its parent. **(D)** The ingrowth of ^{100}Rh with the isolated parent ^{100}Pd. **(E)** Total activity from the isolated ^{100}Pd representing both ^{100}Pd decay and ^{100}Rh growth until transient equilibrium is attained.

If the parent and daughter nuclides were to be separated, the daughter nuclide would decay according to its half-life as indicated by curve C. The isolated parent nuclide would, however, show an increase in activity with time owing to the ingrowth of daughter until transient equilibrium is attained. Curve D of Fig. 1.36 shows the ingrowth of daughter nuclide from a freshly isolated parent. Because $N_B^0 = 0$ at time $t = 0$ (time of separation of parent and daughter), the last term of Eq. 1.168 falls out to give

$$N_B = \frac{\lambda_A N_A^0}{\lambda_B - \lambda_A}\left(e^{-\lambda_A t} - e^{-\lambda_B t}\right) \tag{1.182}$$

The term $\lambda_A N_A^0 /(\lambda_B - \lambda_A)$ describes the rate of production of the daughter divided by the difference between the daughter and parent decay constants, which may be written as

$$N_B = (N_B)_{max}(e^{-\lambda_A t} - e^{-\lambda_B t}) \tag{1.183}$$

similar to the case of Eq. 1.173. Since the activity, A_B, of the daughter atoms is proportional to the number of daughter atoms, or $A_B = k_B \lambda_B N_B$, where k is as defined previously, Eq. 1.183 may also be written as

$$A_B = (A_B)_{max}(e^{-\lambda_A t} - e^{-\lambda_B t}) \tag{1.184}$$

Because the maximum daughter activity in this sample is 100 dpm, Eq. 1.184 may be used to calculate the ingrowth of daughter nuclide with $(A_B)_{max} = 100$.

Curve E of Fig. 1.36 illustrates the activity of the isolated parent nuclide. It is found by summing curves A and D and consequently accounts for the simultaneous decay of the parent nuclide and the ingrowth of the daughter. Notice that the slopes of curves A, B, and E are identical when transient equilibrium is attained, that is, the rates of decay of both the parent and daughter are identical.

E. No Equilibrium

The cases of secular equilibrium and transient equilibrium, which involve decay schemes whereby the parent nuclide is longer lived than its daughter, were just considered. In other cases in which the daughter nuclide is longer lived than its parent, $\lambda_A > \lambda_B$, no equilibrium is attained. Instead, the parent nuclide of shorter half-life eventually decays to a negligible extent, leaving only the daughter nuclide, which decays by its own half-life. The following decay scheme of ^{56}Ni serves as an example:

$$^{56}_{28}\text{Ni} \xrightarrow{t_{1/2}=6.4\,\text{d}} {}^{56}_{27}\text{Co} \xrightarrow{t_{1/2}=77.3\,\text{d}} {}^{56}_{26}\text{Fe} \quad \text{(stable)} \qquad (1.185)$$

The parent nuclide ^{56}Ni decays by electron capture with a half-life of 6.4 d, whereas its daughter ^{56}Co decays with the longer half-life of 77.3 d by electron capture and β^+ emission. Curve A of Fig. 1.37 illustrates the decay of initially pure ^{56}Ni parent nuclide. The decay of ^{56}Ni is followed by the ingrowth (production) of the ^{56}Co daughter nuclide, shown by curve B. The ingrowth of daughter is calculated from Eq. 1.168, of which the last term,

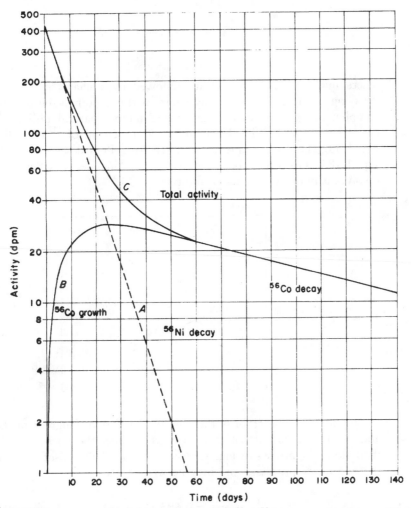

FIGURE 1.37 Growth and decay curves of the ^{56}Ni(^{56}Co) parent–daughter nuclides following the isolation or fresh preparation of the parent nuclide ^{56}Ni. (A) Decay of pure parent nuclide ^{56}Ni. (B) Ingrowth of daughter nuclide ^{56}Co. (C) Total activity representing both ^{56}Ni decay and the simultaneous growth and decay of ^{56}Co daughter.

$N_B^0 e^{-\lambda_B t}$, falls out because $N_B^0 = 0$ at time $t = 0$. The number of daughter atoms N_B of Eq. 1.168 may be converted to activity, A_B, by the term $A_B = k_B \lambda_B N_B$ as discussed previously. The total activity illustrated by curve C of Fig. 1.37 depicts both the simultaneous decay of parent nuclide and the growth and decay of daughter determined by summing curves A and B. Notice from Fig. 1.37 that the parent nuclide activity in this example becomes negligible after around 55 d, after which the total activity, curve C, has a slope corresponding to the decay rate of the daughter nuclide.

F. More Complex Decay Schemes

Other decay schemes exist that involve a chain of numerous nuclides such as

$$A \rightarrow B \rightarrow C \rightarrow \cdots \rightarrow N \tag{1.186}$$

where nuclides A, B, and C are followed by a chain of a number N of decaying nuclides. A long decay chain of this type may be observed in the complex decay schemes of high-atomic-number natural radionuclides such as ^{235}U, ^{238}U, and ^{232}Th. The complex decay scheme of ^{232}Th is illustrated in Fig. 1.38. The decay sequence of $^{232}_{90}Th$ to $^{212}_{83}Bi$ is described by the general Eq. 1.186. However, the continuation of this decay scheme with $^{212}_{83}Bi$ involves a branching decay of the type.

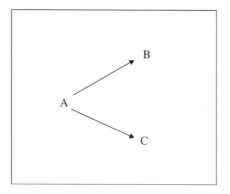

In this example $^{212}_{83}Bi$ is the parent of the two daughter nuclides $^{212}_{84}Po$ and $^{208}_{81}Tl$. The half-life of ^{212}Bi is written under the nuclide symbol rather than along the arrows of Fig. 1.38 because the ^{212}Bi half-life is a function of the two decay processes and may be written as

$$t_{1/2} = 0.693/(\lambda_A + \lambda_B) \tag{1.187}$$

where λ_A and λ_B are the decay constants of the two separate decay processes.

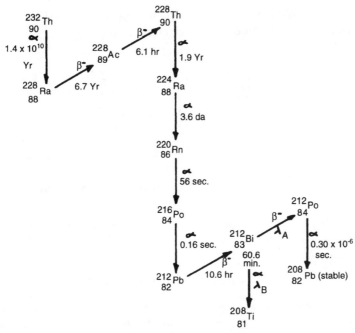

FIGURE I.38 Decay scheme of natural ^{232}Th.

VII. RADIOACTIVITY UNITS AND RADIONUCLIDE MASS

A. Units of Radioactivity

The units used to define radioactivity or, in other words, the activity of a sample are written in terms of the number of atoms, N, disintegrating per unit of time, t. We can use Eq. 1.142 previously discussed in this chapter to calculate the activity of any given mass of radionuclide. The equation, namely $\Delta N/\Delta t = \lambda N$, defines the proportionality between the rate of decay of a radionuclide and the number of atoms of the radionuclide in a sample. As an example, we may use Eq. 1.142 to calculate the activity of 1 g of ^{226}Ra as follows:

$$\Delta N/\Delta t = \lambda N$$
$$\Delta N/\Delta t = (0.693/t_{1/2})(N) \tag{1.188}$$

where $\lambda = 0.693/t_{1/2}$ as derived previously (Eq. 1.152). If we take the half-life, $t_{1/2}$, of ^{226}Ra to be 1599 y and substitute for N, in the preceding equation, the number of atoms per mol of ^{226}Ra, we can write

$$\Delta N/\Delta t = (0.693/1599 \text{ y})(6.022 \times 10^{23} \text{ atoms}/226 \text{ g})$$

where, according to Avogadro's number, there are 6.022×10^{23} atoms per gram mole of substance. If we now convert the half-life of ^{226}Ra from units of years to minutes, we can calculate the number of atoms of ^{226}Ra disintegrating per minute (dpm) per gram according to

$$\Delta N/\Delta t = \left[\frac{0.693}{(1599\,\text{y})(365\,\text{d y}^{-1})(24\,\text{h d}^{-1})(60\,\text{m h}^{-1})}\right]$$
$$\times \left(\frac{6.022 \times 10^{23}\,\text{atoms}}{226\,\text{g}}\right)$$
$$\Delta N/\Delta t = \left(\frac{0.693}{8.404 \times 10^8\,\text{m}}\right)(2.665 \times 10^{21}\,\text{atoms g}^{-1})$$
$$= 2.19 \times 10^{12}\,\text{atoms per minute per gram}$$

The activity of 1 g of ^{226}Ra is the basis of the unit of radioactivity known as the curie (Ci). One curie is almost equal to the activity of 1 g of ^{226}Ra or, by definition,

$$1\,\text{Ci} = 2.22 \times 10^{12}\,\text{dpm} = 3.7 \times 10^{10}\,\text{dps}$$

Therefore, one curie of activity or any multiple of the curie of any radionuclide defines the number of atoms disintegrating per unit of time in minutes or seconds.

The rate of decay in terms of time in seconds gives rise to a more recently adopted Système International d'Unités (SI) unit of activity, which is the becquerel (Bq), where by definition

$$1\,\text{Bq} = 1\,\text{dps}$$

Therefore, we can interrelate the curie and becquerel as follows:

$$1\,\text{Ci} = 2.22 \times 10^{12}\,\text{dpm} = 3.7 \times 10^{10}\,\text{dps} = 37\,\text{GBq}$$

Likewise, smaller units of the curie, namely the millicurie (mCi) and microcurie (μCi), may be interrelated with the becquerel as follows:

$$1\,\text{mCi} = 2.22 \times 10^9\,\text{dpm} = 3.7 \times 10^7\,\text{dps} = 37\,\text{MBq}$$

and

$$1\,\mu\text{Ci} = 2.22 \times 10^6\,\text{dpm} = 3.7 \times 10^4\,\text{dps} = 37\,\text{kBq}$$

Another unit of activity recommended in the early 1960s by the International Union of Pure and Applied Physics, but less frequently used, is the rutherford, where 1 rutherford $= 10^6$ dps and 1 microrutherford would be equivalent to 1 dps or 1 Bq (Buttlar, 1968; Das and Ferbel, 1994).

B. Correlation of Radioactivity and Radionuclide Mass

From Eq. 1.188 and calculations made in the previous Section VII.A, we can see that, for samples of a given level of activity, radionuclides of shorter half-life will contain a smaller number of radioactive atoms than radionuclides of longer half-life.

We can use Eq. 1.188 again to compare two radionuclides of relatively short and long half-lives to see the magnitude of the differences in radionuclide masses we would encounter for any given level of radioactivity. For example, we may take the radionuclide ^{32}P of 14.3-day half-life and the radionuclide ^{14}C of 5730-year half-life and calculate the activity per gram and grams per curie of each radionuclide for comparative purposes. These calculations are as follows.

1. ^{32}P, half-life $= 14.3$ days:

$$\Delta N/\Delta t = (0.693/t_{1/2})(N)$$

$$\Delta N/\Delta t = \left[\frac{0.693}{(14.3\,\text{d})(24\,\text{h}\,\text{d}^{-1})(60\,\text{m}\,\text{h}^{-1})}\right]\left(\frac{6.023 \times 10^{23}}{32\,\text{g}}\right)$$

$$= 6.32 \times 10^{17}\,\text{dpm per gram}\,^{32}\text{P}$$

If, by definition, 1 curie $= 2.22 \times 10^{12}$ dpm, we can convert this activity per gram of ^{32}P to grams ^{32}P per curie as follows:

$$2.22 \times 10^{12}\,\text{dpm Ci}^{-1}/6.32\times 10^{17}\,\text{dpm g}^{-1}\,^{32}\text{P} = 3.51 \times 10^{-6}\,\text{g}\,^{32}\text{P per Ci}$$

$$= 3.51\times 10^{-6}\,\text{mg}^{32}\text{P per mCi}$$

2. ^{14}C, half-life $= 5730$ years:

$$\Delta N/\Delta t = \Delta t = \left[\frac{0.693}{(5730\,\text{y})(365\,\text{d}\,\text{y}^{-1})(24\,\text{h}\,\text{d}^{-1})(60\,\text{m}\,\text{h}^{-1})}\right]\left(\frac{6.022 \times 10^{23}}{14\,\text{g}}\right)$$

$$= 9.90 \times 10^{12}\,\text{dpm per gram}\,^{14}\text{C}$$

This activity per gram of ^{14}C is converted to grams ^{14}C per curie as follows:

$$2.22 \times 10^{12}\,\text{dpm Ci}^{-1}/9.90 \times 10^{12}\,\text{dpm g}^{-1}\,^{14}\text{C} = 0.224\,\text{g}\,^{14}\text{C per Ci}$$

$$= 0.224\,\text{mg}\,^{14}\text{C per mCi}$$

The calculated mass of ^{32}P in 1 curie of activity is almost a million fold less than the calculated mass of ^{14}C in 1 curie of activity. In general, research with radionuclides involves the handling and analysis of lower levels of radioactivity in millicuries, microcuries, and picocuries, and so on. The masses of radioactive atoms in the milli-, micro-, and picocurie levels of radioactivity are obviously much smaller than encountered at the curie level. It is important, therefore, to be aware of the order of magnitude of radioactive atom masses involved, which leads us to the concept of "carrier-free" samples of radionuclides, discussed subsequently.

C. Carrier-Free Radionuclides

A carrier-free radionuclide sample is generally a solution in which all of the atoms of a particular element consist of the radioactive isotope; that is, no stable isotope of that element is present. A stable isotope of the particular element is referred to as carrier. It is common to encounter carrier-free radionuclide samples. Many of the radionuclides procured from commercial producers are supplied carrier free. It is important, therefore, to be aware of the masses of radioactive isotope in the carrier-free sample and any consequences that may be involved when very small quantities (e.g., 10^{-6} to 10^{-12} g or smaller) of radioactive nuclide may be involved.

For example, in Section VII.B we calculated that there was only 3.51×10^{-6} g of ^{32}P per curie of radioactivity. A millicurie of carrier-free ^{32}P, which is a level of activity and form normally procured from a radioisotope supplier, would contain only 3.51×10^{-9} g of ^{32}P and zero grams of stable phosphorus. It is a common procedure to dilute the carrier-free ^{32}P to the microcurie level of activity prior to working with the radionuclide such as in tracer studies. One microcurie of the carrier-free ^{32}P would contain only 3.51×10^{-12} g of phosphorus. Obviously, therefore, we should consider the consequences of working with such small amounts of phosphorus in solution. Over the past 40 years of working with carrier-free radioactive nuclide sources, the author has experienced the absorption of significant quantities of carrier-free radionuclides onto the surface of glassware. If we consider the ionic characteristics of the chemical forms of certain radionuclide sources and the minute quantities these may possess in the carrier-free form, significant quantities of certain carrier-free radionuclides can be lost from solution by absorption on the inner surface of glassware, onto the surface of precipitates, and so forth. For example, when working with carrier-free ^{32}P sources, if a particular experiment calls for the addition of carrier, the author will add carrier to the radionuclide source during the dilution procedure. If carrier is not desired, the procedure recommended by Chase and Rabinowitz (1968) can be utilized. For example, if it is desired to dilute a carrier-free solution of $NaH_2{}^{32}PO_4$ in a volumetric flask, it is best to treat the flask first with a 1% solution of NaH_2PO_4 prior to the addition of the carrier-free solution. The volumetric flask and any other glassware used in the dilution may be rinsed with the 1% NaH_2PO_4. Alternatively, the volumetric flask may be filled with the 1% NaH_2PO_4 solution and allowed to sit for several hours. The flask is then rinsed with deionized water to remove unabsorbed phosphorus. The flask can then be used to prepare a dilution of carrier-free $NaH_2{}^{32}PO_4$. It is important, however, to rinse the flask with a solution of the same chemical form as the radioisotope, if it is desirable to prevent contamination of the radioisotope with another chemical form.

REFERENCES

Abele, H. (2000). The standard model and the neutron β-decay. *Nucl. Instrum. Methods Phys. Res., Sect. A* **440**, 499–510.

Anderson, H. H., Sørensen, H., and Vadja, P. (1969). Excitation potentials and shell corrections for the elements $Z_2 = 20$ to $Z_2 = 30$. *Phys. Rev.* **180**, 373–380.

Argonne National Laboratory (1963). "Reactor Physics Constants." ANL-5800, 2nd ed. United States Atomic Energy Commission, Washington, DC.

Arzumanov, S., Bondarenko, L., Chernyavsky, S., Drexel, W., Fomin, A., Geltenbort, P., Morozov, V., Panin, Yu., Pendlebury, J., and Schreckenbach, K. (2000). Neutron lifetime measured by monitored storing of untra-cold neutrons. *Nucl. Instrum. Methods Phys. Res., Sect. A* **440**, 511–516.

Bacon, G. E. (1969). "Neutron Physics." Wykeham Publications, London.

Balasubramanian, P. S. (1997). Anodically oxidized aluminum layer as a useful substrate for the fabrication of ^{147}Pm sources for beta-ray thickness gauges. *J. Radioanal. Nucl. Chem.* **223**(1–2), 79–81.

Balasubramanian, P. S. (1998). A simple procedure for the fabrication of high activity beta-radiation sources of ^{147}Pm for use in beta-ray thickness gauges. *J. Radioanal. Nucl. Chem.* **229** (1–2), 157–160.

Bambynek, W., Crasemann, B., Fink, R. W., Freund, H. U., Mark, H., Swift, C. D., Price, R. E., and Rao, P. V. (1972). X-ray fluorescence yields, Auger, and Coster-Kronig transition probabilities. *Rev. Mod. Phys.* **44**, 716–813.

Berger, M. J. and Hubbell, J. H. (1997). Photon attenuation coefficients. *In* "Handbook of Chemictry and Physics" (D. R. Lide and H. P. R. Frederikse, Eds.), 77th ed., pp. 10-250–10-254, CRC Press, Boca Raton.

Berger, M. J. and Seltzer, S. M. (1982). Stopping powers and ranges of electrons and positrons. *Nat. Bureau Standards Publ.* NBSIR 82–2550, pp. 168.

Bethe, H. A. (1933). Quantenmechanik der Ein- und Zwei-Electronen-Probleme. In "Handbuch der Physik" (H. Geiger and K. Scheel, Eds.), 2nd ed., Vol. 24, Part I, Springer, Berlin.

Bethe, H. A. and Ashkin, J. (1953). Passage of radiations through matter. In "Experimental Nuclear Physics" (E. Segré, Ed.), Vol. 1, J. Wiley, New York.

Bohr, N. and Lindhard, J. (1954). Electron capture and loss by heavy ions penetrating through matter. *Kgl. Danske Videnskab Selskab, Mat.-Fys. Medd.* **28**(7), 1–30.

Bonn, J., Bornshein, L., Bornshein, B., Fickinger, L., Flatt , B., Kraus, Ch., Otten, E. W., Schnall, J. P., Ulrich, H., Weinheimer, Ch., Kazachenka, O., and Kovalik, A. (2000). The Mainz neutrino mass experiment. *In* "Proceedings of the XIX International Conference on Neutrino Physics and Astrophysics (Neutrino-2000)", Sudbury, Canada, June 16–21, 2000 (in press).

Browne, E., Firestone, R. B., and Shirley, V. S. (1986). "Table of Radioactive Isotopes." John Wiley & Sons, New York.

Bryne, J. (1994). "Neutrons, Nuclei and Matter, an Exploration of the Physics of Slow Neutrons." Institute of Physics, London.

Burcham, W. E. and Jobes, M. (1994). "Nuclear and Particle Physics." Longman Scientific & Technical, Essex.

Buttlar, H. V. (1968). "Nuclear Physics, an Introduction." Academic Press, New York.

Canberra Nuclear. (1996). Neutron detection and counting. *In* "Canberra Nuclear Instruments Catalog", 9th ed., pp. 37–39. Canberra Industries, Meridien, CT.

Casnati, E., Baraldi, C., Boccaccio, P., Bonifazzi, C., Singh, B., and Tartari, A. (1998). The effect of delta rays on the ionometric dosimetry of proton beams. *Phys. Med. Biol.* **43**, 547–558.

Chase, G. D. and Rabinowitz, J. L. (1968). "Principles of Radioisotope Methodology," 3rd ed., pp. 140–143, Burgess Publishing Company, Minneapolis.

Clapp, T. G., Titus, K. J., Olson, L. H., and Dorrity, J. L. (1995). The on-line inspection of sewn seams. *National Textile Center Annual Report (August)*, 221–230.

Compton, A. H. (1923). The spectrum of scattered x-rays. *Phys. Rev.* **22**(5), 409–413.

Coursey, B. M., Calhoun, J. M., and Cessna, J. T. (1993). Radioassays of yttrium-90 used in nuclear medicine. *Nucl. Med. Biol.* **20**(5), 693–700.

Cowan, C. L., Jr., Reines, F., Harrison, F. B., Kruse, H. W., and McGuire, A. D. (1956). Detection of the free neutrino: a confirmation. *Science* **124**, 103–104.

Crookes, W. (1903a). The emanation of Radium. *Proc. Roy. Soc. (London)* **A71**, 405–408.

Crookes, W. (1903b). Certain properties of the emanation of Radium. *Chemical News* **87**, 241.

Cucinotta, F. A., Nikjoo, H., and Goodhead, D. T. (1998). The effects of delta rays on the number of particle-track traversals per cell in laboratory and space exposures. *Rad. Res.* **150**, 115–119.

Das, A. and Ferbel, T. (1994). "Introduction to Nuclear and Particle Physics." John Wiley & Sons, New York.

Dean, J. A. (1995). "Analytical Chemistry Handbook." McGraw-Hill, New York.

Dolan, T. J. (1982). "Fusion Research: Principles, Experience and Technology." Pergamon Press, New York.

Ehman, W. D. and Vance, D. E. (1991). "Radiochemistry and Nuclear Methods of Analysis." John Wiley & Sons, New York.

Einstein, A. (1905). Über einen die Erzeugung und Verwandlung des Lichtes betreffenden heuristischen Gesichtspunkt. *Annalen der Physik, Leipzig* **17**, 132–148.

Elster, J. and Geitel, H. (1903). Über die durch radioactive Emanation erregte scintillierende Phosphoreszenz der Sidot-Blende. *Phys. Z.* **4**, 439–440.

Evans, R. D. (1950). "The Atomic Nucleus." McGraw-Hill, New York.

Evans, R. D. (1972). "The Atomic Nucleus." McGraw-Hill, New York.

Farhataziz and Rodgers, M. A. J. (1987). "Radiation Chemistry, Principles and Applications." VCH Publishers, Inc., New York.

Feather, N. (1938). Further possibilities for the absorption method of investigating the primary β-particles from radioactive substances. *Proc. Cambridge Phil. Soc.* **34**, 599–611.

Fenyves, E. and Haiman, O. (1969). "The Physical Principles of Nuclear Radiation Measurements." Academic Press, New York.

Firestone, R. B., Shirley, V. S., Baglin, C. M., Frank Chu, S. Y., and Zipkin, J. (1996). "Table of Isotopes," Vols. I and II, 8th ed., John Wiley & Sons, New York.

Flammersfield, A. (1946). Eine Beziehung zwischer Energie und Reichweite für Beta-Strahlen kleiner und mittlerer Energie. *Naturwissenschaften* **33**, 280–281.

Friedlander, G., Kennedy, J. W., and Miller, J. M. (1964). "Nuclear and Radiochemistry," 2nd ed., John Wiley & Sons, New York.

Gautier, M. A., Ed. (1995). "Health and Environmental Chemistry: Analytical Techniques, Data Management, and Quality Assurance." Manual LA-10300-M, Vol. III, UC-907, pp. WR190-1–WR190-16, Los Alamos National Laboratory, Los Alamos, NM.

Gautreau, R. and Savin, W. (1999). "Theory and Problems in Modern Physics." McGraw-Hill, New York.

Gibson, J. A. B. and Piesch, E. (1985). "Neutron Monitoring for Radiological Protection." Technical Report Series No. 252, International Atomic Energy Agency, Vienna.

Gibbs, W. W. (1998). A massive discovery. *Sci. Am.* **279**(2), 18–20.

Glendenin, L. E. (1948). Determination of the energy of beta particles and photons by absorption. *Nucleonics* **2**, 12–32.

Gratta, G. and Wang, Y. F. (1999). Towards two-threshold, real-time solar neutrino detectors. *Nucl. Instrum. Methods Phys. Res., Sect. A* **438**, 317–321.

Grau Malonda, A. and Grau Carles, A. (2002). Half-life determination of ^{40}K by LSC. *Appl. Radiat. Isot.* **56**, 153–156.

Halpern, A. (1988). "Schaum's 3000 Solved Problems in Physics." McGraw-Hill, New York.

Hawryluk, R. J., Adler, H., Alling, P., Ancher, C., Anderson, H., Anderson, J. L., Ascroft, D., Barnes, C. W., and Barnes, G. (1994). Confinement and heating of a deuterium-tritium plasma. *Phys. Rev. Lett.* **72**, 3530–3533.

Holden, N. E. (1997a). Table of isotopes. *In* "CRC Handbook of Chemistry and Physics" (D. R. Lide, Ed.), 77th ed., pp. 11-38–11-143, CRC Press, Boca Raton, FL.

Holden, N. E. (1997b). Neutron scattering and absorption properties. *In* "CRC Handbook of Chemistry and Physics" (D. R. Lide, Ed.), 77th ed., pp 11-144–11-158, CRC Press, Boca Raton, FL.

Hubbell, J. H. (1969). "Photon Cross Sections, Attenuation Coefficients, and Energy Absorption Coefficients from 10 keV to 100 GeV." NSRDS-NBS 29, Natl. Stand. Ref. Data Ser., National Bureau of Standards (U.S.), pp. 80.

Isotope Products Laboratories (1995). Californium-252 fission foils and neutron sources. *In* "Radiation Sources for Research, Industry and Environmental Applications," p. 55, Isotope Products Laboratories, Burbank, CA.

Janni, J. F. (1982). Proton range-energy tables, 1 keV–10 GeV. *Atomic Data Nucl. Date Tables* **27**, 147–339.

JET Team (1992). Fusion energy production from a deuterium-tritium plasma in the JET tokamak. *Nucl. Fusion* **32**, 187–201.

Karelin, Y. A., Gordeev, Y. N., Karasev, V. I., Radchenko, V. M., Schimbarev, Y. V., and Kuznetsov, R. A. (1997). Californium-252 neutron sources. *Appl. Radiat. Isot.* **48**(10–12), 1563–1566.

Kearns, E., Kajita, T. and Totsuka, Y. (1999). Detecting massive neutrinos. *Sci. Am.* **281**(2), 64–71.

Kesterbaum, D. (1998). Neutrinos throw their weight around. *Science* **281**(5383), 1594–1595.

Knapp, A. K., Abrams, M. D., and Hulbert, L. C. (1985). An evaluation of beta attenuation for estimating aboveground biomass in a tallgrass praire. *J. Range Manag.* **38**(6), 556–558.

Koch, L. (1995). Radioactivity and fission energy. *Radiochim. Acta* **70/71**, 397–402.

Kostroun, V. O., Chen, M. S., and Crasemann, B. (1971). Atomic radiation transition probabilities to the 1s state and theoretical K-shell fluorescence yields. *Phys. Rev.* **A3**, 533–545.

Krane, K. S. (1988). "Introductory Nuclear Physics." John Wiley & Sons, New York.

Kudo, H. (1995) Radioactivity and fusion energy. *Radiochim. Acta* **70/71**, 403–412.

L'Annunziata, M. F. (1971). Birth of a unique parent-daughter relation: secular equilibrium. *J. Chem. Educ.* **48**, 700–703.

L'Annunziata, M. F. (1987). "Radionuclide Tracers, Their Detection and Measurement." Academic Press, London, pp. 505.

L'Annunziata, M. F. and Passo, C. J., Jr. (2002). Cherenkov counting of yttrium-90 in the dry state; correlations with phosphorus-32 cherenkov counting data. *Appl. Radiat. Isot.* **56**, 907–916.

Lee, T. D. and Yang, C. N. (1956). Question of parity conservation in weak interactions. *Phys. Rev.* **104**, 254–258.

Lindhard, J. and Scharff, M. (1960). Recent developments in the theory of stopping power. I. Principles of the statistical method. *In* "Penetration of Charged Particles in Matter," National Academy of Sciences-National Research Council, Publication 752, p. 49.

Lindhard, J. and Scharff, M. (1961). Energy dissipation by ions in the keV region. *Phys. Rev.* **124**, 128–130.

Lobashev, V. M., Aseev, V. A., Belasev, A. I., Berlev, A. I., Geraskin, E. V., Golubev, A. A., Golubev, N. A., Kazachenko, O. V., Kuznetsev, Yu. E., Ostroumov, R. P., Ryvkis, L. A., Stern, B. E., Titov, N. A., Zadorozhny, S. V., and Zakharov, Yu. I. (2000). Neutrino mass anomaly in the tritium beta-spectrum. *In* "Proceedings the XIX International Conference on Neutrino Physics and Astrophysics (Neutrino-2000)," Sudbury, Canada, June 16–21, 2000 (in press).

Mann, W. B. (1978). A Handbook of Radioactivity Measurements Procedures. National Council on Radiation Protection and Measurements. CCRP Report No. 58, Washington, DC.

Mapleston, P. (1997). Film thickness gauges meet market needs for quality, cost. *Mod. Plastics* **74**, 73–76.

Martin, R. C., Laxon, R. R., Miller, J. H., Wierzbicki, J. G., Rivard, M. J., and Marsh, D. L. (1997). Development of high-activity ^{252}Cf sources for neutron brachytherapy. *Appl. Radiat. Isot.* **48**(10–12), 1567–1570.

Martin, R. C., Knauer, J. B., and Balo, P. A. (2000). Production, distribution and applications of Californium-252 neutron sources. *Appl. Radiat. Isot.* **53**, 785–792.

McLane, V., Dunford, C. L., and Rose, P. F. (1988). "Neutron Cross Sections," Vol. 2, "Neutron Cross Section Curves." Academic Press, San Diego.

Mederski, H. J. (1961). Determination of internal water by beta gauging technique. *Soil Sci.* **92**, 143–146.

Mederski, H. J. and Alles, W. (1968). Beta gauging leaf water status: influence of changing leaf characteristics. *Plant Physiol.* **43**, 470–472.

Michael Lederer, C. and Shirley, V. S., Eds. (1978). "Table of Isotopes." 7th ed., John Wiley & Sons, New York.

Miley, G. H. and Sved, J. (1997). The IEC – A plasma-target-based neutron source. *Appl. Radiat. Isot.* **48**(10–12), 1557–1561.

Morozov, V. A., Churin, I. N., and Morozova, N. V. (1998). Nuclear experimental techniques – three-dimensional delayed coincidence single-crystal scintillation time spectrometer. *Instrum. Exp. Tech.* **41**(5), 609.

Mozumder, A., Chatterjee, A., and Magee, J. L. (1968). Theory of radiation chemistry. IX. Mode and structure of heavy particle tracks in water. *Amer. Chem. Soc. Series Adv. in Chem.* **1**, p. 27.

Murray, R. L. (1993). "Nuclear Energy. An Introduction to the Concepts, Systems, and Applications of Nuclear Processes," 4th Ed. Pergamon Press, Oxford.

Nakahata, M. (2000). Neutrinos underground. *Science* **289**(5482), 1155–1156.

Nielsen, D. R. and Cassel, D. K. (1984). Soil water management. *In* "Isotopes and Radiation in Agricultural Sciences" (M. F. L'Annunziata and J. O. Legg, Eds.), Vol. I, pp. 37–43, Academic Press, San Diego.

Nakayama, F. S. and Ehrler, W. L. (1964). Beta ray gauging technique for measuring leaf water content changes and moisture status of plants. *Plant Physiol.* **39**, 95–98.

Northcliffe, L. C. (1963). Passage of heavy ions through matter. *Ann. Rev. Nucl. Sci.* **13**, 67–102.

O'Leary, G.J. and Incerti, M. (1993). A field comparison of three neutron moisture meters. *Aust. J. Exp. Agric.* **33**, 59–69.

Obregewitsch, R. P., Rolston, D. E., Nielsen, D. R., and Nakayama, F. S. (1975). Estimating relative leaf water content with a simple beta gauge calibration. *Agron. J.* **67**, 729–732.

Ogando, J. (1993). Nuclear web gauging keeps pace with processor needs. *Plastics Technol.* **39**, 46–49.

Ohm, H., Liang, M., Molner, G., and Sistemich, K. (1990). Delayed-coincidence measurement of subnanosecond lifetimes in fission fragments. "The Spectroscopy of Heavy Nuclei 1989 Proceedings of the International Conference on the Spectroscopy of Heavy Nuclei, Agia Pelagia, Crete, June 25–July 1, 1989. Institute Physics Conference Series. No. 105, 323–328. Adam Hilger, Ltd., Bristol.

Passo, C. J., Jr. and Cook, G. T. (1994). "Handbook of Environmental Liquid Scintillation Spectrometry. A Compilation of Theory and Methods." Packard Instrument Company, Meriden, CT.

Patel, S. B. (1991). "Nuclear Physics, an Introduction." John Wiley & Sons, New York.

Paul, W. and Steinwedel, H. (1955). Interaction of electrons with matter. *In* "Beta- and Gamma-ray Spectroscopy" (K. Siegbahn, Ed.) North-Holland, Amsterdam.

Pichlmaier, A., Butterworth, J., Geltenbort, P., Nagel, H., Nesvizhevsky, V., Neumaier, S., Schreckenbach, K., Steichele, E., and Varlamov, V. (2000). MAMBO II: neutron lifetime measurement with storage of untra-cold neutrons. *Nucl. Instrum, Methods Phys. Res., Sect. A* **440**, 517–521.

Rachinhas, P. J. B. M., Simões, P. C. P. S., Lopes, J. A. M., Dias, T. H. V. T., Morgado, R. E., dos Santos, J. M. F., Stauffer, A. D., and Conde, C. A. N. (2000). Simulation and experimental results for the detection of conversion electrons with gas proportional scintillation counters. *Nucl. Instrum. Methods Phys. Res., Sect. A* **441**, 468–478.

Radchenko, V. M., Ryabinin, M. A., Andreytchuk, N. N., Gavrilov, V. D., and Karelin, Ye A. (2000). *Appl. Radiat. Isot.* **53**, 833–835.

Reines, F. (1960). Neutrino interactions. *Ann. Rev. Nucl. Sci.* **10**, 1–26.

Reines, F. (1979). The early days of experimental neutrino physics. *Science* **203**, 11–16.

Reines, F. (1994). 40 years of neutrino physics. *Prog. Part. Nucl. Phys.* **32**, 1–12.

Reines, F. and Cowen, C. L. Jr. (1956). The neutrino. *Nature* **178**, 446–449.

Reines, F. and Cowen, C. L. Jr. (1957). Neutrino physics. *Phys. Today* **10**(8), 12–18.

Robson, J. M. (1950a). Radioactive decay of the neutron. *Phys. Rev.* **77**, 747A.

Robson, J. M. (1950b). Radioactive decay of the neutron. *Phys. Rev.* **78**, 311–312.

Rohrlich, F. and Carlson, B. C. (1954). Positron-electron differences in energy loss and multiple scattering. *Phys. Rev.* **93**, 38–44.

Roy, R.R. and Reed, R.D. (1968). "Interactions of Photons and Leptons with Matter." Academic Press, New York.

Rutherford, E. (1919). Collision of α-particles with light atoms. *Nature (London)* 103, 415–418.

Rutherford, E. (1920). Nuclear constitution of atoms. *Proc. Royal Soc. London* 97A, 374–401.

Rutherford, E. (1936). The development of the theory of atomic structure. In "Background to Modern Science" (J. Needham and W. Pagel, Eds.), pp. 61–74. Macmillan Company, New York.

Schwarzschild, A. (1963). A survey of the latest developments in delayed coincidence measurements. *Nucl. Instrum. Methods* 21, 1–16.

Segré, E. (1968). "Nuclei and Particles." W. A. Benjamin, New York.

Seltzer, S. M. and Berger, M. J. (1982a). Evaluation of the collision stopping power of elements and compounds for electrons and positrons. *Int. J. Appl. Radiat. Isot.* 33, 1189–1218.

Seltzer, S. M. and Berger, M. J. (1982b). Procedure for calculating the radiation stopping power for electrons. *Int. J. Appl. Radiat. Isot.* 33, 1219–1226.

Seltzer, S. M. and Berger, M. J. (1984). Improved procedure for calculating the collision stopping power of elements and compounds for electrons and positrons. *Int. J. Appl. Radiat. Isot.* 35(7), 665–676.

Serway, R. A., Moses, C. J., and Moyer, C. A. (1997). "Modern Physics", 2nd ed., Harcourt College Publishers, New York.

Sheffield, J. (2001). The future of fusion. *Nucl. Instrum. Methods Phys. Res., Sect. A* 464, 33–37.

Snell, A. H., and Miller, L. C. (1948). On the radioactive decay of the neutron. *Phys. Rev.* 74, 1217–1218.

Snell, A. H., *et al.* (1950). Radioactive decay of the neutron. *Phys. Rev.* 78, 310–311.

Snow, W. M., Chowdhuri, Z., Dewey, M. S., Fei, X., Gilliam, D. M., Greene, G. L., Nico, J. S., and Wietfeldt, F. E. (2000). A measurement of the neutron lifetime by counting trapped protons. *Nucl. Instrum. Methods Phys. Res., Sect. A* 440, 528–534.

Sørensen, H. and Anderson, H. H. (1973). Stopping power of Al, Cu, Ag, Au, Pb, and U for 5–18 MeV protons and deuterons. *Phys. Rev.* 8B, 1854–1863.

Spinks, J. W. T. and Woods, R. J. (1990). "An Introduction to Radiation Chemistry." 3rd ed., John Wiley & Sons, Inc., New York.

Strachan, J. D., Adler, H., Barnes, C. W., Barnes, G., *et al.* (1994). Fusion power production from TFTR plasmas fueled with deuterium and tritium. *Phys. Rev. Lett.* 72, 3526–3529.

Sundaresan, M. K. (2001). "Handbook of Particle Physics." CRC Press, Boca Raton, FL.

Tait, W H. (1980). "Radiation Detection." Butterworths, London.

Taylor, L. S., Tubiana, M., Wyckoff, H. O., Allisy, A., Boag, J. W., Chamberlain, R. H., Cowan, E. P., Ellis, F., Fowler, J. F., Fränz, H., Gauwerky, F., Greening, J. R., Johns, H. E., Lidén, K., Morgan, R. H., Petrov, V. A., Rossi, H. H., and Tsuya, A. (1970). "Linear Energy Transfer." ICRU Report 16. International Commission on Radiation Units and Measurements, Washington, D.C.

Titus., K. J., Clapp, T. G., and Zhu, Z. (1997). A preliminary investigation of a beta-particle transmission gauge for seam quality determination. *Textile Res. J.* 67, 23–24.

Tsoulfanidis, N. (1995). "Measurement and Detection of Radiation." 2nd ed., Taylor and Francis, Washington, DC.

Tsybin, A. S. (1997). New physical possibilities in compact neutron sources. *Appl. Radiat. Isot.* 48(10–12), 1577–1583.

Tumul'kan, A. D. (1991). Typical calibration curves for beta thickness gauges. *Measurement Techniques* 34(1), 24.

Turner, J. E. (1995). "Atoms, Radiation and Radiation Protection." 2nd ed., John Wiley & Sons, New York.

Woan, G. (2000). "The Cambridge Handbook of Physics Formulas." Cambridge University Press, Cambridge.

Yi, C. Y., Han, H. S., Jun, J. S., and Chai, H. S. (1999). Mass attenuation coefficients of β^+-particles. *Appl. Radiat. Isot.* 51, 217–227.

2
GAS IONIZATION DETECTORS

KARL BUCHTELA

Atominstitute of the Austrian Universities, A-1020 Vienna, Austria

I. INTRODUCTION: PRINCIPLES OF RADIATION DETECTION BY GAS IONIZATION

When radiation penetrates matter, energy of the radiation is passed on to the matter and the radiation is shielded or even stopped. The atoms or molecules of matter are brought to a state of higher energy, an excited state, or they are ionized if the energy of the radiation is high enough.

Alpha, beta, and gamma rays are known as ionizing radiation. On passing through a gas, these radiations create positive ions and electrons.

123

Those charged particles either cause chemical reactions or recombine, finally producing neutral specimens again. But if an electric field is applied, the positive ions start to migrate to the cathode and the electrons are attracted by the anode. If the field strength, the applied voltage per unit length, is high enough to prevent recombination during migration of the ions and electrons, all of them arrive at the electrodes. They are collected at the electrodes, and by the detection of this electric charge using a suitable electric circuit, an indication of the presence of ionizing radiation is given.

Gas ionization detectors consist of a gas volume in an enclosure that is either sealed or constructed in such a way as to permit a continuous flow of the filling gas. Within that gas volume an electric field is applied across the electrodes. The outer wall frequently serves as one of the electrodes, the cathode, while a wire rod, a grid, or a plate in the middle of the gas volume serves as the anode.

Although there are many different variations in the design of gas ionization counters, a cylindrical system with a central wire or rod, called a "counting tube," is very common. Many designs with different shapes and geometries have been realized. Some of them are suitable for a very wide range of useful applications, some were designed for a very special investigation, and others have been realized only to learn more about the operating principles of ionization detectors in order to improve the performance of this type of radiation detection device.

In this chapter a selection is given from numerous developments in the field of gas ionization detectors. It should be mentioned that radiation measurement methods today place emphasis mainly on radiation spectroscopy. Solid-state and scintillation detectors offer unique advantages in that field of applications. Nevertheless, a great deal of interesting and useful research work is still done using ionization detectors and new developments and applications are reported in the literature.

A very interesting development can be observed in the field of position-sensitive detectors such as micro-strip gas chambers with good localization properties (Sauli, 2001), Bellazzini et al. (2002). Although gas ionization detectors are extremely useful, problems and limitations have to be faced and careful planning of experiments to recognize and deal with those limitations is extremely important (Bateman et al., 1994).

Review articles are available in journals providing information regarding recent developments, achievements, trends, and future perspectives of gas ionization detectors (Sauli, 1998, 2001).

The suitability of gas ionization detector systems for a given kind and energy of radiation depends on the type (composition, pressure) of filling gas to be ionized; the applied field strength; the size, shape, and geometry of the detector volume and electrodes; and the type and thickness of the construction material that surrounds the detector gas volume. Also, environmental factors such as temperature should not be totally neglected. Last but not least, the design of the electric circuit that handles the output signal plays a very important role.

The geometric design of a detector also depends mainly on its application. The size and shape have to be chosen appropriately if small or large areas have to be surveyed by the detector; if it has to be submerged in a liquid; or if, by use of a suitably thin wall, alpha and low-energy beta particles are permitted to enter the detector volume; and finally, if radiation energy has to be determined or if the localization or distribution of the radioactive material in a given specimen is of primary importance.

There are three kinds of gas-filled detectors: ion chambers, proportional counters, and Geiger-Mueller counters. They differ mainly in the strength of the electric field applied between their electrodes. Their common and different characteristics are discussed in this chapter.

II. CHARACTERIZATION OF GAS IONIZATION DETECTORS

A. Ion Chambers

Gas ionization detectors can be characterized by the effects created by different field strengths between the charge-collecting electrodes. The relationship between the pulse size produced and the potential applied across the electrodes of a gas ionization detector is shown in Fig. 2.1. The pulse size depends on the field strength and also on the type of radiation that enters the detector volume and creates ions.

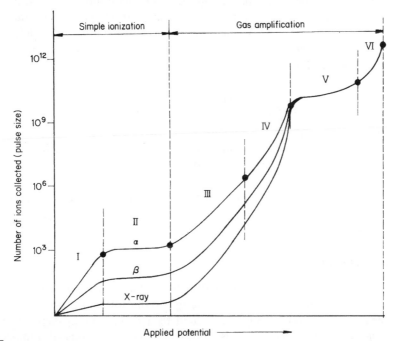

FIGURE 2.1 Relationship between the pulse size produced and the potential applied across the electrodes of a gas ionization chamber exposed to alpha, beta, and x radiation. Various regions are labeled by Roman numerals as follows: region I, recombination region; region II, simple ionization region; region III, proportional region; region IV, limited proportional region; region V, Geiger-Mueller region; region VI, continuous discharge region. (From L'Annunziata, 1987.)

At low field strength, many slowly migrating ion pairs still have the opportunity to recombine. This recombination region is not used for radioactivity detectors. As more voltage is applied, more ions and electrons produced by the ionizing radiation are collected at the electrode. Finally, a field strength is reached at which the now rapidly migrating ions do not have a chance to recombine. Thus, a saturation region is reached where all the ions produced directly by the radiation event, the primary ions, are collected at the electrodes. A further increase of field strength cannot attract more ions because all of them have already been collected. Ion chambers operate in this region. The amount of charge collected at the electrodes directly shows the ionization effects of the incident radiation.

The design of ion chambers can be tailored for a special type of radiation and information about radiation energies can be provided. As the output signal is directly related to the ionization effect, ion chambers are very useful in radiation dosimetry.

Alpha particles produce a great amount of ions along a short path length of travel (high linear energy transfer). They are easily detected because they provide a high output signal. Beta particles and gamma rays produce a very low signal, and rather sophisticated circuits are sometimes needed for amplification of such low-voltage signals.

For a short calculation example and to give an idea about the requirement for electronic circuits combined with ion chambers, it is assumed that a radioactive source emits one alpha particle per second (activity 1 Bq) with an energy of 5 MeV and all the energy of the alpha particles is deposited in the gas volume (air) of the counting chamber. The ionization energy of that gas should be 32.5 eV.

$$5\,\text{MeV}/32.5\,\text{eV} = 1.5 \times 10^5$$

ion pairs are produced by one alpha particle. Thus 1.5×10^5 ion pairs or 1.5×10^5 electrons are produced by one alpha particle per second, corresponding to an electric charge of

$$(1.5 \times 10^5 \text{ electrons/s})(1.6 \times 10^{-19} \text{ coulomb}) = 2.4 \times 10^{-14} \text{ coulomb/s}$$
$$= 2.4 \times 10^{-14} \text{ ampere}$$

B. Proportional Counters

If the field strength is increased further, additional ionization starts to occur because of the higher kinetic energy of the migrating primary ions. These primary ions, now being accelerated to a higher energy than the ionization energy of the detector gas, produce secondary ions by impacts. With increasing field strength, a great number of additionally produced ions are accelerated, the number still being proportional to the number of primary ions. This gas ionization detector region is called the proportional region. In that region, radiation with different abilities to produce primary ions (alpha, beta, or gamma radiation) can still be discriminated, or they are registered by "gross counting" without separation. Also, radiation of the same type but with different energies can be discriminated (García-León *et al.*, 1984).

With further gas multiplication due to higher field strength some nonlinearities will be observed. This effect marks the beginning of the limited proportional region.

C. Geiger-Mueller Counters

As the field strength is increased further, excitations of atoms and molecules are observed that, by the emission of ultraviolet light, can start additional ionization processes. In this region, referred to as the Geiger-Mueller region, the total number of ions produced is independent of the number of primary ions and, therefore, also independent of the type and energy of radiation. A further increase of the field strength causes a continuous discharge (see Fig. 2.1).

In the Geiger-Mueller region all primary ionization effects produce the same maximum response in the detector. Geiger-Mueller counting tubes operate in this region and thus provide no direct information about the type and energy of radiation. Information related to the type and energy of radiation can be provided only by observing shielding effects related to this radiation. Alpha particles are stopped by a thin layer of matter, beta particles show a maximum range in penetrating a shielding material before they enter the detector, and photons show a somehow logarithmic decrease in intensity with increasing thickness of the material. In the earlier days of radiation measurements such experimental setups were frequently used for rough determination of radiation type and energy (Chase and Rabinowitz, 1967).

III. DEFINITION OF OPERATING CHARACTERISTICS OF GAS IONIZATION DETECTORS

In the case of ionization detectors, as well as other detector types, some operating parameters are important for characterizing their capabilities: efficiency, resolution, and resolving time of the detector. For some special detector designs, the position sensitive detectors, also the capability to give precise information regarding the spatial distribution of particles or photons entering the detector volume is of importance.

A. Counting Efficiency

The efficiency refers to the number of particles or photons emitted by a radiation source related to the number of interactions registered by the counting system. This is usually called the absolute efficiency.

$$\text{Absolute efficiency} = \frac{\text{number of signals recorded by the detector}}{\text{number of particles or photons emitted by the source}}$$

Not always are all particles or ions striking the detector volume registered. Therefore another kind of efficiency is used which is called the intrinsic efficiency, defined as:

$$\text{Intrinsic efficiency} = \frac{\text{number of signals recorded by the detector}}{\text{number of particles or photons striking the detector}}$$

With ionization detectors the absolute efficiency of charged particles can go up to nearly 100%. For gamma rays and x-rays the efficiency is frequently much lower because of the relatively poor interaction of the radiation with the gaseous detection volume of the ionization counters. Therefore a higher density of the gaseous volume can sometimes be obtained by using a counting gas of high atomic number (xenon) and by increasing the gas pressure in the ionization detector.

B. Energy Resolution

The energy resolution characterizes the ability of the detector to discriminate between two radiations with energies that are different but rather close to each other. A characteristic figure is given by the full width at half-maximum (FWHM), the width of a peak in a radiation energy spectrum display halfway between the baseline and top of the peak. If E_0 is the energy at the peak maximum and ΔE is the full width at half-maximum, the resolution is given as $R = \Delta E / E_0$, which can also be recorded as a percentage.

Small values of FWHM and of the resolution are a measure of the potential of a detector to provide individual information related to two radiations of approximate energy. Because of the statistical nature of any interaction of radiation with matter, resolution never can be perfect. In addition, electronic noise contributes to the deterioration of resolution.

Not all detectors can provide information about radiation energy.

C. Resolving Time

The resolving time refers to the minimum time interval a detector needs to recover from the interaction with a radiation event and be able to register a following event. For many counting devices, not the resolving time of the detector but the resolving time of the electronic system (e.g., the data handling and processing steps) sets the limits for dealing with high count rates. Counting losses induced by resolving time of a counting system can be a limiting factor in measurements. Several methods for resolving time determination and correction are presented in the literature (Gardner and Liu, 1997; Lee and Gardner, 2000; Vinagre and Conde, 2001).

D. Localization

Some detector designs can give information about the entrance region of particles or photons into the detector or about the distribution of radioactive material in a sample. They can give an image of a radioactive specimen by showing the longitudinal or even two-dimensional distribution of radio-activity.

Position-sensitive detectors based on proportional counting systems were developed by Charpak in the 1960s; and these played a decisive role in many discoveries in particle physics. These types of detectors, providing the opportunity for "nuclear imaging," are nowadays very important also in many experiments in biology and medicine (Charpak, 1970; Charpak and Sauli, 1978; Geltenbort, 1994; Nickles et al., 2002).

IV. ION CHAMBERS

Ionization chambers can be considered as one of the simplest devices for radioactivity measurements. They were used in the very early days of research dealing with the detection of ionizing radiation. But even today new designs for special purposes are developed.

The chamber is made of a nonporous material, the electrodes are usually parallel plates, and the filling gas may have a pressure from a few tenths up to some tens of bars. When ionizing radiation passes through the gas, ion pairs are created. If a sufficiently high voltage gradient prevents recombination, these ions drift toward the electrodes. The output signal registered by the electric circuit can be a flow of current, a charge or voltage pulse, or a total collected amount of electrical charge. Thus three types of ion chambers are known: ion chambers operating in the current mode, ion chambers operating in the pulse mode, and electrostatic or charge integration ion chambers.

With ion chambers operating in the current mode, an electrical current flow is registered, which is initiated by the electrons and ions collected at the electrodes during the time of observation. With the pulse mode type of chamber, single signals, such as voltage pulses created by the ions arriving at the electrodes from a single ionization event, are registered by applying suitable electronic amplification circuits. Electrostatic or charge integration ion chambers are similar to electroscopes. A static electric charge is given to a system consisting of a thin foil or fiber that is suspended parallel to a solid support or to a second fiber or foil. Because of the repulsion of like charges, the fiber or foil will be bent to stay at some distance from the support or the second foil or fiber. Ionizing radiation gradually discharges the system, and this causes the foils or fibers to move back to their original position.

Because of their simple construction and relatively low cost, ion chambers still have many applications. Information related to the type and energy of radiation can be obtained, and the ion chambers can be designed for the detection of low as well as high radioactivity levels. Many kinds of gases can be used to fill the detector volumes.

A. Operating Modes of Ion Chambers

I. Ion Chambers Operating in the Current Mode

One of the most important applications of an ion chamber in everyday radiochemistry is as a portable survey instrument for radiation monitoring purposes. A volume of counting gas, mostly air, is enclosed within walls made of metal-lined plastic or aluminum. These types of walls are "air equivalent." Thus accurate measurements can be made for gamma radiation if the energy of the gamma radiation is high enough to penetrate the walls without significant attenuation, but also low enough to establish electronic equilibrium in these walls. Usually, for gamma radiation with energy lower than 50 keV, attenuation effects have a considerable effect on the efficiency of such detectors. With these instruments the saturated ion current is measured by using an electrometer circuit that is battery powered. Converting the DC signal of an ion chamber to an AC signal provides a more stable amplification, such as with the vibrating-reed electrometer or dynamic capacitor.

2. Charge Integration Ionization Chambers

A frequently used type of ionization counter is operated on the charge integration principle. This type of ionization chamber is charged initially. The drop of charge during exposure to a radiation field can be measured using a charger-reader mechanism and provides information regarding the dose from the radiation field to which the ionization chamber was exposed.

A familiar device is the ionization pocket chamber. These ionization chambers are also charged initially, but they are equipped with a small integral quartz fiber electroscope. An initial charging sets the scale of the electroscope to zero. The total integrated dose can be read periodically by observing the migration of the quartz fiber. This can be done very simply by optical observation, just by holding the pen-shaped pocket chamber up to a source of light and looking at the scale of the fiber electroscope through a small integrated magnifying glass. The accuracy and sensitivity of these devices are limited by leakage current across the insulator material of the ionization chamber.

3. Pulse Mode Ion Chambers

Like other ionization detectors, such as proportional counters and Geiger-Mueller tubes, ionization counters can also be used in pulse mode, in which each separate alpha particle, beta particle, or gamma quantum creates a distinguishable pulse signal. Advantages of pulse mode ionization chambers are their sensitivity and the ability to measure the energy of radiation and thus to be applicable in radiation spectroscopy. Today, such pulse mode ionization chambers have been mostly replaced by semiconductor detectors. Nevertheless, for special applications, such as neutron counting facilities, such chambers are still in use.

Pulse amplitudes from all types of ion chambers are relatively small. In theory, the maximum signal amplitude accumulated from the ion pairs produced by the interaction of, for example, an alpha particle in air along its track within the chamber is of the order of 10^{-5} V. Such a signal can be processed, but rather sophisticated electronic systems are required. Pulses from a single photon interaction are a hundred times smaller, and successful and accurate amplification is difficult and at times even impossible. Internal amplification within the detector volume, which is described in the section of this chapter dealing with proportional counting tubes, helps to overcome these problems.

B. Examples and Applications of Ion Chambers

I. Calibration of Radioactive Sources

Standardization of gamma-emitting radionuclides (e.g., in nuclear medicine applications) is frequently carried out by comparing the ion current from a material with unknown amount of activity with the ion current produced by a standard material of the same radionuclide. In that way one takes advantage of the excellent long-term stability and reproducibility of the ion current produced from the same activity. When operating in the saturation region, the current depends only on the geometry and the activity of a given radioactive material. Chamber volumes can be up to several $1000\,cm^3$ and the walls are made of solid materials, such as steel or brass. The collecting electrode in the inner part is made of a thin metal foil to avoid as much as possible attenuation of the radiation.

High sensitivity can be obtained if pressurized gas is used for the ionization chamber. Of course, this will cause the background current to increase but not be as great as that produced by radiation sources. Pressurized chambers are used for the measurement of gamma-emitting nuclides.

The ion chamber region is usually reached by adjusting the voltage for the electrodes. Saito and Suzuki (1999) used a multi-electrode ion chamber for measuring absolute fluence rate of x-rays. They adjusted the ion chamber region by varying the gas pressure at a given voltage.

2. Measurement of Gases

Many radioactive gases can be incorporated in the filling gas of ionization detectors. Also, in ionization chambers a gas can be sampled on a continuous flow-through basis. The ionization current produced by a gas can be calculated simply and straightforwardly only if the radiation is fully absorbed in the gas volume of the ionization chamber. These types of flow-through ionization chambers are used for monitoring air that contains small amounts of radioactive gas. But a number of difficulties arise if the air is subject to atmospheric changes. Such perturbations of air properties can be due to the content of aerosols, moisture, ions, and so on (Jalbert and Hiebert, 1971; Mustafa and Mahesh, 1978; Waters, 1974).

The change of ionization current due to smoke particles is the operational basis for smoke detectors. In such smoke detectors a built-in alpha source provides a constant ionization current under normal atmospheric conditions. A twin chamber with enclosed air without flow-through capability is used for the reference ion current.

The design of twin chambers can also be used for background compensation. A twin chamber filled with pure air records the background without flow through of the air to be monitored. In that way, compensation for a changing background can easily be achieved, for example, in case of a changing gamma-ray background during air monitoring.

Current mode ion chambers have been very useful in the measurement of radon. The background is low and the counting efficiency high (practically 100%). Experiments have also been reported to provide data for the radon content of groundwater by placing an ion chamber together with a known amount of water for three hours in a leak proof container. The amount of radon in the air can be related to the concentration of radon in the water sample (Amrani *et al.*, 2000).

Tritium measurements using ion chambers present a problem if elastomeric seals are applied. Those elastomeric materials are irreversibly contaminated and the background of the ion chamber is increased. Colmenares (1974) constructed a chamber using ultrahigh-vacuum metallic seals, a metal construction of negligible water adsorption capacity and sapphire as isulator material. The chamber is bakeable up to 450°C and the contamination problems are avoided.

3. Frisch Grid Ion Chambers

Because of the slow ion mobility in gases and the slow drift of ions at the applied field strength in ion chambers, the use of pulse-type ion chambers is restricted to low pulse rates if signals are wanted that are related accurately to the original charge of ions and electrons generated by the radiation. Pulse-shaping circuits designed for low frequencies also make these systems rather susceptible to interference from microphone signals produced by mechanical vibrations.

Therefore pulse-type ion chambers are frequently operated in such a mode that they sense the collected electrons only, not the created positive ions, which migrate much more slowly than the electrons. In that case the pulse amplitude is related to the drift of the electrons only. The signal therefore has a much faster rise time, and higher counting rates can be successfully registered. But because the amplitude of the signal now depends also on the position of the interaction within the ion chamber gas, there is no well-defined information related to the total number of ions created, which means there is no information about the energy of the radiation. However, methods have been developed to overcome the problem of the dependence of the pulse amplitude on the position of the interaction within the chamber. The region of the chamber volume is divided into two parts by a grid. This grid is maintained at a potential between those of the cathode and anode. The mechanical construction of the grid should allow electrons to pass through; it should be as "transparent" to electrons as possible. By suitable positioning of the radiation source outside the chamber or by effective collimation of its radiation, the emitted particles or rays interact with the gas in the ion chamber in a well-defined region between this grid and the negative electrode of the chamber. Thus positive ions simply migrate to the cathode. Electrons are attracted by the transparent grid initially but are further accelerated toward the anode, which is at a much more positive potential than the transparent grid.

Electronic circuits are designed in such a way that, with the electron migration from grid to anode, the voltage between grid and anode drops and a signal is created that depends only on the electron drift and not on the migration of both electrons and cations. Therefore, the slow rise related to ion drift is eliminated. Also, because all electrons are accelerated by the same potential difference, the amplitude of the pulse is independent of the position of the interaction. The amplitude is proportional only to the number of ion

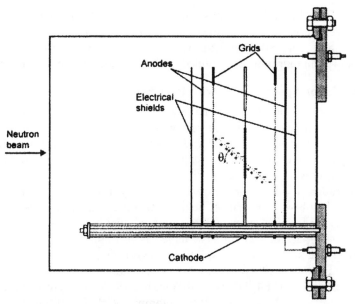

Neutron beam

FIGURE 2.2 Schematic picture of a parallel gridded ionization chamber with one common cathode (From Tutin *et al.*, 2001, reprinted with permission from Elsevier Science.)

pairs, the number of electrons produced along the path of the interacting particle or ray. This type of ion chamber is called a Frisch grid chamber after the scientist who designed the detector (Knoll, 1989). Such Frisch grid chambers have been extremely useful in studies dealing with particle physics.

Gridded ionization chambers are ideally suitable for studies related to nuclear fission because such detectors have not only a practically 100% detection efficiency but they can also provide information about fission fragment properties such as energy, mass, charge, and emission angle. Tutin *et al.* (2001) have designed an efficient device for such investigations, consisting of a combination of two parallel plate ionization chambers with grids and a common central cathode plate (Fig. 2.2). The central aluminum disk cathode has a hole of 78 mm diameter where two thin aluminum foils, covered with uniform layers of fissile materials (^{232}Th, ^{238}U), are mounted back to back. The grids are mounted on aluminum annular disks with an inner diameter of 160 mm and consist of tungsten wires 0.1 mm in diameter spaced by 1.25 mm. The chamber is filled with 90% argon and 10% methane at atmospheric pressure without continuous gas flow. Fission fragments emitted from the cathode are stopped in the space between the cathode and anode, free electrons drift to the anode whereas the slow ions can be treated as being static for a short interval of time. At the end of the electron drift the collected charges can be related to the emission angle of the fission fragments.

There are also applications to be found in the life sciences. Lohmann *et al.* (1998) used a detector system of the Frisch grid chamber type in angiography for the determination of contrast agent (iodine) by "dichromography." According to this method two images with monochromatic x-rays just below and above the absorption edge of the contrast agent are simultaneously obtained and subtracted. Although monochromatic x-rays having suitable intensity to visualize arteries of 1 mm diameter are only provided by synchrotron radiation, the authors concluded that the requirements for application of intravenous coronary angiography are fulfilled with the Frisch gridded detector.

4. Radiation Spectroscopy with Ion Chambers

Pulse-type ion chambers have attracted some interest again, after years during which most of the interest was focused on semiconductor detectors. Ion chambers can be designed and constructed in any shape and size, and for charged particles the pressure can be tailored to an optimum for a desired stopping power (Fulbright, 1979). Also, there is practically no deterioration or degradation due to irradiation, which may adversely affect the application of ion chambers in radiation fields, and ion chambers can be fabricated by using available workshop facilities without high expenses.

Pulse-type ion chambers have been applied in low-level alpha measurements, and good resolutions have been obtained that may even be comparable with the resolution of semiconductor detectors (Gruhn *et al.*, 1982; Bertolini, 1984; Hoetzl and Winkler, 1984; Shenhav and Stelzer, 1985; Kotte *et al.*, 1987; Nowack, 1987; Domnikov *et al.*, 2001).

It was demonstrated that additional information regarding charged particle properties such as atomic number and charge state can be obtained by designing a chamber in such a way that particle pathways are parallel instead of perpendicular to the direction of the electric field. Thus, the drift time of electrons to the grid will be different for electrons created at the beginning of the track and those from the end of the path. The shape of the output pulse will therefore reflect the distribution of ion pairs along the track according to what is called a Bragg curve. With that technique, known as Bragg curve spectrometry, additional information such as atomic number and particle charge can be obtained. For that a detailed analysis of the pulse shape is necessary.

Khriachkov *et al.* (2000) used an alpha-particle spectrometer based on a Frisch grid chamber for studies of (n, α)-reactions induced by fast neutrons. Energy and emission angle of alpha-particles could be detected.

Combinations of ionization chambers with position sensitive ionization detector devices were used by Menk *et al.* (2000) for small-angle x-ray scattering (SAS) investigations. These systems are intended to be used for experiments in some European synchrotron centers.

5. Electret Detectors

Electret types of ion chambers make use of the drop of surface voltage on a plastic material. The plastic specimen is a dielectric material, usually Teflon, which is quasi-permanently charged. It is called an electret and usually has the shape of a disk about 1 mm thick and 10 mm in diameter. Electrets are prepared by being heated and simultaneously exposed to an electric field. Due to this process, many dipoles in the material become oriented in a preferred direction. After the heating, the material is "frozen" and is able to keep the position of its electric dipoles for a long period of time. A voltage gradient of several hundred volts can be maintained between the surfaces of the electret disk.

One surface of the electret is kept in contact with the wall of an ion chamber, which builds up an electric field in the chamber. Ionizing radiation causes a decrease of charge in that system, resulting in a partial neutralization of the charge at the electret. Measurement of the electret voltage difference before and after irradiation allows determination of the amount of ionization. The system has to be calibrated and can be used for determination of environmental radiation doses.

Amrani et al. (2000) used an electret ion chamber for the determination of the radon content of groundwater. They put an electret ion chamber together with a known amount of water in a leak proof container. The reading of the electret ion chamber provides the radon content in the air, and this value could be related to the concentration of radon in the water sample.

6. Fission Chambers

For use in nuclear reactors, miniaturized ion chambers have been constructed that are equipped with stainless steel walls lined with highly enriched uranium. Argon at a pressure of several bars is mainly used as a filling gas. Because of the high pressure, the dimensions of the detector volume can easily be kept larger than the range of the fission products created by the uranium-235 (n, f) reaction.

Long-term operation causes problems because of the burnup of the fissile material (Böck and Balcar, 1975). To compensate for this, so-called regenerative chambers have been designed. These chambers contain a combination of fertile (^{238}U, ^{234}U) and fissile (^{235}U, ^{239}Pu) material as a lining of the inner detector walls. Fission chambers may also show a memory effect after a prolonged period of operation in a reactor core. This is due to a buildup of fission products in the detector volume. Because of the fission product activity, some residual ionization still can be measured even without exposure to a flux of neutrons.

Because of the scarcity of commercially available enriched uranium-235 material, fission detectors have been developed on the basis of uranium-233. Figure 2.3 shows a schematic diagram of the uranium-233 fission chamber designed by Prasad and Balagi (1996). The chambers were filled with argon (97%) and nitrogen (3%) at 1 bar. Low and high sensitivities were obtained by using two kinds of electrode coatings. Low-sensitivity counters have a uranium-233 coating on the anode and high-sensitivity counters have a coating on the cathode. The main disadvantage of uranium-233 is its high specific alpha activity. This can cause pileup effects and spurious counts if the system is applied in pulse mode operation.

Incineration of transuranic elements by neutron induced fission could be a promising way to reduce the long term radiotoxicity of these materials in radioactive waste. In order to measure on line the fission rate of actinide targets a new generation of micro fission chambers have been constructed by Fadil et al. (2002) for their use at the high flux reactor in Grenoble at a flux density of $10^{15}\,n\,cm^{-2}\,s^{-1}$. To avoid pulse pile up the chamber has to operate in current mode. Helium, a gas with high ionization potential, is used under such high flux conditions. Consequently the problem of gas leakage during the operation of the chamber at high temperatures has to be considered.

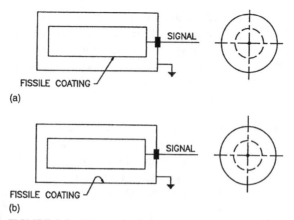

FIGURE 2.3 Schematic diagram of uranium-233 fission chambers. Low-sensitivity counters have a uranium-233 coating on the anode (a) and high-sensitivity counters have a coating
on the cathode (b). The electrode spacings are 4 mm. (From Prasad and Balagi, 1996.)

V. PROPORTIONAL GAS IONIZATION DETECTORS

Proportional gas ionization detectors operate at a higher voltage gradient
than ion chambers. The primary ion pairs created by ionizing radiation are
accelerated much more and acquire a high kinetic energy. By colliding with
other atoms or molecules along their drift, these ions and electrons induce
secondary ionization. More ions and electrons are released with energy for
further ionization of the filling gas. This multiplication process is called a
Townsend avalanche or Townsend cascade. The anode wire must be very
thin to obtain a region of sufficient field strength by applying a reasonable
voltage. The gas amplification takes place mainly in the region of high
voltage gradient near the anode wire. Still, this amplification can be kept
linearly proportional to the original ionization; the number of ions after gas
amplification is proportional to the number of primary ions created by the
ionizing radiation directly. Detailed explanations and descriptions of
phenomena in gas ionization proportional counters are given by Charpak
(1970) and Charpak and Sauli (1978).

Proportional counting tubes can be sealed, with the source of radiation
kept outside the tube. A thin window permits radiation penetration into the
detector volume. Another configuration is designed for flow-through of gas and
the sample can be inserted into the detector volume. These "windowless"
counting systems are useful for the detection of alpha particles and low-energy
beta particles. A maximum counting efficiency of 50%, theoretically for a 2π
counting geometry, is achieved. A 4π geometry can be achieved by using two
flow-through tubes with the sample mounted on a thin foil between the tubes.

Proportional counters usually operate in the pulse mode.

For proportional counters, special gases or mixtures of gases have to be
used. The filling gas should not form anions and should not contain
components that attract electrons. The noble gases meet this requirement
optimally. The formation of secondary Townsend avalanches should also be

avoided. Such secondary avalanches are created by the emission of ultraviolet (UV) photons. This light is produced by the deexcitation of molecules or atoms of the filling gas. To prevent this effect, a component is added to the filling gas that absorbs the energy from the excited species. This additive must get rid of the energy through nonradiative modes, such as dissociation. By this mechanism the ion cascade is localized near its origin and propagates only along the electric field. No other secondary avalanches are created. A frequently used filling gas consists of 90% argon and 10% methane and is called P10 gas (Alkhazov *et al.*, 1967).

Other gas mixtures (Penning gas) consist of a noble gas (neon, argon), the parent gas, with a small amount of an additive (methane, acetylene) of lower ionization energy than the lowest excited state of the parent gas (Järvinen and Sipilä, 1984).

Gas amplification factors of 10^4 can easily be obtained. Therefore, rather simple electric circuits can be used for pulse amplification and pulse handling. Also, the effects of electronic noise can easily be avoided, because the output pulses created by that phenomenon are small.

Gas gain in proportional counting should be an exponential function of the applied high voltage. But in proportional counters filled with mixtures of argon and a low amount of a molecular gas secondary avalanches develop and, as a consequence, gas gain increases faster than exponentially with the applied high voltage (Bronic and Grosswendt, 2001).

Proportional counters, using the fast pulses from electron collection, have a short resolving time of less than $1\,\mu s$. Proportional counters have a high intrinsic efficiency for alpha and beta particles. Photons are detected mainly by Compton effects produced in the walls of the counter. Thus, the intrinsic efficiency for gamma rays is rather low, especially for gamma photons with higher energies.

Counting losses with proportional counters are due to wall effects and to nondetection of very low-energy beta particles. Stanga *et al.* (2002) proposed a calculation model for the correction of counting losses. By means of such calculations the accuracy of internal gas counting methods can be improved, and tedious and time consuming energy calibration procedures can be shortened or even avoided. Proportional counting (PC) is frequently applied to the preparation of reference sources by absolute activity measurements also referred to as radionuclide standardization. Such radionuclide standardization methods involving joint proportional and solid scintillation detector arrangements [i.e., $4\pi\beta(PC)-\gamma Na(Tl)$ counting] are discussed in Chapter 11.

A. Examples and Applications of Proportional Counters

I. Gross Alpha-Beta Counting, Alpha–Beta Discrimination, and Radiation Spectroscopy Using Proportional Gas Ionization Counters

With gross alpha–beta counting no attempt at any discrimination is made. Just the sum of all alpha and beta particles is detected. Gas proportional counting is one of the methods frequently used for gross counting (Passo and Kessler, 1992; PerkinElmer, 1992).

Proportional counters are also frequently used to distinguish between alpha and beta particles from a mixed source. Alpha particles, because of their high linear energy transfer, produce a high number of interactions with the gas in the detector volume. A beta particle produces a much lower number of ions per centimeter along its track than an alpha particle. The gas amplification factor is constant at a given voltage, so the output pulse is much higher for interactions of alpha particles compared with beta particles. With a suitable discriminator level or gate, the pulses created by alpha particles can be detected at a rather low voltage setting. For the detection of beta particles a higher voltage has to be used to overcome the discriminator level.

The alpha particles from a mixed source are registered at a lower voltage, the alpha plateau. At a higher voltage alpha and beta particles are detected and gross alpha–beta counting is accomplished.

Alpha and beta radiation can also be discriminated according to the pulse length. Alpha pulses have a different pulse shape than beta pulses. Semkow and Parekh (2001) could demonstrate that alpha-radioactivity can be measured accurately in the presence of beta-radioactivity but the opposite is not always true due to alpha to beta cross-talk. This cross-talk depends mainly on the alpha-decay scheme and is due to the emission of conversion electrons, Auger electrons, and x-rays.

It is usually assumed that the counting efficiency of a 2π geometry alpha particle detector is 50%. Unfortunately this is not true in practical measurements because of self absorption and backscattering. Several theories have been developed for the calculation of backscattering and self absorption effects. Rodríguez et al. (1997) have presented a review on these topics and also developed new theories. Backscattering depends on the atomic mass of the backing material of the radiation source. Corrections can be found experimentally by preparing samples of various thicknesses and extrapolation to zero sample thickness. But such determinations are only possible with radioactive material with suitable long half-life.

To a limited extent, proportional counters can also be used for radiation spectroscopy (Järvinen and Sipilä, 1984; Jahoda and McCammon, 1988). Pulse height analysis can be applied for radiation spectroscopy for a given type of radiation. To perform pulse height analysis properly, the particles or rays to be analyzed have to release their energy totally within the gas volume of the counter; that is, they must be totally absorbed within the counter. Proportional detectors are used for x-ray spectrometry in the field of x-ray fluorescence analysis if high resolution is not required. Because of the gas amplification process, proportional counters have a poorer resolution than ion chambers. Today, mostly semiconductor detectors are used for x-ray spectroscopy.

Szaloki et al. (2000) have reviewed the essential progress in x-ray spectroscopy, and they point out that, although the gas filled proportional detectors are not superior to semiconductor detectors in resolution, microstrip proportional counters are applied for many investigations including new developments in the field of radioisotope excited XRF-analysis, especially at low energy regions (x-rays below 10 keV).

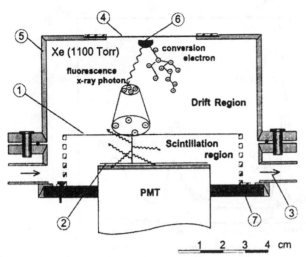

FIGURE 2.4 Design of a gas proportional counter. I. grid mesh with high electron transparency, 2. grid evaporated onto the photomultiplier, 3. to xenon gas purification system, 4. thin aluminized plastic window, 5. stainless steel enclosure, 6. Cd-109 source, 7. insulating material. (From Rachinhas *et al.*, 2000, reprinted with permission from Elsevier Science.)

Xenon gas proportional scintillation counters are used for the detection of x-rays when high detection efficiency and good energy resolution are required (Lopes *et al.*, 2000, 2001; Monteiro *et al.*, 2001; Simões *et al.*, 2001). An excellent example is the detector used by Rachinhas *et al.* (2000) for the identification of internal conversion electrons produced by the decay of Cadmium-109 and Xenon-133m to investigate details of the decay scheme of these radionuclides. The main aim was to selectively detect and identify conversion electrons of Cadmium-109. Figure 2.4 shows the design of the gas proportional scintillation counting device. The stainless steel enclosure holds also a thin plastic window, which is aluminized on the inner side to provide a uniform field strength at the drift region. Drift and scintillation region are separated by a grid mesh with high electron transparency. A second grid is evaporated directly onto the photomultiplier and therefore the scintillation region is in direct contact with the photomultiplier and a high collection efficiency of the UV scintillation photons is guaranteed. Primary electron clouds are produced by ionizing radiation in the absorption region and these drift under the influence of a low electric field towards and through the first grid into the scintillation region where, due to a much higher field strength, scintillations are produced in the xenon filling gas. The electric pulses of the photomultiplier are fed to an amplifier operating with very short shaping times and, as a result, pulse shapes resemble very closely the scintillation light bursts. This produces an efficient pulse shape discrimination and a very detailed interpretation of the pulse height spectra (see Fig. 2.5.).

As reliable detectors proportional counters are frequently used for standardization of radionuclides. García-Toraño *et al.* (2002) compared three methods for the standardization of Cesium-134: absolute counting with a 4π

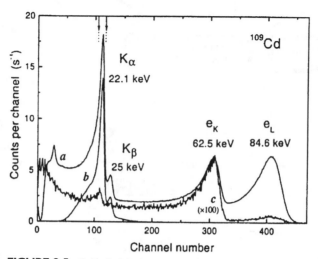

FIGURE 2.5 Pulse height spectrum obtained from a Cd-109 source placed inside the gas proportional scintillation counter. Curve a: raw pulse height distribution, Curve b is obtained when pulses with duration outside the range form 3.6 to 4.0 μs are rejected. Curve c (values multiplied by 100) shows pulses that appear within a 20 μs interval after a 22.1 keV pulse within the range determined by Curve b. (From Rachinhas et al., 2000, reprinted with permission from Elsevier Science.)

NaI(Tl) detector, liquid scintillation, and a proportional counter (argon and methane as counting gas at atmospheric pressure) in coincidence with a NaI(Tl) detector-system. It was shown that all the results of the standardization have been consistent and that any of the three methods that were applied was well suited for this type of application. The theory and principles of 4π counting are provided in Section XI of Chapter 11.

The reliability of 4π pressurized gas proportional counters have been further demonstrated by Altzitzoglou et al. (2002) during their work dealing with the comparison of three methods to standardize a Strontium-89 solution. Correction for self absorption of the samples for gas proportional counting was obtained by plotting the activity concentration of the solution against the mass of radioactive sample. A new half-life value for strontium-89 (50.61 ± 0.05 days) was determined in this work.

International comparison and standardization programs frequently result not only in getting more accurate data of radiation properties but also in improving measurement procedures. Self absorption corrections for beta measurements of solid samples have to be applied depending on the thickness of the specimen. Johansson et al. (2002) demonstrated that the self absorption of beta particles from Thallium-204 show a linear relation to the logarithm of the dry mass of the source. They describe a way to minimize and correct for self absorption in solid sources of Thallium-204 and nuclides with similar decay properties. Also a special device for source drying is described. Warm dry nitrogen jets (60°C) are blown on the rotating source material which is mixed with colloidal silica (Ludox®) to decrease the crystal size of the solid deposit.

2. Position-Sensitive Proportional Counters

a. Single-Wire Proportional Counters

In a proportional counter the position of the avalanche is limited to a small portion of the anode wire length. Some designs of proportional counters are capable of sensing the position of this avalanche and thus providing information about the position of an event taking place within the volume of the proportional counter. If the proportional counting tube is cylindrical with a central wire, electrons drift along the radial field lines. Thus the position of the avalanche indicates the axial position of the initial ion pairs and the position of the entering radiation to be detected. Of course, if the incident radiation extends for some distance along the counting tube, only an approximate region of the incident radiation can be determined. The principle of charge division is most frequently used to determine the position of the ion avalanche. For that purpose, the central anode wire is made of a material having a rather high electric resistance per unit length (Ohsawa *et al.*, 2000). By that means the charge that is collected at the wire electrode is divided between the amplifiers placed at both ends of the anode wire. The charges on those ends are collected in proportions related to the geometric position of the ion avalanche interacting with the wire electrode. A conventional output pulse is provided by summing up the response of the amplifiers and thus getting information about the total charge collected. A signal related to the position is provided by dividing the signal output of one amplifier by the output related to the total charge collected. The pulse height of this new signal indicates the relative position along the length of the central anode wire (Fischer, 1977; Westphal, 1976). Either analog signal handling or digital pulse processing techniques can be applied for this purpose.

Another method for position sensing uses pulse rise time measurements. With this technique the relative rise times of the output pulses of the preamplifiers placed on both ends of the anode wire are determined. Interactions that take place far from one of the preamplifiers result in pulses with a much longer rise time than events close to the preamplifier position. From the rise time difference of the two preamplifiers, a signal can be created that is related to the position of the ion avalanche along the electrode wire. Good results regarding spatial resolution are observed. For well-collimated alpha particles the FWHM can be 0.15 mm for a tube 200 mm long. Such position-sensitive proportional detectors have been applied for x-rays and neutrons, for magnetic spectroscopy of charged particles, and for localization of beta-emitting spots on thin-layer or paper chromatograms (Goulianos *et al.*, 1980).

b. Multiwire Proportional Counters

For many purposes proportional counters with a number of anode wires instead of one central anode wire offer advantages. A grid of anode wires can be placed between two flat cathode plates. Near the cathode plate the field is nearly uniform and electrons drift in that homogeneous field toward the anode wire grid. Near the wires the field strength increases and, as electrons approach this region, they are accelerated toward the nearest anode wire and an ion avalanche is created. Because of this, the signal appears only at a single anode wire and the position of the primary ionizing event can be localized in the dimension perpendicular to the direction of the anode wires. This multiwire proportional counter was developed by Charpak in the 1960s and played a decisive role, not only in many discoveries in particle physics but also in many experiments in biology and medicine (Charpak, 1970; Charpak and Sauli, 1978; Geltenbort, 1994). The technique for position-sensitive counting by using cathode wires of high resistivity has already been discussed. This technique can be used in addition to the plate and multiwire design and a two-dimensional signal pattern can be obtained. Another technique uses a detector construction with the cathode plate divided into narrow strips perpendicular to the anode wires. The induced charge to the nearest strip is recorded. Such position-sensing detectors with large areas are applied in high-energy particle research (Uozumi et al., 1993; Hayakawa and Maeda, 1994).

The relatively low signal amplitude is a disadvantage of these detectors. Therefore, for some applications a hybrid detector system, between proportional and Geiger-Mueller detectors, may be useful because of the much higher signal amplitudes achieved; these are referred to as self-quenched streamer detectors (Knoll, 1989).

c. Microstrip and Micropattern Ionization Counters

Wire proportional chambers were mostly developed at CERN and have been a major step forward in particle detector technology. Even now the field of developing new varieties and improving available designs remains very active. Microstrip gas counters, micromesh designs, nonplanar variants of microstrip gas counters, like the "compteur a trous (CAT)," secondary electron emission gas detectors, and some other varieties have been developed. A description of design with their special features and advantages are given by Fourme (1997). Christophel et al. (1998) present the development of a 2D microgap wire chamber and describe their plans to build large surface detectors. Efforts are also being made toward the use of such position-sensitive detectors to other fields in addition to basic research in particle physics. Ortuño-Prados et al. (1999) describes the use of a multi-wire proportional counter as a potential detector for protein crystallography and other wide-angle diffraction experiments. Fried et al. (2002) present the first results obtained with a large curved 2D position-sensitive neutron detector, which had been constructed for the protein crystallography station at Los Alamos National Laboratory, Babichev et al. (2001) report about their experience in medical radiography. The advantage of using multi-wire proportional counters as high count rate detectors as well as their usefulness for producing dynamic images of high statistical quality is pointed out by Barr et al. (2002). A detailed summary regarding the advances in gas avalanche radiation detectors and their application in biomedical investigations is given by Breskin (2000).

Microstrip gas chambers are ionization counters in which anodes and cathodes are not single plates but are constructed as thin metal strips on a solid insulating support (Barbosa et al., 1992; Bouclier et al., 1992a,b,c,

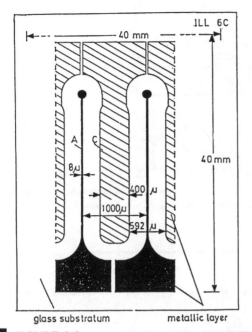

FIGURE 2.6 Structure of a microstrip plate. Electrode spacings are 400 μm and the metallic layers on the glass support, the electrodes, have a thickness of 150 nm. The small strips are the anodes. (From Oed, 1995, reprinted with permission from Elsevier Science.)

1995; Oed, 1995; Pallares *et al.*, 1995). With such a system the spot of the ionization track can be localized because ion production and migration and current flow take place in a well-defined single electrode strip region. Thus, position-sensitive counting can be achieved. Such microstrip gas chambers can be obtained with very small spacing between the electrodes. A small pitch results in good resolution. Even at the beginning of their development results were rather encouraging. At proportional gains above 10^4 good energy resolution (12% for 5.9 keV), position accuracies around 30 μm, and high rate capabilities were obtained. An example is shown in Fig. 2.6. This microstrip chamber was constructed by Oed (1995) using photolithographic techniques. The small strips are the anodes, and the electric field lines between the electrode strips are plotted in Fig. 2.7. An electron that is set free in the gas volume in front of the microstrip plate and reaches the microstrip plate creates an avalanche in a very well defined small region.

A two-dimensional position-sensitive detector was realized by Barbosa *et al.* (1992). Two sets of microstrips are orthogonally oriented, forming a two-dimensional sensitive electrode, which is used in a multiwire proportional configuration as shown in Fig. 2.8. The two cathode systems are isolated by a silicon dioxide layer only 2 μm thick and are therefore at practically the same distance (3 mm) from the anode wires. Therefore the signals induced in both orthogonal electrodes are of the same amplitude. The authors aimed to define a two-dimensional x-ray detecting unit that also could be upgraded to a submillimeter spatial resolution detector.

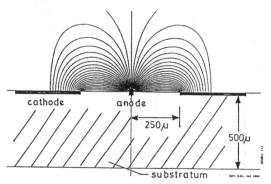

FIGURE 2.7 Plot of the electric field lines between the electrode strips of the microstrip plate. (From Oed, 1995, reprinted with permission from Elsevier Science.)

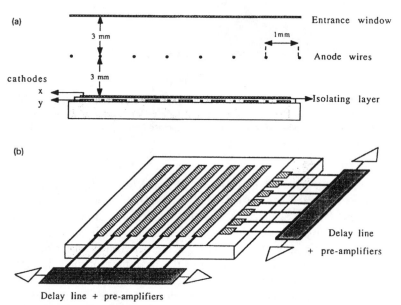

FIGURE 2.8 Two-dimensional position-sensitive detector arrangement. Two sets of microstrips at a distance of 2 μm are orthogonally oriented and connected to delay lines. The anode wires are arranged between the microstrips and the entrance window at a distance of 3 mm. (From Barbosa et al., 1992, reprinted with permission from Elsevier Science.)

There are some limitations to this design of detectors. One has to apply manufacturing techniques such as those used in the field of microelectronics. The total sensitive area of such counters seems to be limited. Also, there are charge buildup effects of the supporting insulating materials. This can have a substantial influence on the gas gain at high fluxes. Ion avalanches can cause accumulation of electric charge on the insulating surface between the strips, which modifies the electric field around the electrodes and changes the gas multiplication characteristics. To avoid this, a surface conductivity of the insulating support can be created, for example, by ion implantation. However, the use of all these sophisticated manufacturing techniques imposes constraints on the size of such radiation detectors. In the beginning, glass and quartz were used for insulating support between the electrodes. Later developments dealt with the application of plastic supports. These materials offer some advantages. They are flexible, and therefore nonplanar detectors can be designed. Cylindrical geometries with very small radii can be realized. Plastic materials not only have the advantage of lower atomic number of their constituents compared with glass but also can be made much thinner. Multiple scattering and photon conversion can be reduced. Plastic materials are also available with a wide range of electrical resistivities, and the design can be tailored to solve the problem of charge buildup. However, plastic materials have to meet the requirements of suitable mechanical stability. Bouclier *et al.* (1995) accomplished microstrip construction on plastic foils by applying a photolithographic etching technique on a layer of aluminum about $0.3\,\mu$m thick on plastic. The distance between the electrodes was about $400\,\mu$m. This is somehow wider than the usually applied $200\,\mu$m and is necessary because of the coarse optical quality of the plastic arrangements compared with glass support microstrips. Also, surface cleaning of plastic before vacuum evaporation of the aluminum cannot be done as perfectly as for glass supports.

Gains close to 10^3 could be reached with the equipment designed by Bouclier *et al.* (1995). Also, good energy-resolution for low-energy x-rays was achieved.

The current tendency in the field of gaseous detectors is the replacement of wire chambers by advanced micropattern electron multipliers to obtain an improvement in spatial accuracy and counting rate capability. Electrode patterns are deposited by microlithographic techniques on insulating substrates. Due to the small distances between cathode and anode (50–$200\,\mu$m) these multipliers offer localization accuracy around a few tens of micrometers. The rapid collection of the ion avalanches considerably reduces space charge buildup which influences the counting rate limitations. Many types of detectors in this family provide 2D localization in a single detector element.

Many new types of gas detectors with additional microstructures like the gas electron multiplier system (GEM) and other designs are currently being developed (Horikawa *et al.*, 2002).

The gas electron multiplier (GEM) was introduced by Sauli (1997). A GEM detector consists of a thin polymer foil ($25\,\mu$m), which is metal clad ($18\,\mu$m) on both sides and perforated to yield a high density of holes ($70\,\mu$m diameter and $100\,\mu$m apart). Photolithographic techniques have been used for manufacturing. A voltage is applied onto the two faces of the metal clad foil and therefore the field is very strong inside the holes. The device is inserted in a gas detector on the path of drifting electrons. Primary electrons produced by ionization of the gas layer above the foil are sucked into the holes where an avalanche process takes place. By that process the charge drifting through the holes is amplified. Most of the secondary electrons produced in the avalanche are transferred to the region below the foil where these electrons are collected by an anode and cause a detection signal. Coupled to other devices like multiwire or micropattern chambers, higher gains are obtained or an operation in less critical field strength conditions are

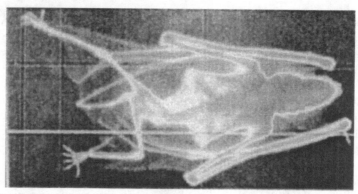

FIGURE 2.9 Absorption radiography of a small mammal recorded with a 2D-GEM detector. (Sauli 2001, reprinted with permission from Elsevier Science.)

permitted. The fast response time generated by electrons are one of the main characteristic advantages of the GEM detector. The GEM detector has been originally developed for application in particle physics. But it has also been applied successfully in other fields of research. Two dimensional GEM detectors have been used to obtain x-ray absorption images to show their applicability in medical diagnostics. Figure 2.9 shows an absorption radiography of a small mammal using 8 keV x-rays. The real size of the image is 3 cm × 6 cm. The position resolution depends on the photoelectron range in the gas. Presently the application of this technique is mostly limited by the readout speed of the electronic system; nevertheless, there are promising developments ongoing in this field (Sauli, 2001).

Photomultiplier tubes are frequently used in instrumentation for medical diagnosis such as with gamma cameras or CT equipment where light from large scintillator arrays has to be recorded. An alternative and probably more economic device for light detection and 2D recording would be the use of a thin solid photocathode combined with gas avalanche multipliers and a micropattern device (Fig. 2.10). It may even be possible to include several GEMs to such a device in cascade. Each GEM operates at a low gain whereby a high total gain is achieved. In addition the photocathode is shielded from photon feedback induced by ion avalanches (Fig. 2.11).

Ongoing work is focused on the improvement of GEM detector performance (Assaf, 2002), and their quality control at the manufacture stage will be needed. Fraga *et al.* (2000) have shown that visible light emitted by the GEM avalanches can be successfully used for quality control of the material, to determine their uniformity and to identify local defects. It is much more effective than the normal optical inspection.

Bellazzini *et al.* (1999) introduced the WELL detector as a new type of position-sensitive gas proportional counter. The basic design is similar to the GEM detector. The main difference between the GEM and the WELL detector is that the GEM alone acts only as an amplifying stage whereas the WELL detector has read-out strips directly placed onto the insulating foil providing a position-sensitive compact system. Printed circuit board

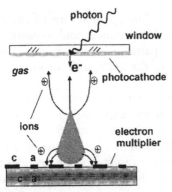

FIGURE 2.10 The principle of the gas avalanche photomultiplier: Photons stimulate the emission of electrons from the photocathode into the gas, avalanche multiplication takes place near the anodes of the micropattern device, ions are collected on neighboring cathodes, some ions may drift to the photocathode. (From Breskin 2000, reprinted with permission from Elsevier Science.)

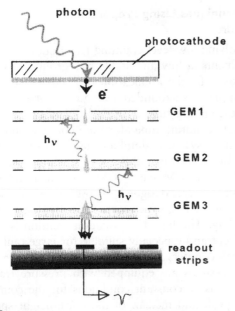

FIGURE 2.11 The multi-GEM phomultiplier concept, providing high total gain and 2D recording by a micropattern device. (From Breskin 2000, reprinted with permission from Elsevier Science.)

technology was employed to fabricate the amplifying structures (Bellazzini et al., 1999; Pitts and Martin, 2001).

Although the development of position-sensitive chambers are mainly dedicated for applications in high energy physics these types of detectors are also instruments of choice for radiation detection and localization in other fields of basic and applied research. Breskin provides many examples of applications in biology and medicine. Among these are (i) the comparison of images obtained by autoradiographic techniques and ionization detectors, (ii) images of ionizing particle track patterns demonstrated as applications in nanodosimetry, and (iii) examples of the application of x-ray imaging and neutron imaging (Breskin, 2000).

Yu *et al.* (1999) designed a position-sensitive x-ray detector with curved electrodes for large angle x-ray diffraction experiments at a synchrotron at Brookhaven National Laboratory. The detector can cover an angle of 45° and has an arc length of 20 cm with a radius of curvature of 25 cm.

Comprehensive reviews regarding the recent developments in the field of micropattern gas detectors are presented by Sauli (1999, 2002) and Bellazzini *et al.* (2001, 2002).

Microstrip and micropattern gas chambers were also filled with ^{3}He to be used as neutron counters (Iguchi *et al.*, 1994; Hayakawa and Maeda, 1996; Radeka *et al.*, 1998).

For additional information on the application of multiwire and multipattern proportional counters, see Chapter 13.

3. Low Level Counting Techniques Using Proportional Gas Ionization Detectors

For Investigations involving low-level counting techniques, e.g., low-level radiocarbon dating experiments, a low and stable background is a necessity. Today this is achieved mostly by the application of "active shielding." The counting tube for the sample is surrounded by "guard tubes," which are combined with the sample counting tube by a anticoincidence circuit. Only the counts due to the sample counting tube alone are counted and not those registered by both counting systems simultaneously which are due to background radioactivity. In earlier times this active shielding was a ring of sometimes more than 20 Geiger-Mueller (GM) counting tubes. Later, umbrella-shaped guard tubes were designed. Those were in some cases displaced by liquid scintillator guards, which were specially designed for low-level anticoincidence shielding. The liquid scintillation solution is frequently based on a mineral oil solvent and especially suitable for large tanks. Within the guard chamber, several counters based on proportional detectors are sometimes installed. Some systems are equipped with pressure transmitters and temperature sensors to ensure constant conditions for the counting gas. Also, measurement of the peak and median of the pulse height spectrum is used to obtain information about the purity of the counting gas.

Several guard counter designs are described in the literature and a remarkable construction has been proposed and tested by Theodórsson and Heusser (1991). They suggest an arrangement of flat guard counters on the external sides of the main shield instead of the inner region of the shield as usual. In this way the weight and space of the inner shield can be reduced. They also claim that the effects of secondary nuclear reactions causing background effects are considerably reduced.

A new detector type for low-level anticoincidence counting is designed and constructed by Zhang *et al.* (2002). A CdTe semiconductor counter is used as a guard detector forming also the wall of the low level proportional counting tube. This equipment is applied successfully to radiocarbon dating investigations, and the authors suggest also other fields of application.

Background reduction by electronic circuit design can be accomplished by pulse shape discrimination (Mäntynen *et al.*, 1987; Äikää *et al.*, 1992). By applying pulse shape discrimination, the background is reduced by more than 70%, and only 20% of the efficiency is lost. Figures of merit are improved by a factor of nearly 2. With a counting time of 44 hours measurable ages up to 56,000 years are achieved.

Carbon dioxide, being a "slower" gas than, for example, methane, is better suited for pulse shape discrimination. On the other hand, purity requirements are much more severe for carbon dioxide. If pulse shape discrimination for background reduction is used, the total length of the rising pulses is measured. The accumulated rise time of an irregular (i.e., a background) pulse is much longer than the rise time of a beta pulse. Yet some background remains, for example, that arising from gamma-emitting radionuclides in the construction material. The factors necessary for all these improvements are provided by Äikää *et al.* (1992).

At present, three measurement methods for radiocarbon dating are available: accelerator mass spectrometry, low-level liquid scintillation counting, and low-level gas proportional counting. During the past several years gas proportional counting methods had become less attractive for radiocarbon dating studies. Some authors are of the opinion that the application of gas proportional counting for radiocarbon dating should be reconsidered, as multidetector gas proportional counting systems offer some advantages.

A modern multidetector system has the advantage of parallel counting, which saves a great deal of time. Also, it takes less time to prepare carbon dioxide from a 1-g carbon sample than to carry out a benzene synthesis from the same amount of sample. This benzene is used as an additive to a liquid scintillation cocktail as discussed in Chapter 6. With parallel counting in a multidetector system based on ionization detectors, one of the samples is always a background sample, thus providing continuous monitoring of the background. Pulse rise discrimination techniques can be used in addition to reduce the number of background counts. If pulse rise analysis techniques are used to reject the slower rising background pulses, the counting efficiency is reduced by 18%, but at the same time the background is reduced by a factor of 3.3. A dramatic reduction of background counts is obtained by anticoincidence shielding. Like anticoincidence systems, liquid scintillation guard detectors are frequently used for active shielding.

According to the investigation of Theodórsson (1991a), a multidetector gas proportional counting system seems to be highly competitive. Of course, the accelerator mass spectrometry technique has clear superiority over radiometric methods, especially for very small samples, but considering the high price of accelerator mass spectrometry equipment, it seems likely that accelerator mass spectrometry systems and gas proportional counting will be used in the future and these will complement each other very well. Because of the potential of accelerator mass spectrometry, scientists hesitated to apply and further improve gas proportional counting. Future developments, especially with respect to computer-assisted gas proportional counting systems, will be of interest.

Proportional counting devices still play an important role in radiocarbon dating investigations. For example, Facorellis *et al.* (2001) report about measurements from 19 sites in Greece (Thessaly) providing information about human presence from 48,000 years before our age with a lot of interesting details also regarding climatic conditions of the past.

Low-level liquid scintillation analyzers with active shielding can provide low background count rates of 0.3 cpm for ^{14}C measurements, making liquid scintillation an attractive method for ^{14}C dating. Chapters 5 and 6 of this book provide detailed information on low-level radiocarbon measurements by liquid scintillation analysis.

At the National Institute of Standards (NBS/NIST) tritium standards are calibrated regularly using liquid scintillation and gas proportional counting methods. Using the available data from measurements over 38 years Unterweger and Lucas (2000) could obtain a more accurate and precise value for the half-life of that radionuclide (4504 ± 9 days).

The available data from international comparison projects have also been used to study the state of art of tritium low-level measurement techniques. The objective was to find a realistic value for the sensitivity which could be demanded in ultra-low-level tritium investigations. Theodórsson (1999) reported that during intercomparison investigations only two laboratories could reach a standard deviation of ≤ 0.03 TU for weak samples. The achievement of a good level of sensitivity and accuracy for tritium measurement is an urgent requirement because otherwise the possibility of obtaining reliable hydrological information that tritium can give as a natural tracer would be severely limited. Improved future counting systems are discussed. It is again mentioned that gas proportional counting systems can be improved significantly by moving the guard counters to the outer surface of the shield as it had been already proposed by Theodórsson and Heusser (1991).

Measurements have been carried out also to verify theoretical aspects, such as the investigations of Kuzimov and Osetrova (2000) on the shape of the carbon-14 beta-spectrum. Their examinations yielded results which are consistent with some of the theoretical predictions but which contradict the prediction of others. These findings may help researchers arrive at more accurate theories.

4. Applications in Environmental Monitoring, and Health Physics

a. Radon in Water

Zikovsky and Roireau (1990) have developed a simple method for the measurement of radon in water using proportional counters. The method is based on the purging of radon from water with argon, which is bubbled through the water sample and then directed to the counting tube. Argon picks up the radon that was dissolved in the water. A gas purification system removes humidity and oxygen. The high voltage is set for the alpha plateau and thus a very low background of less than 0.2 cpm and a counting efficiency of 25% are obtained, giving a detection limit of 0.02 Bq L^{-1}. This detection limit compares favorably with that of other methods developed for the determination of radon in water.

Radon daughters (polonium-218, lead-214, polonium-214) contaminate the detector, and after each measurement a waiting period of at least 1 hour, depending on the activity measured, is necessary because of the decay of these daughter products. Otherwise a correction for the residual activity has to be applied.

b. Measurement of Plutonium-241

Rosner *et al.* (1992) have built a proportional counting system, that is especially suitable for the measurement of ^{241}Pu. Plutonium-241 is the only significant beta-emitting transuranium nuclide in low-level waste from nuclear power plants. Quantitation of plutonium-241 in low-level waste and environmental samples is of interest because ^{241}Pu is a precursor of other transuranium nuclides that have longer half-lives, greater environmental mobility, and greater radiotoxicity. Americium-241, with a half-life of 432 years, is the daughter product of plutonium-241 and has relatively high radiotoxicity.

Plutonium-241 can be determined indirectly by alpha-spectroscopic measurements of its daughter nuclide americium-241. Measurements based on the ingrowth of the daughter radionuclide ^{241}Am can be done only after a long growth period. Even after 4 years the activity ratio ^{241}Am : ^{241}Pu is only 1 : 166. Thus the lower limit of detection for ^{241}Pu by direct measurement using proportional counting is about 10 mBq according to the work of Rosner *et al.* (1992), whereas via ^{241}Am buildup about 200 mBq is needed for detection.

Some authors have applied liquid scintillation counting to the direct measurement of plutonium-241. Because of the rather high background of commonly available liquid scintillation equipment, this method can be applied only for samples with a relatively high content of plutonium-241. Investigations of that type have been carried out in regions with elevated fallout levels such as Scandinavia or with samples from the nuclear industry or weapons test sites. Lower limits of detection of 35–65 mBq have been reported. However, a low-level liquid scintillation analyzer equipped with a BGO detector guard and time-resolved liquid scintillation counting (TR-LSC) background discrimination electronics is capable of counting environmental ^{241}Pu at a low background of 2.4 cpm (M. F. L'Annunziata, personal communication).

However, because of the nonspecific character of beta radiation, liquid scintillation and proportional counting require very pure samples for counting. Therefore the chemical purity of the samples and the self-absorption due to the presence of matrix material in the counting sample are the critical points in the proportional counting procedure.

For proportional counting special equipment is needed. This equipment can be obtained by modification of commercially available systems.

c. Measurement of Iron-55

For some radionuclides that are difficult to detect during radioprotection measures, gas ionization detectors still offer good possibilities. Iron-55 is a possible contaminant around nuclear reactors, and during planned repairs suspension and dispersion of this radioisotope of iron have to be monitored to avoid intake by workers. The low-energy x-rays of iron-55 (5.9 keV) are stopped by most detector windows. This radiation is also difficult to detect in the presence of other contaminating radionuclides. Surette and Waker (1994) have designed a monitoring system based on a sealed xenon-filled proportional counter with a thin beryllium window. The detector is combined with a single-channel analyzer and a shuttle mechanism that permits positioning of air filter or swipe media. For a counting time of 100 seconds the detection limit is around 10 Bq. The thin window of the proportional counting tube (0.05 mm) allows more than 90% of photons with an energy of 3 keV or greater to pass through. The monitoring system is sufficiently sensitive to detect well below the maximum permissible level of surface contamination and also below the maximum permissible concentration in air of the facility for which it was designed.

d. Tritium in Air

Proportional counters can also be used for tritium monitoring in air, as demonstrated by Aoyama (1990). Monitoring of tritium in air is required in the environment of 14-MeV neutron generators, heavy-water reactors, and reprocessing plants and will also be necessary at nuclear fusion reactors. Tritium must be detected separately from other radioactive volatile noble gases and air activation products. For occupational radiation protection and emission control, a real-time measurement and high sensitivity are necessary to meet the legal requirements for radiation protection and emission control. To respond to an accidental release, a wide range of detection is essential. High sensitivity can be obtained by using systems equipped with anticoincidence shielding or pulse shape discrimination. Conventional proportional counters suffer from the disadvantage of requiring a counting gas and have a rather short operation range.

Aoyama (1990) described a method for tritium monitoring in air by the use of flow-through proportional counters with air as a counting gas. The counters need no counting gas other than the sampled air. The electronic equipment attached to the counting system comprises pulse height discrimination, anticoincidence shielding, and background compensation. In that way it is possible to detect and measure tritium in an external gamma background and also in alpha and beta backgrounds originating from other gaseous radioactive materials in the air sample. It was reported that a lower detection limit of $0.005\,\mathrm{Bq\,cm^{-3}}$ in the presence of natural background can be obtained in a counting time of 1 min. Also, a wide range up to $5000\,\mathrm{Bq/cm^3}$ (up to six decades) can be managed by this system. The proportional counting detector is rather complicated, consisting of an arrangement of anode wires and cathode meshes. A schematic picture of this arrangement is shown in Fig. 2.12. Outer layers of the counter were used as guard counters to eliminate gamma background. Gaps between individual arrangements of anode and cathode were kept longer than the maximum range of tritium in air, thereby avoiding coincidence effects caused by tritium. Such coincidence effects were used to exclude other beta rays. The alpha component from radon and its daughter nuclides was eliminated by pulse

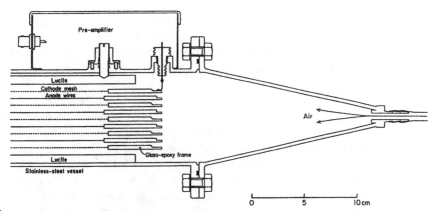

FIGURE 2.12   Cross section of a tritium monitor, which uses air as a counting gas. The detector consists of four layers of multiwire proportional counters. The air flows uniformly through the counter. (From Aoyama, 1990, reprinted with permission from Elsevier Science.)

height discrimination. Data derived from pressure, temperature, and humidity sensors were transferred to a computer and used to control the high voltage and to correct coincidence count rates.

e. Radiostrontium

Low level proportional counters are applied for quantitative radionuclide measurement after radiochemical separation procedures. Mateos *et al.* (2000) have designed and constructed a semiautomatic analysis system for the determination of ^{90}Sr/^{90}Y in aqueous samples using a sequential injection method. The beta-measurements are made twice within 24 hours and from these results the initial activity of strontium-90 and yttrium-90 is calculated. Thus the time consuming yttrium milking method can be overcome.

Vaca *et al.* (2001) compared strontium-90 measurement methods using a Berthold LB770 counter and a Quantulus 1220 liquid scintillation spectrometer. The proportional counter had a passive shield of 20 cm thick lead and an active gas proportional guard counter. The samples can be measured simultaneously by that device and a background from 0.3 to 0.6 cpm, depending on the detector location along the gas flow pathway is obtained. It is surprising that for gas proportional counting a minimum detectable activity of 0.13 Bq/kg is reported, for Cerenkov counting 0.37 Bq/kg.

Crown ether technologies were used by Scarpitta *et al.* (1999) to measure the strontium-90 content of Brookhaven National Laboratory groundwater samples. With gas proportional and liquid scintillation counting minimum detectable levels of $37\ \mathrm{Bq\,m^{-3}}$ were achieved using a processed sample of 1 liter and a counting time of 1 hour.

Proportional counting is also used for the determination of strontium-90 in human bones and teeth in Greece. Measurement was performed on yttrium-90 after equilibration with strontium-90 and liquid extraction using *bis* (2-ethyl-hexyl) hydrogen phosphate. Analyses were performed during 1992–1996 on 108 samples from 896 individuals. Samples were classified according to the age and sex of the donors. In bones an average of 30 mBq strontium-90 per gram calcium was found with only a small variation with respect to age and sex. From the variation of the activity in teeth it can be concluded that the contamination from the atmospheric nuclear weapons test exceeds by far that caused by the Chernobyl accident (Stamoulis *et al.*, 1999).

f. Health Physics

Tissue equivalent proportional counters can be used to evaluate the radiation dose and dose equivalent for gamma rays and neutrons. Techniques to separate out the dose and energy spectra of neutrons and charged particles are necessary for health physics investigations in space shuttle conditions. Braby and Badhwar (2001) used a combination of a tissue equivalent and a hydrogen-free detector. Both have nearly the same response to photons, but the hydrogen free detector is insensitive to neutrons below about 10 MeV. Thus the neutron dose can be obtained by subtraction. Similar considerations are used also for the separation of charged particles and neutrons.

For thermoluminescence dating, the radiation dose the object had been exposed to must be known. Most frequently this radiation dose is due to the content of alpha- and beta-emitting natural radionuclides within the material of the archeological specimen. Proportional counting techniques can be used to determine the activity of the material, and from this analytical result, the radiation dose can be calculated. Troja *et al.* (1995) give an example for this type of activity measurements and dose calculations.

Nano-dosimetry will be of interest for investigations in microbiological radiation effects. Tamboul and Watt (2001) built a gridded parallel plate proportional counter, operating at low pressure (1 Torr). This corresponds to a mean chord diameter of 1.8 nm. The device is designed to have a response to radiation simulating that of a bimolecular target of about the same sensitive volume, e.g., a double-stranded DNA molecule.

VI. GEIGER-MUELLER COUNTERS

As already mentioned, with Geiger-Mueller (GM) tubes much higher electric fields are applied than with ion chambers and proportional counters. Because of the high electric field, the intensity of an individual avalanche is enhanced. As a consequence of the emission of UV photons, which are released during deexcitation of atoms or molecules inside the tube, additional avalanches are created. One avalanche therefore can trigger another at a different position in the detector chamber volume. The number of avalanches grows exponentially. Also, the number of slowly migrating positive ions increases. The increasing number of positively charged ions near the electrode causes the field strength to decrease, and further creation of avalanches is stopped because ion pair multiplication requires a sufficiently high electric field strength. The discharge in a Geiger-Mueller tube is terminated at about the same total produced charge, regardless of the amount of ions initially created by the radiation event. Therefore all output pulses from a Geiger-Mueller tube are of about the same size. The output pulse amplitudes of GM tubes are very large compared with signals of ion chambers and proportional counters, usually of the order of several volts. Simple electronic circuits can be used to register Geiger-Mueller output signals, but no information about the type and energy of the incident radiation can be obtained. Besides this lack of information about the type and energy of radiation, Geiger-Mueller tubes have a rather long resolving time compared with proportional counting tubes. Therefore their use is limited to relatively low count rates, only a few hundred counts per second. Resolving time corrections can be applied, but the resolving time depends not only on the field strength but also on the observed count rate (Jones and Holford, 1981).

The decaying source method is probably the most general and accurate of the methods for measuring the observed and true counting rates over the entire counting rate range of interest. For that a very pure radionuclide source of known half-life is essentially needed. True count rate and observed count rate differ considerably at high activity of the radioactive source however, with time, the background corrected observed count rate will approach the true count rate. This type of experimental determination of dead time is frequently used to test the usefulness of mathematically based models for correction (e.g., Gardner and Liu, 1997; Lee and Gardner, 2000),

Counting losses induced by resolving time of a counting system can be a limiting factor in measurements. Vinagre and Conde (2001) presented an interesting method for the determination of resolving time of a counting system. They added an additional pulse to each pulse of the counting system and varied the delay time of this additional pulse. By observing the total count rate as a function of the delay time good results for the resolving time could be obtained.

It warrants mention that Geiger-Mueller tubes show a remarkable energy dependent response for high energy photons above 3 MeV. This was again pointed out by Neumann *et al.* (2002) to be a relevant factor in accurate dose determinations.

A. Designs and Properties of Geiger-Mueller Counters

I. Fill Gas

The fill gas for Geiger-Mueller counting tubes has to meet requirements similar to those for the fill gas for proportional counters. Argon and helium are most frequently used. The gas pressure is on the order of tenths of bars, and depending on the size and shape of the tubes a voltage on the order of hundreds of volts is applied. Geiger-Mueller tubes are usually permanently sealed and operate at low gas pressure, although designs have been realized using atmospheric pressure and flow-through to replenish the fill gas and flush out impurities.

2. Quenching

After the termination of the discharge, the slowly migrating positive ions of the fill gas finally arrive at the cathode, which is usually the outer wall of the counting tube. At this electrode the cations capture electrons from the cathode surface and a corresponding amount of energy is liberated. If this liberated energy exceeds the ionization energy of the cathode material, additional electrons are set free from this electrode. These newly generated free electrons migrate to the anode and create another avalanche. This finally results in a continuous output of pulses. The probability of this additional electron drift is rather low, but because of the high number of cations at the field strength conditions in a Geiger-Mueller tube, this effect of multiple pulses is observed. With Geiger-Mueller tubes special precautions have to be taken to prevent the formation of additional avalanches. This can be done by reducing the bias voltage after the Geiger discharge. This external quenching can be achieved by using a suitable electronic circuit (resistor and capacitance) that determines the time of restoration of the high voltage following a Geiger discharge. The restoration time is usually on the order of milliseconds and therefore this design is suitable only for low count rates.

It is more common today to use internal quenching, which involves the addition of a suitable compound to the fill gas. The ionization energy for this additive to the fill gas (quench gas) must be lower than the ionization energy for the fill gas. Although confusing, the same expression "quench gas" is used for both the additive to a fill gas of proportional counters, which has to absorb UV photons, and the additive to a fill gas in the Geiger-Mueller tube, which should be able to neutralize the drifting ions of the original filling gas by electron transfer. The ions of the quench gas migrate to the cathode and are also neutralized. But the liberated ionization energy is now consumed by the quench gas and causes dissociation of the quench gas molecules. Some quench gases, such as halogens (e.g., chlorine or bromine), show spontaneous recombination; other quench gases, such as organic compounds (e.g., ethanol), are consumed, and therefore the lifetime of an organic-quenched Geiger-Mueller tube is limited to about 10^9 counts. Quench gases are usually added at an amount of several percent to the fill gas of the Geiger-Mueller tube.

A relatively long time is needed (100–500 μs) to clean the positive ions that are formed during the avalanche propagation.

The transition from proportional mode to Geiger-Mueller mode takes place at increasing field strength. Golovatyuk and Grancagnolo (1999) could demonstrate that this transition also depends on the concentration of a quenching gas. This fact may be of relevance if pulse shape analysis is used for particle identification. If the concentration of quenching gas is low, gas amplification, as a function of high voltage, increases more rapidly and the boundary between the proportional region and the Geiger-Mueller region may be crossed easily. Results of pulse shape analysis may not be interpreted correctly.

3. Plateau

For the simple electronic circuits that are usually designed for use with Geiger-Mueller tubes a minimum pulse amplitude is required for count registering. At a given voltage this minimum pulse amplitude is exceeded by all signals, as soon as that voltage, the Geiger discharge region, is reached. Therefore, on increasing the voltage while exposing the Geiger-Mueller tube to a radioactive source of constant activity, pulse registering starts rather abruptly and the counts per unit of time remain relatively constant (plateau of a Geiger-Mueller counter).

Geiger-Mueller tubes are frequently rated on the basis of the slope of the plateau region. The slope of the plateau region of halogen-quenched tubes is usually less flat than that of organic-quenched tubes (2–3% per 100 volts). But halogen-quenched tubes can usually be operated at a lower voltage than organic-quenched tubes.

4. Applications

The design of Geiger-Mueller tubes is usually similar to that of proportional tubes. But most frequently the end window type is used. Geiger-Mueller tubes can also have the shape of "needle tubes," in which the anode consists of a needle. In the vicinity of the needle point the field strength varies by $1/r^2$ instead of the $1/r$ variation near a wire or rod electrode. Therefore, counters with a very small active volume can be manufactured.

Because a Geiger discharge is created by a single ion pair, alpha and beta particles, once they penetrate the wall or window, are registered with very high efficiency. Gamma rays are detected by the electrons that are observed as a result of interaction of the gamma ray with the walls of the counting tube via the photoelectric effect or Compton effect. The efficiency of Geiger-Mueller tubes for gamma rays is very low and also depends on the atomic number of the material used to make the tubes.

Currently, Geiger-Mueller tubes are used most frequently for radiation monitoring and contamination control in day-to-day radiochemistry work.

Photon doses in mixed fields (neutrons/gamma) are frequently measured with Geiger-Mueller counters. But it has to be mentioned that the response of Geiger-Mueller detectors depends on photon energy, especially for photon energies above 3 MeV. Dealing with the analysis of neutron and photon components during calibration experiments Neumann *et al.* (2002) point out that the knowledge of spectral distribution of the photons is essential for accurate dose determinations.

a. Environmental Radioassay

Radon Bigu (1992) designed a fully automatic system for the unattended quantitation of radon-222 and radon-220 progeny. He used a GM beta particle detector with a pancake configuration. The instrument is a microprocessor-based system that consists of a sampling device, an electronic scaler, and a personal computer. The computer records all sampling and counting routines.

The sampling device consists of a filter about 5 cm in diameter facing the detector at a distance of about 0.5 cm. The air flow rate is $1.4\,L\,min^{-1}$. However, the measurement and data procedure is rather complex and requires a rather sophisticated computer program. Basically, the following steps are required:

- The sampling and counting for a given period provides results for the combined radon-222 and radon-220 progeny contribution.

- A counting period after the sampling records the beta particle activity versus time and thus permits assay of radon-222 daughter products.
- From the results of the preceding steps the contribution of radon-220 progeny can be calculated.

Difficulties may arise if measurements have to be made under transient conditions, for example, when there are rapid changes in the concentration of radon-222, or changes in the aerosol concentration or aerosol size distribution. This effect is related to the half-life of the radionuclides of interest. For radon-220 it is particularly acute because of the long half-life of lead-212. The full effect of any of these changes (perturbations) is felt by the detectors after about one half-life of the dominating radionuclide of the decay chain.

Fluorine-18 Papp and Uray (2002) used a very simple experimental setup for the determination of fluorine-18 attached to aerosol particles in a laboratory where syntheses for positron emission tomography are carried out. Aerosol samples were collected by drawing the air through a glass-fibre filter using a mobile high-volume air sampler. The filter discs were counted under an end-window Geiger Mueller tube (mica window $2\,mg\,cm^{-2}$ thickness and 35 mm diameter, background about 32 cpm). Following this very simple experimental procedure subsequent measurements and a rather complicated computation using Bateman-type differential equations have to be carried out to distinguish between the radioactivity of the airborne natural radionuclides like ^{218}Po, ^{214}Pb ^{214}Bi, ^{212}Pb, ^{212}Bi, ^{208}Tl, and Fluorine-18. Therefore the method cannot provide instantaneous results; however, very low activity concentrations, around $1\,Bq/m^3$, corresponding to 160 atoms/m^3 can be detected. This method can be applied also to the determination of any other airborne beta-emitting radionuclide if its half-life differs sufficiently from those of the progenies of radon and thoron.

Radiostrontium The beta counting of yttrium-90 after growth to equilibrium with strontium-90 had been used during an extensive and remarkable investigation carried out by Russian and Norwegian scientists in the South Ural region near the site of the first weapon grade plutonium production reactor complex in Russia. Geiger-Mueller-counting tubes had been used for the determination of beta particles and Strand *et al.* (1999) reported that they found 720 kBq/kg of strontium-90 in sediments and 8 to 14 kBq/L in water.

Cosma (2000) carried out strontium-90 determinations in Romania without previous chemical separation procedures. He used aluminum plates to absorb low-energy beta particles and thereby detect only the high-energy beta radiation of yttrium-90. He obtained values between 40 and 75 kBq/kg in sediments and soil after the Chernobyl accident in Romania.

Chu *et al.* (1998) compared three methods for the determination of radiostrontium, the nitric acid precipitation method, ion exchange and crown ether separation procedures. They analyzed soil, tea leaves, rice, and milk powder. Their main statement is that by application of the crown ether method the time consuming and hazardous nitric acid precipitation method is avoided. Measurements were carried out using gas flow Geiger-Mueller tubes and Cerenkov counters. Data are given for strontium-90 and strontium-89.

VII. SPECIAL TYPES OF IONIZATION DETECTORS

A. Neutron Detectors

Practically every type of neutron detector consists of a target material that is designed to produce charged particles by interaction with neutrons. Those charged particles can be detected by any suitable detector, such as an ionization detector. The nuclear interactions resulting in the production of charged particles are governed by the reaction cross section. This cross section depends strongly on the energy of the neutrons as described in Chapter 1. In searching for such nuclear reactions one has to consider that the cross section should be as large as possible. Detectors with high efficiency and small dimensions can be designed in this way .

The most popular nuclear interaction for the measurement of neutrons is the $^{10}B(n, \alpha)^7Li$ reaction. It can be used for the measurement of slow neutrons. The cross section decreases rapidly with increasing neutron energy as illustrated in Fig. 1.14 of Chapter 1. This reaction is very useful because of the large cross section for thermal neutrons (3840 barns) and because of the rather high isotopic abundance of the boron isotope with mass number 10 (19.8%). Usually, boron trifluoride is used as an additive to the host gas in proportional counting tubes.

The reaction $^3He(n, p)^3H$ has a significantly higher cross section for thermal neutrons, but the relatively high cost of ^3He has somewhat limited the application of this target material for proportional neutron counting tubes. The ^3He counters can be used for what is usually called a hostile environment, and they find application in well logging investigations (Glesius and Kniss, 1988). Glesius and Kniss provide a review of such applications for borehole measurements.

For the detection of delayed neutrons Loaiza (1999) used an array of Helium-3 counters embedded in polyethylene. High efficiency, low dead time and gamma-insensitivity were the requirements for this counting device. The system was tested using an Am/Li source, the accuracy relative to a standard source embedded in graphite was about 3%, the efficiency 29% and the dead time $0.46\,\mu s$. Most gamma pulses have been suppressed by proper setting of amplifier gain and discriminator. Thus all the necessary requirements for the investigations could be fulfilled.

In several places there are plans to construct spallation neutron sources. For experiments with such neutron sources detectors will be required with two dimensional response, good time resolution and capability for neutron energy determination. Radeka et al. (1998) built multiwire chambers up to $50\,cm \times 50\,cm$ with helium-3 and propane as filling gas mixture and work is in progress to construct a large curved detector for protein crystallography studies at a pulsed spallation source at Los Alamos. The detector will be placed 16 m from the neutron creation point, and thus a single neutron pulse time would act as a monochromator for neutrons.

Also new detector designs, like GEM detectors have been introduced for neutron measurements. Lopes *et al.* (1999) combine the principles of proportional scintillation counters and gas electron multipliers, Fraga *et al.* (2002) applied helium-3 as a filling gas to determine neutrons.

The $^6Li(n, \alpha)^3H$ reaction cannot be used for gas ionization counters because a lithium-containing gas for proportional counters is not available. But ^6Li counting scintillators are quite common as detectors for neutrons as described in Chapter 11.

The cross sections of uranium-233, uranium-235, and plutonium-239 for fission reaction with thermal neutrons are very large, and the fission products that form the "charged particles" to be detected in a proportional counting tube have very high kinetic energy (about 160 MeV). This facilitates discrimination from the alpha emission of the fissile materials that are neutron targets of the counting system. Little success was achieved in trying to produce these neutron targets as a gaseous additive to the host gas of proportional counting tubes. Commonly the surfaces of the electrodes are covered with a deposit of the fissile material. This system is frequently applied, for example, for the fission chambers that are used for reactor as well as nonreactor applications.

As mentioned previously, the BF_3 proportional tube is the most widely used detector for slow neutrons. Somehow the boron trifluoride can serve both purposes, as a target for slow neutrons and also as a proportional counting gas for the reaction products of the $^{10}B(n, \alpha)^7Li$ reaction. Although other boron-containing gases have been investigated, BF_3 offers good properties as a proportional gas and also a high boron content compared with other gaseous boron compounds. Usually boron-10 is highly enriched for use in boron trifluoride counting tubes. This provides a much higher efficiency than is obtained with naturally occurring boron. Tubes with enriched boron-10 have about five times higher efficiency for thermal neutron counting than tubes filled with boron in its natural isotopic abundance.

According to the reaction $^{10}B(n, \alpha)^7Li$, the output signal handling seems to be simple and straightforward for the application of boron-10 to the detection or even spectroscopy of neutrons. However, the energy spectrum and pulse processing for BF_3 tubes can be rather complicated in detail. Recoil ^7Li also contributes to the energy spectrum and the nuclear reaction leads to either a ground state (94%) or excited state (6%) of ^7Li. Also, the volume of the counting tube in general is not sufficiently large compared with the range of the alpha particles or even the range of ^7Li recoil atoms. Therefore, the energy of these reaction products is not deposited totally in the gas volume, but interaction with the walls of the tube occurs. This results in distortion of the energy spectrum recorded from ionizing effects in the gaseous volume. Summarizing, one can say that the BF_3 tube is a detector from which, by differential pulse height analysis, little useful information is obtained about the energy spectrum of the incident radiation. The pulse height spectrum depends mainly on the size and shape of the detector. Therefore counting is done only at a high voltage providing a flat region at a plateau and a discriminator setting is used at which all neutrons are counted but all low-amplitude effects are rejected. Low-amplitude effects are due mainly to gamma rays producing secondary electrons from wall interactions. But at very high gamma radiation fields problems arise because of pileup effects. Also, BF_3 suffers from radiation decomposition in gamma fields of high intensity. Some authors try to absorb decomposition products of BF_3 by applying activated charcoal as an absorbent.

Position-sensitive neutron counters are essential for measuring the neutron flux distribution in critical assemblies. For that purpose neutron counting tubes with unusual dimensions may be constructed, such as a 1.2-m-long and 8-mm-diameter tube designed by Uritani *et al.* (1995). With that device nonuniformities in a critical assembly could be detected and correction measures undertaken.

I. BF_3 Tube Construction

If the dimension of BF_3 counting tubes increases, detection efficiency is improved and wall effects are suppressed. To some extent, increasing the gas pressure has the same effects. Some consideration has to be given to the materials used for tube construction to avoid radioactivation effects due to neutron capture by the materials used for tube wall construction. Because of its rather low neutron interaction, aluminum is frequently the material of choice; however, if a low background is essential, one has to keep in mind that aluminum contains a small amount of alpha-emitting materials. For such low-level investigations stainless steel is preferred as a construction material for BF_3 tubes. Elevated temperature has some adverse effects on counting performance. Above 100–150°C pulse amplitude and pulse height resolutions are decreased because of desorption of impurities from construction materials inside the tube. Extensive studies of the temperature dependence of BF_3 proportional counters were carried out by Sakamoto and Morioka (1994). Some phenomena that depend on temperature were related to impurities in the enclosed gas and also to construction details of the electrodes.

Usually BF_3 tubes are operated at a rather high voltage. Therefore spurious pulses are possible due to leakage current through insulators, especially under conditions of high humidity. Also, detector microphonics have been observed if the counting system is subject to shock or vibrations.

2. Detectors for Fast Neutrons

It has to be kept in mind that the gas ionization detectors previously described, namely BF_3 and ^{3}He detectors, which are based on the conversion of neutrons to directly detectable charged particles, are capable of detecting only slow neutrons. The cross section responsible for the ^{10}B$(n, \alpha)^7$Li and ^{3}He$(n, p)^3$H reactions decreases rapidly for neutrons with higher energies. To use these detectors for the determination of fast neutrons, the high-energy particles have to be slowed down, i.e., moderated. The low detection efficiency for high-energy neutrons of slow neutron detectors can be greatly improved by surrounding the detector volume with a layer of moderating material, for example, hydrogen- and carbon-containing materials such as paraffin. Fast neutrons lose part of their initial high kinetic energy by impacts with the moderator molecules before reaching the sensitive volume of the detector. However, neutrons can escape from the moderator layer by scattering without reaching the detector volume or can be captured by moderator materials. Thus an increase of the thickness of the moderator layer will not proportionally increase the number of thermalized neutrons counted by the detector. A maximum counting efficiency will be observed at a specific moderator thickness. This optimal thickness depends on the initial energy of the fast neutrons to be detected and varies from a few centimeters for neutrons with energies of keV up to several tens of centimeters for neutrons having energies in the MeV range.

There is no general method for neutron spectrometry, especially around the eV region. The "slowing-down time" method can be applied for such investigations and conventional BF_3 tubes are used (Maekawa and Oyama, 1995a,b, 1997).

Toyokawa *et al.* (1995) described a multipurpose neutron counter, applicable to the measurement of fluence, energy distribution, and radiation dose equivalent. This system consists of a spherical polyethylene moderator and three ^3He position-sensitive tubes inserted into the moderator orthogonally to each other. These three position-sensitive tubes provide information about the thermal neutron distribution in the spherical moderator, and from that information the foregoing parameters can be evaluated.

For neutron spectrometry in the MeV range, ^3He ionization chambers can be used. Iguchi *et al.* (1994) carried out investigations dealing with the application of these detectors in neutron spectrometry. Their ^3He detector consists of a cylindrical gridded ionization chamber (Fig. 2.13). Monte Carlo simulation was applied to estimate the detector response. Four kinds of reactions in the detector gas were considered in the calculations: $^3He(n,p)t$, $^3He(n,d)d$, and $^3He(n,n)$, and $^1H(n,n)$ elastic scattering. Corresponding to these calculations, the response functions were measured with monoenergetic neutrons at various energy points.

Pulse height and rise time distribution analysis of signals from neutron proportional counters were used to reject undesirable signals of hydrogen-filled proton recoil counters, ^3He-filled counters, and BF_3 counters. Gamma ray background and wall effect pulses can be reduced by that method (Sakamoto and Morioka, 1993).

Neutron measurements in an environment with high gamma-radiation doses are of interest in the field of nuclear safeguards. Especially neutron-gamma coincidence counting is of particular interest for spent-fuel measurements for burnup verification and in several steps of nuclear fuel reprocessing. The high gamma background has limited the selection of neutron detectors. Neutron fission chambers do not possess sufficient efficiency to be used in coincidence counting and BF_3 tubes suffer from radiation damage. Beddingfield *et al.* (2000) have carried out comprehensive research to optimize the helium-3 neutron proportional counter performance in high gamma ray dose environment. There are many parameters to be observed, such as tube size, gas pressure, gamma-ray dose, gamma-ray pile up, gamma-ray energy, radiation damage to the gas mixture and to the

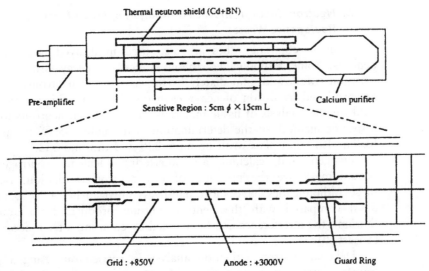

FIGURE 2.13 Schematic view of an ³He gas ionization chamber. The detector consists of a cylindrical gridded ionization chamber. The sensitive volume is fixed to 5 cm in diameter and 15 cm in length by guard rings at both ends. The chamber is filled with helium-3, argon, and methane at a pressure of several bars. A calcium purifier in the chamber is used to remove hydrogen produced from the ³He(n, p)³H reaction from the detector gas. Thermal neutrons are shielded by a boron layer outside the tube. (From Iguchi *et al.*, 1994, reprinted with permission from Elsevier Science.)

preamplifier system, etc. There is no best option of counting tube design for all mixed-field applications; however, from the presented amount of experimental data a good choice for a useful special design can be made.

a. Long Counter

Most neutron detector systems suffer from the disadvantage that the counting efficiency depends strongly on the energy of the neutrons to be detected. The so-called long counters try to avoid that disadvantage. A long counter consists of a neutron detector tube, most frequently a BF_3 tube, that is placed in the central region of a paraffin cylinder. The paraffin cylinder is covered with a layer of B_2O_3 and with an additional layer of paraffin. Only one end of the inner paraffin cylinder is not covered by the boron and additional paraffin. Thus, the device is sensitive only to neutrons coming from the direction of this end. Any neutron arriving from that direction is moderated and has a good chance of arriving at the central BF_3 tube. To give low-energy neutrons a better chance of reaching the tube, holes are drilled in the front end of the inner paraffin layer (Hunt and Mercer, 1978).

Because of the nearly energy-independent response of this type of counting tube, the arrangement is also called a "flat response" detector. Many variations of such flat response detectors have been designed and constructed, some of them using ³He tubes, pressurized filling gas, multiple tube arrangements, and so on. One has to be aware that the counting efficiency of such neutron counting systems is rather low, sometimes much less than 1% (East and Walton, 1982).

b. Neutron Counting in Nuclear Analysis of Fissile Materials and Radioactive Waste

Neutron counting tubes are also used in delayed neutron activation analysis. Some radionuclides follow a beta, n decay process; their beta emission is followed immediately by the emission of neutrons. Such nuclides are produced by fission of heavy elements. Therefore this procedure can be used for the analysis of fissile materials. Using thermal neutrons for fission is a specific method for the determination of uranium-235. With fast neutrons fission also takes place with uranium-238 and thorium-232.

Oxygen and calcium are interfering elements. Nitrogen-17 and potassium-48 are the products of fast neutron irradiation. But because those radionuclides have short half-lives (nitrogen-17, 4.2 s, potassium-48, 5.8 s) compared with the neutron-emitting products from uranium and thorium, the interference can be avoided by counting after a decay period of at least 20 s.

Delayed neutron activation analysis is carried out using a pneumatic transfer system at a neutron source of sufficient flux density, usually a reactor. The samples are first positioned near the reactor core by the transfer system and after a suitable irradiation period ($\sim$60 s) and decay period ($\sim$20 s) samples are counted ($\sim$60 s) at a neutron detector assembly.

Thorium interference due to fast neutron-induced fission is overcome by irradiation with and without cadmium shielding. The delayed neutron activation analysis is used mainly for the determination of uranium and thorium at trace levels in minerals. Fully automatic systems are available, with detection limits on the order of 0.01 μg/g for uranium and 1 μg/g for thorium.

Neutron counters have been applied also to the determination of transuranium elements. A high-sensitivity neutron counting tube arrangement was used successfully for the determination of plutonium in radioactive waste drums at Lawrence Livermore National Laboratory (Hankins and Thorngate, 1993a,b). It was reported that the sensitivity of this equipment is about 10 times better than the sensitivity of x-ray and gamma-ray instruments that are normally used. Helium-3 counting tubes are arranged outside the waste package. These ^{3}He counters are covered with paraffin with an outside lining of cadmium. Fission neutrons passing the cadmium barrier are thermalized in the paraffin layer and detected by the ^{3}He tubes.

Another system uses a pulsed electron beam from a linear accelerator to produce high-energy photon bursts from a metallic converter. The photons induce fission in transuranium elements. When fission is induced in such material, delayed neutrons can be detected by a sensitive neutron counting system (Lyoussi et al., 1996).

Not only transuranium elements are determined in waste using neutron counting; moisture measurements of the radioactive waste are also carried out. The thermalization of neutrons from an isotopic neutron source is detected by a proportional neutron counting tube. The moisture content of the waste is an important parameter that determines the combustibility of waste materials (Lentsch et al., 1996).

c. Moisture Measurements

Moisture measurements are based on the principle of neutron moderation by hydrogen atoms. In neutron water gauges, neutrons are most frequently produced through (α, n) reactions, e.g., $^9Be(\alpha, n)^{12}C$. These neutrons have a spectrum of energies from 0 to about 10 MeV. The neutrons of high energies are moderated (slowed down) by elastic scattering with hydrogen atoms from water. These slow neutrons are detected by a counting device which is only sensitive for slow neutrons e.g., a BF_3 counter. Modeling of the interactions of slow neutrons with different media, like soils, is not easy and therefore a calibration is needed to convert the slow neutron counts to water content. Usually a neutron moisture meter device is combined with a density gauge.

O'Leary and Incerti (1993) have undertaken a study to compare three neutron moisture meters during field experiments. They made measurements in different types of soil and moisture content and discussed also calibration problems which are of prime importance to get reliable results.

The theory and practice of measuring the water content in large volumes of material by neutron thermalization and the measurement of thermal neutrons with BF_3 or ^3He detectors are reviewed by Nielsen and Cassel (1984) and Bacchi et al. (2002).

B. Multiple Sample Reading Systems

In radioassay methods in biochemistry and medicine, a high sample number throughput is frequently essential. Radioactivity quantitation on solid supports, and radioimmune, dot blot, cell proliferation, and receptor binding assays require systems for counting a rather high number of samples in a given time. Simultaneous counting methods for a large number of samples are desirable. For these applications multiple sample reading systems have been designed. Bateman (1994) has constructed a multipin detector. The pins are centred in holes in a metallic collimator system and 60 beta-sensitive positions are obtained.

A system with 96 individual detectors working in the Geiger-Mueller region has also been manufactured (Roessler et al., 1993; Hillman et al., 1993a). A high sample throughput is achieved and the counting procedure is about 40 times faster than single-detector assay procedures. Of course, the counting efficiency for tritium is much lower than that achieved with liquid scintillation counting, but the background is reduced because the GM detectors are very small.

Roessler et al. (1993) compared several methods for receptor binding assays and compared the sample throughputs. Hillman et al. (1993b) applied the 96-sample measurement system for chromium-51 retention assays. Several other application examples can be found in the literature (Alteri, 1992; Hutchins, 1992).

Microplate assays related to investigations using radioactive tracers have attracted great interest during the past decade. For microplate assays radioactivity has to be measured from samples on a solid support that may hold 96 samples in an area of 8×12 cm. Cells or tissues are incubated in the presence of a radiolabeled substrate simultaneously in all positions of such a microplate. After the incubation the nonincorporated components must be separated from the incorporated radioactive substrate at each position of the microplate. Applying conventional techniques, this was usually done by filtration and washing one sample after the other. The radioactive residue on the filters was counted using liquid scintillation counting (LSC) techniques. All this was a rather time-consuming and expensive procedure. Great progress was achieved by developing a sample harvester that could harvest and wash 96 samples simultaneously, saving a great deal of time and work. This harvesting and washing procedure can be performed by a specially constructed fully automatic cell harvester from conventional microplates, or special filter bottom foils for these plates can be used. The application of a radioactivity reading system that can analyze 96 samples simultaneously greatly simplifies the microplate radioassay techniques.

Two different types of microplates can be chosen, those with and those without a removable bottom. The removable bottom consists of a membrane filter material that can be easily stripped from the bottom of the microplate.

These solid support samples can now be measured using either liquid scintillation counting (standard LSC or multidetector LSC) or ionization detector techniques, such as proportional ionization detector counting, position-sensitive proportional counter scanning, or multidetector avalanche gas ionization detector quantitation. See Chapters 5 and 11 of this book for a detailed description of scintillation analysis in the microplate sample format.

For position-sensitive proportional counter scanning, systems similar to those used for scanning thin layer chromatograms (TLC) or paper chromatograms (PC) are used. With a position-sensitive wire detector 12 samples in a single row can be counted simultaneously. This method suffers from some disadvantages. This type of detector has a very low counting efficiency for low-energy beta emitters, and it is subject to high amounts of cross talk when high-energy beta-emitting radionuclides such as phosphorus-32 are analyzed. Also, the efficiency is not uniform across the entire length of the wire. It seems that this technique is rather unsuitable for quantitative simultaneous multicounting applications. Therefore systems with individual detectors in the format of the microplate were designed and manufactured. Open-end gas avalanche detectors are used and the systems are capable of quantitating tritium, carbon-14, phophorus-32, sulfur-35, iodine-125, and many other beta emitters. Of course, the filter mat must be dry but there is no addition of cocktail. The filter is not destroyed and can be used for further investigations. Also, the amount of waste is minimized.

A detailed description and examples of applications are given by Kessler (1991). This technique can be applied to the radioassay of dot blots and labeled cell proliferation assays. With conventional autoradiography and densitometry, the range of radioactivity measurements is much smaller than with a multidetector system, because an x-ray film shows a saturation effect in blackening. Also, the exposure time for x-ray films is much longer than the measurement time for ionization detectors. Other more quantitative imaging methods are described in Chapter 13. The ionization multidetector arrangement seems to be comparable to a liquid scintillation multidetector system

(Kessler, 1991). However, commercially available high-sample-throughput multidetector microplate scintillation analyzers described in Chapter 5 provide higher counting efficiencies and higher sample throughputs.

C. Self-Powered Detectors

Self-powered neutron detectors are fabricated with a material incorporated in the detector volume that has a high cross section for neutrons. By neutron capture, a beta-emitting radionuclide is formed. The detector operates by directly measuring the flow of current produced by the beta particles. No external bias voltage is needed. Other types of self-powered detectors for neutron counting are operated by the current that is produced by ionization due to gamma emission related to neutron capture during fission. The main advantages of these self-powered neutron detectors are their small size and the simple electronics necessary for this type of detector (Knoll, 1989).

Disadvantages are the low levels of the output signals, a slow response time, and sensitivity of the response to the neutron spectrum. Self-powered detectors have to be operated in the current mode, because the signal created by a neutron can be only a single electron.

D. Self-Quenched Streamer

Traditionally, gas ionization detectors are categorized as ion chambers, proportional counters, and Geiger-Mueller tubes. But another type of counting system based on ionization effects has been developed and applied. This is a type of gas multiplication detector that is somewhat different from proportional and Geiger-Mueller counting systems. It is called a self-quenched streamer (SQS) or limited streamer detector and is frequently used in position-sensitive multiwire detector systems. In conventional proportional and Geiger-Mueller counters, UV photons play a significant role in the propagation of an ion avalanche. If the propagation of the avalanche is kept small by the field strength or by absorption of UV, the system works in the proportional mode. If UV photons are able to create additional avalanches that may spread through the entire length of the anode wire and the whole process is terminated only by the creation of a space charge around the anode, the system works in the GM mode and the output signal does not depend on the original ionization effect (e.g., on the number of primary ions produced by the radiation event). In the self-quenched streamer mode the ion avalanches are controlled in a special way. The counting tube is filled with a gas mixture that absorbs UV photons. Therefore, no additional avalanches far from the original avalanche pathway can be created through excitation by photon absorption. Avalanches, therefore, grow and propagate in the shape of a streamer. The streamers have a diameter of about $200\,\mu$m and extend a few millimeters from the anode. They terminate at low field strength at larger radii of the detector.

If the voltage is high enough, a single electron can create a streamer. The streamers have a final length that depends on the voltage applied. The formation of such streamers is supported by anode wires with relatively large diameters (0.1 mm).

SQS detectors have some properties of both proportional and Geiger-Mueller detectors. The rather high internal gas amplification is useful for position-sensitive detectors. Position-sensitive detectors operating in the proportional region have much smaller signal amplitudes. But, as in the case of Geiger-Mueller tubes, the signal amplitudes in SQS detectors no longer provide information about the energy of the primary radiation event. Spreading of the avalanche along the total length of the anode wire is prevented. Thus the resolving time is much shorter than with Geiger-Mueller tubes (Knoll, 1989).

E. Long-Range Alpha Detectors

Traditional alpha detectors suffer from limitations related to the very short range of alpha particles in air. If sample and detector together are kept in a vacuum or sample and detector are operated in close proximity, reasonable efficiency is achieved. An alpha particle produces about 30,000 ions per 1 MeV of its particle energy (see Chapter I). These ions can be transported over significant distances by a moving stream of air to a detector. For that purpose, a current of air can be generated by a small fan and the ions can be transported over a distance of several meters. The current of air is finally monitored by an ion chamber (Garner *et al.*, 1994). By using air as the detector gas, alpha contamination on any complicated surface can be measured (MacArthur *et al.*, 1992, 1993; Allander *et al.*, 1994; Vu *et al.*, 1994). Figure 2.14 shows the principle of a long-range alpha detector. It is shown that the detector is sensitive to the ionized air molecules produced by the passage of an alpha particle rather than to the alpha particle itself. The detector consists mainly of two grids (see Fig. 2.15) across which an electric field is applied. One type of ion is attracted by the high-voltage (HV) grid, the other by the sense grid. Both possible polarities for the grids have been applied with equal sensitivity. The charge collected at the sense grid is

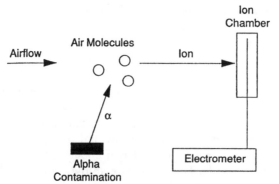

FIGURE 2.14 Principle of a long-range alpha detector operation. Ions created by alpha particles are transported to the detector by air flow. (From MacArthur *et al.*, 1992, with permission ©1992 IEEE.)

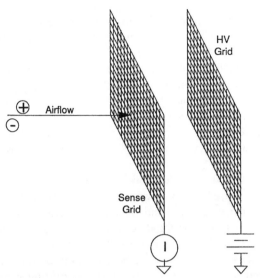

FIGURE 2.15 Construction detail of a long-range alpha detector. An electric field is applied across the grids. One type of ion is attracted by the HV grid, the other one by the sense grid. (From MacArthur et al., 1992, with permission ©1992 IEEE.)

measured by a suitable electronic circuit and used to determine the ionization, the number of alpha particles.

Instead of using air flow, the ions produced by alpha particles can be transported to the detector by an electrostatic field. At Los Alamos National Laboratory long-range alpha detectors have been built for several applications, such as monitoring of soil surface and liquid effluents. A hand monitor has also been constructed. The method was applied to radon measurements (Bolton, 1994). Some effort has been made to use the long-range detectors for the measurement of beta contamination (Johnson et al., 1994).

Real-time alpha activity monitoring is one of the applications for which the ionization detectors show several advantages. A monitoring system for real-time alpha monitoring was developed at Los Alamos and tested at the Radioactive Liquid Waste Treatment Facility as a means for real-time monitoring of liquid waste influent (Whitley et al., 1996). This system determines the alpha activity of the wastewater by measuring the ionization of ambient air above the surface at a rather long distance. The distance to the surface of the liquid described by Whitley et al. (1996) was about 4 inches. Sometimes this type of design causes problems because of changing levels of the surface to be monitored, for example, with liquids.

The ionization counting system consists of a metal enclosure and a signal plate that is maintained at 300 V DC. The box is maintained at ground potential. A highly sensitive electronic circuit is used to detect changes in current to the plate. Changes in alpha activity in the contaminated liquid at the 10 nCi/L level could be detected.

The authors claim that this kind of measurement equipment can be useful for monitoring low-level liquid streams before discharge into the environment. But a more sensitive design will be necessary to ensure regulatory compliance and offer the opportunity for field application.

Ionization monitoring has one great advantage in simultaneous measurement of the entire body of a person. Air currents which are created by high capacity fans can be drawn from all surfaces of the body of a person who is positioned in a monitoring chamber (size 91 cm × 91 cm × 213 cm). High air flow rate is essential because of the rather short "ion lifetime." Koster et al. (1998) describe a contamination measurement facility for alpha particle radionuclides, which is used at Los Alamos National Laboratory for test experiments.

Another example for the application of ionization monitoring is a portable swipe monitor, based on long-range alpha detection (Whitley et al., 1998). This facility consists of two independent detection chambers. The swipe is placed in one chamber for the detection of the alpha contamination and the other chamber records signals due to the presence of radon or other background radionuclides. The response to beta contamination is about 100 times weaker compared to the same activity of alpha-emitting radionuclides. A unit applicable to rapid field measurements is available with dimensions of 28 cm × 13 cm × 14 cm, and weight of 5 kg.

F. Liquid Ionization and Proportional Detectors

Detector materials of high density offer some advantages, particularly for the detection of radiation with low linear energy transfer and high energy. Radiation spectroscopy in many cases can be carried out much more reliably using detector materials of higher density. Consequently, research related to liquid and solid-state ionization detectors is carried out. Noble gases in the liquid or solid phase are dielectric materials where created electrons remain free if all electronegative impurities can be removed.

Among the noble gases, xenon has attracted much interest as a filling medium for ionization-type detectors, such as ion chambers and proportional counting systems. The start of the ion multiplication phenomenon is observed at a field strength of 10^8 V/m. At 10^5 V/m the electron drift velocity is about 3×10^3 m/s. Main obstacles to the construction of such detectors are the requirements for operation at a low temperature and for extensive purification of detector medium. Liquid xenon ionization chambers compared with sodium iodide (NaI) detectors have a similar gamma efficiency and a higher energy resolution (L'Annunziata, 1987). Of course, the energy resolution of semiconductor gamma detectors is still better.

The size of useful liquid or solid noble gas ionization detectors depends on the purity of the filling material. Position sensing by large detectors can be carried out by measuring the electron drift time. Gridded versions of such ion chamber detectors have also been reported.

Liquid ionization chambers (Ar, Xe) are frequently used in basic nuclear physics, e.g., for the search for weakly interacting massive particles (WIMPs), e.g., the neutrinos predicted by supersymmetric theories (Ovchinnikov and Parusov, 1999).

Some information is available related to nonpolar liquids as ionization detectors at room temperature. Here, the purity that can be achieved and maintained for the applied material is extremely important. Research has been carried out using, for example, tetramethylsilane. This material was used for ion chambers working in pulse and current mode (Knoll, 1989).

G. Dynamic Random Access Memory Devices (DRAM)

Soft errors are induced to dynamic random access memory devices (DRAM) and therefore Chou *et al.* (1997) studied their use as radiation detectors. Samples of DRAMs from several manufacturers, just off the shelf products, have been used for that study. Memory content of the DRAM was reset, and then after irradiation, the number of flipped cells was determined. Once counted, the memory content is reset again. Experimental results using alpha particle radiation indicate that the soft error is linearly related to irradiation time as well as the radiation source intensity. This linearity could not be obtained with gamma radiation. Nevertheless, it can be assumed that high density DRAMs may be promising counters for charged particle detections. They could also be used for the counting of neutrons if the DRAMs are coated with a layer of neutron sensitive materials.

REFERENCES

Äikää, O., Mäntynen, P., and Kankainen, T. (1992). High performance ^{14}C gas proportional counting system applying pulse shape discrimination. *Radiocarbon* **34**, 414–419.

Alkhazov, G. D., Komar, A. P., and Vorobev, A. A. (1967). Ionization fluctuation and resolution of ionization chambers and semiconductor detectors. *Nucl. Instrum. Methods* **48**, 1–12.

Allander, K. S., Bounds, J. A., and Mac Arthur, D. W. (1994). Application of the long-range alpha detector (LRAD) to the detection of natural-occurring radioactive materials (NORM). Proc. WM '94 Conf., Tucson, AZ.

Alteri, E. (1992). Measurement of reverse transcriptase activity of an HIV-1 virus stock prepared in A 3.01 lymphblastoid cells. *Matrix Application Note*, PerkinElmer Life and Analytical Sciences, Boston, MA.

Altzitzoglou, T., Denecke, B., Johansson, L., and Sibbens, G. (2002). Standardisation of ^{89}Sr using different methods. *Appl. Radiat. Isot.* **56**, 447–452.

Amrani, D., Cherouati, D. E., and Cherchali, M. E. H. (2000). Groundwater radon measurements in Algeria. *J. Environ. Radioact.* **51**, 173–180.

Aoyama, T. (1990). A tritium in-air monitor with compensation and additional recording of α, β and γ-backgrounds. *IEEE Trans. Nucl. Sci.* **37**, 885–991.

Assaf, J. (2002). High voltages influence on the response of two-stage GEM detector. *Radiat. Meas.* **35**, 7–12.

Babichev, E. A., Baru, S. E., Groshev, V. R., Khabakhpashev, A. G., Krainov, G. S., Leonov, V. V., Neustroev, V. A., Porosev, V. V., Savinov, G. A., and Shekhtman, L. I. (2001). Usage of two types of high-pressure xenon chambers for medical radiography. *Nucl. Instrum. Methods Phys. Res., Sect. A* **461**, 430–434.

Bacchi, O. O., Reichardt, K., Calvache, M., Nielsen, D., Vachaud, G., Eaglesham, A., Chalk, P. M., Urquiaga, S., Zapata, F., Laurent, J.-P., Thony, J. L., Vauchlin, M., and Moutonnet, P. (2002). "Neutron and Gamma Probes: Their Use in Agronomy." International Atomic Energy Agency Training Course Series No. 16 (IAEA-TCS-16), pp. 75, IAEA, Vienna.

Barbosa, A. F., Riekel, C., and Wattecamps, P. (1992). Two dimensional x-ray detector based on microstrip and multiwire design. *Nucl. Instrum. Methods, Sect. A* **323**, 247–251.

Barr, A., Bonaldi, L., Carugno, G., Charpak, G., Iannuzzi, D., Nicoletto, M., Pepato, A., and Ventura, S. (2002). A high-speed, pressurised multi-wire gamma camera for dynamic imaging in nuclear medicine. *Nucl. Instrum. Methods Phys. Res., Sect. A* **477**, 499–504.

Bateman, J. E. (1994). Fundamentals of gas counters. *Med. Radiat. Detect.* **1994**, 89–99.

Bateman, J. E., Sore, J., Knight, S. C., and Bedford, P. (1994). A new gas counter for radio-immunoassay. *Nucl. Instrum. Methods Phys. Res., Sect. A* **348**, 288–292.

Beddingfield, D. H., Menlove, H. O., and Johnson N. H. (1999). Neutron proportional counter design for high gamma-ray environments. *Nucl. Instrum. Methods Phys. Res., Sect. A* **422**, 35–40.

Beddingfield, D. H., Johnson, N. H., and Menlove, H. O. (2000). ^{3}He neutron proportional counter performance in high gamma-ray dose environments. *Nucl. Instrum. Methods Phys. Res., Sect. A* **455**, 670–682.

Bellazzini, R., Bozzo, M., Brez, A., Gariano, G., Latronico, L., Lumb, N., Massai, M. M., Papanestis, A., Raffo, R., Spandre, G., and Spezziga, M. A. (1999). A two-stage, high gain micro-strip detector. *Nucl. Instrum. Methods Phys. Res., Sect. A* **425**, 218–227.

Bellazini, R., Bozzo, M., Brez, A., Gariano, G., Latronico, L., Lumb, N., Papanestis, A., Spandre, G., Massa, M. M., Raffo, R., and Spezziga, M. A. (1999). The WELL detector. *Nucl. Instrum. Methods Phys. Res., Sect. A* **423**, 125–134.

Bellazini, R., Brez, A., Gariano, G., Latronico, L., Lumb, N., Moggi, A., Reale, S., Spandre, G., Massai, M. M., Spezziga, M. A., Toropin, A., Costa, E., Soffitta, P., and Pacella, D. (2001). Micropattern gas detectors: the CMS MSGC project and gaseous pixel detector application. *Nucl. Instrum. Methods Phys. Res., Sect. A* **471**, 41–54.

Bellazzini, R., Spandre, G., and Lumb, N. (2002). Progress with micro-pattern gas detectors. *Nucl. Instrum. Methods Phys. Res., Sect. A* **478**, 13–25.

Bertolini, G. (1984). Alpha particle spectroscopy by gridded ionization chamber. *Nucl. Instrum. Methods* **223**, 285–289.

Bigu, J. (1992). Design and operation of an automated beta-particle counting system for the measurement of ^{220}Rn (and ^{222}Rn) progeny. *Appl. Radiat. Isot.* **43**, 443–448.

Böck, H. and Balcar, E. (1975). Long-time behaviour of regenerative in-core neutron detectors with ^{238}U-^{239}Pu electrodes during power cycling. *Nucl. Instrum. Methods* **124**, 563–571.

Bolton, R. D. (1994). Radon monitoring using long-range alpha detector-based technology. Nuclear Science Symposium, Norfolk, VA.

Bouclier, R., Florent, J. J., Gaudaen, J., Million, G., Ropelewski, L., and Sauli, F. (1992a). Microstrip gas chambers on thin plastic supports. *IEEE Trans. Nucl. Sci.* **39**, 650–653.

Bouclier, R., Florent, J. J., Gaudaen, J., Sauli, F., and Shekhtman, L. (1992b). Development of microstrip gas chambers on thin plastic supports. *Nucl. Instrum. Methods Phys. Res., Sect. A* **315**, 521–528.

Bouclier, R., Florent, J. J., Gaudaen, J., Millon, G., Pasta, A., Ropelewski, L., Sauli, F., and Shekhtman, L. (1992c). High flux operation of microstrip gas chambers on glass and plastic supports. *Nucl. Instrum. Methods Phys Res., Sect. A* **323**, 240–246.

Bouclier, R., Capeans, M., Evans, J., Garabatos, C., Manzin, G., Million, G., Ropelewski, L., Sauli, F., Shekhtman, L. I., Temmel, T., and Fischer, G. (1995). Optimization of design and beam test of microstrip gas chambers. *Nucl. Instrum. Methods Phys. Res., Sect. A* **367**, 163–167.

Braby, L. A. and Badhwar, G. D. (2001) Proportional counter as neutron detector. *Radiat. Meas.* **33**, 265–267.

Breskin, A. (2000). Advances in gas avalanche radiation detectors for biomedical applications. *Nucl. Instrum. Methods Phys. Res. Sect. A* **454**, 26–39.

Bronic, I. K. and Grosswendt, B. (2001). Experimental study of gas mixtures in strong non-uniform electric fields. *Radiat. Phys. Chem.* **61**, 477–478.

Charpak, G. (1970). Evolution of the automatic spark chambers. *An. Rev. Nucl. Sci.* **20**, 195–254.

Charpak, G. and Sauli, F. (1978). Multiwire proportional chambers and drift chambers. *Nucl. Instrum. Methods* **162**, 405–428.

Chase, G. and Rabinowitz, J. (1967). Principles of Radioisotope Methodology. *Burgess Publishing Company, Minneapolis.*

Chou, H. P., Chou, T. C., and Hau, T. H. (1997). Evaluation of high density DRAMs as a nuclear radiation detector. *Appl. Radiat. Isot.* **48**, 1601–1604.

Christophel, E., Dracos, M., and Strub, R. (1998). The 2D-microgap wire chamber. *Nucl. Instrum. Methods Phys. Res., Sect. A* **419**, 515–518.

Colmenares, C. A. (1974). Bakeable ionization chamber for low-level tritium counting. *Nucl Instrum. Methods Phys. Res.* **150**, 549.

East, L. V. and Walton, R. B. (1969). Polythene moderated ^{3}He neutron detectors. *Nucl Instrum. Methods Phys. Res.* 161–166.

Cosma, C. (2000). Strontium-90 measurement after the Chernobyl accident in Romanian samples without chemical separation. *Spectrochimica Acta Part B* **55**, 1165–1171.

Domnikov, V. N., Saltykov, l. S., Slusarenko, L. I., and Shevchenko, S. V. (2001). About the effectiveness of spectrometry in alpha-activity monitoring of industrial air-borne particles. *Appl. Radiat. Isot.* **55**, 543–547.

Evans, A. E., Jr. (1982). Energy dependence of the response of a ^{3}He long counter. *Nucl. Instrum. Methods* **199**, 643–644.

Facorellis, Y., Kyparissi-Apostolika, N., and Maniatis, Y. (2001). The cave of Theopatra, Kalambaka: Radiocarbon evidence for 50,000 years of human presence. *Radiocarbon* **43**, 1029–1048.

Fadil, M., Blandin, Ch., Christophe, S., Déruelle, O., Fioni, G., Marie, F., Mounier, C., Ridikas, D., and Trapp, J. P. (2002). Development of fission micro-chambers for nuclear waste incineration studies. *Nucl. Instrum. Methods Phys. Res. Sect. A* **476**, 313–317.

Fischer, B. E. (1977). A digital processor for position sensitive detectors. *Nucl. Instrum. Methods* **41**, 173–181.

Fourme, R. (1997). Position sensitive gas detectors: MEPCs and their gifted descendants. *Nucl. Instrum. Methods Phys. Res., Sect. A* **392**, 1–11.

Fraga, F. A. F., Fetal, S. T. G., Ferreira Maqes, R., and Policarpo, A. P. L. (2000). Quality control of GEM detectors using scintillation techniques, *Nucl. Instrum. Methods Phys. Res. Sect. A* **442**, 417–422.

Fraga, F. A. F., Margato, L. M. S., Fetal, S. T. G., Fraga, M. M., F. R., Maques, R. F, Policarpo, A. J. P. L., Guerard, B., Oed, A., Manzini, G., and van Vuure, T. (2002). CCD readout of GEM-based detectors. *Nucl. Instrum. Methods Phys. Res., Sect. A* **478**, 357–361.

Fried, J., Harder, J. A., Mahler, G. J., Makowiecki, D. S., Mead, J. A., Radeka, V., Schaknowski, N. A., Smith, G. C., and Yu, B. (2002). A large, high performance, curved 2D position-sensitive neutron detector. *Nucl. Instrum. Methods Phys. Res., Sect. A* **478**, 415–419.

Fulbright, H. W. (1979). Ionization chambers. *Nucl. Instrum. Methods* **62**, 21–28.

García-León, M., García-Montaño, E., and Madurga, G (1984). Characterization of ^{99}Tc by shape of its plateau with a gas-flow proportional counter. *Int. J. Appl. Radiat. Isot.* **35**, 195–200.

García-Toraño, E., Barquerro, R. L., and Roteta, M. (2002) Standardization of ^{134}Cs by three methods. *Appl. Radiat. Isot.* **56**, 211–214.

Gardner, R. P. and Liu, L. (1997). On extending the accurate and useful counting rate of GM counter detector systems. *Appl. Radiat. Isot.* **48**, 1605–1615.

Garner, S. E., Bounds, J. A., Allander, K. S., Caress, R. W., Johnson, J. D., and MacArthur, D. W. (1994). A compendium of results from long-range alpha detector soil surface monitoring: June 1992–May 1994. LA-12861-MS. Los Alamos National Laboratory Document.

Geltenbort, P. (1994). Recent results with microstrip gas chambers. *Nucl. Instrum. Methods Phys. Res., Sect. A* **353**, 168.

Glesius, F. L. and Kniss, T. A. (1988). He-3 neutron detectors for hostile environments. *IEEE Trans. Nucl. Sci.* **35**, 867–871.

Golovatyuk, V. and Grancagnolo, F. (1999). Observation of transition between proportional and Geiger-Müller modes in helium-isobutane gas mixtures. *Nucl. Instrum. Methods Phys. Res., Sect. A* **428**, 367–371.

Goulianos, K., Smith, K. K., and White, S. N. (1980). *Anal. Biochem.* **103**, 64–69.

Gruhn, C. R., Binimi, M., Legrain, R., Loveman, R., Pang, W., Loach, M., Scott, D. K., Shotter, A., Symons, T. J., Wouters, J., Zismon, M., and Devier, R. (1982). Bragg curve spectroscopy. *Nucl. Instrum. Methods* **196**, 33–40.

Hankins, D. E. and Thorngate, J. H. (1993a). A high sensitive neutron counter and waste drum counting with the high sensitivity neutron instrument. UCRL-ID-111750. Document of the Lawrence Livemore National Laboratoy, University of California.

Hankins, D. E. and Thorngate, J. H. (1993b). A neutron counting instrument for low level transuranic waste. Document of the Lawrence Livermore National Laboratoy, University of California, UCRL-ID-115887.

Hayakawa, Y. and Maeda Y. (1994). Performance of a microstirp proportional counter. *Annu. Rep. Res. Reactor Inst. Kyoto Univ.* **27**, 34–41.

Hayakawa, Y. and Maeda Y. (1996). Microstrip gas chamber for x-rays and neutrons. *Jpn. J. Appl. Phys.* **35**, 123–125.

Hillman, G. G., Roessler, N., Fulbright, R. S., Edson Pontes, J., and Haas, G. P. (1993a). Application of the direct beta counter Matrix 96 for cytotoxic assays: simultaneous processing and reading of 96 wells using a ^{51}Cr retention assay. *Cancer Immunol. Immunother* **36**, 351–356.

Hillman, G. G., Roessler, N., Fulbright, R. S., Edson Pontes, J., and Haas, G. P. (1993b). [51]Cr release assay adapted to a 96-well format sample reading. *Biotechniques* **15**, 744–749.

Hoetzl, H. and Winkler, R. (1984). Experience with large-area Frisch grid chambers in low-level alpha spectrometry. *Nucl. Instrum. Methods* **223**, 290–295.

Horikawa, S., Inaba, S., Kawai, H., Matsumoto, T., Nakayama, H., Tajima, Y., Takamatsu, K., Tsurju, T., and Yoshida, H. Y. (2002). Development of micro-gap wire chamber. *Nucl. Instrum. Methods Phys. Res., Sect. A* **481**, 166–173.

Hunt, J. B. and Mercer, R. A. (1978). The absolute calibration of a long counter by the associated activity techniques. *Nucl. Instrum. Methods* **156**, 451–457.

Hutchins, D. (1992). The bioassay of Cytokines. *Matrix Application Note*, PerkinElmer Life and Analytical Sciences, Boston, MA.

Iguchi, T., Nakayamada, N., Takahashi, H., and Nakazawa, M. (1994). Neutron spectrometry using a [3]He gas ionization chamber. *Nucl. Instrum. Methods Phys. Res., Sect. A* **353**, 152–155.

Jahoda, K. and McCammon, D. (1988). Proportional counters as low energy photon detectors. *Nucl. Instrum. Methods Phys. Res., Sect. A* **272**, 800–813.

Jalbert, R. A. and Hiebert, R. D. (1971). Gamma insensitive air monitor for radioactive gases. *Nucl. Instrum. Methods* **96**, 61–66.

Järvinen, M. L. and Sipilä, H. (1984). Improved proportional counters for practical application. *IEEE Trans. Nucl. Sci.*, NS-31, 356–359.

Johansson, L., Sibbens, G., Altzitzoglou, T., and Denecke, B. (2002). Self-absorption correction in standardisation of [204]Tl. *Appl. Radiat. Isot.* **56**, 199–203.

Johnson, J. D., Allander, K. S., Bounds, J. A., Garner, S. E., Johnson, J. P., and MacArthur, D. W. (1994). Long range alpha detector (LRAD) sensitivity to beta contamination and soil moisture. *IEEE Trans. Nucl. Sci.* **41**, 755–757.

Jones, A. R. and Holford, R. M. (1981). Application of Geiger-Mueller counters over a wide range of counting rates. *Nucl. Instrum. Methods* **189**, 503–509.

Kessler, M. J. (1991). A new, rapid analysis technique for quantitation of radioactive samples isolated on a solid support. Proceedings of the International Conference on New Trends in Liquid Scintillation Counting and Organic Scintillators 1989, Lewis Publishers, Chelsea, MI.

Khriachkov, V. A., Ketlerov, V. V., Mitrofanov, V. F., and Semenova, N. N. (2000). Low-background spectrometer for the study of fast neutron-induced (n,α) reactions. *Nucl. Instrum. Methods Phys. Res., Sect. A* **444**, 614–621.

Knoll, G. F. (1989). "Radiation Detection and Measurement," John Wiley & Sons, New York.

Koster, J. E., Bounds, J. A., Kerr, P. L., Steadman, P. A., and Whitley, C. R. (1998). Whole body personnel monitoring via ionization detection. *IEEE Trans. Nucl. Sci.* **45**, 976–980.

Kotte, R., Keller, H. J., Ortlepp H. G., and Strary, F. (1987). Bragg peak spectroscopy of low-energy heavy ions. *Nucl. Instrum. Methods Phys. Res., Sect. A* **257**, 244–252.

Kuzminov, V. V. and Osetrova, N. J. (2000). Precise measurement of ^{14}C beta Spectrum by using a wall-less proportional counter. *Phys. At. Nucl.* **63**, 1292–1296.

L'Annunziata, M. F. (1987). "Radionuclide Tracers, Their Detection and Measurement," Academic Press, New York.

O'Leary, G. J. and Incerti, M. (1993) A field comparison of three neutron moisture meters. *Australian Journal of Experimental Agriculture* 33, 59–69.

Lee, S. H. and Gardner, R. P. (2000). A new G-M counter dead time model. *Appl. Radiat. Isot.* 53, 731–737.

Lentsch, J. W., Babad, H., Stokes, T. I., Hanson, C. E., Vargo, G. F., and Boechler, G. N. (1996). New instruments for characterization of high level waste storage tanks at the Hanford site. Presented at WM'96, Tucson, AZ.

Loaiza, D. J. (1999). High-efficiency ^{3}He proportional counter for the detection of delayed neutrons. *Nucl. Instrum. Methods Phys. Res., Sect. A* 422, 43–46.

Lohmann, M., Besch, H. J., Dix, W. R., Dünger, O., Jung, M., Menk, R. H., Reime, B., and Schildwächter, L. (1998). A high sensitive two-line detector with large dynamic range for intravenous coronary angiography. *Nucl. Instrum. Methods Phys. Res., Sect. A* 419, 276–283.

Lopes, J. A. M., dos Santos, J. M. F., Conde, C. A. N., and Morgado, R. E. (1999). A new integrated photosensor for gas proportional scintillation counters based on the gas electron multiplier (GEM). *Nucl. Instrum. Methods Phys. Res., Sect. A* 426, 469–476.

Lopes, J. A. M., dos Santos, J. M. F., and Conde, C. A. N. (2000). A large area avalanche photodiode as the VUV photosensor for gas proportional scintilllation counters. *Nucl. Instrum. Methods Phys. Res., Sect. A* 454, 421–425.

Lopes, J. A. M., dos Santos, J. M. F., Morgado, R. E., and Conde, C. A. N. (2001). A Xenon gas proportional scintillation counter with a UV-sensitive large-area avalanche photodiode. *IEEE Trans. Nucl. Sci.* 48, 312–319.

Lyoussi, A., Romeyer-Dherey, J., and Buisson, A. (1996). Low level transuranic waste assay system using sequential photon interrogation and on line neutron counting signatures. Presented at WM'96, Tucson, AZ.

MacArthur, D. W., Allander, K. S., Bounds, J. A., Catlett, M. M., and Mcatee, J. L. (1992). Long-range alpha detector (LRAD) for contamination monitoring. *IEEE Trans. Nucl. Sci.* 39, 952–957.

MacArthur, D. W., Allander, K. S., Bounds, J. A., Caress, R. W., Catlett, M. M., and Rutherford, D. A. (1993). LRAD Surface Monitors. *LA-12524-MS. Los Alamos National Laboratory Document*.

Maekawa, F. and Oyama, Y. (1995a). Neutron spectrum measurement in the energy region of eV with the slowing down time method. In Proceedings of the 9th Workshop on Radiation Detectors and Their Uses (M. Miyajima, S. Sasaki, T. Iguchi, N. Nakazawa, and M. Takebe, Eds.) Nat. Inst. Phys. KEK Japan.

Maekawa, F. and Oyama, Y. (1995b). Measurement of low energy neutron spectrum below 10 keV with the slowing down time method. *Nucl. Instrum. Methods Phys. Res., Sect. A* 372, 262–274.

Maekawa F. and Oyama, Y. (1997). Measurement of neutron energy spectrum below 10 keV in an iron shield bombarded by deuterium tritium neutrons and benchmark test of evaluated nuclear data from 14 MeV to 1 eV. *Nucl. Sci. Eng.* 125, 205–217.

Mäntynen, P., Äikää, O., Kankainen, T., and Kaihola, L. (1987). Application of pulse shape discrimination to improve the precision of the carbon-14 gas proportional counting method. *Applic. Radiat. Isot.* 38, 869–873.

Mateos, J. J., Gomez, E., Garcias, F., Casas, M., and Cerdá, V. (2000). Rapid ^{90}Sr/^{90}Y determination in water samples using a sequential injection method. *Appl. Radiat. Isot.* 53, 139–144.

Menk, R. H., Sarvestani, A., Besch, H. J., Walenta, A. H., Amenitsch, H., and Bernstorff (2000). Gas gain operations with single photon resolution using an integrating ionization chamber in small angle x-ray scattering experiment. *Nucl. Instrum. Methods Phys. Res., Sect. A* 440, 181–190.

Monteiro, C. M. B., Lopes, J. A. M., Simoes, P. C. P. S., dos Santos, J. M. F., and Conde, C. A. N. (2001). An argon gas proportional scintillation counter with UV avalanche photodiode scintillation readout. *IEEE Trans. Nucl. Sci.* 48, 1081–1086.

Mustafa, S. M. and Mahesh, K. (1978). Criterion for determining saturation current in parallel plate ionization chambers. *Nucl. Instrum. Methods* 150, 549–553.

Neumann, S., Böttger, R., Guldbakke, S., Matzke, M., and Sosaat, W. (2002). Neutron and photon spectrometry in mono energetic neutron fields. *Nucl. Instrum. Methods Phys. Res., Sect. A* 476, 353–357.

Nickles, J., Bräuning, H., Bräuning-Demian, A., Dangendorf, V., Breskin, A., Chechik, R., Rauschnabel, K., and Schmidt Böcking, H. (2002). A gas scintillation counter with imaging optics and large area UV-detector. *Nucl. Instrum. Methods Phys. Res., Sect. A* **477**, 59–63.

Nielsen, D. R. and Cassel, D. K. (1984). Soil water management, *In* "Isotopes and Radiation in Agricultural Sciences" (M. F. L'Annunziata and L. O. Legg Eds.) Vol. 1, pp. 35–65, Academic Press, London and New York.

Nowack, G. F. (1987). Electrical compensation method of mechanically induced disturbances in gas-filled radiation detectors. *Nucl. Instrum. Methods Phys. Res., Sect. A* **255**, 217–221.

Oed, A. (1995). Properties of micrso-strip gas chambers (MSGC) and recent developments. *Nucl. Instrum. Methods Phys. Res., Sect. A* **367**, 34–40.

Ohsawa, D., Masaoka, S., Katano, R., and Isozumi, Y. (2000). Resolution of a position sensitive proportional counter with a resistive anode wire of carbon fiber. *Appl. Radiat. Isot.* **52**, 943–954.

Ortuño-Prados, F., Bazzano, A., Berry, A., Budtz-Jørgensen, C., Hall, C., Helsby, W., Lewis, R., Parker, B., and Ubertini, P. (1999). A high-pressure MWPC detector for crystallography. *Nucl. Instrum. Methods Phys. Res., Sect. A* **420**, 445–452.

Ovchinnikov, B. M. and Parusov, V. V. (1999). A method for background reduction in an experiment for WIMP search with a Xe (Ar)-liquid ionization chamber. *Astroparticle Physics* **10**, 129–132.

PerkinElmer (1992). High throughput screening of samples containing alpha & beta radionuclides: an overview of methods. *Application Note*, PerkinElmer Life and Analytical Sciences, Boston, MA.

Pallares, A., Barthe, S., Bergtold, A. M., Brom, J. M., Cailleret, J., Christophel, E., Coffin, J., Eberle, H., Fang, R., Fontaine, J. C., Geist, W., Kachelhoffer, T., Levy, J. M., Mack, V., Schunck, J. P., and Sigward, M. H. (1995). Microstrip gas chambers on implanted substrates. *Nucl. Instrum. Methods Phys. Res., Sect. A* **367**, 185–188.

Papp, Z. and Uray, I. (2002). Sensitive method for the determination of ^{18}F attached to aerosol particles in a PET centre. *Nucl. Instrum. Methods Phys. Res., Sect. A* **480**, 788–796.

Passo, C. and Kessler, M. (1992). "The essentials of alpha/beta discrimination," PerkinElmer Life and Analytical Sciences, Boston, MA.

Pitts, W. K. and Martin, M. D. (2001). Experience with laser microfabricated detectors at the University of Louisville. *Nucl. Instrum. Methods Phys. Res. Sect. A* **471**, 268–271.

Prasad, K. R. and Balagi, V. (1996). Uranium-233 fission detectors for neutron flux measurement in reactors. *Rev. Sci. Instrum.* **67**, 2197–2201.

Rachinhas, P. J. B. M., Simoes, P. C. P. S., Lopes, J. A. M., Dias, T. H. V. T., Morgado, R. E., dos Santos, J. M. F., Stauffer, A. D., and Conde, C. A. N. (2000). Simulation and experimental results for the detection of conversion electrons with gas proportional scintillation counters. *Nucl. Instrum. Methods Phys. Res., Sect. A* **441**, 468–478.

Radeka, V., Schaknowski, N. A., Smith, G. C., and Yu, B. (1998). High performance, imaging, thermal neutron detectors. *Nucl. Instrum. Methods Phys. Res., Sect. A* **419**, 642–647.

Rodríguez, P. B., Sánchez, A. M., and Tomé, F. V. (1997). Experimental studies of self-absorption and backscattering in alpha-particle sources. *Appl. Radiat. Isot.* **48**, 1215–1220.

Roessler, N., Englert, D., and Neumann K. (1993). New instruments for high throughput receptor binding assays, *J. Receptor Res.* **13**, 135–145.

Rosner, G., Hötzl, H., and Winkler, R. (1992). Determination of [241]Pu by low level beta proportional counting, application to Chernobyl fallout samples and comparison with the [241]Am build-up method. *J. Radioanal. Nucl. Chem.* **163**, 225–233.

Saito, N. and Suzuki, I. H. (1999). Absolute fluence rates of soft x-rays using a double ion chamber. *J. Electron Spectrosc. Relat. Phenom.* **101–103**, 33–37.

Sakamoto, S. and Morioka, A. (1993). Pulse shape discrimination with proportional counters for neutron detection. Proceedings of the Seventh Workshop on Radiation Detectors and Their Uses (M. Miyajima, S. Sasaki, Y. Yoshimura, T. Iguchi, and N. Nakazawa Eds.) *Nat. Inst. Phys. KEK Japan.*

Sakamoto, S. and Morioka, A. (1994). Temperature dependence of BF_3 proportional counters. *Nucl. Instrum. Methods Phys. Res., Sect. A* **353**, 160–163.

Santos, J. M. F., Dias, T. H. V. T., Reyes Cortes, S. D. A., and Conde, C. A. N. (1989). Novel techniques for designing gas proportional scintillation counters for x-ray spectrometry. *Nucl. Instrum. Methods Phys. Res., Sect. A* **280**, 288–290.

dos Santos, J. M. F., Lopes, J. A. M., Veloso, J. F. C. A., Simoes, P. C. P. S., Dias, T. H. V. T., dos Santos, F. P., Rachinhas, P. J. B. M., Ferreira, L. F. R., and Conde, C. A. N. (2001). Development of portable gas proportional scintillation counters for x-ray spectrometry. *X-Ray spectrom.* **30**, 373–381.

Sauli, F. (1997). GEM: a new concept for electron amplification in gas detectors. *Nucl. Instrum. Methods Phys. Res., Sect. A* **386**, 531–534.

Sauli, F. (1998). Gas detectors: Recent developments and future perspectives. *Nucl. Instrum. Methods Phys. Res., Sect. A* **419**, 189–201.

Sauli, F. (1999). Recent developments and applications of fast position-sensitive gas detectors. *Nucl. Instrum. Methods Phys. Res. Sect. A* **422**, 257–262.

Sauli, F. (2001). Gas detectors: Achievements and trends. *Nucl. Instrum. Methods Phys. Res., Sect.A* **461**,47–54.

Sauli, F. (2002). Micro-pattern gas detectors. *Nucl. Instrum. Methods Phys. Res., Sect. A* **477**, 1–7.

Scarpitta, S., Odin-McCabe, J., Gaschott, R., Meier, A., and Klug, E. (1999). Comparison of four ^{90}Sr groundwater analytical methods. *Health Phys.* **76**, 644–656.

Semkow, T. M. and Parekh, P. P. (2001). Principles of gross alpha and beta radioactivity detection in water. *Health Phys.* **81**, 567–573.

Shenhav, N. J. and Stelzer, H. (1985). The mass dependence of the signal peak height of a Bragg-curve. *Nucl. Instrum. Methods* **228**, 359–364.

Simoes, P. C. P. S., dos Santos, J. M. F., and Conde, C. A. N. (2001). Driftless gas proportional scintillation counter pulse analysis using digital processing techniques. *X-ray Spectrom.* **30**, 342–347.

Stamoulis, K. C., Assimakopoulos, P. A., Ioannides, K. G., Johnson, E., and Soucaco, P. N. (1999). Strontium-90 concentration measurements in human bones and teeth in Greece. *Sci. Total Environ.* **229**, 165–182.

Stanga,.D., Picolo, J. L., Coursol, N., Mitev, K., and Moreau, I. (2002). Analytical calculations of counting losses in internal gas proportional counting. *Appl. Radiat. Isot.* **56**, 231–236.

Strand, P., Brown, J. E., Drozhko, E., Mokrov, Y., Salbu, B., Oughton, D., Christensen, G. C., and Amundsen, I. (1999). Biogeochemical behaviour of ^{137}Cs and ^{90}Sr in the artificial reservoirs of Mayak PA, Russia. *Sci. Total Environ.* **241**, 107–116.

Surette, R. A., and Waker, A. J. (1994). Workplace monitoring of swipes and air filters for ^{55}Fe. *IEEE Trans. Nucl. Sci.* **41**, 1374–1378.

Szalóki, I., Török, S. B., Ro, C. U., Injuk, J., and Van Grieken, R. E. (2000). X-ray Spectrometry. *Anal. Chem.* **72**, 211–233.

Tamboul, J. Y. and Watt, D. E. (2001). A proportional counter for measurement of the bio-effectiveness of ionising radiations at the DNA level. *Nucl. Instrum. Methods Phys. Res., Sect. B* **184**, 597–608.

Theodórsson, P. (1991a). Gas proportional versus liquid scintillation counting, radiometric versus AMS dating. *Radiocarbon* **33**, 9–13.

Theodórsson, P. and Heusser, G. (1991). External guard counters for low-level counting systems. *Nucl. Instrum. Methods Phys. Res., Sect. B* **53**, 97–100.

Theodórsson, P. (1999). A review of low-level tritium system and sensitivity requirements. *Appl. Radiat. Isot.* **50**, 311–316.

Tieh-Chi Chu, Jeng-Jong Wang, and Yu-Ming Lin (1998). Radiostrontium analytical method using crown-ether compound and Cerenkov counting and its application in environmental monitoring. *Appl. Radiat. Isot.* **49**, 1671–1675.

Toyokawa, H., Urotani, A., Mori, C., Takeda, N., and Kudo, K. (1995). A multipurpose spherical neutron counter. *IEEE Trans. Nucl. Sci.* **42**, 644–648.

Troja, S. O., Cro, A., and Picouet, P. (1995). Alpha and beta dose-rate determination using a gas proportional counter. *Radiat. Meas.* **24**, 297–308.

Tutin, G. A., Ryzhov, I. V., Eismont, V. P., Kireev, A. V., Condé, H., Elmgren, K., Olsson, N., and Renberg, P. U. (2001). An ionization chamber with Frisch grids for studies of high-energy neutron-induced fission. *Nucl. Instrum. Methods Phys. Res., Sect. A* **457**, 646–652.

Unterweger, M. P. and Lucas, L. L. (2000). Calibration of the National Institute of Standards and Technology tritiated-water standards. *Appl. Radiat. Isot.* **52**, 527–431.

Uozumi, Y., Sakae, T., Matoba, M., Ijiri, H., and Koori, N. (1993). Semi-microscopic formula for gas gain of proportional counters. *Nucl. Instrum. Methods Phys. Res., Sect. A* **324**, 558–564.

Uritani, A., Kuniya, Y., Takenaka, Y., Toyokawa, H., Yamane, Y., Mori, S., Kobayashi, K., Shiroya, S., and Ichihara, C. (1995). A long and slender position-sensitive helium-3 proportional counter with an anode wire supported by a ladder shaped solid insulator. *J. Nucl. Sci. Technol.* **32**, 719–726.

Vaca, F., Manjón, G., Cuéllar, S., and Garcia-Leon, M. (2001). Factor of merit and minimum detectable activity for ^{90}Sr determinations by gas-flow proportional counting or Cherenkov counting. *Appl. Radiat. Isot.* **55**, 849–851.

Veloso, J. F. C. A., dos Santos, J. M. F., and Conde, C. A. N. (2000). A proposed new microstructure for gas radiation detectors: The microhole and strip plate. *Rev. Si. instrum.* **71**, 2371–2376.

Veloso, J. F. C. A., dos Santos, J. M. F., and Conde, C. A. N. (2001). Gas proportional scintillation counters with a CsI-covered microstrip plate UV photosensor for high-resolution x-ray spectrometry. *Nucl. Instrum. Methods Phys. Res., Sect. A* **457**, 253–261.

Vinagre, F. L. R. and Conde, C. A. N. (2001). Method for effective dead time measurement in counting systems. *Nucl. Instrum. Methods Phys. Res., Sect. A* **462**, 555–560.

Vu, T. Q., Allander, K. S., Bolton, R. D., Bounds, J. A., Garner, S. E., Johnson, J. D., Johnson, J. P., and MacArthur, D. W. (1994). Application of the long-range alpha detector for site-characterization technology. *Proceedings of the WM '94 Conference. Tucson, AZ.*

Waters, J. R. (1974). Precautions in the measurement of tritium concentration in air when using flow-through chambers. *Nucl. Instrum. Methods* **117**, 39–43.

Westphal, G. P. (1976). A high precision pulse-ratio circuit. *Nucl. Instrum. Methods* **134**, 387–390.

Whitley, C. R., Johnson, J. D., and Rawool-Sullivan, M. (1996). Real-time alpha monitoring of a radioactive liquid waste stream at Los Alamos National Laboratorty. WM 1996, Tucson, AZ.

Whitley, C. R., Bounds, J. A., and Steadman, P. A. (1998). A portable swipe monitor for alpha contamination. *IEEE Trans. Nucl. Sci.* **45**, 533–535.

Yu, B., Smith, G. C., Siddons, D. P., Pietraski, P. J., and Zojceski, Z. (1999). Position sensitive gas proportional detectors with anode blades. *IEEE Trans. Nucl. Sci.* **46**, 338–431.

Zhang, L., Takahashi, H., Hinamoto, N., Nakazawa, M., and Yoshida, K. (2002). Design of a hybrid gas proportional counter with CdTe guard counters for ^{14}C dating system. *Nucl. Instrum. Methods Phys. Res., Sect. A* **478**, 431–434.

Zikovsky, L. and Roireau, N. (1990). Determination of radon in water by argon purging and alpha counting with a proportional counter. *Appl. Radiat. Isot.* **41**, 679–681.

3

SOLID STATE NUCLEAR TRACK DETECTORS

RADOMIR ILIĆ
Faculty of Civil Engineering, University of Maribor, Smetanova 17, 2000 Maribor, Slovenia; and Jožef Stefan Institute, Jamova 39, 1000 Ljubljana, Slovenia

SAEED A. DURRANI
School of Physics and Astronomy, University of Birmingham, Birmingham B15 2TT, UK

I. INTRODUCTION

Since its discovery in 1958 (Young, 1958; Silk and Barnes, 1959), the technique now generally known as Solid State Nuclear Track Detection (SSNTD) has, over the last few decades, become a popular and well-established method of measurement in a large number of fields involving different aspects of radioactivity or nuclear interactions. The reasons for its widespread use include the basic simplicity of its methodology and the low cost of its materials, combined with the great versatility of its possible applications—as will become clear in what follows. Other important factors include the small geometry of the detectors, and their ability—in certain cases—to preserve their track record for almost infinite lengths of time (indeed, mineral grains in geological and planetary materials less than a millimeter across can, by suitable treatment, be made to reveal the billions of years old record of their radiation history). The fact that the detectors, in themselves, do not need any electronic/electrical instrumentation means that they can be deployed under field conditions and in remote, fairly inaccessible places for long durations of time without the need of human intervention or backup, except for initial placement and final retrieval; and their ruggedness is of great merit in making this possible.

The basis of this technique lies in the fact that when heavy, charged particles (protons upward) traverse a dielectric medium, they are able to leave long-lived trails of damage that may be observed either directly by transmission electron microscopy (TEM)—provided that the detector is thin enough, viz. some μm across—or under an ordinary optical microscope after suitable enlargement by etching the medium. It is, in our view, not appropriate or necessary in the present Handbook to give a detailed historical and theoretical account of the discovery and the basic mechanisms involved in the phenomena of track production and revelation. The interested reader is referred to standard texts on this subject, e.g. Fleischer *et al.* (1975) or Durrani and Bull (1987). A brief outline of the basic principles and methods is, however, traced out in Section II.

The detecting media most often used in the field of SSNTD applications fall in two distinct categories. In the first category are polymeric—or plastic—detectors. These are most widely used not only for radiation monitoring and measurement, e.g. in health physics/radiation protection, or in environmental research and applications such as measuring radon levels in dwellings or in the field, but also in many other fields involving nuclear physics and radioactivity. In what follows, it is this type of detectors—viz. the polymeric ones—that we shall deal with most extensively. The second category of detectors is natural mineral crystals (and glasses) that have, imprinted within them, a record of their radiation (and thermal) history over the aeons. These find their greatest application in fields such as geology, planetary sciences (especially lunar and meteoritic samples), oil exploration, etc. Some of these minerals (e.g. sheets of mica) can, of course, also be used as custom-made detectors of heavy-ion or induced-fission bombardment. They can, for instance, be used inside reactor cores—since, by and large, they do not record neutron-recoils, and can withstand high temperatures and γ-ray exposures (both of which properties are generally lacking in plastic detectors).

As stated above, by far the most widely used SSNTD detectors today are plastics, which—unlike mineral crystals—do not require special preparation such as grinding and polishing. They are also much more sensitive than crystals and glasses, since some of them can record charged particles from protons upward. Several types of special track-recording polymers are commercially available—offering stable/constant recording efficiencies and good reproducibility of results. (For environmental effects on these properties, e.g. aging processes, storage conditions, etc., see Subsection II.B.2.) At present, the most sensitive and also the most widely used plastic is the CR-39 polymer (a polyallyldiglycol carbonate). It can record all charged nucleons, starting with protons. Cellulose nitrates and acetates can record α-particles upward. The Lexan polycarbonate, one of the earliest plastic SSNTDs to be used, responds to nuclei of charge equal to or greater than $Z = 6$ without special treatment (but also to α's by treating exposed detectors to UV radiation). Given in the footnote below[1] are some useful addresses of manufacturers. All polymer detectors are relatively inexpensive (sheets of $\sim 20\,\text{cm} \times 25\,\text{cm}$ cost less than a hundred US dollars, from which hundreds of detectors can be cut out).

Apart from use in radiation protection and environmental measurements, polymer detectors—often in great numbers and of very large sizes—are also employed in specialist research, e.g. cosmic-ray measurements (in balloons) or in spacecraft (for long-duration exposures); heavy-ion interactions; study of exotic decays; and measurements of life-times of artificially created superheavy elements—for all such applications the reader is referred to specialist texts (Fleischer *et al.*, 1975; Durrani and Bull, 1987; journals such as *Radiation Measurements* (Pergamon Press) and its predecessors) for further details and references.

Earth scientists who use mineral crystals as natural detectors for, e.g., age-determination of rocks, or research workers in the field of planetary science applications, have to utilize specialist machinery and techniques for crystal cutting, grinding and polishing; heavy-liquid and magnetic separation of minerals; special microscopes for mineral identifications; sample-mounting and replication methods, etc. Once again, the interested reader is referred to the sources listed at the end of the preceding paragraph. If mica sheets are used as detectors, no cutting or polishing machinery is required. Glass detectors are also easy to cut and polish (ordinary microscope slides, made of glass, in fact, need no polishing when used as SSNTD detectors).

We close these introductory paragraphs by quoting from the editorial in the very first issue of the journal *Nuclear Track Detection*—now entitled *Radiation Measurements*—written by one of us (SAD)—which stated, apropos of the SSNTD technique: ... "(it) has grown to such an extent that now there is hardly a branch of science and technology where it does not have actual or potential applications. Fields where well-established applications of this technique already exist include fission and nuclear physics; space physics; the study of meteoritic and lunar samples; cosmic rays; particle accelerators and reactors; metallurgy, geology and archaeology; medicine and biology; and many more" (Durrani, 1977). One only needs to add here that, with the passage of time, the above claim has become truer than ever—as will be authenticated by the sections that follow hereunder.

---

[1]CR-39 (*Polyallydiglycol carbonate*): (i) Page Mouldings (Pershore) Ltd, Pershore, Worcs, UK; (ii) American Acrylics and Plastics, Stratford, CT, USA; (iii) Tastrak, c/o H H Wills Physics Laboratory, Bristol, UK; (iv) Intercast Europe SpA, Parma, Italy.
Lexan (*Bisphenol-A polycarbonate*): General Electric Co., Schenectady, NY, USA.
Makrofol (*Bisphenol-A polycarbonate*): Bayer AG, Leverkusen, Germany.
LR 115; CN 85 (*Cellulose nitrate*): Kodak Pathé, Vincenne, France.

II. FUNDAMENTAL PRINCIPLES AND METHODS OF SOLID STATE NUCLEAR TRACK DETECTION

A. Physics and Chemistry of Nuclear Tracks

I. Formation of Latent Tracks

As stated earlier, we shall concentrate in the present treatment, in the main, on polymeric (plastic) materials as the prime example of solid state nuclear track detectors (SSNTDs) most widely used for radiation monitoring and measurement purposes. Some coverage will, however, be extended to other dielectric materials (e.g. mineral crystals, glasses) that are of importance in geological and cosmological fields.

The first thing to state here is that there are no universally accepted models for the formation of latent tracks in dielectric solids. In polymers, two processes are believed to determine the formation of a latent track: (1) defect creation and (2) defect relaxation; these are briefly outlined below.

The defect creation process can be subdivided into the following steps:

(i) The primary interaction between the passing particles and the atoms of the medium which takes place over a very short time (of the order of 10^{-17} s for 1 MeV α-particles).

(ii) The electronic collision cascade process, which spreads out from the particle trajectory: it leaves behind a positively charged plasma zone, and produces chemically activated molecules outside this zone. The process lasts approximately 3 orders of magnitude longer than the primary interaction (i.e. $\sim 10^{-14}$ s).

(iii) The atomic collision cascade is the next process, which occurs owing to the "Coulomb explosion" of the remaining charged plasma. The process takes place within a timescale of $\sim 10^{-12}$ s.

The defect relaxation can be subdivided into two processes:

(i) Aggregation of the atomic defects within the depolymerized zone (track core) into an extended defect over a timescale of about 10^{-10} s.

(ii) Relaxation of molecular defects via secondary reactions of chemically activated species in the partly depolymerized zone (track halo). This process occurs on a timescale of ~ 1 s.

Axial and radial sections through a latent track are shown in Fig. 3.1. The track core, ~ 10 nm in diameter, corresponds to the range of the atomic collision cascade. In this zone the molecular weight is drastically reduced. The track core is surrounded by a track halo, 100–1000 nm in diameter, corresponding to the electronic collision cascade, with modified chemical properties.

a. Factors Determining the Production of 'Stable'/Etchable Tracks

The following conclusions have been drawn from extensive studies in the field of SSNTD, although modifications of these criteria are always possible

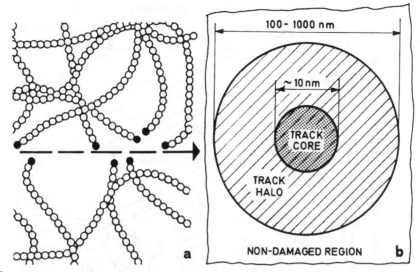

FIGURE 3.1 Axial (a) and radial (b) sections through a particle track in a polymer (not drawn to scale). Chain breaks allow preferential etching at a lower damage density (Ilić, 1990).

as new discoveries are made in the future.

1. Only "heavy" charged particles (protons upward) are capable of producing stable latent tracks—i.e. not electrons, etc.
2. There is generally a lower limit or critical value of the primary ionization, or of the linear rate of energy loss (dE/dx; or LET) by the particle concerned, which must be exceeded for the tracks to be "registered" (or revealed by appropriate etching). Thus, very fast (and hence low-LET) charged particles fail to leave etchable tracks until they have slowed down sufficiently in a medium to attain the critical value of LET. But when, in terms of the Bragg curve, they become very slow (i.e. fall on the low-energy side of the Bragg peak) toward the end of their range, they again become incapable of producing etchable tracks by virtue of picking up electrons from the medium and gradually losing their effective positive charge.
3. Different dielectrics have different "sensitivity" or "threshold" for recording heavy ions of a given energy per nucleon (or v/c, where v is the velocity of the particle, and c the speed of light); see Fig. 3.2. Thus, most mineral crystals are unable to "register" lightly charged ions, while most polymers are capable of recording naturally emitted alpha particles (though some may require special treatment, e.g. UV exposure, before yielding etchable tracks); and some (especially CR-39 plastic) can even record protons of moderately high energies (up to $\sim 70\,\mathrm{MeV}$). Note that fast neutrons can produce "intrinsic" tracks through proton recoils of the hydrogen content of the plastic detector.
4. If sufficient heat is applied to the dielectric prior to etching, it may partially or totally lose the latent track by the "healing" of the

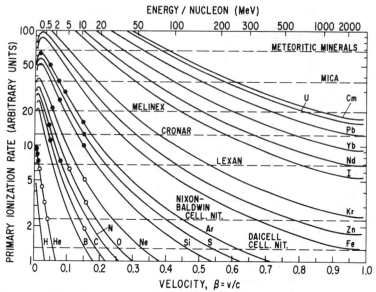

FIGURE 3.2 Primary ionization as a measure of the intensity of track damage in various nonconducting solids is given as a function of velocity β (relative to that of light) and of energy per nucleon, for a number of nuclei. The damage density increases with increasing charge, or atomic number; it also generally increases as the particle slows down. The horizontal lines represent the thresholds for track recording in materials ranging from sensitive plastics (bottom) to typical constituents of meteorites (top). The experimental points for accelerator ions in Lexan polycarbonate are given as open circles for zero registration and as filled circles for 100% registration. Note that the registration threshold of the most sensitive plastic detector (CR-39) lies below the x-axis of the figure (Fleischer, 1998).

etchable damage in the medium. In mineral crystals, a temperature of some hundreds of °C applied for an hour or so may result in producing substantial "fading" of the tracks; while in plastics, a temperature of ~100–200°C for an hour can produce a similar degree of fading (leading to the nonrevelation of tracks by subsequent etching); see Table 3.1 (from Durrani and Bull, 1987).

Based on, or by incorporating, the above properties and factors, a number of theories or hypotheses have been put forward by different authors to explain the basic mechanisms of track production in dielectric solids. None of these, however, have been able to yield verifiable quantitative predictions or data for track production in the various media or for charged particles of given energies and types. The reader is referred to standard texts (e.g. Fleischer *et al.*, 1975; Durrani and Bull, 1987; Spohr, 1990) for an in-depth understanding of the various theories—which range from the ion-explosion spike model (where ionization by the charged particle leads to sufficient lattice damage in crystals to yield etchable tracks), through the concepts of point defects and extended defects produced in single crystals by high fluences of energetic heavy ions, to the scission of polymeric chains by ionizing particles and δ-rays leading to etchability of the plastic detector.

TABLE 3.1 **Track Retention Characteristics of Some Common Detectors. Typical Temperatures for 100% Loss of Fission Tracks (FT) in 1 h of Annealing are Shown (Durrani and Bull, 1987)**

Material	100% FT loss in 1 h (°C)*
Plastics	
Cellulose nitrate	80–100
CR-39	~250**
Lexan	>185
Makrofol	165
Glasses	
Soda-lime glass	350–400
Tektite glass	~500
Mineral Crystals	
Apatite	350–400
Clinopyroxene	500–600
Epidote	625–725
Feldspar (Plagioclase)	700–800
Merrillite (Whitlockite)	~450
Mica	500–600
Olivine	400–500
Orthopyroxene	450–500
Quartz	1000
Sphene	650–800
Zircon	750–850

*These temperatures should be regarded only as rough guides. The retention temperatures for both minerals and plastics depend on their exact composition as well as on the etching conditions employed. Many of the mineral names, in particular, cover a wide range of compositions.

**At this temperature, CR-39 develops extensive cracks and becomes discolored.

2. Visualization of Tracks by Chemical and Electrochemical Etching

a. Chemical Etching (CE)

Chemical etching of plastic detectors is straightforward; that of mineral crystals, nearly so. The etching is usually carried out in thermostatically controlled baths (kept constant to ~±0.5 °C). Some useful etchants for nuclear track detectors are summarized in Table 3.2.

For **plastics**, the most frequently used etchant is the aqueous solution of NaOH (or KOH), with concentrations ranging from a molarity of 1–12 (~6 M being the most popular). The temperatures usually employed range from ~40 to 70 °C. In some cases, ethyl alcohol is added to the etchant to increase sensitivity and speed of etching. A large (glass or plastic) beaker is

TABLE 3.2   **Some Useful Etchants for Nuclear Track Detectors* (adapted from Durrani and Bull, 1987)**

Material	Etchant
Polycarbonate plastics	Aqueous NaOH solution; typically 1–12 M. Temperature: 40–70°C
	Alternatively, 'PEW' solution: 15 g KOH + 45 g H_2O + 40 g C_2H_5OH. Temperature: 70°C
Cellulose nitrate plastics	NaOH; 1–12 M. Temperature: 40–70°C
CR-39 plastic (allyldiglycol carbonate)	NaOH, KOH solutions; 1–12 M. Temperature: 40–70°C
Orthopyroxenes and clinopyroxenes	6 g NaOH + 4 g H_2O. Boiling, under reflux
Mica	48% HF. Temperature: 20–25°C**
Glasses	1–48% HF. Temperature: 20–25°C
Feldspars	1 g NaOH + 2 g H_2O. Boiling, under reflux
Apatite, Whitlockite[+]	0.1–5% HNO_3. Temperature: 20–25°C
Zircon	11.5 g KOH + 8 g NaOH (eutectic). Temperature: 200–220°C
Olivine	1 ml H_3PO_4 + 1 g oxalic acid + 40 g disodium salt of EDTA + 100 g H_2O; NaOH added to bring pH to 8.0 (the 'WN solution'). Boiling, under reflux
Sphene	1HF : 2HNO_3 : 3HCl : 6H_2O. Temperature: 20°C

*For a more extensive list see Fleischer *et al.* (1975). Note that etching times will vary according to the exact etching conditions (temperature and concentration of etchant) and the nature of the track-forming particle. In most cases they are a few hours (but of the order of a few seconds or minutes for some glasses or micas etched in 48 vol% HF). They should be determined by trial and error for each detector type. M stands for the molarity of the etching solution.

**Note that muscovite needs ∼20–30 min, but biotite only a few minutes, of etching.

[+]In current mineralogical usage, whitlockite is termed "merrillite."

usually placed inside the temperature-controlled bath, and it is this beaker that contains the etching solution. Into this are suspended, by means of springs, etc., several detectors that are to be etched simultaneously, with a lid covering the top of the beaker to reduce the evaporation—and the resulting increase of the solute concentration—of the etchant solution. Sometimes a stirring mechanism is incorporated.

The transformation of a latent into a visible track is brought about by the simultaneous action of two etching processes: chemical dissolution along the particle track at a (quasi-) linear rate V_T, and the dissolution of the bulk material at a lower rate, V_B. In accordance with the basically different properties of etched tracks, the detectors can be classified into two categories: (i) thin detectors, where the majority of etched tracks are etched-through holes, and (ii) thick detectors, where the residual foil thickness is greater than the etched-track depth. A simple schematic model for track etching in thick and thin detectors is shown in Figs. 3.3(a,b). Figure 3.3c depicts the important concept of Θ_c, the critical angle of etching.

On the basis of a comprehensive study (Somogyi, 1980, 1990), it was found that the etch-rate ratio V (= V_T/V_B) as a function of the residual range

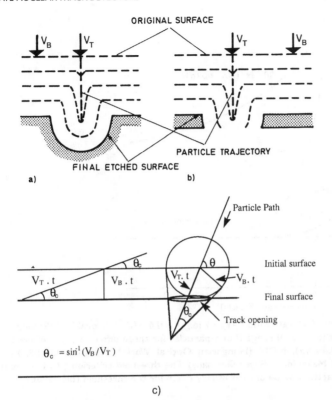

ORIGINAL SURFACE

PARTICLE TRAJECTORY

FINAL ETCHED SURFACE

a)

b)

Particle Path

Initial surface

$V_T \cdot t$

$V_B \cdot t$

$V_T \cdot t$

$-V_B \cdot t$

Final surface

Track opening

$\theta_c = \sin^{-1}(V_B/V_T)$

c)

FIGURE 3.3 Schematic representation of track etching in thick (a), and thin (b), detector foils. Development of track profile during the etching process for a particle entering detector surface at 90° is illustrated. Bulk etch rate and track etch rate are denoted by V_B and V_T respectively (Ilić, 1990). (c) Concept of the critical angle of etching. When an irradiated detector is treated with an appropriate etchant, the velocity of etching along the latent track (V_T) is larger than the bulk velocity of etching (V_B) elsewhere in the medium. In the figure shown, the ratio $V_T/V_B = 3$. There is an angle Θ_c for each medium and a given heavy ion such that, by the time that the etchant travels a distance $V_B t$ vertically into the body of the detector, it reaches the end of the range of the particle proceeding along that "dip angle" Θ_c at the same instant – i.e. $V_B t/V_T t = \sin\Theta_c$. Only tracks making dip angles with the detector surface, such that $\Theta > \Theta_c$, will thus leave observable track openings. The half-cone angle of all such etch pits is also $\Theta_c = \sin^{-1}(V_B/V_T)$ (Durrani, 1997).

R of the particles in polymers can be described by

$$V = 1 + e^{-aR+b}, \text{ for polycarbonates and cellulose nitrate;} \quad (3.1)$$

$$\text{and } V = aR^{-b}, \text{ for allyldiglycol carbonate.} \quad (3.2)$$

Here a and b are fitting parameters. The function $V(R)$ for three commonly used detectors for α-particles is shown in Fig. 3.4. The threshold criterion ($V=1$) is marked in this figure. In practice, a value of $V=1.2$ is usually taken for the registration threshold. Etching conditions that remove a layer whose maximum thickness is equal to the range of the particles in the detector are recommended.

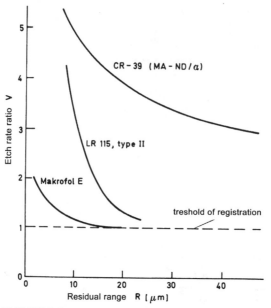

FIGURE 3.4 Etch rate ratio $V_T/V_B = V$ (e.g. $V = 11.6 \times R^{-0.464}$, for CR-39 (Somogyi, 1990)) as a function of the residual range R of α-particles for three different types of plastic detectors (CR-39 (MA-ND/α), MOM (Hungarian Optical Works), Hungary; LR 115, Kodak Pathé, France; and Makrofol E, Bayer, Germany). The threshold criterion ($V = 1$) is marked in the figure. In practice a value of 1.2 is usually taken for the threshold (Ilić, 1990).

Glass detectors are normally etched in aqueous HF solutions (usually diluted downward from the 48 vol.% maximum strength of HF) at room temperature. Teflon beakers have to be used for containing the hydrofluoric acid (which would attack a glass beaker).

Mineral crystals—which have been appropriately ground and polished, either as found in nature or prior to artificial irradiation—are etched by a variety of etching reagents of different molarities and at different temperatures. Detailed etching recipes may be found in Fleischer *et al.* (1975), and Enge (1980); an abbreviated table (Table 3.2) has been given above (from Durrani and Bull, 1987). Figure 3.5 shows some typical shapes of etched tracks in (a) plastics, (b) crystals, and (c) glasses.

b. Electrochemical Etching (ECE)

If track density is not high (i.e. is less than $\sim 10^3$ tracks cm^{-2}), it is often helpful to enlarge the tracks for ease of counting. This can be done by electrochemical etching (ECE)—first proposed by Tommasino (1970)—which enlarges the chemically etched tracks ($r \sim 1\,\mu m$) a hundredfold or so. The principle of the ECE method is to apply a high-frequency (several kHz) high electrical field ($\sim 30–50\,kV\,cm^{-1}$) across two compartments of an etching cell, filled with a conducting etchant solution (e.g. NaOH), and separated by a plastic detector containing etchable tracks on its surface. After a period of chemical pre-etching (Fig. 3.6), which produces sharp-tipped tracks, the electric field at the tip builds up to a value equalling the breakdown limit of

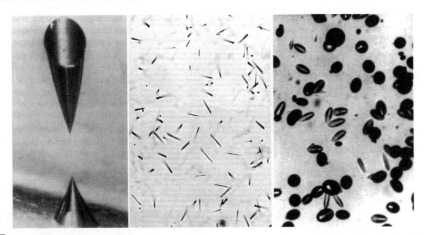

FIGURE 3.5 (a) Etched track of a cosmic ray (Argon) ion that penetrates an Apollo electrophoresis device made of Lexan (Fleischer, 1998); (b) etched spontaneous-fission tracks (^{238}U) in Durango apatite (Durrani and Bull, 1987); and (c) etched neutron-induced fission tracks (^{235}U) in obsidian glass (Fleischer, 1998). Note that different magnifications have been used for these images.

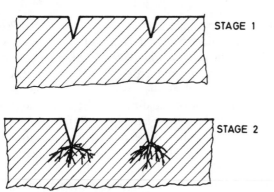

FIGURE 3.6 Formation of electrochemical etch spots. Stage I: formation of track pits due to (early or pre-) etching process. Stage 2: treeing at the tip of the track pit due to electrical breakdown of the dielectric medium (Durrani and Bull, 1987; Ilić, 1990).

the dielectric medium (i.e. the plastic detector). At this point, "treeing" takes place resulting in large Lichtenberg-type figures surrounding the track-tip (see Durrani and Bull, 1987, for details; and Matiullah *et al.*, 1987, for the design of an electrochemical etching cell and its electronic circuitry).

Figure 3.7 shows a picture of typical CE and ECE track-spots produced on plastic detectors by radon (Ilić, 1990).

B. Track Detector Types and Properties

I. General Properties

Etched tracks have been observed in hundreds of materials. These materials include, in particular, polymers, inorganic glasses and mineral crystals (see Fleischer *et al.*, 1975; Durrani and Bull, 1987). Tracks have also

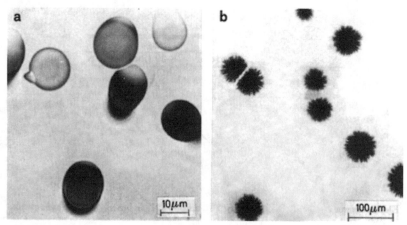

FIGURE 3.7 Tracks of α-particles emitted by ^{222}Rn and its decay products in (a) CR-39 (chemically etched), and (b) in Makrofol-E (electrochemically etched). Note that electrochemically etched tracks are usually one to two orders of magnitude bigger than chemically etched tracks (Ilić, 1990).

been observed in some oxide semiconductors ($Bi_2Sr_2CaCu_2O_x$) (Provost *et al.*, 1995); intermetallic compounds ($NiZr_2$; NiTi); and, most recently, in metals (Ti) (Barbu *et al.*, 1995). In general, the SSNTDs may be considered to be mainly dielectric solids, i.e. poor conductors of heat and electricity. A value of $\sim 2000\,\Omega\,cm$ has been quoted as the lower limit for the resistivity, and $\sim 0.06\,cm^2\,s^{-1}$ as the upper limit of the diffusivity of a medium, for tracks to be formed in it (Fleischer *et al.*, 1975; Fleischer, 1981). Various authors have suggested that track formation should be related to a number of different parameters, such as total energy loss rate, primary ionization, restricted energy loss, thermal conductivity, etc. In practice, the track formation criteria may be tested heuristically by irradiating a given material with different ions at various energies and recording those cases for which etchable tracks are formed.

As already mentioned, polymers are the most sensitive detectors. Being made of long-chain molecules, they are susceptible to effects of chain breaks, which can be created at considerably lower energy transfers to electrons (2–3 eV) than are needed in inorganic solids for the lowest-energy ionization processes (10–15 eV). Chain scission, in turn, lowers the molecular weight and allows more rapid chemical attack at the increased number of chain ends (Fleischer, 1998).

Characteristics of some of the most widely used polymer detectors are given in Table 3.3. Because of its good sensitivity, stability against various environmental factors, and high degree of optical clarity, CR-39 has become the most favored SSNTD.

2. Aging and Environmental Effects

The durability of tracks in some solids is noteworthy, since it allows them to persist under adverse conditions of temperature, pressure, etc. This stability has permitted primordial tracks to be identified that were formed not

TABLE 3.3 Useful Characteristics of Some Plastic Detectors (adapted from Ilić, 1990)

Material	Composition	Trade Name	Density ($g\,cm^{-3}$)	Refractive Index
Cellulose nitrate	$C_6H_8O_9N_2$	CN 85[1]	1.52	
		CA 8015[1]	1.52	1.51
		LR 115[1]		
		Daicel[2]	1.42–1.45	1.505
		DNC[3]	1.4	1.5
Bisphenol-A polycarbonate	$C_{16}H_{14}O_3$	Makrofol[4]	1.29	
		Lexan[5]		
Allyldiglycol-carbonate	$C_{12}H_{18}O_7$	CR-39[6]	1.32	1.45
		MA-ND[7]		
		TASTRAK[8]		

[1]Kodak Pathé
[2]Dai Nippon Co., Japan
[3]Cellulose nitrate produced in Russia
[4]Bayer AG, Germany
[5]General Electric Co., USA
[6]American Acrylics, USA; Homalite, USA; Baryotrack, Japan; Pershore, UK
[7]MOM (Hungarian Optical Works), Hungary
[8]Track Analysis Systems Ltd, UK

long after the end of nucleosynthesis of our solar system in meteoritic minerals. Similarly, dosimetry measurements of charged products from neutron interactions can be made in an intense background of more sparsely ionizing radiations, e.g. γ-rays (Fleischer, 1998). Tracks in minerals and glasses can withstand vast doses of electrons and of UV radiation, and show no effects of exposure to external oxygen. However, the stability of response and the sensitivity of plastics is dependent on environmental conditions. The dependence of the response of a polymer on the manufacturing process, as well as on the amount and duration of exposure to UV radiation, oxygen, humidity, temperature and storage conditions (the "aging" process) has been studied by a number of investigators (Homer and Miles, 1986; Tidjani, 1990, 1991; Khayrat and Durrani, 1995; Tsuruta, 1997; Miles, 1997). However, the physics of these processes is still not fully understood (Durrani and Ilić, 1997). Large variations in the efficiency for alpha-particle detection were observed in some detectors exposed to solar light. Fading effects, which change the track revelation properties of polymers, such as the etching rate, "etch induction time," and track revelation efficiency (which is adversely affected by the application of high temperatures before or during the etching process), have also been observed. Exposure to O_2 distorts the surface quality of un-doped CR-39, and creates poor transparency, thus resulting in decreased accuracy of measurement when using transmitted illumination. The beneficial effect of antioxidant doping of CR-39 on the stability of the material and of latent tracks in it has been reported. It was observed that the "etch induction time" (etching time before track revelation starts) increases, and the etch rate ratio V_T/V_B decreases, if the detector is irradiated in a vacuum, because of the outgassing of oxygen from the detector itself (Csige *et al.*, 1988). This has important implications for plastic detectors exposed in space missions (e.g. long-duration exposures) or in balloons for cosmic-ray studies. It becomes important to keep an accurate record of variations of temperature during such explorations in space, since they can affect the track recording and retention efficiencies of the detectors (see review paper by Durrani, 1991; and also O'Sullivan *et al.*, 1984).

Ultraviolet exposure can also dramatically change the properties of polymeric track detectors (see, e.g., Khayrat and Durrani, 1995). If the UV irradiation is too pronounced, V_B can drastically increase, so that the detector sensitivity is reduced. Since the fading effects of latent tracks are enhanced at higher temperatures, the response of a given detector may be somewhat different in different geographical regions under natural environment. Careful work is still required to quantify these possible factors and to establish control of at least some of the parameters involved, with reference to particular plastics. The interested reader is referred to Ilić and Šutej (1997), Miles (1997), and Fleischer (1998), and to references cited in those texts.

C. Track Evaluation Methods

I. Manual/Ocular Counting

Manual (or more accurately, ocular: eye-) counting denotes non-automatic counting of etched tracks generally using an optical microscope, with a moving stage, and two eyepieces (which range between $\sim 8\times$ and $16\times$). The choice of objectives employed depends on the track density, etch-pit size, and the degree of resolution required. The objectives used most often for counting purposes are 20 (or 25)$\times$ and 40 (or 45)$\times$. If pit size needs to be measured, then $63\times$ or 95 (or 100)$\times$ may have to be used. For better resolution, oil-immersion objectives may be employed—but dry objectives are easier to use. Usually, fields of view (fov's) are chosen in an unbiassed manner such that contiguous fields are brought into view by linear movement of the stage along an arbitrarily chosen x-axis, followed by counting in the next parallel line by moving the stage along the y-axis by the width of one fov. What is important is to ensure that no tracks are counted more than once and none are left out through any bias. Any lower limit on the size of acceptable etch pits must be consistently imposed by a given observer. Criteria for genuine tracks (whose pits have regular shapes—whether circular or conic sections in the case of glasses and the CR-39 plastic, and whose conical bottoms appear as pinpoints of light by moving the objective up and down; or needle-like in the case of mineral crystals or certain plastic detectors and particle types) as against defects, scratches and other artifacts, have usually got to be learnt by new workers, who should first familiarize themselves with detectors artificially irradiated with α-particles or fission fragments, and etched with care. Track densities are expressed either in relative terms (i.e. tracks per field of view) or in absolute terms (in which case the area of each fov for a given objective is determined, once for all, using graticules supplied by the manufacturer). If a given track density (say, tracks cm^{-2}) is then to be converted into a dose (e.g. $Bq\,m^{-3}\,h$), one requires a standard source for exposure, followed by etching under identical conditions—or one may use a theoretical approach (see Subsection III.A.1). For statistical errors, see Subsection II.D.3.

To count electrochemical etch spots (usually several tens of μm in radius) in plastic detectors, it is normally sufficient to employ $\sim 20\times$ objective.

2. Spark Counting

A spark counter (Cross and Tommasino, 1970) is a semiautomatic device occasionally used to count low track densities (10^2–$10^3\,cm^{-2}$)—e.g. those encountered in radon monitoring or personnel neutron dosimetry. Here, a thin plastic detector foil (~ 10–$20\,\mu m$) containing through-holes (produced by over-etching of the film exposed to alpha particles, etc.), is interposed between two electrodes: a cathode, and an anode which is effectively in the form of an aluminized plastic foil (e.g. Mylar); see Fig. 3.8 and Durrani

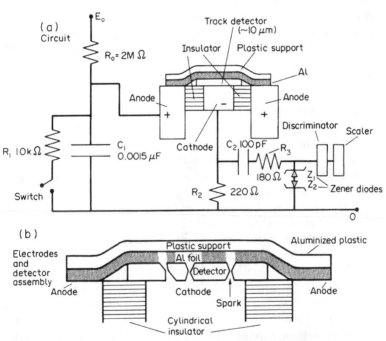

FIGURE 3.8 Details of the circuit (a), and electrodes and detector assembly (b), for a spark counter. The anode and the cathode of the detector consist of two coaxial cylindrical conductors separated by an insulator. The irradiated plastic detector foil (~ 10–$20\,\mu m$ thick), etched so as to produce through-holes, is placed on the cathode, and covered by another plastic foil $\sim 100\,\mu m$ thick (essentially, for support as a backing), which is thinly aluminized on the lower face to offer a conducting path. When the switch is opened, the capacitor C_l is raised in potential toward the applied voltage E_0, and a voltage appears across the electrodes and hence across the etched detector. Eventually, a discharge takes place between the anode and the cathode across an etched track. Sparks jump through different holes in the detector foil in random sequence; but only once per through-hole, since each spark destroys the conducting Al element in its vicinity. The sparks are counted by a scaler via a discriminator. After each spark, the capacitor C_l needs to be recharged by the applied voltage E_0 to provide sufficient potential for the next spark (Durrani and Bull, 1987).

and Bull (1987) for details. When a sufficiently high voltage is applied across the detector film ($\sim 500\,\mathrm{V}$), sparks jump across the detector via the through-holes, one by one—electronic circuitry ensuring that the flow of current after each spark causes the potential across a capacitor temporarily to fall below the breakdown value required for sparking. Each breakdown (or the spark) may be counted electronically. Since the spark burns out a hole in the thin ($\sim 1\,\mu\mathrm{m}$) aluminum coating, exposing the non-conducting plastic backing, the same through-hole is not counted again. At the end of the counting sequence, the aluminized foil shows a visible pattern of holes corresponding to the original track-holes in the detector. The spark holes can be easily counted either by naked eye or under a low-power microscope.

3. Advanced Systems for Automatic Track Evaluation

Fully automatic analysis systems for track evaluation are currently available on the market.[2] A number of other automated systems have been developed in-house by various research laboratories by upgrading the conventional optical microscope operation with additional hardware and/or suitable software (e.g. Rusch et al., 1991; Fews, 1992; Skvarč, 1993). Reports on recent developments of such systems may be found in the Proceedings of the two latest conferences on Nuclear Tracks in Solids (Chambaudet et al., 1999; Ilić et al., 2001). With such advanced systems, simultaneous measurements are made of the track size parameters (area, minor and major axes); of the grey levels inside the track; or of the average greyness (i.e. brightness) of a single track. A typical hardware configuration of a modern microprocessor system is shown in Fig. 3.9 (after Lengar, 2001). The main components of such a system are: an optical microscope equipped with autofocus and an X–Y moving stage; a CCD video camera; a digitizer; and a personal computer. The image of the detector surface is thus produced by a conventional optical microscope, and transmitted by the CCD camera to the computer. The stored image comprises a number (e.g. 512×512) of pixels, each of them with a number (say, 256) of grey levels. The X–Y stage is capable of moving over large areas (e.g. $30\,\mathrm{cm} \times 10\,\mathrm{cm}$) in steps of, e.g. $1\,\mu\mathrm{m}$. The detector foil can be usually scanned at a rate of up to four frames per second. The magnification used is typically $0.5\,\mu\mathrm{m}$/pixel, giving a resolution of $0.2\,\mu\mathrm{m}$ by interpolation along a line of pixels. The setup is fitted with an autofocus system, capable of focusing to within about $1\,\mu\mathrm{m}$. With the help of appropriate software, these systems can carry out many tasks such as: measuring the spatial density of etched tracks; determining their two-dimensional coordinates; areas; grey levels; statistical distribution, etc. Because of their speed, such automatic systems are not only becoming popular for routine work but they are currently also revolutionizing the possibilities for more advanced research work in fields such as high-energy heavy-ion interactions; exotic decays; cosmic-ray and monopole investigations, etc.,

---

[2]For example: ELBEK Bildanalyse GmbH, Siegen, Germany.
AUTOSCAN Systems Pty. Ltd., Brighton, Victoria, Australia.

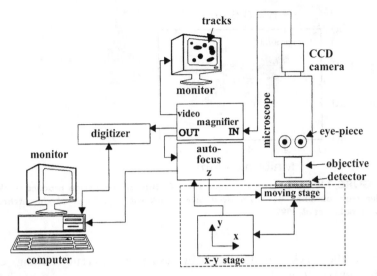

FIGURE 3.9 Automatic system for track analysis, **TRACOS** (Skvarč, 1993). The block diagram is adapted from Lengar, 2001.

where myriads of fields of view need to be examined in search of rare events. These systems are able to detect particle tracks at great speed as well as to discriminate against all types of nontrack defects. The detector foil is scanned in successive horizontal passes along the x-axis, alternating in plus and minus directions.

The principal phases of the scanning are (cf. Fews, 1992):

1. At the microscope stage position, a new frame is digitized into the computer frame buffer;
2. The stage is instructed to move exactly one image frame to the position of the next frame;
3. The image frame is searched for candidate regions where tracks may be located. This procedure generates a list of candidate events, and typically takes ~ 100 ms;
4. The perimeter of each event is then calculated by a special procedure of image processing, which is the most critical stage of the analysis;
5. Selection criteria are applied, which enable one to discriminate between particle tracks and background events;
6. The final orientation of the track is determined, and parametric measurements performed;
7. Detailed track calculations are performed, as required, on individual tracks, either during the scan or off-line.

To illustrate the capabilities of the representation of the measured data, an example obtained by the TRACOS system (Skvarč, 1993) is presented in Fig. 3.10 (see also Subsection III.B.1). Here the grey level is plotted against the major axis of the ^{10}B$(n, \alpha)^{7}$Li reaction-product tracks in CR-39. The tracks formed by the ^{10}B$(n, \alpha)^{7}$Li reaction products are easily separated

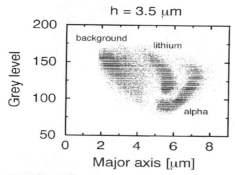

FIGURE 3.10 Grey level vs major axis of ^{10}B(n, α)^{7}Li reaction product tracks in CR-39 detector, obtained at a removed layer thickness $h = 3.5\,\mu$m of the etched detector (Izerrouken et al., 1999).

from the background tracks by using appropriate etching conditions (Izerrouken *et al.*, 1999).

New techniques have been developed for the selective enhancement and evaluation of radiographic images in track detectors on the basis of image processing of individual "image element units"—viz., etched tracks. Such selective radiographs, based on the assessment of the size and optical properties of individual tracks, have been obtained by Skvarč *et al.* (1999) by using their TRACOS advanced track analysis system. The applicability of such techniques is illustrated in Subsection III.C.3.

Using such a system, a new method, which enhances the measurements of charge-changing and other reactions by tracing the trajectories of charged particles through a stack of nuclear track detectors, has recently been developed (Skvarč and Golovchenko, 2001). Here, a complex software was developed in order to allow:

1. Matching of the tracks of the same particle on successive foils;
2. Connecting successfully matched tracks into trajectories;
3. Recognizing charge-changing reactions, and calculating reaction points and fragment emission angles.

All data structures generated as above are stored in a postgreSQL database, which allows flexible development of further compound data structures and complex data queries.

D. Basics of Measurement Procedures

I. Revelation Efficiency

When a detector—whether a polymer or a crystal or glass—is immersed in an etchant, the etching process starts at its top (or external) surface, and proceeds inward by etching away the detector, layer by layer, at a general or bulk velocity of etching, say V_B. It is known that when the etchant comes across a trail of damage produced by a charged particle, it proceeds along that track at a greater velocity, say V_T, the enhanced value depending on the nature and energy of the charged particle. It can be shown (see Fleischer *et al.*, 1975, or Durrani and Bull, 1987, for details) that this results in **not** revealing those tracks that make a shallower (i.e. smaller) angle with the surface of the detector than a "critical angle of etching," $\Theta_c = \sin^{-1}(V_B/V_T)$: see Fig. 3.3c. If all the tracks encountered have emanated from points within the body of the detector (e.g. from uranium fission within a mineral crystal), then it has been shown that the "revelation efficiency" for such tracks is $\cos^2 \Theta_c$ (rather than 1). If, however, the tracks have been made by an external thin source of track-forming particles placed in contact with the detector, the registration efficiency can be shown to be $1 - \sin \Theta_c$ (further details of these, and more complex, geometries may be seen in Durrani and Bull, 1987, pp. 64–72). These "revelation" (or "registration"/"detection"/"etching") efficiencies represent the ratio of the number of observed etched tracks to the number of latent damage trails crossing a unit area of the "original" surface of the detector (i.e. where the etching first starts). In the case of crystals exposed to fission fragments, $V_T \gg V_B$, so that $\Theta_c \sim 0$, and hence both $\cos^2 \Theta_c$ (the revelation efficiency for internal tracks) and $1 - \sin \Theta_c$ (for external tracks) tend to unity. For glasses (with $\Theta_c \sim 30°$ for fission tracks), and in CR-39 (with $\Theta_c \sim 20°$ for alpha particles), the detection efficiencies may differ considerably from unity. Thus, for $\Theta_c = 30°$, $\cos^2 \Theta_c = 0.75$ and $(1 - \sin \Theta_c) = 0.5$; and for $\Theta_c = 20°$, $\cos^2 \Theta_c = 0.883$ and $(1 - \sin \Theta_c) = 0.658$. It must be remembered that, strictly speaking, V_T, and hence Θ_c, changes along a charged particle's path all the time as it traverses a medium, continuously losing its energy, and possibly its charge, and consequently its rate of ionization. It must be emphasized that V_T—and hence Θ_c and the revelation efficiency—varies from particle to particle, even for the same detecting medium.

2. Sensitivity

By "sensitivity" we mean here the ratio of the number of revealed tracks to that of the incident particles ultimately responsible for the tracks; in other words, tracks per incident particle, or track density per unit fluence of the incident particles. For heavily charged particles of moderate energies per nucleon (e.g. fission fragments, which have energies ~ 0.5–1 MeV per nucleon), the sensitivity is close to 1 in most detectors. However, for fast neutrons, which can produce tracks only (or mostly) through the recoil of hydrogen nuclei contained in the material of, say, CR-39, the sensitivity (for chemically etched "intrinsic" tracks) is only $\sim 10^{-3}$ to 10^{-4} tracks/neutron; for electrochemical etching, it may fall by a further factor of 10 depending on the energy and the fluence of the neutrons (see Al-Najjar *et al.*, 1979, for details). Of course, in many dielectrics (e.g. mineral crystals or glasses), charged particles whose LET is below a critical value for the detecting material (e.g. cosmic rays or accelerated heavy ions), or whose (effective) charge is below the registration threshold of the detector, fail to leave an etchable track until they have slowed down sufficiently (cf. Fig. 3.2). CR-39 plastic is so far the only known polymeric detector that can register proton tracks.

3. Statistical Errors

In common with other radioactive measurements, Poisson statistics apply to the counting of nuclear tracks (see Chapter 7 of this Handbook for details). Other factors affecting the reproducibility of results may, in this context, be considered to be systematic errors—which include etching conditions; aging and environmental effects on the detecting material; criteria adopted by the observer for track identification and acceptance, etc.

For good statistics, it is necessary to actually count $\sim 400\text{--}1000$ tracks (yielding *ca.* 5–3% errors, respectively). Here the expression "actually count" is used advisedly. If one actually counts, say, only 4 tracks in one field of view that is $10^{-2}\,\text{cm}^2$, yielding a track density of 400 tracks cm^{-2}, the statistical error remains equivalent to 4 ± 2 tracks, i.e. $\pm 50\%$, and does not become 400 ± 20 tracks cm^{-2} i.e. 5%! (In other words, it remains $(4 \pm 2)/10^{-2} = 400 \pm 200$ tracks cm^{-2}.) Thus, in the case of low track density, one needs to count tracks over a large number of fields of view in order to gather statistically reliable results. (For instance, in the above-cited case, one needs to count ~ 100 fields of view, i.e. accumulate ~ 400 tracks, in order to yield a $\sim \pm 5\%$ error.)

4. Background Measurement

Most detectors have a natural background of tracks, which become revealed upon etching. If the background is negligible in comparison with the tracks deliberately produced by irradiation, it may be just ignored without producing a perceptible difference in the expressed error. If, however, the background is significant, there are two alternative procedures. In the first approach, the background is eliminated by an appropriate method. For instance, in the case of fission track dating of rocks (see Subsection III.A.2), if the crystal in question has a high background of natural fission tracks from its ^{238}U content, it needs to be given a suitable high-temperature treatment (e.g. heating it at $\sim 500°\text{C}$ for 1 h) to remove the background prior to reactor irradiation for inducing fission in the ^{235}U content of the crystal. In the second approach, for instance with α-tracks produced in a plastic detector exposed to environmental radon, (i) the pre-existing background is minimized by keeping the detector appropriately shielded from atmospheric radon—e.g. by keeping the detector, prior to exposure, under a peelable thin layer of protective plastic foil or under some (Al) wrapping that is only removed just before exposing the detector to Rn; and (ii) a sufficiently large number of background tracks are counted in the un-irradiated detector in order to get good statistics for the background tracks. The irradiated detector, too, then has to be counted over a sufficiently large number of fields of view (fov's). For instance, if the genuine track density is 200 tracks cm^{-2} and the background track density is 40 tracks cm^{-2}, then we get the following situation, when each fov $= 10^{-2}\,\text{cm}^2$:

A. Count 200 fov's ($= 2\,\text{cm}^2$) for the irradiated detector, yielding total tracks (genuine + background) $= 480 \pm 480^{1/2}$. Then suppose that we count only 10 fov's of the un-irradiated detector, yielding a background of 4 ± 2 tracks; then this is equivalent to 80 ± 40 tracks over 200 fov's. In such a case the genuine tracks are found to be $(480 \pm 480^{1/2}) - (80 \pm 40) = 400 \pm (480 + 1600)^{1/2} = 400 \pm 45.6$, i.e. a statistical error of 11.4%.

B. Count 200 fov's for the irradiated detector as before. But then count 400 fov's ($= 4\,\text{cm}^2$) for background (un-irradiated detector); this would yield $160 \pm 160^{1/2}$ tracks, i.e. $80 \pm 40^{1/2}$ tracks over 200 fov's. Hence the genuine tracks are $(480 \pm 480^{1/2}) - (80 \pm 40^{1/2}) = 400 \pm 520^{1/2} = 400 \pm 22.8$, i.e. a statistical error of $\pm 5.7\%$.

Thus, it is obvious that if the background is a significant fraction of the genuine tracks (viz., 20% in the above case), then the background, too, has to be counted over a large number of fov's (viz., twice as many (that is 400 fov's) as for the irradiated detector) if a final error in the background-corrected value is to be comparable to the percentage error from the raw data.

5. Calibration and Standardization

One of the major drawbacks of SSNTDs is the strong variability of their sensitivity—which may vary from batch to batch and even from sheet to sheet in the same batch supplied by a manufacturer.

Calibration is performed with beams of known ions; or with known neutron flux and spectrum; or with known radon concentration, etc. If at all possible, one should rely on direct calibration with ions whose charge and energy (Z and E) are similar to those of the particles being studied. In practice, for a given ion, a particular etchant, and a given detector, a response curve V_T vs R has to be generated, where V_T is along-the-track velocity of the etchant, and R the residual range of the ion in that detector. Further information on the calibration of radon, neutron and cosmic-ray dosimeters may be found in relevant literature (e.g. Miles *et al.*, 1996; Tommasino, 2001; Benton *et al.*, 2001). In 1984, the European Radiation Dosimetry Group (EURADOS) initiated a program on the use of SSNTDs for neutron dosimetry in cooperation with the Commission of the European Communities. The major aim of this series of experiments was to provide standardized irradiation for laboratories from Europe and elsewhere in the world, which use SSNTDs routinely or in particular fields of research. Since then, a number of neutron or proton irradiation exercises have been conducted. Similarly, in order to ensure that radon measurements made by different laboratories are mutually compatible and consistent, an outstanding program of intercomparison of passive radon monitors has been carried out periodically by the National Radiological Protection Board (UK) since 1982 (Miles *et al.*, 1996).

III. MEASUREMENTS AND APPLICATIONS

A. Earth and Planetary Sciences

I. Radon Measurements

Radon measurements are one of the most widely used applications of SSNTDs today. Radon is a naturally occurring radioactive gas that constitutes

TABLE 3.4 The Decay Products of $^{222}_{86}$Rn, a Gaseous Member of the Naturally Occurring Radioactive Series $^{238}_{92}$U $\rightarrow$ $^{206}_{82}$Pb (adapted from Durrani and Bull, 1987)

Isotope	Atomic No. Z	Half-life	Radiations emitted	α-particle decay energy (MeV)
^{226}Ra	88	1600 y	α	4.78
^{222}Rn	86	3.825 d	α	5.49
^{218}Po	84	3.05 min	α	6.00
^{214}Pb	82	26.8 min	β, γ	–
^{214}Bi	83	19.9 min	β, γ	–
^{214}Po	84	164 μs	α	7.69
^{210}Pb	82	22.3 y	β, γ	–
^{210}Bi	83	5.01 d	β	–
		$(+3.0\times10^6$ y	α	4.95)
^{210}Po	84	138.4 d	α	5.30
^{206}Pb	82	Stable	–	–

both a hazard—e.g. lung cancer, especially in confined spaces such as uranium mines—and a helpful resource—e.g. means for uranium exploration and, putatively, for earthquake prediction (Fleischer, 1997a).

Radon ($Z = 86$) is a chemically inert, noble element, which is quite mobile at normal temperatures. It is a decay product found in each of the three naturally occurring radioactive chains headed, respectively, by ^{238}U, ^{232}Th and ^{235}U; each of these radon isotopes decays by α-emission. Of these three radioisotopes, ^{222}Rn (from ^{238}U; usually called simply radon), because of its relatively long half-life ($\tau_{1/2} = 3.82$ d; $E_\alpha = 5.49$ MeV) and natural abundance, is the most important isotope. ^{220}Rn (from ^{232}Th—sometimes also called thoron) is of less importance, owing largely to its relatively short half-life ($\tau_{1/2} = 55.6$ s; $E_\alpha = 6.29$ MeV). The role of ^{219}Rn (a descendant of ^{235}U)—because of its very low natural abundance as well as the very short half-life ($\tau_{1/2} = 3.96$ s; $E_\alpha = 6.82$ MeV) is usually considered to be entirely negligible.

A vast literature exists on radon and its measurements—the most widely used technique for the measurement of radon being, in fact, the SSNTD method (see the book by Durrani and Ilić, 1997, for a general survey of this subject area). In what follows, we summarize the methods and applications of SSNTDs in the field of radon measurements. In this description we shall concentrate our attention mostly on the long-lived isotope ^{222}Rn. It should be remembered, however, that the short-lived solid daughters of ^{222}Rn also play an important role—by getting "plated out" on solid surfaces (including those of human lungs) and then decaying by (health-damaging) α-emission. Table 3.4 gives the decay chain of ^{222}Rn, together with the half-lives of the product isotopes as well as the types and energies of the radiations emitted.

TABLE 3.5 Upper Theoretical Limit and Measured Values of the Response of Some Radiometers for ^{222}Rn, Used in Large-scale Radon Surveys (adapted from Nikolaev and Ilić, 1999)

Type of Radiometer	Detector	Response (tracks cm^{-2}/kBq m^{-3} h)
Upper theoretical limit[1]	CR-39	14
	LR 115	3
Diffusion[2]	CR-39	2.5
Membrane permeation[3]	CR-39	1.2–6.2
	LR 115	1.7
	Makrofol E	0.67–1.64
Bag permeation[4]	LR 115	0.49
Charcoal collection[5]	CR-39	545
Electret collection[6]	CR-39	2486
Electrostatic collection[7]	CR-39	5000

[1]Calculated for open ('bare') detectors (CR-39 and LR 115)
[2]A tube with a detector located at one end of the diffusion zone formed by the tube
[3]An enclosure (cup-type) that allows ^{222}Rn to enter through a permeable membrane (Fig. 3.11b)
[4]A bag-type permeation sampler, formed from a heat-sealed plastic bag (filter) made of polyethylene
[5]Charcoal acts as a collector of radon from the air.
[6]An electret acts as a collector of radon decay products.
[7]Here the Rn daughter products are collected by an electrostatic field on a thin metal foil placed on the detector.
Note that incorporation of an electret, etc. (in the last three entries), makes the radiometers/dosimeters vastly more efficient in collecting radon and its daughters – and hence far exceed the theoretical limit shown above.

a. Response of Detectors to Radon and Radon Daughters

In deriving the response of a plastic detector (e.g. CR-39) to the decay products of Rn, let us follow the simplified first-order model calculations of Durrani (1997); a more precise calculation may be found in Fleischer and Mogro-Campero (1978) and Ilić and Šutej (1997). The measured response of some commonly used dosimeters is given in Table 3.5. Imagine a detector of area $1\,\text{cm}^2$ lying at the bottom of a cylinder of air, R cm high (Fig. 3.11a), where R is the range of the radon-decay α-particle ($E_\alpha = 5.49\,\text{MeV}$; range in air, 4 cm). The cylinder thus represents a "thick source" of α-particles, its top being the maximum height from which a radon-α can reach the detector. The volume of this cylinder is R cm^3 ($= R \times 10^{-6}\,\text{m}^3$). Assume the radon activity concentration to be C_a (Bq m^{-3}), so that the total activity of the cylinder is $C_a R \times 10^{-6}$ Bq. If the exposure time for the detector is t_e seconds, the total number of disintegrations of ^{222}Rn during that time will be $C_a R \cdot 10^{-6} t_e$ in number (a Becquerel being 1 disintegration per second).

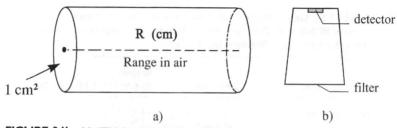

FIGURE 3.11 (a) "Thick-source" geometry of an α-emitter. It is assumed that the source (say radon) is uniformly distributed in the air, in which the α range is R (cm). Then the maximum distance from which an emitted α can reach the detector (taken to be 1 cm² in area) is R cm. The energy of the α-particles arriving at the base of the cylinder (of height R) will then vary from E_{max} (the full energy), in the case of particles contiguous to the detector at the base, to 0 for α's emitted at the top of the cylinder of air. Of the α-particles emitted in the decay of the source-nuclei contained in the cylinder, only 1/4 will reach the detector (Durrani, 1997); (b) A passive (filter-type) dosimeter for long-term radon monitoring (Ilić and Šutej, 1997). Air, containing radon gas (i.e. both isotopes, ^{222}Rn and ^{220}Rn), enters at the bottom of the dosimeter, which incorporates a permeable membrane. The time taken for the gas to diffuse through the membrane, effectively discriminates against the entry of ^{220}Rn owing to the very short half-life: 55.6 s vs 3.82 d for ^{222}Rn. Typical size of the (CR-39) detector, attached to the "ceiling" of the dosimeter, is 1 cm × 1 cm, with a thickness ranging from ∼ 100–1000 μm; typical sensitive volume of the dosimeter is 100 cm³.

Now it can be shown (see Durrani and Bull, 1987, pp. 64–69) that the particles reaching the base of a "thick source" are ¼ of the total emissions from all heights up to R (instead of ½ of all emanations in a 4π geometry). Also, if the critical angle of etching for the α-particles is Θ_c, the fraction actually revealed by etching is $\cos^2\Theta_c$. Hence, from all the above considerations, the revealed track density per cm² (i.e. the area of the base of the cylinder) is given by

$$\rho(\text{cm}^{-2}) = \frac{1}{4}C_aR \cdot 10^{-6}t_e\cos^2\Theta_c \qquad (3.3)$$

Now, Θ_c depends on the nature of the detector as well as the energy of the α-particle. In the case of a "thick source," E_α varies from the full energy of an α-particle at 0 height (viz. 5.49 MeV) to 0 energy for an α arriving from a height R cm. An average value of Θ_c for this spectrum of α-energies has, thus, to be used. For CR-39, an average value of $\Theta_c \sim 15°$ may be assumed to be reasonable, so that $\cos^2\Theta_c \sim 0.93$. The full range in air for the 5.49 MeV α's is $R \simeq 4$ cm. It is also customary to consider a radon concentration activity of $C_a = 1$ kBq m^{-3}, and an exposure time of $t_e = 1$ h $= 3600$ s.

On substituting the above values, Eq. 3.3 yields the value (for CR-39):

$$\rho \simeq \frac{1}{4} \cdot (10^3) \cdot (4 \cdot 10^{-6}) \cdot (3.6 \cdot 10^3) \cdot 0.93 = 3.35 \text{ (tracks cm}^{-2})/(\text{kBq m}^{-3}\text{ h})$$

Another popular detector for radon measurements is LR 115 (a cellulose nitrate). This, however, is sensitive to α-particles only between 2 and 4 MeV. Hence the relevant value of the range in air $R \simeq 1.9$ cm; the corresponding

$\bar{\Theta}_c$ for α's of this energy range in LR 115 is approximately $40°$, so that $\cos^2 \Theta_c = 0.59$. With these modifications, Eq. 3.3 now yields a value (for LR 115):

$$\rho \simeq \tfrac{1}{4} \cdot 1.9 \cdot 3.6 \cdot 0.59 = 1 \text{ (track cm}^{-2})/(\text{kBq m}^{-3}\text{ h})$$

The above is the track density from the decay of ^{222}Rn itself. In addition, if the daughter products, ^{218}Po and ^{214}Po, are in full secular equilibrium with the progeny (i.e. if the 'equilibrium factor' $F = 1$—which is true for a fully closed system), then one would get two further α's from these daughter products per radon decay, leading to a total track density three times the values given in Eq. 3.3—provided the daughter products are assumed to remain uniformly distributed in the cylinder of air. If, however, $F = 0.5$ (which is often the case in practice), then one would get only 1 α from the daughter products per ^{222}Rn decay—leading to a total track density twice the value given in Eq. 3.3 and the numerical values quoted above for CR-39 and LR 115.

In a practical case, if the radon activity concentration in a typical home is 50 Bq m^{-3}, but the detector exposure is for, say, 90 days ($= 2160$ h), so that the total disintegrations are equivalent to 108 kBq m^{-3} h, the track density in CR-39 from ^{222}Rn alone would be $3.35 \times 108 = 362$ tracks cm^{-2}—a very moderate value to count.

b. Types of Measurement

Homes Over the last twenty or thirty years an intensive effort has been made globally to measure radon levels in dwellings and workplaces in view of the perceived hazard to human health posed by high radon environments. Various national regulatory bodies have promulgated "action levels" for radon concentration in homes (typical values being around 200 Bq m^{-3}), beyond which remedial action becomes mandatory (see parts of Chapter 3 in Durrani and Ilić, 1997, for details).

Since radon levels in homes fluctuate with weather conditions (pressure, temperature)—e.g. low barometric pressure causes higher exhalation rates of radon from the ground—it is common to leave radon dosimeters *in situ* for, say, 3 months at a time to smooth out such variations. The radon levels in homes greatly depend on the rate of ventilation of air (greater ventilation reduces the Rn concentration); the height of a given room above the ground level (the higher the room, lower the Rn level as a rule); the building materials and structural characteristics, etc. A number of national authorities, as well as industrial firms, have produced simple "passive dosimeters" for radon measurements in buildings—some providing a service by mail (the home-owner receives a few dosimeters by post; places them at various (undisturbed) positions around the house for 6–12 weeks; and then posts them to the national authority for the etching and counting of the plastic detectors). Most of these dosimeters really aim at the counting of the Rn tracks; the radiation dose resulting from both ^{222}Rn and its daughters is then simply inferred from the observed track density, in view of the standardized geometry and characteristics of the dosimeter. Figure 3.11b shows a typical passive dosimeter. For other, improved, dosimeters—e.g. those incorporating a membrane; an electret or activated charcoal (both of which enhance the radon collection efficiency), etc.—the reader is referred to reviews by Ilić and Šutej (1997) and by Nikolaev and Ilić (1999): for the topic is too vast to be adequately covered in the present chapter. For epidemiological, biological, health-related, etc., consequences of radon exposure, the interested reader is again referred to monographs on the subject (see, e.g., Nazarof and Nero, 1988, as well as relevant chapters and references cited therein and in Durrani and Ilić, 1997, for guidance. One example of the latter is the section contributed by Muirhead (1997) in the last-mentioned book).

Field   Another important area of radon measurements in recent decades has been to study radon in the geological context. Here, there are several distinct branches of activity. The first is the determination of radon emission values in the field as an aid to uranium prospecting. Here, the reader is referred to the chapter by Khan *et al.* (1997) in the book by Durrani and Ilić (1997). Another branch is covered by chapters contributed by Fleischer (1997b) and by Monnin and Seidel (1997a) in the above-cited book on the subjects of radon-based earthquake prediction and volcanic surveillance, respectively.

Finally, we might mention the subject area of geological correlation of radon levels in the field, measured by implanting a network of plastic tubes (~ 1–1.5 m long and ~ 10–12 cm in diameter) into the soil, with SSNTD-containing cans sitting inside the tubes at the bottom of the holes. These cans usually incorporate a filter to impede the passage of the thoron gas (^{220}Rn) and thus almost to eliminate this very short-lived ($\tau_{1/2} = 55.6$ s) component of radon during its ingress. After leaving the cans *in situ* for a period of ~ 30 days, they are removed and all the plastic detectors etched and track-counted in the laboratory.

Elaborate analytical procedures have been developed, based on geostatistical methodologies of sampling (e.g. "unbalanced nesting") and working out of correlation coefficients (e.g. by using intersample distance as a variable, and plotting out a "variogram"), to establish any correlation between the localized geology/lithology and the measured radon level (in $Bq\,m^{-3}$) in the can at that sampling point (see, e.g., Badr *et al.*, 1993, for details). The present conclusion of such measurements is (cf. Durrani, 1999) that, while there is some correlation between the radon concentration levels on the ground and the underlying geology/lithology on a medium-distance scale (some hundreds of meters), the correlation on a localized scale (1–10 m) is highly erratic. Thus, for any epidemiological/environmental purpose—e.g. to determine what is likely to be the radon level inside a house built at point x in a given area—it is necessary to measure the surface radon level at that exact point x. Otherwise, we can only make general estimates of the radon levels expected over the area concerned. This has called into question the validity of generalized statements sometimes made by epidemiologists/environmentalists, etc., regarding radon levels in a geographical region or area and the expected incidence of, say, leukaemias in that region.

It must be pointed out here that, at the time of writing (autumn 2002), while there is general agreement among the experts that elevated radon levels in homes do lead to significantly enhanced incidence of lung cancer in the dwellers, the case for enhanced leukaemias from this source is still open to controversy. Further sharply focused studies of all aspects of such epidemiological correlations are, therefore, highly desirable.

2. Fission Track Dating

Fission track dating (FTD) is one of the earliest applications of the SSNTD technique (Price and Walker, 1962). The idea is to use mineral crystals themselves—which are to be dated—as the natural track detectors. The dating is based on the fact that all mineral crystals contain some uranium as a trace element—ranging from parts per billion (ppb) to several thousand parts per million (ppm) by weight—a typical value being a few ppm. The ^{238}U component (natural abundance, 99.3%) of the U-content undergoes natural fission at a fixed rate (with a fission half-life of $\sim 10^{16}$ year, i.e. a fission decay constant λ_f of $\sim 7 \times 10^{-17}\,year^{-1}$). This leaves latent fission tracks—produced by the energetic fission fragments—in the body of the crystal at a known time-rate, which can be easily revealed by etching the crystal (after grinding and polishing its surface to eliminate scratches, etc.) in an appropriate reagent. If, then, one knew the uranium content of the crystal, it would be easy to calculate the time elapsed (since the crystal had last solidified) that had resulted in the number of fission tracks actually observed in the crystal. The uranium content is actually determined by irradiating the crystal—after having eliminated the pre-existing natural tracks by heating it to a high temperature—with a known fluence (i.e. total neutrons incident per unit area—in other words, the time-integrated flux) of thermal neutrons in a reactor. The thermal neutrons produce induced fission in the ^{235}U component of the uranium content; and since the thermal fission cross section is known, this would reveal the ^{235}U content and hence the ^{238}U content (viz., 139 times the ^{235}U content).

A detailed derivation of the equations given below may be seen in Durrani and Bull (1987, pp. 200–202), but upon using the values of the natural constants involved one arrives at the following expressions:

(i) For relatively young rocks ($A \ll 4.5 \times 10^9$ year, the (α-decay) half-life of ^{238}U), the age A is given by

$$A = 6 \cdot 10^{-8}(\rho_s/\rho_i)\Phi\,year \qquad (3.4)$$

where ρ_s is the natural (or spontaneous) fission track density (cm^{-2}) on the surface of the etched crystal; and ρ_i is the induced-fission track density (cm^{-2}) resulting from a thermal-neutron fluence $\Phi(cm^{-2})$. For instance, if the natural track density is $2 \times 10^3\,cm^{-2}$ and that induced by a fluence of 10^{16} thermal neutrons cm^{-2} is $2 \times 10^4\,cm^{-2}$, then, from Eq. 3.4, $A = 6 \times 10^7$ year, i.e. 60 Myear.

(ii) For rocks of ages non-negligible compared to the (α-decay) half-life of ^{238}U (viz. 4.5×10^9 years), one needs to use a more complex age equation (since the initial quantity of ^{238}U was significantly greater than that found today, α-decay having continuously reduced the number of fissioning ^{238}U nuclei), viz.

$$A' = 6.49 \cdot 10^9 \ln\{[0.924 \cdot 10^{-17}(\rho_s/\rho_i)\Phi] + 1\} \text{ year} \qquad (3.5)$$

For instance, if $\rho_s = 4 \times 10^4 \text{ cm}^{-2}$; $\rho_i = 2 \times 10^4 \text{ cm}^{-2}$ from $\Phi = 10^{16} \text{ cm}^{-2}$ then, by substituting these values in Eq. 3.5,

$$A' = 6.49 \cdot 10^9 \cdot \ln(1.1848) = 6.49 \cdot 10^9 \cdot 0.1696 \text{ year} = 1.10 \cdot 10^9 \text{ year.}$$

As mentioned above, this is the "age of solidification" of the crystal (whether A or A'). If a severe thermal episode has intervened, which has annealed out the pre-existing fission tracks, then, as the crystal gradually cools down once again, a "closure temperature" is eventually reached after which track retention again sets in; the age given by A or A' is thus the "track-retention age" of the crystal. Actually, the fact that a thermal episode results not only in total elimination of tracks but also in the partial shrinking or shortening of other fission tracks produced as the crystal gradually cools down, means that one would obtain a histogram of lengths of tracks. Such a histogram can be used not only to infer the thermal history of a rock (Wagner, 1981) but this approach has also been used as a pointer in important geological operations such as search for oil (the temperature-cum-pressure regime over some millions of years that may produce light hydrocarbons such as petroleum in a geological formation, happens to correspond to the same temperature window—~ 55–$120°C$—that can produce partial shortening of tracks in associated apatite crystals. The latter may, thus, act as pointers to oil reservoirs (see, e.g., Green *et al.* (1989)). Other applications of fission tracks are studies of orogenesis, uplift rates of rocks, movements of geological faults, etc. (see Wagner and Van den haute, 1992).

The fission track dating of archaeological materials such as glasses—provided that they have sufficient U-content and are reasonably old—can be done on the same principles as geological samples. Interesting applications have included the dating of an obsidian dagger (Fleischer *et al.*, 1965) used in prehistoric times which had been burnt in a fire (thus resetting the fission-track clock); identifying the original source of obsidian glass found in a Mesolithic cave on the mainland of Greece—where no volcanic sources of glass exist (Durrani *et al.*, 1971); and the dating of the use of fire by the Peking Man, in whose "hearth" sphene crystals had been discovered with partially annealed fission tracks (Guo, 1982).

3. Planetary Science

Under the title "Planetary Science," we shall briefly cover some topics of research on lunar and meteoritic samples.

a. Lunar Samples

In the heyday of lunar research (early 1970s), the SSNTD technique played a prominent role in helping the scientists unravel the radiation history of the moon. The method could work wonders with minuscule quantities of lunar material: grains some hundreds of μm across—and weighing merely some tens of μg—allowed information stored in them over hundreds of millions of years to be decoded. We shall cite here only one example of the type of results that this technique could yield. The fact that near-perfect vacuum prevailed on the moon means that even low-energy (solar) cosmic rays had been able to reach the surface of the moon. Unetched lunar grains, subjected to transmission electron microscopy, showed up enormous track densities (up to 10^{10}–10^{11} tracks cm^{-2}). One important and unexpected result was the finding—made on samples collected from the surface of the moon as well as from varying depths down to ~ 3 m below the surface by both manned (US) and unmanned (Russian) missions, using drilling devices—that the track density histograms in lunar grains at **all** depths (down to ~ 3 m) were roughly the same. This— combined with the fact that even grains $\sim 400\,\mu$m across, found at depths of up to ~ 3 m, sometimes showed track density gradients across their surface (indicating that at one time they must have lain right at the top of the moon for these low-energy cosmic rays to undergo appreciable attenuation over such tiny—viz., some hundreds of μm—distances)—gave rise to the concept of "cosmic gardening" on the surface of the moon. Thus, it was postulated that a churning and mixing of the soil in the top several meters—probably caused by micrometeoritic bombardment—took place, such that over a timescale of some hundreds of millions of years, the top soil was completely turned over (Comstock *et al.*, 1971; Bhandari *et al.*, 1973; Durrani *et al.*, 1980). It is doubtful that such a phenomenon could have been discovered by any other technique. Other studies and results, presented in thousands of pages of Proceedings of Lunar Science Conferences, are too numerous to be summarized here.

b. Meteoritic Samples

Meteoritic crystals, unlike the lunar material, are still available for study in many laboratories of the world. They, too, however, constitute too specialist a field to warrant extensive coverage in this Handbook. We shall restrict ourselves to just three examples of the use of the SSNTD technique in this subject area.

Age determination   Here, some modifications have to be made to the analysis leading to Eq. 3.5 given above in Subsection III.A.2 for age A'. Meteorites are known to be probably older than any other constituents of the solar system that we have access to at present. At the time of the formation of meteorites ($\sim 4.6 \times 10^9$ years ago), there used to be a lot of ^{244}Pu (now almost entirely extinct), whose half-life is 82 Myear; most of it therefore underwent decay over the first $\sim 10\ \tau_{1/2}$, viz. the first ~ 800 Myear. A part of this decay was through fission, and most of the tracks found in meteoritic crystals are, in fact, those from the fission of ^{244}Pu rather than ^{238}U. A complicated re-iterative procedure has to be used—starting with an assumed abundance ratio of ^{244}Pu/^{238}U at a reference time t_0 years ago—to obtain the optimum value of Δt (the track **nonretention** interval immediately following the reference time t_0, when the crystals were still too hot to record tracks). Then $t = t_0 - \Delta t$ gives the track-retention, or the "fission track" age of the meteorite (see Durrani and Bull, 1987, pp. 230–232).

A further complication is that it is not just the spontaneous fission of ^{244}Pu and ^{238}U that has left tracks in the meteoritic crystals, but so have ancient cosmic rays: both by themselves and also by generating spallation recoil tracks and those resulting from cosmic-ray-induced fission in the above two isotopes. For details, see the above reference as well as relevant chapters in Fleischer *et al.* (1975).

Cooling-down Rate of the Early Solar System   Pellas and Storzer (1981) have developed an ingenious method of estimating the cooling down rate of the solar nebula, following the end of nucleosynthesis—when temperatures of the constituents of the system, including meteorite "parent bodies," were too high to allow tracks to be retained by the crystals. These authors used (i) the decay rate of the ^{244}Pu content ($\tau_{1/2} = 82$ Myear) as the "palaeo-clock"—each half-life reducing the track-production rate to one-half; and (ii) the track-retention temperature of the various constituent crystals as a "paleo-thermometer": e.g., if zircons had begun to retain the fission tracks, the meteoritic material must have cooled down to $\sim 700°$C; and if olivines had done so, the temperature must be down to $\sim 500°$C. The cooling down rates of the early solar system were calculated from such considerations to be $\sim 1°$C per million years (within a factor of ~ 10 either way). For further details see the above reference (and also Durrani, 1981; Durrani and Bull, 1987, pp. 232–235).

Determination of Pre-atmospheric Size of Meteorites   Fleischer and coworkers (e.g. Fleischer *et al.*, 1967a,b) have pioneered methods of calculating the pre-ablation (i.e. in-space, prior to atmospheric entry) size of meteorites. The main principle of the method is that as galactic cosmic rays enter a meteorite from outside, they undergo attenuation in such a way that the lower-energy (softer) components fall off first with a high attenuation coefficient (i.e. with a shorter attenuation length); the surviving (harder) cosmic-rays then attenuate with ever-increasing attenuation lengths. By measuring the fall-off rate of the cosmic-ray tracks from the present (i.e. post-ablation) top surface of the meteorite as a function of distance from that surface it is then possible, in principle, to estimate how much thickness of the outer layers of the meteorite must have ablated away to leave behind the *present* top surface. For finer details of the procedure, see the references above (also, e.g., Bull and Durrani, 1976).

4. Cosmic Ray Measurements: Particle Identification

The application of SSNTD in the field of charged-particle identification was initiated in 1967 (Price *et al.*, 1967). The ability to extract quantitative information about individual particles soon led to its use in cosmic ray measurements. The principles of such measurements, and the results obtained thereby, are outlined in reference books such as Fleischer *et al.* (1975):

Durrani and Bull (1987); Fleischer (1998); and Marenny (1987). The interested reader may find further information in a number of special issues of the journal *Radiation Measurements* (and its predecessors), dealing with topics relevant to cosmic rays, space radiation, and space missions (Benton, 1992, 1994; 1996a,b; Benton and Adams, 1992; Benton and Panasyuk, 1999; Benton and Badhwar, 2001; Benton *et al.*, 1996, 2001). Here we shall only briefly describe the main procedures used for particle identification in cosmic ray measurements.

The so-called "multiple-sheet method" is illustrated in Fig. 3.12, in which a particle that crosses five detector sheets comes to rest in the sixth sheet. After exposure to cosmic rays in space, all six sheets are etched. Since the rate of ionization increases downward, i.e. along the direction of the particle's progress, the cone-shaped etch-pits steadily lenghthen; the final etched shape (in sheet 6) is cylindrical or test-tube like because preferential etching (with a velocity V_T) ended at the site where the particle came to rest. The length of each of the ten cones gives the localized value of the ionization rate; and the distance from each cone to the final rounded-out location gives the 10 residual ranges of the particle—providing, in this case, a tenfold redundancy that improves the quality of the measurements of the cosmic-ray charge and energy. A plot of $V_T(R)$—i.e. a curve depicting the change of track-etch

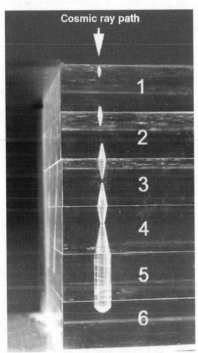

FIGURE 3.12 Photograph of a 3D model of the track of a cosmic ray slowing down in a stack of six plastic sheets. Note that the rate of change of the etched cone length with distance, in a given medium, is a unique function of the atomic number and mass of the cosmic ray particle. The length of the etched cone increases from top to bottom through sheets 1, 2, 3, 4, 5 as the velocity of the particle decreases, until finally it stops in sheet 6 (The model was made by the group headed by W. Enge at Kiel University, Germany).

velocity V_T as a function of R, the residual range of the particle—when used in conjunction with an appropriate calibration based on accelerator irradiation with known heavy ions of known energies, thus provides a high-resolution method of identifying extremely heavy cosmic rays in a polycarbonate stack. The latest results obtained from the "long duration exposure facility" (providing a 6-year exposure in a spacecraft in Earth orbit) have been published by O'Sullivan *et al.* (2001a). An instructive example of establishing ancient cosmic-ray spectra and identification of intergalactic particles leaving tracks in meteorites is found in a paper by Green *et al.* (1978).

B. Physical Sciences

The SSNTD technique has been used in a variety of nuclear physics and related studies comprising: neutron-induced fission; charged particle induced fission; photofission and electrofission; ternary fission; high energy reactions; spontaneously fissioning isomers; quest for superheavy elements; search for new materials and for exotic modes of decay; development of accelerator-driven systems; hunt for monopoles; detection of neutron quantum states, etc.

For each of the above-mentioned topics, substantial numbers of papers are cited in reference books (Fleischer *et al.*, 1975; Durrani and Bull, 1978; Fleischer, 1998; Marenny, 1987). The latest results can be found in the Proceedings of Conferences on Nuclear Tracks in Solids, published in the last decade (Brandt *et al.*, 1991; Guo *et al.*, 1993; Perelygin *et al.*, 1995; Ilić *et al.*, 1997, 2001; Chambaudet *et al.*, 1999); in special issues of *Radiation Measurements* (Benton and Panasyuk, 1999; Benton and Badhwar, 2001); and in recently published review papers (Khan and Qureshi, 1999; Ditlov, 2001; Brandt, 2001; Durrani, 2001; Benton *et al.*, 2001; Poenaru *et al.*, 2002). In the following Subsection, principles of the measurements involved and some of the main applications are outlined.

I. Particle Spectrometry

SSNTDs do not offer very fine energy resolution to allow them to be used for accurate spectrometric purposes. One reason for poor resolution is that the etching procedure introduces a good deal of statistical variability or "spread" in the measured track parameters. For instance, if one measures the diameter of the etch-pit mouth opening of an α-track in a CR-39 plastic, corresponding to monoenergetic α-particles, the diameters (when plotted as size vs frequency) will be found to possess a "histogram" of sizes rather than a sharp single-value ('δ-function') peak. The reason for this spread is twofold. The first relates to the etching being a statistical process. Secondly, unless the α's are strictly collimated, those incident on the detector surface at different angles will penetrate to different depths below the surface and thus produce different etch-pit openings. The resolution $\Delta d/d$—where Δd is the full width of the histogram peak at half-height—is usually 10–20%. In the case of "thick-source geometry"—e.g. radon α's arriving at the detector surface after having traversed different thicknesses of air—the incident particles will, of course, have residual energies ranging from 0 to E_{max} (full α-energy) at the

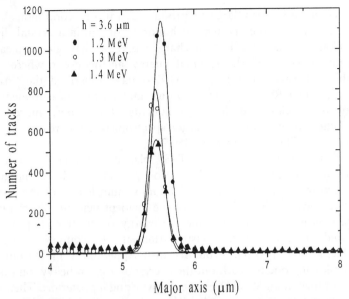

FIGURE 3.13 Etched track size distributions (measured by TRACOS, cf. Skvarč, 1993) of α-particles in CR-39 detector. Irradiations were carried out using ^{239}Pu as the α-particle source ($E_\alpha = 5.156$ MeV). Low-energy (1.2, 1.3, and 1.4 MeV) α-particles were obtained by varying the source-detector distance in air. The removed layer thickness of the etched detector is denoted by h. Since the incidence angle of the α-particles was 90°, the major axis = minor axis = track diameter (Izerrouken et al., 1999).

point of incidence at the detector, thus producing etch-pits of vastly different diameters. An example of the distribution of track sizes (diameters) for normally incident α-particles is shown in Fig. 3.13.

Continuing with the theme of using the diameter of an α-particle etch-pit in a plastic detector such as CR-39, Khayrat and Durrani (1999) have shown that the relationship between diameter-size and α-energy may have two opposite modes of dependence. If the etching is carried out until the end of the particle range is reached, then the higher the α-energy, the larger the diameter (since the diameter corresponds to the full, i.e. integrated, damage imparted to the detector material by the dissipation of the particle's energy). If, however, a "short-etching" is carried out, then—since at high particle energies, dE/dx, i.e. the linear rate of energy deposition or LET, is generally **smaller** than at lower energies—the diameter corresponding to a **high**-energy particle will, in fact, be smaller than that for a lower-energy particle for equal durations of etching. It is, thus, necessary to bear in mind which mode of etching is being employed.

Another approach is to use degrading foils (Al, plastic, etc.) of different thicknesses to filter out lower-energy α's and register only the higher-energy α's. The thickness of the degrading foil will also give an indication of the α-particle's energy.

If one is interested in using the length of the particle range in, say, a mineral crystal, then one has to remember that very high-energy particles will produce etchable tracks only toward the end of their range (when the dE/dx

has become sufficiently large to produce a trail of "continuous" damage in the medium)—viz. the last tens or hundreds of μm in a crystal. In principle, the etchable range of a known charged particle may give a measure of its minimum energy—or the residual energy at the point where it becomes etchable. Al-Najjar and Durrani (1984) have described the "track profile technique" in CR-39 to perform range and energy measurements on high-energy α-particles and fission fragments. For other, more specialized, approaches to energy (and charge) spectrometry, the reader is referred to Chapter 6 of Durrani and Bull (1987).

Fast neutrons, incident on CR-39 plastic, are able to produce proton-recoil tracks in the detector. The recoil protons can have energies ranging from zero to full neutron energy. The maximum length of the proton tracks can, thus, give us the full energy of the incident neutron. From range-energy tables, it is possible to determine the energy of the (recoil) proton from its measured range in a given plastic. Alternatively, computer programs exist (e.g. Henke and Benton, 1968; see also Appendix 1 in Durrani and Bull, 1987) allowing one to work out the energy of a given heavy ion (or a proton) from its range in an SSNTD plastic, by working *a posteriori*. Then the highest energy found for proton-recoil track might be taken to be the energy of the incident neutron.

2. Heavy Ion Measurements

Among early examples of the application of SSNTDs to the study of *low-energy* heavy ions is the work by Gottschalk and coworkers, initiated in 1983. Since then, they have published an extensive review paper (Gottschalk *et al.*, 1996). Results obtained from studies over the last few years may be found in the Proceedings of the three latest conferences on Nuclear Tracks in Solids (Chambaudet *et al.*, 1999; Ilić *et al.*, 1997, 2001). A large number of *low-energy* heavy-ion nuclear reactions have been studied, and extensive data compiled. The data comprise: total and partial cross sections; elastic-scattering angular distributions; and determination of reaction mechanisms as well as masses, kinetic energies, and angular distributions of the reaction products. This technique offers possibilities for detailed investigations of reaction Q-values, kinetic energy losses, mass transfer functions, etc.

Investigations of kinematical analyses of heavy ion reactions have been extended over the years to the *high-energy* region by workers such as Brechtmann and Heinrich (1988), and continued by several groups in Europe, USA, Russia, and elsewhere. The problems investigated include the search for projectile fragments with fractional charges; mean-free paths of relativistic heavy ion fragments; charge correlation and transfer momenta for heavy ion fragmentation, etc. Advanced methods, based on the utilization of advanced automatic systems for track analysis (see Subsection II.C.3), are a good alternative to electronic measuring systems. Excellent charge resolution can be obtained with these advanced techniques. Such studies have made useful contributions to the understanding of the basic phenomena in question.

As an example of the application of the technique, an experiment for the measurement of the total charge-changing and partial cross-sections in

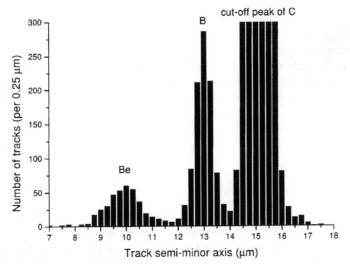

FIGURE 3.14 Spectrum of tracks produced from ^{12}C bombardment, measured at a depth of 5.5 cm in water/CR-39 stack. The initial and the exit energies of the ^{12}C beam were 275 and 207 MeV per nucleon, respectively. The total charge-changing cross section, and the cross sections for the production of B and Be fragments, were directly measured (Golovchenko et al., 2002).

the interaction of a 207 MeV/nucleon ^{12}C beam with water is illustrated in Fig. 3.14 (after Golovchenko *et al.*, 2002). Here a stack made of CR-39 detectors, with a water target, was exposed at right angles to a ^{12}C beam of initial energy 275 MeV/nucleon in the biology port of the HIMAC facility (at the National Institute of Radiological Sciences, Chiba, Japan). The detector plates were $\sim 600\,\mu$m in thickness, interleaved with the water target. The fragments produced in the target were measured along the stack plates, as were the primary ions and the product particles. After chemical etching of the detectors, track evaluations were performed by the automatic system (TRACOS, cf. Skvarč, 1993).

3. Neutron Measurements

Ever since their discovery, SSNTDs have been extensively applied to the study of the complex problems of neutron dosimetry. A number of different approaches have been used by various research groups in performing neutron dosimetry with SSNTDs around nuclear facilities and in space as well as for the study of basic physics. Recent review papers (e.g. Benton *et al.*, 2001) summarize neutron dosimetry measurements in spacecraft over the past 20 years. The results achieved so far in personnel monitoring of neutrons in workplaces are summarized in a recent review paper by Tommasino (2001). Most recently, a neutron spectrometer and a method of measurement of quantum states of neutrons with SSNTD have been developed (Nesvizhevsky *et al.*, 2000, 2002).

Generally speaking, there are two approaches: either one observes direct neutron effects in the detector such as the ^{1}H(n, p) intrinsic reaction; or one observes induced reaction products, from a "converter screen" placed in close contact with the detector, using a reaction such as ^{6}Li(n, α)^{3}H. In the following paragraphs the principles of neutron measurements are briefly outlined.

Neutrons, being electrically uncharged, cannot produce etchable tracks directly, and therefore are usually detected via charged nuclear-reaction products, using an appropriate neutron converter. There exists a quasi-linear relation between the observed track density ρ and the neutron fluence Φ, such that

$$\Phi = \frac{\rho - \rho_o}{K} \tag{3.6}$$

where ρ_o is the background track density and K is the detector response (tracks/neutron).

Consider an example: ^{10}B(n, α)^{7}Li reaction commonly used for thermal neutrons (Fig. 3.15). We would like to derive the correlation between the ρ and Φ. Suppose the range of one of the reaction products in the medium of the converter is denoted by R; then we can detect the reaction product emitted from depths in the converter ranging from 0 to R. Since in our case two particles (α and ^{7}Li), are emitted, two values of the range, R_α and R_{Li} in the medium, have to be taken into account. It can then be easily shown (see, e.g., Durrani and Bull, 1987, p. 69) that the track density registered at the top surface of the converter, and hence of the detector which is in contact with it, is given by

$$\rho = \frac{1}{4}n(R_\alpha \cos^2 \Theta_c + R_{Li} \cos^2 \Theta_c) \tag{3.7}$$

where Θ_c is the critical angle of etching (assumed to be the same for both types of particles) and n is the reaction density per unit volume of the source,

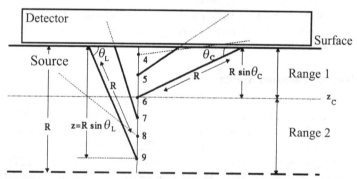

FIGURE 3.15 Detection efficiency of α-particles generated within a thick source. The body of the source material is subdivided into two regions: Region I and Region II. Region I stretches from depth $z = 0$ to $z = z_c = R \sin \Theta_c$, where Θ_c is the critical angle of etching. In this region Θ_c is the governing factor: if the latent track makes an angle $> \Theta_c$ (e.g. for track 5), it will be revealed by etching; if $\Theta < \Theta_c$ (e.g. for track 4), it will fail to be revealed; track 6 just makes it (with $\Theta = \Theta_c$). Region II extends from depth z_c to $z = R$. Here the direction of emission, i.e., angle Θ, is the governing factor. Thus from depth 9, a track making a minimum angle $\Theta_L = \sin^{-1}(z/R)$ sets the limit: all tracks contained within the angles Θ_L to $\pi/2$ with the surface will be revealed by etching. Latent tracks 7 and 8 represent those that will, and will not, etch out, respectively (after Durrani and Bull, 1987).

viz. $\Phi\Sigma_a$. Here Σ_a is the macroscopic cross section (cm^{-1}) for the reaction on boron atoms, and Φ is the number of incident neutrons (cm^{-2}).

In the case of fast neutrons, a more complex, energy-dependent expression has to be used for the interaction

$$n = \int_0^E \Sigma(E)\Phi(E)\,dE \qquad (3.8)$$

where $\Sigma(E)$ and $\Phi(E)$ are the values of the two parameters in the neutron energy interval E to $(E+dE)$.

An important parameter is the (lower) limit of detection. This value decreases with decrease in the background, although it never vanishes. The smallest detectable neutron fluence Φ_d is defined as

$$\Phi_d = \frac{L_d}{K} \qquad (3.9)$$

where L_d is the lowest track density detectable, and K is the response (tracks/neutron).

A new fast neutron detection technique called coincidence counting of tracks, by which the background signal can be greatly reduced, has recently been proposed. The essence of the coincidence counting method is the measurement of the $^1\text{H}(n, p)$ reaction product tracks with a pair of SSNTD foils placed in close contact during the irradiation. For details, see Lengar *et al.* (2002). Track density vs neutron fluence for single and coincidence tracks is shown in Fig. 3.16.

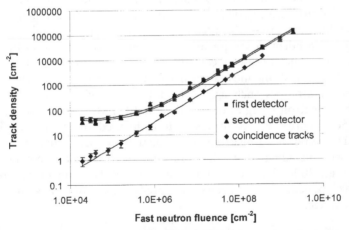

FIGURE 3.16 Track density vs neutron fluence for single and coincidence tracks. Here the detection of fast neutrons is performed with a pair of **CR-39** detector foils (via $^1\text{H}(n, p)$ reaction). After subsequent chemical etching, the evaluation of the etched tracks is performed by automatic track analysis system (**TRACOS**). Only tracks produced by the same recoil nuclei in the surface layers of both detector foils are taken into account as 'coincident tracks'. The lower limit for neutron detection by the coincidence detector was found to be two orders of magnitude lower than that obtained with a detector based on counting tracks in a single foil of **CR-39** (Lengar *et al.*, 2002).

a. Thermal Neutrons

Thermal and epithermal neutrons can be measured by using α-sensitive detectors (e.g., cellulose nitrates, allyldiglycol carbonate) in combination with neutron converter screens containing ^{6}Li ($\sigma_a = 950$ b) or ^{10}B ($\sigma_a = 3840$ b), where b stands for barn: $1 \, b = 10^{-28} \, m^2$. For this purpose, a number of commercially available neutron converters (metal B; ^{10}B; LiF; $Li_2B_4O_7$; B_4C) have been developed. Detector response for, e.g., CR-39/B detector/convertor system, has been found to be $\sim 8 \times 10^{-3}$ tracks/n_{th} under the etching conditions: 6.25 M NaOH at 70°C for 30–180 min (Ilić *et al.*, 1986). ^{235}U(n_{th}, f) reaction ($\sigma_f = 586$ b), with detectors such as Makrofol KG; LG-760 or LG-750 phospate glasses, etc., has also been used for the measurement of thermal neutrons. Similarly, the ^{239}Pu(n, f) reaction can also be used for such measurements. Details about the construction and characteristics of dosimeters which contain all three fissionable isotopes ^{237}Np, ^{235}U, and ^{238}U are described in Fleischer *et al.* (1975) and references therein.

b. Fast Neutrons

Fast neutrons leave recoil-proton tracks from the H content of polymeric detectors; some contribution also comes from (n, p) and (n, α) reactions with the C, N, and O constituents of different plastic detectors. All such tracks are termed "intrinsic tracks"—though the vast majority ($\sim 95\%$) are proton-recoil tracks. These tracks can be both chemically and electrochemically etched. In a series of papers published by Matiullah, Durrani and coworkers in the 1980s (see, e.g., Matiullah and Durrani, 1987a,b; Matiullah *et al.*, 1988; Durrani and Matiullah, 1988; James *et al.*, 1987), these authors have described the construction of 3-dimensional dosimeters consisting of layers of plastic detectors with varying hydrogen contents—some acting as "radiators" of recoiling protons—which can act as direction-independent as well as energy-independent dose-equivalent dosimeters (see Subsection III.C.1).

If one wants to use a converter for fast neutrons (as one does with thermal neutrons), the most attractive converters are those based on the (n, f) reaction in isotopes such as ^{232}Th and ^{237}Np, which have thresholds for fission at neutron energies of ~ 1 MeV and 0.1 MeV, respectively.

The majority of recent research is focused upon characterization of SSNTDs' response to neutrons as a function of energy and the direction of incidence—including the development of predictive computer codes (Peurrung, 2000; Luszik-Bhadra *et al.*, 2001). The aim of this research is to develop neutron dosimeters that are accurate over a sufficiently large energy range.

4. Nuclear and Reactor Physics

SSNTDs have been used in about 100 nuclear laboratories worldwide, many of which have their own accelerators and/or nuclear reactors. SSNTDs are particularly widely used in a variety of nuclear physics experiments, e.g. for the recording of rare events (such as spontaneous fission; search for monopoles). The earlier landmarks in the history of nuclear applications of

SSNTDs are surveyed in the book by Fleischer *et al.* (1975). The interested reader may find new results in a number of contributions in Proceedings of conferences on Nuclear Tracks in Solids published in the 1990s or so (Brandt, *et al.*, 1991; Guo *et al.*, 1993; Perelygin *et al.*, 1995; Ilić *et al.*, 1997, 2001; Chambaudet *et al.*, 1999).

Radioactive decay with spontaneous emission of particles heavier than α's was predicted in 1980, and such an "exotic" nuclear decay mode was first observed by Rose and Jones (1984). Experimental difficulties are caused mainly by low yield in the presence of a strong background of α-particles. Here it is an advantage to use those types of detectors (such as polyethylene terephthalate or phosphate glass) which are not sensitive to alphas and other low-Z particles. Price *et al.* (1985) were the first to use the SSNTD technique to study the spontaneous emission of heavy ions from certain high-Z radionuclides (^{222}Ra, ^{224}Ra). This research was continued by other groups from the former Soviet Union, Europe, USA, China, and Japan, etc. Systematics of experimental results obtained by SSNTDs and other detectors (until now 19 nuclides have been known to have heavy-fragment radioactivity with the emission of ^{14}C, ^{20}O, ^{19}F, ^{24,25,26}Ne, ^{28,30}Mg and ^{32,34}Si); comparison of theory with experiments; and identification of possible candidates for future experiments are presented in a recently published review paper (Poenaru *et al.*, 2002). As an example, two recently obtained results are given below.

Tretyakova *et al.* (2001) have studied the cluster decay of ^{242}Cm$\rightarrow^{34}$Si$+^{208}$Pb, and measured its partial half-life using phosphate glass detectors. The corresponding partial half-life was found to be $(1.4\pm0.3)\times10^{23}$ s. The branching ratios relative to α-decay and relative to spontaneous fission were found to be 1.0×10^{-16} and 1.6×10^{-9}, respectively. The exotic nuclear decay of ^{230}U$\rightarrow^{22}$Ne$+^{208}$Pb was investigated with a polyester track detector (Qiangyan *et al.*, 2002), and the preliminary branching ratio for the emission of heavy ions to α-particles was found to be $(1.3\pm0.8)\times10^{-14}$.

SSNTDs have been used to measure cross sections down to 10^{-35} cm^2 as well as to visualize a number of interesting nuclear processes (such as ternary fission). Beginning with the earliest observation of fission tracks in mica, SSNTDs have been used to generate new data on spontaneous fission half-lives, life-times of compound nuclei, fission cross sections and fission barrier heights (Fleischer *et al.*, 1975; Fleischer, 1998; Gangrskij *et al.*, 1992; Khan and Qureshi, 1999; Durrani, 2001).

The search for superheavy elements (SHE) is an ongoing activity, and SSNTDs are playing an important role to verify theories such as those predicting that there should be an "island of stability" for elements around $Z=114$, where half-lives could go up to 10^3 years (Brandt, 2001; Durrani, 2001). In the past, search for the tracks of superheavy elements in meteorites has been conducted by Flerov and his coworkers (see e.g. Perelygin and Stetsenko, 1977). Besides these applications, SSNTDs are useful for studying properties of new man-made heavy elements with Z values beyond 104.

In the context of reactor physics, several laboratories are involved in the research to transmute long-lived poisonous radioactive materials (e.g. ^{239}Pu) into shorter-lived fission fragments or stable nuclides. SSNTDs play an important role in the determination of the energy-dependent neutron fluence in small volumes (a few cm^3), or in the exact profile determination of the primary proton beams (Brandt, 2001). Neutron flux distributions in and around the core of research reactors have also been studied using reference glasses containing known amounts of uranium (Durrani, unpublished data).

5. Radiography

An important property of SSNTDs is their ability to register and localize individual radiation events. Thus, an image of the objects emitting or transmitting radiations is formed on the SSNTDs exposed to them. A number of radiographic techniques have been developed for the physical and chemical characterization of materials. The application of SSNTDs as a research tool in the laboratory as well as for large-scale analytical purposes (e.g. for non-destructive imaging for industrial use) is being explored on an on-going basis. Such applications can be classified into two categories: (i) autoradiography, and (ii) transmission radiography. According to the nature of the detected radiation and/or experimental setup, autoradiography can be subdivided into: (i) autoradiography based on natural radioactivity; (ii) neutron-induced autoradiography; (iii) ion-induced autoradiography; (iv) photon-induced autoradiography; and (v) ion or neutron activation autoradiography. Transmission radiography can also be subdivided into: (i) neutron radiography; (ii) ion radiography; (iii) ion lithography; and (iv) ion channelography. The basic principles of the techniques are given in various reference books (e.g. Fleischer *et al.*, 1975; Flerov and Bersina, 1979; Harms and Wyman, 1986; Spohr, 1990; Rusov *et al.*, 1991). Further information may also be found in the Proceedings of the conferences on Nuclear Tracks in Solids (Brandt *et al.*, 1991; Guo *et al.*, 1993; Perelygin *et al.*, 1995; Ilić *et al.*, 1997, 2001; Chambaudet *et al.*, 1999). Schematic representation of tracks in "thin" (i.e. of thickness less than the range R of the particle) and "thick" ($> R$) detectors is shown in Fig. 3.17.

A detailed physical model of image formation in SSNTDs was formulated by Ilić and Najžer (1990a). On the basis of this model, the following types of calculations were carried out: large-area signal transfer function (Ilić and Najžer, 1990a); space-dependent transfer functions in thin (Ilić and Najžer, 1990b) and thick (Ilić and Najžer, 1990c) detectors; and the relevant image quality factors (Ilić and Najžer, 1990d). The theoretical calculations were verified experimentally for a number of SSNTDs (Ilić and Najžer, 1990a–d; Pugliesi and Pereiria, 2002). Large-area signal transfer function relates the detector's optical density D to the exposure ε. Here D is defined as

$$D = \log \frac{I_o}{I} = -\log T \tag{3.10}$$

where I_0 is the intensity of incident light, I is the intensity of transmitted light, and T is the fraction of light transmitted by the detector. On the basis of the above-mentioned model it was found that

$$D = -\log[(T_f e^{-\varepsilon} + T_t(1 - e^{-\varepsilon})], \text{ for thin detectors; and} \qquad (3.11)$$

$$D = -\log[(T_f e^{-\varepsilon} + T_{t1}(1 - e^{-\rho\pi r_{t1}^2}) + T_{t2}(e^{-\rho\pi r_{t1}^2} - e^{-\varepsilon})], \qquad (3.12)$$
for thick detectors.

Here the exposure (a dimensionless quantity) is defined as

$$\varepsilon = S_t \rho \qquad (3.13)$$

where S_t is the average track mouth-opening area and ρ is the track density. The meanings of the symbols T_f, T_t, T_{t1}, T_{t2}, and r_{t1} are explained in the caption of Fig. 3.17.

Autoradiographic image quality expressed in terms of three image-quality factors is characterized by:

1. The spatial resolution quoted in terms of the image unsharpness for $\varepsilon < 1$ and a track size smaller than the range R of the particle is approximately equal to $0.77R\cos\Theta_c$. The track size influence on image unsharpness begins to predominate when the average track size is approximately equal to the range of the particle.

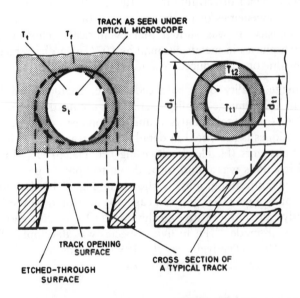

a) THIN DETECTOR   b) THICK DETECTOR

FIGURE 3.17 Schematic representation of tracks in thin (a), and thick (b), detectors. Light transmission through the track-free area is denoted by T_f. Light transmission through the area covered by tracks in the thin detector is denoted by T_t, whereas S_t is the track area in the thin detector. In the thick detector, an inner circle with a track diameter $d_{tI} = 2r_{tI}$, and light transmission T_{tI}, is surrounded by an external ring (responsible for the darkening of the image) with light transmission T_{t2}. Track diameter is denoted by $d_t = 2r_t$ (Ilić and Najžer, 1990a).

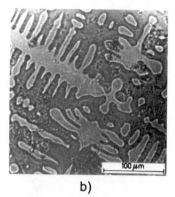

a)   b)

FIGURE 3.18 Neutron-induced radiographs of boron-containing carbon steel. Autoradiographs of the same area taken with CR-39 detector (a), and gelatine (b), are presented. The autoradiograph (a) was obtained at a thermal neutron fluence of 9.5×10^{11} cm^{-2}, and the autoradiograph (b) at the thermal neutron fluence of 5×10^{15} cm^{-2}. The boron is concentrated in the dark interdendritic areas (Najžer et al., 1982).

2. The maximum value of the detector contrast attainable with SSNTDs was found to be 1/ln10. Contrast sensitivity (minimum recognizable fractional change in concentration) as small as 10% can be obtained with some detectors and/or etching conditions.

3. The smallest detail discernible in a radiographic image is determined by the inhomogeneities of the image caused by statistical fluctuations in track density. It was shown that cylindrical inclusions of light-isotope-rich phases in metals as small as 10^{-15} g can be detected using some neutron- or ion-induced reactions.

Optimum image quality of a balanced image, characterized by equal importance of all three image quality factors, is obtained at $\varepsilon = 1$. In the Subsection III.B.6 (Elemental Analysis and Mapping), the use of some radiographic techniques is presented in detail. Concentration sensitivity of the method for light elements (H, He, Li, B) using ^{2}H(^{3}He, p)^{4}He, ^{6}Li(n, α)^{3}H, and ^{10}B(n, α)^{7}Li reactions was found to be down to ppm range. An example of neutron-induced autoradiography is shown in Fig. 3.18. Uranium concentration in the ppb range (10^{-9} g/g) can easily be measured by the ^{235}U(n, f) reaction. Recently an ultrasensitive technique (10^{-14}–10^{-15} g/g) for the determination of man-made ^{239}Pu in living species was developed (Perelygin and Churburkov, 1997; Perelygin et al., 1999) by the use of combined neutron and gamma ray activation techniques.

6. Elemental Analysis and Mapping

An interesting use of the SSNTD technique has been its application in measuring the amount and spatial distribution of certain types of elements in a sample.

Here there are two possibilities. First is where the element in question is radioactive in itself—giving out, say, α-particles or fission fragments. The second is that exposure to, say, thermal neutrons can produce a reaction in the given isotope, leading to the emission of charged particles such as α's or fission fragments.

Geologists who may, for instance, wish to determine the uranium content of a rock, routinely use the first approach. Here the only requirement is to place an α-sensitive plastic detector (e.g. CR-39, CN 85) in contact with a roughly polished surface of the U-containing rock for an appropriate length of time. The detector is then removed, etched, and α-counted under an optical microscope. A simple equation then gives the U content, as shown below.

Suppose the range of the α-particle in the rock sample is R cm, then it can be shown rigorously (e.g. see Durrani and Bull, 1987, pp. 64–69) that of all the α's emanating from all depths down to R cm below the top surface of a thick sample, only 1/4 will manage to reach the top surface (and hence the detector). It is usual to convert the linear range R cm to mass per unit area of the rock sample by multiplying R cm by the density (g cm^{-3}) of the material, yielding say m_s (g cm^{-2}), where s refers to the rock sample. Suppose also that the weight-fraction of the element (or isotope) x of interest—^{238}U in our case—is a fraction C_x of the bulk matrix; and the α-decay constant of element x is λ_x (s^{-1}). Then, remembering that the number of atoms per gram of x is given by N/A_x [where N is Avogadro's number (6×10^{23}), and A_x the atomic weight of the element (or isotope) x], we obtain the following expression for the etched-track density of the α-particles after an exposure time of t (s)

$$\rho(tracks\,cm^{-2}) = \tfrac{1}{4}m_s\lambda_x(C_x/A_x)\,Nt\cos^2\Theta_c \qquad (3.14)$$

where Θ_c is the critical angle of etching for α-particles incident on the detector, which leads to the etching efficiency $\cos^2\Theta_c$ in the case of a "thick source."

Let us take the case of ^{238}U (which is nearly the whole of the element U) in a silicate rock. The range of α-particles in silicates is ~ 4 mg cm^{-2} ($\sim 15\,\mu$m, the density being ~ 2.7 g cm^{-3}). With a $\tau_{1/2} = 4.47 \times 10^9$ year, the α-decay constant is found to be $(\ln 2 = 0.693)/((4.47 \times 10^9)(3.15 \times 10^7))s = 0.492 \times 10^{-17}\,s^{-1}$. But it must be remembered that, normally, the ^{238}U content is in secular equilibrium with all its descendants down to ^{206}Pb in the rock, so that a total of 8 α's are produced per decay of ^{238}U (all at the same rate). If, then, the U content of the sample is taken as $C_x = 1000$ ppm ($= 0.1\%$ by weight), the time of exposure as $t = 24$ h $= 8.64 \times 10^4$ s, and the critical angle for α's in a CR-39 detector, $\Theta_c = 15°$ (with $\cos^2\Theta_c = 0.933$), Eq. 3.14 yields the following value for ρ:

$$\rho = (8/4)\,(4 \cdot 10^{-3})\,(0.492 \cdot 10^{-17})\,[10^{-3} \cdot 6 \cdot 10^{23}/238]\,(8.64 \cdot 10^4)\,0.933$$
$$= 8 \cdot 10^3 \text{ tracks cm}^{-2}$$

which is an easy track density to measure. If all the other values are known except C_x, the observed value of ρ will then immediately yield the value of C_x, i.e. the U content of the rock.

As an alternative to the above—where spontaneous production of α's is taking place from the decay of an element (^{238}U and its descendants in our case)—one could induce fission in the ^{235}U content by thermal neutron bombardment in a reactor.

Here, one replaces C_x of Eq. 3.14 by IC_x (where $I = 7 \times 10^{-3}$ is the ratio of ^{235}U atoms to the total U atoms). One also has to replace the total number of α's produced per ^{238}U atom over time t (viz. $\lambda_x t$) by the expression $\Phi \sigma_f$, where Φ (cm^{-2}) is the time-integrated flux (i.e. the fluence) of thermal neutrons, and σ_f (cm^2) is the induced-fission cross section per ^{235}U atom. And, of course, N/A_x is now replaced by N/A_5, where A_5 is the atomic weight of ^{235}U. Finally, one must remember that there are two fission fragments (going in opposite directions) per fission (so that the fraction arriving at the top source is 1/2 rather than 1/4).

With all the above changes, the formula in the case of induced fission of the ^{235}U content of a rock resulting from a fluence Φ of thermal neutrons (cm^{-2}), we obtain the following expression for the track density (cm^{-2}) of fission fragments in the detector placed in contact with the top surface of the sample:

$$\rho \, (\text{tracks cm}^{-2}) = \tfrac{1}{2} m_s (IC_x N/A_5) \Phi \sigma_f \cos^2 \Theta_c \qquad (3.15)$$

Small changes need to be made (compared to Eq. 3.14) in the numerical values of m_s and Θ_c; for fission fragments the range of (a single) fission fragment is closer to ~ 5 mg cm^{-2} in silicates; and Θ_c for fission fragments is closer to $\sim 10°$ in CR-39 (so that $\cos^2\Theta_c \sim 0.97$).

As an example, if we use $\Phi = 10^{13}$ thermal neutrons cm^{-2} with $\sigma_f \sim 5.86 \times 10^{-22}$ cm^2; and changing C_x of total U to 10 ppm ($= 10^{-5}$), we obtain the following value for ρ from Eq. 3.15:

$$\rho = (1/2) \, (5 \cdot 10^{-3}) \, (7 \cdot 10^{-3} \cdot 10^{-5}) \, (6 \cdot 10^{23}/235) \, (10^{13}) \, (5.86 \cdot 10^{-22}) \, 0.97$$
$$= 2.54 \cdot 10^3 \text{ tracks cm}^{-2}$$

Note that if one does not use an external detector in the case of induced fission, but simply counts the number of fission tracks reaching the (well-polished) top surface of the rock sample after appropriate etching, then—since the critical angle Θ_c for silicates ~ 0—the $\cos^2\Theta_c$ term in Eq. 3.14 may be replaced by 1. (Since α-tracks are not revealed by etching in silicate crystals, the rock sample cannot be utilized as a detector in the first approach above, using ^{238}U.)

A third scenario for elemental analysis is when, for instance, thermal neutrons are used to produce an (n, α) reaction in a given element (or isotope) distributed in the main matrix of the sample. Here, the situation is analogous to that of induced fission—except for the fact that σ_a now is the cross section for the (n, α) reaction, and only one α is emitted per reaction. With these modifications, Eq. 3.15 now becomes

$$\rho \, (tracks \, \text{cm}^{-2}) = \tfrac{1}{4} m_s (C_x N/A_x) \Phi \sigma_a \cos^2 \Theta_c \qquad (3.16)$$

A plastic detector must, of course, now be used, and Θ_c refers to the etching of the α-tracks in the detector.

The last approach has been used for studying the elemental mapping and content-estimation of elements such as Li, B, Pb, Po, Th, U, Pu, and many others (see examples in Fleischer *et al.*, 1975; Flerov and Bersina, 1979; Durrani and Bull, 1987; Fleischer, 1998; Durrani, 2001).

It may be worth emphasizing the fact here that, in the case of all the above-described approaches, one not only obtains an estimate of the quantity of the element or isotope under examination in a given sample but also a replica of the distribution pattern of that element in the sample; hence the word "mapping" in the title of this Subsection.

C. Biological and Medical Sciences

I. Radiation Protection Dosimetry/Health Physics

Measuring doses of radiation to which humans have been exposed is important for their biological safety. Among topics related to the application of SSNTDs in radiation protection (or health physics) are: (i) radon dosimetry (in homes, workplaces, mines); (ii) neutron dosimetry (especially around nuclear or accelerator facilities); and (iii) heavy ion dosimetry (space missions; supersonic air travel; personnel dosimetry of regular crew members of high-altitude aircraft). These subjects are briefly covered below.

a. Radon Dosimetry

Exposure to radon gas, which is present naturally in the environment, constitutes over half of the radiation dose received by the general public annually. The deleterious effects of high radon levels on human health—especially in regard to lung cancer, though less so in regard to leukaemias—are well documented. At present, the most widely used method of measuring radon concentration levels is based on the use of SSNTDs (see Subsection III.A.1, which covers many aspects of radon measurements). For the coverage of dosimetric and health physics aspects of radon, the reader is referred to Jönsson (1997a,b), Miles and Ball (1997), Muirhead (1997), Pineau (1997), and Sohrabi (1997). For a review of radon as a health hazard at home see Durrani (1993). The use of SSNTDs, whether bare or placed in special chambers (passive dosimeters—produced by many national regulatory bodies as well as commercially: see Fig. 3.11b) is quite simple and cheap, and provides the possibility of large-scale surveys with many simultaneous measurements in dwellings, etc. The information, integrated over a long enough time (several days to several months, in order to smooth out diurnal and seasonal variations), gives reliable average values of the biological dose. The activity concentration of ^{222}Rn as small as $1\,\mathrm{Bq\,m^{-3}}$ may be measured with some of these dosimeters. Radon levels in homes vary greatly from country to country, and even from region to region in a country; but the average global values are around $50\,\mathrm{Bq\,m^{-3}}$. Regulatory bodies in a number of countries have laid down "action levels" for radon activity concentration in homes (e.g. $200\,\mathrm{Bq\,m^{-3}}$ in both new and existing homes in the UK), beyond which remedial action becomes mandatory. Exposed dosimeters—usually both in living rooms and in bedrooms—may be sent, even by post, to a laboratory where the SSNTD detectors are etched and evaluated, usually by automatic track counting devices.

A variety of radon dosimeters have been developed worldwide. The most widely used are those based on the work of Fleischer and Mogro-Campero (1978) in the USA; Urban and Piesch (1981) in Germany; Bartlett and Bird (1987) and Hardcastle et al. (1996) in the UK; Doi et al. (1994) in Japan; Tommasino (1988) in Italy; and Vorobyev et al. (1991) in Russia. Calibration and standardization of such detectors have been described by Miles (1997); their utilization in radon monitoring devices by Ilić and Šutej (1997); and their comparison with other radon monitoring devices by Monnin and Seidel (1997b). Properties of SSNTDs suitable for Rn measurements and transformation of latent to visible tracks are described by Durrani (1997) and Tommasino (1997). For the coverage of the applications of these dosimeters in geophysical science the reader is refered to: Åkerblom and Mellander (1997); Balcázar (1997); Fleischer (1997a,b); Hakl et al. (1997); Khan et al. (1997); and Monnin and Seidel (1997a). The interested reader may find more information in the recently published review by Tommasino (2001). In a recent article in the American Scientist, Fleischer (2002) has presented a lucid account of "serendipitous radiation monitors", including a description of retrospective monitoring of radon exposure by examining α-particle tracks recorded by the CR-39 lenses of spectacles worn by their subjects.

b. Neutron Dosimetry

Several aspects of neutron measurements have been covered in Subsection III.B.3 above. Neutron dosimetry has been of importance ever since nuclear reactors came into operation round the world (i.e. since the 1940s and 1950s)—for it was recognized early on that the exposure of reactor personnel to fast and slow neutrons must be kept under tight surveillance in view of the health hazards involved—not least in the case of criticality accidents. Since the 1970s a number of SSNTD-based dosimeters have been evolved. One of the earliest was described by Walker et al. (1963). The basic details of such dosimetric systems may be seen in Chapter 7 of Durrani and Bull (1987); see also review paper by Tommasino (2001).

For thermal neutrons, (n, α) reactions are generally utilized, incorporating converter screens containing compounds of boron and/or lithium, e.g. $Li_2B_4O_7$ (the (n, α) cross section of ^{10}B—which is $\sim 20\%$ of natural B—being 3840 barns $(10^{-28} \, m^2)$; and of ^6Li (7.5% of natural Li) being 940 barns), placed in contact with α-sensitive detector foils such as LR 115, CN 85 or CR-39. Occasionally the (n, f) reaction is also employed—but the fissile materials incorporated in the converter screen, such as ^{235}U, can give the wearer an unacceptably large γ-ray dose from the fission reactions produced.

Fast neutron doses can be measured either by examining the "intrinsic" recoil proton tracks, produced through interactions with the hydrogen content, say, of the CR-39 detector, or (cf. Harrison, 1978) by producing "fast fission" in ^{238}U, ^{232}Th or ^{237}Np (which have thresholds ranging from neutron energies of $\sim 1 \, MeV$ to $\sim 100 \, keV$); but, again, background γ-radiations would present a health hazard. For very-high-energy neutrons, reactions such as $^{12}C(n, n')3\alpha$ have been used, which have a threshold of ~ 10 MeV (see, e.g., Balcázar and Durrani, 1980; Al-Najjar, *et al.*, 1986).

An important concept in neutron dosimetry is the determination of the "equivalent dose," which takes into account the "quality factor" of neutrons that depends on their energy. The quality factor (a dimensionless quantity) for fast neutrons is taken to be 5–20, depending on the energy (Dörshel *et al.*, 1996). The unit of dose equivalent is sievert: 1 Sv = (Absorbed dose in Gy) × Quality factor; where the SI unit of absorbed dose, gray, is defined as: $1\,Gy = 1\,J\,kg^{-1}$. It so happens that the energy-dependent fission cross section of ^{237}Np mimics the variation of the quality factor of neutrons versus their energy. Hence a nearly energy-independent track production rate can be obtained in a dosimeter incorporating ^{237}Np as a converter.

Much work has also been done to develop direction-independent and energy-independent neutron dosimeters incorporating, for instance, layers of "radiators" containing different proportions of hydrogen content and hence yielding different quantities of recoiling proton tracks. Some of these dosimeters are based on the electrochemical etching (with energy-dependent efficiency of revelation) of the tracks (see, e.g., papers by Matiullah, Durrani and their coworkers: Matiullah and Durrani, 1987a,b; Matiullah *et al.*, 1988; Durrani and Matiullah, 1988; James *et al.*, 1987). More complex systems have also been evolved, e.g. "albedo dosimeters", which incorporate a CR-39 detector, a ^{6}LiF or $Li_2B_4O_7$ radiator, and a Cd cover. Albedo dosimeters respond not only to the incident fast neutrons but also to those reflected by the wearer's body and thus thermalized (see, e.g., Gomaa *et al.*, 1981). The aim of all such dosimeters is for their dose equivalent response to cover the whole energy spectrum of the incident neutrons.

c. Heavy Ion Dosimetry

Reference has been made in Subsection III.B.2 above to heavy ion measurements, and in Subsection III.A.4 to cosmic-ray measurements. In recent years, increasing attention has been paid to the heavy ion and cosmic-ray dose received, in particular by the crew members, but also by the travelling public, in high-altitude and supersonic aircraft. At such heights (~ 10000 m and above)—and during space flights—solar flares as well as solar and galactic cosmic rays may present a non-negligible health hazard to humans (Spurný, 2001); at ground level, these radiations are severely curtailed by the Earth's atmosphere. Surveys of aircrew exposure to such radiations have been carried out by Curzio *et al.* (2001a,b); O'Sullivan *et al.* (2000, 2001b); Donnelly *et al.* (2001), and by others, using arrays of both active and passive detectors, including SSNTDs. Fluences of, and doses imparted by, high-energy and high-charge particles at such high altitudes have been successfully measured by these devices.

2. Environmental Sciences

The best applications of the SSNTD technique to environmental studies are obviously those that exploit its strongest suits, namely where integration of the effects in question is advantageous (e.g. when the signal is weak in terms of intensity or temporal frequency); the phenomenon contains charged-particles—be it in the presence of more intense but weakly ionizing radiations; and where field studies less amenable to electronic gadgetry are of importance. For illustration, we treat below some representative areas of successful SSNTD applications.

a. Measurement of Uranium and Radium Concentrations in Water, Milk, Soil and Plants, etc.

Such measurements have been successfully carried out, among others, by: Gamboa *et al.*, (1984); Ramola *et al.* (1988); Fleischer and Raabe (1977). The methods are straightforward. Plastic detector foils are either left immersed in water or in contact with the samples in question, or implanted in the soil and left undisturbed for a period of days or weeks (depending on the intensity of the signal). After exposure, the detectors are retrieved, etched (chemically or electrochemically) and counted for α-particles. Sohrabi and coworkers (1993) have, in particular, carried out SSNTD studies on high natural radiation levels in homes and schools in the Ramsar area of Iran. Results of such studies—especially in areas of high natural radiation levels in India, Brazil, and China—may be found in the Proceedings of the conferences on this subject (Sohrabi *et al.*, 1993; Wei *et al.*, 1997; Burkart *et al.*, 2002). A posthumous review paper by Somogyi (1990) gives a useful account of the environmental behavior of radium.

The U-content of plants, soil, etc., can also be carried out by inducing thermal-neutron fission in the ^{235}U isotope, followed by autoradiography of the leaves, etc., which may have assimilated U either from the soil or as a result of deposition of U-bearing dust particles (see, e.g., Bersina *et al.*, 1995).

b. Plutonium in the Environment

Environmental hazards of the long-lived ($\tau_{1/2} = 24100$ year) radioisotope ^{239}Pu, forming a part of the nuclear waste generated all over the world by nuclear power plants from their ^{238}U-containing fuel, have highlighted the need for strict surveillance of plutonium in the environment. Perelygin has been a strong proponent for the need of such surveys using the SSNTD method (see, e.g., Perelygin and Churburkov, 1997; Perelygin *et al.*, 1999). The methods proposed by these authors—entailing thermal-neutron fission of ^{239}Pu—aim at attaining a measurement sensitivity of 10^{-14} to 10^{-15} g of Pu per g of human tissue. The benefits of being able to quantify such health hazards to all living species by relatively inexpensive methods on a large scale are obvious.

c. "Hot Particle" Measurements

Our last illustrative topic in this subject area is the measurement of "hot particles" released, in particular, in the meltdown of the Chernobyl nuclear power plant in Ukraine in 1986. The nuclear fallout covered vast areas not only in the former Soviet Union and the rest of Europe but also in many other parts of the world. The worst affected areas were, of course, in Ukraine and the nearby Belarus, and in the surrounding regions. Vast amounts of radioactivity were carried by the blast and the accompanying plume of active debris; these eventually settled on forests, plants, crops, and soil in both

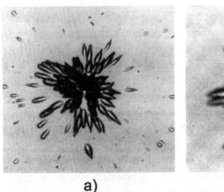

a)   b)

FIGURE 3.19 Hot-particle detection. Images of hot particles, absorbed by filters in the working zones of plutonium reprocessing plant Mayak, in Ukraine, were obtained on CR-39 detectors, characterizing hot particles of relatively high (a), and low (b), activity (Bondarenko et al., 1999).

inhabited and uninhabited areas. Winds and rains carried them indiscriminately in all directions in the following weeks and months. Huge quantities of "hot particles" were found, in subsequent measurements over the years, to have been deposited on tree leaves; and those buried in the soil kept migrating sideways, upwards, and downwards by the movement of groundwater and through the action of weather, rains, and other natural forces.

The SSNTD technique has proved to be a very suitable method of measuring the effects of hot particles in the environment (Fig. 3.19)—both their activity and their temporal and spatial movements. Some useful papers here are those by Boulyga et al. (1999); Bondarenko et al. (1999); and Sajo-Bohus et al. (1998). Here, SSNTD radiography was applied to identify the aerosol-contained hot particles from the Chernobyl fallout. Fission was induced in the transuranium elements deposited on aerosol filters, using the (n, f) and (γ, f) interactions produced by thermal neutrons and energetic gamma rays. The resulting clusters of fission fragments were then detected and mapped by track detectors; so were the α-emissions from the heavy radioisotopes involved.

In another representative paper using the SSNTD technique, Badr and Durrani (1993) measured the α-activity of human hair and charred sheep lungs collected from subjects around the epicenter of the Chernobyl accident after the lapse of several months (and possibly years). Only one of the five samples measured showed a significant excess of α-radioactivity.

Quantitative measurements of the environmental effects of nuclear accidents are, obviously, of great importance; and the SSNTD technique provides a means for simple, inexpensive, and widespread surveys of such effects.

3. Cancer Diagnostics and Therapy

Studies of the structure of latent tracks that have led to predicting certain effects in physical, chemical, and biological systems have recently been reviewed (Hill, 1999; Katz and Cucinotta, 1999). From the many examples of the medical applications of etched tracks, we shall only consider here cancer therapy as an important illustration. Nuclear therapy has encompassed the use of photons, electrons, pions as well as neutrons and protons. Recent additions to this list are: (i) radiotherapy with light ions (carbon, oxygen, neon); and (ii) boron neutron capture therapy. These approaches appear to be quite promising—as they open up further fields of selective treatment of cancer with radiation.

In general, in comparison with the earlier types of radiation listed above (r, e, etc.), light ions exhibit more suitable physical and biological properties for cancer treatment owing to: (a) excellent depth-dose profile—based on the Bragg curve (i.e., an increase of energy deposition with penetration distance, culminating in a sharp and high peak followed by rapid fall-off in dose beyond it); and (b) increased biological efficiency and reduced oxygen effects at the end of the particle range. All these allow a greater dose to be delivered at the tumor location, avoiding unwanted exposure of neighboring healthy tissues (Petti and Lennox, 1994). However, techniques for hadron (especially light-ion) therapy are far from standardized at present (Lennox, 2001). Before performing an actual treatment, all the physical properties of the particular ion beam should be carefully determined, since they are altered as the ions pass through a tissue. To achieve this goal, a number of useful experiments for planning cancer therapy with ion beams of ^{12}C, ^{16}O, ^{19}F, and ^{20}Ne in the energy range 40–200 MeV/n have been performed with SSNTDs during the last decade (see, for instance, Golovchenko et al., 2002, and references therein). The tissue to be irradiated has been simulated by water, plexiglass, and CR-39. From these experiments the following parameters have been obtained: (i) the partial cross sections and yields of primary beam fragmentation leading to the production of lower-Z ions; (ii) fluences and linear energy transfer (LET) values along the penetration path; (iii) beam ranges; and (iv) complete depth-dose profiles, including range stragglings and residual ionization formed due to longer-range fragments.

Boron neutron capture therapy (BNCT) has been revitalized during the past few years, in the wake of the termination of clinical trials in the USA (around 1961) and the continued clinical application in Japan since 1968. The treatment relies on the selective accumulation or retention of boron compounds in tumor tissue, and the subsequent exposure to thermal neutrons. During the latter phase, the tumor tissue gets irradiated by the ^{10}B $(n, \alpha)^7Li$ reaction products. The accurate measurement of the ^{10}B distribution in the tumor is essential for evaluating the potential usefulness of various ^{10}B-delivery compounds. For this purpose, the neutron-induced autoradiography with SSNTDs has been found to be the most powerful technique (Skvarč et al., 1999; Ogura et al., 2001; Durrani, 2001). The reasons for this include: (i) high concentration sensitivity (average boron concentrations down to the ppm range can be measured. Local concentration in structural detail (cells) as small as 100–1000 μm^2 can be measured with statistical errors of about 10%); (ii) high spatial resolution (a few μm); and (iii) ability to selectively image boron distribution in a whole-body section.

An example is given in Fig. 3.20 (Skvarč et al., 1999), where a thermal neutron radiograph of the whole-body section of a mouse is shown. Here, a

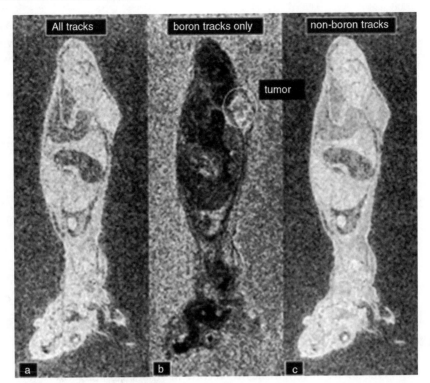

FIGURE 3.20 Thermal neutron induced autoradiograph of the whole-body section of a mouse, obtained by selective radiography with SSNTDs, using the TRACOS system. Bright regions correspond to areas with high α-track densities. The mouse was given an intravenous injection of ^{10}BSH solution (116 ppm), sacrificed, and frozen 6 h after the injection. (a) thermal neutron induced autoradiograph in CR-39 detector obtained at a neutron fluence of 1.7×10^9 n cm^{-2} showing all tracks produced; (b) selected boron autoradiograph; and (c) the difference between the two images, which represents the non-boron tracks (Skvarč et al., 1999).

compound, ^{10}BSH, entrapped in a polyethylene glycol (PEG) binding liposome (116 ppm), was prepared and intravenously injected into a tumor-bearing mouse. The mouse was sacrificed, frozen, and cut into 40 μm thick sections; it was then monted on a 3M scotch adhesive tape. The whole-body section of the mouse, suffering from a pancreatic cancer tumor, was put in close contact with a CR-39 detector and irradiated with thermal neutrons at a fluence of 1.7×10^9 n cm^{-2}. A selective radiograph (boron-generated tracks only) was produced by recently developed image-enhancement techniques (Skvarč et al., 1999), based on the utilization of their advanced systems for track evaluation, TRACOS.

IV. CONCLUSION

The contents of this chapter will, it is hoped, have demonstrated what a versatile and powerful technique the Solid State Nuclear Track Detection (SSNTD) method is. As one of us wrote in a recent review article (Durrani,

2001): "The spectrum of information revealed by the technique extends from delineating the history of the cosmos over billions of years to observing exotic decays lasting a minute fraction of a second." The method also covers topics such as the measurement of radon levels in dwellings; radiation protection dosimetry in nuclear energy establishments and hospitals; the pinpointing of "hot particles" from a nuclear fallout; elemental mapping in industrial materials; prospecting for oil and uranium deposits; fission-track dating of rocks—and many more. One could almost say that the limit of the applications of the SSNTD technique is the limit of one's imagination—although there are, of course, limitations! Despite its versatility, the technique is relatively simple and, at its basic level, inexpensive—which makes it particularly attractive for the Third World laboratories. Presently, over 300 papers per year are being published globally in this discipline, covering the various topics touched upon in this chapter as well as many other applications in science and technology, both on Earth and in space. However, we leave it to our readers—whether in the First World or the Third!—to identify future research and development areas where they can fruitfully apply the SSNTD method: and to extend its use to domains not yet dreamt of.

ACKNOWLEDGMENTS

One of us (SAD) wishes to record his thanks to the Jožef Stefan Institute, Ljubljana, for its hospitality over a period of two weeks during August–September 2002, which enabled the coauthors of this chapter to finalize its contents in an intensive and sharply targeted effort. Our thanks are also due to Ms Uršula Turšič, who typed most of the text and cheerfully incorporated almost endless amendments proposed by the coauthors; Mr. Bojan Žefran, who prepared all the figures and tables; and other staff and colleagues (in particular Dr. Jure Skvarč and Igor Lengar MSc) of RI's Group at the Institute for their concerted and highly skilled support of our activities. Finally, we wish to record our sincere thanks to the Editor of this Handbook, Dr. Michael L'Annunziata, for his unfailing, courteous, and prompt help and support of our work at all stages of organizing and writing this chapter.

REFERENCES

Åkerblom, G. and Mellander, H. (1997). Geology and radon. *In* "Radon Measurements by Etched Track Detectors: Applications to Radiation Protection, Earth Sciences and the Environment" (S. A. Durrani and R. Ilić, Eds.), pp. 21–49. World Scientific, Singapore.

Al-Najjar, S. A. R., Abdel-Naby, A., and Durrani, S. A. (1986). Fast neutron spectrometry using the triple-α reaction in the CR-39 detector. *Nucl. Tracks* **12**(1–6), 611–615.

Al-Najjar, S. A. R., Bull, R. K., and Durrani, S. A. (1979). Electrochemical etching of CR-39 plastic: applications to radiation dosimetry. *Nucl. Tracks* **3**, 169–183.

Al-Najjar, S. A. R. and Durrani, S. A. (1984). Track profile technique (TPT) and its applications using CR-39. I: Range and energy measurements of alpha-particles and fission fragments. *Nucl. Tracks* **8**, 45–50.

Bartlett, D. T. and Bird, T. V. (1987). Technical specification of the NRPB radon personal dosimeter. NRPB-R208, HMSO, London.

Badr, I. and Durrani, S. A. (1993). Alpha-activity measurements of biological samples from areas around the site of the Chernobyl nuclear disaster using the SSNTD technique. *Nucl. Tracks Radiat. Meas.* **22**(1–4), 849–850.

Badr, I., Oliver, M. A., Hendry, G. L., and Durrani, S. A. (1993). Determining the spatial scale of variation in soil radon values using a nested survey and analysis. *Rad. Prot. Dosim.* **49**, 433–442.

Balcázar, M. (1997). Radon and geothermal energy production. *In* "Radon Measurements by Etched Track Detectors: Applications to Radiation Protection, Earth Sciences and the Environment" (S. A. Durrani and R. Ilić, Eds.), pp. 345–362. World Scientific, Singapore.

Balcázar, M. and Durrani, S. A. (1980). High-energy spectrometry with plastic SSNTDs. *Nucl. Instrum. Meth.* **173**, 131–135.

Barbu, A., Dammak, H., Dunlop, A., and Leseuer, D. (1995). Ion tracks in metals and intermetallic compounds. *MRS Bull.* **20**, 29–34.

Benton, E. V., Ed. (1992). "Space Radiation". *Nucl. Tracks Radiat. Meas.* **20**(1).

Benton, E. V., Ed. (1994). "Special Section on Space Radiation". *Radiat. Meas.* **23**(1).

Benton, E. V., Ed. (1996a). "Space Radiation". *Radiat. Meas.* **23**(1).

Benton, E. V., Ed. (1996b). "Space Radiation: Results of the Long Duration Exposure Facility (LDEF)". *Radiat. Meas.* **26**(1–6).

Benton, E. V. and Adams, J.H. Jr., Eds. (1992). "Galactic Cosmic Radiation: Constraints on Space Exploration". *Nucl. Tracks Radiat. Meas.* **20**(3).

Benton, E. V. and Panasyuk, M., Eds. (1999). "Space Radiation Environment". *Radiat. Meas.* **30**(5).

Benton, E. V. and Badhwar, G.D., Eds. (2001). "Prediction and Measurements of Secondary Neutrons in Space". *Radiat. Meas.* **33**(3).

Benton, E. V., Adams, J. H. Jr., and Panasyuk, M., Eds. (1996)."Space Radiation Environment. Empirical and Physical Models". *Radiat. Meas.* **26**(3).

Benton, E. R., Benton, E. V., and Frank, A. L. (2001). Neutron dosimetry in low-earth orbit using passive detectors. *Radiat. Meas.* **33**, 255–263.

Bersina, I. G., Brandt, R., Vater, P., Hinke, K., and Schutze, M. (1995). Fission track autoradiography as a means to investigate plants for their contamination with natural and technogenic uranium. *Radiat. Meas.* **24**, 277–282.

Bhandari, N., Goswami, J. N., and Lal, D. (1973). Surface irradiation and evolution of the lunar regolith. *In* "Proc. Fourth Lunar Sci. Conf.", pp. 2275–2290. Pergamon Press, New York.

Bondarenko, O. A., Korneev, A. A., Onishchuk, Yu. N., Berezhnoy, A. V., Aryasov, P. B., Antonyuk, D., and Dmitrienko, A. V. (1999). Application of SSNTD for maintenance of radiation and nuclear safety of the Sarcophagus. *Radiat. Meas.* **30**, 709–714.

Boulyga, S. F., Kievietskaja, A. I., Kievets, M. K., Lomonosova, E. M., Zhuk, I. V., Yaroshevich, O. I., Perelygin, V. P., Petrova, R., Brandt, R., and Vater, P. (1999). Nuclear track radiography of hot aerosol particles. *Radiat. Meas.* **31**(1–6), 191–196.

Brandt, R. (2001). Some contributions from SSNTD towards nuclear science: From multifragmentation towards accelerator driven systems (ADS). *Radiat. Meas.* **34**(1–6), 211–219.

Brandt, R., Spohr, R., and Vater, P., Eds. (1991). "Proc. 15th Int. Conf. on Particle Tracks in Solids". Marburg, 1990, *Nucl. Tracks Radiat. Meas.* **19**(1–4).

Brechtmann, C. and Heinrich, W. (1988). Measurements of elemental fragmentation cross-section for relativistic heavy-ions using CR-39 plastic nuclear track detectors. *Nucl. Instrum. Meth.* **B29**, 675–679.

Bull, R.K. and Durrani, S.A. (1976). Cosmic-ray tracks in the Shalke meteorite. *Earth Planet. Sci. Lett.* **32**, 35–39.

Burkart, W., Sohrabi, M., and Bayer, A., Eds. (2002). "Proc. 5th Int. Conf. on High Levels of Natural Radiation and Radon Areas". Munich, 2000, Elsevier, Amsterdam, 322 pp.

Chambaudet, A., Fromm, M., Ilić, R., Vater, P., Dubois, C., and Rebetez, M., Eds. (1999). "Proc. 19th Int. Conf. on Nuclear Tracks in Solids". Besançon, 1998. *Radiat. Meas.* **31**(1–6).

Comstock, G. M., Evwaraye, A. O., Fleischer, R. L., and Hart, H. R. Jr. (1971). The particle track record of lunar soil. *In* "Proc. Second Lunar Sci. Conf." pp. 2569–2582. MIT Press, Cambridge, Mass.

Cross, W. G. and Tommasino, L. (1970). A rapid reading technique for nuclear particle damage tracks in thin foils. *Radiat. Effects* **5**, 85–89.

Csige, I., Hunyadi, I., Somogyi, G., and Fujii, M. (1988). Vacuum effect on the etch induction time and registration sensitivity of polymer track detectors. *Nucl. Tracks Radiat. Meas.* **15**(1–4), 179–182.

Curzio, G., Grillmaier, R. E., O'Sullivan, D., Pelliccioni, M., Piermattei, S., and Tommasino, L. (2001a). The Italian national survey of aircrew exposure: I. Characterisation of advanced instrumentation. *Radiat. Prot. Dos.* **93**, 115–123.

Curzio, G., Grillmaier, R. E., O'Sullivan, D., Pelliccioni, M., and Tommasino, L. (2001b). The Italian national survey of aircrew exposure: II. On-board measurements and results. *Radiat. Prot. Dos.* **93**, 125–133.

Ditlov, V. (2001). The evaluation of track theory throughout the history of the international solid state detector conferences. *Radiat. Meas.* **34**(1–6), 19–26.

Doi, M., Fujimoto, K., Kobayashi, S., and Yonehara, H. (1994). Spatial distribution of thoron and radon concentrations in the indoor air of a traditional Japanese wooden house. *Health Phys.* **66**, 43–49.

Donnelly, J., Thomson, A., O'Sullivan, D., Drury, L. O'Cis, and Wenzel, K.-P. (2001). The cosmic ray actinide charge spectrum derived from a $10\,m^2$ array of solid state nuclear track detectors in earth orbit. *Radiat. Meas.* **34**(1–6), 273–276.

Dörshel, B., Schuricht, V., and Stener, J. (1996). "The Physics of Radiation Protection". Nucl. Techn. Publ., Ashford, 309 pp.

Durrani, S. A. (1977). Editorial. *Nucl. Track Detection* **1**, 1–2.

Durrani, S. A. (1981). Track record in meteorites. *Proc. Roy Soc. London* **A374**, 239–251.

Durrani, S. A. (1991). The effect of irradiation temperature on the response of track recording crystalline and polymeric media: A brief review. *Nucl. Tracks Radiat. Meas.* **19**(1–4), 61–70.

Durrani, S. A. (1993). Radon as a health-hazard at home – what are the facts? *Nucl. Tracks Radiat. Meas.* **22**(1–4), 303–317.

Durrani, S. A. (1997). Alpha-particle etched track detectors. *In* "Radon Measurements by Etched Track Detectors: Applications to Radiation Protection, Earth Sciences and the Environment" (S. A. Durrani and R. Ilić, Eds.), pp. 77–101. World Scientific, Singapore.

Durrani, S. A. (1999). Radon concentration values in the field: Correlation with underlying geology. *Radiat. Meas.* **31**(1–6), 271–276.

Durrani, S. A. (2001). Nuclear Tracks: A success story of the 20th century. *Radiat. Meas.* **34**(1–6), 5–13.

Durrani, S. A. and Bull, R. K. (1987). "Solid State Nuclear Track Detection". Pergamon Press, Oxford, 304 pp.

Durrani, S. A. and Matiullah (1988). Development of thermal and fast-neutron dose equivalent dosimeter using CR-39 detector with radiators-converters. *Radiat. Prot. Dosim.* **23**, 183–186.

Durrani, S. A. and Ilić, R., Eds. (1997). "Radon Measurements by Etched Track Detectors: Applications to Radiation Protection, Earth Sciences and the Environment". World Scientific, Singapore, 387 pp.

Durrani, S. A., Khan, H. A., Taj, M. A., and Renfrew, C. (1971). Obsidian source identification by fission track analysis. *Nature* **233**, 242–245.

Durrani, S. A., Bull, R. K., and McKeever, S.W.S. (1980). Solar flare exposure and thermoluminescence of Luna 24 core material. *Phil. Trans. Roy. Soc. London* **A297**, 41–50.

Enge, W. (1980). Introduction to plastic nuclear track detectors. *Nucl. Tracks Radiat. Meas.* **4**, 283–308.

Fews, A. P. (1992). Fully automated image analysis of etched tracks in CR-39. *Nucl. Instrum. Meth.* **B71**, 465–478.

Fleischer, R. L. (1981). "Nuclear Track Production in Solids". Progress in Material Sciences, Chalmers Anniversary Volume, pp. 98–128. Pergamon Press, Oxford.

Fleischer, R. L. (1988). Radon in the environment – opportunities and hazards. *Nucl. Tracks Radiat. Meas.* **14**, 421–435.

Fleischer, R. L. (1997a). Radon: Overview of properties, origin and transport. *In* "Radon Measurements by Etched Track Detectors: Applications to Radiation Protection, Earth Sciences and the Environment" (S. A. Durrani and R. Ilić, Eds.), pp. 1–20. World Scientific, Singapore.

Fleischer, R. L. (1997b). Radon and earthquake prediction. *In* "Radon Measurements by Etched Track Detectors: Applications to Radiation Protection, Earth Sciences and the Environment" (S. A. Durrani and R. Ilić, Eds.), pp. 285–299. World Scientific, Singapore.

Fleischer, R. L. (1998). "Tracks to Innovation". Spinger Verlag, New York, 193 pp.

Fleischer, R. L. (2002). Serendipitous radiation monitors. *American Scientist* **90**, 324–331.

Fleischer, R. L. and Raabe, O. G. (1977). Fragmentation of respirable PuO_2 particles in water by alpha-decay-mode of dissolution. *Health Phys.* **32**, 253–257.

Fleischer, R. L. and Mogro-Campero, A. (1978). Mapping of integrated radon emanation for detection of long-distance migration of gases within the earth: Techniques and principles. *J. Geophys. Res.* **83**, 3539–3549.

Fleischer, R. L., Price, P. B., Walker, R. M., and Leakey, L. S. B. (1965). Fission track dating of a mesolithic knife. *Nature* **205**, 1138.

Fleischer, R. L., Price, P. B., Walker, R. M., Maurette, M., and Morgan, G. (1967a). Tracks of heavy primary cosmic rays in meteorites. *J. Geophys. Res.* **72**, 331–353.

Fleischer, R. L., Price, P. B., Walker, R. M., and Maurette, M. (1967b). Origin of fossil charged-particle tracks in meteorites. *J. Geophys. Res.* **72**, 355–366.

Fleischer, R. L., Price, P. B., and Walker, R. M. (1975). "Nuclear Tracks in Solids: Principles and Applications". University of California Press, Berkeley, 605 pp.

Flerov, G. N. and Bersina, I. G. (1979). "Radiography of Minerals and Ores" (in Russian). Atomizdat, Moscow, 223 pp.

Gamboa, I., Jacobson, I., Golzarri, J. I., and Espinosa, G. (1984). Uranium contents determination in commercial drinkable milk. *Nucl. Tracks Radiat. Meas.* 8(1–4), 461–463.

Gangrskij, Yu. P., Markov, B. N., and Perelygin, V. P. (1992). "Registration and Spectrometry of Fission Fragments" (in Russian). Energoatomizdat, Moscow, 312 pp.

Golovchenko, A. N., Skvarč, J., Yasuda, N., Giacomelli, M., Tretyakova, S. P., Ilić, R., Bimbot, R., Toulemonde, M., and Murakami, T. (2002). Total charge-changing and partial cross-section measurements in the reactions of ~ 100–$250\,MeV$/nucleon ^{12}C in carbon, paraffin, and water. *Phys. Rev.* **C66**, 014609, 1–7.

Gomaa, M. A., Eid, A. M., Griffith, R. V., and Davidson, K. J. (1981). CR-39 carbonate plastic as a neutron albedo and threshold dosimeter. *Nucl. Tracks* **5**, 279–284.

Gottschalk, P. A., Vater, P., Brandt, R., Qureshi, I. E., Khan, H. A., and Fiedler, G. (1996). Kinematical analysis of heavy ion induced nuclear reactions using solid state nuclear track detectors. *Phys. of Particles and Nuclei* **27**, 154–178.

Green, P. F., Bull, R. K., and Durrani, S. A. (1978). Particle Identification from track–etch rates in minerals. *Nucl. Instrum. Meth.* **157**, 185–193.

Green, P. F., Duddy, I. R., Gleadow, A. J. W., and Lovering, J. F. (1989). Apatite fission–track analysis as a paleotemperature indicator for hydrocarbon exploration. *In* "Thermal History of Sedimentary Basins" (N. D. Naeser and T. H. McCulloh, Eds.), pp. 181–195. Springer-Verlag, New York.

Guo, S. L. (1982). Some methods in fission track dating of Peking man. *In* "Proc. 11th Int. Conf. on Solid State Nuclear Track Detectors" (P. H. Fowler and A. M. Clapham, Eds.), pp. 371–374. Pergamon Press, Oxford.

Guo, S. -L., Zhao, C. -D., and Sun, H., Eds. (1993). "Proc. 16th Int. Conf. on Solid State Nuclear Track Detectors". Beijing, 1992, *Nucl. Tracks Radiat. Meas.* **22**(1–4).

Hakl, J., Hunyadi, I., and Várhegyi, A. (1997). Radon monitoring in caves. *In* "Radon Measurements by Etched Track Detectors: Applications to Radiation Protection, Earth Sciences and the Environment" (S. A. Durrani and R. Ilić, Eds.), pp. 259–283. World Scientific, Singapore.

Hardcastle, G. D., Howarth, C. B., Naismith, S. P., Algar, R. A., and Miles, J. C. H. (1996). NRPB etched-track detectors for area monitoring of radon. NRPB-R283 HMSO, London.

Harrison, K. G. (1978). A neutron dosimeter based on ^{237}Np fission. *Nucl. Instrum. Meth.* **157**, 169–178.

Harms, A. A. and Wyman, R. (1986). "Mathematics and Physics of Neutron Radiography". Reidel, Dordrecht, 163 pp.

Henke, R. P. and Benton, E. V. (1968). Charged particle tracks in polymers: No. 5–A computer code for the computation of heavy ion range-energy relationships in any stopping material. US Naval Radiological Defense Laboratory, Report USNRDL-TR-67-122, San Francisco.

Hill, M. A. (1999). Radiation damage to DNA: The importance of track structure. *Radiat. Meas.* **31**(1–6), 15–23.

Homer, J. B. and Miles, J. C. H. (1986). The effects of heat and humidity before, during and after exposure on the response of PADC (CR-39) to alpha particles. *Nucl. Tracks* **12**(1–6), 133–136.

Ilić, R. (1990). Damaged track detectors for alpha particle registration – track formation and detector processing. *In* "Proc. Int. Workshop on Radon Monitoring in Radioprotection, Environmental Radioactivity and Earth Sciences" (L. Tommasino *et al.*, Eds.). Trieste, 1989, pp. 133–144. World Scientific, Singapore.

Ilić, R. and Najžer, M. (1990a). Image formation in track-etch detectors – I. The large area signal transfer function. *Nucl. Tracks Radiat. Meas.* **17**, 453–460.

Ilić, R. and Najžer, M. (1990b). Image formation in track-etch detectors – II. The space dependent transfer function in thin detectors. *Nucl. Tracks Radiat. Meas.* **17**, 461–468.

Ilić, R. and Najžer, M. (1990c). Image formation in track-etch detectors – III. The space dependent transfer function in thick detectors. *Nucl. Tracks Radiat. Meas.* **17**, 469–473.

Ilić, R. and Najžer, M. (1990d). Image formation in track-etch detectors – IV. Image quality. *Nucl. Tracks Radiat. Meas.* **17**, 475–481.

Ilić, R. and Šutej, T. (1997). Radon monitoring devices based on etched track detectors. *In* "Radon Measurements by Etched Track Detectors: Applications to Radiation Protection, Earth Sciences and the Environment" (S.A. Durrani and R. Ilić, Eds.), pp. 103–128. World Scientific, Singapore.

Ilić, R., Rant, J., Humar, M., Somogyi, G., and Hunyadi, I. (1986). Neutron radiographic characteristics of MA-ND type (allyl-diglycol-carbonate) nuclear track detector. *Nucl. Tracks* **12**(1–6), 933–936.

Ilić, R., Vater, P., and Kenawy, M. A., Eds. (1997). "Proc. 18th Int. Conf. on Nuclear Tracks in Solids". Cairo, 1996, *Radiat. Meas.* **28**(1–6).

Ilić, R., Jenčič, I., and Vater, P., Eds. (2001). "Proc. 20th Int. Conf. on Nuclear Tracks in Solids". Portorož, 2000. *Radiat. Meas.* **34**(1–6).

Izerrouken, M., Skvarč, J., and Ilić, R. (1999). Low energy alpha particle spectroscopy using CR-39 detector. *Radiat. Meas.* **31**(1–6), 141–144.

James, K., Matiullah, and Durrani, S. A. (1987). A computer-program for calculation of the dose-equivalent response of fast-neutron dosimeter based on electrochemically etched CR-39 detector. *Nucl. Tracks Radiat. Meas.* **13**, 3–24.

Jönsson, G. (1997a). Indoor radon surveys. *In* "Radon Measurements by Etched Track Detectors: Applications to Radiation Protection, Earth Sciences and the Environment" (S. A. Durrani and R. Ilić, Eds.), pp. 157–177. World Scientific, Singapore.

Jönsson, G. (1997b). Soil radon surveys. *In* "Radon Measurements by Etched Track Detectors: Applications to Radiation Protection, Earth Sciences and the Environment" (S. A. Durrani, and R. Ilić, Eds.), pp. 179–188. World Scientific, Singapore.

Katz, R. and Cucinotta, F. A. (1999). Tracks to therapy. *Radiat. Meas.* **31**(1–6), 379–388.

Khan, H. A., Qureshi, A. A., and Qureshi, I. E. (1997). Radon and mineral exploration. *In* "Radon Measurements by Etched Track Detectors: Applications to Radiation Protection, Earth Sciences and the Environment" (S. A. Durrani and R. Ilić, Eds.), pp. 319–343. World Scientific, Singapore.

Khan, H. A. and Qureshi, I. E. (1999). SSNTD applications in science and technology – A brief review. *Radiat. Meas.* **31**(1–6), 25–36.

Khayrat, A. H. and Durrani, S. A. (1995). The effect of UV exposure on the track and bulk etching rates in different CR-39 plastics. *Radiat. Meas.* **25**(1–4), 163–164.

Khayrat, A. H. and Durrani, S. A. (1999). Variation of alpha-particle track diameter in CR-39 as a function of residual energy and etching conditions. *Radiat. Meas.* **30**, 15–18.

Lengar, I. (2001). Measurements of the neutrons with a pair of CR-39 etched track detectors. M.Sc. Thesis, University of Ljubljana, Ljubljana, 2001.

Lengar, I., Skvarč, J., and Ilić, R. (2002). Fast neutron detection with coincidence counting of recoil tracks in CR-39. *Nucl. Instrum. Meth.* **B192**, 440–444.

Lennox, A. J. (2001). Accelerators for cancer therapy. *Radiat. Phys. Chem.* **61**, 223–226.

Luszik-Bhadra, M., Matzke, M., and Schuhmacher, H. (2001). Development of personal neutron dosimeters at the PTB and first measurements in the space station MIR. *Radiat. Meas.* **33**, 305–312.

Marenny, A. M. (1987). "Dielectric Track Detectors in Radiation Physics and Radiobiology Experiments" (in Russian). Energoatomizdat, Moscow, 184 pp.

Matiullah and Durrani, S. A. (1987a). Chemical and electrochemical registration of protons in CR-39 – Implications for neutron dosimetry. *Nucl. Instrum. Meth.* **B29**, 508–514.

Matiullah and Durrani, S. A. (1987b). A cubical fast neutron dosimeter based on electrochemically etched CR-39 detectors with polymeric front radiators. *Radiat. Prot. Dosim.* **10**, 77–80.

Matiullah, Taylor, C., and Durrani, S. A. (1987). An integrated-circuit based variable power supply for electrochemical etching. *Nucl. Tracks Radiat. Meas.* **13**, 67–70.

Matiullah, Durrani, S. A., and Khan, G. A. (1988). A practical fast-neutron dosimeter based on electrochemically etched CR-39 detector with angle-independent response. *Nucl. Instrum. Meth.* **34**, 499–504.

Miles, J. C. H. (1997). Calibration and standardization of etched track detectors. *In* "Radon Measurements by Etched Track Detectors: Applications to Radiation Protection, Earth Sciences and the Environment" (S. A. Durrani and R. Ilić, Eds.), pp. 143–154. World Scientific, Singapore.

Miles, J. C. H. and Ball, T. K. (1997). Mapping of the probability of high radon concentration. *In* "Radon Measurements by Etched Track Detectors: Application to Radiation Protection, Earth Sciences and the Environment" (S. A. Durrani and R. Ilić, Eds.), pp. 209–223. World Scientific, Singapore.

Miles, J. C. H., Algar, R. A., Howarth, C. B., Hubbard, L., Risica, S., Kies, A., and Poffijn, A. (1996). "Results of the 1995 European Commission Intercomparison of Passive Radon Detectors". European Commission, Directorate-General XII, EUR 16949 EN, Brussels.

Monnin, M. M. and Seidel, J. L. (1997a). Radon and volcanic surveillance. *In* "Radon Measurements by Etched Track Detectors: Applications to Radiation Protection, Earth Sciences and the Environment" (S. A. Durrani and R. Ilić, Eds.), pp. 301–318. World Scientific, Singapore.

Monnin, M. M. and Seidel, J. L. (1997b). Radon measurement techniques. *In* "Radon Measurements by Etched Track Detectors: Applications to Radiation Protection, Earth Sciences and the Environment" (S. A. Durrani and R. Ilić, Eds.), pp. 51–74. World Scientific, Singapore.

Muirhead, C. R. (1997). Radon-induced health effect. *In* "Radon Measurements by Etched Track Detectors: Applications to Radiation Protection, Earth Sciences and the Environment" (S. A. Durrani and R. Ilić, Eds.), pp. 243–257. World Scientific, Singapore.

Najžer, M., Humar, M., and Ilić, R. (1982). Microautoradiography with gelatine. *In* "Proc. 11th Int. Conf. on Solid State Nuclear Track Detectors" (P. H. Fowler and A. M. Clapham, Eds.), 1981, pp. 77–80. Bristol.

Nazaroff, W. W. and Nero, A. V., Eds. (1988). "Radon and its Decay Products in Indoor Air". John Wiley and Sons, New York, 518 pp.

Nesvizhevsky, V. V., Börner, H., Gagarski, A. M., Petrov, G. A., Petuhkov, A. K., Abele, H., Bäßler, S., Stöferle, T., and Soloviev, S. M. (2000). Search for quantum states of the neutron in gravitational field: Gravitational levels. *Nucl. Instrum. Meth.* **A440**, 754–759.

Nesvizhevsky, V. V., Börner, H. G., Petuhkov, A. K., Abele, H., Baeßler, S., Rueß, F. J., Stöferle, T., Westpahal, A., Gagarski, A. M., Petrov, G. A., and Strelkov, V. (2002). Quantum states of neutrons. *Nature* **415**, 297–299.

Nikolaev, V. A. and Ilić, R. (1999). Etched track radiometers in radon measurements: A review. *Radiat. Meas.* **30**, 1–13.

Ogura, K., Yamazaki, A., Yanagie, H., Eriguchi, M., Lehman, E. H., Küehne, G., Bayon, G., and Kobayashi, H. (2001). Neutron capture autoradiography for study on boron neutron capture therapy. *Radiat. Meas.* **34**(1–6), 555–558.

O'Sullivan, D., Thompson, A., Adams, J. A., and Beahm, A. (1984). New results on the investigation of nuclear track detectors response with temperature. *Nucl. Tracks* **8**, 143–146.

O'Sullivan, D., Bartlett, D., Grillmaier, R., Heinrich, W., Lindborg, L., Schraube, H., Silari, M., Tommasino, L., and Zhou, D. (2000). Investigation of radiation fields at aircraft altitudes. *Radiat. Prot. Dosim.* **92**(1–3), 195–198.

O'Sullivan, D., Thompson, A., Donnelly, J., Drury, L. O., and Wenzel, K. P. (2001a). The relative abundance of actinides in the cosmic ray radiation. *Adv. in Space Res.* **27**, 785–789.

O'Sullivan, D., Zhou, D., and Flood, E. (2001b). Investigation of cosmic rays and their secondaries at aircraft altitudes. *Radiat. Meas.* **34**(1–6), 277–280.

Pellas, P. and Storzer, D. (1981). [244]Pu fission track thermometry and its applications to stony meteorites. *Proc. Roy. Soc. London* **A374**, 253–270.

Perelygin, V. P. and Stetsenko, S. C., (1977). Search for tracks of galactic cosmic nuclei with $Z \geq 110$ in meteorite olivines. *JETP Lett.* **32**, 608–610.

Perelygin, V. P. and Churburkov, Y. T. (1997). Man-made plutonium in environment – Possible serious hazard for living species. *Radiat. Meas.* **28**(1–6), 385–392.

Perelygin, V. P., Vater, P., Ilić, R., and Durrani, S. A., Eds. (1995). "Proc. 17th Int. Conf. on Solid State Nuclear Track Detectors". Dubna, 1994, *Radiat. Meas.* **25**(1–4).

Perelygin, V. P., Churburkov, Yu, T., Dimitriev, S. N., Oganesjian Yu., Ts, Petrova R. I., and Drobina, T. P. (1999). On problem of ultrasensitive determination of man-made plutonium in living species. *Radiat. Meas.* **31**(1–6), 407–412.

Petti, P. L., and Lennox A. J. (1994). Hadronic radiotherapy. *Annu. Rev. Nucl. Part. Sci.* **44**, 155–197.

Peurrung, A. J. (2000). Recent developments in neutron detection. *Nucl. Instrum. Meth.* **A443**, 400–415.

Pineau, J. F. (1997). Radon dosimetry and monitoring in mines. *In* "Radon Measurements by Etched Track Detectors: Applications to Radiation Protection, Earth Sciences and the Environment" (S. A. Durrani and R. Ilić, Eds.), pp. 189–208. World Scientific, Singapore.

Poenaru, D. N., Nagame, Y., Gherghescu, R. A., and Greiner, W. (2002). Systematics of cluster decay modes. *Phys. Rev.* **C65**, 054308, 1–6.

Price, P. B. and Walker, R. M. (1962). Observations of fossil particle tracks in natural micas. *Nature* **196**, 732–734.

Price, P. B., Fleischer, R. L., Peterson, D. D., O'Ceallaigh, C., O'Sullivan, D., and Thomson, A. (1967). Identification of isotopes of energetic particles with dielectric track detectors. *Phys. Rev.* **164**, 1618–1620.

Price, P. B., Stevenson, J. D., Barwick, S. W., and Ravn, H. L. (1985). Discovery of radioactive decay of Ra-222 and Ra-224 by C-14 emission. *Phys. Rev. Lett.* **54**, 297–299.

Provost, J., Simon, Ch., Hervievu, M., Groult, D., Hardy, V., Studer, F., and Toulemonde, M. (1995). Swift heavy ions in insulating and conducting oxides: Tracks and physical properties. *MRS Bull.* **20**(129), 2–28.

Pugliesi, R. and Pereira, M. A. S. (2002). Study of neutron radiography characteristics for the solid state nuclear track detector Makrofol-DE. *Nucl. Instrum. Meth.* **A484**, 613–618.

Qiangyan, P., Wifean, Y., Shuanggui, Y., Zongwei, L., Taotao, M., Yixiao, L., Dengming, K., Jimin, Q., Zihua, L., Mutian, Z., and Shuhonh, W. (2002). Search for heavy ion emission from the decay of [230]U. *Phys. Rev.* **C62**, 044612, 1–2.

Ramola, R. C., Singh, S., and Virk, H. S. (1988). Uranium and radon estimation in some water samples from Himalayas. *Nucl. Tracks Radiat. Meas.* **15**(1–4), 791–793.

Rose, H. J. and Jones, G. A. (1984). A new kind of natural radioactivity. *Nature* **307**, 245–247.

Rusch, G., Winkel, E., Noll, A., Heinrich, W. (1991). The Siegen automatic measuring system for track detectors: New developments. *Nucl. Tracks Radiat. Meas.* **19**(1–4), 261–266.

Rusov, V. D., Babikova, Yu. F., and Yagola, A. G. (1991). "Image Reconstruction in Electron – Microscopic Autoradiography" (in Russian). Energoatomizdat, Moscow, 216 pp.

Sajo-Bohus, L., Pálfalvi, J., and Greaves, E. D. (1998). Hot particle spectrum determination by track image analysis. *Radiat. Phys. Chem.* **51**, 467–468.

Silk, E. C. H. and Barnes, R. S. (1959). Examination of fission fragment tracks with an electron microscope. *Phil. Mag.* **4**, 970–972.

Spurný, F. (2001). Radiation doses at high altitudes and during space flights. *Radiat. Phys. Chem.* **61**, 301–307.

Skvarč, J. (1993). Automatic image analysis system TRACOS. *Inf. MIDEM (J. Microelectronics, Electronic Components and Materials)* **23**, 201–205.

Skvarč, J. and Golovchenko, A. N. (2001). A method of trajectory tracing of $Z \leq 10$ ions in the energy region below 300 MeV/u. *Radiat. Meas.* **34**(1–6), 113–118.

Skvarč, J., Ilić, R., Yanagie, H., Ogura, K., Rant, J., and Kobayashi, H. (1999). Selective radiography with etched track detectors. *Nucl. Instrum. Meth.* **B152**, 115–121.

Sohrabi, M. (1997). High radon levels in nature and in dwellings: Remedial actions. *In* "Radon Measurements by Etched Track Detectors: Applications to Radiation Protection, Earth Sciences and the Environment" (S. A. Durrani and R. Ilić, Eds.), pp. 225–242. World Scientific, Singapore.

Sohrabi, M., Ahmed, J. U., and Durrani, S. A., Eds. (1993). "Proc. 3rd Int. Conf. on High Levels of Natural Radiation". Ramsar, 1990, IAEA, Vienna, 618 pp.

Somogyi, G. (1980). Development of etched nuclear tracks. *Nucl. Instrum. Meth.* **173**, 1–14.

Somogyi, G. (1990). Methods for measuring radium isotopes: track detection. *In* "The Environmental Behaviour of Radium", IAEA Technical Report Series 310, IAEA, Vienna, pp. 229–256.

Spohr, R. (1990). "Ion Tracks and Microtechnology". Vieweg, Braunschweig, 272 pp.

Tidjani, A. (1990). Effects of UV-light on the efficiency of alpha-particle detection of CR-39, LR 115 type-II and CN 85. *Nucl. Tracks Radiat. Meas.* **17**, 491–495.

Tidjani, A. (1991). Property modifications in UV irradiated polymeric track detectors. *Nucl. Instrum. Meth.* **B58**, 43–48.

Tommasino, L. (1970). Electrochemical etching of damage tracks detectors by H.V. pulse and sinusoidal waveform. Intern. Rept. Lab. Dosimetria e Standardizzazione, CNEN Casaccia, Rome.

Tommasino, L. (1988). Assessment of natural and man made alpha emitting radionuclides. *Nucl. Tracks Radiat. Meas.* **15**(1–4), 555–565.

Tommasino, L. (1997). Track registration: Etching and counting methods for nuclear tracks. *In* "Radon Measurements by Etched Track Detectors: Applications to Radiation Protection, Earth Sciences and the Environment" (S. A. Durrani, and R. Ilić, Eds.), pp. 129–141. World Scientific, Singapore.

Tommasino, L. (2001). Personal dosimetry and area monitoring for neutron and radon workplaces. *Radiat. Meas.* **34**(1–6), 449–456.

Tretyakova, S. P., Bonetti, R., Golovchenko, A., Guglielmetti, A., Ilić, R., Mazzochi, Ch., Mikheev, V., Oglobin, A., Ponomarenko, V., Shigin, V., and Skvarč, J. (2001). Study of cluster decay of ^{242}Cm using SSNTD. *Radiat. Meas.* **34**(1–6), 241–243.

Tsuruta, T. (1997). Effects of heat-treatment and gamma-ray irradiation on the etch-pit formation in allyl-diglycol carbonate resin. *J. Nucl. Sci. Techn.* **34**, 1015–1021.

Urban, M., and Piesch, E. (1981). Low level environmental radon dosimetry with a passive track etch detector device. *Radiat. Prot. Dosim.* **1**, 97–109.

Vorobyev, I. B., Krivokhatskiy, A. S., Krisyuk, E. M., Nikolaev, V. A., Pautov, V. P., and Terentiev, M. V. (1991). Structure and application of equipment assembly for measurement of radon volume activity in houses and soils. *Nucl. Tracks Radiat. Meas.* **19**(1–4), 431–432.

Wagner, G. A. (1981). Fission-track ages and their geological interpretation. *Nucl. Tracks* **5**, 15–25.

Wagner, G. A. and Van den haute, P. (1992). "Fission-Track Dating". Enke, Stuttgart, 285 pp.

Walker, R. M., Price, P. B., and Fleischer, R. L. (1963). A versatile disposable dosimeter for slow and fast neutrons. *Appl. Phys. Lett.* **3**, 28–29.

Wei, L., Sugahara, T., and Tao, Z., Eds. (1997). "Proc. 4th Int. Conf. on High Levels of Natural Radiation and Radon Areas". Beijing, 1995, Elsevier, Amsterdam, 438 pp.

Young, D. A. (1958). Etching of radiation damage in lithium fluoride. *Nature* **182**, 375–377.

4

SEMICONDUCTOR DETECTORS

PAUL F. FETTWEIS AND JAN VERPLANCKE
Canberra Semiconductor N.V., B-2250 Olen, Belgium

RAMKUMAR VENKATARAMAN, BRIAN M. YOUNG AND HAROLD SCHWENN
Canberra Industries, Inc., Meriden, CT, USA

I. INTRODUCTION

A. The Gas-Filled Ionization Chamber

A semiconductor detector can be best compared to a classical ionization chamber described elsewhere in this book (Chapter 2). A schematic diagram of such an ionization chamber is given in Fig. 4.1. It consists essentially of a gas-filled (Kr, Xe, ...) capacitor to which a bias (H.T.) is applied. An ionizing particle (alpha, p, d, beta, ...) will create a certain number N of pairs of positive ions and electrons, where N is given by

$$N = \frac{E}{\varepsilon} \tag{4.1}$$

E represents the kinetic energy of the particle and ε the energy necessary to create one ion–electron pair.

Handbook of Radioactivity Analysis, Second Edition

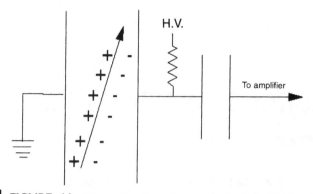

FIGURE 4.1 Schematic view of a gas-ionization chamber. The positive and negative charges formed by the ionizing particle are integrated and the resulting pulse, whose height is proportional to the deposited energy, is sent to the amplifier for further treatment.

In order to act as a spectrometer (i.e., an instrument able to count the number of entering particles and to measure their energy), an ionization chamber must fulfill three basic conditions:

1. The ionizing particle must lose all its kinetic energy inside the sensitive volume of the detector.
2. All created charges must be collected by the applied bias and contribute to the pulse formation.
3. In the absence of any ionizing particles, no charges may be collected by the electric field.

B. The Semiconductor Detector

A reverse-biased p-n or p-i-n Ge or Si diode fulfills all three of these basic conditions to function as a solid state ionization chamber (Fig. 4.2). Indeed, the intrinsic or depleted region of the junction acts as the sensitive volume and the whole may be regarded as a capacitor having a (small) leakage current between the p+ and n+ contacts in the absence of any ionizing radiation. From an electronic point of view, it may be regarded as a capacitor in parallel with a direct current (DC) source.

The detector capacitance depends on the detector dimensions. Its magnitude is determined by the area of the p+ and n+ contacts, their separation, and the dielectric constant of the semiconductor. The p+ contact carries a negative space charge and the n+ contact a positive space charge. In the intrinsic region an electrical field exists due both to the space charges and the applied reverse bias. In Ge detectors this intrinsic region may be very large (up to 60 mm), typical values for silicon detectors are 150–1000 μm, and Si(Li) detectors have a thickness of 3–4 mm.

An ionizing particle entering (or created in) the intrinsic region will lift a certain number of electrons from the valence band, into the conduction band, generating a certain number of pairs consisting of positive holes and negative electrons swept away to the p+ and n+ contacts, respectively, by the existing electric field.

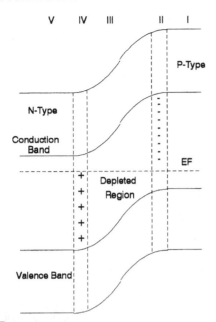

FIGURE 4.2 Band structure of a p-n junction. The probability $P(E)$ of occupation of an electronic level E in a solid is given by $P(E) = 1/\{[\exp(E - E_F)/kT + 1\}$, where E_F represents the Fermi level, k the Boltzmann constant, and T the temperature. Note that $P(E_F) = 0.5$. In a p-type semiconductor E_F lies close to the valence band, in a n-type semiconductor close to the conduction band, and in an intrinsic semiconductor approximately halfway between both bands. In an unbiased p-n junction the height of the Fermi level E_F depends only on the temperature. Five regions are distinguished in a p-n junction: the p-region, the negative space charge region, the intrinsic region, the positive space charge region, and the n-region. In a reversed biased p-n junction the potential barrier is enhanced and the p-n junction acts as a diode. In a p-i-n junction, E_F is no longer constant and the intrinsic region is increased.

Variations in shape and rise time make the amplitude of the current pulse unsuitable for spectroscopic aims, as its intensity is not proportional to the deposited energy (Kröll *et al.*, 1996). What is important for spectroscopic applications is the integral of the current pulse. Therefore a charge-sensitive (integrating) preamplifier (Fig. 4.3) has to be used, which transforms the current pulse, i_{in}, into a step voltage V_0. The latter is proportional to the incident energy, if the amplification factor, A, is very large:

$$V_0 = \int i_{in} dt / C_F = Q/C_F = Nq/C_F = Eq/(\varepsilon C_F) \qquad (4.2)$$

where Q represents the total charge Nq and ε the energy necessary to excite an electron hole pair. This energy may not be confused with the forbidden energy gap, which is much smaller (Table 4.1). It means that about 33% of the available energy is actually converted into electron hole pairs. The rest serves to excite lattice vibrations and is lost in the pulse formation (Leo, 1987; Goulding and Landis, 1982). Protracted accumulation of charges on the feedback capacitor C_F must be avoided. Therefore C_F has to be

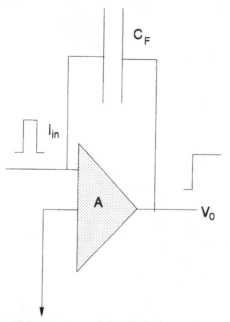

■ **FIGURE 4.3** **Charge sensitive preamplifier.**

■ **TABLE 4.1** **Some Important Ge and Si Properties**

	Ge (at 77 K)	Si (at 300 K)
Electron mobilty: μ_e in cm^{2}V s	3.6E4	1350
Hole mobility: μ_h in cm^{2}V s	4.2E4	480
Energy ε needed to create 1 e$^-$–h$^+$ pair.	2.96 eV	3.62 eV
Atomic number Z	32	14
Forbidden energy gap	0.746 eV	1.115

discharged in time, most commonly by a resistor R_F or by pulsed reset techniques (see Fig. 4.4).

Apart from the integrator stage, a (resistive feedback) preamplifier may have a second stage. A differentiation and a pole-zero cancellation circuit couple the two stages (Fig. 4.4). The rise time of the signal is determined by the output signal of the detector along with the preamplifier speed. A typical fall time is of the order of 50 μs. For digitization, the signal has to be further transformed by a shaping amplifier. The task of the shaping amplifier is complex. It transforms the shape and amplitude from the preamplifier signal in order to:

1. Improve the signal to noise ratio via adjustable shape form and width.
2. Make the signal suitable for digitization in an analog-to-digital converter.
3. Make the output independent of the signal rise time.
4. Facilitate calibration of the spectrum.

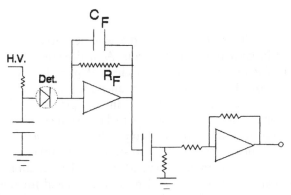

FIGURE 4.4 Schematic drawing of a resistive feedback preamplifier.

A well-chosen peaking time, that is, the time needed for the signal to reach its maximum amplitude, is important to reduce the electronic noise and thus to improve the detector resolution. The output pulse should not be too long in order to prevent spurious summation of independent pulses separated by very small time intervals. On the contrary, for very short peaking times ($\sim 1 \, \mu$s), the peak shaping is ended before completion of the integration, which would mean important loss of information. This so-called ballistic deficit is particularly important in large γ detectors. Analog ("Gated Integrator") or digital techniques may be incorporated in the pulse processing to minimize spectrum broadening due to ballistic deficit. A comparative study of different ballistic deficit correction methods versus input count rate has been carried out (Duchêne and Moszynski, 1995). Finally, the signal is sent to an analog-to-digital converter (ADC) and a multichannel analyzer (MCA), which measures the pulse height and constructs a spectrum, that is a histogram of pulses classified as a function of their pulse height. The analog amplifier and ADC can be replaced by a digital signal processing (DSP) module. DSP is a technique whereby the detector signal is digitized directly as it comes from the preamplifier, with only some minor preconditioning. The digitized data are then filtered and optimized using digital processing algorithms and finally transferred to the MCA for storage, view, and analysis. DSP allows implementation of signal filtering functions that are not possible through traditional analog signal processing. Benefits include higher throughput, reduced sensitivity to ballistic deficit, adaptive processing, improved resolution, and improved temperature stability for repeatable performance.

C. Fundamental Differences Between Ge and Si Detectors

In Table 4.1 three important differences between Ge and Si are given. These are the energy gap, the atomic number Z, and the mobilities μ_e and μ_h of the majority carriers. Together with the purity and charge-carrier lifetime, they influence the thickness of the depletion region of a biased p-n junction.

I. The Energy Gap

There is a 50% difference between the energies needed to create an electron–hole pair in germanium and silicon. A Si detector may be used at room temperature for the spectroscopy of charged particles. A Ge detector has to be cooled below 100 K in order to reduce the leakage current due to thermal generation of charge carriers to an acceptable level. This has important consequences: a Ge detector has to be operated inside a vacuum chamber and cooled to liquid nitrogen temperatures. The sensitive detector surfaces are thus protected from moisture and other condensable contaminants. That means that, independent of the junction itself, an entrance window exists that makes Ge detectors less suited for the detection of charged particles and also affects the efficiency for low-energy photons.

2. The Atomic Number

In Chapter 1, the three typical interactions of electromagnetic radiation with matter have been detailed. The electrons scattered (photoelectric effect or Compton scattering) or generated (pair production) by one of the three basic interactions excite a certain number of electron–hole pairs and are responsible for the peak formation. For γ-spectroscopy, the photoelectric effect contributes directly to the full energy peak. Indeed, as the total energy of a γ-ray is transferred to an electron, the kinetic energy of the electron will be proportional to the energy of the incoming γ-ray. For the efficiency of a γ-spectrometer preference should thus be given to a semiconductor material having a high photoelectric cross section. Figure 4.5 shows the photoelectric cross section of Si and Ge as a function of energy. One sees immediately that Ge beats Si by one to two orders of magnitude. This is expected, as the photoelectric cross section depends roughly on the fifth power of the atomic

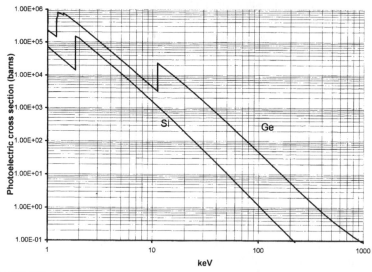

FIGURE 4.5 Photoelectric cross section (barns) of Si (lower curve) and Ge (higher curve) as a function of energy (keV).

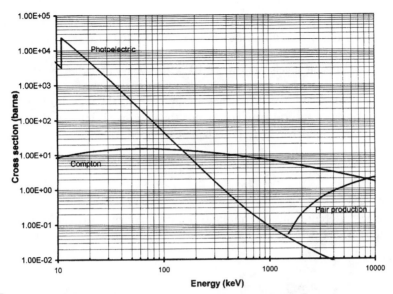

FIGURE 4.6 Compton, photoelectric, and pair production cross section of Ge for high-energy γ-rays.

number Z. Figure 4.6 shows the Compton, photoelectric, and pair production cross section of Ge for γ-ray energies up to 10 MeV.

The Compton cross section is the dominant one for all energies except the very lowest ($E_\gamma \leq 150$ keV) and the very highest ($E_\gamma = 8.5$ MeV). The Compton effect too contributes strongly to the full energy peak by multiple Compton scattering under the condition that the last interaction is a photoelectric one and that all the preceding Compton interactions take place in the Ge crystal. In large-volume detectors the probability of multiple Compton scattering increases. If the last interaction does not occur by the photoelectric effect or if one of the multiple Compton interactions takes place outside the sensitive volume of the detector, the pulse will contribute to the Compton continuum.

The threshold of 1022 keV for the pair production process (see Chapter 1) is clearly seen. It is remarkable that the pair production cross section of a 10 MeV γ-ray equals that of the photoelectric cross section at about 300 keV. It plays an important role in the spectroscopy of high-energy γ-rays. Full absorption of two, one, or none of the 511 keV annihilation lines will contribute to the "full energy," the "single escape," or the "double escape" peak. All three peaks carry full spectroscopic information and are discussed in some detail in Section II.B.2.

3. The Purity or Resistivity of the Semiconductor Material

It is known (Knoll, 1989) that the thickness of the depletion region of a planar semiconductor is given by

$$d = \sqrt{\frac{2\varepsilon V}{qN}} \tag{4.3}$$

where V represents the applied bias and N the net concentration level of electrically active impurities in the bulk, q the electronic charge, and ε the energy gap needed to excite one e^-–h^+ pair. For Ge, active impurity concentration levels as low as 10^{10} atoms/cm^3 of either p or n type can be achieved. This corresponds roughly to 1 impurity atom per 1012 atoms! The application of a reversed bias of up to 5000 V thus leads to a depletion thickness of several cm. This is not the case in Si. Indeed, the resistivity ρ of the semiconductor material can be expressed as

$$\rho = \frac{1}{q\mu N} \qquad (4.4)$$

where μ represents the mobility of the majority carrier. Equation 4.4 may thus be written as

$$d = \sqrt{2\varepsilon V \rho \mu} \qquad (4.5)$$

From Table 4.1 one sees that μ is much smaller for Si than for Ge. If d is expressed in μm, V in volts and ρ in ohm-cm, Eq. 4.5 reduces for Si to

$$d = 0.562\sqrt{\rho V} \qquad (4.6)$$

A thickness of up to 315 μm can be obtained for typical resistivities of $\approx$3000 ohm-cm and a bias of $\sim$100 V. It is thus not possible to realize high-volume detectors with Si. Except for x-rays or low-energy γ-rays, Si detectors are used mainly for charged particles. Since Si detectors may be used at room temperature they may be placed in a vacuum chamber together with the source. The absence of any supplementary entrance window allows the particles to reach the sensitive volume of the detector.

In Si(Li) detectors, the excess acceptor ions in p-type Si may be compensated by Li donor ions. This way, a thickness of up to 5 mm of the active p-i-n region can be obtained. These detectors are predominantly used in x-ray spectroscopy.

4. Charge Carrier Lifetime τ

The charge carrier lifetime τ is the time that the carriers (electrons in the conduction and holes in the valence band) remain free. Trapping centers reduce this lifetime. The maximum signal height V_0 (Eq. 4.2) from the preamplifier after interaction of the detector with the ionizing radiation is given by

$$V_0 = \frac{Eq}{\varepsilon C_{\mathrm{F}}}\left(1 - \frac{d}{\mu\mathcal{E}\tau}\right) \qquad (4.7)$$

where d is the distance traveled by the charge, E the energy deposited in the detector, $\mathcal{E}$ the electric field, μ the mobility of the charges, q the elementary charge, ε the energy needed to excite one $e^- h^+$ pair, τ the charge carrier lifetime, and C_F the feedback capacitor value.

In order to have good charge collection and thus to avoid tailing, $\mu \mathcal{E} \tau \gg d$ where the minimum value for τ for detector grade semiconductor material is 5 ms for Si at 300 K and 20 μs for Ge at 77 K.

II. Ge DETECTORS

A. High-Purity Ge Detectors

The depletion layer of a p-i-n Ge detector must have a thickness of several centimeters in order to enhance the probability of an interaction of a γ-ray with the sensitive detector material and thus be useful as a γ-ray spectrometer.

Today large Ge crystals of either p or n type are grown with the low impurity levels needed. The detectors fabricated from these crystals are called intrinsic or high-purity detectors. They can be stored indefinitely at room temperature.

Detectors of different size or geometry are available, such as planar detectors, coaxial detectors, and well-type detectors. Others differ in the choice of contacts, of the choice of the entrance window (Al, Be, . . .), the selection of the cryostat construction materials, and so on. In Section II.E they will be briefly described together with their main applications. However, before doing so, it is important to analyze the main features of a γ-spectrum, to understand the influence of the parameters that are used to characterize a germanium detector and to know the different sources of background. Only a clear understanding of these features will allow the user to choose the right detector for a specific application.

B. Analysis of Typical γ-Spectra

1. Spectrum of a Source Emitting a Single γ-ray with $E\gamma < 1022$ keV

Figure 4.7 shows the decay scheme of ^{137}Cs, one of the important long-lived $(T_{1/2} = 30.17$ y) fission products and a common contaminant. It emits two β-rays of 1176 (6%) and 514 keV (94%) exciting a 2.55-minute isomeric level of ^{137}Ba. This isomeric level de-excites itself by the emission of a single γ-ray of 661.66 keV. The M4 isomeric transition is highly converted $(\alpha_{total} = 0.11)$; that is the de-excitation can take place through the emission of a γ-ray but also by the ejection of an atomic electron (a conversion electron) with subsequent delayed emission of the characteristic ^{137}Ba x-rays. Even though ^{137}Cs generates one of the simplest spectra possible (Fig. 4.8), it is worthwhile to take a closer look at it. The spectrum was taken with a 25% n-type Ge ("REGe") detector placed in an RDC low-background cryostat and a ULB Pb castle. The most striking is the full energy peak at 661.66 keV carrying the full spectroscopic information. The x-rays of the daughter element ^{137m}Ba are clearly seen: two doublets at $31.82 \sim 32.19$ keV and $36.4 \sim 37.3$ keV. For most other γ transitions the intensity of the x-rays of

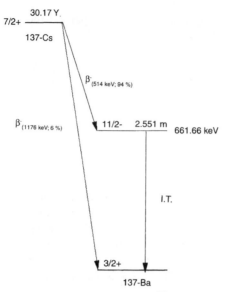

FIGURE 4.7   Decay scheme of ¹³⁷Cs; I.T. stands for isomeric or internal transition.

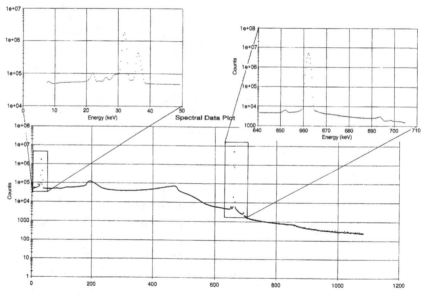

FIGURE 4.8   Gamma-spectrum of ¹³⁷Cs emitting a single γ-ray at 661.66 keV taken with a 25% n-type Ge detector placed in an RDC low background cryostat and a ULB Pb castle. Besides the photopeak at 661.66 keV, the x-rays of the daughter element ¹³⁷ᵐBa are seen: two doublets at 31.82–32.19 keV and 36.4–37.3 keV. The weaker lines at 22.11, 26.3, and 651.8 keV correspond to the Ge escape peaks, and the 693.9 keV peak is a random sum peak between the Ba x-rays and the 661.66 keV main peak.

the daughter element will be less pronounced, as most γ transitions have a much smaller total conversion coefficient. The small peak at 693.9 keV corresponds to the random sum peak between the Ba x-rays and the 661.66-keV photopeak. The weak peaks at 651.8, 22.31, and 26.52 keV correspond to the Ge escape peaks (see Section II.B.3) of the 661.66 keV-line and of the Ba x-rays.

Besides these well-defined peaks, two broad peaks are seen, namely the Compton edge and the backscatter edge. The first has an energy of about 478 keV. It is due to a 180° Compton scattering inside the active volume of the detector with subsequent escape of the Compton γ-ray from the detector's active volume. The second is due to a 180° Compton scattering in the detector surroundings with subsequent detection in the detector of the escaped Compton-scattered γ-ray having an energy of about 184 keV. The broadness of these peaks is due to the fact that the scattering angle of 180° is only approximately fulfilled. Finally, the broad elevation in the continuum around 845 keV is due to the summation of the backscatter edge with the photopeak. In Fig. 4.9 the energies of the Compton edge and the backscatter edge are given as a function of energy of the primary gamma-ray. Notice that the backscatter edge tends toward a saturation value of about 200 keV. The energy of the Compton edge is given by

$$E_{CE} = E_\gamma - E_{BS} \tag{4.8}$$

Note that both curves cross at about 250 keV. For γ-rays with $E_\gamma < 250$ keV the positions of the Compton edge and the backscatter edge are thus reversed. The continuum at the lower energy side from the Compton edge is due to Compton scattering inside the active volume of the detector with subsequent escape of the Compton-scattered γ-ray and to the bremsstrahlung emitted during the interaction of the betas and electrons with the detector surroundings. The maximum of this bremmsstrahlung-continuum is equal to that of the emitted beta, 1176 keV in the case of ^{137}Cs.

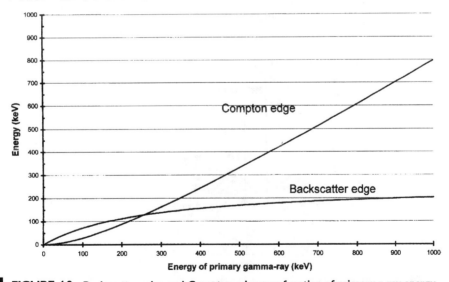

FIGURE 4.9 Backscatter edge and Compton edge as a function of primary γ-ray energy.

The continuum between the Compton edge and the full-energy peak is due to multiple Compton scattering, where the last interaction is a Compton event rather than a photoelectric effect. This leads to the fact that the continuum to the left of the full-energy peak is generally higher than to its right.

The background above the full-energy peak is due to the bremsstrahlung of the 1176-keV β transition and to origins not related to the source, as discussed in greater detail later in Section D.

2. Spectrum of a Multiple-γ-ray Source Emitting at Least One γ-ray with an Energy $\geq$ 1022 keV

The spectrum can be complicated even when only a small number of γ-rays are emitted during the radioactive decay. The case in which one or several γ-rays surpass the energy of 1022 keV is especially interesting. This will be illustrated with the help of the γ-spectrum of ^{24}Na formed for example by the ^{23}Na(n, γ)^{24}Na reaction and decaying with a half-life of 15.03 h to ^{24}Mg. This decay takes place in >99% of all cases by a β-transition of 1.389 MeV and in 0.06% by a β-transition of 276 keV. From the decay scheme shown in Fig. 4.10 one sees that two strong (>99%) coincident γ-rays of 1368.9 and 2754.2 keV exist, as well as a weak γ transition of 3867.2 keV (0.06%), also in coincidence with the 1368.9-keV line. The total of only three γ-rays, all surpassing the threshold for a possible pair production, leads to the quite complex spectrum of Fig. 4.11. It shows, besides the backscatter peak at about 200 keV, a total of 13 well-defined peaks.

The three full-energy peaks at 1368.9, 2754.2, and 3867.2 keV are clearly seen. The intense first two are accompanied by a well-pronounced Compton edge at approx. 1100 and 2400 keV; whereas the Compton edge of the weak 3867.2 keV line is almost lost in the general background.

If pair production takes place, two annihilation quanta of 511 keV are emitted at 180°. When the two are fully absorbed, they contribute to the

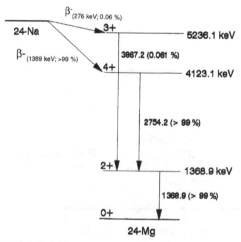

FIGURE 4.10   Decay scheme of ^{24}Na.

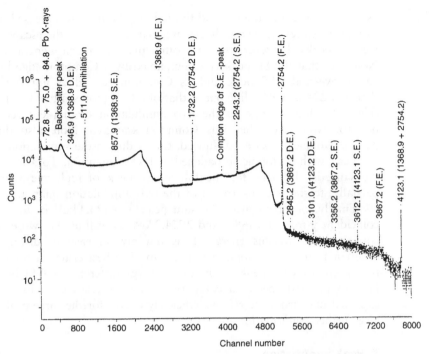

FIGURE 4.11 γ-Spectrum of ^{24}Na emitting two strong (>99%) coincident γ-rays at 1368.9 and 2754.2 keV and a weak γ-ray at 3867.2 kev (0.06%). Thirteen well-defined peaks are observed. Their origin is explained in the text.

full-energy peak. When one escapes from the detector without interaction a discrete single escape (SE) peak is generated having an energy of

$$E_{SE} = E_\gamma - 511\,\text{keV} \tag{4.9}$$

When both annihilation quanta escape, the double escape peak (DE) is generated with an energy of

$$E_{DE} = E_\gamma - 1022\,\text{keV} \tag{4.10}$$

Besides the Doppler-broadened 511-keV line (see Section C.1.c), all six escape peaks can be recognized in the spectrum. Those of the 1368.9-keV line are weak, as the energy is too close to the threshold energy of 1022 keV (Fig. 4.6). It is worthwhile to take a closer look at the strong DE-peak at 2754.2 − 1022 = 1732.2 keV and the corresponding SE-peak at 2243.2 keV. The peak shape of the first one is a mirror image of the full-energy peak. The background to the right of the peak is higher than that to the left! This is due to multiple Compton scattering of one or both annihilation quanta (the last interaction not being a photoelectric effect), whereby the energy of the Compton electron adds to the energy of the DE peak, increasing the continuum to its right. On the contrary, the SE peak is perfectly symmetric as the continuum to the left and the right of the peak is increased by multiple Compton scattering. These shapes are characteristic for the escape peaks. For weaker peaks, however, they are often masked by the general continuum. Note also that the SE peak is accompanied by a Compton edge but that the gap between the SE peak and its Compton edge does not correspond to that of a 2243.2 keV-γ-line but to that of a Compton scattered 511 keV-γ-line (Fig. 4.9). Indeed, one of the two annihilation quanta escaped from the detector while the other was Compton scattered. Note also that the SE peak is Doppler-broadened. Indeed, this is due to the summation of a sharp DE peak with a Doppler-broadened 511-keV quantum. To a lesser amount, the same is also true for the full-energy peak of high-energy γ-rays due, partially, to the full absorption of the two annihilation quanta.

Finally, one recognizes the sum peak at 4123.1 keV of the two strong coincident γ-rays at 1368.9 and 2754.2 keV as well as the two escape peaks corresponding to this energy. It is not always easy to distinguish these different peaks in an unknown spectrum. The best criterion for recognizing the different escape peaks is the exact energy difference of 511 or 1022 keV. For complicated spectra a comparison of their relative intensities with the expected ones from the relative efficiency curves for the three peak types can give further confirmation.

3. Peak Summation

In Fig. 4.11 different sum peaks have been seen. They merit further attention. Real sum peaks have to be distinguished from random sum peaks. Real sum peaks are due to coincident γ-rays simultaneously detected. Their energy equals the sum of the two individual energies. The interpretation can be confirmed by their intensity if measured with the same detector at a different source to detector distance. Indeed, the probability P of a real sum peak is given by:

$$P = I \cdot p \cdot \varepsilon_1 \cdot \varepsilon_2 \tag{4.11}$$

where I is the intensity (Bq) of the source, ε_1, ε_2 are the counting efficiencies for γ_1 and γ_2, respectively, and p is the intensity of the less abundant of the two coincident γ-rays summing up.

If $\varepsilon_1 \approx \varepsilon_2$ one sees that the intensity of the sum peak varies roughly as the square of the efficiency. The phenomenon of True Coincidence Summing, also referred to as Cascade summing and its impact on γ-ray full energy peaks is discussed in Section II.B.4 of this chapter.

In addition to real sum peaks, spurious sum peaks due to the finite time resolution can occur. Their probability is given by

$$P = 2 \cdot \tau \cdot I^2 \cdot \varepsilon_1 \cdot p_1 \cdot \varepsilon_2 \cdot p_2 \tag{4.12}$$

where τ is the time resolution of the detection system and p_1 and p_2 are the branching ratios of the two γ-rays summed up accidentally.

The intensity of random sum peaks depends, therefore, thus on the square of the source intensity and on the time resolution. An illustration of a random sum peak is given in Fig. 4.8. Avoid, if possible, the use of intense sources in order to minimize random summation. If the radioactive source decays with a certain transition probability λ, the intensity I is given by

$$I = I_0 \cdot e^{-\lambda \cdot t} = I_0 \cdot e^{(-0.693/T_{1/2}) \cdot t} \tag{4.13}$$

I_0 representing the initial intensity and $T_{1/2}$ the half-life. Inserting Eq. 4.13 into Eq. 4.12, one sees that the probability P of occurrence of a spurious sum peak depends on $I^2 = I_0^2 \cdot e^{-2 \cdot \lambda \cdot t}$. The probability of occurrence of a spurious sum peak decays thus with a transition probability of $2 \cdot \lambda$ rather than λ, or a half-life $T_{1/2}/2$ rather than $T_{1/2}$, i.e. twice as fast as the isotope itself. This is a firm criterion for their recognition. Pulse pile-up rejection in modern processing electronics can reduce random summing to a great degree.

4. True Coincidence Summing Effects

In most cases of radioactive decay, a parent nuclide decays to an excited energy level of a daughter nuclide by emitting an alpha or a beta particle, or via electron capture. The transition from the excited state to the ground state of the daughter nuclide may then occur by the emission of two or more gamma-rays in a cascade. Since the excited states have life times on the order of pico-seconds, it is highly probable that the γ-rays emitted in a cascade are detected within the resolving time of a gamma-ray spectrometer. The γ-rays are then said to be detected in true coincidence. In the case of a nuclear decay occurring via electron capture, X-rays will be emitted which may also be detected in true coincidence with a gamma-ray. As a result of True Coincidence Summing or Cascade Summing, the detector accumulates the sum total of the energy deposited by the cascading gammas from a given nuclear decay. Therefore, events are lost (summing-out) or gained (summing-in) from the Full Energy Peak (FEP) of the gamma-ray of interest, and any activity determination based on the FEP will be in error. It is therefore, necessary to correct for true coincidence effects.

Figure 4.12 gives an example of a radioactive decay where cascade summing occurs. In the above example, the parent nuclide undergoes a beta decay to the excited energy state E_1 of the daughter nucleus. The de-excitation to the ground state of the daughter nuclide occurs via the emission of gamma-rays γ_1 and γ_2 in a cascade or via the emission of gamma-ray γ_3 directly to the ground state. Assuming that the gamma-rays γ_1 and γ_2 are detected in true coincidence, a FEP measurement of γ_1 or γ_2 suffers from cascade summing losses and the FEP measurement of γ_3 suffers from cascade summing gains. It must be noted that cascade summing losses are not just limited to the counts appearing in the sum peak. Rather, the detector may accumulate the full energy deposition from one of the gamma-rays (say γ_1) and a partial energy deposition from the second gamma-ray (say γ_2), resulting in a count being lost from the FEP of γ_1. In the pulse height spectrum, these events will appear in the continuum between the energy of γ_1

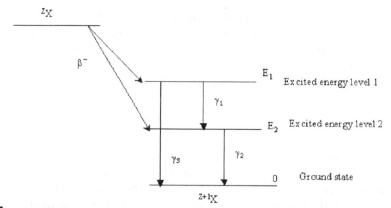

FIGURE 4.12   An example decay scheme showing cascading gamma-rays.

and the energy of the sum peak. The same argument can be made for γ_2 as well. Therefore, the cascade summing losses depend on *the total efficiency* of the detector for the gamma-rays in the cascade. In the case of cascade summing gains, the two gamma-rays γ_1 and γ_2 in the above example deposit their full energy in the detector. The resulting event appears at the same energy as that of γ_3. Cascade summing gains are dependent on the peak efficiencies of the detector at the gamma-ray energies of interest. Besides the detection efficiencies, cascade summing losses or gains also depend on the gamma-ray emission probabilities and any angular correlations involved in the gamma-ray emission. It is important to note that the magnitude of cascade or *true* coincidence summing losses or gains is dependent on the counting geometry, and not on the source activity.

Angular correlation between two gamma-rays emitted in a cascade is defined as the relative yield of γ_2 about the $0°$ direction defined by the detector position, given that γ_1 is emitted in the same direction. Angular correlations arise because the direction of emission of the first gamma-ray is related to the orientation of the angular momentum of the intermediate state. If the lifetime of the intermediate state is short, the orientation of the angular momentum will persist. The direction of the second gamma-ray will be related to the angular momentum of the intermediate state, and hence to the direction of the first gamma-ray (Evans, 1955). Angular correlation effects in general are not very significant when correcting for cascade summing effects. But for measurements requiring a high degree of accuracy (few tenths of a percent), it is indeed necessary to take angular correlation effects into account.

Detailed discussions on the subject of true coincidence summing can be found in several standard text books (Debertin and Helmer, 1988; Knoll, 1989).

a. True Coincidence Correction for a Simple Case

For a simple decay scheme such as the one shown in Fig. 4.12, it is straight-forward to derive a correction factor for cascade summing losses or gains. For γ, the full energy peak rate in the absence of cascade summing can be written as follows.

$$\dot{N}_{10} = A \cdot p_1 \cdot \varepsilon_1 \qquad (4.14)$$

The quantity A is the source activity, p_1 is the emission probability of γ_1, and ε_1 is the full energy peak efficiency at the energy of γ_1. Since γ_1 and γ_2 are emitted and detected in true coincidence, the energy deposited in the detector may be the sum of the full energy from γ_1 and a part of the energy (up to the full energy) from γ_2. This results in events being lost from the full energy peak of γ_1. Since any type of interaction involving γ_2 will result in a loss of count from the FEP of γ_1, the *total detection efficiency* of γ_2 is used in determining the cascade summing loss.

The peak rate of γ_1 in the presence of cascade summing is written as follows.

$$\dot{N}_1 = A \cdot p_1 \cdot \varepsilon_1 - A \cdot p_1 \cdot \varepsilon_1 \cdot \varepsilon_{t2} \qquad (4.15)$$

The correction factor is derived from Eqs. 4.14 and 4.15.

$$COI = \frac{\dot{N}_1}{\dot{N}_{10}} = 1 - \varepsilon_{t2} \qquad (4.16)$$

In deriving a correction factor for γ_1, one has to keep in mind that not all emissions of γ_2 are preceded by γ_1. A fraction of γ_2 is preceded by beta decay. The peak rate of γ_2 in the presence of cascade summing is given in Eq. 4.17.

$$\dot{N}_2 = A \cdot p_2 \cdot \varepsilon_2 - A \cdot p_2 \cdot \varepsilon_2 \cdot \left(\frac{p_1}{p_2}\right)\varepsilon_{t1} \qquad (4.17)$$

The cascade summing correction factor for γ_2 is therefore,

$$COI = \frac{\dot{N}_2}{\dot{N}_{20}} = 1 - (p_1/p_2)\varepsilon_{t1} \qquad (4.18)$$

From Eqs. 4.16 and 4.18 it is evident that total efficiency ε_t is should be known in order to determine the correction factor for cascade summing losses.

In the case of γ_3, one has to correct the full energy peak for cascade summing gains. The peak count rate in the absence of cascade summing is written as,

$$\dot{N}_{30} = A \cdot p_3 \cdot \varepsilon_3 \qquad (4.19)$$

The full energy peak of γ_3 will gain events only when γ_1 and γ_2 deposit their full energies in the detector.

$$\dot{N}_3 = A \cdot p_3 \cdot \varepsilon_3 + A \cdot p_1 \cdot \varepsilon_1 \cdot \varepsilon_2 \qquad (4.20)$$

The correction factor is therefore,

$$COI = \frac{\dot{N}_3}{N_{30}} = 1 + (p_1 \varepsilon_1 \varepsilon_2 / p_3 \varepsilon_3) \qquad (4.21)$$

For decay schemes involving three or more gamma-rays in a cascade, the analytical formulae for summing out probabilities especially, become quite cumbersome to calculate. Over the last two decades, several authors have generalized these formulae for complex decay schemes and have reported them in the literature (Andreev *et al.*, 1972; Moens *et al.*, 1982; De Corte and Freitas, 1992).

b. True Coincidence Correction Using Canberra's Genie2000 Software

Cascade summing losses could be as high as 30–40% at close-in geometries, depending on the type of detector used and the specific nuclide that is being measured. If the detector is calibrated with a standard source identical in shape and size to that of the sample, and the nuclide(s) under study are the same in the standard and sample, then no correction need be applied for true coincidence summing. In all other cases correction factors must be applied if measurements are required to be performed at close-in geometries. Canberra Industries has developed and patented a technique for calculating correction factors for true coincidence or cascade summing losses and gains [U.S. Patent 6,225,634]. The algorithms that perform the calculations have been incorporated into Canberra's Genie2000 Gamma Analysis software package (version 2.0 and later). Genie2000 can calculate the true coincidence correction factors for a wide variety of counting geometries and for an exhaustive list of nuclides and gamma-ray lines. To compute the correction factors, Canberra's method requires a single intrinsic peak-to-total efficiency curve and a so-called spatial response characterization or ISOCS characterization for each detector. Canberra's ISOCS (In Situ Object Calibration Software) is a powerful mathematical tool to calculate HPGe full energy peak efficiencies for practically any source geometry (Bronson and Young, 1997; Venkataraman *et al.*, 1999).

The Genie2000 algorithms for calculating the true coincidence correction factors for voluminous sources are based on the work done by V. P. Kolotov *et al.* (1996). In this method, the voluminous source is first divided into a large number of equal volume sub-sources. A point location is selected within each sub-source using a pseudo-random sequence. The true coincidence correction factor at each of these point locations is calculated and then integrated to determine the overall correction factor for the entire source.

It was previously noted that the total efficiency of the detector, ε_t, is required to compute the correction factor for true coincidence losses. For a point source at a location "i" the total efficiency at a given gamma-ray energy may be determined, provided the full energy peak efficiency ε_p and the peak-to-total ratio (P/T) are known at the given energy.

$$\varepsilon_{t,i} = \varepsilon_{p,i}/(P/T) \tag{4.22}$$

The full energy peak efficiency is calculated using the ISOCS characterization for the given detector. The P/T ratio is obtained from the intrinsic peak-to-total efficiency curve determined for the detector. The true coincidence correction for the gamma-ray of interest, g, is given by the equation,

$$COI_{g,i} = (1 - L_g) \times (1 + S_g) \tag{4.23}$$

where L_g is the probability of summing out and S_g is the probability of summing in. These probabilities are the sum of the partial probabilities calculated for individual decay chains involving the gamma line of interest.

$$L_g = \sum_{j=1}^{n} L_{g,j} \tag{4.24}$$

$$S_g = \sum_{j=1}^{m} S_{g,j} \tag{4.25}$$

The calculation of summing out probability L_g requires the knowledge of nuclear data such as the gamma-ray yields, branching ratios, and internal conversion coefficients, as well as total detection efficiencies. Summing in probability S_g requires the knowledge of nuclear data and full energy peak efficiencies. The generalized formulae reported in the literature to compute the summing out and summing in probabilities for complex decay schemes have been incorporated into the methodology developed by Kolotov *et al.* By calculating the coincidence correction factors ($COI_{g,i}$) for a large number of infinitesimally small sub-sources and then integrating, the correction factor for the whole voluminous source is obtained.

It is desirable to use the spatial characterization for the specific HPGe detector, if available. However, it is not a necessary condition. Koskelo *et al.* (2001) have shown that it is sufficient to use an approximate detector characterization in order to obtain good cascade summing results with Genie2000. Venkataraman and Moeslinger have demonstrated the feasibility of employing a discrete number of generic detector response characterizations for carrying out cascade summing corrections on gamma-ray spectra obtained with non-characterized HPGe detectors (Venkataraman and Moeslinger, 2001). A set of generic detector characterizations has therefore been made available within Genie2000.

c. True Coincidence Correction Using Ortec's Gamma Vision Software

The technical details given in this section are from a paper published by Ron Keyser *et al.* (2001). The method for true coincidence correction implemented in Ortec's GammaVision software is based on the work by Blaauw *et al.* (1993). In this method, the probability of recording a count in the full energy peak is given by,

$$P_{Ei} = g_i \varepsilon_{\text{fullenergy}, Ei} \prod_{j \neq i} (1 - g_j \varepsilon_{\text{total}, Ej}) \tag{4.26}$$

where, P_{Ei} is the probability of a count in the full energy peak, $\varepsilon_{\text{fullenergy}, Ei}$ is the full energy efficiency at an energy E_i, $\varepsilon_{\text{total}, Ej}$ is the total efficiency at an energy E_j, and g_i and g_j are the transition probabilities for gamma-rays with energies E_i and E_j, respectively. Thus, the determination of the correction factor is reduced to knowing the full energy efficiency, the total efficiency, and the decay scheme of the nuclides in question. In addition to the full energy efficiency, the total efficiency includes the peak-to-total ratio, an absorption correction, and terms that correct the efficiency for an extended source.

5. Ge-Escape Peaks

For low-energy γ-rays or for extremely thin detectors, when the interaction takes place close to the detector border, a certain probability exists that a Ge x-ray escapes from the detector. This probability is thus particularly important in detectors having thin windows (see Fig. 4.20). The parasitic peaks are observed at energies of

$$E_\gamma - 9.88\,\text{keV} \text{ (escape of the } K_\alpha \text{ line)}$$

and

$$E_\gamma - 10.98\,\text{keV} \text{ (escape of the } K_\beta \text{ line)}$$

The latter is five times less probable than the former. An illustration of several Ge escape peaks can be found in Fig. 4.8.

With these general aspects of a γ-spectrum in mind, it is time now to take a closer look at the characteristics of a Ge detector such as resolution and efficiency, which play an important role in the choice of an appropriate detector.

C. Standard Characteristics of Ge Detectors

I. Energy Resolution

From the spectra discussed in Section B, it is clear that the observed peaks have a finite width. Peak broadening is due to the statistical fluctuations in the number of electron–hole pairs created in the active detector volume (FWHM)$_{\text{det}}$ and to the electronic noise of the different elements of the amplification chain. The resolution is expressed by full width at half-maximum

(FWHM), and it can be readily obtained from the spectra. The different noise contributions add quadratically according to the equation

$$\text{FWHM} = \sqrt{(\text{FWHM})^2_{\text{det}} + (\text{FWHM})^2_{\text{elect}}} \qquad (4.27)$$

$(\text{FWHM})_{\text{det}}$ and $(\text{FWHM})_{\text{elect}}$ represent the detector and the electronic contribution in the total FWHM.

The energy E released in the detector is shared by two processes, namely direct ionization and lattice vibrations. Both processes may lead to the generation of

$$N = \frac{E}{\varepsilon} \qquad (4.28)$$

electron–hole pairs according to Eq. 4.1 described in Section I.A. The second process obeys a Gaussian distribution and, if direct ionization would be negligible, the variance σ_N of the number of charge carriers N would be given by the equation

$$\sigma_N = \sqrt{N} = \sqrt{\frac{E}{\varepsilon}} . \qquad (4.29)$$

When the variance σ is expressed in energy units (eV), Eq. 4.29 becomes

$$\sigma = \varepsilon\sqrt{N} = \sqrt{E\varepsilon} \qquad (4.30)$$

and the intrinsic FWHM_{det} is calculated as

$$(\text{FWHM})_{\text{det}} = 2.35\sqrt{E\varepsilon} \qquad (4.31)$$

where the factor 2.35 is a statistical property of the Gaussian distribution and gives the ratio between FWHM and the variance of a Gaussian distribution. In practice, however, direct ionization is not negligible at all, justifying the introduction of a correction factor F, the so-called Fano factor:

$$\text{FWHM}_{\text{det}} = 2.35\sqrt{FE\varepsilon} \qquad (4.32)$$

The Fano factor has an approximate value of 0.1 for Ge and Si. In Fig. 4.13 the approximate intrinsic FWHM is given as a function of γ-ray energy.

a. The Electronic Noise Contribution (FWHM)elect and Its Time Behavior

Depending on the detector type, resolutions (FWHM) lower than 1.8 keV at 1332 keV, 0.50 keV at 122 keV, and 0.15 keV at 5.9 keV are common. This implies an electronic noise contribution of <0.8, 0.22, and 0.10 keV, respectively. The electronic noise depends, amongst others, on the capacitance of the detector and thus on the detector dimensions and geometry.

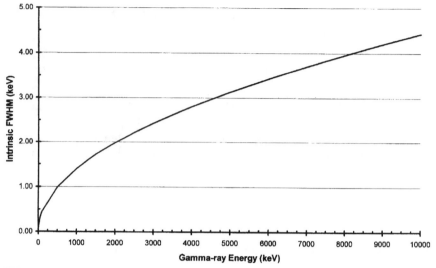

FIGURE 4.13 Approximate intrinsic FWHM as a function of γ-ray energy.

Electronic noise is any undesired fluctuation that is superimposed on the signal. It contributes to the finite resolution of the detector. In electrical circuits it stems from random processes such as the random collection of electrons or the arbitrary thermal movement of electrons in a resistor (Goulding and Landis, 1982). It can be represented as a voltage or current generator with zero average value and random positive and negative peaks. Noise is a statistical phenomenon and can be described as a time average of the squares of all positive and negative values. One has to realize that a counting rate of one 1-MeV γ-ray per second losing its complete energy in the active volume of the detector generates a current of only 5.41×10^{-14} C/s, and this has to be registered with a precision of better than 0.2% if a resolution (FWHM) of 1.8 keV is desired. This is a very difficult task for any electronic measuring chain.

In a detector amplifier system, three different noise contributions may be distinguished as functions of their time behavior.

The Step Noise or Parallel Noise $(FWHM)_S$ arises from the discrete character of any current i_n flowing in the input circuit of the preamplifier. This current is integrated on the capacitor C_f (see Figs. 4.3 and 4.4). The two main sources of step noise are the detector leakage current and the thermal noise of the feedback resistor. It can be represented by a current generator, generating current-pulses at the input of the preamplifier. It is proportional to

$$(FWHM)_S \propto \sqrt{\left(I_l + 2\frac{kT}{R_f}\right)\tau} \qquad (4.33)$$

where I_l represents the total current of the detector (leakage current plus current generated by the detected radiation), k the Boltzmann constant, T the temperature of the feedback resistor, and τ the shaping time (measuring time of the amplifier).

Step noise can be reduced by:

1. Measuring at shorter shaping times
2. Reducing the current through the detector (e.g., by measuring at a lower counting rate)
3. Choosing a feedback resistor with a high resistance, or avoiding it by using a different reset mechanism

The Delta Noise or Series Noise $(FWHM)_D$ is mainly associated with the shot noise in the first stage of the preamplifier (FET). Delta noise is proportional to

$$(FWHM)_D \propto C \sqrt{\frac{T}{g_m \tau}} \qquad (4.34)$$

where g_m represents the transconductance of the FET and C is the total capacitance at the input of the preamplifier.

Delta noise can thus be reduced by:

1. Measuring at longer shaping time
2. Miniimizing the detector and stray capacitance
3. Selecting a low-noise FET with large transconductance.

The Flicker-Noise or 1/f Noise $(FWHM)_F$ is independent of the detector capacity and exists only in association with a direct current. It is independent of the shaping time τ and is thus less relevant for the present discussion.

All these different noise contributions sum up quadratically with the intrinsic noise discussed in Section C.1. The total noise is thus given by

$$(FWHM)_{tot} = \sqrt{(FWHM)_{det}^2 + (FWHM)_D^2 + (FWHM)_S^2 + (FWHM)_F^2} \quad (4.35)$$

It is particularly instructive to look at the dependence of the noise on the shaping time τ. Figure 4.14 gives a schematic view of the square of the total FWHM. For most detectors, measured with a Gaussian shaper at low counting rates, the optimum shaping time lies between 3 and $8\,\mu$s, corresponding to a peaking time between 6 and $16\,\mu$s. It is important to realize that, at high counting rates, the average DC current through the detector will increase and consequently also the step noise. The optimum shaping time will thus tend to lower values at high counting rate! In either case the optimum shaping time for a given measurement condition should be determined experimentally.

The importance of step noise and delta noise also depends on the actual shape of the amplifier signal. A semi-Gaussian shaper gives one of the best compromises between both step noise and delta noise. For high-count-rate measurements with large coaxial detectors (see Section II.E) a gated

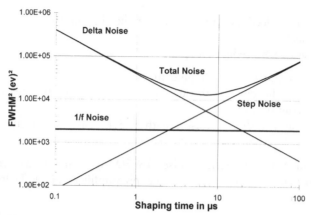

FIGURE 4.14 Noise as a function of the shaping time τ.

integrator (e.g., the Canberra model 2024 Spectroscopy Amplifier) or longer rise times with DSP processors may be used in order to minimize the ballistic deficit.

b. Interference with Mechanical Vibrations and with External RF Noise

Vibrations of the detector cryostat, or even audible noise, may also lead to spectrum broadening. This has to do with the fact that the germanium detector crystal and some leads are at high voltage while they are closely surrounded by conductors on ground potential. This way, effective capacitors are formed. Their value can change when the crystal, the leads or cryostat parts vibrate. Since a capacity, C, can be written as the ratio of a charge, Q, over a voltage, V, a changing capacity can be interpreted by the detector's electronics circuit as being due to a changing charge, in the same way as a detected photon gives rise to a change of charges. Provided that the frequency of the mechanical vibration or noise is not filtered out by the RC-filter network of the amplifier, this noise can sum up with real photon events and show up as peak broadening.

To minimize this sensitivity for "microphony," the user should avoid excessive audible noise and vibrations in the vicinity of the detector, e.g., by placing the detector on some damping material. Detectors are also less sensitive for microphony at lower shaping times. For extreme applications, e.g., for use on board of helicopters or air planes, detector manufacturers can change the mechanical construction of the detector so that its eigen-frequencies do not correspond with the characteristic noise or vibration frequencies of the plane or helicopter.

The spectroscopy system can also behave as an effective antenna for strong RF signals from the environment. Pick up of these may also lead to peak broadenings. Sensitivity for pick-up depends strongly on details such as orientation, grounding and bundling of cables, contact resistance between the various components of the spectroscopy system components, etc. For extreme RF noise, detector manufacturers can change the cryostat and preamplifier hardware to render them virtually immune for pick up of RF signals.

Finally, peak degradation resulting from more fundamental physical processes can also occur, including detector temperature change, Doppler broadening, recoil shift, and recoil broadening. These will be discussed briefly.

c. Other Sources of Peak Degradation

Temperature Change of the Detector. A small temperature dependence of the energy gap and the energy ε necessary for creating an electron–hole pair (Table 4.1) of Si and Ge exists and is given by

$$\frac{\Delta\varepsilon}{\varepsilon} \approx \frac{0.00015}{K} \tag{4.36}$$

The peak position of a 1.3-MeV transition thus changes by about 0.2 keV per kelvin. This effect can cause some peak broadening, especially at higher γ-ray energies, when the cryostat temperature is not stable.

Doppler Broadening. This is observed when the γ-ray is emitted by an "object" that is not at rest. The most commonly known Doppler-broadened line is the 511 keV annihilation line. When a positron (e^+) comes to rest it combines ("annihilates") with an electron (e^-) to form a positronium that decays almost immediately into two quanta of 511 keV. Due to momentum conservation these are emitted at $180°$ (Chapter 1). As the annihilating electrons may have a rest kinetic energy, the annihilation lines are Doppler broadened.

Doppler broadening can also occur when a γ-ray is emitted by an isotope "in flight" after a nuclear reaction. An example is provided by the well-known 479.9-keV γ-line emitted in the reaction

$$^{10}_{5}\text{B}(\text{n}, 2\alpha)t$$

As the cross section for this reaction is very high (3837 barns), boron constitutes a very effective thermal neutron shield but adds an intense Doppler-broadened background line if no special shielding is used.

Note that because of the almost isotropic emission of γ-rays, a Doppler broadened line is always symmetric.

Recoil Broadening. This must not be confused with Doppler broadening. A fast neutron can transfer a large amount of its energy to a recoiling nucleus (see Chapter 1). This recoil-energy is generally not seen by the detector as it takes place in the target. However, Stelson *et al.* (1972) showed that, in the case in which the target is the detector itself, the supplementary energy from the recoil contributes to the formation of electron–hole pairs in the detector. These supplementary charge carriers add to those due to the γ transition, resulting in an odd-shaped peak, having a normal low-energy and a long high-energy slope. Bunting and Kraushaar (1974) detected this phenomenon. It was further described by Verplancke (1992) and Heusser (1993). In particular, it is seen for certain background lines induced by (n, n') reactions in the Ge crystal itself, such as the 691.0-keV line due to the $^{72}\text{Ge}(\text{n}, \text{n}')^{72}\text{Ge}$ (Table 4.3). Recoil broadening always resuslts in a right side asymmetric peak.

Recoil Energy Shift. A γ-ray is emitted by a nucleus when it passes from an excited state to a lower energy state. The latter can be an excited state or the ground state. However, depending on the nature (multipolarity) of the transition, the excited state can de-excite alternatively through the emission of a conversion electron. Furthermore, it shares its energy with the energy of the recoiling nucleus. This recoil energy E_R is given by

$$E_R = \frac{E_\gamma^2}{2Mc^2} \tag{4.37}$$

where E_γ represents the energy of the emitted γ-ray, M the mass of the nucleus, and c the speed of light. If the energy is given in keV and the mass in atomic units (M of $^{12}C = 12$), the corresponding energy difference E_g of the nuclear states involved is given by

$$E_g = E_\gamma + E_R = E_\gamma \left(1 + \frac{E_\gamma}{1,862,300M}\right). \tag{4.38}$$

This effect is completely negligible in general and therefore, $E_g \approx E_\gamma$. However, if M is small and E_γ high, the difference between E_g and E_γ can be significant, as pointed out by Greenwood and Chrien (1980), among others. For instance, the γ-ray of 10,829.1 keV produced in the $^{14}N(n,\gamma)^{15}N$ reaction populates the ground state and is issued from an excited level at 10,833.3 keV. This energy corresponds to the binding energy of the neutron. Note that medium- and high-efficiency Ge detectors perform excellently even at this very high energy, the most useful peak being the double escape peak, which suffers no Doppler broadening as outlined in Section II.B.2.

Radiation Damage. Low-energy tailing can be due to electronics but also to the presence of trapping centers or "deep levels" in the detector. These trap electrons or holes for periods longer than the time needed for pulse formation (Eq. 4.7). Trapping centers may be created by radiation damage in the detector induced by fast charged particles and/or fast neutrons. A 16-MeV neutron creates four times more trapping centres than a 1.6-MeV neutron. Charged particles are easy to shield. This is not true for fast neutrons. They have to be thermalized by a large, hydrogen-rich layer and must be subsequently absorbed by a high-cross section material. The effects of deep-level defects in a high radiation environment have been studied by Lutz (1996).

d. The Gaussian Peak Shape

The peak shape is closely related to the resolution. In principle, the peak shape follows Poisson statistics. If the number of counts is ≥ 20, the shape of the full-energy peak is given by a Gaussian distribution of the values x around the energy channel E according to the equation

$$f(x) = e^{-(x-E)^2/2\sigma^2} \tag{4.39}$$

TABLE 4.2   Theoretical Gaussian Peak Widths

A = Fraction from the maximum	B = Width at fraction	B/FWHM
1/2 (FWHM)	$2.35 \cdot \sigma$	1
1/10 (FW1/10M)	$4.29 \cdot \sigma$	1.82
1/20	$4.9 \cdot \sigma$	2.08
1/50	$5.59 \cdot \sigma$	2.37

The maximum of the distribution lies at $x = E$. Table 4.2 gives the width of the distribution for a few points. Few detectors, if any, exhibit the theoretical peak shape. Most modern detectors have a ratio FWTM (full width at tenth-maximum) to FWHM (full width at half-maximum) of better than 1.9, but a ratio of 2 is common for larger detectors. Mainly detectors made of n-type germanium may show higher FWTM/FWHM ratios.

A FWTM/FWHM ratio smaller than 1.82 may be indicate that the peak has shifted during the accumulation of the spectrum or that the peak is actually a doublet.

2. The Peak-to-Compton Ratio

Following the IEEE standards (ANSI/IEEE Standard, 1986) the peak-to-Compton ratio is defined as the ratio between the maximum number of counts in the channel at the top of the 1332.5-eV peak of ^{60}Co and the average channel count between 1040 and 1096 keV. It depends on the resolution and efficiency, and also on the presence of material in the vicinity of the active detector region, as these materials may backscatter γ-rays into the detector. It plays a role in the "background due to the presence of the source," as will be discussed in Section II.D.

3. The Detector Efficiency

The efficiency ε_γ is a measure of the probability (expressed in absolute values or in per cent) that a γ-ray of energy E_γ is fully absorbed in the active volume of the detector or, in other words, the probability that it contributes to the full-energy peak. It depends basically on the solid angle Ω under which the source is seen by the detector and on intrinsic factors characteristic of the detector.

a. Geometrical Efficiency Factor

In the case of a point source situated on the axis of a circular detector with a flat surface facing the source, the geometric efficiency η is given by a simple analytical formula

$$\eta = \frac{\Omega}{4\pi} = \frac{1 - \cos[\arctan(r/d)]}{2} \tag{4.40}$$

where r represents the radius of the detector, d the distance between detector and source including the distance between detector and endcap, and Ω the solid angle under which the source sees the detector. Moens *et al.* (1981) and Moens and Hoste (1983) proposed an extension of this formula for the case of an extended source or a non-axial source. More realistic computations are based on numerical approaches (e.g., Monte Carlo, Canberra's ISOCS/LabSOCS software).

b. The Intrinsic Efficiency ε_i and the Transmission T_γ

The overall efficiency may be given by

$$\varepsilon_\gamma = \eta \varepsilon_i T_\gamma \tag{4.41}$$

Fundamental effects such as the photoelectric effect and multiple Compton scattering discussed in Section II.B.1 are included in an intrinsic factor ε_i. Other factors such as the thickness of the different entrance windows, the p^+ or n^+ layers and the encapsulation of the source itself are included in the transmission factor T_γ, and η represents the geometric efficiency (Eq. 4.40). Figure 4.15 gives the transmission through different endcaps and dead Ge layers. The importance of T_γ is illustrated in Fig. 4.16, which shows two experimental efficiency curves for the same low-energy detector germanium detector (LEGE) of 200 mm^2 surface and 10 mm thickness obtained with a mixed ^{241}Am^{137}Cs^{60}Co source placed at 5 cm. The only difference is the entrance window, 0.15-mm Be in the first case and 0.5-mm Al in the second. A big difference in efficiency is observed below 20 keV. The increase in efficiency at the very low energy side is due to the beginning of the influence of the K-absorption edge of Ge. It is clearly seen in the upper curve (Be window) but is strongly reduced by the higher absorption of Al as shown by the lower curve.

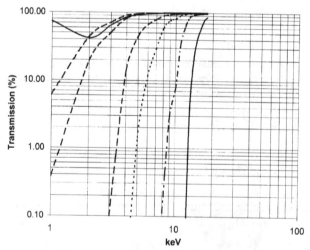

FIGURE 4.15 Some typical transmission curves. Plain curves: Ge dead layer of 0.3 μm (implanted window for *REGe* detector or "thin window" for Canberra's *XtRa* or *BEGe* detectors) and 0.5 mm (Li diffused layer). Dashed curves: cryostat Be window of 0.05, 0.1, and 0.5 mm. Dotted curve: cryostat Carbon window of 0.5 mm. Dashed-dotted curve: cryostat Al window of 0.5 mm.

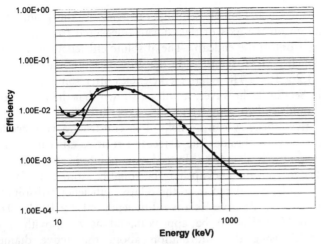

FIGURE 4.16 Superposition of two efficiency curves for the same low-energy detector mounted with two different endcaps: 0.15 mm Be (upper curve) an 0.5 mm Al (lower curve). The source–detector distance was 5 cm. The curves are polynomial fits through the experimental points.

The transmission T_γ is given by

$$T_\gamma = e^{-\mu d}. \tag{4.42}$$

Here μ represents the total absorption coefficient for the γ-ray and d the thickness of the specific absorber. The absorption coefficient μ can be expressed in g/cm^2 (the mass absorption coefficient μ_m) or in cm^{-1} (the linear absorption coefficient μ_l) whether d is expressed in g/cm^2 or in cm. The following relation exists between both:

$$\mu_m = \frac{\mu_l}{\rho} \tag{4.43}$$

where ρ represents the density of the absorber in g/cm^3. It is sometimes convenient to express the thickness d as half-thickness $d_{1/2}$. The relation between both is given by:

$$d_{1/2} = \frac{\ln 2}{\mu} \tag{4.44}$$

where μ stands for either μ_l or μ_m.

c. Relative Efficiency

The efficiency cited by the manufacturer of a Ge detector following the IEEE standards (ANSI/IEEE Standard 325-1986) does not represent the absolute efficiency of the detector. It represents the ratio of the absolute detector efficiency at 1332.5 keV (^{60}Co) to that of the same γ-ray obtained with a 3 inch × 3 inch NaI(Tl) scintillation detector, the point source being placed at 25 cm on the axis of the endcap (measured from the center of the source to the front of the endcap). The absolute efficiency at 1.3 MeV of such a 3 inch × 3 inch NaI(Tl) scintillation detector measured at the same distance of 25 cm is 1.2×10^{-3}. The relative efficiencies offer some means to compare detectors. In Section E, however, it will be shown that the notion of "relative efficiency" could lead to completely false expectations.

The relative efficiency can be helpful to construct a very crude absolute efficiency curve when, besides the detector's relative efficiency, the diameter of its active volume is also given. Indeed, two points of this curve for a point source at 25 cm from the endcap may be roughly estimated. The first point, at 1.3 MeV, can be calculated from the relative efficiency (Absolute efficiency at 1.3 MeV and at 25 cm = Relative Efficiency × 1.2×10^{-3}). The second point, at 100–150 keV, can be approximated as $\varepsilon = \eta$ with η calculated from Eq. 4.25 using the information about the active diameter. This approximation is based on the assumption that, around 100–150 keV, ε_i is near 100% and the transmission correction negligible (at this precision). These two points may be joined by a straight line on a log–log-scale. It is evident that such a rough estimate may not be used for actual measurements as large errors are introduced by the application of the oversimplified assumptions.

d. The Experimental Efficiency Curve

In order to analyze a gamma-ray spectrum to obtain a source activity or gamma emission rate, it is necessary to know the detection efficiency for each peak observed in the spectrum. This can be accomplished by mapping the detection efficiency curve versus gamma-ray energy over a range of energies. Such a curve can be established by the use of one or more calibration sources. If N_0 represents the number of radioactive atoms present in the calibration source at the starting moment of the measurement and $\lambda = 0.693/T_{1/2}$ its transition probability, where $T_{1/2}$ represents the half-life,

$$N_d = N_0(1 - e^{-\lambda \Delta t}) \tag{4.45}$$

atoms decay during the measuring time Δt. If Δt is small with respect to $T_{1/2}$ as is generally the case, Eq. 4.45 reduces to

$$N_d = N_0 \lambda \Delta t = I_0 \Delta t \tag{4.46}$$

I_0 being the activity in Bq. The number N_r of registered counts in the full-energy peak is thus given by

$$N_r = I_0 \Delta t \varepsilon p \tag{4.47}$$

where ε contains all efficiency factors discussed and p represents the branching ratio of the γ-ray measured. The efficiency can thus be readily calculated. Evidently, I_0 has to be corrected for the decay during the time elapsed when the calibration source was certified, according to the equation

$$I_0 = I_{\text{cert}} \, e^{-\lambda t} \tag{4.48}$$

where I_{cert} represents the certified intensity of the source for a given day and hour and t the time elapsed until the actual measurement.

To establish the complete efficiency curve, a great number of efficiencies ε_γ for various energies should be obtained experimentally, either by the use of several calibrated standard mono-energetic γ-sources such as ^{137}Cs or, depending on the energy range desired, by the use of one or several multi-gamma sources such as ^{241}Am (11.959 keV), ^{60}Co (1173 ~ 1333 keV), ^{56}Co (1 ~ 3 MeV), ^{152}Eu (121 ~ 1408 keV), or ^{133}Ba (53 ~ 383 keV). For still lower energies, x-ray sources such as the Mn x-rays [5.88765 keV (50.5%), 5.89875 keV (100%), and 6.49 keV (20.3%)] emitted during the EC decay of ^{55}Fe can be useful. This calibration work is straightforward using modern software such as GENIE (Canberra Industries). Nevertheless, attention should be paid to the following points:

1. Ensure that all standard sources have the same form and are placed at the same distance from the detector. These must be identical to those of the samples to be measured.
2. Ensure that the encapsulations used for all calibration sources and the samples to be measured are the same, especially when (very) low energy measurements have to be performed.
3. Do not use intense sources if multi-gamma-ray sources are used emitting two or more coincident gammas. This will lead to losses due to random summing (see II.B.3).
4. If the standard source used in the calibration emits multiple gamma-rays in true coincidence (^{152}Eu, ^{60}Co ^{88}Y etc.), then one has to be cognizant of true coincidence summing (or cascade summing) losses or gains affecting the full energy peaks. True coincidence summing losses or gains lead to an underestimation of the measured efficiencies while the summing gains lead to an overestimation. These effects become worse with high efficiencies (small source–detector distances and/or large detectors). A correction factor may have to be used to correct for the effects of true coincidence summing (see II.B.4) before the efficiency calibration curve can be used in the analysis.

Figure 4.17 shows a typical efficiency curve of a 25% p-type coaxial detector. A mixed ^{133}Ba^{137}Cs^{60}Co source was placed at a distance of 5 cm. The full line represents a fourth order polynomial fit in $1/E$, E being the energy in keV of the experimental points.

e. Mathematical Efficiency Calculations

As described in the previous section, the detection efficiency curve may be obtained by measuring one or more calibration standards. These standards should emit gamma-rays that span the range of energies expected to be

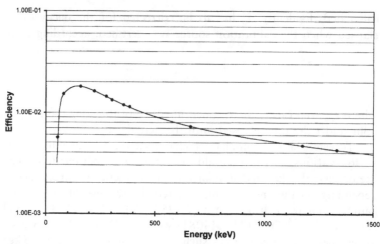

FIGURE 4.17 Experimental efficiency curve of a 25% p-type coaxial detector. A mixed ^{133}Ba–^{137}Cs–^{60}Co source was placed at a distance of 5 cm. The full line represents a 4th order polynomial fit in I/E, E being the energy in keV of the experimental points.

present in the actual samples to be measured. In addition, it is essential that the calibration measurements have the same geometry as the actual samples. Specifically, this means that the source dimensions, source material and density, container wall thickness and density, and source-to-detector positioning must all be same for the calibration standards versus the samples. In many cases it is not practical to obtain calibration standards that match the samples to be counted. Examples of this situation are

- the samples to be counted span a large variety of shapes and sizes and densities (e.g., decommissioning activities), thereby requiring an unacceptably large number of calibration standards to buy, count, and dispose of.
- the samples to be counted are too large to fabricate equivalent calibration standards (e.g., railroad cars full of soil and debris).

In such cases it becomes necessary to utilize mathematical methods to obtain detection efficiency curves. One common approach is to use Monte Carlo computer modeling techniques. These techniques derive their name from the use of computer-generated random numbers to mimic the random processes that take place in real-life gamma-ray emission, scattering, absorption, and detection events. Three computer packages that are in common use are

- MCNP (Monte Carlo N-Particle). See Briesmeister, 2002.
- EGS (Electron Gamma Shower). See Nelson *et al.*, 1985.
- GEANT. See CERN Applications Software Group, 1994.

These codes allow for description of the counting geometry as well as features of the source gamma-ray emission spectrum and features of the detector. These packages are extremely flexible and, consequently, extremely complex. To obtain accurate results it is particularly important to provide detailed information about the structure of the detector. The best results are obtained by benchmarking the calculations from a detector model against measurements with calibration standards; in effect, calibrating the mathematical model. An example of this sort of approach, including rather sophisticated details of the detector structure, is given in Friedman *et al.*, 2001. After developing an accurate model of the detector (i.e. a model proven to be able to reproduce measured efficiencies), it can be used to calculate efficiencies for other source geometries. Clearly this is a very complex process. Development of the geometry model and execution of the calculation take a lot of time and require a high level of sophistication and experience on the part of the user. It is well rewarded, however – a good Monte Carlo model can reproduce detection efficiencies with accuracies that rival those obtainable with calibration standards.

To reduce the amount of time and expertise necessary to obtain reliable efficiencies from Monte Carlo techniques, other approaches have been developed. Typically these are simplified discrete-ordinates calculations that divide the source region into small volume pixels ("voxels") that can be approximated as point sources. The detection efficiency for a given voxel is obtained by approximating the efficiency for the equivalent naked point source and then accounting for the attenuation losses through any absorbing materials between the point source and the detector. The detection efficiency for the entire source is simply obtained by summing the efficiencies for all the voxels.

Examples of such computer codes are Canberra's ISOCS and LabSOCS software packages (Venkataraman *et al.*, 1999). At the Canberra factory, a given detector is "characterized" by developing an MCNP model that reproduces several standard source measurements made with the detector. From there the MCNP code is used to map out the efficiency of the detector for a naked (i.e. unattenuated) point source at any location within 500 meters of the detector. This map of efficiency versus position and energy (referred to as the "characterization" for that detector) is then provided in the form of a look-up table to be used with the ISOCS/LABSOCS software. The software itself provides a simple interface for users to select a basic geometry template (e.g., box, cylinder, planar source, etc.) and to specify the details of the source and possible passive absorbers (e.g., dimensions, densities, etc.) as well as the details of the source-to-detector vector. Utilizing the characterization to obtain the naked point source efficiencies, the software divides the source into pointlike voxels to calculate the detection efficiency as described in the previous paragraph. Typical calculation times for common geometries are on the order of seconds, and rarely more than one minute; and the calculated results are typically accurate to better than 10% depending on the complexity of the geometry. For detectors that have been characterized by Canberra, this is an extremely flexible and powerful technique.

D. Background and Background Reduction

It seems natural to define the background as all pulses registered by the detection system when no source is present. However, it was seen earlier (see Section II.B) that the Compton effect and bremsstrahlung give rise to an important continuum that also has to be regarded as a real background. It is thus useful to distinguish between the background with and without a source.

I. Background in the Presence of a Source

The background due to the source itself is essentially the continuum generated by Compton scattering and by bremsstrahlung. These effects have been analyzed in Section II.B and no further discussion is needed. However, it is this background that very often governs the detection limits.

2. Background in the Absence of the Source

Background in the absence of a source has three different origins: man-made isotopes, natural isotopes, and cosmic radiation. In contrast to the background from a source, it contains, besides a continuum due to cosmic interactions in the crystal (Verplancke, 1992), many discrete γ lines. The most common discrete lines are summarized in Table 4.3. The first column of Table 4.3 gives the energy, the second column the isotope in which the nuclear transition responsible for the emission of a γ-ray takes place, and the third column the decaying isotope and/or the reaction responsible for the formation of the isotope. The fourth column gives the intensity of the γ-ray. A fifth column is reserved for various remarks, such as the origin of the γ-ray, the principal decay mode, the half-life, and, when possible, the intensity of the prompt reaction γ-rays.

a. Man-Made Isotopes

Here we find essentially fission isotopes such as ^{137}Cs due to fallout from the former bomb-testing in the atmosphere, nuclear accidents, or isotopes formed by man-made nuclear reactions such as those of ^{60}Co.

b. Natural Isotopes

Here we find ^{40}K and the isotopes belonging to the natural decay chains: ^{238}U (Table 4.4), ^{235}U (Table 4.5), and ^{232}Th (Table 4.6). These tables give the decay mode, the half-life, and the main γ-rays of the various isotopes. The parent nuclei, ^{238}U, ^{235}U, and ^{232}Th, are very long lived. Their half-lives are several orders of magnitude longer than those of the longest lived daughter elements. They reach a secular equilibrium meaning that the intensities of the various γ-rays may be compared directly with each other after correction for α or β branching (see Chapter 1). However, the equilibrium may be disturbed if physical or chemical separation took place. Just two examples:

1. The "emanation" of noble gases (^{222}Rn, ^{220}Rn, ^{219}Rn), daughters from natural U and Th. In particular, U is often found underground or in the construction materials of buildings. Consequently, ^{222}Rn (radon) may concentrate in closed rooms. The characteristic γ-rays of its daughters ^{214}Pb and ^{214}Bi are very common background lines. The intensity of Rn lines in the background spectrum may fluctuate a lot with the weather conditions.
2. Separation in geological time due to different solubilities of the various elements.

TABLE 4.3 **Background Lines Observed in Ge-Spectra (This List is Neither Complete nor Should all Lines be Present in Each Spectrum)**

γ-line(keV)	Isotope[a]	Reaction[b]	$I_\gamma{}^c$(%)	Remarks
13.26	^{73m}Ge	^{72}Ge$(n,\gamma)^{73m}$Ge	0.09	$T_{1/2}=0.5$ s: isomeric transition produced continuously by thermalized neutrons from Cosmic origin (see also 66.7 keV line).
14.41	^{57}Fe	^{57}Fe$(p,n)^{57}$Co ^{56}Fe$(p,\gamma)^{57}$Co ^{56}Fe$(d,n)^{57}$Co	8.8	EC-decay ($T_{1/2}=271.3$ d): particles from Cosmic origin.
46.5	^{210}Bi	^{210}Pb	3.65	β^--decay ($T_{1/2}=22.28.3$ h): ^{238}U series
49.9	^{223}Ra	^{227}Th	0.52	α-decay ($T_{1/2}=11.43$ d): ^{235}U series
50.1			7.28	
53.2	^{230}Th	^{234}U	0.12	α-decay ($T_{1/2}=1.47$E5 y): ^{238}U series.
53.4	^{73m}Ge	^{72}Ge$(n,\gamma)^{73m}$Ge	10.5	$T_{1/2}=0.5$ s: is produced continuously by thermalized neutrons from cosmic origin.
63.32	^{234}Pa	^{234}Th	4.49	β^--decay ($T_{1/2}=24.1$ d): ^{238}U series.
66.7	^{73m}Ge	^{72}Ge$(n,\gamma)^{73m}$Ge	0.5	$T_{1/2}=0.5$ s: is produced continuously by thermalized neutrons from cosmic origin. Sum peak $53.4+13.26$ and individual line. As the lines are produced inside the detector, the probability for summation is almost 100%.
67.7	^{226}Ra	^{230}Th	0.38	α-decay ($T_{1/2}=8$E4 y): ^{238}U series.
68.7	^{73}Ge	^{73}Ge$(n,n')^{73}$Ge		Prompt γ-line produced by inelastic scattering of fast neutrons from cosmic origin.
72.80 74.97 84.45 84.94 87.3	Pb	Pb X-Ray		Mainly due to external conversion in the Pb-shield.
81.23	^{231}Pa	^{231}Th	0.89	β^--decay ($T_{1/2}=25.5$ h); ^{235}U series.
82.09	^{231}Pa	^{231}Th	0.4	β^--decay ($T_{1/2}=25.5$ h); ^{235}U series.
84.21	^{231}Pa	^{231}Th	6.6	β^--decay ($T_{1/2}=25.5$ h); ^{235}U series.
84.37	^{224}Ra	^{228}Th	1.9	α-decay ($T_{1/2}=1.91$ y); ^{232}Th series.
92.6	^{234}Pa	^{234}Th	5.16	β^--decay ($T_{1/2}=24.1$ d): ^{238}U series.
93.32	^{67}Zn	^{65}Cu$(\alpha,2n)^{67}$Ga	48.0	EC-decay ($T_{1/2}=78.3$ h): α-particles from Cosmic origin. See also 184.5 and 194.25 keV lines
99.6	^{228}Th	^{228}Ac	1.37	β^--decay ($T_{1/2}=6.15$ h); ^{232}Th series.
109.89	^{19}F	^{19}F$(n,n')^{19}$F		Prompt γ-line produced by inelastic scattering of fast neutrons from cosmic origin.
122.4	^{57}Fe	^{57}Fe$(p,n)^{57}$Co ^{56}Fe$(d,n)^{57}$Co ^{56}Fe$(p,\gamma)^{57}$Co ^{57}Co$(n,n')^{57}$Co		EC-decay ($T_{1/2}=271.3$ d): particles from Cosmic origin.
122.4	^{219}Rn	^{223}Ra	1.19	α-decay ($T_{1/2}=11.43$ d): ^{235}U series.

(continued)

TABLE 4.3 (Continued)

γ-line(keV)	Isotope[a]	Reaction[b]	I_γ^c(%)	Remarks
129.6	^{228}Th	^{228}Ac	2.45	β^--decay ($T_{1/2} = 6.15$ h); ^{232}Th series.
131.2	^{234}U	^{234}Pa	20	β^--decay ($T_{1/2} = 6.7$ h): ^{238}U series.
136.47	^{57}Fe	^{57}Fe(p, n)^{57}Co ^{56}Fe(d, n)^{57}Co ^{56}Fe(p, γ)^{57}Co ^{57}Co(n, n′)^{57}Co	11.0	EC-decay ($T_{1/2} = 271.3$ d): particles from Cosmic origin.
139.7	^{75m}Ge	^{74}Ge(n, γ)^{75m}Ge	39.0	$T_{1/2} = 48$ s: isomeric transition produced continuously by thermalized neutrons from Cosmic origin.
143.58	^{57}Fe	^{57}Fe(p, n)^{57}Co ^{56}Fe(d, n)^{57}Co ^{56}Fe(p, γ)^{57}Co ^{57}Co(n, n′)^{57}Co	1.0	See also 14.12, 122.4, and 136.47 eV lines.
143.8	^{231}Th	^{235}U	10.9	α-decay ($T_{1/2} = 7.05$E8 y): ^{235}U series.
143.9	^{226}Ra	^{230}Th	0.05	α-decay ($T_{1/2} = 8$E4 y): ^{238}U series.
144.2	^{219}Rn	^{223}Ra	3.26	α-decay ($T_{1/2} = 11.43$ d): ^{235}U series.
154.1	^{219}Rn	^{223}Ra	3.26	α-decay ($T_{1/2} = 11.43$ d): ^{235}U series.
159.7	^{77m}Ge	^{76}Ge(n,γ)^{77m}Ge	11.0	$T_{1/2} = 52.9$ s: isomeric transition produced continuously by thermalized neutrons from cosmic origin.
163.3	^{231}Th	^{235}U	5.0	α-decay ($T_{1/2} = 7.05$E8 y): ^{235}U series.
174.9	^{71m1}Ge	^{70}Ge(n, γ)^{71m1}Ge	1.0	$T_{1/2} = 73$ ns: isomeric transition produced continuously by thermalized neutrons from Cosmic origin.
184.59	^{67}Zn	^{65}Cu(α, 2n)^{67}Ga	62.0	EC-decay ($T_{1/2} = 78.3$ h): isomeric transition; α-particles from Cosmic origin. See also 93.32 and 194.24 keV lines.
185.7	^{231}Th	^{235}U	57.5	α-decay ($T_{1/2} = 7.05$E8 y): ^{235}U series.
185.91	^{66}Cu	^{65}Cu(n, γ)^{66}Cu		Prompt neutron capture γ-line produced by thermalized neutrons from Cosmic origin.
186.1	^{222}Rn	^{226}Ra	3.57	α-decay ($T_{1/2} = 1601$ y): ^{238}U series.
194.25	^{67}Zn	^{65}Cu(α, 2n)^{67}Ga	1.0	β^+-decay ($T_{1/2} = 78.3$ h): α-particles from Cosmic origin.
198.4	^{71m2}Ge	^{70}Ge(n, γ)^{71m2}Ge	99.0	$T_{1/2} = 22$ ms: is produced continuously by thermal. neutrons from cosmic origin. Sum peak 23.5 + 174
203.1	Cu	^{63}Cu(n, γ)^{64}Cu		Prompt neutron capture γ-ray, $I = 6.64\%$ in nat. isotope-mixture; is produced continuously by thermalized neutrons from cosmic origin.
205.3	^{231}Th	^{235}U	5.0	α-decay ($T_{1/2} = 7.05$E8 y): ^{235}U series.
209.3	^{228}Th	^{228}Ac	3.88	β^--decay ($T_{1/2} = 6.15$ h); ^{232}Th series.
215.5	^{77}As	^{76}Ge(n, γ)^{77m}Ge	21.0	$T_{1/2} = 52.9$ s: β^--decay of isomeric level excited continuously by therm. neutrons from Cosmic origin.

(*continued*)

■ **TABLE 4.3** (Continued)

γ-line(keV)	Isotope[a]	Reaction[b]	I_γ^c(%)	Remarks
215.99	^{224}Ra	^{228}Th	0.3	α-decay ($T_{1/2} = 1.91$ y): ^{232}Th series.
226.4	^{234}U	^{234}Pa	5.9	β^--decay ($T_{1/2} = 6.7$ h): ^{238}U series.
227.2	^{234}U	^{234}Pa	5.5	β^--decay ($T_{1/2} = 6.7$ h): ^{238}U series.
236.0	^{223}Ra	^{227}Th	11.2	α-decay ($T_{1/2} = 11.43$ d): ^{235}U series.
238.6	^{212}Bi	^{212}Pb	43.6	β^--decay ($T_{1/2} = 10.64$ h); ^{232}Th series.
241.0	^{220}Rn	^{224}Ra	3.97	α-decay ($T_{1/2} = 11.43$ d): ^{235}U series.
241.98	^{214}Bi	^{214}Pb	7.5	β^--decay ($T_{1/2} = 26.8$ m): ^{238}U series.
256.0	^{223}Ra	^{227}Th	7.6	α-decay ($T_{1/2} = 11.43$ d): ^{235}U series.
269.2	^{219}Rn	^{223}Ra	13.6	α-decay ($T_{1/2} = 11.43$ d): ^{235}U series.
270.2	^{228}Th	^{228}Ac	3.43	β^--decay ($T_{1/2} = 6.15$ h); ^{232}Th series.
271.2	^{215}Po	^{219}Rn	9.9	α-decay ($T_{1/2} = 3.96$ s): ^{235}U series.
277.4	^{208}Pb	^{208}Tl	6.31	β^--decay ($T_{1/2} = 3.05$ m); ^{232}Th series.
278.3	^{64}Cu	^{63}Cu(n, γ)^{64}Cu		Prompt neutron capture γ-ray, $I = 30.12\%$ in nat. isotope-mixture; is produced continuously by thermalized neutrons from cosmic origin.
283.7	^{227}Ac	^{231}Pa	1.6	α-decay ($T_{1/2} = 4243$): ^{235}U series.
288.1	^{208}Tl	^{212}Bi	0.34	α-decay ($T_{1/2} = 1.01$ h): ^{232}Th series.
295.2	^{214}Bi	^{214}Pb	18.5	β^--decay ($T_{1/2} = 26.8$ m): ^{238}U series.
300.0	^{227}Ac	^{231}Pa	2.39	α-decay ($T_{1/2} = 4243$ y): ^{235}U series.
300.1	^{212}Bi	^{212}Pb	3.34	β^--decay ($T_{1/2} = 10.64$ h); ^{232}Th series.
302.7	^{227}Ac	^{231}Pa	2.24	α-decay ($T_{1/2} = 4243$ y): ^{235}U series.
323.3	^{219}Rn	^{223}Ra	3.9	α-decay ($T_{1/2} = 11.43$ d): ^{235}U series.
328.3	^{228}Th	^{228}Ac	2.95	β^--decay ($T_{1/2} = 6.15$ h); ^{232}Th series.
330.1	^{227}Ac	^{231}Pa	1.31	α-decay ($T_{1/2} = 4243$ y): ^{235}U series.
338.3	^{219}Rn	^{223}Ra	2.789	α-decay ($T_{1/2} = 11.43$ d): ^{235}U series
338.3	^{228}Th	^{228}Ac	1.25	β^--decay ($T_{1/2} = 6.15$ h); ^{232}Th series.
351.0	^{207}Tl	^{211}Bi	2.76	α-decay ($T_{1/2} = 2.14$ m): ^{235}U series.
351.92	^{214}Bi	^{214}Pb	38.5	β^--decay ($T_{1/2} = 19.9$ m): ^{238}U series.
367.94	^{200}Hg	^{199}Hg(n, γ)^{200}Hg		Prompt neutron capture γ-ray, $I = 81.35\%$ in nat. isotope-mixture; is produced continuously by thermalized neutrons from cosmic origin. Its observation is mainly due to the high reaction yield and the enormous thermal cross-section of ^{199}Hg of 2000 barn.
401.7	^{215}Po	^{219}Rn	6.64	α-decay ($T_{1/2} = 3.96$ s): ^{235}U series.
404.8	^{211}Bi	^{211}Pb	3.83	β^--decay ($T_{1/2} = 36.1$ m: ^{235}U series
409.5	^{228}Th	^{228}Ac	1.94	β^--decay ($T_{1/2} = 6.15$ h): ^{232}Th series.
426.99	^{211}Bi	^{211}Pb	1.72	β^--decay ($T_{1/2} = 36.1$ m: ^{235}U series.
427.89	^{125}Te	^{124}Sn(p, γ)^{125}Sb	29.4	β^--decay ($T_{1/2} = 2.77$ a): protons from Cosmic origin.

(*continued*)

TABLE 4.3 (Continued)

γ-line(keV)	Isotope[a]	Reaction[b]	I_γ^c(%)	Remarks
444.9	^{219}Rn	^{223}Ra	1.27	α-decay ($T_{1/2} = 11.43$ d): ^{235}U series.
452.83	^{208}Tl	^{212}Bi	0.31	α-decay ($T_{1/2} = 1.01$ h): ^{232}Th series.
463.0	^{228}Th	^{228}Ac	4.44	β^--decay ($T_{1/2} = 6.15$ h): ^{232}Th series.
463.38	^{125}Te	^{124}Sn(p,γ)^{125}Sb	0.15	β^--decay ($T_{1/2} = 2.77$ a): protons from Cosmic origin.
510.8	^{208}Pb	^{208}Tl	22.6	β^--decay ($T_{1/2} = 3.05$ m); ^{232}Th series.
511.0	Anni.			This very common Doppler broadened line finds its origin in the annihilation of β^+-particles occurring in the β^+-decay or the pair production process induced by high energy γ-rays ($E_\gamma > 1022$ keV) of Cosmic origin and/or due to nuclear decay or various nuclear reactions. The many possible origins allow no prediction of its intensity. It may not be used to estimate the intensity of a β^+-decay branching. Is also produced by muon-induced pair production.
549.7	^{216}Po	^{220}Rn	0.1	α-decay ($T_{1/2} = 55.6$ s): ^{232}Th series.
558.2	^{114}Cd	^{113}Cd(n,γ)^{114}Cd		Prompt neutron capture γ-ray, $I = 79.71\%$ in nat. isotope-mixture; is produced continuously by thermalised neutrons of Cosmic origin.
562.9	^{76}Ge	^{76}Ge(n,n′)^{76}Ge		Prompt γ-line produced by inelastic scattering of fast neutrons from Cosmic origin. Right asymmetric line-shape due to recoil of the Ge-atoms induced by (n,n′) reaction.
563.3	^{134}Ba	^{133}Cs(n,γ)^{134}Cs	8.38	β^--decay ($T_{1/2} = 2.06$ a). This isotope is found in reactor waste (Chernobyl fallout) but not in the fall-out of bomb testing. This is due to the fact that it is no fission product, as it is screened by the stable ^{134}Xe. It is however found among the reactor fission products, as ^{133}Cs is the stable end product of the A $= 133$ fission chain having a yield of 7.87%.
568.7	^{234}U	^{234}Pa	3.3	β^--decay ($T_{1/2} = 6.7$ h): ^{238}U series.
569.5	^{234}U	^{234}Pa	10.0	β^--decay ($T_{1/2} = 6.7$ h): ^{238}U series.
569.79	^{207}Pb	^{207}Pb(n,n′)^{207}Pb ^{206}Pb(n,γ)^{207}Pb		Prompt γ-line produced by inelastic scattering of fast neutrons from cosmic origin or by thermal neutron capture.
583.2	^{208}Pb	^{208}Tl	84.5	β^--decay ($T_{1/2} = 3.05$ m); ^{232}Th series.

(continued)

████ **TABLE 4.3** (Continued)

γ-line(keV)	Isotope[a]	Reaction[b]	$I_\gamma{}^c$(%)	Remarks
595.9	^{74}Ge	^{73}Ge(n, γ)^{74}Ge ^{74}Ge(n, n')^{74}Ge		Prompt neutron capture γ-ray, $I = 34.65\%$ in nat. isotope-mixture; is produced continuously by thermalized neutrons from cosmic origin. Prompt γ-line produced by inelastic scattering of fast neutrons from Cosmic origin. Right asymmetric line-shape due to recoil of the Ge-atoms induced by (n, n')-reaction.
604.7	^{134}Ba	^{133}Cs(n, γ)^{134}Cs	97.6	See comments 563.3 keV line.
651.0	^{114}Cd	^{113}Cd(n, γ)^{114}Cd		Prompt neutron capture γ-ray, $I = 15.23\%$ in nat. isotope-mixture; is produced continuously by thermalized neutrons from cosmic origin.
600.55		^{124}Sn(p, γ)^{125}Sb	17.78	β^--decay ($T_{1/2} = 2.77$ y): protons from Cosmic origin.
606.64	^{125}Te	^{124}Sn(p, γ)	5.02	β^--decay ($T_{1/2} = 2.77$ y): protons from Cosmic origin
609.3	^{214}Po	^{214}Bi	44.8	β^- decay ($T_{1/2} = 19.9$ m): ^{238}U series.
635.9	^{125}Te	^{124}Sn(p, γ)^{125}Sb	11.32	β^--decay ($T_{1/2} = 2.77$ y): protons from Cosmic origin.
661.66	^{137m}Ba	^{137}Cs	85.0	Fission isotope β^--decay ($T_{1/2} = 30.17$ y): bomb testing + Chernobyl fallout.
669.6	^{63}Cu	^{63}Cu(n, n')^{63}Cu		Prompt γ-line produced by inelastic scattering of fast neutrons from cosmic origin
671.40	^{125}Te	^{124}Sn(p, γ)^{125}Sb	1.8	β^--decay ($T_{1/2} = 2.77$ y): protons from Cosmic origin.
691.0	^{72}Ge	^{72}Ge(n, n')^{72}Ge		Prompt γ-line produced by inelastic scattering of fast neutrons from cosmic origin. This line is a 0^+–0^+ and can thus only take place by internal conversion as electrical monopol transitions are strictly forbidden. The asymmetric rightside shape is due to imperfect transformation of the recoil energy due to the neutron scattering and is observed—in contrary to other (n, n') reactions, due to the fact that the recoil takes place in the Ge and thus inside to the detector.
727.3	^{212}Po	^{212}Bi	6.25	β^--decay ($T_{1/2} = 1.01$ h): ^{232}Th series.
751.8	^{65}Zn	^{63}Cu(α2n)^{65}Ga	50.7	β^+-decay ($T_{1/2} = 15$ m): continuously formed by α-particles from Cosmic origin.
766.0	^{234}U	^{234m}Pa	0.21	β^--decay ($T_{1/2} = 1.17$ m); ^{238}U series.
768.4	^{214}Po	^{214}Bi	4.88	β^--decay ($T_{1/2} = 19.9$ m): ^{238}U series.
769.7	^{73}As	^{73}Ge(p, nγ)^{73}As		Prompt γ-line produced by p,n-reaction with protons from Cosmic origin.

(*continued*)

■ **TABLE 4.3** (Continued)

γ-line(keV)	Isotope[a]	Reaction[b]	$I_\gamma{}^c$ (%)	Remarks
770.8	^{65}Cu	^{65}Cu(n, n′)^{65}Cu		Prompt γ-line produced by inelastic scattering of fast neutrons from Cosmic origin.
772.4	^{228}Th	^{228}Ac	1.58	β^--decay ($T_{1/2} = 6.15$ h): ^{232}Th series.
785.6	^{212}Po	^{212}Bi	1.11	β^--decay ($T_{1/2} = 1.01$ h): ^{232}Th series.
794.9	^{228}Th	^{228}Ac	4.34	β^--decay ($T_{1/2} = 6.15$ h): ^{232}Th series.
795.8	^{134}Ba	^{133}Cs(n, γ)^{134}Cs	85.4	See 563.3 keV line.
801.9	^{134}Ba	^{133}Cs(n, γ)^{134}Cs	8.73	See 563.3 keV line
803.3	^{206}Pb	^{206}Pb(n, n′) ^{206}Pb^{210}Po	0.001	Prompt γ-line produced by inelastic scattering of fast neutrons from cosmic origin. α-decay ($T_{1/2} = 138.4$ d): ^{238}U series.
805.7	^{114}Cd	^{113}Cd(n, γ)^{114}Cd		Prompt neutron capture γ-ray, $I = 5.1$% in nat. isotope-mixture; is produced continuously by thermalized neutrons from cosmic origin.
810.80	^{58}Fe	^{59}Co(γ, n)^{58}Co ^{59}Co(n, 2n)^{58}Co ^{58}Fe(p, n)^{58}Co ^{57}Fe(p, γ)^{58}Co ^{57}Fe(d, n)^{58}Co ^{58}Fe(n, p)^{58}Mn	≈ 100 82.2	β^--decay ($T_{1/2} = 63$ s): is produced continuously by fast γ's and particles of cosmic origin.
831.8	^{211}Bi	^{211}Pb	3.83	β^--decay ($T_{1/2} = 36.1$ m): ^{235}U series.
833.95	^{77}Ge	^{72}Ge(n, n′)^{72}Ge		Prompt γ-line produced by inelastic scattering of fast neutrons from cosmic origin. Right asymmetric line-shape due to recoil of the Ge-atoms induced by (n, n′)-reaction.
834.6	^{54}Cr	^{54}Cr(p, n)^{54}Mn ^{53}Cr(d, n)^{54}Mn ^{53}Cr(p, γ)^{54}Mn	100	EC-decay ($T_{1/2} = 312.2$ d): protons from cosmic origin.
835.7	^{228}Th	^{228}Ac	1.68	β^--decay ($T_{1/2} = 6.15$ h): ^{232}Th series.
846.8	^{56}Fe	^{56}Fe(n, n′)^{76}Fe	19.0	Prompt γ-line produced by inelastic scattering of fast neutrons from cosmic origin. The absence of the 1,282.6 keV line allows to distinguish it from the same line excited in the decay of ^{56}Co.
846.8	^{56}Fe	^{56}Fe(p, n)^{56}Co		β^+-decay ($T_{1/2} = 78.76$ d): The presence of the 1,238.2 keV line allows to distinguish it from the same line excited in the ^{56}Fe(n, n′)^{76}Fe reaction.
860.6	^{208}Pb	^{208}Tl	12.42	β^--decay ($T_{1/2} = 3.05$ m); ^{232}Th series.
868.1	^{73}Ge	^{72}Ge(n, γ)^{73}Ge		Prompt neutron capture γ-ray, $I = 30.12$% in nat. isotope-mixture; is produced continuously by thermalized neutrons from cosmic origin.
880.51	^{234}U	^{234}Pa	9	β^--decay ($T_{1/2} = 6.7$ h): ^{238}U series.
883.24	^{234}U	^{234}Pa	15	β^--decay ($T_{1/2} = 6.7$ h): ^{238}U series.
897.6	^{207}Bp	^{207}Tl	0.24	β^--decay ($T_{1/2} = 4.79$ m): ^{235}U series.

(*continued*)

TABLE 4.3   (Continued)

γ-line(keV)	Isotope[a]	Reaction[b]	I_γ^c(%)	Remarks
911.2	^{228}Th	^{228}Ac	26.6	β^--decay ($T_{1/2}=6.15$ h): ^{232}Th series.
925.0	^{234}U	^{234}Pa	2.9	β^--decay ($T_{1/2}=6.7$ h): ^{238}U series.
926.0	^{234}U	^{234}Pa	11.0	β^--decay ($T_{1/2}=6.7$ h): ^{238}U series.
927.1	^{234}U	^{234}Pa	11.0	β^--decay ($T_{1/2}=6.7$ h): ^{238}U series.
934.1	^{214}Po	^{214}Bi	3.03	β^--decay ($T_{1/2}=19.9$ m): ^{238}U series.
946.0	^{234}U	^{234}Pa	12	β^--decay ($T_{1/2}=6.7$ h): ^{238}U series.
962.1	^{65}Cu	^{63}Cu(n, n')^{63}Cu		Prompt γ-line produced by inelastic scattering of fast neutrons from cosmic origin.
964.8	^{228}Th	^{228}Ac	5.11	β^--decay ($T_{1/2}=6.15$ h): ^{232}Ht series.
969.0	^{228}Th	^{228}Ac	16.20	β^--decay ($T_{1/2}=6.15$ h): ^{232}Ht series
1001.0	^{234}U	^{234m}Pa	0.59	β^--decay ($T_{1/2}=1.17$ m); ^{238}U series.
1039.5	^{70}Ge	^{70}Ge(n, n')^{70}Ge		Prompt γ-line produced by inelastic scattering of fast neutrons from cosmic origin. Right asymmetric line-shape due to recoil of the Ge-atoms induced by (n, n') reaction.
1063.64	^{207}Pb	^{207}Pb(n, n')^{207}Pb ^{206}Pb(n, γ)^{207}Pb		Prompt γ-line produced by inelastic scattering of fast neutrons from cosmic origin or by thermal neutron capture. See also 569.79 keV line.
1077.41	^{68}Zn	^{65}Cu(α, n)^{68}Ga	3.0	β^+-decay ($T_{1/2}=68.3$ m): α-particles of Cosmic origin.
1097.3	^{116}Sn	^{115}In(n, γ)^{116m1}	55.7	β^--decay ($T_{1/2}=54.1$ m): formed by thermalized neutrons from cosmic origin.
1115.5	^{65}Cu	^{65}Cu(n, n')^{65}Cu ^{65}Cu(p, n)^{65}Zn	50.75	Prompt γ-line produced by inelastic scattering of fast neutrons from cosmic origin. EC+β^+-decay ($T_{1/2}=244$ d) Formed by fast neutrons or protons from cosmic origin.
1120.4	^{214}Po	^{214}Bi	14.8	β^--decay ($T_{1/2}=19.9$ m): ^{238}U series.
1124.51	^{65}Cu	^{70}Ge(n, α2n)^{65}Zn	50.75	EC+β^+-decay ($T_{1/2}=244$ d). Formed by fast neutrons from cosmic origin. Note that it is the same line as the above mentioned 1115.5 keV transition. As the reaction takes place inside the Ge-detector itself, its energy sums up with the Kα−X-ray of Cu. It is thus possible to distinguish the formation reaction of ^{65}Zn.
1173.2	^{60}Ni	^{59}Co(n, γ)^{60}Co	100	β^--decay ($T_{1/2}=5.172$ y): This isotope is a common contamination in modern steel and is introduced at the high furnace level.
1204.1	^{74}Ge	^{74}Ge(n, n')^{74}Ge		Fast neutrons from cosmic origin. Right asymmetric line-shape due to recoil of the Ge-atoms induced by (n, n') reaction.
1238.26	^{56}Fe	^{56}Fe(p, n)^{56}Co	13.4	β^+-decay ($T_{1/2}=78.76$ d): See 846.8 keV line.
1238.8	^{214}Po	^{214}Bi	5.86	β^--decay ($T_{1/2}=19.9$ m): ^{238}U series.

(continued)

TABLE 4.3   (Continued)

γ-line(keV)	Isotope[a]	Reaction[b]	$I_\gamma{}^c$(%)	Remarks
1291.65	^{59}Co	^{58}Fe(n, γ)^{59}Fe	57.0	β^--decay ($T_{1/2} = 45.1$ d). Is produced continuously by thermalized neutrons of cosmic origin.
1293.5	^{116}Sn	^{115}In(n, γ)^{116m}In	85.0	In β^--decay ($T_{1/2} = 54.1$ m): formed by thermalized neutrons from cosmic origin.
1293.64	^{41}K	^{40}Ar(n, γ)^{41}Ar	99.16	β^--decay ($T_{1/2} = 1.83$ h); is produced continuously by thermalized neutrons from cosmic origin. Is a common B.G. line near air-cooled fission reactors.
1327.0	^{63}Cu	^{63}Cu(n, n′)^{63}Cu		Prompt γ-line produced by inelastic scattering of fast neutrons from cosmic origin.
1332.5	^{60}Ni	^{59}Co(n, γ)^{60}Co	100	See 1173.2 keV-line.
1377.6	^{57}Co	^{58}Ni(γ, n)^{57}Ni ^{58}Ni(n, 2n)^{57}Ni	30.0	β^+ + EC-decay ($T_{1/2} = 36.0$ h)
1377.6	^{214}Po	^{214}Bi	3.92	β^--decay ($T_{1/2} = 19.9$ m): ^{238}U series.
1408.0	^{214}Po	^{214}Bi	2.48	β^--decay ($T_{1/2} = 19.9$ m): ^{238}U series.
1412.1	^{63}Cu	^{63}Cu(n, n′)^{63}Cu		Prompt γ-line produced by inelastic scattering of fast neutrons from cosmic origin.
1460.8	^{40}Ar	^{40}K	99.16	EC and β^{+-}decay. ($T_{1/2} = 1.277\text{E}+8$ y) Widespread natural radioactive isotope. The modal human body contains about 4000 Bq of this isotope.
1481.7	^{65}Cu	^{65}Cu(n, n′)^{65}Cu		Prompt γ-line produced by inelastic scattering of fast neutrons from Cosmic origin.
1547.0	^{63}Cu	^{63}Cu(n, n′)^{63}Cu		Prompt γ-line produced by inelastic scattering of fast neutrons from cosmic origin.
1588.2	^{228}Th	^{228}Ac	3.27	β^--decay ($T_{1/2} = 6.15$ h): ^{232}Th series.
1620.6	^{212}Po	^{212}Bi	1.6	β^--decay ($T_{1/2} = 1.01$ h): ^{232}Th series.
1729.6	^{214}Po	^{214}Bi	2.88	β^--decay ($T_{1/2} = 19.9$ m): ^{238}U series.
1764.5	^{214}Po	^{214}Bi	15.96	β^--decay ($T_{1/2} = 19.9$ m): ^{238}U series.
2204.1	^{214}Po	^{214}Bi		β^--decay ($T_{1/2} = 19.9$ m): ^{238}U series.
2223.2	^{2}H	^{1}H(n, γ) ^{2}H		Prompt neutron capture γ-ray, $I = 100\%$ in nat. isotope- mixture; is produced continuously by thermalized neutrons from cosmic origin.
2614.6	^{208}Pb	^{208}Pb(n, n′) ^{208}Pb^{208}Tl	99.2	Prompt γ-line produced by inelastic scattering of fast neutrons from cosmic origin. β^--decay ($T_{1/2} = 3.05$ m); ^{232}Th series.

[a]The isotope in which the transition takes place is mentioned. The reaction responsible or the parent nucleus are given in the second column.

[b]The isotopes formed by (n, γ)-reaction can also be formed by (d, p)-reaction or even by (n, 2n)-reaction if the isotope wit N + 2 neutrons is stable.

[c]Intensity of the γ-line in% and per decay. If possible the intensity of reaction γ's is given in the column reserved for the remarks.

TABLE 4.4 ^{238}U Natural Decay Chain

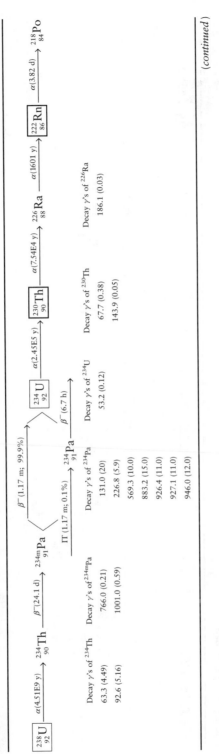

$^{238}_{92}$U $\xrightarrow{\alpha(4.51\text{E}9\ y)}$ $^{234}_{90}$Th $\xrightarrow{\beta^-(24.1\ d)}$ $^{234\text{m}}_{91}$Pa $\xrightarrow{\beta^-(1.17\ m;\ 99.9\%)}$ $^{234}_{92}$U $\xrightarrow{\alpha(2.45\text{E}5\ y)}$ $^{230}_{90}$Th $\xrightarrow{\alpha(7.54\text{E}4\ y)}$ $^{226}_{88}$Ra $\xrightarrow{\alpha(1601\ y)}$ $\boxed{^{222}_{86}\text{Rn}}$ $\xrightarrow{\alpha(3.82\ d)}$ $^{218}_{84}$Po

$\xrightarrow{\text{IT}(1.17\ m;\ 0.1\%)}$ $^{234}_{91}$Pa $\xrightarrow{\beta^-(6.7\ h)}$

Decay γ's of ^{234}Th	Decay γ's of $^{234\text{m}}$Pa	Decay γ's of ^{234}Pa	Decay γ's of ^{234}U	Decay γ's of ^{230}Th	Decay γ's of ^{226}Ra
63.3 (4.49)	766.0 (0.21)	131.0 (20)	53.2 (0.12)	67.7 (0.38)	186.1 (0.03)
92.6 (5.16)	1001.0 (0.59)	226.8 (5.9)		143.9 (0.05)	
		569.3 (10.0)			
		883.2 (15.0)			
		926.4 (11.0)			
		927.1 (11.0)			
		946.0 (12.0)			

(continued)

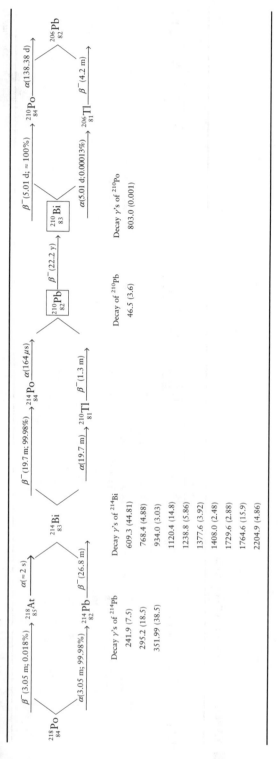

TABLE 4.4 (Continued)

$^{218}_{84}\text{Po}$

$\xrightarrow{\beta^- (3.05 \text{ m}; 0.018\%)}$ $^{218}_{85}\text{At}$ $\xrightarrow{\alpha(\approx 2 \text{ s})}$

$\xrightarrow{\alpha(3.05 \text{ m}; 99.98\%)}$ $^{214}_{82}\text{Pb}$ $\xrightarrow{\beta^- (26.8 \text{ m})}$ $^{214}_{83}\text{Bi}$

$\xrightarrow{\beta^- (19.7 \text{ m}; 99.98\%)}$ $^{214}_{84}\text{Po}$ $\xrightarrow{\alpha(164 \,\mu s)}$

$\xrightarrow{\alpha(19.7 \text{ m})}$ $^{210}_{81}\text{Tl}$ $\xrightarrow{\beta^- (1.3 \text{ m})}$

$^{210}_{82}\text{Pb}$ $\xrightarrow{\beta^- (22.2 \text{ y})}$ $^{210}_{83}\text{Bi}$

$\xrightarrow{\beta^- (5.01 \text{ d}; \approx 100\%)}$ $^{210}_{84}\text{Po}$ $\xrightarrow{\alpha(138.38 \text{ d})}$ $^{206}_{82}\text{Pb}$

$\xrightarrow{\alpha(5.01 \text{ d}: 0.00013\%)}$ $^{206}_{81}\text{Tl}$ $\xrightarrow{\beta^- (4.2 \text{ m})}$

Decay γ's of ^{214}Pb

241.9 (7.5)
295.2 (18.5)
351.99 (38.5)

Decay γ's of ^{214}Bi

609.3 (44.81)
768.4 (4.88)
934.0 (3.03)
1120.4 (14.8)
1238.8 (5.86)
1377.6 (3.92)
1408.0 (2.48)
1729.6 (2.88)
1764.6 (15.9)
2204.9 (4.86)

Decay of ^{210}Pb

46.5 (3.6)

Decay γ's of ^{210}Po

803.0 (0.001)

TABLE 4.5 ^{235}U Natural Decay Chain

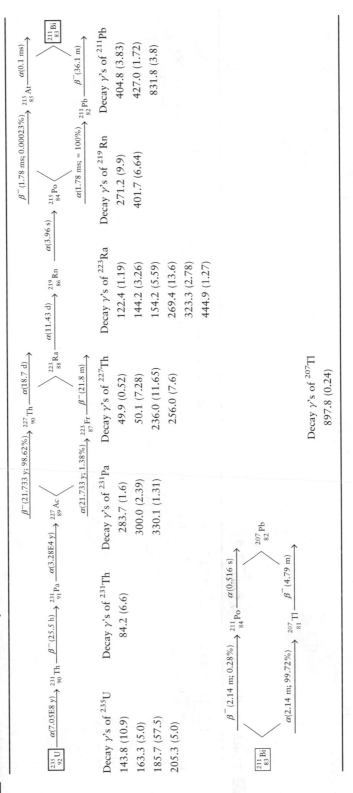

TABLE 4.6 ^{232}Th Natural Decay Chain

$$\boxed{^{232}_{90}\text{Th}} \xrightarrow{\alpha(1.4\text{E}10\ y)} {}^{228}_{88}\text{Ra} \xrightarrow{\beta^-(6.79\ y)} {}^{228}_{89}\text{Ac} \xrightarrow{\beta^-(6.15\ h)} {}^{228}_{90}\text{Th} \xrightarrow{\alpha(1.91\ y)} {}^{224}_{88}\text{Ra} \xrightarrow{\alpha(3.66\ d)} \boxed{^{220}_{86}\text{Rn}} \xrightarrow{\beta^-(55.6\ s)} {}^{216}_{84}\text{Po} \xrightarrow{\alpha(0.156\ s)} {}^{212}_{82}\text{Pb}$$

Decay γ's of ^{228}Ac

129.6(2.45)
209.3(3.88)
270.24 (3.43)
328.0(2.95)
338.3(11.3)
463.0(4.44)
772.4(1.5)
794.9(4.34)
835.7(1.68)
911.2(26.6)
964.8(5.11)
969.0(16.2)
1588.2(3.27)
1630.6(1.6)

Decay γ's of ^{228}Th

84.37 (1.27)
216.0 (0.26)

Decay γ's of ^{224}Ra

241.0 (3.97)

Decay γ of ^{220}Rn

549.7 (0.1)

(*continued*)

TABLE 4.6 (Continued)

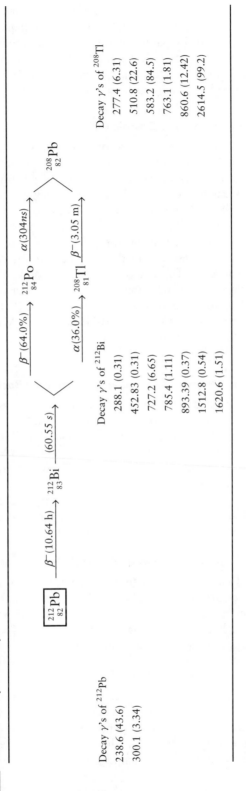

$^{212}_{82}\text{Pb}$ $\xrightarrow{\beta^- (10.64\text{ h})}$ $^{212}_{83}\text{Bi}$ $\xrightarrow{(60.55\text{ s})}$

$\xrightarrow{\beta^- (64.0\%)}$ $^{212}_{84}\text{Po}$ $\xrightarrow{\alpha (304ns)}$ $^{208}_{82}\text{Pb}$

$\xrightarrow{\alpha (36.0\%)}$ $^{208}_{81}\text{Tl}$ $\xrightarrow{\beta^- (3.05\text{ m})}$

Decay γ's of ^{212}Pb

238.6 (43.6)
300.1 (3.34)

Decay γ's of ^{212}Bi

288.1 (0.31)
452.83 (0.31)
727.2 (6.65)
785.4 (1.11)
893.39 (0.37)
1512.8 (0.54)
1620.6 (1.51)

Decay γ's of ^{208}Tl

277.4 (6.31)
510.8 (22.6)
583.2 (84.5)
763.1 (1.81)
860.6 (12.42)
2614.5 (99.2)

This can be used for dating of geological formations based on the $^{234}U/^{230}Th$ ratio. When there is a chance that the equilibrium of the daughter isotope with its parent is disturbed, only the intensities of the γ-rays belonging to the partial decay chain of this daughter can be compared directly. These long-lived isotopes, whose equilibrium with the parent can be disturbed, are marked by a frame in Tables 4.4–4.6.

3. Background of Cosmic Origin

Cosmic rays comprise primarily very high-energy (up to $10^8 \sim 10^9\,GeV$) protons and α particles originating from stellar processes in supernovas with a mean energy between 5 and $10\,GeV$ per nucleon. These particles undergo collisions in the stratosphere, where they give rise to various π and K mesons as well as to muons, neutrinos, electrons, neutrons, and photons. Typical fluxes at sea level are $10^{-2}\,particles/cm^2 \cdot s \cdot steradian$ distributed according to a $\cos^2\theta$ law, θ being the polar angle. About 75% of all particles at sea level are π-mesons and the absolute proton flux is of the order of 0.1% of all particles. These cosmic rays constitute a very important part of the background in the absence of a source and contribute to the continuum as well as to the activation of various nuclei. These effects continue to attract the attention of various researchers, such as Wordel *et al.* (1996) and Heusser (1996). A comprehensive overview of the origin of cosmic rays has been given by Celnikier (1996).

a. "Prompt," Continuously Distributed Background

Charged particles can penetrate the sensitive volume of the detector, giving rise to a continuous background in coincidence with the primary particle. The energy loss per collision of 10-MeV electrons, 100-MeV mesons or 1000-MeV protons is approximately $1.8\,MeV/g\,cm^2$, generating in a Ge detector a signal of about 10 MeV per cm traversed. Cosmic particles also produce showers of secondary particles (p, e^-, e^+), mainly in the detector shielding. In turn, these secondary charged particles produce bremsstrahlung and annihilation lines. This secondary radiation contributes more specifically to the background in the lower energy region of a shielded detector.

b. Neutron-Induced "Prompt" Discrete γ-Rays

Fast neutrons can induce prompt γ-rays by the (n, n') reaction. This is particularly important when the reaction takes place in the Ge itself or in other materials in the vicinity of the detector such as Cu, Fe, Pb, and Cd. These lines are summarized in Table 4.3. The Compton scattering of these γ-rays also adds to the continuum.

c. "Delayed" γ-Rays

Delayed γ-rays are due to the de-excitation of isotopes formed either in the Ge itself or in the material surrounding the detector. They are due to capture of thermalized neutrons or to more exotic nuclear reactions also mentioned in Table 4.3. These isotopes also contribute to the continuum by bremsstrahlung and Compton scattering.

4. Background Reduction

Background reduction is a difficult and delicate operation. The optimum shielding should take the isotopes to be measured into account as well as the energy range and the lower limit of detection desired. But as local conditions can vary strongly, no off-the-shelf solution can be given for all cases. Some general rules remain valid under all conditions and will be discussed subsequently.

a. Passive Background Reduction

Passive background reduction is based on the absorption of undesired γ-rays by an absorber placed between the detector and the source of the background. The transmitted intensity is given by Eq. 4.49. If the absorption coefficient is expressed in half-thickness (cm), the transmitted intensity I is given by

$$I = I_0 e^{-0.693d/d_{1/2}} \tag{4.49}$$

where I_0 is the initial flux, $d_{1/2}$ the half-thickness, and d the actual thickness of the shielding. In Fig. 4.18 the half-thickness for Cu, Sn, Pb, and Si is given as a function of energy. Good shielding should be sufficiently thick; for example 10 times the half-thickness, in order to reduce the background by a factor of 1000. This would translate into a Pb thickness of 8.8 cm for a 1000 keV γ-ray. In practice a thickness of 10 or 15 cm is often chosen. It must be remembered here that lead contains ^{210}Pb $(T_{1/2} = 21\,\text{y})$, as the Pb ores and the coke used in the melting process contain U traces that continuously form ^{210}Pb. The ^{210}Pb content in lead varies according to its origin and age. Specific lead is available for Ultra-Low Background shielding.

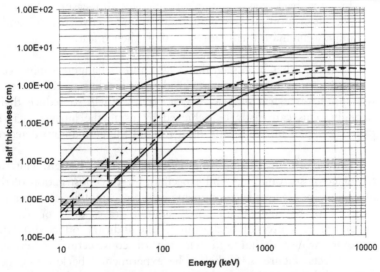

FIGURE 4.18 Half-thickness in cm for Pb (lower plain curve), Sn (dashed curve), Cu (dotted curve) and Si (upper plain curve) as a function of energy.

On the other hand, the Pb shield should not be too thick, in order to reduce the production of fast neutrons by cosmic particles with the subsequent production of n-induced background lines summarized in Table 4.3. Fast neutrons are difficult to stop. Several tens of centimeters of hydrogen-rich material is needed to thermalize them. Once thermalized, they can be stopped by high-cross section materials such as B or Cd. However, the absorption process generates new γ-rays, including a Doppler-broadened 480 keV for B and a whole spectrum of neutron capture γ-rays for Cd. It is possible to stop thermal neutrons without the production of new γ-rays by the ^{6}Li$(n, \alpha t)$ reaction. However, the limited availability of ^{6}Li does not make this a usable alternative.

Fluorescent Pb x-rays can be reduced by a supplementary lining of the Pb shield by lower Z material such as Cd or Sn. A 1-mm thickness of Sn stops 95% of all Pb x-rays, and a supplementary lining of 1.5 mm of Cu stops most of the Sn x-rays and raises the total absorption of lead x-rays to 98.5%. Once again, thick linings should not be used, otherwise the continuous background due to the backscattering of the source γ-rays will increase. This is due to the fact that the Compton effect responsible for the backscattering varies with the atomic number Z while the photoelectric effect is proportional to Z^5. Also, the plastic inner layer often used to prevent contamination of the shield should be as thin as possible. Lining a 10-cm Pb shield with 1-mm Cd, 2-mm Cu, and 10-mm Plexiglas increases the background by as much as 30% at 25 keV and 15% at 1000 keV.

Cosmic background and the associated activation and reaction lines can be adequately reduced by placing the detector deep underground. At a depth of 1000 m-water-equivalent the neutron flux induced by cosmic particles is less than 1% of the flux observed at sea level.

b. Active Background Reduction

In active background reduction we cover all measures that are not based on the absorption of the undesired γ-rays. Active background reduction techniques eliminate their causes or limit their effect.

Venting. It has been seen that ^{222}Rn and its daughters may accumulate in a closed area and, in particular, inside the shielding of the detector. Venting with an Rn-free gas such as N_2 or Ar may help to reduce their presence strongly. This can easily be achieved by using the nitrogen gas boiling off from the liquid nitrogen tank used to cool the germanium detector.

Choice of Construction Materials and Cryostat Design. Materials used in the detector surroundings and especially for the construction of the cryostat as well as the electronic components of the preamplifier can contain elements such as Al, Be, and Sn, that may contain traces of U or Th, constituting an undesired source of background. The industry offers different types of cryostats whose design and selection of construction materials minimize these effects. Figure 4.19 shows the experimental background per keV per hour (cph/keV) for the same detector element mounted in different Canberra cryostats: the classical vertical dipstick cryostat and the same cryostat with

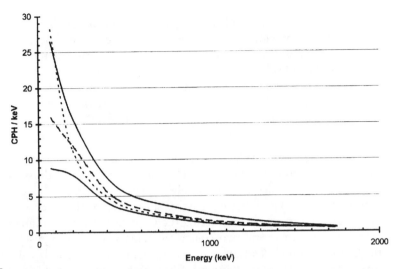

Background in counts per keV and per hour for the same detector placed in different cryostats: Upper plain curve: standard Canberra vertical dipstick cryostat; dashed curve: same cryostat with the building materials close to the crystal replaced by low background alternatives; dotted curve: same cryostat but with an additional low-background lead disk between the crystal and preamplifier; lower plain curve: Ultra-low-background Canberra cryostat model 7500SL RDC-ULB.

the materials close to the detector element selected for low background. The third spectrum is taken with the same cryostat but with an additional low-background lead disk between the crystal and preamplifier. It is seen that this additional lead disk reduces the background from the preamplifier and from the floor at higher energies only. The lowest background at all energies is obtained with the 7500Sl-RDC-ULB cryostat. In this cryostat, the low-background detector chamber is separated from the preamplifier chamber and from the rest of the cryostat with a thin tube holding the cold finger. This part of the cold finger is off-set from the lower part in the dipstick cryostat, preventing a direct line of view between the detector element and the floor (Ceuppens *et al.*, 1996).

The Compton Suppression Spectrometer. The Compton continuum is observed when the Compton-scattered γ-ray escapes from the detector. When a large scintillation detector [NaI(Tl), plastic, or BGO)] surrounds the Ge detector and the source, a coincidence signal between this shield and the Ge detector can be used to suppress the Compton pulse. However, the following two important remarks have to be formulated here:

1. Above 200 keV Compton scattering occurs predominantly in the forward direction. The optimum active shield should be designed with this property and the actual source-detector geometry in mind. By Compton suppression, the remaining continuum after adequate passive shielding, can be reduced by a factor of 5 or more.
2. A Compton suppression active shield also rejects coincident lines such as the 1173.2- and 1332.5-keV lines of ^{60}Co. For isotopes having

two or more coincident lines, strong spectrum deformation will occur, and the previously established efficiency curve will no longer be valid. The application of this technique is limited therefore to specific applications. This effect is similar to that observed in a well-type detector with one important difference. In both cases the intensity of coincident γ lines is reduced, but in the case of a well-type detector the intensity of the coincident lines is found back in the sum peaks, as will be discussed in Section E.

The Cosmic Veto Shield. High-energy charged cosmic particles contribute to the continuum in the spectrum. This background can be drastically reduced by an active veto detector, generally a plastic scintillator, placed above, or surrounding, the Ge detector's passive shield (see for instance Semkow *et al.*, 2002). Any cosmic particle entering the Ge detector will also interact with the veto detector generating two coincident signals that can be used to suppress the Ge pulse. Long dead times of a few tens of microseconds will also reduce the "delayed" cosmic background. With the help of a veto shield, background reductions of 99% have been obtained in the 10-MeV region, where the background is solely due to cosmic events (Müller *et al.*, 1990).

E. The Choice of a Detector

1. General Criteria

The industry offers a large variety of germanium detector models each of which is tailored for a particular application or energy range. Figure 4.20, for instance, summarizes the various models offered by Canberra. This table is self-explanatory. More details of each of the models can be found in the catalogues from the various manufacturers.

In this chapter, we will focus on a few models only, namely the well type detector, the coaxial detectors and the Broad-Energy-Germanium or "BEGe" detector.

2. The Germanium Well-Type Detector

A well-type detector is designed to surround the sample so that close to 4π geometry is obtained. The source sees the negligible thin ion-implanted p^+ contact. It is thus the ideal detector when small samples (test tube sized) have to be measured routinely. It must be emphasized, however, that with this detector type, coincident γ-rays are subject to intense summation, leading to strong spectrum deformation, namely a reduction of the individual peaks and an enlargement of the sum peaks. Consequently, for each sample (or at least for each isotope mixture) a specific and carefully established efficiency calibration is needed.

3. Limitations to the "Relative Efficiency" Quoted for Coaxial Detectors

Traditionally, the shapes and geometries of most HPGe coaxial detectors that are offered on the market today, are designed to optimize resolution and relative efficiency as defined in Section II.C.3. The notion of "relative efficiency," however, does not tell the spectroscopist anything about the

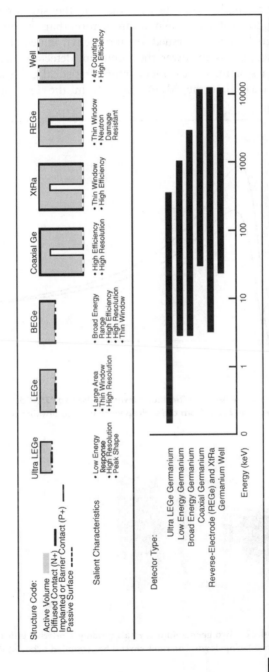

FIGURE 4.20 Summary of the various Ge detector models offered by Canberra, with the energy range they cover and their salient performance characteristics. (From Canberra catalogue ed. 12.)

real behavior of this detector at energies other than 1.3 MeV or in real measurement situations with sources different from a point source at 25 cm distance. Figure 4.21, for instance, compares the absolute efficiencies of the two detectors from Fig. 4.22. Both detectors have the same "relative efficiency" of 35%, but the absolute efficiencies, even for a point source at 25 cm, are very different for all energies different from 1.3 MeV. It is seen that the detector with the best absolute efficiency in the energy range considered has a large diameter, a shorter length, and sharper edges at the side of the entrance window (this window is facing down in this picture) than the other detector.

A goal that is pursued by most environmental and low-level gamma spectroscopists is to lower the minimum detectable activity (MDA) of their detection system, i.e. to obtain more statistical evidence in less time. It has been shown that the MDA depends in the first place on the detection

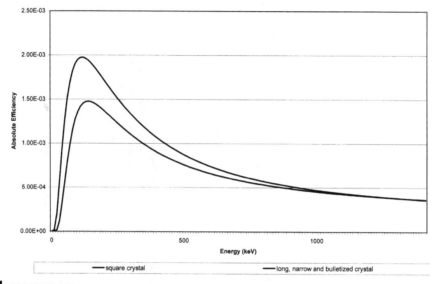

FIGURE 4.21 Absolute efficiency curves for the two *"35%"* detectors shown in Figure 4.22. Point sources at 25 cm from the endcap were used to obtain these curves.

FIGURE 4.22 Two germanium crystals yielding a relative efficiency of 35%. The entrance windows are facing down. The right crystal has rounded edges at the window side.

TABLE 4.7 Performance of a Long, 70% n-type Detector with Rounded Edges vs. a 50% Thin Window (XtRa) p-type Ge Detector. The Source is a Disk Source on the Detector Window

| Energy (keV) | Net count rate (cps) | | Background (cps) | | FWHM (keV) | | Relative MDA | |
	70% n-type	50% XtRa	70% n-type	50% XtRa	70% n-type	50% XtRa	70% n-type	50% XtRa
59	16.3	18.9	2.06	1.81	1.04	0.82	1	0.72
88	16.3	18.6	0.82	0.65				
122	16.2	17.5	1.11	0.77				
165	11.7	13.1	1.08	0.79				
392	6.84	7.42	0.74	0.56				
514	5.45	5.87	0.78	0.34				
662	4.63	4.89	0.61	0.46	1.63	1.34	1	0.75
898	3.24	3.33	0.56	0.45				
1173	2.63	2.66	0.33	0.26				
1333	2.38	2.36	0.16	0.14	2.23	1.81	1	0.85
1836	1.87	1.83	0.11	0.06				

efficiency and in the second place on number of background counts and peak-width. It thus appears that it pays the most to increase the detection efficiency. This, however, does not always mean that one needs to choose a bigger detector or a detector with a higher relative efficiency. More important is to select a detector that is better matched with the source to be measured.

This principle is dramatically demonstrated with the example summarized in Table 4.7. Table 4.7 shows the net count rates, the number of background counts, energy-resolutions, and relative MDAs obtained with a cylindrical multigamma source positioned on the endcaps of two different detectors. The "50% XtRa" detector is a Canberra thin window p-type coaxial detector with relatively sharp edges at the window side, a diameter of 65.5 mm and a length of 65 mm, similar to the shape of the crystal on the left side in Fig. 4.22. The "70% n-type" or "REGe-type" detector has strongly rounded edges – beyond the diameter of the source, a diameter of 69.7 mm and a length of 80.3 mm. It is seen in Table 4.7 that the "smaller" detector for this particular detector-source geometry, yields a higher counting rate at all energies below 1.2 MeV, a lower background, better energy resolutions and thus lower MDAs than the "bigger" detector!

4. The Broad Energy Germanium, or "BEGe" Detector

Observations like those described in Section II.E.3 led some detector manufacturers to build detectors that are optimized for certain specific applications (Verplancke, 1999; Keyser *et al.*, 1998). The Broad Energy Germanium or "BEGe"-detector from Canberra is developed to give a detector that is best adopted for low level applications with extended sources and energies ranging from 5 keV to 2 MeV. It makes use of the best available and selected germanium material (generally of p-type), has relatively sharp edges, a very thin window that is 100% transparent for energies of 3 keV and up, a low capacitance (and thus very low electronic noise), very high resolutions (low FWHM) at lower energies, large active surfaces (up to 5000 mm^2) and fixed dimensions. The cryostat is equipped with a carbon window that has a transmission of more than 85% at 10 keV and close to 100% for energies higher than 20 keV.

III. Si DETECTORS

A. Si(Li) X-ray Detectors

Si(Li) detectors are made by compensating the excess acceptor ions of a p-type crystal with Li donor ions by a process called lithium drifting. The nominal thickness varies between 2 and 5 mm and the active area between 12 and 80 mm^2 and resolutions between 140 and 190 eV are achieved. Like Ge detectors, they are operated in a liquid nitrogen cryostat. They find their application mainly in x-ray analysis. From this point of view they should be compared with low-energy or ultra-low-energy Ge detectors. These latter have better resolutions for reasons explained in Section I.C. All generalities mentioned in the Section II related to Ge detectors can be transposed readily to Si(Li) detectors. As with low-energy Ge detectors, the efficiency for low energy γ-rays is governed by the various entrance windows. The efficiency for high-energy γ-rays drops drastically above 20 keV and reaches nearly zero at 100 keV, whereas a low-energy Ge detector still has appreciable efficiency at 1000 keV as illustrated in Fig. 4.16. What seems to be a disadvantage may turn out to be an advantage in many applications. Indeed, the low efficiency for higher energy γ-rays reduces not only the full energy peaks but also the continuous background due to the presence of the source and more particularly to Compton scattering decreasing the lower limit of detection in the x-ray region. Finally, γ-spectra or x-ray spectra taken with Si(Li) detectors are less disturbed by the escape of Si x-rays than Ge detectors by the escape of Ge x-rays (see Section II.B.3). Indeed Si K$_\alpha$ x-rays have an energy of only 1.74 keV. The choice between a low-energy Ge detector and an Si(Li) detector is thus governed solely by the projected application. At room temperature, Si(Li) detectors are sometimes used as high-energy particle detectors.

B. Si Charged Particle Detectors

Silicon charged particle detectors such as diffused junction detectors (DJD) or silicon surface barrier detectors (SSBs) have served the scientific and industrial community for several decades (Knoll, 1989). In the gold–silicon detector, the n-type silicon has a gold surface barrier as the front contact and deposited aluminum at the back of the detector as the ohmic contact. Current applications, however, require detectors having lower noise, better resolution, higher efficiency, greater reliability, more ruggedness, and higher stability than older technologies could produce. Modern ion-implanted detectors such as the Passivated Implanted Planar Silicon (PIPS) detectors are now recommended as charged particle detector. They surpass the older detector types in almost every respect.

Salient advantages of PIPS technology include the following:

- Buried ion implanted junctions. No epoxy edge sealant is needed or used. This increases the detector stability. Ion implantation ensures thin, abrupt junctions for good α resolution.
- SiO_2 passivation. It allows long-term stability and low leakage currents.
- Low leakage current, typically 1/10 to 1/100 of an SSB (surface barrier detector).
- Low noise.
- Thin windows ($\leq 500\,\text{Å}$ equivalent of Si). This results in less straggling in the entrance windows and thus in better α resolution.
- Ruggedness (cleanable surface).
- Bake-able at high temperatures.
- Long lifetime.

In the detection process the particle is stopped in the depletion region, forming electron–hole pairs. The energy necessary to form a single electron–hole pair depends on the energy gap ε (Table 4.1) of the detector material, but it is essentially independent of the energy of the incoming particle. Consequently, the number N of electron–hole pairs ultimately formed is directly proportional to the energy of the stopped particle as expressed in Eq. 4.1. This eventually results in a pulse proportional to the energy of the charged particle. The thickness d (Eq. 4.6) of the depletion region depends on the applied bias voltage. Partial or full depletion with or without over-voltage is possible as illustrated in Fig. 4.23. The capacitance in pF is given by

$$C = \frac{1.05A}{d} \tag{4.50}$$

where A represents the surface area of the junction in cm^2 and d its thickness in cm. The surface seen by the charged particles is called the active area of the detector. It is required for the calculation of the efficiency. The junction area is typically 20% larger than the active area.

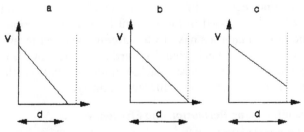

FIGURE 4.23 Thickness d of the depletion layer as a function of applied bias: (a) partially depleted detector, (b) fully depleted detector and (c) fully depleted detector with overvoltage.

The noise level of charge-sensitive preamplifiers is usually given by the manufacturer for zero input capacitance. It increases with capacitance (Eq. 4.34), and the rate of increase is also specified by the manufacturer. The detector capacitance is reduced at higher bias voltages as long as the detector is not fully depleted. The lowest noise and best resolution are thus obtained at higher voltages within the recommended range. At voltages above that recommended by the manufacturer, the reverse leakage current is likely to increase, causing excessive noise and loss of resolution.

I. Alpha Detectors

Alpha spectroscopy finds applications in widely different disciplines such as:
- Radiochemical analysis
- Environmental studies and surveys
- Health physics
- Survey of nuclear sites through the off-line detection of emitted actinides.
- Geological and geomorphologic studies (such as U–Th dating).

It requires high resolution, high sensitivity, and low background.

1. High resolution is ensured by the thin entrance window over the detector surface. It reduces energy straggling in the entrance window. Energy straggling is due to the random nature of the interaction of a charged particle with the detector material. This leads to a spread in energy if a beam of charged particles passes through a certain thickness of absorber and, consequently results in an increase of the peak width (Knoll, 1989). A thin window means less straggling and better resolution. Furthermore, the low leakage current ensures a low electronic noise contribution. Both properties together allow high α resolution. Values $\leq 18\,\text{keV}$ (FWHM) are routinely achieved for a detector with an active area of $450\,\text{mm}^2$. Note, however, that the obtainable resolution depends not only on the detector but also on external factors such as vacuum and source preparation described later in this chapter. Table 4.8 shows some typical specifications and operating characteristics for modern, ion-implanted α detectors.

2. High sensitivity is enhanced by good resolution, which reduces the background below the peak. A depletion depth of $140\,\mu\text{m}$ is enough to absorb α particles of up to 15 MeV covering the complete range of all α emitting radionuclides. For larger detector diameters ($1200\,\text{mm}^2$), absolute efficiencies $\geq 40\%$ can be achieved. This is illustrated in Fig. 4.24 and discussed in more detail later in this section. Packaging and mounting materials have to be carefully selected to avoid possible contaminants. Low background is further ensured by clean manufacturing and testing procedures. Backgrounds of $\leq 0.05\,\text{cts/(h cm}^2)$ in the energy range 3–8 MeV are achieved routinely.

a. Factors Influencing Resolution and Efficiency

Detector-Source Distance. All α particles reaching the active area of the detector will be counted. The counting efficiency is thus given by the

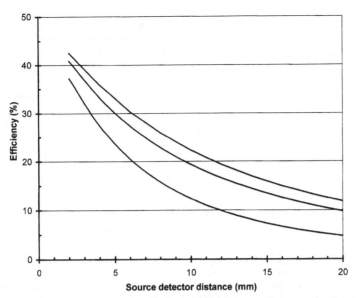

FIGURE 4.24 Calculated efficiencies for a $1200 \, mm^2$ (upper curve), a $900 \, mm^2$ (middle curve) and a $300 \, mm^2$ (lower curve) detector as a function of the source–detector distance h; the source diameters where respectively 35, 32, and 15 mm.

geometric efficiency, $\eta = \Omega/4\pi$, where Ω is the solid angle under which the detector subtends the source. For the case of a circular detector on axis with a circular isotropic source disk, this solid angle can be computed by Monte Carlo calculations (Williams, 1966; Carchon *et al.*, 1975) and is available in tabulated form (Gardner *et al.*, 1980). Figure 4.24 gives the calculated efficiencies for 1200-, 900-, and 300-mm^2 detectors as a function of source to detector distance. The source diameters are 35, 32, and 15 mm. Actual efficiencies may be slightly different, especially at small source detector distances, because of factors such as self-absorption in the source. Efficiencies of $\geq 40\%$ are obtainable.

In Table 4.8 alpha resolutions (FWHM) for the 5.486-MeV alpha line of ^{241}Am are given in the case of a detector source distance of $d = 15 \, mm$, using standard Canberra electronics. When the source approaches the detector, line broadening is expected, as the mean slope of the α particles entering the detector is increased, resulting in an effectively increased thickness of the entrance window and subsequent higher energy straggling (Aggarwal *et al.*, 1988). For ion-implanted detectors this energy straggling is minimized because of the very thin entrance window of 500 Å. For comparison, the entrance window in equivalent Si is $\cong 800$ Å for an SSB with a gold window and > 2000 Å with an aluminum window. Empirically, it has been proven, that for a 300- to 600-mm^2 detector the increase in FWHM stays below 50% for distances as small as 2 mm. Consequently, for a 300-mm^2 detector the increase of the α resolution at a source–detector distance of 2 mm with respect to that at 15 mm is thus expected to be $\leq 17 \times 0.50$ or $\leq 8.5 \, keV$. This results in an FWHM $\leq 26 \, keV$. The increase in FWHM decreases to 10% at $d = 8 \, mm$ and is practically negligible for distances $> 10 \, mm$.

TABLE 4.8   Some Examples of Operating Characteristics for α-Detectors

Active Area (mm²)	300	450	600	900	1200
Thickness (μm min/max)	150/315	150/315	150/315	150/315	150/315
Recommended Bias (V)	+20/80 V	+20/80 V	+20/80 V	+20/80 V	+20/80 V
Si-Resistivity (min Ω-cm)	2000	2000	2000	2000	2000
Operating Temp (min/max)	−20/+40	−20/+40	−20/+40	−20/+40	−20/+40
Leakage current (at 25°C) (typical/max in nA)[a]	15/70	25/100	30/120	40/200	60/300
α-resolution (keV)[b]	17/19	18/20	23/25	25/30	30/37
Absolute efficiency (%)[c]					
at 2 mm spacing	36.8	40.2	41.0	43.6	44.5
at 5 mm spacing	23.5	28.3	31.2	34.8	36.9
at 15 mm spacing	7.3	10.1	12.4	16.1	18.9

[a]These values are 5–10 times smaller than those of corresponding surface barrier detectors.
[b]For the 5.486 MeV alpha line of ^{241}Am at 15 mm detector–source spacing using standard Canberra electronics. Beta resolution is 5 keV less than alpha resolution and is approximated by pulser line width.
[c]With a source diameter of 15 mm.

Source Radius. It is interesting to take a closer look at the influence of the source diameter on the efficiency. Figure 4.25 shows the geometric efficiency of a 450-mm² and a 1700-mm² detector as a function of the source radius for a source to detector distance of 5 mm. One sees immediately that the efficiency of the bigger detector is much greater, whatever source radius is chosen. Note, however, the existence of an inflection point for $R_s = R_d$ as well as the sharp decrease in efficiency beyond this point. R_s and R_d represent the source and detector radii. The diameter of the source should thus never exceed the diameter of the detector. If a uniform specific source activity A_s (Bq/cm²) is assumed, the total number of counts registered in a time t is proportional not only to the efficiency but also to the total activity of the source deposited on the surface area or, in other words, the efficiency multiplied by A_s. Figure 4.26 gives this number as a function of the source radius in arbitrary units. Note that when the source radius exceeds that of the detector, the gain in source surface is exactly compensated by the loss in efficiency. The optimum source radius thus equals the radius of the detector. This general rule is independent of the source to detector distance.

Source thickness. Sources must be homogeneous and thin in order to avoid energy straggling due to self-absorption (Burger *et al.*, 1985). Self-absorption is proportional to the thickness of the source and inversely proportional to the specific activity. For typical values of specific activities on the order of 100 Bq/cm², the self-absorption is generally negligible for carrier-free sources. However, the effect of thickness of the carrier-free source depends on the transition probability of the isotope in question, which increases with increasing half-life. Expressed in energy loss, it is on the order of 0.03 keV for "short"-lived isotopes such as ^{239}Pu ($T_{1/2} = 2.4 \times 10^4$ y) and ^{230}Th ($T_{1/2} = 7.5 \times 10^4$ y), while

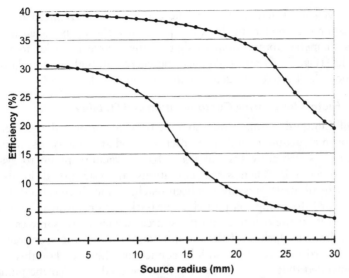

FIGURE 4.25 Geometrical efficiency of a 1700 mm² (upper curve) and a 450 mm² (lower curve) α-detector as a function of the source diameter given in mm for a source detector distance of 5 mm.

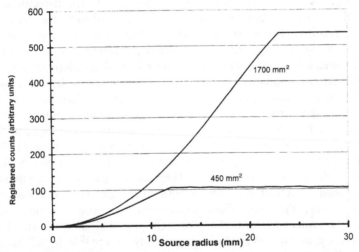

FIGURE 4.26 Number of counts registered during a certain time Δt (arbitrary units) for a 1700 mm² and a 450 mm² detector, as function of the source radius in mm.

for "long"-lived isotopes such as ^{238}U ($T_{1/2} = 4.7 \times 10^9$ y) it is on the order of 5 keV. Indeed, a 10^5 times smaller transition probability requires the presence of 10^5 times more source material in order to reach the same activity (see Chapter 1). When estimating the source thickness of a non-carrier-free source all isotopes deposited together with the isotope of interest must be considered. This can be due either to a different isotope of the same element or to the simultaneous deposition of other elements during source preparation.

Problems can also arise with very intense sources, as the source thickness and, therefore, the self-absorption is proportional to the total source activity.

For a given total activity the specific activity can be reduced by choosing a larger source diameter. In this case, preference should be given to a detector with a diameter about equal to that of the source in order to increase the efficiency (Fig. 4.26) and to reduce the energy straggling, as relatively fewer α particles will strike the detector at an acute angle.

b. Factors Influencing Contamination and Stability

Oil Contamination. Alpha sources have to be placed together with the detector in a vacuum chamber in order to avoid any energy loss in the air gap. Typical α-spectroscopy systems use a rotary vacuum pump to evacuate the α-spectrometer(s). When static conditions are established in the vacuum system (the ultimate pressure has been reached) and there is no substantial gas flow toward the pump, oil particles can back-stream toward the spectrometer and deposit on the detector and the source surfaces. The same can happen in a more dramatic fashion if the pump is disabled and the spectrometer draws air backward toward the manifold connecting the two. For this reason it is recommended that a back-streaming filter be used between the pump and the detector source vacuum chamber to prevent oil contamination.

Particulate and Recoil Contamination. Contamination of detectors can take place when particles from sources gravitate to the detector surface and stick there or are splattered, sputtered, or splashed onto the detector surface by the recoil energy imparted to the nucleus of an α-emitting atom. In the latter case the energy of the particles may be sufficient to implant themselves in the detector so that they cannot be removed nondestructively. Much of the casual contamination can be removed from PIPS detectors by cleaning with a cotton ball saturated with isopropanol. Vigorous scrubbing will not harm the PIPS detector. Recoil contamination is almost never 100% removable. It is best avoided by careful sample preparation, avoiding hot samples, or using the techniques reported by Sill and Olson (1970), which involve operating the spectrometer with an air barrier and/or a bias voltage between the detector and source. They show that recoil contamination can be reduced by a factor of up to 1000 if an air layer of about $12\,\text{mg/cm}^2$ exists between the detector and source and if the source is negatively biased by a few volts. By straggling, the air gap will increase the FWHM of α-peaks by a few keV, which is probably acceptable in all but the most demanding of applications.

c. Stability of the Detection System

Both long-term and temperature stability are important in detectors used for α-spectroscopy because count times are often many hours or days and gain shifts during data accumulation lead to erroneous or unusable spectra.

Long-Term Stability. Long-term stability is affected by the impact of the environment on the detector junctions. SSB detectors sometimes fail with prolonged exposure to room atmosphere and at other times fail when operated for prolonged periods under high vacuum. This instability is caused by the epoxy edge encapsulation that is required for this type of detector. The

PIPS detector has junctions that are buried in the silicon bulk and no epoxy encapsulation is needed or used which ensures intrinsic long-term stability.

Temperature Stability. The leakage current of silicon diodes doubles for every 5.5–7.5°C change in ambient temperature. Since the preamplifier HV bias resistor is a noise contributor, it is necessarily of high value, typically 100 MΩ. With an SSB detector having a leakage current of 0.5 μA, the change in bias voltage at the detector for a 2°C change in ambient temperature can be as much as 13 V. This is enough bias change to affect the overall gain of the preamplifier by a substantial amount. Modern PIPS detectors have a typical leakage current of less than one-tenth that of SSB detectors or DJD. Consequently, system gain change as a function of temperature is proportionally less, so that for operational temperatures of up to 35°C no significant peak shifts are observed.

d. The Minimum Detectable Activity (MDA)

The minimum detectable activity (MDA) at the 95% confidence level is given by

$$MDA = \frac{2.71 + 4.65\sqrt{b}}{t\eta P} \tag{4.51}$$

where t is the counting time, η the counting efficiency, P the yield of the α measured, and b the background counts.

The two detector-bound parameters, background (b) and efficiency (η), are particularly favorable in the case of an α-PIPS detector. For a 450-mm^2 detector ($\eta = 0.40$, $b = 6$ counts/d) and for an overnight run ($t = 15\,h = 54,000\,s$) one has thus $MDA = 0.54\,mBq$ if a 100% yield for the α ray is assumed, as well as the worst-case condition that all background counts are in the peak or region of interest. The limiting factor is often not the absolute MDA expressed in Bq, but rather the specific minimum detectable activity SMDA expressed in Bq/cm^2:

$$SMDA = \frac{MDA}{S_s} \tag{4.52}$$

where S_s represents the area of the source in cm^2.

The background in practical applications is often compromised by the presence of higher energy α lines that produce counts in the spectrum at lower energies. PIPS detectors are notably free of these tailing effects in comparison with SSB detectors of equivalent efficiency, in part because of their thin entrance window. Comparisons between the two types of detectors have shown a difference of as much as a factor of 3 in this background tailing or continuum. This translates into an improvement in MDA by a factor $\sqrt{3}$.

2. Electron Spectroscopy and β-Counting

PIPS detectors can also be used for electron spectroscopy and β counting. The thin entrance window of the PIPS detector provides little attenuation even for weak beta particles. In the β ray and (conversion) electron energy region (< 2000 keV) the resolution (FWHM) is approximated by the pulser line width. Canberra provides special β PIPS detectors fabricated from higher ohmic material, having a minimum thickness of 475 μm and allowing full absorption of electrons of up to 400 keV. Note, however, that higher energy electrons can be fully absorbed too. This is due to the fact that high-energy electrons do not follow a straight path inside the detector but rather change direction, so the real path of the electrons inside the detector is much greater than the detector thickness d. For example, the conversion electrons of the 661-keV γ line of ^{137}Cs at 624.8 and 629.7 keV are clearly seen. If only β counting is needed, the efficiency is uncompromised as long as the detector absorbs enough energy from the β ray to exceed the noise level.

As in the case of α-spectroscopy, the main factor influencing detector efficiency for electron spectroscopy (e.g., spectroscopy of low-energy conversion electrons) is governed by the geometric efficiency η. Note, however, that in the calculations the junction area and not the active area has to be taken into account, as the detector-mount is partially transparent for electrons. Furthermore, backscattering of low-energy electrons from the detector surface may cause significant loss of efficiency. By analogy with the experimental values of the fraction of normally incident electrons backscattered from thick slabs of aluminum (see, e.g., Knoll, 1989), it can be inferred that between 10 and 13% of the electrons whose energies lie between 50 and 700 keV are backscattered by thick slabs of Si, and the backscattered fraction drops sharply for higher energies. On the other hand, if backscattering occurs in the source, it may increase the apparent number of β particles, as electrons emitted outside the solid angle sustained by the detector can be scattered inside this solid angle. Efficiency calibration for electron spectroscopy must be done, therefore, with multi-energy standards, prepared in the same way as the unknowns. Source backings should be of low-Z materials to minimize source backscattering effects. Conversion electrons show up most clearly if they are not in coincidence with nuclear β particles. This is the case if the decay takes place through an isomeric level of the daughter such as in the decay of ^{137}Cs (Fig. 4.7) or if it takes place by almost pure electron capture (such as with ^{207}Bi, often used as standard). If the conversion electrons are in coincidence with the β particles, they can sum up with the nuclear electrons (Eq. 4.11). The resulting sum peak will be continuously distributed as the β particles. If β activities have to be measured, these conversion electrons can furnish supplementary counts. This is the case for example, if the sum peak surpasses the energy of the β threshold. This effect is isotope specific. The β threshold is not given by the thickness of the entrance window, which is negligible for all practical cases, but by the noise of the detector and electronics. In practice, a value of three times the electronic noise (FWHM) is taken. As nuclear β rays have a continuous energy distribution, this effect has to be taken into account when source intensities have to be measured. Indeed, part of the emitted electrons can lie under the threshold. This depends essentially on the form of the β spectrum and has to be considered individually.

Beta particles can "channel" between crystal planes of the detector and lose energy at a lesser rate than if they cross planes. To minimize this effect,

β-PIPS detectors [(as well as continuous air monitoring (CAM) PIPS] are made from silicon wafers that are cut off-axis. Small errors in calculated efficiency, however, remain possible.

Finally, it has to be noted that β detectors with an active thickness of 475 μm have small sensitivity for γ-rays. Indeed, from Fig. 4.18 it follows that the half-thickness of Si for the total absorption of γ-rays of 50 and 100 keV is 0.631 cm and 1.63 cm, respectively, so that for these energies 4.72 or 2.00% of all γ-rays falling on the detector will undergo an interaction. This can lead to a supplementary pulse or a sum pulse (Eq. 4.11).

3. Continuous Air Monitoring

The increasing demand for safety of nuclear installations calls for continuous survey of airborne radioactive particles inside and around nuclear sites, and the potential for nuclear accidents calls for a worldwide survey of the atmosphere. In particular, it is important to know whether, instantaneously or over a certain time, β and/or α activities remain below imposed limits. For a judicious choice of a continuous air monitoring system, the influence of the detector on the system performance should be understood.

Airborne radioactive particle concentration limits are expressed in Derived Air Concentration (DAC) units and are isotope specific. One DAC corresponds to an isotope concentration of 1 Bq/m^3. For certain α emitters these limits are extremely low. For example, for ^{239}Pu in soluble form, the DAC limit corresponds to a value of 0.08 Bq/m^3. The exposure is expressed in DAC-hours, that is, the concentration in Bq/m^3 multiplied by the exposure time in hours. In order to detect these activities, air is pumped through a filter at a speed of about 1 m^3/h. A detector continuously measures the accumulated activity. An instrument should be able to detect an activity concentration of 8 DAC-hours, that is, 1 DAC in 8 hours, 2 DAC in 4 hours, and so on.

This requirement is further complicated by the fact that the α background varies due to simultaneous collection and counting of the α activity from ^{222}Rn progeny, which can be significantly higher than the desired MDA. The β background also varies but, unlike the cause of the α background, this is mainly due to cosmic events.

For off-line measurements of filter samples, standard α or β detectors can be used under certain conditions. On-line measurements, however, require special characteristics, in particular, light-tightness, moisture resistance, and corrosion protection. Figure 4.27 shows an exploded view of a Canberra CAM PIPS detector. Depletion layers between 120 and 325 μm are possible. Their main characteristics are:

1. Operable in light to 5000 lumens
2. Corrosion resistant varnish coated
3. Moisture resistant varnish coated
4. Low bias voltage (10–90 V)
5. β and α discriminated by energy
6. Wide temperature range and low leakage current
7. High β sensitivity, 300 μm active thickness

CAM PIPS DETECTOR - SERIES CAM

- Alpha ,Beta Counting In Harsh Environments
- Wide Operating Temperature Range
- Cleanable Detector Surface

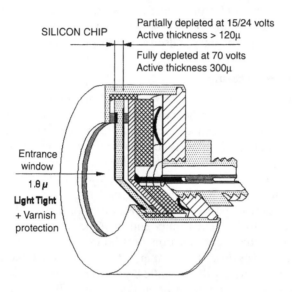

SILICON CHIP

Partially depleted at 15/24 volts
Active thickness > 120μ

Fully depleted at 70 volts
Active thickness 300μ

Entrance
window

1.8μ

Light Tight
+ Varnish
protection

ALPHA and BETA Counting

FIGURE 4.27 Exploded view of a **CANBERRA CAM** Detector (Continuous Air Monitoring) detector.

a. Light-tightness and Resistance to Harmful Environments

Silicon detectors are fundamentally light sensitive. In continuous air monitoring, the detector is not protected by a vacuum chamber and light may reach the detector in some cases. CAM PIPS detectors are made with a front surface coating of 0.5-μm-thick aluminum, which blocks the light. Furthermore, because of the nature of continuous air monitoring, detectors are often used in a harmful environment, such as a humid and/or dusty atmosphere charged with corrosive gases. In order to extend the usable lifetime of the detectors, CAM detectors are covered with a 1 μm varnish coating, providing mechanical and chemical resistance against abrasion, solvents, and corrosion. This varnish corresponds to a supplementary absorption layer of about 0.6 μm silicon equivalent. In vacuum, these supplementary windows cause roughly a doubling of the α resolution. However, one has to take into account the energy straggling in the air gap between filter and detector and in the filter itself, which makes straggling in the entrance window relatively unimportant. This is illustrated by Fig. 4.28, showing the empirical resolution (FWHM) of a CAM450 and a CAM1700 detector for the 5499.2-keV α line of ^{238}Pu as a function of the source–detector distance. The FWHM decreases with the distance, contrary

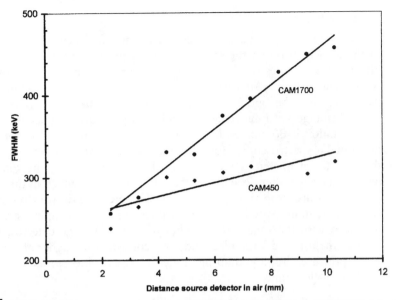

FIGURE 4.28 Empirical resolution (FWHM) of a CAM450 and CAM1700 detector for the 5499.2 keV α-line of ^{238}Pu as a function of the source detector distance.

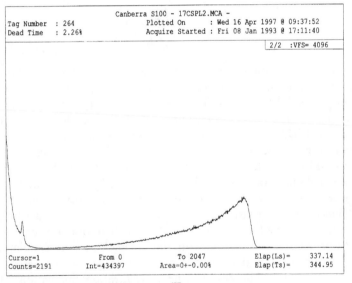

FIGURE 4.29 Beta-spectrum of ^{137}Cs in the presence of a Alpha source of ^{239}Pu, taken in air with a CAM1700 detector.

to the situation for a detector in a vacuum. Note the quasi-linear increase of the FWHM for distances of up to 10 mm. This degraded resolution is, however, still good enough to separate completely the α and β activity as illustrated in Fig. 4.29, showing the β spectrum of ^{137}Cs (Fig. 4.7) in the presence of a ^{239}Pu alpha-source taken with a CAM1700 detector with a source–detector distance of 4.3 mm.

b. Efficiency

In normal continuous air measurements, no efficiency loss is expected due to the air gap between the source and the detector. Indeed, the range in air of typical α-particles of about 5 MeV is several centimeters, and the air gap is normally < 1 cm. All earlier remarks on the efficiency remain valid, in particular that the optimum source diameter equals the detector diameter. The advantage of a big detector and a large source radius is evident, as the resolution is dominated by the air gap and not by the detector radius as in normal α-spectroscopy. However, the source diameter should never exceed the detector diameter as seen in Fig. 4.25. Furthermore, the total activity deposited on the filter depends on the pumping speed, which in turn is limited by the pressure drop through the filter. The pressure drop increases linearly with the pumping speed. For a given throughput the pumping speed needed decreases with the square of the filter diameter. A large detector, therefore, permits the use of a large filter and, as a consequence, higher air flow for the same pressure drop, permitting larger total activities to be deposited on the filter in less time.

c. Background and MDA Problems in Continuous Air Monitoring

In the case of continuous air monitoring, Eq. 4.51 can be written in the form

$$\text{MDA} = \frac{2.71 + 4.65\sigma_b}{t\eta 3600} \tag{4.53}$$

where t is the pumping and measuring time expressed in hours, σ_b the standard deviation of the background, and η the fractional counting efficiency. Besides the measuring time t, the most important parameter is the standard deviation of the background, which is quite different in the α and β region. The MDA, therefore, must be examined separately for α and β emitters.

For α emitters, the background b is no longer given by the proper background of the detector but rather by the activity of the ^{222}Rn progeny accumulated simultaneously on the filter, which can be higher than the α activity of concern. Whether or not the air in the laboratory is filtered, values of 4–40 Bq/m^3 can be regarded as quite normal, and DAC-values of 0.08 Bq/m^3 have to be detected for soluble ^{239}Pu. Furthermore, the concentration of the Rn-progeny in air varies with time. Therefore, the standard deviation σ_b is determined not only by the square-root of the registered number of background counts but also by the concentration fluctuations. Indeed, all α lines due to ^{222}Rn and its progeny lie above the α energies of ^{239}Pu. Consequently, due to tailing effects, these peaks contribute to the background beneath the ^{239}Pu peaks. The energy discrimination shown in Fig. 4.29 is good enough to ensure complete α and β separation despite the tailing effects inherent in continuous air measurements.

If a counting efficiency of $\eta = 40\%$, a pumping speed of 1 m^3/h, a pumping time of 8 hours, and a constant background of 40 Bq/m^3 are assumed, a total number of $0.5 \times 8 \times 3600 \times 40$ disintegrations occur due to the background accumulated on the filter. This leads to an MDA (Eq. 4.51) of 0.3 Bq in $4\,\text{m}^3$ of air or $0.08\,\text{Bq/m}^3$. Up to four times better results can be obtained by using background subtraction based on stripping methods, that is by subtracting the independently determined contribution of higher energy background peaks under the peak of interest.

The background in the β region ($2.1\,\text{counts/min cm}^2$) is largely of cosmic and γ-ray origin. Let us assume that a 450-mm^2 detector is used close to a filter of almost equal size. The background in a 8-hour run is thus $8 \times 60 \times 2.1 \times 4.5 = 4536$ counts. Assuming an 8-hour accumulation on the filter, this leads to an MDA of $0.03\,\text{Bq}$ for the mean $4\,\text{m}^3$ counted during 8 hours or to $0.01\,\text{Bq/m}$. It must be stressed that the actual MDA will depend on the experimental setup.

IV. SPECTROSCOPIC ANALYSES WITH SEMICONDUCTOR DETECTORS

Semiconductor detectors [e.g., SSB, PIPS, Ge(Li), HPGe, and Si(Li) detectors] are typically operated in a pulse mode and the pulse amplitude is taken as a measure of the energy deposited in the detector. Typically, the pulse amplitude data are presented as a differential pulse height spectrum. Because of real-world effects (such as electronic noise and the various interactions that can occur within the sample, between the sample and the detector, and within the detector itself), the peaks that result in alpha spectra have a very different shape from those that result in gamma spectra. The peak shapes that occur in gamma-ray spectra have been studied and described extensively in the literature (e.g., Gunnink and Niday, 1972; Helmer and Lee, 1980). Figure 4.30 shows a detailed analysis of a gamma-ray peak and the shape

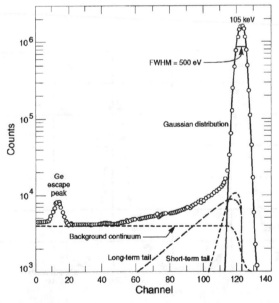

FIGURE 4.30 The detailed shape of an observed peak from a Ge(Li) detector with the principal shape components indicated. (Adapted from Gunnick and Niday, 1972.)

components proposed by Gunnink and Niday (1972) to model the peak. The shape of a peak in an alpha-particle spectrum has been described and modeled by Wätzig and Westmeier (1978). Other models of alpha-particle peaks have been proposed by Garcia-Toraño and Aceña (1981), Amoudry and Burger (1984), and Kirby and Sheehan (1984). Representative examples of a gamma-ray peak and alpha-particle peaks are presented in Figs. 4.31 and 4.32, respectively.

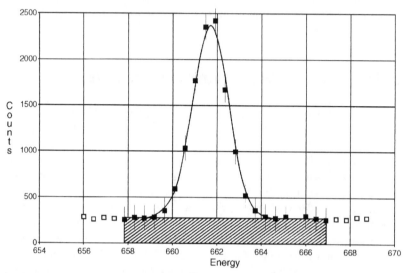

FIGURE 4.31   A 661.6 keV peak from ^{137}Cs as observed at approximately 0.5 keV/channel.

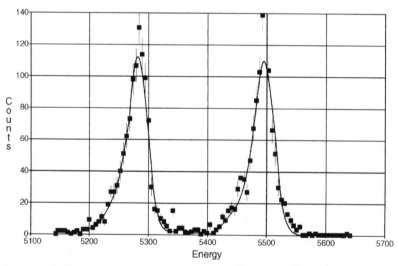

FIGURE 4.32   The 5.276 MeV alpha peaks of ^{243}Am and ^{241}Am, respectively. Note the asymmetry (tailing) of the peaks.

A. Sample Preparation

Warning: The chemical procedures discussed in this section involve the use of strong acids, caustic solutions, and very high temperatures. Appropriate precautions should be observed when handling such materials or working with such conditions. Particular caution should be exercised when working with perchloric acid, as the addition of perchloric acid to a solution containing any organic (carbon) material can result in a very vigorous reaction or EXPLOSION!

The very different interaction mechanisms and thus attenuation characteristics of gamma-rays and alpha particles demand very different considerations in their sample preparations. As alpha particles lose energy virtually continuously along their track, they have a distinct range. In the energy range of interest (typically 4–7 MeV), alpha particles can be stopped by a sheet of paper or approximately 2–8 cm of air (at STP). Thus, encapsulating the sample is out of the question. In fact, even minimal amounts of material between the emitting nuclide and the detector can degrade the energy of the alpha particles to the point that spectroscopic identification becomes difficult, if not impossible. On the other hand, gamma-rays can penetrate relatively long distances in a material without interaction (and concomitant loss of energy), so containment of volumetric (thick) samples of γ emitters is not only possible but routinely employed.

I. Sample Preparation for Alpha Spectrometry

Sample preparation must convert the raw sample into a form that is suitable for alpha spectrometry. This implies two requirements for the preparation:

1. Produce a thin sample
2. Chemically separate elements that would produce chemical or radiochemical interferences.

In addition, the final form of the source should be rugged enough to be handled safely, chemically stable, and free of all traces of acid and solvent to prevent damage to the counting chambers and detectors. Proper sample preparation is essential to ensure an accurate quantitative assay as well as high resolution. In general, sample preparation requires three steps:

1. Preliminary treatment
2. Chemical separation
3. Sample mounting

These three steps will now be reviewed, starting with sample mounting and ending with preliminary treatment. This order has been chosen because it is easier to understand why certain things are required in the early steps after one understands the requirements of the later steps.

a. Sample Mounting

In addition to energy straggling, there are geometric effects that alter the energy resolution of alpha spectra. The need for a thin sample is demonstrated in Figs. 4.33 and 4.34. Figure 4.33 demonstrates that the variation in the energy of (originally) monoenergetic particles escaping the sample is proportional to the sample thickness, and Fig. 4.34 illustrates the fact that particles leaving the sample or entering the detector at angles other than perpendicular have a longer path length in the energy-degrading materials of the sample matrix and detector dead layer. This variation of the track length in energy-degrading material causes a variation in the observed energy, which contributes to increased line breadth and tailing. Referring to the notation

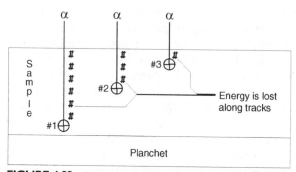

FIGURE 4.33 **Three atoms of an alpha-emitting nuclide (labelled 1, 2, and 3) are deposited at different depths within the thickness of the sample. The energy of the alpha particle from the atom labelled #1 will be degraded more than that of #2, which in turn is degraded more than that of #3. Thus the observed energy of the alpha particles from a thick (monoenergetic) sample will have a distribution of energies reflecting the thickness of the sample (as well as due to straggling).**

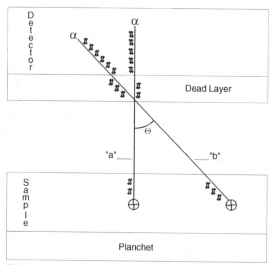

FIGURE 4.34 Since the sample (may) and the detector (definitely does) have a finite radius, alpha particles can leave the sample and enter the detector at angles other than perpendicular. Track *a* leaves the sample and enters the detector at right angles while Track *b* leaves the sample and enters the detector at an angle θ from perpendicular. Track *b* has a path length (through the sample and through the detector dead layer) that is greater than track *a* by a factor of $1/\cos \theta$. Thus an alpha particle emitted along track *b* will have a greater energy degradation than a particle emitted along Track *a*.

of Fig. 4.34, the difference in the track length of a particle traveling along path *a* versus path *b* is given by

$$\Delta = \text{difference in track length} = (d + t)\left(\frac{1}{\cos\theta} - 1\right) \qquad (4.54)$$

Thus, one can reduce the line breadth by

1. Minimizing *d*, the dead layer (window) on the detector
2. Minimizing *t*, the thickness of the sample
3. Minimizing *θ*, the acceptance angle of the detector

These items were introduced in a general sense in Sections III.B.1.a and III.B.1.b.

From the preceding discussion, it is clear that (all else being equal) the thinner the sample, the better the resolution. Thus the optimum sample mount in terms of resolution would be a monatomic layer of sample atoms; however, in practice, thicker mounts are typical. Methods that have been used to mount sources for alpha spectrometric measurements using semiconductor detectors include

1. Vacuum sublimation
2. Electrospraying
3. Electrodeposition (a) from an aqueous solution and (b) from an organic solution—also referred to as molecular plating
4. Hydroxide or fluoride co-precipitation and filtration as a thin source
5. Evaporation from an organic solvent
6. Evaporation from an aqueous solution

An excellent review of the various sample mounting methods is given by Lally and Glover (1984).

Vacuum sublimation. If the overriding concern is to achieve the highest possible resolution, one should consider mounting the sample by vacuum sublimation. Although the method is capable of producing very good resolution, it is not quantitative, and it is more appropriate to metrology applications (such as the precise measurement of alpha-particle energies) than to general radiochemical assay. Vacuum sublimation requires an apparatus in which the sample is heated to a sufficiently high temperature in a vacuum that the sample is vaporized and then sublimed onto a substrate. Samples mounted by this method have produced resolutions of 4–5 keV with magnetic spectrographs and approximately 11 keV (FWHM) with a surface barrier detector.

Electrospraying. Sample mounting by electrospraying can produce extremely thin sources as well as deposits of up to 1 mg/cm^2 with high efficiency. The method requires an apparatus in which the sample is dissolved in an organic solvent and sprayed from a fine capillary tube or hypodermic needle (with the tip squared off) against a substrate that forms the cathode of the apparatus. An electrode may be placed in the solution, or the needle itself can be made the anode of this device. With a potential of up to 8 kV applied between the needle and substrate, the organic solution is ejected as a fine spray so that the organic solvent evaporates before reaching the cathode. In this manner, only solid particles reach the cathode. To ensure a uniform deposit, the cathode is typically rotated during the spraying. Electrosprayed sources have produced resolutions of approximately 17 keV (FWFIM).

Electrodeposition. Samples may be mounted by electrodeposition from an aqueous solution or an organic solution, in which case the method is generally referred to as molecular plating. The method produces rugged sources that may be kept in the laboratory indefinitely and is frequently used for preparing rugged alpha sources. Electrodeposition is applicable to a wide range of work from metrology measurements to radiochemical assays. Resolutions of < 20 keV are possible with semiconductor detectors. In addition to the production of thick sample deposits, impurities can affect the yield of the technique. Thus steps to chemically separate the element of interest and place this element in an appropriate solution for electrodeposition must precede the electrodeposition.

Electrodeposition from Organic Solutions. This technique involves passing a low current at high voltage through an organic solution. It is reasonably rapid and virtually quantitative; near-quantitative recoveries of many of the actinides have been reported in about 1 hour. The method requires the use of reasonably pure solutions. As little as $100 \mu g$ of iron or aluminum (which deposits on the cathode along with the actinides) in solution can cause the deposit to be thick and produce degraded resolution. One precaution concerning molecular plating that should be noted is the use of high voltages and volatile organic solvents. This combination can present a hazard, particularly in confined areas such as a glove box.

Electrodeposition from Aqueous Solutions. In contrast to molecular plating, electrodeposition from aqueous solutions is usually performed at voltages of approximately 12–20 V with sufficient current capacity to provide a few hundred mA/cm^2. The method can produce quantitative yields from pure actinide solutions; however, impure solutions may produce less than quantitative yields. The use of a complexing agent, such as hydrofluoric acid, sodium bisulfate, tri/diethylenetriaminepentaacetic acid (DTPA), or ethylenediaminetetraacetic acid (EDTA), can make the electrolyte more tolerant of impurities. One drawback of electrodeposition from an aqueous solution is that it is time consuming, taking up to several hours to complete a deposition. Figure 4.35 shows the amount of Pu remaining in the plating solution as a function of time for electrodeposition of Pu from a 1 M H$_2$SO$_4$ solution. It is apparent from this figure that, to achieve high recovery, one must commit a substantial amount of time to the electrodeposition step.

Electrodeposition is applicable to many elements including the actinides (Talvitie, 1972). Procedures for electrodepositing radium (Roman, 1984); thorium (Roman, 1980); and uranium, thorium, and protactinium (McCabe *et al.*, 1979; Ditchburn and McCabe, 1984) have been presented.

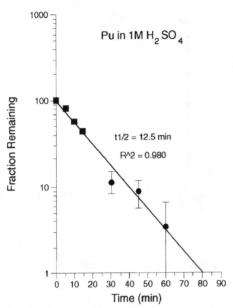

FIGURE 4.35 The results of an experiment designed to evaluate the rate of electrodeposition of Pu from I M H_2SO_4 (pH = 3.2). Such experiments are used to optimize conditions and evaluate plating times. (From Burnett, 1992.)

Cable *et al.* (1994) have investigated the optimization of the chemical and physical parameters affecting the electrodeposition for alpha spectrometry of uranium, thorium, protactinium, americium, and plutonium using a custom-designed electrodeposition unit.

Electrodeposition cells vary from very simple to rather elaborate. Schematic drawings of two designs are presented in Figs. 4.36 and 4.37. The sample is electrodeposited on a metallic substrate, typically a disk of stainless steel, nickel, or copper (although other materials have been used successfully), which functions as the cathode of the electrodeposition cell. Only one side of the disk should be exposed to the plating solution. The anode is normally made of platinum.

In general, the actinide elements thorium through curium can be electro-deposited as hydrous oxides from a buffered, slightly acidic aqueous solution without prior oxidation. Following electrodeposition, the cathode disk is often heated to convert the deposited actinide compound to the anhydrous state or flamed to convert it to an oxide. The high temperature will also volatilize the spontaneously volatile component of any polonium that may have inadvertently deposited on the disk. Sill and Olson (1970) report that heating the disk on an uncovered hot plate for 5 minutes reduces the spontaneously volatile component of poloniurn to a generally acceptable level without loss of lead or polonium. As the volatility of polonium produces a pseudo-recoil effect, by which the detector can become contaminated, it is desirable to eliminate the spontaneously volatile component of polonium to prevent contamination of the detector (see Section IV.B.1) and counting interferences. Care should be taken in heating the disk, as ignition at red heat

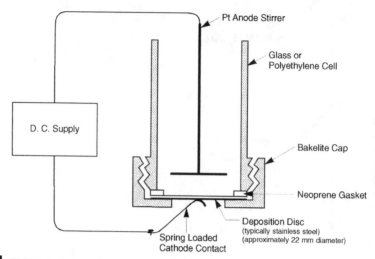

FIGURE 4.36 **Schematic drawing of a simple electrodeposition cell.**

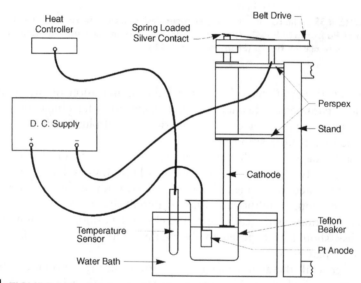

FIGURE 4.37 **Schematic drawing of a rotating disc electrodeposition unit. The disc upon which the sample is to be deposited (cathode) is mounted on the end of the spindle which rotates at 3600 rpm. (From Burnett, 1990.)**

can volatilize lead (if present), which can carry other nonvolatile components with it, resulting in a loss of material. The volatility of polonium is highly dependent upon the disk material and the conditions of the deposition. Contamination of samples by polonium has been shown to occur via a variety of pathways including spontaneous deposition from the air and from acid baths used to clean recycled deposition disks. For a more complete discussion of polonium's role as a contaminant and interference, see Sill and Olson (1970) and Sill (1995).

Co-precipitation and Filtration as a Thin Source. Co-precipitation and filtration is a fast, inexpensive method used to mount samples for alpha spectrometry. Basically, the method provides for the co-precipitation of the nuclides of interest as either a hydroxide or a fluoride using either cerium or neodymium[1] as a carrier to produce an extremely finely divided precipitate, which is deposited by filtration over a substrate of ceric hydroxide, cerous fluoride, or neodymium fluoride. The substrate is prepared by filtering the substrate solution (typically ceric hydroxide or neodymium fluoride) through a 0.1 μm membrane filter. These very finely divided precipitates plug the filter and provide a very smooth and nearly impenetrable surface upon which the co-precipitated (with cerous hydroxide or neodyinium fluoride) nuclides of interest lie.

The method as presented by Sill and Williams (1981) uses cerium carrier and substrate (in both the oxide and fluoride forms) and it was proposed that lanthanum and neodymium could be used equally well. Subsequent extensions of this procedure tended to focus on the chemical separations that allow the elements of interest to be separated from each other and placed in a chemical form that permits them to be coprecipitated (typically as a fluoride or hydroxide). Hindman (1986) presented a method by which the actinides (thorium, uranium, plutonium, and americium) are separated from each other by coprecipitation and mounted as fluorides on a neodymium fluoride substrate, and Sill (1987a) presented methods for the precipitation of actinides as fluorides or hydroxides for high-resolution alpha spectrometry.[2]

The method continues to evolve, being combined with many separation procedures to handle a wide variety of sample types (Sill, 1987b; Sill and Sill, 1989). More recently, it has been demonstrated that satisfactory resolutions can be obtained even with the sample mounted directly upon the filter, that is, without first depositing a substrate on the filter (Sill and Sill, 1994).

The method can produce excellent resolution[3] provided the total mass of the sample layer (nuclides of interest, carrier, and any impurities) is kept below approximately 100 μg.[2] Using an SSB detector, Sill and Williams (1981) found an FWHM for ^{239}Pu of about 65 keV when mounted by this method, compared with an FWHM of about 50 keV for a ^{239}Pu source

---

[1]Lanthanum has also been proposed as a carrier (Sill and Williams, 1981); however, later work (Hindman, 1986) indicated that there are certain disadvantages associated with the use of lanthanum: the purity of available lanthanum reagents is a problem, lanthanum is not as soluble as neodymium in the small pyrosulfate fusions of this procedure, and the precipitation characteristics of lanthanum are not as advantageous as those of neodymium.

[2]The 100-μg limit applies when deposited in a 7/8-inch-diameter circle (on a 25-mm filter) producing a thickness of $\sim$25 μg/cm^2. Sill and Williams (1981) warm against attempting to distribute the sample over an area greater than that of the detector in an attempt to decrease the sample thickness, as the large entry angle of alpha particles into the detector produces unacceptable amounts of tailing in the spectrum.

[3]Today, possibly because of improved filters and detectors, one can expect to achieve routinely a resolution of 40–50 keV with samples mounted by coprecipitation and filtration, while electrodeposited samples typically produce a resolution of 20–40 keV.

electrodeposited on polished stainless steel. Noting the limitation on the size of the mounted sample, some care should be exercised in selecting the initial size of the sample. If the sample contains more than $100\,\mu$g of the nuclide of interest, the resolution will suffer. Obviously, a smaller initial sample size should be chosen.

In addition, it should be noted that certain sample types (e.g., soils) frequently contain trace quantities of elements that produce chemical interferences with the elements of interest. This can lead to degraded resolution if the total mounted mass exceeds approximately $100\,\mu$g. In this case, one has two choices:

1. If the mass of the interfering elements exceeds $100\,\mu$g, then a more specific separation is required.
2. If the mass of the interfering elements is less than approximately $75\,\mu$g, one might be able to use them in place of the carrier in the coprecipitation of the sample fraction in which these interfering elements occur. For example, 1 g of an average soil contains approximately $75\,\mu$g of the light lanthanides (lanthanum, praseodymium, neodymium, etc.), which can be used in place of the cerium carrier to coprecipitate the (actinide) element of interest that occurs in the same fraction as the light lanthanides. (As the light lanthanides are typically trivalent, they typically end up in the americium fraction.)

This method of sample mounting is not limited to chemical separations by coprecipitation. Any separation scheme that produces purified fractions of the elements of interest (e.g., ion exchange, extraction) can precede this method of sample mounting.

Direct Evaporation of an Organic Solution. Direct evaporation of an organic solution produces sources with reasonable resolution by generating nearly solid-free deposits of some alpha-emitting elements. Basically, the method requires extraction of the elements of interest into an organic solution followed by the evaporation of this solution on a stainless steel disk. Examples of organic solutions that may be used include thenoyltrifluoroacetone (TTA) in benzene or xylene to complex uranium and thorium and TTA in toluene to chelate plutonium.

The method typically starts with reasonably pure fractions of the elements of interest obtained by ion exchange or solvent extraction. This solution is then evaporated to dryness and treated with a small volume[4] of perchloric and nitric acids to oxidize any residual organic matter. Following the dissolution of the sample, the pH is adjusted to about 3.0 by the addition of 1.0 M NaOH, and the elements of interest are extracted into approximately 1 mL of an approximately 0.4 M TTA solution. Small stoppered centrifuge tubes may be used to avoid the introduction of excessive

---

[4]As the organic solution will eventually have to be evaporated, it is expedient to keep the volume to a minimum. Since the chemical yield of the extraction increases as the ratio of the volume of the aqueous phase to that of the organic phase decreases, it follows that the extraction should be carried out from small volumes ($\sim$5 mL) to maximize the recovery.

amounts of air during the mixing of the organic and aqueous phases. A small Pasteur pipette may then be used to transfer the separated organic phase dropwise onto the stainless steel disk for evaporation. To promote uniform drying, the disk can be placed on a heated brass cylinder or common iron washer. As a final step, the disk may be flamed to a dull red to ensure the removal of all residual organic material.

Direct Evaporation of an Aqueous Solution. This method is typically not used for the preparation of high-resolution sources as the material does not deposit uniformly. Any salts in the aqueous solution, including the active material itself, tend to deposit as crystals and aggregates. The resulting self-absorption causes a decrease in resolution. Although spreading agents (such as tetraethylene glycol) can be added to the solution to reduce the crystallization problem during the evaporative deposition, they tend to leave substantial quantities of organic material in the deposit that must later be burned off, causing poor adherence of the nuclide to the disk.

b. Chemical Separation

As the initial sample may be rather large (on the order of 1 g or more) and the mounted sample needs to be very small (in the microgram range), it is necessary to separate the elements of interest from the bulk of the sample. Once the alpha-emitting elements are separated from the bulk sample, it may not be necessary to separate the various alpha-emitting elements from each other before counting; see, for instance, Sill and Sill (1994). However, as the alpha-particle energies of many nuclides differ by as little as 15–30 keV (which is comparable to the energy resolution of the detectors used in alpha spectrometry), chemical separation of such nuclides is required to eliminate these radiochemical interferences and make quantitative analysis possible.

Unlike cold chemistry, in which standard methods abound, there are no standard (prescribed) methods for radiochemical procedures other than for drinking water as given in the EPA 900 series. The trend in the United States in recent years has been for the acceptability of a radiochernical procedure to be performance based. That is, there is no one mandatory procedure with which to perform a given analysis. Rather, a procedure is considered acceptable if one can demonstrate acceptable performance in cross-checks, analysis of knowns, and so forth.

To perform the necessary chemical separations, one must get the elements of interest into solution. This will be discussed in Section IV.A.1.c. Assuming the elements of interest have been dissolved, numerous separation procedures are available. A brief overview of the various methods is presented in the following with references to the scientific and commercial literature from which the detailed procedures may be obtained.

Separation by Precipitaiton/Co-precipitation. This technique has been documented extensively in the literature (Sill, 1969, 1977, 1980; Sill and Williams, 1969; Sill *et al.*, 1974). The method is frequently used in conjunction with sample mounting by the method of coprecipitation and filtration as a thin source (Sill and Williams, 1981; Hindman, 1986; Sill and Sill, 1994), but it can also be used preparatory to electrodeposition.

Separation by Ion Exchange. This is probably still the most common method of chemical separation for the preparation of samples for alpha spectrometry. The method depends on the selective adsorption and desorption of ionic species on ion exchange resins and thus requires that the element of interest be in a form that may be adsorbed by the resin.

Numerous procedures for chemical separations by ion exchange have been presented. Quantitative separation of uranium, thorium, and protactinium by ion exchange has been demonstrated by McCabe *et al.* (1979). An improved method for the purification of protactinium was later presented by Ditchburn and McCabe (1984). Numerous other procedures for chemical separations by ion exchange have also been presented in the literature. In addition to the numerous texts on the subject (e.g., Small, 1989), manufacturers of ion exchange resins[5] are often excellent sources of resource material.

Chemical Extraction. Chemical extraction is a separation technique that relies on the difference in the solubility of the element of interest in an organic solvent versus an aqueous solution. Traditionally, the two components of the system were maintained in the liquid phase and the method was referred to as liquid–liquid extraction. However, an innovative application of solvent extraction has been developed at the Argonne National Laboratory in which the solvent extraction system is adsorbed on a macroporous polymeric support that immobilizes the extractant and diluent to form the stationary phase of an extraction chromatographic system.

Separation by Liquid–Liquid Extraction. This method of extraction requires that the element of interest be in true ionic solution in an aqueous medium and not complexed (chelated or bound) in any manner. That is, liquid–liquid extraction will not extract the element of interest from suspended solid or colloidal material. In addition, the presence of organic (and in some cases inorganic) complexing materials in the aqueous phase will, in many cases, cause the extraction to be unsuccessful.

The difference in the solubility of the element of interest in the organic solvent versus the aqueous solution is expressed in terms of the distribution coefficient, K_d, which is defined as

$$K_d = \frac{C_{org}}{C_{aq}} \tag{4.55}$$

From this definition, it follows that the percent recovery of an extraction is given by

$$\% \text{ recovery} = \frac{K_d V_{org}}{K_d V_{org} + V_{aq}} \times 100 \tag{4.56}$$

[5]For example, Bio-Rad Laboratories, Inc., 2000 Alfred Nobel Drive, Hercules, CA 94547 and The Dow Chemical Company, P.O. Box 1206, Midland, MI 48641-1206.

The principles of liquid–liquid extraction and the derivation of Eq. 4.56 are provided by L'Annunziata (1979).

In general, organic acids, ketones, ethers, esters, alcohols, and organic derivatives of phosphoric acid have all been used for extraction. The Purex process, which is generally used for the reprocessing of nuclear fuel, makes use of tributyl phosphate (TBP) in an inert hydrocarbon diluent to extract both uranium and plutonium. Methyl isobutyl ketone (MIBK) has also been used for the extraction of U and Pu from spent fuel. Thenoyltrifluoroacetone (TTA) can be used to extract some actinides. Sill *et al.* (1974) have presented a procedure by which the actinides are extracted into Aliquat 336, followed by stripping of these elements from the organic extracts. Although liquid–liquid extraction can be used as a precursor to further separations, samples are frequently mounted directly from the organic phase by evaporation of the organic solvent.

Extraction Chromatographic Systems. This extraction system is used much like ion exchange resins. One advantage of these materials is their high specificity. They are marketed by Eichrom Industries, Inc.[6] Procedures are available for the separation of americium, plutonium, and uranium in water (Eichrom Industries, 1995a); uranium and thorium in water (Eichrom Industries, 1995b); uranium and thorium in soil (Eichrom Industries, 1994); and thorium and neptunium in water (Eichrom Industries, 1995c). A method for screening urine samples for the presence of actinides using these extraction chromatographic materials has been presented by Horwitz *et al.* (1990).

c. Preliminary Treatments

Preliminary treatments typically vary with the objectives of the experiment and the sample matrix. Basically, they are performed to attain one or more of the following objectives:

1. To separate the component(s) of interest from the remainder of the sample
2. To ensure that the sample is representative of the bulk sample
3. To ensure that the sample remains representative of the bulk sample
4. To preconcentrate the component(s) of interest
5. To introduce chemical tracers and ensure equilibration with analyte isotopes
6. To prepare the sample for the chemical procedures that are to follow, that is, dissolve the sample

Variable and/or incomplete sample dissolution is a major cause of inaccurate radiochemical analyses. To ensure accurate and reproducible results, it is essential that all of the element of interest be brought into solution. A variety of methods have been suggested and used to prepare samples for alpha spectrometry, including high-temperature fusions, acid leaching, and a variety of "digestions" typically involving acid bombs at

[6]Eichrom Industries, Inc., 8205 S. Cass Avenue, Suite 107, Darien, IL 60559.

elevated temperature and pressure. Sill and Sill (1995) provide convincing arguments for the use of high-temperature fusions, citing examples of the failure of other methods to place selected elements into solution. For a more complete discussion of decomposition methods, see Bock (1979) or Sulcek and Povondra (1989).

For liquid samples, one must first decide what is of concern. Is it the dissolved material, the particulate material, or the total (both together)? If the sample is to be separated into soluble and particulate components, the first step should be to filter the sample. Following filtration, the liquid portion should be acidified to prevent biological growth as well as to keep trace elements in solution (at $6 < pH < 8$, many metallic elements form insoluble hydroxides, which can then "plate" onto the walls of the sample container). Acidification of a liquid sample before filtration can introduce a bias in the individual components as the acidification of the sample will leach the particulate matter. Radiochemical tracers, if they are to be used, should be added immediately following acidification. Sufficient time for isotopic equilibration should be allowed before any further chemical procedures are performed. Burnett (1990) suggests that "24 hours appears to be sufficient for equilibration of most radiotracers with uranium-series isotopes in natural waters."

Following isotopic equilibration, one can perform a preconcentration step if desired. Preconcentration is frequently used to obtain sufficient material when the concentration of the material of interest is very low. Common methods of preconcentration include ion exchange, coprecipitation, and the use of adsorptive filters such as manganese-coated acrylic fibers, which have high adsorptive capacities and can be used to preconcentrate elements such as radium, thorium, protactinium, and actinium.

To ensure total dissolution of the element(s) of interest in a total water sample or even the liquid phase when there is a possibility that the element(s) of interest is chelated with organic material or otherwise bound in a form that would interfere with its separation, a high-temperature fusion may be employed. Such a procedure is described by Sill and Sill (1994).

A simple flowchart for the preliminary treatment of liquid samples might appear as follows:

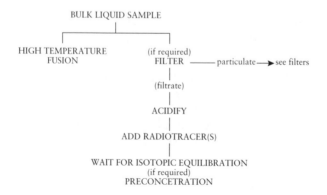

The particulate fraction can be treated in the same way as solid samples (soils, etc.) once the presence of the filter is addressed. Typically, the filter is "digested" either by ashing or by dissolving in strong acid and treating the residue as one would a solid sample; however, polycarbonate (membrane) filters are resistant to acids and do not submit to acid dissolution.

A simple flowchart for the preliminary treatment of filtrates might appear as follows:

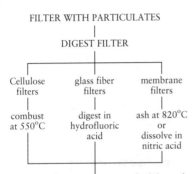

FILTER WITH PARTICULATES

DIGEST FILTER

Cellulose filters	glass fiber filters	membrane filters
combust at 550°C	digest in hydrofluoric acid	ash at 820°C or dissolve in nitric acid

proceed to preliminary treatment of solid samples
(radiotracers added during fusion/digestion/acid leach)

For solid samples, one must first decide what is of concern. Is it the total sample or some fraction thereof? Typically, for soil samples, one is concerned with the sample in total; however, if the sample is to be separated according to particle size, then the first step should be to fractionate the sample according to particle size. Following fractionation, the sample should be ground and mixed well to ensure homogeneity. Finally, the subsample, on which the chemical separations will be performed, should be measured.

As with any sample being prepared for alpha spectrometry, the elements of interest need to be brought into solution before their separation. Although in some cases it may be possible to remove the element(s) of interest from the bulk of the sample by leaching in strong acid and separating the liquid and solid phases by centrifuging or filtration, it is generally recommended that a total dissolution of the sample be performed to ensure that the element(s) of interest is indeed brought into solution.

Typically, a high-temperature fusion (e.g., pyrosulfate or potassium fluoride fusion) is used to ensure the total dissolution of a solid sample (Sill and Williams, 1981; Hindman, 1984; Sill and Sill, 1994). Detailed procedures for this technique have been presented by Hindman (1984) and Sill and Sill (1994). One drawback of this method is the expense of the required platinum dish and its limitations in terms of compatibility with certain chemicals and processes. Acknowledging this drawback, Sill and Sill (1995) have presented a procedure for performing a pyrosulfate fusion in borosilicate glassware.

A simple flowchart for the preliminary treatment of solid samples might appear as follows:

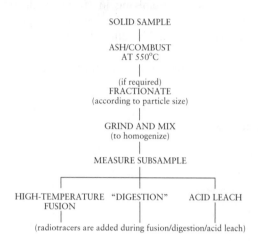

2. Sample Preparation for Gamma Spectrometry

Before the advent of high-resolution spectrometers, radiochemical separations were often required prior to counting. Although such procedures are still useful in some cases, they are not covered in this section. Rather, this section focuses on the preparation of samples that do not require extensive chemical preparation.

The first step in sample preparation is to collect the sample. Care should be taken during sample collection to ensure that the sample is representative of the bulk material. For example, air sampling for particles should employ isokinetic sampling. For soil sampling, care should be taken to prevent cross-contamination of samples by the collection tools.

Assuming one is analyzing bulk samples (e.g., there is no chemical separation or preconcentration), the basic function of the sample preparation is to make the sample look like the standard that was assumed for the efficiency calibration. Whether the calibration standard is an actual source or a mathematical model such as is used for Monte Carlo calibrations, the standard is prepared with or assumed to have certain properties (e.g., dimensions, density, distribution). The sample must be prepared in a manner that reproduces these properties. For example, if it is assumed that the active material is uniformly distributed in a liquid sample, then plating of the active material on the walls of the container must be avoided. To this end, liquid samples may be acidified.

A significant difference between alpha and gamma spectrometry is that in gamma spectrometry, the nuclides of interest are not removed from the bulk sample, so the properties of the bulk sample (density, homogeneity, etc.) become important.

In other ways, the sample preparation considerations for gamma spectrometry are similar to those for alpha spectrometry. For example, for liquid samples, one must still decide which component is of concern. Is it the dissolved material, the particulate material, or the total (both together)? If the sample is to be separated into soluble and particulate components, the first step should be to filter the sample. Following filtration, the filtrate should be acidified both to prevent biological growth and to keep trace elements in solution (at $6 < pH < 8$, many metallic elements form insoluble hydroxides, which can plate onto the walls of the sample container). In alpha spectrometry, one wants to keep the active material in solution so that it might be chemically separated; however, in gamma spectrometry, one tries to keep the active material in solution to ensure that the geometric distribution of the active material is consistent with the assumed distribution of the calibration standard. Acidification of a liquid sample before filtration can introduce a bias in the individual components, as the acidification of the sample will leach the particulate matter. If the sample is to be analyzed in total (without regard to which fraction contains what activity), acidification as the first step in sample preparation is appropriate as it keeps trace elements in solution. The fact that acidification also leaches the particulate matter is not critical in this case, as the dissolved material will then be uniformly distributed in the liquid, which is most likely the distribution assumed for the efficiency calibration. In fact, if a sample is to be analyzed in total and it contains particulate matter, one should pay particular attention to ensuring that the material is distributed as assumed. That is, if one prepares the calibration standard assuming the active material is uniformly distributed, then one should attempt to ensure that the active material in the sample is also uniformly distributed. In other words, shake it up. Particulate matter in liquid samples can present a difficulty, particularly with long sample counts, as the particulate matter can settle during the counting period, causing a bias to develop.

B. Analysis—Analytical Considerations

I. Analytical Considerations in Alpha Spectrometry

One can use a peak search program to identify peaks in an alpha spectrum, but it is more typical to use a library-driven and/or user-defined search, as the separations that are typically performed in the preparation of the sample severely limit the nuclides that could be found in any given fraction. Thus, one simply analyzes the regions of the spectrum where the nuclides of interest could be. In addition, library-driven routines are more suited to the analysis of small, poorly defined peaks that are frequently encountered in low-level (environmental) alpha spectrometry.

If the peaks are fully resolved from one another, a simple summation of the counts in each peak provides an accurate value for the peak area. If there is any overlap of peaks, one should use an algorithm (typically implemented in a computer program) that is capable of calculating the areas of peaks that overlap. The algorithm should use a peak model that includes a low-energy tail, which is typical of alpha peaks. A variety of mathematical models and methods have been used in various computer codes designed to analyze complex alpha spectra. Examples of these include ALFUN

(Wätzig and Westmeier, 1978), NOLIN (Garcia-Toraño and Aceña, 1981), DEMO (Amoudry and Burger, 1984), and GENIE-PC (Koskelo *et al.*, 1996).

These programs assume no continuum distribution under the peaks, as alpha particle interactions do not provide a mechanism by which a continuum distribution could be generated. Rather, any alpha particle interaction in the detector would necessarily contribute to the peak (or tail of the peak). This should not be interpreted as meaning that there is no "background" contribution, but rather that the background contributions also form or contribute to a peak (or tail). Background contributions may result from:

1. Contamination of the counting chamber and/or detector, which can be determined by counting the empty chamber.
2. Contaminants in the process reagents and/or mounting materials, which can be determined by counting a method blank.

For radiochemical analyses (assays), these background contributions must be subtracted from the observed spectrum to determine the sample (only) count rate to determine accurately the sample activity. Some commercially available alpha spectrometry software packages differentiate between these two contributions as an area correction (item 1 above) and a reagent correction (item 2 above). The reagent correction is often implemented by scaling the contribution of one (reagent) nuclide to another (reagent nuclide), in which case the area correction should be implemented before the reagent correction. For example, if a ^{242}Pu tracer is used that contains trace levels of ^{239}Pu, one would need to subtract the tracer's contribution to the ^{239}Pu peak from the observed spectrum to determine the sample's contribution to the ^{239}Pu peak. The tracer's contribution to the ^{239}Pu peak may be determined as a fraction of the ^{242}Pu that is present due to the tracer. However, if ^{242}Pu contamination is present in the counting chamber, the chamber's contribution to the observed ^{242}Pu peak must be subtracted from the observed ^{242}Pu peak before the ^{242}Pu peak can be used to determine the quantity of reagent present and thereby the reagent's contribution to the ^{239}Pu peak.

As discussed in Section III.B.1.a, the efficiency for the detection of alpha particles is independent of energy or emitter and is strictly a function of the geometric efficiency of the source–detector configuration. Thus the question becomes, what is the ideal source–detector configuration.

As discussed in Section III.B.1.a, for a given specific source activity, A_s (Bq/cm^2), the optimum source diameter (from efficiency considerations only) is equal to the detector diameter. However, for a fixed amount of activity (as one would obtain from a given sample), the count rate depends only on the geometric efficiency, which, as Fig. 4.25 shows, increases with decreasing source diameter. The practical ramification of this is that one can increase the counting efficiency by depositing the sample in a smaller diameter. However, as the sample diameter decreases, the sample thickness increases and can cause a decrease in resolution. Thus the minimum sample diameter is constrained by the effect of sample thickness on resolution. For example, a 100-μg sample deposited in a diameter of 1 inch ($\sim$5 cm^2) results in a sample thickness of 20 μg/cm^2, as does a 25-μg sample deposited in a diameter of

1/2 inch ($\sim$1.25 cm^2). With approximately equal sample thicknesses, the resolution of the two samples will be roughly equivalent,[7] but the smaller diameter sample will have a higher counting efficiency. Thus the counting efficiency for low-activity (actually low-mass) samples can be increased by depositing them in a smaller diameter within the constraints imposed by the effects of sample thickness on resolution. The resolution of samples mounted by the fluoride precipitation method is generally acceptable if the sample thickness is kept at 25 μg/cm^2.

However, in practice, the diameter of the sample deposition is not adjusted from sample to sample but rather specified by the sample mounting procedure. The size of the sample deposition, as well as the initial mass of the sample, called for in a procedure should be based on the anticipated sample quantities (concentrations), maximum desired sample thickness, diameter of the detector, and available sizes of commercially available filters, filter holders, electrodeposition disks, and so on. Thus, instead of altering the diameter of the deposition to increase the count rate of low-activity (low-mass) samples, one typically increases the size of the initial sample. Applying these considerations to the practical problem of achieving the lowest possible minimum detectable activity (MDA) (in terms of Bq/g) for a given sample analysis produces a protocol that requires:

1. The diameter of the sample mount should be approximately equal to the diameter of the detector.
2. The amount of initial raw sample to be used in the analysis should be maximized within the constraint that the final mounted sample thickness does not exceed 25 μg/cm^2 (or whatever thickness is demanded by resolution considerations).

Although it has been recommended by Sill and Olson (1970) that "sources should be placed at least 1.5 diameters from the detector to obtain optimum resolution," one needs to appreciate that there is a trade-off between efficiency and resolution. While placing the source closer to the detector causes a decrease in resolution (larger FWHM), it also increases the counting efficiency. As discussed in Section III.B.l.a, the increase in the FWHM at distances as close as 2 mm can be expected to be no greater than 50% (for 300- to 600-mm^2 detectors). Such a decrease in resolution may be deemed tolerable in light of the increase in efficiency so obtained. For low-level counting, it is not unusual to sacrifice resolution in order to increase the counting efficiency.

Another important consideration in alpha counting is the problem of recoil contamination, which can occur when the progeny of the alpha-emitting nuclides being observed are ejected from the sample (due to the kinetic energy of recoil from the initial alpha emission) and become attached to the detector. The short-lived alpha-emitting progeny then contribute to the alpha spectrum.

---

[7]Actually, the smaller diameter sample should produce a slight advantage in terms of spectral tailing, as the maximum entry angle (of alpha particles into the detector) is less for the smaller diameter sample.

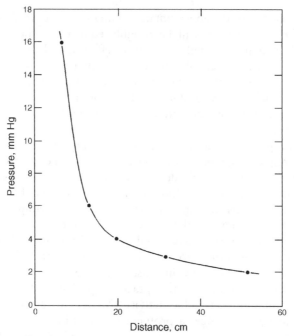

FIGURE 4.38 Relationship between distance and pressure required to stop recoiling ^{221}Fr atoms. (From Sill and Olson, 1970, reprinted with permission of the American Chemical Society, ©1970.)

Sill and Olson (1970) have demonstrated a reduction in recoil contamination "by a factor of at least 10^3 with a loss in resolution of only 1 or 2 keV by leaving enough air in the counting chamber to produce $12\,\mu g/cm^2$ of absorber between the source and detector, and applying a negative potential of 6 volts to the source plate." Figure 4.38 summarizes the relationship between distance and pressure required to stop recoiling ^{221}Fr atoms as described by Sill and Olson (1970), who observed that "the range of the recoiling atoms was between 12 and $16\,\mu g/cm^2$ for all distances checked."

Neither the air layer nor the negative bias on the sample plate individually is sufficient to prevent recoil contamination of the detector. Both the air layer and the negative bias together are required to prevent recoiling daughter atoms from reaching the detector. Today, most commercially available alpha spectrometers provide a readout of the chamber pressure and a negative bias on the sample plate relative to the detector.[8]

In summary, the resolution is improved with

1. A thinner entrance window on the detector.
   Trade-off: None.
2. Thinner sample deposition.
   Trade-off: Larger diameter samples (of the same total activity) have a lower counting efficiency and greater low-energy tailing due to large angle entry of the alpha particles into the detector.

[8]Equivalently, the detector can be biased positive relative to the sample plate.

3. A thinner air column (absorber) as measured in g/cm^2 between the sample and detector.
   Trade-off: Increased recoil contamination if the air column is too thin ($< 12 \, \mu$g/cm^2).
4. A longer (straighter) path between the sample and detector.
   Trade-off: Lower counting efficiency.

2. Analytical Considerations in Gamma Spectrometry

The American National Standards Institute has published a standard that provides guidance in the calibration and use of germanium spectrometers:

> Methods for the calibration and use of germanium spectrometers for the measurement of gamma-ray energies and emission rates over the energy range from 59 keV to approximately 3000 keV and for the calculation of source activities from these measurements are established. Minimum requirements for automated peak finding are stated. Methods for measuring the full-energy peak efficiency with calibrated sources are given. Performance tests that ascertain the proper functioning of the Ge spectrometer and evaluate the limitations of the algorithms used for locating and fitting single and multiple peaks are described. Methods for the measurement of and the correction for pulse pileup are suggested. Techniques are recommended for the inspection of spectral-analysis results for large errors resulting from summing of cascade gamma-rays in the detector. Suggestions are provided for the establishment of data libraries for radionuclide identification, decay corrections, and the conversion of gamma-ray rates to decay rates.[8a]

Typically, an automated gamma spectral analysis requires the following steps:

1. Peak location
2. Calculation of peak areas
3. Correction of peak areas (if required; e.g., subtraction of system background or reference peak correction for random summing losses)
4. Calculation of the efficiency at the peak energies
5. Calculation of activity

Generally, the activity of a sample is calculated in gamma spectrometry from the following equation:

$$\text{Activity(in Bq at } t = 0) = \frac{\text{Net Peak Area} (E)}{T_\text{L} \, \text{Efficiency} \, (E) \, \text{Intensity}_\gamma(E)} \frac{\lambda \, T_R}{1 - e^{-\lambda \, T_R}} \quad (4.57)$$

[8a]ANSI N42.14-1991, copyright © 1991, IEEE. All rights reserved.

where $t = 0$ is the starting time of data acquisition, T_L the live time of the count, T_R the real time of the count, and λ the decay constant:

$$\lambda = \frac{\ln(2)}{T_{1/2}} \tag{4.58}$$

where $T_{1/2}$ is the radionuclide half-life and

$$\frac{\lambda T_R}{1 - e^{-\lambda T_R}} = \text{decay correction for decay during the counting period}^9 \tag{4.59}$$

$$\text{Intensity}_\gamma(E) = \frac{\text{gamma emission rate (at energy } E)}{\text{disintegration rate}} \tag{4.60}$$

and

$$\text{Efficiency}(E) = \frac{\text{full-energy deposition rate (at energy } E)}{\text{gamma emission rate (at energy } E)} \tag{4.61}$$

The efficiency is supposed to express the relationship between the full-energy deposition rate and the emission rate of a sample. Typically, the full-energy deposition rate is approximated by the net count rate, so that the emission rate of a sample is determined by a simple proportional scaling of the observed count rate from a standard of known emission rate as follows:

$$\frac{\text{Emission rate of sample}}{\text{Count rate of sample}} = \frac{\text{(known) emission rate of standard}}{\text{observed count rate of known standard}} \tag{4.62}$$

so that

$$\text{Emission rate of sample} = \frac{\text{count rate of sample}}{\left(\begin{array}{c} \text{observed count rate of known standard}/ \\ \text{(known) emission rate of standard} \end{array} \right)}$$

$$= \frac{\text{count rate of sample}}{\text{efficiency}}$$

In other words, the efficiency is supposed to specify the number of full-energy depositions (of energy E) in the detector per gamma-ray (of energy E) emitted by a source of a given geometry.[10] The efficiency may be determined

---

[9]This correction factor [which essentially converts the nominal count rate (Peak Area/T_L) to the count rate at time $t = 0$] is derived assuming the dead time is constant during the counting period. As such, it is an approximation that is valid only for materials whose half-life is long relative to the count time.

[10]The term geometry is used to indicate the geometric distribution of a source (or sample) relative to the detector, the materials between the source and detector, the size and configuration of the detector, and so forth.

by mathematical methods such as Monte Carlo calculations.[11] More traditionally, it is determined by dividing an observed count rate for a calibration standard of a given geometry by the known emission rate of the standard as indicated above. The assumption inherent in this methodology is that the observed count rate (or total net counts in a given time period) is equal to the full-energy deposition rate of the standard (or total number of full-energy depositions of energy E in a given time period).

Most problems in the quantification of gamma spectra can be traced to a discrepancy between the observed count rate and the full-energy deposition rate. There are many potential causes for such a discrepancy, among which are the following:

- Failure to correct for dead time causes full-energy depositions to be unrecognized.
- Pulse pileup causes full-energy depositions to be unrecognized.
- Random summing causes full-energy depositions to be unrecognized.
- Coincidence summing causes full-energy depositions to be unrecognized.
- Incorrect assessment of peak area produces an incorrect count rate that does not properly represent the full-energy deposition rate.

The basic steps involved in a gamma spectral analysis will now be reviewed.

a. Peak Location

The first step in the analysis is to locate the peaks in the spectrum. This can be accomplished by either a library-driven routine or a search-driven routine.

The library-driven routine uses a list of energies (a library[12]) of peaks for which one wishes to search. It then calculates the net area of the region over which each listed peak, if present, would exist. The area so calculated may then be reported (even if it is negative), whereas other programs first determine whether the net area is statistically significant. The region over which the peak is assumed to exist is usually determined from a "shape calibration," that is, a relationship of the FWHM versus energy and possibly a tailing parameter versus energy.

The ability of a library-driven routine to identify peaks (and nuclides) is limited to the entries in the library. One should also be aware that spectral artifacts (such as the backscatter peak) can produce false-positive peak identifications.

On the positive side, library-driven routines provide the following advantages:

1. The ability to identify small peaks
2. The ability to identify poorly shaped peaks
3. The ability to unfold complex multiplets

[11]With the increas in computer power that has become available in recent years, mathematical calibrations have become more practical and more widely available. Mathematical calibrations are particularly well suited to in situ counting and other geometries for which the production of a calibration standard would be impractical if not impossible.

[12]As the library is typically used again during the analysis process for nuclide identification and activity calculation, it generally includes the nuclide name, the nuclide half-life, and the gamma-ray intensities, in addition to the gamma-ray energies.

This last strength is used to great advantage in a program known as MGA (Gunnink, 1990), which quite arguably epitomizes the capabilities of library-driven routines. It was designed to determine the isotopic abundances of a plutonium sample. A mixture of plutonium isotopes (238–242) produces a spectrum that is too complex to analyze by traditional means. However, knowing that the sample was pure plutonium at one time, all of the potential component nuclides are known and a library of all their energy lines and relative abundances can be specified. By knowing the detailed line shape of gamma and x-ray peaks, one can generate the envelope function for a given mixture of the component nuclides. Essentially, a least-squares fit of the envelope function to the observed spectrum using (among other parameters) the relative abundances of the plutonium isotopes as independent variables then yields the relative abundances of the plutonium isotopes.

A search-driven routine applies some mathematical methodology to the spectral data to distinguish peaks from the continuum distribution. A method that is often employed is to apply a symmetric zero-area transform (often referred to as a sliding transform, sliding filter, digital filter, or filter) to the spectral data. The method was proposed by Mariscotti (1967) and employed in the programs SAMPO (Routti and Prussin, 1969) and HYPERMET (Phillips and Marlow, 1976) and several commercially available programs that followed (e.g., Canberra Industries' GENIE family of spectrometry systems). The transformed spectrum (which can be thought of as a response function) will be zero where the spectrum is constant, nearly zero where the spectrum is slowly varying, and large (either positive or negative, depending on the definition of the transform) in the region of a peak. Thus one merely needs to scan the response for regions that exceed some threshold value to find peak locations in the spectrum. The response will be strongest when the width of the feature in the spectrum (ideally a peak) most closely matches the width of the filter. This has two implications:

- The width of the filter should be chosen to match the expected width of the peaks. This is typically accomplished by use of a shape calibration.
- This algorithm tends to discriminate against features that are both wider and narrower than the filter width. Hence, spectral artifacts in Ge and Si(Li) detector spectra whose width differs significantly from the expected peak width (such as the backscatter peak) tend to be filtered out.

A common error in the use of search-driven routines is failure to match the sensitivity of the peak search routine to the detection limit assumed in the calculation of the minimum detectable activity (MDA) or the lower limit of detection (LLD). These calculations assume that peaks of a given size (relative to the background) can be detected at a given confidence level. This is not necessarily true if the sensitivity (response threshold) is not selected appropriately. This has been recognized and addressed in ANSI standard

N42.14 (1991), which states in Section 5

> If an automated peak-finding routine is used in the spectral analysis, it should be able to find small well-formed single peaks whose areas are statistically significant (above background).

and provides a test for automatic peak-finding algorithms in Section 8.1, which

> ...has been designed to determine how well singlet peaks on a flat baseline that are at or above an "observable" level can be found (i.e., detected) with the peak-finding algorithm.

The standard goes on to state that

> The peak-finding algorithm is expected to find a peak in a spectrum whose area, $A, = L_P \sqrt{[(2.55)(\text{FWHM})y_i]}$, where 2.55 is based on $\pm 3\sigma$ for a Gaussian peak, FWHM is the full width in channels at half maximum of the peak, y_i are the average counts in each baseline channel, and $L_P = 2.33$ corresponds to the value L_P initially suggested for this test.[12a]

The value of 2.33 for L_P was chosen to correspond to the critical level, L_C as defined by Currie (1968) to be the decision limit at which a count is assumed to be detectable with 95% confidence (i.e., $\alpha =$ probability of a false positive = 5%).[13]

Advantages of search-driven routines include the following:

1. The ability to locate peaks even if one did not anticipate their presence.
2. The ability to differentiate between peaks and other spectral features.

A search-driven algorithm should also be able to locate the individual components of a multiplet (two or more peaks that overlap). Depending on the methodology employed, this functionality can be incorporated in the peak-locating routine, however, some other methods can be implemented only during the peak (area) analysis routine. Therefore, the discussion of multiplet deconvolution will be taken up in the next section on peak area analysis.

[12a]ANSI N42.14-1991, copyright © 1991, IEEE. All rights reserved.

[13]Currie's derivation of the limit of detection was based on single-channel (gross counting) considerations and, as such, is not strictly applicable to multichannel analysis; however, it has become common practice to apply the equations and concepts from his derivation to multichannel analysis even though there are additional considerations and uncertainties in multichannel analysis that are not incorporated in these equations. For instance, the probability of a false-positive identification is not strictly a function of the size of the background, as the peak-locating algorithm will not (falsely) identify a peak if the region of the null spectrum under consideration is reasonably flat (regardless of size). Furthermore, the uncertainty associated with the ability to detect a peak as a function of peak shape is not included. As an example, consider two peaks with equal net area at (or slightly above) the critical level. A peak-locating algorithm may detect one and not the other, because its ability to recognize a peak is dependent on the shape of the spectral distribution.

b. Peak Area Analysis

The next step in the analysis is to calculate the net area of the observed peaks. This is typically accomplished either by a summation method or by fitting a function that represents the assumed peak shape to the observed data and reporting the area under the peak function as the net area:

$$\text{Summation:} \quad \text{Area} = \sum_{i=\text{left}}^{\text{right}} y_i - b_i \qquad (4.64)$$

$$\text{Fit:} \quad \text{Area} = \int P(\alpha_1, \alpha_2, \alpha_3, \ldots, x)\, dx \qquad (4.65)$$

where i = channel number, left = the leftmost channel of peak region (to be fitted), right = the rightmost channel of peak region (to be fitted), y_i = number of (gross) counts in channel i, b_i = continuum contribution to channel i, and $P(\alpha_1, \alpha_2, \alpha_3, \ldots, x)$ = the "best fit" mathematical function that models the assumed peak shape. The best fit is typically determined by the method of least squares, which requires that χ^2 be minimized, where χ^2 is defined as

$$\chi^2 = \sum_i w_i [y_i - b_i - P(\alpha_1, \alpha_2, \alpha_3, \ldots, x_i)]^2 \qquad (4.66)$$

where w_i = the weighting applied to the ith point and the α_k are the free parameters of the model.

The fit method is applicable to both singlets and multiplets; the summation method (by itself) cannot assess the contributions from the individual components of a multiplet. Thus, multiplet analysis requires some sort of fit to be performed. Notice that both the summation and fit methods require that the continuum contribution (background) under the peak, b_i, be specified. One way to estimate this contribution is to assume a particular mathematical model for the background and determine its parameters from the channels immediately to either side of the peak.

Two commonly used background models are the linear background and step background (Gunnink, 1979),[14] which may be determined from the spectral data as follows:

$$\text{Linear model:} \quad b_i = \frac{B_\text{L}}{n} + \frac{B_\text{R} - B_\text{L}}{n(N+1)} i \qquad (4.67)$$

$$\text{Step model:} \quad b_i = \frac{B_\text{L}}{n} + \frac{B_\text{R} - B_\text{L}}{nG} \sum_{i=\text{left}}^{i} y_i \qquad (4.68)$$

---

[14]Mathematically, a step function can be expressed by a variety of functions. The function presented here is the one proposed by Gunnink (1979) and used in Canberra Industries' GENIE software.

where

$$B_L = \sum_{i=\text{left}-n}^{\text{left}-1} y_i \tag{4.69}$$

$$B_R = \sum_{i=\text{right}+1}^{\text{right}+n} y_i \tag{4.70}$$

where n is the number of channels to be averaged on each side of the peak to determine the background, N is the number of channels in the peak region, and

$$G = \text{integral of the peak region} = \sum_{i=\text{left}}^{\text{right}} y_i \tag{4.71}$$

For single well-resolved peaks, the linear and step backgrounds produce approximately equivalent results for the area calculation. However, for multiplets, the linear and step backgrounds can produce different results. As shown in Figs. 4.39 and 4.40, the step background is greater than the linear background on the left side of the multiplet and less than the linear background on the right side. For multiplets containing a large component and a small component, the step background places the major portion

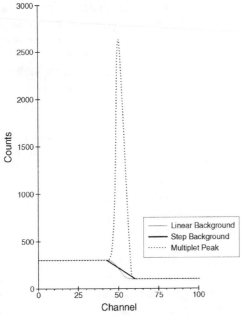

FIGURE 4.39 **A comparison of a step background versus a linear background for a multiplet with the small component on right side.**

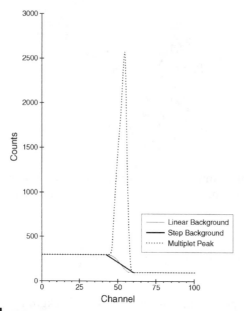

FIGURE 4.40 **A comparison of a step background versus a linear background for a multiplet with the small component on left side.**

of the background change under the major component, and the linear background changes linearly across the peak region. This results in the linear approximation understating (relative to the step approximation) the background under the minor component (thus overstating the net area) when the minor component is on the left side of the multiplet and overstating the background (relative to the step approximation) under the minor component (thus understating the net area) when the minor component is on the right side of the multiplet.

The accuracy of the analysis of multiplet components and the ability to detect the presence of these components are also addressed in ANSI standard N42.14 (1991), as Section 5 states:

> ... The peak fitting routine should be able to find multiplet peaks that meet the peakarea criteria for a singlet, are approximately the same intensity, and are separated by 1 FWHM. (see Section 8 for the test procedures). Optimization of the peak search parameters of the peak-finding algorithm is left to the user or software vendor. These should be adjusted so that statistically significant peaks are found with a minimum number of false peaks being reported (see performance tests in Section 8).[14a]

Section 8.3 of the standard (ANSI N42.14, 1991) provides performance tests for "the Doublet-Peak Finding and Fitting Algorithms." These tests can be used to determine which background function provides the most accurate

---

[14a]ANSI N42.14-1991, copyright © 1991, IEEE. All rights reserved.

multiplet analysis as well as to evaluate an algorithm's overall accuracy in determining the areas of the components of a multiplet. Investigations using such tests have been presented by Koskelo and Mercier (1990) and Mercier and Koskelo (1992), in which several commercially available analysis programs were evaluated.

As stated previously, the ability to detect a multiplet can be incorporated in the peak locate routine or the peak (area) analysis routine. Peak search algorithms that employ a symmetric zero-area transform (and determine the presence of the peak from a large response in the transformed spectrum) to detect the individual components of a multiplet at the peak search stage can generally resolve components of equal size separated by one FWHM or more. As the difference in size of the individual components increases, the separation required to be able to detect the smaller component becomes greater. The individual components of a multiplet can also be detected during the peak fit stage by inspecting the residuals of the fit. After the initial fit, the residuals are examined. The presence of an unresolved component causes the residuals to deviate significantly from zero. Thus, one can add another component (peak) to the fitting function and fit the data again. This algorithm (particularly when used in conjunction with shape information) tends to be more sensitive to the detection of multiplet components than the symmetric zero-area transform; however, some care should be taken in the application of this method, as one needs to distinguish between normal statistical fluctuations and significant deviations.

c. Peak Area Corrections

The next step in the analysis is to correct the observed net area of the peaks for any systematic errors. Such errors can be caused by (but are not limited to)

- Environmental background
- Pulse pileup
- Random summing
- Coincidence summing

Correcting for the presence of environmental background can be done by subtracting the count rate of the environmental background peak from the observed count rate of the sample peak. The count rate of the environmental background peak should be established by counting a sample blank (as opposed to an empty shield), as the sample itself can shield the detector from the source of the environmental background. Counting the empty shield rather than a sample blank biases the environmental background count rate high, resulting in a low bias for the corrected sample count rate.

Pulse pileup, random summing, and coincidence summing all cause an event of energy E to increment a channel that corresponds to an energy $E' > E$. Thus the net area of the peak at energy E does not accurately represent the number of full-energy events of energy E.

For purposes of this discussion, random summing is defined as two independent depositions occurring within AT of one another, where AT is less than the discrimination time of the fast discriminator used by the pileup rejector (PUR) (typically, approximately 500 ns), so that the amplifier pulse appears to be a single event.

Pulse pileup is defined as two depositions occurring within ΔT of one another where ΔT is greater than the discrimination time of the fast discriminator used by the pileup rejector but less than the amplifier pulse width, so that one amplifier pulse starts before the preceding pulse has ended. The resulting pulse is thus distorted in that it has an amplitude and/or width that differs from the first event alone. The distinction between pulse pileup and random summing is that the pileup rejector can tell that the pileup pulse is the result of two events (and thus discriminate against these pulses, whereas it cannot discriminate against pulses resulting from random summing).[15]

Coincidence summing is defined as two depositions originating from a single event within ΔT of one another, where ΔT is less than the discrimination time of the fast discriminator used by the pileup rejector, so that the amplifier pulse appears to be a single event. The distinction between coincidence summing and random summing is that coincidence summing originates from a single event. Examples are, cascade gammas (such as the 1332-keV gamma that follows the 1173-keV gamma following the decay of ^{60}Co), coincident x and γ rays following electron capture (such as the 14-keV x-ray and 1836-keV γ ray following the electron capture decay of ^{88}Y), and coincident emission of γ rays along with the 511-keV annihilation photons following positron emission (such as in the decay of ^{58}Co in which the daughter emits the 810-keV gamma in coincidence with the annihilation photons). A further distinction between random summing and coincident summing is that coincident summing is independent of count rate, but random summing is a function of count rate.

The discrepancy (bias) caused by pulse pileup between the number of full-energy counts (net peak area) and the number of full-energy events can be reduced by the use of a pileup rejector or corrected for in software by use of a reference peak correction. PURs can keep the observed full-energy count rate (net area divided by live time) within approximately 1% of the true full-energy deposition rate for input count rates[16] below approximately 20,000 counts/s. A PUR is ineffective against random summing, as the summed pulse appears to be a single event, against which the PUR cannot discriminate. With a discrimination time of 500 ns, the probability of random summing approaches 1%[17] at rates of approximately 20,000 counts/s.

An alternative to PUR is reference peak correction, which can be used to correct for both pileup and random summing (even dead time if the count

[15]With a further distinction that $\Delta T <$ the linear gate time (LGT) of the ADC in the case of "leading-edge pileup," and LGT $< \Delta T <$ pulse width in the case of "trailing-edge pileup."

[16]Where the input count rate must be defined as the full-spectrum input rate, because an event of any size summing with an event of interest will remove the count from the peak of interest.

[17]Note that this application of a PUR does not imply an overall accuracy or precision of 1%, but rather a limitation on the *bias* caused by just pulse pileup. All of the normal considerations associated with the measurement of a nuclear count rate still apply.

times are measured in real time rather than live time). The corrected peak area is given by

$$A_C = A_O \frac{P_{\text{NOM}}}{P_O} \qquad (4.72)$$

where A_O is the observed net peak area, P_{NOM} the nominal (net) count rate of the reference peak, and P_O, the observed (net) count rate of the reference peak.

The reference peak can be produced by either a pulser or a radioactive source in a fixed position. It should be noted that when a pulser is used to inject a signal into the preamplifier, the preamplifier output may have a shape slightly different from that of the output that results from charge Injection by the detector (i.e., the detection of a gamma-ray). The difference in preamplifier pulse shapes makes it impossible to pole zero the amplifier properly for both pulse types. This will cause a slight degradation in the system resolution.

Both pileup rejection and reference peak corrections are ineffective for correcting for coincidence summing. However, correction factors of the form

$$C = \frac{\text{full-energy events (of energy } E)}{\text{full-energy counts (at energy } E) \text{ (i.e., net peak area)}} \qquad (4.73)$$

can be calculated as presented by Andreev *et al.* (1972, 1973) and McCallum and Coote (1975). Debertin and Schötzig (1979) have extended these equations and incorporated them in the computer program KORSUM, which allows calculation of coincidence summing corrections for arbitrary decay schemes. This program does not include effects due to angular correlations or coincidence with β rays or bremsstrahlung, as the authors considered contributions from these effects to the total summing correction to be low and smaller than the uncertainty of the correction. The authors used the program to calculate correction factors for a point source geometry and a beaker geometry and obtained good agreement with experimental values.

Note that to correct for coincidence summing, the correction factors must be calculated and applied on a line-by-line basis and their calculation requires knowledge of the peak and total efficiencies for the particular detector–geometry combination. The specificity of the correction factors to be determined made it difficult to incorporate the techniques into a commercial analysis package, until now. Within the past few years, however, manufacturers such as Canberra Industries, and Ortec, have developed their own analysis packages that include algorithms to correct for true coincidence summing (or cascade summing) effects. The effect of true coincidence summing and the techniques employed to correct for it are discussed in Section II.B.4. of Chapter 4.

Virtually all commercially available spectrometers include live-time correction and pileup rejection, and some systems also include reference peak correction capability in their software. To date, no commercial supplier of gamma spectrometers has incorporated a coincidence summing correction in any commercially available system.

d. Efficiency Calculation

The next step in the analysis is to calculate the efficiency at the energy of each observed peak. Because it would not be practical to try to measure the efficiency at every possible energy that might be encountered in a sample, most software uses an efficiency calibration to calculate the efficiency for any given energy. Typically, these calibrations consist of a functional expression of efficiency as a function of energy:

$$\text{Efficiency}\,(E) = f(E) \tag{4.74}$$

The functional dependence is typically determined by the method of least-squares fitting or sometimes spline fitting. The energy-efficiency coordinates from which these functions are calculated can be determined by measuring a source of known emission rate or by mathematical calculation, such as with the Monte Carlo method. A typical efficiency calibration from a commercial spectroscopy package is shown in Fig. 4.41.

Care should be taken to ensure that the calibration standard represents the samples to be counted. That is, the calibration standard and samples should be identical in size, shape, density, spatial distribution of active material, and so on. Source position is relatively more critical for close geometries. These precautions apply to mathematical calibrations as well as to those performed with radioactive standards.

e. Nuclide Identification and Activity Calculation

The next step in the analysis is to identify the nuclides that are present in the sample and to calculate their activity. Some simple schemes allow the nuclide identification to be independent of the activity calculation, whereas other schemes depend on the activity calculation to identify a nuclide positively. This will be clarified subsequently by example.

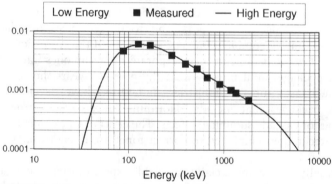

FIGURE 4.41 A typical efficiency calibration as displayed by a Canberra **GENIE-PC** system.

A variety of nuclide identification algorithms have been used in commercial software packages. Following is a list of some of the nuclide identification algorithms that have been used in commercial packages.

1. The simplest algorithm requires that for a particular nuclide to be identified, a peak must exist in the spectrum (within some user-defined energy tolerance) at every gamma-ray energy listed in the library for that nuclide.
2. Another algorithm requires that only a given (user-defined) fraction of the listed gamma-rays of a particular nuclide be observed (as peaks within some user-defined energy tolerance) to identify that nuclide positively.
3. Yet another algorithm requires that only a given (user-defined) fraction of the listed gamma-rays of a particular nuclide be observed (as peaks within some user-defined energy tolerance) to identify the nuclide potentially, its positive identification being dependent on a nonzero activity being calculated for it.

The equation for activity is obtained from the following expression:

Observed net peak area (E)

$$= \int_{t=0}^{t=T_R} A_O\, e^{-\lambda t}\, \text{Intensity}_\gamma\,(E)\, \text{Efficiency}\,(E)(1-DT)\,dt \quad (4.75)$$

where DT is the fractional dead time. If the dead time is assumed to be constant for the duration of the acquisition, one can solve this equation for A_O to obtain Eq. 4.76 (as stated previously and restated here):

$$\text{Activity (in Bq at } t=0) = \frac{\text{Net Peak Area } (E)}{T_L \text{Efficiency}\,(E)\, \text{Intensity}_\gamma (E)} \frac{\lambda T_R}{1-e^{-\lambda T_R}} \quad (4.76)$$

where $T_L = T_R\,(1-DT)$.

This equation has been applied in a variety of ways in commercial packages. One of the simplest nuclide identification–activity calculation methodologies calculates the activity of a nuclide from a single "key line." Following an independent nuclide identification routine (typically, algorithm 1 or 2 from the preceding list of nuclide identification algorithms), the activities of the identified nuclides are calculated from Eq. 4.76.

The most serious deficiency of these simple independent nuclide identification algorithms occurs in spectra that have an interference (i.e., two or more nuclides contributing to a single peak). In particular, these algorithms tend to produce both false-positive identifications and grossly inaccurate quantifications due to the presence of interfering nuclides. This is easily demonstrated by considering the case of ^{75}Se. If one were to analyze a sample containing only ^{75}Se, there would be (at least) five peaks in the spectrum at 121, 136, 264, 279, and 400 keV. With a 1-keV energy tolerance, these peaks would also be identified as ^{57}Co and ^{203}Hg in addition to ^{75}Se.

^{75}Se		^{57}Co		^{203}Hg	
Energy	Intensity	Energy	Intensity	Energy	Intensity
121.117	17.14%	122.061	85.60%		
136.001	58.27%	136.474	10.68%		
400.660	58.5%				
279.544	24.79%			279.197	81.46%
400.660	11.37%				

If one were to calculate the activities of the "identified nuclides" by assigning the entire area of a peak to each nuclide without regard to the interferences, then for every 100 Bq of ^{75}Se the following activities (within statistical fluctuations) would be reported:

^{57}Co: 20.0 Bq based on the 121-keV peak
545.6 Bq based on the 136-keV peak
X Bq where $10.0 < X < 545.6$

If a weighted mean activity is employed

^{203}Hg: 30.4 Bq based on the 279-keV peak
^{75}Se: 100.0 Bq based on any of the lines

Of course, for mixtures the situation is even worse, as none of the nuclides are calculated correctly. To remedy this situation, one of two algorithm is generally employed. The first makes use of an interference library in which one explicitly identifies which lines of which nuclides interfere with what other nuclides. The activity of the nuclide that has lines without interference is then calculated from the peaks for which there is no interference (e.g., the 264- and 400-keV lines of ^{75}Se in the preceding case). This nuclide's contribution to the peaks with which it interferes is then calculated and subtracted from the observed area of these peaks. In the preceding example, the contribution of ^{75}Se to the 121- and 136-keV peaks is calculated from the calculated activity of ^{75}Se and subtracted from the observed areas of the 121- and 136-keV peaks. The balance of the peak areas is then attributed to ^{57}Co, and its activity is calculated from these "corrected" peak areas. Similarly, the contribution of ^{75}Se to the 279-keV peak is calculated from the calculated activity of ^{75}Se and subtracted from the observed area of the 279-keV peak. The remaining area is then attributed to ^{203}Hg and used to calculate the activity of ^{203}Hg. One limitation of this algorithm is that the user must explicitly identify all of the potential interferences before the analysis for this method to idenjtify the interferences that are present.

Another algorithm, originally proposed by Gunnink and Niday (1972) and implemented in SAMP080 (Koskelo *et al.*, 1981) and GENIE-PC (Koskelo and Mercier, 1995), that can resolve this situation involves setting up a simultaneous set of equations that express each potentially identified nuclide's contribution to each observed peak. Using the previous example, in which the observed peaks (at 121-, 136-, 264-, 279-, and 400-keV) cause

^{75}Se, ^{57}Co, and ^{203}Hg to be identified, one would obtain the following set of equations:

$$\text{Int}_{121\text{-Se}} * A_{\text{Se}} + \text{Int}_{122\text{-Co}} * A\text{Co} = \frac{\text{Peak Area (121)}}{\varepsilon(121) * T_{\text{L}}} \qquad (4.77a)$$

$$\text{Int}_{136\text{-Se}} * A_{\text{Se}} + \text{Int}_{136\text{-Co}} * A_{\text{Co}} = \frac{\text{Peak Area (136)}}{\varepsilon(136) * T_{\text{L}}} \qquad (4.77b)$$

$$\text{Int}_{264\text{-Se}} * A_{\text{Se}} = \frac{\text{Peak Area (264)}}{\varepsilon(264) * T_{\text{L}}} \qquad (4.77c)$$

$$\text{Int}_{279\text{-Se}} * A_{\text{Se}} + \text{Int}_{279\text{-Hg}} * A_{\text{Hg}} = \frac{\text{Peak Area (279)}}{\varepsilon(279) * T_{\text{L}}} \qquad (4.77d)$$

$$\text{Int}_{400\text{-Se}} * A_{\text{Se}} = \frac{\text{Peak Area (400)}}{\varepsilon(400) * T_{\text{L}}} \qquad (4.77e)$$

in which A_{Se} is the (unknown) activity of ^{75}Se; A_{Co} the (unknown) activity of ^{57}Co; A_{Hg} the (unknown) activity of ^{203}Hg; $\text{Int}_{121\text{-Se}}$ the gamma intensity of the 121-keV emission from ^{75}Se, and so on; Peak Area(121) is the observed peak area of the 121-keV peak; $\varepsilon(121)$ is the peak efficiency at 121-keV; and T_{L} is the live time of the data acquisition.

This set of five equations in only three unknowns is obviously overdefined; however, a best fit solution can be obtained by minimizing chi squared (the sum of the squares of the residuals[18]). Because each observed emission rate has some uncertainty associated with its measurement, we might prefer to place more weight on the lines that have the smallest uncertainty and less weight on the lines that have the greatest uncertainty. This weighting can be accomplished by multiplying the residual (for each line) by a weighting factor which is the inverse of the variance of the measured emission rate (for that line) so that a weighted chi squared is given by

$$\chi^2 = \sum_{i=\text{lines}} \sum_{j=\text{nuclides}} w_i \left[\frac{\text{Peak Area}(i)}{\varepsilon(i)T_{\text{L}}} - \text{Int}_{ij} A_j \right]^2 \qquad (4.78)$$

This quantity is minimized when the activities, A_j, satisfy the condition

$$\frac{\partial \chi^2}{\partial A_j} = 0 \quad \text{for all } Aj \qquad (4.79)$$

The solutions obtained with this formalism are the weighted average activities for each nuclide. One advantage of this algorithm is that the user

---

[18]The residuals being defined as the differences between the observed emission rate and the emission rate implied by the solution.

is not required to identify explicitly all of the potential interferences before the analysis for this method to identify the interferences that are present. As long as the nuclides that need to be included in the interference set satisfy the identification criteria of the nuclide identification algorithm, their interferences will be recognized by this method.

To take fullest advantage of this algorithm, one should have a very complete nuclide library—in terms of both nuclides and lines of each nuclide-to ensure that all potential nuclides and interferences can be identified. To illustrate the need for a complete library, consider the analysis of a natural soil containing ^{228}Ac, which has a gamma emission of 835.7 keV with an intensity of approximately 1.7%. Because of this low intensity, people often omit it from their library. As a result, ^{54}Mn (which has a gamma emission of 834.8-keV) is reported because there is nothing else to which the 835-keV peak can be attributed. If the 835.7-keV line is included in the library for ^{228}Ac, the 835 keV peak can be (correctly) attributed to the ^{228}Ac and the simultaneous solution can eliminate the (false) identification of the ^{54}Mn.

Another precaution that should be taken when setting up libraries is related to the half-life that is entered for each nuclide. Most commercial software implements corrections for decay during data acquisition, decay back to some sample time, and, if the sampling occurred over an extended period, decay during sample collection. These corrections produce erroneous (high) values for short-lived material in equilibrium with a long-lived parent if the true half-life of the short-lived material is used for the decay correction. If the short-lived material is in equilibrium with the long-lived parent, one can substitute the halflife of the long-lived parent in the decay corrections by using the parent's halflife in the short-lived daughter's library entry.

REFERENCES

Aggarwal, S. K., Duggal, R. K., Shah, P. M., Rao, R., and Jain, H. C. (1988). Experimental evaluation of the characteristic features of passivated ion implanted and surface barrier detectors for alpha spectrometry of plutonium. *J. Radioanal. Nucl. Chem.* **120**, 29–39.

Amoudry, F. and Burger, P. (1984). Determination of the ^{239}Pu/^{240}Pu isotopic ratio by high resolution alpha spectrometry. *Nucl. Instrum. Methods Phys. Res.* **223**, 360–367.

Andreev, D. S., Erokhina, K. I., Zvonov, V. S., and Lemberg, I. Kh. (1972). *Instrum. Exp. Tech.* **25**, 1358.

Andreev, D. S., Erokhina, K. L, Zvonov, V. S., and Lemberg, I. Kh. (1973). *Izv. Akad. Nauk. SSR Ser. Fiz.* 37(8), 1609.

ANSI N42.14–1991. "American National Standard Calibration and Use of Germanium Spectrometers for the Measurement of Gamma-Ray Emmission Rates of Radionuclides," Copyright C) 1991 by the Institute of Electrical and Electronics Engineers, Inc (IEEE). The IEEE disclaims any responsibility or liability resulting from the placement and use in the described manner. Information is reprinted with the permission of the IEEE.

Blaauw, M., (1993). The use of sources emitting coincident γ-rays for determination of absolute efficiency curves of highly efficient Ge detectors. *Nucl. Instrum. Methods Phys. Res., Sect. A* **332**, 493–500.

Bock, R. (1979). "A Handbook of Decomposition Methods in Analytical Chemistry." International Textbook Company.

Briesmeister, J. (2002). MCNP—A general Monte Carlo N-particle transport code. Version 4C. Los Alamos National Laboratory Publication LA-13709-M.

Bronson, F.L. and Young, B.M. (1997). Mathematical Calibrations of Germanium Detectors and Instruments that use them. Proceedings of the 5th Annual NDA/NDE Waste Characterization Conference, Salt Lake City, Utah, January, 1997.

Bunting, R. L. and Kraushaar, J. J. (1974). Short lived radioactivity induced in Ge(Li) gamma-ray detectors by neutrons. *Nucl. Instrum. Methods* **118**, 565–572.

Burger, P., De Backer, K., and Schoemnaeckers, W. (1985). 2nd International Technical Symposium on Optical and Electro-optical Science and Engineering, Nov. 25–29 and Dec. 2–6, 1985, Cannes, France.

Burnett, W. C. (1990). "Alpha Spectrometry: A Short Course Emphasizing the Practical Applications of Alpha Spectrometry." Canberra Industries, Meriden, CT.

Burnett, W. C. (1992). "Advanced Alpha Spectrometry: A Short Course Emphasizing Advanced Techniques in Alpha Spectrometry." Canberra Industries, Meriden, CT.

Cable, P., Burnett, W. C., Hunley, D., Winnie, J., McCabe, W., and Ditchburn, R. (1994). Investigating the chemical and physical controls on electrodeposition for alpha spectrometry. Proceedings, 40th Conference on Bioassay, Analytical and Environmental Radiochemistry, Nov. 13–17, 1994, Cincinnati, OH.

Carchon, R., Van Camp, E., Knuyt, G., Van De Vijver, R., Devos, J. and Ferdinande, H. (1975). A general solid angle calculation by a Monte Carlo method. *Nucl. Instrum. Methods* **128**, 195–199.

Celnikier, L. M., (1996). Cherche source des rayons cosmiques … désespérément. *Bull. Soc. Fr. Phys.* **108**, 6–10.

Ceuppens, M., Verplancke, J., and Tench, O. (1996). Low background germanium detectors; environmental laboratory to underground counting facility. Presented at the workshop on Methods and Applications of Low Level Radioactivity Measurements, Nov. 7–8, D-Rossenclorf-Dresden.

CERN Applications Software Group (1994). GEANT: Detector Description and Simulation Tool. CERN Program Library Long Writeup W5013.

Currie, L. A. (1968) Limits for qualitative detection and quantitative determination: Application to radiochemistry. *Anal. Chem.* **40**, 586–593.

Debertin, K. and Helmer, R. G. (1988). "Gamma- and X-Ray Spectrometry with Semiconductor Detectors." North-Holland, Amsterdam.

Debertin, K. and Schötzig, U. (1979). Coincidence summing corrections in Ge(Li)-spectrometry at low source-to-detector distances. *Nucl. Instrum. Methods* **158**, 471–477.

De Corte, F. and Freitas, C. (1992). The correction for γ-γ, γ-KX and γ-LX true-coincidences in k_o-standardized NAA with counting in a LEPD. *J. Radioanal. Nucl. Chem.* **160**, 253–267.

Ditchburn, R. G. and McCabe, W. J. (1984). An improved method for the purification and electrodeposition of protactinium for application to the INS uranium-series dating project. Institute of Nuclear Sciences (New Zealand) R-325.

Duchêne, G. and Moszynski, M. (1995). Ballistic deficit correction method for large Ge detectors. High counting rate study. *Nucl. Instrum. Methods Phys. Res., Sect* A **357**, 546–558.

Eichrom Industries, Inc. (1994). ACS06: uranium and thorium in soil. Illinois.

Eichrom Industries, Inc. (1995a). ACW03: americium, plutonium and uranium in water. Illinois.

Eichrom Industries, Inc. (1995b). ACW01: uranium and thorium in water. Illinois.

Eichrom Industries, Inc. (1995c). ACW08: thorium and neptunium in water. Illinois.

Ejiri, H. and de Voight, M. J. A. (1989). "Gamma-ray Electron Spectroscopy in Nuclear Physics." Oxford Studies in Nuclear Physics, Clarendon Press, Oxford.

Evans, R. D. (1955). "The Atomic Nucleus." McGraw-Hill, New York.

Friedman, R. J., Reichard, M. C., Blue, T. E., and Brown, A. S. (2001). Evaluation of scatter contribution from shielding materials used in scatter measurements for calibration range characterization. *Health Physics* **80**, 54–61.

Garcia-Toraño, E. and Aceña, M. L. (1981). NOLIN: Nonlinear analysis of complex alpha spectra. *Nucl. Instrum. Methods* **185**, 261–269.

Gardner, R., Verghese, K., and Lee, H. M. (1980). The average solid angle subtended by a circular detector coaxial to a isotopic source. *Nucl. Instrum. Methods*, **176**, 615–617.

Gimore, G. and Hemingway, J. D. (1995). "Practical Gamma-Ray Spectrometry." John Wiley and Sons, New York.

Goulding, F. S. and Landis, D. A. (1982). Signal processing for semiconductor detectors. *IEEE Trans. Nucl. Sci.* **29**, 1125–1141.

Greenwood, R. C. and Chrien, R. E. (1980). Precise γ-ray energies from the ^{14}N(n,γ)^{15}N and ^{23}Na(n,γ)^{24}Na reactions. *Nucl. Instrum. Methods* **175**, 515–519.

Gunnink, R. (1979). "Computer techniques for analysis of gamma-ray spectra." Proceedings, ANS Topical Conference, Computers in Activation Analysis for Gamma-Ray Spectroscopy. CONF-780421, Mayaguez, Puerto Rico, pp. 109–138.

Gunnink, R. (1990). "MGA: A Gamma-Ray Spectrum Code for Determining Plutonium Isotopic Abundances, Vol. 1, Methods and Algorithms." UCRL-LR-103220, Lawrence Livermore National Laboratory.

Gunnink, R. and Niday, J. B. (1972). "Computerized Quantitative Analysis by Gamma-Ray Spectrometry." Lawrence Livermore National Laboratory Rept. UCRL-51061, Vol. 1, 1972.

Helmer, R. G. and Lee, M. A. (1980). Analytical functions for fitting peaks from Ge semiconductor detectors. *Nucl. Instrum. Methods* **178**, 499–512.

Heusser, G. (1993). Cosmic ray induced background in Ge-spectrometry. *Nucl. Instrum. Phys. Res., Sect B* **83**, 223–228.

Heusser, G. (1996). Cosmic ray interaction study with low-level Ge-spectrometry. *Nucl. Instrum. Pbys. Res., Sect A* **369**, 539–543.

Hindman, F. D. (1986). Actinide separations for alpha spectrometry using neodymium fluoride coprecipitation. *Anal. Cbem.* **58**, 1238–1241.

Horwitz, E. P., Dietz, M. L., Nelson, D. M., LaRosa, J. J., and Fairman, W. D. (1990). Concentration and separation of actinides from urine using a supported bifunctional organophosphorus extractant. *Anal. Chim. Acta* **238**, 263–271.

Keyser, R.M., Twomey T.R., and Sangsingkeow, P. (1998). Matching Ge detector element geometry to sample size and shape: One does not fit it all. Proceedings of the 1998 Winter Meeting of the ANS, Nov. 1998.

Keyser, R.M., Haywood, S.E., and Upp, D.L. (2001). Performance of the True Coincidence Correction Method in Gamma Vision, Proceedings of American Nuclear Society 2001 Annual Meeting, Milwaukee, WI, June 2001.

Kirby, H. W. and Sheehan, W. E. (1984). Determination of ^{238}Pu and ^{241}Pu in ^{239}Pu by alphaspectrometry. *Nucl. Instrum. Methods Phys. Res.* **223**, 356–359.

Knoll, G. F. (1989). "Radiation Detection and Measurement." John Wiley and Sons, New York.

Kolotov, V. P., Atrashkevich, V. V., and Gelsema, S. J. (1996). Estimation of true coincidence corrections for voluminous sources. *J. Radoanal. Nucl. Chem.* **210**(1), 183–196.

Koskelo, M. J. and Mercier, M. T. (1990). Verification of gamma spectroscopy programs: A standardized approach. *Nucl. Instrum. Methods Phys. Res., Sect. A* **299**, 318–321.

Koskelo, M. J. and Mercier, M. T. (1995). Verification of gamma spectroscopy programs: Multiple area problems and solutions. *J. Radioanal. Nucl. Chem. Articles* **193**, 211–217.

Koskelo, M. J., Aarnio, P. A., and Routti, J. T. (1981). SAMPOSO: An accurate gamma spectrum analysis method for microcomputers. *Nucl. Instrum. Methods* **190**, 89–90.

Koskelo, M. J., Burnett, W. C., and Cable, P. H. (1996). An advanced analysis program for alpha-particle spectrometry. *Radioact. Radiochem.* **7**(1) 18–27.

Koskelo, M. J., Venkataraman, R., and Kolotov, V. P. (2001). Coincidence summing corrections using alternative detector characterization data. *J. Radioanal. Nucl. Chem.* **248**(2), 333–337.

Kröll, Th., Peter, L., Elze, Th. W., Gerl, J., Happ, Th., Kaspar, M., Schaffner, H., Schremmer, S., Schubert, R., Vetter, K., and Wollerrsheim, H. J. (1996). Analysis of simulated and measured pulse shapes of closed-ended HPGe detectors. *Nucl. Instrum. Methods Phys. Res., Sect A* **371**, 489–496.

Laborie, J.M., Le Petit, G., Abt, D., and Girard, M. (2002). Monte Carlo calibration of the efficiency response of a low-background well-type HPGe detector. *Nucl. Instrum. Methods Phys. Res., Sect. A* **479**, 618–630.

Lally, A. E. and Glover, K. M. (1984). Source preparation in alpha spectrometry. *Nucl. Instrum. Methods Phys. Res.* **223**, 259–265.

L'Annunziata, M. F. (1979). "Radiotracers in Agricultural Chemistry," pp. 345–359. Academic Press, New York.

Leo, W. R. (1987). "Techniques for Nuclear and Particle Physics Experiments." Springer-Verlag, New York.

Lutz, G. (1996). Effects of deep level defects in semiconductor detectors. *Nucl. Instrum. Methods Phys. Res., Sect A* **377**, 234–243.

Mariscotti, M. (1967). A method for automatic identification of peaks in the presence of background and its application to spectrum analysis. *Nucl. Instrum. Methods* **50**, 309–320.

McCabe, W. J., Ditchburn, R. G., and Whitehead, N. E. (1979). "The Quantitative Separation, Electrodeposition and Alpha Spectrometry of Uranium, Thorium and Protactinium in Silicates and Carbonates." R-262, DSIR, Institute of Nuclear Sciences, New Zealand, 29 p.

McCallum, G. J. and Coote, G. E. (1975). Influence of source-detector distance on relative intensity and angular correlation measurements with Ge (Li) spectrometers. *Nucl. Instrum. Methods* **130**, 189–197.

Mercier, M. T. and Koskelo, M. J. (1992). Verification of gamma-spectroscopy programs: accuracy and detectability. *J. Radioanal. Nucl. Chem. Articles* **160**(1), 233–243.

Moens, L. and Hoste, J. (1983). Calculation of the peak efficiency of high-purity germanium detectors. *Int. J. Appl. Radiat. Isot.* **34**, 1085–1095.

Moens, L., De Donder, J., Lin, Xi-lei, De Corte, F., De Wispelaere, A., Simonits, A., and Hoste, J. (1981). Calculation of the absolute peak efficiency of gamma-ray detectors for different counting geometries. *Nucl. Instrum. Methods* **187**, 451–472.

Moens, L., De Corte, F., Simontis, A., Lin Xilei, De Wispelaere, A., De Donder, J., and Hoste, J. (1982). Calculation of the absolute peak efficiency of Ge and Ge(Li) detectors for different counting geometries. *J. Radioanal. Nucl. Chem.* **70**, 539–550.

Müller, G., Wissmann, F., Schröder, F., Mondry, G., Brinkmann, H. J., Smend, F., Schumacher, M., Fettweis, P., and Carchon, R. (1990). Low-background counting using Ge(Li) detectors with anti-muon shield. *Nucl. Instrum. Methods Phys. Res., Sect. A* **295**, 133–139.

Nelson, W.R., Hirayama, H., and Rogers, D.W.O. (1985). The EGS4 code system. Stanford Linear Accelerator, Stanford University, SLAC-265.

Philips, G. W. and Marlow, K. W. (1976). Automatic analysis of gamma-ray spectra from germanium detectors. *Nucl. Instrum. Methods* **137**, 526–536.

Roman, D. (1980). The electrodeposition of thorium in natural materials for alpha spectrometry. *J. Radioanal. Chem.* **60**, 317–322.

Roman, D. (1984). Electrodeposition of radium on stainless steel from aqueous solutions. *Appl. Radiat. Isot.* **35**, 990–992.

Routti, J. T. and Prussin, S. G. (1969). Photopeak method for the computer analysis of gamma-ray spectra from semiconductor detectors. *Nucl. Instrum. Methods* **72**, 125–142.

Semkow, T.M., Parekh, P.P., Schwenker, C.D., Khan, A.J., Bari, A., Colaresi, J.F., Tench, O.K., David, G., and Guryn, W. (2002). Low background gamma spectrometry for environmental radioactivity. *Appl. Radiat. Isot.* **57**, 213–223.

Sill, C. W. (1969). *Health Phys.* **17**, 89–107.

Sill, C. W. (1977). Determination of thorium and uranium isotopes in ores and mill tailings by alpha spectrometry. *Anal. Chem.* **49**, 618–621. (See *Anal. Chem.* **49**, 1648, for correction.)

Sill, C. W. (1980). Determination of gross alpha-strontium, neptunium and/or uranium by gross alpha counting on barium sulfate. *Anal. Chem.* **52**, 1452–1459.

Sill, C. W. (1987a). Precipitation of actinides as fluorides or hydroxides for high-resolution alpha spectrometry. *Nucl. Chem. Waste Manage.* **7**, 201–215.

Sill, C. W. (1987b). Determination of radium-226 in ores, nuclear wastes and environmental samples by high-resolution alpha spectrometry. *Nucl. Chem. Waste Manage.* **7**, 239–256.

Sill, C. W. (1995). Rapid monitoring of soil, water, and air dusts by direct large-area alpha spectrometry. *Health Phys.* **69**, 21–33.

Sill, C. W. and Olson, D. G. (1970). Sources and prevention of recoil contamination of solid-state alpha detectors. *Anal. Chem.* **42**, 1596–1607.

Sill, C. W. and Sill, D. S. (1989). Determination of actinides in nuclear wastes and reference materials for ores and mill tailings. *Waste Manage.* **9**, 219–229.

Sill, C. W. and Sill, D. S. (1994). Simultaneous determination of actinides in small environmental samples. *Radioact. Radiochem.* **5**, (2), 8–19.

Sill, C. W. and Sill, D. S. (1995). Sample dissolution. *Radioact. Radiochem.* **6**(2), 8–14.

Sill, C. W. and Williams, R. L. (1969). Radiochemical determination of uranium and the transuranium elements in process solutions and environmental samples. *Anal. Chem.* **41**, 1624–1632.

Sill, C. W. and Williams, R. L. (1981). Preparation of actinides for alpha spectrometry without electrodeposition. *Anal. Chem.* **53**, 412–415.

Sill, C. W., Puphal, K. W., and Hindman, F. D. (1974). Simultaneous determination of alpha-emitting nuclides of radium through californium in soil. *Anal. Chem.* **46**, 1725–1737.

Small, H. (1989). "Ion Chromatography." Plenum, New York.

Stelson, P. H., Dickens, J. K., Raman, S., and Tramell, R. C. (1972). Deterioration of large Ge(Li) diodes caused by fast neutrons. *Nucl. Instrum. Methods* **98**, 481–484.

Sulcek, Z. and Povondra, P. (1989). "Methods of Decomposition in Inorganic Analysis." CRC Press, Boca Raton, FL.

Talvitie, N. A. (1972). Electrodeposition of actinides for alpha spectrometric determination. *Anal. Chem.* **44**, 280–283.

Turner, J. E. (1986). "Atoms, Radiation and Radiation Protection" Pergamon Press, New York.

Verplancke, J. (1992). Low level gamma spectroscopy: Low, lower, lowest. *Nucl. Instrum. Methods Phys. Res., Sect. A* **312**, 174–182.

Venkataraman, R., Bronson, F., Atrashkevich, V., Young, B.M., and Field, M. (1999). Validation of in situ object counting system (ISOCS) mathematical efficiency calibration software. *Nucl. Instrum. Methods Phys. Res., Sect. A* **442**, 450–454.

Venkataraman, R., and Moeslinger, M. (2001). Using generic detector characterization templates for Cascade Summing Correction, Proceedings of American Nuclear Society 2001 Annual Meeting, Milwaukee, WI, June 2001.

Verplancke, J. (1999). About shapes and geometries of high purity germanium detectors. Presented at the Nuclear Physics Conference in Madrid, Aug. 1999.

Wätzig, W. and Westmeier, W. (1978). ALFUN-a program for the evaluation of complex alpha-spectra. *Nucl. Instrum. Methods* **153**, 517–524

Williams, I. R. (1966). Monte Carlo calculation of source-to-detector geometry. *Nucl. Instrum. Methods* **44**, 160–162.

Wordel, R., Mouchel, D., Altzitzoglou, T., Heusser, G., Quintana, A. B., and Meynendonckx, P. (1996). Study of neutron and muon background in low-level germanium gamma-ray spectrometry. *Nucl. Instrum. Methods Phys. Res., Sect. A* **369**, 557–562.

5

LIQUID SCINTILLATION ANALYSIS: PRINCIPLES AND PRACTICE

MICHAEL F. L'ANNUNZIATA
The Montague Group, P.O. Box 5033 Oceanside, CA 92052–5033, USA

MICHAEL J. KESSLER (DECEASED)[1]

[1]This chapter is dedicated to the memory of Michael J. Kessler, Ph.D. who contributed to the First Edition of the Handbook of Radioactivity Analysis in 1997. Dr. Kessler also provided the author with much encouragement during the planning of that First Edition. His sudden passing in April of 1997 was a great loss to all who knew him and to the world scientific community. He was a dear friend and esteemed colleague.

I. INTRODUCTION

Liquid scintillation counting (LSC) or liquid scintillation analysis (LSA) has been a very popular technique for the detection and quantitative measurement of radioactivity since the early 1950s. The technique has been most useful in studies of the life sciences and the environment. Many of the principles of LSA overlap in the fields of low-level environmental radioactivity monitoring and the measurement of higher levels of radioactivity used in research, radioisotope applications, and nuclear power. However, the techniques and principles used in the LSA of environmental radioactivity per se will not be covered in detail in this chapter. The reader is directed to Chapter 6 for additional information on the use of LSA for the measurement of either natural levels of radionuclides or low levels of man-made radionuclides found in the environment.

Applications of LSA in the measurement of radionuclides used as tracers has lead to a large number of cutting edge and Nobel prize-winning discoveries in the life sciences over the past 40 years. The LSA technique in scientific research remains one of the most popular experimental tools used for the quantitative analysis of radionuclides. These include principally alpha- and beta-particle-emitting atoms; but may also include weak gamma-, x-ray, and Auger electron emitters. Recent advances have been made in the application of liquid scintillation to the analysis of neutrons, gamma radiation, and high-energy charged particles, and a treatment of these advances will be included in this chapter.

The wide popularity of LSA is a consequence of numerous advantages, which are high efficiencies of detection, improvements in sample preparation techniques, automation including computer data processing, and the spectrometer capability of liquid scintillation analyzers permitting the simultaneous assay of different nuclides.

II. BASIC THEORY

A. Scintillation Process

The discovery of scintillation in organic compounds was documented in a thesis by Hereforth (1948) under the leadership of Kallmann as related in a historical account by Niese (1999). In her thesis presented on September 13, 1948 at the Technical University Berlin – Charlottenburg, Hereforth reported that aromatic compounds could convert absorbed energy of nuclear radiation into light photons. Her thesis was followed by papers authored by Kallmann (1950) and Reynolds *et al.* (1950) on liquid scintillation counting (LSC) that demonstrated certain organic compounds in solution-emitted fluorescent light when bombarded by nuclear radiation. Certainly the origin of LSA as a technique for the quantification of radioactivity is attributed to the original papers by Kallmann and Reynolds in 1950. The fluorescence or emission of photons by organic compounds (fluors) as a result of excitation can be readily converted to a burst of electrons with the use of a photomultiplier tube (PMT), and subsequently measured as an electric pulse.

The technique of LSC involves placing the sample containing the radioactivity into a glass or plastic container, called a scintillation vial, and adding a special scintillation cocktail. Samples may also be analyzed by high-sample-throughput LSA in plastic microplates containing 24, 96, or 384 sample wells per microplate, which accept sample–fluor cocktail volumes in the range of 20–150 μL. High-sample-throughput microplate LSC is described in Section XI of this chapter. Common capacities of scintillation vials that can be easily accommodated in conventional automatic liquid scintillation analyzers vary from 4 to 20 mL capacity; however, microfuge tubes of 0.5–1.5 mL capacity can also be counted directly in a conventional LSA with the use of special microtube holders.

Both plastic and glass LSC vials have certain advantages and disadvantages in terms of background, solvent permeability, fragility, and transparency, etc. Polyethylene plastic vials are permeable when stored containing the old traditional fluor solvents such as benzene, toluene, and xylene; however, these vials do not display solvent diffusion when the more environmentally safe commercial fluor cocktails are used (e.g. Ultima Gold[TM], Pico-fluor[TM], Opti-fluor[TM], etc.), which use diisopropylnaphthalene (DIN), pseudocumene, or linear alkylbenzene solvents. The plastic vials are also unbreakable, less expensive, and display lower backgrounds than the glass vials. Glass vials, however, provide the advantage of transparency to visualize the sample and fluor cocktail solution to permit inspection for undesirable properties such as color, residue, or sample inhomogeneity.

The scintillation cocktail is composed of a solvent such as DIN, or a linear alkylbenzene together with a fluor solute such as 2,5-diphenyloxazole (PPO) dissolved in a concentration of approximately $2-10 \, g \, L^{-1}$. The liquid fluor cocktails are available commercially, and these are made to be compatible and mixable with radioactive samples dissolved in either organic solvents or aqueous media. When samples are dissolved in aqueous media, three different chemical components are required in the fluor cocktail solution: the organic solvent, organic scintillator, and surfactant (emulsifier). The choice of solvent, scintillator, and surfactant for the preparation of contemporary fluor cocktails is dictated by the need for efficient energy transfer and light output in the scintillation process even under very high aqueous sample loads exceeding 50% water as well as the need for environmentally safe solutions with low toxicity, high flash point, and low disposal costs. To meet these needs, some commercially available formulations use diisopropylnaphthalene or a linear alkylbenzene solvent. A few of these commercial fluor cocktails were noted in the previous paragraph. The properties and performance of the modern environmentally safe solvents and some of the commercially available cocktails made from these solvents have been reviewed and tested by Takiue *et al.* (1990a), Neumann *et al.* (1991), and Thomson (1991). Chapter 8 provides detailed information on the composition of liquid scintillation fluor cocktails and sample preparation techniques.

The radioactivity in the form of the sample is placed into the scintillation cocktail to form a homogeneous counting solution. The liquid scintillation process that occurs in a scintillation cocktail is shown in Fig. 5.1. The first step in the process is the interaction of the radioactivity with the solvent molecules of the liquid scintillation cocktail. These solvent molecules, as seen in Fig. 5.1, are organic in nature and contain at least one aromatic ring. Because the solvent molecules are in greater concentration than the solute fluor molecules in the fluor cocktail, the solvent molecules will absorb the major portion of the nuclear radiation emitted by the sample in the cocktail solution. The result is the formation of activated organic solvent molecules, which transfer their energy to the organic scintillator or fluor. Organic scintillators are chosen because they are soluble in the organic solvent, they can easily accept the energy from the activated solvent molecule, and they produce an activated or excited scintillator molecule. These excited

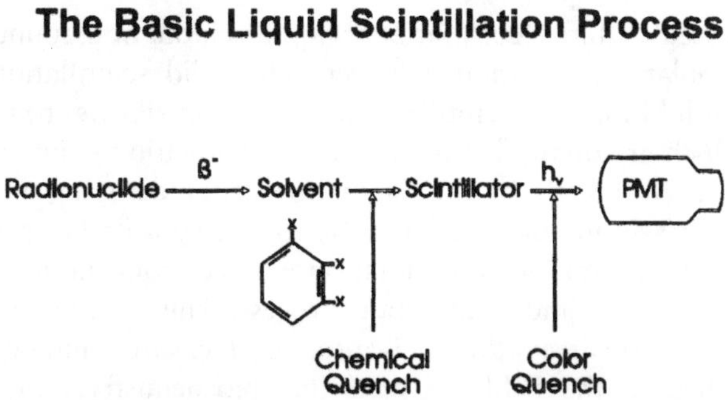

The Basic Liquid Scintillation Process

FIGURE 5.1 An illustration of the sequence of events in the basic liquid scintillation process. A radionuclide will dissipate its energy of decay (e.g., β^--particle energy) in the liquid scintillation cocktail containing solvent and fluor. The aromatic solvent absorbs most of the energy of the beta particle. The energy of excitation of the solvent is then transferred to the scintillator (fluor) molecules, which upon deexcitation emit photons of visible light. The light photons are detected by a photomultiplier tube (PMT), which converts the light photons into a flow of electrons and further amplify the current pulse. Points of interference caused by chemical and color quench are also indicated.

scintillator molecules rapidly lose their energy and return to their original ground state by way of a fluorescent mechanism. The energy is released as a flash of light in the wavelength range of 375–430 nm for each radioactive decay process occurring in the fluor cocktail. The wavelength of emission depends on the scintillator dissolved in the fluor cocktail. The intensity or brightness of the light flash that is produced is a function of the energy and the type of nuclear decay.

The entire process of liquid scintillation counting can be described by using the following analogy. The original nuclear decay energy absorbed in the fluor cocktail can be thought of as a battery; and the fluor cocktail itself can be considered as a light source or lamp fueled by the battery. The amount of energy in the battery cannot be determined by sight, touch, taste, or smell; however, the battery energy will govern the light intensity emitted by the lamp. This is the scintillation cocktail's purpose. It converts the original nuclear decay energy to flashes of light by way of the process shown in Fig. 5.1. The intensity of the light flashes is directly proportional to the original nuclear energy dissipated in the fluor cocktail. The higher the energy, the brighter the resultant light flash. For example, tritium, which is a low-energy beta-particle emitter ($E_{max} = 18.6$ keV), would produce relatively very low intensity light flashes for each beta-particle absorbed in the fluor cocktail, such as dim light from a lamp. However, ^{32}P, which is a high-energy beta-particle emitter ($E_{max} = 1710$ keV) would produce a light intensity approximately 100 times brighter in the fluor cocktail (like a large spotlight). Thus, the light intensity emitted by a scintillation fluor cocktail reflects the original nuclear decay energy, and the number of light flashes per unit time is proportional to the number of nuclear decays in that time unit or, in other words, the sample radioactivity (e.g., disintegrations per minute or DPM).

A liquid scintillation analyzer may also be used to measure the fluorescence produced when radioactive nuclides are adsorbed onto or in close proximity to the surface of a plastic or glass scintillator (solid scintillator) located within a conventional liquid scintillation counting vial or well of a microplate scintillation analyzer. The solid scintillation counting process uses a solid inorganic scintillator (e.g., yttrium silicate) to produce the light flashes, which are quantified by the liquid scintillation counter. The light flashes are produced directly by the interaction of the decaying nuclear event with the inorganic scintillator. The intensities of the light flashes produced are proportional to the energies of the radiation emitted from the nuclear decay, similar to that for the liquid scintillation process. This technique, known as scintillation proximity assay (SPA), is used to measure binding reactions, commonly studied in the fields of medicine, biochemistry, and molecular biology, without the need to separate bound from free fractions. It uses glass or plastic solid scintillation microspheres together with a low-energy-emitting isotope-labeled (^{3}H or ^{125}I) ligand. The method is described briefly in Section XI of this chapter and in more detail in Chapter 11, "Solid Scintillation Analysis," as it is a solid scintillation technique which utilizes a liquid scintillation counter.

B. Alpha-, Beta-, and Gamma-Ray Interactions in the LSC

The scintillation process and light that is produced are different for the alpha, beta, and gamma decay processes. These decay processes are described in detail in Chapter 1. Only a brief treatment is provided here. The alpha decay process is illustrated by Eq. 1.1 and Fig. 1.1 of Chapter 1. During the alpha decay process, a helium nucleus, which is composed of two protons and two neutrons, is released with a specific energy (monoenergetic) from the atomic nucleus. The general decay energy range for alpha particles is 2–8 MeV. When alpha decay occurs in a liquid scintillation cocktail, the alpha particles interact with the fluor cocktail to produce light (approximately 1 photon/keV of original decay energy). The light intensity is converted into an electric pulse of magnitude proportional to the light intensity via a photomultiplier tube described in Section III of this chapter.

If we compare the linear range (R_{cm}) in centimeters of a 5.5 MeV alpha particle from ^{241}Am in water ($R_{cm} = 0.0048$ cm) to the range of a 0.55 MeV beta particle from ^{10}Be in water ($R_{cm} = 0.178$ cm), we see that the alpha particle travels a much shorter distance, only 2.7 hundredths (0.0048/0.178) that of the beta particle, regardless of the fact that the alpha particle possessed ten times the energy of the beta particle (see Chapter 1 for calculations of range and energy for alpha and beta particles). The higher charge and mass of the alpha particle compared with the beta particle are responsible for the reduced range of the alpha particle (see Chapter 1) and less efficient excitation energy transfer to solvent and fluor. Alpha particles produce light in the liquid scintillation cocktail at about one-tenth the light intensity per unit of particle energy of beta particles (Horrocks, 1974). Therefore, in the case of alpha particles, which are monoenergetic, a single pulse height peak is seen for each alpha decay, at a pulse height equivalent to approximately one-tenth its original nuclear decay energy. A 5-MeV alpha particle, therefore, would be detected at approximately 500 keV in a liquid scintillation cocktail. Consequently, the pulse heights of alpha particles and

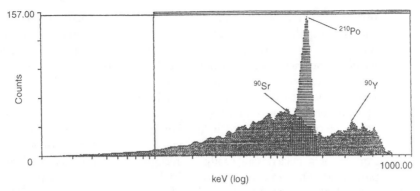

FIGURE 5.2 The overlapping pulse height spectra produced by a mixture of the 5.30 MeV alpha particles of ^{210}Po and the 0.55 MeV beta particles of ^{90}Sr in a PerkinElmer 2750TR liquid scintillation analyzer. The sample contains a mixture of ^{210}Po + ^{90}Sr(^{90}Y) in a scintillation cocktail having 50% water (1:1 mixture of water and Ultima Gold AB^{TM} fluor cocktail) and displayed a tSIE of 277. Notice the relatively sharp peak of the 5.3 MeV alpha-particle pulse height spectrum overlapping with the 0.55 MeV beta-particle pulse height spectrum. The ^{90}Y daughter is in secular equilibrium with its parent ^{90}Sr. The α-peak of ^{210}Po and the β peaks of ^{90}Sr(^{90}Y) appear in a separate α-MCA and β-MCA. The two pulse height spectra of the MCAs are overlapped to demonstrate the overlapping pulse heights produced by the ^{210}Po and ^{90}Sr. (L'Annunziata, 1997, unpublished work.)

beta particles in the same sample often overlap even when the alpha particles emitted from certain radioactive nuclides are of energy about 10 times greater than the E_{max} of beta particles emitted by other radionuclides in the same sample. The overlapping liquid scintillation pulse height spectra of ^{210}Po and ^{90}Sr(^{90}Y) in the same sample are illustrated in Fig. 5.2.

The pulse decay times of the light emissions from alpha and beta events are also different. An alpha pulse in the scintillation process can be about 35–40 ns longer than a pulse event produced by beta particles. Using this characteristic, simultaneous analysis of alphas and betas in the same pulse height energy range can be performed. The discrimination of alpha and beta particles, which produce overlapping pulse height intensities, will be explained briefly in Section XI of this chapter and in greater detail in Chapter 6.

The counting efficiency (how efficiently the nuclear decay is detected) is approximately 100% for almost all alpha decays using a liquid scintillation cocktail. Because of the unique pulse height spectral characteristics of alpha detection in an LSC (see Fig. 5.2) and their slower pulse decay times, alpha particles can be distinguished easily from most other nuclear decay radiations with the liquid scintillation analyzer.

The second and most common radionuclide decay process is the production of a beta particle. Beta decay can take place by either negatron (β^-) or positron (β^+) emission. The production of a negative beta particle (negatron) is described by Eq. 1.16 of Chapter 1 and several examples are given in Eqs. 1.17–1.23 of that chapter. During the common beta decay process, a neutron is converted to a proton and an electron (negative beta particle) and a antineutrino. The beta particle (negatron) is equivalent to an electron in property, and the antineutrino is a particle of zero charge and nearly zero mass. The total decay energy that is released in the beta decay process is shared between the beta particle and the neutrino. This total decay energy is usually expressed as the E_{max}, which is the maximum energy that is released in the decay process. The decay energy is shared between the beta particle and neutrino, but only the beta particle can be detected by the scintillation process. Thus, the resultant spectrum for all beta decays starts at zero and goes to the maximum decay energy (E_{max}) as illustrated in Fig. 1.4 of Chapter 1. Approximately 10 photons of light per keV of beta-particle decay energy are produced in the liquid scintillation process. Because of the broad spectrum of beta-particle energies emitted by a given radionuclide sample, beta decays can easily be recognized by this distinct broad spectral pattern as illustrated in Fig. 1.4 of Chapter 1 on the linear energy scale or Fig. 5.2 of this chapter illustrating the pulse height spectra of $^{90}Sr(^{90}Y)$ on a logarithmic energy scale.

The second type of beta decay produces a positron or positive beta particle. This beta decay process converts a proton to a neutron and a positively charged electron (positron) accompanied by the emission of a neutrino. Positron emission is described by Eq. 1.29 of Chapter 1, and an example is provided by Eq. 1.30 of the same chapter. The positron is an antiparticle of an electron; it possesses an opposite charge and a spin in the opposite direction to that of the electron. The total energy released in the positron decay process is shared between the positron and the neutrino similar to the negatron decay process. The positron will lose its kinetic energy in matter via ionization. When it comes to a near stop, it comes into contact with an electron, its antiparticle, and is annihilated with the simultaneous production of two gamma-ray photons of 0.51 MeV energy equivalent to the two annihilated electron masses. See Chapter 1 for more detailed information on both the negatron and positron decay processes.

The liquid scintillation counting efficiency for beta particles (negatrons or positrons) is dependent on the original energy of the beta decay. For most beta particles with a decay energy above 100 keV, the counting efficiency is 90–100%, but for lower energy beta decays the efficiency is normally in the range of 10–60% depending upon the degree of quench in the sample. The phenomenon of quench and its effect on liquid scintillation counting efficiency are described in Sections IV and V of this chapter.

Another common nuclear decay process is gamma-ray emission. In this process, a gamma ray is emitted from the nucleus of the decaying atom. The gamma ray is electromagnetic radiation or, in other respects, a photon particle. The general energy range for gamma rays is 50–1500 keV. Gamma-ray emission often accompanies alpha, beta, or electron capture (EC) decay processes. Bremsstrahlung or x-radiation, which is electromagnetic radiation originating from electron energy transitions, also accompany the EC decay process. When gamma-emitting radionuclides are detected by the liquid scintillation counter, it is not the gamma ray that is detected to a very significant degree, but rather the alpha particles, beta particles, or atomic electrons (Auger and internal-conversion electrons) that may be produced during decay process occurring in the liquid scintillation fluor cocktail. Gamma rays from sample radionuclides in the scintillation cocktail can produce Compton electrons, although these interactions are less significant in magnitude in the liquid fluors. In general, electromagnetic radiation makes only a minor contribution to excitation in liquid scintillation fluor compared to charged-particle radiation. For example, if we consider the liquid scintillation analysis of ^{125}I, which decays by electron capture with the emission of gamma rays and daughter x-radiation, liquid scintillation counting efficiencies as high as 85% are reported. However, the excitations in the liquid scintillation fluor are due mainly to the absorption of Auger and internal-conversion electrons and only a minor contribution ($\sim$8%) is the result of x-rays produced during the decay process (L'Annunziata, 1987).

C. Cherenkov Photon Counting

Beta particles of energy in excess of 263 keV can be detected and quantified in water or other liquid medium using the liquid scintillation analyzer without the use of scintillation fluor cocktail. The sample is simply placed in a clear liquid solution (often aqueous) and detected by the light produced by the Cherenkov effect. Charged particles, such as beta particles, that possess sufficient energy can travel at a velocity exceeding the speed of light in media such as water, organic solvents, plastic, and glass. When this occurs, the charged particle will produce Cherenkov photons, which extend from the ultraviolet into the visible wavelengths. The light that is produced is low intensity and is normally detected in the low-energy counting region of 0–50 keV. High-energy beta-particle emitters, which emit a significant number of beta particles in excess of 263 keV, can be analyzed by counting the Cherenkov photons in the liquid scintillation analyzer without fluor cocktail. Some examples are ^{32}P ($E_{max} = 1710$ keV), ^{90}Sr(^{90}Y) where the E_{max} of ^{90}Y beta particles is 2280 keV, ^{86}Rb ($E_{max} = 1770$ keV occurring at an 88% intensity (probability per decay) or 680 keV at a 12% intensity) and ^{89}Sr ($E_{max} = 1490$ keV). The Cherenkov counting efficiency of these radionuclides is in the range of approximately 35–70% depending on color quench. The process of Cherenkov counting is treated in detail in Chapter 9. In general, it is important to remember that when quantifying radionuclides by Cherenkov counting, the counting region should be set to a lower energy (0–50 keV) to encompass only the low pulse height spectra produced by Cherenkov photons, and no fluor cocktail is required.

III. LIQUID SCINTILLATION COUNTER (LSC) OR ANALYZER (LSA)

As described previously, the scintillation process involves the conversion of nuclear decay energy into light flashes. Therefore, to quantify the nuclear decay event and to satisfy needs for automation and multiple user programs, an LSC must be able to perform the following functions: (1) it must be able to detect light flashes that occur in the scintillation vial with fluor cocktail or solid scintillator (SPA) and be able to determine the number of light flashes and their intensity; (2) it must be able to hold a large number of scintillation vials (>400) of various sizes (e.g. 20, 8, 7, and 4 mL and microfuge or Eppendorf tubes); (3) it must have the ability to process automatically various types of samples using different counting conditions

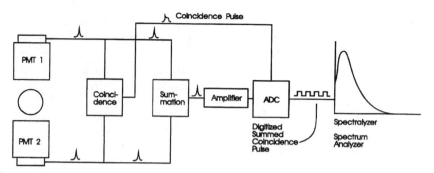

FIGURE 5.3 Schematic diagram of the components of a contemporary liquid scintillation analyzer (from Kessler, 1989).

and counting programs (e.g., single radionuclides, multiple radionuclides, quench corrections, direct DPM, Cherenkov counting) using programmable counting setups or counting protocols; (4) it must be able to process the data from flashes of light per minute to counts per minute (CPM) and then convert these count rates to actual nuclear disintegration rates or nuclear decay events per minute (DPM) using a quench correction method or direct DPM method, and (5) it must perform data analysis and reduction, special computer-managed data application programs, and instrument performance assessment and be able to assist in the diagnosis of instrumentation problems.

The first and most important task of the LSA is the detection and quantification of the number of light flashes and their corresponding intensities. This is accomplished by the heart of the LSC, the light detection and quantification components. A simple block diagram of the LSC is illustrated in Fig. 5.3. Three basic components are found in this part of the LSC, namely, the detector(s), a counting circuit, and a sorting circuit.

In order to quantify the radioactivity in the sample, the sample is loaded into the counting chamber using either an up- or downloading elevator mechanism. The downloading mechanism has the basic advantage of being able to prevent any external light from entering the counting chamber by using a double light seal mechanism. The double light seal is implemented by automatic loading of the sample vial from the sample chamber deck to a holding area, where the sample is sealed from external light. The sample is subsequently moved into the counting chamber, which is below the holding area. Because of this unique downloading mechanism, the photomultiplier tube (PMT) high voltage can remain on at all times and the PMT background stabilized. Once the sample has been loaded into a light-tight chamber, the light is detected using two photomultiplier tubes.

The PMTs convert the light photons emitted from the liquid scintillation vial to electrons when the light photons hit a bialkalie photocathode located inside the face of the PMT as illustrated in Fig. 5.4. The electrons produced at the PMT photocathode are amplified through a series of positively charged dynodes, each dynode having an increasing positive voltage along the series. The increasing voltage accelerates the initial photoelectrons produced at the PMT photocathode to yield an avalanche of electrons, resulting in a pulse amplification. In the PMT in Fig. 5.4 a series of 12 dynodes are illustrated,

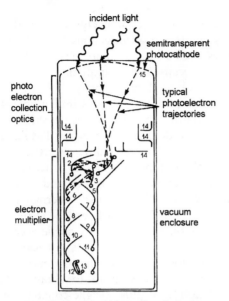

1 - 12 dynodes 14 focussing electrodes
13 anode   15 photocathode

FIGURE 5.4 Schematic of a photomultiplier tube (courtesy of **SCIONIX HOLLAND B.V.,
3981 LA Bunnik, The Netherlands).**

whereby the final electron amplification is collected at the anode of the PMT. The light, which is produced in the scintillation vial, is thereby converted to a corresponding electronic signal. Because the amount of light produced in the scintillation vial is normally very low (10 photons per keV energy absorbed in the liquid fluor cocktail), the PMT must be able to amplify the light by a large factor. This amplification factor is approximately 10 million times for the standard PMT used in the LSC.

As illustrated in the block diagram of Fig. 5.3, two PMTs are used for the measurement of the light intensity from the nuclear decay processes in the sample vial. The two PMTs permit coincidence light detection and coincidence pulse summation required for the LSC to be able to detect low-energy radionuclides such as tritium ($E_{max} = 18.6$ keV) and to distinguish instrument background from true nuclear events. If only a single PMT were used in the LSC, the background level would be approximately 10,000 CPM for a 0–2000 keV counting region. This high background is normally due to the large amplification factor from the PMT that is applied to the signal resulting from any light flashes emitted from the scintillation vials. This high background count rate mainly occurs in the 0–10 keV region (thermal and electronic background noise). In the LSC, two PMTs and a coincidence circuit are used to help differentiate background signals from true nuclear decay events in the scintillation vial, which is referred to as coincidence counting. The principle behind coincidence counting is based upon the fact that, when a nuclear decay event occurs in the scintillation vial, light is produced which is isotropic (i.e., is emitted equally in all directions). Since the decay process and resultant scintillation process produce multiphoton events (about 10 photons per keV of nuclear energy dissipated in a liquid scintillation cocktail), light is emitted in all directions from the scintillation vial. The decay process and resultant scintillation are very rapid (approximate light decay time is 2–10 ns). Because the scintillation process produces multiphoton events and the events decay rapidly, we can distinguish most background from true nuclear decay in the scintillation vial. If light is produced in the scintillation vial inside the analyzer detection area, it will be emitted in all directions and be detected by the two PMTs in the very short pulse decay time of 2–10 ns. If a signal is detected in both the PMTs within a coincidence resolving time of 18 ns, it is accepted as a true nuclear decay event. If on the other hand, a background event occurs in one of the PMTs or in the electronic circuitry (e.g., thermal or electronic noise), it will produce only a single event, which will be detected by only one of the two PMTs in the 18 ns time frame. Such a single event is rejected as occurring external to the sample or, in other words, a background event. By using two PMTs and the coincidence circuit, the instrument background is reduced from 10,000 CPM with a single PMT to about 30 CPM with two PMTs for a wide-open 0–2000 keV pulse height counting region. The PMT signal that is sent to the coincident circuit is an analog signal with a pulse height that reflects and is proportional to the original nuclear decay energy.

The next part of the detection area illustrated in Fig. 5.3 is the summation circuit. This circuit has a dual purpose. The first is to reassemble the original two coincident signals into an individual signal with the summed intensity. This helps to optimize the signal-to-noise ratio in the instrument. The second purpose is to compensate for the light intensity variations due to the position of the nuclear decay in the vial that would occur when samples containing color are counted. If only one of the two PMT signals were used in counting a colored sample, the signal height would be dependent on where in the scintillation vial the light was produced. If the light was produced near the edge of the scintillation vial, a brighter flash of light would be detected by the PMT that is in closer proximity to that edge of the vial. However, with two PMTs and a summed signal, the final pulse height produced by the PMT is not affected by the position of the nuclear decay in the presence of color in the sample counting vial.

Subsequent to pulse summation in the LSC, the signal is further amplified and sent to the analog-to-digital converter (ADC). The ADC converts the signal from an analog signal, which is a pulse with a certain height, to a single number that represents its pulse height or intensity. The digital pulses are finally sorted on the basis of their magnitude or pulse height number. The sorting can be accomplished by one of two methods: pulse height analysis (PHA) or multichannel analysis (MCA).

PHA, which is the older of the two methods, utilizes only two discriminators, an upper and a lower energy discriminator. An upper level discriminator is set such that all of the pulses with a certain energy of interest are always lower than this upper level. A lower level discriminator is also set to help reduce background and other counting interferences of low magnitude. When an event is detected, its pulse height is measured; if it has an intensity lower than the upper discriminator and higher than the lower discriminator, it is accepted as a true nuclear event. If any of the pulses fall outside this range, they are rejected and lost by the counting circuitry. All of the pulses that fall into the accepted range are counted, hence the term liquid scintillation counting (LSC).

The second and more contemporary method of sorting pulses is MCA. The MCA is a series of bins or slots, where different pulse height magnitudes are placed once they have been detected. Two types of MCAs are commonly used: linear and logarithmic. The linear MCA provides data with pulse heights calibrated to represent decay energy in keV on a linear scale. For a common 4000 channel linear MCA, each channel may represent approximately 0.5 keV of energy. The logarithmic MCA displays the pulse heights in channels plotted along a logarithmic scale as illustrated in Fig. 5.2.

All of the pulses collected in MCAs are not only counted but analyzed in terms of their number and height; therefore, the liquid scintillation counter is now more often referred to as the liquid scintillation analyzer (LSA). A linear MCA output with a typical beta-particle pulse height spectrum is illustrated in Fig. 5.5.

The second function of the modern LSA is to move and count various types and sizes of sample vials containing scintillation fluor cocktail. Most modern LSAs are cassette based. This means that sample vials are placed in racks holding between 12–18 individual scintillation vials or samples. Specific cassettes are available for holding scintillation vials of different sizes. With a large-vial (20 mL) scintillation counter, vials and/or sample holders of most other sizes can be counted.

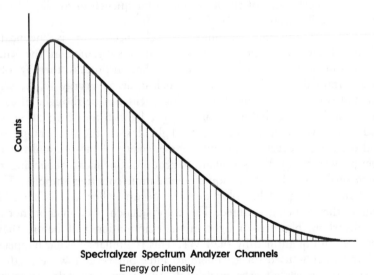

Spectralyzer Spectrum Analyzer Channels
Energy or intensity

FIGURE 5.5   Illustration of a typical liquid scintillation beta-particle pulse height spectrum collected in the many channels of a multichannel analyzer (MCA). A typical linear MCA will have as many as 4000 channels calibrated over the energy range of 0–2000 keV. (Courtesy of PerkinElmer Life and Analytical Sciences, Boston, MA.)

Many persons can use the same instrument by establishing counting protocols to analyze different radionuclides under different counting conditions and sample sizes. The key functions of the LSA — analyzing sample data by determining sample quench levels, converting count rate (e.g., counts per minute or CPM) to disintegration rate (e.g., disintegrations per minute or DPM) for unknown samples, and automatic monitoring of the performance of the instrument (instrument performance assessment) — will be described in detail later in this chapter.

IV. QUENCH IN LIQUID SCINTILLATION COUNTING

In scintillation counting the sample is either dissolved in a liquid scintillation cocktail or adsorbed onto a solid scintillator in a sample vial or microplate well. In order to quantify the nuclear events as activity in terms of DPM, the LSA counts the number of flashes of light in a preselected time period to provide a count rate (CPM) of the sample. The sample count rate is dependent on how efficiently the nuclear decay events are converted to light flashes that are detected and quantified by the LSA. Because the sample solution is always present, it can absorb nuclear decay energy thereby preventing this energy from being absorbed by the chemical fluor molecules, or the solution can absorb photons of light that are emitted by the scintillation cocktail. This causes the phenomenon called quench. We can define quench as interference with the conversion of decay energy to photons emitted from the sample vial. Quench can be the result of two common causes: (1) the presence of chemicals in the fluor cocktail that are mixed with the sample and (2) a colored substance that comes from the sample. The points of interference of chemical and color quench in the liquid scintillation process are illustrated in Fig. 5.1.

The first and most common quench mechanism is chemical quench. Chemical quench is caused by a chemical substance in the sample that absorbs nuclear decay energy in the scintillation process, thereby, obstructing to a certain degree the transfer of nuclear decay energy to the scintillation cocktail solvent. A chemical quenching agent can be thought of as a sponge that absorbs energy before it can produce light in the scintillation process. In addition to reducing the number of light flashes, the quenching process can and often does decrease the apparent intensity of the original nuclear decay energy as seen by the scintillation process. Chemical quenching occurs to some degree in almost all liquid scintillation counting samples. The second mechanism of quench, color quench, occurs when color is visible in the sample that is being counted. The color quench phenomenon normally acts by absorbing photons of light in the scintillation vial before they can be detected and quantified by the PMT. This is similar to what happens when a colored filter is used on a camera to filter out certain wavelengths of light. Chemical quench absorbs nuclear decay energy and color quench absorbs photons of light. These quenching phenomena reduce the number of counts per minute (CPM) of the sample that are detected by the LSA. In order to compensate for quench and determine the sample activity or DPM (nuclear decay rate), it is necessary to know the counting efficiency defined by the following equation:

$$\% \text{ efficiency} = \text{CPM/DPM} \times 100. \qquad (5.1)$$

where CPM is the count rate of the sample determined by the LSA and DPM is the actual sample disintegration rate. The relationship between CPM and DPM of the sample varies according to the energy of the original nuclear decay at a given degree of quench. The lower the energy of the decay, the greater is the effect of quench on the counting efficiency for beta-emitting radionuclides. This is illustrated in Fig. 5.6, which shows the liquid scintillation pulse height spectra of seven tritium samples ($E_{max} = 18.6$ keV) that were prepared with the same activity (DPM), but with different amounts of 0.5 M HNO_3, which acts as a chemical quenching agent. The LSA determined the CPM for each sample by summing the area under the pulse height spectrum of each sample. As illustrated in Fig. 5.6, the least quenched sample is that which contains no HNO_3. The area under the pulse height spectrum of the first sample had 126,287 CPM and the highest pulse heights with a maximum equivalent to approximately 18.6 keV. The counting efficiency for this sample is calculated as

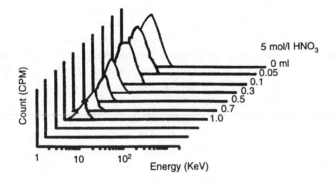

HNO_3 (0.5M) (mL)	^{3}H (CPM)	^{3}H (DPM)	^{3}H (%E)
0.00	126,287	210,000	60.1
0.05	115,834	210,000	55.1
0.10	102,218	210,000	48.7
0.30	61,211	210,000	29.1
0.50	39,846	210,000	19.0
0.70	25,239	210,000	12.0
1.00	16,091	210,000	7.7

FIGURE 5.6 Pulse height spectra of seven samples of ^{3}H of equal activity containing varying amounts of 0.5M HNO_3 quenching agent. The pulse height spectra are plotted on a logarithmic scale with pulse height calibrated to equivalence keV energy. The liquid scintillation counting (detection) efficiencies for each sample are listed as percentages. (Courtesy of PerkinElmer Life and Analytical Sciences, Boston, MA.)

126,287 CPM/210,000 DPM or 60.1%. The second sample is quenched by the added 0.05 mL of HNO_3, and as a result (1) the CPM is reduced to 115,834, (2) the endpoint or maximum intensity of the pulse height spectrum of this sample is reduced, and (3) the counting efficiency of the sample is reduced to 55%. As illustrated in Fig. 5.6, when the sample is quenched more and more, the maximum observed pulse height is reduced further and the CPM collected under the pulse height spectrum is reduced. For example, the last sample listed, which contains the highest amount of quenching agent, gave the lowest count rate of 16,091 CPM and a calculated counting efficiency of only 7.7%. Thus, as the quench increases for tritium, both the maximum pulse height and the total CPM are reduced significantly. We can conclude that chemical quenching agents, although dilute and small in quantity, can have a significant effect on the counting efficiency of tritium.

On the other hand, an isotope such as carbon-14, which emits beta particles of energy almost 10 times higher ($E_{max} = 156$ keV) than tritium, quenching agents cause a significant reduction in the maximum pulse heights but have a less significant effect on the pulse counts collected than was observed in the case of tritium. Table 5.1 shows the effect of the quenching agent on five samples of carbon-14. The five samples contained the same activity (100,000 DPM), but increasing amounts of quenching agent. The quenching agent is not given here, but a common quenching agent used for these studies is nitromethane over the range of 0–100 μL per 20 mL of fluor cocktail. The endpoint of the pulse height spectra (maximum pulse height expressed in keV) of each sample listed in Table 5.1 changed significantly from sample 1 to sample 5 as chemical quench increased; however, the sample count rates (total counts collected under the pulse height spectra per given period of time) did not change as drastically as for tritium. As can be seen from Table 5.1, pulse height spectral intensity (maximum pulse height), changes as the sample is quenched, but the efficiency or CPM value (area of energy spectrum) changes only slightly. The overall conclusion is that for beta-particle emitters, the lower the energy (E_{max}) of the beta decay, the greater is the effect of quench on the counting efficiency of the radionuclide.

For alpha-emitting radionuclides the phenomenon of quench does not significantly effect the counting efficiency as shown in Fig. 5.7. As the quench of the sample is increased, the monoenergetic alpha peak is simply shifted to

TABLE 5.1 Effect of Quench on Carbon-14 Counting Efficiency in Liquid Scintillation Analysis

Sample	Maximum pulse height (keV)	CPM	DPM	Efficiency (%)
1	156	95,000	100,000	95.0
2	112	94,500	100,000	94.5
3	71	92,500	100,000	92.5
4	43	90,500	100,000	90.5
5	32	87,000	100,000	87.0

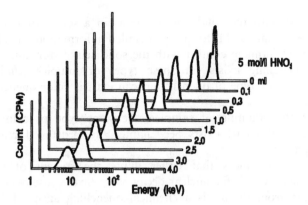

HNO$_3$ (0.5M) (mL)	^{241}Am (CPM)	^{241}Am (DPM)	^{241}Am (%E)	^{241}Am %FWHM
0.00	46,660	46,700	100	42
0.05	46,632	46,700	100	a
0.10	46,692	46,700	100	a
0.30	46,149	46,700	99	47
0.50	46,234	46,700	99	a
0.70	46,371	46,700	99	a
1.00	46,394	46,700	99	a
1.50	46,448	46,700	99	53
2.00	46,256	46,700	99	a
3.00	46,148	46,700	99	a
4.00	46,080	46,700	99	a

^{a}% FWHM not calculated.

FIGURE 5.7 Pulse height spectra of 11 samples of ^{241}Am of equal activity containing varying amounts of 0.5M HNO$_3$ quenching agent. The pulse height spectra are plotted on a logarithmic scale with pulse height calibrated to equivalence keV energy. The liquid scintillation counting (detection) efficiencies for each sample are listed as percentages. The alpha peak resolutions are measured as percent full width at half-maximum. (Courtesy of PerkinElmer Life and Analytical Sciences, Boston, MA.)

lower pulse heights, but the total area under the pulse height spectrum or detection efficiency (equivalent to counting efficiency) is not changed significantly. Also, as illustrated in Fig. 5.7, the alpha-particle pulse height spectrum in the LSA undergoes some peak broadening (reduced resolution) proportional to the level of quench; although this will have no significant effect on detection efficiency. The resolution is determined by the percent full width at half maximum according to Eq. 11.27 of Chapter 11. Vera Tomé *et al.* (2002) studied alterations in alpha-peak shape in liquid scintillation with the potential of utilizing LSA for alpha spectrometry.

For gamma emitters, the quenching phenomenon is very similar to that observed with beta emitters (Ishikawa and Takiue, 1973). See Section VI of this chapter for a treatment on the liquid scintillation analysis of gamma-emitting radionuclides.

The effect of quench using solid scintillators in an LSA is shown in Figure 11.36 of Chapter 11. When using solid scintillators, the sample is normally placed directly on the solid scintillator and dried or the sample is counted adsorbed onto the solid scintillator as in a scintillation proximity assay (SPA). See Chapter 11 for more detailed information on SPA. The sample is in direct intimate contact with the solid scintillator, and therefore no chemical quenching exists for these types of samples. Under these circumstances the only type of quench that can exist is color quench from a colored sample.

For Cherenkov counting of samples, the same type of color quench without any chemical quench exists. See Chapter 9 for a detailed treatment of Cherenkov counting.

All chemical substances that either dilute the solvent of the fluor cocktail or compete with it for nuclear decay energy will cause quench. Even dissolved oxygen from the air is a chemical quenching agent (Takiue and Ishikawa, 1974); its effect can be seen in the LSA of weak (low-energy) beta particle-emitting radionuclides such as tritium. More information on chemical quenching agents and their classifications can be obtained in a previous book (L'Annunziata, 1987). As chemical quenching agents in the samples we analyze generally cannot be avoided and the effect of quench on detection efficiency is significant with many radionuclides, it is important to correct for quench when necessary. This will permit accurate measurement of sample activities in disintegration rate (e.g., DPM).

V. METHODS OF QUENCH CORRECTION IN LIQUID SCINTILLATION COUNTING

Because some type of quenching exists in almost all types of samples that are quantified by the scintillation counting process, it is important to understand the methods that can be used to correct for quench. These methods allow us to relate and even convert the count rate (CPM) to the actual number of nuclear decays or disintegration rate (DPM) of a sample. This can be accomplished by one of the following methods: (1) internal standard method, (2) sample spectrum method, (3) external standard method, and (4) direct DPM method. Each of these methods can be used for quench correction and DPM determination. Each has distinct advantages for various sample types and/or radionuclides. These will be discussed subsequently together with explanations of the when and why of using these techniques.

A. Internal Standard (IS) Method

The internal standard (IS) method is the oldest and most tedious, and it can be the most accurate method if great care is taken in its implementation. The technique involves a series of steps for each sample. The first step is to count each sample and obtain an accurate count rate (CPM) value for each. Then the samples are removed from the LSC, and a known activity (DPM) of a radionuclide standard is added to each sample; hence, the term internal standard is applied to this technique. After the addition of the internal standard and thorough mixing of the standard and sample, the samples are recounted to obtain the CPM of the sample plus the internal standard. Once the CPM of the sample and the CPM of the sample plus internal standard are obtained, the following equation is applied to determine the counting efficiency of the sample:

$$E = \frac{C_{s+i} - C_s}{D_i} \qquad (5.2)$$

where C_{s+i} is the count rate of the sample after the addition of the internal standard, C_s is the count rate of the sample before the addition of the internal standard, and D_i is the disintegration rate of the added aliquot of internal standard. The disintegration rate of the sample, D_s, may then be calculated as follows:

$$D_s = C_s/E \qquad (5.3)$$

For example, if the counting efficiency for a given sample was found to be 0.25 according to Eq. 5.2 and the sample count rate was found to be 25,000 CPM, the activity of the sample can be calculated to be 25,000 CPM/ 0.25 = 100,000 DPM.

Several assumptions and restrictions are made for the internal standard method some of which may be intuitively obvious. These are described as follows: (1) The same radionuclide must be used for the internal standard as the sample radionuclide; for example, a tritium-labeled standard must be used with samples containing tritium. Hendee *et al.* (1972) showed that [^{3}H]toluene and [^{3}H]hexadecane are good internal standards for organic-compatible fluor cocktails and [^{3}H]water or [^{3}H]hexadecane serve well for aqueous-compatible fluor cocktails when assaying for tritium. The organic standards labeled with ^{14}C are good internal standards for counting efficiency determinations of samples containing ^{14}C. (2) The internal standard added to the sample must have a count rate at least 100 times that of the sample. (3) The addition of the internal standard to the sample must not alter the quench in the sample to any significant degree. (4) The activity (DPM) of the internal standard must be accurately known, as with a National Institute of Standards and Technology (NIST) traceable standard. The ^{3}H and ^{14}C standards noted above are available from PerkinElmer Life and Analytical Sciences, Boston, Massachusetts for use as liquid scintillation internal standards. (5) This method of determining sample activities requires accurate sample transfer procedures, which can be tedious when working with many samples and small volumes of internal standard. Dobbs (1965) and Thomas *et al.* (1965) have investigated syringe dispensing techniques for the addition of internal standards to samples in scintillation counting vials.

The internal standard method is used most often for environmental samples (low-count-rate samples) where the counting times of samples are long compared to counting times of the samples with internal standard. This method, if performed properly, is the most accurate of all the quench correction methods. The major disadvantages of this technique are the time and the number of sample-handling steps required for each sample.

B. Sample Spectrum Characterization Methods

Sample spectrum characterization methods of quench correction involve the use of some characteristics of the sample spectrum as a measure of quench in the sample. Some of these methods are described subsequently.

I. Sample Channels Ratio (SCR)

The sample channels ratio (SCR) method was applied often during the early generations of liquid scintillation counters that were equipped with only the PHA or single channel analyzer for data storage and analysis. Nevertheless, the method is applicable with most of the commercial LSAs today. It also remains as a useful method for modern LSAs not equipped with external standards, and the SCR method has other applications described further on in this chapter. The method involves counting the sample in two counting regions defined by lower-level (LL) and upper-level (UL) pulse discriminator settings. The count rate in each counting region varies according to the level of quench in the sample due to the pulse height spectral shift from higher to lower magnitudes caused by sample quench. An example of the pulse height shift according to quench level is illustrated in Fig. 5.8. As illustrated, a sample that is more highly quenched will produce pulse events of lower magnitude (height) than a sample that is lesser quenched.

The SCR quench correction method requires firstly defining the widths of two counting regions also referred to as counting channels or windows. The lower and upper discriminator levels of one region are selected so as to provide a narrow counting region, which can register pulses of only low magnitude (e.g., Channel 1, 0–300 of Fig. 5.8). The discriminator levels of the second counting region are set to provide a wider counting region, which can register most of the pulses of both low and high magnitude (e.g., Channel 2, 0–700 of Fig. 5.8).

A shift in pulse height due to quenching produces a change in the ratio of the pulses registered (counts) by the two regions. The degree of spectral shift and magnitude of change in the sample channels ratio (SCR), such as CPM_1/CPM_2 or sample count rate in Channel 1 over the sample count rate in Channel 2, are dependent on the severity of quench. Consequently, if a series of quenched standards consisting of scintillation vials each containing the same amount of radioactive standard but increasing amounts of quenching agent were counted, they would show a variation in the channels ratio and counting efficiency, such as that illustrated in Fig. 5.9. The procedures used to prepare sets of quenched standards are described in Section V.D of this chapter. The quench correction curve, once prepared for a given radionuclide and fluor cocktail, may be used as a standard curve for determining the counting efficiency of a sample from its channels ratio. The values of counting efficiency for the standard curve are calculated according to

$$E = C_{std}/D_{std} \tag{5.4}$$

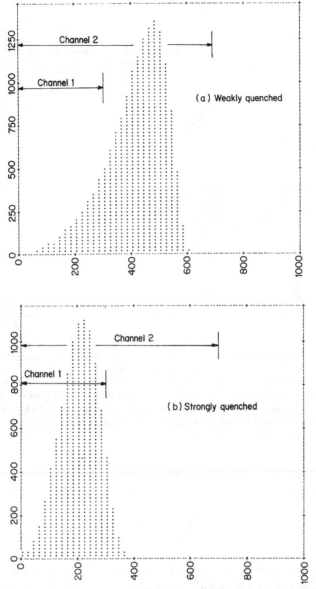

FIGURE 5.8 a) Weakly-quenched and (b) strongly-quenched pulse height spectra produced by ^{33}P in relation to two overlapping counting regions (Channels 1 and 2) of a Beckman LQ 7800 liquid scintillation analyzer. The counting channels 1 and 2 are defined by lower- and upper-level discriminator settings of 0–300 and 0–700, respectively. (L'Annunziata, 1986, unpublished work.)

where E is the counting or detection efficiency with values between 0 and 1.0, C_{std} is the count rate of the quenched standard in units of counts per minute (CPM) or counts per second (CPS), and D_{std} is the disintegration rate of the quenched standard in units of disintegrations per minute (DPM) or disintegrations per second (DPS). The activities of the unknown samples are

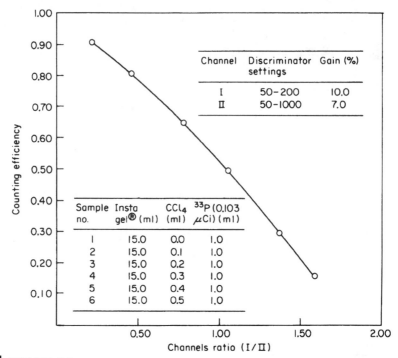

FIGURE 5.9 Typical channels ratio quench correction curve. The channels ratio (I/II) represents the count rate of ^{33}P from Channel I divided by the count rate from channel II. The discriminator and gain settings for each channel are given. The data were obtained from six samples each containing 15 mL of commercial scintillation cocktail (Insta Gel) and 1.0 mL of [^{33}P]orthophosphate of known activity (0.103 μCi equivalent to 3.81 kBq). Each sample contained increasing amounts of quenching agent (CCl$_4$) as described in the table inset. (L'Annunziata, 1986, unpublished work.)

determined from the count rate of the sample in the wider channel divided by the detection efficiency obtained from the SCR quench correction curve or

$$D_s = C_s/E \qquad (5.5)$$

where D_s is the disintegration rate of the sample, C_s is the count rate of the sample in the wide-open channel (i.e., the wider channel from which the detection efficiencies of the quenched standards were determined), and E is the detection efficiency obtained from the SCR quench correction curve.

A more detailed treatment of this method can be found in reviews by L'Annunziata (1984, 1987). The method is less often used with modern LSAs due to the advent of MCAs in commercial LSA instrumentation, which utilize sample spectrum quench-indicating parameters or external standard quench correction methods. Also the SCR technique is generally not useful with samples of low count rate or high quench, because the counts in one or both of the channels may be so low that a channels ratio becomes meaningless, or long periods of counting time would be required to achieve acceptable levels of statistical accuracy.

2. Combined Internal Standard and Sample Channels Ratio (IS-SCR)

Dahlberg (1982) devised a combination of the IS and SCR methods (IS-SCR), which ameliorates the disadvantages of the two techniques. The high dependence on accurate dispensing of internal standards in the IS technique and the high errors encountered at low count rates in the SCR technique have been eliminated in the combined IS-SCR method for quench correction. In the combined IS-SCR method, the disadvantage of the SCR method at low count rates is avoided by the addition of internal standard (IS) to low radioactivity samples. The SCR values are then taken for quench correction, instead of calculating the efficiency by the "classical" IS method of measuring the contribution to count rate by the known amount of standard added. As only an SCR value is required after adding an internal standard, the dependence of the "classical" IS method on accurate dispensing of standard to sample is also eliminated.

A similar combined IS-SCR technique was devised by McQuarrie and Noujaim (1983) for the counting efficiency determinations of either ^{3}H, ^{14}C, or both nuclides as a mixture. The unique characteristic of this method is the use of ^{67}Ga as the internal standard for either ^{3}H, ^{14}C, or the dual nuclide mixture. The liquid scintillation pulse-height spectrum of ^{67}Ga is characterized by two peaks (Fig. 5.10) corresponding to 8 keV Auger electrons and 93 keV conversion electrons, which are similar in energy to the average beta-particle energy of 5.7 keV for ^{3}H and 49 keV for ^{14}C. A ratio of the measured activity of the two ^{67}Ga peaks is used to reflect the degree of quenching in the sample. The sample is easily recovered after the internal standard ^{67}Ga decays ($t_{1/2} = 78$ h) and accurate dispensing of the internal standard to sample is not required, because only the ratio of activity between the two peaks is used to monitor quench.

3. Sample Spectrum Quench-Indicating Parameters

With the development of the multichannel analyzer (MCA), sample spectrum quench-indicating parameters (QIPs) have become more sophisticated, as all of the channels of the MCA can be used simultaneously to measure quench. Examples of quench-indicating parameters that measure quench by sample spectrum characterization are the spectral index of the

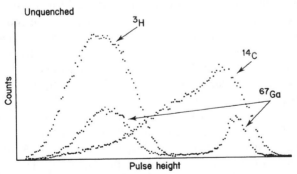

FIGURE 5.10 Liquid scintillation spectra of ^{3}H, ^{14}C, and ^{67}Ga (From McQuarrie and Noujaim, 1983.)

sample (SIS), the spectral quench parameter of the isotope, SQP(I), and the asymmetric quench parameter of the isotope, AQP(I).

a. Spectral Index of the Sample (SIS)

The SIS is a measure of the mean pulse height or center of gravity of the sample pulse height spectrum, which is utilized in the Tri-Carb LSAs of PerkinElmer Life and Analytical Sciences. The pulses produced from photon events are linearly amplified, digitized, and stored in an MCA to produce a complete sample pulse height spectrum in a region of pulse heights calibrated to represent the energy scale from 0 to 2000 keV. The SIS is a measure of the first moment of the pulse height spectrum proportional to the average energy of the beta spectrum times a factor K or

$$ \text{SIS} = K \frac{\sum_{X=L}^{U} X \cdot n(x)}{\sum_{X=L}^{U} n(x)} \tag{5.6} $$

where X is the channel number (see the beta-particle pulse height spectrum with respect to the numerous channels of the MCA in Fig. 5.5), $n(x)$ is the number of counts in Channel X, L and U are the lowest and uppermost limits of the pulse height spectrum, and K is a factor, which fixes the SIS of unquenched ^{3}H and ^{14}C at 18.6 and 156, respectively, corresponding to the maximum beta-particle energies of ^{3}H and ^{14}C in keV. Therefore, the SIS reflects the endpoint or maximum energy of the sample pulse height spectrum as well as the magnitude and shape of the spectrum. From Eq. 5.6 we see that the value of SIS is (1) unitless, (2) always greater than 1.0, (3) becomes smaller in magnitude as quench increases for a given radionuclide, and (4) at a given level of quench, beta emitters of higher E_{max} will produce higher values of SIS.

An example of count rate (CPM) and quench-indicating parameter (SIS) data collected for a series of ten quenched tritium standards is given in Table 5.2. This data was collected by the LSA when the instrument counted each tritium quenched standard to provide a count rate (CPM) for each standard, which is listed in column 2 of Table 5.2. After the count rate of each standard is obtained, the LSA measures the QIP of each standard, in this case SIS, according to Eq. 5.6. The next step required for the preparation of the quench correction curve is the calculation of the percent counting efficiency for each standard according to Eq. 5.1. The instrument makes this calculation by taking the CPM (column 2) and dividing by the DPM (column 3) of each quenched standard and multiplying by 100 to obtain the percent counting efficiency.

The data of counting efficiency and quench-indicating parameter, SIS, listed in Table 5.2 is then taken automatically by the instrument to plot the quench correction curve for tritium illustrated in Fig. 5.11. Another quench correction curve for ^{14}C is also plotted in Fig 5.11. The ^{14}C quench correction curve was prepared in a fashion similar to the procedure described using ^{14}C quenched standards. Figure 5.11, therefore, illustrates plots of the

TABLE 5.2 Data Collected for the Preparation of a ³H Quench Correction Curve of Counting Efficiency versus the Quench-indicating Parameter SIS

Standard	CPM	DPM	Efficiency (%)[a]	SIS
1	68,000	100,000	68	18.6
2	64,000	100,000	64	16.0
3	58,000	100,000	58	14.8
4	52,000	100,000	52	13.6
5	48,000	100,000	48	12.0
6	38,000	100,000	38	11.0
7	29,000	100,000	29	10.5
8	23,000	100,000	23	9.2
9	18,000	100,000	18	8.5
10	13,000	100,000	13	8.0

[a]The % efficiency here refers to the % counting efficiency calculated according to Eq. 5.1. For the calculation of sample activities from count rate the decimal equivalent of % counting efficiency is used (e.g., 0.68 for 68%).

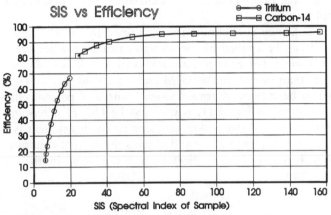

FIGURE 5.11 Quench correction curves for ³H and ¹⁴C based on the quench-indicating parameter SIS, a sample pulse height spectrum characterization method.

quench correction curves for two radionuclides on the same graph. Several observations can be made from these two curves. The first observation is that for ¹⁴C, which is a beta-particle-emitting radionuclide of intermediate energy ($E_{max} = 156$ keV), quench has a marked effect on the endpoint or maximum energy, as the SIS decreases from 156 to 25). However, the count rate (area under the pulse height spectrum of each standard) or counting efficiency (CPM/DPM) decreases only slightly (0.95–0.83) as illustrated in Fig. 5.11. Therefore, for midrange to higher-energy beta-particle-emitting radionuclides, quench does not have a marked effect on the counting efficiency of the sample as on the apparent endpoint energy. The second observation is related to the tritium quench correction curve. In the case of tritium both the pulse height spectrum endpoint and the counting efficiency are dramatically reduced as a result of quench. The curve of percent counting efficiency versus SIS is very steep for tritium. This dramatic slope can result in a rather large error in DPM values, if accurate SIS values are not obtained. Also, it is intuitively obvious that the spectrum characterization method of determining the quench-indicating parameter is dependent on the counts in the sample. The larger the number of counts, the more accurate is the measurement of sample spectrum quench parameter (e.g., SIS). From these observations, it is clear that the sample spectrum characterization method of determining the QIP should be used only when mid- to high-energy radionuclides are being quantified and when the count rate of the sample is well above background (> ~1000 CPM). We shall see further on in this chapter that quench-indicating parameters derived from an external standard are more versatile and applicable to samples of both low and high activity (Section V.C). However, quench-indicating parameters derived from the sample spectrum are particularly useful when external standards cannot be applied such as in color quench correction for Cherenkov counting as demonstrated by L'Annunziata and coworkers (see Noor *et al.*, 1996a). The SIS is also a valuable tool in spectrum unfolding for the analysis of a mixture of two beta-particle-emitting radionuclides (L'Annunziata, 1997b) described further on in this chapter.

Once a quench correction curve is plotted by the LSA and stored in its memory, it can be applied by the LSA to calculate the activity (DPM) of an unknown sample. For example, an unknown sample is counted and the LSA provides a count rate of 36,000 CPM and a SIS value of 12. The radionuclide is known to be tritium. A tritium quench curve of percent efficiency versus SIS, as illustrated in Fig. 5.11, is used by the LSA to determine the percent counting efficiency of that unknown sample. The instrument is programmed to read the stored quench curve and obtains the percent counting efficiency of 48% from the curve. The sample activity is calculated by the LSA according to the equation

$$DPM_s = \frac{CPM_s}{E} \qquad (5.7)$$

where DPM_s is the sample activity in disintegrations per minute, CPM_s is the count rate of the unknown sample, and E is the counting efficiency obtained from the quench correction curve as a decimal, not as a percent. The value of E should generally be in the range between 0 and 1.0, as the decimal representation of the percent counting efficiency over the range of 0–100%. Therefore, in this example, the instrument calculates the activity of the unknown sample as 36,000 CPM/0.48 and the resultant value of 75,000 DPM is obtained. The LSA can perform this type of analysis for all samples of unknown activities.

b. Spectral Quench Parameter of the Isotope Spectrum or SQP(I)

The spectral quench parameter of the isotope or SQP(I) is also referred to as the mean pulse height of the isotope spectrum (Rundt, 1991). It is utilized

quench indicating parameter common to all external standard sources, namely, the external standard channels ratio (ESCR) technique.

I. External Standard Channels Ratio (ESCR)

The external standard channels ratio (ESCR) technique for the determination of counting efficiencies is similar to the SCR method described previously. The principal difference is that the channels ratio produced by the external standard Compton pulse height spectrum is utilized rather than that produced by the sample pulse height spectrum. The ESCR was once a very popular quench-indicating parameter before the advent of modern LSAs equipped with more versatile QIPs, such as $H\#$, SQP(E), and tSIE; however, the ESCR method is still used with the older generation instruments.

The ESCR technique offers the advantage that the optimum channel widths (counting regions) and gains for the channels ratio determinations are often factory set to monitor the scintillation events produced by Compton electrons, that result from the interaction of gamma rays of the external standard with the scintillation cocktail and vial wall. In the previously described SCR technique, the channel widths and gain settings which produce the best (most linear) quench correction curve must be determined experimentally and will differ from radionuclide to radionuclide.

In practice, the ESCR quench correction curve is prepared firstly by counting a series of variable quenched radionuclide standards in a preselected counting region from which counting efficiency values are obtained. After each of the above counts are obtained, an additional count is made for each sample exposed to the external gamma-ray source. The external standard counts are collected in two other preselected counting channels, and the net count rate in these two channels due to the external standard is computed by subtracting from both channels those pulses or counts contributed by the sample nuclide. The ESCR is obtained from the external standard count rates in these two channels. A plot of counting efficiency versus ESCR is then made as illustrated in Fig. 5.12.

Since the channels ratio in the ESCR method arises from count produced by an external source, the ratio determination does not suffer from poor statistical accuracy for samples with low count rates, as does the SCR method. However, the ESCR technique has certain disadvantages among which are (1) the quench correction curves are dependent on sample volume, (2) the quench correction curves display a greater difference for color and chemical quenching, and (3) at least one minute additional counting time is required to count each sample exposed to the external standard. Wigfield and Cousineau (1978) found excellent agreement between counting efficiencies and various combinations of chemical and color quenching. They advise that the user investigate his/her own counting system and scintillation cocktail to evaluate the acceptability of this technique within acceptable error for samples with chemical and color quenching before discarding the use of this technique.

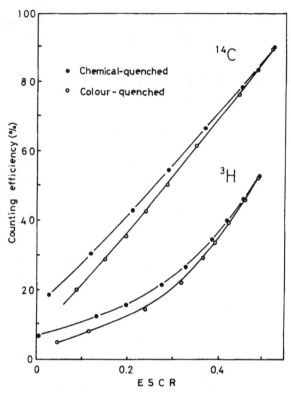

FIGURE 5.12 Color and chemical quench correction curves based on the external standard channels ratio (From Takiue *et al.*, 1983, reprinted with permission from Elsevier Science).

2. H-number (H#)

The *H*-number (*H*#) as a quench-indicating parameter was first proposed by Horrocks (1976a,b, 1977, 1978a), and it remains today a popular method for quench correction in liquid scintillation analysis with Beckmann Instruments (L'Annunziata, 1987).

The technique involves the irradiation of liquid scintillation counting vials containing standards in scintillation cocktail, varying in their degree of quench, with an external radionuclide source $^{137}Cs(^{137m}Ba)$ emitting monoenergetic gamma radiation. The radiation reaching the scintillation vials and samples consists exclusively of 0.662 MeV gamma rays, as the 0.032 MeV x-rays from ^{137m}Ba are absorbed by the source container. Through the Compton effect, the gamma radiation produces a spectrum of Compton scatter electrons of varying energies between zero and E_{max} in the scintillation fluor cocktail.

The spectrum of energies of the Compton electrons are constant from sample to sample. However, the scintillation photon intensities and concomitant pulse heights produced by the Compton electrons will vary depending on the amount and type of quenching agent in each sample. The Compton scatter electrons produce a spectrum of pulse events as illustrated in

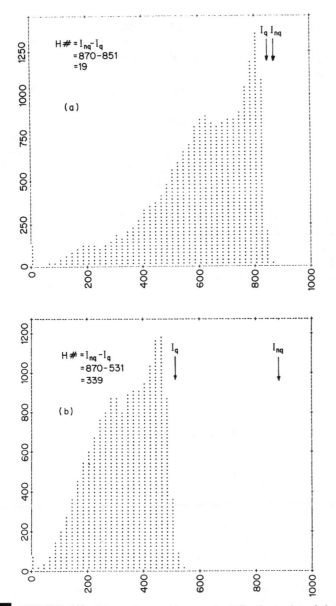

FIGURE 5.13 **Effect of quenching on the inflection point of the external standard** ^{137}**Cs Compton edge for (a) a weakly quenched sample** ($H\# = 19$) **and (b) a strongly quenched sample** ($H\# = 339$) **in a Beckman LS 7800 liquid scintillation analyzer (L'Annunziata, 1986, unpublished work).**

Fig. 5.13. If only those Compton scatter electrons with energy E_{max} are considered, these would produce pulse heights of maximum magnitude as a peak referred to as the Compton edge. The magnitude of the pulse height spectrum at the Compton edge is maximum for a sample free of quenching agents and saturated with nitrogen gas (nonquenched sample). The Compton edge of quenched samples is encountered at lower pulse heights than that of the nonquenched sample, and the degree of spectral shift is a function of the amount of quench in the sample. A measure of the degree of spectral shift or difference in E_{max} pulse response is called the H#, and it is a measure of the amount of quench in a sample.

In practice, the H# concept is applied to the quench correction of samples counted with LSAs with logarithmic pulse height conversion. Such systems convert the initial pulse-height response to the logarithm of the pulse height. Thus, initial pulse responses which may differ by a factor of 1000, may be handled by a single amplifier and pulse-height analyzer and, as reported by Horrocks (1978b), there is a constant logarithmic difference between response relationships at different quench levels. For example, a 50 percent reduction in photon yield or an increase in quench by a factor of two represents a constant difference of 0.301 or the logarithm of two between the logarithmic response relationships of different quench levels. This is illustrated in Fig. 5.14.

With logarithmic response relationships, the measured pulse height, H, may be defined using the notation of Horrocks as

$$H = a + b \log E \qquad\qquad (5.9)$$

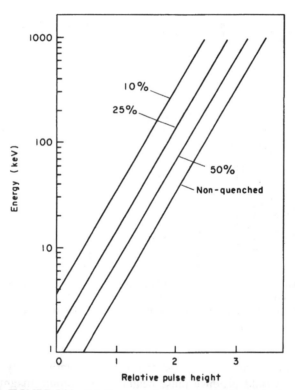

FIGURE 5.14 Relative quench effect on the logarithmic response for different electron energies. Curves are marked to indicate logarithmic response at quench levels of 50, 25, and 10% compared to the pulse-height response from nonquenched scintillation media. Curves marked '25%', '50%' and 'nonquenched' are separated by a constant value of 0.301 relative pulse height (From Horrocks, 1980).

where H is commonly expressed in discriminator division, a is the pulse-height response for a $1 \, keV$ electron, and b is the slope of the energy response curve. For different levels of quench, the slope b will remain constant, but the value a will differ. For a nonquenched sample the measured pulse height, H_o, may be defined as

$$H_o = a_o + b \log E \tag{5.10}$$

Likewise, for a quenched system the measured pulse height, H_q, may be written as

$$H_q = a_q + b \log E \tag{5.11}$$

The difference between the measured pulse-height responses for electrons of the same energy (e.g., E_{max} from ^{137}Cs Compton edge) in nonquenched and quenched systems is defined as $H\#$ or

$$H\# = H_o - H_q \tag{5.12}$$
$$= a_o - a_q + b \log E - b \log E \tag{5.13}$$
$$= a_o - a_q \tag{5.14}$$

The $H\#$ is determined by taking the difference between the relative pulse heights at the inflection points of ^{137}Cs Compton spectra of nonquenched and quenched samples, as illustrated in Fig. 5.13. In the examples presented in Fig. 5.13, $H\#$ values of 19, and 339 are illustrated. Greater and lesser degrees of quench will result in a corresponding variation in the magnitude of the $H\#$. The inflection points of the quenched and nonquenched samples are determined automatically by multichannel analysis while exposing the samples to an external $^{137}Cs(^{137m}Ba)$ source. The MCA divides the pulse height scale into narrow channels and accumulates the counts in each channel over a given period of time. A microprocessor then compares the counts in each channel to define the Compton spectra and precisely locate the inflection points.

With certain liquid scintillation spectrometers, for example, those of Beckmann Instruments, the inflection point of the external standard Compton edge produced by a nonquenched fluor cocktail is factory set at 870 discriminator units. Quenched samples produce Compton edges with inflection points at lower discriminator levels, and the magnitude of the difference defines the $H\#$ (see the examples in Fig. 5.13).

In practice, a standard curve is prepared to relate counting efficiency to $H\#$. This requires the preparation of a set of standards in liquid scintillation vials containing the same and known activity (DPM) of radionuclide and increasing amounts of quenching agent in scintillation fluor cocktail (see Section V.D for procedures for preparing quenched standards). These standards are then counted in optimal region settings (LL and UL discriminator settings) and the counting efficiency ($E = cpm/dpm$ or cps/Bq) for each quenched standard is plotted against the $H\#$ as illustrated in

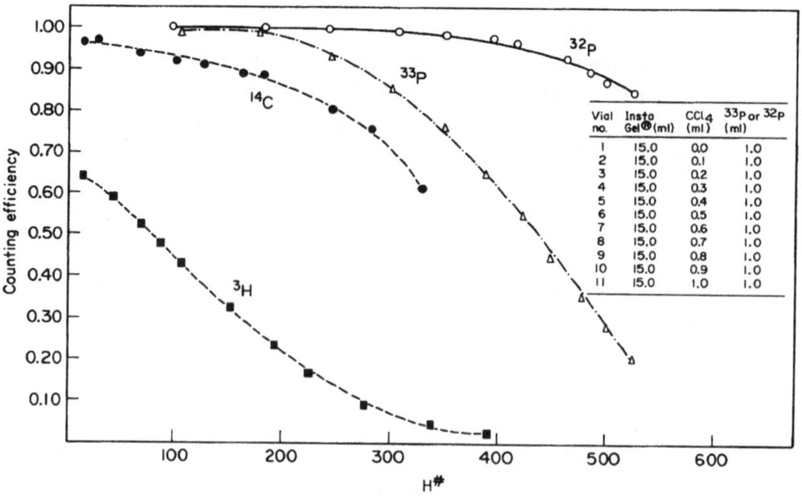

Vial no.	Insta Gel® (ml)	CCl_4 (ml)	^{33}P or ^{32}P (ml)
1	15.0	0.0	1.0
2	15.0	0.1	1.0
3	15.0	0.2	1.0
4	15.0	0.3	1.0
5	15.0	0.4	1.0
6	15.0	0.5	1.0
7	15.0	0.6	1.0
8	15.0	0.7	1.0
9	15.0	0.8	1.0
10	15.0	0.9	1.0
11	15.0	1.0	1.0

FIGURE 5.15 Quench correction curves for ^{3}H, ^{14}C, ^{33}P, and ^{32}P based on the quench-indicating parameter *H#*. The ^{3}H and ^{14}C plots were obtained from commercially obtained quenched standards. Those of ^{33}P and ^{32}P were obtained from standards prepared with Insta Gel scintillation cocktail and CCl_4 quenching agent according to the table inset (L'Annunziata, 1986, unpublished work).

Fig. 5.15. The counting efficiency of specific radionuclide samples of unknown activity are determined from their *H#* and the standard curve for that radionuclide.

The *H#* technique offers certain advantages over the "classical" quench correction methods such as SCR and ESCR. These advantages are (1) any sample can have only one *H#* value, contrary to channels ratio techniques, (2) the *H#* technique results in less variable quench correction curves over a wider range of counting efficiency, (3) if the *H#* of a nonquenched standard is properly calibrated, the *H#* of any given sample would be constant from instrument to instrument, although the counting efficiency may not necessarily be constant.

3. Relative Pulse Height (RPH) and External Standard Pulse (ESP)

The relative pulse height (RPH) and external standard pulse (ESP) quench correction techniques are similar in concept to the previously described *H#* quench monitor procedure (L'Annunziata, 1987).

In the ESP technique reported by Laney (1976, 1977) and evaluated by McQuarrie *et al.* (1980), the liquid scintillation spectrometer measures the degree of quench in a sample by the shift in the average pulse height, P_s, originating from Compton electrons produced by an external ^{133}Ba gamma-ray source as compared to the average pulse height, P_r, produced in a sealed nonquenched reference vial stored in the elevator mechanism of the counter. The shift in the average pulse heights is defined by the ratio

$$\mathrm{ESP} = \frac{P_r - P_\infty}{P_s - P_\infty}, \tag{5.15}$$

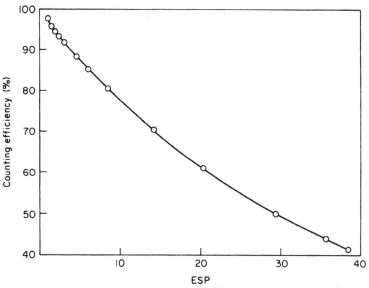

FIGURE 5.16 Variations in counting efficiency with ESP using quenched ^{14}C standards (From data of McQuarrie et al., 1980, reprinted with permission from Elsevier Science).

where P_∞ is a correction term corresponding to Compton electrons produced by the external gamma-ray source under "infinite" quenching. In ESP determinations, the entire external standard pulse-height spectrum of quenched and nonquenched samples is stored in a multichannel analyzer from which the average pulse height is determined.

As described in the case of $H\#$, a quench correction curve is prepared relating counting efficiency with ESP, using a set of quenched standards containing the same activity of radionuclide and varying amounts of quenching agent in fluor cocktail. A quench correction curve is obtained by plotting counting efficiency against ESP as illustrated in Fig. 5.16. Laney (1976) defined the relative pulse height (RPH) as the reciprocal of the ESP. Both can serve as quench indicating parameters, but ESP produces quench correction curves with more linearity (Grau Malonda, 1999).

4. Spectral Quench Parameter of the External Standard or SQP(E)

The quench-indicating parameter SQP(E) is measured with ^{226}Ra or ^{152}Eu (Günther, 1998) as the external standard source in LSAs of the LKB and Wallac instruments (Kouru, 1991; Grau Malonda, 1999). These instruments are now manufactured by PerkinElmer Life and Analytical Sciences. An MCA with 1024 logarithmic channels is used to determine the position of 99.5% of the endpoint of the external standard spectrum to define SQP(E). The gamma source for the SQP(E) determination is positioned below the counting vial containing scintillation fluor cocktail. The SQP(E) defines the uppermost channel number (endpoint) that comprises 99.5% of the total counts of the external standard pulse height spectrum. Only a small portion of the endpoint, the remaining 0.5% of the total counts or area under the pulse height spectrum are excluded (Kessler, 1989). As described by Grau Malonda (1999) the SQP(E) for a nonquenched sample is defined as

$$SQP(E) = P - 400 \qquad (5.16)$$

where SQP(E) corresponds to the i value of the equation

$$\sum_{j=1}^{n} N_j \geq (1-r) \sum_{j=400}^{n} N_j > \sum_{j=i+1}^{n} N_j \qquad (5.17)$$

where N_j is the total number of external standard counts in channel j, $r = 0.995$, $n = $ the total number of channels $= 1024$, and $\sum_{j=400}^{n} N_j = N_{tot}$ is the total number of external standard counts or area under the external standard pulse height spectrum above channel 400. The above formula indicates that the first 400 channels of the external standard pulse height spectrum are excluded from the calculations. The 400 channels at the lower end of the pulse height spectrum corresponds to approximately 0–20 keV events; and the objective of the exclusion is to reduce that portion of the spectrum that could vary from any "wall effect" that would occur whenever scintillation fluor solvent penetrates into the plastics wall of the counting vial. The wall effect is enhanced scintillation resulting from organic solvents such as benzene or toluene penetrating into the plastic vial wall. However, this effect generally does not occur to any appreciable extent with the modern solvents based on linear alkyl benzene and diisopropylnaphthalene (see Chapter 8). In the elucidation by Grau Malonda (1999) the value of P is obtained from the equation

$$P = i + \frac{1}{N_i} \left[\sum_{j=i}^{n} N_j - (1-r)N_{tot} \right] \qquad (5.18)$$

where N_i is obtained from the equation

$$N_i = \frac{1}{3} \left[\sum_{j=i-1}^{i+1} N_j \right] \qquad (5.19)$$

As required with other quench-indicating parameters it is necessary to count a set of quenched standards all having a known and constant activity of radionuclide but varying levels of quench. From the count rates of each standard and SQP(E) value measured by the LSA, a standard curve of counting efficiency versus SQP(E) is plotted as illustrated in Fig. 5.17. When a sample of unknown activity is analyzed in the LSA, the instrument will determine the SQP(E) value of the sample, and from the standard curve extract the counting efficiency.

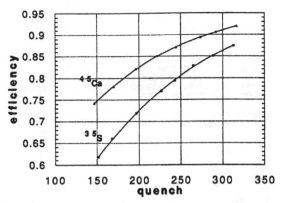

FIGURE 5.17 Calibration (quench correction) curves for ^{45}Ca and ^{35}S. The quench-indicating parameter is the SQP(E) (From Grau Carles et al., 1993f).

5. Transformed Spectral Index of the External Standard (tSIE)

The external standard quench correction methods previously described define specific characteristics of the external standard pulse height spectrum as quench-indicating parameters such as, (1) the magnitude of the average pulse heights of the external standard Compton spectrum (e.g., ESP and RPH), (2) the inflection point at the Compton edge (e.g., $H\#$), and (3) the endpoint of the external standard pulse height spectrum (e.g., SQP(E)). Another popular external standard quench-indicating parameter was first reported by Everett et al. (1980) and Ring et al. (1980) under the designation of spectral index of the external standard (SIE). The SIE is similar to the SIS described previously with the exception that the SIE characterizes the external standard pulse height spectrum in the same fashion as the SIS characterizes the sample pulse height spectrum (Kessler, 1989). The objective of SIE is to characterize the external standard pulse height spectrum to the extent of quantifying the various features of the pulse height distribution and any changes in these features, which could occur as a result of quench. Features such as the spectral peak, slope at various points of the spectrum, and maximum pulse height will govern the center of gravity of the pulse height spectrum, which will obviously change according to quench level (L'Annunziata, 1987). The SIE is calculated as

$$SIE = k \frac{\sum_{x=L}^{u} X \cdot n(x)}{\sum_{x=L}^{u} n(x)} \tag{5.20}$$

where k is a factor assigned to provide a maximum value to the SIE of a non-quenched standard, X is the channel number, $n(x)$ is the number of counts or pulse events in channel X, and L and U are lower and upper limits that encompass the pulse height spectrum. The lower limit L is set above zero sufficient to eliminate changes in pulse events of low magnitude produced by the "wall effect." That could occur if fluor cocktail solvent were to penetrate the plastic wall of the scintillation counting vial. Notice the close similarity of

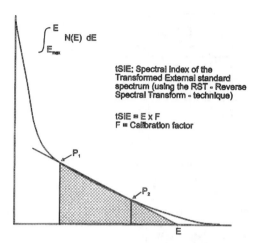

FIGURE 5.18 Transformed liquid scintillation pulse height spectrum of an external ^{133}Ba standard. The tSIE is calculated by the extrapolated value E times a calibration factor F to provide a quench-indicating parameter in the range of 0–1000. The highest value of tSIE = 1000 is set using an unquenched ^{14}C standard. (Courtesy of PerkinElmer Life and Analytical Sciences, Boston, MA.)

the above equation for SIE to that used to calculate SIS (Eq. 5.6). The values of SIE are unitless, always greater than 1.0, and of magnitude that will vary according to quench (i.e., the higher the quench level in the sample, the lower will be the SIE value).

A further development based on the SIE is the transformed spectral index of the external standard (tSIE) introduced by Packard Instruments and now an integral division of PerkinElmer Life and Analytical Sciences. The tSIE method of quench correction uses ^{133}Ba as the gamma-ray source. The Compton spectrum of this external standard is obtained in an MCA, such as the SIE described previously and transformed by performing a reverse back sum on the spectrum to obtain the transformed spectrum as illustrated in Fig. 5.18. From the transformed spectrum, an endpoint energy is determined by a reverse spectral transform (RST) technique using two points on the spectrum and extrapolating to the energy axis (Kessler, 1989). The simplified mathematical expression of the reversed spectral transform is

$$\int_{\hat{E}_{max}}^{\hat{E}} N(\hat{E})d\hat{E} = \left(\int_{\hat{E}_{max}}^{\hat{E}_2} N(\hat{E})d\hat{E} - \int_{\hat{E}_{max}}^{\hat{E}_1} N(\hat{E})d\hat{E} \right) \frac{\hat{E} - tSIE}{\hat{E}_2 - \hat{E}_1} \qquad (5.21)$$

where $\hat{E}$ is the transformed energy and tSIE is calculated as one of the parameters of the RST function as

$$tSIE = \hat{E} - (\hat{E}_2 - \hat{E}_1) \frac{\displaystyle\int_{\hat{E}_{max}}^{\hat{E}} N(\hat{E})d\hat{E}}{\displaystyle\int_{\hat{E}_{max}}^{\hat{E}_2} N(\hat{E})d\hat{E} - \int_{\hat{E}_{max}}^{\hat{E}_1} N(\hat{E})d\hat{E}} \qquad (5.22)$$

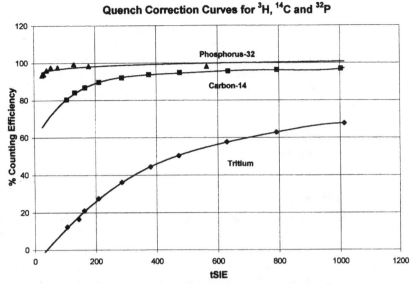

FIGURE 5.19   Quench correction curves for ^{3}H ($E_{max} = 18.6$ keV), ^{14}C ($E_{max} = 156$ keV), and ^{32}P ($E_{max} = 1710$ keV) using tSIE as the external standard quench-indicating parameter with a Packard 2700TR LSA. The optimum counting efficiencies for the unquenched samples of ^{3}H, ^{14}C, and 5% water-quenched ^{32}P were 67.6, 96.9, and 98.1%, respectively. (L'Annunziata, 1996, unpublished work.)

The final tSIE (extrapolated endpoint times calibration factor) is calculated on the basis of the tSIE being equal to 1000 for a nonquenched ^{14}C sample, that is used for calibration and normalization. Additional information on the measurement of tSIE and its applications is given by Kessler (1991). The ^{133}Ba gamma-ray source for the measurement of the tSIE is positioned below the sample vial. The positioning of the external standard under the sample produces a quench measurement that compensates for variations of sample volume. The value of tSIE, therefore, can be accurately determined even for small (< 1 mL) sample–fluor cocktail mixtures.

The major advantages of using the external standard quench-indicating parameter tSIE rather than the QIP based on the sample spectrum (SIS) can be ascertained from Fig. 5.19, which shows quench correction curves for tritium, ^{14}C, and ^{32}P plotted on the same graph. These plots of percent efficiency versus tSIE were obtained with a PerkinElmer 2700TR LSA using three sets of quenched standards, one set of standards for each radionuclide. The plots illustrate the dynamic range of the quench-indicating parameter from 1000 for the nonquenched cocktail mixtures to less than 100 for the highly quenched samples of the three radioisotopes. Also, we can note from Fig. 5.19 that, for a given level of quench, the counting efficiencies are higher for radionuclides that emit beta particles of higher energy; and quench has less effect on the counting efficiencies of radionuclides that emit beta particles of higher energy. It is important to recall that tSIE is radioisotope independent; it is a function of the quality or quench level of the fluor cocktail. The second advantage of the external standard method over the sample spectrum characterization method for the determination of QIP is that the external standard method is sample count rate independent. The external standard quench-indicating parameter (QIP) does not depend on the count rate of the sample, but depends on the counts created by the external gamma-ray source and the resultant Compton electrons produced within the scintillation fluor cocktail. The only disadvantage of the external standard method is that each sample must be counted alone and then counted again with the external standard present. This extra counting step usually requires about 6–120 s additional counting time depending on the sample volume and quench level. This disadvantage is of little significance; sample counting with and without the presence of the external standard including the measurement of the QIP is fully automated.

6. G-number (G#)

The G# is a quench-indicating parameter (QIP) first described in detail by Grau Carles and Grau Malonda (1992) as a patent and subsequently reported by Grau Carles et al. (1993a). The method was designed to provide an accurate QIP regardless of the level of quench in a sample, even when the quench level is so high that the counting efficiency of the beta-emitting radionuclide is reduced to less than 1%. The idea behind this approach of Grau Carles and Grau Malonda is based on the use of an external standard which emits considerable quantities of high-energy gamma radiation sufficient to produce appreciable numbers of Compton electrons in the sample scintillation cocktail that have energies above the Cherenkov threshold of 263 keV. The LKB Rack Beta liquid scintillation analyzers (PerkinElmer Life and Analytical Sciences) are equipped with such a gamma source, namely ^{226}Ra; and the development of this technique was therefore demonstrated with the LKB instrument. The Compton electrons produced by the ^{226}Ra external standard will produce a pulse height spectra, which is a result of photons emitted by the sample originating from a combination of scintillation and Cherenkov events. As explained by Grau Carles and Grau Malonda (1992), both scintillation and Cherenkov photons are detected simultaneously within the 18 ns time window of the coincidence circuitry of the LSA. However, when samples are very highly quenched in the liquid scintillation cocktail, the liquid scintillation diminishes considerably to the point that the pulse height spectrum produced by the ^{226}Ra external standard becomes the result of mostly Cherenkov photons produced by the portion of Compton electrons of energy in excess of 263 keV. In this fashion, regardless of the level of quench, a characteristic of the pulse height spectrum of the ^{226}Ra can be measured to provide an accurate QIP even when the scintillation process is quenched to the extent that beta particle-emitting radionuclides are counted at an efficiency of less than 1%.

Like all other methods of quench correction, this technique requires the preparation of a set of quenched standards of the radionuclide of interest. The quenched standards are counted in a suitable counting region defined by lower-level and upper-level discriminator settings. The count rate of each quenched standard is determined in this counting region, and the counting efficiency is calculated. The quenched standards are also exposed to the ^{226}Ra external standard gamma-ray source, and the resulting ^{226}Ra pulse height

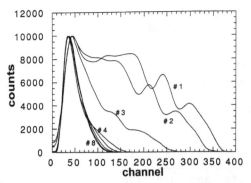

FIGURE 5.20 Compton spectra from ^{226}Ra adjusted and normalized for different quench levels (From Grau Carles and Grau Malonda, 1992).

spectrum is produced in the channel region between 10 and 500. Figure 5.20 illustrates pulse height spectra produced by eight quenched standards exposed to ^{226}Ra external standard. The spectra show how the pulse heights shift from higher to lower magnitudes (higher channels to lower channels) as quench increases. At very high levels of quench (standards #4 to #8 of Fig. 5.20), the pulse height spectra from ^{226}Ra are the consequence of mainly Cherenkov photons produced by the Compton electrons in excess of 263 keV energy.

The G-number is based on the analysis of the displacement of the final part (endpoint) of the ^{226}Ra external standard pulse height spectra as a function of quench. According to Grau Carles *et al.* (1993a) Fourier series are fitted to the pulse height spectra produced by the ^{226}Ra external standard to enable their normalization to the number of counts, y_N, of the first peak in the pulse height spectra due to Cherenkov light created by the most energetic Compton electrons (i.e., the left-most peaks in Fig. 5.20). The spectral interval within the limits $y_N/10$ and $y_N/500$ is taken from the final part of each external standard pulse height spectrum. A linear relationship in the selected interval is obtained by raising the number of counts y to the power α or

$$y \rightarrow y^\alpha \quad 0 < \alpha < 1 \tag{5.23}$$

where α is the value, that provides the best regression coefficient in the line

$$y^\alpha = ax + b \tag{5.24}$$

For the channel with a number of counts $y = y_N/100$, the G-number is given by

$$G = \frac{(y_N/100)^\alpha - b}{a} \tag{5.25}$$

Examples of typical quench correction curves obtained with quenched standards of ^{35}S, ^{14}C, ^{45}Ca, and ^{89}Sr are illustrated in Fig. 5.21, where it is clearly illustrated that the G-number serves as an excellent quench-indicating

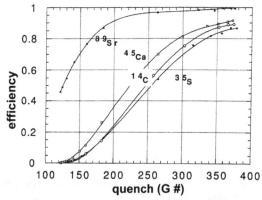

FIGURE 5.21 Calibration or quench correction curves of counting efficiency as a function of the G-number for ^{35}S, ^{14}C, ^{45}Ca, and ^{89}Sr (From Grau Carles and Grau Malonda, 1992).

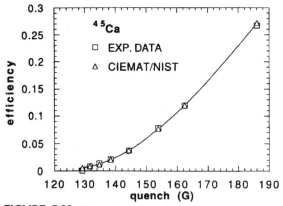

FIGURE 5.22 Experimental and **CIEMAT/NIST** computed efficiencies for ^{45}Ca (From Grau Carles et al., 1993a, reprinted with permission from Elsevier Science).

parameter over the widest possible range of counting efficiency from the highest detection efficiency to the lowest ($< 1\%$). This is more clear from the expanded quench correction curve for highly quenched ^{45}Ca illustrated in Fig. 5.22 over the counting efficiency range of $< 1\%$ to approximately 25%.

There is no documented rationale for the selection of the name "G-number" for the identification of this quench-indicating parameter. However, the writer can only assume that the letter "G" calls to mind the first letter of the family names Grau Carles and Grau Malonda, who devised this technique. Consequently, this QIP could likewise be called the "Grau-number." There exists a similar corollary in the previous development of the H-number by D.L. Horrocks described previously in Section C.2 of this chapter. In recognition of its founder, the H-number is also referred to as the Horrocks-number.

D. Preparation and Use of Quenched Standards and Quench Correction Curves

A quench correction curve is a calibration curve of percent efficiency versus a quench-indicating parameter (e.g., $H\#$, SQP(E), tSIE, and G#). Examples of quench correction curves are found in Figs. 5.9, 5.11, 5.12, 5.15–5.17, 5.19, 5.21, and 5.22. The quench correction curve is prepared from a set of quenched standards, which is a series of samples containing the same radionuclide in scintillation cocktail, all with the same DPM but different levels of quench.

I. Preparation of Quenched Standards

There are two methods used for obtaining a set of quenched standards. The first method is to purchase a set of sealed standards for the radionuclide of interest and scintillation cocktail type that one plans to use; and with the quenched standards, prepare a quench correction curve of percent counting efficiency versus a suitable QIP. Quenched standards of ^{3}H and ^{14}C are available commercially, because of their relatively long half-lives. Some suppliers (e.g., PerkinElmer Life and Analytical Sciences, Downers Grove, IL) can provide the quenched standards according to specification including variables such as (1) radionuclide standard activity, (2) scintillation cocktail, (3) quenching agent, (4) counting vial size and type, and (5) sample volume. Sets of quenched standards for ^{3}H and ^{14}C are sold on the market without customer specifications. If a user is interested in procuring these, it is important that he or she procure the set of quenched standards with vial size, sample volume, and scintillation cocktail compatible with their experimental samples. The second method is to prepare a series of quenched standards in the laboratory and to prepare the quench correction curve from these quenched standards. This method can be the most reliable when very accurate DPM values are required, because the user can control all aspects of the preparation of the quenched standards to most closely represent the chemistry of his or her experimental samples. A detailed description of the procedure for the preparation of quenched standards and a quench correction curve from the quenched standards is provided next. An outline of the steps is as follows:

1. **Choose the type and size of counting vial and sample volume that will be used.** Counting vials come in various sizes (e.g., 4, 6, 8, 20 mL) and as glass and plastic. Although glass and plastic vials may perform similarly (Elliott, 1984), there can be differences depending on the cocktail used and radionuclide analyzed. The size of the vial can have a significant effect on counting geometry and the quench correction curve would vary significantly according to this variable. The vial size and type, scintillation cocktail-sample volume, and cocktail composition of the quenched standards should be the same as the experimental samples (Collé, 1997a,b).

2. **Choose the scintillation cocktail that will be used.** Commercial scintillation cocktails come in various chemical compositions with differing properties, some miscible with organic solvents and others with aqueous sample solutions (see Chapter 8). Scintillation cocktails that use solvents such as toluene, xylene, pseudocumene, or linear alkylbenzene may be used with samples in organic solvents while cocktails using diisopropylnaphthylene (DIN) or phenylxylylethane

(PXE) as solvents may be mixed with aqueous samples. Most accurate results are obtained when the chemistry of the quenched standards are the same as the experimental samples. Detection efficiencies can vary between different scintillation cocktail compositions. Error can be introduced when determining the activity (e.g., DPM) of an experimental sample mixed in a scintillation cocktail different from that from which the quenched standards were prepared. The error is particularly pronounced in the case of the LSA of low-energy radionuclides such as tritium (Collé, 1997a,b). It is best therefore to prepare quenched standards with the same scintillation cocktail to be used with the experimental samples.

3. **Prepare a stock solution to contain the radionuclide standard of interest of known activity (DPM) in the scintillation cocktail.** The radionuclide standard should have an activity that is accurately known such as one traceable to a primary standard (e.g., NIST traceable standard) or a radionuclide standardized according to a known method of standardization, such as the CIEMAT/NIST method described in Section IX.A of this chapter. The standard used must be compatible and thoroughly miscible with the cocktail. The stock solution should be of sufficient volume to prepare more vials than standards that are needed to allow for the possibility of discarding some standard vials for reasons described subsequently. For example, if 10 quenched standards each containing 10 mL of scintillation cocktail are desired, then prepare a stock solution of more than 150 mL of scintillation cocktail containing radionuclide standard to allow the testing of 15 standards with the possibility of discarding 5 as described in step (5) below. The level of the radioactivity in the stock scintillation cocktail should be high enough to require counting the standards for only a short period of time and still achieve good or acceptable counting statistics. Activities of approximately 200,000 DPM per vial of low-energy emitting radionuclide, such as ^{3}H, or approximately 100,000 DPM per vial of higher-energy emitting radionuclide, such as ^{14}C, should be adequate.

4. **Transfer the exact aliquot of radionuclide standard – scintillation cocktail stock solution into each of the vials to be used for the preparation of the quench correction curve.** The stock solution may be added by pipette or gravimetrically. The volume of the quenched standards should be similar to the volume of the experimental samples against which the quench correction curve will be used. Counting geometry due to volume differences can affect the counting efficiency.

5. **Count each of the standards and determine whether the count rate (CPM) of each is within acceptable counting statistics.** A counting region that encompasses the entire radionuclide pulse height spectrum can be used. The standards are of high activity and background counts can be ignored. Replicate counts of each standard (e.g., count each standard from 5 to 10 times). Obtain a mean count for all of the standards. As an excess number of standards are prepared, any of the standards that deviate more than 2% from the mean count rate

can be discarded. This provides us with a set of standards of equal activity in scintillation cocktail.

6. **Select a suitable quenching agent and add increasing amount of the quenching agent to each standard.** In this step the amounts of quenching agent do not have to be added with precision. Only the amount of radionuclide standard in each vial, prepared in the previous step, must be exact. With respect to the quenching agent added, it is only important that each vial have increasing amounts of quenching agent so that a quench correction curve of counting efficiency versus QIP can be established over a broad range of quench levels. Nitromethane is a popular quenching agent, because it is a strong quencher and only small increments are required. For example, if the scintillation cocktail uses toluene as a solvent and there are 10 quenched standards of either ^{3}H or ^{14}C, then the following increments of nitromethane can be added to the vials: 0, 1, 5, 10, 18, 25, 35, 45, 55, and 65 μL to provide a broad range of quench levels. Notice that the first vial contains no added quenching agent. It would be the least-quenched standard. Radionuclide standards that emit higher-energy beta particles (higher E_{max}) generally require larger increments of quenching agent to provide a quench curve, that displays a significant reduction in counting efficiency against a QIP. For example, Fig. 5.15 shows ^{33}P ($E_{max} = 249$ keV) and ^{32}P ($E_{max} = 1710$ keV) to undergo little change in counting efficiency over a wide range of quench level. Quenching agents have less effect on counting efficiency as the E_{max} of the beta emitter increases as illustrated in Figs. 5.15 and 5.19 using ^{3}H, ^{14}C, ^{33}P, and ^{32}P as examples. Other scintillation cocktails respond differently to nitromethane. For example, cocktails containing DIN or PXE solvent may require larger increments of nitromethane quenching agent. In such cases the following increments of nitromethane for a set of 10 quenched standards may be appropriate: 0, 5, 10, 15, 28, 45, 70, 110, 150, and 230 μL. Other popular quenching agents are CCl_4 (see Fig. 5.15), chloroform, acetone, and water, which are not as strong quenchers as nitromethane and also display differing degrees in their relative strength of quench. The amounts of quenching agent required will differ, because not all agents quench equally. Classification of quenching agents according to their quenching power is given by L'Annunziata (1987). The quenching agent used should be soluble and not react with the scintillation cocktail. A color dye can be selected as a quenching agent for a set of quenched standards when color quench is expected in the experimental samples. Water is a relatively weak quenching agent; and when analyzing for tritium, different water loads in scintillation cocktail, such as 1 : 9, 2 : 8, 3 : 7, 4 : 6, and 5 : 5 (water : cocktail ratio), may suffice for the preparation of a tritium quench correction curve, particularly when quenching agents other than water are not expected to be present in the experimental samples. If other chemical constituents are expected to be present in water samples to be analyzed, these can be added to the quenched standards in increasing amounts

to fully simulate the chemistry of the experimental samples. The most ideal case is where the quenching agents in the quenched standards are identical to those encountered in the experimental samples, although many researchers ignore this for reasons of convenience.

7. **If in doubt concerning the amounts of quenching agents to add, predetermine this experimentally.** Whenever the amounts of quenching agent required are in doubt, it is easy to predetermine this by adding various amounts of quenching agent to scintillation cocktail in counting vials without radionuclide standard. The vials can then be counted without radionuclide only to determine the external standard quench-indicating parameter (QIP) for each vial. Adjustments can be made with more or less quenching agent in order to achieve a wide range of quench levels according to the QIP.

8. **Label the quenched standards by number of letter (e.g., 1 to 10 or A to J) and isotope, date, etc.** The quenched standards can be labeled only on the top of the counting vial cap. A round self-adhesive label may be placed on the cap or the information written on the top of the cap with a fine-tipped indelible pen.

9. **Store the quenched standards in the dark for a day or more for stability against any possible photo- or chemiluminescence.** Photo- and chemiluminescence are possibilities that can occur when either the counting vials are open or when quenching agent is added, respectively. The samples can be counted on a daily basis to observe if there is any reduction in count rate with time as evidence of luminescence. Once stability is confirmed the quenched standards can be used to prepare a quench correction curve.

A procedure alternative to the above for the preparation of a set of quenched standards would be to dispense the desired volume of scintillation cocktail to a set of counting vials. The volume would depend on the combined volume of sample and scintillation cocktail planned for the experimental counting vials (e.g., 10, 15 or even 20 mL depending on vial size and capacity). The radionuclide standard is then added to each vial in equal amounts using a very precise microliter syringe with an adapter to help assure the addition of the same amounts to each vial. A Hamilton syringe equipped with a Cheney adapter (Hamilton Company, Reno, NV 89502, USA or CH-7402 Bonaduz, GR, Switzerland) may be suitable. The writer finds it easier to prepare standards of equal activity by preparing a stock solution of radionuclide standard in scintillation cocktail and dispensing this solution into counting vials as described in steps 3 and 4 above.

If stored under refrigeration (5–10°C), sets of quenched standards may be stable for two to three years. It is best to keep records of the quench correction curves prepared from a given set of quenched standards from time to time (e.g., monthly basis) to check their stability.

2. Preparation of a Quench Correction Curve

The quench correction curve or plot of counting efficiency versus quench-indicating parameter (QIP) must be determined with a set of quenched standards, and this curve stored in the memory of the liquid scintillation analyzer. The preparation of the quench correction curve is described as follows:

1. **Set up a counting protocol on the LSC to plot a quench correction curve of percent counting efficiency versus a quench-indicating parameter.** One of the QIPs that uses a sample spectrum characterization method may be used, although an external standard QIP is more often preferred for reasons described in Section V.C of this chapter. Some modern LSAs store the entire pulse height spectrum of the sample counted onto the memory of the hard disk. With these instruments it is often not necessary to set the counting region defined by lower level (LL) and upper level (UL) discriminator settings. Once the pulse height spectrum of each quenched standard is stored in the instrument, the count rate in any counting region defined by LL and UL discriminator settings can be extracted. However, in older generation instrumentation it is necessary to first define the counting region according to LL and UL discriminator settings prior to counting the set of quenched standards. The counting efficiency can vary according to counting region settings, and consequently quench correction curves of counting efficiency versus QIP can also differ according to counting region settings.

2. **Count the quenched standards at a statistical accuracy of at least 0.5% 2s.** Due to high activities of standards in each vial (100,000–200,000 DPM) the counting time required to reach the 0.5% 2s statistics should not exceed 5 min per quenched standard.

3. **Obtain a plot of the percent counting efficiency of the radionuclide standard versus the QIP.** Modern LSAs will store this data in computer memory. When experimental samples are counted the instrument should use the QIP measured for the sample and determine the counting efficiency from the correction curve.

3. Use of a Quench Correction Curve

The objective of the quench correction curve (also referred to as a calibration curve) is to determine the counting efficiency of experimental samples. From the counting efficiency the count rate (e.g., CPM) of the sample is converted to radionuclide activity (e.g., DPM) according to Eq. 5.7 described previously. When using a quench correction curve it is important to keep in mind certain rules, some of which may be intuitively obvious. These are the following:

1. A quench correction curve is good for only one radionuclide.
2. The quench correction curve is dependent on counting region defined by lower-level and upper-level pulse height discriminator settings.
3. The quench correction curve is scintillation cocktail dependent (Collé, 1997a,b). It is important to be certain that the scintillation cocktail

used for the experimental samples is the same as that used for the preparation of the quench correction curve. If a different cocktail is used for experimental samples compared to that used to prepare the quench correction curve, it is necessary to confirm that there is no significant difference between the cocktails in performance for the given radionuclides. Differences in performance of cocktails are more pronounced when analyzing for relatively low-energy beta emitters such as ^{3}H and ^{14}C.

4. Quench correction curves using an external standard QIP are preferred, because these are independent of sample activity.

5. Quench correction curves that utilize a sample spectrum QIP are dependent on sample activity and limited to sample count rates well above background ($>$ approximately 1000 CPM).

E. Combined Chemical and Color Quench Correction

When there are significant quantities of chemical or color quench in scintillation cocktail there can be a significant difference in the quench correction curve obtained. This is particularly the case when there is a high degree of either chemical or color quench (high quench level) and for the liquid scintillation analysis of relatively weak beta particle-emitting radionuclides such as ^{3}H ($E_{max} = 18.6$ keV) or ^{14}C ($E_{max} = 156$ keV). Examples of differences between chemical and color quench correction curves can be seen in Figs. 5.12 and 5.23. Such differences can be observed regardless of the quench-indicating parameter used. The differences in the two curves is based on the two different mechanisms of quench; namely chemical quench that entails the inhibition of energy transfer from cocktail solvent to fluor molecules and color quench that entails the absorption of light photons emitted by the scintillation cocktail (see Fig. 5.1). As noted by Takiue *et al.* (1991b) the liquid scintillation pulse height distribution of a color-quenched sample is different from that of a chemical-quenched sample, even if both the samples have the same activity and counting efficiency (see Fig. 5.24). Therefore, the pulse height distribution of the external standard will be different for either color- or chemical-quenched producing different QIPs and different quench correction curves. The difference is more significant at high levels of quench, and becomes less significant as the beta-particle energy (E_{max}) of the radionuclide increases.

The best alternative when color exists in the sample is to decolorize (e.g., sample bleaching or oxidation of organic samples to CO_2 and H_2O, see Chapter 8), thereby eliminating the problem of two mechanisms of quench leaving behind only chemical quench, which is present in all samples with the exception of the artifical nonquenched argon-purged standards. If decolorization is not possible, most modern LSAs are equipped with color correction programs or algorithms that will correct for the difference between chemical and color quench. When both color and chemical quench are significant and cannot be avoided, it is recommended that the color correction program provided with the instrumentation be utilized, if available, at high quench levels (e.g., tSIE < 400).

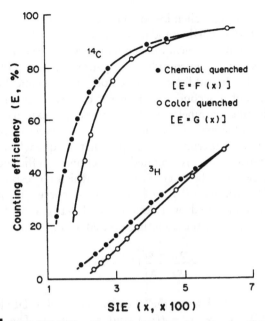

FIGURE 5.23 Chemical and color quench correction curves based on an external standard (From Takiue *et al.*, 1991a, reprinted with permission from Elsevier Science).

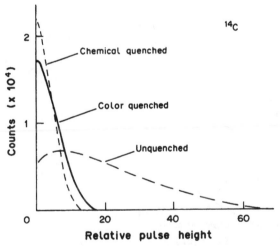

FIGURE 5.24 Liquid scintillation pulse height distributions of chemical- and color-quenched ^{14}C samples. Both samples have identical activity and counting efficiencies (From Takiue *et al.*, 1991a, reprinted with permission from Elsevier Science).

An example of a practical program for correction of the difference between color and chemical quench correction curves was formulated by Takiue *et al.* (1991a). This method entails the preparation of two sets of quenched standards of a given radionuclide. One set of standards is prepared with a color-quenching agent (e.g., bromothymol blue, methyl red, or bromocresol green) that produces minimal chemical quench, and another set of standards is prepared using a chemical-quenching agent (e.g., nitromethane or CCl₄). The sets of quenched standards are used to plot two curves consisting of a color- and chemical-quench correction curve of counting efficiency versus any external quench-indicating parameter (e.g., H#, SQP(E), or tSIE, see Fig. 5.23). In addition, the color- and chemical-quenched standards are used to plot a second set of curves consisting of the external standard quench-indicating parameter plotted against the external standard counts, referred to as double external standard relation curves (DESR curves, see Fig. 5.25). When an experimental sample that is quenched by both chemical and color constituents, is counted, the counting efficiency of the unknown sample has the value between E_1 and E_2, which corresponds to the external standard (ES) counts of u_o of the DESR curves illustrated in Fig. 5.25. Hence, as demonstrated by Takiue *et al.* (1991a), according to the geometry depicted in Fig. 5.25, the counting efficiency (E_o) is defined as

$$E_o = E_1 + \frac{x_o - x_1}{x_2 - x_1}(E_2 - E_1) \tag{5.26}$$

where x_1 and x_2 are obtained from the chemical and color DESR curves, respectively (Fig. 5.23), where $x_1 = f(u_o)$ and $x_2 = g(u_o)$. The efficiency values E_1 and E_2 are obtained from the chemical and color quench correction curves, respectively (Fig. 5.23), where $E_1 = F(x_1)$ and $E_2 = G(x_2)$. Equation 5.26 is then written as

$$E_o = F(x_1) + \frac{x_o - f(u_o)}{g(u_o) - f(u_o)}[G(x_2) - F(x_1)]. \tag{5.27}$$

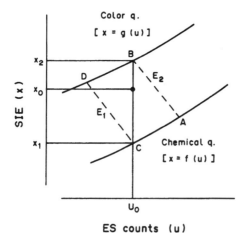

FIGURE 5.25 DESR curves for chemical- and color-quenched radionuclide used for the calculation of the counting efficiency of an experimental combined chemical- and color-quenched sample as described in Section V.E, where x_o and u_o are the quench-indicating parameter and external standard (ES) counts for the experimental combined chemical- and color-quenched sample (From Takiue et al., 1991a, reprinted with permission from Elsevier Science).

Takiue *et al.* (1991a) used polynomial curve fitting with the least squares method to define the coefficients of the quench correction and DESR curves. This color correction method is easily applied with the computer application programs of most modern LSAs. Nevertheless, the problem of combined color and chemical quench in samples is best averted by decolorization of samples prior to the addition of scintillation cocktail. Also, the difference between chemical and color quench correction curves is more pronounced with beta-emitting radionuclides of relatively low beta-particle E_{max} such as ^3H and ^{14}C combined with high levels of quench.

F. Direct DPM Methods

The Direct DPM methods entail the LSA measurement of the absolute activities or disintegration rates of radionuclides, particularly beta emitters, under various levels of quench without the use of quench correction curves. These methods are described subsequently.

I. Conventional Integral Counting Method (CICM)

During the early years of liquid scintillation counting it was discovered that an extrapolation of integral counting curves to zero discriminator bias could be used to determine the absolute activities (DPM) of alpha emitters (Basson and Steyn, 1954) and beta emitters (Steyn, 1956) without interference from gamma emission. The method applied to alpha emitters received little attention, because the LSA counting efficiency of alpha emitters was close to 100% even in these early years of liquid scintillation development. As far as beta emitters are concerned, this technique, known as integral counting, received some popular attention and applications in the late 1950s and during the 1960s. Some recent developments have made this technique a practical and accurate method for the absolute activity measurement of beta-emitting radionuclides.

The work of Goldstein (1965) demonstrated the broad range of radionuclides that may be analyzed by integral counting as well as the simplicity of the procedure involved. In the development and testing of integral counting, Goldstein (1965) used the first and only commercial LSA available at that time, which was a Packard 314 liquid scintillation spectrometer. The procedure involved three pulse height discriminators labeled AA', B, and C. The AA' discriminator was set just above the noise level to reject noise pulses. The C discriminator (upper level discriminator) was turned off or disengaged so that all of the pulses of magnitude above the B discriminator would be registered and counted. The height of the B discriminator was varied in the range of 10–30 volts in 5-volt increments. The count rates for a given beta-emitting sample in scintillation cocktail were collected for each setting of the B discriminator. With the B discriminator at its lowest setting the count rate is highest. With each incremental increase in the height of the B discriminator, the count rate diminishes, because fewer and fewer pulses are detected. The resulting plot of count rate on a logarithmic scale versus the B discriminator bias (volts) setting on a linear scale would be linear with negative slope, which could be extrapolated back to zero bias volts. At this point of extrapolation the count rate (CPM) at zero bias would be the disintegration rate (DPM) of the sample. This extrapolation method is currently referred to as the conventional integral counting method (CICM). It is an effective method for extrapolating to sample DPM for beta emitters or beta–gamma emitters, which emit beta particles with an $E_{max} > 200$ keV, including ^{147}Pm, ^{45}Ca, ^{99}Tc, ^{36}Cl, ^{204}Tl, ^{89}Sr, ^{90}Sr(^{90}Y), ^{91}Y, ^{32}P, ^{131}I, ^{85}Kr, ^{131m}Xe, and ^{60}Co, among others, regardless of quench level. Corrections for quench in the sample scintillation cocktail mixtures are not necessary, because the quench level in the sample affects only the slope of the integral curve, and extrapolation of the curve to zero discriminator bias ends at the same count rate for all quench levels with expected statistical deviations ($< 2\%$ error). Homma and Murakami (1977) also applied the conventional integral counting method to determine the activity of ^{226}Ra after separating the equilibrated ^{222}Rn into a liquid scintillator. The disintegration rates of ^{222}Rn and its daughters, which include both alpha and beta emitters were determined by this method at various quench levels.

The conventional integral counting method for sample DPM determinations generally cannot be applied to the measurement of beta-emitting radionuclides of $E_{max} <$ about 200 keV. However, Homma et al. (1994a) developed the technique into the modified integral counting method (MICM), which can be used to determine the activities of all beta-particle emitters including ^{3}H of very low energy ($E_{max} = 18.6$ keV) and with higher accuracy.

2. Modified Integral Counting Method (MICM)

The modified integral counting method was reported by Homma and coworkers (1993a,b) who modified the CICM by extrapolating the integral counting curves, not to the zero pulse height as described above for the CICM, but to the zero detection threshold of the liquid scintillation spectrometer, which refers to the average energy required to produce a measurable pulse. They applied the new method to analyze the activity of alpha and beta emitters including ^{222}Rn and its daughters as well as to the low-energy beta emitters ^{3}H, ^{14}C, ^{35}S, and ^{45}Ca with 100% detection efficiency. The method is described subsequently in more detail.

The modified integral counting method as was determined by Homma et al. (1994a) is carried out by the following procedure:

1. The first step requires the determination of the zero detection threshold of the particular LSA utilized for the analysis. This is carried out by measuring a standardized nonquenched ^{3}H sample according to the integral counting method described earlier. The observed integral count rates of the ^{3}H standard are plotted at several pulse heights and the curve is then extrapolated to the count rate, which is equivalent to the disintegration rate (DPM) of the ^{3}H standard. The keV value (pulse height) at this count rate represents the zero detection threshold. The zero detection threshold was found by Homma et al. (1994b) to vary from instrument to instrument over the range of $2.4–3.5 \pm 0.2$ keV.

2. Once the zero detection threshold is determined for the particular instrument, the absolute disintegration rate of any low-energy beta emitter ($E_{max} <$ about 200 keV) including ^{3}H as well as high-energy beta emitter ($E_{max} >$ about 200 keV) is determined by extrapolating the integral pulse height spectrum of the radionuclide of interest to the previously determined zero detection threshold.

Examples of results obtained from the modified integral counting method applied to the activity determination of ^{35}S and ^{45}Ca are illustrated in Fig. 5.26. As noted by Homma *et al.* (1994a,b) it is obvious from the plots illustrated that extrapolation of the integral pulse height spectrum to only the zero pulse height leads to an intercept value that is lower than the actual DPM of the radionuclide. However, extrapolation of the integral counting curve to the zero detection threshold leads to the actual disintegration rate of the sample.

The modified integral counting method was reported also by Homma *et al.* (1993a,b, 1994c) for the determination of ^{222}Rn and its daughters ^{218}Po, ^{214}Pb, ^{214}Bi, and ^{214}Po. Total α and β activity was determined with 100% counting efficiency. The MICM can be applied to the activity measurements of

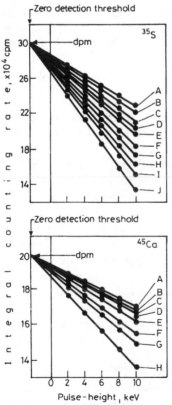

FIGURE 5.26 Extrapolation plots of the integral count rates of quenched ^{35}S and ^{45}Ca to the zero detection threshold for determination of the radionuclide disintegration rates. Letters A, B, C... denote samples with increasing quench levels. Deviations from actual DPM values were < 1% for all plots (From Homma *et al.*, 1994b).

α- and β-emitters as single radionuclide samples or mixtures; and γ emission does not interfere in most cases (Homma *et al.*, 1994a). Measurements of ^{222}Rn with activity ranges of 0.2–22.9 Bq/L in natural water samples obtained from private wells and springs were carried out by Murase *et al.* (1998) with the MICM applying 100-minute counting times, which provided activities with an overall uncertainty of 3.1%. The modified integral counting method has a practical simplicity similar to the efficiency tracing (ET) method described next. The ET method is most often used to determine the activity of single and multiple β and $\beta-\gamma$ emitters; it can be applied also to the activity measurements of mixtures of α- and β-emitters (see Fujii and Takiue, 1988b, and Section VIII of Chapter 9, and Fig. 9.11).

3. Efficiency Tracing with ^{14}C (ET)

Efficiency tracing (ET) with ^{14}C is another practical and simple extrapolation method applied generally to the absolute activity measurements of β-emitting radionuclides with the exception of tritium. This method should not be confused with the CIEMAT/NIST efficiency tracing method described in Section IX of this chapter. The ET method was demonstrated by Takiue and Ishikawa (1978) to provide accurate DPM values for 14 radionuclides. A subsequent study by Ishikawa *et al.* (1984) showed that the technique provided accurate DPM measurements of 11 additional β- and $\beta-\gamma$-emitting radionuclides, namely, ^{14}C, ^{32}P, ^{36}Cl, ^{46}Sc, ^{59}Fe, ^{60}Co, ^{63}Ni, ^{86}Rb, ^{90}Sr(^{90}Y), ^{131}I, ^{134}Cs, and ^{147}Pm regardless of quench level.

The efficiency tracing with ^{14}C (ET) technique involves the following steps:

1. A ^{14}C nonquenched standard is counted in six separate counting regions, such as 0, 2, 4, 6, 8, and 10 to the upper limit of the pulse height scale. Counting regions, such as 0–2000, 2–2000, 4–2000, 6–2000, 8–2000, and 10–2000 keV for lower level (LL) to upper level (UL) pulse height discriminator settings on a keV equivalent scale, may serve as one example of workable counting regions. However, other counting regions may be used. See L'Annunziata (1997) and L'Annunziata and coworkers (Noor *et al.*, 1996a).
2. The percent counting efficiency values of the nonquenched ^{14}C standard in each of the six counting regions are calculated according to Eq. 5.1.
3. An unknown sample is subsequently counted in the same six regions as the nonquenched ^{14}C standard.
4. The six CPM values of the unknown sample are plotted against the six percent counting efficiency values of the nonquenched ^{14}C standard.
5. The curve is then extrapolated to 100% counting efficiency, where the CPM of the unknown sample is equal to its DPM. Extrapolation may require a linear or multilinear regression least-squares best fit of the data points and definition of the equation to the line or curve to most accurately determine the point of intersection at 100% counting efficiency. An example of eight efficiency tracing curves for

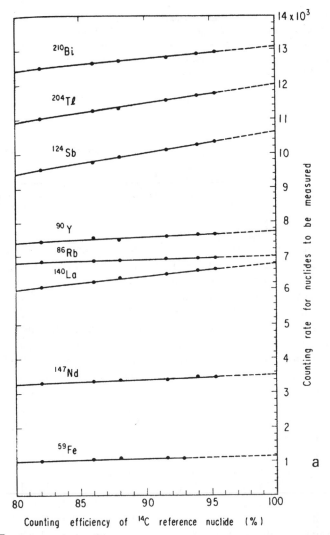

FIGURE 5.27 Efficiency tracing curves of eight radionuclide samples. Extrapolated values (dashed portion of the curves) indicate the counting rates at 100% counting efficiency or DPM of each nuclide (From Takiue and Ishikawa, 1978, reprinted with permission from Elsevier Science).

the DPM determination of eight radionuclides is illustrated in Fig. 5.27. The dashed portion of the plots are the extrapolated segments to 100% counting efficiency.

The efficiency tracing direct DPM method is unaffected by sample scintillation cocktail volumes over the range of 1–20 mL, composition of the scintillation cocktail, amount or kind of quenching agent, or size and material of the counting vial. These variables affect the slope and possibly even the shape (curvature) of the efficiency tracing curve of the sample of unknown activity; however, the extrapolated value of CPM at 100% counting efficiency remains

TABLE 5.3 Percent Recoveries of Calculated Activities of Five Composite Mixtures of ^{86}Rb–^{35}S–^{33}P Determined by the Efficiency Tracing (ET) Technique[a]

Sample DPM (in hundreds) ^{86}Rb : ^{35}S : ^{33}P	Total DPM (actual)	Total DPM (ET)	Total DPM recovery (%)
4326:7294:7194	18,814	18,671	99.2
2146:3620:3424	9,190	9,185	99.9
1042:1794:1550	4,386	4,408	100.5
3113:5510:4620	13,243	13,237	100.0
432.3:742.6:646.1	1,821	1,819	99.9

[a]From L'Annunziata and coworkers (see Noor *et al.*, 1996a), reprinted with permission from Elsevier Science).

constant. In other words, regardless of these sample variables, the extrapolated ET curve provides the DPM of the sample. The constant DPM values obtained, regardless of these variables, were demonstrated by Fujii *et al.* (1986) in a study of the efficiency tracing DPM measurements of ^{14}C, ^{35}S, ^{32}P, ^{36}Cl, ^{45}Ca, and ^{131}I on filter disks in LSA counting vials.

An additional attribute of the efficiency tracing technique is the possibility of determining the total DPM of mixtures of β-emitting radionuclides and α–β emitters as demonstrated by Fujii and Takiue (1988a,b) and L'Annunziata and coworkers (see Noor *et al.*, 1996a). Table 5.3 illustrates the excellent recoveries obtained for total DPM measurements of mixtures of ^{86}Rb $+$ ^{35}S $+$ ^{33}P. Tests were also performed in this same work to demonstrate that the total DPM measurements of mixtures provided constant recoveries regardless of quench level. It should be noted, however, that the technique provides the total DPM of the mixture and not the activities of the individual radionuclide components of the mixture.

This direct DPM method is extremely useful for the determination of activities of radionuclides of relatively short half-life for which NIST-traceable standards are not available commercially. The ET-DPM method is an automatic radionuclide activity analysis option available with some liquid scintillation analyzers of PerkinElmer Life and Analytical Sciences (Boston, MA). With these LSAs, the DPM of any β-emitting radionuclide of $E_{max} > 70$ keV is determined with a homogeneous radionuclide sample in scintillation cocktail using a preprogrammed ET-DPM counting mode. The instrument automatically determines the DPM of the sample including a plot of the efficiency tracing curve irrespective of the sample quench level.

The efficiency tracing DPM technique is reviewed by Kessler (1991). The method was tested by L'Annunziata (1997a) for β-emitting radionuclide samples over a wide range of quench and levels of sample activity. On the basis of these tests the following conclusions and recommendations were made:

1. The efficiency tracing DPM (ET-DPM) technique is an accurate method for determining the total activity (DPM) of β-emitting and β–γ-emitting radionuclide samples with the exception of ^{3}H.

2. The ET-DPM method can be used to determine the total activity (DPM) of a mixture of β-emitting radionuclides.

3. If the direct DPM measurement of ^{3}H is wanted, most commercial LSAs have a Direct-DPM mode, which determines the DPM of ^{3}H via coded assay method, that likely involves the use of a quench correction curve. For precise work it is best to analyze for ^{3}H activity using a quench correction curve prepared by the user with a scintillation cocktail equivalent to the cocktail containing the sample of unknown ^{3}H activity.

4. No quench correction curves are needed, either for chemical or color quench, when using the ET-DPM method.

5. For samples of low count rates (< 1500 CPM), it is recommended that one use a conventional DPM method, which utilizes a quench correction curve. A quench correction curve may be made from a higher level of the radionuclide in question after standardization by the ET-DPM method.

6. For best results samples should be counted for a duration sufficient to achieve a % 2 sigma standard deviation (% 2s) of 1% or lower of the count rate.

7. The ET-DPM method is very useful for the determination of activities (DPM) of radionuclides of short half-life for which quenched standards are not available. However, if a very precise activity of the nuclide is required within the limits established by a national bureau of standards (e.g., NIST) for that of a primary standard the CIEMAT/NIST or other comparable method of radionuclide standardization is recommended. This method is described in Section IX of this chapter.

8. The ET-DPM method may be used to determine with certitude the activity of a source radionuclide prior to initiating a tracer experiment with that nuclide. Before beginning an experiment with a radionuclide as a tracer, it is best not to accept blindly the cited activity provided by the radioisotope supplier on the label of the source container. It is best to prepare replicate samples of the radionuclide source and use the ET-DPM method to determine the absolute activity (DPM) of the radionuclide source before beginning an experiment with that source.

4. Multivariate Calibration

The principles of multivariate calibration, including the multivariate methods of multiple linear regression (MLR), principal component regression (PCR), and partial least squares (PLS), among others are described in detail by Beebe and Kowalski (1987) and Thomas and Haaland (1990). The practical applications of MLR are indisputable, as this statistical method is of widespread use. The authors noted that PCR and PLS is gaining acceptance in chemistry, as the laboratory computer can facilitate data collection and processing required for multivariate calibration. As a statistical mathematical tool, multivariate calibration can be applied to a chemical or physical analysis when more than one measurement is acquired for each sample. Mathematical data matrices are written according to the numbers of samples and variables implicated in an analytical result. As explained by Beebe and Kowalski (1987), PLS is a factor-based modeling procedure in which factors are defined for any linear combination of the variables in the data matrices. The PLS algorithm estimates the factors in these matrices to provide a prediction of the observation on an unknown sample. Information on the mechanics of calibration and prediction using the PLS algorithm can be obtained from Geladi and Kowalski (1986) and a rigorous treatment of the methodology from Lorber et al. (1986).

García et al. (1996) applied partial least squares (PLS) as a multivariate calibration to determine (predict) ^{14}C activities in samples over the activity range of 1.48–15.16 DPM per sample–scintillation cocktail mixture. They used a Packard Tri-Carb® 2000 CA/LL liquid scintillation analyzer (now PerkinElmer Tri-Carb LSA) and samples with variable quench levels using a quenching agent in the concentration range of 0–0.6% CCl_4. For the multivariate calibration they applied PLS regression using the PLS-Toolbox package for MATLAB devised by Wise (1992). The factors considered in two models constructed by García et al. (1996) consisted of ^{14}C content, quenching, blank, and spectral variability in one model, with blank omitted from the second model. They obtained slightly improved results omitting the blank (background) among the factors considered. Among 16 samples tested over the activity range of 1.48–15.16 DPM per sample, they obtained predicted activities with a relative error in the range of 0–5.4% (average = 1.09% relative error).

This is the first application of multivariate calibration to single-label sample activity determinations. The multivariate calibration approach has been applied by Toribo et al. (1995, 1996, and 1997) for the simultaneous liquid scintillation analysis of a mixture of alpha emitters. The multivariate calibration approach to the analysis of ^{14}C is new and not yet applied routinely. However, it has the advantage of being a time-saving approach to low-level liquid scintillation counting, because background information is not needed and, therefore, total counting time is reduced.

5. Other Direct DPM Methods

Other direct DPM methods exist. These methods such as the triple-to-double coincidence ratio (TDCR) efficiency calculation technique and the CIEMAT/NIST efficiency tracing with ^{3}H technique are employed specifically to the standardization of radionuclides by liquid scintillation analysis rather than routine radioactivity measurements. A treatment of these methods are found in Section IX of this chapter concerning radionuclide standardization.

VI. ANALYSIS OF X-RAY, GAMMA-RAY, ATOMIC ELECTRON AND POSITRON EMITTERS

Liquid scintillation analysis (LSA) is used principally for the analysis of beta- and alpha particle-emitting nuclides. However, liquid scintillation is also applied to the analysis of certain gamma-ray emitters and nuclides decaying by electron capture and emitting x-rays, Auger electrons, and internal conversion electrons (Grau Malonda, 1999). This broad potential of LSA was illustrated previously in this chapter (see Fig. 5.10) where the 8 keV Auger electrons and 93 keV conversion electrons emitted from ^{67}Ga produce pulse height spectra with peaks that coincide closely to those produced by ^{3}H and ^{14}C (McQuarrie and Noujaim, 1983).

Vanadium-49 is another radionuclide that decays by pure electron capture. It emits x-rays and Auger electrons of very low energy (< 5 keV). Rodríguez Barquero *et al.* (1998) and Rodríguez Barquero and Los Arcos (2000) demonstrate that LSA is the preferred method, because LSA is not affected by self-absorption problems that would otherwise be prevalent with such low-energy electron emitters. They report liquid scintillation counting efficiencies between 8 and 25% for ^{49}V.

In certain cases the counting efficiencies of radionuclides that decay by electron capture emitting x-rays and Auger electrons may be higher with the liquid scintillation technique than those attainable with the thin-walled NaI(Tl) solid scintillation crystal detector. The LSA with its automatic sample changer and computer is more commonly encountered in laboratories than its solid scintillation counterpart. This reflects the driving force behind finding broader ranges of application of liquid scintillation counting.

The interaction of x- and gamma-rays with liquid scintillation cocktail is principally the result of the Compton effect whereby part of the energy of the x- or gamma-ray photon is imparted to orbital electrons. An ejected electron (Compton electron) imparts its energy in material in a fashion similar to that of a beta particle. The absorption of its energy by the liquid scintillation cocktail results in fluorescence with the emission of photons of visible light. In liquid scintillation cocktail, the photoelectric effect usually does not occur over 30 eV. However, the photoelectric interaction is significant for radionuclides decaying by electron capture and emitting low-energy x-rays (Grau Malonda and Grau Carles, 2000). Bransome (1973) reports that the photoelectric effect can be evident at higher gamma-ray energies in the glass vial walls or, in the sample–scintillation cocktail mixture, if the scintillator is loaded with heavy elements.

Cherenkov photons will be produced in liquid scintillators to a significant extent if gamma-ray energies are high enough to produce Compton electrons of sufficient energy to cause the Cherenkov effect (Grau Carles and Grau Malonda, 1992 and Grau Carles *et al.*, 1993). The Cherenkov effect is discussed in detail in Chapter 9.

Numerous studies have been undertaken on the liquid scintillation analysis of ^{55}Fe, which decays exclusively by electron capture emitting x-rays and Auger electrons of low energy, 0.6–6.5 keV. Some examples that can be cited are Dern and Hart (1961a,b), Perry and Warner (1963), Eakins and Brown (1966), Cosolito *et al.* (1968), Miller *et al.* (1969), Cramer *et al.* (1971), Horrocks (1971), Grau Malonda (1982), Ortiz *et al.* (1993), Günther (1998), Ceccatelli and De Felice (1999), and Grau Malonda and Grau Carles (2000). Electron capture decay gives rise to the emission of x-rays, Auger electrons, and internal conversion electrons, which interact with the liquid scintillation cocktail to cause fluorescence. Recent papers report counting efficiencies of 50–62% for [55]Fe (Ortiz *et al.*, 1993; Günther, 1998; Grau Malonda and Grau Carles, 1999, 2000).

Another popular radionuclide, which decays exclusively by electron capture, is [125]I. The radionuclide is currently very popular as a tracer in the biological sciences (e.g., Malkov *et al.*, 2000; Teresa *et al.*, 2000; Larsson *et al.*, 2001) and liquid scintillation is a very convenient and efficient means of analysis. The electron-capture decay results in the emission of 35 keV gamma radiation in 6.7% of the transitions and the emission of converted electrons in 93.3% of the transitions (Grau Malonda, 1999). Early studies reported a liquid scintillation counting efficiency of 56% for ^{125}I by standard liquid scintillation counting techniques (Rhodes, 1965), and later yet higher counting efficiencies of over 80% were reported by Jordan *et al.* (1974), Horrocks (1976c), Ring *et al.* (1980), Chandrasekaran (1981), Kits *et al.* (1985), and Grau Carles *et al.* (1994c). For example, the quench calibration curve for ^{125}I of counting efficiency versus the sample spectrum quench indicating parameter SIS in Fig. 5.28 illustrates an optimum counting efficiency of 80%. More recent studies by Grau Malonda and Grau Carles (2000) report counting efficiencies of over 88% for ^{125}I in Insta Gel Plus and Ultima Gold liquid scintillation cocktail.

As discussed in Chapter 1, either the electron-capture decay process or the emission of an internal conversion electron leaves an orbital electron vacant. For the case of ^{125}I this vacancy may be filled by electrons from outer shells giving rise to the emission of x-rays of the Te daughter

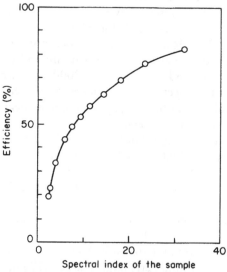

FIGURE 5.28 Quench correction curve for ^{125}I based on the sample pulse height quench-indicating parameter, **SIS.** The photon emissions from ^{125}I in scintillation cocktail are due to cocktail interactions with 35 keV gamma-ray emissions in 6.7% of the transitions, internal conversion electrons in 93.3% of the transitions and abundant Auger electron and Te K x-ray emissions (see Table of Radioactive Isotopes in the Appendix). [From **Ring et al., 1980,** reprinted with permission from Elsevier Science.]

nuclide and Auger electrons. Horrocks (1976c) explained that the electron capture process in ^{125}I involves capture of a K-shell electron in 80% of the decay transitions and an L-shell in the remaining 20%. The Te L x-ray is 3.5 keV in energy and totally absorbed by the liquid scintillation cocktail, whereas, the Te K x-ray is emitted with 27.7 keV energy and has a high probability of escape. He concluded that the excitations in the liquid scintillation cocktail are due mainly to the absorption of Auger and internal conversion electrons, and only a minor contribution (about 8% of the fluor excitations) is the result of x-rays produced during the decay process.

Zinc-65 is another radionuclide used in the biological sciences as a tracer (Wolterbeek *et al.*, 2002). It decays by electron capture and positron emission (see Table of Radioactive Isotopes in the Appendix). Günther (1998) and Sandhya and Subramanian (1998) report liquid scintillation counting efficiencies up to 76%. Only 1.5% of the ^{65}Zn nuclide transitions to stable ^{65}Cu occur via β^{+} emission ($E_{max} = 325$ keV). Consequently positron interaction with scintillation cocktail contributes only a small portion to the overall detection efficiency. About 50.5% of the transitions occur via electron capture (EC) to the ground state of ^{65}Cu and the remaining 48% by EC with accompanying gamma emission (Günther, 1998). Consequently, the abundant atomic electron and x-radiation, that accompany the EC decay process of ^{65}Zn, are the emissions that generate significant liquid scintillator excitation and light emission.

Barosi *et al.* (1980) reports the liquid scintillation assay of ^{51}Cr with a maximum counting efficiency of 87%. Chromium-51 decays by electron capture and 10% of the excited nuclei simultaneously undergo further decay with the emission of gamma radiation of 320 keV energy or the emission of internal conversion electrons of 315 keV. Internal conversion electron emission compete with the emission of gamma radiation, and the conversion electrons are always slightly lower in energy than the gamma radiation. The energy difference is equivalent to the binding energy of the atomic electron (see Chapter 1, Eq. 1.43). X-ray and Auger-electron emission, which accompany electron-capture decay, also must be considered among the processes that generate scintillation fluor excitation. Chromium-51 decays with the emission of 5 keV x-rays and 4.5 keV Auger electrons in 91% of the transitions (see Barosi *et al.*, 1980). The double radionuclide tracer ^{59}Fe–^{51}Cr was assayed by Barosi *et al.* (1980) in red blood-cell kinetic studies. If the double label is assayed by NaI(Tl) solid scintillation counting of the gamma-ray photopeaks, optimum counting efficiencies of 15% and 3% are reported for ^{59}Fe and ^{51}Cr, respectively. However, if liquid scintillation counting is used, optimum counting efficiencies of 20% and 15% are reported for the ^{59}Fe and ^{51}Cr double label, respectively. The fivefold increase in the counting efficiency of ^{51}Cr is due mainly to the liquid scintillation cocktail absorption of x-ray and Auger electron energy.

Positron-emitting nuclides can be assayed by LSA with a high detection efficiency when positron emission is the principal mode of decay. As described in Chapter 1, positrons have similar interactions, ranges, and stopping powers as negatrons of similar energy; however, in addition positrons will produce annihilation radiation (0.511 keV gamma rays) when the positrons come to rest in the proximity of atomic electrons. As is the case of negatrons, counting efficiencies of positrons as high as 100% can be obtained with the use of conventional liquid scintillation cocktails, and, when the E_{max} of the positrons are sufficiently above the Cherenkov threshold in water (see Chapter 9), these may be analyzed by Cherenkov counting (Table 5.4).

Wiebe *et al.* (1980) and McQuarrie *et al.* (1981) measured the Compton electron contribution to the count rate resulting from the interactions of 511-keV annihilation gamma rays with the liquid scintillation cocktail. Thus, in addition to positron–scintillation cocktail interactions, they point out that as much as 24% of the observed count rate is due to scintillation cocktail interactions with Compton electrons originating from annihilation radiation. Wiebe *et al.* (1980) showed that the largest amount of energy that may be deposited in a liquid scintillation cocktail is equivalent to the sum of the highest positron energy and the energy deposited by the annihilation gamma rays or

$$E_{max} = E_{\beta^+} + 2E_{e^-} \tag{5.28}$$

where E_{max} is the maximum energy deposited by the positron in the scintillation cocktail, E_{β^+} is the maximum positron energy, and E_{e^-} is the energy of the Compton edge associated with 511-keV annihilation gamma radiation (341 keV). For example, in the case of ^{18}F, $E_{max} = 635\,keV + 2(341\,keV) = 1317\,keV$. Two times the energy of the annihilation radiation Compton edge must be accounted for, because a positron annihilates with the simultaneous emission of two gamma rays of 511 keV energy.

Other radionuclides analyzed by LSA that decay by electron capture with the emission of gamma radiation, x-rays, and atomic electrons are ^{54}Mn, ^{85}Sr, ^{88}Y, ^{109}Cd, and ^{133}Ba among others (Los Arcos *et al.*, 1991; Grau Carles *et al.*, 1994c; Grau Malonda and Grau Carles, 1999; Wolterbeek and

---

TABLE 5.4 Liquid Scintillation (LS) and Cherenkov Counting Efficiencies of a Few Positron-emitting nuclides[a]

Nuclide	$E_{\beta^+\,max}$	Half-life	LS Counting efficiency (%)	Cherenkov counting efficiency in water (%)
^{18}F	0.635 (96.9%)[b]	109.7 m	100	3.7
^{68}Ga	0.80 (1.3%) 1.889(89%)	68.3 m	100	47
^{34m}Cl	1.35(24%)	32.0 m	100	57
^{34}Cl	2.47(28%) 4.50(47%)[c]	1.5 s		

[a]From McQuarrie *et al.* (1981).
[b]Energy values are in MeV and the intensities of the decay mode are given alongside in parenthesis.
[c]Of ^{34m}Cl.

van der Meer, 2002). Liquid scintillation counting efficiencies of over 60% and 70% are reported for ^{85}Sr and ^{109}Cd, respectively (Grau Carles *et al.*, 1994c).

VII. COMMON INTERFERENCES IN LIQUID SCINTILLATION COUNTING

The counting interferences most commonly found in LSA and how each can be recognized and/or corrected to obtain accurate and reproducible DPM values must be considered. Six major counting interferences exist in the scintillation counting of samples: (1) background radiation, (2) quench (color, chemical, and ionization), (3) multiple radionuclides in the same sample, (4) luminescence, (5) static, and (6) wall effect. Each of these interferences will be considered here and in other parts of this chapter with special attention given to their identification, elimination, or means of correcting for any errors that these may generate.

A. Background

Background is defined as counts arising from sources external to the sample, such as cosmic or environmental radiation, and from instrument noise and PMT crosstalk. When determining sample count rates (CPM_s) from which sample disintegration rates (DPM_s) will be determined according to procedures described previously in Section V, it is necessary to obtain an accurate measure of the background count rate (CPM_{bkg}) whenever background counts are significant relative to the sample counts. Background count rates are determined by counting a blank counting vial containing the scintillation cocktail plus all other chemical constituents used in the preparation of samples with the exception of the radionuclide of interest. In other words, the blank should have the same quench level as the radioactive samples to be analyzed. For example, if the radioactive samples are measured in a sample–scintillation cocktail mixture of 50% water (1 : 1 water load), the background count rate should be determined in the blank sample 1 : 1 water–scintillation cocktail mixture. Ideally any other quenching agents that may be present in the sample should also be present in the background blank counting vial. Such a blank can be obtained by preparing a sample containing no radionuclide of interest in a fashion similar to the preparation of the experimental samples. Once a blank is prepared, it must be counted for a sufficient period of time to get an accurate measurement of its count rate. The time required for counting background blanks can be decided by using statistical criteria presented in Chapter 7.

Once the background count rate is determined, most modern LSAs store the background pulse height spectrum in computer memory. The background counts for any given counting region of the pulse height spectrum can then be subtracted automatically from the sample count rates to provide a net count rate according to the following equation:

$$CPM_{net} = CPM_s - CPM_{bkg}. \qquad (5.29)$$

If the background count rate (CPM_{bkg}) is significant compared to the sample count rate (CPM_s), it is necessary to subtract the background contribution according to the above equation. The net count rate (CPM_{net}) is then used to determine disintegration rates according to Eq. 5.7 described in Section V.B.a of this chapter. The majority of the background counts are found in the low end of the liquid scintillation pulse height spectrum, such as 0–5.0 keV for instruments that have pulse height spectra calibrated over the energy range of 0–2000 keV.

Methods of reducing background to optimize instrument performance are provided in Section XVII.C of this chapter and in Chapter 6.

B. Quench

Color and chemical quench are described in detail in Section IV of this chapter. In brief, quench affects the scintillation photon intensity and efficiency of detection of radionuclides in the liquid scintillation cocktail. The lower the radiation energy of the radionuclide, the greater is the effect of quench on the counting efficiency of the sample. Four common methods of quench correction are (1) internal standardization, (2) sample spectrum quench correction, (3) external standard quench correction, and (4) Direct DPM methods. These are treated in detail in Section V. A third mechanism of quench not yet described is ionization quench, which is the reduction in the number of radiation-excited scintillation cocktail molecules as a consequence of ionization generated by the nuclear radiation with the associated reduction in photon intensity. Ionization quench is corrected for generally in the quench correction techniques described in Section V of this chapter, and therefore, it may be of little concern to many experimentalists. However, persons involved in the standardization of radionuclides, particularly radionuclides emitting low-energy beta particles or those decaying by electron capture must include ionization quench correction in the calculation of counting efficiency utilized in the CIEMAT/NIST and TDCR procedures (Grau Malonda and Grau Carles, 1999, 2000; Grau Carles and Grau Malonda, 2001; García and Grau Malonda, 2002) described in Section IX of this chapter.

C. Radionuclide Mixtures

Multiple radionuclides in samples can present an interference when the energy spectra of the two radionuclides overlap. This is due to the fact that all beta-emitting radionuclides produce a continuum spectrum of beta-particle energies from zero to the E_{max} as illustrated in Fig. 1.4 of Chapter 1. If two radionuclides are present in the same sample (e.g., ^{3}H and ^{14}C of E_{max} 18.6 and 156 keV, respectively), the lower-energy beta particles of tritium

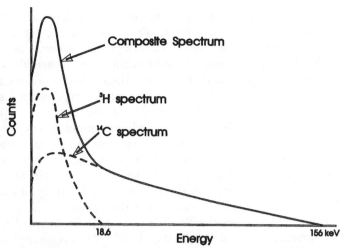

FIGURE 5.29 Overlapping liquid scintillation pulse height spectra of ^{3}H and ^{14}C with characteristic beta-particle E_{max} of 18.6 and 156 keV, respectively. An example of a composite spectrum of the two radionuclides as a mixture is also illustrated.

produces a spectrum of pulse heights equivalent to the range of zero to at most 18.6 keV depending upon the quench in the sample as illustrated in Fig. 5.29. The second radionuclide ^{14}C, which emits beta particles of higher energy, produces a pulse height spectrum overlapping that produced by tritium and extend to a maximum of 156 keV. Figure 5.29 is a graphic composite spectrum of a sample with both ^{3}H and ^{14}C. In order to quantify the separate radionuclide activities (DPM) of such a dual mixture in a sample, the count rates and counting efficiency of each radionuclide must be determined. Several methods may be employed for the measurement of two, three, or even more beta-emitting radionuclides in a mixture by LSA. These methods are described in detail in Section VIII of this chapter.

D. Luminescence

Luminescence in liquid scintillation fluor cocktails or in aqueous buffer media refers to the emission of light photons as a consequence of energy absorption and concomitant molecular excitation from origins other than nuclear radiation. Luminescence can be a practical tool in the study of biochemical reactions or an interference in LSA. This section will provide a treatment of the various types of luminescence encountered in LSA and recommendations on how to avoid or minimize any interference that some types of luminescence can present.

I. Bioluminescence

Bioluminescence occurs when a biochemical reaction produces photons, which is a desirable reaction when used as a tool to study certain biochemical assays. Some examples are the biochemical reactions catalyzed by avidin-alkaline phosphatase, horseradish peroxidase, β-galactosidase, luciferase, xanthine oxidase, and ATP assays via luciferin–luciferase. The LSA can be utilized to count all luminescent events when these are used to study certain biochemical reactions. Under such circumstances, the LSA counting protocol should be set to count the experimental samples in the Single Photon Counting (SPC) mode. In this counting mode only a single PMT is used without the coincidence counting circuitry. SPC is used because all luminescent events are single photon in nature and would be eliminated by the two PMTs and the coincidence counting circuitry. Therefore, in the SPC mode the coincidence counting circuit is disabled and only one PMT is operational. In addition, because bioluminescence normally produces a large amount of light compared to a radioactive event, the high voltage on the PMT is automatically lowered in the SPC mode to prevent saturation of the PMT. The counting region for bioluminescent samples in the LSA is generally 0–10 keV for pulse height scales calibrated in keV energy equivalence.

2. Photoluminescence and Chemiluminescence

Luminescence as an interference can occur in the assay of radioactive samples in scintillation cocktail. There are primarily two types of luminescence: photo- and chemiluminescence.

Photoluminescence is the result of the exposure of the sample–scintillation cocktail mixture to ultraviolet light. Photoluminescence is normally single photon in nature and decays in a matter of minutes. Therefore, letting freshly prepared samples in scintillation cocktail remain in the dark of the LSA for 10–15 min completely eliminates any photoluminescence that may occur. A precount delay time can also be used with counting protocols of certain LSAs.

The second type of luminescence, chemiluminescence, is a frequent interference in the liquid scintillation assay of radioactive samples. This is the production of light within the scintillation cocktail due to a chemical reaction. Chemical reactions that cause chemiluminescence often occur when scintillation cocktail is added to the sample solution in the liquid scintillation counting vial. A chemical reaction can occur, for example, when scintillation cocktail is added to a basic sample solution (pH 8–14) or when a chemical substance, such as hydrogen peroxide, is present in the sample. The pH effects and chemical interactions with some component of the scintillation cocktail cause molecular excitation and light emission. Some types of samples that can produce a considerable chemiluminescence are tissue or cell digests with inorganic bases (e.g., NaOH, KOH, Solvable^{TM}) or organic bases (e.g., Soluene 350^{TM}).

During the chemiluminescent reaction, single photons are produced in the scintillation cocktail, and because of their large number they may bypass the coincidence circuit and be registered as counts together with counts produced by the radionuclides in the sample. The counting of single-photon chemiluminescence events by the coincidence (dual PMT) circuitry can be demonstrated from the equation of Horrocks and Kolb (1981), which is written as

$$N_C = 2_{\tau R} N_1 N_2, \tag{5.30}$$

where N_C is the coincidence count rate, τR is the coincidence resolving time (e.g., 30 ns), and N_1 and N_2 are the single-event count rates from photomultiplier tubes 1 and 2, respectively. As the single-photon events from chemiluminescence increase in frequency, the probability that they produce a coincidence count rate correspondingly increases according to Eq. 5.30. This is illustrated by the data of Table 5.5. Luminescence count rates can be very high, depending on the chemical constituents and/or pH of the sample. From Table 5.5, it is obvious that if sample count rates are low, luminescence can be a source of serious error. In any event, luminescence should be identified, if it occurs, and it should be eliminated before counting to avoid error in the activity determinations of experimental samples.

Luminescence, which may occur in sample–scintillation cocktail mixtures, can be detected easily when counting relatively high-energy β-emitting radionuclides ($E_{max} > 70$ keV). For example, Fig. 5.30 illustrates the pulse height spectrum of a luminescent sample, a tritium sample, and a ^{14}C sample. As illustrated in Fig. 5.30, the luminescence spectrum occurs at the very low energy portion of the pulse height spectrum occurring generally in the 0–6 keV region for pulse height spectra on an energy equivalent scale. Luminescence in a sample can be recognized easily by one of three methods, namely, (1) spectral display, (2) counting region settings, and (3) instrumental detection and measurement. As illustrated in Fig. 5.30, a sample containing appreciable chemiluminescence produces a pulse height spectral peak in the 0–6 keV region on top of the main radioactivity pulse height peak. The figure illustrates the chemiluminescence pulse height spectrum overlapping with a dual ^{3}H and ^{14}C pulse height spectra. When counting single-radionuclide samples two counting regions can be used to detect luminescence. For example, when counting ^{14}C the following counting

TABLE 5.5 **Random Coincidence Count Rate N_C, as a Function of Single Photon Levels, N_1 and N_2**

$N_1 = N_2$(CPM)[a]	N_C(CPM)[b]
10000	0.1
50000	2.5
100000	10.0
500000	250
1000000	1000
5000000	25000

[a]From Horrocks and Kolb (1981).
[b]$N_C = [2(30 \times 10^{-9})/60](N_1/N_2) = 10^{-9}(N_1/N_2)$.

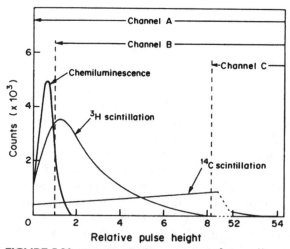

FIGURE 5.30 Pulse height distributions of ^3H and ^{14}C scintillation pulses and of chemiluminescence pulses, and channel settings for the analytical measurement of each activity and chemiluminescence count rate (From Takiue *et al.*, 1985, reprinted with permission from Elsevier Science).

regions may be used: region A: 0–156 keV and region B: 4.0–156 keV. If the counts in these regions are similar, then little if any luminescence would be expected. If, on the other hand, the counts in region A (0–156 keV) are much higher (25–500%), luminescence may be predicted. In addition, many modern LSAs are able to determine the magnitude of luminescence in a sample and print the percent luminescence on the same page as the CPM and DPM values of each sample. The percent luminescence is calculated by the instrument according to the equation

$$\% \text{luminescence} = \frac{\text{chance coincidence events}}{\text{true coincidence events}} \times 100 \qquad (5.31)$$

When luminescence is detected or even suspected in experimental samples, it can be controlled, corrected for, or even eliminated as discussed subsequently.

3. Luminescence Control, Compensation, and Elimination

Once luminescence is recognized as a problem with a particular set of samples, it can be controlled by using proper sample preparation procedures and even eliminated or compensated for by certain liquid scintillation counter controls. The four most common methods of reducing or correcting sample luminescence are: (1) chemical methods, (2) temperature control, (3) counting region settings, and (4) delayed coincidence counting also referred to as random coincidence counting.

a. Chemical Methods

Chemical methods used to avoid or suppress chemiluminescence are reviewed by Peng (1976). Among these, neutralization of alkaline sample solutions with a nonoxidizing acid is recommended, as basic sample solutions are often the main cause of chemiluminescence. This can be accomplished by: (1) adding 10 mL of acetic acid to a gallon of scintillation cocktail providing this does not alter the performance of the fluor cocktail, (2) neutralizing the basic sample solution before adding the scintillation cocktail to the sample counting vial, or (3) using a special scintillation cocktail designed to reduce or suppress luminescence such as Insta-Fluor, which contains a chemiluminescence inhibitor or Hionic-Fluor, which displays a very fast chemiluminescence decay property. See Chapter 8 for scintillation cocktail properties and characteristics. This normally reduces the amount of chemiluminescence. A problem associated with sample neutralization may occur when counting large macromolecules found in certain biological samples, in which acidification of samples may cause precipitation. Precipitation of sample most often includes precipitation of the radioactive material and, therefore, loss of counting efficiency and incorrect DPM values.

b. Temperature Control

Acidification of sample solution as already described followed by heating to 40°C is often recommended to reduce chemiluminescence, or acidification may be omitted when it causes precipitation and the sample solution only heated. Heating the sample to 40°C helps drive the chemiluminescence reaction to its endpoint. This is possible because the reaction is a chemical reaction and every 10°C increase in temperature doubles the reaction rate. An alternative procedure of temperature control of luminescence is to cool the reaction using an LSA with temperature control. Cooling slows the reaction rather than accelerating its termination. Cooling reduces counts from chemiluminescence, which can be eliminated altogether by counting region setting or delayed coincidence counting, discussed subsequently in this section.

c. Counting Region Settings

This is the recommended method for all radionuclides with the exception of tritium. This method is relatively simple but not often used. If the counting region for a mid- to high-energy β-emitting radionuclide (e.g., ^{14}C and higher energy β emitters) is set at approximately 10 keV or above, no luminescence of any kind will be observed in the sample counts and, therefore, no correction will be necessary. The only precaution when performing DPM determination is to set the same counting region for the quench correction curve and for the experimental samples. A counting region for ^{14}C to avoid error from luminescence would be, for example, 10.0–156 keV for lower level and upper level discriminator settings when using instruments that have pulse height discriminator settings calibrated to keV energy equivalence. In the case of tritium, a counting region cannot be set to avoid luminescence, because tritium emits very low-energy beta particles ($E_{max} = 18.6$ keV) producing a pulse height spectrum that greatly overlaps that of luminescence. In the case of tritium, chemiluminescence must be eliminated or measured and subtracted from the sample counts for accurate activity calculations (Takiue et al., 1984). Figure 5.30 illustrates other discriminator settings to define three counting regions, proposed by Takiue *et al.* (1985), which permit the simultaneous counting of chemiluminescence, ^{3}H, and ^{14}C, when the DPM analysis of dual-radionuclide samples containing chemiluminescence is required. Equations for calculating the activities (DPM) of the dual-radionuclide samples and the count rates due to chemiluminescence are given by Takiue *et al.* (1985, 1986). Detailed information on multiple-radionuclide analysis is provided in Section VIII of this chapter.

d. Delayed Coincidence Counting

Delayed coincidence counting, also referred to random coincidence counting, is a method for the elimination of error resulting from luminescence that can be applied to the liquid scintillation analysis of all radionuclides including tritium. For this method to work, a circuit with a 20-ns delay is added to one of the two PMTs. The sample containing the radioactivity is counted simultaneously with and without the delay coincidence circuit enabled. If the coincident circuit is used without the delay or random coincidence circuit, both the luminescence and radioactive decays will be detected. The radioactive decay events are accepted by the coincidence counting circuit because they are isotropic multiphoton events; and luminescent single-photon events are also accepted by the coincidence counting circuit because they occur at such a high count rate in the sample, and are detected within the resolving time of the counting circuit. Now, if the coincident circuit is used with a delay mode added to one of the PMTs, only the single-photon luminescence events will be detected, because of their high count rate of occurrence. Radioactive decay, which produces isotropic multiphoton events, will not be detected by the coincidence circuit due to a 20-ns delay in one of the two PMTs. Finally, the counts collected from the two readings with and without the delay circuit enabled are subtracted channel by channel of the MCA over the pulse height region equivalent to 0–6.0 keV. The resultant spectrum will be a product of the actual nuclear decay events without chemiluminescence. This special delay method is known as the luminescent detection and correction method. It can be applied to all radionuclides independent of radioisotope decay energy. Luminescence detection and correction are available with most state-of-the-art LSAs (Kessler, 1989).

E. Static

Electrostatic discharge is a photon-producing interference in liquid scintillation counting. Static electricity may be generated by friction or pressure between two materials. When nonconductive materials are separated, one material develops a positive and the other a negative charge. Static consists of charged ions, positive or negative, which are atoms electrically out of balance due to the removal or addition of electrons. The intensity of static electricity can be measured as positive or negative voltage on the surface of matter in magnitudes of tens of thousands of volts. The discharge of static electricity is a random event; but the phenomenon commonly occurs with many materials we may come into contact with when preparing samples for counting, including scintillation counting vials. Static electricity is produced easily in low-humidity rooms during the time of the year when dry heat is used to warm laboratories. A common characteristic of static electricity is its stability; it can remain on the surface of scintillation counting vials for relatively long periods of time. When the scintillation vial is placed in the counting chamber and electrostatic discharges occur, it is like an electrical lightning storm occurring in or on the surface of the scintillation vial producing random pulse events. The static charge buildup can have many causes, including shipping, handling, use of plastic gloves, and low humidity in the sample preparation area. Plastic vials tend to build up more of a static charge than glass vials.

There are primarily four methods of reducing or eliminating static from sample vials for liquid scintillation counting: (1) the use of an electrostatic controller, (2) selection of vial type, (3) antistatic wipes, and (4) humidification of the sample preparation and counting area. A brief description of each method is given.

The electrostatic controller is a circular donut-shaped device located in the elevator tube through which the counting vial must pass before it is moved into the counting chamber. In certain instruments it contains eight geometrically located electrodes, which generate a 360° field of electrically produced ions. When the counting vial passes through the electrostatic controller, it enters the field of electrically produced counterions, which can neutralize static electricity on the counting vial surface in a matter of 2 s. This process occurs just before the robotic positioning of the vial into the counting chamber located between the two PMTs. Contemporary state-of-the-art LSAs are equipped with an electrostatic controller. Although the electrostatic controller offers no guarantee of removing all static from the counting vial surface, there may be no need to take any other steps to control static on the surface of counting vials before placing them in the LSA sample changer.

Further measures may be taken to guard against static charge collection and discharge from counting vial surfaces. One step is to select a type of counting vial that would tend to collect less static electricity. Because plastic tends to hold a static charge more than glass, the use of glass vials helps to reduce the static charge for most samples. The disadvantages of using glass vials is that they are more expensive and more difficult to dispose of. The alternative to glass vials is to use special "antistatic" plastic vials. These vials are manufactured with a special plastic treatment that greatly reduces the amount of static on the vial surfaces compared with standard plastic vials. Another technique is to wipe each vial with an "antistatic" wipe or with a moist cloth just before placing the vial into the sample changer of the liquid scintillation counter. This readily removes the static charge on the surface of the vial just before counting. The final step that may be taken to reduce static is to increase the humidity in the room where the samples are prepared as well as in the counting area.

F. Wall Effect

When samples are counted in plastic vials with traditional cocktails, the organic scintillator from the cocktail can penetrate the wall of the plastic vial.

Traditional cocktails are those made with solvents such as toluene, xylene, and pseudocumene. This can cause a problem when external standard quench correction methods are used, because plastic vials with solvent penetration can scintillate causing a distortion of the external standard pulse height spectrum. The result would be inaccurate quench-indicating parameters, which would give erroneous counting efficiencies and, as a consequence, error in the DPM measurements of samples. This problem can be overcome easily by always using environmentally safer cocktails. The newer environmentally safer cocktail solvents, such as diisopropylnaphthalene and linear alkylbenzene will not penetrate into the wall of the plastic vial and cause the wall effect. As a general rule, always use glass vials when using traditional scintillation cocktails and use either plastic or glass vials with the environmentally safer cocktails.

VIII. MULTIPLE RADIONUCLIDE ANALYSIS

There is a wide-ranging need in the scientific community to analyze multiple radionuclides as mixtures. These include dual-, triple-, and multiple radionuclide mixtures stemming from research in the chemical, biological, nuclear power, and environmental sciences, among others, as reviewed by L'Annunziata (1984b), Takiue et al. (1991b,c, 1992, 1995, 1999), Toribio et al. (1995, 1996, 1997, 1999), Fujii et al. (2000), Kashirin et al. (2000), and Nayak (2001). Several methods are available for analyzing the activity of more than one β-emitting radionuclide in the same sample. Because of the broad spectrum of β particle energies emitted by any given radionuclide anywhere between zero and E_{max}, we always observe a broad pulse height spectrum in the LSA from zero to a maximum pulse height. Therefore, all liquid scintillation pulse height spectra from different β-emitting radionuclides overlap. Because of this spectral overlap and the very broad characteristics of β particle pulse height spectra, it was traditionally considered feasible to analyze by LSA at most three β-emitting radionuclides in the same sample provided their β particle energy maxima differed by a factor of three or four (L'Annunziata, 1979, 1984, and 1987). However, advances in LSA during 1990–1997 have revealed new regionless spectral unfolding and deconvolution methods capable of analyzing several β-emitting radionuclides in the same sample with the aid of computer processing. Even radionuclide mixtures of ^{14}C ($E_{max} = 156\,keV$) and ^{35}S ($E_{max} = 167\,keV$), which for decades were thought to be impossible to resolve because of their similar β particle energies can now be identified and quantified as mixtures. A description of the techniques used to resolve and quantify mixtures of β- and α-emitting radionuclides by LSA is provided in this section.

A. Conventional Dual- and Triple-Radionuclide Analysis

The conventional methods described in detail in this section refer, for the most part, to the analysis of two β-emitting radionuclides in a mixture.

However, the same principles may apply to the analysis of three radionuclides as a mixture, and reference to this will be given below, where appropriate.

I. Exclusion Method

The exclusion method is one of the original methods applied to the analysis of a dual-radionuclide mixture by LSA. It is described in detail by Okita *et al.* (1957), Kobayashi and Maudsley (1970), and L'Annunziata (1979). The technique is rarely used today, because of the availability of more efficient methods of dual- or triple-radionuclide analysis. Nevertheless, it is presented briefly here, as the reader will encounter occasional reference to this method in the current literature.

The dual radionuclide mixture of ^{3}H ($E_{max} = 18.6$ keV) and ^{14}C ($E_{max} = 156$ keV) will be taken as an example to describe this method, which requires a relatively large difference in β particle energies between the two radionuclides. It is recommended generally that the E_{max} of the two radionuclides differ by a factor of 3 or 4. Two counting regions are defined using lower level (LL) and upper level (UL) pulse height discriminators such that in counting region 1, also referred to as region A and in this case, we can refer to it as the tritium region (e.g., LL $= 0.0$ and UL $= 18.6$ keV), where both the tritium spectrum and the spilldown of the carbon-14 spectrum into the tritium region are found. In region 2, also referred to as region B and in this case the carbon-14 region, pulse height discriminators are set (e.g., LL $= 18.6$ and UL $= 156$ keV) to allow only pulses from the carbon-14 spectrum. The name exclusion method is derived from the fact that region 2 excludes all tritium pulses. When counting a mixture of ^{3}H and ^{14}C, count rates (CPM) will be collected in each region, and the following equations are used to calculate the DPM for ^{3}H and ^{14}C:

$$H = \frac{N_1 - C_{C_1}}{h_1} \qquad (5.32)$$

and

$$C = \frac{N_2}{c_2} \qquad (5.33)$$

where H and C are the activities or DPM of ^{3}H and ^{14}C, respectively, in the mixed radionuclide sample, N_1 and N_2 are the count rates (CPM) in regions 1 and 2 of the LSA, h_1 and c_1 are the counting efficiencies of ^{3}H and ^{14}C in counting region 1 and c_2 is the counting efficiency of ^{14}C in counting region 2.

From these equations it is clear that five parameters are needed to calculate the DPM for both the tritium and ^{14}C dual-labeled samples. The CPM in regions 1 and 2 are determined automatically by the LSC. The three efficiency factors are determined using three quench correction curves, which consist of plots made from a series of tritium and ^{14}C quenched standards. The three curves are constructed by plotting % counting efficiency for tritium and ^{14}C in region 1 and ^{14}C in region 2 versus an external standard quench-indicating parameter.

2. Inclusion Method

In this method the counting regions are set such that there are spillup and spilldown of pulse events in each region from both radionuclides. Again, we will use ^{3}H and ^{14}C as an example of a typical dual mixture, keeping in mind that the procedure presented here and equations used will work for any dual mixture of β-emitting radionuclides provided the β particle energies (E_{max} values) of the two radionuclides differ significantly. Also, in this discussion we will refer to lower energy and higher energy radionuclides. In this example, ^{3}H ($E_{max} = 18.6\,keV$) is the lower energy radionuclide and ^{14}C ($E_{max} = 156\,keV$) the higher energy radionuclide of the mixture.

For this method to work it is necessary that two counting regions (regions A and B) are established by setting the lower level and upper level pulse height discriminators to assure that there will be significant spillup of pulse events from ^{3}H into region B and the unavoidable spilldown of pulse events from ^{14}C into region A. An illustration of pulse height discriminator settings, which establish the counting regions for the analysis of two radionuclides by this inclusion method is given in Fig. 5.31.

Some modern LSAs are available with preset dual counting regions for the activity analysis of dual mixtures such as ^{3}H–^{14}C, ^{3}H–^{32}P, and ^{3}H–^{125}I. For other radionuclide combinations, it is necessary to establish the LL and UL discriminator settings for the appropriate spillup and spilldown of pulse events from the two radionuclides. The procedure used to establish these regions will be discussed later on in this section. For the case of the ^{3}H–^{14}C mixtures, counting region A is normally set by discriminators LL = zero and UL = 12.0 keV, while region B is defined by the discriminator settings LL = 12.0 and UL = 156 keV, when the pulse height spectra are displayed on a linear scale in β particle energy equivalents (Fig. 5.31).

After the two counting regions are defined, it is necessary to prepare quench correction curves, which can be used to determine the counting efficiencies of the ^{3}H and ^{14}C (or lower energy and higher energy radionuclides) in the two counting regions. Two sets of quench standards are required, one set of ^{3}H and one set of ^{14}C quenched standards. If two other radionuclides need to be analyzed, a set of quenched standards of the lower energy radionuclide and a set of quenched standards of the higher energy radionuclide are required. The procedure for preparing a series of quenched standards was given in Section V.D of this chapter. Each series of quenched standards is counted in regions A and B and, as a result, four quench correction curves are created, as illustrated in Fig. 5.32.

When using quench curve correction, determination of the radionuclide activities becomes more difficult as spillover of ^{14}C into the tritium region (region A) increases with quench or, in other words, when the ratio of ^{14}C to tritium increases in region A. If the number of counts from ^{14}C into the tritium region becomes large, the correction of the tritium counts can result in a small number that is less accurate. Likewise, as quench increases, the spillover of ^{3}H pulse events into region B (^{14}C region) diminishes and can even

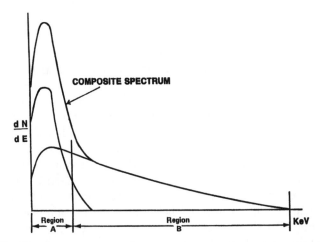

FIGURE 5.31 Typical component and composite pulse height spectra observed for two β-emitting radionuclides (e.g., ^{3}H and ^{14}C) in an approximately 1:1 mixture and having significantly different β-particle energy maxima. Two counting regions are illustrated (regions A and B) for use in the inclusion method, which are set to allow spillup of the lower energy radionuclide from region A into region B, whereby counts from both radionuclides appear in both regions. (Courtesy of PerkinElmer Life and Analytical Sciences, Boston, MA.)

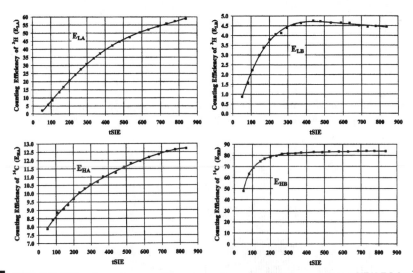

FIGURE 5.32 Quench correction curves of counting efficiency versus tSIE/AEC for ^{3}H and ^{14}C in region A (LL–UL: 0–12.0 keV) and region B (LL–UL: 12.0–156 keV) for the dual radionuclide analysis of ^{3}H–^{14}C. The notations E_{LA}, E_{LB}, E_{HA} and E_{HB} are the counting efficiency factors defined in Eqs. 5.36 and 5.37. The quench correction curves were obtained with ^{3}H and ^{14}C quenched standards counted in regions A and B with a PerkinElmer 2770TR/SL liquid scintillation analyzer.

disappear, which makes the calculations for the ^{3}H and ^{14}C activities invalid. Therefore, to maintain optimal counting conditions, it is necessary to keep the amount of spillover of the ^{14}C pulse events in the tritium region A at a fairly constant level as well as the spillup of tritium events into region B. This is accomplished using an automatic windows tracking method called

AEC (automatic efficiency control) or AQC (automatic quench compensation). As the sample is counted, the LSA determines the level of quench of the experimental sample using a quench-indicating parameter (e.g., $H\#$ or tSIE), and the counting regions are adjusted automatically so that the spillover of the ^{14}C into the tritium region (region A) is kept at a fairly constant level (10–15%), and the spillup of ^3H in the ^{14}C region (region B) is also preserved. Figure 5.33 illustrates that the liquid scintillation pulse height spectrum of a radionuclide diminishes with quench, and AEC automatically moves a discriminator setting according to the degree of quench in the sample. If dual label samples of tritium and ^{14}C are counted using quench curves and automatic window tracking methods (e.g., AEC or AQC) are employed, the resultant quench correction curves shown in Fig. 5.32 can be used to determine the counting efficiency of each radionuclide in each counting region. As illustrated, four curves are created, one for each of the two nuclides in each of the two regions. The major feature to note is that the amounts of spilldown of ^{14}C into the ^3H region and the spillup of ^3H into the ^{14}C region are kept constant.

The equations used to calculate the DPM for each radionuclide are derived from the following equations, which describe the count rate in the two counting regions:

$$CPM_A = D_L E_{LA} + D_H E_{HA} \qquad (5.34)$$

and

$$CPM_B = D_L E_{LB} + D_H E_{HB} \qquad (5.35)$$

where CPM_A and CPM_B are the count rates of a dual radionuclide sample in regions A and B, respectively; D_L and D_H are the disintegration rates

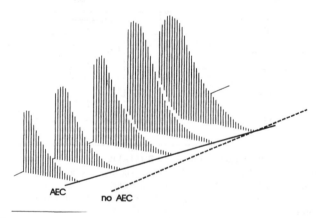

AEC   no AEC

FIGURE 5.33 Illustration of automatic region tracking. Using automatic efficiency control (AEC), the liquid scintillation analyzer automatically moves the upper level discriminator of a counting region from pulse heights of higher magnitude to those of lower magnitude (right to left) according to the degree of quench in a sample. The pulse height spectra of five samples are illustrated, each at different levels of quench. The samples of higher quench level are those of smallest pulse number and magnitude. (Courtesy of PerkinElmer Life and Analytical Sciences, Boston, MA.)

(DPM) of the lower energy (e.g., ^{3}H) and higher energy (e.g., ^{14}C) radionuclides, respectively; E_{LA} and E_{LB} are the counting efficiencies of the lower energy radionuclides in regions A and B, respectively; and E_{HA} and E_{HB} are the counting efficiencies of the higher energy radionuclide in regions A and B, respectively. The four counting efficiency factors in Eqs. 5.34 and 5.35 are obtained automatically by the LSA from the four quench correction curves (Fig. 5.32) stored in the computer memory of the LSA. Therefore, the preceding two equations still have two unknowns, namely, D_L and D_H, which are solved for simultaneously to obtain

$$D_L = \frac{CPM_A E_{HB} - CPM_B E_{HA}}{E_{LA} E_{HB} - E_{LB} E_{HA}} \tag{5.36}$$

and

$$D_H = \frac{CPM_B E_{LA} - CPM_A E_{LB}}{E_{LA} E_{HB} - E_{LB} E_{HA}} \tag{5.37}$$

For the activity determinations of a dual mixture, the LSA determines the count rates of the sample in regions A and B, and the quench-indicating parameter of the sample. From the value of the quench-indicating parameter, the LSA automatically extracts the needed four counting efficiency factors from the quench correction curves (e.g., Fig. 5.32) and then automatically calculates the disintegration rates of the two radionuclides in the mixture according to Eqs. 5.36 and 5.37. All the calculations are done by the software of the liquid scintillation analyzer.

Because the descriminator settings defining the counting regions A and B will move automatically according to the amount of quench in the sample, as measured by the QIP, it is important to note that varying the counting region settings changes the quench correction curves (efficiency vs QIP). This is generally of little concern to the analyst because modern LSAs save, on the hard disk of the computer, the entire pulse height spectra of the quenched radionuclide standards. Consequently, as the counting regions are changed automatically by the instrument according to quench level, so are the resultant new quench correction curves automatically redetermined by the instrument.

Often it is necessary to analyze radionuclide mixtures for which no preset counting regions have been established in the LSA. In such a case, it is necessary to find and select the optimum LL and UL discriminator settings to define counting regions A and B. The procedure used to obtain the proper discriminator settings is as follows, using the ^{33}P–^{32}P dual radionuclide mixture as an example:

1. To find the appropriate LL and UL discriminator settings for the inclusion method for dual-radionuclide activity analysis, we must count first a known activity (DPM) of the lower-energy radionuclide

(e.g., ^{33}P) as a pure radioisotope sample in a wide range of counting regions starting at a LL discriminator setting of zero and progressively increasing the UL discriminator. A count rate (CPM) for each region setting is recorded.

2. This procedure is repeated with a known activity (DPM) of the higher-energy radioisotope (e.g., ^{32}P), which is counted in the same regions selected in the above step 1. The count rates (CPM) of the higher-energy radionuclide in the counting regions are recorded. Both samples of the low-energy radionuclide (e.g., ^{33}P, $E_{max} = 249$ keV) and the higher-energy radionuclide (e.g., ^{32}P, $E_{max} = 1700$ keV) used for this exercise should be at similar and low levels of quench, that is, the lowest level of quench expected for any given unknown mixture.

3. The counting efficiencies of the separate low- and high-energy radionuclides (e.g., ^{33}P and ^{32}P) standards in the various counting regions are then calculated according to the equation %E = (CPM/ DPM)(100).

4. The counting efficiencies of the low-energy radionuclide in the individual counting regions are plotted against the counting efficiencies of the high-energy radionuclide in the same counting regions as illustrated in Fig. 5.34.

5. The objective of this exercise is to find discriminator settings at which there will be significant overlap (spillover) of counts from both radionuclides in the two counting regions required for the inclusion method. These conditions are found in the "knee" section of the curve. Hence, a counting region is selected arbitrarily from the knee section of the curve as the counting region A for the dual-label radionuclide analysis. In the example using the ^{33}P and ^{32}P radionuclide standards a counting region (LL–UL) of 0–100 keV from the knee of the curve was selected from the eighth data point counting from left to right of Fig. 5.34.

6. Having defined in this way one of the counting regions (region A), we can then proceed to select the discriminator settings for the second counting region (region B). The discriminator settings for region B are defined by selecting its LL discriminator setting to be equivalent to the upper limit of region A (e.g., 100 keV for the ^{33}P–^{32}P double label) and selecting the UL discriminator setting to encompass all pulses of highest magnitude arising from the higher energy radionuclide. In this example the UL discriminator setting of 1700 keV was selected, because no pulses from this double-radionuclide mixture can reach beyond 1700 keV. The second counting region (region B) for the case of the ^{33}P–^{32}P dual label was therefore defined by the LL and UL discriminator settings of 100–1700 keV.

Figure 5.35 illustrates the composite pulse height spectrum of a dual-isotope sample of ^{33}P–^{32}P with an approximate 1:1 activity ratio as seen on the LSA computer screen. The discriminator settings established for regions A and B as required for the analysis of these two isotopes in the same sample are also seen in the figure. The inclusion method has been found to provide

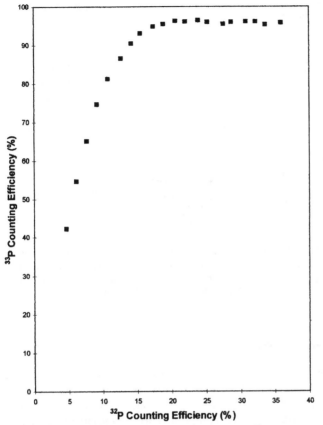

FIGURE 5.34   Effect of region settings on ^{33}P and ^{32}P counting efficiencies over the range of LL–UL, 0–30 to 0–220 keV. The data points represent increasing counting region widths in increments of 10 keV (From L'Annunziata, 1997b).

absolute recoveries of the ^{3}H–^{14}C, ^{33}P–^{32}P, ^{55}Fe–^{59}Fe, ^{90}Sr–^{90}Y, and other double radionuclide mixtures, for a wide range of activity ratios of the two radionuclides (L'Annunziata, 1984b, 1987, 1997b; Kessler, 1989; Viteri and Kohaut, 1997; Benitez-Nelson and Buessler, 1998; Lee *et al.*, 2002). Fujii and Takiue (2001) report the unique analysis of airborne ^{3}H and ^{14}C in activity concentrations as low as 0.01 Bq cm^{-3} by suspension of the radionuclides in a "foggy scintillator" created with an ultrasonic wave generator. The radionuclides could be analyzed as single or dual-radionuclide mixtures.

If we apply the inclusion method to a triple-radionuclide analysis (e.g., a mixture of ^{3}H, ^{14}C, and ^{32}P), three counting regions and three sets of quench standards would be required to prepare quench correction curves for each radionuclide in each counting region. Consequently three CPM values, one for each of the three counting regions, are obtained and three equations solved to calculate the DPM of each of the three radionuclides. For an excellent example of a triple-radionuclide analysis of this type and the calculations and quench curves involved, the reader should refer to Schneider and Verbrugge (1993).

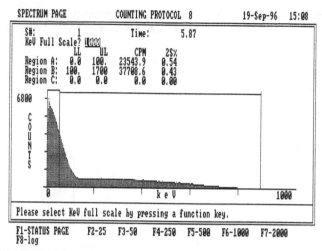

FIGURE 5.35 A composite pulse height spectrum of a $^{33}P-^{32}P$ dual-radionuclide sample as displayed on the computer screen of a PerkinElmer 2300TR liquid scintillation analyzer. Counting regions A and B are set for the dual-radionuclide analysis by the conventional inclusion method. (L'Annunziata, 1996, unpublished work.)

B. Digital Overlay Technique (DOT)

This method uses an external standard to measure quench and a specimen overlay to obtain the DPM for multiple-radionuclide samples. The shape of the sample spectrum is used to resolve dual or triple radionuclide samples by fitting the spectrum of each component to the measured composite spectrum. Spectral fitting is accomplished by the instrument (e.g., Wallac RackBeta LSA of PerkinElmer Life and Analytical Sciences), which maintains a spectrum library that covers a large quench region of both chemical and color quench for the radionuclides. The technique is reviewed and tested by Kouru (1991) and Kouru and Rundt (1991) and described in patents by Rundt and Kouru (1989, 1992). They demonstrate the method to perform as well as the counting region method previously described. Other spectral deconvolution methods are described subsequently in detail.

C. Full Spectrum DPM (FS-DPM)

Full Spectrum DPM is a user-friendly method available with PerkinElmer liquid scintillation analyzers for the measurement of many dual radionuclide combinations including $^3H-^{14}C$, $^3H-^{32}P$, $^3H-^{35}S$, $^{14}C-^{32}P$, $^{33}P-^{32}P$, $^{35}S-^{32}P$, $^3H-^{125}I$, $^{125}I-^{131}I$, $^{51}Cr-^{14}C$, $^{67}Ga-^{68}Ga$, $^{55}Fe-^{59}Fe$, $^{125}I-^{14}C$, $^{59}Fe-^{51}Cr$, and $^{89}Sr-^{90}Sr$. The analysis protocol for FS-DPM is easy to set up, because no counting regions need to be defined. The full-spectrum DPM method utilizes the spectral index of the sample (SIS) of the double radionuclide sample to "unfold" the separate pulse height spectra of the composite spectrum. This is possible because the SIS of the composite spectrum is a function of the individual distributions and the fractional counts of each radionuclide as well as the level of quench in the sample. The direct proportionality of SIS to the radionuclide composition in the sample was demonstrated by L'Annunziata and coworkers (see Noor *et al.*, 1995, 1996b), who determined the activity ratios of ^{3}H:^{14}C and ^{35}S:^{32}P from only the spectral index of the sample.

The key to spectrum unfolding used in full-spectrum DPM is the SIS. At a given level of quench, each radionuclide has a defined pulse height-energy distribution and hence a unique SIS value. When two radionuclides are combined into a single sample, the resultant pulse height is the sum of the two individual distributions. The SIS of the total distribution is a function of the SIS of the individual distributions and the fractional counts of each radionuclide. If the SIS_L and the SIS_H are the spectral index values of the low- and high-energy radionuclides of a dual-radionuclide sample, then the SIS of the total distribution SIS_T can be calculated as

$$SIS_T = \frac{(SIS_L)(\sum N_L E + SIS_H)(\sum N_H E)}{\sum N_L E + \sum N_H E} \tag{5.38}$$

where $\sum N_L E$ = accumulated counts from the low-energy radionuclide (e.g., ^{3}H) and $\sum N_H E$ = accumulated counts from the high-energy radionuclide (e.g., ^{14}C).

From Eq. 5.38 the following equations are derived, which define the count rates of the low-energy radionuclide (CPM_L) and the high-energy radionuclide (CPM_H) of a composite sample:

$$CPM_L = \frac{SIS_H - SIS_T}{SIS_H - SIS_L}(CPM_T) \tag{5.39}$$

and

$$CPM_H = \frac{SIS_T - SIS_L}{SIS_H - SIS_L}(CPM_T). \tag{5.40}$$

For a detailed treatment on the derivation of Eqs. 5.39 and 5.40, the reader may refer to Kessler (1989) and van Cauter and Roessler (1991).

To determine the count rates and disintegration rates of unknown dual-radionuclide samples, four quench correction curves must be prepared with two sets of quenched standards. Series of quenched standards of the low-energy radionuclide and of the high-energy radionuclide are needed. The quenched standards are counted in a regionless environment; that is, no counting region discriminator settings need to be established. The data collected by the LSA from the counting of each of the quenched standards are the quench-indicating parameter (tSIE), the SIS values of the low- and high-energy radionuclide standards (SIS_L) and (SIS_H), respectively, and the percent counting efficiency of the low- and high-energy radionuclide standards. The LSA then plots automatically the four quench correction curves such as the curves illustrated in Fig. 5.36.

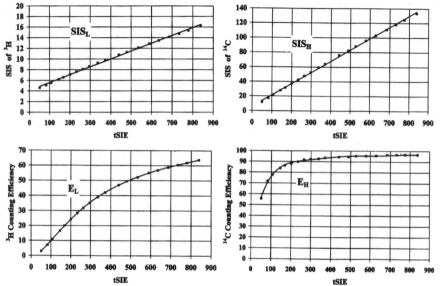

FIGURE 5.36   Quench correction curves for Full Spectrum DPM (FS-DPM) analysis of ^{3}H and ^{14}C mixtures. From the tSIE of a given sample the liquid scintillation analyzer obtains from these curves the SIS of the lower- and higher-energy radionuclides, SIS$_L$ and SIS$_H$, respectively, and the percent counting efficiency of the lower- and higher-energy radionuclides, E$_L$ and E$_H$, respectively. The above curves were obtained from ^{3}H and ^{14}C quenched standards with a PerkinElmer 2770TR/SL liquid scintillation analyzer.

For the analysis of unknown activities of the radionuclide components of a dual-radionuclide sample, the LSA will first determine the total count rate (CPM_T) and the tSIE of the sample. From the one value of tSIE the instrument extracts automatically the values of SIS_L and SIS_H for the composite sample using two of the quench correction curves (Fig. 5.36), which are stored in computer memory. The LSA then calculates the count rate values of the low- and high-energy radionuclides CPM_L and CPM_H respectively, according to Eqs. 5.39 and 5.40. The instrument then automatically converts these count rates to the disintegration rates of the low- and high-energy radionuclide DPM_L and DPM_L respectively, according to the equations

$$DPM_L = CPM_L/E_L \tag{5.41}$$

and

$$DPM_H = CPM_H/E_H \tag{5.42}$$

where E_L and E_H, the counting efficiencies of the low- and high-energy radionuclides respectively, are obtained from the respective quench correction curves illustrated in Fig. 5.36.

This DPM method provides accurate DPM values for dual radionuclides when the endpoint energies (E_{max}) of the two radionuclides differ by a factor of 3 and the activity ratios are in the range of $1:25$ and $25:1$. For additional information on this technique see Kessler (1989), De Filippis (1991), van Cauter and Roessler (1991), and L'Annunziata (1997b). The method was utilized for the analysis of ^{89}Sr–^{90}Sr mixtures by Hong *et al.* (2001) within 4 h after strontium separation from liquid waste. For low-level counting they report a counting efficiency of 95% for ^{89}Sr and 92% for ^{90}Sr with the full-spectrum DPM method and lower limits of detection of 37 mBq/L for ^{90}Sr and 32 mBq/L for ^{89}Sr with a 60-min counting time. Altzitzogluo *et al.* (1998) and Lee *et al.* (2002) also report the use of the method for the analysis of ^{90}Sr from the ^{90}Sr–^{90}Y parent–daughter mixture. Lee *et al.* (2002) compared the full-spectrum DPM method and inclusion method (Section VIII.A.2) for the ^{89}Sr–^{90}Y analysis and found equal performance for both. The full-spectrum DPM method is easier to carry out for the less experienced analyst, as no counting regions need to be established.

D. Recommendations for Multiple Radionuclide Analysis

The following are a few suggestions to follow when performing DPM determinations for multiple-radionuclide samples by one of the methods previously described: (1) for the preparation of quench correction curves prepare the sets of quenched standards in the same cocktail, vial size, and total sample volume as the samples to be analyzed. (2) Try to have an excess of the lower energy radionuclide compared to the higher energy radionuclide, because these methods favor more accurate DPM determinations for the lower energy radionuclide. (3) Always use the automatic region tracking (AEC or AQC) for dual- or triple-radionuclide samples when using the conventional DPM methods (inclusion methods) to maintain constant spillover of the radionuclides in the required counting regions. (4) Try to maintain a minimum activity (DPM) of approximately 1000 for each of the two radionuclides. This latter recommendation on sample activities is important for samples analyzed by the DOT or Full Spectrum DPM methods, as sample count rates must be well above background to obtain statistical accuracy in the determination of sample spectra. Full-spectrum DPM and DOT DPM are count rate dependent and should not be used for low-count-rate samples. If all of these suggestions are observed, then DPM for multiple-labeled samples can be accurate and reproducible ($\pm 3\%$ or better).

E. Statistical and Interpolation Methods

The multiple-radionuclide analysis methods previously discussed are limited to the analysis of not more than three β-emitting radionuclides in the same sample. From the beginning of liquid scintillation analysis in the 1950s up to about 1990 the regionless analysis of more than three β-emitting radionuclides in the same sample was considered to be impracticable or not feasible. The broad pulse height spectra produced by β particles made the task of deconvoluting the pulse height spectra of more than three radionuclides in the same sample appear daunting. However, with the advent of technological advances including applications of multichannel analyzers in LSA and direct computer processing of LSA data, it has become easy to analyze simultaneously numerous α-, α-β-, or α-β-γ-emitting radionuclides in a mixture, evidenced by the research reported by Takiue et al. (1990b, 1991b,c, 1992, 1995, 1999), Matsui and Takiue (1991) and Fujii et al. (1999, 2000) on the application of the most-probable-value theory to simultaneous multiple-radionuclide (as many as seven) analysis and the work of Grau Carles et al. (1993c), Grau Malonda et al. (1994) and Grau Carles (1996) on the use of spectral deconvolution and interpolation methods to multiple-radionuclide (as many as six) activity analysis. These techniques can be applied also to mixtures of α- and β-emitters in one multichannel analyzer by spectral unfolding without α-β discrimination as demonstrated by Grau Carles et al. (1996).

It was traditionally believed that ^{14}C and ^{35}S in the same sample could be neither identified nor analyzed by liquid scintillation counting, because of the very close similarities of the β particle endpoint energies (E_{max}) of these two radionuclides. However, work first reported by Grau Carles and Grau Malonda (1991) demonstrated the accurate analysis of these two radionuclides in a mixture using spectral dilatation-interpolation and least-squares fitting. Activity ratios of ^{14}C/^{35}S were analyzed with an accuracy within about 3%. Grau Carles with Rodríguez Barquero and Grau Malonda (1993a) reported further improvements to this methodology.

Another advancement is the application of multivariate calibration (MVC) to the deconvolution of α- and α-β-emitting radionuclides in the same sample as reported by Toribio et al. (1995, 1996, 1997, 1999) without the need for α-β discrimination

Studies by Kashirin et al. (2000) demonstrated the use of a Windows-based computer analysis of spectra library created with complex radionuclide mixtures in different ratios to permit the activity analysis of mixtures of several α- and β-emitting radionuclides in a mixture. A spectral deconvolution technique was developed also by Verrezen and Hurtgen (2000) to determine the activities of low-energy β activity (e.g., ^{3}H) in the presence of several high-energy β emitters.

These advances in the development of techniques for the analysis of several radionuclides in the same sample will be discussed here. The reader is invited to refer to the literature cited for more details on the techniques involved.

I. Most-Probable-Value Theory

Takiue et al. (1990a,b) and Matsui and Takiue (1991) reported the application of the most-probable-value theory as a new technique to the simultaneous liquid scintillation analysis of four β-emitting radionuclides in the same sample. This technique was expanded to the simultaneous analysis of six β-emitting radionuclides by Takiue et al. (1991c, 1992) and even seven β- and β-γ-emitting radionuclides (Takiue et al., 1995).

The technique requires only a contemporary LSA equipped with a multichannel analyzer, sets of quenched standards for the radionuclides to be measured, and a personal computer for data processing. The approach to this technique, as it was first devised for the analysis of samples containing mixtures of ^{3}H, ^{14}C, ^{32}P, and ^{45}Ca, calls for more counting regions than the number of nuclides to be measured. According to Takiue et al. (1990b) and Matsui and Takiue (1991), the method may be described using the four radionuclide composite sample consisting of ^{3}H, ^{14}C, ^{32}P, and ^{45}Ca, as an example. The count rates of a sample observed in each counting region are defined by the following equations:

$$n_1 = Aa_1 + Bb_1 + Cc_1 + Dd_1 \tag{5.43}$$

$$\vdots$$

$$n_i = Aa_i + Bb_i + Cc_i + Dd_i \tag{5.44}$$

$$\vdots$$

$$n_m = Aa_m + Bb_m + Cc_m + Dd_m \tag{5.45}$$

where $n_1, \ldots, n_i, \ldots, n_m$ are the count rates of a sample in different counting regions ($m >$ number of nuclides in the sample), $A, B, C,$ and D are the activities of ^{3}H, ^{14}C, ^{32}P, and ^{45}Ca, respectively, and $a_i, b_i, c_i,$ and d_i are the respective counting efficiencies in the ith counting region.

The counting efficiencies of a radionuclide in each counting region are determined by means of external standard quench correction curves plotted using sets of quenched standards for each radionuclide. Sets of quenched standards are not commercially available for many radionuclides. These can be prepared in the laboratory in advance by first determining the activities of separate pure β- and β–γ-emitting radionuclides using a direct DPM method, such as efficiency tracing DPM (ET-DPM) as described in Section V.F or a more precise radionuclide standardization according to the CIEMAT/NIST efficiency tracing technique described in Section IX.A of this chapter. Even pure radionuclides procured commercially should be analyzed to confirm their exact activities by a direct DPM method. Once the DPM values of the pure radionuclides are known, sets of quenched standards can be prepared according to the procedure described in Section V.D of this chapter.

From the series of Eqs. 5.43–5.45 and following the derivations of Takiue et al. (1990b) and Matsui and Takiue (1991), the most probable values of A, B, C, and D, that is, the activities of ^{3}H, ^{14}C, ^{32}P, and ^{45}Ca must be determined, and the following equation is derived to search a minimum value (S):

$$S = \sum_{i=I}^{m} w_i \{n_i - (Aa_i + Bb_i + Cc_i + Dd_i)\}^2 \tag{5.46}$$

where w_i is the arithmetic weight of the measurement in the ith counting region, which is calculated by $1/N_i$ where N_i is the total number of counts in the ith counting region.

The most probable value, that is, the activities of each radionuclide A, B, C, and D should satisfy the following condition:

$$\frac{\partial S}{\partial A} = \frac{\partial S}{\partial B} = \frac{\partial S}{\partial C} = \frac{\partial S}{\partial D} = 0. \qquad (5.47)$$

The following normal equations are then derived:

$$A \sum w_i a_i^2 + B \sum w_i a_i b_i + C \sum w_i a_i c_i + D \sum w_i a_i d_i = \sum w_i a_i n_i, \quad (5.48)$$

$$A \sum w_i b_i a_i + B \sum w_i b_i^2 + C \sum w_i b_i c_i + D \sum w_i b_i d_i = \sum w_i b_i n_i, \quad (5.49)$$

$$A \sum w_i c_i a_i + B \sum w_i c_i b_i + C \sum w_i c_i^2 + D \sum w_i c_i d_i = \sum w_i c_i n_i, \quad (5.50)$$

$$A \sum w_i d_i a_i + B \sum w_i d_i b_i + C \sum w_i d_i c_i + D \sum w_i d_i^2 = \sum w_i d_i n_i. \quad (5.51)$$

These equations can be solved for radionuclide activities A, B, C, and D, representing DPM of ^{3}H, ^{14}C, ^{32}P, and ^{45}Ca, respectively, using the determinant calculated by a personal computer. For example, as given by Matsui and Takiue (1991) the determinant for the calculation of the ^{3}H activity from the above equations is the following:

$$H = \frac{1}{K} \begin{vmatrix} \sum w_i a_i n_i & \sum w_i a_i b_i & \sum w_i a_i c_i & \sum w_i a_i d_i \\ \sum w_i b_i n_i & \sum w_i b_i^2 & \sum w_i b_i c_i & \sum w_i b_i d_i \\ \sum w_i c_i n_i & \sum w_i c_i b_i & \sum w_i c_i^2 & \sum w_i c_i d_i \\ \sum w_i d_i n_i & \sum w_i d_i b_i & \sum w_i d_i c_i & \sum w_i d_i^2 \end{vmatrix} \qquad (5.52)$$

where

$$K = \begin{vmatrix} \sum w_i a_i^2 & \sum w_i a_i b_i & \sum w_i a_i c_i & \sum w_i a_i d_i \\ \sum w_i b_i a_i & \sum w_i b_i^2 & \sum w_i b_i c_i & \sum w_i b_i d_i \\ \sum w_i c_i a_i & \sum w_i c_i b_i & \sum w_i c_i^2 & \sum w_i c_i d_i \\ \sum w_i d_i a_i & \sum w_i d_i b_i & \sum w_i d_i c_i & \sum w_i d_i^2 \end{vmatrix} \qquad (5.53)$$

For this case six counting regions were used in the multichannel pulse height analyzer for the measurement of the four nuclides. The discriminator settings of counting regions 1, 2, 3, and 4 are set to receive significant pulses from ^{3}H, ^{14}C, ^{45}Ca, and ^{32}P, respectively, with overlapping pulse height distributions. Channels 5 and 6 were set to receive pulses mostly from the medium-energy β-emitting radionuclides ^{14}C and ^{45}Ca, as the use of double channel settings for the medium-energy β emitters produces more accurate data. The counts of the quenched standards collected in the various counting regions should exceed 10^4 to keep error at a minimum. The mean percent recovery was 2.4% for eight samples containing different proportions of ^3H, ^{14}C, ^{45}Ca, and ^{32}P at different quench levels, which represent 32 radionuclide analyses (8 samples $\times$ 4 radionuclides). Matsui and Takiue (1991) modified the technique by using only three counting regions for the analysis of the four radionuclides in a mixture. This required counting the unknown sample at two quench levels determined by the quench-indicating parameter tSIE. The sample was counted twice, that is, before and after the addition of quench agent. Mean recoveries by this modified approach for 28 radionuclide analyses (7 samples $\times$ 4 radionuclides) was 3.6%.

The approach used by Takieu et al. (1990b) can be applied to the simultaneous liquid scintillation analysis of six different β-emitting radionuclides in a mixture as demonstrated by Takiue et al. (1991c, 1992). They demonstrated the activity analysis of ^3H–^{63}Ni–^{14}C–^{45}Ca–^{36}Cl–^{32}P by application of the most-probable-value theory. A PerkinElmer Tri-Carb Model 4000 was used, and the samples were counted in 12 counting regions as illustrated in Fig. 5.37. Thus, the measurement of an unknown sample requires 12 observation equations of the general type described in Eqs. 5.43–5.45. The 12 equations are written according to the following:

$$n_i = Aa_i + Bb_i + Cc_i + Dd_i + Ee_i + Ff_i \quad (i = 1\text{–}12) \qquad (5.54)$$

where n_i is the count rate of a sample in the ith counting region. A, B, C, D, E, and F are the activities of the six radionuclides, namely ^3H, ^{63}Ni, ^{14}C, ^{45}Ca, ^{36}Cl, and ^{32}P, and a_i, b_i, c_i, d_i, e_i, and f_i are the respective radionuclide

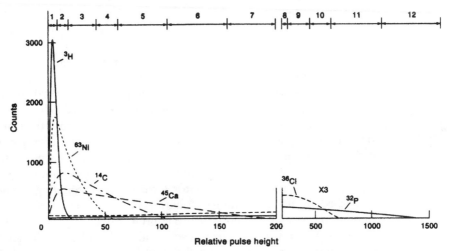

FIGURE 5.37 Liquid scintillation pulse height distributions of six pure beta emitters and region settings for analytical measurements. (From Takiue et al., 1992, reprinted with permission from Elsevier Science.)

counting efficiencies in the ith channel. Because six radionuclide activities must be determined simultaneously, a six-by-six matrix is derived and written similarly to the case of a four-by-four matrix (Eq. 5.52) written for the analysis of four radionuclides. The mean recovery for 60 analyses (i.e., 10 samples × 6 radionuclides per sample) was 3.9%.

The technique can be applied to the liquid scintillation analysis of even low-level β–γ-emitting radionuclides together with β emitters as demonstrated by Takiue et al. (1995). In this case, mixtures of the following seven radionuclides were determined by application of the most-probable-value theory: ^{51}Cr–^{3}H–^{125}I–^{14}C–^{45}Ca–^{22}Na–^{32}P. In this case 14 counting regions were used. The lower limits of detection based on the analysis of 30 samples was calculated as $0.01\,\mathrm{Bq\,mL^{-1}}$ for higher energy radionuclides and $0.05\,\mathrm{Bq\,mL^{-1}}$ for lower energy radionuclides in the mixtures.

A further development of this technique is its application to the analysis of radionuclide combinations with similar pulse height distributions, such as ^{3}H–^{125}I and ^{3}H–^{51}Cr, regardless of the different decay modes of these radionuclides, that is, β decay with ^{3}H and electron capture (EC) decay with ^{125}I and ^{51}Cr. Takiue et al. (1991b) demonstrated the successful application of this procedure to the analysis of combinations ^{3}H–^{14}C–^{125}I and ^{3}H–^{14}C–^{51}Cr in a wide range of activity ratios and quench levels.

More recent work by Takiue et al. (1999) and Fujii et al. (1999, 2000) include the combined use of liquid scintillation and NaI(Tl) spectrometers to permit the simultaneous determination of the activities of many more nuclides in only one calculation process and, at the same time, enhance the accuracy of the radionuclide activity determinations. This method is referred to as a hybrid radioassay technique, because both liquid and solid scintillation spectrometers are used for a given sample. The NaI(Tl) solid scintillation detector provides additional sensitivity, as it would be particularly sensitive to x- and gamma-ray emitters and high-energy beta emitters, which produce considerable Bremsstrahlung radiation. For example, Takiue et al. (1999) and Fujii et al. (2000) analyzed a mixture of seven radionuclides (i.e., ^{3}H, ^{14}C, ^{22}Na, ^{32}P, ^{45}Ca, ^{51}Cr, and ^{125}I using 12 counting regions defined as illustrated in Figs. 5.38 and 5.39. The hybrid radioassay technique was applied by Fujii et al. (2000) to the analysis of the seven radionuclides in waste solutions with a detection limit of approximately $0.005\,\mathrm{Bq\,mL^{-1}}$.

The procedures outlined in this section are carried out easily with any LSA equipped with a MCA or, for the hybrid radioassay, data provided by conventional liquid and solid scintillation analyzers without any modification of the equipment and with a personal computer for data processing. If the counting protocols of the LSA allow for only three counting regions with upper- and lower-level discriminator settings, more counting regions can be established by using additional counting protocols. However, this calls for counting the samples in more than one protocol. The age has come when the LSA can be called definitely also a multiple-radionuclide spectrometer, because numerous x-ray, β- and β–γ-emitting radionuclides can be identified and analyzed simultaneously. Beta spectrometry applications of the LSA now play a role of increased importance.

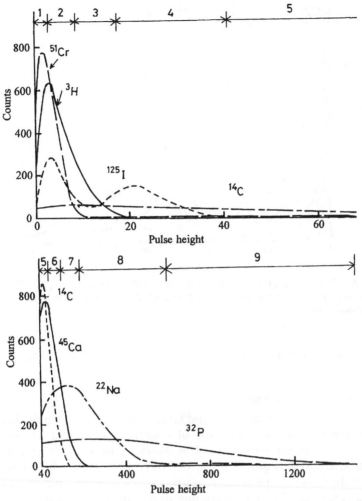

FIGURE 5.38 Liquid scintillation pulse height distributions of seven nuclides and channel settings for analysis (From Takiue *et al.*, 1999, reprinted with permission from Elsevier Science).

2. Spectral Deconvolution and Interpolation

Over the period of 1991–1996 a new technique was reported and developed by researchers at CIEMAT, Madrid, which is a powerful spectral unfolding method for the simultaneous activity analysis of numerous β-emitting radionuclides, including β-emitting nuclides of very similar energy maxima (e.g., ^{14}C and ^{35}S) and even some α–β-emitting radionuclides in the same MCA without α–β discrimination. A description of these techniques is provided here. References are cited for additional information.

The procedure involved is described by Grau Carles (1993) for two types of LSAs, namely, the LSA that analyzes pulse height spectra on a logarithmic scale and the LSA that uses a linear pulse height scale. As explained by Grau Carles (1993), the method has three key components: spectral fitting,

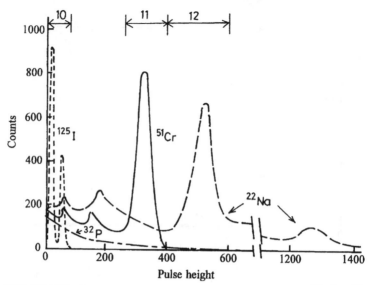

FIGURE 5.39   NaI(Tl) scintillation pulse height distributions of ^{125}I, ^{51}Cr, ^{22}Na, and ^{32}P and channel settings for analysis (From Takiue *et al.*, 1999, reprinted with permission from Elsevier Science).

spectrum unfolding, and spectral interpolation. These components to the analysis are described next.

a. Spectral Fitting

As noted by Grau Carles and Grau Malonda (1991), pulse height spectra are generally defined in terms of discrete pulse height or energy values; however, they are histograms. It is therefore necessary to have a continuous mathematical function defining the spectra. For pulse height spectra on a logarithmic scale, the spectral function is obtained by fitting Fourier series to the experimental spectra according to the following function:

$$f_F(\omega) = \begin{cases} a + b\omega + \displaystyle\sum_{k=1}^{n} c_k \sin\dfrac{k\pi\omega}{M} & 0 \le \omega \prec \omega^* \\[2mm] 0 & \omega \succ \omega^* \end{cases} \qquad (5.55)$$

where $\omega = 0$ and $\omega = \omega^* = M$ are the first and the last values of the spectrum and N is the number of harmonics. The coefficients a, b, and c_k are:

$$a = y_0$$

$$b = \frac{y_M - y_0}{M}$$

$$c_k = \frac{2}{M}\sum_{j=0}^{M} y_j' \sin\frac{\pi k\omega_j}{M} \qquad (5.56)$$

where

$$y'_j = y_j - (a + b\omega_j)$$
$$\omega_j = j = 1, 2, \ldots, M.$$

For the case of an LSA that uses a linear pulse height scale the spectral function is obtained by fitting the Chebyshev series to the experimental spectra determined by the following:

$$f_c(\omega) = \begin{cases} \left[\displaystyle\sum_{k=1}^{N} c_k T_{k-1}(\omega) \right] - \tfrac{1}{2} c_1 & 0 \leq \omega \prec \omega^* \\[2ex] 0 & \omega \succ \omega^* \end{cases} \qquad (5.57)$$

where $\omega = 0$ and $\omega = \omega^* = M$ are the first and the last values of the spectrum, $T_k(\omega)$ are the Chebyshev functions defined in the interval $[0, \omega^*]$ and c_k are the coefficients given by

$$c_k = \frac{2}{N} \sum_{j=1}^{N} f(\lambda_j) T_{k-1}(\lambda_j) \qquad (5.58)$$

where the values λ_j are the zeros of the function $T_k(\omega)$.

b. Spectrum Unfolding

According to descriptions by Grau Carles (1993) and Grau Carles *et al.* (1993b), the spectral deconvolution method is a simultaneous standardization technique, providing radionuclide activities, based on spectral shape analysis of the component nuclides in the mixture. The spectrum unfolding is based on minimizing the expression:

$$\min \left\{ \sum \left(y_i(X + Y) - a y_i(X) - b y_i(Y) \right)^2 \right\} \qquad (5.59)$$

for a dual-radionuclide mixture, where y_i is the number of counts in channel i for the nuclide in brackets, X and Y are the radionuclides in the mixture, and a and b are the parameters obtained from the least-squares fit. This minimum condition, as explained by Grau Carles (1993) and Grau Carles *et al.* (1994a), can be applied only when all spectra $y_i(X+Y)$, $y_i(X)$ and $y_i(Y)$ have the same quench value, that is, the same quench-indicating parameter. Therefore, it is necessary to obtain the spectra $y_i(X)$ and $y_i(Y)$, for the same quench value of the mixture. This is achieved by the spectral interpolation described subsequently in the next part of the analysis procedure. The activities in DPM for the two nuclides X and Y are obtained from the following:

$$A(X) = \frac{ay_i(X)}{t\varepsilon(X)} \tag{5.60}$$

and

$$A(Y) = \frac{by_i(Y)}{t\varepsilon(Y)} \tag{5.61}$$

where t is the counting time in minutes and $\varepsilon(X)$ and $\varepsilon(Y)$ are the counting efficiencies of radionuclide X and Y, respectively.

When a mixture of more than two radionuclides is analyzed, Eq. 5.59 is written as

$$\min\left\{ \sum_i \left(y_i \left(\sum_j X_j \right) - \sum_j a_j y_i(X_j) \right)^2 \right\} \tag{5.62}$$

where X_j are the component nuclides, a_j the coefficients that make the condition minimum, and y_i the number of counts in channel i.

c. Spectral Interpolation

Spectra at different quench levels have different end points or maxima. Therefore, as noted earlier, all spectra are interpolated to the same end point and maxima to validate the application of Eqs. 5.59 or 5.62 for spectral unfolding of dual or multiple (more than three) radionuclides, respectively. Intepolation of spectra, as explained by Grau Carles *et al.* (1993c), is carried out in the following steps: (1) a mathematical transformation is found that makes all spectra pass through common maxima, inflection, and endpoints; (2) the transformed spectra are then interpolated channel by channel; and (3) the required spectrum is found by inverse transformation.

As explained by Grau Carles (1993), the spectral function, such as the Chebyshev fitting described previously by Eq. 5.57, is a mathematical manipulation of the spectra that smoothes the spectra and eliminates all superfluous statistical fluctuations. If $f_i(w)$ ($i = 1, 2, 3$) are the spectral functions, these may be divided into two regions, such as

$$f_i(\omega) = \begin{cases} g_i(\omega) & 0 \leq \omega \prec \omega_{i\alpha} \\ h_i(\omega) & \omega_{i\alpha} \leq \omega \leq \omega_{i\beta} \end{cases} \tag{5.63}$$

where $\omega_{i\alpha}$ is the position of the maximum and $\omega_{i\beta}$ is the endpoint for spectrum i. As explained by Grau Carles (1993), the following mathematical transformation:

$$\omega' = a\omega \tag{5.64}$$

$$a = \frac{a\omega_{1\alpha}}{\omega_{i\alpha}} \tag{5.65}$$

allows the transformation of all $g_i(\omega)$ functions, making them pass through the maximum of the spectral function $f_i(\omega)$. In the same way, the transformation

$$\omega' = b\omega + c \tag{5.66}$$

where

$$b = \frac{\omega_{1\alpha} - \omega_{1\beta}}{\omega_{i\alpha} - \omega_{i\beta}} \tag{5.67}$$

$$c = \frac{\omega_{i\alpha}\omega_{1\beta} - \omega_{1\alpha}\omega_{i\beta}}{\omega_{i\alpha} - \omega_{i\beta}} \tag{5.68}$$

takes the maximum and the endpoint of the function $f_1(\omega)$ as common points for each function $h_i(\omega)$.

When the spectral functions are transformed, a channel-by-channel interpolation provides a spectrum y_i^*. As explained by Grau Carles (1993), the following step determines the inverse equation, which transforms the spectrum y_i^* into the spectrum y_i:

Let $f_p^*(\omega)$ be the spectral function for y_i^* distribution and divide the function into the two regions:

$$f_p^*(\omega) = \begin{cases} g_p^*(\omega) & 0 \le \omega \prec \omega_{1\alpha} \\ h_p^*(\omega) & \omega_{1\alpha} \le \omega \le \omega_{i\beta} \end{cases} \tag{5.69}$$

The positions of the maximum $\omega_{p\alpha}$ and the end point $\omega_{p\beta}$ of the spectrum y_i for Q_p (quench value or quench-indicating parameter) can be found by interpolation in quench correction curves of channel number versus quench-indicating parameter, such as one illustrated for ^{14}C in Fig. 5.40. An example of interpolated spectra of ^{14}C and ^{35}S with overlapping spectral maxima, inflection points and endpoints is illustrated in Fig. 5.41.

Then the inverse transformation

$$\omega' = a'\omega \tag{5.70}$$

where

$$a' = \frac{\omega_{p\alpha}}{\omega_{1\alpha}} \tag{5.71}$$

transforms $g_p^*(\omega)$ into $g_p(\omega)$, and

$$\omega' = b'(\omega) + c' \tag{5.72}$$

where

$$b' = \frac{\omega_{p\alpha} - \omega_{p\beta}}{\omega_{1\alpha} - \omega_{1\beta}} \tag{5.73}$$

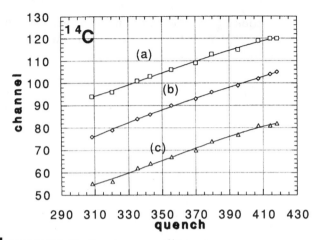

FIGURE 5.40 Fitted curves of ^{14}C local parameters versus quench: (a) end points of the spectra, (b) inflection points, and (c) position of the spectral maxima. The quench-indicating parameter is SQP(E). (From Grau Carles *et al.*, 1993a, reprinted with permission from Elsevier Science.)

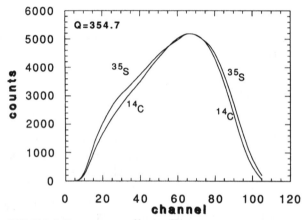

FIGURE 5.41 Computed ^{14}C and ^{35}S spectra for the quench value Q = 354.7, where Q is the SQP(E). (From Grau Carles *et al.*, 1993a, reprinted with permission from Elsevier Science.)

$$c' = \frac{\omega_{1\alpha}\omega_{p\beta} - \omega_{p\alpha}\omega_{1\beta}}{\omega_{1\alpha} - \omega_{1\beta}} \qquad (5.74)$$

transforms $h_p^*(\omega)$ into $h_p(\omega)$. The functions $g_p(\omega)$ and $h_p(\omega)$ define the spectral function $f_p(\omega)$, the spectrum y_i for the particular quench level Q_p.

The power of this method can be evidenced in the spectral deconvolution of ^{14}C ($E_{max} = 156\,keV$) and ^{35}S ($E_{max} = 167\,keV$), which have very close β-particle energy maxima with a ratio of the energy maxima of only 1.07. The spectral unfolding of a 2.2:1 mixture of ^{14}C and ^{35}S is illustrated in Fig. 5.42. The difference spectrum illustrated is a result of subtracting the actual ^{14}C–^{35}S pulse height spectrum from the fitted Fourier spectral function

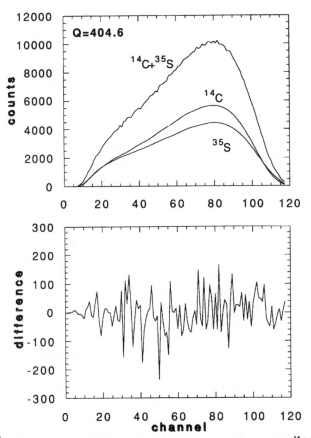

FIGURE 5.42   Spectrum unfolding of a sample containing $^{14}C/^{35}S$ with an activity ratio of 2.2 : I and the difference spectrum (From Grau Carles et al., 1993a, reprinted with permission from Elsevier Science).

$f_F(\omega)$. As explained by Grau Carles (1993), the difference spectrum shows that the fluctuations are statistical, that is, Fourier spectral function fits the empirical data well. For the deconvoluted ^{14}C–^{35}S sample spectrum illustrated in Fig. 5.42, Grau Carles et al. (1993b) determined the ^{14}C and ^{35}S activities with a 1.3 and 3.6% discrepancy, respectively, from the true DPM values. The ^{14}C–^{35}S dual mixture was considered traditionally inseparable spectrometrically by LSA. Another radionuclide pair with very similar pulse height spectral endpoints is ^{3}H–^{55}Fe. Grau Carles et al. (1993c) again demonstrated the potential of this spectrum unfolding method with this nuclide pair. The power of the method is also demonstrated in the analysis of the traditional ^{14}C–^{3}H radionuclide mixtures, as low discrepancies from the true activity values are reported by Grau Carles et al. (1991) at various levels of quench and high-count-rate ratios of ^{14}C–^{3}H up to 100 : 1. Grau Carles et al. (1993f) demonstrated that this spectral deconvolution method will also provide good recoveries for multiple-radionuclide activity determinations in scintillation cocktails consisting of aqueous gel suspensions, such as aqueous ^{45}Ca–^{35}S in Insta-Gel. Mean recoveries expressed as percent discrepancy between computed and actual radionuclide DPM values for a wide range of ^{45}Ca–^{35}S activity ratios from 10.3 : 1 to 1 : 8.1 were 2.3 and 3.1% for ^{45}Ca and ^{35}S, respectively.

In addition to dual-radionuclide mixtures, this spectral unfolding method is applicable to the activity analysis of numerous β-emitting radionuclides. When numerous radionuclides are in the mixture, the previously described Eq. 5.62 is the general expression upon which spectrum unfolding is based. The spectral unfolding and activity analysis of up to six different β-emitting radionuclides in the same mixture are reported by Grau Carles et al. (1993d,e), Grau Malonda et al. (1994), and Grau Carles (1996). Among the radionuclides analyzed by this technique are mixtures of ^{89}Sr, ^{90}Sr, and ^{90}Y even before secular equilibrium was reached between the ^{90}Sr and ^{90}Y. The percent recoveries, that is, percent discrepancy or uncertainty, of the analyzed activities are amazingly low in many circumstances. An example is given in Fig. 5.43, which illustrates the spectral unfolding of a ^{3}H–^{14}C–^{125}I–^{131}I–^{32}P mixture and the computer output with activity analysis and percent uncertainties of the activity determinations for each radionuclide.

The spectrum unfolding technique has been tested by Grau Carles et al. (1996) with a mixture of α- and β-emitting radionuclides consisting of ^{234}Th + ^{234m}Pa + ^{230}Th. The activities of the α- and β-emitting radionuclides could be determined by the spectrum unfolding method without pulse shape analysis (PSA) for α–β separation. Percent discrepancies for activity determinations of the radionuclides in the mixture by spectrum unfolding were much lower than discrepancies obtained by the traditional PSA method.

The computer program MLOG is described by Grau Carles (1996) for the spectrum unfolding methods previously described. The MLOG program was applied by Grau Carles et al. (1998) to determine the separate activities of the parent–daughter nuclides, ^{125}Sb and ^{125m}Te and thereby, determine the ^{125}Sb to ^{125m}Te beta branching ratio.

3. Multivariate Calibration

The application of multivariate calibration for the determination of ^{14}C activities was discussed in Section V.F.4 of this chapter. It has been applied by Toribio et al. (1995, 1996) in the liquid scintillation analysis of mixtures containing the three main α-emitting isotopes of plutonium, ^{238}Pu, ^{239}Pu, and ^{240}Pu. The multivariate methods used were principal component regression (PCR) and partial least squares regression (PLSR). As explained by Toribio et al. (1995), these methods can estimate the concentration of radionuclide of interest without the need to consider the other radionuclides present in the system that affect the experimental signal, provided the isotopes are present in both the unknown samples and the calibration set. The advantage here is that the methods are suitable for the analysis of samples with unknown interferences, thereby avoiding the need of a separation step.

For the PCR and PLSR analysis, Toribio et al. (1995) used a program written in FORTRAN 77 run on a personal computer with a DOS operating system described by Tauler et al. (1991), and the programs were validated by comparison with the PLS-Toolbox package for MATLAB written by Wise (1992). The α emissions of ^{239}Pu and ^{240}Pu have overlapping energies, and

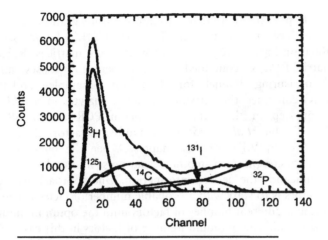

Local Quench Parameters and Efficiency

Nuclide	Quench	Maximum	End	Efficiency
P32	280.8	112.0	136.8	0.9964
I131	280.8	76.6	113.4	0.9620
I125	280.8	29.1	54.1	0.6115
C14	280.8	40.5	71.9	0.8647
H3	280.8	13.7	35.5	0.2129

Nuclide	Counting rate (cpm)
P32	27291.9
I131	12298.1
I125	11669.4
C14	21525.8
H3	28965.3

Nuclide	Activity (dpm)	Uncertainty (%)
P32	27391.1	0.5
I131	12784.0	2.3
I125	19083.2	3.4
C14	24893.0	1.9
H3	136049.3	2.2

FIGURE 5.43 Separated components of a composite mixture of $^3H + {}^{14}C + {}^{125}I + {}^{131}I + {}^{32}P$ by spectrum unfolding and the underlying computer output of activity analysis (From Grau Carles, 1996).

the predicted error for the PCR or PLSR separation of these two nuclides was about 35%. These two nuclides are determined together as one component or analyte, $^{240+239}Pu$, which is similar to other established procedures using surface barrier detectors. Samples containing standards and unknowns of low-level activity (3.42–174 DPM) were counted for 5 h and data analyzed by relating known plutonium activities to alpha disintegrations registered in energy channels of a PerkinElmer Tri-Carb CA/LL liquid scintillation analyzer. Relationships were sought between two matrices $R(M,N)$ and $C(M,K)$. Matrix $R(M,N)$ contained the instrumental responses, number of counts at N measuring channels for M calibration samples, and the other matrix $C(M,K)$ contained the activities of the K analytes in the M samples. The important step in the data treatment in PCR and PLSR models, emphasized by Toribio *et al.* (1995), is the number of factors chosen as noted previously in Section V.F.4 of this chapter. When the number is chosen correctly, the data compression step filters out noise and useless data without sacrificing significantly the desired information. One approach to the proper selection of factors is to start with a probable number of factors. The analysis is repeated using a different number of factors until the optimum number that fits the data is found. The optimum number of factors in this case was found to be four, all attributable to sample activities. No factor was attributed to background, as predictions did not improve when background was subtracted.

The values obtained for $^{240+239}$Pu and ^{238}Pu activities for samples containing varying amounts of these radionuclides in mixtures over the range of 2.43–174 DPM had errors of about 7 and 10% for the high- and low-activity samples, respectively. Background (blank) subtraction is not required. This approach to radionuclide analysis is still at its early stages of development; however, Toribio *et al.* (1995) conclude that the multivariate calibration technique provides a detection limit one order of magnitude higher than that obtained with a surface barrier detector. However, the detection limit would be expected to drop with the use of α–β-discrimination and active shielding in LSA. With α–β-discrimination in many LSAs the β-emitting isotope ^{241}Pu could also be included in the mixture, and activity measurements of mixtures containing the α- and β-emitting isotopes of plutonium in the same sample, as ^{238}Pu $+ ^{239+240}$Pu $+ ^{241}$Pu, would be possible.

Organic extracting agents are often used to improve the alpha spectral resolution in the liquid scintillation analysis of alpha emitters. However, subsequent to extraction, alpha peaks are reported to display a continuous shift along their axis with time for periods in excess of 200 h (Toribio *et al.*, 1997). To overcome error due to the instability of alpha spectral peak positions, Toribio *et al.* (1997, 1999) report the application of moving curve fitting (MCF) to the multivariate calibration analysis of mixtures of the four alpha-emitting isotopes ^{242}Pu, $^{239+240}$Pu, and ^{238}Pu with <15% relative errors of predicted values and an activity quantification threshold of 15 DPM.

A different approach to multiple radionuclide analysis involves the establishment of a library of radionuclide spectra constituting different mixtures of radionuclides at varying levels of quench. The spectral library, which is stored as computer data, must be prepared from radionuclide standards in various mixtures and activity ratios including also spectral distortions caused by the wall effect. Kashirin *et al.* (2000) describe this approach, which they have tested for complex mixtures of up to eight radionuclides including ^{3}H, ^{14}C, ^{60}Co, ^{63}Ni, ^{90}Sr $+ ^{90}$Y, ^{125}I, ^{137}Cs, and ^{241}Am. They applied algorithms described in patents by Belanov *et al.* (1997, 1998) to minimize deviations between real and model spectra regardless of linear or logarithmic scales used for the presentation of spectra. The spectra are analyzed after dividing the spectra into groups of channels with boundaries defined by the equation

$$N_k = N_k - 1 + [(k+1)/2], \quad k = 1, \ldots, K \qquad (5.75)$$

where k is the group index and K the number of groups. The method is limited to samples that are not highly quenched (tSIE > 100) and to radionuclides in the established library of spectra.

Often the LSA of radionuclide mixtures is complicated by the need to analyze low-energy beta emitters, such as ^{3}H, contaminated by greater activities of high-energy beta impurities. Verrezen and Hurtgen (1996, 2000) describe a multiple-window spectrum unfolding technique for the analysis of low-energy ^{3}H in the presence of up to 10-fold higher activities of high-energy beta emitters over a wide range of quench (e.g., tSIE $\geq$ 100). The method is based on the fact that the ratio of the net count rates in two fixed counting regions for a single radionuclide is a constant at a given quench level. Consequently, any spectral shape (e.g., impurity spectrum) can be reconstructed using the established ratio and a pure reference spectrum. On the basis of this hypothesis Verrezen and Hurtgen (2000) calculate the spectral contribution of the impurity in the low-energy part of the liquid scintillation pulse-height spectrum and correct for the impurity interference on the measured low-energy sample spectrum.

IX. RADIONUCLIDE STANDARDIZATION

The authors include radionuclide standardization here as an altogether different specialty of LSA, as it pertains to the purest form of measurement of the activity or disintegration rates (DPM) of radionuclides. Therefore, it includes the methods used for the absolute activity of radionuclides as reference sources from which other radionuclide activities may be calibrated. The standardization of radionuclides is important in many fields, among which one of the most demanding has been radionuclide standardization for applications and research in therapeutic nuclear medicine (Coursey *et al.*, 1991, 1994). This section will include some of the most popular techniques applied to the standardization of radionuclides.

A. CIEMAT/NIST Efficiency Tracing

This method of radionuclide standardization, as related in a historical account, was first conceived by Augustín Grau Malonda in 1978 with the objective of developing a procedure applicable to any liquid scintillation analyzer, any scintillation cocktail, and any radionuclide (Grau Malonda, 1999). As the procedure could not be based solely on theoretical calculations of counting efficiency, Grau Malonda devised a procedure that combined theoretical calculations based on the radionuclide to be analyzed and experimental data provided by the LSA. Early descriptions of the method applied to the activity analysis of pure beta-particle emitters are provided by Grau Malonda (1982a), and Grau Malonda and García-Toraño (1982), and for radionuclides decaying by electron capture by Grau Malonda (1982b).

The name CIEMAT/NIST currently attributed to the method is the result of collaboration on the method that started in 1984 between the Centro de Investigaciones Energéticas Medioambientales y Tecnológicas (CIEMAT), Madrid (the affiliation of Agustin Grau Malonda) and the National Institute of Standards and Technology (NIST), Gaithersburg, which at that time was called the National Bureau of Standards (the affiliation of Bert M. Coursey). Collaboration between the two institutions was intended to develop the method into one that could be used as a reference method for the standardization of radionuclides by any laboratory in the world with a conventional LSA. The principles of the method and many of the major references to this technique will be presented here. Comprehensive treatments on the procedures and calculations involved are available from books by Grau Malonda (1995, 1999). A paper by García-Toraño et al. (1991) also provides a good summary of the method.

I. Theory and Principles (^{3}H as the Tracer)

As noted by Grau Malonda (1995, 1999) the original plan, conceived in 1978, was to develop a procedure for the standardization of radionuclides, which is applicable to any commercial LSA available in most laboratories, any liquid scintillation cocktail, and any radionuclide. To achieve this goal the method had to be based on a combination of theoretical calculations related to the particular radionuclide under investigation and its emissions as well as theoretical calculations related to a primary standard tracer nuclide (e.g., ^{3}H) and its emissions together with experimental data from the LSA which provides information concerning the instrument, scintillation cocktail, and radionuclide (e.g., count rate and quench indicating parameter).

The method is centered on the theoretical computer-based calculations of the counting efficiency of the radionuclide to be analyzed and that of a reference primary standard, such as tritium, for different values of figure of merit (M) and an experimental quench correction curve obtained from a set of ^{3}H quench standards. The figure of merit (M) is a term used in this technique, which is defined as the β particle energy in keV required to produce one photoelectron by the photocathode or, in other words, one photoelectron that reaches the first dynode of the photomultiplier tube. The relationship of radionuclide counting efficiency and figure of merit for pure β emitters and liquid scintillation systems comprised of two photomultiplier tubes in coincidence is defined by

$$\varepsilon = \int_0^{E_{\max}} N(E) \left\{ 1 - \exp\left[\frac{-E \cdot X(E)}{2 \cdot M} \right] \right\}^2 dE \qquad (5.76)$$

where $E_{\max}$ is maximum β particle energy, $N(E)$ is the theoretical β particle energy distribution, $X(E)$ is the correction for ionization quenching and wall losses (see Section IX.A.5 for a treatment on ionization quenching), and M is the figure of merit (Grau Malonda, 1982b; Grau Malonda and García-Toraño, 1982a; Grau Malonda and Los Arcos, 1983; Grau Malonda et al., 1985; Coursey et al., 1986, 1989).

The objective of the method is to obtain the counting efficiency of a nuclide under investigation (e.g., ^{14}C) from the counting efficiency of an absolute or primary standard of ^{3}H. Hence, the term 'efficiency tracing' is used in naming the technique. Since the ^{3}H is a primary standard, its efficiency is taken as experimental data, while that of the nuclide under investigation (e.g., ^{14}C) is obtained via a theoretical calculation. It was essential, therefore, that the calculation model for the counting efficiency, as envisaged by Grau Malonda, be so complete and accurate that the final results of the counting efficiency of the nuclide under investigation would be better than the precision limits of the experiment and instrumentation.

The term 'figure of merit' in this section has a totally different meaning than that used in other parts of this book, where the reader will find reference to figure of merit as relating counting efficiency to background in low activity samples. In this section figure of merit refers exclusively to the quantitative yield of the photomultiplier dynode and its relation to counting efficiency. To avoid confusion, the term 'free parameter' or λ has been adopted instead of figure of merit when relating photomultiplier dynode yield and counting efficiency in this radionuclide standardization technique. Consequently Eq. 5.76 for the expression of the theoretical counting efficiency is often written as

$$\varepsilon = \int_0^{E_{\max}} N(E) \left\{ 1 - \exp\left[\frac{-EQ(E)}{2\lambda} \right] \right\}^2 dE \qquad (5.77)$$

where $Q(E)$ is the ionization quenching correction factor or $X(E)$ of Eq. 5.76, and λ is the free parameter or figure of merit (M) of Eq. 5.76.

The calculation model relies on the decay schemes of the nuclide of interest, and for this purpose, tables of beta-particle and positron spectra were published by Grau Malonda and García-Toraño (1978 and 1981a, respectively) from which the spectral area over any given energy range may be obtained. Also, tables of calculated counting efficiencies as a function of the free parameter were published for numerous radionuclides decaying by pure beta-particle emission (Grau Malonda and García-Toraño, 1981b; Grau Malonda et al., 1985), electron-capture decay (Grau Malonda, 1982b), positron emission (Grau Malonda and García-Toraño, 1982b), beta–gamma-ray emission (García-Toraño and Grau Malonda,

1988), and electron-capture decay with gamma-ray emission (Grau Malonda and Fernández, 1985).

As the calculation of the theoretical counting efficiency must account for all beta transitions, Eq. 5.77 is modified accordingly. For example, ^{125}Sb undergoes 8 beta-decay transitions to the daughter nuclide ^{125m}Te. Consequently, to account for all beta-decay transitions, Grau Carles et al. (1998) used the following expression, according to Grau Malonda (1995, 1999), for the calculation of ^{125}Sb counting efficiency

$$\varepsilon = \sum_{i=1}^{8} p_i \int_0^{(E_m)_i} N_i(E) \left\{ 1 - \exp\left[\frac{-EQ(E)}{2\lambda}\right] \right\}^2 dE \qquad (5.78)$$

where p_i, $(E_m)_i$, and $N_i(E)$ are respectively the 8 beta-particle transition intensities, end-point energies (E_{max}), and Fermi (beta-particle energy) distributions; E, $Q(E)$, and λ are as previously described. The calculated counting efficiency of ^{125}Sb as a function of the free parameter is therefore, a sum of the calculated efficiencies for each of the 8 beta-decay transitions.

The calculation of the theoretical counting efficiency can be yet more complex when numerous beta- and gamma-transitions are part of the radionuclide decay scheme. The standardization of ^{110m}Ag described by García-Toraño et al. (2000) is an interesting example. It decays to ^{110}Cd by β^- emission (98.65%) and isomeric deexcitation by γ emission (1.35%) to ^{110}Ag, which decays to ^{110}Cd by β emission. The decay scheme comprises more than 12 β branches and 50 gamma rays. García-Toraño et al. (2000) used the CIEMAT/NIST method to standardize samples of ^{110m}Ag, which required the calculation of the Fermi spectra of all β branches, as well as the Compton spectra produced by the γ-ray interactions with the scintillator. A total of 128 decay pathways are possible, and the researchers calculated their probabilities and counting efficiencies to obtain the overall counting efficiency. The expression used for the calculation of the theoretical counting efficiency, in this case, was that described by García-Toraño et al. (1991) for a radionuclide that undergoes β decay followed by a cascade of n γ transitions

$$\varepsilon = \int_0^{E_\beta} \int_0^{E_{\gamma 1}} \cdots \int_0^{E_{\gamma n}} N(E) S_1(E_1) \cdots S_n(E_n)$$
$$\times \left\{ 1 - \exp\left[\frac{-E_\beta Q(E_\beta) - \sum_{i=1}^{n} E_{\gamma i} Q(E_{\gamma i})}{2\lambda}\right] \right\}^2 dE \, dE_1 \cdots dE_n \qquad (5.79)$$

where $N(E)$ and $S_i(E)$ are the β spectrum and Compton spectrum of a given γ transition respectively, $Q(E)$ is the ionization quenching factor, λ is the free parameter or figure of merit, and E_β, $E_{\gamma 1}, \ldots, E_{\gamma n}$ are the maximum energies of the β and γ transitions, respectively. García-Toraño et al. (2000) factorized the integral expressions of Eq. 5.79 so that single integrals could be solved and the results combined to yield the complete counting efficiency.

Computer programs or codes have been developed for the rapid computation of counting efficiency as a function of the figure of merit (i.e., free parameter) as described by Grau Malonda (1995, 1999) and Grau Malonda *et al.* (1987). The computer program BETA3 (Grau Malonda, 1999) and EFFY has been written for computations involving pure beta (negatron and positron) emitters, and the programs EFFY2, EFFY4 (García-Toraño and Grau Malonda, 1981, 1985), and CN2001A (Günther, 2001 and Zimmerman *et al.*, 2002) are available for the computation of counting efficiency as a function of free parameter for complex-decay beta emitters. The programs calculate the beta spectra using the theory of radioactive beta decay, taking into account the shape factors. The total detection probability is obtained by dividing the spectrum into bands, calculating the partial detection probabilities of each of them, and summing. The average and mean energies in the spectra are also calculated. Because of the different decay processes involved, special programs have been written to compute the counting efficiency as a function of free parameter such as EFYGA for x-ray and gamma emitters (García-Toraño and Grau Malonda, 1987); EBEGA for beta–gamma emitters (García-Toraño *et al.*, 1988); VIASKL (Los Arcos *et al.*, 1987); VIAS1 (Grau Malonda, 1999) for $K–L$ shell electron capture nuclides; EMI (Grau Carles *et al.*, 1994b) for electron-capture gamma and isomeric transition nuclides; EMIS (Grau Malonda, 1999) for radionuclides decaying by electron capture, electron capture coincident with a gamma transition, and pure isomeric transition; and ADDI (Grau Malonda, 1999) for computing the counting efficiency of a radionuclide decaying by electron capture and beta-decay transition. A paper by Los Arcos *et al.* (1991) provides a detailed treatment of the method applied to multigamma electron-capture radio-nuclides. The computer programs are available from CIEMAT, Avda. Complutense 22, 28040 Madrid, Spain. Also, the program CIENIST99 written in Visual Basic 5 reported by Günther (2000) combines the CIEMAT/NIST codes BETA, EBEGA, and EMI to accommodate multi-beta/multi-gamma transitions.

The procedure integrates theoretical calculations with experimental measurements. The free parameter (figure of merit) is an essential component in the theoretical calculation of the counting efficiency, while the quench-indicating parameter (QIP) of a particular liquid scintillation analyzer (e.g., tSIE, SQP(E), or H#) is required for the experimental aspect of this method. However, as explained by Grau Malonda and García-Toraño (1982a), the figure of merit cannot be obtained directly from experiment and the quenching parameter is not theoretically computable. Therefore, the method makes use of a plot that relates the figure of merit (free parameter) with the quench-indicating parameter. Such a plot is obtained by integrating an experimental quench correction curve of a primary standard, such as ^{3}H (e.g., counting efficiency vs. a QIP), with a theoretically computed curve of counting efficiency versus figure of merit.

Tritium is the most suitable standard for this method when the analysis of beta emitters is required. It is readily available as an absolute standard in tritiated water or in organic form, it has a relatively long half-life, and as noted by Coursey *et al.* (1998), it provides for more sensitive extrapolations to the low-energy portions of the beta-particle spectra than higher-energy standards. Radioactive standards of tritiated water disseminated by the Laboratoire Primaire des Ionizants (LPRI), France and the National Institute of Standards and Technology (NIST), USA were intercompared by Zimmerman and Collé (1997a) and demonstrated an apparent mean disagreement between standards of $< 0.4\%$ on a relative basis. As a standard for the CIEMAT/NIST method ^{3}H has the advantage that any uncertainties on the efficiency curve of this radionuclide are reduced for the efficiency calculations of the more energetic β emitters. See Grau Malonda and García-Toraño (1982a) and Grau Malonda (1995, 1999) for calculations demonstrating the uncertainty propagations for tritium.

The objective is to obtain a counting efficiency quench correction curve for any radionuclide, which may decay by β-, β–γ, x-ray and γ-photon emissions, EC decay, and isomeric transition. The counting efficiency quench correction curve is then used to determine the absolute activity (DPM) of a radionuclide sample under investigation from its experimental quench-indicating parameter.

2. Procedure

The procedure for obtaining the counting efficiency quench correction curve for any nuclide under investigation with ^{3}H as the standard is outlined next. The sequence of steps in the procedure is illustrated in Fig. 5.44 and the curves established as a result of each step are illustrated in Fig. 5.45:

1. A set of ^{3}H quenched standards is counted in the LSA and a quench correction curve of tritium counting efficiency versus the quench-indicating parameter is plotted. See curve (a) of Fig. 5.45.
2. The theoretical counting efficiency of ^{3}H (ε_T) as a function of figure of merit is computed. See curve (b) of Fig. 5.45.
3. From the preceding two relationships, the figure of merit as a function of the quench-indicating parameter is obtained. Although these two relationships are determined from ^{3}H standards, the figure of merit as a function of quench-indicating parameter is independent of any radionuclide. It is a "universal" curve, which may be applied to any radionuclide, but is suitable only for the particular LSA and scintillator used. See curve (c) of Fig. 5.45.
4. The theoretical counting efficiency of the nuclide under investigation (ε_{nuc}) as a function of figure of merit is calculated. See curve (d) of Fig. 5.45.
5. Consequently, the relationship between the counting efficiency of the nuclide under investigation (ε_{nuc}) and the quench-indicating parameter can be obtained. This plot is obtained from the two preceding relationships, that of the universal curve of figure of merit versus the quench-indicating parameter and the computed curve of ε_{nuc} versus figure of merit. See curve (e) of Fig. 5.45.

The final curve obtained in step 5 above can be used to determine the counting efficiencies of experimental samples. The experimental sample of the

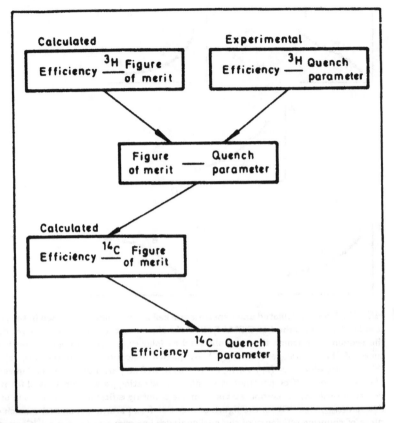

FIGURE 5.44 Radionuclide standardization method of CIEMAT/NIST ^{3}H efficiency tracing. The diagram illustrates the sequence of calculated and experimental relationships, which are defined to yield the quench correction curve of any radionuclide under investigation. Tritium is the standard (or tracer nuclide) in all cases, while ^{14}C, illustrated here as an example, is the radionuclide under investigation. (From Grau Malonda and García-Toraño, 1982a, reprinted with permission from Elsevier Science.)

nuclide under investigation is counted and the count rate (CPM) and quench-indicating parameter are recorded. From the quench-indicating parameter and the quench correction curve of ε_{nuc} versus QIP, the counting efficiency of the experimental sample ε_{nuc} is obtained. The activity (DPM) of the experimental sample is then calculated as

$$A = \frac{CPM}{\varepsilon_{nuc}} \qquad (5.80)$$

The computer calculations of counting efficiency as a function of figure of merit (M) were reported by Grau Malonda *et al.* (1985) for 35 pure β-emitting radionuclides. These fitted data of counting efficiency as a function of figure of merit are reported as polynomial coefficients to the fifth degree according to the equation

$$\ln \varepsilon = A_0 + A_1 \ln M + A_2 (\ln M)^2 + \cdots \qquad (5.81)$$

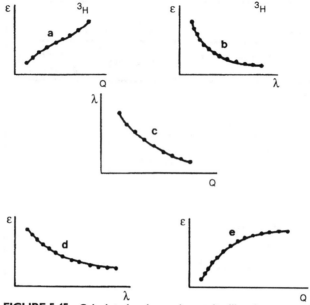

FIGURE 5.45 Calculated and experimental calibration curves obtained in the radionuclide standardization method of **CIEMAT/NIST** ^{3}H efficiency tracing. The curves are described in the sequence by which they are obtained as follows: (a) experimental quench correction curve of ^{3}H counting efficiency (ε_T) vs. quench indicating parameter (Q); (b) computed ^{3}H counting efficiency (ε_T) vs. figure of merit or free parameter (λ); (c) "universal" curve of figure of merit (free parameter) vs. quench indicating parameter derived from the above two relationships; (d) computed curve of the counting efficiency of the nuclide under investigation (ε_{nuc}, e.g., ^{14}C) vs. the figure of merit or free parameter; and (e) quench correction curve of counting efficiency of the nuclide under investigation (ε_{nuc}, e.g., ^{14}C) vs. the instrumental quench-indicating parameter derived from curves (c) and (d). (From Grau Malonda, 1995, 1999).

These coefficients are listed in Table 5.6. The computer programs for calculating these coefficients and those of many other radionuclides are described in the publications cited in the previous section under theory and principles, and the entire programs are available from Grau Malonda (CIEMAT, Avda. Complutense 22, 28040 Madrid, Spain) or Grau Malonda (1999). The figure of merit of the standard ^{3}H is obtained from the relationship of (ε_T) versus M. The good fit for these data is reported by Grau Malonda *et al.* (1985) according to the following fifth-degree polynomial:

$$\ln M = B_0 + B_1(\ln \varepsilon) + B_2 (\ln \varepsilon)^2 + \cdots \qquad (5.82)$$

where the values of B_i for ^{3}H are reported to be $B_0 = 3.11669$, $B_1 = -0.585619$, $B_2 = 0.137378$, $B_3 = -0.148844$, $B_4 = 0.0540364$, and $B_5 = -0.00705037$. Finally, the calculated quench curve of ε_{nuc} versus a QIP for 34 of the radionuclides listed in Table 5.6 can be obtained from the experimental quench correction curve for ^{3}H (i.e., ε_T versus QIP) and the data obtained using Eqs. 5.81 and 5.82 and their respective polynomials.

TABLE 5.6 Polynominal Coefficients for Fitting Counting Efficiency as a Function of the Figure of Merit

Nuclide	Polynominal Coefficients						Maximum disc (%)
	A_0	A_1	A_2	A_3	A_4	A_5	
^{3}H	0.405213 + 1	−0.519852	−0.238284	−0.542988 − 1	0.233709 − 1	−0.229258 − 2	0.2
^{10}Be	0.460257 + 1	−0.833326 − 3	−0.14108 − 2	−0.369043 − 2	0.239038 − 2	−0.675201 − 3	0.07
^{14}C	0.456074 + 1	−0.498280 − 1	−0.212707 − 1	0.599262 − 2	−0.114070 − 1	0.955890 − 3	0.3
^{31}Si	0.460386 + 1	−0.996311 − 3	−0.651159 − 3	−0.102653 − 2	0.592699 − 3	−0.174146 − 3	0.02
^{32}Si	0.456478 + 1	−0.405983 − 1	−0.175982 − 1	0.923926 − 3	−0.493393 − 2	−0.792020 − 4	0.1
^{32}P	0.460414 + 1	−0.759601 − 3	−0.494713 − 3	−0.795852 − 3	0.460529 − 3	−0.133668 − 3	0.01
^{33}P	0.457255 + 1	−0.319945 − 1	−0.144253 − 1	−0.128012 − 2	−0.230966 − 2	−0.452069 − 3	0.07
^{35}S	0.454759 + 1	−0.580615 − 1	−0.236917 − 1	0.373332 − 2	−0.928844 − 2	0.654008 − 3	0.2
^{39}Ar	0.459712 + 1	−0.619246 − 2	−0.340035 − 2	−0.367905 − 2	0.187337 − 2	−0.608778 − 3	0.06
^{45}Ca	0.457021 + 1	−0.338821 − 1	−0.151187 − 1	−0.174736 − 2	−0.212018 − 2	−0.456534 − 3	0.06
^{63}Ni	0.441361 + 1	−0.166749	−0.657592 − 1	−0.176547 − 1	−0.139047 − 1	0.280594 − 2	0.07
^{66}Ni	0.455387 + 1	−0.504729 − 1	−0.212871 − 1	0.588624 − 3	−0.573001 − 2	0.103022 − 3	0.2
^{69}Zn	0.459965 + 1	−0.455061 − 2	−0.239135 − 2	−0.231221 − 2	0.110588 − 2	−0.369529 − 3	0.03
^{79}Se	0.457489 + 1	−0.375947 − 1	−0.161750 − 1	0.961543 − 2	−0.109656 − 1	0.656409 − 3	0.3
^{85}Kr	0.459722 + 1	−0.682161 − 2	−0.340506 − 2	−0.288481 − 2	0.123660 − 2	−0.433468 − 3	0.04
^{87}Rb	0.459557 + 1	−0.777543 − 2	−0.575931 − 2	−0.465487 − 2	0.291824 − 2	−0.128978 − 2	0.06

^{89}Sr	0.460272 + 1	−0.237735 − 2	−0.119424 − 2	0.438239 − 3	−0.154756 − 3	0.01
^{90}Sr	0.459429 + 1	−0.892563 − 2	−0.451710 − 2	0.172480 − 2	−0.597759 − 3	0.06
^{90}Y	0.460413 + 1	−0.106759 − 2	−0.558493 − 3	0.242267 − 3	−0.808739 − 4	0.008
^{91}Y	0.460279 + 1	−0.218860 − 2	−0.110022 − 2	0.409650 − 3	−0.143794 − 3	0.01
^{99}Tc	0.457235 + 1	−0.312754 − 1	−0.140836 − 1	−0.891300 − 3	−0.555615 − 3	0.03
^{106}Ru	0.426331 + 1	−0.279785	−0.120473	0.389097 − 2	0.928245 − 3	0.4
^{109}Pd	0.446763 + 1	−0.156553	−0.109538	0.902461 − 2	0.629885 − 3	0.6
^{113m}Cd	0.459211 + 1	−0.114637 − 1	−0.560371 − 2	0.133895 − 2	−0.579840 − 3	0.04
^{115}In	0.460051 + 1	−0.234863 − 2	−0.216760 − 2	0.277809 − 2	−0.799807 − 3	0.08
^{121}Sn	0.458112 + 1	−0.214372 − 1	−0.100221 − 1	0.702033 − 3	−0.677667 − 3	0.03
^{123}Sn	0.460229 + 1	−0.283529 − 2	−0.139369 − 2	0.430745 − 3	−0.161408 − 3	0.02
^{135}Cs	0.457944 + 1	−0.268855 − 1	−0.129479 − 1	−0.278706 − 2	−0.546387 − 3	0.1
^{143}Pr	0.459885 + 1	−0.551151 − 2	−0.275820 − 2	0.930999 − 3	−0.342315 − 3	0.03
^{185}W	0.458369 + 1	−0.191761 − 1	−0.898716 − 2	0.842663 − 3	−0.645690 − 3	0.03
^{187}Re	0.167732 + 1	−0.165469 + 1	−0.152546	−0.572250 − 2	0.368647 − 3	0.01
^{188}W	0.457681 + 1	−0.256581 − 1	−0.117439 − 1	0.558416 − 4	−0.620994 − 3	0.01
^{209}Pb	0.459299 + 1	−0.109936 − 1	−0.529640 − 2	0.112054 − 2	−0.511350 − 3	0.04
^{241}Pu	0.394623 + 1	−0.535624	−0.217935	0.206078 − 1	−0.184730 − 2	0.2
^{249}Bk	0.450771 + 1	−0.938177 − 1	−0.365791 − 1	−0.141435 − 1	0.176016 − 2	0.3

[a]From Grau Malonda *et al.* (1985) reprinted with permission from Elsevier Science.

The CIEMAT/NIST ^{3}H efficiency tracing has become a very popular method for radionuclide standardization, as no special equipment is required. A common modern LSA and personal computer is all that is needed. The principal advantage of the CIEMAT/NIST ^{3}H efficiency tracing method is that only standards of ^{3}H are required, that is, no standards of the nuclide under investigation are needed. Also, the calculated uncertainties reported in the activity analysis of a broad range of radionuclides are low. The following are some examples of radionuclides standardized by this method and the calculated uncertainties expressed as a percent in parenthesis: ^{90}Sr(^{90}Y) ($\pm 2\%$) reported by Grau Malonda and Los Arcos (1983); ^{14}C ($\pm 0.20\%$) reported by Coursey et al. (1986); ^{99m}Tc ($\pm 3.4\%$) reported by Grau Malonda and Coursey (1987); ^{36}Cl ($\pm 0.3\%$) reported by Rodríguez Barquero et al. (1989); ^{241}Pu ($\pm 3.0\%$) and ^{63}Ni ($\pm 1.1\%$) reported by Coursey et al. (1989); ^{186}Re ($\pm 1.61\%$) reported by Coursey et al. (1991); ^{35}S ($\pm 1.0\%$) reported by Calhoun et al. (1991); ^{54}Mn ($\pm 0.71\%$) reported by Rodríguez et al. (1992); ^{93m}Nb ($\pm 0.56\%$) reported by Günther and Schötzig (1992); ^{45}Ca ($\pm 0.25\%$) and ^{55}Fe ($\pm 0.6\%$) reported by Ortiz et al. (1993); ^{59}Fe ($\pm 0.4\%$) and ^{131}I ($\pm 0.4\%$) reported by Günther (1994); ^{125}I ($\pm 1.5\%$), ^{85}Sr ($\pm 1.5\%$), and ^{109}Cd ($\pm 1.5\%$) reported by Grau Carles et al. (1994c); ^{45}Ca ($\pm 0.4\%$) reported by Rodríguez Barquero et al. (1994); ^{41}Ca ($\pm 2.4\%$) reported by Rodríguez Barquero and Los Arcos (1996); ^{125}Sb(^{125m}Te) ($\pm 1.5\%$) reported by Grau Carles et al. (1997, 1998); ^{89}Sr ($\pm 0.48\%$) reported by Coursey et al. (1998) and Cruz et al. (2002); and ^{89}Sr ($\pm 0.2\%$) reported by Altzitzoglou et al. (2002); ^{90}Sr ($\pm 1.1\%$) reported by Altzitzoglou et al. (1998) for low-level activities (~ 45 Bq kg^{-1}) in dry bone ash; ^{62}Cu ($\pm 0.01\%$) reported by Zimmerman and Cessna (1999); ^{153}Sm ($\pm 0.19\%$) reported by Schötzig et al. (1999);^{110m}Ag ($\pm 0.29\%$) reported by García-Toraño et al. (2000); ^{237}Np ($\pm 0.2\%$) reported by Günther (2000); ^{204}Tl ($\pm 0.3\%$) reported by Hult et al. (2000) and Johansson et al. (2002); ^{177}Lu ($\pm 0.8\%$) reported by Zimmerman et al. (2001); ^{134}Cs ($\pm 0.19\%$) reported by García-Toraño et al. (2002); ^{188}W/^{188}Re ($\pm 0.4\%$) reported by Zimmerman et al. (2002); and ^{40}K ($\pm 0.1\%$) reported by Grau Carles and Grau Malonda (1997) and Grau Malonda and Grau Carles (2002).

3. Cocktail Physical and Chemical Stability

Liquid scintillation cocktail volume can have a significant effect on the count rate of weak β emitters such as ^{63}Ni ($E_{max} = 67$ keV). A study by Zimmerman and Collé (1997b) demonstrated that a 7% variation in the liquid scintillation counting efficiency of ^{63}Ni can be observed over the scintillation cocktail volume range of 1–20mL. However, they demonstrated that the CIEMAT/NIST method is able, nevertheless, to trace the observed ^{63}Ni activity to about 0.1%. To assure precise standardizations when employing the CIEMAT/NIST efficiency tracing method some researchers will centrifuge the liquid scintillation counting vial to assure that no significant amount of nuclide and cocktail remain adhered to the inner surface of the counting vial cap. Gravimetric dispensing of radionuclide solutions is also a common practice.

The physical and chemical stability of samples in scintillation cocktail is a priority. Terlikowska *et al.* (1998) identified the causes of instability of samples which include (1) degradation of the scintillator or cocktail, (2) adsorption of the nuclide on the walls of the counting vial, (3) settlement of the aqueous phase of the nuclide source, (4) change in the physical characteristics (e.g., size) of the aqueous micelles in cocktail mixtures, and (5) evolution of quenching in the cocktail due to a chemical change in its composition (e.g., an increase of dissolved oxygen in the cocktail with time). A quench-indicating parameter measured with an external radiation source may not provide any evidence of a problem in the cases of the above instabilities 2 and 4. Consequently, the stability of cocktails must be studied thoroughly for each nuclide standardized. It is recommended also that standardizations of any particular nuclide be carried out in accord with previously tested and recommended cocktails and procedures cited in the literature. Carrier is used by some researchers in radionuclide standardization to avoid error that might occur due to adsorption of nuclide on the walls of counting vials. For example, the following carriers have been added to nuclide samples for this purpose: 0.25 mL of Sr carrier solution containing $4 \, mg \, Sr \, g^{-1}$ in $1 \, mol \, L^{-1}$ HCl in the standardization of ^{89}Sr (Coursey *et al.*, 1998), 0.1 mg Mn or Zn carrier in 1 mL of 0.05 mol/L of EDTA solution (Günther, 1998), vials previously saturated with a carrier solution containing $100 \, \mu g/g$ of $NiCl_2$ in 0.1 M HCl in the standardization of ^{63}Ni (Terlikowska *et al.*, 1998), and ^{177}Lu solution in carrier containing approximately 0.06 g of $LuCl_3$ per mL of $1 \, mol \, L^{-1}$ HCl (Zimmerman *et al.*, 2001).

4. Potential Universal Application

In the preceding treatment on the theory and experimental procedure involved in the CIEMAT/NIST method it was established that the counting efficiency as a function of the 'figure of merit' (M) or 'free parameter' (λ) is found for both the standard (tracer) nuclide (ε_T) and the radionuclide under investigation (ε_{nuc}). Under such circumstances, as noted by Grau Malonda *et al.* (1985), the ratio of the counting efficiencies ($\varepsilon_T/\varepsilon_{nuc}$) is also known as a function of M. Consequently, as the counting efficiency of the tracer nuclide (e.g., ^{3}H) is known, the other can be computed. Tritium is considered, therefore, not only as a standard nuclide, but also a tracer nuclide, because the counting efficiency of the nuclide of interest is found against that of the reference nuclide, ^{3}H in this case. (Most often ^{3}H is used as the tracer for the standardization of beta emitters, but we will see in the next section that a new tracer, ^{54}Mn is recommended for the standardization of radionuclides that decay by pure electron capture.) With this in mind and in view of the curves illustrated in Fig. 5.45, we can see that the curves ε_T versus QIP (obtained from quenched standards) and ε_{nuc} versus QIP (obtained from the relationships of ε_T vs. M, M vs. QIP, and ε_{nuc} vs. M) can be integrated to provide a curve directly relating the counting efficiency of the nuclide under investigation (ε_{nuc}) and that of the tritium standard or tracer nuclide (ε_T). Such a curve is illustrated in Fig. 5.46. With this relationship the counting efficiency of a given nuclide (ε_{nuc}) at a certain QIP could be determined from the counting efficiency of tracer nuclide (ε_T) at the same QIP.

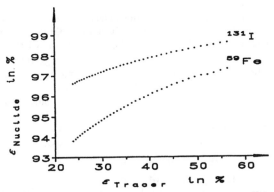

FIGURE 5.46 Detection (counting) efficiency of ^{59}Fe and ^{131}I vs. efficiency of the tracer nuclide ^{3}H (From Günther, 1994, reprinted with permission from Elsevier Science.)

To facilitate the application of the CIEMAT/NIST efficiency tracing method Günther (1996) proposed the following procedure:

1. Calculation of the efficiencies. Calculating the efficiencies of radionuclides versus the corresponding tritium efficiency can be done in a national standards laboratory. The results should be parameterized as coefficients k_i of the polynomial in Eq. 5.83. Tables 5.7 and 5.8 summarize the results obtained at the Physikalisch-Technische Bundesanstalt (PTB) for a variety of pure β emitters and β–γ emitters.

$$\varepsilon_{\text{nuc}} = \sum_{i=0}^{3} k_i \, \varepsilon_{\text{T}}^i \tag{5.83}$$

2. Verifying the results. This step is also best done at one of the national standards laboratories, because they have a stock of standard solutions. An experimental comparison of the results with those obtained with other methods is necessary. Tables 5.7 and 5.8 show some results.

3. Laboratory procedure. The user must determine the quench-indicating parameter Q automatically with each single measurement, calculating the corresponding tritium efficiency ε_{T} of the individual measurement using his/her own tritium quench curve, and calculating the efficiency ε_{nuc} with this efficiency and the coefficients of Table 5.7 or 5.8. This can be shortened by a computer program. The tritium efficiency measurements obtained with the individual LSC device as well as the polynomial coefficients k_i can be combined and included in a program (LSCAL, available from E. W. Günther, PTB, Bundesallee 100, D-38116 Braunschweig, Germany) to obtain and use an efficiency curve

$$\varepsilon_{\text{nuc}} = \sum_{i=0}^{n} m_i \, Q^i \tag{5.84}$$

with new coefficients m_i.

As explained by Günther (1996), with this procedure, only the count rate and the quench parameter Q (same as the quench-indicating parameter, QIP)

TABLE 5.7 For β Emitters, Coefficients k_i of the Polynomial Eq. 5.83 Correlating the Nuclide Counting Efficiencies (ε_{nuc}) with Tritium Counting Efficiencies (ε_T), and a Comparison of the Standardization by Efficiency Tracing with Tritium to Other Methods[a]

Nuclide	Coefficents of the least-squares fit[b]					Efficiency[c] (%) ($\varepsilon_T = 50\%$)	Comparison[d]	
	k_0	k_1	k_2	k_3	k_4		Method	Δ (%)
^{14}C	72.21	0.9538	−1.4846E-2	0.917E–4	0	94.25	I	<0.5
^{32}P	99.39	0.0145	−0.0102E-2	0	0	99.86	IC,B	<0.5
^{33}P	80.97	0.6279	−0.9657E-2	0.598E–4	0	95.70	–	
^{35}S	69.35	0.9633	−1.4284E-2	0.868E–4	0	92.65	–	
^{45}Ca	80.43	0.6377	−0.9737E-2	0.602E–4	0	95.50	–	
^{63}Ni	22.76	2.4709	−4.8562E-2	5.603E–4	−2.633E–6	78.48	–	
^{89}Sr	96.97	0.0685	−0.0472E-2	0	0	99.22	B	<0.5
^{90}Sr	90.20	0.3099	−0.4652E-2	0.286E–4	0	97.64	–	<0.5
^{90}Y	98.39	0.0378	−0.0263E-2	0	0	99.62	B	<0.5
^{90}Sr/^{90}Y	190.09	0.2222	−0.1567E-2	0	0	197.29	IC,B	<0.5
^{99}Tc	81.10	0.6123	−0.9312E-2	0.575E–4	0	95.62	I	<0.5
^{147}Pm	74.15	0.8224	−1.2363E-2	0.7596E–4	0	93.86	ET	1.4
^{204}Tl	92.96	0.1436	−0.0914E-2	0	0	97.86	B	~1

[a]From Günther (1996).
[b]The coefficients are from Eq. 5.83. If k_4 is not listed, the value is 0.
[c]A typical efficiency of unquenched samples with a tritium efficiency of 50%.
[d]The methods used to determine the activity of the standard solutions and the difference Δ from the result obtained with these methods. C, $4\pi\beta-\gamma$ coincidence counting; IC, calibrated ionization chamber; B, $4\pi\beta$ proportional counting; I, intercomparison; G, $4\pi\gamma$ counting; ET, efficiency tracing with coincidence counting.

are necessary to obtain the sample activity. If the liquid scintillation counter is equipped with a self-normalization procedure, the repetition of the determination of the tritium quench correction curve need not be carried out except over longer intervals of time. The procedure proposed by Günther (1996) is summarized in Fig. 5.47.

All of the previous treatment on the CIEMAT/NIST radionuclide standardization method dealt with the use of standard ^{3}H as the tracer nuclide. Standard tritium has proven to be the best tracer for the standardization of β- or $\beta-\gamma$-emitting nuclides, and it has been the standard of choice since the inception of this technique. However, for the standardization of ^{55}Fe and ^{65}Zn, Günther (1998) has demonstrated ^{54}Mn to be a more robust tracer. Iron-55 decays by pure electron capture and ^{65}Zn decays 98.6% by electron capture, and the remaining 1.4% by positron emission. Günther (1998) demonstrates that the use of standard ^{54}Mn, which decays by pure electron capture, as the tracer nuclide provides radionuclide activities for ^{55}Fe with a total relative uncertainty of only 0.44% whereas the uncertainty when using tritium as the tracer was 1.67%. Likewise, the total uncertainty in the standardization of ^{65}Zn using ^{54}Mn as the tracer nuclide was only 0.44%, while the uncertainty when using ^{3}H as the tracer increased

TABLE 5.8 For $\beta-\gamma$ Emitters, Coefficiencts k_i of the Polynomial Eq. 5.83 Correlating the Nuclide Counting Efficiencies (ε_{nuc}) with Tritium Counting Efficiencies (ε_T), and a Comparison of the Standardization by Efficiency Tracing with Tritium to Other Methods[a]

Nuclide	Coefficients of the least-squares fit				Efficiency (%) ($\varepsilon_T = 50\%$)	Comparison	
	k_0	k_1	k_2	k_3		Method	Δ (%)
^{59}Fe	87.30	0.3778	−0.5006E-2	0.2580E-4	6.9	IC,C	< 0.5
^{60}C	86.06	0.4489	−0.6756E-2	0.4126E-4	96.78	IC,C,I	< 0.5
^{129}I	79.90	0.7893	−1.2180E-2	0.6958E-4	97.61	IC,I	< 0.5
^{131}I	93.86	0.1408	−0.1015E-2	0	98.36	IC,C,I	< 0.5
^{137}Cs/^{137}Ba	107.90	0.179	−0.1284E-2	0	113.65	IC,G	< 0.5
^{134}Cs	82.81	0.5048	−0.7181E-2	0.4313E-4	95.49	IC,C,I	< 0.5
^{141}Ce	93.13	0.1424	−0.0995E-2	0	97.76	IC,C,I	< 0.5
^{203}Hg	79.43	0.6568	−0.9905E-2	0.6100E-4	95.13	IC,I	0.6

[a]See footnotes, Table 5.7 from Günther (1996).

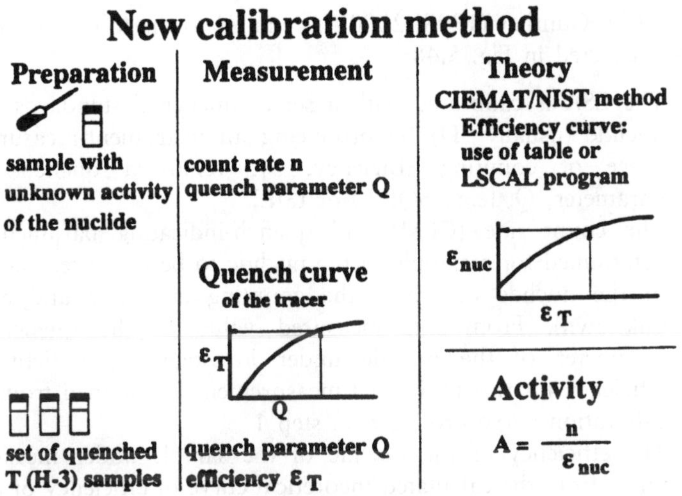

New calibration method

FIGURE 5.47 The proposed calibration method. The counting efficiency of the nuclide to be measured (ε_{nuc}) is obtained with the quench parameter (Q) of the measurement, the tritium quench curve and the tabulated coefficients of an efficiency curve of the nuclide [from Günther (1996)].

to 0.93%. The use of ^{54}Mn as a tracer nuclide is discussed further in the next section of this chapter.

Further streamlining of the CIEMAT/NIST efficiency tracing standardization method is described by Günther (1998, 2000, 2002b). As described previously the methodology involves the calculated theoretical efficiency curve of the efficiency of the nuclide under investigation (ε_{Nuc}) versus the theoretical efficiency curve of the tracer nuclide (ε_T). The procedure as

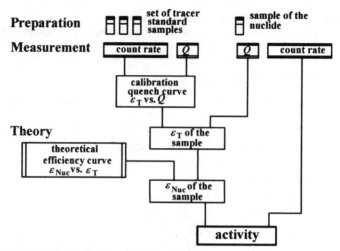

FIGURE 5.48 Principle of the **CIEMAT/NIST** method. Q is the quench parameter measured, ε_T is efficiency of the tracer, ε_{nuc} the efficiency of the nuclide to be measured. (From Günther, 1998, reprinted with permission of Elsevier Science.)

portrayed by Günther (1998, 2000) is reduced to the following steps, which are also depicted in Fig. 5.48:

1. The LSA is calibrated with a set of quenched standards of a tracer nuclide (usually ^{3}H) by producing an instrument-measured quench curve of counting efficiency, ε_T, versus a quench- indicating parameter, Q, [e.g., SQP(E) or tSIE].

2. The count rate (CPM) and quench-indicating parameter, Q, are determined for a sample of the nuclide to be measured (i.e., a sample of the nuclide of which the counting efficiency and activity are unknown). From this measured value of the quench-indicating parameter of the nuclide under investigation, a fictitious tracer efficiency for this individual measurement is obtained from the tracer calibration curve produced in step 1.

3. The efficiency of the nuclide of the sample under measurement is taken from the calculated theoretical curve of efficiency of the nuclide under investigation (ε_{Nuc}) versus the theoretical efficiency curve of the tracer nuclide (ε_T) and/or taken from a database containing the polynomial equation for the curve.

The experimentally determined quench correction curve prepared with the standard tracer nuclide in step 1 should be useful for a long period of time (several months or more) with modern LSAs that are properly cared for in terms of normalization and calibration, and monitored frequently with automated instrument performance assessments (IPAs). The monitoring of instrument stability by regular automated IPAs is discussed further on in this chapter. The theoretically calculated counting efficiencies of the nuclide under investigation and the tracer nuclide are carried out via computer programs (codes) already elaborated for specific nuclide decay schemes available as previously described in this chapter or available from databases containing the polynomial for the relationship of the theoretical calculated nuclide efficiency and the tracer efficiency described by the equation.

$$\varepsilon_{\text{nuc}} = \sum_{i=0}^{n} k_i \, \varepsilon_{\text{T}}^i \qquad (5.85)$$

which is the general equation for the polynomial of the least-squares best fit for the relationship, such as that of Eq. 5.83 with n number of coefficients. With the above two curves established, it becomes a simple matter to determine the counting efficiency of a sample and thereby standardize a radionuclide sample simply from its count rate (CPM) and its quench-indicating parameter (Q).

This approach has been tested by other researchers with excellent results using ^{3}H as the tracer nuclide including the standardization of ^{125}Sb(^{125m}Te) in equilibrium with $< 1.5\%$ uncertainty by Grau Carles et al. (1998), ^{153}Sm with 0.19% uncertainty by Schötzig et al. (1999), ^{237}Np with $<0.2\%$ uncertainty by Günther (2000), ^{204}Tl with $< 0.3\%$ uncertainty by Hult et al. (2000), and ^{177}Lu and ^{188}Re with 0.22 and 0.3% uncertainty, respectively, by Schötzig et al. (2001). Günther (2002b) reports excellent agreement between the CIEMAT/NIST and $4\pi\beta$-γ standardization method with the following deviations between the two methods expressed as a percent: ^{134}Cs ($< 0.3\%$), ^{153}Sm ($< 0.2\%$), ^{169}Er ($< 0.2\%$), ^{177}Lu (0.6%), and ^{188}Re (0.2–1.0%). A unique application of the CIEMAT/NIST method is reported by Günther (2002c) for the analysis of ^{32}P and ^{33}P activities in angioplastic balloons where ^{33}P is an impurity. The relative standard uncertainties for the determination was 0.32% for ^{32}P and 1.6% for ^{33}P impurity.

5. Ionization Quenching and Efficiency Calculations (^{3}H or ^{54}Mn as the Tracer)

The theoretical calculation of the counting efficiency of the nuclide under investigation and of the tracer nuclide (usually ^{3}H for β and β-γ emitters or ^{54}Mn for nuclides of low atomic number that decay by electron capture) is carried out using the basic equation of Grau Malonda and coworkers, Eq. 5.77, and variations of the equation depending on the decay scheme of the radionuclide under measurement (e.g., Eqs. 5.78 and 5.79). A key factor in the theoretical calculation is the computation of the ionization quench function $Q(E)$, which has been revisited by several researchers including Los Arcos and Ortiz (1997), Cassette et al. (1998), Günther (1998), Rodríguez Barquero et al. (1998), Ceccatelli and DeFelice (1999), Grau Malonda (1999), Grau Malonda and Grau Carles (1999, 2000), Grau Carles and Grau Malonda (2001), and García and Grau Malonda (2002).

Ionization quenching is a reduction in the intensity of the flash of light (photon intensity) as a result of ionization caused by the nuclear radiation as it travels through the scintillator solvent. As described by Grau Malonda (1999) the fluorescence yield or number of photons $L(E)$ produced by an ionizing particle traveling through a liquid scintillator is a nonlinear function of the particle energy (E). The nonlinearity increases with the stopping power or ionization power of the particle. Therefore, a low-energy beta particle will cause higher ionization quenching than a higher-energy beta particle.

According to Birks (1964) the specific fluorescence or number of photons produced per particle path length of travel, dL/dx, is defined by

$$\frac{dL}{dx} = \frac{\eta_0 \, dE/dx}{1 + kB \, (dE/dx)} \qquad (5.86)$$

where η_0 is the scintillation efficiency or number of fluorescence photons emitted per unit of absorbed energy, dE/dx is the radiation stopping power (i.e., stopping power of the scintillator for one beta particle of energy E) in units of $\mathrm{MeV\,cm^{-1}}$ or $\mathrm{MeV\,cm^2\,g^{-1}}$ (see Chapter 1, Section V.A) and kB is the ionizing quench parameter or constant with units of $\mathrm{cm\,MeV^{-1}}$ or $\mathrm{g\,MeV^{-1}\,cm^{-2}}$.

In the absence of ionization quench the specific fluorescence would be defined as

$$\frac{dL}{dx} = \eta_0 \frac{dE}{dx} \qquad (5.87)$$

and the term $[1 + kB(dE/dx)]^{-1}$ of Eq. 5.85 is the reduction in fluorescence due to ionization quenching. The formula usually used for the ionization quench function in the CIEMAT/NIST calculation of counting efficiency is the formula of Birks (1964)

$$Q(E) = \frac{1}{E} \int_0^{E_{max}} \frac{dE}{1 + kB(dE/dx)} \qquad (5.88)$$

According to Grau Malonda and Grau Carles (1999) the ionization quench function $Q(E)$ is determined by applying the Birks formula (Eq. 5.88) of which the stopping power is approximated to a polynomial equation. Also, the ionization quench function is evaluated analytically whereby $Q(E)$ is expressed in terms of kB. An empirical equation for $Q(E)$ is expressed by Grau Malonda and Grau Carles (1999) as a quotient of polynomial equations, that are fitted for different kB values between 0.001 and 0.020 g $\mathrm{MeV^{-1}\,cm^{-2}}$. They also analyzed the influence of the tracer nuclide (i.e., ^{3}H or [54]Mn) and demonstrated that ^{3}H remains the recommended tracer for the analysis of nuclides that undergo beta decay while [54]Mn provides a more robust standardization of certain nuclides that decay by electron capture and possess a low atomic number close to that of iron. Also they note that the calculated formula for $Q(E)$ and kB value must be the same for the tracer and radionuclide under measurement. Computer programs BETAKB and EMISKB are provided by Grau Malonda (1999) for the calculation of the theoretical counting efficiencies of radionuclides that undergo beta decay or electron-capture decay, respectively, with subroutines for the computation of the ionization quench function $Q(E)$ for a given Birks parameter kB (see also Los Arcos and Ortiz, 1997). Günther (1998) also demonstrated the improved accuracy of standardization of the electron-capture nuclides ^{55}Fe and ^{65}Zn when ^{54}Mn is used as the tracer.

In a review of the CIEMAT/NIST and TDCR methods (see Section IX of this chapter) Grau Carles and Grau Malonda (2001) present new concepts to improve the robustness of the radionuclide standardization methods. Currently, in all of the models, the counting efficiency is computed on the basis of the photoelectron output of the photomultiplier (PM) photocathode. Although the results of the model calculations agree well with the experimental results, they propose that the correct calculated theoretical counting efficiency should be based on the final PM anode output rather than on the photocathode or any dynode outputs. Consequently modifications to the theoretical counting efficiency calculations and determinations of the optimum kB parameter are proposed (Grau Carles and Grau Malonda, 2001 and García and Grau Malonda, 2002). Günther (2002a) provides a review of the prospects for the universal application of the CIEMAT/NIST method. In general, the Visual Basic CN2001 program has been useful for the calculation of theoretical counting efficiencies of the nuclide under measurement versus the efficiency of the ^{3}H tracer of nuclides decaying by β, β–γ, EC, EC–γ and for isomers and nuclides with mixed decay schemes. It is available from the author at the Physikalisch-Technische Bundesanstalt (PTB), Bundesalee 100, 38116 Braunschweig, Germany, and it is based on the CIEMAT programs EFFY (García Toraño and Grau Malonda, 1985), EMI (Grau Carles *et al.*, 1994b), CEGA2 and KB (Los Arcos and Ortiz, 1997). In his review Günther (2002a) summarizes that pure beta emitters present no problem with activities determined with uncertainties within 0.2 and 0.5%. The standardization of some pure gamma emitters (e.g., ^{93m}Nb) are determined with 100% efficiency. Beta/gamma emitters are determined with as high precision as pure beta emitters. The EC nuclides do present a problem with higher levels of uncertainty; however, better models such as those proposed by Grau Malonda and Grau Carles (2000), the EMI2 program of Grau Malonda *et al.* (1999), and the use of ^{54}Mn as a tracer for some EC nuclides (not mentioned in the review) may reduce uncertainties. The CIEMAT/NIST method is, in most cases, not called for when the activities of alpha emitters are needed, as the liquid scintillation counting efficiency of alpha emitters are generally close to unity. However, when the alpha emission is part of a complex decay scheme, such as ^{237}Np/^{233}Pa, alpha transitions are included in the CN2001 program of the CIEMAT/NIST method (Günther, 2000, 2002a).

B. $4\pi\beta - \gamma$ Coincidence Counting

$4\pi\beta - \gamma$ coincidence counting is a direct method of activity determination, as it is independent of any quench indicating parameters and no reference standards are required for counting efficiency determinations. An additional advantage of this method is that the activity of a nuclide can be determined from experimental counting data only without the need for detector efficiencies.

As may be construed from the name applied to this method, it involves the coincidence detection of two types of radiations from a given radionuclide. The methods can be applied, therefore, to radionuclides that emit in coincidence more than one distinguishable type of radiation. Two types of detectors are required in coincidence to distinguish exclusively two types of radiation emissions, such as a liquid scintillation detector (or gas proportional counter) to measure a beta emission and a solid scintillation (or semiconductor detector) capable of measuring a gamma emission. Since the technique involves, in most cases, the use of a solid scintillation counter as one of the detectors, this method of radionuclide standardization is discussed in Chapter 11. The reader is invited to refer to Chapter 11 on Solid Scintillation Analysis for detailed information on this topic.

C. Triple-to-Double Coincidence Ratio (TDCR) Efficiency Calculation Technique

The first triple photomultiplier liquid scintillation counting system with an electronic circuit having two different coincident outputs was reported originally by Schwerdtel (1965, 1966a,b). A new triple PMT detector design and the original version of the method of radionuclide standardization now referred to as the triple-to-double coincidence ratio (TDCR) efficiency calculation technique were published in a national report in 1979 by Pochwalski and Radoszewski and in the international journals by Pochwalski et al. (1988). The integration of theoretical calculations of counting efficiency based on the radiation emissions of the radionuclide under investigation with the experimental data obtained from the LSA, somewhat similar to the approach taken in the CIEMAT/NIST efficiency tracing method described earlier was reported by Broda et al. (1988) and Grau Malonda and Coursey (1988), and Grau Carles and Grau Malonda (1989). These papers formed the basis for subsequent research and development in this method, which is described subsequently.

I. Principles

The liquid scintillation analyzer (LSA), in this case, is different from the conventional commercially available equipment, described previously in this chapter. The TDCR efficiency calculation technique requires an LSA equipped with three photomultiplier tubes and two different coincident outputs. Such equipment is generally designed and manufactured generally by those laboratories dedicated to radionuclide standardization. An example of such an instrument, referred to as the TDCR liquid scintillation counter, is that used by the Laboratoire Primaire des Rayonnements Ionisants (IPRI), France and reported by Cassette and Vatin (1992). A diagram of the three-PMT LSA is provided in Fig. 5.49. The optical chamber of the LSA accommodates a standard 20-mL liquid scintillation counting vial, and facing the vial are three photomultiplier tubes at 120° angles to each other. With such an arrangement, dual and triple phototube coincidence outputs are possible as a requirement for this method. For example,

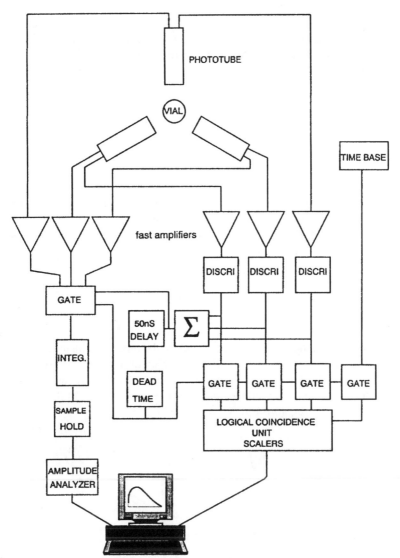

FIGURE 5.49   The counting system of a three-phototube liquid scintillation analyzer with link to a personal computer. The components are similar to those found in a conventional dual-phototube LSA including amplifiers, pulse height discriminators, summation circuitry (Σ), electronic gates, coincidence circuit, and scalers. (From Cassette and Vatin, 1992, reprinted with permission Elsevier Science.)

the two coincident outputs used by Pochwalski *et al.* (1988) from photomultiplier tubes A, B, and C were the count rate from the triple coincidences (ABC) and the count rate from the logic sum of three different double coincidences (AB+BC+CA). Since this initial report, many developments have been made. Advances in the method are described by Simpson and Meyer (1992b and 1994b), from which the subsequent treatment and mathematical notation used are taken.

In the case of a pure β-emitting source on which are viewed simultaneously three normalized photomultiplier tubes the double-tube coincidence count rate N_d and the triple-tube coincidence count rate N_t are defined as

$$N_d = N_0\,\varepsilon_d \qquad\qquad (5.89)$$

and

$$N_t = N_0\,\varepsilon_t \qquad\qquad (5.90)$$

where N_0 is the disintegration rate of the radionuclide source and ε_d and ε_t are the true double-tube and triple-tube counting efficiencies, respectively. From Eqs. 5.89 and 5.90, the ratios of the counting efficiencies and count rates can be written as

$$\frac{\varepsilon_t}{\varepsilon_d} = \frac{N_t}{N_d} = R. \qquad\qquad (5.91)$$

As in the previously described CIEMAT/NIST radionuclide standardization method of efficiency tracing, theoretical counting efficiencies are calculated and these are combined with experimental data. Therefore, the actual counting efficiencies ε_d and ε_t for a pure β emitter can be represented by the theoretical counting efficiencies described by Grau Malonda and Coursey (1988) and Simpson and Meyer (1992a–c) as

$$\varepsilon_2 = \int_0^{E_m} N(E)[1 - \exp\{-P \cdot F(E) \cdot E\}]^2 dE, \qquad\qquad (5.92)$$

and

$$\varepsilon_3 = \int_0^{E_m} N(E)\left[1 - \exp\{-P \cdot F(E) \cdot E\}\right]^3 dE, \qquad\qquad (5.93)$$

where ε_2 and ε_3 are the double- and triple-tube efficiencies, respectively, $N(E)$ represents the number of β particles of energy between $E + dE$ of the β particle spectrum calculated according to the Fermi theory of decay (Fermi spectrum), E_m is the maximum β particle energy, $F(E)$ is the relative scintillation efficiency (Gibson and Gale, 1968) and P is the figure of merit or free parameter, the only unknown quantity of the theoretical efficiency equations. The figure of merit (free parameter) was defined previously in this chapter in Section IX.A concerning the CIEMAT/NIST efficiency tracing technique where a "system" figure of merit M was used, whereby $P = 1/2M$ for a double-phototube system and $P = 1/3M$ for a triple-phototube system (Simpson and Meyer, 1992b). The efficiencies, therefore, are a function of a single unknown parameter P, which are written as $\varepsilon_2 = f(P)$ and $\varepsilon_3 = g(P)$. The two functions of P can replace the efficiencies noted in Eq. 5.91 as follows:

$$\frac{\varepsilon_t}{\varepsilon_d} = \frac{N_t}{N_d} \quad \text{or} \quad \frac{g(P)}{f(P)} = R. \tag{5.94}$$

The ratio R is experimentally measured by the LSA and the value of P is computed directly. As noted by Simpson and Meyer (1994b), the figure of merit P should be single-valued for a given value of R. With the calculated value of P for a given experimental sample, the counting efficiency values ε_2 and ε_3 can be calculated according to Eqs. 5.92 and 5.93. The activity of the radionuclide source can then be calculated using Eqs. 5.89 and 5.90 or

$$N_0 = \frac{N_d}{\varepsilon_2} = \frac{N_t}{\varepsilon_3}. \tag{5.95}$$

The calculation of the theoretical efficiencies according to Eqs. 5.92 and 5.93 are similar to Eq. 5.77 previously described in Section IX.A. According to Grau Carles and Grau Malonda (2001) the equations applicable to the theoretical counting efficiency calculations for beta-particle emitters for a three-photomultiplier system as used in the TDCR method are the following irrespective of the number of active photomultipliers:

$$\varepsilon_2 = \int_0^{E_m} N(E) \left\{ 1 - \exp\left[\frac{-EQ(E)}{3\lambda}\right] \right\}^2 dE \tag{5.96}$$

for two photomultipliers working in coincidence, and

$$\varepsilon_3 = \int_0^{E_m} N(E) \left\{ 1 - \exp\left[\frac{-EQ(E)}{3\lambda}\right] \right\}^3 dE \tag{5.97}$$

for three photomultipliers in coincidence. The variables are the same as those defined previously for Eq. 5.77 in Section IX.A. The above calculations are based on a definition of the figure of merit (free parameter) as a computation β particle energy in keV required to produce one photoelectron at the output of the photocathode or photoelectron yield at the first dynode. However, in a review of the CIEMAT/NIST and TDCR methods Grau Carles and Grau Malonda (2001) proposed a new concept with modification to the efficiency calculations to improve the robustness of the radionuclide standardization methods. According to the review, the counting efficiency calculations should be based not on the photocathode or any dynode outputs, but rather on the final photomultiplier anode output.

2. Experimental Conditions

Simpson and Meyer (1994b) demonstrated that any change in the counting efficiency due to any altered chemical quenching state of the sample will manifest itself in the measured triple and double counting rates, thereby producing a different value of R. They demonstrated, therefore, that the figure of merit P extracted by this method for a given sample includes effects due to chemical quenching so that the calculated sample activity N_0 is independent of chemical quench effects. Using wide ranges of $CHCl_3$ as a quenching agent (18–2287 mg $CHCl_3$ load), they demonstrated no significant quench effect on the calculated activity of ^{14}C. Also, smaller $CHCl_3$ quenching agent loads (13–409 mg) had no effect on the calculated activity of ^{63}Ni, and chloroform loads of up to 60 and 270 μL produced no effect on the calculated activities of ^{99}Tc and ^3H, respectively.

The calculation of the counting efficiencies ε_2 and ε_3 are carried out with the CIEMAT computer programs, such as, EFFY2 written by García-Toraño and Grau Malonda (1985) previously described in this chapter. Simpson and Meyer (1994b) report the use of a modified EFFY2 program to operate on a personal computer, which calculates the ratio R of the triple-to-double efficiencies as a function of figure of merit. They note that in practice, Eq. 5.94 is solved by incrementally varying the figure of merit, P, in the computer calculation, which in effect reads the counting efficiencies corresponding to the measured ratio R from plots of efficiency curves such as those illustrated in Fig. 5.50. Broda et al. (2000) describe a computer program, TDCRB-1, to calculate the radioactive concentration of a solution based on the TDCR method. The computer programs such as those described previously in Section IX.A of this chapter may include subroutines for the calculation of the ionization quench function $Q(E)$ for any given Birks ionizing quench parameter kB (see Eq. 5.88 of this chapter). Cassette et al. (2000)

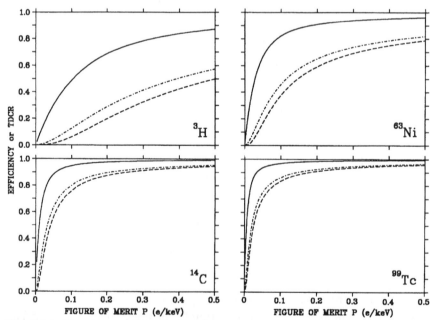

FIGURE 5.50 Calculated two-phototube efficiency ε_2 (dashed-dotted curves), three-phototube efficiency ε_3 (dashed curves) and the ratio $\varepsilon_3/\varepsilon_2$ (solid curves) for each of the pure beta-particle emitters indicated. (From Simpson and Meyer, 1994b, reprinted with permission from Elsevier Science.)

demonstrated that the Birks kB factor is independent of the cocktail chemistry and optical design of the TDCR system, the influence of the kB factor is negligible for high-energy nuclides (e.g., ^{14}C and higher-energy beta emitters), and the optimum value of the kB factor is indepentent of the nuclide. Broda et al. (2002) also, could not find any dependence of the kB factor on the solute or solvent; however, they point out that the proper fitting of the kB ionization quench parameter in the TDCR model is vital to improving the accuracy of this standardization method.

The uncertainties of the calculated activities by this TDCR efficiency calculation technique was estimated by Simpson and Meyer (1994b) to be $\pm 0.9\%$ for ^{14}C, $\pm 2.5\%$ for ^{63}Ni, and $\pm 1.62\%$ for ^{99}Tc. Broda and Pochwalski (1993) report the TDCR standardizations of the electron-capture nuclides ^{55}Fe and ^{54}Mn with ± 1.5 and $\pm 1.6\%$ uncertainties, respectively. Studies by Simpson and Meyer (1996) demonstrate the use of the TDCR efficiency calculation technique for the activity analysis of ^{204}Tl to within 1% uncertainty. Broda et al. (1998) report the standardization of ^{139}Ce with $\pm 0.5\%$ uncertainty using the TDCR method. In a unique approach to radionuclide standardization, Hwang et al. (2002) developed a multichannel timescaling method to enable the direct measurement of accidental coincidences in a triple-photomultipler counting system. This permitted the determination of the double and triple coincidences via corrections due to dead time. They report the measurement of ^{204}Tl with $\pm 0.84\%$ uncertainty, and the limitations to this new method.

The TDCR efficiency calculation technique is a proven method for radionuclide standardization; however, special liquid scintillation equipment is required. The CIEMAT/NIST method previously described in this chapter has had more appeal and currently shows more promise for universal application in many laboratories because the method does not require any special LSA.

X. NEUTRON/GAMMA-RAY MEASUREMENT AND DISCRIMINATION

The same year that Hereforth (1948) under the leadership of H. Kallmann, documented in her thesis the discovery that beta and gamma radiation produced scintillation in certain organic compounds, Bell (1948) demonstrated that fast neutrons could be detected in an anthracene scintillator by the proton recoils produced by neutron–proton collisions. Although neutrons have no charge and consequently cannot produce ionization or excitation directly in a scintillation fluor, the essentially equal mass of the neutron and protons (hydrogen atoms) in organic compounds facilitate transfer of energy from the neutron to protons via elastic scattering as described in Chapter 1, Section II.G.3a. Because of the almost equal mass of the neutron and proton, the maximum fraction of the kinetic energy that a fast neutron can lose in a single head-on collision with a proton is calculated as 0.999 or 99.9% (see Table 1.5, Chapter 1). Although, partial energy transfers also occur through less direct neutron-proton collisions, it is clear that fast neutron energy is absorbed more easily by a substance rich in protons (hydrogen atoms) compared to another substance possessing atoms of higher atomic numbers. The recoil protons with their acquired kinetic energy produce ionization until their energy is dissipated in the medium and they come to a full stop. A liquid scintillation cocktail, rich in protons (high H:C ratio), can be used to measure fast neutron radiation. The energy of the recoil protons dissipated in the scintillation cocktail solvent will excite the solute fluor molecules resulting in the emission of visible light similar to the effect of alpha- and beta-particle interactions discussed previously.

Another radiation type that does not produce direct ionization in a liquid scintillation fluor is gamma radiation. Like x-radiation described previously, gamma rays will interact with liquid scintillation cocktail via principally Compton interactions with atomic electrons. The transfer of gamma-ray photon energy to atomic electrons will result in the liberation of Compton electrons. The energy of these electrons will be absorbed by the liquid scintillation solvent in the same way as beta particles described previously. Compton-electron energy absorption will result in fluor excitation and the emission of visible light of intensity proportional to the amount of energy absorbed in the scintillation cocktail.

The two mechanisms of fluor excitation via recoil proton or Compton electron energy absorption forms the basis for the discrimination of neutron and gamma radiation, respectively, by liquid scintillation. The liquid scintillation discrimination of the two radiation types will be discussed further on in this section.

Only a brief description of the scintillation detection of neutrons and neutron/gamma-ray discrimination will be presented here. More detailed information can be obtained from references specific to the subject (Harvey and Hill, 1979; Rausch, *et al.*, 1993; Brooks, 1993, 1997; Poenaru and Greiner, 1997; and Peurrung, 2000) and from Chapter 11.

A. Detector Characteristics and Properties

The liquid scintillation cocktails used for the detection and measurement of neutrons are organic solutions high in hydrogen content to enhance the efficiency of trapping the neutron kinetic energy through neutron-proton collisions, which results in the production of recoil protons. Since high gamma-ray backgrounds are often present the ability of discriminating between incident neutron and gamma radiation through pulse shape discrimination (PSD) becomes important and some liquid scintillation cocktails are most suitable for this purpose. Also, some organic scintillation cocktails are devoid of hydrogen, such as BC-509 (NE226) or BC-537 (NE230), that contain hexafluorbenzene (C_6F_6) or deuterated benzene (C_6D_6), respectively. Such cocktails have reduced sensitivity for neutrons and can be used for the measurement of gamma radiation in the presence of neutron radiation. Some examples of commercially available cocktails for the detection of neutrons and gamma radiation or for neutron/gamma-ray discrimination are provided in Table 5.9. These liquid scintillators are used devoid of oxygen gas. Dissolved atmospheric oxygen reduces their light output by about 30%, and eliminates any pulse shape discrimination

TABLE 5.9 Properties and Applications of Liquid Scintillators Used for Neutron Radiation Measurements or Neutron/Gamma Radiation Discrimination[a]

Scintillator	Properties	Applications
BC-501A[b] (NE213)[c]	Xylene[d] (>90%) Aromatic fluors (<10%)[e] Atomic ratio, H/C: 1.212 λ_{max}: 425 nm[f] Light output: 78% anthracene[g] Decay time: 3.2 ns[h] Density: 0.874 g cm^{-3} Refractive index, n$_D$: 1.505 Flash point: 26°C	Yields excellent n/γ pulse shape discrimination (PSD) over a wide energy range up to ~100 MeV (Horwáth et al., 2000, and Nakao et al., 2001)
BC-505 (NE224)	1,2,4-trimethylbenzene[d] (97.5%) Aromatic fluors (< 0.5%) Atomic ratio, H/C: 1.331 λ_{max}: 425 nm Light output: 80% anthracene Decay time: 2.5 ns Density: 0.977 g cm^{-3} Refractive index, n$_D$: 1.505 Flash point: 48°C	High light output and transmission; High flash point makes it safer to use and transport than xylene or toluene-based scintillators Suitable for large-volume detectors including anti-Compton and anticoincidence shields
BC-509 (NE226)	Hexafluorobenzene[d] Formula: C$_6$F$_6$ λ_{max}: 425 nm Light output: 20% anthracene Decay time: 3.1 ns Density: 1.61 g cm^{-3} Refractive index, n$_D$: 1.38 Boiling point: 80°C Flash point: 10°C	Essentially free of hydrogen atoms (with exception of fluor molecules) only 0.18 atom% of hydrogen atoms; Low sensitivity to fast and moderated neutrons, insensitivity increased with pulse shape discrimination to enhance gamma-ray detection in n/γ discrimination Can be used to detect neutrons when it is desireable to avoid thermalizing properties of hydrogen
BC-517H (NE235H)	1,2,4-trimethylbenzene (< 30%) Mineral oil (> 70%) Aromatic fluors (< 0.3%) Atomic ratio, H/C: 1.89 λ_{max}: 425 nm Light output: 52% anthracene Mean free path: > 5 meters[i] Decay time: 2.5 ns Density: 0.86 g cm^{-3} Refractive index, n$_D$: 1.476 Flash point: 81°C	Mineral oil-based scintillator where long mean free paths are required and high light output important; Compatible with acrylic plastics such as Plexiglas® and Perspex®, as well as many metals and reflective coatings. Pulse shape discrimination capability

(continued)

TABLE 5.9   (Continued)

Scintillator	Properties	Applications
BC-519	1,2,4-trimethylbenzene ($<40\%$) Mineral oil ($<60\%$) Fluors (0.5%) Density 0.86 g cm^{-3} Atomic ratio, H/C: >NE213 Flash point: >62°C Boiling point: > 204°C	Provides excellent n/γ pulse shape discrimination for neutron energies up to 100 MeV (Horwáth *et al.*, 2000).
BC-521 (NE323)	1,2,4-trimethylbenzene ($>85\%$) Gadolinium (0.5–1.5% w/w) Light output: 68% anthracene[j] 57% anthracene[k] Atomic ratio, H/C: 1.314 λ_{max} : 424 nm Decay time: 3.6 ns Density: 0.89 g cm^{-3} Refractive index, n_D: 1.50 Flash point: 44°C	Utilized for neutron spectrometry with pulse shape discrimination and neutrino research; Neutrons moderated by hydrogen of the scintillator can be captured by the Gd yielding beta particles and gamma rays;
BC-523 (NE311)	1,2,4-trimethylbenzene ($>30\%$) Methyl borate ($<60\%$) 1-methylnaphthylene (10%) Light output: 65% anthracene Atomic ratio, H/C: 1.74 Natural ratio ^{10}B/^{11}B: 0.245 Enriched ratio ^{10}B/^{11}B: 9.0 λ_{max} : 424 nm Decay time: 3.7 ns Density: 0.916 g cm^{-3} Refractive index, n_D: 1.4 Boiling point: 68.9°C	Available at natural ^{10}B isotope abundance or enriched; Useful for total neutron absorption spectrometry, fast neutrons produce a prompt recoil proton pulse with initial scatterings in the cocktail; thermalized neutrons may undergo the ^{10}B(n,α)^{7}Li capture; the capture pulse is in delayed coincidence with the prompt pulse, useful in identifying neutron events (Yen *et al.*, 2000); Useful for n/γ separation by PSD
BC-537 (NE230)	Deuterated benzene ($>98\%$) Formula: (C_6D_6) Aromatic fluors ($<2\%$) Light output: 61% anthracene Atomic ratio, D/H: 114:1 Atomic ratio, D/C: 0.99 λ_{max} : 425 nm Decay time: 1.8 ns Density: 0.954 g cm^{-3}	Useful for n/γ separation as sensitivity to fast neutrons is reduced by deuterium.

(*continued*)

◼ **TABLE 5.9** (Continued)

Scintillator	Properties	Applications
	Refractive index, n_D: 1.50	
	Boiling point: 79.1°C	

[a]Data from Bicron, Newbury, Ohio, USA & NE Technology Ltd (Saint-Gobain Crystals and Detectors UK Ltd. Beenham, Reading, Berkshire, England.
[b]Bicron scintillator.
[c]NE Technology scintillator type equivalent.
[d]Solvent.
[e]Solute (fluor).
[f]Wavelength of fluorescence peak.
[g]Percentage of anthracene light output.
[h]Short component of de-excitation light.
[i]Light transmission.
[j]For 0.5% Gd w/w.
[k]For 1.0% Gd w/w.

properties that the cocktails might possess. These cocktails are purged thoroughly with nitrogen gas prior to use.

Liquid scintillation neutron detectors vary in size from small to medium-sized (4–12 cm in diameter by 4–12 cm in length) up to large volume detectors comprising several cubic meters depending on the research objectives (see Table 5.9 and Zecher *et al.*, 1997). The efficiency and energy resolution of the detector will be governed by its size. Energy is deposited by fast neutrons in the liquid scintillation cocktail by collisions with hydrogen atoms (n–p collisions) and, as a consequence, recoil proton excitation of cocktail solvent and fluor molecules. However, gamma radiation will deposit energy mainly via Compton interactions resulting in Compton-electron excitation of cocktail and fluor. Since both neutrons and gamma rays are generally intermixed, both radiations should be considered. The neutrons will encounter a wide range of collisions with protons including direct head-on collisions and complete transfer of energy to less-direct collisions even just a glancing with a proton resulting in various scattering interactions and only partial energy transfers to protons. Likewise, Compton electrons will vary in energy depending on the energy of the incident gamma radiation.

If the liquid scintillation detector is very small or the neutron or γ-ray energies very high, a considerable number of recoil protons or Compton electrons can escape from the detector. This results in an incomplete deposition of either proton-recoil or Compton-electron energy in the cocktail referred to as the 'wall effect'. When the wall effect is significant, the event (n-p or γ-e) recorded by the photomultiplier tube as a pulse height would be smaller than what would be expected had the recoil proton or Compton electron deposited all its energy in the cocktail. Consequently, the 'response function', which is the pulse height registered per neutron or gamma-ray energy, would be shifted to lower pulse heights. For example, in the analysis of γ-ray photons of energy over the range of 7–20 MeV, Novotny *et al.* (1997) chose NE213 scintillation detectors large enough (e.g. 5 cm in diameter by 10 cm in length, 206 cm³ in volume) taking into account the maximum range of Compton electrons to be detected to avoid a distortion of the pulse height response by wall effects. Likewise, for the measurement of neutron radiation Nakao *et al.* (2001) demonstrated, that for a NE213 liquid scintillator of 12.7 cm diameter by 12.7 cm long the proton-escape probability ranges from approximately 1% to more than 50% of the recoil protons generated for neutrons over the energy range of 25 to over 300 MeV. The neutron detection efficiencies were determined by Nakao *et al.* (2001) by integrating the pulse height response functions at various calibrated pulse height thresholds. At a lower-limit discriminator settings or pulse height threshold of 1.15 MeV (calibrated with a ^{60}Co source) the neutron detection efficiencies varied from 24 to 10% over the neutron energy range of 23–132 MeV. Satoh *et al.* (2001) developed a Monte Carlo code, designated SCINFUL-QMD, for the calculation of neutron detection efficiencies for neutron energies up to 3 GeV in liquid scintillators such as NE213.

It is important to note here that for heavy particles (e.g., protons and α particles) the light output from the liquid scintillator is nonlinear with energy. Unlike electrons or β particles, which produce a relatively linear light output with energy over the range of 0.100–1.6 MeV (Novotny *et al.*, 1997 and Horváth *et al.*, 2000) heavy particles exhibit high values of stopping power (dE/dx) in scintillator and thereby produce higher specific ionization and a saturation effect that yields less light per particle energy loss. The nonlinearity of light output per particle energy loss for protons results in recoil proton energies occurring at lower pulse heights in the pulse height spectrum. Consequently, the use of computer codes used to unfold neutron and photon energy spectra require accurate measurement of the light output response as a function of energy deposited in the scintillator (Klein and Neumann, 2002, Neumann *et al.*, 2002, Reginatto *et al.*, 2002, and Schmidt *et al.*, 2002). Radionuclide sources with well-separated γ-ray energies are used to establish the pulse height scale in terms of electron energies. Gamma-photon sources produce Compton edges in the scintillator where the Compton-electron energies can be used to convert light outputs into electron equivalent energies or MeVee (Horwáth *et al.*, 2000, Nakao *et al.*, 2001, and Klein and Neumann, 2002).

The detector size will govern the number of multiple neutron scatterings. Neutron scattering can result in a broad range of energy transfers to recoil protons, from very little neutron energy when a neutron is simple glanced off a proton (near hit) up to the maximum neutron energy resulting from a head-on collision with a hydrogen atom. All recoil protons from a given neutron after multiple collisions and scatterings will produce photons all of which are detected by the photomultiplier tube as they will all occur with the coincidence time gate of the scintillation analyzer. The sum of the photons produced by the recoil protons from the multiple scatterings of one neutron will produce a pulse height from the photomultiplier proportional to the light output. Due to the summation of photons from multiple scatterings the resultant pulse heights will be higher than what would occur at reduced scattering.

Neutrons are also scattered by carbon atoms in the scintillator; however, carbon recoils do not produce and significant light output in liquid scintillator (Klein and Neumann, 2002). The H:C atomic ratios in liquid scintillator may vary from 1.2 to 1.9 (Table 5.9). The higher the ratio or greater the atom% H in the scintillator, the higher will be the probability of n–p collisions and the production of proton recoils. However, carbon scattering will cause neutron energy loss and consequently a reduction in pulse height response as a function of energy. As illustrated in Table 1.5 of Chapter 1, the maximum fraction of kinetic energy that may be lost by a neutron collision with ^{12}C is 28.6% resulting from a head-on n-^{12}C collision. A neutron may lose, therefore, anywhere between 0 and 28.6% of its energy in a single collision with carbon. The scattered neutron may then collide with a proton, but the maximum energy of the recoil proton in this case would be within the range of 100 and 71.4% of the original neutron energy. Consequently, carbon atoms in the scintillator will reduce the pulse height response as a function of neutron energy.

At lower neutron energies, particularly where neutrons may become thermalized in the liquid scintillator, special detectors may be employed, such as BC-523, that contain natural or enriched ^{10}B in the scintillation cocktail (Table 5.9) to enhance detection efficiencies. The incident neutrons are moderated by n–p collisions in the cocktail and then undergo $^{10}B(n,\alpha)^7Li$ capture according to Eq. 1.74 of Chapter 1 and more specifically as follows:

$$_0^1n + {}_5^{10}B \rightarrow \alpha + {}_3^7Li(g.s.) + 2.790\,\text{MeV (6.3\%)} \tag{5.98}$$

$$\rightarrow \alpha + {}_3^7Li^* + \gamma(0.478\,\text{MeV}) + 2.312\,\text{MeV (93.7\%)} \tag{5.99}$$

The branching ratios for the two reactions are 6.3 and 93.7%, respectively. The branch described by Eq. 5.98 leaves the ^7Li at the ground state, whereas the other branch (Eq. 5.99) leaves the ^7Li at an excited state; and the α particles plus $^7Li^*$ ions create scintillation light equivalent to light produced by 65-keV electrons or beta particles. Yen *et al.* (2000) report the use of 4 cm diameter detectors of ^{10}B-enriched scintillator joined together in a circular array of 55 detectors. They report detector efficiencies of 95, 85, and 71% at neutron energies of 10, 100, and 1000 eV, respectively.

B. Neutron/Gamma-ray (n/γ) Discrimination

The detection of neutrons is often complicated with background gamma radiation. The neutron detector can be shielded from radiation background arising from charged particles by use of absorber material between the source and detector or via magnetic field deflection. This leaves only gamma rays, which remain undeflected by magnetic fields and relatively unattenuated by absorbers.

I. Pulse Shape Discrimination (PSD)

Neutron and γ-ray interactions in a liquid scintillator may be separated by pulse shape discrimination (PSD) according to principles that are similar to alpha- and beta particle (α–β) discrimination in liquid scintillation described further on in this chapter and in Chapter 6. The principles behind (n/γ) discrimination rest on different degrees of excitation caused by heavy charged particles versus lighter charged particles in the liquid scintillation cocktail. Neutron interactions produce the relatively heavy recoil protons, whereas γ rays produce much light Compton electrons in scintillation cocktail. The heavier charged particles produce greater specific ionization (higher LET) in cocktail than electrons and, consequently, cocktail fluors reach higher excited states that take longer for deexcitation to occur. Deexcitation is manifested as fluorescence. The fluorescence is analyzed by the electric pulse it produces in the photomultipler tube (PMT). The shape of the pulse is a function of the specific ionization of the particle (heavy particle versus lighter particle), and the shape can be defined in two components, namely, a fast (prompt) component or rise time or a slow component or tail. The prompt component of fluorescence has a decay time of only a few nanoseconds (Table 5.9), while the delayed component or tail may last a few hundred nanoseconds. A heavy charged particle, such as a proton, will produce a large fraction of its fluorescence light in the slow component, while a much lighter charged particle will cause fluorescence with a much smaller portion of the light in the slow component.

To exploit the phenomena of differing pulse shapes the method of Heltsley *et al.* (1988) is commonly used. As described by Zecher *et al.* (1997) the method entails the use of two analog-to-digital converters (ADCs), one that integrates the total charge of the pulse, and the other that integrates the charge from some fixed time fraction (portion) of the pulse (i.e., either the first prompt component or the tail). By comparing the magnitude of the charge collected in the fraction of the pulse to the total charge in the pulse, one can determine which species created the pulse, proton or electron, and consequently, neutron or γ ray. A two-dimensional plot is made of either the prompt component (first 30–40 nanoseconds) or the tail, which is the remaining pulse event following the first 30–40 nanoseconds, that is, the remaining part excluding the prompt portion of the pulse event. Zecher *et al.* (1997) called the fraction of charge collected from the prompt event as QFAST, that collected from the tail portion as QTAIL, and the total integrated charge from the PMT as QTOTAL. The signals and gates used to collect total charge from a fluorescence decay pulse event and the fraction QTAIL or QFAST (by diference) is illustrated in Fig. 5.51.

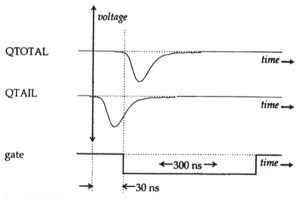

FIGURE 5.51 Signals and gate used in n/γ pulse-shape discrimination. (From Heltsley et al., 1988, reprinted with permission from Elsevier Science.)

The charge collected from either the prompt component QFAST or slow (tail) component QTAIL is plotted against QTOTAL. Fig. 5.52 illustrates QFAST plotted against QTOTAL. Since QTOTAL is a charge proportional to the light output, it can be calibrated in MeV equivalents using γ-ray sources that produce Compton electrons in liquid scintillation. The Compton edge of each of the γ-ray sources are used to calibrate the energy scale rather than protons, because the light output in scintillator is linearly dependent to electron energy, whereas, the linearity does not hold for protons as discussed previously. As illustrated in Fig. 5.53, the PSD causes the pulse events (points)

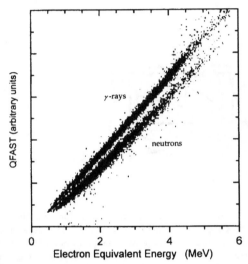

FIGURE 5.52 A PSD spectrum from a cell (liquid scintillation detector). The neutrons and γ-rays are from a **Pu-Be** source placed I m perpendicularly from the center of the cell. (From **Zecher** *et al.*, 1997, reprinted with permission from **Elsevier Science.**)

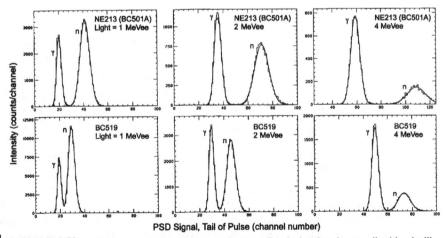

FIGURE 5.53 Comparison of neutron/gamma-ray discrimination in two liquid scintillators – **NE2I3 (BC50IA)** and **BC5I9** for three values of light output – I, 2, and 4 MeVee. (From **Horváth** *et al.*, 2000, reprinted with permission from **Elsevier Science.**)

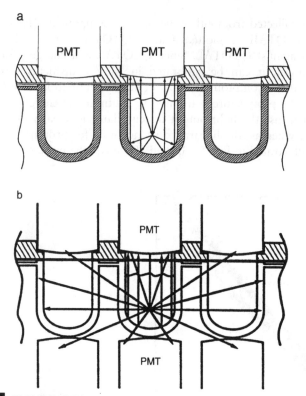

FIGURE 5.54 Two detector designs for microplate scintillation counting. (a) a cross section of a microplate segment illustrating three white opaque sample wells and photo-multiplier tubes (PMTs) aligned above each sample well using only one PMT per sample well and up to 12 PMTs to analyze simultaneously as many as 12 samples. (e.g., PerkinElmer TopCount Scintillation and Luminescence Counter); (b) cross section of the detector design that requires two PMTs per sample and clear sample microplates. Optical crosstalk by photon escape and travel from one sample well to other wells is illustrated (solid lines with arrows) [Courtesy of PerkinElmer Life and Analytical Sciences, Boston, MA.]

to fall into one of two groups, namely, a group for γ-ray-induced events (more light in the prompt component and less in the tail) and another group for neutron-induced events (less light in the prompt component and more in the tail). To evaluate n/γ discrimination at different light outputs Horwáth et al. (2000) utilized data points within a narrow verticle slice of the two-dimensional plot of the type illustrated in Fig. 5.52 with the slice centered at a value of light output to produce a PSD spectrum. Fig. 5.53 illustrates the results of taking the verticle slices of data points from the two-dimensional plot for light outputs of 1.0, 2.0, and 4.0 MeV equivalents. To calibrate the integrated charge (proportional to channel number in Fig. 5.53) of the PMT pulses to light output, Horwáth et al. (2000) used the Compton edge of three γ-ray sources, those of ^{60}Co ($E_\gamma = 1.17$ and 1.33 MeV), ^{228}Th ($E_\gamma = 2.66$ MeV), and Pu-Be ($E_\gamma = 4.44$ MeV). Gamma-ray sources are used to calibrate the energy scale of the PMT because the pulse height response to electron-induced scintillation is a linear function of electron energy.

The calibration is usually referred to as electron equivalent energy or MeVee as noted in Fig. 5.53.

2. Time-of-Flight (TOF) Spectrometry

Another method used to discriminate neutrons and γ-ray photons is via time-of-flight (TOF) spectrometry. In liquid scintillation cells sufficiently large, such as the NE213 detectors measuring 10 cm in diameter by 5 cm in length used by Neumann et al. (2002), a significant fraction of neutrons and γ-ray photons would be expected to undergo multiple scattering. A γ-ray travels at the speed of light, whereas neutrons travel slower. For example, a 2.0-MeV neutron travels at only $0.065c$ or 6.5% of the speed of light (see Fig. 1.10 of Chapter 1). Consequently, a γ-ray that undergoes multiple Compton scattering in a scintillation cocktail would produce a pulse event in the PMT before a neutron that undergoes multiple scattering. As noted by Peurrung (2000) the success of this method hinges upon the fact that the average interval between proton recoil events in neutron scattering is > 3 ns, whereas γ-ray interactions occur significantly more quickly and can thus be discriminated against. Neumann et al. (2002) coupled the scintillation detector to the PMT at distances of up to 6.305 m from the neutron source (accelerator target) via a lightguide. TOF spectrometry is possible only at accelerator facilities with pulsed beam technique. TOF was recorded using the constant fraction timing of the anode signal of the PMT and the beam pickup signal of the recorder. Gamma photon spectrometry was carried out by setting a window on the prompt photon peak in the TOF spectrum to measure the pulse height spectra that was subsequently submitted to computer code for unfolding to provide spectral fluences. For neutron analysis spectral unfolding of deconvoluted pulse height spectra was possible without TOF selection.

XI. MICROPLATE SCINTILLATION AND LUMINESCENCE COUNTING

Many applications use radioactivity as a method of screening a series of compounds or reactions in an effort to find a new therapeutic agent, drug candidate, or to assess a molecular biology function. In these types of applications, many thousands of samples are counted each week. Because of the numerous samples involved, large amounts of scintillation vials and cocktails are required for traditional LSA, which add cost to the sample analysis. This large number of samples also creates a large amount of radioactive waste. Consequently, microplate scintillation counting instruments have been developed to reduce waste, increase sample throughput via the simultaneous analysis of several samples with multiple detectors, provide automation capabilities, and reduce the cost of consumable materials and waste disposal.

The microplate scintillation counter uses a microplate sample format, which was developed to provide for the analysis of large numbers of samples in a relatively small format. A microplate is a small container with many sample wells. It is about the size of a postal card measuring approximately

$5 \times 3.4 \times 0.75$ inches (L × W × H). The microplate comes in three sample well formats, namely, 24-well plates (6×4 wells), 96-well plates (12×8 wells) and 384-well plates (24×16 wells). The 24-well microplate sample wells hold up to 375 and $1500 \, \mu L$ of sample plus scintillation cocktail per well for shallow- and deep-well plates, respectively. The 96-well microplate sample wells hold 75 and $350 \, \mu L$ of sample per well for shallow- and deep-well plates, respectively, and the 384-well microplate sample wells hold up to $80 \, \mu L$ of sample per well. It is clear that only relatively small samples can be analyzed in these microplates, but the footprint is small, compact, and easy to automate. As a result of the small sample size and small amount of scintillation cocktail required, the activity analysis of many biological and molecular biology samples can be carried out without producing a large amount of radioactive waste. The second major feature of the microplate scintillation counter is the ability to count up to 12 samples simultaneously directly in the microplate.

Only a brief description of the microplate scintillation analyzer is provided here. More information on this instrumentation is available in Chapter 11, Section V.B.

A. Detector Design

The microplate scintillation and luminescence instrument can be of single- or multiple-detector design (up to 12 detectors). The microplate instrument can count samples directly in microplate wells using one of the two mechanisms: single or dual PMT technology. In a microplate the sample wells are all packed in close proximity in a very small area; therefore, two primary issues must be addressed in microplate counting in terms of detector design, namely crosstalk and background reduction.

B. Optical Crosstalk

The normal microplate is made of clear polystyrene. If radioactivity or luminescence is to be detected in the microplate with multiple photomultiplier tube detectors, light created in one sample well of the microplate can be "seen" by a phototube viewing an adjacent well. This is defined as optical crosstalk. Two approaches may be taken to resolve the optical crosstalk problem: its prevention or its correction.

Figure 5.54 shows two detector designs. The detector arrangement, illustrated in Fig. 5.54a, utilizes only one phototube per sample well and as many as 12 phototubes to analyze 12 samples simultaneously, as employed in the TopCount microplate scintillation and luminescence counter. A white reflective and opaque microplate is utilized whereby light produced by the radioactivity in the liquid scintillator is reflected back toward the phototube and no optical crosstalk occurs. Therefore, the best method to prevent optical crosstalk is to use an opaque microplate with a single phototube per sample detector. A white microplate is used for the liquid scintillation analysis of radioactive samples and a black microplate for single photon counting of luminescent samples. The optical crosstalk that occurs with the use of clear microplates and two phototube detectors per sample is illustrated in Fig. 5.54b. This detector design utilizes two photomultiplier tubes per sample, one above and the other below the sample. A clear microplate is required in this case, and even in the measurement of weak beta particle-emitting radionuclides, such as ^{3}H, scintillation photons from one sample well can be seen by neighboring phototubes intended for other sample wells. Optical crosstalk of this kind can be significant in phototubes located more than two wells distant from the radioactive sample. To correct for optical crosstalk in this type of detector design using clear microplates, a mathematical algorithm must be used. The correction depends on the type of microplate, the radioisotope, the sample quench level, and the detection method used.

In summary, it is best to use a microplate scintillation counter that employs opaque microplates and a single phototube per sample well, with as many as 12 phototubes to analyze simultaneously up to 12 samples, whereby the problem of optical crosstalk is avoided altogether. For further reading on crosstalk interferences in microplate scintillation counting see Effertz *et al.* (1993).

C. Background Reduction

The two detector designs of microplate scintillation counters, illustrated in Fig. 5.54, must take different approaches to the problem of background reduction. The design with two photomultiplier tubes per sample well uses conventional coincidence counting and additional lead shielding to reduce background, whereas the single-phototube design in TopCount microplate scintillation analyzers uses a time-resolved liquid scintillation counting (TR-LSC) technology for background reduction.

The single-PMT time-resolved scintillation counting technique uses pulse counting to distinguish between true scintillation pulse events and background noise. This obviates the need for a second PMT per sample and reduces lead shielding requirements. The elimination of a second PMT per sample also facilitates close physical alignment of PMTs for the simultaneous counting of up to 12 samples on a single microplate. Single-PMT counting with TR-LSC uses scintillators with relatively long decay periods. A scintillator with a long decay constant emits photons, after β-particle excitation, over a longer period of time. Each scintillation pulse produces a photon packet followed by a series of pulses as illustrated in Fig. 5.55, whereas, PMT noise creates single pulse events. In TopCount TR-LSC the characteristics of a pulse are determined over a period of time (e.g., 200 ns) after the initial photon packet is detected. If it is followed by one or more additional pulses within the resolving time period of 200 ns, the pulse is accepted as a true scintillation event. If no additional pulses are detected within the resolving time period, the initial pulse is rejected as background noise. The resolving time circuit to initiate pulse decay discrimination is triggered when a pulse height exceeds the single photon event (SPE) threshold. The number of pulses above the SPE threshold is counted during the resolving time period. Multiple pulses detected in the resolving time are accepted as a valid event, which is further analyzed in the

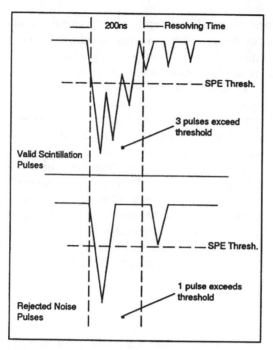

FIGURE 5.55 Single PMT time-resolved pulse discrimination employed in the PerkinElmer TopCount Microplate Scintillation and Luminescence Counter. (Courtesy of PerkinElmer Life and Analytical Sciences, Boston, MA.)

pulse height analyzer of the LSA. If multiple pulses are not detected in the resolving time interval, the triggering pulse is rejected as background noise.

The single-phototube design in scintillation counters with TR-LSC background reduction utilizes opaque sample microplates to prevent optical crosstalk from sample to sample. All types of samples, including liquid scintillator, solid scintillator, luminescent samples, filters, and membranes, can be counted using this technique. An external standard can also be used for DPM determinations when the single-phototube design is employed with the phototube detector located above the microplate sample well. The external standard can be positioned automatically by the instrument below the sample well of the microplate when needed for automatic quench correction determinations. This external standard DPM method is useful particularly in the measurement of dual radionuclides and samples with low count rates. For further information on single-PMT technology in microplate scintillation analyzers, see Effertz *et al.* (1993).

D. Applications

A microplate scintillation and luminescence counter has broad applications, and with these are distinct sample preparation methods that are used for radioactivity analysis or luminescence counting. All of these applications are intended for small sample volume and high sample throughput analysis. These are discussed briefly below.

I. Liquid Scintillation Analysis (LSA)

When liquid scintillation counting is required with the microplate analyzer, scintillation cocktail and sample are added to each well of the microplate. This method is used for counting organic or aqueous samples directly in the microplate wells. An example of this can be taken from enzyme activity studies in which sample is extracted or filtered directly into the liquid scintillation cocktail and counted. Many examples can be taken from the literature. The work of Harwood Jr. (1995) demonstrates enzyme activity measurements with the TopCount microplate scintillation counter for ^{3}H measurements. In this case the enzyme activity is measured with only a single transfer from reaction vessels or wells of a 96-well reaction plate to the filtration wells of a 96-well PerkinElmer Unifilter GF/B filtration plate following acid precipitation, allowing liquid scintillation counting to be conducted directly in the filtration plate without the need for either removal of the filter from the plate or transfer of the filter to liquid scintillation counting vials prior to radioactivity analysis.

Microplate liquid scintillation counting applications are numerous. Among these a few will be cited including papers dealing with general biological applications (Pye *et al.*, 1998; Xu *et al.*, 1998; Craddock *et al.*, 1999; Underwood and Blumenthal, 1999; Upham, 1999; Yin *et al.*, 2000), and wipe test applications (Hill, 1996).

2. Solid Scintillator Microplate Counting

Solid scintillation counting in microplates involves depositing a sample into the well of a special plastic microplate (e.g., PerkinElmer LumaPlate and Wallac Scintiplate), which contains a solid scintillator at the bottom of the well. The material is dried and then counted in a microplate scintillation counter. By this technique the activities (DPM) of small volumes of any nonvolatile samples labeled with beta particle or Auger electron-emitting nuclide may be determined. The small samples involved are particularly applicable to cytotoxicity, immunoassay, receptor binding, enzyme activity, and other metabolic studies. In certain circumstances the technique can replace conventional LSA in counting vials when nonvolatile forms of the radionuclides and small volumes need to be analyzed. See Chapter 11, Section V.B.1 for detailed information on this solid scintillation counting technique with microplate scintillation analyzers.

3. Scintillation Proximity Assay (SPA)

Scintillation proximity assay (SPA) is a technology for the analysis of binding reactions, commonly studied in the medical and biochemical sciences, which circumvents the need to separate bound from free fractions. Glass or plastic solid scintillator microspheres are used in this assay together with an isotope-labeled (^{3}H or ^{125}I) ligand. With the use of modern liquid handling equipment and an automatic scintillation counter for samples in a microplate format, hundreds of samples may be prepared and analyzed in a single day, because traditional processes for the separation of bound and free fractions are not required.

See Chapter 11, Section V.B.2 for detailed information on the use of microplate scintillation counters for scintillation proximity assays.

4. Luminescence Assays

The microplate scintillation and luminescence counter is a state-of-the-art automated microplate luminometer suitable for use with all glow-type or enhanced flash-type luminescent chemistries. Bio- and chemiluminescence assays are carried out in the 96- or 384-well microplate format via single photon counting (SPC). Applications include immunoassays, DNA probe assays, cell growth assays, and reporter gene assays. High-throughput automated and unattended assays are carried out with multiple-phototube detectors as previously described, while 96- or 384-well microplates are stacked in the instrument in quantities of 30 or more. Automated liquid handling systems and other robotic systems can allow unattended high-throughput screening utilizing glow luminescence as reported by Walton (1996). The background is approximately 20 counts per second (CPS) and accurate measurements of over 20×10^6 CPS are possible.

Some examples of typical assays include reporter gene assays (Scheirer *et al.*, 1994, and Plautz *et al.*, 1996), chemiluminescence assay of polymerase chain reaction products (Brillanti *et al.*, 1991, Kaneko *et al.*, 1992, and Garson and Whitby, 1994), neutrophil activation studies (Lieberman, 1995) and cell proliferation and cytotoxicity assays (Roelant and Burns, 1995).

5. Receptor Binding and Cell Proliferation Assays

These techniques require filtration of the bound or reacted radionuclide from the free. There are two methods of performing this and both involve using a cell harvester. The cell harvester is a special instrument that can filter samples from each well of the microplate, deposit them onto a filter or membrane, and wash away the nonreacted or unbound radiolabeled material. Harvesters are available that can filter harvest from 6 to 96 samples simultaneously in a microplate format. The harvesting can be performed directly on a filter or membrane and cocktail added to the filter in a special "bag" or directly onto the filter in a special holder. The second method is to use a special filter-integrated microplate (e.g., UniFilter plate) onto which harvesting can be performed directly using this microplate and cocktail added to each well prior to counting in a microplate scintillation counter. The most common assays of this type include receptor binding, which are used routinely for the evaluation of pharmaceutical agents by assessing their ability to interfere with the specific binding of a radiolabeled ligand to its receptor. See Chapter 11, Section V.B.2 for detailed information on scintillation proximity assay.

E. DPM Methods

The methods of determining sample activities as DPM in a microplate are very similar to those used for a conventional LSA. Two primary methods are used, which require the measurement of a quench indicating parameter (QIP) derived from either the sample spectrum or the external standard spectrum.

As described previously in Section V.B of this chapter, the sample spectrum method uses some feature of the actual spectrum of the sample to measure quench. Two different sample spectrum quench-indicating parameters are used, namely, the transformed spectral index of the sample (tSIS) and the spectral endpoint energy [SQP(I)] depending on the make of the instrument. In order to use these quench-indicating parameters to correct for quench or other QIPs based on external standard pulse height spectra [e.g., tSIE or SQP(E)], a quench curve must be prepared. This curve can be prepared by measuring the counting efficiencies of quenched standards directly in the sample wells of a microplate, or the quenched standards can be prepared in vials or test tubes and thereafter pipetted into the wells of a microplate.

Once the quench correction curve has been prepared, it is stored in the instrument memory and applied to the unknown sample analysis to determine the DPM of each sample. The sample spectrum methods of quench correction have the same limitations that were discussed in Section V.B of the chapter for conventional LSC. These limitations include (1) sample spectrum quench-indicating parameters are count rate dependent, (2) the tritium quench curve has a sharp slope, and (3) DPM determinations are less accurate.

The alternative and preferred method of quench correction is the use of an external standard (e.g., ^{133}Ba, ^{226}Ra, or ^{152}Eu) gamma source and the preparation of a quench correction curve based on the external standard pulse height spectrum. The external standard quench indicating parameter [e.g., tSIE or SQP(E)] is used as previously discussed in detail in Section V.D of this chapter. This method is sample count rate independent, and it has a large dynamic range even for tritium (see Fig. 5.56). This method of quench correction can be used only with single-PMT microplate scintillation counters, which have the PMT aligned above the sample well of the microplate. With microplate counters having two PMTs aligned above and

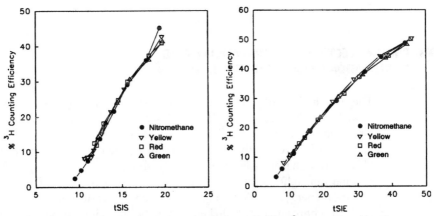

FIGURE 5.56 TopCount microplate scintillation counter ^{3}H quench correction curves using the 96-well microplate, PicoPlate-96, and MicroScint-20 scintillation cocktail. Left curves: tSIS is the quench-indicating parameter. Right curves: tSIE is the quench-indicating parameter. (From Effertz *et al.*, 1993.)

below the sample wells there remains no space to position the external standard appropriately within the instrument.

F. Advantages and Disadvantages

The microplate scintillation and luminescence counter has the basic advantage of high sample throughput, as up to 12 samples can be analyzed simultaneously for radioactivity or luminescence. Hands-off fully automated analysis of thousands of samples is possible, particularly in the simplest case of a microplate analyzer with 32 microplates stacked for counting, each microplate containing 96 sample wells, which provides for 3072 unattended analyses. Microplates containing 384 sample wells each with $80\,\mu L$ capacity can be stacked up to 40 microplates at one time, which provides for over 15,000 samples of unattended analysis. The cost reduction in microplate counting compared with conventional liquid scintillation counting due to the reduced cocktail consumption and consequent reduction in radioactive waste expense and vial cost is a definite advantage. Microplates offer also a format that is easy to handle, leaving less chance for error caused from placing samples out of order, as sample wells cannot be mixed up or intermixed. Once samples are prepared on a microplate, they are kept in place on the microplate as a fingerprint. Microplates are easily portable and stored without occupying much space.

A disadvantage of the microplate scintillation analyzer is the small volumes of sample that can be accommodated in the sample wells. A 20-mL liquid scintillation vial can accept approximately 10 mL of aqueous sample mixed with 10 mL of suitable liquid scintillation cocktail. However, the largest sample well of a 24-well microplate has a capacity of only 1.5 mL $(1500\,\mu L)$, of which approximately 0.75 mL at most can be an aqueous sample and 0.75 mL a suitable cocktail. That is a 13-fold difference in sample size between the two methods of analysis. Therefore, sample activities in microplate scintillation counting must be higher, or longer counting times are required.

XII. PHOTON ELECTRON REJECTING ALPHA LIQUID SCINTILLATION (PERALS) SPECTROMETRY

The interaction of alpha particles with liquid scintillation cocktail was described earlier in Section II.B of this chapter. Of particular interest is that alpha particles are generally detected at a very high counting efficiency of about 100% even with high degrees of quench in the scintillation cocktail. However, alpha particles produce only about one-tenth the light intensity (i.e. scintillation light output) as beta particles or gamma radiation per unit of radiation energy. A 5-MeV alpha particle will produce a pulse height spectral peak at approximately 500 keV when liquid scintillation pulse heights are calibrated on a scale equivalent to particle energy in keV. (Such pulse height energy scales are calibrated with beta particles or electrons, as electrons yield a linear light output per particle energy dissipated in the liquid scintillator as discussed previously in Section X.A of this chapter.) This posed a problem in the liquid scintillation analysis of α emitters when found in the presence of β emitters or β–γ emitters, because their pulse height spectra would overlap even though the energy of the α particle radiation was 10 times that of the β-particle radiation. For example, the overlapping liquid scintillation pulse height spectrum of the α emitter ^{210}Po ($E = 5.30$ MeV) and the β emitter ^{90}Sr ($E_{max} = 546$ keV) are illustrated in Fig. 5.2 in Section II.B of this chapter. Therefore, since the early 1970s research has gone into the development of liquid scintillation spectrometers that could reject the pulses produced by β and γ radiation and provide an instrument that could measure α particle radiation with good resolution, high counting efficiency and low background superior to the surface barrier detectors used for measurements of α particle radiation. This lead to the development of PERALS spectrometry. The history of the work that led to the development of PERALS spectrometry, is described by McDowell and McDowell (1991). A description of the instrumentation, its performance, and applications will be provided here.

The development of PERALS spectrometry had to solve two problems of α particle liquid scintillation spectrometry: (1) the poor energy resolutions obtained with α emitters in quenching aqueous-accepting scintillation cocktails in conventional LSAs and (2) the interference caused by overlapping spectra from β- and γ-emitters in the same sample. The problem of resolution of α pulse height spectra was solved by modifying the following elements of the liquid scintillation system as described by McDowell and McDowell (1993): (1) the α emitters of interest had to be placed into a nonquenching, organophilic complex in a completely organic and highly efficient scintillator; (2) the detector assembly had to be modified with an efficient reflector arrangement and light-coupling optics to transmit the maximum amount of scintillation light to the photocathode of the PMT, and (3) a diffuse reflector had to be used in the detector assembly so that the light from an α event, anywhere in the sample would appear the same to the PMT, because of the nonuniform response of PMTs.

The design of the high-resolution α liquid scintillation spectrometer, known as the PERALS spectrometer, is illustrated in Fig. 6.12 of Chapter 6. The PERALS spectrometer is produced by ORDELA, Inc, Oak Ridge, TN, USA. The α emitters are firstly placed into an organic "extractive scintillator" often via some solvent extraction process that enables the radionuclides to pass from the aqueous sample phase into the organic scintillator phase. The extractive scintillator contains an organic extractant (extractive molecule) capable of forming selective complexes with certain alpha emitters to facilitate their removal from aqueous solutions and transfer into the organic scintillator phase. The organic extractants chosen must be pure, exhibit a minimum of quenching, and be high in the extractive power for the target element(s) as described by McDowell and McDowell (1993) and McDowell (1996). Popular extractive scintillators consist of a combination of scintillator (e.g., PPBO), solvent (e.g., toluene), and an extractant. Some commercial extractive scintillators used for PERALS spectrometry are provided by ETRAC, Knoxville, TN, USA, among which are ALPHAEX, THOREX, URAEX, and POLEX, which contain, as extractant, di(2-ethyl-hexyl)-phosphoric acid (HDEHP), 1-nonyldecylamine

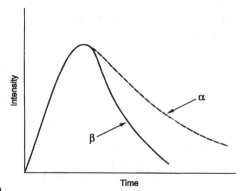

FIGURE 5.57 **Alpha and beta pulse decay in a liquid scintillator. The decay time of light produced by the α particle in certain scintillators is about 35–40 ns longer than that of the faster decaying light produced by the β particle.**

sulfate, tri-*n*-octylamine sulfate (TNOA), and tri-octyl-phosphine oxide (TOPO), respectively. A culture tube measuring $10 \times 75\,mm$ containing 1 mL of the radionuclide sample in the extractive scintillator is placed in the detecting chamber of the PERALS spectrometer. The instrument contains only one photomultiplier tube. Dual photomultiplier tubes are not needed, in contrast to conventional LSA, because the thermal electron noise in the phototubes is always well below the pulse heights produced by α particles (McKlveen, and McDowell, 1984). Light-coupling silicone oil and a white reflecting surface made of barium sulfate and binder are important to meet the light transfer requirements from sample tube to the photocathode for optimum resolutions of α particle pulse height spectra.

   To provide for clean α particle pulse height spectra without the interference from β particle smear, the PERALS spectrometer discriminates the α- from β particle events in the liquid scintillator. The instrument makes use of pulse shape discrimination (PSD), which can take advantage of the longer decay time of light produced by α particle interactions in scintillation fluor and reject the faster decaying light produced by β particle or γ-ray interactions. Fig. 5.57 illustrates the light pulse shapes of alpha and beta events in a liquid scintillator. Depending on the scintillator used, the decay times of α particle interactions can be 30–40 ns longer than β particle interactions. As explained by Cadieux (1990), each input pulse to the system generates an amplified voltage pulse for pulse height analysis and a pulse shape signal with a voltage proportional to the time length of the incoming pulse. A discriminator generates a gating signal for only the alpha particles after analyzing the pulse shape voltages. The proper discriminator gating signal is set by use of a microcurie γ-ray source outside of the PERALS sample chamber, which generates Compton electrons that simulate β particle interactions inside the scintillation fluor cocktail, whereby the discriminator can be adjusted to reject pulse events from the Compton electrons. Following pulse shape discrimination, a multichannel analyzer (MCA) collects and analyzes pulse height spectra from the α particle interactions.

   PERALS spectrometry has proven to be a valuable tool in the measurement of ^{226}Ra and ^{222}Rn in drinking water with detection limits of

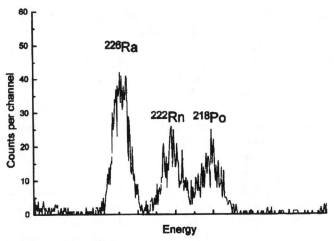

FIGURE 5.58 Radium spectrum of La Bourbouls-Choussy thermal water (3000 ± 200 mBq activity and 88650 s counting time. (From Aupiais *et al.*, 1998, reprinted with permission© 1998 American Chemical Society.)

0.006 Bq L^{-1} of ^{226}Ra using only 6 mL of sample (Aupiais *et al.*, 1998) and 0.68 Bq L^{-1} of ^{222}Rn with the extraction of 1 L of water (Hamanaka *et al.*, 1998). An example of the α particle discrimination achievable by PERALS with low activity water samples is provided by Fig. 5.58 illustrating the pulse height spectra of ^{226}Ra, ^{222}Rn, and ^{218}Po with α particle emissions at 4.78, 5.49, and 6.00 MeV respectively. PERALS spectrometry has demonstrated to be a practical method for the analysis of the actinides with high resolution and at low levels of activity. For example, the following studies can be cited: the analysis of ^{232}U-^{234}U-^{238}U at resolutions of 266, 233, and 200 keV FWHM reported by Dacheux *et al.* (2000), limits of detection as low as 1 mBq L^{-1} for the extraction of uranium, thorium, plutonium, americium, and curium in aqueous solutions of up to 250 mL (Dacheux and Aupiais, 1997 and Dacheux and Aupiais, 1998). Measurements of ^{237}Np by PERALS spectrometry at levels as low as 23.7 and 9.5 pg L^{-1} after counting times of 3 and 10 days were demonstrated by Aupiais *et al.* (1999). Selective extraction of ^{210}Po for PERALS spectrometry by Véronneau *et al.* (2000) yielded detection limits below 1 mBq L^{-1} for 200 mL of solution and 1000 min counting times. Because of the relatively high spectral resolutions achieved by PERALS spectrometry the normal Gaussian shape of alpha peaks sometimes display a high-energy tail in addition to the pure Gaussian function. Aupiais and Dacheux (2000) demonstrated that this asymmetry is due to internal conversion. They conclude that for the accurate PERALS analysis of some radionuclides, such as isotopes of thorium, it is necessary to account for *L*- and *M*-shell internal conversion contributions in the activity measurements.

The PERALS spectrometer can be considered the most sensitive α-spectrometric method available. As reported by McDowell and McDowell (1993) the PERALS spectrometer rejects 99.99% of ambient and sample β and γ counts from the α spectrum providing ambient radiation backgrounds of 0.001 CPM or less. An alpha pulse height energy resolution of 4.2% FWHM is achievable, which is limited by the quality of contemporary PMTs. The lower limits of detection are estimated at 0.17 mBq (0.01 DPM). The α counting efficiency is 99.7%, as a small 0.3% is lost through α particle collisions with the sample tube.

The major disadvantage of PERALS spectrometry is that it is specific for the spectrometric analysis of α emitters only, that is, this instrument or method is not able to determine α and β activities simultaneously as reviewed by Pates *et al.* (1996a). Other conventional LSAs equipped with automatic sample changers are available commercially, that provide α–β separation by pulse decay analysis or pulse shape analysis and spectrometric analysis of both α and β emitters in the same sample, albeit at a lower α particle energy resolution than obtained by PERALS spectrometry. The simultaneous liquid scintillation measurement of α and β emitters in the same sample by conventional LSA is described in the following and in Chapter 6.

XIII. SIMULTANEOUS α–β ANALYSIS

As described in Sections II.B and XII of this chapter, α and β pulse height spectra often overlap. The pulse height spectra of the α and β emitters ^{210}Po + ^{90}Sr(^{90}Y), respectively, illustrated in Fig. 5.2, are a good example of this spectral overlap. However, many contemporary LSAs are equipped with the circuitry capable of analyzing α- and β-emitting radionuclides in the same sample. These instruments may use pulse decay analysis (PDA) or pulse shape analysis (PSA), first reported by Buchtela *et al.* (1974) and Thorngate *et al.* (1974), to differentiate alpha from beta decay events. In PDA a pulse decay time discriminator, referred to as the pulse decay discriminator (PDD), monitors the length of time for a pulse event originating from a photon emission to decay. As described in Section XII and illustrated in Fig. 5.57, fluor excitation events originating from α-particle interactions have 35–40 ns longer decay lifetimes than events originating from β-particle interactions. This is due to the longer deexcitation and light emission processes in scintillation fluors after α particle interactions. In PDA, the differences in the lengths of pulse decay events resulting from α- and β-particle interactions can be measured by the charge differences collected at the photomultiplier output at the tail portion of the pulse events. (See also Section X.B.1 on the similarity with pulse shape discrimination used to discriminate n/γ interactions with liquid scintillators). The fluorescence decay after an α particle interaction, displays a longer tail (Fig. 5.57), and consequently produces a higher charge at the PMT output for the tail portion of the pulse event when compared to a fluorescence decay pulse event produced by a β particle. The actual difference in decay lifetimes between the α and β events depends on the chemical composition of scintillation fluor cocktail used and particle energies (Pujol and Sanchez-Cabeza, 1997, Pates *et al.*, 1998, and Rodriguez Barquero and Grau Carles, 1998).

Instruments that use PSA to discriminate between α and β events also take advantage of the greater length of the α-produced pulse compared to the

β-produced pulse. In PSA, the ratio of the area of the tail of a pulse, beginning at 50 ns after pulse initiation, is compared to the total pulse area (McDowell, 1966 and Rodriguez Barquero and Grau Carles, 1998), which provides a method of assigning a pulse to a β event (short pulse) or α event (long pulse). This is also referred to as pulse shape discrimination (PSD).

The LSA, which is equipped with pulse decay discrimination or pulse shape discrimination (PDD) or PSD for α–β analysis, is also equipped with two MCAs. All of the pulse events originating from scintillation photon emissions with a decay time longer than the PDD setting are sent to the α-MCA and those events which have a shorter lifetime are sent to the β-MCA. It is necessary, therefore, to find the optimum PDD setting to get the best separation of α- and β-radionuclide activities from mixtures into separate MCAs.

A. Establishing the Optimum PDD Setting

The optimum pulse decay discriminator setting is found by counting a pure α emitter source and a pure β emitter source in the LSA in the same fluor cocktail, sample composition, sample volume, and type of vial used as the experimental samples to be measured. The optimum PDD setting is affected by sample quench level, the specific quenching agents in the sample, and the E_{max} of the β emitter (Pates *et al.*, 1998). The β-particle energy is an influencing factor, because the PMT pulse event has a specific length for a given β particle energy, and the pulse length increases with increasing event energy. Also, as noted by Pates *et* al. (1998) the delayed component (tail) of pulse events is a function of the amount of π-electron singlet and triplet excitation states in the scintillation solvent produced by the ionizing radiation. Alpha particles (or heavy charged particles with high LET) will produce more triplet-state excitation than the lighter beta particles. The triplet states take longer to undergo deexcitation fluorescence. However, the solvents as well as quenching agents present in the sample can affect the delayed component of deexcitation. Some components in cocktails can inhibit deexcitation and prolong triplet-state excitation, while other chemicals in the cocktail (e.g., O_2, CCl_4, $CHCl_3$, etc.) are electron scavengers, and able therefore, to interact with higher energy states and reduce the delayed component. When gross α and gross β are needed and the radionuclides in the experimental samples may not be known, one uses an α- and β-standard of similar energy to that of the α and β radionuclides in the samples (Passo and Cook, 1994). The following calibration steps are required:

1. The pure β emitter standard dissolved in a suitable liquid scintillation cocktail is placed into the LSA for counting. The chemical composition of this standard should be the same as those of the unknown experimental samples. During this calibration procedure the instrument automatically counts the pure β emitter standard at various PDD settings. At each PDD setting the instrument selects and sends all pulses of duration longer than the PDD setting to an α-multichannel analyzer (α-MCA) and pulses of shorter duration to

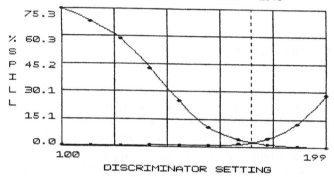

ALPHA/BETA SPILLOVER CURVE FOR: SR90P021

	%SPILLOVER	
SETTING	ALPHA	BETA
100	0.64	75.33
111	0.60	68.56
122	0.58	59.09
133	0.74	42.92
144	0.85	24.95
155	0.97	10.74
166	1.69	4.17
177	4.51	1.17
188	12.55	0.34
199	28.50	0.18

| | | %Spillover | |
		ALPHA	BETA
SETTING IN USE:	171	2.52	2.45
COMPUTED SETTING:	171	2.52	2.45

FIGURE 5.59 Alpha/beta misclassification curve for an aqueous sample of ^{210}**Po + **^{90}**Sr(**^{90}**Y) in Ultima Gold AB scintillation cocktail, measured by a PerkinElmer 2700TR liquid scintillation analyzer. This crossover curve was made from the same sample, which produced the pulse height spectra in Fig. 5.2. (From L'Annunziata 1997, unpublished work.)**

the β-multichannel analyzer (β-MCA). For each PDD setting the LSA calculates the percent spillover of the β events into the α-MCA. A curve of the percent spillover versus the PDD setting as illustrated in Fig. 5.59 is then plotted automatically.

2. The pure α emitter standard dissolved in a suitable cocktail of the same chemistry as the previously described β emitter (step 1) is placed into the LSA for counting. Its chemical composition should also be the same as the unknown experimental samples. As in step 1, the instrument automatically counts the pure α emitter standard at various PDD settings. At each PDD setting the instrument selects and sends all pulses of duration longer than the PDD setting to an α-multichannel analyzer (α-MCA) and pulses of shorter duration to the β-multichannel analyzer (β-MCA). For each PDD setting the LSA calculates the percent spillover of the α events into the β-MCA. A curve of the percent spillover versus the PDD setting as illustrated in Fig. 5.59 is then plotted automatically by the LSA.

3. The PDD setting at which the two spillover curves cross is selected as the optimum PDD setting for simultaneous α–β analysis in the same

sample (e.g., PDD = 171 of Fig. 5.59). The two spillover curves plotted in this fashion are referred to as crossover plots. At the crossover PDD setting there is minimum spillover of α events into the β-MCA and β events into the α-MCA, when the activities of both α and β events must be determined. However, if only the α emitter activity is of interest in a sample containing both α and β emitters, a higher PDD setting can be selected, well above the crossover setting (e.g., PDD = 199 of Fig. 5.59), where the spillover of β events into the α-MCA approaches 0. Alternatively, if only the β-emitter activity is of interest in a sample containing both α and β emitters, a lower PDD setting can be selected, well below the crossover setting (e.g., PDD = 122 of Fig. 5.59), where the spillover of α events into the β-MCA is lowest.

B. α–β Spillover Corrections and Activity Calculations

Once the optimum PDD setting is determined with, the detection efficiency of the α particles in the α-MCA and that of the β particles in the β-MCA must be determined. This may be best determined by the use of internal standardization as described in Section V.A of this chapter. The α-MCA and β-MCA will provide net (background-subtracted) count rates (CPM) for α and β particles, respectively. The net count rates must be corrected for spillover or misclassification and the net count rates converted to α and β activities (disintegration rates). The calculations required for the determination of α and β count rates corrected for spillover are derived in Chapter 6. The following equations described by Bakir and Bem (1996) provide the sample gross α and β activity concentrations in units of Bq L^{-1}:

$$A_\alpha = \frac{I_\beta E_\beta - I_\alpha S_{\beta/\alpha}}{\left(E_\alpha E_\beta - S_{\alpha/\beta} S_{\beta/\alpha}\right) VC60} \qquad (5.100)$$

$$A_\beta = \frac{I_\alpha E_\alpha - I_\beta S_{\alpha/\beta}}{\left(E_\alpha E_\beta - S_{\alpha/\beta} S_{\beta/\alpha}\right) VC60} \qquad (5.101)$$

where A_α and A_β are the gross α particle and gross β particle activities, respectively, in units of Bq L^{-1}, I_α and I_β are the net (background subtracted) count rates in the α and β MCAs, respectively, $S_{\alpha/\beta}$ and $S_{\beta/\alpha}$ are the percent spillover of α events into the β MCA and β events into the α MCA, respectively, E_α and E_β are the detection efficiencies for α particles in the α MCA and β particles in the β MCA, respectively, V is the volume of analyzed sample in liters, C is the concentration factor or the degree of concentration of the sample prior to analysis, and 60 is a conversion factor to change count rate from CPM to CPS.

The backgrounds, count rates, and detection efficiencies in the α MCA and β MCA will depend on the pulse height discriminator settings used in the α and β channels, respectively, and count mode used (e.g., LLCM or NCM) as described in Section XVII.C. The optimum PDD setting must be

determined at the same discriminator and count mode settings as those used
for the experimental sample analysis.

C. Optimizing α–β Discrimination in PDA

The chemistry of the sample, which includes the scintillation cocktail used
and all of the quenching agents present in the sample, plays a crucial role in
α–β discrimination in PDA. The following are some guidelines to follow to
enhance α–β separation performance

1. The cocktail-sample mixture must be homogeneous. The chemistry of
   the sample including the scintillation cocktail and kinds and amounts
   of quenching agents present should be the same as those of the α and
   β standards used to plot the crossover curves for determination of the
   optimum PDD setting. The sample must be completely solubilized in
   the liquid scintillation cocktail. Acids are often necessary to keep
   α emitter salts in solution in the cocktail. Organic samples and filters
   are often processed through ashing followed by acid dissolution of salt
   residues (Yang and Guo, 1995). Table 5.10 provides some guidelines
   on the acid loading capacity (mL acid/10 mL cocktail) at room
   temperature (20–22°C). The minimum α–β misclassification (at
   crossover point from spillover curves) for small sample sizes
   (≤ 0.5 mL acid) is also provided. Water-soluble paper for smear
   tests is a sample type that might be used to avoid excessive acids in
   cocktail (Takiue et al., 1989a), although glass fiber filter is also
   demonstrated as a good swipe material for α–β contamination assays.
2. Minimize the sample quench. The sample volume (e.g., aqueous
   solution) will dilute the scintillation fluor cocktail and also introduce
   quenching agents. Minimum α/β spillover is obtained generally at
   lowest levels of quench.
3. Use organic acceptor cocktails rather than cocktails that have
   surfactants or emulsifiers. Table 5.10 lists liquid scintillation cocktails
   that provide excellent α–β discrimination performance. They all use

TABLE 5.10 Acid Loading Performance of α/β Compatible Liquid Scintillation Cocktails

Cocktail (10 mL)	H_2O (mL)	1 M HCl (mL)	2 M HCl (mL)	1 M HNO$_3$ (mL)	2 M HNO$_3$ (mL)	% Misclassification (minimal sample loads)
Ultima Gold	3.50	0.25	0.10	0.70	0.30	0.6
Ultima Gold XR	10.0	2.00	0.90	3.00	1.75	2.2
Ultima Gold AB	10.0	5.50	2.25	3.25	2.25	0.5
Insta-Gel XF + 20% w/v naphthalene	1.5	1.3	1.1	1.4	1.4	0.5

[a]The loading capacity of the acids are mL acid per 10 mL of cocktail. From PerkinElmer Life
and Analytical Sciences, Boston.

diisopropylnaphthalene (DIN) as a solvent with the exception of Insta-Gel XF, which has pseudocumene as a solvent. Pseudocumene and the alkylbenzenes are "fast" solvents, which are less efficient for α–β separation. Naphthalene is added to Insta-Gel XF in a proportion of 20% w/v to improve α–β separation by acting as an intermediate in the energy transfer process between solvent and fluor increasing the production of π-electron triplet excitation states produced by α-particle interactions (Passo and Cook, 1994). Naphthalene is not required with the Ultima Gold Cocktails, as the DIN solvent serves the same purpose (Thomson, 1991). The DIN solvent-based cocktails, particularly Ultima Gold AB, have been demonstrated to provide excellent α–β separations, while increasing surfactant concentration degrades the cocktail performance for α–β analysis (Pates *et al.*, 1993, 1996b).

4. Use organic extractive scintillators that are selective for the α emitters of interest. The use of organic (water immiscible) extractive scintillators was described in Section XII of this chapter. This is recommended when only the α-activity is desired, as after extraction of the dissolved aqueous sample with the selective organic scintillator, the α emitters transfer from the aqueous phase to the organic scintillator leaving most of the β emitters behind in the aqueous phase. Organic extractive scintillators are useful when only α emitters are of interest, when there is an overwhelming β emitter activity with α emitters of interest, and to improve the energy resolution of α pulse height spectra. The work of Yang *et al.* (1992, 1994) and the information provided are recommended as additional resources on this topic.

5. Use specific extractive resins to concentrate the radionuclides of interest. For example, actinide extractive resin, available from Eichrom Technologies Inc., Darien IL, USA or Eichrom Europe, S.A.R.L., Paris, France, is used to extract elements of high atomic number (e.g., Th, U, Pu, Am, Cm, etc) from aqueous solutions. The extraction of the actinides together with α–β pulse shape discrimination provide improved limits of detection (Horwitz *et al.*, 1997 and Eikenberg *et al.*, 1999).

6. Purging the dissolved oxygen from the cocktail with argon gas will improve α–β separation by reducing quench. This is not commonly performed except in research to obtain unquenched samples, as the additional step is time consuming and the gain in α–β separation is relatively small.

7. Time-resolved pulse decay analysis (TR-PDA) significantly improves α/β separation or, in other words, reduces misclassification of α as β and β as α (Passo, 1996). TR-PDA, available with PerkinElmer LSAs, is PDA combined with time-resolved background discrimination (TR-LSC), which is described briefly in Section XVII.C.5.d of this chapter and in more detail in Chapter 6. This effect, as described by Passo (1996), is due to enhanced α rejection from the β-MCA and associated lowering of the β spillover into the α-MCA due to a change in the shape of the α misclassification curve.

D. Quenching Effects in α–β Discrimination

As noted previously in Section C the α–β discrimination is improved by minimizing quench in the sample, which is accomplished by reducing the sample volume as the amount of water and acids among other substances in the sample can cause quench. The following are important effects of quench on α–β separation and guidelines to follow in controlling the effects of quench or even taking advantage of these effects, when they occur:

1. Alpha misclassification increases with quench. Once the optimum PDD setting is determined for a given sample quench level as described in Section B above, any increase in quench level causes an increase in the misclassification of α events into the β-MCA. This can be seen from the alpha into beta curve of Fig. 5.60, which was prepared by measuring firstly the optimum PDD setting for a given value of quench (e.g., tSIE = 620 in this example) and then measuring the percent spillover at various quench levels after adding incremental amounts of quenching agent. The degree of change in misclassification is a function of the type of quenching agent present in the sample. Some quenching agents are strong electron scavengers (e.g., O_2, $CHCl_3$, CCL_4), and these are able to interact more easily with π-electron triplet states reducing the delayed component of fluorescence deexcitation (Pates *et al.*, 1998). Therefore, it is important that the chemistry of the α and β standards used to determine the optimum PDD setting be identical to the chemistry of samples under investigation.

2. Counting efficiency of α emitters is expected to remain at 100% (Aburai *et al.*, 1981; Takiue *et al.*, 1989a; Parus *et al.*, 1993); however, loss of α events into the β-MCA occurs with increased quench. Caution should be taken in assuming 100% detection efficiency, and the detection efficiency should be determined in all cases. Reduced counting efficiencies can occur with instruments that count afterpulses to discriminate against background as described in Section XVII.C.

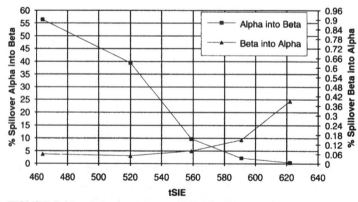

FIGURE 5.60 Percent misclassification curves of alpha and beta events as a function of quench level (tSIE) measured at a constant PDD setting determined at tSIE 620 (Courtesy of PerkinElmer Life and Analytical Sciences, Boston, MA.)

3. Beta misclassification decreases with an increase in quench as illustrated in Fig. 5.60. However, the percent misclassification and counting efficiency are a function of the E_{max} of the β particle emitter (Pates *et al.*, 1998). Therefore, to assure accurate measurement of misclassification and optimum PDD setting, it is best to plot a crossover curve using the same β emitter or one of similar E_{max} as that of the β emitter expected to be in the samples to be analyzed.

4. The best α/β separation performance is maintained by keeping the quench level (sample chemistry), sample volume, and vial type the same as the pure α and β standards used to determine the optimum PDD setting. This is done by maintaining the sample chemical composition, sample size, fluor cocktail type and volume, and vial type the same for the standards as the experimental samples. If it is not possible to predict the quench level of the experimental samples, the optimum PDD settings for a wide range of quench levels should be determined using quenching agents that are identical to those found in the samples to be analyzed. Yang (1996) clearly demonstrates this by plotting a curve of optimum PDD setting versus tSIE.

For additional information on quench effects on α–β separation see Pates *et al.* (1996c, 1998).

XIV. SCINTILLATION IN DENSE (LIQUID) RARE GASES

Liquid rare gases, such as argon and xenon, have properties of high scintillation yield as a medium of detection. In particular, liquid xenon is considered most suitable for γ-ray detection because of its high atomic number and short decay constants of the scintillation light ($\tau_1 = 4.3$ ns, $\tau_2 = 22$ ns for short and long components, respectively), which comes close to that of plastic scintillators (Hitachi *et al.*, 1983; Tanaka *et al.*, 2001). The efficiency of liquid xenon as a scintillator is reported by Aprile *et al.* (1990) to be similar to that of NaI(Tl) solid scintillation detector, but with a shorter scintillation decay time. Both liquid argon (LAr) and liquid xenon (LXe) are widely studied as efficienct liquid scintillation detector for alpha particles and internal conversion electrons (Aprile *et al.*, 1990; Miyajima, *et al.*, 1992; Tanaka *et al.*, 2001), γ radiation and neutron spectrometry from the scintillation efficiency of electron and nuclear recoil events (Arneodo *et al.*, 2000; Akimov *et al.*, 2002), and β–γ discrimination (Benetti *et al.*, 1993; Arneodo *et al.*, 2000). Liquid rare gas detectors are currently being studied for their utility in nuclear medicine as imaging detectors in Positron Emission Tomography or PET (Chepel, 1993; Chepel *et al.*, 1999; Solovov *et al.*, 2000), tomography imaging telescopes (Time Projection Chambers) for research in medium-energy gamma-ray astrophysics (Aprile *et al.*, 1990; Cennini *et al.*, 1999), galactic Dark Matter Search consisting of weakly interacting massive particles (Arneodo *et al.*, 2000; Akimov *et al.*, 2002), and relativistic heavy ion detection (Tanaka, *et al.*, 2001). Applications will expand in light of the advantageous detector characteristics of high atomic number, scintillation efficiency, and short fluorescence decay times.

The mechanisms of scintillation in liquid argon and liquid xenon are similar and have been investigated thoroughly (Kubota et al., 1978a,b, 1979; Doke, 1981; Doke et al., 1990). As described by Cennini et al. (1999) scintillation light in liquid argon is produced by ionizing radiation either by direct ionization of an Ar atom followed by excimer formation and deexcitation according to

$$\text{Ar}^* + \text{Ar} \rightarrow \text{Ar}_2^* \rightarrow 2\text{Ar} + h\nu \ (128\,\text{nm}) \qquad (5.102)$$

or via ionization, recombination, excimer formation, and deexcitation as

$$\text{Ar}^\oplus + \text{Ar} \rightarrow \text{Ar}_2^\oplus + \text{e} \rightarrow \text{Ar}_2^* \rightarrow 2\text{Ar} + h\nu \ (128\,\text{nm}) \qquad (5.103)$$

Deexcitation from the Ar_2^* to the dissociated ground state 2Ar gives rise to fluorescence displaying fast and slow decay components of about 5 ns and 1 μs, respectively with relative yields of 23 and 77% at an emission of approximately 128 nm wavelength. The light emission is abundant with approximately 1 photon emitted per 25 eV of energy dissipated in LAr by ionizing radiation or 40,000 photons per MeV. The wavelength of light emission is in the vacuum ultraviolet (VUV) region. Consequently, glass-faced photomultipliers cannot be used. More costly photomultipliers with MgF_2 or CaF_2 windows are required. The addition of liquid xenon in small concentrations (≤ 100 ppm) into the LAr serves as a wavelength shifter from 128 to 173 nm permitting photomultipliers with quartz windows (Cennini et al., 1999). The scintillation mechanism for liquid xenon is similar to that described above for LAr with an average of 1 photon emitted per 14.7 eV of energy dissipated by ionizing α particles (Tanaka et al., 2001). By applying a wavelength shifter consisting of 0.4 mg cm^{-2} of sodium salicylate onto the inner surface of the LXe scintillation detector, Tanaka et al. (2001) were able to use a standard photomultiplier with Pyrex glass face. The avalanche photodiodes (APD) described in Chapter 11 are currently under investigation as a possible replacement of the photomultiplier tube for liquid rare gas scintillation detectors as the APD could be placed directly into the LXe scintillation detector. Solovov et al. (2000) demonstrated the feasibility of applying large area APD at the LXe temperature of $-105°$C as the scintillation photosensor.

Arneodo et al. (2000) demonstrated the excellent linear response of scintillation fluorescence versus energy deposited in liquid xenon using several γ-ray sources. The linearity is illustrated in Fig. 5.61. Typical energy distribution in liquid Xe with ^{57}Co and ^{137}Cs γ-ray sources placed outside the LXe scintillation detector measuring 5.0 cm and 4.6 cm for the top and bottom diameters and 4.0 cm in height are illustrated in Fig. 5.62. The width of each energy distribution is dominated by a light collection efficiency that varies from 5 to 12% depending on the point of interaction of the γ-ray within the LXe detector volume. The overall detector efficiency was estimated

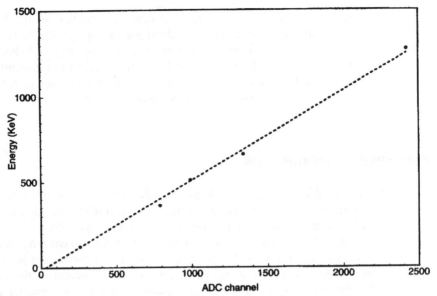

FIGURE 5.61 Linearity of deposited energy vs. scintillation in liquid xenon with [57]Co (122 keV), [133]Ba (356 keV), [22]Na (511, 1275 keV) and [137]Cs (662 keV) gamma-ray sources placed outside the detectors. (From Arneodo *et al.*, 2000, reprinted with permission from Elsevier Science.)

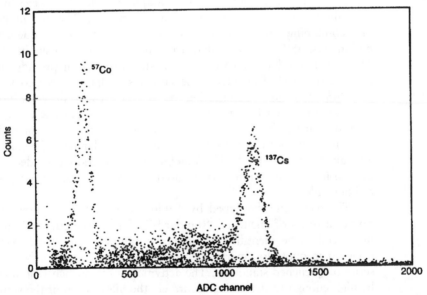

FIGURE 5.62 Scintillation light spectrum of [57]Co and ^{137}Cs in liquid xenon. (From Arneodo *et al.*, 2000, reprinted with permission from Elsevier Science.)

by Arneodo *et al.* (2000) to be approximately 1.5% for the 175 nm UV photons from the LXe scintillation taking into account the average light collection efficiency of 8%, PMT quantum efficiency of 19%, and photo-extraction potential of LXe of 15 eV.

Recent studies have demonstrated infrared scintillation in LAr and LXe detected with an InGaAs photodiode (Bressi *et al.*, 2000, 2001). The LXe infrared scintillation light is emitted in a narrow region below 1200 nm wavelength with a relatively 2-fold lower light yield than gaseous xenon. It appears, therefore, that so far only the ultraviolet emissions of LAr and LXe have demonstrated applications in the detection and measurement of ionizing radiation.

XV. RADIONUCLIDE IDENTIFICATION

It is possible to use an LSA to identify automatically an unknown pure β-emitting radionuclide sample; once it is identified, the LSA can automatically determine the activity (DPM) of that same sample.

One of the obstacles encountered in identifying β-emitting radionuclides by LSA is that the pulse height spectrum of all β emitters are broad over the range from 0 keV to the end point energy corresponding to the E_{max} of the β emissions. Also, the pulse height spectrum of a given β emitter changes in size and endpoint energy according to the level of quench in a sample. Consequently, the pulse height distribution of a high-energy β emitter under a high degree of quench could resemble that of a low-energy β emitter under a low degree of quench.

By means of a double-ratio technique Takiue *et al.* (1989b) developed a method for identifying unknown β emitters. The technique is based on combining the information provided by two quench-indicating parameters, tSIE and SIS, that were described previously in this chapter. The tSIE is produced by the external standard, and it provides information defining the level of quench in the sample-scintillation fluor cocktail mixture. The SIS is a quench-indicating parameter, which depends not only on the degree of quench in the sample but also the β particle energy spectrum. The two quench-indicating parameters together, therefore, provide information, that reflects the size and shape of a particular pulse height spectrum at any given quench level. The combined information provided by the two quench-indicating parameters is used to identify pure β or β–γ emitting radionuclides.

The technique developed by Takiue *et al.* (1989) required the preparation of a series of quenched "standards" of all of the pure radionuclides that may need to be identified as unknown samples. The word "standards" is enclosed in quotation marks, because the activity need not be known in the series of quenched standards. The activities of the quenched standards should be high enough to determine accurately the SIS of each of the standards, and they are used only once to prepare a double-ratio curve for a particular radionuclide.

Once the series of quenched standards has been prepared for each radionuclide of interest, the SIS and tSIE of each quenched standard are determined and plotted as illustrated in Fig. 5.63. The family of curves illustrated in the figure becomes separated from the lowest curve for ^{3}H to the uppermost curve for ^{32}P according to the maximum energy (E_{max}) and

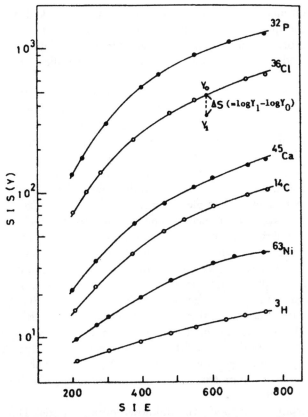

FIGURE 5.63   Double ratio curves for identifying an unknown radionuclide. Each curve is constructed from the SIS and SIE of a series of quenched "standards" of each nuclide. (From Takiue et al., 1989b, reprinted with permission from Elsevier Science.)

average energy (E_{av}) of each radionuclide, because the SIS plotted along the ordinate reflects both E_{max} and E_{av}. The curves, therefore, provide a definition of the magnitudes of the pulse height spectral distributions as a function of quench, which are radionuclide specific.

To apply this technique, an unknown sample is monitored for its SIS and tSIE value and then compared with the double-ratio curves, such as those illustrated in Fig. 5.63, that were prepared in advance. For the tSIE and SIS of an unknown sample the radionuclide can be identified using the double-ratio curves provided the deviation $\Delta S/S$ of the experimental plotted point of the unknown sample is less than ± 0.02, or

$$\frac{\Delta S}{S} = \frac{(\log Y_1 - \log Y_0)}{S} < \pm 0.02 \qquad (5.104)$$

where ΔS is the difference in ordinate ($\log Y_1 - \log Y_0$) between the plotted point Y_1 and the corresponding value Y_0 on the double-ratio curve, and S is the logarithm of the SIS value ($\log Y_0$) on the curve. Takiue et al. (1989b) demonstrated that the activities of the unknown samples can also be determined automatically, if the double-ratio curves are linked by a computer program to the quench correction curve of the appropriate radionuclide. In essence, therefore, an unknown pure β emitter in scintillation fluor cocktail could be placed in an LSA for both automatic identification and activity analysis.

The above procedure was demonstrated to work for several radionuclides with the exception of ^{14}C and ^{35}S. The close similarities of the β particle energy spectra of ^{14}C ($E_{max} = 156$ keV) and ^{35}S ($E_{max} = 167$ keV) produce overlapping double-ratio curves, which prevents the identification of these two radionuclides in unknown samples. However, Natake $et\ al.$ (1996) showed that pulse height analysis could be used to identify ^{14}C and ^{35}S in unknown samples. The slight difference in the pulse height distributions could be distinguished using the sample channels ratio (SCR), which is the ratio of two counts in two measurement channels. The optimum channel settings were selected to obtain as great a difference between the two radionuclides as possible for a given quench level. The SCR for a series of quenched "standards" of ^{14}C and ^{35}S are then plotted against the H number in what is referred to as a double quench parameter curve (DQP) curve. The H number is an external standard quench-indicating parameter (QIP) described previously in this chapter. Other external standard QIPs, such as tSIE or SQP(E), can also be used in this technique. To apply the technique, the SCR and the H number are determined for the unknown sample and the two QIP values are plotted on the previously prepared DQP curve. The radionuclide is then identified when the plot of that point lies in the neighborhood of either the ^{14}C or ^{35}S DQP curve.

In another approach to identify unknown pure β emitters Dodson (1996) developed the technique of characterizing the pulse height spectrum according to the following three properties: E_{max}, the isotope center number I, and the spectral resolution (FWHM or F). The isotope center number I is defined as

$$I = \frac{\sum (N_i E_i)}{\sum (N_i)} \tag{5.105}$$

where E_i is the energy of the N_i electron. For quenched sets of radionuclide standards, both I and F are found to be linearly dependent on E_{max}. The slopes and intercepts of the linear equations defining I and F as a function of E_{max} over a range of quench levels are tabulated for each radionuclide. The measurement of I, E_{max}, and FWHM of an unknown nuclide's sample spectrum coupled with a relevant algorithm permit the identification of the unknown nuclide within the parameters of the defined reference set.

Alpha-particle emitters were identified by Kaihola (2000) of Wallac Instruments (now PerkinElmer Life and Analytical Sciences) who demonstrated that the external standard quench-indicating parameter SQP(E) could be used to correlate the alpha-particle sample pulse height distribution with its emission energy as a function of quench level. Two known energy calibration lines as a function of quench are required to verify the emission energy and identify the radionuclide in liquid scintillation alpha spectroscopy.

XVI. AIR LUMINESCENCE COUNTING

A novel approach to α particle counting is the use of a liquid scintillation counter to detect luminescence events caused by the α particle excitation of N_2 in air. The technique was devised by Takiue and Ishikawa (1979). As explained by these investigators, the alpha particle passing through air dissipates its energy, forming excited nitrogen molecules N_2^*, excited nitrogen molecular ions N_2^{+*}, and excited nitrogen atomic ions N^{+*} as follows with relative energy consumptions as given in parenthesis:

$$
N_2 + W_\alpha
\begin{cases}
\longrightarrow N_2^* & (50\%) \\
\longrightarrow N_2^{+*} + e^- & (45\%) \\
\longrightarrow N^{+*} + N + e^- & (5\%)
\end{cases}
\tag{5.106}
$$

where W_α is the α particle energy dissipated in air, e^- is an electron, and N is the nitrogen atom. The emission spectra are derived from entirely N_2^* and N_2^{+*} bands.

The oxygen in air (21%) does not contribute to the air luminescence but rather acts as a quencher by forming a complex with N_2^*, leading to a radiationless deactivation of excited molecular nitrogen. Argon, whose concentration in air is only 1%, undergoes luminescence. However, its photon emissions make little contribution to the overall nitrogen luminescence because of the low concentration of argon in air compared to that of nitrogen (78%).

The sample preparation and counting procedure recommended by Takiue and Ishikawa (1979) is simple and relatively inexpensive. Approximately 0.1 g of a solution of α particle-emitting nuclide solution is deposited onto a 0.84 mg cm^{-2} thick Tetoran film (Toray, Inc.), dried with an infra-red lamp, and inserted into an empty liquid scintillation glass counting vial so that the film and sample remain suspended in the air around the center of the vial.

The sample may be counted as such without further preparation, or the air in the vial may be purged with pure nitrogen as a gas scintillator.

When the nuclides are counted in air, counting efficiencies of 72 and 74% for ^{210}Po and ^{241}Am, respectively, are obtained, and purging with pure nitrogen leads to an increase of 10–15% in counting efficiency. The method offers several advantages: (1) there is no concern about quenching except for that of oxygen, which remains constant from sample to sample; (2) there is no need to be concerned about the preparation and handling of costly scintillators, as air is the only medium of detection; (3) the technique offers excellent reproducibility from sample to sample because of the absence of variable quenching agents; (4) the spectral emissions of nitrogen overlap well with the spectral response of photomultiplier tubes used in LSAs, obviating the need for wavelength shifters; and (5) normal air containing 78% nitrogen may be used as the detection medium.

The main disadvantages of the air luminescence method are (1) reduced counting efficiency, as liquid scintillation with fluor cocktail provides 100% counting efficiency for α emitters even under high levels of quench; (2) the fact that the pulse height spectrum is a peak produced by the air luminescence photons, so that there are no α-radionuclide spectrometry capabilities; (3) the limitation of the sample size is to about 0.1 mL of radioactive solution that is deposited and dried on the transparent film; and (4) there is no possibility of analyzing β emitters in the sample, that is, simultaneous $\alpha-\beta$ analysis is not feasible.

These original findings of Takiue and Ishikawa were expanded by Takiue (1980) for the routine analysis of α emitters on smear (swipe) paper in air without scintillation cocktail. The technique involves swiping a 100 cm^2 contaminated area with a smear paper (25 mm diameter, 30 mg cm^{-2}) at a pressure of about 0.2 kg cm^{-2}. The contaminated smear sample thus obtained is placed directly in an empty glass scintillation counting vial with the contaminated side of the paper facing upward and measured without adding a liquid scintillation cocktail as illustrated in Fig. 5.64.

The α particle counting efficiency is reduced because of the 2π counting geometry with the smear sample at the bottom of the counting vial. Takiue (1980) observed variation in smear removal efficiencies from different contaminated surfaces. The counting efficiencies of ^{210}Po or ^{241}Am as smear samples on paper varied between 10 and 20% from sample to sample. The following advantages of the air luminescence method for monitoring α-emitter radioactive contamination can be cited: (1) sample preparation for α-contamination surface monitoring is simple and inexpensive, because no scintillation cocktail is required or cocktail waste produced; and (2) low backgrounds are achieved (~ 9 CPM with the instrumentation used by Takiue), providing a detection limit of $1 \times 10^{-7} \mu$Ci cm^{-2} for loose contamination. Disadvantages of the air luminescence method are that (1) much lower and more variable counting efficiencies (10–20%) are

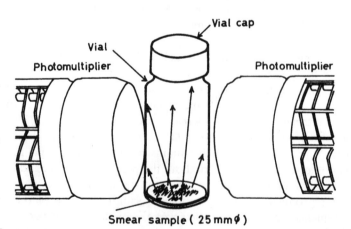

FIGURE 5.64 Measurement geometry for a smear sample in air (contaminated side of paper is facing upward) and detector photomultiplier tubes of a liquid scintillation spectrometer (From Takiue, 1980.)

obtained in air luminescence analysis of smear samples, compared to a constant 100% counting efficiency obtained by LSA, and (2) β-emitters cannot be analyzed by air luminescence, that is, simultaneous α/β analysis with air luminescence is not possible.

Caution must be taken to assure that there are no significant activities of β particles with energies in excess of 180 keV, which is the threshold energy for the production of Cherenkov radiation by β particles (electrons) in glass. For samples containing high-energy β emitters, Takiue (1980) demonstrates that a double-region counting technique could be used to discriminate between pulses produced by Cherenkov photons and air luminescence.

Further studies by Homma et al. (1987) demonstrated the possibilities of measuring ^{222}Rn and its daughter nuclides in air in the gaseous phase by air luminescence counting. A counting efficiency of 42% was achieved. The endpoint of the air luminescence spectrum produced by ^{222}Rn is about 18 keV, hence a preset counting region normally used for ^{3}H liquid scintillation analysis, could be used for the air luminescence analysis.

In the LSA of ^{222}Rn, Murase et al. (1989a) demonstrated that air luminescence produced by ^{222}Rn and its daughters in the gaseous phase above the liquid scintillator can give rise to significant air luminescence counts. Therefore, to avoid error caused by air luminescence in the LSA of ^{222}Rn, currently approved methods require filling the liquid scintillation vial to a total volume of 20 mL with sample and fluor cocktail. For example, the procedure used by the US EPA (EPA/EERF-Manual-78–1) for the analysis of radon in water (see Passo and Cook, 1994) calls for the injection of 10 mL of freshly sampled water into 10 mL of an organic accepting liquid scintillation cocktail (e.g., Opti-Fluor O). The ^{222}Rn partitions into the organic fluor phase, and insignificant gaseous space remains in the standard 20-mL liquid scintillation counting vial for any significant air luminescence interference. To avoid any interference from luminescence by ^{222}Rn alpha particles in the air above the liquid scintillation cocktail Murase et al. (1999) selected a lower-level discriminator setting of 25 keV. Thus, all pulse events that may originate from air luminescence would be rejected, as these are found in the 0–18 keV region only.

Murase et al. (1989b) demonstrated the air luminescence analysis of ^{210}Po, ^{238}U, and ^{241}Am. The technique involves spotting small amounts (0.1 mL) of aqueous solutions of the α emitters on the bottom of glass liquid scintillation vials and evaporating the samples to dryness with an infrared lamp. The counting efficiencies obtained were between 33.3 and $33.7 \pm 0.23\%$. The air luminescence pulse height spectra correspond to the region of 0–18 keV for pulse height calibrated on an energy scale. Therefore, a counting region preset for ^{3}H LSA can be used for the air luminescence counting. Following air luminescence counting, the samples can be recovered from the vials for other studies. The counting efficiencies are easily determined by liquid scintillation analysis of samples that can be carried out at random or for as many vials as desired. LSA after air luminescence was carried out by Murase et al. (1989b) after dissolving the radionuclide residue by wetting the bottom of the vial with 0.5 mL H_2O followed with liquid scintillation cocktail. The LSA of α emitters is 100%. Therefore, simply the difference between the count rates obtained by LSA and air luminescence analysis can be used to determine the air luminescence counting efficiency, as the count rate by LSA is also the disintegration rate. A disadvantage of the air luminescence analysis is the very small sample size of 0.1 mL that can be analyzed. Larger samples might leave too thick a residue at the bottom of the scintillation vial after drying, which would produce samples with significant self-absorption.

Air luminescence counting was used by Homma et al. (1996) to standardize ^{222}Rn samples that are applied to calibrate detectors utilized for the determination of ^{222}Rn in air. The standardization of ^{222}Rn as described by Homma et al. (1996) involves collecting a sample of ^{222}Rn from approximately 20 mL of sample of a 0.1 M HCl solution of 20 kBq ^{226}RaCl$_2$ with a syringe and then transferring the ^{222}Rn gas to a standard liquid scintillation counting vial, which was filled with air and closed with a silicon rubber stopper. The ^{222}Rn in the vial is allowed to sit for 3.5 h before measurement to allow the daughters ^{218}Po, ^{214}Pb, ^{214}Bi, and ^{214}Po to reach transient equilibrium with the ^{222}Rn. This sample is referred to as the ^{222}Rn "standard". The ^{226}Ra solution from which the ^{222}Rn is taken does not have to be a standardized solution, because the ^{222}Rn determination is not based on secular equilibrium between ^{226}Ra and ^{222}Rn. It is not even necessary to know how much ^{222}Rn from the ^{226}Ra solution was actually removed. It is necessary only to standardize the ^{222}Rn that is transferred into the rubber stopper-sealed scintillation counting vial. Standardization of the gaseous ^{222}Rn is done by air luminescence counting. The air luminescence counting efficiency of the ^{222}Rn was determined by using the counting efficiency of a standardized ^{210}Po source. The ^{210}Po counting efficiency (E) was determined by calculating $E = \alpha/A$, where α is the air luminescence count rate of a ^{210}Po source and A is the α particle emission rate of the ^{210}Po source expressed in α particles per second in a solid angle of 2π steradians. Homma et al. (1996) underscore the importance of measuring the air luminescence counting efficiency of ^{222}Rn, as instrument models may vary in optical properties, PMT properties, and circuit properties. Morita-Murase et al. (2001) have pursued further the application of air luminescence counting to the measurement of ^{222}Rn. They standardized a ^{222}Rn sample collected from the airspace above a ^{226}RaCl$_2$ solution with a syringe. The ^{222}Rn was injected into a small quartz tube (7 mm in diameter by 40 mm in length) sealed at both ends with silicon rubber stoppers. After 3.5 h transient equilibrium of ^{222}Rn with its daughters ^{218}Po, ^{214}Pb, ^{214}Bi, and ^{214}Po is reached. The activity of the ^{222}Rn was then determined via analysis of the daughter ^{214}Pb γ-rays with a Ge semiconductor well detector. The small quartz tube with known ^{222}Rn activity was placed into a standard empty liquid scintillation vial, and the vial agitated until the quartz tube was broken liberating the ^{222}Rn into the air of the vial. The ^{222}Rn was counted with an LSA to provide the air luminescence counting efficiency of ^{222}Rn for the particular LSA used. An air luminescence counting efficiency of 42.6 $\pm$0.2% for ^{222}Rn was reported.

XVII. LIQUID SCINTILLATION COUNTER PERFORMANCE

Now that the basic principles and procedures of LSA have been described, it is important to review the methods used to optimize the performance of the LSA. This is particularly important when sensitivity is key to our measurements and we need methods that can be used to assess the performance of the LSA to insure that the instrument is providing reliable and reproducible data.

A. Instrument Normalization and Calibration

The first item to be addressed is that of assessing performance and calibration of the LSC to be sure that the activity data (DPMs) are accurate and reproducible. The first area to review is the calibration methods that are used for LSCs. One method is to use a flame-sealed standard in scintillation cocktail, usually a ^{14}C source as the calibration normalization source for the LSC. This sealed standard should be an absolute calibrated standard such as a NIST traceable standard so that its DPM is accurately known and it produces light pulses similar to those found in real samples.

The calibration procedure of LSAs varies according to the manufacturer of commercial instrumentation. One method of calibration [e.g., Packard (now PerkinElmer) Instruments] involves loading an unquenched ^{14}C standard source into the counting chamber, the high voltage for each of the PMTs is set automatically to obtain the optimum ^{14}C efficiency, and the ^{14}C endpoint energy of the standard is fixed at 156 keV with the minimum background. Such a calibration should be performed at least weekly as long as the instrument is being used. Another calibration method used with other liquid scintillation counters utilizes a flashing light-emitting diode (LED). This is normally used on instrument types with which the high voltage is turned off during sample loading and unloading, for example, uploading instruments. This method uses a flashing LED to calibrate the PMT high voltage each time a sample is loaded into the counting chamber.

B. Assessing LSA Performance

Assessing the performance of the LSA on a routine basis is important to have assurance that the instrument is operating within acceptable parameters and to have a standing record over time of the instrument performance. To maintain good laboratory practice and satisfy regulatory agencies, it is often necessary to have records of the instrument performance on a routine basis, such as daily, weekly, or monthly, to provide proof that deviations in instrument performance do not have any effect on the analytical results. Instrument performance assessment (IPA) is carried out generally with a set of standards that are NIST traceable. These standards can be the same as those used for calibrating the instrument.

The IPA is normally carried out on a regular basis and the results stored and printed in a tabular and/or graphic form with a time and date stamp. Eight parameters should be assessed using a set of 3 standards (tritium,

^{14}C, and background): (1) tritium efficiency, (2) tritium background, (3) tritium figure of merit, (4) tritium chi-square, (5) ^{14}C efficiency, (6) ^{14}C background, (7) ^{14}C figure of merit, and (8) ^{14}C chi-square. The tritium and ^{14}C efficiencies are determined by the LSA using two sealed standards counted in predefined counting regions. This requires approximately 1 min for each sample and should be performed daily. The backgrounds are determined using a sealed background standard with two preset counting regions for tritium and ^{14}C. This requires approximately 60 min because an accurate CPM value needs to be obtained to assure reproducibility. Background assessment should be performed at least weekly. The ^{3}H and ^{14}C figure of merit performance values are determined using the E^2/B calculation, which is the percent counting efficiency squared divided by the background count rate in CPM. This is an excellent indicator of any change in performance for the LSC; it can be calculated using the 3 standards, ^{3}H, ^{14}C, and background, and it should be performed daily. The final performance parameters are the ^{3}H and ^{14}C chi-square. This is a monitor of short-term stability in the LSA and involves counting the ^{3}H and ^{14}C sealed radionuclide standards 20 times for 0.5 min each. The chi-square tests are used to confirm that the count rate measurements vary within the normal and expected deviations; that is, the count rates are neither too consistent nor too variable to be true. The chi-square tests should be performed weekly. By monitoring and recording these parameters, Good Laboratory Practices (GLP) can be satisfied.

The instrument performance assessment (IPA) can be either performed manually by the LSC users with the data placed into a spreadsheet or automatically by an LSC that has automatic IPA and instrument calibration. Figure 5.65 shows a plot of the data obtained with an automatic IPA program on a Packard (PerkinElmer) LSA. This type of graph can be displayed, stored in computer memory, and printed for each of the 8 IPA parameters described previously. In any given graph for any of the 8 parameters, such as the example in Fig. 5.65a, an average line is calculated and the 1, 2, and 3 sigma values are provided so that the user can evaluate any trends or outlying points. If an outlying point is observed, this information is printed on the final data output. In addition, a series of suggestions can be provided so that the user can assess the problem further and take preventative action if any parameters are too frequently outside of recommended limits. An example, is provided in Fig. 5.65b. Possible problems that might be indicated are counting contamination, PMT failure, dirty PMTs, and electronic failures.

C. Optimizing LSC Performance

The procedures for determining sample activities in DPM in routine sample analysis are dealt with in detail in previous parts of this chapter. This section will describe methods used to optimize performance particularly for low-count-rate samples when LSA sensitivity is most important. There are several methods of optimizing performance, and these will be described in detail in this section.

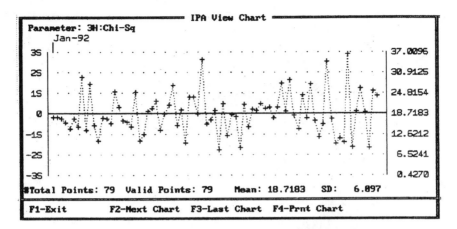

FIGURE 5.65 Chi-square instrument performance assessment (IPA) data displayed on the computer monitor of a liquid scintillation analyzer (a) as a chart and (b) as a table with instructions if measured parameters are obtained too frequently outside of recommended limits (Courtesy of PerkinElmer Life and Analytical Sciences, Boston, MA.)

I. Counting Region Optimization

If an isotope such as ^{14}C is being quantified and a low level of detection is required, the counting region, defined by lower level (LL) and upper level (UL) pulse height discriminators, must be optimized for best performance measured by the figure of merit (E^2/B). The figure of merit is calculated as the square of the percent counting efficiency of the radionuclide of interest divided by the background count rate expressed in CPM. Optimizing the counting sensitivity (E^2/B) with spectrum analysis of sample and background is achieved by the following procedure:

1. Count a known activity (DPM) of the radionuclide of interest and store the sample spectrum, as illustrated in Fig. 5.66 (upper plot), in the computer memory. The sample size, sample chemistry, and amount of cocktail used should be similar to those of the experimental samples.

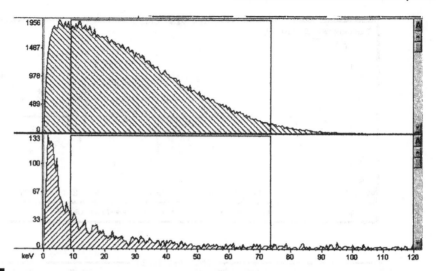

FIGURE 5.66 Pulse height spectra of a ^{14}C sample (above) and a background or blank sample (below). For this particular sample and blank, chemical composition, sample load, cocktail, and vial type the **LL** (left) and **UL** (right) discriminator settings of 9.0–73.5 were found to provide region optimization. In this example, the reduced counting region produced a sacrifice of 18% in the ^{14}C counting efficiency but a reduction of 54% in the background counts. Region optimization is specific to radionuclide, sample-cocktail chemistry or quench level, and vial type and size (L'Annunziata, 1996, unpublished work).

2. Prepare a blank sample (no radionuclide) plus cocktail in a counting vial to measure background. Count the blank sample and store the background spectrum, as illustrated in Fig. 5.66 (lower plot). It is important to prepare the sample and background (blank) counting vials in steps 1 and 2 with the same cocktail and any other chemical components in identical amounts to provide the same chemistry and the same quench levels in both counting vials. For example, if the experimental samples will have a 50% water load (e.g., 10 mL H_2O + 10 mL cocktail), then the blank background vial should be prepared similarly, but without radioisotope. Ideally, the background counting vial should have the same chemical composition and QIP, that is, the same tSIE, SQP(E) or H#, as the sample vial with the exception that the background vial will not contain any added radionuclide.

3. Start with a wide-open counting region defined with the LL discriminator at 0 keV and the UL discriminator at the sample pulse height spectrum endpoint, and measure the count rates from the sample and background vials in the wide-open counting region. From the count rate data from the sample and background calculate the figure of merit E^2/B.

4. Repeat step 3 by increasing the LL discriminator by suitable increments and calculating the E^2/B after each change of the LL discriminator setting.

5. Decrease the UL discriminator by suitable increments and measure the E^2/B for each setting of the LL discriminator already tested. Record the E^2/B for each of the LL–UL counting regions tested.

      6. Select the optimum counting region as that region defined by the
LL–UL settings that gave the highest E^2/B.

The above procedure can be carried out using a "Replay" option that
is included with many LSAs. This allows the variation of the LL and
UL discriminator settings to obtain new count rates from computer-stored
pulse height spectra without having to recount the sample or background.
Also, certain LSAs are equipped with SpectraWorks or similar computer
program that will (1) process automatically the sample and background
spectra, (2) calculate the optimum LL–UL discriminator settings for a
counting region, and (3) provide a value of the E^2/B for that optimum region.
SpectraWorks will also calculate the lower limits of detection (LLD)
or minimal detectable activity (MDA) for a particular radionuclide, counting
efficiency, and background obtained. The LLD or MDA is defined further
on in this chapter. For more information on SpectraWorks see Anonymous
(1995).

      An example of results obtained with region optimization can be taken
from the following case:

> A sample of ^{14}C and a background blank were prepared similarly in
> liquid scintillation counting vials with the exception that the background
> vial had no radionuclide added. For a wide-open counting region of
> 0–156 keV defined where LL and UL discriminators are calibrated in keV,
> the instrument gave a 95% counting efficiency for the ^{14}C, a background
> count rate of 25.0 CPM, and a calculated E^2/B of 361. After region
> optimization according to the foregoing procedure, an optimum counting
> region of 18–102 keV was found, which gave a ^{14}C counting efficiency
> of only 63.1% but a background of 3.38 CPM, and a calculated E^2/B
> of 1178. Region optimization caused a reduction by more than 30% in
> the counting efficiency from 95 to 63.1% but an even greater reduction of
> more than 10-fold in the background. The counting region of 18–102 keV
> is the optimum counting region for the particular sample and instrument
> used in this example. Region optimization is important when sample count
> rates are low (close to background) and low levels of detection are
> required.

      It is important to emphasize here that an optimum counting region so
determined is good only for the particular radionuclide, sample size, vial type,
cocktail-sample chemistry (i.e., quench level), and LSA used. Therefore, we
should keep in mind that when samples vary in quench, the optimized
counting regions (window) vary as well. If the quench level between samples
vary only slightly, one can widen slightly the optimum counting region to
accommodate small changes in spectral shifts.

2. Vial Size and Type

      Optimization of performance for the LSC is achieved also by selection of
the proper size of vial and vial material (e.g., glass or plastic). The data in

TABLE 5.11 Tritiated Water in Ultima Gold LLT Scintillation Cocktail in Glass and Plastic, Large and Small Vials, Performance Comparison

Vial size (mL)	Vial material	Water (%)	Counting Region 0–18.6 keV			Counting Region 0.5–5.0 keV		
			Efficiency[a] (%)	Background (CPM)	E^2/B	Efficiency[a] (%)	Background (CPM)	E^2/B
20	Glass	0.5	30.95	16.87	56.8	28.91	8.04	103.9
20	Plastic	0.5	31.71	10.15	99.1	29.31	3.88	221.1
7	Glass	0.5	29.55	12.81	68.1	27.75	5.92	130.1
7	Plastic	0.5	30.10	7.82	115.8	28.10	2.58	306.1

[a]% Counting efficiency = (CPM/DPM)(100).

Table 5.11 provide ^3H counting efficiencies, backgrounds, and figures of merit (E^2/B) for a standard (wide-open) counting region and an optimized counting region for two counting vial types and sizes. The data in this table provide evidence of some well-known generalities that should be considered when optimizing counting performance. First, plastic vials produce a lower background count than the corresponding glass vials. In this example a 15–30% improvement in performance (E^2/B) was obtained with plastic vials. Second, small vials for low-volume samples provide a lower background with about the same counting efficiency as large vials. Third, the optimized counting region gives a better performance than the standard counting region. From the example taken in Table 5.11 the total performance was enhanced with small plastic vials over large glass vials in an optimized counting region by almost a factor of 3 (E^2/B 103.9–306.1). The only problem with using smaller vials is that the sample capacity is lower, and if maximum sample size is required, the larger vials should be used.

3. Cocktail Choice

Choosing the proper liquid scintillation fluor cocktail optimizes counting performance. When sample counts are low, it is best to choose a cocktail that can hold as much sample as possible and still provides the maximum counting efficiency with a minimum amount of background. Chapter 8 provides a great deal of information that can be helpful in selecting a cocktail as a function of sample chemical properties to optimize counting performance.

4. Counting Time

The error associated with the measurement of sample counts and count rate is a function of counting time. As demonstrated in Chapter 7, we can calculate that the standard deviation of the counts collected and the calculation of the associated count rate are reduced according to the length of the counting time. This is particularly relevant when the sample counts are very close to background, and accurate measurements of sample counts and background counts must be made to distinguish the difference within reasonable error limits. Therefore, liquid scintillation counting performance can be optimized by increasing the sample and background counting times. The longer the counting time the lower is the detection level or minimal detectable activity (MDA). The MDA also referred to as the lower limits of detection (LLD) in units of $Bq\,L^{-1}$ can be calculated by the following equation from Prichard *et al.* (1992) as described by Passo and Kessler (1993), Passo and Cook (1994), and Biggen *et al.* (2002):

$$LLD = \frac{L_D}{60\,EVTX} \tag{5.107}$$

where LLD is the lowest activity concentration in Bq L^{-1} that yields a net count above background with a 95% probability, L_D is the detection limit according to Currie (1968) and Hurtgen *et al.* (2000), which is the true net signal that can be detected with a given probability, 60 is a factor for conversion of DPM to Bq (i.e., 60DPM/1Bq), E is the fractional detection efficiency (CPM/DPM), V is the sample volume in liters, T is the count time in minutes, and X is any factor that is relevant, such as, decay correction, chemical yield, etc. As described by Passo and Cook (1994), when the background and sample are counted as pairs, that is, the same counting time, and the background counts collected are >70, the detection limit may be written as

$$L_D = 4.65\sqrt{B} \tag{5.108}$$

When the lower limit of detection (LLD) is expressed in the less common units of pCi L^{-1}, the conversion factor 60 of Eq. 5.107 is replaced with the factor 2.22 for the conversion of DPM to pCi (i.e., 2.22 DPM/pCi).

To illustrate the application and interpretation of calculated lower limits of detection (LLD) and detection limits (L_D) let us take the following example to calculate the LLD and, from these calculations, determine how many sample counts must be acquired to reach a calculated LLD:

If in a particular LSA counting region and a cocktail chemistry the analysis of ^3H yielded a total of 1300 background counts in a 500-minute counting time, that is, $B = 1300$ and background $= 2.60$ CPM, the counting efficiency for ^3H was 35% or 0.35, and the sample volume analyzed was 10 mL, the lower limit of detection according to Eq. 5.107 would be

$$LLD = \frac{4.65\sqrt{1300\,\text{counts}}}{(60\,\text{DPM/Bq})\,(0.35)\,(0.01\text{L})\,(500\,\text{min})}$$
$$= 1.60\,\text{Bq}\,\text{L}^{-1} \tag{5.109}$$

In the above calculation the limit of detection is

$$L_D = 4.65\sqrt{1300\,\text{counts}} = 168\,\text{counts} \tag{5.110}$$

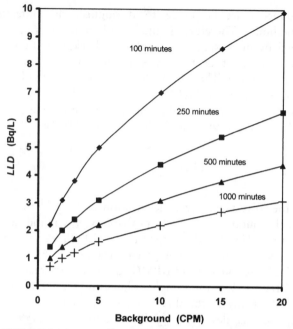

FIGURE 5.67   The calculated *LLD* or *MDA* in Bq L^{-1} as a function of sample background in CPM for 4 different sample and background counting times: 100, 250, 500, and 1000 minutes for ^{3}H tritium analysis with a counting efficiency of 35% or 0.35 and sample volume of 10 mL.

Therefore, to achieve the calculated LLD of 1.60 Bq L^{-1} a minimum number of sample counts defined by

$$L_D + B = 168 + 1300 = 1468 \text{ counts} \tag{5.111}$$

must be registered with a 10 mL sample in a 500-min counting time to conclude that 1.60 Bq L^{-1} are detected with a 95% confidence level.

From Eqs. 5.107 and 5.108 and the calculation 5.109 we can see that increasing the counting time and reducing the background will reduce the LLD or MDA. This is illustrated by Fig. 5.67 where we can see that, by reducing the background, we can achieve a desired LLD with a shorter counting time.

5. Background Reduction

Often the easiest method of optimizing LSC performance is to reduce the background. As described previously in this chapter, region optimization, vial size, and vial material are important variables to control in the reduction of background. This section will discuss 5 other methods that can be used to reduce background and enhance performance for LSA.

a. Temperature Control

The original purpose of temperature control or cooling during the early years of liquid scintillation counting was to reduce background counts from thermal noise of the PMTs. Modern bialkalie photomultiplier tubes are manufactured to have a low noise even at room temperature. The purpose of temperature control in contemporary liquid scintillation counting is to control and maintain the sample temperature, sometimes required for optimizing sample chemistry even at extremes of external temperatures. The sample chemistry in some cocktails requires lower temperature to accept more sample, and as a result temperature control is necessary.

b. Underground Counting Laboratory

This method of background reduction can be used for very low level counting and this extreme measure has been taken more often in the past for the measurement of very low levels of naturally occurring radionuclides such as ^{3}H in ground water and ^{14}C dating. This involves building a special counting room below the ground to reduce high-energy cosmic radiation that may cause background counts in the liquid scintillation counter. However, modern pulse discrimination electronics, described briefly in Part d following and in more detail in Chapter 6, will reduce instrument backgrounds to levels at which underground laboratories are generally not needed.

c. Shielding

The last two areas that are used to reduce background will be described in a little more detail here but will be covered more thoroughly in Chapter 6. Shielding the instrument from external radiation can reduce background. There are two types of shielding, passive and active shielding.

Passive shielding is the "brute force" approach. This requires more and more lead (up to 1000 kg) to reduce background. The extra lead reduces background from environmental gamma photons associated with building materials and instrument construction materials. In addition, this extra lead can reduce the "soft" cosmic muon components of background. The lead shielding is usually lined internally with cadmium and/or copper to absorb any secondary x-rays and cosmic components.

Active shielding can be one of the two types, a guard detector with or without anticoincidence circuitry and components. The guard detector is an external detector [e.g., bismuth germanate (BGO)] that surrounds the area between the PMTs and the sample. The BGO has special properties described in detail in Chapter 11. BGO is a solid crystal scintillator with a high density and high atomic number. The high density and atomic number give BGO excellent "stopping power" for the detection of highly penetrating radiation. BGO can be used to discriminate against gamma and muon components of cosmic background. This property of BGO in combination with TR-LSC pulse discrimination electronics (described in the following) can discriminate more background events.

Another approach to active shielding to reduce background is to manufacture a detector shield above, below, and surrounding the sample vial and PMTs consisting of an external counting chamber of a liquid or solid scintillator, such as mineral oil scintillator, plastic scintillator (e.g., NE110), or even a NaI(Tl) crystal. In addition to the guard material, two PMTs and an anticoincidence circuit are added. The detector guard rejects much of the environmental gamma radiation, as well as soft and hard cosmic radiation.

d. Pulse Discrimination Electronics

Three different pulse discrimination methods used, namely, pulse shape analysis (PSA), pulse amplitude comparison (PAC), and time-resolved LSC (TR-LSC).

Pulse shape analysis (PSA) relies on the fact that a true nuclear decay signal produces a very prompt pulse in the LSC, whereas a background event produces a pulse with a longer decay time. By setting up the instrument to discriminate pulses on the basis their shape, background can be discriminated from true nuclear decay signal. This discrimination is specific and must be optimized for each vial material, vial size, and sample-cocktail chemistry.

The second method is PAC, which compares the amplitude of the pulse from each of the two PMTs. If the event occurs in the scintillation vial, photons of approximately the same intensity will be seen by both PMTs, but if the event originates from outside the sample, it will produce pulses in each of the PMTs of different pulse heights. If the PAC deviates significantly from unity, the pulse is discriminated as a background.

The third method, TR-LSC, is a patented method of reducing background by discriminating against pulses on the basis of the number of afterpulses that occur following an initial pulse event over a given period of time. These afterpulses are most commonly created by background events. Afterpulses are generally more numerous among nonquenchable background events and less numerous with events originating from the scintillation cocktail. With the TR-LSC technique the LSA can be manufactured or programmed to count afterpulses, known as pulse index discrimination, which can identify background events by counting afterpulses with varying degrees of sensitivity. Therefore, among various LSAs, the amount of pulse index discrimination for background rejection can be factory set or programmed to provide different sensitivities for the reduction of background (e.g., high-sensitivity count mode, low-level count mode, and ultra low-level count mode with BGO detector guard). Tables 5.12 and 5.13 illustrate how significantly background can be reduced and performance enhanced using this latter technique of background reduction. The data shown in these two tables might typically be found in a low-level counting laboratory. The samples are placed in the optimum counting geometry, optimum vial size and vial material, optimum cocktail, and counted in the optimum counting region with the TR-LSC method of electronic discrimination at various levels of pulse index discrimination with and without an active guard.

The data of Table 5.12 illustrate that the use of TR-LSC alone and in combination with an active BGO guard detector greatly increases the performance of the LSC for ^{14}C samples. This is also the case for many other radionuclide samples with low count rates. For the case of ^{14}C about 15–20 times better performance is obtained with the ultra low-level TR-LSC and the BGO guard detector.

The data of Table 5.13 illustrate that the use of TR-LSC alone and in combination with a BGO active guard detector greatly increases the

TABLE 5.12 Radiocarbon Dating Samples as 3.5 mL Benzene with PPO/POPOP Scintillator[a]

Amount of pulse index discrimination	Detector guard	^{14}C efficiency (%)	Background (CPM)	Figure of Merit (E^2/B)
None	No	83.45	9.67	720
Normal TR-LSC	No	81.87	7.07	948
High-sensitivity TR-LSC	No	78.51	4.74	1,300
Low-level TR-LSC	No	70.10	1.38	3,560
Ultra-low-level TR-LSC + BGO Active Guard	Yes	65.01	0.33	12,675

[a]The optimized counting region is 10–102 keV with various levels of TR-LSC pulse index discrimination applied to the pulse events. Courtesy of PerkinElmer Life and Analytical Sciences, Boston.

TABLE 5.13 Environmental Tritium Water Samples as 10 mL Water in 10 mL Ultima Gold LLT Scintillation Cocktail[a]

Amount of pulse index discrimination	Detector guard	^{3}H efficiency (%)	Background (CPM)	Figure of Merit (E^2/B)
None	No	26.50	18.45	38
Normal TR-LSC	No	26.24	12.75	54
High-sensitivity TR-LSC	No	24.68	9.25	66
Low-level TR-LSC	No	22.59	3.33	153
Ultra-low-level TR-LSC + BGO Active Guard	Yes	20.01	1.00	400

[a]The optimized counting region is 0.5–5.0 keV with various level of TR-LSC pulse index discrimination applied to the pulse events. Courtesy of Perkin Elmer Life and Analytical Sciences, Boston.

performance of the LSC for tritium samples in general with low count rates. For the case of tritium about a 10-fold better performance is achieved with the use of ultra-low-level TR-LSC in combination with the BGO guard detector.

For additional information on time-resolved liquid scintillation counting (TR-LSC) see papers by Roessler *et al.* (1991) and Passo and Roberts (1996).

XVIII. CONCLUSIONS

In conclusion, the best ways of optimizing performance of the LSC for low-level samples are to (1) optimize counting geometry, such as optimum sample size, vial size, and vial material; (2) increase counting time; (3) optimize counting region; (4) optimize cocktail selection, and (5) use electronic discrimination and an active guard detection method. See also Chapter 6 for additional information on optimizing performance in liquid scintillation analysis.

REFERENCES

Aburai, T., Takiue, M., and Ishikawa, H. (1981). Quantitative measurement method of α-emitters by using liquid scintillation counter connected with multichannel pulse-height analyzer. *Radioisotopes* **30**, 579–583.

Akimov, D., Bewick, A., Davidge, D., Dawson, J., and Howard, A. S. *et al.* (2002). Measurements of scintillation efficiency and pulse shape for low energy recoils in liquid xenon. *Phys. Lett. B* **524**, 245–251.

Altzitzoglou, T., Larosa, J. L., and Nicholl, C. (1998). Measurement of ^{90}Sr in bone ash. *Appl. Radiat. Isot.* **49**(9–11), 1313–1317.

Altzitzoglou, T., Deneke, B., Johansson, L., and Sibbens, G. (2002). Standardization of ^{89}Sr using three different methods. *Appl. Radiat. Isot.* **56**, 447–452.

Anonymous (1995). SpectraWorks^{TM} for windows spectrum analysis software. *Cronical Software Notes* **SC-003**, PerkinElmer Life and Analytical Sciences, Boston, pp. 4.

Aprile, E., Mukherjee, R., and Suzuki, M. (1990). A study of the scintillation light induced in liquid xenon by electrons and alpha particles. *IEEE Trans. Nucl. Sci.* **37**(2), 553–558.

Arneodo, F., Baiboussinov, B., Badertscher, A., Benetti, R., and Bernardini, E. *et al.* (2000). Scintillation efficiency of nuclear recoil in liquid xenon. *Nucl. Instrum. Methods Phys. Res., Sect. A* **449**, 147–157.

Attie, M. R. P., Koskinas, M. F., Dias, M. S., and Fonseca, K. A. (1998). Absolute disintegration rate measurements of Ga-67. *Appl. Radiat. Isot.* **49**(9–11), 1175–1177.

Aupiais, J., Fayolle, C., Gilbert, P., and Dacheux, N. (1998). Determination of ^{226}Ra in mineral drinking waters by α liquid scintillation with rejection of β-γ emitters. *Anal. Chem.* **70**, 2353–2359.

Aupiais, J., Dacheux, N., Thomas, A. C., and Matton, S. (1999). Study of neptunium measurement by alpha liquid scintillation with rejection of β-γ emitters. *Anal. Chim. Acta* **398**, 205–218.

Aupiais, J., and Dacheux, N. (2000). Understanding the peak asymmetry in alpha liquid scintillation with β/γ discrimination. *Radiochim. Acta* **88**, 391–398.

Baerg, A. P. (1973). The efficiency extrapolation method in coincidence counting. *Nucl. Instrum. Methods* **112**, 143–150.

Bakir, Y. Y. and Bem, H. (1996). Application of pulse decay discrimination liquid scintillation counting for routine monitoring of radioactivity in drinking water. *In* "Liquid Scintillation Spectrometry 1994" (G.T. Cook, D.D. Harkness, A.B. MacKenzie, B.F. Miller, and E.M. Scott, Eds.) pp. 293–299, Radiocarbon 1966, Tucson.

Barosi, G., Cazzola, M., and Perugini, S. (1980). *In* "Liquid Scintillation Counting, Recent Applications and Development", Vol. II, pp. 517–523, Academic Press, New York and London.

Basson, J. K., and Steyn, J. (1954). Absolute alpha standardization with liquid scintillators. *Proc. Phys. Soc. A* **67**, 297–298.

Beebe, K. R., and Kowalski, B. R. (1987). An introduction to multivariate calibration and analysis. *Anal. Chem.* **59**, 1007A–1017A.

Belanov, S. V., Kashirin, I. A., Malinovskiy, S. V., Egorova, M. E., Efimov, K. M., Tikhomirov, V. A., and Sobolev, A. I. (1997). The method of identifying radionuclides with the use of a liquid scintillation counter. RF Patent No. 2,120,646.

Belanov, S. V., Kashirin, I. A., Malinovskiy, S. V., Ermakov, A. I., Efimov, K. M., Tikhomirov, V. A., and Sobolev, A. I. (1998). The method of identifying radionuclides with the use of a liquid scintillation counter. RF Patent No. 98,106,407.

Bell, P. R. (1948). The use of anthracene as a scintillation counter. *Phys. Rev.* **73**(11), 1405-

Benetti, P., Calligarich, E., and Dolfini, R. *et al.* (1993). Detection of energy deposition down to the keV region using liquid xenon scintillation. *Nucl. Instrum. Methods Phys. Res., Sect. A* **327**, 203–206.

Benitez-Nelson, C. R., and Buessler, K. O. (1998). Measurement of Cosmogenic ^{32}P and ^{33}P activities in rainwater and seawater. *Anal. Chem.* 70, 64–72.

Biggen, C. D., Cook, G. T., MacKenzie, A. B., and Pates, J. M. (2002). Time-efficienct method for the determination of ^{210}Pb, ^{210}Bi, and ^{210}Po activities in seawater using liquid scintillation spectrometry. *Anal. Chem.* 74, 671–677.

Birks, J. B. (1964). The Theory and Practice of Scintillation Counting. Pergamon Press, Oxford.

Bransome Jr., E. D. (1973). *Semin. Nucl. Med.* 3, 389–399.

Bressi, G., Carugno, G., Conti, E., Iannuzzi, D., and Meneguzzo, A. T. (2000). Infrared scintillation in liquid Ar and Xe. *Nucl. Instrum. Methods Phys. Res., Sect. A* 440, 254–257.

Bressi, G., Carugno, G., Conti, E., Del Noce, C., and Iannuzzi, D. (2001). Infrared scintillation: a comparison between gaseous and liquid xenon. *Nucl. Instrum. Methods Phys. Res., Sect. A* 461, 378–380.

Brillanti, S. *et al.* (1991). Effect of alpha interferon therapy on hepatitis C viremia in community-acquired chronic non-A non B hepatitis A quantitative polymerase chain reaction study. *J. Med. Virol.* 34, 136–141.

Broda, R., and Pochwalski, K. (1993). The ETDCR method of standardizing ^{55}Fe and ^{54}Mn. *In* "Liquid Scintillation Spectrometry 1992" (J. E. Noakes, F. Schönhofer, and H. A. Polach, Eds.), pp. 255–260. Radiocarbon, Tucson.

Broda, R., Pochwalski, K., and Radoszewski, T. (1988). Calculation of liquid-scintillation detector efficiency. *Appl. Radiat. Isot.* 39, 159–164.

Broda, R., Péron, M. N., Cassette, P., Terlikowska, T., and Hainos, D. (1998). Standardization of ^{139}Ce by the liquid scintillation counting using the triple to double coincidence ratio method. *Appl. Radiat. Isot.* 49(9–11), 1035–1040.

Broda, R., Cassette, P., Maletka, K., and Pochwalski, K. (2000). A simple computing program for application of the TDCR method to standardization of pure-beta emitters. *Appl. Radiat. Isot.* 52, 673–678.

Broda, R., Maletka, K., Terlikowska, T., and Cassette, P. (2002). Study of the influence of the LS-cocktail composition for the standardization of radionuclides using the TDCR model. *Appl. Radiat. Isot.* 56, 285–289.

Brooks, F. D. (1993). Developments in neutron detection. *In* "Nuclear Techniques for Analytical and Industrial Applications" (G. Vourvopoulos and T. Raradellis, Eds.), pp.151–170. Western Kentucky University, Bowling Green.

Brooks, F. D. (1997). Neutron detectors and spectrometers. Proc. SPIE 2867, 538–549.

Buchtela, K., Tschurlovits, M., and Unfried, E. (1974). Eine Methode zur Unterschiedung von α- und β-Strahlen in einem Flüssigszintillationsmessgerät. *Int. J. Appl. Radiat. Isot.* 25, 551–555.

Cadieux, J. R. (1990). Evaluation of a photoelectron-rejecting alpha liquid-scintillation (PERALS) spectrometer for the measurement of alpha-emitting radionuclides. *Nucl. Instrum. Methods Phys. Res., Sect. A* 299, 119–122.

Calhoun, J. M., Coursey, B. M., Gray, D., and Karam, L. (1991) The standardization of ^{35}S methionine by liquid scintillation efficiency tracing with ^{3}H. *In* "Liquid Scintillation Counting and Organic Scintillators" (Harley Ross, John E. Noakes and Jim D. Spaulding, Eds.), pp. 317–323, Lewis Publishers, Chelsea, MI 48118.

Cassette, P., and Vatin, R. (1992). Experimental evaluation of TDCR models for the 3 PM liquid scintillation counter. *Nucl. Instrum. Methods Phys. Res., Sect. A* 312, 95–99.

Cassette, P., Altzitzoglou, T., Broda, R., Collé, R., Dryak, P., De Felice, P., Günther, E., Los Arcos, J. M., Ratel, G., Simpson, B., and Verregen, F. (1998). Comparison of activity concentration measurement of ^{63}Ni and ^{55}Fe in the framework of the EUROMET 297 project. *Appl. Radiat. Isot.* 49, 1403–1410.

Cassette, P., Broda, R., Hainos, D., and Terlikowska, T. (2000). Analysis of detection-efficiency variation techniques for the implementation of the TDCR method in liquid scintillation counting. *Appl. Radiat. Isot.* 52, 643–648.

Ceccatelli, A., and De Felice, P. (1999). Standardisation of ^{90}Sr, ^{63}Ni and ^{55}Fe by the 4πβ liquid scintillation spectrometry method with ^{3}H-standard efficiency tracing. *Appl. Radiat. Isot.* 51, 85–92.

Cennini, P., Revol, J. P., Rubbia, C., Sergiampietri, F., and Bueno, A. *et al.* (1999). Detection of scintillation light in coincidence with ionizing tracks in a liquid argon time projection chamber. *Nucl. Instrum. Methods. Phys. Res., Sect. A* 432, 240–248.

Chandrasekaran, E. S. (1981). Measurement of iodine-125 by liquid scintillation counting method. *Health Phys.* **40**(6), 896–898.

Chepel, V. (1993). A new liquid xenon scintillation detector for positron emission tomography. *Nucl. Tracks Rad. Meas.* **21**(1), 47–51.

Chepel, V., Solovov, V., van der Marel, J., Lopes, M. I., Crespo, P., Jeneiro, L., Santon, D., Ferreira Marques, R., and Policarpo, A. J. P. L. (1999). The liquid xenon detector for PET: recent results. *IEEE Trans. Nucl. Sci.* **NS-46**(4), 1038–1044.

Collé, R. (1997a). Systematic effects of total cocktail mass (volume) and H_2O fraction on $4\pi\beta$ liquid scintillation spectrometry of ^{3}H. *Appl. Radiat. Isot.* **48**(6), 815–831.

Collé, R. (1997b). Cocktail mismatch effects in $4\pi\beta$ liquid scintillation spectrometry of ^{3}H: implications based on the systematics of ^{3}H detection efficiency and quench indicating parameter variations with total cocktail mass (volume) and H_2O fraction. *Appl. Radiat. Isot.* **48**(6), 833–842.

Cosolito, F. J., Cohen, N., and Petrow, H. G. (1968). Simultaneous determination of iron-55 and stable iron by liquid scintillation counting *Anal. Chem.* **40**(1), 213–215.

Coursey, B. M., Gibson, J. A. B., Heitzmann, M. W., and Leak, J. C. (1984). Standardization of technetium-99 by liquid scintillation counting. *Appl. Radiat. Isot.* **35**, 1103–1112.

Coursey, B. M., Mann, W. B., Grau Malonda, A., Garcia-Toraño, E., Los Arcos, J. M., Gibson, J. A. B., and Reher, D. (1986). Standardization of carbon-14 by $4\pi\beta$ liquid scintillation efficiency tracing with hydrogen-3. *Appl. Radiat. Isot.* **37**, 403–408.

Coursey, B. M., Lucas, L. L., Grau Malonda, A., and Garcia-Toraño, E. (1989). The standardization of plutonium-241 and nickel-63. *Nucl. Instrum. Methods Phys. Res., Sect. A* **279**, 603–610.

Coursey, B. M., Cessna, J., Garcia-Toraño, E., Golas, D. B., Grau Molanda, A., Gray, D. H., Hoppes, D. D., Los Arcos, J. M., Martin-Casallo, M. T., Schima, F. J., and Unterweger, M. P. (1991). The standardization and decay scheme of Rhenium-186. *Appl. Radiat. Isot.* **42**, 865–869.

Coursey, B. M., Calhoun, J. M., Cessna, J., Golas, D. B., and Schima, F. J. (1994). Liquid-scintillation counting techniques for the standardization of radionuclides used in therapy. *Nucl. Instrum. Methods Phys. Res., Sect. A* **339**, 26–30.

Coursey, B. M., Schima, F. J., Golas, D. B., Palabrica, O. T., Suzuki, A., and Dell, M. A. (1998). Measurement standards for strontium-89 for use in bone palliation. *Appl. Radiat. Isot.* **49**(4), 335–344.

Craddock, B. L., Orchiston, E. A., Hinton, H. J., and Welham, M. J. (1999). Dissociation of apoptosis from proliferation, protein kinase B activation, and BAD phosphorylation in interleukin-3-mediated phosphoinositide 3-kinase signaling. *J. Biol. Chem.* **274**(15), 10633–10640.

Cramer, C. F., Nicholson, M., Moore, C., and Teng, K. (1971). Computer programs for calculation of liquid scintillation data from long-lived and short-lived radionuclides used in single or dual labeling. *Int. J. Appl. Radiat. Isot.* **22**, 17–20.

Cruz, P. A. L., Loureiro, J. S., and Bernardes, E. M. O. (2002). Standardization of ^{89}Sr solution from a BIPM intercomparison using a liquid scintillation method. *Appl. Radiat. Isot.* **56**, 457–459.

Currie, L. A. (1968). Limits for qualitative detection and quantitative determination – application to radiochemistry. *Anal. Chem.* **40**, 586–593.

Dacheux, N., and Aupiais, J. (1997). Determination of uranium, thorium, plutonium, americium, and curium ultratraces by photon electron rejecting α liquid scintillation. *Anal. Chem.* **69**, 2275–2282.

Dacheux, N., and Aupiais, J. (1998). Determination of low concentrations of americium and curium by photon/electron rejecting alpha liquid scintillation. *Anal. Chim. Acta* **363**, 279–294.

Dacheux, N., Aupiais, J., Courson, O., and Aubert, C. (2000). Comparison and improvement of the determinations of actinide low activities using several α liquid scintillation spectrometers. *Anal. Chem.* **72**, 3150–3157.

Dahlberg, E. (1982). Quench correction in liquid scintillation counting by a combined internal standard-samples channels ratio technique. *Anal. Chem.* **54**(12), 2082–2085.

De Filippis, S. (1991). ^{55}Fe and ^{59}Fe: a qualitative comparison of four methods of liquid scintillation activity analysis. *Radioactivity Radiochem.* **2**, 14R&R–21R&R.

Dern, R. J., and Hart, W. L. (1961a). Doubly labeled iron. I. Simultaneous liquid scintillation counting of isotopes ^{55}Fe and ^{59}Fe as ferrous perchlorate. *J. Lab. Clin. Med.* **57**, 322–330.

Dern, R. J., and Hart, W. L. (1961b). Doubly labeled iron. II. Separation of iron from blood samples and preparation of ferrous perchlorate for liquid scintillation counting. *J. Lab. Clin. Med.* **57**, 460–467.

Dobbs, H. E. (1965). "Dispersing Solutions for Liquid Scintillation Counting", Memorandum No. AERE-M1574, UK Atomic Energy Research Establishment, Harwell.

Dodson, C. L. (1996). Radionuclide identification in liquid scintillation spectrometry. *In* "Liquid Scintillation Spectrometry 1994" (G. T. Cook, D. D. Harkness, A. B. MacKenzie, B. F. Miller and E. M. Scott, Eds.), *Radiocarbon* 361–364.

Doke, T. (1981). Fundamental properties of liquid argon, krypton, and xenon as radiation detector media. *Port. Phys.* **12**(1–2), 9–48.

Doke, T., Masuda, K., and Shibamura, E. (1990). Estimation of absolute photon yields in liquid argon and xenon for relativistic electrons. *Nucl Instrum. Methods, Sect. A* **291**, 617–620.

Eakins, J. D., and Brown, D. A. (1966). An improved method for the simultaneous determination of iron-55 and iron-59 in blood by liquid scintillation counting. *Int. J. Appl. Radiat. Isot.* **17**, 391–397.

Elliott, J. C. (1984). Effect of vial composition and diameter on determination of efficiency, background, and quench curves in liquid scintillation counting. *Anal. Chem.* **56**, 758–761.

Effertz, B., Neuman, K., and Englert, D. (1993). Single photomultiplier technology for scintillation counting in microplates. *In* "Liquid Scintillation Spectrometry 1992" (J. E. Noakes, F. Schönhofer, and H. A. Polach, Eds.), *Radiocarbon* 37–42.

Eikenberg, J., Zumsteg, I., Rüthi, M., Bajo, S., Fern, M. J., and Passo, C. J. (1999). Fast radiochemical screening of transuranium radionuclides in urine using actinide extractive resin and low-level α/β LSC. *Radioact. Radiochem.* **10**, 19–30.

Everett, L. J., Ring, J. G., and Nguyen, D. C. (1980). *In* "Liquid Scintillation Counting, Recent Applications and Development", Vol. I "Physical Aspects" (C.-T. Peng, D. L. Horrocks, and E. L. Alpen, Eds.), pp. 119–128. Academic Press, New York and London.

Fujii, H., and Takiue, M. (1988a). Radioassay of dual-labeled samples by sequential Cherenkov counting and liquid scintillation efficiency tracing technique. *Nucl. Instrum. Methods Phys. Res., Sect. A* **273**, 377–380.

Fujii, H., and Takiue, M. (1998b). Radioassay of alpha- and beta-emitters by sequential Cherenkov and liquid scintillation counting. *Appl. Radiat. Isot.* **39**, 327–330.

Fujii, H., and Takiue, M. (2001). Foggy scintillation counting technique. *Appl. Radiat. Isot.* **55**, 517–520.

Fujii, H., Takiue, M., and Ishikawa, H. (1986). Activity determination of disc samples with liquid scintillation efficiency tracing technique. *Appl. Radiat. Isot.* **37**, 1147–1149.

Fujii, H., Matsuno, K., and Takiue, M. (1999). Construction of analytical beta ray monitor for liquid waste. *Radioisotopes* **48**, 465–471.

Fujii, H., Matsuno, K., and Takiue, M. (2000). Hybrid radioassay of multiple radionuclide mixtures in waste solutions by using liquid and NaI(Tl) scintillation monitors. *Health Phys.* **79**(3), 294–298.

Funck, E., and Nylandstedt Larsen, A. (1983). The influence from low energy x-rays and Auger electrons on $4\pi\beta$-γ coincidence measurements of electron-capture-decaying nuclides. *Int. J. Appl. Radiat. Isot.* **34**, 565-

García, G., and Grau Malonda, A. (2002). The influence of stopping power on the ionization quench factor. *Appl. Radiat. Isot.* **56**, 295–300.

García, J. F., Izquierdo-Ridorsa, A., Toribio, M., and Rauret, G. (1996). Classical versus multivariate calibration for a beta emitter (^{14}C) activity determination by liquid scintillation counting. *Anal. Chim. Acta* **331**, 33–41.

García-Toraño, E., and Grau Malonda, A. (1981). EFFY, a program to calculate the counting efficiency of beta particles in liquid scintillators. *Comput. Phys. Commun.* **23**, 385–391.

García-Toraño, E., and Grau Malonda, A. (1985). EFFY2, a new program to compute the counting efficiency of beta particles in liquid scintillators. *Comput. Phys. Commun.* **36**, 307–312.

García-Toraño, E., and Grau Malonda, A. (1987). EFYGA, a Monte Carlo program to compute the interaction probability and the counting efficiency of gamma rays in liquid scintillators. *Comput. Phys. Commun.* **47**, 341–347.

García-Toraño, E., and Grau Malonda, A. (1988). Cálculo de la eficiencia de recuento de nucleidos que experimentan desintegración beta y desexcitación gamma simple. *Report CIEMAT* 616, Madrid.

García-Toraño, E., Grau Malonda, A., and Los Arcos, J. M. (1988). EBEGA—the counting efficiency of a beta-gamma emitter in liquid scintillators. *Comput. Phys. Commun.* **50**, 313–319.

García-Toraño, E., Martin Cassallo, M. T., Rodríguez. L., Grau, A., and Los Arcos, J. M. (1991). On the standardization of beta-gamma-emitting nuclides by liquid scintillation counting. *In* "Liquid Scintillation Counting and Organic Scintillators" (Harley Ross, John E. Noakes, and Jim D. Spaulding, Eds.), pp. 307–316, Lewis Publishers, Chelsea, MI.

García-Toraño, E., Roteta, M., and Rodríguez Barquero, L. (2000). Standardization of ^{110m}Ag by liquid scintillation and $4\pi\beta$-γ coincidence counting. *Appl. Radiat. Isot.* **52**, 637–641.

García-Toraño, E., Rodríguez Barquero, L., and Roteta, M. (2002). Standardization of ^{134}Cs by three methods. *Appl. Radiat. Isot.* **56**, 211–214.

Garson, J., and Whitby, K. (1994). Nucleic acid quantification by chemiluminescence assay of polymerase chain reaction products. *TopCount Topics*, **TCA-020**, Packard BioScience Company, Meriden, CT 06450, pp. 4.

Geladi, P., and Kowalski, B. R. (1986). Partial least-squares regression: a tutorial. *Anal. Chim. Acta* **185**, 1–17.

Gibson, J. A. B., and Gale, H. J. (1968). Absolute standardization with liquid scintillation counters. *J. Sci. Instrum.* **1**(2), 96–106.

Goldstein, G. (1965). Absolute liquid-scintillation counting of beta emitters. *Nucleonics* **23**, 67–69.

Grau Carles, A. (1993). A new linear spectrum unfolding method applied to radionuclide mixtures in liquid scintillation spectrometry. *Appl. Radiat. Isot.* **45**, 83–90.

Grau Carles, A. (1996). MLOG, the simultaneous standardization of multi-nuclide mixtures. *Comput. Phys. Commun.* **93**, 48–52.

Grau Carles, A., and Grau Malonda, A. (1989). Electron-capture standardization with a triple phototube system. *Anales de Física, Ser. B* **85**, 160–176.

Grau Carles, A., and Grau Malonda, A. (1991). A new procedure for multiple isotope analysis in liquid scintillation counting. *In* "Liquid Scintillation Counting and Organic Scintillators" (Harley Ross, John E. Noakes and Jim D. Spaulding, Eds.), pp. 295–306, Lewis Publishers, Chelsea, MI 48118.

Grau Carles, A., and Grau Malonda, A. (1992). "Precise System for the Determination of the Quench Parameter of Radioactive Samples in Liquid Phase", Patent P. 9202639, December 29, 1992, Registry of Industrial Property, Madrid, Spain.

Grau Carles, A., and Grau Malonda, A. (1997). Calibración del ^{40}K por centelleo líquido. Determinación del periodo de semidesintegración. Report CIEMAT 831, Madrid.

Grau Carles, A., and Grau Malonda, A. (2001). Free parameter, figure of merit and ionization quench in liquid scintillation counting. *Appl. Radiat. Isot.* **54**, 447–454.

Grau Carles, A., Martin-Casallo, M. T., and Grau Malonda. (1991). Spectrum unfolding and double window methods applied to standardization of ^{14}C and ^{3}H mixtures. *Nucl. Instrum. Methods Phys. Res., Sect. A* **307**, 484–490.

Grau Carles, A., Grau Malonda, A., and Rodríguez Barquero, L. (1993a). Cherenkov radiation effects on counting efficiency in extremely quenched liquid scintillation samples. *Nucl. Instrum. Methods Phys. Res., Sect. A* **334**, 471–476.

Grau Carles, A., Rodríguez Barquero, L., and Grau Malonda, A. (1993b). Standardization of ^{14}C and ^{35}S mixtures. *Nucl. Instrum. Methods Phys. Res., Sect. A* **335**, 234–240.

Grau Carles, A., Rodríguez Barquero, L., and Grau Malonda, A. (1993c). A spectrum unfolding method applied to standardization of ^{3}H and ^{55}Fe mixtures. *Appl. Radiat. Isot.* **44**, 581–586.

Grau Carles, A., Rodríguez Barquero, L., and Grau Malonda, A. (1993d). Standardization of multi-nuclide mixtures by a new spectrum unfolding method. *J. Radioanal. Nucl. Chem., Letters* **176**, 391–403.

Grau Carles, A., Rodríguez Barquero, L., and Grau Malonda, A. (1993e). Simultaneous standardization of ^{90}Sr-^{90}Y and ^{89}Sr mixtures. *Appl. Radiat. Isot.* **44**, 1003–1010.

Grau Carles, A., Rodríguez Barquero, L. and Grau Malonda, A. (1993f). Double-label counting of heterogeneous samples. *In* "Liquid Scintillation Spectrometry 1992" (J. E. Noakes, F. Schönhofer, and H. A. Polach, Eds.), pp. 239–249, Radiocarbon, Tucson.

Grau Carles, A., Rodríguez Barquero, L., and Grau Malonda, A. (1994a). Deconvolution of ^{204}Tl/^{36}Cl and ^{147}Pm/^{45}Ca dual mixtures. *Nucl. Instrum. Methods Phys. Res., Sect. A* **339**, 71–77.

Grau Carles, A., Grau Malonda, A., and Grau Carles, P. (1994b). EMI, the counting efficiency for electron capture, electron capture-gamma and isomeric transition. *Comput. Phys. Commun.* **79**, 115–123.

Grau Carles, A., Grau Malonda, A., and Rodríguez Barquero, L. (1994c). Standardization of ^{125}I, ^{85}Sr and ^{109}Cd by CIEMAT/NIST method. *Appl. Radiat. Isot.* **45**, 461–464.

Grau Carles, A., Grau Malonda, A., and Gómez Gil (1996). Standardization of U(X$_1$ + X$_2$): the ^{234}Th + ^{234m}Pa + ^{230}Th mixture. *Nucl. Instrum. Methods Phys. Res., Sect. A* **369**, 431–436.

Grau Carles, A., Rodriguez Barquero, L., and Jimenez De Mingo, A. (1997). ^{125}Sb to ^{125m}Te branching ratio. ICRM '97, 1–12. (preprint manuscript).

Grau Carles, A., Rodriguez Barquero, L., and Jimenez De Mingo, A. (1998). ^{125}Sb to ^{125m}Te branching ratio. *Appl. Radiat. Isot.* **49**(9–11), 1377–1381.

Grau Malonda, A. (1982a). Measurement of beta radioactivity by liquid scintillation counting. (in Spanish), Doctoral Thesis No. 171/82. Universidad Complutense de Madrid, Madrid.

Grau Malonda, A. (1982b). Counting efficiency for electron-capturing nuclides in liquid scintillator solutions. *Int. J. Appl. Radiat. Isot.* **33**, 371–375.

Grau Malonda, A. (1995). "Modelos de Parámetro Libre en Centelleo Líquido". Editorial CIEMAT, Avda. Complutense 22, 28040 Madrid, pp. 387.

Grau Malonda, A. (1999). "Free Parameter Models in Liquid Scintillation Counting". Editorial CIEMAT, Madrid, pp. 416.

Grau Malonda, A., and Coursey, B. M. (1987). Standardization of isomeric-transition radionuclides by liquid-scintillation efficiency tracing with hydrogen-3: application to Technetium-99m. *Appl. Radiat. Isot.* **38**, 695–700.

Grau Malonda, A., and Coursey, B. M. (1988). Calculation of beta-particle counting efficiency for liquid-scintillation systems with three phototubes. *Appl. Radiat, Isot.* **39**, 1191–1196.

Grau Malonda, A., and Fernández, A. (1985). Cálculo de la eficiencia de detección de nucleidos que se desintegran por captura electrónica y emission gamma. *Informe Junta de Energía Nuclear*, JEN, 575, Madrid.

Grau Malonda, A., and García-Toraño, E. (1978). Espectros beta, I. Espectros simples de negatrons. *Informe Junta de Energía Nuclear*, JEN, 427, Madrid.

Grau Malonda, A., and García-Toraño, E. (1981a). Cálculo de la eficiencia de detección en liquidos centelleadores. I. Nucleidos que se desintegran por emission simple de negatrons. *Informe Junta de Energía Nuclear*, JEN, 488, Madrid.

Grau Malonda, A., and García-Toraño, E. (1981b). Espectros beta, II. Espectros simples de positrons. *Informe Junta de Energía Nuclear*, JEN, 489, Madrid.

Grau Malonda, A., and García-Toraño, E. (1982a). Evaluation of counting efficiency in liquid scintillation counting or pure β-ray emitters. *Int. J. Appl. Radiat, Isot.* **33**, 249–253.

Grau Malonda, A., and García-Toraño, E. (1982b). Cálculo de la eficiencia de detección en liquidos centelleadores. II. Nucleidos que se desintegran por emission simple de positrons. *Informe Junta de Energía Nuclear*, JEN, 518, Madrid.

Grau Malonda, A., and Grau Carles, A. (1999). The ionization quench factor in liquid-scintillation counting standardizations. *Appl. Radiat. Isot.* **51**, 183–188.

Grau Malonda, A., and Grau Carles, A. (2000). Standardization of electron-capture radionuclides by liquid scintillation counting. *Appl. Radiat. Isot.* **52**, 657–662.

Grau Malonda, A., and Grau Carles, A. (2002). Half-life determination of ^{40}K by LSC. *Appl. Radiat. Isot.* **56**, 153–156.

Grau Malonda, A., and Los Arcos, J. M. (1983). Un nuevo procedimiento para la Calibración del ^{90}Sr y del ^{90}Y mediante centelleo en fase liquido. *Anales de Física B* **79**, 5–9.

Grau Malonda, A., García-Toraño, E., and Los Arcos, J. M. (1985). Liquid-scintillation counting efficiency as a function of the figure of merit for pure beta particle-emitters. *Int. J. Appl. Radiat, Isot.* **36**, 157–158.

Grau Malonda, A., García-Toraño, E., and Los Arcos, J. M. (1987). Free parameter codes to compute the counting efficiency in liquid scintillators. *Trans. Amer. Nucl. Soc.* **55**, 55–56.

Grau Malonda, A., Rodriguez Barquero, L., and Grau Carles, A. (1994). Radioactivity determination of ^{90}Y, ^{90}Sr and ^{89}Sr mixtures by spectral deconvolution. *Nucl. Instrum. Methods Phys. Res., Sect. A* **339**, 31–37.

Grau Malonda, A., Grau Carles, A., and Grau Carles, P. (1994b). EMI, the counting efficiency for electron capture, electron capture-gamma and isomeric transitions. *Comput. Phys. Commun.* **79**, 115–123.

Grau Malonda, A., Grau Carles, A., Grau Carles, P., and Galiano Casas, G. (1999). EMI2, the counting efficiency for electron capture by a $KL_1L_2L_3M$ model. *Comput. Phys. Commun.* **123**, 114–122.

Grigorescu, L. (1963). Mesure absolue de l'activité des radionucléides par la méthode des coïncidences beta-gamma. Corrections des coincidences instrumentals et de temps morts. MESUCORA Congress, Paris.

Grigorescu, L. (1973). Accuracy of coincidence measurements. *Nucl. Instrum. Methods* **112**, 151-

Grigorescu, E. L., Sahagia, M., Razdolescu, A., Luca, A., and Radwan, R. M. (1998). *Appl. Radiat. Isot.* **49**(9–11), 1165–1170.

Günther, E. W. (1994). Standardization of ^{59}Fe and ^{131}I by liquid scintillation counting. *Nucl. Instrum. Methods Phys. Res., Sect. A* **339**, 402–407.

Günther, E. W. (1996). A simple method for transferring the tritium calibration of an LSC system to other radionuclides. *In* "Liquid Scintillation Spectrometry 1994" (G. T. Cook, D. D. Harkness, A. B. MacKenzie, B. F. Miller and E. M. Scott, Eds.), pp. 373–379. Radiocarbon, Tucson.

Günther, E. (1998). Standardization of the EC nuclides ^{55}Fe and ^{65}Zn with the CIEMAT/NIST LSC tracer method. *Appl. Radiat. Isot.* **49**(9–11), 1055–1060.

Günther, E. (2000). Standardization of ^{237}Np by the CIEMAT/NIST LSC tracer method. *Appl. Radiat. Isot.* **52**, 471–474.

Günther, E. (2001). Computer program CN2001A. Physikalisch-Technische Bundesanstalt, Braunschweig, Germany.

Günther, E. (2002a). What can we expect from the CIEMAT/NIST method? *Appl. Radiat. Isot.* **56**, 357–360.

Günther, E. (2002b) Determination of the activity of β-γ radionuclides usine the CIEMAT/NIST method. Proceedings of the International Symposium LSC 2002, Karlsruhe, May 7–11, 2001. Radiocarbon, Tucson (in press).

Günther, E. (2002c). Determination of the ^{32}P activity in angioplastic balloons by LSC. *Appl. Radiat. Isot.* **56**, 291–293.

Günther, E., and Schötzig, U. (1992). Activity determination of ^{93m}Nb. *Nucl. Instrum. Methods Phys. Res., Sect. A* **312**, 132–135.

Hamanaka, S., Shizuma, K., Wen, X., Iwatani, K., and Hasai, H. (1998). Radon concentration measurement in water by means of α liquid scintillation spectrometry with a PERALS spectrometer. *Nucl. Instrum. Methods Phys. Res., Sect. A* **410**, 314–318.

Harvey, J. A., and Hill, N. W. (1979). Scintillation detectors for neutron physics research. *Nucl. Instrum. Methods* **162**, 507–529.

Harwood Jr., H. J. (1995). Protein farnesyltransferase: measurement of enzymatic activity in a 96-well format using TopCount microplate scintillation counting technology. *Anal. Biochem.* **226**, 268–278.

Heilgeist, M. (2000). Use of extraction chromatography, ion chromatography and liquid scintillation spectrometry for rapid determination of strontium-89 and strontium-90 in food in cases of increased release of radionuclides. *J. Radioanal. Nucl. Chem.* **245**(2), 249–254.

Heltsley, J. H., Brandon, L., Galonsky, A., Heilbronn, L., Remington, B. A., Vandes Molen, A., Yurkon, J., and Kasagi, J. (1988). *Nucl. Instrum. Methods Phys. Res., Sect. A* **263**, 441–445.

Hendee, W. R., Ibbott, G. S., and Crisha, K. L. (1972). ^{3}H-toluene, ^{3}H-water and ^{3}H-hexadecane as internal standards for toluene- and dioxane-based liquid scintillation cocktails. *Int. J. Appl. Radiat. Isot.* **23**, 90–95.

Hereforth, L. (1948). "Die Fluoreszensanregung organischer Substanzen mit Alphateilchen, schnellen Elektronen und Gammastrahlen." Thesis presented Sept. 13, 1948, Technical University, Berlin-Charlottenburg.

Hill, W. A. G. (1996). Use of TopCount[TM] for radioactivity determinations in wipe test. *TopCount Topics*, **TCA-027**, Packard BioScience Company, Meriden, CT 06450, pp. 4.

Hino, Y., and Ohgaki, H. (1998). Absolute measurement of ^{192}Ir. *Appl. Radiat. Isot.* **49**(9–11), 1179–1183.

Hitachi, A., Takahashi, T., Funayama, N., Masuda, K., Kikuchi, J., and Doke, T. (1983). Effect of ionization density on the time dependence of luminescence from liquid argon and xenon. *Phys. Rev. B* **27**(9), 5279–5285.

Homma, Y., and Murakami, Y. (1977). Study on the applicability of the integral counting method for the determination of ^{226}Ra in various sample forms using a liquid scintillation counter. *J. Radioanal. Chem.* **36**, 173–184.

Homma, Y., Murase, Y., and Takiue, M. (1987). Determination of ^{222}Rn by air luminescence method. *J. Radioanal. Nucl. Chem., Letters* **119**, 457–465.

Homma, Y., Murase, Y., and Handa, K. (1993a). Comparison of a modified integral counting method and efficiency tracing method for the determination of ^{222}Rn by liquid scintillation counting. *In* "Liquid Scintillation Spectrometry 1992" (J. E. Noakes, F. Schönhofer, F. and H. A. Polach, Eds.), *Radiocarbon* 59–62.

Homma, Y., Murase, Y., and Handa, K. (1993b). Determination of atmospheric radioactivity using a membrane filter and liquid scintillation spectrometry. *In* "Liquid Scintillation Spectrometry 1992" (J. E. Noakes, F. Schönhofer, F. and H. A. Polach, Eds.), *Radiocarbon* 63–67.

Homma, Y., Murase, Y., and Handa, K. (1994a). The zero detection threshold of a liquid scintillation spectrometer and its application to liquid scintillation counting. *Appl. Radiat. Isot.* **45**, 341–344.

Homma, Y., Murase, Y., and Handa, K. (1994b). Absolute liquid scintillation counting of ^{35}S and ^{45}Ca using a modified integral counting method. *J. Radioanal. Nucl. Chem.* **187**, 367–374.

Homma, Y., Murase, Y., and Handa, K. (1994c). A modified integral counting method and efficiency tracing method for measuring ^{222}Rn by liquid scintillation counting. *Appl. Radiat. Isot.* **45**, 699–702.

Homma, Y., Murase, Y., Handa, K., Koyama, S., Suzuki, N., and Horiuchi, K. (1996). Rapid calibration of detectors for determining ^{222}Rn using air luminescence counting. *In* "Liquid Scintillation Spectrometry 1994" (G. T. Cook, D. D. Harkness, A. B. MacKenzie, B. F. Miller, and E. M. Scott, Eds.), *Radiocarbon* 111–116.

Hong, K. H., Cho, Y. H., Lee, M. H., Choi, G. S., and Lee, C. W. (2001). Simultaneous measurement of ^{89}Sr and ^{90}Sr in aqueous samples by liquid scintillation counting using the spectrum unfolding method. *Appl. Radiat. Isot.* **54**, 299–305.

Horrocks, D. L. (1971). Obtaining the possible maximum of 90 percent efficiency for counting of ^{55}Fe in liquid scintillator solutions. *Int. J. Appl. Radiat. Isot.* **22**, 258–260.

Horrocks, D. L. (1974). "Applications of Liquid Scintillation Counting". Academic Press, New York, pp. 340.

Horrocks, D. L. (1976a). The mechanism of the liquid scintillation process. *In* "Liquid Scintillation Science and Technology" (A. A. Noujaim, C. Ediss, and L. I. Wiebe, Eds.), pp. 1–16. Academic Press, New York and London.

Horrocks, D. L. (1976b). Absolute disintegration rate determination of beta-emitting radionuclides by the pulse height shift extrapolation method. *In* "Liquid Scintillation Science and Technology" (A. A. Noujaim, C. Ediss, and L. I. Wiebe, Eds.), pp. 185–198. Academic Press, New York and London.

Horrocks, D. L. (1976c). Measurement of ^{125}I by liquid scintillation methods. *Nucl. Instrum. Methods* **133**, 293–301.

Horrocks, D. L. (1977). "The H-Number Concept", Publication No. 1095 NUC-77-IT. Beckman Instruments Inc., Irvine, pp. 26.

Horrocks, D. L. (1978a). A new method of quench monitoring in liquid scintillation counting: the H number concept. *In* "Liquid Scintillation Counting" (M. A. Crook and P. Johnson, Eds.), Vol. 5, pp. 145–168. Heyden, London.

Horrocks, D. L. (1978b). A new method of quench monitoring in liquid scintillation counting: the H number concept. *J. Radioanal. Chem.* **43**, 489–521.

Horrocks, D. L. (1980). Effect of quench on the pulse height distribution for tritium-containing samples – high quench levels. *In* "Liquid Scintillation Counting, Recent Applications and Development" (C.-T. Peng, D. L. Horrocks, and E. L. Alpen, Eds.), Vol. I, pp. 199–210, Academic Press, New York and London.

Horrocks, D. L., and Kolb, A. J. (1981). Instrumental methods for detecting some common problems in liquid scintillation counting. *Lab. Pract.* **30**, 485–487.

Horváth, Á., Ieki, K., Iwata, Y., Kruse, J. J., Seres, Z., Wang, J., Weiner, J., Zecher, P. D., and Galonsky, A. (2000). Comparison of two liquid scintillators used for neutron detection. *Nucl. Instrum. Methods Phys. Res., Sect. A* **440**, 241–244.

Horwitz, E. P., Chiarizia, R., and Dietz, M. L. (1997). DIPEX: a new extraction chromatographic material for the separation and preconcentration of actinides from aqueous solution. *Reactive & Functional Pol.* **33**, 25–36.

Houtermans, H., and Miguel, M. (1962). 4π-β-γ coincidence counting for the calibration of nuclides with complex decay schemes. *Int. J. Appl. Radiat. Isot.* **13**, 137–142.

Hughes, K. T., Ireson, J. C., Jones, N. R. A., and Kivela, P. (2001). Color quench correction in scintillation proximity assays using Paralux Count Mode. *Application Note*, pp. 12, PerkinElmer Life Sciences, Boston, MA.

Hult, M., Altzitzoglou, T., Denecke, B., Persson, L., Sibbens, G., and Reher, D. F. G. (2000). Standardization of ^{204}Tl at IRMM. *Appl. Radiat. Isot.* **52**, 493–498.

Hurtgen, C., Jerome, S., and Woods, M. (2000). Revisiting Curie – how low can you go? *Appl. Radiat. Isot.* **53**, 45–50.

Hwang, H.-Y., Park, J. H., Park, T. S., Lee, J. M., Cho, Y. H., Byun, J. I., Choi, O., Jun, J.-S., Lee, M. H., and Lee, C. W. (2002). Development of MCTS technique for 3-PM liquid scintillation counting. *Appl. Radiat. Isot.* **56**, 307–313.

Ishikawa, H., and Takiue, M. (1973). Liquid scintillation measurement for β-ray emitters followed by γ-rays. *Nucl. Instrum. Methods* **112**, 437–442.

Ishikawa, H., Takiue, M., and Aburai, T. (1984). Radioassay by an efficiency tracing technique using a liquid scintillation counter. *Int. J. Appl. Radiat. Isot.* **35**, 463–466.

Johansson, L., Sibbens, G., Altzitzoglou, T., and Denecke, B. (2002). Self-absorption correction in standardization of ^{204}Tl. *Appl. Radiat. Isot.* **56**, 199–203.

Jordan, W. C., Spiehler, V., Haendiges, R., and Helman, E. Z. (1974). Evaluation of alternative counting methods for radioimmunoassay of hepatitis-associated antigen (HB Ag). *Clin. Chem.* **20**(7), 733–737.

Kaihola, L. (2000). Radionuclide identification in liquid scintillation alpha-spectroscopy. *J. Radioanal. Nucl. Chem.* **243**(2), 313–317.

Kallman, H. (1950). Scintillation counting with solutions. *Phys. Rev.* **78**(5), 621–622.

Kaneko, S. *et al.* (1992). Quantification of hepatitis C RNA by competitive polymerase chain reaction. *J. Med. Virol.* **37**, 278–282.

Kashirin, I.A., Ermakov, A. I., Malinovskiy, S. V., Belanov, S. V., Sapozhnikov, Yu. A., Efimov, K. M., Tikhomirov, V. A., and Sobolev, A. I. (2000). Liquid scintillation determination of low level components in complex mixtures of radionuclides. *Appl. Radiat. Isot.* **53**, 303–308.

Kessler, M. J., Ed. (1989). "Liquid Scintillation Analysis, Science and Technology". Publ. No. 169–3052, PerkinElmer Life and Analytical Sciences, Boston.

Kessler, M. J. (1991). Applications of quench monitoring using transformed external standard spectrum (tSIE). *In* "Liquid Scintillation Counting and Organic Scintillators" (Harley Ross, John E. Noakes and Jim D. Spaulding, Eds.), pp. 343–364, Lewis Publishers, Chelsea, MI 48118.

Kessler, M. J. (1991). Absolute activity liquid scintillation counting: an attractive alternative to quench-corrected DPM for higher energy isotopes. *In* "Liquid Scintillation Counting and Organic Scintillators" (Harley Ross, John E. Noakes and Jim D. Spaulding, Eds.), pp. 647–653, Lewis Publishers, Chelsea, MI 48118.

Kits, J., Látalová, M., Látal, F., and Zich, O. (1985). Determination of the activity of ^{125}I by liquid scintillation measurement. *Int. J. Appl. Radiat. Isot.* **36**, 320.

Kobayashi, Y., and Maudsley, D. V. (1970). Practical aspects of double isotope counting. *In* "Current Status of Liquid Scintillation Counting" (E. D. Bransome Jr., Ed.), pp. 76–78. Grune and Stratton, New York.

Klein, H., and Neumann, S. (2002). Neutron and photon spectrometry with liquid scintillation detectors in mixed fields. *Nucl. Instrum. Methods Phys. Res., Sect. A* **476**, 132–142.

Kouru, H. (1991). A new quench curve fitting procedure: fine tuning of a spectrum library. *In* "Liquid Scintillation Counting and Organic Scintillators" (Harley Ross, John E. Noakes and Jim D. Spaulding, Eds.), pp. 247–255, Lewis Publishers, Chelsea, MI 48118.

Kouru, H., and Rundt, K. (1991). Multilabel counting using digital overlay technique. *In* "Liquid Scintillation Counting and Organic Scintillators" (Harley Ross, John E. Noakes and Jim D. Spaulding, Eds.), pp. 239–246, Lewis Publishers, Chelsea, MI 48118.

Kubota, S., Nakamoto, A., Takahashi, T., Hamada, T., Shibamura, E., Miyajima, M., Masuda, K., and Doke, T. (1978a). Recombination luminescence in liquid argon and in liquid xenon. *Phys. Rev. B* **17**(6), 2762–2765.

Kubota, S., Hishida, M., and Ruan, J.-Z. (1978b). Evidence for a triplet state of the self-trapped excitation states in liquid argon, krypton and xenon. *J. Phys. C* **11**, 2645–2651.

Kubota, S., Hishida, M., Suzuki, M., and Ruan, J.-Z. (1979). Dynamical behavior of free electrons in the recombination process in liquid argon, krypton, and xenon. *Phys. Rev. B* **20**(8), 3486–3496.

Laney, B. H. (1976). External standard method of quench correction: advanced techniques. *In* "Liquid Scintillation Science and Technology" (A. A. Noujaim, C. Ediss, and L. I. Wiebe, Eds.), pp. 135–152. Academic Press, New York and London.

Laney, B. H. (1977). Two-parameter pulse height analysis in liquid scintillation. *In* "Liquid Scintillation Counting", (M. A. Crook and P. Johnson, Eds.), Vol. 4, pp. 74–84. Heyden, London.

L'Annunziata, M. F. (1979). "Radiotracers in Agricultural Chemistry". Academic Press, London, pp. 536.

L'Annunziata, M. F. (1984a). The detection and measurement of radionuclides. *In* "Isotopes and Radiation in Agricultural Sciences" (M. F. L'Annunziata and J. O. Legg, Eds.), Vol 1, pp. 141–231. Academic Press, London.

L'Annunziata, M. F. (1984b). Agricultural biochemistry: reaction mechanisms and pathways in biosynthesis. *In* "Isotopes and Radiation in Agricultural Sciences" (M. F. L'Annunziata and J. O. Legg, Eds.), Vol 2, pp. 105–182. Academic Press, London.

L'Annunziata, M. F. (1987). "Radionuclide Tracers, Their Detection and Measurement". Academic Press, San Diego. pp. 505.

L'Annunziata, M. F. (1997a). Efficiency tracing DPM (ET-DPM) and Direct-DPM – Instrument performance data. *Counter Intel. Tri-Carb LSC Application Note*, PerkinElmer Life and Analytical Sciences, Boston, pp. 8.

L'Annunziata, M. F. (1997b). Comparison of conventional and full spectrum DPM (FS-DPM) analysis of $^{33}P-^{32}P$ double labels – Instrument performance data. *Counter Intel. Tri-Carb LSC Application Note*, PerkinElmer Life and Analytical Sciences, Boston, pp. 6.

Larsson, J., Wingårdh, K., Berggård, T., Davies, J. R., Lögdberg, L., Strand, S.-E., and Åkerström, B. (2001). Distribution of iodine-125-labeled α_1-microglobulin in rats after intravenous injection. *J. Lab. Clin. Med.* **137**(3), 165–175.

Lee, M. H., Chung, K. H., Choi, G. K., and Lee, C. W. (2002). Measurement of ^{90}Sr in aqueous samples using liquid scintillation counting with full spectrum DPM method. *Appl. Radiat. Isot.* **57**(2), 257–263.

Lieberman, M. M. (1995). Determination of neutrophil activation by chemiluminescence using the Packard TopCount^{TM} microplate scintillation and luminescence counter. *TopCount Topics*, **TCA-022**, Packard BioScience Company, Meriden, CT 06450, pp. 2.

Lorber, A., Wangen, L. E., and Kowalski, B. R. (1986). A theoretical foundation for the PLS algorithm. *J. Chemom.* **1**(1), 19–31.

Los Arcos, J. M., and Ortiz, F. (1997). kB: a code to determine the ionization quench function $Q(E)$ as a function of the kB parameter. *Comp. Phys. Commun.* **103**, 83–94.

Los Arcos, J. M., Grau Molanda, A., and Fernandez, A. (1987). VIASKL: a computer program to evaluate the liquid scintillation counting efficiency and its associated uncertainty for K-L-atomic shell electron-capture nuclides. *Comput. Phys. Commun.* **44**, 209–220.

Los Arcos, J. M., Grau, A., and García-Toraño, E. (1991). LSC standardization of multigamma electron-capture radionuclides by the efficiency tracing method. *In* "Liquid Scintillation Counting and Organic Scintillators" (Harley Ross, John E. Noakes, and Jim D. Spaulding, Eds.), pp. 611–622, Lewis Publishers, Chelsea, MI 48118.

Malkov, V. A., Panyutin, I. G., Neumann, R. D., Zhurkin, V. B., and Camerini-Otero, R. D. (2000). Radioprobing of a recA-three-stranded DNA complex with iodine-125: evidence for recognition of homology in the major groove of the target duplex. *J. Molec. Biol.* **299**(3), 629–640.

Mann, W. B., Rytz, A., Spernol, A., and McLaughlin, W. L. (1988). "Radioactivity Measurements, Principles and Practice". Pergamon Press, Oxford, pp. 937.

Matsui, Y., and Takiue, M. (1991). Liquid scintillation radioassay of multi-labeled beta-emitters. *Appl. Radiat. Isot.* **42**, 841–845.

McDowell, W. J. (1996). Recent applications of PERALS® spectrometry. *In* "Liquid Scintillation Spectrometry 1994" (G. T. Cook, D. D. Harkness, A. B. MacKenzie, B. F. Miller and E. M. Scott, Eds.), pp. 157–165. Radiocarbon, Tucson.

McDowell, W. J., and McDowell, B. L. (1991). Liquid scintillation alpha spectrometry: a method for today and tomorrow. *In* "Liquid Scintillation Counting and Organic Scintillators" (Harley Ross, John E. Noakes, and Jim D. Spaulding, Eds.), pp. 105–122, Lewis Publishers, Chelsea, MI 48118.

McDowell, W. J., and McDowell, B. L. (1993). The growth of a radioanalytical method: alpha liquid scintillation spectrometry. *In* "Liquid Scintillation Spectrometry 1992" (J. E. Noakes, F. Schönhofer, and H. A. Polach, Eds.), pp. 193–200. Radiocarbon, Tucson.

McKlveen, J. W., and McDowell, W. J. (1984). Liquid scintillation alpha spectrometry techniques. *Nucl. Instrum. Methods Phys. Res.* **223**, 372–376.

McQuarrie, S. A., Wiebe, L. I., and Ediss, C. (1980). *In* "Liquid Scintillation Counting, Recent Applications and Development" (C.-T. Peng, D. L. Horrocks, and E. L. Alpen, Eds.), pp. 291–299. Academic Press, New York and London.

McQuarrie, S. A., and Noujaim, A. A., Ediss, C., and Wiebe, L. I. (1981). *Trans. Amer. Nucl. Soc.* **39**, 27–28.

McQuarrie, S. A., and Noujaim, A. A. (1983). ⁶⁷Ga: a novel internal standard for LSC. *In* "Advances in Scintillation Counting" (S. A. McQuarrie, C. Ediss and L. I. Wiebe, Eds.), pp. 57–65, University of Alberta Printing Services, Edmonton.

Miller, M., Kereiakes, J. G., and Friedman, B. I. (1969). Determination of iron-59 and iron-55 in [blood] plasma using liquid scintillation counting. *Int. J. Appl. Radiat. Isot.* **20**(2), 133–135.

Miyajima, M., Sasaki, S., Tawara, H., and Shibamura, E. (1992). Absolute number of scintillation photons in liquid xenon by alpha-particles. *IEEE Trans. Nucl. Sci.* **39**(4), 536–540.

Morita-Murase, Y., Murakami, I., and Homma, Y. (2001). The air luminescence count for the rapid determination of ²²²Rn in a liquid scintillation spectrometer. *Chem. Lett.* **2001**, 238–239.

Murase, Y., Homma, Y., and Takiue, M. (1989a). Effect of air luminescence counts on determination of ²²²Rn by liquid scintillation counting. *Appl. Radiat. Isotopes* **40**, 295–298.

Murase, Y., Homma, Y., Takiue, M., and Aburai, T. (1989b). Determination of air luminescence spectra for alpha-emitters with liquid scintillation spectrometers. *Appl. Radiat. Isotopes* **40**, 291–294.

Murase, Y., Homma, Y., Murakami, I., and Handa, K. (1998). Assay of ²²²Rn in water samples by a modified integral counting method. *Appl. Radiat. Isot.* **49**(7), 861–865.

Murase, Y., Homma, Y., Murakami, I., Handa, K., and Horiuchi, K. (1999). Indoor ²²²Rn measurements using an activated charcoal detector. *Appl. Radiat. Isot.* **50**, 561–565.

Natake, T., Takiue, M., and Fujii, H. (1996). Nuclide identification for pure beta-emitting radionuclides with very similar beta end-point energies using a liquid scintillation spectrometer. *Nucl. Instrum. Methods Phys. Res., Sect. A* **378**, 506–510.

Nayak, D. (2001). Multitracer techniques: applications in chemical and life sciences. *Appl. Radiat. Isot.* **54**, 195–202.

Nakao, N., Kurosawa, T., Nakamura, T., and Uwamino, Y. (2001). Absolute measurements of the response function of an NE213 organic liquid scintillator for the neutron energy range up to 206 MeV. *Nucl. Instrum. Methods Phys. Res., Sect. A* **463**, 275–287.

NCRP (1985). "A Handbook of Radioactivity Measurements Procedures". Report No. 64. National Council on Radiation Protection and Measurements, Washington, DC 20014.

Neumann, K. E., Roessler, N., and ter Wiel, J. (1991). Safe scintillation chemicals for high efficiency, high throughput counting. In "Liquid Scintillation Counting and Organic Scintillators" (Harley Ross, John E. Noakes, and Jim D. Spaulding, Eds.), pp. 35–41, Lewis Publishers, Chelsea, MI 48118.

Neumann, S., Böttger, R., Guldbakke, S., Matzke, M., and Sosaat, W. (2002). Neutron and photon spectrometry in monoenergetic neutron fields. Nucl. Instrum. Methoods Phys. Res., Sect. A **476**, 353–357.

Niese, S. (1999). The discovery of organic solid and liquid scintillators by H. Kallmann and L. Herforth 50 years ago. J. Radioanal. Nucl. Chem. **241**(3), 499–501.

Noor, A., Kasim, N., and L'Annunziata, M. F. (1995). Application of pulse height spectral analysis to double-label counting of ^{35}S-^{32}P. Appl. Radiat. Isot. **46**, 791–797.

Noor, A., Zakir, M., Burhanuddin, R., Maming, and L'Annunziata, M. F. (1996a). Cerenkov and liquid scintillation analysis of the triple label ^{86}Rb-^{35}S-^{33}P. Appl. Radiat. Isot. **47**, 659–668.

Noor, A., Zakir, M., Burhanuddin, R., Kasim, N., Nurr, L. A., Anthony, Maming, Agung, M., and L'Annunziata, M. F. (1996b). Pulse height spectral analysis of ^{3}H:^{14}C ratios. Appl. Radiat. Isot. **47**, 767–775.

Novotny, T., Büermann, L., Guldbakke, S., and Klein, H. (1997). Response of nE213 liquid scintillation detectors to high-energy photons (7 MeV $< E_\gamma <$ 20 MeV). Nucl. Instrum. Methods Phys. Res., Sect. A **400**, 356–366.

Okita, G. T., Kabara, J. J., Richardson, F., and LeRoy, G. V. (1957). Assaying compounds containing ^{3}H and ^{14}C. Nucleonics **15**, 111–114.

Ortiz, J. F., Los Arcos, J. M., and Grau Malonda, A. (1993). CIEMAT/NIST standardization method extended to anode outputs for beta and electron-capture nuclides. In "Liquid Scintillation Spectrometry 1992" (J. E. Noakes, F. Schönhofer and H. A. Polach, Eds.), pp. 261–267. Radiocarbon, Tucson.

Park, T. S., Hwang, H. Y., and Lee, J. M. (1998). An improved coincidence counting technique for standardization of radionuclides. Appl. Radiat. Isot. **49**(9–11), 1147–1151.

Parus, J. L., Raab, W., and Radoszewski, T. (1993). Liquid scintillation counting of plutonium and/or americium concentrations. In "Liquid Scintillation Spectrometry 1992" (J. E. Noakes, F. Schönhofer, and H. A. Polach, Eds.), pp. 233–237. Radiocarbon, Tucson.

Passo, Jr., C. J. (1996). Time-resloved pulse decay analysis (TR-PDA): a refinement for alpha/beta discrimination. In "Liquid Scintillation Spectrometry 1994" (G. T. Cook, D. D. Harkness, A. B. MacKenzie, B. F. Miller, and E. M. Scott, Eds.), pp. 37–41. Radiocarbon, Tucson.

Passo, Jr., C. J., and Cook, G. T. (1994). "Handbook of Environmental Liquid Scintillation Spectrometry, A Compilation of Theory and Methods". Publ. No. PMC0387. PerkinElmer Life and Analytical Sciences, Boston.

Passo, Jr., C. J., and Kessler, M. J. (1993). Selectable delay before burst – a novel feature to enhance low-level counting performance. In "Liquid Scintillation Spectrometry 1992" (J. E. Noakes, F. Schönhofer, and H. A. Polach, Eds.), pp. 51–57. Radiocarbon, Tucson.

Passo, Jr., C. J., and Roberts, D. J. (1996). Expanded energy for time-resolved liquid scintillation counting: an enhancement for programmable TR-LSC®. In "Liquid Scintillation Spectrometry 1994" (G. T. Cook, D. D. Harkness, A. B. MacKenzie, B. F. Miller, and E. M. Scott, Eds.), pp. 67–74. Radiocarbon, Tucson.

Pates, J. M., Cook, G. T., MacKenzie, A. B., and Thomson, J. (1993). The development of alpha/beta separation liquid scintillation cocktail for aqueous samples. J. Radioanal. Nucl. Chem., Articles **172**, 341–348.

Pates, J. M., Cook, G. T., and MacKenzie, A. B. (1996a). Alpha/beta separation liquid scintillation spectrometry: current trends. In "Liquid Scintillation Spectrometry 1994" (G. T. Cook, D. D. Harkness, A. B. MacKenzie, B. F. Miller, and E. M. Scott, Eds.), pp. 267–281. Radiocarbon, Tucson.

Pates, J. M., Cook, G. T., MacKenzie, A. B., and Thomson, J. (1996b). The effect of cocktail fluors on pulse shapes and alpha/beta separation liquid scintillation spectrometry. In "Liquid Scintillation Spectrometry 1994" (G. T. Cook, D. D. Harkness, A. B. MacKenzie, B. F. Miller, and E. M. Scott, Eds.), pp. 317–326. Radiocarbon, Tucson.

Pates, J. M., Cook, G. T., MacKenzie, A. B., and Passo, Jr., C. J. (1996c). Quenching and its effect on alpha/beta separation liquid scintillation spectrometry. *In* "Liquid Scintillation Spectrometry 1994" (G. T. Cook, D. D. Harkness, A. B. MacKenzie, B. F. Miller, and E. M. Scott, Eds.), pp. 75–85 Radiocarbon, Tucson.

Pates, J. M., Cook, G. T., MacKenzie, A. B., and Passo, Jr., C. J. (1998). Implications of beta energy and quench level for alpha/beta liquid scintillation spectrometry calibration. *Analyst* **123**, 2201–2207.

Peng, C. T. (1976). Chemiluminescence. *In* "Liquid Scintillation Science and Technology" (A. A. Noujaim, C. Ediss, and L. I. Wiebe, Eds.), pp. 313–329. Academic Press, New York and London.

Perry, S. W., and Warner, G. T. (1963). A method of sample preparation for the estimation of 55Fe in whole blood by the liquid scintillation counting technique. *Int. J. Appl. Radiat. Isot.* **14**, 397–400.

Peurrung, A. J. (2000). Recent development in neutron detection. *Nucl. Instrum. Methods Phys. Res., Sect. A* **443**, 400–415.

Plautz, J. D., Strayer, C. A., and Kay, S. A. (1996). Automated recording of luciferase-reported gene transcription in living seedlings and fruit flies. *TopCount Topics*, **TCA-025**, Packard BioScience Company, Meriden, CT 06450, pp. 4.

Pochwalski, K., Broda, R., and Radoszewski, T. (1988). Standardization of pure beta emitters by liquid-scintillation counting. *Appl. Radiat. Isot.* **39**, 165–172.

Poenaru, D. N., and Greiner, W., Eds. (1997). Experimental Techniques in Nuclear Physics. Walter de Gruyter, Berlin.

Prichard, H. M., Venso, E. A., and Dodson, C. L. (1992). Liquid scintillation analysis of ^{222}Rn in water by alpha-beta discrimination. *Radioact. Radiochem.* **3**(1), 28–36.

Pujol, L., and Sanchez-Cabeza, J.-A. (1997). Role of quenching on alpha/beta separation in liquid scintillation counting for several high capacity cocktails. *Analyst* **122**, 383–385.

Pye, D. A., Vives, R. R., Turnbull, J. E., Hyde, P., and Gallagher, J. T. (1999). Heparan sulfate oligosaccharides require 6-O-sulfation for promotion of basic fibroblast growth factor mitogenic activity. *J. Biol. Chem.* **273**(36), 22936–22942.

Rausch, C., Bucherl, T., Gaehler, R., Von Seggern, H., and Winnaker, A. (1993). Recent developments in neutron detection. *Proc. SPIE* **1737**, 255–263.

Razdolescu, A. C., Grigorescu, M., Sahagia, M., Luca, A., and Ivan, C. (2000). Standardization of ^{169}Yb by the $4\pi\beta$-γ method. *Appl. Radiat. Isot.* **52**, 505–507.

Reginatto, M., Goldhagen, P., and Neumann, S. (2002). Spectrum unfolding, sensitivity analysis and propagation of uncertainties with the maximum entropy deconvolution code MAXED. *Nucl. Instrum. Methods Phys. Res., Sect. A* **476**, 242–246.

Reynolds, G. T., Harrison, F. B., and Salvini, G. (1950). Liquid scintillation counters. *Phys. Rev.* **78**(4), 488.

Rhodes, B. A. (1965). Liquid scintillation counting of radioiodine. *Anal. Chem.* **37**(8), 995–997.

Ring, J. G., Nguyen, D. C., and Everett, L. J. (1980). *In* "Liquid Scintillation Counting, Recent Applications and Development", "Physical Aspects" (C.-T. Peng, D. L. Horrocks, and E. L. Alpen, Eds.), Vol. I, pp. 89–104. Academic Press, New York and London.

Rodríguez Barquero, L., and Grau Carles, A. (1998). The influence of the primary solute on alpha/beta discrimination. *Appl. Radiat. Isot.* **49**(9–11), 1065–1068.

Rodríguez Barquero, L., and Los Arcos, J. M. (1996). ^{41}Ca standardization by the CIEMAT/NIST LSC method. *Nucl. Instrum. Methods Phys. Res., Sect. A* **369**, 353–358.

Rodríguez Barquero, L., and Los Arcos, J. M. (2000). Measurement of ^{49}V half-life. *Appl. Radiat. Isot.* **52**, 569–571.

Rodríguez Barquero, L., Grau Malonda, A., Los Arcos Merino, J. M., and Suarez Contreras, C. (1989). Preparación y calibración por centelleo liquido de una muestra de Cl-36. *Anales de Física B* **85**, 55–69.

Rodríguez, L., Los Arcos, J. M., and Grau, A. (1992). LSC standardization of ^{54}Mn in inorganic and organic samples by the CIEMAT/NIST efficiency tracing method. *Nucl. Instrum. Methods Phys. Res., Sect. A* **312**, 124–131.

Rodríguez, L., Los Arcos, J. M., and Grau Malonda, A. and Garcia-Toraño, E. (1994). LSC standardization of ^{45}Ca by the CIEMAT/NIST efficiency tracing method. *Nucl. Instrum. Methods Phys. Res., Sect. A*, **339**, 6–13.

Rodríguez, L., Los Arcos, J. M., Ortiz, F., and Jiménez, A. (1998). ^{49}V standardization by the CIEMAT/NIST LSC method. *Appl. Radiat. Isot.* 49(9–11), 1077–1082.

Roeland, C. H., and Burns, D. A. (1995). Introduction to CytoLite^{TM}; a chemiluminescent reagent system for cell proliferation and cytotoxicity assays. *Lite-Guides* LGA-001, Packard BioScience Company, Meriden, CT 06450, pp. 6.

Roessler, N., Valenta, R. J., and van Cauter, S. (1991). Time-resolved liquid scintillation counting. *In* "Liquid Scintillation Counting and Organic Scintillators" (Harley Ross, John E. Noakes, and Jim D. Spaulding, Eds.), pp. 501–511, Lewis Publishers, Chelsea, MI 48118.

Rundt, K. (1991). The effect of quench curve shape of the solvent and quencher in a liquid scintillation counter. *In* "Liquid Scintillation Counting and Organic Scintillators" (Harley Ross, John E. Noakes, and Jim D. Spaulding, Eds.), pp. 257–268, Lewis Publishers, Chelsea, MI 48118.

Rundt, K., and Kouru, H. (1989). Liquid scintillation counter for measuring the activity of radioactive samples containing a multiple of radioactive isotopes. US Patent No. 4,918,310.

Rundt, K., and Kouru, H. (1992). Apparatus and a method for measuring the activity of radioactive samples containing a multiple of radioactive isotopes. US Patent No. 5,134,294.

Sandhya, D., and Subramanian, M. S. (1998). Radiometric determination of trace amounts of zinc using liquid scintillation counting. *Talanta* 46, 921–926.

Satoh, D., Shigyo, N., Iwamoto, Y., Kitsuki, H., and Ishibashi, K. (2001). Study of neutron detection efficiencies for liquid organic scintillator up to 3 GeV. *IEEE Trans. Nucl. Sci.* 48(4), 1165–1167.

Scheirer, W., Roelant, C., and Burns, D. A. (1994). Introduction to LucLite^{TM}; a bioluminescent reagent system for reporter gene assays. *Lite-Guides* LGA-002, Packard BioScience Company, Meriden, CT 06450, p. 5.

Schmidt, D., Asselineau, B., Böttger, R., Klein, H., Lebreton, L., Neumann, S., Nolte, R., and Pichenot, G. (2002). *Nucl. Instrum. Methods Phys. Res., Sect. A* 476, 186–189.

Schneider, E. W., and Verbrugge, M. W. (1993). Radiotracer method for simultaneous measurement of cation, anion and water transport through ion-exchange membranes. *Appl. Radiat. Isot.* 44, 1399–1408.

Schötzig, U., Schönfeld, E., Günther, E., Klein, R., and Schrader, H. (1999). Standardization and decay data of ^{153}Sm. *Appl. Radiat. Isot.* 51, 169–175.

Schötzig, U., Schrader, H., Schönfeld, E., Günther, E., and Klein, R. (2001). Standardization and decay data of ^{177}Lu and ^{188}Re. *Appl. Radiat. Isot.* 55, 89–96.

Schwerdtel, E. (1965). *G.I.T. Fachzeitschrift für das Laboratorium* 9, 881.

Schwerdtel, E. (1966a). Recent developments in liquid scintillation counting. *Kerntechnik* 8, 517–520.

Schwerdtel, E. (1966b). Simple method for an exact efficiency determination in liquid scintillation counting of low-energy β-emitters. *Atomkernenergie* 11, 324–325.

Simpson, B. R. S. (2002). Radioactivity standardization in South Africa. *Appl. Radiat. Isot.* 56, 301–305.

Simpson, B. R. S., and Meyer, B. R. (1988). A multiple-channel 2- and 3-fold coincidence counting system for radioactivity standardization. *Nucl. Instrum. Methods Phys. Res., Sect. A* 263, 436–440.

Simpson, B. R. S., and Meyer, B. R. (1992a). The standardization of ^{201}Tl by liquid scintillation coincidence counting. National Accelerator Centre Report. NAC/92–01, PO Box 72, Faure, 7131 South Africa.

Simpson, B. R. S., and Meyer, B. R. (1992b). Direct determination of the activity of non-gamma-emitting radionuclides by the TDCR efficiency calculation technique: a review of the present status. Report NAC/92–02, September 1992, National Accelerator Centre, PO Box 72, Faure, 7131 South Africa. pp 18.

Simpson, B. R. S., and Meyer, B. R. (1992c). Further investigations of the TDCR efficiency calculation technique for the direct determination of activity. *Nucl. Instrum. Methods Phys. Res., Sect. A* 312, 90–94.

Simpson, B. R. S., and Meyer, B. R. (1994a). Standardization and half-life of ^{201}Tl by the $4\pi(x,e)-\gamma$ coincidence method with liquid scintillation counting in the 4π-channel. *Appl. Radiat. Isot.* 45, 669–673.

Simpson, B. R. S., and Meyer, B. R. (1994b). Direct activity measurement of pure beta-emitting radionuclides by the TDCR efficiency calculation technique. *Nucl. Instrum. Methods Phys. Res., Sect. A* **339**, 14–20.

Simpson, B. R. S., and Meyer, B. R. (1996). Activity measurement of ^{204}Tl by direct liquid scintillation measurements. *Nucl. Instrum. Methods Phys. Res., Sect. A* **369**, 340–343.

Simpson, B. R. S., and van Oordt, E. J. (1997). Data acquisition program for the NAC radioactivity standards laboratory. NAC Report NAC/97–03, National Accelerator Centre, PO Box 72, Faure, 7131 South Africa.

Solovov, V. N., Chepel, V., Lopes, M. I., Ferreira Marques, R., and Policarpo, A. J. P. L. (2000). Study of large area avalanche photodiode for detecting liquid xenon scintillation. *IEEE Trans. Nucl. Sci.* **47**(4), 1307–1310.

Steyn, J. (1956). Absolute standardization of beta-emitting isotopes with a liquid scintillator counter. *Proc. Phys. Soc. A* **69**, 865–867.

Suárez, J. A., Rodríguez, M., Espartero, A. G., and Piña, G. (2000). Radiochemical analysis of ^{41}Ca and ^{45}Ca. *Appl. Radiat. Isot.* **52**, 407–413.

Takiue, M. (1980). Simple and rapid measurement of α-rays on smear samples using air luminescence. *Health Phys.* **39**, 29–32.

Takiue, M., and Ishikawa, H. (1974). Quenching analysis of liquid scintillation. *Nucl. Instrum. Methods* **118**, 51–54.

Takiue, M., and Ishikawa, H. (1978). Thermal neutron reaction cross section measurements for fourteen nuclides with a liquid scintillation spectrometer. *Nucl. Instrum. Methods* **148**, 157–161.

Takiue, M., and Ishikawa, H. (1979). α-ray measurement due to air luminescence employing a liquid scintillation spectrometer. *Nucl. Instrum. Methods* **159**, 139–143.

Takiue, M., Natake, T., and Hayashi, M. (1983). Double ratio technique for determining the type of quenching in liquid scintillation measurement. *Int. J. Appl. Radiat. Isot.* **34**, 1483–1485.

Takiue, M., Hayashi, M., Natake, T., and Ishikawa, H. (1984). Elimination of chemiluminescence in liquid scintillation measurement. *Nucl. Instrum. Methods Phys. Res.* **219**, 192–195.

Takiue, M., Natake, T., Hayashi, M., and Yoshizawa, Y. (1985). Analytical subtraction of chemiluminescence counts for dual-labeled samples in liquid scintillation measurement. *Int. J. Appl. Radiat. Isot.* **36**, 285–289.

Takiue, M., Natake, T., Fujii, H., and Ishikawa, H. (1986). Modification of a dual-label analysis data processing system for chemiluminescence corrections in liquid scintillation counting. *Nucl. Instrum. Methods Phys. Res., Sect. A* **247**, 395–398.

Takiue, M., Fujii, H., Aburai, T., and Ishikawa, H. (1989a). Usefulness of water-soluble paper for smear test of low-energy β- and α-emitters using a liquid scintillation technique. *Health Phys.* **56**, 367–371.

Takiue, M., Natake, T., and Fujii, H. (1989b). Nuclide identification of β-emitter by a double ratio technique using a liquid scintillation counter. *Nucl. Instrum. Methods Phys. Res., Sect. A* **274**, 345–348.

Takiue, M., Fujii, H., and Homma, Y. (1990a). Reliability of the activity determined by the quenching correction method for two groups of emulsion scintillators. *Appl. Radiat. Isot.* **41**, 195–198.

Takiue, M., Matsui, Y., Natake, T., and Yoshizawa, Y. (1990b). A new approach to analytical radioassay of multiple beta-labeled samples using a liquid scintillation spectrometer. *Nucl. Instrum. Methods Phys. Res., Sect. A* **293**, 596–600.

Takiue, M., Matsui, Y., and Fujii, H. (1991a). Activity determination of simultaneously chemical and color quenched samples using a liquid scintillation counter. *Appl. Radiat. Isot.* **42**(3), 241–244.

Takiue, M., Matsui, Y., and Fujii, H. (1991b). Liquid scintillation radioassay for multiple radionuclide mixtures by the most probable value theory. *J. Radioanal. Nucl. Chem.* **152**, 227–236.

Takiue, M., Fujii, H., Natake, T., and Matsui, Y. (1991c). Analytical measurements of multiple beta-emitter mixtures with a liquid scintillation spectrometer. *J. Radioanal. Nucl. Chem., Letters* **155**, 183–193.

Takiue, M., Matsui, Y., Natake, T., and Fujii, H. (1992). Nuclide identification of pure-beta emitter mixtures with liquid scintillation spectrometry. *Appl. Radiat. Isot.* **43**, 853–857.

Takiue, M., Natake, T., and Fujii, H. (1995). Liquid scintillation radioassay for low-activity beta-emitter mixtures by the method of least squares. *J. Radioanal. Nucl. Chem., Letters* **200**, 247–258.

Takiue, M., Natake, T., and Fujii, H. (1999). A hybrid radioassay technique for multiple beta-emitter mixtures using liquid and NaI(Tl) scintillation spectrometers. *Appl. Radiat. Isot.* **51**, 429–434.

Tanaka, M., Doke, T., Hitachi, A., Kato, T., Kikuchi, J., Masuda, K., Murakami, T., Nishikido, F., Okada, H., Ozaki, K., Shibamura, E., and Yoshihira, E. (2001). LET dependence of scintillation yields in liquid xenon. *Nucl. Instrum. Methods Phys. Res., Sect. A* **457**, 454–463.

Tauler, R., Izquierdo-Ridorsa, A., and Casassas, E. (1991). Comparison of multivariate calibration methods applied to the spectrophotometric study of mixtures of purine and pyrimidine bases. *An. Quím.* **87**, 571–579.

Teresa, M., Robela, C. P., Camargo, I. M. C., Oliveira, J. E., and Bartolini, P. (2000). Single-step purification of recombinant human growth hormone (hGH) directly from bacterial osmotic shock fluids, for the purpose of ^{125}I-hGH preparation. *Protein Expression Purif.* **18**(2), 115–120.

Terlikowska, T., Cassette, P., Péron, M. N., Broda, R., Hainos, D., Tartes, I., and Kempisty, T. (1998). Study of the stability of 63Ni sources in Ultima Gold® liquid scintillation cocktail. *Appl. Radiat. Isot.* **49**(9–11), 1041–1047.

Thomas, R. C., Judy, R. W., and Harpvotlian, H. (1965). Dispenser for addition of internal standard in liquid scintillation counting. *Anal. Biochem.* **13**, 358–360.

Thomas, E. V., and Haaland, D. M. (1990). Comparison of multivariate calibration methods for quantitative spectral analysis. *Anal. Chem.* **62**, 1091–1099.

Thomson, J. (1991). Di-isopropylnaphthalene - a new solvent for liquid scintillation counting. *In* "Liquid Scintillation Counting and Organic Scintillators" (Harley Ross, John E. Noakes and Jim D. Spaulding, Eds.), pp. 19–34, Lewis Publishers, Chelsea, MI 48118.

Thorngate, J. H., McDowell, W. J., and Christian, D. J. (1974). An application of pulse shape discrimination to liquid scintillation counting. *Health Phys.* **27**, 123–126.

Toribio, M., García, J. F., Izquierdo-Ridorsa, A., Tauler, R., and Rauret, G. (1995). Simultaneous determination of plutonium alpha emitters by liquid scintillation counting using multivariate calibration. *Anal. Chim. Acta* **310**, 297–305.

Toribio, M., García, J. F., Izquierdo-Ridorsa, A., and Rauret, G. (1996). Optimization of counting conditions and simultaneous determination of ^{238}Pu, ^{239}Pu and ^{240}Pu by liquid scintillation counting. *In* "Liquid Scintillation Spectrometry 1994" (G. T. Cook, D. D. Harkness, A. B. MacKenzie, B. F. Miller, and E. M. Scott, Eds.), *Radiocarbon* 157–165.

Toribio, M., García, J. F., Izquierdo-Ridorsa, A., and Rauret, G. (1997). Multivariate calibration and spectrum position correction for simultaneous determination of alpha and beta emitting plutonium isotopes by liquid scintillation. *Anal. Chim. Acta* **356**, 41–50.

Toribio, M., Padró, A., García, J. F., and Rauret, G. (1999). Determination of mixtures of alpha emitting isotopes (^{242}Pu, $^{239+240}$Pu, ^{238}Pu) by using liquid scintillation-moving curve fitting. *Anal. Chim. Acta* **380**, 83–92.

Underwood, C. J., and Blumenthal, D. K. (1999). Adaptation of the protein kinase filter paper assay to a 96-well microtiter format. *Anal. Biochem.* **267**, 235–238.

Upham, L. (1999). Conversion of 96-well assays to 384-well homogeneous assays. *Biomed. Prod.* (*January*), 10–11.

Van Cauter, S., and Roessler, N. (1991). Modern techniques for quench correction and DPM determination in windowless liquid scintillation counting: a critical review. *In* "Liquid Scintillation Counting and Organic Scintillators" (Harley Ross, John E. Noakes, and Jim D. Spaulding, Eds.), pp. 219–237, Lewis Publishers, Chelsea, MI 48118.

Vera Tomé, F., Gómez Escobar, V., and Martín Sánchez, A. (2002). Study of the peak shape in alpha spectra measured by liquid scintillation. *Nucl. Instrum. Methods Phys. Res., Sect. A* (in press).

Véronneau, C., Aupiais, J., and Dacheux, N. (2000). Selective determinatioin of polonium by photon electron rejecting alpha liquid scintillation (PERALS system). *Anal. Chim. Acta* **415**, 229–238.

Verrezen, F., and Hurtgen, C. (1996). Radioassay of low-level, low-energy beta activity in multilabeled samples containing high-energy beta impurities using liquid scintillation spectrometry. *Radiocarbon* **1996**, 381–389.

Verrezen, F., and Hurtgen, C. (2000). A multiple window deconvolution technique for measuring low-energy beta activity in samples contaminated with high-energy beta impurities using liquid scintillation spectrometry. *Appl. Radiat. Isot.* **53**, 289–296.

Viteri, F. E., and Kohaut, B. A. (1997). Improvement of the Eakins and Brown method for measuring ^{59}Fe and ^{55}Fe in blood and other iron-containing materials by liquid scintillation counting and sample preparation using microwave digestion and ion-exchange column purification of iron. *Anal. Biochem.* **244**(1), 116–123.

Walton, T. D. (1996). Implementation of a robotics system for high throughput screening utilizing glow luminescence. *TopCount Topics*, **TCA-026**, Packard BioScience Company, Meriden, CT 06450, pp. 6.

Wiebe, L. I., McQuarrie, S. A., and Ediss, C. (1980). *In* "Liquid Scintillation Counting, Recent Applications and Development, (C.-T. Peng, D. L. Horrocks, and E. L. Alpen, Eds.), Vol. 1, pp. 81–87. Academic Press, New York and London.

Wigfield, D. C., and Cousineau, C. M. E. (1978). Some empirical observations concerning the liquid scintillation counting of coloured solutions. *Can. J. Chem.* **56**, 2173–2177.

Wise, B. M. (1992). PLS-Toolbox for use with MATLAB^{TM}. Center for Process Analytical Chemistry (CPAC), University of Washington, Seattle.

Wolterbeek, H. Th., and van der Meer, A. J. G. M. (2002). Transport rate of arsenic, cadmium, copper and zinc in *Potamogeton pectinatus L.*: radiotracer experiments with ^{76}As, ^{109,115}Cd, ^{64}Cu and ^{65,69m}Zn. *Sci. Total Environ.* **287**, 13–30.

Xu, X., Nakano, T., Wick, S., Dubay, M., and Brizuela, L. (1999). Mechanism of Cdk2/Cyclin E inhibition by p27 and p27 phosphorylation. *Biochem.* **38**, 8713–8722.

Yan, C. G. (1996). Improvement of accuracy of efficiency extrapolation method in 4π β-γ coincidence counting *Nucl. Instrum. Methods Phys Res., Sect. A* **369**, 383–387.

Yang, D. (1996). Calibration and quench correction for alpha liquid scintillation analysis. *In* "Liquid Scintillation Spectrometry 1994" (G. T. Cook, D. D. Harkness, A. B. MacKenzie, B. F. Miller, and E. M. Scott, Eds.), pp. 339–344. Radiocarbon, Tucson.

Yang, D., and Guo, Y. (1995). Determination of alpha radioactivity in vegetable ashes with liquid scintillation analysis. *Alpha Beta Application Note*, **ABA-007**, Packard BioScience Company, Meriden, CT 06450, pp. 4.

Yang, D., Zhu, Y., and Mobius, S. (1992). Rapid method for alpha counting with extractive scintillator and pulse decay analysis. *Alpha Beta Application Note*, **ABA-002**, Packard BioScience Company, Meriden, CT 06450, pp. 8.

Yang, D., Zhu, Y., and Jiao, R. (1994). Determination of Np, Pu and Am in high level radioactive waste with extraction-liquid scintillation counting. *J. Radioanal. Nucl. Chem.* **183**, 245–260.

Yi-Fen, Y., Bowman, J. D., Bolton, R.D., and Crawford, B. W. *et al.* (2000). A high-rate ^{10}B-loaded liquid scintillation detector for parity-violation studies in neutron resonances. *Nucl. Instrum. Methods Phys Res., Sect. A* **447**, 476–489.

Yin, H., Greenberg, G. E., and Fischer, V. (2000). Application of Wallac MicroBeta radioactivity plate counter and Wallac Scintiplate in metabolite profiling and identification studies. *Application Note* H78423, pp. 15, PerkinElmer Life Sciences, Boston.

Zecher, P. D., Galonsky, A., Kruse, J. J., Gaff, S. J., Ottarson, J., Wang, J., Deák, F., Horváth, A., Kiss, A., Seres, Z., Ieki, K., Iwata, Y., and Schelin, H. (1997). A large area, position-sensitive neutron detector with neutron/γ-ray discrimination capabilities. *Nucl. Instrum. Methods Phys. Res., Sect. A* **401**, 329–344.

Zimmerman, B. E., and Collé, R. (1997a). Comparison of the French and U. S. National ^{3}H (tritiated H$_2$O) standards by $4\pi\beta$ liquid scintillation spectrometry. *Appl. Radiat. Isot.* **48**, 521–526.

Zimmerman, B. E., and Collé, R. (1997b). Cocktail volume effects in $4\pi\beta$ liquid scintillation spectrometry with ^{3}H-standard efficiency tracing for low-energy β-emitting radionuclides. *Appl. Radiat. Isot.* **48**, 365–378.

Zimmerman, B. E., and Cessna, J. T. (1999). The standardization of ^{62}Cu and experimental determinations of dose calibrator settings for generator-produced ^{62}CuPTSM. *Appl. Radiat. Isot.* **51**, 515–526.

Zimmerman, B. E., Unterweger, M. P., and Brodack, J. W. (2001). The standardization of ^{177}Lu by $4\pi\beta$ liquid scintillation spectrometry with ^{3}H-standard efficiency tracing. *Appl. Radiat. Isot.* **54**, 623–631.

Zimmerman, B. E., Cessna, J. T., and Unterweger, M. P. (2002). The standardization of ^{188}W/^{188}Re by $4\pi\beta$ liquid scintillation spectrometry with the CIEMAT/NIST ^{3}H-standard efficiency tracing method. *Appl. Radiat. Isot.* **56**, 315–320.

6

ENVIRONMENTAL LIQUID SCINTILLATION ANALYSIS

GORDON T. COOK
Scottish Universities Research and Reactor Centre, East Kilbride, Glasgow G75 0QF, Scotland

CHARLES J. PASSO, JR.
PerkinElmer Life and Analytical Sciences, Downers Grove, Illinois 60515

BRIAN CARTER
Ontario Power Generation Inc., Whitby, Ontario L1N 1E4, Canada

Packard Bioscience and Wallac Oy, referred to in this chapter, are now divisions within PerkinElmer Life and Analytical Sciences.

Handbook of Radioactivity Analysis, Second Edition
Copyright © 2003 Elsevier Science (USA). All rights reserved.

I. INTRODUCTION

Environmental liquid scintillation counting (LSC) is the measurement of both natural and anthropogenic radionuclides in the natural environment. In the vast majority of environmental applications, radionuclide concentrations are low and the contribution of instrument background to the precision of the measurement is significant. During the 1980s, a new generation of commercially manufactured liquid scintillation analyzer (LSA) was introduced. These instruments were classified as "low level" because the background reduction features that were incorporated into them enabled the quantification of much lower radionuclide activities than were previously possible, i.e., the E^2/B (Figure of Merit) factor is increased (where E = efficiency and B = background) and this results in an improvement in the minimum detectable activity (MDA). Well-documented examples of such improvements include: (i) the extension of the effective age limit for radiocarbon dating from approximately 40,000 to 55,000 years and even > 60,000 years (Polach *et al.*, 1988; McCormac *et al.*, 1993; Cook, 1995; Plastino *et al.*, 2001) and (ii) the detection of approximately 1 Bq l^{-1} of ^{3}H in natural waters without electrolytic enrichment (Rank, 1993). In addition to the introduction of background reduction features, the incorporation of multichannel analyzers (MCAs) and microcomputer technology into modern LSAs has added spectrometric and sophisticated data handling capabilities that have made LSC a much more attractive analytical tool for environmental analyses. Because of these advances, low level LSC is now commonly used for the following applications:

1. Measurement of natural series radionuclides, in many cases with respect to health issues. Many are now routinely measured at natural environmental levels in a range of sample matrices including waters, sediments, soils, air, etc. These include isotopes of radium (Ra), uranium (U), thorium (Th), ^{210}Pb, ^{222}Rn, and ^{231}Pa (Mobius *et al.*, 1993; Saarinen and Suksi, 1993; Venso *et al.*, 1993; Al-Masri and Blackburn, 1995; Pates *et al.*, 1996a,b; Tinker and Smith, 1996; Salonen and Hukkanen, 1997; Aupiais *et al.*, 1998; Sato *et al.*, 2000; Wallner, 2001).

2. Monitoring the environment for radionuclide releases associated with nuclear fuel cycle activities (fuel enrichment, fuel fabrication, power generation, and fuel reprocessing facilities). This would principally be the analysis of beta emitting radionuclides without significant gamma emissions including ^{3}H, ^{14}C, ^{36}Cl, ^{89,90}Sr, ^{90}Y, ^{99}Tc, and ^{241}Pu but could also include analysis of alpha emitting radionuclides (Amano

and Kasai, 1988; Wilken and Joshi, 1991; Verrezen and Hurtgen, 1992; Ryan et al., 1993; Cook et al., 1995; Lee and Lee, 1999; Spry et al., 2000; Heilgeist, 2000; Tolmachyov et al., 2001). In addition, the analyses of effluent waters, urine, and other matrices for alpha activity is now widely reported (Cadieux et al., 1994; Eikenberg et al., 1996; Dacheux and Aupiais, 1997).

3. Studying the rates of processes in the environment. This would mainly be carried out using radionuclides of natural origins and would include (i) ^{14}C dating (Thomson et al., 1995; Zaretskaia et al., 2001), (ii) ground water movement and dating using ^{3}H (Rank, 1993), (iii) marine sediment mixing, productivity, and particle flux studies using ^{234}Th (Pates et al., 1996b; Kersten et al., 1998; Nour et al., 2002) and ^{210}Pb/^{210}Bi/^{210}Po (Biggin et al., 2002).

The general theory of LSC is discussed in Chapter 5 while the focus of this chapter is confined to a discussion of the features that have enhanced the performance of LSAs and the application of these instruments to the measurement of environmentally significant radionuclides.

II. LOW LEVEL LIQUID SCINTILLATION COUNTING THEORY

A. Sources of Background

Instrument background is derived from many sources including natural radioactivity in the instrument's construction materials, e.g. the photo-multiplier tubes (PMTs), natural radioactivity in the vial walls or cap, cosmic ray interactions with the vial walls and vial contents, and thermal ionisations in the PMTs, etc. These can be divided broadly into two categories termed quenchable and unquenchable background (Horrocks, 1985). Quenchable background comprises approximately one-third of the total and is primarily due to cosmic rays and other high energy radiations that penetrate the lead shielding, the reflector, and the glass wall of the vial before interacting with the scintillation cocktail (Horrocks, 1985; van Cauter, 1986). In addition, radioactive contamination of the scintillation cocktail can add to the quenchable component. Quenchable background increases with increasing cocktail volume and is reduced by quenching agents (Horrocks, 1985). The pulse height spectrum extends beyond the ^{3}H and ^{14}C energy ranges and the light pulses so produced are similar to those of true beta events.

Background events originating outside the cocktail are unaffected by quenching agents and thus are referred to as unquenchable (Horrocks, 1985). Unquenchable background events contribute around two-thirds of the total LSA background (van Cauter, 1986). The primary source is high energy cosmic radiation and natural radioactivity in the glass vials, which interact with the vial and PMT glass surfaces to produce low photon yield Cerenkov events and fluorescence that are primarily confined to the ^{3}H energy region. Unquenchable background can be distinguished electronically from valid scintillation events because the pulse pattern that results is characterized by a series (burst) of trailing low-photon-yield afterpulses that follow the initial

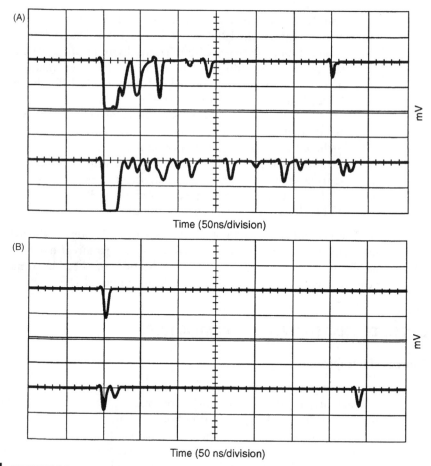

FIGURE 6.1 Typical PMT pulse shapes from **(A)** unquenchable and **(B)** quenchable events. The top halves of A and B are the pulses from the right hand side **PMT** and the bottom halves are the pulses from the left hand side **PMT** (Passo and Roberts, 1996). Reprinted by permission of *Radiocarbon*, University of Arizona, Tucson.

prompt pulse (van Cauter, 1986). Typical pulse patterns due to quenchable and unquenchable events are illustrated in Fig. 6.1.

B. Background Reduction Methods — Instrument Considerations

Until the introduction of commercially manufactured low level LSAs in the 1980s, almost all background reduction methods were attempted by the instrument users. They consisted mainly of modifications to existing instrumentation and included the following: increasing the amount of passive shielding, reducing the voltage applied to the PMTs, masking the PMTs or the vials to minimize PMT crosstalk, or using alternative low background materials for vial (mainly quartz or Teflon^{TM}) or vial holder (e.g., delrin) construction (Tamers, 1965; Calf and Polach, 1974; Noakes, 1977; Haas, 1979; Gupta and Polach, 1985). In addition, a number of experimental devices incorporating cosmic guard detectors were fabricated (Pietig and

Scharpenseel, 1964; Alessio *et al.*, 1976; Punning and Rajamae, 1977; Iwakura *et al.*, 1979; Jiang *et al.*, 1983). However, it was not until background reduction features were combined with modern microprocessor technology that the full potential of low level LSA instrumentation was realized. Modern low level instruments employ one or more of the following features to reduce background:

I. Enhanced Passive/Graded Shielding

Some instruments employ very large amounts of lead shielding. For example, the Wallac 1220 Quantulus has a shield of approximately 630 kg. This is composed primarily of low specific activity lead with a graded cadmium and copper inner lining (Kojola *et al.*, 1984). Also, the Beijing Nuclear Instrument Factory manufactures an LSA with a passive shield comprising of a 7.5 cm thickness of lead and 4 cm of stainless steel, which has a total weight of 1500 kg (Guan Sanyuan and Xie Yuanming, 1992). These shields reduce the background from environmental gamma photons associated with building and instrument construction materials, as well as the "soft" cosmic muon component of background and the graded cadmium and copper lining in the Quantulus shield absorbs secondary x-rays and thermal neutrons. However, passive shielding alone does not remove high energy gamma photons and the hard cosmic ray component. Anticoincidence shielding (an active guard detector) or a quasi-active guard in combination with some type of pulse shape analysis are the most effective ways to reduce their contribution to background (Kojola *et al.*, 1984; Guan Sanyuan and Xie Yuanming, 1992; Noakes and Valenta, 1996).

2. Active Guard Detectors

An active shield or guard detector consists of a volume of scintillating material (e.g., a plastic scintillator such as NE-110, a mineral oil-based scintillator, or even a large sodium iodide crystal) and either one or two additional PMTs. This guard surrounds the detector assembly and its PMTs are in anticoincidence with the sample PMTs (e.g. Kojola *et al.*, 1984). Events detected only by the guard detector and events simultaneously detected by the guard and sample PMTs are rejected. Typically, the guard will reject much of the residual environmental gamma radiation, the soft cosmic component, and the harder cosmic muon flux. This has the effect of reducing PMT crosstalk from Cerenkov light-induced pulses, which are primarily caused by the interaction of cosmic muons with the PMTs. Several types of active shields are employed in commercially available LSAs. For example, the Wallac 1414 Guardian employs a plastic scintillator, the Wallac 1220 Quantulus uses a mineral oil-based scintillator (Fig. 6.2), and the Kvartett (Einarsson, 1992) uses a large sodium iodide crystal.

3. Pulse Discrimination Electronics

a. Pulse Shape Analysis (PSA)

A beta scintillation event is comprised of a fast (prompt) and a delayed component. The majority of the light (derived from excited singlet states)

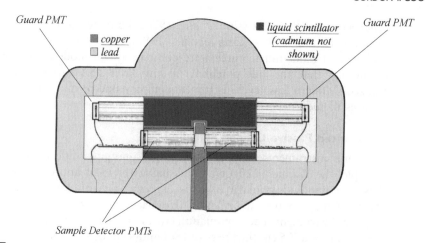

Guard PMT ■ copper ■ liquid scintillator Guard PMT
□ lead (cadmium not shown)

Sample Detector PMTs

FIGURE 6.2 Schematic representation of the active guard, passive shield and detector chamber in the Wallac Quantulus 1220 LSA. Reprinted by permission of PerkinElmer Life and Analytical Sciences.

is within the prompt component (typically 2–8 ns in duration). The delayed component (produced largely by the annihilation of triplet states) may persist for several hundred nanoseconds. The relative amounts of light in the prompt and delayed components have long been known to be dependent on the specific ionization induced by different types of particles. The pulse shape analyzer in the Quantulus is an analog device that integrates the tail of each pulse and compares this with the total pulse integral to produce an amplitude independent parameter that relates to the pulse shape (Kaihola, 1991). It is on this basis that PSA can be used for particle detection, which includes alpha–beta discrimination. Although the most effective use of PSA is for alpha counting, the technique has also been used for background discrimination (Kaihola, 1991). The background of glass vials is charac- terized by long scintillation light pulses due to the interaction of cosmic or other environmental radiation with the vial (much of the latter originates from the inherent radioactivity of the vial itself). PSA is used to discriminate these long scintillation events from the shorter scintillations of true beta events (Fig. 6.3). PSA values are adjustable within a finite range and the optimum value (in which E^2/B is maximized) can be determined automatically or empirically determined by the user. This optimum will be specific for vial type and chemistry. As in all pulse shape discrimination schemes, there is a finite probability of a slight decrease in counting efficiency due to the misclassification of events. Several of the leading LSA manufacturers, including Wallac and Beckman Instruments Inc., use some form of PSA that is capable of background reduction.

b. Pulse Amplitude Comparison (PAC)

This technique compares the ratio of the pulse amplitudes produced by the two PMTs (Kaihola, 1991). Photons produced in the cocktail from a radioactive decay event are evenly distributed and the number impinging

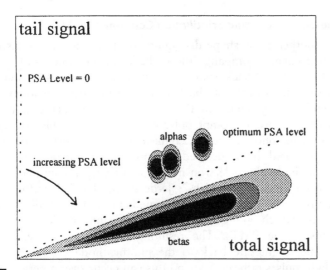

FIGURE 6.3 Schematic representation of the pulse shape analysis (PSA) feature used by Wallac (Pates *et al.*, 1996). Reprinted by permission of *Radiocarbon*, University of Arizona, Tucson.

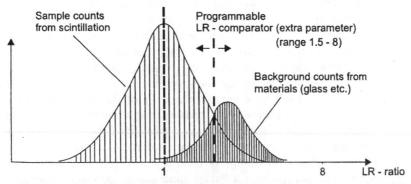

FIGURE 6.4 Schematic representation of the pulse amplitude comparator (PAC) feature used by Wallac. Reprinted by permission of PerkinElmer Life and Analytical Sciences.

on each photocathode will be similar. Thus, the ratio of the pulse amplitudes will vary around a value of one. Photons produced in the vial wall or in the PMTs will show greater variation, e.g., if a scintillation is produced through the interaction of a cosmic muon with a PMT, the amplitude of the pulse observed by one PMT will differ from that observed by the other and the PAC ratio will deviate significantly from unity. Therefore, the pulse amplitude ratios will be less evenly distributed. Effectively, this is a PMT crosstalk reduction feature and is used in low level LSA instruments manufactured by Wallac (Fig. 6.4). The PAC values are adjustable within a defined range and the optimum is empirically determined. Again, the use of PAC can lead to a small reduction in counting efficiency due to misclassification of events.

c. Time-Resolved Liquid Scintillation Counting (TR-LSC®)

This is another pulse shape discrimination technique and is used by the Packard Instrument Company (now PerkinElmer Life and Analytical Sciences) to reduce background count rates by discriminating the unquenchable component of background from true beta events and the quenchable background, on the basis of the number and energies of afterpulses that follow a prompt pulse event. Unquenchable background is the result of low photon yield Cerenkov events consisting of a prompt component followed by a delayed component comprising of a burst of afterpulses (of single photoelectron magnitude) that can continue for as long as $5\,\mu s$ (Dressler and Spitzer, 1967; Jerde *et al.*, 1967; Roodbergen *et al.*, 1972). The prompt pulses derived from true beta events will also be followed by afterpulses, which are caused by the collision of electrons with gas molecules inside the PMTs. The resulting ionized gas molecules take a significant time to migrate back to the photocathode where they produce secondary light pulses (afterpulses) and this can occur over a period of several microseconds, thereby emulating a background event. In addition, the higher the energy of the initial event, the more electrons are liberated and the higher the probability of afterpulse production. Nevertheless, a greater number of afterpulses will usually result from an unquenchable background event than from a true scintillation event, for a given energy. Therefore, the solution is to incorporate a system that allows some afterpulses and in addition, this acceptance criterion is scaled with event energy. The acceptance criteria are stored in the form of a "burst look-up table" that correlates the acceptable number of afterpulses versus event energy. Valenta (1987) designed burst counting circuitry to discriminate unquenchable background events from true beta events on the basis of the number of burst afterpulses that follow the prompt pulse, relative to the amplitude of this prompt pulse. The total number of afterpulses is termed the pulse index and this is used to produce three dimensional information that is comprised of energy and time-resolved information (number of afterpulses) for each detected event. Figure 6.5A demonstrates that a background sample produces a significant number of afterpulses or bursts at the low energy end of the spectrum. In contrast, an unquenched ^{3}H sample generates few afterpulses and only at the high energy end of the spectrum (Fig. 6.5B) while an air quenched ^{3}H sample prepared in the laboratory (Fig. 6.5C) generates almost no afterpulses. Graded shielding is not required on Packard (PerkinElmer) liquid scintillation analyzers because TR-LSC is effective at eliminating the background that is caused by the interaction of secondary x-rays with the counting chamber. In addition, background due to Cerenkov radiation and fluorescence, which is the result of the interaction of beta radiation from ^{40}K in the glass of the PMTs and counting vial, is also effectively eliminated. Because true beta events are followed by afterpulses, there is a finite probability that valid events may be rejected, thus, TR-LSC may reduce counting efficiency. In some applications, rejection of valid events is more probable, e.g. (1) in deoxygenated samples, (2) when using long fluorescence life-time (slow) scintillators that produce a significant delayed component, i.e. those

A

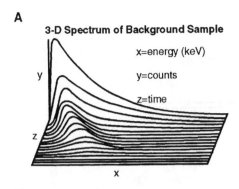

B

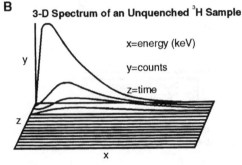

C

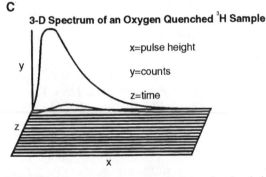

FIGURE 6.5 Three dimensional plots of pulse height spectra for: (A) a background sample, (B) an unquenched ³H sample and (C) a laboratory prepared (quenched) ³H sample. Reprinted by permission of PerkinElmer Life and Analytical Sciences.

cocktails that contain di-isopropyl naphthalene (DIN) and phenyl-ortho-xylylethane (PXE), and (3) with higher energy beta emitters (energies greater than or equal to ^{14}C) where the delayed component persists, even in the presence of oxygen. The ability to optimize TR-LSC for all three of the above conditions was made possible by the introduction of programmable TR-LSC. TR-LSC has a preset delay of 75 ns between the onset of the prompt pulse and the commencement of afterpulse or burst counting. Programmable TR-LSC, on the other hand, provides the ability to adjust the delay between 75 and 800 ns (Passo and Kessler, 1993). Graphic representations of nonprogrammable and programmable TR-LSC are shown in Fig. 6.6.

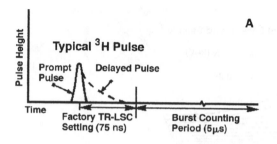

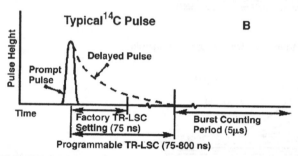

FIGURE 6.6 Schematic representations of (A) nonprogrammable TR-LSC and (B) programmable TR-LSC (Passo and Roberts, 1996). Reprinted by permission of *Radiocarbon*, University of Arizona, Tucson.

4. TR-LSC Quasi-active Detector Guards

a. Slow Scintillating Plastic

An additional feature that reduces background was incorporated into earlier Packard Tri-Carb XL models employing TR-LSC. This feature was in the form of a modified detector assembly that consisted of a long fluorescence lifetime scintillating plastic that surrounded the sample vial in the counting chamber (Valenta and Noakes, 1989). This plastic detector guard was optically coupled to the face of each PMT and was coated with a reflective material. It had very good optical transmission in the 400–500 nm range and the decay time of the slower component of the scintillating plastic was approximately 250 ns. When the slow fluor in the plastic (external to the sample) was excited by a high energy gamma or muon event that satisfied the instrument coincidence requirement, the phosphorescence of the slow fluor provided an extended burst of afterpulses that aided the background discrimination. Thus, the plastic acted as an active guard (but without dedicated PMTs). One factor that had to be stressed when using the plastic guard was that the emission spectrum of the cocktail should not significantly overlap the absorption spectrum of the guard. An appropriate secondary scintillator (e.g., POPOP, *bis*-MSB) was used to ensure that significant spectral overlap was avoided.

b. Bismuth Germanate (BGO)

A quasi-active detector guard, based on BGO ($Bi_4Ge_3O_{12}$), was developed by Packard Instrument Company (Noakes and Valenta, 1996)

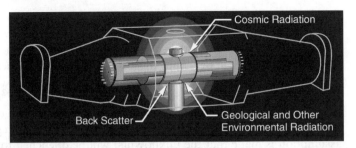

FIGURE 6.7 Schematic representation of the quasi active BGO guard, passive shield and detector chamber in the Packard 2770TR/SL. Reprinted by permission of PerkinElmer Life and Analytical Sciences.

and incorporated into the Tri-Carb 2770TR/SL and the current 3170TR/SL models (Fig. 6.7). BGO is a nonhygroscopic scintillator that exhibits low afterglow and its efficient gamma photon stopping power ($Z = 83$; density $= 7.13 \, \text{g cm}^{-3}$) makes it an effective cosmic guard. BGO gamma detectors were first introduced commercially (circa 1979) for use in computerized tomography (CT) applications and because of its high gamma cross-section, BGO has also been used in positron emission tomography (PET), high energy calorimetry, and medium energy physics (Bicron Corporation, 1988). As a detector guard for LSA, BGO is used to discriminate against the gamma and muon components of cosmic background. Events that interact with BGO produce pulses that are of much longer duration (at least a factor of 10) than beta and alpha pulses in the cocktail. If an event interacts with the detector and satisfies the coincidence requirement, TR-LSC is then used to discriminate the longer cosmic background pulses from beta pulses. Similar to the slow scintillating plastic guard already discussed, the BGO guard completely surrounds the sample vial in the modified detector assembly. However, this detector assembly is not optimal for analysis of beta–gamma emitting radionuclides where the gamma photons are coincident with the beta particle and sufficiently energetic to interact with the BGO guard as this increases the probability of misclassification of beta events as background.

5. Counting Region Optimization

The limit of detection of a measurement is determined by the signal-to-background ratio. In liquid scintillation counting, the signal-to-background ratio is expressed as the square of the detection efficiency divided by the background (E^2/B) and is termed the Figure of Merit (FOM). Counting region optimization is used to maximize the E^2/B and thereby improve the limit of detection. Thus, even without the use of background reduction features, limits of detection can be improved by simply optimizing the counting regions.

a. Region Optimization Procedures and Requirements Under Constant Quench Conditions

A blank (background) and a standard sample containing the radionuclide of interest are required to optimize the counting region for that radionuclide.

The requirements for proper region optimization include:

1. The standard containing the radionuclide of interest should have sufficient known activity and must be in the same cocktail and quenched to the same degree as the unknown samples. Equivalent quench ensures that the region chosen gives the optimum E^2/B for the sample. Different quench levels will shift the counts in or out of the region and thus affect the counting efficiency in that region. The assumption is made that the unknown(s) will be quenched to the same degree. In reality, there will be slight differences in quench and for this reason, once the optimum region is determined, quench correction should be applied to calculate the absolute activity. The quench curve should relate counting efficiency in the optimized region (for a series of known activity standards that are quenched to different degrees) to the quench indicating parameter (H-number, SQP(E), t-SIE, etc). Production of this type of quench curve will account for the minor quench variations in the samples. Wallac employs a digital overlay technique (DOT) where reference spectra, stored in the memory of the counter, are fitted to the sample spectra. Rundt (1992) gives a comprehensive review of this technique.

2. The blank should be quenched to the same degree as the standard and the unknown samples. Minor fluctuations in the degree of quenching will result in only small changes in the blank activity. Therefore, it is generally not necessary to construct a quench curve for the blank. Small changes in the blank spectrum will not adversely affect the optimum window determination.

By using a representative standard and an appropriate blank, a reasonable approximation to the optimum counting window can be determined for unknown samples near background levels. In some cases, it may be possible to use the unknown sample as the standard if the activity is several times the background. In this case, the count time should be long enough to minimize the error associated with the measurement. However, counting efficiency must still be determined. In many modern LSAs, region optimization can be performed automatically.

b. Region Optimization Under Variable Quench Conditions

For samples that have highly variable quench, it may not be practical to maintain a fixed counting region and still maintain maximum counting sensitivity. A practical solution is to determine the equivalent, unquenched spectral endpoint by counting the least quenched sample in the batch of unknowns or by counting the least quenched standard of the quench set. Quench standards and samples are counted using this region setting with Automatic Quench Compensation (AQC is a Beckman patent), Automatic Efficiency Compensation (AEC is a derivative of AQC and is a Packard Feature), or an equivalent technique. AQC adjusts the upper limit of the counting region to the observed spectral endpoint of the quenched sample. By adjusting the region in this manner, background is reduced and high counting efficiency is maintained. For variably quenched samples, the use of AQC represents the best choice for maintaining high sensitivity provided the sample activity is sufficiently high to produce a good spectral endpoint. The DOT technique employed by Wallac can also be used for a variable quench situation, its accuracy will be reduced when count rates are very low.

6. Process Optimization

The term "optimize" or "optimizing" is used in many areas to simply describe the process of determining the best or optimum windows (see Counting Region Optimization above) on an LSA, for a given sample geometry. While the counter is certainly an integral part of producing results, it is just that, a part. It is one step in a "process" and it is this process which should always be considered for "optimization". The process would include collection, storage and preparation/extraction of the sample, the preparation for counting and the actual sample counting. The goal of the optimization would be to attain the required results (precision, accuracy, and sensitivity) with the lowest cost per sample in terms of technician time, materials/equipment, and counter time. If the equipment is new, the counter performance must be determined prior to evaluating the process. The first step is to consider the process requirements:

1. What is the required lower limit of detection (LLD)?
2. Is waste production an important consideration?
3. Does the sampling process have restrictions (size, time, etc)?
4. Is equipment time restricted?
5. Is sample preparation/extraction time limited?
6. Is sample storage limited?
7. Is cocktail/vial cost per sample important?

The second step is to define the current performance (LLD) of existing resources. Typical performance data for three counters is presented in Table 6.1. The third and final step is to consider all the factors from step 1 and employing the performance data from step 2, to arrive at the most efficient (optimum) procedure for the application. Given all of the process variables this can become quite a challenge. By using a decision tree, such as the example in Fig. 6.8, the operation can be arranged into a fairly straightforward exercise. Counter setup and sample preparation can be widely varied and still meet the same sensitivity requirements. For example, assume that a laboratory currently uses the first counter from Table 6.1 (Counter A) and

TABLE 6.1 Instrument Performance Summary for ^{3}H Analysis Using 10 mL Water and 10 mL Scintillation Cocktail

Counter	Opt. Window (keV)	Efficiency (%)	Bkgd (cpm)	Count Time (min)	LLD (Bq L^{-1})
A	0.0–3.5	21.82	2.08	200 (2 × 100)	3.7
B	0.0–5.5	20.03	1.53	200 (2 × 100)	3.4
C	0.0–4.0	23.71	0.95	200 (2 × 100)	2.3

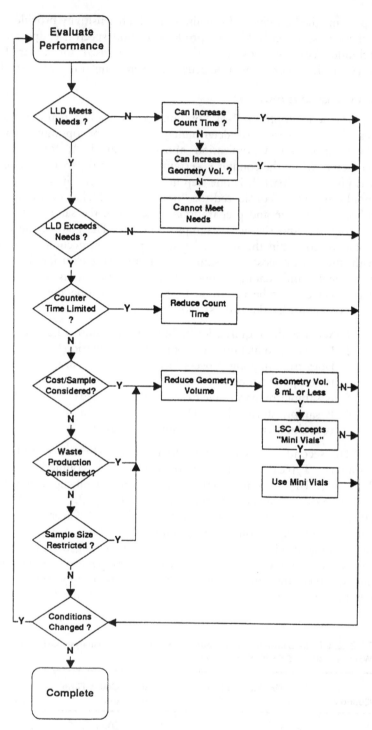

FIGURE 6.8   Process optimizing decision tree.

TABLE 6.2   Example of Optimizing to Performance Requirements for ^{3}H

Counter	Sample Vol. (mL)	Cocktail Vol. (mL)	Efficiency (%)	Bkgd (cpm)	Count Time (min)	LLD (Bq L^{-1})
A	10	10	21.82	2.08	200 (2 × 100)	3.7
C^a	3	3	24.11	0.41	400 (2 × 200)	3.7
C	10	10	23.71	0.95	90 (2 × 45)	3.7

^{a}8 mL vial.

that the required LLD for the process is 3.7 Bq L^{-1} ^{3}H. A new counter arrives (Counter C) and must be integrated into the system. As shown, the performance of the new counter is significantly better and if the same counting protocol is maintained, the samples will be "over counted". This additional performance can be exploited as shown in Table 6.2, to reduce either the required count time or sample geometry.

C. Background Reduction Methods — Vial, Vial Holder, and Cocktail Considerations

I. Vials

Any natural radioactivity in the walls or cap of the counting vial will obviously increase the background count rate and for this reason, glass vials with a low ^{40}K content are desirable. Plastic vials display lower background than glass, partly due to their lower radioactivity and also to their lower density, which results in a lower probability of particle/photon interactions with the vial walls. However, care must be exercised over the choice of scintillation cocktail when using plastic vials. Typically, cocktails based on di-isopropylnaphthalene (DIN) and phenyl-ortho-xylylethane (PXE) will not attack the plastic vial whereas cocktails containing solvents such as benzene, pseudocumene, and toluene are of very limited use due to solvent diffusion into the plastic. Low diffusion plastic vials that have a thin Teflon coating reduce solvent diffusion to some extent but do not eliminate it. Static charge may build up on plastic vials and this can lead to spurious high count rates (the typical static symptom). To prevent any buildup of static charge, it is suggested that vials with an antistatic coating should be used. The antistatic vials are slightly more expensive, but the costs associated with recounting samples usually more than offsets any price difference. Vials manufactured from silica (Hogg *et al.*, 1991; Hogg and Noakes, 1992; Wilson *et al.*, 1996) and Teflon (Polach *et al.*, 1983; Kalin and Long, 1989) are quite commonly used for radiocarbon dating and have been demonstrated to produce significant background reductions. Their main drawback is the comparatively high cost compared to low ^{40}K glass while a lack of uniformity of response has also been noted by Einarsson (1992) for his silica vials. Additionally, great care must be taken in their cleaning between sample additions, however, under certain circumstances, e.g. where samples are very small or their ages are approaching the limit of detection, they can prove very beneficial.

2. Vial Holders

Inert vial holders made from materials such as delrin have been used to enable instruments such as the Quantulus to accept nonstandard-sized vials (particularly those made from silica) and indirectly, their use brings about background reductions. The slow scintillating plastic used as the material for construction of the quasi-active guard in the Packard Tri-Carb XL models was also used to produce vial holders that would accommodate 7 mL vials. The vial holders enhanced the background reduction even further when used in conjunction with the plastic guard, however, they were expensive and could be damaged by some solvents. They were most useful for applications such as radiocarbon dating where the total sample volume is often < 7 mL. Where the sample size was not limited, the use of 20 mL vials (and therefore no vial holder) was preferable as the MDA from the smaller sample would be poorer. In addition, these vial holders were also used effectively in TR-LSC instruments that did not contain the plastic guard (Cook *et al.*, 1989). Similarly, vial holders manufactured from BGO and designed to accommodate 7 mL vials were also used to provide additional background discrimination although, again they were expensive. These vial holders worked best in an instrument that had the BGO guard detector and electronics, but also performed quite well in other TR-LSC systems operating in the low level count mode (Cook, 1995). Owing to their limited use and high cost, neither type of vial holder is now commercially available.

3. Cocktail Choice and Optimization

Overall, liquid scintillators should be prepared with reagents that are low in natural radioactivity. Of course, when dealing with aqueous samples, the limit of detection will also be influenced by the loading capacity. The selected cocktail(s) should be evaluated to determine the sample to cocktail mixing ratio that will give the best performance. The cocktail should be tested at various sample loads through its rated capacity, while holding the total geometry constant. Each mixing ratio is tested using a background/standard pair to determine optimum windows and evaluate the LLD. In the example presented in Fig. 6.9, Ultima Gold LLT (Packard) is evaluated from 10–60% sample loading, with a constant counting geometry of 20 mL. The best performance for this example occurs at a 50% sample load (ratio = 1 : 1). Regardless of the total sample/cocktail volume, the best counting performance (LLD) will be achieved using an optimum 1 : 1 mixing ratio. The detailed legend box in Fig. 6.9 outlines the optimum counting windows for each percent sample loading.

D. Background Reduction Methods — Environment

To achieve optimum performance from an LSA, it is important to ensure that the instrument is stable. Lack of stability may be manifested by changes in

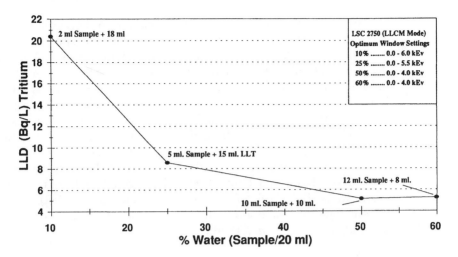

All tests performed using 20ml. Antistatic Vials

FIGURE 6.9   Example of cocktail performance evaluation.

efficiency, background and the quench indicating parameter, etc. Stability is generally aided by placing the LSA in a stable counting environment, where temperature and humidity are controlled, the atmosphere is not corrosive, and the power supply to the instruments is of good quality. As a general guide, samples should be allowed to stand for several hours (4–6) in the counter prior to counting as this minimizes chemiluminescence and/or photoluminescence. The effects of chemiluminescence can also be reduced by the use of a chilling unit that cools the sample counting chamber.

Static is generated in the counter by the rubbing together of nonconducting materials in a dry environment, hence the requirement for controlling humidity. If the vial entering the counting chamber has a static charge, spurious photons will be generated as the charge dissipates. Static can be minimized by humidifying the counter or the counting room, by using antistatic vials and by employing the static controlling device that is common to LSAs that have an automated sample changer.

Temperature played an important role in the operation of early liquid scintillation counters. The noise rates (singles rates) of early PMTs were much greater than today's and very dependent on operating temperature. For this reason, older instruments were refrigerated to reduce this noise level and reductions in background count rates were significant beyond the effect of the coincidence circuitry. Refrigeration, although beneficial, is not as critical for background reduction in modern instruments because of the introduction of lower-noise bialkali tubes and improvements in tube manufacturing. Today, refrigeration is often used for reasons other than cooling the PMTs. Temperature control may be necessary for special sample preparation (e.g. where a gel cocktail is required and gelling may be impaired within certain temperature ranges), severe operating conditions, minimizing chemiluminescence, or reducing static buildup.

As previously discussed, the contribution to background from high energy radiation, that can penetrate passive shielding and interact with the vial walls and PMTs, etc, can be minimized by the use of active guard detectors and pulse shape analysis techniques. An alternative method for the reduction of high energy radiation, which is particularly effective for instruments without the above features, is to choose a location that provides additional shielding against cosmic or environmental radiation. Examples of this include (i) siting the British Museum's LSAs, used for radiocarbon dating, in the London Underground system (Bowman, 1989), (ii) a specially designed counting laboratory at the University of Arizona that has 10 m of overburden (Kalin and Long, 1989), (iii) a counting laboratory within the Warragamba Dam wall in Australia (Calf and Airey, 1982), and Wallac's surface low level laboratory that was constructed from low background materials (Kaihola *et al.*, 1986).

III. ALPHA/BETA DISCRIMINATION

A. Alpha/Beta Separation Theory

The ability to measure the activity of alpha emitting radionuclides by LSC has been recognized for many years. It has also long been established that alpha and beta emitting radionuclides produce different pulse shapes at the PMT anode and may be separated on this basis (Buchtela *et al.*, 1974; Thorngate *et al.*, 1974; McKlveen and Johnson, 1975). Pulse shape discrimination techniques are well known and have been extensively applied to the rejection of gamma background in neutron spectrometry (Horrocks, 1970; Winyard *et al.*, 1971) and to the isolation of fission events in liquid scintillation counting (Horrocks, 1963). These techniques are often referred to in the literature by different instrument manufacturers as either pulse shape discrimination (PSD), pulse shape analysis (PSA), or pulse decay analysis (PDA). Different names arise because the techniques of pulse shape discrimination can vary, but all are based on electronic circuits that measure some aspect of pulse decay time.

Alpha/beta separation is a comparatively recent feature on commercially available, conventional LSAs. The 4π counting geometry of LSC results in approximately 100% counting efficiency for alpha emitting radionuclides, which, coupled with the low alpha background count rates, provides an alternative to gross alpha–beta by gas flow proportional counting. For certain applications, alpha–beta LSC can also be a very useful alternative to conventional alpha spectrometry using passivated implanted planar silicon (PIPS) and silicon surface barrier (SSB) detectors, however, the much poorer resolution of LSC relative to semiconductor detectors can be a major disadvantage. This is brought about principally because of the relatively large amount of energy required to produce a single photoelectron at the PMT photocathode and, to a much lesser extent, because of the inefficient light production by alpha particles relative to betas.

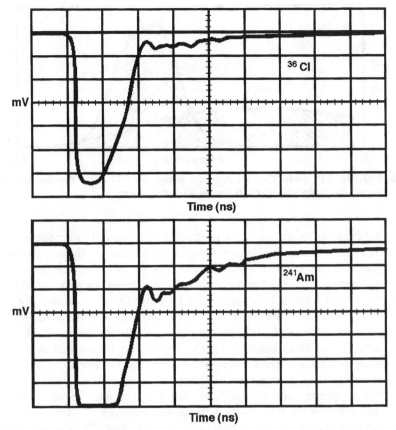

FIGURE 6.10 PMT pulse shapes for ^{36}Cl (beta) and ^{241}Am (alpha). Reprinted by permission of PerkinElmer Life and Analytical Sciences.

To understand how the separation of alpha from beta events is accomplished, it is necessary to examine the processes at a molecular level. Alpha and beta events may be distinguished from one another in a liquid scintillator by examining the electronic pulses that are produced at the PMT anode of the detector. As described in the previous section, these pulses are made up of two components: the prompt and the delayed components, which occur in different proportions in alpha and beta pulses (Horrocks, 1974a). Figure 6.10 demonstrates the difference in pulse shape between ^{36}Cl (beta emitter) and ^{241}Am (alpha emitter) in Insta-Gel cocktail (PerkinElmer Life and Analytical Sciences) containing 20% weight/volume naphthalene. These pulse shapes were measured at the PMT anode using a storage oscilloscope set on averaging mode to produce an average pulse shape. The photons incident on the cathode of the PMT originate from the radioactive decay of excited singlet and triplet states of the fluor molecules in the cocktail. The prompt component arises from the fast, exponential decay of excited singlet states while triplet states can produce photons only upon collision with another molecule in the triplet state, resulting in a longer lifetime of several

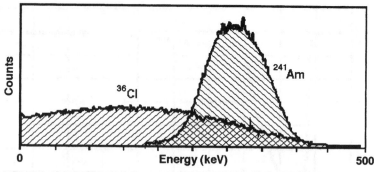

FIGURE 6.11 Multi Channel Analyzer display for ^{36}Cl and ^{241}Am spectra. Reprinted by permission of PerkinElmer Life and Analytical Sciences.

hundred nanoseconds, thus producing the delayed component of the pulse (Brooks, 1979). The higher specific ionization of alpha particles causes a greater proportion of excited molecules to be in triplet states and, hence, alpha pulses have a longer duration. The longer duration of alpha pulses is the basis of alpha–beta separation by pulse shape discrimination.

Most of the environmentally significant alpha emitting radionuclides emit particles in the 4–6 MeV energy region, while the betas of interest typically have E_{max} values below 2.5 MeV. Separation of alpha from beta events is necessary because the energy to light conversion yield from alpha particles is approximately a factor of 10 lower than from betas, with the result that alpha spectra are typically in the beta region. Figure 6.11 illustrates the spectral overlap that is observed between ^{241}Am, which produces alpha particles in the region 5.4–5.5 MeV and ^{36}Cl, which has an E_{max} of 710 keV.

B. Alpha/Beta Instrumentation

Two types of LSC instruments can distinguish α from β events. These are the PERALS® spectrometer (photon/electron-rejecting alpha liquid scintillation) and conventional LSAs with pulse-shape discrimination (PSD) electronics. Although both can simultaneously produce resolved α and β spectra, they are designed with different purposes in mind and, hence, are generally best suited to different types of applications.

I. The PERALS® Spectrometer

This instrument was designed by the late W. Jack McDowell, G. N. Case and coworkers (McDowell and McDowell, 1994) as a method of α spectrometry in addition to conventional solid state detection, and has its own characteristics, advantages, and disadvantages. Optics, sample preparation techniques and sample geometry are geared to producing the maximum α–α energy resolution by enhancing light output, as well as reducing interference from β and γ emissions. A cross section of the PERALS® detector is illustrated in Fig. 6.12. As a result, the optics and geometry are not optimized for β spectrometry, and the lack of passive shielding and

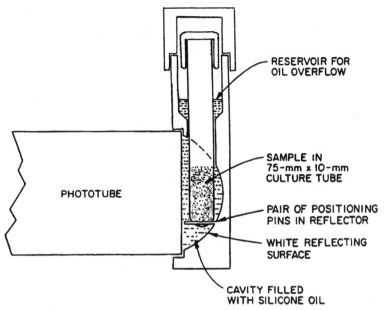

FIGURE 6.12 Cross-section of the PERALS$^©$ spectrometer (McDowell and McDowell, 1993). Reprinted by permission of *Radiocarbon*, University of Arizona, Tucson.

coincidence circuitry result in increased background in the β spectrum. McDowell and McDowell (1994) suggest that this instrument is really best suited for α measurements alone. A key element of the PERALS® system is extractive scintillators, whereby an organophilic complexing agent, usually specific to a single element, is incorporated into the scintillation cocktail. This serves two purposes: (1) specific radionuclide extraction may be accomplished, and (2) quenching caused by water, acid, or other sample impurities is minimized, thus further improving the energy resolution. A series of extractive scintillators for the analysis of a range of actinides and other naturally occurring radionuclides has been developed (McDowell and McDowell, 1994) and used successfully both with PERALS® and other LSAs utilizing PSD. The authors also recommend purging all samples with nitrogen or an alternative oxygen-free gas to improve energy resolution even further by reducing oxygen quenching.

The advantages of this system are (i) high alpha–beta discrimination, producing >99.9% rejection of β–γ events while maintaining a > 99% counting efficiency for the α events, (ii) improved alpha resolution (200–300 keV FWHM) compared with a standard LS counter, (iii) the observed backgrounds in the α spectra are typically lower than in a conventional counter.

2. Conventional LS Spectrometers with Pulse-Shape Discrimination

The major manufacturers of LSAs produce instruments capable of PSD for simultaneous determination of α and β activity. The α–α energy resolution that can be achieved by conventional LS spectrometers is generally poorer than with the PERALS® spectrometer (300 to > 500 keV FWHM), however, these instruments have definite advantages for gross α–β analysis. Also, automatic sample changers are standard features, which are advantageous for routine analyses or for relatively short counting times. In some instances, the use of extractive scintillators is not practical or adds unnecessary complications to the sample preparation. For example, radon can be prepared for analysis by simply shaking a sample of water with a water-immiscible cocktail. However, the use of conventional LSAs does not preclude McDowell's methods. Finally, these instruments also allow the use of an α-emitting isotope as an yield tracer in the analysis of a β-emitting isotope and vice versa (e.g., Pates *et al.*, 1996b).

As stated before, each manufacturer has its own method of PSD and these methods are referred to by different terminologies. This can be confusing, so we include here a discussion of various methods of PSD used with each instrument and explain the terms used, as we understand them.

a. Wallac (now PerkinElmer Life and Analytical Sciences)

Wallac (now PerkinElmer Life and Analytical Sciences) has manufactured several α–β LSAs: Wallac 1409, Wallac Quantulus 1220™, Wallac 1414 WinSpectral™, and the Wallac1414 Guardian™ DSA systems. The Quantulus 1220 is the only system skill in production. Wallac terms its version of PSD as pulse shape analysis (PSA), and time discriminator (TD) settings are PSA levels. PSA integrates the tail of the scintillation pulse for a long enough period to enable differentiation between short and long pulses. The pulse length information is normalized to the pulse height to achieve amplitude independence (i.e., energy-independence). The optimum PSA level is a numeric parameter that is represented schematically by a line that separates the alpha from beta events. Events falling above the line are classified as α events, and those below, as β events (Fig. 6.3). Both the 1414 Guardian and the 1220 Quantulus™ have active guard counters, in addition to other features that contribute to background reduction (see Section II for details).

b. Packard Instrument Co. (now PerkinElmer Life and Analytical Sciences)

The Tri-Carb® 2900TR and 3100TR series of LSAs have an optional PSD feature called pulse decay analysis (PDA), with a TD called the pulse decay discriminator (PDD). PDA distinguishes between α and β events using the zero-crossing method of PSD, which is so named because each scintillation pulse is doubly differentiated and the time at which this new curve crosses the zero line is compared to the PDD value (Fig. 6.13). Pulses that cross in a shorter time are classified as β events and those that cross in a longer time are classified as α events. Again, this method should be amplitude-independent. These instruments can also reduce background by activation of the burst-counting circuitry (BCC) feature (see Section II for details). Time resolved pulse decay analysis (TR-PDA) is the combined use of TR-LSC background discrimination and pulse decay analysis. By applying TR-LSC to PDA, alpha events that are misclassified as beta are discriminated out by TR-LSC since the long delayed component of alpha pulses resemble background events. Thus, alpha events

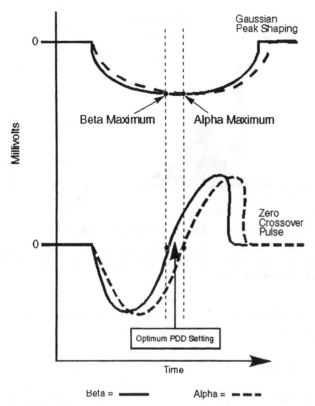

FIGURE 6.13 Schematic representation of the zero-crossing method of pulse shape discrimination used by Packard Instrument Co. Reprinted by permission of PerkinElmer Life and Analytical Sciences.

misclassified as betas will be eliminated from the beta multichannel analyzer (MCA). In addition, the misclassification of beta events will also be reduced since the TR-PDA causes a shift to a higher optimum PDD value. At higher PDD values, fewer beta events can satisfy the time requirement for an alpha pulse and this reduces the beta spill into the alpha MCA. The net result of TR-PDA is a reduction in the misclassification of beta and alpha events. An added benefit of TR-PDA is that TR-LSC background discrimination allows simultaneous low level beta counting.

c. Beckman Coulter Inc.

The Beckman LS 6500 series include the optional capability for PSD, in which α and β scintillation pulses are distinguished by a ratio technique. The area of each pulse is found at times t_1 and t_2, corresponding to the prompt and delayed components respectively. The ratio of these areas is compared with the ratio of known α and β pulses, resulting in a time discrimination called the R-value (Dodson, 1991). The R-values are plotted versus cpm response and the optimum TD setting is the R-value that provides the minimum spillover of alpha and beta events. Typically, this minimum is represented graphically as the minimum value between the alpha and beta

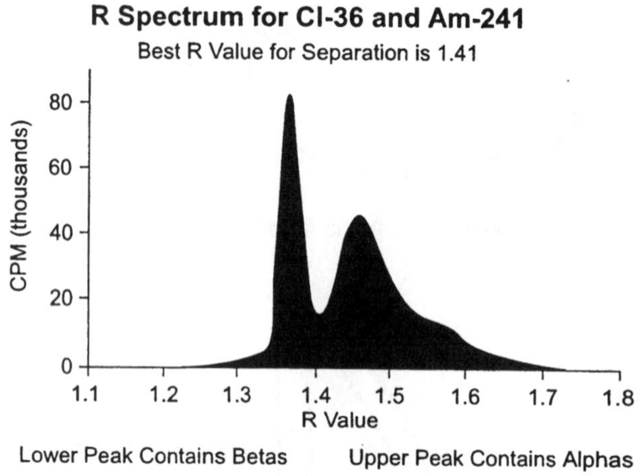

R Spectrum for Cl-36 and Am-241
Best R Value for Separation is 1.41

Lower Peak Contains Betas   Upper Peak Contains Alphas

FIGURE 6.14 Schematic representation of the determination of the optimum "R" value setting for pulse shape discrimination used by Beckman Instrument Inc. Reprinted by permission of Beckman Instrument Inc.

R-value spectra. The determination of the optimum R-value for pure ^{241}Am and ^{36}Cl standards is illustrated in Fig. 6.14.

C. Cocktail and Vial Considerations

1. Cocktail Choice

a. Aqueous-Accepting Cocktails

Aqueous-accepting cocktails are typically used in standard β-spectrometers and are quite suitable for gross alpha–beta measurements and where alpha–alpha resolution is not critical. They are infrequently used in the PERALS® system. The reasons for this are as follows. The presence of surfactants and other additives in these cocktails (and, of course, the aqueous sample itself) causes a significant degree of quenching and therefore a reduction in light output. This in turn has an adverse effect on the degree of alpha–beta separation and alpha–alpha resolution. Thus, the degree of separation is always a trade-off against sample loading. In addition, standard cocktails for beta counting applications employing fast solvents such as xylene, pseudocumene, toluene, and alkylbenzenes do not result in pulse shapes that are sufficiently different that they produce optimum separation of alpha from beta events. To overcome the poor separation, 20% naphthalene has been added to standard cocktails for alpha–beta separation applications (Oikari *et al.*, 1987). Naphthalene improves alpha–beta separation by acting as an intermediate in the energy transfer process between the solvent and fluor (Brooks, 1979; McDowell and McDowell, 1994). This more energetically favourable route increases the production efficiency of excited fluor molecules. The production rate of fluor triplet states is especially enhanced because energy transfer to triplet states relies on a physical approach and is affected by the concentration of both the fluors and the intermediate (naphthalene) in the cocktail. As more triplet states become occupied, the delayed component of the PMT anode pulse increases and has the effect of stretching the alpha pulses relative to the beta pulses. The ability of naphthalene to act as an intermediate in this way is the result of the extensive delocalization of its electrons. During the late 1980s and early 1990s, a range of cocktails based on the solvent di-isopropylnaphthalene (DIN) was introduced (Thomson, 1991). DIN has many advantages over the traditional cocktails (see Chapter 8), including the fact that it is nontoxic, nonflammable, biodegradable, and is similar to naphthalene in its ability to stretch alpha pulses relative to betas and therefore to enhance alpha–beta separation. Cocktails based on DIN, PXE, or DIN together with a second "safe" solvent have now almost completely taken over from fast cocktails containing naphthalene. Furthermore, experiments to assess the suitability of (i) secondary fluors with a range of fluorescence lifetimes and (ii) some alternative scintillating cocktail solvents have failed to improve upon the alpha–beta separation efficacy provided by commercially available cocktail formulations (Pates *et al.*, 1996c; Passo *et al.*, 2002).

b. Extractive Scintillators

Extractive scintillators are an elegant combination of two technologies (liquid scintillation and liquid–liquid extraction) in which an organophilic extractant is incorporated in an organic solvent in the presence of a fluor and naphthalene. Extractive scintillators enable extraction of the nuclide of interest from an aqueous sample directly into the scintillator, and since virtually nothing except the radionuclide of interest is transferred, there is reduced quenching compared with that observed in aqueous-accepting cocktails. This results in improved separation of β–γ from α events and improved α–α resolution. There is also a reduction in the variations in quenching between samples, and this facilitates nuclide identification. Products for the extraction of uranium, thorium, plutonium, polonium, radium, and so on are all marketed by ETRAC Laboratories Inc., Knoxville and Oak Ridge, Tennessee (Table 6.3). A typical formulation for extraction of uranium would comprise 96.3 g of Adogen 364 (tertiary amine), 160 g of

TABLE 6.3 **Properties of the Commercially Available Extractive Scintillators**

Extractive Scintillator	Elements Extracted	Applications
ALPHAEX$_\alpha$™	Th, Pa, U, Np, Pu, Am, Cm, Bk, Cf, Es, Fm, Md, No	Any original matrix after conversion to: 1 M HNO_3 for Th, U(IV, VI); pH 2–3, 1 M nitrate for others
URAEX$_\alpha$™	Uranium selectivity	From 1 M sulfate at pH 2
THOREX$_\alpha$™	Thorium and uranium	From 1 M sulfate at pH 2
POLEX$_\alpha$™	Polonium selectively	From 7.5 M H_3PO_4, 0.01 M HCL
RADAEX$_\alpha$™	Radium selectively	From 0.3–0.5 M $NaNO_3$ at pH 11–12
RADONS$_\alpha$™	Radon selectively	Radon from water or aqueous solution
STRONEX$_\alpha$™	Strontium selectively	From 0.3–1.0 M $NaNO_3$ at pH 9–11

purified naphthalene, 4 g of the scintillator PBBO [2-(4'-biphenylyl-6-phenylbenzoxazole)] made up to 1 L with toluene (Bouwer *et al.*, 1979). A comprehensive guide to the PERALS® and extractive scintillators can be found in McDowell and McDowell (1994).

2. Vial Choice

In experiments that we have conducted, no significant improvements in the separation of alpha from beta events have been observed when using plastic vials, although, peak resolution is improved. In addition, a progressive deterioration in separation with time (commencing within 48 h of sample preparation) was observed when using plastic vials, apparently caused by a stretching of the beta pulses due to an interaction between the cocktail and the plastic vial. This effect was not observed when using glass vials and there was also no significant difference in separation between 7-mL and 20-mL vials. Alpha/beta separation with glass vials can be improved significantly by etching the outside wall of the vial to improve the light output. This also has the effect of improving peak resolution.

D. Alpha/Beta Calibration

To optimize alpha–beta separation performance, it is essential to determine the optimum TD setting at which there is equal and minimum spill of alpha pulses into the beta MCA and beta pulses into the alpha MCA. Figure 6.15 illustrates the percent spillover or percent misclassification of ^{241}Am and ^{36}Cl samples in a Packard instrument. The determination of an optimum TD requires two standards, preferably, one of the pure alpha emitter of interest and one of the pure beta emitter of interest. For the most accurate results, the standards must be as nearly identical as possible to the unknown samples in their chemistry, volume, vial type, and so on. To arrive at the optimum setting, each standard is counted individually at a range of TD settings. Some instruments perform this operation automatically while others perform it semiautomatically. It is possible to favor counting either alpha or beta events by adjusting the TD value appropriately to minimize interference from the other radionuclide.

I. Misclassification Calculations

The calculation required to determine the actual cpm is essentially the same as that used in dual label studies and can be defined as follows:

$X_\alpha =$ alpha misclassification as beta
$X_\beta =$ beta misclassification as alpha
$A_T =$ true count rate due to alpha disintegrations
$B_T =$ true count rate due to beta disintegrations
$A_O =$ observed count rate in alpha MCA
$B_O =$ observed count rate in beta MCA

Alpha into beta misclassification (X_α) is defined as the ratio of counts accumulated in the beta MCA to counts accumulated in both the alpha and

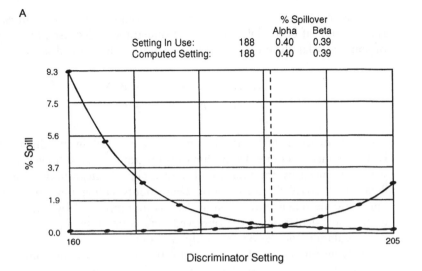

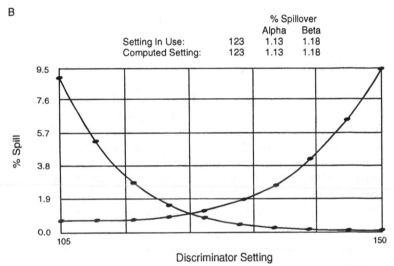

FIGURE 6.15   Percentage misclassification of ^{36}Cl and ^{241}Am in a Packard 2550TR/AB under (A)normal laboratory prepared levels of quenching and (B) artificially high levels of quenching. Reprinted by permission of PerkinElmer Life and Analytical Sciences.

beta MCAs as measured with the alpha standard.

$$X_\alpha = B_O/(A_O + B_O)$$

Similarly, beta into alpha misclassification (X_β) is defined as:

$$X_\beta = A_O/(A_O + B_O)$$

as measured with the beta standard.

It is important to understand that the count rate observed in each MCA is a function of both alpha and beta disintegrations because of the occurrence of some misclassification. This relationship can be defined as:

$$A_O = A_T - A_T X_\alpha + B_T X_\beta$$
$$B_O = B_T - B_T X_\beta + A_T X_\alpha$$

The first equation states that the observed count rate in the alpha MCA (A_O) is due mainly to counts from alpha disintegrations (A_T); however, this value will be reduced by the total number of alpha disintegrations counted in the beta MCA ($A_T X_\alpha$). Furthermore, beta counts falling into the alpha MCA must also be taken into account. This is accomplished by adding the $B_T X_\beta$ term. The calculation of the observed beta count rate (B_O) is the reverse of the calculation for the observed alpha count rate.

Solving for A_T and B_T and substitution of the A_T expression into the B_T expression and vice versa, we obtain the final equations for A_T and B_T:

$$A_T = \frac{A_O - A_O X_\beta - B_O X_\beta}{1 - X_\beta - X_\alpha}$$

and similarly

$$B_T = \frac{B_O - B_O X_\alpha - A_O X_\alpha}{1 - X_\alpha - X_\beta}$$

Example of Misclassification Calculation. The calculation is further illustrated by counting mixtures of ^{36}Cl and ^{241}Am ranging from approximately 500 DPM of ^{36}Cl and 50,000 DPM of ^{241}Am to approximately 50,000 DPM of ^{36}Cl and 500 DPM of ^{241}Am in a total volume of 10 mL of cocktail (with appropriate background samples) at the optimum TD value.

The results were as follows:

Misclassification of alphas into the beta MCA (as a fraction) = $0.0071 = X_\alpha$
Misclassification of betas into the alpha MCA (as a fraction) = $0.0063 = X_\beta$
Background in alpha-MCA = 0.69 CPM (0–2000 keV)
Background in beta-MCA = 39.6 CPM (0–2000 keV)

Table 6.4 shows the gross and background-corrected raw instrument data. Table 6.5 compares the actual alpha and beta activities in the vial with the calculated spill-corrected activities derived from the raw data.

2. Quenching and Quench Correction of Percentage Misclassification

In the PERALS® system, all aspects of sample preparation (but principally the use of extractive scintillants) are geared towards minimising the quenching and maximising the constancy of composition of samples. In standard LS counters, when aqueous-accepting cocktails are employed,

▉ **TABLE 6.4** **Gross and Net (Background Corrected) Raw Data**

Sample No.	Observed Gross Counts		Net Counts (Background Subtracted)	
	Beta-MCA	Alpha-MCA	Beta-MCA	Alpha-MCA
1	48545	684	48506	683
2	46533	2731	46494	2730
3	44254	5313	44215	5312
4	24435	25323	24395	25332
5	5208	46131	5168	46130
6	2833	47160	2793	47159
7	811	50775	771	50774

▉ **TABLE 6.5** **Results of Performing Misclassification Correction on the Net Count Rates**

Sample No.	Actual (vial) dpm		Spill Corrected dpm	
	^{36}Cl	^{241}Am	^{36}Cl	^{241}Am
1	48684	449	48811	378
2	46694	2496	46771	2453
3	44187	5004	44459	5068
4	24315	24847	24369	25348
5	4961	45118	4869	46429
6	2425	47626	2472	47481
7	507	49595	411	51135

variations in quenching are much more likely and have to be accounted for in any calculations.

Figures 6.15A and B represent misclassification plots determined in a Tri Carb LSA instrument for the same activities of standards but under different quench conditions. Figure 6.15A represents relatively unquenched conditions compared with Fig. 6.15B. Both the optimum time discrimination (setting) TD value and the percentage misclassification of events change under these different conditions. It should be noted, however, that this effect will vary according to the isotopes being measured and may also vary, in a more limited fashion, from instrument to instrument. When the degree of quenching for a particular set of samples varies, there are two possible approaches. The first approach is to produce a single pair of alpha–beta standards that are similar in their quench-indicating parameter value to the least-quenched sample and then to progressively quench them and reoptimize the TD at each quench level. This produces a series of optimum TD values and percentage misclassifications for a range of quenching. The degree of sample quenching in each sample may then be measured by making a short count on each sample to determine the value of the quench-indicating parameter. Subsequently, the samples have to be counted at their individual optimum PDD conditions; however, this could conceivably require a separate counting protocol for each sample. The alternative approach is again to prepare a pair of standards that are equivalent in quenching to the least-quenched samples, determine the optimum TD and then progressively quench the standards. However, in this approach, the misclassification is always determined at the original TD setting. This allows the construction of a plot of percentage misclassification versus quench-indicating parameter for a single TD setting. All samples may then be counted within a single protocol and a correction for misclassification as a function of quench is applied. This approach would be in addition to a quench curve that relates quenching to detection efficiency.

IV. ANALYSIS OF BETA EMITTING RADIONUCLIDES

A. Tritium (^{3}H)

I. Environmental Occurrence

Tritium is a pure, low-energy beta emitter ($E_{max} = 18.6$ keV) with a half-life of 12.43 years. Present-day tritium has three major sources of origin: (1) natural production in the upper atmosphere through cosmic ray-induced spallations of nuclides and particle capture reactions of nitrogen and oxygen (Geyh and Schleicher, 1990), (2) residual activities from atmospheric nuclear weapons testing (Libby, 1963), and (3) on-going nuclear fuel cycle operations. Since the partial atmospheric nuclear test ban treaty in 1963, the worldwide levels of tritium in the environment have been decreasing at a rate approximately equal to its half-life (Okada and Momoshima, 1993). Generally, nuclear fuel cycle operations are of concern to the immediate locale only, with relatively minor contributions to more general environmental tritium levels. Regulated monitoring of specific sites and hydrological studies are the principal applications of environmental tritium analysis, and, of course, as the "background" levels decrease, there is a corresponding requirement for the measurement of lower and lower activities.

Tritium activities are commonly described in terms of tritium units (TU), where 1 TU = 1 atom of tritium per 10^{18} atoms of hydrogen or 1 TU = 7.19 DPM L^{-1} of water = 0.118 Bq L^{-1} of water. The natural activities in precipitation vary from about 25 TU at high latitudes to about 4 TU in the equatorial zone (Geyh and Schleicher, 1990).

2. Sample Preparation and Analysis

a. Sample Handling

Samples taken for the purpose of tritium analysis should be sealed in airtight containers at the point of sampling (glass or high-density plastic is preferred), as the hydrogen from water or water-bearing samples is readily exchanged with atmospheric hydrogen in the form of water vapor. In the case of biological samples and some water-bearing biologically active samples (sludge and silt) that cannot be processed immediately, freezing is a good way to preserve the sample composition. Frozen samples should be thawed completely and the external areas of the container dried prior to opening in order to minimize sample exposure to condensation of atmospheric water (onto the sample) or isotopic exchange with water vapor in the air. When samples are exposed, it should be assumed that contamination is a possibility, and steps to prevent or monitor this are required.

b. Sample Preparation

For the purposes of this section, the term "sample preparation" refers to the processes required to render the sample into a pure state (water or benzene), that is, ready to be mixed with the scintillation cocktail or fluor in preparation for liquid scintillation counting. Generally, three methods of preparation are employed for environmental tritium analysis of water. These are direct addition, electrolytic enrichment, and benzene synthesis, and the method of sample preparation depends on the type of analysis to be performed. Samples requiring extraction of water, such as biological and soil samples, always require a purification or extraction step regardless of the analysis method to be employed. Water samples require purification (distillation is the norm) for electrolytic enrichment and benzene synthesis analysis, but this step can often be omitted with the direct addition method. Tritium can exist in biological samples as part of the water component (FWT, free water tritium) or as part of the organic structure (BT, bound tritium). To analyze for the BT component, the sample is dried (freeze drying or low-temperature oven drying at $\sim$60–80°C) and then combusted. The water of combustion is collected, purified as required, and analyzed by the direct addition method.

Direct Addition. Some water samples are relatively pure as received (e.g., municipal drinking water) and can be counted without any purification steps. For routine operations this can be a significant consideration, because the purification step greatly increases the sample preparation time. Purification is recommended if (1) color is visible (includes cloudiness from suspended solids), (2) organics are present, or (3) the sample history or source is unknown (may contain interfering chemicals).

Once the requirements of a reasonably pure sample are met, it is ready to be mixed with a scintillation cocktail and counted.

Electrolytic Enrichment. This procedure is based on the principle of selective isotopic enrichment using electrolysis. Because of the slightly higher binding energies, molecules of THO are not decomposed to form H_2 and O_2 as readily as H_2O or DHO (T = tritium, D = deuterium). Water samples are distilled, then made slightly alkaline (Ascione *et al.*, 1995) and placed in an electrolysis cell. A constant current is applied to the cell, and electrolytic decomposition of the water reduces the sample to about 5% of the initial volume. Isotopic fractionation, due to the higher binding energy, typically concentrates over 90% of the tritium in the remaining water. The concentrated sample is purified by distillation or vacuum distillation and counted by the direct addition method. This procedure usually takes 5–7 days to complete;

therefore, multicell designs containing 20–40 cells are very common. Modern systems include temperature control of samples (~2–4°C) and electronic regulation of the electrolysis process. Each cell in the system must be characterized for isotopic separation performance, which is done by adding a known quantity of tritium to the cell to determine the "recovery efficiency." Each time an electrolysis run (batch) is carried out, different cells are spiked with a known activity. In this way, the recovery efficiency (RE) of each cell is checked against a known standard after every few uses. The enrichment factor (EF) is simply the RE multiplied by the initial weight (W_i) of sample divided by the final weight (W_f) of sample and is expressed as a numerical value (e.g., enrichment factor = 15). The electrolysis cells are generally designed to handle 150–250 mL samples, with enrichment factors of 15 to > 30.

$$EF = W_i \div W_f \times RE$$

For example, a 250 g water sample is enriched and 10 g of sample remains; the recovery efficiency is 91%. EF = 250 ÷ 10 × 0.91 = 22.75.

Benzene Synthesis. The final method of preparation is the synthesis of high-purity benzene from water samples (De Filippis and Noakes, 1991). Sample water is added to calcium carbide in an evacuated reaction vessel and acetylene (C_2H_2) is generated. The acetylene is then cyclotrimerized to high-purity benzene using a chromium or vanadium catalyst. For liquid scintillation counting of benzene, there is no requirement for a scintillation cocktail, as the fluors can be added in solid form directly to the counting vial. One sample per day per vacuum apparatus can be handled, with the production of 8–15 mL of high-purity benzene. The method is only marginally more sensitive than direct addition and is not generally used. It is included for completeness and will not be discussed further.

c. Sample Purification/Extraction Techniques

Low-level counting of tritium requires that water samples be reasonably pure and free from interfering substances. Given the wide variety of sample matrices that environmental monitoring programs may encounter, various cleanup techniques are employed. If there is any doubt about the purity of a sample it is always best to take proactive action, as repeating the analysis invariably costs more (manpower, equipment, and time) than the initial sample cleanup.

Filtering. Simple filtering of a "dirty" water sample often produces a clear water that is suitable for counting. This should be used only when the source of the sample is well known, such as a routinely sampled stream or well, and there is little chance of the presence of interferences. This type of cleanup should be used only for samples in which mud or silt is suspended. An alternative would be to allow the suspended material to settle out and then decant the clear sample.

Ion Exchange. Mixed-bed ion exchange columns designed specifically for tritium analysis are commercially available. A typical column (Eichrom Industries, Tritium Column) is composed of a prefilter and beds of anion, cation, and organic specific resins. It is designed to give results comparable to those for simple distillation. The column is prerinsed with reference background water and then the sample is passed through for "cleanup." This type of column is a useful tool but should be used only with sample media that have been pretested and the cleanup performance verified. Simple water samples (ground, lake, etc.) are likely candidates for this method. More complex samples, such as rainwater exposed to plant leaves, can become highly colored and require a more aggressive procedure.

Freeze Drying. Freeze drying is the process whereby a sample is frozen and then exposed to high vacuum. The water contained in the sample goes from the solid phase directly to a vapor (sublimation), and if the vacuum is drawn through a cold trap, this vapor can be collected as pure water. Routine application of this process, for the purposes of extracting pure water for tritium analysis, is generally accomplished using custom glassware and cryogenic (liquid nitrogen, LN_2) cooling. The sample is first frozen solid using a dewar of LN_2; the dewar is then moved to a collecting flask (cold trap) and a high vacuum applied. The ambient room temperature supplies sufficient heat to accelerate the sublimation process, which may take several hours (depending on the volume of water required). The water sample is collected in the cold trap (as deposited ice) and is simply melted and recovered for analysis.

Azeotropic Distillation. An azeotrope is a liquid mixture that has a constant boiling point and whose vapor has the same composition as the liquid. Several compounds, such as toluene, benzene, and cyclohexane, form suitable azeotropes with water. Each is compatible with scintillation cocktails, immiscible in water, and cyclic in structure, making them resistant to exchange with external hydrogen. Cyclohexane is the preferred compound as it is the least hazardous and provides a suitable level of performance. The composition of the azeotrope formed by cyclohexane (boiling point 81°C) and water is 91.5 and 8.5%, respectively, and has a boiling point of 69.8°C. Azeotropic distillation (Bretthauer *et al.*, 1973) provides a simple method for extracting water from a wide variety of media (e.g., honey, milk, vegetation, soil, fish), for tritium analysis using liquid scintillation counting. Because these materials are largely composed of free water, an azeotropic distillation will yield, from a mixture of sample and cyclohexane, not necessarily azeotropic in proportion, an azeotropic vapor and, hence, distillate. The distillate separates into two layers with water forming the bottom layer. Using suitable glassware, the distillate is collected and the water portion removed for analysis.

Distillation. Plain or vacuum distillation is useful for water samples that do not contain volatile materials but may have chemical contaminants that would render them unsuitable for analysis.

d. Reference Background Water

The quality of the "reference background water" is the cornerstone of any program undertaking environmental tritium analysis as it provides the reference point against which all samples will be compared; therefore, it is critical that this water is tritium-free, otherwise an analytical bias will result. Deep wells and glacial aquifers are likely sources of tritium-free water due to their long period of isolation from atmospheric exposure. Identifying a suitable source of background water can be a challenging task, and communicating with laboratories with existing environmental tritium monitoring programs can save considerable effort and time in locating these sources. To evaluate the suitability of a reference background water it is simplest to use the electrolytic enrichment technique. A sample of the water is enriched by 15–30 times and analyzed by LSC, as a sample, using the original water as the background. This allows detection of any tritium present, at a much lower limit of detection (typically $< 0.1 \, \text{Bq} \, \text{L}^{-1}$). Suggestions on handling the background reference water supply are as follows: (1) obtain large batches (1–2 years supply) as it is more efficient and economical to validate, (2) store in airtight glass containers, and (3) implement procedures for handling, validation testing, and maintenance of testing and usage records.

e. Standards

Standards are used to ensure consistent counter performance and to check sample preparation materials and counting conditions for each batch of samples to be counted. A traceable tritium standard should be purchased in the form of HTO and aliquots accurately dispensed and diluted to give solutions of known activity. Common practice is to prepare stock solutions of high activity, which are then volumetrically diluted to working standards. As a guideline, the working standard should contain sufficient activity to yield $< 0.1\%$ counting error (at 1 sigma), when counted using the sample protocol. Keeping the tritium standard close to the guideline offers several advantages: (1) since a minimum of activity is used, potential cross contamination to samples is reduced; (2) some counters employ special background enhancing features that are sensitive to high count rates, and if the count rate of the standard exceeds the limit, inconsistent results can occur; (3) the lower the activity level employed, the more cost effective the use of the purchased tritium standard will be.

Suggestions for handling tritium standards are as follows: (1) store and prepare in airtight glass containers (i.e., dedicated volumetric flasks); (2) use stock and working solutions; (3) keep records of all certificates and preparation and testing of all stock and working solutions; (4) dispense standards in a properly ventilated area, well away from any samples.

f. Quality Control

The purpose of the quality control program is to demonstrate that the parameters involved in the production of results are within expected tolerances. The counting efficiency and background count rate are confirmed for each batch of samples by including standards and background samples. These are satisfactory for demonstrating the equipment performance but do not validate the sample handling or preparation. Two other sample types are used to demonstrate that sample handling does not adulterate or contaminate the sample and that the preparation of samples for counting is acceptable:

1. Process control samples. Samples with low levels of tritium can easily be contaminated by external sources. A process control sample is basically a sample of reference background water that is exposed to the same processes as the sample. This sample is then counted along with the rest of the sample batch and compared with the regular background to demonstrate the absence of contamination during the process. For example, if a batch of milk samples is being freeze dried to extract water for tritium analysis, a process control sample would also be freeze dried and included in the sample batch.

2. Quality control (QC) samples. The purpose of the QC sample is to introduce a sample with known activity into the handling process. Whenever a purification or extraction technique is performed, a QC sample is added to the batch and processed as a regular sample. Because the QC sample has known activity, this demonstrates the performance of the entire analytical and counting process. The QC sample water is prepared by adding a measured amount of standard stock or working standard solution to a quantity of reference background water. The activity level required varies with the count conditions but should be calculated to yield a 2 sigma counting error of the order of 10%. Charting the data from QC samples gives an historical baseline against which current results can be judged. This aids in detecting trends, poor sample preparation, and counting or counter problems.

g. Quality Assurance

Quality assurance tests are independent tests performed to provide confidence in the counting system and generally involve the analysis of known and unknown (blind) samples from external agencies. A good example of this for tritium would be the Laboratory Intercomparison Program administered by the US Environmental Protection Agency (US EPA). Water samples are distributed to the participating laboratories as "blind" samples for analysis. The analysis results are reported back to the US EPA and are tabulated for all program participants. This serves as a check on the accuracy of the analysis process and gives insight into the performance of other laboratories. Intercomparisons with other laboratories using known and unknown samples are also useful when they can be arranged.

B. Radiocarbon (^{14}C)

I. Environmental Occurrence

Carbon-14 has a half-life of 5730 years and decays by beta emission ($E_{max} = 156\,keV$) to the stable nuclide ^{14}N. It is ubiquitous in our environment and is also one of the most common radionuclides measured by LSC. Its occurrence in the environment is from three main sources; (1) natural production via the interaction of cosmically produced neutrons with atmospheric nitrogen in the upper atmosphere, (2) atmospheric nuclear weapons testing (principally during the 1950s and 1960s), and (3) discharges from facilities associated with the nuclear fuel cycle (principally nuclear power plants and fuel reprocessing plants). Additional minor sources include (1) discharges from facilities that synthesize radio-labelled compounds for biomedical, biological, and agricultural tracer studies, (2) discharges from hospitals, universities, research centers, and so on, of the labelled materials used in tracer studies, and (3) natural production by spallation reactions in surface rocks and soils.

The environmental applications for ^{14}C measurements can be divided into three main categories as follows:

1. Radiocarbon dating. The relatively constant natural production rate of ^{14}C in the upper atmosphere, its uniform uptake as ^{14}CO$_2$ into living plant material, conversion to plant carbohydrates, and subsequent transfer through the food chain are the basis of the radiocarbon dating technique. The end result of the food chain transfer is almost uniform labelling of all living organisms. ^{14}C dating is used in a wide range of scientific disciplines including archeology, geology, soil science, climate reconstruction, and oceanography, and LSC is a widely used measurement technique for ^{14}C dating (Thomson *et al.*, 1995; Pessenda *et al.*, 2001; Walker *et al.*, 2001).

2. Environmental measurements. Typically, these measurements are performed on samples from the environment around facilities associated with the nuclear fuel cycle and facilities that produce labelled compounds. Although ^{14}C is not among the most abundant of the anthropogenic radionuclides that enter our environment as a result of nuclear fuel cycle activities, its long half-life, high environmental mobility, and ability to enter the food chain mean that it delivers one of the highest collective effective dose equivalents to the global population, hence the level of interest in its measurement in the environment (Buzinny *et al.*, 1998; Gulliver *et al.*, 2001; Isogai *et al.*, 2002).

3. Food adulteration studies. Petroleum derivatives are occasionally used to adulterate natural food and drink products without the buyer's knowledge. Because petroleum-based products are sufficiently old that they contain no ^{14}C, depletions in the natural ^{14}C content are normally indicative of adulteration. Examples of this include the adulteration of wines, spirits, wine or cider vinegars, and other natural products (Schönhofer, 1989; Schönhofer, 1992).

2. Sample Preparation and Analysis

a. Sample Preparation

Benzene synthesis. This is the preferred technique for radiocarbon dating because of its higher precision, although it is also used for environmental and food adulteration studies. Benzene synthesis involves the following steps: Sample conversion to CO_2 is by combustion in a pure oxygen atmosphere for organic carbon samples or acid hydrolysis for inorganic carbon samples (e.g., shells and biogenic deep ocean sediments). The CO_2 is subsequently converted to lithium carbide by reaction with molten lithium, and on cooling, the addition of water causes the production of acetylene (Barker, 1953). The acetylene is then cyclotrimerized to benzene using a chromium- or vanadium-based catalyst (Noakes et al., 1963). Benzene is an ideal counting medium for the following reasons:

1. It is a clear aromatic solvent capable of dissolving sufficient fluor concentrations.
2. It has excellent energy transmission properties.
3. It has a high carbon content (92.3%).
4. It is relatively easy to synthesize.
5. It has reasonable resistance to quenching.

In most instances, solid fluors would be added directly, rather than a scintillation cocktail. This minimizes volume additions and therefore any increase in quenchable background. The trend in the radiocarbon dating discipline has been to use only a primary fluor and to use it at relatively high concentrations (12–15 mg per gram of benzene) to facilitate weighing. This procedure would be common for researchers using instruments such as the Wallac Quantulus 1220. This is now possible because the wider wavelength response of modern PMTs has negated the requirement for a secondary fluor (wavelength shifter). A primary and a secondary fluor (bis-MSB) are recommended for users of Packard instruments with nonprogrammable TR-LSC. The recommended concentrations are 2.8 mg of butyl-PBD and 3.0 mg of bis-MSB per gram of sample benzene, and highly accurate techniques for dispensing such small weights are available (Cook et al., 1990a,b; Anderson and Cook, 1991; Cook and Anderson, 1992). A primary fluor alone can be used with programmable TR-LSC provided the burst delay is increased beyond 75 ns.

CO_2 Absorption. Direct absorption of CO_2 using quaternary amines is a widely used technique, particularly where the activities are enriched relative to natural production. Carbo-Sorb (Packard), Optisorb '1' (Wallac), and Carbamate-1 (National Diagnostics) are examples of high-capacity carbon dioxide absorbers that are compatible with LSC cocktails. Direct absorption into inorganic bases is an alternative means of absorbing the CO_2 in which aqueous solutions of up to 1 M sodium or potassium hydroxide or methanolic solutions of 2 M potassium hydroxide have been used. However, these have three distinct disadvantages: they are strong quenching agents, they have low trapping capacities compared with the amines, and they produce severe chemiluminescence.

For CO_2 samples absorbed into base solutions, cocktails that can accept high-ionic-strength solutions are added. For low-precision measurements, the advantages of the absorption technique are that sample preparation is less time consuming than for benzene synthesis and the sample preparation apparatus is relatively simple. The major disadvantage of the CO_2 absorption technique is the amount of carbon that can be absorbed. For example, 10 mL of Carbo-Sorb is approximately the maximum quantity that would be miscible with a scintillation cocktail in a 20-mL vial. This is capable of absorbing 58 mmol of CO_2, which is equivalent to 0.7 g of carbon. In comparison, it is possible to add approximately 19 g of carbon into a vial via benzene synthesis, although 3–7 g is more typical. Detailed information on the absorption technique may be obtained from Qureshi et al. (1989).

Direct counting. On occasion, when adulteration of alcoholic drinks is being studied, direct counting of the alcohol may be performed. Schönhofer (1989) has demonstrated that direct counting of uncolored spirits with an ethanol concentration of about 40% v/v is feasible. A 1:1 ratio of spirit to cocktail (Quickszint 400, Zinsser) was found to be suitable. For coloured spirits and wine, distillation was required to concentrate and purify the ethanol. A 1:1 mix of cocktail to 85–90% ethanol was used.

For ^{14}C measurements in vinegar, acetic acid has been isolated by continuous extraction with di-isopropyl ether and subsequent distillation. Subsequently, 4 mL of concentrated acetic acid in 6 mL of Quickszint 400 were analyzed (Schönhofer, 1992).

b. Standards (primarily for ^{14}C dating)

The Primary Modern Reference Standard for radiocarbon measurements is wood from the AD 1890 tree ring (and therefore uncontaminated by fossil fuel CO_2 – Suess Effect) whose activity was measured in the 1950s and corrected for ^{14}C decay to the year of growth to determine the absolute activity. Radiocarbon ages are expressed in years B.P. (before present – where present is the year 1950). In practice, all laboratories use a Secondary Modern Reference Standard against which sample activities are determined. The secondary standard most commonly in use is Oxalic Acid SRM 4990C which is distributed by the National Institute of Standards and Technology, Gaithersburg, USA. This reference material consists of a 1000 lb lot of oxalic acid prepared by fermentation of French beet molasses using *Aspergillus niger*; 0.7459 times the specific activity of this material (normalized to a $\delta^{13}C$ value of −25‰) is equal to the activity of the AD 1890 wood, which in turn equates to 226 Bq kg^{-1} of carbon.

c. Quality Assurance

There is an inevitable diversity of experimental approaches within radiocarbon dating facilities and in this situation, the issue of comparability of results amongst laboratories becomes paramount. Over the past 20 years, a number of international intercalibration programs have taken place (ISG, 1982; Scott et al., 1990; Rozanski et al., 1992; Gulliksen and Scott, 1995). These intercomparisons are of direct benefit to the participating laboratories as an independent check of laboratory performance and provide indirect benefit to the user communities. As a result of these studies, a range of natural samples (e.g., dendrochronologically dated wood, marine sediment, peat, barley, and so on) are available, albeit in limited quantities. These materials have known ages or consensus activities that cover the entire applied radiocarbon timescale.

d. Calculation of Results and Radiocarbon Conventions

Most radiocarbon laboratories, in performing their calculations, use an activity value A_{ON}, which is 74.59% of the measured net Oxalic Acid SRM 4990C activity (A_{OX}), normalized for ^{13}C fractionation according to:

$$A_{ON} = 0.7459 A_{OX}[1 - \{2(25 + \delta^{13}C)/1000\}]$$

(Measurements of $\delta^{13}C$ are made by stable isotope mass spectrometry and are relative to the VPDB standard, which is a Cretaceous belemnite, *Belemnita americana*, from the Peedee formation in South Carolina).

The term *Percent Modern* equates to the activity ratio $A_{SN}/A_{ON} \times 100\%$ where A_{SN} is the sample activity (A_S) normalized for sample fractionation to $-25‰$. However, the activity A_{ON} depends on the year of measurement (y) and has to be corrected for decay between 1950 and the year (y) of actual measurement. The *absolute international standard activity* is given by

$$A_{abs} = A_{ON} e^{\lambda(y-1950)}$$

where y is the year of oxalic acid measurement and $\lambda = 1/8267 \, y^{-1}$ is based on the 5730 yr half-life.

For geochemical and environmental ^{14}C studies, the term *absolute percent modern* (pM) is more commonly used. This is defined as:

$$pM = A_{SN}/A_{abs} \times 100\%$$

When one refers to 100% modern, this effectively is the specific activity that prevailed in 1890, prior to the onset of the Suess Effect; 100% modern is equivalent to $226 \, Bq \, kg^{-1} \, C$.

For radiocarbon age calculations the equation is:

$$t = \frac{1}{\lambda} \ln \frac{A_{ON}}{A_{SN}}$$

where t = time (age), $\lambda = \ln 2/5568$ (where 5568 years is the Libby half-life of ^{14}C), A_{ON} is 74.59% of the measured net Oxalic Acid SRM 4990C activity (A_{OX}), normalized for ^{13}C fractionation, and A_{SN} is the sample activity (A_S) normalized for sample fractionation to $-25‰$.

A comprehensive discussion of the reporting of ^{14}C data is given in Stuiver and Polach (1977) and recently updated in Mook and van der Plicht (1999).

C. Nickel-63 (^{63}Ni)

I. Environmental Occurrence

^{63}Ni is a weak beta-emitting radionuclide of $E_{max} = 67 \, keV$ and has a half-life of 9.2 years. It exists in the coolant water of nuclear power reactors (Lo *et al.*, 1993) and is formed by neutron capture of nickel released from the steel piping and so on due to corrosion. It is included in the list of low-level, long-lived radioactive wastes from nuclear power reactors specified in the US Nuclear Regulatory Commission (NRC) Regulation 10CFR Part 61 published by the NRC in 1982.

2. Sample Preparation and Analysis

The measurement of ^{63}Ni in environmental samples by liquid scintillation counting (LSC) requires separation and purification from the original sample matrices because of its weak beta radiation and interference from coexisting radionuclides. Procedures for measuring ^{63}Ni have been reported for many sample matrices, including liquid effluents, water, seawater, vegetation, urine, and resins (Harvey and Sutton, 1970; Kramer, 1981; Radwan *et al.*, 1981; Kojima and Furukawa, 1985; NUREG, 1985; Russow and Dermietzel, 1990; Lo *et al.*, 1993).

Because it exists in the coolant water of nuclear power reactors, water and liquid effluents are the most common sample matrices to be assayed. Chemical separation of nickel by precipitation with dimethylglyoxime (DMG) followed by quantitative analysis by LSC is the method commonly used for assay, although precipitation as the dipyridine nickel dithiocyanate complex $[Ni(C_2H_4N)_4](CNS)_2$ and solubilization in a dioxane-based liquid scintillation cocktail has also been reported (Harvey and Sutton, 1970).

The sample matrix dictates whether any specific sample collection or preservation procedures are required. Although the scientific literature does not emphasize specific collection procedures, the addition of HCl or HNO_3 to lower the pH to <2 and addition of a stable nickel carrier and/or a complexing agent have all been documented (Harvey and Sutton, 1970; Terlikowska-Drozdziel and Radoszewski, 1992; DOE, 1993). Filtering of water samples is commonly performed (Lo *et al.*, 1993) and samples such as sludges, resins, or organic material require wet oxidation with strong acids to enable dissolution of the sample (NUREG, 1985; Russow and Dermietzel, 1990).

Sample processing includes chemical separation methods which are required to isolate the ^{63}Ni for LSC. An $Fe(OH)_3$ scavenge has been used and is necessary to separate the nickel from organic complexing agents such as ethylenediaminetetraacetic acid (EDTA) and other contaminants in order to precipitate the nickel quantitatively as nickel dimethylglyoxime. Nickel remains in the precipitate at a pH of 6–8, but at higher pH, $Ni(OH)_2$ starts to dissolve. Anion exchange has also been used to pretreat reactor effluent samples to remove heavy metals that are likely to interfere with the precipitation (Harvey and Sutton, 1970), and a preconcentration procedure for enrichment of ^{63}Ni from seawater by adsorption on hydrous magnesium oxide has also been described (Lo *et al.*, 1993). Dewberry *et al.* (1999) used a Ni-DMG precipitation followed by Eichrom Industries' Ni-selective extraction chromatography resin to determine ^{63}Ni in the presence of fission product and plutonium alpha–beta activities at activities up to 10^3 times the observed ^{63}Ni activity.

A representative chemical separation method involving DMG has been published in the US Department of Energy (DOE) Methods Compendium manual (DOE, 1993). The method is used to determine the activity of ^{59}Ni and ^{63}Ni in a solution and was developed to assay drainable liquids and acid-dissolved sludges. In this method, a nickel sample solution is loaded onto a column of DMG, which is prepared by mixing an ethanol slurry of DMG with Microthene (Quantum Inc., Tuscola, Illinois), a 50-mesh polyethylene powder. The LSA was calibrated for efficiency by counting the standards and blanks that were prepared with the analytical batch. Quench correction was not performed in this procedure because chemical quench was assumed to be relatively constant between the standards and samples, as the standards were processed identically. Details of the method and possible interferences are published as radiochemistry procedure RP300 in the March 1993 addendum of the DOE Methods Compendium manual. No detection limit was reported for the DOE procedure, however, various limits of detection have been reported by other investigators. Russow and Dermietzel (1990) reported a minimum detectable activity for ^{63}Ni in plant tissue of 80 Bq g^{-1} dry matter, assuming a count time of 10 min and a standard deviation of $\pm 3\%$ at the 95% confidence level. Harvey and Sutton (1970) spiked samples with varying known amounts of ^{63}Ni and stable nickel carrier and could detect as little as 0.099 Bq per sample with counting efficiencies in excess of 60% and a background of 20–25 CPM using an older generation LSA. Kojima and Furukawa (1985) reported a detection limit of 0.06 Bq g^{-1} at 95% confidence assuming a counting time of 1000 min in their method, and Kramer (1981) reported a limit of 0.055 Bq per sample in urine. None of these measurements was made with instruments capable of low-level performance, since most of the reported backgrounds were in the range 20–25 CPM. Modern instrumentation equipped with background reduction features as discussed in Section II would produce significantly lower detection limits.

D. Strontium-89 and Strontium-90/Yttrium-90 (^{89}Sr and ^{90}Sr/^{90}Y)

I. Environmental Occurrence

^{90}Sr ($E_{max} = 546$ keV and half-life $= 28.5$ y) is a fission product and ^{89}Sr ($E_{max} = 1492$ keV and half-life $= 50.55$ d) is part of the mass 89 fission chain, so their main sources in the environment are atmospheric nuclear weapons testing and releases from the nuclear fuel cycle. ^{90}Sr is normally in equilibrium with its ^{90}Y daughter ($E_{max} = 2282$ keV and half-life $= 2.67$ d). Fallout from nuclear weapons testing is primarily responsible for the ^{89}Sr and ^{90}Sr/^{90}Y concentrations found globally in soils. ^{90}Sr is important because of its relatively high fission yield and long physical and biological half-lives. ^{89}Sr shares the same biological significance but has a much shorter physical half-life and thus will not have a long-term environmental impact. Water, milk, soil, vegetation, and urine are the typical sample matrices that have been analyzed.

2. Sample Preparation and Analysis

The quantitative analysis of radiostrontium is based on three major considerations: sample pretreatment (to bring the sample into a suitable form), radiochemical separation and nuclear counting (Wilken and Joshi, 1991). Much of the emphasis on the assay of radiostrontium has been focused on the separation chemistry, as it mimics calcium in its behavior and this makes separation difficult in environmental samples where, inevitably, calcium is abundant.

a. Early LSC Methods

Several methods of measuring strontium using a liquid scintillation counter have been reported (Uyesgi and Greenberg, 1965; Piltingsrud and Stencel, 1972; Carmon, 1979; Reynolds and Eldridge, 1980; Martin, 1987; Tait *et al.*, 1997; Tait *et al.*, 1998; De Vol *et al.*, 2001). These are categorized as liquid scintillation and/or Cerenkov counting methods.

The measurement of ^{90}Sr alone by LSC is complicated because ^{89}Sr may also be present and interferes with the ^{90}Sr measurement. Typically, preconcentration and chemical separation by extraction and chromatographic techniques have been applied because of the complexity of the sample matrix and the low levels of ^{89}Sr and ^{90}Sr and other environmental radionuclides in the sample. The exact separation scheme depends on the sample matrix and whether individual activities of ^{89}Sr and ^{90}Sr are required. Cerenkov measurements usually require two or more measurements for ^{89}Sr and ^{90}Sr in the presence of ^{90}Y (Carmon, 1979; Reynolds and Eldridge, 1980). Early attempts at strontium assay by LSC involved precipitation of strontium as insoluble salts and suspension in a suitable cocktail containing a gelling agent. Others report a method of concentrating inorganic ions from large aqueous samples with a scintillating ion exchange resin (Heimbuck *et al.*, 1963). All of these early methods suffered from several limitations, such as failure to correct for yield recovery, quenching, and nonuniformity of samples. In addition, the methods required that strontium and yttrium be in secular equilibrium (except for Piltingsrud and Stencel, 1972). This requirement added considerably to the time needed to complete an assay. Methods were then attempted to assay for ^{90}Sr under nonequilibrium conditions. One such method also employed the use of a gelling agent to suspend an insoluble strontium precipitate containing ^{90}Y, ^{90}Sr, and ^{89}Sr and a spectrum unfolding computer program to determine the quantity of each of the three isotopes (Piltingsrud and Stencel, 1972). Another method involved precipitating as strontium carbonate, dissolving in glacial acetic acid, and adding 2-ethylhexanoic acid to form strontium and yttrium 2-hexanoates which are then soluble in a toluene-based cocktail (Uyesgi and Greenberg, 1965).

b. Recent LSA Methods

Currently, the most popular separation methods for radiostrontium involve the use of ion exchange chromatography (Amano and Yanase, 1990), liquid–liquid (solvent) extraction (Dietz and Horowitz, 1993; Tait and Wiechen, 1993), and extraction chromatography (Dietz *et al.*, 1991; Vajda *et al.*, 1992).

Liquid–liquid (solvent) extraction has been used for process-scale separation and preconcentration of radionuclides, and recent research has focused on the development of a workable acid-side extraction process for the removal and recovery of ^{90}Sr from nuclear waste streams (Dietz and Horowitz, 1993). Macrocyclic polyethers (crown ethers) have been proposed to overcome the unique problems of the low charge and large ionic radius of the Sr ion. This low charge density made it difficult to strip away associated water molecules that must accompany the strontium into the organic-phase to maintain electrical neutrality. The net result was poor extraction efficiency. An improvement was developed by Dietz and Horowitz (1993) when they combined a crown ether [*bis*-4,4'(5')-*tert*-butylcyclohexano-18-crown-6] in 1-octanol. This improved extraction specificity and the ability to accommodate a substantial amount of water. Liquid–liquid extraction methods using crown ethers have been applied to analytical work, but they are time consuming and may generate a substantial amount of radioactive waste. A more manageable approach that has been adopted is to coat a small portion of the selective extractant onto an inert solid support. Sr•Spec resin (strontium specific), that has become commercially available, (Eichrom Industries, Darien, Illinois) is such a chromatographic material. The use of an extraction agent on an inert support is known as extraction chromatography.

Separation methods using extraction chromatography are very popular for assaying radiostrontium, and the technique has been applied to bioassay as well as environmental samples. Dietz *et al.* (1991) employed the technique to determine ^{89}Sr and ^{90}Sr in urine. The separation scheme involved acid digestion of the urine sample (600 mL) to remove organics, and precipitation of basic calcium phosphate. The residue was redissolved in nitric acid and loaded onto an Sr•Spec column to isolate radiostrontium.The strontium-containing sample was ultimately eluted from the column with 5 mL of water and mixed with 17 mL of scintillation cocktail. The authors reported that the advantage of this method is that samples can be counted immediately after strontium separation. Yields from the coprecipitation step and column recoveries were reported to be around $95 \pm 5\%$ overall. The limit of detection was reported as about $25\,mBq\,L^{-1}$. Hong *et al.* (2001) report limits of detection (60-min count time) of 37 and $32\,mBq\,L^{-1}$ for ^{90}Sr and ^{89}Sr, respectively, for their analysis of aqueous samples, following Sr separation using Sr•Spec resin while Heilgeist (2000) reports a limit of detection of approximately $100\,mBq\,kg^{-1}$ in fresh produce for the analysis of ^{90}Sr and ^{89}Sr following separation with Sr•Spec resin.

Tait *et al.* (1998) have developed a solid-phase extraction method for strontium in milk, using cryptand 222 bound to Dowex 50W × 8 cation exchange resin. This method does not require drying, ashing, acid leaching, and precipitation and under optimum conditions, using this technique, 95% of the Sr was removed from 1 L of milk in 4 h. The Sr was subsequently eluted from the resin using hot 1 M NaCl in 1 M HCl and then precipitated as strontium carbonate which was subsequently dissolved and then added to a scintillation cocktail for ^{89}Sr and ^{90}Sr analysis. An earlier version of this method (Tait *et al.*, 1997) employed Sr•Spec resin to isolate the Sr in a suitable form for LSC analysis following extraction with cryptand 222/Dowex 50W × 8.

c. Cerenkov Counting Methods

Several attempts have been made to determine radiostrontium by Cerenkov counting (Randolph, 1975; Buchtela and Tschurlovits, 1975; Carmon, 1979; Reynolds and Eldridge, 1980; Martin, 1987; Rucker, 1992; Chu *et al.*, 1998; Vaca *et al.*, 2001). These methods are based on the fact that both ^{89}Sr and ^{90}Y can be detected with high counting efficiency (>40%) by Cerenkov counting in an aqueous solution. Interference from ^{90}Sr is at a minimum because of its low Cerenkov counting efficiency (< 1.4%). This fact has been the reason for the appeal of Cerenkov counting, because it allows immediate determination of ^{89}Sr in a fresh mixture of ^{89}Sr and ^{90}Sr. In all of these references, conventional separation techniques, including precipitation, ion exchange, or liquid extraction, were used to isolate radiostrontium. After separation, the ^{90}Y component is used to quantify the ^{90}Sr activity. Martin (1987) radiochemically separated the ^{90}Y component, and Carmon (1979) mathematically calculated the amount of ^{90}Y ingrowth within 2 h after the separation. Reynolds and Eldridge (1980) waited for ^{90}Y ingrowth (^{89}Sr levels were found to be insignificant). Rucker (1992) used Cerenkov counting for ^{89}Sr and then added scintillation cocktail to determine the total ^{89}Sr and ^{90}Sr activity with a mathematical correction for ^{90}Y ingrowth. Banavali *et al.* (1992) proposed a method that involves the use of carbonate precipitation and extraction chromatography to isolate total radiostrontium, which is determined by Cerenkov counting (primarily ^{89}Sr). The sample is Cerenkov counted again 2–3 days later, when there has been sufficient ^{90}Y ingrowth, to determine the activity of ^{90}Sr via ^{90}Y. Cerenkov counting techniques allow the use of ^{85}Sr as a yield tracer because this isotope is not detected by Cerenkov counting. This can make the process simpler, because evaporative steps and desiccation steps to determine a gravimetric yield can be avoided. Calculated detection limits of 0.35 Bq L^{-1} for ^{89}Sr and 0.29 Bq L^{-1} for ^{90}Sr were reported by Rucker (1992) for a 1-L sample, 20-min count time, and 80% chemical recovery.

In summary, much of the emphasis on the assay of radiostrontium has been on separation chemistry. Recently, extraction chromatography methods have gained in popularity. The analytical techniques for the determination of ^{89}Sr, ^{90}Sr, and ^{90}Y are dependent on the decay characteristics of the three radionuclides and the desired detection limits. LSC offers the ability to measure radiostrontium at very high counting efficiencies after separation. Cerenkov counting of the ^{90}Y daughter can be combined with cocktail counting of radiostrontium to determine the radiostrontium component of a sample. LSA detection limits are reported to be <1 Bq L^{-1}. While significant advances have been made in the analysis of radiostrontium, its measurement in solid matrices still presents a problem to many analysts. The IAEA recently organized an intercomparison exercise involving 158 laboratories, worldwide, to analyze a mineral sample spiked at 3 different levels with ^{90}Sr. The major sources of bias leading to underestimation and overestimation of ^{90}Sr activities are discussed in De Regge *et al.* (2000).

E. Technetium-99 (^{99}Tc)

1. Environmental Occurrence

^{99}Tc is a pure beta emitter with an E_{max} of 292 keV and a half-life of 2.12×10^5 years. It is a fission product, and its presence in the environment is due to nuclear weapons tests and nuclear fuel cycle operations, although the contribution to soils from weapons fallout is small (6.2 mBq kg^{-1}) (Rouston and Cataldo, 1978). Substantial quantities are produced from the running of nuclear reactors although releases to the environment are primarily from fuel reprocessing operations. The principal reasons for the level of interest in environmental measurements of this nuclide are its long half-life, high mobility and solubility as the pertechnetate ion (TcO$_4^-$), and high transfer rate from soil to edible vegetation.

2. Sample Preparation and Analysis

Several methods have been used to measure ^{99}Tc in the environment. Separation techniques include anion exchange, organic extraction, selective precipitation, or combinations of these (Boyd and Larson, 1960; Golchert and Sedlet, 1969; Holm *et al.*, 1984; Cattarin *et al.*, 1985; Silva *et al.*, 1988; Paviet *et al.*, 1991; Wigley *et al.*, 1999), and quantitative analysis has typically been by liquid scintillation or gas proportional counting.

A representative ion exchange separation method is described by Silva *et al.* (1988). The method involves separating ^{99}Tc from samples of archived Nevada Test Site (NTS) ground water using AG 1-X8 (100–200 mesh) anion exchange resin with ^{95m}Tc as the chemical yield tracer. A spectral stripping technique was used to determine concentrations of ^{99}Tc and ^{95m}Tc from standard spectra for these isotopes. The ^{99}Tc and ^{95m}Tc were finally identified from both their characteristic spectra and column elution patterns. The detection limit was reported to be approximately 0.5 DPM L^{-1}.

The procedures based on liquid–liquid extraction or a combination of liquid–liquid extraction and ion exchange use various extractants, including DB18-C6 crown ether (dibenzo-18-crown-6), carbon tetrachloride, and an aqueous solution of sodium thiosulfate and tetraphenyl arsenium chloride followed by tri-*n*-butyl phosphate (Walker *et al.*, 1980; Luxenburger and Schuttelkopf, 1984; Paviet *et al.*, 1991; Verrezen and Hurtgen, 1992). Wigley *et al.* (1999), used a combination of ion exchange and liquid–liquid extraction to effect separation. In this method, ^{99m}Tc was used as the yield tracer for analysis of a range of environmental samples. These were combusted, the residue extracted into HNO$_3$, purified by ion exchange and then a final separation of the Tc from Ru was carried out by extracting the Tc into 5% tri-*n*-octylamine in xylene. The xylene fraction containing the Tc was then mixed with a commercial scintillation cocktail. Recovery of the ^{99m}Tc was determined immediately by gamma spectrometry. The sample was then left for 2 weeks to allow the ^{99m}Tc to decay and subsequently counted by LSC to determine the ^{99}Tc activity.

Selective extraction methods using TEVA•Spec resin (Eichrom Industries, Darien, Illinois) and liquid scintillation counting or inductively coupled plasma mass spectrometry (ICP-MS) have been reported for measurement of technetium concentrations in environmental samples (Sullivan *et al.*, 1991; Beals, 1992; Scarpitta, 1993). A consolidation of the extraction chromatography method for water analysis has recently been included in the US Department of Energy (DOE) Methods Compendium manual (DOE/EM 0089T, Rev. 1, 1993). The method proposed analyses by either ICP-MS or LSC. Quantitative analysis by LSC is viewed as a very practical approach, because access to an LSC is more universal than access to an ICP-MS. There are several possible variations on this technique. For example, the resin can be flushed into a vial containing a gel cocktail or the ^{99}Tc can be eluted with 7–9 M HNO_3 and an aliquot submitted for LSC. In addition, ^{95m}Tc or ^{99m}Tc can be used as a yield tracer. The LSC detection limit is reported to be approximately $37\,mBq\,L^{-1}$ for the extraction technique.

F. Lead-210 (^{210}Pb) [Bismuth-210 (^{210}Bi) and Polonium-210 (^{210}Po)]

I. Environmental Occurrence

^{210}Pb (half-life = 22.3 years), ^{210}Bi (half-life = 5.01 days) and ^{210}Po (half-life = 138.38 days) are all members of the ^{238}U decay series (Fig. 6.16). ^{210}Pb decays via combined β^- emission ($E_{max} = 61\,keV$), γ emission ($E_\gamma = 46.5\,keV$), and internal conversion to the β^--emitting ^{210}Bi ($E_{max} = 1161\,keV$). In turn, ^{210}Bi decays to the α-emitting ^{210}Po ($E_\alpha = 5.3\,MeV$). A detailed decay scheme is illustrated in Fig. 6.17. The main reasons for carrying out ^{210}Pb/^{210}Bi/^{210}Po analyses are two-fold: (1) Radiological considerations. Much attention has been paid to their analysis in drinking waters and in air (e.g., Katzlberger *et al.*, 2001; Wallner, 2001) and (2) they can be used as tracers of rates and mechanisms of processes in the environment. ^{210}Pb, ^{210}Bi, and ^{210}Po are all particle reactive species and this property has been used to advantage in the study of a range of processes. For example, ^{210}Pb has been used to study bioturbation in slowly accumulating sediment systems, e.g., deep ocean sediments (Thomson *et al.*, 1993), and accumulation rates in rapidly

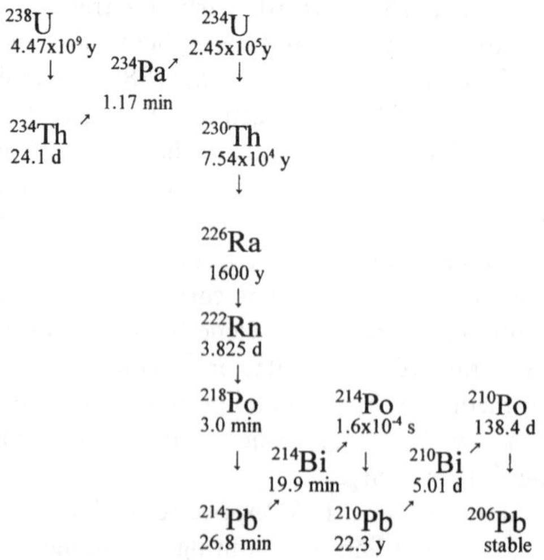

FIGURE 6.16 ^{238}U natural decay series.

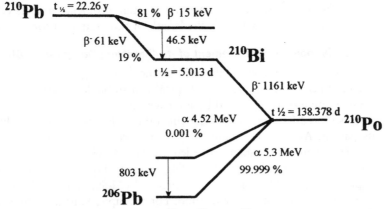

FIGURE 6.17 Decay scheme for ^{210}Pb, ^{210}Bi and ^{210}Po.

accumulating sediment systems, e.g., some peat profiles and lake sediments (MacKenzie *et al.*, 1998). Typically, there is sufficient ^{210}Pb activity in such systems to allow direct counting by gamma spectrometry to be carried out.

They have also been used in the study of a range of processes occurring in the marine water column. For example, ^{210}Po/^{210}Pb activity ratio measurements have been used to study biological productivity and boundary scavenging (Nozaki *et al.* 1976; 1997; 1998; Smith and Ellis, 1995) while Bacon *et al.* (1988) and Wu and Boyle (1997) have used ^{210}Pb as an analog for contaminants delivered to the oceans from the atmosphere. However, the activities present in seawater are extremely low and preconcentration steps are required to allow large sample volumes to be processed.

2. Sample Preparation and Analysis

There are three traditional ways by which ^{210}Pb is typically analyzed:

a. Direct Counting by Gamma Spectrometry

This can be accomplished by counting the 46.5 keV gamma photon emissions using an N-type HPGe gamma photon detector with a beryllium or carbon epoxy thin window. The difficulty in this method is the comparatively low counting efficiency of HPGe detectors combined with the low absolute intensity of the 46.5 keV gamma photon (approx. 4%). However, this technique is feasible when analyzing peat and sediment samples as described above.

b. Indirectly by Measurement of its α-emitting Grand-daughter (^{210}Po)

In sediments, ^{210}Pb and ^{210}Po are often found in equilibrium and the activity of ^{210}Pb can be estimated from that of the ^{210}Po. The ^{210}Po is typically analyzed by alpha spectrometry following preconcentration and spontaneous deposition onto silver foils. However, in the upper water column, equilibrium between ^{210}Pb and ^{210}Po is rarely achieved and separate ^{210}Pb and ^{210}Po determinations are required. To achieve this, a polonium yield tracer is added to the sample (^{208}Po or ^{209}Po), ^{210}Po must then be quantitatively removed for analysis after which the sample is respiked with the polonium yield tracer and stored for several months to allow ingrowth of

^{210}Po. However, there are several potential problems associated with this method, not least of which is the length of time required for ^{210}Po ingrowth.

c. By Indirect Measurement of its β¯emitting Daughter (^{210}Bi)

This is not as commonly used as the ^{210}Po ingrowth method but it has been applied in marine studies. Typically, a stable lead carrier is added to the sample, the lead is separated by ion exchange and precipitated using $Pb(SO_4)$. This is stored for 1 month, after which the ^{210}Bi approaches equilibrium with its parent. Analysis is usually by gas proportional counting.

The major advantage afforded by liquid scintillation is high counting efficiency. Potentially, ^{210}Pb, ^{210}Bi, and ^{210}Po can all be analyzed at approaching 100% counting efficiency by LSC. Direct determination of ^{210}Pb by LSC is not a new method. For example, Fairman and Sedlet (1968) demonstrated the potential of this technique using standardized solutions of ^{210}Pb, ^{210}Bi, and ^{210}Po in secular equilibrium and a modification of the ion exchange technique, developed by Raby and Hyde (1952), to separate them prior to counting. However, the advances have mainly occurred since the introduction of MCA-based, microcomputer-controlled systems in the 1980s and the development of more sophisticated solid and liquid phase extraction systems. Several methods have been used to effect Pb separation including (i) precipitation (e.g., as $PbSO_4$), (ii) solid-phase extraction (e.g., Sr•Spec resin), (iii) liquid-phase extraction (e.g., Polex extractive scintillator), (iv) ion exchange, and (v) combinations of these techniques. Al-Masri *et al.* (1997) determined ^{210}Pb in environmental water samples by adding stable lead carrier as the yield tracer, followed by precipitation as a sulphate and dissolution in alkaline EDTA. They then allowed ^{210}Bi to grow in and determined the activity by Cerenkov counting, thereby avoiding potential interferences from alpha and soft beta emitters, although, the counting efficiency by this method was comparatively low (20%). Peck *et al.* (1998) demonstrated an increase in Cerenkov counting efficiency form 17–75% through the use of Triton X-100 and sodium salicylate as enhancers. Under these conditions, ^{210}Po caused some interference, which was not present in the absence of the enhancers. Kim *et al.* (2001) used a similar $PbSO_4$ precipitation technique to determine ^{210}Pb in groundwater and soil but in this instance, they counted the precipitate directly in a gel scintillation cocktail and determined the ^{210}Pb from the total spectrum of ^{210}Pb and the ingrown ^{210}Bi. Vajda *et al.* (1997) determined ^{210}Pb and ^{210}Po in a range of matrices including soils, sediments, and biological tissues using Sr•Spec resin. This retains both ^{210}Po and ^{210}Pb which can then be eluted sequentially using HNO_3 and HCl respectively. Analysis of the ^{210}Po was by alpha spectrometry, using ^{208}Po as the yield tracer while the ^{210}Pb was determined by LSC following gravimetric yield determination of the stable lead carrier. Wallner (1997) determined ^{210}Pb, ^{210}Bi, and ^{210}Po in aerosol samples by dissolution and extraction of Bi and Po into Polex, leaving Pb in the aqueous phase. ^{210}Bi and ^{210}Po were determined by α–β LSC after short-lived ^{212}Pb and its progeny had been allowed to decay. ^{210}Pb was determined by 2 methods (i) direct measurement by adding cocktail to the aqueous phase and (ii) carrying out a second Polex extraction on the aqueous phase after allowing the ^{210}Bi to grow in for 2 weeks, followed by LSC determination. A similar approach was also adopted by Katzlberger *et al.* (2001) for the analysis of natural drinking waters. Most recently, Biggin *et al.* (2002) have developed a method for determining each of ^{210}Pb, ^{210}Bi, and ^{210}Po in seawater using ^{212}Pb, ^{207}Bi, and ^{208}Po, respectively, as yield tracers. This offers the opportunity to examine the geochemical cycling of the ^{210}Pb/ ^{210}Bi/ ^{210}Po system in the marine environment and to use them as tracers of scavenging processes operating on timescales ranging from a few days to tens of years. Separation of the 3 radionuclides was accomplished using Sr•Spec resin. Both the ^{210}Pb and ^{210}Bi were determined by LSC while the ^{210}Pb was determined by alpha spectrometry. Figure 6.18 is a schematic representation of the method.

G. Thorium-234 (^{234}Th)

I. Environmental Occurrence

Thorium has an average crustal abundance of $8.1\,\mathrm{mg\,kg^{-1}}$ but this is comprised almost entirely of ^{232}Th (half-life $= 1.41 \times 10^{10}$ years). ^{234}Th (half-life $= 24.1$ days) is the immediate daughter of ^{238}U (half-life $= 4.47 \times 10^{9}$ years). It is a low-energy beta emitter ($E_{max} = 190\,\mathrm{keV}$) which also produces gamma photons at low probabilities (63 keV at 3.8%, 92.8 keV at 2.7%, and 92.4 keV at 2.7%). Its daughter, ^{234m}Pa (half-life $= 1.17\,\mathrm{min}$) occurs in secular equilibrium with its parent within the time frame of most analytical procedures and because ^{234m}Pa decays by the emission of a high-energy beta particle ($E_{max} = 2.33\,\mathrm{MeV}$), it can also be used to determine the ^{234}Th activity of a sample.

Thorium occurs only in the 4+ oxidation state and is a highly insoluble, highly particle reactive and of low environmental mobility. In contrast, under oxidising conditions uranium occurs in the 6+ oxidation state and exists as the soluble uranyl ion ($UO_2{}^{2+}$). These contrasting geochemical behaviors have enabled any disequilibrium between the ^{234}Th/^{238}U daughter/parent pair to be used to study a number of short term processes occurring in the oceans. In the water column, the major applications for ^{234}Th/^{238}U disequilibrium have focused on either elucidating oceanic carbon fluxes or modeling reactive pollutant behaviors in coastal environments. In the sediment this disequilibrium is typically used to study reworking processes (Martin and Sayles, 1987).

2. Sample Preparation and Analysis

The two methods traditionally used to determine ^{234}Th activities are: (1) gamma spectroscopy via the low probability gamma emissions, which results in the requirement for very large sample volumes (>200 L) (Buesseler *et al.*, 1992, 2001) and (2) determination of the beta emissions from either ^{234m}Pa or both ^{234}Th and ^{234m}Pa by gas proportional counting which requires sample volumes in the range 2–100 L (Coale and Bruland, 1987; Moran and Buesseler, 1993; Buesseler *et al.*, 2001). However, Anderson *et al.* (1991) introduced a liquid scintillation method that utilized the alpha-emitting ^{230}Th isotope as the yield tracer for the combined beta emissions

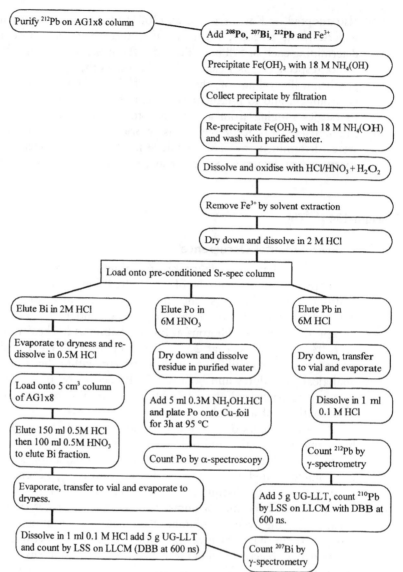

FIGURE 6.18 Flowchart of the analytical procedure for analysis of ^{210}Pb, ^{210}Bi and ^{210}Po in filtered seawater samples. (From Biggen *et al.*, 2002, reprinted with permission from *Anal. Chem.* (2002) 74, 671–677. Copyright 2002 American Chemical Society.

from ^{234}Th and ^{234m}Pa. Briefly, this method was as follows:

1. Filtered seawater (20 L) was spiked with ^{230}Th yield tracer.
2. Thorium was scavenged from the sample by coprecipitation with ferric hydroxide.
3. The ferric hydroxide precipitate was redissolved and thorium was purified by anion exchange.

Since alpha particles produce about one-tenth of the light per keV that would be produced by a beta particle, there is interference between the ^{230}Th

spectrum, from its 4.7 MeV alpha particles and the ^{234m}Pa/^{234}Th beta spectrum. However, Anderson *et al.* (1991) produced samples of constant quench and under these conditions, the extent of the interference remains constant. Through the use of separate ^{230}Th and ^{234}Th/^{234m}Pa spectra, produced at the same quench level as samples, they estimated the degree of interference in suitable counting windows. Under their conditions, these counting windows were 10–70 keV or 0–80 keV for ^{234m}Pa/^{234}Th and 100–220 keV for ^{230}Th. These gave counting efficiencies of approx. 100% for ^{230}Th and 137% (0–80 keV) relative to ^{234}Th for the combined ^{234m}Pa/ ^{234}Th. More recently, Pates *et al.* (1996b) refined this method significantly through the use of an LS spectrometer with α–β separation capability to overcome the spectral interference problem. Figure 6.19 illustrates the ^{230}Th yield tracer spectrum and the underlying ^{234m}Pa/^{234}Th continuum when counted using a Packard LSC with alpha–beta separation capability. These spectra were derived from the analysis of a 20 L seawater sample. They achieved a limit of detection of 0.4 mBq L^{-1} for a 20 L sample and a counting time of 400 min. Preparation of the ^{230}Th yield tracer from urananite ore requires a significant degree of preparation, the full details of which are given in Pates *et al.* (1996b).

Nour *et al.* (2002) developed a similar method for the determination of ^{234}Th in sediments using the ^{230}Th yield tracer. They used extraction chromatography (TEVA•Resin) rather than ion exchange to separate and purify the thorium and determined the ^{234}Th via the ^{234m}Pa by Cerenkov counting. Subsequent to this, cocktail was added and the sample counted at a PSA level that gave low beta misclassification, in order to optimize the LSC for ^{230}Th measurement to determine thorium yields.

Kersten *et al.* (1998) used a thorium specific diatomite adsorption technique in which 100 L samples of seawater were filtered and spiked with

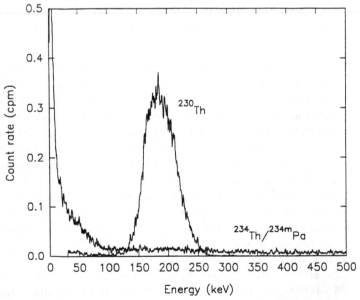

FIGURE 6.19 ^{230}Th yield tracer spectrum and underlying ^{234m}Pa/ ^{234}Th spectrum.

an α-emitting ^{229}Th yield tracer. The water was acidified to pH 4 using HNO$_3$ and 20 g of diatomite (MERCK 159241) were stirred in and the pH raised to 8–9 using NaOH. Following further stirring and then settling, the diatomite was removed by decanting and filtration, and the adsorbed thorium was eluted using 9M HCl. The eluate was then prepared for α–β liquid scintillation spectrometry by taking it to dryness, adding 1 mL of 1M HCl and 20 mL of cocktail.

H. Plutonium-241 (^{241}Pu)

I. Environmental Occurrence

^{241}Pu is a low-energy beta emitter with an E_{max} of 21 keV and a half-life of 14.4 years. It would be considered to have relatively low radiotoxicity; however, it decays to ^{241}Am (half-life = 433 years), which is an alpha particle emitter and is highly radiotoxic. The fallout from past nuclear weapons testing and releases associated with nuclear fuel cycle operations (in particular from reprocessing) as gaseous or liquid effluents are the two predominant sources of environmental contamination (Pimpl, 1992). In nuclear power facilities, the production of ^{241}Pu begins with neutron capture by ^{238}U followed by a succession of beta decays and neutron captures to yield ^{240}Pu and ^{241}Pu (Martin, 1986). In fact, ^{241}Pu is the only significant beta-emitting transuranic nuclide in low-level radioactive waste. The fact that it is a pure, low-energy beta emitter means that LSC is the ideal analytical technique for its measurement.

2. Sample Preparation and Analysis

Pretreatment and processing procedures typically involve ashing the samples at 500°C, leaching with strong acids, purification by solvent extraction and/or ion exchange, and alpha spectroscopy to measure the alpha plutonium activity. Thereafter, several methods can be employed to determine the ^{241}Pu activity:

1. Allowing sufficient time for ingrowth of the ^{241}Am daughter. The electrodeposited source is then recounted and the increase in count rate within a window that includes both the ^{238}Pu and the ^{241}Am is determined. From this, the ^{241}Am is calculated, followed by a back calculation to determine the original ^{241}Pu content (Day and Cross, 1981). The major drawback of this method is the slow ingrowth of the ^{241}Am daughter, which greatly extends the analysis time.

2. Leaching the disk with nitric acid, again after a considerable time has elapsed from the initial electrodeposition. This is followed by radiochemical separation of the ^{241}Am, electrodeposition, and α spectrometry (Holm and Persson, 1977). Again, back calculation allows determination of the original ^{241}Pu activity, but this technique suffers from the drawback already discussed.

3. Several researchers have recovered the deposited plutonium, typically by boiling the disk with nitric acid and then purifying it, for example, by liquid–liquid extraction or ion exchange. An aliquot of the sample,

in a suitable matrix, is then added to a scintillation cocktail and the activity determined (Schell *et al.*, 1981; Pimpl, 1992).

4. Direct measurement of the electrodeposited ^{241}Pu has also been carried out (Horrocks and Studier, 1958; Cook and Anderson, 1991; Ryan *et al.*, 1993). In this technique, the disk with the electrodeposited plutonium is placed, face up, in the base of a scintillation vial. A few milliliters of scintillation cocktail are added and the sample is counted. The cocktail should preferably be one designed for a non-aqueous medium as it will contain no surfactants and will produce less quenching and therefore a higher counting efficiency. Ryan *et al.* (1993) listed several advantages for this technique over the others: (a) no need to strip the plutonium from the disk, (b) no need to wait for ^{241}Am ingrowth, (c) no detectable chemiluminescence, (d) no spectral degradation provided the plated disks are specular, (e) sufficient sensitivity for measurement of environmental levels despite this being a 2π counting geometry (compared with the 4π geometry of technique 3 above), and (f) easy retrieval and storage of disks.

The alpha spectroscopy approach has largely been replaced by liquid scintillation analysis because a quick, direct analysis of ^{241}Pu after separation is possible. However, initial alpha spectroscopy is still required if measurement of all plutonium isotopes is necessary. LSC with pulse shape analysis has been used to reduce interferences in the energy region of the ^{241}Pu (Lee and Lee, 1999). A typical ^{241}Pu (beta) spectrum with an associated alpha plutonium peak is shown in Fig. 6.20, which was the result of a TOPO/heptane extraction from nitric acid solution (Fliss and Enge, 1993). Cook and Anderson (1991), in their study of cocktails, reported a minimum detection limit of 44 mBq per sample based on 2.83 times the background error and a 100-min count. Pimpl (1992) reported a detection limit of 50 mBq for a 100-min count using a formula based on 3.29 (95% confidence) times the background error. Hands and Conway (1977) reported a detection limit of 63 mBq per sample based on their procedure for a 100-min count time and a significance of three times the background error.

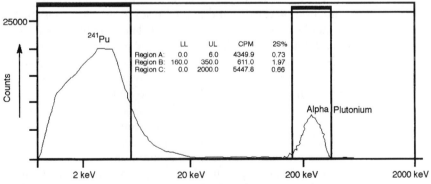

FIGURE 6.20 Composite LSA spectrum of ^{241}Pu and alpha emitting Pu isotopes. Reprinted by permission of PerkinElmer Life and Analytical Sciences.

V. ANALYSIS OF ALPHA-EMITTING RADIONUCLIDES USING CONVENTIONAL LS SPECTROMETERS WITH PULSE SHAPE DISCRIMINATION

The much poorer resolution of LSC compared with semiconductor detection methods is a fundamental limitation in the use of this technique for alpha spectrometry. Even the resolution of the PERALS system (which is optimized for alpha spectrometry), in combination with extractive scintillators, is poorer by a factor of about 10; nevertheless, there are many applications in alpha spectrometry where resolution is not critical. Furthermore, software programs for processing and deconvoluting alpha spectra are being developed that make it possible to carry out a wider range of analyses (Egorov *et al.*, 2001).

The range of extractive scintillators produced by McDowell and co-workers has enabled the development of analytical techniques for a number of elements that have alpha-emitting isotopes. These include thorium, polonium, radium, radon, uranium, and the transuranics. In some cases, the energies of the individual isotopes are sufficiently different that resolution is feasible (e.g., thorium); in others (e.g., uranium), resolution will be a limiting factor in determining individual isotopic activities. A comprehensive discussion of the applications of PERALS is beyond the scope of this chapter and we would advise anyone interested in this system to consult McDowell and McDowell (1994); however, Table 6.3 summarizes the commercially available extractive scintillators, the elements that they extract, and their applications (from McDowell and McDowell, 1993). In addition, the development of extraction chromatography resins (e.g., TRU•Spec and U/TEVA•Spec resins) has afforded a greater ability to carry out the type of separation required for analysis of alpha-emitting radionuclides by LSC but again, resolution will be the limiting factor.

The following discussion is largely limited to conventional LS systems, however, it should be borne in mind that virtually all of the analytical techniques for alpha emitters were either initially designed for the PERALS system or can be adapted for it.

A. Gross Alpha Measurements

Detection of alpha activity on glass fiber "swipe" samples, to check for surface contamination, is feasible and limited resolution is also possible, provided the paper is not heavily loaded with material. The glass fiber paper is simply placed in the scintillation vial with a small volume of extractive scintillator (Alphaex) or a DIN-containing cocktail suitable for nonaqueous media, to give optimum discrimination of the beta–gamma activity. The degree of pulse shape discrimination is sufficient to eliminate most of the beta–gamma activity.

In a similar manner, gross alpha activities (excluding radium and radon) may be determined for a range of environmental materials. The most commonly analyzed matrix is water, in particular well water. Here, the sample can be dried down and taken up in a minimum of dilute acid and then mixed with a DIN-based cocktail that will accept an aqueous medium. If the sample is in solid form, it is dissolved using a suitable method that leaves it in the chloride, nitrate, or perchlorate form and then extracted using Alphaex.

A recent approach to the measurement of gross alphas in liquid samples, which combines extraction technology with liquid scintillation analysis technology, has been developed. The use of extraction resins (Eichrom Industries, Darien IL, USA) and alpha liquid scintillation counting (Thakkar, *et al.*, 1997) has resolved sample preparation problems associated with drying the sample on a planchet for traditional gas flow proportional counting (GPC). Many of the inherent limitations associated with the traditional GPC method are overcome by using this approach and results demonstrate faster sample preparation, lower detection limits and shorter counting times. This approach has also been applied to the counting of urine samples for bioassay (Eikenberg, *et al.*, 1998).

B. Radium-226 (^{226}Ra)

I. Environmental Occurrence

^{226}Ra (half-life $= 1602$ years) is a naturally occurring radioisotope of the ^{238}U decay series. Earth, marine, and environmental scientists often require analysis of ^{226}Ra in natural waters because of public health concerns (e.g., Aupiais *et al.*, 1998) and because it has proved a useful tracer of geochemical processes, particularly in the marine environment (Burnett and Tai, 1992). The measurement of radium in public water supplies has become a matter of interest because it is one of the most hazardous elements with respect to internal radiation exposure (Higuchi *et al.*, 1984), indeed, regulations have made the analysis of ^{226}Ra and ^{228}Ra very common in US ground water. Natural waters and drinking water are by far the most common sample matrices assayed for radium by liquid scintillation methods, although the ^{226}Ra content of tissue, soil, sediment, and rock samples has also been determined by liquid scintillation methods (Higuchi, 1981; Blackburn and Al-Masri, 1992; Saarinen and Suksi, 1992).

2. Sample Preparation and Analysis

Many procedures for determining the ^{226}Ra activity in water involve liquid scintillation counting of the ^{222}Rn daughter product either alone or together with its daughter nuclides; therefore, any measurement of ^{226}Ra will also be relevant to ^{222}Rn. A typical method for the analysis of ^{226}Ra in water involves a volume reduction and purging with helium followed by sealing of the sample, which is then set aside for approximately 3 weeks to allow radon ingrowth. The high solubility of radon in organic solvents is exploited for making these measurements, as the water sample forms a separate phase when mixed with organic-accepting cocktails and the radon contained in the water sample fully partitions into the organic cocktail phase in about 3 h. The radon content can then be analyzed quantitatively by a conventional LSC using a wide window. Typical liquid scintillation cocktails that are used in these procedures are based on toluene, xylene, or mineral oil solvents.

A common alternative method is to coprecipitate the radium with barium sulfate and/or lead sulfate. The precipitate can then be decomposed using phosphoric acid and the sample mixed with a toluene-based cocktail.

The high solubility of radon in nonpolar solvents such as toluene enables counting of this two phase system. Rapid ingrowth of ^{222}Rn daughters means that the gross counting efficiency is close to 500% (3 α particles and 2 β particles), however, background count rates can be high, thus limiting the limit of detection. Alpha/beta separation by pulse shape discrimination can be used to reduce the background by basing the measurement only on the alpha spectrum (Satoh *et al.*, 1984). Alternatively, Sato *et al.* (2000) have used LSC combined with α–β coincidence for low-level measurements. This method is based on the coincidence counting of the β particle from ^{214}Bi and the α-particle from ^{214}Po (half-life $= 163\,\mu$s). These paired pulses can be separated electronically from background events and single pulses from other radionuclides, giving significant reductions in the background. A compact, single photomultiplier tube LSC with automatic sample changing of up to 7 samples that is dedicated to analysis of these paired pulses has been developed by Gudjonsson and Theodorsson (2000).

Specific chemical extractants are now also used commonly for radium separation. These include sym-di[4(5)-*tert*-butylbenzo]-16-crown-5-oxyacetic acid (DTBDB16C5-OAcH) (Chu and Lin, 2001) and 2-methyl-2-heptylno-nanoic acid (HMHN) together with dicyclohexano-21-crown-7 [Cy(2)21C7] (Aupiais *et al.*, 1998), which are the active constituents of RADAEX that was developed by McDowell and coworkers.

C. Radon-222 (^{222}Rn)

1. Environmental Occurrence

^{222}Rn (half-life $= 3.8$ days) is an inert noble gas and the immediate daughter product of ^{226}Ra. Natural radiation accounts for the majority of human exposure to radiation, and ^{222}Rn and its short-lived daughter products are the largest contributors to this radiation dose (Tso and Li, 1987).

2. Sample Preparation and Analysis

a. ^{222}Rn Measurements in Air

Radon has long been identified as a serious health hazard in uranium mines and has also been cited as a hazard in indoor air in certain regions (e.g., the Cornwall region of the United Kingdom). Radon gas emanates from the soil into homes from cracks in flooring or through basement floors by molecular diffusion or pressure-driven flow (Tso and Li, 1987) and may also be present in building materials. Because of the radon health concern, several methods have been developed to monitor for radon and its daughters in air. These include alpha track, activated charcoal adsorption, continuous radon monitoring, grab radon sampling, and radon progeny integrated sampling (Passo and Floeckher, 1991). As a screening method, the charcoal adsorption technique is popular because it is inexpensive and easy to use. Typically, radon gas is trapped by activated charcoal contained in a circular canister, approximately 10×3 cm in dimension, and the trapped radon daughters are detected by gamma spectrometry (George, 1984). An attractive alternative to the gamma spectrometry method of detection has been to use liquid scintillation counting to measure the radon and daughters trapped by activated charcoal (Prichard *et al.*, 1980; Prichard and Marien, 1985). The liquid scintillation charcoal technique offers greater sensitivity, simplicity, and the ability to automate the counting of large numbers of samples. This has made the LSC method of detection ideal for radon screening. The Niton Corporation markets a radon gas collection device that is a specially designed plastic scintillation vial containing a porous canister held securely near the top of the vial (Passo and Floeckher, 1991). The canister contains a bed of activated charcoal (1.3 g) and silica desiccant (0.9 g) and a removable cap to prevent moisture or radon from entering the vial during storage or after exposure. These detectors are passive collection devices requiring no power, and exposure is initiated by removing the cap to allow the radon in the air to diffuse passively into the charcoal. The typical exposure time is 48 h, at which point the radon accumulation has reached 95% of equilibrium with the air. The exposure period is ended by replacing the cap on the detector. Eluting the radon from the charcoal is accomplished by adding 10 mL of a xylene-based cocktail to the bottom of the vial. The desorption of radon takes place through the vapor phase because the cocktail is not in direct contact with the charcoal. Desorption is about 80% complete after 3 h, which is the time for full equilibrium of the decay products. The maximum count rate is achieved in 8 h. Final radon concentrations are calculated by taking the observed CPM due to radon and daughters counted in a 25–900 keV region of interest and applying an empirically determined calibration factor, as well as factors to correct for the decay of radon, adsorption time, and elution time. A database computer program that contains calculation routines based on the factors for these detectors is also available through Niton Corporation. A minimum detection limit for radon measurements in air by the Niton system is of the order of 15 mBq L^{-1}.

b. ^{222}Rn Measurements in Water

The ^{222}Rn concentration in water is due to the decay of ^{226}Ra associated with the surrounding rock and soil. Radon gas percolates through the soil and rock and dissolves in the water; therefore, the concentration of radon in water is often higher than one would expect if the activity were due only to that supported by ^{226}Ra dissolved in the water. As previously stated, the popularity of using liquid scintillation for radon analysis is due to the high solubility of radon in organic solvents such as toluene and xylene, which are used in LSC cocktails. Properly collected water samples can be added directly to a nonaqueous-accepting scintillation cocktail and form a two-phase aqueous/organic system. The radon will be partitioned between the water, scintillation cocktail, and the air space in the vial and will be available for measurement by LSC methods. Much of the detail given earlier for ^{226}Ra analysis applies to ^{222}Rn, apart from the volume reduction, which would cause the ^{222}Rn to be lost from the sample so that subsequent ingrowth would be from the supported component only (i.e., from ^{226}Ra dissolved in the water). Much information has been published concerning the proper method for sampling water for radon measurement in domestic premises, and these methods can be adapted for other situations.

The following are the basic steps:

1. Attach a sampling funnel and tubing to the faucet.
2. Turn on the water and allow a steady flow for 2 min.
3. Slow the water flow, invert the funnel (mouth up), and adjust the flow so that the pool water in the funnel cavity is not turbulent.
4. Insert the needle of a 20-mL hypodermic syringe below the water surface and withdraw several milliliters of water and discard. Repeat this rinse several times.
5. Withdraw 12–15 mL of water slowly to minimize air bubbles. Invert syringe to eject any air bubbles and retain 10 mL of water.
6. Place the syringe needle under the surface of 10 mL of an appropriate organic-accepting liquid scintillation cocktail contained in a glass scintillation vial and slowly eject the water from the syringe into the cocktail.
7. Slowly withdraw the syringe and tightly cap the vial.
8. Measure the sample by LSC.

Obviously, steps 1, 2, and 3 can be modified if collection is not from a faucet. If taking water from a lake or stream, a direct sample can be taken into a standard EPA-type water collection bottle with a volume of at least 20 mL. These bottles have rubber–Teflon septa and prevent radon leakage from the bottle. There is some advantage in using alpha–beta LSC rather than standard beta LS counting for radon measurements in water. This derives from the low alpha background, which makes it possible to achieve lower detection limits. Alpha/beta LSC methods for radon counting have taken two approaches: the two-phase system and homogeneous counting. The approach used by Salonen (1993) was homogeneous counting with a safe cocktail. Since this method involves direct mixing of the water sample and cocktail, interferences from other radionuclides in surface and ground waters must be considered. However, the author points out that typically the concentrations of interfering radionuclides in surface and ground waters are, on an average, 2–3 orders of magnitude lower than the radon concentrations. The best limit of detection obtained was $0.03\,Bq\,L^{-1}$ for a 400-min count using the Wallac Quantulus and Teflon vials. The author reported better reproducibility with this homogeneous counting approach than with the more common two-phase counting approach. Spaulding and Noakes (1993) used the two-phase approach to obtain a limit of detection of $<0.3\,Bq\,L^{-1}$ for a 60-min count time using a Packard 2550TR/AB and standard glass vials.

D. Uranium

I. Environmental Occurrence

Isotopes of uranium are naturally occurring within the two uranium decay series, either as the parents (i.e., ^{238}U, ^{235}U) or as a daughter within the series (^{234}U). Natural decay series radionuclides occur ubiquitously in rocks, soil, and water, and, together with ^{40}K, they contribute most of the natural radioactivity occurring in our environment. NCRP 50 (1976) provides more details on the occurrence and the concentrations of individual uranium isotopes.

2. Sample Preparation and Analysis

Although ICP-MS and semiconductor alpha spectrometry are important techniques for low-level uranium measurement, liquid scintillation methods have been used to monitor for uranium in a variety of matrices including water, urine, phosphate containing materials, and air (Horrocks, 1974b; Hinton *et al.*, 1990; Miller, 1991; Prichard and Cox, 1991; Venso *et al.*, 1993; Sanchez-Cabeza and Pujol, 1998; Forte *et al.*, 2001). Many of the reported analyses of uranium involve extractive scintillator techniques. For example, Prichard and Cox (1991) and Venso *et al.* (1993) combined an extractive agent, *bis*(2-ethylhexyl)phosphoric acid (HDEHP), with either a toluene-based or a DIN-based cocktail. If alpha–beta discrimination techniques are used, cocktails containing mixtures of phenyl-*ortho*-xylylethane (PXE) and DIN or DIN alone give good alpha–beta separation without purging the sample of oxygen. Fueg *et al.* (1997) used extraction chromatographic resins as part of their separation procedure for analyzing ^{234}U and ^{238}U (and Th isotopes) in soil samples without the use of a U yield tracer. MDAs varied between 0.2 and $0.8\,\mathrm{Bq\,kg^{-1}}$ for 1g of sample and 80,000 s counting time.

The simple extraction procedure used by Prichard and Cox (1991) is shown in Fig. 6.21. The weight of the organic phase (scintillation cocktail) is recorded by weighing the volume in a preweighed scintillation vial to correct for recovery of scintillation solution. Coating the bottom of translucent vials

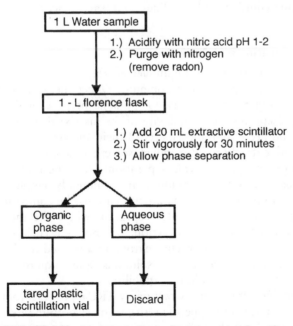

FIGURE 6.21 Analytical scheme for extraction of uranium from water. Reprinted by permission of Perkin Elmer Life and Analytical Sciences.

with a reflective material improved energy resolution to the degree that some nuclide identification among ^{234}U, ^{238}U, and other alphas was possible. Prichard and Cox (1991) reported a minimum detection limit of $0.0377\,Bq\,L^{-1}$ for the HDEHP extraction method and counting with a conventional LSC, assuming 90% combined counting efficiency and recovery, an average background of 3.2 CPM and a counting time of 10 min.

A minimum detectable limit for alpha–beta discriminating LSC has been reported as $0.01\,Bq\,L^{-1}$ because of the lower alpha background (Venso *et al.*, 1993) for a 1-L sample counted for 20 min.

E. Transuranium Elements (Np, Pu, Am, Cm)

1. Environmental Occurrence

The transuranic elements, neptunium (Np), plutonium (Pu), americium (Am), and curium (Cm), are products of nuclear weapons testing and the nuclear fuel cycle. With the exception of plutonium and to a lesser extent americium, which have been widely studied as part of the global fallout following weapons testing, the others are of environmental importance only around fuel reprocessing and disposal sites. Their disposal is of particular concern because most are long lived and have high radiotoxicity. Because of their toxicity, strict restrictions govern the limits allowed for transuranium elements in the environment; therefore, the determination of these elements in high-level radioactive waste is important as it will influence the method of disposal. However, the measurement of transuranics in high-activity waste is not trivial because concentrations are low and activities of beta- and beta–gamma-emitting radionuclides and salt concentrations are high (Yang, 1990).

2. Sample Preparation and Analysis

The measurement of transuranium elements in high-activity waste and in nuclear fuel reprocessing cycles is common. Direct measurement of radionuclides in these sample matrices is not possible because of the number of radionuclides present and the complexity of the matrix; therefore, separation and concentration are important aspects of their analyses.

Liquid–liquid extraction (also called solvent extraction) is usually the method of choice for process-scale separation and concentration of radionuclides. Because extractive techniques are commonly employed in transuranic separation schemes, the cocktail chosen must be compatible with the separation method. For process-scale separation where the radionuclide of interest is separated by solvent extraction, organophilic cocktails constitute the best choice. When extraction chromatography is used as the separation method, samples are aqueous acid solutions and require an aqueous-accepting cocktail with acid-holding capability.

Transuranic element solvent extraction by TRU•Spec (Eichrom Industries, Darien, Illinois), a transuranic-specific resin, involves both the isolation of the elements of interest from the sample matrix and the separation of the transuranic elements from one another on the resin. This involves the use of the organophosphorus extractant octyl(phenyl)-N,N-di-isobutylcarbamoylmethyl-phosphine oxide (CMPO) in a paraffinic hydrocarbon containing a moderate amount of tri-n-butyl phosphate (TBP) as a second extractant.

Yang (1990) proposed two extraction schemes for the isolation of various transuranic elements in high-activity waste streams. One such method is proposed for Np, Pu, and Am. This scheme involves the use of two extractants, di-(2-ethylhexyl)phosphoric acid (HDEHP) and trialkyl phosphine oxide (TRPO). The other scheme was developed for the assay of Np in the presence of relatively high Pu and Am contents. After extraction, the samples were analyzed using an alpha–beta LSC to eliminate any beta interferences that may have been coextracted. By using a combination of solvent extraction and alpha–beta discrimination, Yang observed that alpha activity can be quantitatively analyzed in the presence of a 10^5 excess of beta activity in high-activity waste. Detection limits depend on the degree of concentration achieved in the separation scheme and the amount of beta interference that is present. Detection limits with conventional LSCs equipped with alpha–beta discrimination were of the order of 0.2–$0.5\,\mathrm{Bq\,L^{-1}}$ for a 60-min count time without concentration of the sample.

Recent research by Dacheux, Aupiais and coworkers (Dacheux and Aupiais, 1997; Dacheux and Aupiais, 1998; Aupiais *et al.*, 1999) has focused on the potential of the PERALS system in combination with extractive scintillators for the analysis of a number of transuranic radionuclides. They have demonstrated limits of detection that are often better than were achieved by both inductively coupled plasma mass spectrometry and alpha spectrometry.

REFERENCES

Alessio, M., Allegri, L., Bella, F., and Improta, S. (1976). Study of background characteristics by means of high efficiency liquid scintillation counter. *Nucl. Instrum. Methods Phys. Res.* **137**, 537–543.

Al-Masri, M. S. and Blackburn, R. (1995). Application of Cerenkov radiation for the assay of ^{226}Ra in natural water. *Sci. Total Environ.* **173/174**, 53–59.

Al-Masri, M. S., Hamwi, A., and Mikhlallaty, H. (1997). Radiochemical determination of lead-210 in environmental water samples using Cerenkov counting. *J. Radioanal. Nucl. Chem.* **219**, 73–75.

Amano, H. and Kasai, A. (1988). The transfer of atmospheric HTO released from nuclear facilities during normal operation. *Journal of Environ. Radioact.* **8**, 239–253.

Amano, H. and Yanase, N. (1990). Measurement of ^{90}Sr in environmental samples by cation-exchange and liquid scintillation counting. *Talanta* **37**, 585–590.

Anderson, R. and Cook, G. T. (1991). Scintillation cocktail optimization for ^{14}C dating using the Packard 2000 CA/LL and 2260 XL. *Radiocarbon* **33**, 1–7.

Anderson, R., Cook, G. T., MacKenzie, A. B., and Harkness, D. D. (1991). The determination of ^{234}Th in water column studies by liquid scintillation counting. *In* "Liquid Scintillation Counting and Organic Scintillators, 1989" (H. Ross, J. E. Noakes and J. D. Spaulding, Eds.), pp. 461–470. Lewis Publishers, Michigan.

Ascione, G., Elwood, S. M., Frankenfield, R. A., Griesbach, O. A., and Stencel, J. R. (1995). Practical aspects of environmental analysis for tritium using enrichment by electrolysis. *Radioact. Radiochem.* **6**, 40–49.

Aupiais, J., Dacheux, N., Thomas, A. C., and Matton, S. (1999). Study of neptunium measurement by alpha liquid scintillation with rejection of beta-gamma emitters. *Anal. Chim. Acta* **398**, 205–218.

Aupiais, J., Fayolle, C., Gilbert, P., and Dacheux, N. (1998). Determination of Ra-226 in mineral drinking waters by alpha liquid scintillation with rejection of beta emitters. *Anal. Chem.* **70,** 2353–2359.

Bacon, M. O., Belastock, R. A., Tecotsky, M., Turekian, K. K., and Spencer, D. W. (1988). Lead-210 and polonium-210 in the ocean water profiles of the continental shelf and slope south of New England. *Continental Shelf Res.* **8,** 841–853.

Banavali, A. D., Moreno, E. M., and McCurdy, D. E. (1992). Strontium-89, 90 analysis by Eichrom column chemistry and Cerenkov counting. Abstract in *Proceedings of the 38th Annual Conference on Bioassay, Analytical, and Environmental Radiochemistry.* Santa Fe, New Mexico.

Barker, H. (1953). Radiocarbon dating: Large scale production of acetylene from organic material. *Nature* **172,** 631–632.

Beals, D. (1992). Measuring technetium-99 in aqueous samples by ICP-MS. *In* "Proceedings of the Third International Conference on Nuclear and Radiochemistry", Vienna, Austria.

Bicron Corporation. (1988). BGO ($Bi_4Ge_3O_{12}$) Product Bulletin SC-103A, Newbury, Ohio.

Biggin, C. D., Cook, G. T., MacKenzie, A. B., and Pates, J. M. (2002). Time-efficient method for the determination of ^{210}Pb, ^{210}Bi and ^{210}Po activities in seawater using liquid scintillation spectrometry. *Anal. Chem.* **74,** 671–677.

Blackburn, R., and Al-Masri, M. S. (1992). Determination of radium-226 in aqueous samples using liquid scintillation counting. *Analyst,* **117,** 1949–1951.

Bouwer, E. J., McKlveen, J. W., and McDowell, W. J. (1979). A solvent extraction-liquid scintillation method for assay of uranium and thorium in phosphate-containing material. *Nucl. Technol.* **42,** 102–111.

Bowman, S. (1989). Liquid scintillation counting in the London Underground. *Radiocarbon* **31,** 393–398.

Boyd, G. E. and Larson, Q. V. (1960). Solvent extraction of heptavalent technetium. *J. Phys. Chem.* **64,** 988–996.

Bretthauer, E. W., Compton, E. H., and Moghissi, L. A. (1973). Separation of water from biological and environmental samples for tritium analysis. *Anal. Chem.* **45,** 1565–1566.

Brooks, F. D. (1979). Development of organic scintillators. *Nuclear Instrum. Methods Phys. Res.* **162,** 477–505.

Buchtela, K. and Tschurlovits, M. (1975). A new method for determination of Sr-89, Sr-90, and Y-90 in aqueous solution with the aid of liquid scintillation counting. *Appl. Radiat. Isot.* **26,** 333–338.

Buchtela, K., Tschurlovits, M., and Unfried, E. (1974). Eine Methode zur Unterschiedung von Alpha- und Beta-Strahlen in einem Fluessigszintillationsmessgeraet. *Int. J. Appl. Radiat. Isot.* **25,** 551–555.

Buesseler, K. O., Bacon, M. P., Cochran, J. K., Livingston, H. D., Casso, S. A., Hirschberg, D., Hartman, M. C., and Fleer, A. P. (1992). Determination of thorium isotopes in seawater by non-destructive and radiochemical procedures. *Deep-Sea Res.* **39,** 1103–1114.

Buesseler, K. O., Benitez-Nelson, C., Rutgers van der Loeff, M., Andrews, J., Ball, L., Crossin, G., and Charette, M. A. (2001). An intercomparison of small- and large-volume techniques for thorium-234 in seawater. *Marine Chem.* **74,** 15–28.

Burnett, W. C. and Tai, Wei-Chei. (1992). Determination of radium in natural waters by alpha liquid scintillation. *Anal. Chem.* **64,** 1691–1697.

Buzinny, M., Likhtarev, I., Los, I., Talerko, N. and Tsigankov, N. (1998). ^{14}C analysis of annual tree rings from the vicinity of the Chernobyl NPP. *Radiocarbon* **40,** 373–379.

Cadieux, J. R., Clark, S., Fjeld, R. A., Reboul, S., and Sowder, A. (1994). Measurement of actinides in environmental samples by photon-electron rejecting alpha liquid scintillation. *Nucl. Instrum. Phys. Res. Sect. A* **353,** 534–538.

Calf, G. E. and Airey, P. L. (1982). Liquid scintillation counting of C-14 in a heavily shielded site. *In* "Archaeometry: an Australian Perspective" (W. Ambrose and P. Duerden, Eds.), pp. 351–356, Australian National University Press, Canberra, Australia.

Calf, G. E. and Polach, H. A. (1974). Teflon vials for liquid scintillation counting of carbon-14 samples. *In* "Liquid Scintillation Counting: Recent Developments" (P. E. Stanley and B. Scoggins, Eds.), pp. 224–234. Academic Press, New York.

Carmon, B. (1979). The use of Cerenkov radiation for the assay of radiostrontium in aqueous solutions. *Appl. Radiat. Isot.* **30**, 97.

Cattarin, S., Doretti, L., and Mazzi, U. (1985). Determination of ^{99}Tc in urine by liquid scintillation counting to evaluate internal contamination. *Health Phys.* **11**, 795–804.

Chu, T. C., Wang, J. J., and Lin, Y. M. (1998). Radiostrontium analytical method using crown-ether compound and Cerenkov counting and its applications in environmental monitoring. *Appl. Radiat. Isot.* **49**, 1671–1675.

Chu, T. C. and Lin, C. C. (2001). The solvent extraction of radium using sym-Di[4(5)-tert-butylbenzo]-16-crown-5-oxyacetic acid. *Appl. Radiat. Isot.* **55**, 609–616.

Coale, K. H. and Bruland, K. W. (1987). Oceanic stratified euphotic zone as elucidated by ^{234}Th:^{238}U disequilibria. *Limnol. Oceanogr.* **32**, 189–200.

Cook, G. T. (1995). Enhanced low-level LSC performance for Carbon-14 dating using a bismuth germanate ($Bi_4Ge_3O_{12}$) quasi-active guard. *Radioact. Radiochem.* **6**, 10–15.

Cook, G. T. and Anderson, R. (1991). The determination of ^{241}Pu by liquid scintillation spectrometry using the Packard 2250CA. *J. Radioanal. Nucl. Chem. (Letters)* **154**, 319–330.

Cook, G. T. and Anderson, R. (1992). A radiocarbon dating protocol for use with Packard scintillation counters employing burst-counting circuitry. *Radiocarbon* **34**, 381–388.

Cook, G. T., Begg, F. H., Naysmith, P., Scott, E. M., and McCartney, M. (1995). Anthropogenic ^{14}C marine geochemistry in the vicinity of a nuclear fuel reprocessing plant. *Radiocarbon* **37**, 459–467.

Cook, G. T., Harkness, D. D., and Anderson, R. (1989). Performance of the Packard 2000CA/LL and 2250CA/XL liquid scintillation counters for ^{14}C dating. *Radiocarbon* **31**, 352–358.

Cook, G. T., Hold, A. G., Naysmith, P. and Anderson, R. (1990b). Applicability of 'new technology' scintillation counters (Packard 2000 CA/LL and 2260 XL) for ^{14}C dating. *Radiocarbon* **32**, 233–234.

Cook, G. T., Naysmith, P., Anderson, R., and Harkness, D. D. (1990a). Performance optimization of the Packard 2000 CA/LL liquid scintillation counter for ^{14}C dating. *Nucl. Geophys.* **4**, 241–245.

Dacheux, N. and Aupiais, J. (1997). Determination of uranium, thorium, plutonium, americium and curium ultratraces by photon electron rejecting alpha liquid scintillation. *Anal. Chem.* **69**, 2275–2282.

Dacheux, N. and Aupiais, J. (1998). Determination of low concentrations of americium and curium by photon/electron rejecting alpha liquid scintillation. *Anal. Chim. Acta* **363**, 279–294.

Day, J. P. and Cross, J. E. (1981). ^{241}Am from the decay of ^{241}Pu in the Irish Sea. *Nature* **292**, 43–45.

Dewberry, R. A., Bibler, N. E., and DiPrete, D. P. (1999). Measured ^{63}Ni contents in Savannah River Site high level waste and Defense Waste Processing Facility glass product by Ni-selective ion exchange purification and beta-decay counting. *J. Radioanal. Nucl. Chem.* **242**, 81–89.

De Filippis, S. and Noakes, J. E. (1991). Sample preparation of environmental samples using benzene synthesis followed by high-performance LSC. *Transactions of the American Nuclear Society*, **64**, 79–80.

De Regge, P., Radecki, Z., Moreno, J., Burns, K., Kis-Benedek, G., and Bojanowski, R. (2000). The IAEA proficiency test on evaluation of methods for ^{90}Sr measurement in a mineral matrix – Preliminary evaluation of sources of bias and measurement uncertainties. *J. Radioanal. Nucl. Chem.* **246**, 511–519.

De Vol, T. A., Duffey, J. M., and Paulenova, A. (2001). Combined extraction chromatography and scintillation detection for off-line and on-line monitoring of strontium in aqueous solutions. *J. Radioanal. Nucl. Chem.* **249**, 295–301.

Dietz, M. L. and Horowitz, E. P. (1993). Novel chromatographic materials based on nuclear waste processing chemistry. *LC-GC* **11**, 424–436.

Dietz, M. L., Horowitz, E. P., Nelson, D. M., and Wahlgren, N. (1991). An improved method for determining ^{89}Sr and ^{90}Sr in urine. *Health Phys.* **61**, 871–877.

Dodson, C. (1991). *In* "Alpha/Beta Discrimination on Liquid Scintillation Counters", Beckman Instruments Inc. Publication, Fullerton, California.

DOE (1993). *In* "DOE methods for evaluating environmental and waste management samples", DOE/EM-0089T, Rev 1, National Technical Information Service, U.S. Department of Commerce, Springfield, Virginia.

Dressler, K., and Spitzer, L. (1967). Photomultiplier tube pulses induced by rays. *Rev. Sci. Instrum.* **38**, 436–438.

Egorov, V. N., Chirin, N. A., and Kolomeitsev, G. Y. (2001). Liquid-scintillation alpha spectrometry. *Atomic Energy* **90**, 308–313.

Eikenberg, J., Fiechtner, A., Ruethi, M., and Zumsteg, I. (1996). A rapid screening method for determining gross alpha activity in urine using α/β LSC. *In* "Advances in Liquid Scintillation Spectrometry, 1994" (G. T. Cook, D. D. Harkness, A. B. MacKenzie, B. F. Miller and E. M. Scott, Eds.), pp. 283–292. Radiocarbon Publishers, University of Arizona, Tucson.

Eikenberg, J., Zumsteg, I., Rüthi, M., Bajo S., Passo, C. J., and Fern, M.J. (1998). A Rapid Procedure for Screening Transuranium Nuclides in Urine Using Actinide Resin and Low Level a/b-LSC. Perkin Elmer Life and Analytical Sciences Alpha/Beta Application Note ABA-009.

Einarsson, S. (1992). Evaluation of a prototype low-level liquid scintillation multisample counter. *Radiocarbon* **34**, 366–373.

Fairman, W. D., and Sedlet, J. (1968). Direct determination of lead-210 by liquid scintillation counting. *Anal. Chem.* **40**, 2004–2008.

Fliss, P. and Enge, R. (1993). Detection of alpha emitters in solidified radioactive waste. *Nucl. Eng. Design* **147**, 111–114.

Forte, M., Rusconi, R., Margini, C., Abbate, G., Maltese, S., Badalamenti, P., and Bellinzona, S. (2001). Determination of uranium isotopes in food and environmental samples by different techniques: a comparison. *Radiat. Prot. Dosim.* **97**, 325–328.

Fueg, B., Tschachtli, T., and Krahenbuhl, U. (1997). Alpha liquid scintillation spectrometry used for the measurement of uranium/thorium-disequilibria in soil samples. *Radichim. Acta* **78**, 47–51.

Geyh, M. A. and Schleicher, H. (1990). *In* "Absolute Age Determination", Springer-Verlag, Berlin.

George, A. C. (1984). Passive integrated measurement of indoor radon using activated carbon. *Health Phys.* **46**, 867–872.

Golchert, N. W. and Sedlet, J. (1969). Radiochemical determination of technetium-99 in environmental water samples. *Anal. Chem.* **41**, 669–671.

Guan Sanyuan and Xie Yuanming (1992). Instrumentation and software for low-level liquid scintillation counting radiocarbon dating. *Radiocarbon* **34**, 374–380.

Gudjonsson, G. I. and Theodorsson, P. (2000). A compact automatic low-level liquid scintillation system for radon-in water measurement by pulse pair counting. *Appl. Radiat. Isot.* **53**, 377–380.

Gulliksen, S. and Scott, E. M. (1995). TIRI report. *Radiocarbon* **37**, 820–821.

Gulliver, P., Cook, G. T., MacKenzie, A. B., Naysmith, P., and Anderson, R. (2001). Transport of Sellafield-derived ^{14}C from the Irish Sea through the North Channel. *Radiocarbon* **43**, 869–877.

Gupta, S. K. and Polach, H. A. (1985). *In* "Radiocarbon Dating Practices at ANU", Australian National University Press, Canberra, Australia.

Haas, H. (1979). Specific problems with liquid scintillation counting of small benzene volumes and background count rate estimations. *In* "Proceedings of the 9th International ^{14}C Conference" (R. Berger and H. E. Suess, Eds.), pp. 246–255. University of California Press, Berkeley.

Hands, G. C. and Conway, B. O. B. (1977). Simultaneous determinations of plutonium alpha and beta activity in liquid effluents and environmental samples. *Analyst* **102**, 934–937.

Harvey, B. R. and Sutton, G. A. (1970). Liquid scintillation counting of nickel-63. *Appl. Radiat. Isot.* **21**, 519–523.

Heilgeist, M. (2000). Use of extraction chromatography, ion chromatography and liquid scintillation spectrometry for rapid determination of strontium-89 and strontium-90 in food in cases of increased release of radionuclides. *J. Radioanal. Nucl. Chem.* **245**, 249–254.

Heimbuck, A. H., Gee, H., Bould, H., and Schwarz, W. J. (1963). *USAEC Report*, NYO-10405.

Higuchi, H. (1981). Analytical methods for radium in environmental samples. *Radioisotopes* **30**, 618–627.

Higuchi, H., Uesugi, M., Satoh, K., and Ohashi, N. (1984). Determination of radium in water by liquid scintillation counting after preconcentration with ion exchange resin. *Anal. Chem.*, **56**, 761–763.

Hinton, Jr. E. R., Adams, T. T., and Burchfield, L. A. (1990). Gross alpha counting of air filters using a pulse-shape discriminating alpha liquid scintillation counter. *In* "36th Annual Conference on Bioassay, Analytical, and Environmental Radiochemistry", Oak Ridge, Tennessee

Hogg, A. and Noakes, J. E. (1992). Evaluation of high-purity synthetic silica vials in active and passive vial holders for liquid scintillation counting of benzene. *Radiocarbon* **34**, 394–401.

Hogg, A., Polach, H., Robertson, S., and Noakes, J. E. (1991). Application of high purity synthetic quartz vials to liquid scintillation low-level ^{14}C counting of benzene. *In* "Liquid Scintillation Counting and Organic Scintillators, 1989" (H. Ross, J. E. Noakes and J. D. Spaulding, Eds.), pp. 123–132. Lewis Publishers, Michigan.

Holm, E. and Persson, R. B. R. (1977). Radiochemical studies of ^{241}Pu in Swedish reindeer lichens. *Health Phys.* **33**, 471–473.

Holm, E., Rioseco, J., and Garcia-Leon, M. (1984). Determination of Tc-99 in environmental samples. *Nucl. Instrum. Methods Phys. Res.* **23**, 204–207.

Hong, K. H., Cho, Y. H., Lee, M. H., Choi, G. S., and Lee, C. W. (2001). Simultaneous measurement of ^{89}Sr and ^{90}Sr in aqueous samples by liquid scintillation counting using the spectrum unfolding method. *Appl. Radiat. Isot.* **54**, 299–305.

Horrocks, D. (1963). Interaction of fission fragments with organic scintillators. *Rev. Sci. Instrum.* **34**, 1035.

Horrocks, D. (1970). Pulse shape discrimination with organic scintillation solutions. *Appl. Spectrosc.* **24**, 397–404.

Horrocks, D. L. (1974a). *In* "Applications of Liquid Scintillation Counting", Academic Press, New York.

Horrocks, D. L. (1974b). Measurement of low levels of normal uranium in water and urine by liquid scintillation alpha counting. *Nucl. Instrum. Methods Phys. Res.* **117**, 589–595.

Horrocks, D. L. (1985). Studies of background sources in liquid scintillation counting. *Appl. Radiat. Isot.* **36**, 609–617.

Horrocks, D. L. and Studier, M. H. (1958). Low-level ^{241}Pu analysis by liquid scintillation techniques. *Anal. Chem.* **30**, 1747–1750.

ISG. (1982). An inter-laboratory comparison of radiocarbon measurements in tree-rings. *Nature* **198**, 619–623.

Isogai, K., Cook, G. T. and Anderson, R. (2002). Reconstructing the history of ^{14}C discharges from Sellafield: 1. Atmospheric discharges. *J. Environ. Radioact.* **59**, 207–222.

Iwakura, T., Kasida, Y., Inoue, Y., and Tokunaga, N. (1979). A low-background liquid scintillation counter for measurement of low level tritium. *In* "Behavior of Tritium in the Environment", IAEA Proceedings Series, pp. 163–171. IAEA, Vienna.

Jerde, R. L., Paterson, L. E., and Stein, W. (1967). Effect of high energy radiations on noise pulses from photomultiplier tubes. *Rev. Sci. Instrum.* **38**, 1387–1395.

Jiang, H., Luu, S., Fu, S., Zhang, W., Zhang, T., Ye, Y., Li, M., Fu, P., Wang, S., Peng, Ch., and Jiang, P. (1983). Model DYS low level liquid scintillation counter. *In* "Advances in Scintillation Counting" (S. A. McQuarrie, C. Ediss and L. I. Wiebe, Eds.), pp. 478–493. University of Alberta, Alberta Press.

Kaihola, L. (1991). Liquid scintillation counting performance using glass vials in the Wallac 1220 Quantulus. *In* "Liquid Scintillation Counting and Organic Scintillators, 1989" (H. Ross, J. E. Noakes and J. D. Spaulding, Eds.), pp. 495–500. Lewis Publishers, Michigan.

Kaihola, L., Kojola, H., and Kananen, R. (1986). Low level liquid scintillation performance in a low level surface laboratory. *Nucl. Instrum. Methods Phys. Res.* **B17**, 509–510.

Kalin, R. M. and Long, A. (1989). Radiocarbon dating with the Quantulus in an underground counting laboratory: performance and background sources. *Radiocarbon* **31**, 359–367.

Katzlberger, C., Wallner, G., and Irlweck, K. (2001). Determination of ^{210}Pb, ^{210}Bi and ^{210}Po in natural drinking water. *J. Radioanal. Nucl. Chem.* **249**, 191–196.

Kersten, M., Thomsen, S., Priebsch, W., and Garb-Schonberg, C. D. (1998). Scavenging and particle residence times determined from ^{234}Th/^{238}U disequilibria in the coastal waters of Mecklenburg Bay. *Appl. Geochem.* **13**, 339–347.

Kim, Y. J., Kim, C. K., and Lee, J. I. (2001). Simultaneous determination of ^{226}Ra and ^{210}Pb in groundwater and soil samples by using the liquid scintillation counter – suspension gel method. *Appl. Radiat. Isot.* **54**, 275–281.

Kojima, S. and Furukawa, M. (1985). Liquid scintillation counting of low activity ^{63}Ni. *J. Radioanal. Nucl. Chem.* **95**, 323–329.

Kojola, H., Polach, H., Nurmi, J., Oikari, T., and Soini, E. (1984). High resolution low-level liquid scintillation β-spectrometer. *Appl. Radiat. Isot.* **35**, 949–952.

Kramer, G. (1981). Optimization of the analysis of nickel-63 in urine. *In* "Internal report AECL-7248", Atomic Energy of Canada Ltd., Chalk River, Ontario.

Lee, M. H. and Lee, C. W. (1999). Determination of low level Pu-241 in environmental samples by liquid scintillation counting. *Radiochim. Acta* **84**, 177–181.

Libby, W. F. (1963). Moratorium tritium geophysics. *Journal of Geophysical Research*, 4485–4494.

Lo, J. M., Cheng, B. J., Tseng, C. L., and Lee, J. D. (1993). Preconcentration of nickel-63 in sea water for liquid scintillation counting. *Anal. Chim. Acta.* **281**, 429–433.

Luxenburger, H. J. and Schuttelkopf, H. (1984). *In* "Proceedings of the 5th International Conference on Nuclear Methods in Environmental and Energy Research, CONF/TIC-840408, NTIS, Springfield, Virginia. p. 376.

MacKenzie, A. B., Logan, E. M., Cook, G. T., and Pulford, I. D. (1998). A historical record of atmospheric depositional fluxes of contaminants in west-central Scotland derived from an ombrotrophic peat core. *Sci. Total Environ.* **222**, 157–166.

Martin, J. E. (1986). Determination of ^{241}Pu in low level radioactive wastes from reactors. *Health Phys.* **51**, 621–631.

Martin, J. E. (1987). Measurement of ^{90}Sr in reactor wastes by Cerenkov counting of ^{90}Y. *Appl. Radiat. Isot.* **38**, 953–957.

Martin, W. R. and Sayles, F. L. (1987). Seasonal changes of particle and solute transport processes in nearshore sediments: ^{222}Rn/^{226}Ra and ^{234}Th/^{238}U disequilibrium at a site in Buzzards Bay, MA. *Geochim. Cosmochim. Acta* **51**, 927–943.

McCormac, F. G., Kalin, R. M., and Long, A. (1993). Radiocarbon dating beyond 50,000 years by liquid scintillation counting. *In* "Advances in Liquid Scintillation Spectrometry, 1992" (J. E. Noakes, F. Schönhofer and H. A. Polach, Eds.), pp. 125–133. Radiocarbon Publishers, University of Arizona, Tucson.

McDowell, W. J. and McDowell, B. L. (1993). The growth of a radioanalytical method: alpha liquid scintillation spectrometry. *In* "Advances in Liquid Scintillation Spectrometry, 1992" (J. E. Noakes, F. Schönhofer and H. A. Polach, Eds.), pp. 193–200. Radiocarbon Publishers, University of Arizona, Tucson.

McDowell, W. J. and McDowell, B. L. (1994). "Liquid Scintillation Alpha Spectrometry", CRC Press, Boca Raton.

McKlveen, J. W. and Johnson, W. R. (1975). Simultaneous alpha and beta particle assay using liquid scintillation counting with pulse shape discrimination. *Health Phys.* **28**, 5–11.

Miller, T. J. (1991). Development of a rapid, economical and sensitive method for the routine determination of excreted uranium in urine. *Analytical Letters* **24**, 657–664.

Mobius, S., Kamolchote, P., Ramamonjisoa, T.-L., and Yang, M. (1993). Rapid determination of Ra, Rn, Pb and Po in water using extractive liquid scintillation. *In* "Advances in Liquid Scintillation Spectrometry, 1992" (J. E. Noakes, F. Schönhofer and H. A. Polach, Eds.), pp. 413–416. Radiocarbon Publishers, University of Arizona, Tucson.

Momoshima, N. and Okada, S. (1993). Overview of tritium: characteristics, sources and problems. *Health Phys.*, **65**, 595–609.

Mook, W. G. and van der Plicht, J. (1999). Reporting ^{14}C activities and concentrations. *Radiocarbon* **41**: 227–239.

Moran, S. B. and Buesseler, K. O. (1993). Size-fractionated ^{234}Th in continental shelf waters of New England: implications for the role of colloids in oceanic trace metal scavenging. *J. Marine Res.* **51**, 893–922.

NCRP (1976). Environmental Radiation Measurements. *In* "National Council on Radiation Protection and Measurements", Report No. 50 Washington, DC.

Noakes, J. E. (1977). Considerations for achieving low level radioactivity measurements with liquid scintillation counters. *In* "Liquid Scintillation Counting" (M. A. Crook and P. Johnson, Eds.), **4**, pp. 189–206. Heyden, London.

Noakes, J. E. and Valenta R. J. (1996). The role of $Bi_4Ge_3O_{12}$ as an auxiliary scintillator for $\alpha/\beta/\gamma$ liquid scintillation counting and low level counting. *In* "Advances in Liquid Scintillation Spectrometry, 1994" (G. T. Cook, D. D. Harkness, A. B. MacKenzie, B. F. Miller and E. M. Scott, Eds.), pp. 283–292. Radiocarbon Publishers, University of Arizona, Tucson.

Noakes, J. E., Isbell, A. F., Stipp, J. J., and Hood, D. W. (1963). Benzene synthesis by low temperature catalysis for radiocarbon dating. *Geochim. Cosmochim. Acta* **27**, 797–804.

Nour, S., Burnett, W. C., and Horwitz, E. P. (2002). ^{234}Th analysis of marine sediments via extraction chromatography and liquid scintillation counting. *Appl. Radiat. Isot.* **57**, 235–241.

NUREG (1985). *In* "Assay of long-lived radionuclides in low level wastes from power reactors", NUREG/CR-4101/Part 1, Nuclear Regulatory Commission, Washington, DC.

Nozaki, Y., Thomson, J., and Turekian, K. K. (1976). The distribution of ^{210}Pb and ^{210}Po in the surface waters of the Pacific Ocean. *Earth Planet. Sci. Lett.* **32**, 304–312.

Nozaki, Y., Zhang, J., and Takeda, A. (1997). ^{210}Pb and ^{210}Po in the Equatorial Pacific and the Bering Sea: the effect of biological productivity and boundary scavenging . *Deep-Sea Res. II* **44**, 2203–2220.

Nozaki, I., Dobashi, F., Kato, Y., and Yamamoto, Y. (1998). Distribution of Ra isotopes and the ^{210}Pb balance in the surface seawaters of the mid Northern Hemispher. *Deep-Sea Res. I* **45**, 1263–1284.

Oikari, T., Kojola, H., Nurmi, J., and Kaihola, L. (1987). Simultaneous counting of low alpha and beta particle activities with liquid scintillation spectrometry and pulse shape analysis. *Appl. Radiat. Isot.* **38**, 875–878.

Passo Jr, C. J. and Floeckher, J. M. (1991). The LSC approach to radon counting in air and water. *In* "Liquid Scintillation Counting and Organic Scintillators, 1989" (H. Ross, J. E. Noakes and J. D. Spaulding, Eds.), pp. 375–384. Lewis Publishers, Michigan.

Passo Jr, C. J., and Kessler, M. J. (1993). Selectable delay before burst – a novel feature to enhance low-level counting performance. *In* "Advances in Liquid Scintillation Spectrometry, 1992" (J. E. Noakes, F. Schönhofer and H. A. Polach, Eds.), pp. 51–57. Radiocarbon Publishers, University of Arizona, Tucson.

Passo Jr, C. J., Cook, G. T. and Thomson, J. (2002). The effects of substituted anthracenes and novel scintillating solvents in cocktail formulations for α/β separation in liquid scintillation spectrometry. *In* "Advances in Liquid Scintillation Spectrometry, 2001" (Mobius, S. Eds.), Radiocarbon Publishers, University of Arizona, Tucson. (IN PRESS).

Pates, J. M., Cook, G. T., and MacKenzie, A. B. (1996a). Alpha/beta separation liquid scintillation spectrometry: current trends. In "Advances in Liquid Scintillation Spectrometry, 1994" (G. T. Cook, D. D. Harkness, A. B. MacKenzie, B. F. Miller, and E. M. Scott, Eds.), pp. 267–281. Radiocarbon Publishers, University of Arizona, Tucson.

Pates, J. M., Cook, G. T., MacKenzie, A. B., Anderson, R., and Bury, S. J. (1996b). Determination of ^{234}Th in marine samples by liquid scintillation spectrometry. *Anal. Chem.* **68**, 3783–3788.

Pates, J. M., Cook, G. T., and MacKenzie, A. B. (1996c). The effect of cocktail fluors on pulse shapes and alpha/beta separation liquid scintillation spectrometry. In "Advances in Liquid Scintillation Spectrometry, 1994" (G. T. Cook, D. D. Harkness, A. B. MacKenzie, B. F. Miller, and E. M. Scott, Eds.), pp. 317–326. Radiocarbon Publishers, University of Arizona, Tucson.

Paviet, P., Raymond, A., and Metcalf, R. (1991). A rapid method for the analysis of technetium-99 in liquid radioactive wastes. *In* "Monograph of Second International Conference on Methods and Applications of Radioanalytical Chemistry", p. 44. American Nuclear Society, Washington DC.

Peck, G. A., Smith, J. D., and Cooper, M. D. (1998). Enhanced counting efficiency of Cerenkov radiation from bismuth-210. *J. Radioanal. Nucl. Chem.* **238**, 163–165.

Pessenda, L. C. R., Gouveia, S. E. M., and Aravena, R. (2001). Radiocarbon dating of total soil organic matter and humin fraction and its comparison with ^{14}C ages of fossil charcoal. *Radiocarbon* **43**, 595–601.

Pietig, F. von. and Scharpenseel H. W. (1964). Alterbestimmung mit flussigheits-scintillations-spectrometer, uber die wirksamkeit von abschirmungsabnahmen. *Atompraxis*, 7, 1–3.

Piltingsrud, H. V. and Stencel, J. R. (1972). Determination of ^{90}Y, ^{90}Sr, and ^{89}Sr in samples by use of liquid scintillation beta spectroscopy. *Health Phys.* 23, 121–122.

Pimpl, M. (1992). Increasing the sensitivity of ^{241}Pu determination for emission and immission control of nuclear installations by aid of liquid scintillation counting. *J. Radioanal. Nucl. Chem.* 161, 429–436.

Plastino, W., Kaihola, L., Bartolomei, P., and Bella, F. (2001). Cosmic background reduction in the radiocarbon measurements by liquid scintillation spectrometry at the underground laboratory of Gran Sasso. *Radiocarbon* 43, 157–161.

Polach, H., Calf, G., Harkness, D. D., Hogg, A., Kaihola, L., and Robertson, S. (1988). Performance of new technology liquid scintillation counters for ^{14}C dating. *Nucl. Geophys.* 2, 75–79.

Polach, H., Gower, J., Kojola, H., and Heinonen, A. (1983). An ideal vial and cocktail for low level scintillation counting. *In* "Advances in Scintillation Counting" (S. A. McQuarrie, C. Ediss and L. I. Weibe, Eds.), pp. 508–525. University of Alberta Press, Alberta.

Prichard, H. M. and Cox, A. (1991). Liquid scintillation screening method for isotopic uranium in drinking water. *In* "Liquid Scintillation Counting and Organic Scintillators" (H. Ross, J. E. Noakes and J. D. Spaulding, Eds.), pp. 385–397. Lewis Publishers, Michigan.

Prichard, H. M. and Marien, K. (1985). A passive diffusion ^{222}Rn sampler based on activated carbon adsorption. *Health Phys.*, 48, 797–803.

Prichard, H. M., Gesell, T. F., and Meyer, C. R. (1980). Liquid scintillation analyses for radium-226 and radon-222 in potable waters. *In* "Liquid Scintillation Counting – Recent Appplications and Development, Vol. II, Sample Preparation and Applications" (C.-T. Peng, D. L. Horrocks and E. L. Alpen, Eds.), Academic Press, London.

Punning, J. M. and Rajamae, R. (1977). Some possibilities for decreasing the background of liquid scintillation beta-ray counter. *In* "Low Radioactivity Measurements and Applications" (P. Povinec and S. Usacev, Eds.), pp. 169–171. Slovenske Pedagogicke Nakladatelstvo, Bratislava.

Qureshi, R. M., Aravena, R., Fritz, P., and Drimmie, R. (1989). The CO_2 absorption method as an alternative to benzene synthesis method for ^{14}C dating. *Appl. Geochem.* 4, 625–633.

Raby, A. and Hyde, E. K. (1952). *US Atom. Energy Comm. Rep.* L-2069, 23–32

Radwan, M., Przybylska, A., Mykowska, E., and Gibas, K. (1981). Measurement of nickel-63 low activity in samples of metals and alloys with the help of liquid scintillators. *Appl. Radiat. Isot.* 32, 97–99.

Randolph, R. B. (1975). Determination of strontium-90 and strontium-89 by Cerenkov and liquid scintillation counting. *Appl. Radiat. Isot.* 26, 9–16.

Rank, D. (1993). Environmental tritium in hydrology: present state (1992). *In* "Advances in Liquid Scintillation Spectrometry, 1992" (J. E. Noakes, F. Schönhofer and H. A. Polach, Eds.), pp. 327–334. Radiocarbon Publishers, University of Arizona, Tucson.

Reynolds, S. A. and Eldridge, J. S. (1980). *In* "Liquid Scintillation Counting-Recent Applications and Development" (C.-T. Peng, D. L. Horrocks and E. L. Alpen, Eds.), pp. 397–405. Academic Press, New York.

Roodbergen, S., Kroondijk, R., and Verhuel, H. (1972). Cerenkov radiation in photomultiplier windows and the resulting time shift in delayed coincidence time spectra. *Nucl. Instrum. Methods Phys. Res.* 105, 551–555.

Rouston, R. C. and Cataldo, D. A. (1978). Accumulation of ^{99}Tc by tumbleweed and cheatgrass grown on arid soils. *Health Phys.* 34, 685–690.

Rozanski, K., Stichler, W., Gonfiantini, R., Scott, E. M., Beukens, R. P., Kromer, B., and Van der Plicht, J. (1992). The IAEA ^{14}C intercomparison exercise 1990. *Radiocarbon* 34, 506–559.

Rucker, T. L. (1992). Calculational method for the resolution of ^{90}Sr and ^{89}Sr counts from Cerenkov and liquid scintillation counting. *In* "Liquid Scintillation Counting and Organic Scintillators" (H. Ross, J. E. Noakes and J. D. Spaulding, Eds.), pp. 529–535. Lewis Publishers, Michigan.

Rundt, K. (1992). Digital overlay technique in liquid-scintillation counting – a review. *Radioact. Radiochem.* 3, 14–25.

Russow, R. and Dermietzel, J. (1990). Determination of ^{63}Ni-activity in plant materials. *Isotopenpraxis* **26**, 502–504.

Ryan, T. P., Mitchell, P. I., Vives I Batlle, J., Sanchez-Cabeza, J. A., McGarry, A. T., and Schell, W. R. (1993). Low-level ^{241}Pu analysis by supported-disk liquid scintillation counting. *In* "Advances in Liquid Scintillation Spectrometry, 1992" (J. E. Noakes, F. Schönhofer and H. A. Polach, Eds.), pp. 75–82. Radiocarbon Publishers, University of Arizona, Tucson.

Saarinen, L. and Suksi, J. (1992). Determination of uranium series radionuclides Pa-231 and Ra-226 using liquid scintillation counting (LSC). *In* "Report of the Nuclear Waste Commission of Finnish Power Companies".

Saarinen, L. and Suksi, J. (1993). Determination of uranium series radionuclide ^{231}Pa using liquid scintillation counting. *In* "Advances in Liquid Scintillation Spectrometry, 1992" (J. E. Noakes, F. Schönhofer and H. A. Polach, Eds.), pp. 209–215. Radiocarbon Publishers, University of Arizona, Tucson.

Salonen, L. and Hukkanen, H. (1997). Advantages of low background liquid scintillation alpha-spectrometry and pulse shape analysis in measuring Rn-222, uranium and Ra-226 in groundwater samples. *J. Radioanal. Nucl. Chem.* **226**, 67–74.

Salonen, L. (1993). Measurement of low levels of ^{222}Rn in water with different commercial liquid scintillation counters and pulse shape analysis. *In* "Advances in Liquid Scintillation Spectrometry, 1992" (J. E. Noakes, F. Schönhofer and H. A. Polach, Eds.), pp. 361–371. Radiocarbon Publishers, University of Arizona, Tucson.

Sanchez-Cabeza, J. A. and Pujol, L. (1998). Simultaneous determination of radium and uranium activities in natural water samples using liquid scintillation counting. *Analyst* **123**, 399–403.

Sato, K., Hashimoto, T., Noguchi, M., Nitta, W., Higuchi, H., Nishigawa, N., and Sanada, T. (2000). A simple method for determination of Ra-226 in environmental samples by applying alpha-beta coincidence liquid scintillation counting. *J. Environ. Radioact.* **48**, 247–256.

Satoh, K., Noguchi, M., Higuchi, H., and Kitamura, F. D. (1984). Low level α activity measurements with pulse shape discrimination, application to the determination of α-nuclides in environmental samples. *Radioisotopes* **33**, 841–846.

Scarpitta, S. (1993). Technetium-99 in water and vegetation by liquid scintillation analysis. *In* "39th Annual Conference on Bioassay, Analytical, and Environmental Radiochemistry", Colorado Springs, Colorado.

Schell, W. R., Vick, C. E., and Wurtz, E. A. (1981). A low level laboratory for alpha and gamma counting of environmental samples. *In* "Methods of Low-Level Counting and Spectrometry", pp. 125–149. IAEA-SM-252/22, IAEA, Vienna.

Schönhofer, F. (1989). Determination of ^{14}C in alcoholic beverages. *Radiocarbon* **31**, 777–784.

Schönhofer, F. (1992). ^{14}C in Austrian wine and vinegar. *Radiocarbon* **34**, 768–771.

Scott, E. M., Aitchison, T. C., Harkness, D. D., Cook. G. T., and Baxter, M. S. (1990). An overview of all three stages of the international radiocarbon intercomparison. *Radiocarbon* **32**, 309–319.

Silva, R. J., Evans, R., Rego, J. H., and Buddemeier, R. W. (1988). Methods and results of Tc-99 analysis of Nevada test site ground waters. *J. Radioanal. Nucl. Chem.* **124**, 397–405.

Smith, J. N. and Ellis, K. M. (1995). Radionuclide tracer profiles at the CESAR ice station and Canadian Ice Island in the Western Arctic Ocean. *Deep-Sea Research II*, **42**, 1449–1470.

Spaulding, J. D. and Noakes, J. E. (1993). Determination of ^{222}Rn in drinking water using an alpha/beta liquid scintillation counter. *In* "Advances in Liquid Scintillation Spectrometry, 1992" (J. E. Noakes, F. Schönhofer and H. A. Polach, Eds.), pp. 373–381. Radiocarbon Publishers, University of Arizona, Tucson.

Spry, N., Parry, S. and Jerome, S. (2000). The development of a sequential method for the determination of actinides and Sr-90 in power station effluent using extraction chromatography. *Appl. Radiat. Isot.* **53**, 163–171.

Stuiver, M. and Polach, H. A. (1977). Reporting of ^{14}C data. *Radiocarbon* **19**, 355–363.

Sullivan, T., Nelson, D., and Thompson, E. (1991). Determination of ^{99}Tc in borehole waters using extraction chromatographic resin. *In* "37th Annual Conference on Bioassay, Analytical, and Environmental Radiochemistry", Ottawa, Canada.

Tait, D., Haase, G., and Wiechen, A. (1997). A fast method for the determination of ^{90}Sr in liquid milk by solid phase extraction with cryptand 222 on cation exchange resin. *Kerntechnik* **62**, 96–98.

Tait, D., Haase, G., and Wiechen, A. (1998). Solid phase extraction of strontium by cryptand 222 bound to Dowex $50W \times 8$ for the determination of ^{90}Sr in liquid milk. *Kieler Milchwirtschaftliche Forschungsberichte* **50**, 211–223.

Tait, D. and Weichen, A. (1993). Use of liquid scintillation for fast determination of ^{89}Sr and ^{90}Sr in milk. *Sci. Total Environ.* **130/131**, 447–457.

Tamers, M. A. (1965). Routine carbon-14 dating using liquid scintillation techniques. *In* "Proceedings of the 6th Conference on Radiocarbon and Tritium Dating" (R. M. Chatters and E. A. Olson, Eds.), pp. 53–67.

Terlikowska-Drozdziel, T. and Radoszewski, T. (1992). Investigation of radioactive counting of selected radionuclides in various liquid scintillators. *Nucl. Instr. Methods Phys. Res., Sect. A* **312**, 100–103.

Thakkar, A., Fern, M., Kupka, T., Harvey, J., and Passo, C. J. (1997). A New Approach to Gross Alpha Measurements in Aqueous Samples Using Extraction Chromatography and Liquid Scintillation Counting. Perkin Elmer Life and Analytical Sciences Alpha/Beta Application Note ABA-008.

Thomson, J. (1991). Di-Isopropylnaphthalene – a new solvent for liquid scintillation counting. *In* "Liquid Scintillation Counting and Organic Scintillators" (H. Ross, J. E. Noakes and J. D. Spaulding, Eds.), pp 19–34, Lewis Publishers, Michigan.

Thomson, J., Colley, S., Anderson, R., Cook, G. T., and MacKenzie, A. B. (1993). ^{210}Pb in the sediments and water column of the Northeast Atlantic from 47 to 59°N along 20°W. *Earth Planet. Sci. Lett.* **115**, 75–87.

Thomson, J., Cook, G. T., Anderson, R., MacKenzie, A. B., Harkness, D. D., and McCave, I. N. (1995). Radiocarbon age offsets in different-sized carbonate components of deep-sea sediments. *Radiocarbon* **37**, 91–101.

Thorngate, J. H., McDowell, W. J., and Christian, D. J. (1974). An application of pulse shape discrimination to liquid scintillation counting. *Health Phys.* **27**, 123–126.

Tinker, R. A. and Smith, J. D. (1996). Simultaneous measurement of ^{226}Ra and ^{133}Ba using liquid scintillation counting with pulse shape discrimination. *Anal. Chim. Acta* **332**, 291–297.

Tolmachyov. S., Ura, S., Momoshima, N., Yamamoto, M., and Maeda, Y. (2001). Determination of Cl-36 by liquid scintillation counting from soil collected at the Semipalatinsk nuclear test site. *J. Radioanal. Nucl. Chem.* **249**, 541–545.

Tso, M. W. and Li, C. (1987). Indoor and outdoor ^{222}Rn daughters in Hong Kong. *Health Phys.*, **53**, 175–180.

Uyesgi, G. S. and Greenberg, A. E. (1965). Simultaneous assay of Strontium-90 and Yttrium-90 by liquid scintillation spectrometry. *Appl. Rad. Isot.* **16**, 581–587.

Vaca, F., Manjon, G., Cuellar, S., and Garcia-Leon, M. (2001). Factor of merit and minimum detectable activity for ^{90}Sr determinations by gas flow proportional counting or Cherenkov counting. *Appl. Radiat. Isot.* **55**, 849–851.

Vajda, N., Ghods-Esphahani, A., Cooper, E., and Danesi, P. R. (1992). Determination of radiostrontium in soil samples using a crown ether. *J. Radioanal. Nucl. Chem.* **162**, 307.

Vajda, N., La Rosa, J., Zeisler, R., Danesi, P., and Kis Benedek, G. (1997). A novel technique for the simultaneous determination of ^{210}Pb and ^{210}Po using a crown ether. *J. Environ. Radioact.* **37**, 355–372.

Valenta, R. J. (1987). Reduced background scintillation counting. U.S. Patent No. 4,651,006.

Valenta, R. J. and Noakes, J. E. (1989). Liquid scintillation measurement system with active guard shield. U.S. Patent No. 4,833,326.

van Cauter, S. (1986). Three dimensional spectrum analysis: a new approach to reduce background of liquid scintillation counters. *Packard Instrument Company Application Bulletin* 006, 1–8, PerkinElmer Life and Analytical Sciences, Boston.

Venso, E. A., Prichard, H. M., and Dodson, C. L. (1993). Measurement of isotopic uranium in Texas drinking water supplies by liquid scintillation with alpha-beta discrimination. *In* "Advances in Liquid Scintillation Spectrometry, 1992" (J. E. Noakes, F. Schönhofer and H. A. Polach, Eds.), pp. 425–430. Radiocarbon Publishers, University of Arizona, Tucson.

Verrezen, F. and Hurtgen, C. (1992). The measurement of Technetium-99 and Iodine-129 in waste water from pressurized nuclear-power reactors. *Appl. Radiat. Isot.* **43**, 61–68.

Walker, C. R., Short, B. W., and Spring, H. S. (1980). Determination of technetium-99 by liquid scintillation counting. *In* "Proceedings of the Conference on Analytical Chemistry in Energy

Technology", Gatlinburg, Tennessee, Ann Arbor Scientific Publications Inc, Michigan, pp. 101–110.

Walker, M. J. C., Bryant, C., Coope, G. R., Harkness, D. D., Lowe, J. J., and Scott, E. M. (2001). Towards a radiocarbon chronology of the late-glacial: sample selection strategies. *Radiocarbon* **43**, 1007–1019.

Wallner, G. (1997). Simultaneous determination of ^{210}Pb and ^{212}Pb progenies by liquid scintillation counting. *Appl. Radiat. Isot.* **48**, 511–514.

Wallner, G. (2001). Distribution of ^{212}Pb, ^{214}Pb and ^{210}Pb with its daughter products on aerosol fractions from Vienna and Badgastein (Austria). *Radiochim. Acta* **89**, 791–798.

Wigley, F., Warwick, P. E., Croudace, I. W., Caborn, J., and Sanchez, A. L. (1999). Optimised method for the routine determination of Technetium-99 in environmental samples by liquid scintillation counting. *Anal. Chim. Acta* **380**, 73–82.

Wilken, R.-D. and Joshi, S. R. (1991). Rapid methods for determining ^{90}Sr, ^{89}Sr and ^{90}Y in environmental samples: a survey. *Radioact. Radiochem.* **2**, 14–27.

Wilson, J. E., McCormac, F. G., and Hogg, A. G. (1996). *In* "Advances in Liquid Scintillation Spectrometry, 1994" (G. T. Cook, D. D. Harkness, A. B. MacKenzie, B. F. Miller and E. M. Scott, Eds.), pp. 59–65. Radiocarbon Publishers, University of Arizona, Tucson.

Winyard, R., Lutkin, J. E., and McBeth, G. W. (1971). Pulse shape discrimination in inorganic and organic scintillators. *Nucl. Instrum. Methods Phys. Res.* **95**, 141–154.

Wu, J. and Boyle, E. A. (1997). Lead in the western North Atlantic Ocean: completed response to leaded gasoline phaseout. *Geochim. Cosmochim. Acta* **61**, 3279–3283.

Yang, D. (1990). Study on determination of Np, Pu, and Am with extraction-liquid scintillation counting and its application to assay of transuranium elements in high-level waste. Ph.D. Thesis, Tsinghua University, Beijing, Peoples Republic of China.

Zaretskaia, N. E., Ponomareva, V. V., Sulerzhitsky, L. D., and Zhilin, M. G. (2001). Radiocarbon studies of peat bogs: and investigation of South Kamchatka volcanoes and Upper Volga archaeological sites. *Radiocarbon* **43**, 571–580.

7

RADIOACTIVITY COUNTING STATISTICS

AGUSTÍN GRAU MALONDA

Instituto de Estudios de la Energía, CIEMAT, Avda. Complutense 22, 28040 Madrid, Spain

AGUSTÍN GRAU CARLES

Departamento de Fusión y Física de Partículas, CIEMAT, Avda. Complutense 22, 28040, Madrid, Spain

I. INTRODUCTION

In the measurement process, the object to be observed is always affected by an undetermined interaction between the observer and the observed. As a result, the measured magnitudes are always reproduced with a certain inherent uncertainty caused by the instrument. This uncertainty in the measurements makes the use of error theory essential. When we measure radioactive substances the situation becomes even more complicated, because the radioactivity decay is a random process. In radioactivity counting two types of fluctuations are basically generated, one related to the activity of the sample, when the half-life of the radionuclide is short; and another caused by the random nature of radioactivity decay, which modifies the disintegration rates with time. Since the measurement of radioactivity involves values with different degrees of reliability and validity, the principles of counting statistics must be applied.

II. STATISTICAL DISTRIBUTIONS

In this section we shall study all basic characteristics of both the Poisson and the normal (or Gaussian) distributions, and their relation to radioactivity counting statistics. Although the Poisson distribution involves all processes of radioactivity decay, and therefore the detection of particles and radiation, the normal distribution is applied more often by far and is better known in the majority of cases. Since the application of the Poisson distribution to counting statistics may be tedious and time consuming, and considering that both distributions give identical results when the total number of counts becomes large; our final objective will be the determination of some characteristic parameters, which will allow one to exchange the Poisson distribution by the normal one.

A. The Poisson Distribution

The Poisson distribution describes a random process for which the occurrence probability of a certain event is constant and small. This distribution not only concerns radioactivity counting statistics or nuclear decay, but is also applied to evaluate, in a more or less approximate way, many other processes. Some examples of daily life that verify Poisson statistics are for example: the number of phone calls received at a phone switchboard several minutes before noon, the number of annual strikes in a factory, the number of misprints on a book page, the number of times a piece of a machine fails in a given period of time, the number of fatal traffic accidents each week in a city. Przyborowski and Wilenski (1935) present an application of Poisson law and construct rules to minimize the chance of errors in tests and samples.

In radioactivity decay the following four aspects are fulfilled: (i) all radioactive nuclei have the same decay probability for a given time period, (ii) the decay process of one nucleus is not affected by the decay of other nuclei, (iii) the total number of nuclei and measurement time intervals are sufficiently large, and (iv) the nuclei half-life is long compared with the detection pulse. Therefore, the radioactivity decay is a random process, in which a discontinuous random variable is defined as the number of times a decay event takes place in a continuous period of time t. Additionally, the probability of one decay event occurring in a time increment Δt must be asymptotically proportional to Δt, independently of the time value in the interval Δt and all previous decay events. Under these conditions the Poisson distribution takes the form:

$$P_x(t) = \frac{e^{-at}(at)^x}{x!} \qquad x = 0,1,2,\ldots \tag{7.1}$$

where $P_x(t)$ is the probability that a number of x decay processes take place in time t and a a constant to be determined. Equation 7.1 may be derived in three different ways: by approximation from the binomial distribution (Hoel, 1984 and Eadie *et al.*, 1971), considering first principles (Evans, 1972) or basing all calculations on a Markov process (Feller, 1968; Rozanov, 1977).

Three important properties derived from Eq. 7.1 are the following:

$$\sum_{x=0}^{\infty} P_x = 1 \tag{7.2}$$

$$\sum_{x=0}^{\infty} x P_x = at \tag{7.3}$$

$$\sum_{x=0}^{\infty} (x - at)^2 P_x = at \tag{7.4}$$

The normalization condition, Eq. 7.2, states that the summation of all possible probabilities P_x of the Poisson distribution is one. Equation 7.3 defines the parameter a as the average value of the distribution divided by t, i.e., the number of decay processes per unit of time. The third expression, Eq. 7.4, refers to the identity relation between the variance and the mean, when the Poisson distribution is applied. The most important consequence derived from the three properties, Eqs. 7.2–7.4, is that the single parameter a enables one to determine the first and second order moments (i.e., $\sum_{x=0}^{\infty} x P_x$ and $\sum_{x=0}^{\infty} x^2 P_x$) of the Poisson distribution. By defining the average number of decay processes in time t as the product $\mu = at$, the Poisson distribution can be expressed as

$$P_x = \frac{e^{-\mu} \mu^x}{x!} \tag{7.5}$$

The Poisson distribution is asymmetric, and its shape depends on the parameter μ, which is a real value greater than zero.

Example 7.1 The measurement of a radioactive sample of ^{238}U ($T_{1/2} = 4.47 \times 10^9$ years) during three days gives a total number of counts 133,747. (a) Evaluate the average counting rate. Compute the probability of counting: (b) less than one count in one second, (c) exactly one count, (d) less than 3 counts, (e) a number of counts between 1 and 4, (f) more than 3 counts.

(a) The average counting rate is:

$$\mu = \frac{133,747}{3 \times 24 \times 60 \times 60} = 0.516 \text{ counts/s}$$

(b) The probability of counting less than one count is computed as follows:

$$P(X < 1) = P_0 = e^{-0.516} = 0.597$$

(c) The probability of counting exactly one count:

$$P(X = 1) = \frac{\mu^1 e^{-\mu}}{1!} = 0.516 \times e^{-0.516} = 0.3$$

(d) The probability of counting less than 3 counts is the addition of the probabilities of counting none, 1 and 2 counts

$$P(X < 3) = \sum_{x=0}^{2} P_x = P_0 + P_1 + P_2 = e^{-\mu} + \mu e^{-\mu} + \frac{\mu^2}{2} e^{-\mu}$$

$$= e^{-\mu}\left(1 + 0.516 + \frac{0.516^2}{2}\right) = 0.984$$

(e) The probability of counting a number of counts between 1 and 4 is the addition of the probabilities of counting exactly 2 and 3 counts

$$P(1 < X < 4) = P_2 + P_3 = \frac{\mu^2 e^{-\mu}}{2!} + \frac{\mu^3 e^{-\mu}}{3!} = e^{-\mu}\left(\frac{\mu^2}{2!} + \frac{\mu^3}{3!}\right)$$

$$= e^{-0.516}\left(\frac{0.516^2}{2} + \frac{0.516^3}{6}\right) = 0.093$$

(f) The probability of counting more than 3 counts is

$$P(X > 3) = \sum_{x=4}^{\infty} P_x = 1 - \sum_{x=0}^{3} P_x = 1 - (P_0 + P_1 + P_2 + P_3)$$

$$= 1 - e^{-\mu}\left(1 + \mu + \frac{\mu^2}{2!} + \frac{\mu^3}{3!}\right)$$

$$= 1 - e^{-0.516}\left(1 + 0.516 + \frac{0.516^2}{2} + \frac{0.516^3}{6}\right) = 0.00196$$

It should be remarked that a linear combination of the two random variables X and Y, which both follow the Poisson distribution, gives a third random variable, which does not necessarily follow the Poisson distribution. For the following three random variables $Y = aX$, $Z = X + Y$ and $W = X - Y$, only Z follows the Poisson distribution.

B. The Gaussian Distribution

In general, it is assumed that the random variable X follows a normal distribution of mean μ and variance σ^2, when the probability density function of a set of measured values x yield the function

$$p(x) = \frac{1}{\sigma\sqrt{2\pi}} \exp\left[-\frac{(x-\mu)^2}{2\sigma^2}\right] \qquad (7.6)$$

where $(-\infty < x < \infty)$, $(-\infty < \mu < \infty)$ and $(0 < \sigma^2 < \infty)$.

Frequently, in the course of a number of measurements, we are interested in evaluating the probability that a new measurement of the random variable X verifies a certain condition relative to the value x. This probability is computed in terms of the accumulated distribution function, which is defined as follows:

$$P(X \le x) = \frac{1}{\sigma\sqrt{2\pi}} \int_{-\infty}^{x} \exp\left[-\frac{(t-\mu)^2}{2\sigma^2}\right] dt \qquad (7.7)$$

Geometrically, the integral Eq. 7.7 is interpreted as the area under the curve between the limits of the interval $-\infty \le X \le x$. Since each pair of values (μ, σ^2) defines a different distribution, there is no simple way to tabulate the accumulated distribution function. This inconvenience is partially resolved by defining the normalized random variable Z

$$Z = \frac{X - \mu}{\sigma} \qquad (7.8)$$

for which the probability density and the accumulated distribution functions can be defined as follows:

$$p(z) = \frac{1}{\sqrt{2\pi}} \exp\left(-\frac{z^2}{2}\right) \qquad (7.9)$$

$$P(Z \le z) \equiv \Phi(z) = \frac{1}{\sqrt{2\pi}} \int_{-\infty}^{z} \exp\left(-\frac{t^2}{2}\right) dt \qquad (7.10)$$

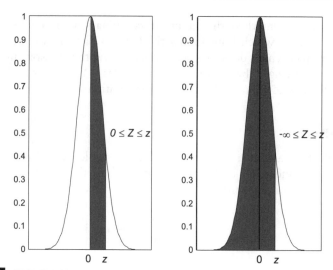

FIGURE 7.1 Integration intervals corresponding to two different criteria for tabulation.

Although the tabulated values of the accumulated distribution function Eq. 7.10 are found in the majority of statistics manuals, the symmetric character of the distribution makes two different criteria of tabulation possible, which may cause confusion. For instance, Triola (1983) assumes the integration interval $0 \leq Z \leq z$, while Newbold (1995) and Abramowitz and Stegun (1972) consider $-\infty < Z \leq z$. These intervals are shown in Fig. 7.1. Since the normal distribution is normalized to one, both criteria of tabulation make the values simply differ in 0.5. Bevington (1969) defines the normalized random variable Z as the absolute value of the difference between the means $|X - \mu|$ divided by the standard deviation

$$Z = \frac{|X - \mu|}{\sigma}$$

The integration interval is $-\infty < Z < \infty$ and the values of $\Phi(Z)$ are two times the values for the interval $0 < Z < \infty$.

Example 7.2 By searching in Table 7.1, calculate the accumulated distribution function Φ for the following limits of the standard random variable Z: (a) $0 \leq Z \leq 1.75$, (b) $-1.75 \leq Z \leq 0$, (c) $Z \geq 1.85$, (d) $Z \leq 1.10$, (e) $Z \leq -1.45$.

(a) The second column in Table 7.1 shows the accumulated distribution function Φ for the values of z listed in the first column. The remaining columns, from third to eleventh, list the intermediate values of Φ when z is incremented in 0.01. Therefore, the sought value of $\Phi(1.75) = 0.4599$ is given in column seven.

(b) Taking into account the symmetry of the curve with respect to the origin we have

$$\Phi(-1.75 \leq Z \leq 0) = \Phi(0 \leq Z \leq 1.75) = 0.4599$$

TABLE 7.1 The Standard Normal Distribution

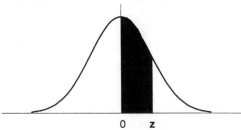

z	0.00	0.01	0.02	0.03	0.04	0.05	0.06	0.07	0.08	0.09
0.0	0.0000	0.0040	0.0080	0.0120	0.0160	0.0199	0.0239	0.0279	0.0319	0.0359
0.1	0.0398	0.0438	0.0478	0.0517	0.0557	0.0596	0.0636	0.0675	0.0714	0.0753
0.2	0.0793	0.0832	0.0871	0.0910	0.0948	0.0987	0.1026	0.1064	0.1103	0.1141
0.3	0.1179	0.1217	0.1255	0.1293	0.1331	0.1368	0.1406	0.1443	0.1480	0.1517
0.4	0.1554	0.1591	0.1628	0.1664	0.1700	0.1736	0.1772	0.1808	0.1844	0.1879
0.5	0.1915	0.1950	0.1985	0.2019	0.2054	0.2088	0.2123	0.2157	0.2190	0.2224
0.6	0.2257	0.2291	0.2324	0.2357	0.2389	0.2422	0.2454	0.2486	0.2517	0.2549
0.7	0.2580	0.2611	0.2642	0.2673	0.2704	0.2734	0.2764	0.2794	0.2823	0.2852
0.8	0.2881	0.2910	0.2939	0.2967	0.2995	0.3023	0.3051	0.3078	0.3106	0.3133
0.9	0.3159	0.3186	0.3212	0.3238	0.3264	0.3289	0.3315	0.3340	0.3365	0.3389
1.0	0.3413	0.3438	0.3461	0.3485	0.5308	0.3531	0.3554	0.3577	0.3599	0.3621
1.1	0.3643	0.3665	0.3686	0.3708	0.3729	0.3749	0.3770	0.3790	0.3810	0.3830
1.2	0.3849	0.3869	0.3888	0.3907	0.3925	0.3944	0.3962	0.3980	0.3997	0.4015
1.3	0.4032	0.4049	0.4066	0.4082	0.4099	0.4115	0.4131	0.4147	0.4162	0.4177
1.4	0.4192	0.4207	0.4222	0.4236	0.4251	0.4265	0.4279	0.4292	0.4306	0.4319
1.5	0.4332	0.4345	0.4357	0.4370	0.4382	0.4394	0.4406	0.4418	0.4429	0.4441
1.6	0.4452	0.4463	0.4474	0.4484	0.4495	0.4505	0.4515	0.4525	0.4535	0.4545
1.7	0.4554	0.4564	0.4573	0.4582	0.4591	0.4599	0.4608	0.4616	0.4625	0.4633
1.8	0.4641	0.4649	0.4656	0.4664	0.4671	0.4678	0.4686	0.4693	0.4699	0.4706
1.9	0.4713	0.4719	0.4726	0.4732	0.4738	0.4744	0.4750	0.4756	0.4761	0.4767
2.0	0.4772	0.4778	0.4783	0.4788	0.4793	0.4798	0.4803	0.4808	0.4812	0.4817

(c) Since the total area under the curve is 1, we obtain

$$P(Z \geq 1.85) = 0.5 - P(0 \leq Z \leq 1.85) = 0.5 - 0.4678 = 0.0322$$

(d) Since the area under the curve for the interval $-\infty < Z \leq 0$ is 0.5, we have

$$P(Z \leq 1.10) = 0.5 + P(0 \leq Z \leq 1.10) = 0.5 + 0.3643 = 0.8643$$

(e) By applying both the normalization and symmetry properties of the curve, we obtain

$$P(Z \leq -1.45) = 0.5 - P(-1.45 \leq Z \leq 0) = 0.5 - P(0 \leq Z \leq 1.45)$$
$$= 0.5 - 0.4265 = 0.0735$$

In practice, the random variable Z has no physical meaning. For this reason, the notation $N(\mu, \sigma^2)$ is adopted preferably in those statistical problems that apply the normal distribution. However, to perform calculations, the tabulated values of the standard normal distribution are still valid. The following examples make it clear how to manage Table 7.1, when the notation $N(\mu, \sigma^2)$ is adopted.

Example 7.3 Consider the random variable X that follows the normal distribution $N(16, 16)$. Compute the accumulated distribution probability $P(X \geq 17)$ from Table 7.1.

The average and the standard deviation are $\mu = 16$ and $\sigma = 4$, respectively. The limits of the interval for the standard random variable Z are obtained from Eq. 7.8.

$$Z = \frac{X - \mu}{\sigma} = \frac{17 - 16}{4} = 0.25$$

Hence,

$$P(X \geq 17) = P(Z \geq 0.25) = 0.5 - P(0 \leq Z \leq 0.25) = 0.5 - 0.0987 = 0.4013$$

Example 7.4 The normal distribution $N(20, 4)$ has the accumulated distribution probability 0.30 for $X \leq x_0$. Compute x_0.

Since $P(Z \leq z_0) = 0.30 < 0.5$, z_0 must be negative. By making use of Table 7.1, we find $z_0 = -0.84$. Now, from Eq. 7.8, $-0.84 = (x_0 - 20)/2$. Thus, $x_0 = 20 - 1.68 = 18.32$.

In contrast to the Poisson distribution, the linear combination of the two random variables X and Y, which are characterized by the normal distribution, generates a new random variable $W = aX + bY$, which always follows the normal distribution. This important property clearly indicates how advantageous the application of the normal distribution compared to the Poisson distribution may be. For large average values of μ, the Poisson distribution is generally approximated to the normal distribution $N(\mu, \mu)$. In such a case the accumulated distribution satisfies the equation

$$P(X \leq x) = e^{-\mu} \sum_{n=0}^{x} \frac{\mu^n}{n!} \approx \Phi\left(\frac{x + 1/2 - \mu}{\sqrt{\mu}}\right) \tag{7.11}$$

TABLE 7.2 Probability for Different Intervals

Interval	Probability
$\mu \pm 0.67\sigma$	0.50
$\mu \pm \sigma$	0.68
$\mu \pm 1.64\sigma$	0.90
$\mu \pm 2\sigma$	0.95
$\mu \pm 3\sigma$	0.997

This approximation sets the standard deviation of the normal distribution equal to the standard deviation of the Poisson distribution, and yields results commonly denoted as $\mu \pm \sigma$, where $\sigma = \sqrt{\mu}$. Since the area under the curve for the interval $(\mu - \sigma, \mu + \sigma)$ is 68% of the total area, the probability that new measurements give results lying in the interval $(\mu - \sigma, \mu + \sigma)$ is 68%. Table 7.2 shows such a probability for different intervals.

III. ANALYSIS OF A SAMPLE OF RESULTS

Both the Poisson and normal distributions are applied to analyze the results of experiments with a large number of observed events. In practice, however, that is not the case, and the observed events in an experiment are frequently limited in number. All statistical information of one experiment is contained in the population distribution (or limit distribution). When a radioactive sample is analyzed, the amount of accumulated results depends on the counting rate and the measuring time, but what about the true value? This section will try to answer this and other questions relative to the analysis of a statistical sample of results.

A. Best Estimate of the True Value

For a symmetric distribution, the best estimate of the true value is given by the center of symmetry of the distribution. This center of symmetry is by definition the single value that agrees with its symmetrical position in the distribution. Since the mean, median, and mode all represent the center of symmetry of the distribution, nothing can be concluded about which one is the best to describe the true value.

For an asymmetric distribution, the mean, median, and mode take on different values. Therefore, the best estimate of the true value may be selected among these three possibilities by considering the requirements of the problem. Certain special cases may lead one to choose the median or the mode as the best estimate, but the mean of the population distribution is the one preferred in the majority of situations.

Let us assume the average of a sample $\bar{x}$ as the best estimate of the true value for a sample of results $x_1, x_2, \ldots, x_n$ that follow an unknown distribution

$$\bar{x} = \frac{1}{n}\sum_{i=1}^{n} x_i \tag{7.12}$$

Obviously, when the total number of measurements $n \to \infty$, the sample of results conforms to the population distribution, and the true value X agrees with the mean value of the distribution μ. Eisenhart (1963) analyzes the concepts of true value and uncertainty in measurement instruments.

B. Best Estimate of Precision

The analysis of the results of an experiment makes it necessary to define the concept of uncertainty. The width of the distribution shows graphically the uncertainty in symmetrical distributions. The two simplest expressions that represent the uncertainty are the absolute mean deviation τ and the standard deviation σ

$$\tau = \int_{-\infty}^{\infty} |x - X|\, p(x)\, dx \tag{7.13}$$

$$\sigma^2 = \int_{-\infty}^{\infty} (x - X)^2 p(x)\, dx \tag{7.14}$$

where X is the true value of the distribution, commonly associated with the mean of the distribution.

To define the absolute average deviation, the absolute differences between x and the true value X are multiplied by the relative frequencies of x in the distribution, and integrated for all possible values of x. On the other hand, Eq. 7.14 squares the difference instead of taking the absolute value. In both cases, the result of integration is a positive number that becomes larger with the distribution width.

The difference between the terms error and uncertainty have been widely discussed in the literature (Burns *et al.*, 1973; Campion *et al.*, 1973; Rabinovich, 2000). An easy description of error theory is given in Beers (1957). The term error is generally applied to describe the difference between the measured and the true values, while the term uncertainty is related to the inaccuracy of the measurements, and is always accompanied with the corresponding confidence probability. The International Vocabulary (1993) defines uncertainty as "an interval having a stated level of confidence." We illustrate the difference between the two definitions with an example, such that the error can be 1%, whereas the uncertainty is ±1% with a level of confidence of 95% (observe how the error can be +1% or −1%, but never ±1%).

The simplest counting statistics is when we have only one counting result and we are interested in associating with it an uncertainty. This is possible because the radioactive emission follows Poisson statistics and, if the measurement does not perturb the process, the standard deviation of the total counts is its square root.

When the counting statistics includes n measurements $x_1, x_2, \ldots, x_n$, the sample variance is given by the expression

$$s_n^2 = \frac{1}{n-1} \sum_{i=1}^{n} (x_i - \bar{x})^2 \tag{7.15}$$

which coincides with the variance Eq. 7.14 for $n \to \infty$.

C. Error Propagation

In radioactivity counting statistics, the Poisson distribution gives the uncertainty in the total number of counts. Other related quantities, e.g., counting rate or the number of counts without background are obtained by error propagation (Natrella, 1963 and Nicholson, 1966).

Suppose the triple measurement of one radioactive sample gives the three independent results x, y, z for the total number of counts, then the respective standard deviations are $\sigma_x = \sqrt{x}$, $\sigma_y = \sqrt{y}$, $\sigma_z = \sqrt{z}$, and the standard deviation of a function of these three variables x, y, z, when these are independent, is given by

$$\sigma_u^2 = \left(\frac{\partial u}{\partial x}\right)^2 \sigma_x^2 + \left(\frac{\partial u}{\partial y}\right)^2 \sigma_y^2 + \left(\frac{\partial u}{\partial z}\right)^2 \sigma_z^2 \tag{7.16}$$

On the other hand, when the three variables x, y, z are correlated, we have

$$\begin{aligned} \sigma_u^2 &= \left(\frac{\partial u}{\partial x}\right)^2 \sigma_x^2 + \left(\frac{\partial u}{\partial y}\right)^2 \sigma_y^2 + \left(\frac{\partial u}{\partial z}\right)^2 \sigma_z^2 + 2\rho_{xy}\left(\frac{\partial u}{\partial x}\right)\left(\frac{\partial u}{\partial y}\right)\sigma_x\sigma_y \\ &+ 2\rho_{yz}\left(\frac{\partial u}{\partial y}\right)\left(\frac{\partial u}{\partial z}\right)\sigma_y\sigma_z + 2\rho_{xz}\left(\frac{\partial u}{\partial x}\right)\left(\frac{\partial u}{\partial z}\right)\sigma_x\sigma_z \end{aligned} \tag{7.17}$$

where ρ_{xy}, ρ_{xz}, ρ_{yz}, are the correlation coefficients for the variables xy, xz, yz, respectively. These coefficients are defined by

$$\rho_{xy} = \frac{1}{(n-1)\sigma_x\sigma_y} \left(\sum_{i=1}^{n} x_i y_i - n\bar{X}\bar{Y} \right) \tag{7.18}$$

$$\rho_{xz} = \frac{1}{(n-1)\sigma_x\sigma_z} \left(\sum_{i=1}^{n} x_i z_i - n\bar{X}\bar{Z} \right) \tag{7.19}$$

$$\rho_{yz} = \frac{1}{(n-1)\sigma_y\sigma_z}\left(\sum_{i=1}^{n} y_i z_i - n\bar{Y}\bar{Z}\right) \tag{7.20}$$

The corresponding correlation coefficient for two independent variables is $\rho = 0$, while for two completely correlated variables it is given by $\rho = \pm 1$. Table 7.3 shows several common functions and their respective partial derivatives. To compute the standard deviations for the counting parameters listed in column 1 of Table 7.4, the application of the error propagation equations in Table 7.3 is required.

Example 7.5 Two students are discussing how to apply the error propagation equations of Table 7.3 on function $u = 2x$ to obtain the standard deviation. Student A considers the standard deviation $\sigma_u = 2\sigma_x$ the best, while student B, after transforming the function into $u = x + x$, obtains the standard deviation $\sigma_u = \sqrt{2}\sigma_x$. Then a third student C joins the discussion arguing that the standard deviation is $\sigma_u = 2\sqrt{5}\sigma_x$ by transforming the function into $u = z - v$, where $z = 4x$ and $v = 2x$. Explain the discrepancies in the standard deviations obtained by the three students.

TABLE 7.3 **Error Propagation**

Equation	Standard Deviation
$u = ax$	$\sigma_u = a\sigma_x$
$u = x \pm y$	$\sigma_u = \sqrt{\sigma_x^2 + \sigma_y^2}$
$u = a + bx + cy$	$\sigma_u = \sqrt{b^2\sigma_x^2 + c^2\sigma_y^2}$
$u = xy/z$	$(\sigma_u/u)^2 = (\sigma_x/x)^2 + (\sigma_y/y)^2 + (\sigma_z/z)^2$
$u = \log x$	$\sigma_u = \sigma_x/x$

TABLE 7.4 **Application of Transmission Errors to Different Counting Results**

Result	Equation	Standard Deviation
Counting rate	$b = C_B/t_B$	$\sigma(b) = \sigma(C_B)/t_B = \sqrt{C_B}/t_B$
Net gross counting	$C_S = C_{S+B} - C_B$	$\sigma(C_s) = \sqrt{\sigma^2(C_{S+B}) + \sigma^2(C_B)}$
		$= \sqrt{C_{S+B} + C_B}$
Net counting rate	$c_S = C_{S+B}/t_{S+B} - C_B/t_B$	$\sigma(c_s) = \sqrt{\sigma^2(C_{S+B})/t_{S+B}^2 + \sigma^2(C_B)/t_B^2}$
		$= \sqrt{C_{S+B}/t_{B+S}^2 + C_B/t_B^2}$
Channels ratio	$Q = N_A/N_B$	$(\sigma(Q)/Q)^2 = (\sigma(N_A)/N_A)^2 + (\sigma(N_B)/N_B)^2$
		$= 1/N_A + 1/N_B$

Only the standard deviation given by student A is correct. The assumption of student B is not valid, because the two terms into which the function u has been separated are correlated. Therefore, from Eq. 7.17 we have

$$\sigma_u^2 = \sigma_x^2 + \sigma_x^2 + 2\sigma_x\sigma_x = (\sigma_x + \sigma_x)^2 = 4\sigma_x^2$$

so

$$\sigma_u = 2\sigma_x$$

in agreement with the result given by student A. The argument of student C is also devoid of the correlation concept. To be correct, the standard deviation should be expressed as

$$\sigma_u^2 = \sigma_z^2 + \sigma_v^2 - 2\sigma_z\sigma_v = 16\sigma_x^2 + 4\sigma_x^2 - 2 \times 4\sigma_x \times 2\sigma_x = (4\sigma_x - 2\sigma_x)^2 = 4\sigma_x^2$$

so

$$\sigma_u = 2\sigma_x$$

which also agrees with the result obtained by student A.

Error propagation is of great interest in planning experiments. Optimization of experiments allows one to reduce to a minimum the uncertainties of experimental results. A good example is the optimization in isotopic dilution as is shown by Angoso *et al.* (1973). Optimization of the figure of merit to obtain the best compromise between sample and background counts can be seen in the papers of Loevinger and Berman (1951), Thomas (1954), Jaffey (1960), Reynolds (1964), Donn and Wolke (1976), and Wyld (1970).

D. Accuracy of the Mean Value

In the previous sections, the mean value μ of a distribution of infinite events was taken as the best estimate of the true value X. In practice, however, the number of available measurements is finite, and frequently low. Since this number varies depending on the experiment, one question that immediately arises is: How can we approximate the mean value to the true value? In other words, how many measurements are necessary to obtain an accuracy better than a certain value?

The verification of one experiment by repeating the same sequence of measurements should provide similar average values X_n, and be in good agreement with the true value X, when the total number of measurements is large. The average value for a sequence of n independent measurements is given by

$$X_n = \frac{1}{n}(x_1 + x_2 + \cdots + x_n) \tag{7.21}$$

By applying the error propagation theory to Eq. 7.21, the following standard deviation of the mean is obtained

$$\sigma(X_n) = \left(\frac{\sigma^2(x_1)}{n^2} + \frac{\sigma^2(x_2)}{n^2} + \cdots + \frac{\sigma^2(x_n)}{n^2} \right)^{1/2} \tag{7.22}$$

Since $x_1, x_2, \ldots, x_n$ come from measuring the same quantity, we have

$$\sigma^2(x_1) = \sigma^2(x_2) = \cdots = \sigma^2(x_n) = \sigma^2(x) \tag{7.23}$$

thus

$$\sigma(X_n) = \frac{\sigma(x)}{\sqrt{n}} \tag{7.24}$$

By taking the average quadratic deviation Eq. 7.15 as the best estimate of the measurements, we obtain

$$S_n = \sigma(X_n) = \frac{\left(\sum_{i=1}^{n} (x_i - X)^2 \right)^{1/2}}{\sqrt{n(n-1)}} = \frac{\sigma_n(x)}{\sqrt{n-1}} = \frac{s_n(x)}{\sqrt{n}} \tag{7.25}$$

which indicates the degree of accuracy of the mean value X_n. The standard deviation S_n in Eq. 7.25 is called the **standard error**, and gives the best estimate of the true value when written as

$$X = X_n \pm S_n \tag{7.26}$$

where S_n is a function of the total number of counts and measurements. The increment of the number of measurements makes the accuracy improve proportionally with the square root.

Frequently the best estimate of one experiment is computed from the mean values instead of directly from the measurements. Suppose X_n and Y_n are the corresponding mean values of two sets of n and m measurements, then the mean value of the new variable $z = x \pm y$ is $Z_{nm} = X_n \pm Y_m$, and its standard deviation is given by

$$\sigma(Z_{nm}) = \sqrt{\sigma^2(X_n) + \sigma^2(Y_m)} \tag{7.27}$$

so

$$S_{nm} = \sqrt{S_n^2 + S_m^2} = \sqrt{\frac{\sigma_n^2(x)}{n-1} + \frac{\sigma_m^2(x)}{m-1}} \tag{7.28}$$

E. Combination of Measurements

We are by now quite familiar with the computation of the uncertainty of non-measured quantities from the uncertainty of directly measured quantities by error propagation. However, how can we combine results from experiments in which we have applied different methods?

Suppose that the successive measurements carried on one radioactive sample give $X_n \pm S_n$ by applying one procedure and $X_m \pm S_m$ by applying another. If both methods were of similar precision, a good estimate of the true value could be

$$X_{n,m} = \frac{X_n + X_m}{2} \tag{7.29}$$

In practice, however, one procedure may be more precise than the other, and the best estimate of the true value is given by the weighted mean

$$X_{n,m} = \alpha X_n + (1 - \alpha)X_m \quad \text{where} \quad 0 \le \alpha \le 1 \tag{7.30}$$

The numerical value of the weight α is computed from S_n and S_m. By applying error propagation to Eq. 7.30, we obtain the equation

$$S_{n,m}^2 = \alpha^2 S_n^2 + (1 - \alpha)^2 S_m^2 \tag{7.31}$$

The error function $S_{n,m}$ becomes minimum when the derivates vanishes

$$\frac{dS_{n,m}^2}{d\alpha} = 2\alpha S_n^2 - 2(1 - \alpha)S_m^2 = 0 \tag{7.32}$$

That is, when

$$\alpha = \frac{S_n^{-2}}{S_n^{-2} + S_m^{-2}} \quad \text{and} \quad 1 - \alpha = \frac{S_m^{-2}}{S_n^{-2} + S_m^{-2}} \tag{7.33}$$

Hence, the best estimate of the true value will be given by

$$X_{n,m} = \frac{1}{S_n^{-2} + S_m^{-2}} \left(\frac{X_n}{S_n^2} + \frac{X_m}{S_m^2} \right) \tag{7.34}$$

and the standard deviation

$$S_{n,m}^2 = \frac{S_n^{-2} + S_m^{-2}}{(S_n^{-2} + S_m^{-2})^2} = \frac{S_n^2 S_m^2}{S_n^2 + S_m^2} \tag{7.35}$$

so

$$S_{n,m}^{-2} = S_n^{-2} + S_m^{-2} \tag{7.36}$$

The subsequent generalization of Eqs. 7.34 and 7.36 to more than two procedures generates the following equations:

$$X_{n,m,p,\ldots} = \frac{1}{S_n^{-2} + S_m^{-2} + S_p^{-2} + \cdots} \left(\frac{X_n}{S_n^2} + \frac{X_m}{S_m^2} + \frac{X_p}{S_p^2} + \cdots \right) \tag{7.37}$$

and

$$S_{n,m,p,\ldots}^{-2} = S_n^{-2} + S_m^{-2} + S_p^{-2} + \cdots \tag{7.38}$$

Frequently, these equations are written applying the following notation:

$$X = \sum_{n=1}^{N} \frac{x_n}{S_n^2} \bigg/ \sum_{n=1}^{N} \frac{1}{S_n^2} \tag{7.39}$$

$$S_{in}^2(X) = 1 \bigg/ \sum_{n=1}^{N} \frac{1}{S_n^2} \tag{7.40}$$

the subscript 'in' denotes an inner type standard deviation. On the other hand, the subscript 'out' is used for outer standard deviations that verify

$$S_{out}(X) = \left[\frac{\sum_{i=1}^{N} (x_i - X)^2 / S_i^2}{(n-1) \sum_{i=1}^{N} 1/S_i^2} \right]^{1/2} \tag{7.41}$$

Both types of standard deviation, S_{out} and S_{in}, are equal for samples taken from the same normal population. However, in many situations, the ratio S_{out}/S_{in} differs from unity. Adopting the expression defined by Topping (1972):

$$Z = \frac{S_{out}}{S_{in}} = \left[\frac{\sum_{i=1}^{N} (x_i - X)^2 / \sigma_i^2}{n-1} \right]^{1/2} \tag{7.42}$$

The observed data can only be considered consistent when the value of Z does not differ significantly from unity. In such a case, the best value for the error of the weighted average is given by the greatest of the two values S_{out} or S_{in}. On the other hand, when Z differs significantly from unity, the conclusion is that systematical errors are present in the measurements. This leads to weights, which do not verify the Eq. 7.33, and depend more on other conditions such as the right performance of the experiment.

Example 7.6 The measurement of the energy emitted by one of the gamma-ray transitions of ^{60}Co was carried out by five different methods. The resulting energies and their respective uncertainties are shown in Table 7.5. Find the mean value and the inner and outer standard deviations. Also compute the value of Z and analyze the consequences of its proximity to unity.

The computation of the weighted average gives

$$X = \frac{\sum_{n=1}^{5} x_n/S_n^2}{\sum_{n=1}^{5} 1/S_n^2} = \frac{1.3840098 \times 10^7}{1.0386478 \times 10^4} = 1332.511 \text{ keV}$$

For the inner and outer type standard deviations we have

$$S_{in} = \frac{1}{\sqrt{\sum_{n=1}^{5} 1/S_n^2}} = \frac{1}{\sqrt{1.0386478 \times 10^4}} = 0.0098$$

and

$$S_{out} = \left[\frac{\sum_{n=1}^{5} (x_n - X)^2/S_n^2}{(5-1)\sum_{n=1}^{5} 1/s_n^2} \right]^{1/2} = \left(\frac{1.914313}{4 \times 1.0386 \times 10^4} \right)^{1/2} = 0.0068$$

respectively.

For the value of Z we obtain

$$Z = \frac{\sigma_{out}}{\sigma_{in}} = \frac{0.0068}{0.0098} = 0.69$$

By applying the F-test in the way shown in Example 7.10, we conclude that we have no reasons to think of different standard deviations.

TABLE 7.5 Energy and Uncertainty of ^{60}Co Gamma-ray

E_T[keV]	Uncertainty
1332.483	0.046
1332.560	0.050
1332.540	0.040
1332.509	0.015
1332.508	0.015

F. The Statement of the Results

The concept of uncertainty is commonly applied to evaluate the degree of feasibility in the results of one experiment. The uncertainty considers two limits to determine how the best estimate may deviate from the true value in terms of probability. The uncertainty of the results of a measurement is generally determined by several components, which may be grouped in two categories, according to the procedure used to determine the numerical values:

- Uncertainties evaluated following statistical procedures (Type A evaluation).
- Uncertainties evaluated by means of other procedures (Type B evaluation).

There is no correspondence between the classification of the uncertainty components in the two categories A and B and the usual classification into random and systematic errors. A detailed discussion on systematic uncertainties is given by Eisenhart (1968). The nature of an uncertainty is conditioned by the use of the corresponding quantity or more clearly, the way in which the quantity appears in the model describing the measuring process. When the corresponding quantity is used in different ways, the random component may be transformed into systematic or vice versa. Therefore, the terms accidental and systematic uncertainties may be misleading when they are used in practice. An alternative nomenclature sometimes used is:

"uncertainty component arising from a random effect"

In these statements the random effect generates a possible random error in the process of measurement, while a systematic effect may generate a possible systematic error. A clear description of the different uncertainties involved in sample standardization is presented in the paper of Zimmerman *et al.* (2001).

Each uncertainty component, which contributes to the resultant uncertainty of a measurement by means of an estimated standard deviation, is referred to as the **standard uncertainty** and denoted by u_i. The u_i value is the positive square root of the estimated variance u_i^2.

The uncertainty component of category A is represented by means of statistics: The estimated standard deviation s_i, which is equal to the positive value of the square root of the estimated variance s_i^2 and the number v_i of the associated degrees of freedom. For this component the standard uncertainty is $u_i = s_i$. The computation of the uncertainty by means of the statistical analysis of an observed set is referred to as **Type A uncertainty** computation, which may be considered as an approximation to the corresponding standard deviation.

In a similar way the **Type B uncertainty** component, denoted by u_j, may be considered as an approximation to the corresponding standard deviation. To understand the background of the statement of uncertainty, we refer to the paper of Müller (1979).

1. Combined Standard Uncertainty

The combined standard uncertainty, u_c, of a measurement is obtained from the individual standard uncertainties, obtained from Type A and B evaluations. The rules given in the error transmission section will be applied here.

The combined standard uncertainty is used in the determination of fundamental constants, in fundamental metrology research and in comparisons.

On the other hand, for commercial, industrial or regulatory applications, the required uncertainties must define an interval with a high probability that the result falls within it. In this case, it is convenient to introduce the concept of **expanded uncertainty**, U, obtained from multiplying u_c by a **coverage factor**, k, so that $U = ku_c$. Generally, the value of the coverage factor k is chosen so that we select a confidence level associated with the interval defined by $U = ku_c$. Two frequently used coverage factors, when the quantity is described by a normal distribution, are $k = 2$ and $k = 3$, which define intervals of confidence of 95.5 and 99.7%, respectively. Evaluation of precision and accuracy in instrument calibration systems is given by Eisenhart (1968).

2. Rules for Expressing Results

In this section we follow the NIST recommendations given by Taylor and Kuyatt (1994). The result of a measurement may be followed by the sign $\pm$ and the corresponding uncertainty. The U value and the k factor or the applied u_c must always be specified. A complete uncertainty description for activity concentration for three measurement systems: liquid scintillation, ionization chambers, and $4\pi\gamma$ spectrometers are presented in the work of Zimmerman *et al.* (2001). It is recommended that the following information be included:

A list of all the components of the standard uncertainty with the appropriate degrees of freedom and the resulting value u_c. The uncertainty components should be identified in agreement with the method used to estimate the numerical values: Type A or B. A detailed description of how each component of the standard uncertainty has been evaluated must be presented. When we take $k \neq 2$, a description of how k has been chosen should be given. As an example we present three different forms of writing the energy and the associated uncertainty for the gamma-ray energy of ^{51}Cr:

(1) $E = (320.08419 \pm 0.00042)$ keV. The number after $\pm$ symbol is the numerical value of the expanded uncertainty $U = ku_c$, with $u_c = 0.00021$ keV, and a coverage factor $k = 2$. We assume that the estimated values of a sample follow a normal distribution with a standard deviation u_c. The unknown energy value is believed to be in the interval defined by U with a level of confidence of 95%.

(2) $E = (320.08419 \pm 0.00052)$ keV. Where the number following the $\pm$ symbol is the numerical value of the expanded uncertainty, $U = ku_c$, and a coverage factor based into a t distribution for $\nu = 9$ degrees of

freedom. U defines an interval containing the unknown energy value with a confidence level of 95%.

(3) $E = 320.08419 \, \text{keV}$ with a combined standard deviation $u_c = 0.00021 \, \text{keV}$. We assume that the possible estimated energy values are distributed following a normal distribution with a standard deviation u_c. The unknown energy value is in the interval $E \pm u_c$ with an approximated confidence level of 68%.

IV. STATISTICAL INFERENCE

From a very general point of view we can consider statistical inference as a form of decision based on probability. In a more limited sense, however, one can say that the statistical inference allows estimation or prediction from the data. Statistical inference gives responses to three types of problems: first, it permits one to test if a statement about the value of a parameter is correct. Second, it allows one to calculate the value of a parameter. Finally, it allows one to see whether some relationship exists between two or more variables and in the affirmative case, to determine this relationship. In this section we analyze the hypothesis testing and the estimation techniques.

A. Hypothesis Testing

A statistical hypothesis is an assertion about the probability density function of a random variable. Thus, the assertion that a random variable possesses a Poisson distribution is an example of a statistical hypothesis. The statement that the mean of a normal distribution is 5 is also a statistical hypothesis.

A test of a statistical hypothesis is a procedure for deciding whether to accept or reject the hypothesis. Before carrying out the test we define a hypothesis called a **null hypothesis,** denoted by H_0. A null hypothesis is a statement about a population parameter that is being tested by the use of sample results and a decision making process. This hypothesis must be contrasted with another called the **alternative hypothesis,** denoted by H_1. An alternative hypothesis is a statement to be accepted if the null hypothesis is rejected. The alternative hypothesis utilizes the same parameter as the null hypothesis but gives motivation for the rest of the test procedure. This is the statement we want to accept. Both hypotheses define complementary regions. The **critical region** is the portion of a distribution that provides values for the sample results causing the rejection of H_0. In other words, the critical region is the acceptance region for the alternative hypothesis. The critical value is the value that determines the critical region. Once the null and the alternative hypotheses have been established, we proceed to carry out the experiment and obtain the statistical sample to quantify the statistics of the test. If this one belongs to the critical region we reject the null hypothesis and accept the alternative.

TABLE 7.6 Possible Consequences of a Final Decision

| Decisions | Reality | |
	H_0 true	H_1 true
Accept H_0	Correct decision	Type II error
Reject H_0	Type I error	Correct decision

The result of our decision implies the possibility to guess correctly or make a mistake. We do not know with certainty what the true result is. To understand the meaning of a hypothesis test and the decision table established before carrying out the experiment, we consider the particular case of a sample, which may be radioactive. Only after carrying out the measurement experiment, may the analyst decide if the sample is radioactive or is not radioactive. Two correct and two incorrect decisions can be taken as is shown in Table 7.6

The interpretation of Table 7.6 is the following:

- If the null hypothesis is certain and it is accepted, the decision is correct.
- If the null hypothesis is certain and it is rejected, the decision is incorrect and the error made is Type I. Its probability is represented by α.
- If the alternative hypothesis is certain and it is accepted, the decision is correct.
- If the alternative hypothesis is certain and it is rejected, the decision is incorrect and the error made is Type II. Its probability is represented by β.

Therefore:

$$\alpha = P \text{ (Type I error)} = P \text{ (reject } H_0 \text{ being certain)}$$
$$= P \text{ (accept } H_1 \text{ when } H_0 \text{ is certain)}$$

$$\beta = P \text{ (Type II error)} = P \text{ (accept } H_0 \text{ being uncertain)}$$
$$= P \text{ (accept } H_0 \text{ when } H_1 \text{ is certain)}$$

The power of the test is defined as $1 - \beta$, the probability of rejecting a false null hypothesis:

$$1 - \beta = P \text{ (accept } H_1 \text{ when it is certain)}$$

Example 7.7 To decide if a sample is radioactive we accept the following null hypothesis: "$H_0: \mu_s = 0$ (the sample is not radioactive)". a) What cases can be given? b) Which is the Type 1 error? c) Which is the Type II error?

d) Which error is more dangerous? e) What do you prefer: α small and β large or α large and β small?

   a) The following cases can be given:
      – The activity of the sample is null and the analyst decides that it is not radioactive. The decision is correct.
      – The activity of the sample is null and the analyst decides that it is radioactive. The decision is incorrect and a Type I error is made.
      – The sample is radioactive and the analyst decides that the sample has radioactivity. The decision is correct.
      – The sample is radioactive and the analyst decides that the sample has no radioactivity. The decision is incorrect and a Type II error is made.
   b) A Type I error is made when the analyst decides that the sample is radioactive while it is not radioactive.
   c) A Type II error is made when the analyst decides that the sample is not radioactive while it is radioactive.
   d) In this case it is more serious to make a Type II error than a Type I error. A Type I error obliges one to take unnecessary precautions with the sample. On the other hand, a Type II error leads us to consider the sample as non-radioactive when in fact it is radioactive. That produces a situation with radiation risk.
   e) It is preferable to have α large and β small than the reverse.

It is not convenient to make α or β equal to zero. The most advisable is to choose a value for α equal to 0.01 or 0.05; when we examine a situation where H_0 is true, 1% or 5%, respectively, of the times we make incorrect decisions. The selected value for α will be determined by the severity of the circumstances of making an error. We have centered our attention on α because the objective of hypotheses testing is to test the truth of the statement contained in H_0 and to make a decision related to the rejection of this hypothesis.

The value of α is equal to the area under the curve corresponding to the rejection region. It is usual to nominate α as the significant value or the size of the critical region. Therefore, the level of significance is the probability α of a Type I error. When we say that the results of a study are statistically significant, we indicate that a value α has been chosen and the hypothesis testing leads to the rejection of the null hypothesis.

When we give α, the area of the rejection region for the distribution is known. For one-tail tests the application of α is immediate. In two-tail tests we must take into account that $\alpha/2$ corresponds to each tail, so that the sum of the areas of the two regions is α. When one knows the area of a region of the normal or t distributions, the value of z or t is obtained directly from tables.

Up to here we have chosen a level of significance before solving the problem. This procedure allows one to obtain a decision in a clear and fast way, defining previously a value for α. Another strategy consists of using what is known as the p-value to express the level of significance achieved by the data. The p-value is defined as the probability of obtaining a statistical test more extreme than z^* if H_0 is true; in other words, the p-value $=$ $P(z > |z^*|)$ where $|z^*|$ is the absolute value of z^*.

B. Confidence Intervals

Interval estimators very probably contain the unknown population parameter. To formalize these statements it is necessary to express them in terms of probability. Let us suppose that θ is the parameter to be estimated. Let us also suppose that we have extracted a random sample and from the sample information it is possible to find two random variables X_1 and X_2, such that $X_1 < X_2$. If these random variables have the property that $1 - \alpha$ is the probability that X_1 is smaller than θ and that X_2 is larger than θ we can write

$$P(X_1 < \theta < X_2) = 1 - \alpha \tag{7.43}$$

where α is a number between 0 and 1. Then the interval X_1 and X_2 is an interval estimator of θ at $100(1 - \alpha)\%$ of confidence. If we call x_1 and x_2 the realizations of both random variables, the interval x_1 and x_2 is the confidence interval at $100(1 - \alpha)\%$ for θ. The quantity $1 - \alpha$ is the level of confidence for the interval. In other words, if we take random samples from the population a large number of times, the parameter θ will be contained in $100(1 - \alpha)\%$ of the calculated intervals. Confidence intervals are written as $x_1 < \theta < x_2$.

C. Statistical Inference

I. Variance of a Population

A problem that frequently appears in statistics is to determine if the standard deviation of a sample corresponds to the standard deviation of a population. If we take random samples of size n from a normal distribution with variance σ^2, we know that the random variable χ^2 allows us to study σ or σ^2 from the probabilities of the curve χ^2.

In radioactivity measurements with pulse counters, it is assumed that a good counter does not perturb measurements and therefore the standard deviation of the total counting is equal to the standard deviation of radioactivity disintegration. To test the reliability of a counter it is very useful to apply the χ^2 test. It allows one to check if a set of experimental data follows a preset statistical law. More details about the application of the χ^2 test in radioactivity measurements is given by Evans (1972). The χ^2 value is defined by the equation:

$$\chi^2 = \frac{\sum_{i=1}^{n} \left[\text{(observed value)}_i - \text{(expected value)} \right]^2}{\text{expected value}} \tag{7.44}$$

where n is the total number of independent classifications i in which the data have been grouped. The expected value is computed from Poisson frequency distribution and corresponds to the mean $\mu \approx \bar{x}$. The measured values x_i are the results of the counting; they should be at least 5. The previous expression is now

$$\chi^2 = \frac{\sum_{i=1}^{n} (x_i - \bar{x})^2}{\bar{x}} \qquad (7.45)$$

We will compare $\sigma = \sqrt{\bar{x}}$ with $s = \sqrt{(\sum_{i=1}^{n} (x_i - \bar{x})^2)/(n - 1)}$, where n is the number of measurements. The null hypothesis is:

$$H_0: \sigma = s \qquad (7.46)$$

and the alternative is:

$$H_1: \sigma \neq s \qquad (7.47)$$

From the alternative hypothesis we conclude that the test is bilateral. Therefore, we will reject the null hypothesis if we obtain

$$\chi^2 > \chi^2_{n-1,\alpha/2} \quad \text{or} \quad \chi^2 < \chi^2_{n-1,1-\alpha/2} \qquad (7.48)$$

TABLE 7.7 **Chi-square (χ^2) Distribution**

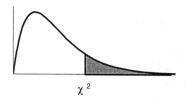

F	0.995	0.99	0.975	0.95	0.90	0.10	0.05	0.025	0.01	0.005
1	–	–	0.001	0.004	0.016	2.706	3.841	5.024	6.635	7.879
2	0.010	0.020	0.050	0.103	0.211	4.605	5.991	7.378	9.210	10.60
3	0.072	0.115	0.216	0.325	0.584	6.251	7.815	9.348	11.34	12.84
4	0.207	0.297	0.484	0.711	1.064	7.779	9.488	11.14	13.28	14.86
5	0.412	0.554	0.831	1.145	1.610	9.236	11.07	12.83	15.09	16.75
6	0.676	0.872	1.237	1.635	2.204	10.64	12.59	14.45	16.81	18.55

TABLE 7.8   Counts of a Radioactivity Sample

Measure	Counts	Δ	Δ^2
1	214	−10	100
2	222	−2	4
3	217	−7	49
4	210	−14	196
5	243	19	361
6	238	14	196

Example 7.8 A radioactive sample was measured with a Geiger counter 6 times. The duration of each measurement was 5 min. Check if the counter works well taking the measurements for Table 7.8.

The average value of the number of counts accumulated in 5 min is:

$$\bar{x} = \frac{\sum_{i=1}^{6} x_i}{n} = \frac{1344}{6} = 224$$

The distribution standard deviation is:

$$\sigma = \sqrt{\bar{x}} = \sqrt{224} = 15.0$$

The observed standard deviation is:

$$s = \sqrt{\frac{\sum_{i=1}^{n}(x_i - \bar{x})^2}{n-1}} = \sqrt{\frac{906}{5}} = 13.5$$

The null hypothesis is:

$$H_0 : \sigma = s$$

The alternative hypothesis is:

$$H_1 : \sigma \neq s$$

The rejection region is obtained from Table 7.7 for $f = n - 1 = 5$ and $\alpha = 0.05$. The critical values are $\chi^2_{5,\,0.025} = 12.83$ and $\chi^2_{5,\,0.975} = 0.831$. From the experimental data we have

$$\chi^2 = \frac{\sum_{i=1}^{n}(x_i - \bar{x})^2}{\bar{x}} = \frac{906}{224} = 4.04$$

The rejection region is $\chi^2 > 12.83$ and $\chi^2 < 0.831$. As χ^2 is outside the rejection region we do not reject the null hypothesis and we conclude that the counter works correctly. Other examples for the application of the χ^2 test to counters with anomalies are available from Grau Carles and Grau Malonda (2000).

2. Variance of Two Populations

Comparing variances requires the introduction of a new statistical test, the F ratio. Two population variances are compared by forming a ratio of their corresponding sample variance. The null hypothesis is $H_0 : \sigma_1^2 = \sigma_2^2$ and the statistical test is

$$F = \frac{s_1^2}{s_2^2} \tag{7.49}$$

The F-statistics is described by the F-distribution depending on the degrees of freedom. Now we have two samples and one degree of freedom for each sample: $f_1 = n_1 - 1$ is the degree of freedom for the numerator and $f_2 = n_2 - 1$ for the denominator. We write these as

$$f = (n_1 - 1, n_2 - 1) \tag{7.50}$$

Since s_1^2 and s_2^2 can never be negative, the F curves start at 0 and are skewed to the right as is shown in Table 7.9. The total area under the curve is equal to 1.

The null hypothesis does not establish an order for σ_1^2 and σ_2^2. When we write $\sigma_1^2 = \sigma_2^2$, we could also write $\sigma_2^2 = \sigma_1^2$. It does not matter which sample variance goes in the numerator of F. When $H_0 : \sigma_1^2 = \sigma_2^2$ is true, under ideal circumstances, $F = s_1^2/s_2^2 = 1$. Thus, to carry out a hypothesis test we must see how far the computed values of F deviate from 1.

If we observe the sketch of Table 7.9 we can appreciate that this table gives critical values larger than 1 for the right hand tail. This decision simplifies our work but it forces us to compromise such that the larger values of s must be in the numerator. If we accept this criterion we can ignore the F values on the left hand tail.

Let us examine in detail Table 7.9 corresponding to $\alpha = 0.025$. The rows of the table indicate the number of degrees of freedom of the numerator and the columns indicate the number of degrees of freedom of the denominator. The critical value is in the intersection of the appropriate row and column. When the value of the degrees of freedom is not explicitly in the table, the following approach must be followed: 1) choose the next degree of freedom, 2) if the degree of freedom in the table is halfway with that of the sample the largest critical value will be taken.

Example 7.9 In example 6 the internal $(S_{in} = 0.0098)$ and external $(S_{ex} = 0.0068)$ uncertainties were obtained. Find if these uncertainties are different at the 95% level of confidence.

TABLE 7.9 *F* Distribution Values of $F_{0.025}$

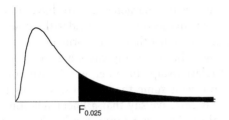

$F_{0.025}$

F deno.	F numerator								
	1	2	3	4	5	6	7	8	9
1	647.79	799.50	864.16	899.58	921.85	937.11	948.22	956.66	963.28
2	38.51	39.00	39.17	39.25	39.30	39.33	39.36	39.37	39.39
3	17.44	16.04	15.44	15.10	14.89	14.74	14.62	14.54	14.47
4	12.22	10.65	9.98	9.60	9.36	9.20	9.07	8.98	8.90
5	10.00	8.43	7.76	7.39	7.15	6.98	6.85	6.76	6.68
6	8.81	7.26	6.60	6.23	5.99	5.82	5.70	5.60	5.52
7	8.07	6.54	5.89	5.52	5.29	5.12	4.99	4.90	4.82
8	7.57	6.06	5.42	5.05	4.82	4.65	4.53	4.43	4.36
9	7.21	5.71	5.08	4.72	4.48	4.32	4.20	4.10	4.03
10	6.94	5.46	4.83	4.47	4.24	4.07	3.95	3.85	3.78

We assume that there is not any difference between the variances:

$$H_0 : S_{in}^2 = S_{ex}^2$$

The alternative hypothesis indicates that the variances are different

$$H_1 : S_{in}^2 \neq S_{ex}^2$$

This alternative hypothesis indicates the contrast is bilateral with $\alpha = 0.05$. The degrees of freedom are $f_{in} = f_{ex} = n - 1 = 4$. From Table 7.9 we obtain the critical value $F_{\alpha/2} = F_{0.025} = 9.60$. The rejection region is then $F_{0.025} > 9.60$. The computed value for $F_{0.025}^*$ is:

$$F_{0.025}^* = \frac{S_{in}^2}{S_{ex}^2} = \frac{0.0098^2}{0.0068^2} = 2.08$$

As the computed value is less than the critical one ($F_{0.025}^* < F_{0.025}$), the null hypothesis cannot be rejected. Therefore, we conclude that there is no reason to think that S_{ex} and S_{in} are different.

V. REGRESSION

The problems analyzed so far were characterized to have one or two independent random variables. In this section we consider the case of two or more random variables related to each other. The form of the relationship may be very varied and unknown, but on many cases it is possible to guess, as a near approach, a linear relationship. In other words, we consider two random variables X and Y, and we assume that the observations tend to be grouped in a straight line. In this case, we say that a linear relationship exists between the two random variables. Once a relationship has been established, we must obtain the function relating the random variables by means of a regression.

A. Linear Regression

The degree of association between two random variables is obtained by applying the correlation between them. This correlation is symmetrical since it is indifferent to the correlation between X and Y or between Y and X. In this section we study the effect on the random variable Y when the random variable X takes a specific value. We limit our analysis to the simplest mathematical structure relating X and Y: the linear relationship. Variations of the problem of fitting a function to a set of data: curvilinear relationships, weighted least squares, nonlinear squares, etc. are analyzed by Draper and Smith (1966). Since we are working with random variables, over time for each value of X a distribution of Y values is obtained; therefore, we will use the concept of conditional distribution. An essential characteristic of this distribution is the mean or the expected value. We denote the expected value of the random variable Y with $E[Y|X = x]$, when the random variable X takes the specific value x. Our assumption of linearity implies that the conditional expected value has a linear dependence on x,

$$E[Y|X = x] = \alpha + \beta x \tag{7.51}$$

where α and β determine the correct line. The interpretation of each one of these constants is immediate. When $x = 0$ we have

$$E[Y|X = 0] = \alpha \tag{7.52}$$

where α is the expected value for the dependent variable Y when the independent variable X takes the value 0.

Let us suppose now that X is increased by 1 unit so that x becomes $x + 1$, then

$$E[Y|X = x + 1] = \alpha + \beta(x + 1) \tag{7.53}$$

and

$$E[Y|X = x + 1] - E[Y|X = x] = \alpha + \beta(x + 1) - (\alpha + \beta x) = \beta \qquad (7.54)$$

Therefore, β, the slope of the line, is the expected increase in Y when X increases by a unit.

In fact, the equations given previously are not verified exactly. Let us suppose that the independent variable takes the value x_i. If we represent by Y_i the corresponding value of the dependent random variable, the expected value is

$$E[Y_i|X = x_i] = \alpha + \beta x_i \qquad (7.55)$$

But, in practice, the value of Y_i will deviate from the expected value. If the difference between the observed and the expected value is denoted by ε_i, we can write

$$\varepsilon_i = Y_i - E[Y_i|X = x_i] = Y_i - (\alpha + \beta x_i) \qquad (7.56)$$

so that

$$Y_i = \alpha + \beta x_i + \varepsilon_i \qquad (7.57)$$

where the random variable ε_i has a mean of 0. The last equation is known as the **population regression line** of data (x_i, Y_i).

We have just described the regression model illustrated in Figure 7.2. For each possible value of the independent variable, the value of the dependent variable may be represented by means of a random variable whose mean is on the regression line. The regression line is drawn through the means of the distributions. For a value of x_i, the independent variable, the deviation of the dependent variable Y_i from the regression line is the error term ε_i.

The regression line is an interesting theoretical construction but, in practice, as we always work with samples of observations, we will never be

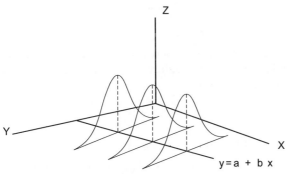

FIGURE 7.2   Probability density functions of the dependent variable for given values of x.

able to obtain this one exactly. Instead of α and β we obtain their estimators: the number a and b. The estimated line has an equation

$$y = a + bx \tag{7.58}$$

Let us suppose that we have a sample of n pairs of observations $(x_1, y_1), (x_2, y_2), \ldots, (x_n, y_n)$. We are interested in obtaining the line that fits the data best. We know that the value x_i produces the value y_i obtained from equation $a + bx_i$, but the exact value for the dependent variable is y_i. The difference between the two is

$$e_i = y_i - (a + bx_i) \tag{7.59}$$

The values of e_i may be positive or negative. If we want to give the same weight to the positive and negative values of the same quantity, a possibility is to work with the square of e_i. The sum of squared differences from the point to the line is

$$SQ = \sum_{i=1}^{n} e_i^2 = \sum_{i=1}^{n} (y_i - a - bx_i)^2 \tag{7.60}$$

The least squares method allows one to estimate the line of a population regression for which the sum of the squares is a minimum. The parameter b can be obtained from the equation

$$b = \frac{\sum_{i=1}^{n} x_i y_i - n\bar{x}\bar{y}}{\sum_{i=1}^{n} x_i^2 - n\bar{x}^2} \tag{7.61}$$

and a with the equation

$$a = \bar{y} - b\bar{x} = \frac{\bar{y} \sum_{i=1}^{n} x_i^2 - \bar{x} \sum_{i=1}^{n} x_i y_i}{\sum_{i=1}^{n} x_i^2 - n\bar{x}^2} \tag{7.62}$$

where $\bar{x}$ and $\bar{y}$ are the respective means of the sample. The line

$$y = a + bx \tag{7.63}$$

is the **sample regression** line of Y on X.

The least squares method is a good procedure to estimate the regression line for the population. This procedure is the most appropriate when the regression line for the population

$$Y_i = \alpha + \beta x_i + \varepsilon_i \tag{7.64}$$

is required. This must fulfil the following conditions:

- Each value of x_i is a fixed number. That is equivalent to saying that the realization of a random variable X_i is independent of the error term ε_i.

- Errors are random variables with an expected value equal to zero

$$E[\varepsilon_i] = 0 \quad (i = 1, 2, \ldots, n) \tag{7.65}$$

- All the random variables ε_i have the same variance σ_ε^2

$$E[\varepsilon_i^2] = \sigma_\varepsilon^2 \quad (i = 1, 2, \ldots, n) \tag{7.66}$$

- The random variables ε_i are not correlated

$$E[\varepsilon_i \varepsilon_j] = 0 \quad \text{for all} \quad i \neq j$$

Bacon (1953) describes the least squares method of fitting a line for different conditions and analyzes the goodness of fitting results from different experiments.

I. Confidence Intervals and Hypothesis Testing

We will analyze the problems of interval construction and the hypothesis testing for the regression parameters of a population. Suppose that the regression line is

$$Y_i = \alpha + \beta x_i + \varepsilon_i \tag{7.67}$$

and the conditions of the previous section are fulfilled. If σ_ε^2 is the common variance for the error terms ε_i, an unbiased estimator of σ_ε^2 is

$$s_e^2 = \frac{\sum_{i=1}^{n} e_i^2}{n - 2} \tag{7.68}$$

where e_i are the residuals of the least squares. These residuals substitute the error terms ε_i which are unknown. We divide by $n - 2$ because we lose two degrees of freedom when estimating the parameters α and β.

If we designate with b the least squares estimate of the slope of the population regression line, the estimator of β is unbiased and the variance is

$$\sigma_b^2 = \frac{\sigma_\varepsilon^2}{\sum_{i=1}^{n} (x_i - \bar{x})^2} = \frac{\sigma_\varepsilon^2}{\sum_{i=1}^{n} x_i^2 - n\bar{x}^2} \tag{7.69}$$

An unbiased estimator of σ_b^2 is provided by

$$s_b^2 = \frac{s_\varepsilon^2}{\sum_{i=1}^{n} (x_i - \bar{x})^2} = \frac{s_\varepsilon^2}{\sum_{i=1}^{n} x_i^2 - n\bar{x}^2} \tag{7.70}$$

In both cases we assume that the conditions of the previous section are fulfilled.

Although the slope is the most interesting parameter, we also give the equation to compute the estimator of the variance of the ordinate on the origin. We substitute β, b, and s_b^2 for α, a, and s_a^2 to have

$$s_a^2 = s_\varepsilon^2 \left(\frac{1}{n} + \frac{\bar{x}^2}{\sum_{i=1}^n x_i^2 - n\bar{x}^2} \right) \tag{7.71}$$

VI. DETECTION LIMITS

Radioactivity measurements are characterized by a variable zero level due to background. This situation obliges one to work with detection and determination limits when the radioactivity of the source is very low.

In this section we analyze the problem of obtaining the detection limits for very low radioactivity measurements. A complete discussion of the detection limits, in a measurement process, requires the introduction of two specific levels: i) a **decision limit** that allows one to deduce whether the result of the analysis indicates that the sample is radioactive or is not radioactive, and ii) a **detection limit** that indicates if an analytical process leads to a quantitative detection. The relationships between these limits and the equations to compute them are also given.

In a general way, two types of devices are considered: counters, characterized to accumulate the information in one channel and spectrometers, where information is distributed in numerous channels. In the latter case, we distinguish between high-resolution detectors, like Ge, and low-resolution detectors, like NaI(Tl).

A. Critical Levels

We will distinguish two fundamental problems in the measurement of very low radioactivity sources: i) given an observed net signal, S, to decide whether a real signal has been detected or, in other words, to decide whether the sample is indeed radioactive. Is $\mu_S > 0$? This question can be addressed by the statistical theory of hypothesis testing, in which one first formulates a test hypothesis. In our case, the null hypothesis H_0 for μ_S is $\mu_S = 0$. This hypothesis and the alternative hypothesis H_1 ($\mu_S > 0$) are mutually exclusive. Together they cover all possible values of μ_S. As a consequence of the intrinsic statistical variation in the counting rates, we can arrive at one of the following two types of judgment errors: i) The error of the first kind or Type I error states that true activity is greater than zero when, in fact, it is zero. ii) The error of the second kind or Type II error states that the true activity is zero when, in fact, it is greater than zero. The probability of making a Type I error is denoted by α and depends on the test procedure. The maximum value of α and the standard deviation of the net signal σ_0, when $\mu_S = 0$, allows one to establish the critical level L_C. An observed signal, S, must exceed L_C to yield the decision "detected." The probability distribution of possible outcomes, when the true net signal is zero, intersects L_C such that the factor

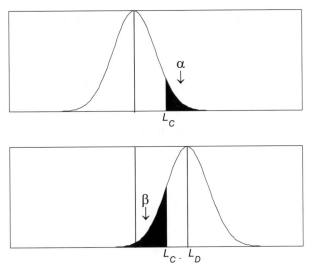

FIGURE 7.3 Type I and II errors.

$1 - \alpha$ corresponds to the correct decision "non-detected." Whereas the probability of making a Type II error, denoted by β, also depends on the size of the measured quantity; in the case of radioactive measurements it depends on the amount of radioactivity of the tested material. The most relevant papers about low-level detection limits are Altshuler and Pasternack (1963), Currie (1968), Currie (1995), and Donn and Wolke (1977).

When the critical level L_C has been established, an a priori detection limit L_D may be established by specifying L_C, the acceptable level β, for a Type II error and the standard deviation σ_D, characterizing the probability distribution of the net signal when its true value μ_S is equal to L_D. Figure 7.3 shows the Type I and II error curves and the critical levels L_C and L_D. The mean μ_S may be between 0 and L_D. When μ_S is between 0 and L_C, we agree that there is no radiation from the sample. When μ_S is between L_C and L_D, there may be radioactivity but when $\mu_S = L_D$ the Type II error is minimum. Therefore, for $L_C < \mu_S < L_D$ although we can have detection, such detection cannot be considered reliable given that the Type II error is not a minimum.

The critical level L_C is given by

$$L_C = k_\alpha \sigma_0 \tag{7.72}$$

and the detection limit by

$$L_D = L_C + k_\beta \sigma_D \tag{7.73}$$

where $k_\alpha \equiv z_\alpha$ and $k_\beta \equiv z_\beta$ are the $k \equiv z$ scores of the standardized normal distribution corresponding to probability levels $1 - \alpha$ and $1 - \beta$, respectively.

When we analyze the pulses due to radioactivity emission, we can assume that the distributions of background and background + source follow the

Poisson distribution. When the count number is sufficiently large, the distributions are approximately normal. Under such circumstances, the variance of the net counting is given by

$$\sigma^2 = \sigma_{S+B}^2 + \sigma_B^2 = \mu_S + \mu_B + \frac{\mu_B}{n} \qquad (7.74)$$

where σ_B is obtained from n measurements without the source. Note than σ depends on the signal level.

If σ_0^2 is the variance when $\mu_S = 0$, and σ_D^2 is the variance for $\mu_S = L_D$, we have

$$L_C = k_\alpha \sigma_0 = k_\alpha (\mu_B + \sigma_0^2)^{1/2} \qquad (7.75)$$

and

$$\sigma_D^2 = \sigma_{S+B}^2 + \sigma_B^2 = \mu_S + \mu_B + \sigma_0^2 = L_D + \sigma_0^2 \qquad (7.76)$$

From Eq. 7.73 we get

$$L_D = L_C + k_\beta (L_D + \sigma_0^2)^{1/2} \qquad (7.77)$$

Solving Eqs. 7.75 and 7.77 we obtain

$$L_D = L_C + \frac{k_\beta^2}{2} \left\{ 1 + \left[1 + \frac{4L_C}{k_\beta^2} + \frac{4L_C^2}{k_\alpha^2 k_\beta^2} \right]^{1/2} \right\} \qquad (7.78)$$

The mean value and the standard deviation without the source allows one to compute L_C and L_D for selected values of α and β by means of Eqs. 7.75 and 7.78. If $k_\alpha = k_\beta = k$ we obtain a considerable simplification of Eq. 7.73, which is reduced to the form

$$L_D = k^2 + 2L_C \qquad (7.79)$$

Example 7.10 Background and a source-plus-background are measured and the counting rates obtained are $C_B = 203$ counts/h and $C_{B+S} = 235$ counts/h. Previously, the background was measured for 200 h, accumulating a total of 40,000 counts (counting rate $B = 200$ counts/h). Compute the values of L_C and L_D when $\alpha = 0.025$ and $\beta = 0.050$, in the following two cases: a) when we know that the background does not change and we can use the value of B, b) when we cannot apply B because the background changes.

The value $k_\alpha = 1.96$ is obtained from Table 7.1 when $\alpha = 0.025$. When $\beta = 0.050$ we have $k_\beta = 1.645$.

The background counting rate is $C_B = 203$ counts/h. As the expected background counting rate is $B = 200$ counts/h, it seems that the background has not changed. The net counting rate is

$$C_S = C_{B+S} - B = 235 - 200 = 35 \text{ counts/h}$$

and the critical counting rate is

$$L_C = k_\alpha\sqrt{\mu_B + \sigma_B^2} \approx k_\alpha\sqrt{B} = 1.96 \times \sqrt{200} = 27.7 \, \text{counts/h}$$

since 35 is greater than 27.7 our decision is that there is activity in the sample. The minimum significant measured counting rate is 27.7 counts/h.

The minimum detectable counting rate from Eq. 7.78 is

$$L_D = L_C + \frac{k_\beta^2}{2}\left\{1 + \left[1 + \frac{4L_C}{k_\beta^2} + \frac{4L_C^2}{k_\alpha^2 k_\beta^2}\right]^{1/2}\right\}$$

$$= 27.7 + \frac{1.645^2}{2}\left\{1 + \left[1 + \frac{4 \times 27.7}{1.645^2} + \frac{4 \times 27.7^2}{1.96^2 \times 1.645^2}\right]^{1/2}\right\}$$

$$= 53.9 \, \text{counts/h}$$

and applying the approximate equation

$$L_D = (k_\alpha + k_\beta)\sqrt{B} = (1.96 + 1.645)\sqrt{200} = 51.0 \, \text{counts/h}$$

When the background changes, we cannot use the mean number B. The net counting rate is obtained applying the following expression

$$C_S = C_{S+B} - C_B = 235 - 203 = 32 \, \text{counts/h}$$

and

$$L_C = k_\alpha\sqrt{\mu_B + \sigma_B^2} = k_\alpha\sqrt{\mu_B + \mu_B} = k_\alpha\sqrt{2\mu_B} \approx k_\alpha\sqrt{2C_B}$$

$$= 1.96\sqrt{2 \times 203} = 39.49 \, \text{counts/h}$$

Since the net counting rate is less than the critical net counting rate, our conclusion is that there is no significant radioactivity in the sample. The minimum detectable counting rate is

$$L_D = L_C + \frac{k_\beta^2}{2}\left\{1 + \left[1 + \frac{4L_C}{k_\beta^2} + \frac{4L_C^2}{k_\alpha^2 k_\beta^2}\right]^{1/2}\right\}$$

$$= 39.49 + \frac{1.645^2}{2}\left\{1 + \left[1 + \frac{4 \times 39.49}{1.645^2} + \frac{4 \times 39.49^2}{1.96^2 \times 1.645^2}\right]^{1/2}\right\}$$

$$= 75.6 \, \text{counts/h}$$

and applying the approximate equation

$$L_D = (k_\alpha + k_\beta)\sqrt{2C_B} = (1.96 + 1.645)\sqrt{2 \times 203} = 72.6 \, \text{counts/h}$$

B. Gamma Spectra

Ge detectors allow the experimental spectroscopists to obtain gamma- and x-ray spectra with high resolution compared with that obtained with NaI(Tl) detectors. This excellent resolution facilitates the qualitative and quantitative analysis of radionuclide mixtures for high and medium activities. However, for low-level samples the low efficiency and the similarity of the small peaks to the fluctuations of the background can make the discrimination between true and false peaks difficult.

Since NaI(Tl) detectors have larger volume and lower resolution than Ge detectors, the detection of low activity peaks is different for each detector. Each one of the Ge peaks can be analyzed taking the two independent constituent parts: the peak and the Compton contribution. On the other hand, the NaI(Tl) peaks force us to consider the overlap contribution of the different spectral components in each of the peaks.

In this section, we analyze the determination and detection limits for Ge detectors. Based upon the acceptable risk of committing a Type I error, a minimum significant measured area for a peak is defined and for a Type II error, a minimum detectable true area is introduced. The study of complete response of the NaI(Tl) detector is carried out taking into account the contribution of the different spectral components.

I. High-Resolution Gamma Spectra

In an experimental Ge gamma spectrum, the minimum detectable area of a peak is the minimum number of photopeak counts that make it detectable. The value of this minimum depends on the spectral background under the peak, but the background usually does not coincide with the detector background. The minimum is predetermined by the statistical risk of including an observed peak when it is not a real peak or to conclude that a real peak is not present when it is really there.

The procedures to discern between real and false peaks are based on the assumption that real peaks show a Gaussian shape. This procedure gives good results when the peaks are sufficiently defined and the Gaussian hypothesis is valid; however, when the peaks are small they can be taken as statistical fluctuations with Gaussian appearance. Therefore, a computational procedure dedicated to determining very small peaks must be able to detect false peaks.

There is a probability that a false peak is accepted as a true peak. This is called a Type I error. Based upon the acceptable risk of committing a Type I error, we define a **minimum significant measured area**. It is assumed that all the peaks in the spectrum whose measured areas are smaller than this limit are discarded and considered as false peaks. This is the Type II error. Consequently, we can define a minimum detectable real peak area such that if the actual photopeak area is at least this large, the risks of committing Type I or Type II errors are less than some preselected values.

An application of this analysis consists of determining the minimum time required to be sure that the risk of making a Type I or II error does not exceed acceptable values. This allows one to predict the time required to assure the detection of small peaks.

a. False Peaks Distribution

Following Head (1972), to know the distribution of false peaks it is necessary to carry out an experiment and have a peak identification computer program. In essence, the experiment consists of preparing "background" spectra with an average measure of 50–50,000 counts/channel. For the purpose of the experiment, the background in the energy range of 80–500 keV can be simulated by the Compton tails of the two ^{60}Co gamma lines, in order to determine false peaks. By means of another gamma source with peaks in the range given above we can produce the real peaks. ^{133}Ba can be used to generate real peaks. The objective is clear: the ^{60}Co Compton distribution produces the "background" and ^{133}Ba gives the true peaks for each background. For a certain background level and several resolution values (1.0, 1.5, 2.0, 2.5, and 3.0 channels, for a real resolution of 1.5 channels), we obtain the number of false and acceptable peaks. From the different analyzed backgrounds we obtain the curves shown in Figure 7.4, where the average area of the false peaks detected $\bar{A}$ and the widths $\sigma_{\bar{A}}$ of these areas as a function of the background B under the peak areas (due to the Compton tails of ^{60}Co gamma lines) are represented.

b. Minimum Significant Area

The probability of committing a Type I error or of concluding that a peak corresponds to a gamma-ray, when in fact it is a false peak, depends upon the peak selection procedure and upon the level of the background under the peak. A minimum significant measured area A_I can be defined by the expression

$$\alpha = P\{\text{Measured area of a false peak} \geq A_I\} \qquad (7.80)$$

A peak whose area is equal or larger than A_I is retained and all peaks whose area is less than A_I are rejected. If we do not have any additional information about the peak, the probability that it is false is α or in other words, the risk of accepting a false peak as true is α.

If we assume that the false peak area distribution is normal, the minimum significant area for committing a Type I error, for a given risk, is given by the equation

$$A_I = \bar{A} + z_\alpha \sigma_{\bar{A}} \qquad (7.81)$$

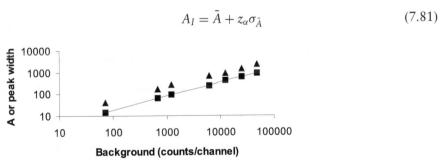

FIGURE 7.4 Average of the peak count area (triangles) and width of the peaks (squares) as a function of background (From Fig. I, *Nucl. Instrum. Methods*, **98**, J.H. Head, Minimum detectable photopeak areas in Ge(Li), 419–428, Copyright (1972), with permission from Elsevier Science).

where z_α is related to α by the equation

$$\alpha = \int_{z_\alpha}^{\infty} \frac{1}{\sqrt{2\pi}} \exp(u^2/2)\, du \qquad (7.82)$$

$\bar{A}$ and $\sigma_{\bar{A}}$ may be obtained by interpolation in Figure 7.4 for a given background B, the value of the minimum significant area $\bar{A}_I$, for a risk α, may be computed from Eq. 7.81.

c. Minimum Detectable Area

Assume that we have obtained A_T counts in the photopeak of a given gamma-ray. When we compute the area by means of a fitting program, the fitted area A_F will seldom be equal to A_T. Due to statistical fluctuations, the measured area for a peak will usually be distributed about the real area ($\bar{A}_F = A_T$) with a width σ_F. As we have seen in the last section, all peaks with mean areas lower than A_I will be discarded as not significant. A Type II error is committed when a real peak $A_T \geq A_I$, due to the spread in the fitted area, gives $A_F < A_I$ and consequently the peak is discarded.

The probability of committing a Type II error depends upon the value of A_I and the fitted area A_F. We define the minimum detectable area A_{II} as

$$\beta = P\{\text{Fitted area } A_F < A_I, \text{ when the true area is } A_T = A_{II}\} \qquad (7.83)$$

As the fitted area is normally distributed with variance σ_F^2 we have

$$A_{II} = A_I + k_\beta \sigma_F \qquad (7.84)$$

where the relationship between β and k_β is the same as that between α and k_α. In this case, σ_F^2 is the variance of the fitted area distribution when $\bar{A}_F = A_T = A_{II}$.

We introduce a hypothesis that allows us to solve the problem with the data we have. We suppose that the variance of the fitted area distribution is similar to the variance of the false peak area distribution ($\sigma_F^2 \approx \sigma_{A_{II}}^2$). As A_{II} is very close to A_I we assume that $\sigma_F^2 \approx \sigma_{A_{II}}^2 \approx \sigma_{A_I}^2 \approx \sigma_{\bar{A}}^2$, consequently

$$A_{II} = A_I + k_\beta \sigma_{\bar{A}} \qquad (7.85)$$

or

$$A_{II} = \bar{A} + (k_\alpha + k_\beta)\sigma_{\bar{A}} \qquad (7.86)$$

$\bar{A}$ and σ_A may be obtained interpolating in Figure 7.4.

Example 7.11 The background spectrum has $B = 10{,}000$ counts/channel. Obtain the values of $\bar{A}$ and $\sigma_{\bar{A}}$. Compute A_I and A_{II} for $\alpha = 10\%$ and $\beta = 1\%$.

For a background $B = 10{,}000$ we obtain from Figure 7.4

$$\bar{A} = 840 \text{ counts} \quad \text{and} \quad \sigma_{\bar{A}} = 320 \text{ counts}$$

The values of k_α and k_β, for $\alpha = 0.10$ and $\beta = 0.01$ are

$$k_\alpha = 1.28 \quad \text{and} \quad k_\beta = 2.33$$

Therefore

$$A_I = \bar{A} + k_\alpha \sigma_{\bar{A}} = 840 + 1.28 \times 320 = 1250 \, \text{counts}$$

and

$$A_{II} = \bar{A} + (k_\alpha + k_\beta)\sigma_{\bar{A}} = 840 + (1.28 + 2.33) \times 320 = 1995 \, \text{counts}$$

The relationship between A_I, A_{II}, and B is

$$A_{II} = K_{\alpha\beta} B^{0.619} \tag{7.87}$$

The exponent is obtained from the graphical representation of A_I and A_{II} as a function of B. The factor $K_{\alpha\beta}$ is a parameter independent of B containing all the information related to the Type I and II error risks that influence A_{II}. In fact, $K_{\alpha\beta}$ is a risk factor depending functionally on α, β, and B. From Eqs. 7.86 and 7.87 we obtain

$$K_{\alpha\beta} = \frac{A_{II}}{B^{0.619}} = \frac{\bar{A} + (k_\alpha + k_\beta)\sigma_{\bar{A}}}{B^{0.619}} \tag{7.88}$$

It is observed that $K_{\alpha\beta}$ is symmetric in α and β.

d. Minimum Counting Time

We have just concluded that we may calculate a minimum area, A_{II}, so that a peak with this area cannot be false with a risk α and a real peak cannot be discarded as false with a risk β. It has been supposed that the background and the peak grow evenly during data accumulation by the spectrometer. In this situation we may compute the counting duration to assure that the peak is detectable with the risks α and β.

Let us suppose that a is the true counting rate for the peak, expressed in counts/h. If $t_{\alpha\beta}$ is the time required, in hours, assuming that the risks for the Type I and II errors do not exceed α and β, respectively, we can write

$$A_{II} = a t_{\alpha\beta} \tag{7.89}$$

$$B = b t_{\alpha\beta} \tag{7.90}$$

taking these expressions to Eq. 7.87 we have

$$a t_{\alpha\beta} = K_{\alpha\beta}(b t_{\alpha\beta})^{0.619} \tag{7.91}$$

the minimum counting time is given by

$$t_{\alpha\beta} = (K_{\alpha\beta}/a)^{2.62} b^{1.62} \tag{7.92}$$

TABLE 7.10 Experimental Values of B, Ā, and $\sigma_{\bar{A}}$. (From Table 2, *Nucl. Instrum. Methods*, 98, J. H. Head, Minimum detectable photopeak areas in Ge(Li), 419–428, Copyright (1972), with permission from Elsevier Science)

B	Ā	$\sigma_{\bar{A}}$
100	50	18.8
500	135	50.9
1000	208	78.1
5000	563	211.6
10000	865	324.9
20000	1328	499.0
40000	2040	766.4

Example 7.12 Table 7.10 shows the experimental values of B, $\bar{A}$ and $\sigma_{\bar{A}}$. Find the relationship between A_{II} and B when $\alpha = 0.10$ and $\beta = 0.01$ or 0.5. Show that $A_{II} = K_{\alpha\beta}B^c$ and compute $K_{\alpha\beta}$ and c. It is assumed that $K_{\alpha\beta}$ is a parameter independent of B.

Equation $A_{II} = K_{\alpha\beta}B^c$ may be fitted taking logarithms

$$\log A_{II} = \log K_{\alpha\beta} + c \log B$$

This is a linear equation

$$y = a + cx$$

where

$$y = \log A_{II}$$

and

$$x = \log B$$

Applying least square fitting to the quantities of Table 7.11
when $\alpha = 0.10$, $\beta = 0.01$, we get $K_{\alpha\beta} = 6.81$ and $c = 0.619$
when $\alpha = 0.10$, $\beta = 0.50$, we get $K_{\alpha\beta} = 4.28$ and $c = 0.619$
As c takes the same values, we may compute $K_{\alpha\beta}$ applying the equation

$$K_{\alpha\beta} = \frac{\bar{A} + (k_\alpha + k_\beta)\sigma_{\bar{A}}}{B^{0.619}}$$

The values of B, $\bar{A}$, and $\sigma_{\bar{A}}$ may be values of Table 7.10 or interpolated logarithmically from Table 7.11; e.g., for $\alpha = 0.01$ and $\beta = 0.025$, when $B = 1000$, $\bar{A} = 208$, and $\sigma_{\bar{A}} = 78$, we have

$$K_{\alpha\beta} = \frac{208 + (2.33 + 1.96) \times 78}{1000^{0.619}} = 7.55$$

▮▮▮ **TABLE 7.11** Values of B and $A_{\parallel}$ for Different α and β Values.

$\alpha = 0.10, \beta = 0.01$		$\alpha = 0.10, \beta = 0.50$	
B	$A_{\parallel}$	B	$A_{\parallel}$
100	118	100	74
500	319	500	201
1000	490	1000	308
5000	1327	5000	834
10000	2038	10000	1281
20000	3130	20000	1967
40000	4806	40000	3021

Example 7.13 Compute the values of $t_{\alpha\beta}$ when $\alpha = 0.01$, $\beta = 0.025$, the background counting rate is $b = 100$ counts/h and $a = 20$ counts/h.

From Eq. 7.92, taking into account that $K_{\alpha\beta} = 7.55$ from the last example we have

$$t_{\alpha\beta} = \left(\frac{7.55}{20}\right)^{2.62} 100^{1.62} = 135 \text{ h}$$

2. Low-Resolution Gamma Spectra

NaI(Tl) detectors generate spectra with very poor resolution compared with the spectra obtained with Ge detectors. Consequently, the procedure described in the previous section is not generally applicable to low-resolution spectra. Isolated peaks cannot be analyzed due to interference from other spectral components. In this section, we introduce a more general procedure based on considering the complete spectral response. This procedure introduces a larger complication in calculations but it is inevitable when we have radionuclide mixtures and the spectrum of one of the radionuclides overlaps the other spectra and vice versa. In this situation, the background must be considered as an independent spectrum. A standard procedure for estimating the radionuclide concentration from gamma-ray spectrometer data is the method of weighted least squares fitting. In this case, it is assumed that the net spectrum of a radionuclide mixture is equivalent to some linear combination of the net characteristic spectra of the radionuclides existing in the sample. The concentration of these radionuclides is represented by the coefficients of the linear combination estimate. Consequently, the estimation of these coefficients is equivalent to the determination of their concentrations.

The solution of least squares is a function $\boldsymbol{\theta}' = (\theta_1, \theta_2, \ldots, \theta_m)$, which minimizes the sum of the squares of the count differences between observed and fitted channels. This sum of squares is denominated a residual variation. We define the following n-dimensional vectors:

$\mathbf{x} = (x_1, x_2, \ldots, x_n)$ net counting rate corresponding to channels $1, 2, \ldots, n$.
$\mathbf{y} = (y_1, y_2, \ldots, y_n)$ gross counts (sample + background) in channels $1, 2, \ldots, n$.
$\mathbf{b} = (b_1, b_2, \ldots, b_n)$ background counts in channels $1, 2, \ldots, n$.

If t is the counting time of the sample and r the counting time of the background, we may define the elements w_{jj} of the diagonal matrix $\mathbf{W}$ as

$$w_{jj} = \frac{y_j}{t^2} + \frac{b_j}{r^2} \tag{7.93}$$

The solution obtained by weighted least squares method is given by the estimator

$$\hat{\theta} = (\mathbf{A}\mathbf{W}^{-1}\mathbf{A}')^{-1}\mathbf{A}\mathbf{W}^{-1}\mathbf{x} \tag{7.94}$$

where $\mathbf{A}$ is an $m \times n$ matrix called the calibration matrix. The dimension m corresponds to the number of calibration spectra and n to the number of channels of each spectrum.

The variance of the estimated concentration parameter for the ith radionuclide is the ith diagonal element of the matrix $(\mathbf{A}\mathbf{W}^{-1}\mathbf{A}')^{-1}$. In order to test the goodness of fit we use the residual mean square statistic

$$s^2 = \frac{1}{n-m}(\mathbf{x} - \mathbf{A}'\hat{\theta})'\mathbf{W}^{-1}(\mathbf{x} - \mathbf{A}'\hat{\theta}) \tag{7.95}$$

which is distributed as a chi-square random variable. A review of methods currently used to unfold particle spectra for measured pulse height distributions and uncertainties propagation is presented by Matzke (2002).

a. Sample with a Single Radionuclide

To illustrate the procedure described in the previous section we consider a simple case: the limit of detection determination when the sample has only one radionuclide and we use a two channel counter. Equation 7.94 becomes

$$\hat{\theta}_1 = \left\{ \frac{a_{11}^2}{y_1/t^2 + b_1/r^2} + \frac{a_{12}^2}{y_2/t^2 + b_2/r^2} \right\}^{-1} \times \left\{ \frac{a_{11}(y_1/t - b_1/r)}{y_1/t^2 + b_1/r^2} + \frac{a_{12}(y_2/t - b_2/r)}{y_2/t^2 + b_2/r^2} \right\} \tag{7.96}$$

the standard error is given by

$$S(\hat{\theta}_1) = \left\{ \frac{a_{11}^2}{y_1/t^2 + b_1/r^2} + \frac{a_{12}^2}{y_2/t^2 + b_2/r^2} \right\}^{-1/2} \tag{7.97}$$

and the detection limit is given by

$$L_D = (k_\alpha + k_\beta)\left\{ \frac{a_{11}^2}{y_1/t^2 + b_1/r^2} + \frac{a_{12}^2}{y_2/t^2 + b_2/r^2} \right\}^{-1/2} \tag{7.98}$$

If the counter has only one channel, the estimate is

$$\hat{\theta}_1 = \left\{ \frac{a_{11}^2}{y_1/t^2 + b_1/r^2} \right\}^{-1} \times \left\{ \frac{a_{11}(y_1/t - b_1/r)}{y_1/t^2 + b_1/r^2} \right\} = \frac{y_1/t - b_1/r}{a_{11}} \tag{7.99}$$

and

$$S(\hat{\theta}_1) = \sqrt{\frac{y_1/t^2 + b_1/r^2}{a_{11}^2}} \tag{7.100}$$

The detection limit is

$$L_D = (k_\alpha + k_\beta)\sqrt{\frac{y_1/t^2 + b_1/r^2}{a_{11}^2}} \tag{7.101}$$

b. Sample with Two Radionuclides

Consider now a radioactive sample with two radionuclides and a counter with two channels. The standard error of the estimated concentrations for radionuclide 1 is $S(\hat{\theta}_1)$ and $S(\hat{\theta}_2)$ for the second radionuclide. It is straightforward to demonstrate the following equations

$$S(\hat{\theta}_1) = \left\{ \frac{1}{d} \left(\frac{a_{21}^2}{y_1/t^2 + b_1/r^2} + \frac{a_{22}^2}{y_2/t^2 + b_2/r^2} \right) \right\}^{-1/2} \tag{7.102}$$

and

$$S(\hat{\theta}_2) = \left\{ \frac{1}{d} \left(\frac{a_{11}^2}{y_1/t^2 + b_1/r^2} + \frac{a_{12}^2}{y_2/t^2 + b_2/r^2} \right) \right\}^{-1/2} \tag{7.103}$$

where d denotes the determinant of the matrix $\mathbf{AW^{-1}A'}$.

The individual estimate of concentrations is impossible when the shape of pulse distributions in both channels is similar. In this situation $d \to 0$.

The detection limit for radionuclide 1 is approximated by

$$L_D(1) = (k_\alpha + k_\beta)\left\{ \frac{1}{d} \left(\frac{a_{21}^2}{y_1/t^2 + b_1/r^2} + \frac{a_{22}^2}{y_2/t^2 + b_2/r^2} \right) \right\}^{-1/2} \tag{7.104}$$

and for radionuclide 2 by:

$$L_D(2) = (k_\alpha + k_\beta)\left\{ \frac{1}{d} \left(\frac{a_{11}^2}{y_1/t^2 + b_1/r^2} + \frac{a_{12}^2}{y_2/t^2 + b_2/r^2} \right) \right\}^{-1/2} \tag{7.105}$$

c. Sample with Several Radionuclides

From the above discussion it is evident that the standard error of the parameter to estimate the concentration of radionuclides, depends on the following factors: counting time of the sample and background, relationship between the spectral shape included in the library and the concentration of the radionuclides present in the sample.

According to the description of Pasternack and Harley (1971), in the multi-radionuclide and multi-channel situation we may consider three different detection limits:

A radionuclide is assumed to be in the sample and the library contains only this radionuclide.

The sample contains only one radionuclide but the library contains this and other radionuclides.

The sample and the library each contain the same radionuclides. The procedure to obtain the detection limits in case 1 is as follows: First we obtain the background spectrum of the system $\mathbf{b}' = (b_1, b_2, \ldots, b_n)$ and the spectrum for a mock sample $\mathbf{y}' = (y_1, y_2, \ldots, y_n)$. Generally we can use the background distribution $\mathbf{b}'$ in the place of $\mathbf{y}'$. Then we apply the least squares analysis and compute the standard error $S(\theta_1)$ from the square root of $(\mathbf{a}'_1 \mathbf{W}^{-1} \mathbf{a}_1)^{-1}$, where $\mathbf{a}_1 = (a_{11}, a_{12}, \ldots, a_n)$ denotes the radionuclide library spectrum. Thus

$$S(\hat{\theta}_1) = \left\{ \sum_{j=1}^{n} \frac{a_{1j}^2}{y_j/t^2 + b_j/r^2} \right\}^{-1/2} \tag{7.106}$$

and

$$L_D = (k_\alpha + k_\beta) \left\{ \sum_{j=1}^{n} \frac{a_{1j}^2}{y_j/t^2 + b_j/r^2} \right\}^{-1/2} \tag{7.107}$$

When we take $\mathbf{b}'$ in place of $\mathbf{y}'(t = r)$ the detection limit is

$$L_D = (k_\alpha + k_\beta) \left\{ \sum_{j=1}^{n} \frac{a_{1j}^2}{2b_j/r^2} \right\}^{-1/2} \tag{7.108}$$

For case 2, the procedure is the same but the matrix $\mathbf{A}$ contains all the spectra of the library. For case 3, the procedure is again the same. It is recommended that the mock sample adequately simulates the sample absorption. When the library does not contain all the sample radionuclides, the estimate may be unacceptable. When the library contains more spectra than the sample, a reduction in precision is observed and the standard error increases; however, the estimates remain unbiased. Explicit mathematical expressions for the bias and the loss of precision when using inadequate calibration matrices are given by Pasternack and Liuzzi (1965).

REFERENCES

Abramowitz, M. and Stegun, I. A. (1972). "Handbook of Mathematical Functions with Formulas, Graphs and Mathematical Tables". NBS. Applied Mathematical Series. 55. Washington.

Altshuler, B. and Pasternack, B. (1963). Statistical measures of the lower limit of detection of a radioactivity counter. *Health Phys.* **9**, 293–298.

Angoso, M., Gimeno, F., Grau Malonda, A., and Domínguez, G. (1973). Isotopic dilution determination of lebaycid in oranges. *J. Radional. Chem.* **13**, 149–154.

Bacon, R. M. (1953). The best straight line among the points. *Am. J. Phys.* **28**, 428–440.

Beers, Y. (1957). "Introduction to the Theory of Errors". Addison-Wesley, Massachusetts.

Bevington, P. R. (1969). "Data Reduction and Error Analysis for the Physical Sciences". McGraw. New York.

Burns, J. E., Campion, P. J., and Williams, A. (1973). Error and Uncertainty. *Metrologia* **9**, 101–102.

Campion, P. J., Burns, J. E., and Williams, A. (1973). "A Code of Practice for Detailed Statement of Accuracy". National Physical Laboratory, London.

Currie, L. A. (1968). Limits for quantitative detection and quantitative determination. *Anal. Chem.* **40**, 586–593.

Currie, L. A. (1995). Nomenclature in evaluation of analytical methods including detection and quantification capabilities. *Pure Appl. Chem.* **67**, 1699–1723.

Donn, J. J., and Wolke, R. L. (1976). The practical design and statistical interpretation of background-dominant counting experiments. *Radiochem. Radioanalyt. Letters* **25**(2), 57–66.

Donn, J. J., and Wolke, R. L. (1977). The statistical interpretation of counting data from measurements of low-level radioactivity. *Health Phys.* **32**, 1–14.

Draper, N. H. and Smith, H. (1966). "Applied Regression Analysis". John Wiley, New York. (Variations to the problem of fitting a function to a set of data. curvilinear relationships, weighted least squares, non linear least squares etc.)

Eadie, W. T., Drijard, D., James, F. E., Ross, M., and Sadoulet, B. (1971). "Statistical Methods in Experimental Physics". Amsterdam, North Holland.

Eisenhart, C. (1963). Realistic evaluation of precision and accuracy of instrument calibration systems. *Journal of Research* **67C**, 161–187.

Eisenhart, C. (1968). Expression on the uncertainties of final results. *Science* **160**, 1201. (Reprinted in Ku, 1969). (A detailed discussion on systematic errors).

Evans, R. E. (1972). "The Atomic Nucleus". Chap. 26–28 and Appendix G. McGraw Hill, New York.

Feller, W. (1968) "An Introduction to Probability Theory and its Applications. Vol I. John Wiley and Sons, Inc., New York.

Grau Carles, P., and Grau Malonda, A. (2000). "Probabilidad Estadística y Errores", Editorial Ciemat, Madrid.

Head, J. H. (1972). Minimum detectable photopeak in Ge(Li) spectra. *Nucl. Instrum. Methods* **98**, 419–428.

Hoel, P. C. (1984). "Introduction to Mathematical Statistics". John Wiley, New York.

International Vocabulary of Basic and General Terms in Metrology. 2nd ed., ISO (1993).

Jaffey, A. H. (1960). Statistical tests for counting. *Nucleonics* **18**, No. 11, 180–184.

Loevinger, R., and Berman, M. (1951). Efficiency criteria in radioactivity counting. *Nucleonics* **9**, 26–39.

Matzke, M. (2002). Propagation of uncertainties in unfolding procedures. *Nucl. Intrum. Methods* **A476**, 230–241.

Müller, J. W. (1979). Some second thoughts on error statements. *Nucl. Instrum. Methods* **163**, 241.

Natrella, M. G. (1963). "Experimental Statistics". National Bureau of Standards, Washington.

Newbold, P. (1995). "Statistics for Business and Economics". Prentice Hall, New Jersey.

Nicholson, W. L. (1966). Statistics of net-counting-rate estimation with dominant background corrections. *Nucleonics* **24**, 118–121.

Pasternack, B. S., and Liuzzi, (1965). Patterns in residuals: a test for regression model adequacy in radionuclide assay. *Technometrics* **7**, 603–621.

Pasternack, B. S., and Harley, N. H. (1971). Detection limits for radionuclides in the analysis of multi-component gamma ray spectrometer data. *Nucl. Instrum. Methods* **91**, 533–549.

Przyborowski, J. and Wilenski, H. (1935). Statistical principles of routine work in testing clover seed for dodder. *Biometrika* **27**, 273–292.

Rabinovich, S. G. (2000). "Measurement Errors and Uncertainties. Theory and Practice". Springer, New York.

Reynolds, S. A. (1964). Choosing optimum counting. *Nucleonics* **22**(8), 104–105.

Rozanov, Y. A. (1977). "Probability Theory: A Concise Course". Dover, New York.

Taylor, B. N., and Kuyatt, C. E. (1994). Guidelines for Evaluating and Expressing the Uncertainty of NIST Measurement Results. NIST Technical Note 1297. Washington.

Thomas, A. (1950). How to compare counters. *Nucleonics* **6**(2), 50–53.

Topping, T. (1972). "Errors of Observation and Their Treatment". Chapman and Hall, London.

Triola, M. F. (1983). "Elementary Statistics". Benjamin/Cummings, Menlo Park, California.

Wyld, G. E. A. (1970). Statistical confidence in liquid scintillation counting. In "The current status of liquid scintillation counting". Grune and Stratton, New York.

Zimmerman, B. E., Unterweger, M. P., and Brodack, J. W. (2001). The standardization of ^{177}Lu by $4\pi\beta$ liquid scintillation spectrometry with ^{3}H-standard efficiency tracing. *App. Radiat. Isot.* **54**, 623–631.

RELEVANT STATISTICAL REFERENCE TABLES

Zwillinger, D. and Kokoska, S. (1999). "Standard Probability and Statistics Tables and Formulae". CRC Press, Boca Raton, Florida.

Murdoch, J. and Barnes, S. A. (1998). "Statistical Tables". Palgrave Macmilan, London.

Lindley, D. V. and Scott, W. F. (1995). "New Cambridge Statistical Tables". Cambridge University Press, Cambridge.

Neave, H. R. (1981). "Elementary Statistical Tables". Routledge, London.

White, J., Yeats, A. and Skipworth, G. (1979). "Tables for Statisticians". Nelson Thornes, London.

Beyer, W. H. (1968). "CRC Handbook of Tables for Probability and Statistics". CRC Press, Boca Raton, Florida.

Owen D. B. (1962). "Handbook of Statistical Tables". Addison-Wesley, Reading, Massachusetts.

James Rohlf, F. and Sokal, R. R. (1994). "Statistical Tables". Freeman, New York.

Abramowitz, M. and Stegun, I. A. (1967). "Handbook of Mathematical Functions with Formulas, Graphs and Mathematical Tables". Dover Publications, New York.

Pearson, K. (1930). "Tables for Statisticians and Biometricians". 3rd ed. Cambridge University Press, London.

Fisher, R. A. and Yates, F. (1974). "Statistical Tables for Biological, Agricultural and Medical Research" Longman, London.

8

SAMPLE PREPARATION TECHNIQUES FOR LIQUID SCINTILLATION ANALYSIS

JAMES THOMSON

PerkinElmer Life and Analytical Sciences, Groningen, The Netherlands

I. INTRODUCTION

It is a sad but very true observation that sample preparation is considered an inconvenient step on the way to obtaining results. Consequently, it is only when the results obtained are different to those expected that the cry "There's something wrong with the cocktail" is heard. While this may or may not be true, the technique of sample preparation in general is only rarely considered responsible for the unexpected results. Correct sample preparation in liquid scintillation analysis (LSA) is essential for both accurate and reproducible analysis, and no amount of instrumental sophistication can ever fully compensate for the problems attendant to a badly prepared sample. Good sample preparation guarantees that the sample will be stable during the analysis, and therefore provides a solid foundation for accurate results. Sample preparation encompasses a wide variety of methods and techniques, and includes dissolution, distillation, extraction, solubilization, digestion, suspension, and combustion. All of these methods of sample preparation hold pitfalls for the unwary and some expertise is generally required. Common to all sample preparation methods is the liquid scintillation counting (LSC) cocktail, and this is the medium that holds the sample during the analysis process. The LSC cocktail is both fundamental to and necessary for the analysis; therefore correct cocktail selection is a critical step in sample preparation.

II. LSC COCKTAIL COMPONENTS

To appreciate the significance of correct cocktail selection, it is useful to explain some fundamentals about the components of LSC cocktails, and thus gain some insight into not only what components are present, but also what properties they impart to the final cocktail. Cocktails can be divided into two main groups: emulsifying cocktails (sometimes called aqueous cocktails) and organic cocktails (sometimes called non-aqueous or lipophilic cocktails). Organic cocktails have only two major components: organic aromatic solvent and scintillators. Emulsifying cocktails have three major components: organic aromatic solvent, emulsifier, and scintillators.

A. Solvents

The traditional or classical aromatic solvents used in LSC are toluene, xylene, and pseudocumene (1,2,4-trimethylbenzene), and these are effective LSC solvents due to the high density of π electrons associated with these solvents. The increase in monoalkyl substitution causes increased electron donation and thus a higher ring electron density; this results in higher counting efficiency:

$$\text{benzene} \rightarrow \text{toluene} \rightarrow \text{xylene} \rightarrow \text{pseudocumene}$$
$$\underrightarrow{\text{increasing efficiency}}$$

Although still popular, the use of toluene, xylene and pseudocumene has, and continues to, decrease due to problems with toxicity, flammability, vapor pressure, smell, and permeation through plastics. These solvents have generally been replaced by the newer generation of "safer" solvents:

<table>
<tr><td>di-isopropylnaphthalene</td><td>phenylxylylethane</td><td>dodecylbenzene</td></tr>
<tr><td>(DIN)</td><td>(PXE)</td><td>(LAB)</td></tr>
</table>

These solvents are characterized by high flash point ($>145°C$), low vapor pressure (<1 mm Hg at $20°C$), low toxicity ($LD_{50} >3000$ mg/kg), low odor and no permeation through plastics. In addition both DIN (Thomson, 1987, 1991) and PXE (Wunderley, 1986), exhibit higher counting efficiency than the classical solvents, and the overall order is:

$$\text{benzene} \rightarrow \text{LAB} \rightarrow \text{toluene} \rightarrow \text{xylene} \rightarrow \text{pseudocumene} \rightarrow \text{PXE} \rightarrow \text{DIN}$$
$$\underrightarrow{\text{increasing efficiency}}$$

Commercial mixtures (e.g., petroleum distillates) have been used instead of pseudocumene but these contain many species other than alkyl benzenes, and include indane, indene, methyl indane, and ethyl indane. These impurities have a detrimental effect on counting efficiency and background, and can be particularly strong chemiluminescers. Indane and indene, even at 50–100 ppm level, will produce severe chemiluminescence when contacted with alkaline materials such as tissue solubilizers. This effect is so severe that the mixture can turn brown or purple and produce backgrounds $>1 \times 10^6$ CPM. These

▬ **TABLE 8.1**   Safety Characteristics of Scintillation Solvents

Solvent	Boiling point	Flash point	Vapour pressure mm Hg at 25°C	Classification (International)	Hazards
Toluene	110°C	4°C	28	Flammable	Inhalation Skin absorption Irritating to skin/eyes
Xylene	138°C	25°C	8	Flammable	Inhalation Skin absorption Irritating to skin/eyes
Cumene	152°C	31°C	5	Flammable	Inhalation Irritating to skin/eyes
Pseudocumene	168°C	50°C	2	Flammable	Inhalation Irritating to skin/eyes
LAB	300°C	149°C	< 1	Harmless	None (as classified)
PXE	305°C	149°C	< 1	Harmless	None (as classified)
DIN	300°C	140°C	< 1	Harmless	None (as classified)
Benzyl toluene	290°C	135°C	< 1	Harmless	None (as classified)
Diphenyl ethane	290°C	135°C	< 1	Harmless	None (as classified)

petroleum distillates can be purified by e.g., distillation, column chromato-graphy, reactive distillation, solvent extraction, and mild sulphonation, but the final product never really approaches the pure solvent as a completely acceptable LSC solvent. An earlier approach at producing a safer solvent system was to use paraffinic and naphthenic solvents in conjunction with a secondary LSC solvent. The paraffinic and naphthenic solvents have poor scintillation properties but the addition of this secondary LSC solvent (sometimes referred to as an energy transfer medium) can boost the counting efficiency up to within 80% of the pure LSC solvent. This approach was largely abandoned due to the large-scale availability of the new safer solvents. A summary of the important characteristics of LSC solvents is shown in Table 8.1.

B. Scintillators

As explained earlier, the fluor or scintillator is a light transducer that converts nuclear energy into light photons. All LSC cocktails contain at least one and usually two scintillators. The primary scintillator is responsible for the initial or primary energy exchange, but the wavelength of the emitted light does not match the optimum detection wavelength of the photo-multiplier tube. Hence a secondary scintillator (or wavelength shifter) is added to shift the wavelength of the emitted light to match the photomultiplier tube. An efficient scintillator should have a high fluorescence quantum efficiency (or high photon yield), a spectrum maximally matched to the response of the photocathode, short fluorescence decay time, a large stokes shift, sufficient solubility, and a low sensitivity to quenching agents. Derivatives of oxazoles, oxadiazoles, phenylenes, *p*-oligophenylenes, styryl-benzenes, benzoxazoles, benzoxazolyl thiophenes, pyrazolines and nitriles have scintillation properties and are potential scintillators. The concentration of the scintillator determines the photon yield of a liquid scintillator. At optimum scintillator concentration the light output is maximum, but beyond this point self- or concentration-quenching occurs due to the formation of excimers. The probability of excimer formation is greater in scintillators with planar configuration than those with bulky substituent groups offering steric hindrance; e.g., PPO (2,5-diphenyloxazole) exhibits a reduced quantum yield value at high concentrations. Substituted polyaryls and *p*-oligophenylenes currently used as scintillators can be rendered more soluble by substitution of hydrogen with alkyl or alkoxy groups. The alkyl groups tend to enhance the solubility in toluene while the alkoxy groups favor an increased solubility in more polar solvents. Primary fluors are generally substituted fluorescent polyaryls and include:

PPO	2,5-diphenyloxazole
PPD	2,5-diphenyl-1,3,4-oxadiazole
PBO	2-(4-biphenylyl)-5-phenyloxazole
PBD	2-phenyl-5-(4-biphenylyl)-1,3,4-oxadiazole
BBD	2,5-di-(4-biphenylyl)-1,3,4-oxadiazole
Butyl-PBD	2-(4-*t*-butylphenyl)-5-(4-biphenylyl)-1,3,4-oxadiazole
BBOT	2,5-bis-2-(5-*t*-butyl-benzoxazoyl) thiophene
TP	*p*-terphenyl

The most popular and widely used is PPO and this preference is due to performance, purity, cost, and availability on a large scale. Butyl-PBD is also occasionally used but it suffers from the drawback that it yellows when in contact with alkaline species such as tissue solubilizers. The situation with secondary scintillators is similar in that there are many scintillators that are useful but only a few are in common use. The important secondary scintillators include:

bis-MSB	*p-bis*-(*o*-methylstyryl)benzene
POPOP	1,4-*bis*-2-(5-phenyloxazolyl)benzene
Dimethyl POPOP	1,4-*bis*-2-(4-methyl-5-phenyloxazolyl)benzene
NPO	2-(1-naphthyl)-5-phenyloxazole
NPD	2-(1-naphthyl)-5-phenyl-1,3,4-oxadiazole
BBO	2,5-di(4-biphenylyl)oxazole
PBBO	2-(4-biphenylyl)-6-phenylbenzoxazole

The most popular and widely used is *bis*-MSB and this preference is due to performance, solubility, purity, cost, and availability on a large scale. Dimethyl POPOP is also occasionally used but it suffers from limited solubility in organic aromatic solvents. Virtually all commercially available LSC cocktails are based on the combination of PPO and *bis*-MSB, and this has been the case now for many years.

C. Surfactants

The most efficient energy transfer takes place in aromatic organic solvents. Since the majority of radioisotopes are present in an aqueous environment that is not miscible with the aromatic solvents, detergents are used to bring the aqueous phase into close contact with the organic phase by forming a microemulsion. Such close contact allows efficient energy transfer and means that radioisotopes in aqueous solutions can be successfully measured both qualitatively and quantitatively. In an LSC cocktail, a microemulsion is a dispersion of very small droplets (radius of approximately 10 nm) of "water-in-oil" and is achieved by the use of detergents. Many detergents have been investigated but only a few specific types have been successfully used in LSC cocktails. The major groups are nonionics, anionics, cationics, and amphoterics, and within each group there are certain useful or favored detergents which are found in many LSC cocktails.

1. Nonionics

This group includes the ethoxylates; one particular type, the alkyl phenol ethoxylates, are found in virtually all LSC cocktails. These are the building blocks upon which cocktail performance is based. Ethoxylates are produced by the vapor or liquid phase ethoxylation of the free hydroxyl group of the alkyl phenol in the presence of a catalyst (usually a carboxylic acid). Such ethoxylates are described as having a specific ethylene oxide or EO length, but this number which refers to the length of the ethoxylate polymer chain is in fact an indication based on properties, not on chemical structure. These ethoxylates have a range of polymer chain lengths present in a normal distribution and the EO length is an indication of the average distribution. Commonly encountered ethoxylates include alkyl phenol ethoxylates, alcohol ethoxylates, amine ethoxylates, acid ethoxylates, castor oil ethoxylates, and ester ethoxylates, but as stated above only the alkyl phenol ethoxylates are useful in LSC cocktail formulation. These other ethoxylates are not used because they fail to produce a stable microemulsion, are generally colored, have limited solubility in aromatic solvents, produce unacceptable chemical quench, and have poor counting efficiency properties. Historically, in the 1960s Patterson and Greene (1965) examined a few nonionic detergents of the alkyl phenol type and concluded that Triton X-100 was the detergent of choice. Unfortunately factors such as the variability of feedstock and ethoxylation conditions were not taken into account, and this resulted in a severe batch-to-batch variations in the original Triton X-100 product. When used in LSC cocktails, the result was microemulsion instability and chemiluminescence; therefore, this meant that detergent batch selection was necessary. Refinements in production methods significantly improved the quality of the alkyl phenol ethoxylates to the extent that standard production material can now be used in LSC cocktails without reverting to batch selection of raw materials. Many manufacturers have alkyl phenol ethoxylates available but since there are a variety of manufacturing methods, there is no guarantee that alkyl phenol ethoxylates from different sources will perform similarly. Sourcing acceptable raw material alkyl phenol ethoxylate can only be achieved through a lengthy trial and error process. When mixed with aromatic LSC solvents alkyl phenol ethoxylates are able to produce microemulsions when aqueous samples are added. The choice of solvent and ethoxylate are critical in that there is a limited range of EO chain lengths which will work with each individual aromatic solvent, and desirable properties such as clarity of microemulsion or gelling can be chosen by careful EO selection. In addition, altering the chain length of the alkyl group of the ethoxylate affects performance. To extend or enhance performance, other detergents are needed and these are normally selected from the other groups.

2. Anionics

This group includes among others alkyl and alkylaryl sulphonates, alcohol sulfates, alkyl suphosuccinates, and phosphate esters (PEs). Of these, the sulphosuccinates, have been found useful due primarily to their synergistic effect when combined with ethoxylates. Sulphosuccinates, even in relatively small quantities, are able to extend the microemulsion region of ethoxylates without affecting the other phases, and have a secondary effect in that they tend to stabilize the gel region. Various sulphosuccinates have been examined but the most popular is sodium di-octylsulphosuccinate. This popularity is due to both its availability in a pure, water-white solid form and its ability to extend the microemulsion range of most ethoxylates. Another group of anionics which are of exceptional benefit are the phosphate esters. The use of neutralized phosphate esters has been patented (ter Wiel and Hegge, 1983), and the benefits resulting from their use include chemiluminescence resistance and microemulsion formation with difficult sample types. In small quantities, the free acid or neutralized phosphate esters act as hydrotropes and can remove the microemulsion instability often encountered with ethoxylates. When used as the free acid the disadvantages are that they can react with certain metals and form colored compounds; they will protonate PPO; and they can cause cocktail degradation through reaction with the solvents. In general, the detergent properties of PEs are enhanced by neutralization but the use of the wrong alkaline material has certain previously unreported drawbacks. If the phosphate esters are neutralized with inorganic alkalis the salt of the phosphate ester has a tendency to precipitate slowly out of the LSC cocktail, especially if the concentration is significant. If ammonia has been used, the presence of an alkaline sample will cause a reaction, resulting in the release of ammonia gas within the vial. Both of these situations are undesirable since they can produce variable quench conditions either in the original cocktail or within the counting vial. Other anionics that are known to have been investigated include sarcosinates, taurates, isothionates, and sulphonates, but little is reported on their benefits. If anionics are used in cocktails either alone or in conjunction with ethoxylates, then care is needed when adding cationics, such as quaternary ammonium hydroxides. There is a potential for reaction between the anionic and cationic, and insoluble compounds can form, either immediately or slowly upon standing. Thus far the formation of an insoluble product has only been seen when sulphosuccinates are the anionic species.

3. Cationics

This class of detergents carries a positive charge on the hydrophobic portion; a property rendering it substantive to negatively charged surfaces, e.g., metals and organic surfaces in general. As stated previously the fact that cationics form salts with anionics virtually precludes their use in cocktails since anionics are commonly present. One area that was tentatively researched was the deliberate reaction between free acid PEs and quaternary ammonium hydroxides, to see if the products had any beneficial detergent properties. The results were a little inconclusive but suggested that further investigation might be worthwhile. In LSC, cationics, such as the quaternary ammonium hydroxides that are powerful bases, are used to solubilize/dissolve a host of animal fats and tissues. Although these cationics have only a limited use in LSC, they nevertheless affect many of the cocktails in use. Being very basic in nature they have the potential to initiate chemiluminescence in many cocktails.

4. Amphoterics

This class of detergents has both positive and negative centers in the molecule; depending upon the condition prevailing, cationic, anionic, or nonionic type properties may be exhibited. Examples of molecules peculiar to this class include betaines, imidazolines, and carboxylates. The commercially available products are at present not suitable due to problems with low concentration availability, high color, and poor solubility in aromatic solvents.

Although most cocktails contain alkyl phenol ethoxylates, it is possible to use a mixed anionic system effectively and one such cocktail was OptiPhase RIA, which was based on sodium di-octyl sulphosuccinate and neutralized phosphate ester in pseudocumene solvent.

D. Cocktails

LSC cocktails can be categorized either as "Classical" or "Safer". The classical cocktails encompass all those based on either toluene, xylene or pseudocumene, while the safer cocktails include those based on DIN, PXE, LAB, benzyl toluene, and 2-phenylethane. Other high flash point solvents such as 2-ethylnaphthalene, C_5-C_{10} alkylbenzenes, hydrogenated terphenyls (Actrel 400), high flash point aromatic fractions (Solvesso 150 & 200), ethylbiphenyl, butylbiphenyl, phenylcyclohexane, phenylnaphthalene, iso-propylbiphenyl, diphenylpropane, di-xylylethane (Freiberg, 1990), and di-cumylethane (Mirsky, 1990), have been investigated but for various reasons have not been used commercially in LSC cocktails. The main reasons include purity, scintillation efficiency, cost, and availability on a large scale. Among the safer cocktails, di-isopropylnaphthalene (DIN) based cocktails, such as the Ultima Gold™ range from PerkinElmer will give high ^{3}H efficiency, and better color and chemical quench resistance. Other safer cocktails based on DIN include some of the Optiphase Hi-Safe range from Wallac, and the AquaSafe, QuickSafe and Quickszint cocktails from Zinsser. Some cocktail manufacturers use PXE as the safer solvent base; examples include the ScintiSafe range from Fisher, the Ecoscint range from National Diagnostics and ReadySafe from Beckman. Finally there are many cocktails based on linear alkyl benzene (LAB) and these include Opti-Fluor® and Emulsifier Safe™ from PerkinElmer, Bio-Safe from RPI, Ecolume from ICN, and LumaSafe from Lumac. Most LSC cocktail suppliers feature both types in their ranges and a full table of the known equivalents is shown in Table 8.2.

III. DISSOLUTION

The primary objective of all sample preparation procedures for liquid scintillation counting (LSC) is to obtain a homogeneous solution for efficient energy transfer from the sample to the LS cocktail. Aqueous solutions are some of the simplest and are most commonly found in liquid scintillation analysis. In general, they provide the environment necessary for many assays and separations, and include the most encountered solvent media for the numerous radioisotopes used in LSC. The main methods for producing aqueous samples are by dissolution, extraction, and distillation. Dissolution simply involves dissolving the sample in water. Extraction can be extraction of the sample from a solid matrix by water (solid/liquid extraction), or extraction of the sample from a liquid matrix by water (liquid/liquid extraction). Distillation involves separation of the aqueous component by evaporation. A variety of LSC cocktails have evolved over the years to accommodate the diverse types, volumes, and concentrations of aqueous samples presented for analysis by LSC and some are considered by Kobayashi and Maudsley (1974), ter Wiel and Hegge (1991), and Peng (1983).

A. Anions

Different types of anions encountered in LSC include chlorides, nitrates, phosphates, acetates, and formates, with sample volumes ranging from less than $100 \mu L$ to greater than $10 mL$, and concentrations varying from less than $10 mM$ to greater than $2 M$. For the purpose of cocktail selection, these aqueous samples can be roughly divided into the following categories:

1. Buffers (e.g., sodium chloride, PBS, potassium phosphate, etc.)

   Low ionic strength (less than $0.1 M$)
   Medium ionic strength ($0.1 M$ to $0.5 M$)
   High ionic strength ($0.5 M$ to greater than $1 M$)

2. Acids (e.g., hydrochloric acid, nitric acid, etc.)
3. Alkalis (e.g., sodium hydroxide, potassium hydroxide, etc.)
4. Other types (e.g., urea, sucrose, imidazole, etc.)

By using this list of categories, it is now possible to assign cocktails for each category and therefore present a simpler and more comprehensive method of cocktail selection than was previously possible. For each category, cocktails will be recommended based on sample acceptance, performance, and safety.

TABLE 8.2 Cocktail Equivalents

LSC cocktails for aqueous samples

Amersham	Baker	Beckman	Fisher	ICN	Lumac	N.D.	NEN	Perkin-Elmer	Perkin-Elmer	Roth	RPI	Zinsser
Saf-T-cocktail		Ready Safe	ScintiSafe 30%	Ecolite	LumaSafe	Ecoscint A		Ultima Gold	Hi-Safe 2			AquaSafe 300+
												Unisafe 1
					Lumagel Safe							Quick-Safe A
			ScintiSafe Plus 50%	Universol ES	LumaSafe Plus			Ultima Gold XR	Hi-Safe 3	Rotiszint Eco-Plus		AquaSafe 500+
			ScintiSafe Gel						Hi-Load			Quickszint 401
				Cytoscint ES		Ecoscint H		Ultima Gold MV	Supermix			
								Ultima Gold AB				
			ScintiSafe Plus 50%					Ultima Gold LLT	Tri-Safe			QuickSafe 400
BCS			ScintiSafe Econo-2		Safefluor S	Uniscint BD	Formula 989	Opti-Fluor	Hi-Safe		Bio-Safe II	
			ScintiSafe Econo-1	Ecolume	Safefluor	Ecoscint		Emulsifier Safe			Econo-Safe	
PCS	Dynagel/ SBS	Ready Gel	Scintiverse I	Universol	Lumagel Plus	Hydrofluor	Aquassure	Insta-Gel Plus	Optiphase Safe	Szintigel Roth	3a70	Quickszint 1
							Aquasol					
ACS/ACSII/	Hydrocount	Ready Value	Scintiverse E	Aquamix		Liquiscint	Aquasol II	Emul. Scint Plus	Optiphase MP	Rotiszint 22/22X	Budget-Solve HFP	Quickszint 212

PCSII	Aqualyte Plus	Ready Protein*	Scintiverse Bio-HP	Cytoscint	Aqualuma Plus	Fluorodyne	Biofluor	Pico-Fluor 15			BioCount	Quickszint 402
			CytoScint			Ultrafluor					Safety-Count	Quickszint 1000
											Safety-Solve	
	Aqualyte	Ready Micro		Betablend	Rialuma	Monofluor	Atomlight	Pico-Fluor 40	Optiphase RIA	Rotiszint 2200	RiaSolve II	Quickszint 2000
	Maxifluor				Aqualuma	Dimiscint/ Bioscint / Soluscint A	Formula 963	Hionic-Fluor		Rotiszint Mini	Neutralizer	
						Filtron X		Filter-Count			Cocktail	Quickszint 361
LSC cocktails for nonaqueous samples/organic samples												
		Ready Organic	ScintiSafe Econo-F	Betamax ES	Safefluor O	Ecoscint O	Mineral Oil	Ultima Gold F	Scint Hi-Safe	Rotiszint 11	Bio-Safe NA	Quickszint-N
							Scintillator	Opti-Fluor O	Betaplate Scint	Secure		
BCS-NA				Betamax								
OCS	Lipofluor		Scintilene		Lipoluma Plus	Betafluor	Econofluor-2	Insta-Fluor Plus	OptiScint Safe	Rotiszint 1100	4a20 and 3a20	
LSC cocktails for flow detectors												
	Bakerflow I	Ready Organic			Lipoluma Plus			Insta-Fluor Plus				Quicksafe Flow 301
					LumaFlow II			Flo-Scint II				Quicksafe Flow 302
	Bakerflow III	Ready Flow III	Scintiverse LC		LumaFlow III	Monoflow	Atomflow	Flo-Scint III	Opti-Flow	Rotiszint 2211		Quicksafe Flow 303

(continued)

TABLE 8.2 Continued

Amersham	Baker	Beckman	Fisher	ICN	Lumac	N.D.	NEN	Perkin-Elmer	Perkin-Elmer	Roth	RPI	Zinsser
					LumaFlow IV			Flo-Scint IV				Quicksafe Flow 306
					LumaFlow A							
					LumaSafe Flow M	Uniscint BD		Ultima-Flo M	Opti-Flow Safe I			Quicksafe Flow 2
					LumaSafe Flow P			Ultima-Flo AP				
					LumaSAfe Flow F			Ultima-Flo AF				
Oxidizer reagents												
	Dynacount					Oxosol H-3		Monophase S	Optisorb-4	Rotiszint OPH		Oxysolve-T
	14C-Fluor					Oxosol-306		Permafluor E+	Optisorb-S	Rotiszint OPC		Zintol-X
	Carbo-Trap				Carbomax Plus	Carbamate-2		Carbosorb E	Optisorb-1			Zintol-II
Tissue and gel solubilizers												
NCS II	Solulyte	BTS-450	ScintiGest		Lumasolve	Solusol	Solvable	Soluene-350	Optisolve	Tissue-sol-Roth	TS-1 and TS-2	Biolute-S
Other Reagents												
				Hyamine hydroxide		Hyamine hydroxide		Hyamine hydroxide				Hyamine hydroxide

TABLE 8.3 Cocktail Selection for Low Ionic Strength Samples (Based on the Use of 10 mL of Cocktail)

	Safer cocktail	Classical cocktail
Low sample volume (0–2.5 mL)	Ultima Gold, MV, AB, LLT, Opti-Fluor, Emulsifier Safe (Poly-Fluor), and other equivalent cocktails.	Insta-Gel Plus (Insta-Gel XF), Pico-Fluor 15 and 40, and other equivalent cocktails.
High sample volume (> 2.5 mL)	Ultima Gold, XR, AB, LLT, Opti-Fluor, and other equivalent cocktails.	Insta-Gel XF, Pico-Fluor 40 and other equivalent cocktails.

B. Low Ionic Strength Buffers

Buffers encountered in this group include 0.01 M PBS (phosphate buffered saline), 50 mM Tris-HCl [Tris(hydroxymethyl)aminoethane hydrochloride], 0.1 M NaCl (sodium chloride), 0.01 M Na_2SO_4 (sodium sulfate), etc. Since these aqueous buffers are relatively dilute, there are comparatively few problems. Both di- and trivalent anions such as SO_4^{2-} and SO_4^{3-} are potentially problematic, due in part to their charge and to their relative size [e.g., chlorides (Cl^-) are much smaller than sulfates (SO_4^{2-})]. These characteristics can impede the formation of a stable microemulsion and can cause phase instability, especially with high concentrations and large volumes. Surprisingly, similar problems can occur with small volumes, particularly within the range of 0.1–0.5 mL of sample in 10 mL cocktail (1–5% sample load). The only other area of concern is color quench problems when using certain metallic salts which are intrinsically colored [e.g., $FeCl_3$ (ferric chloride)]. Any phase instability problem can usually be resolved by decreasing the sample volume or by increasing the cocktail volume. If the problem persists, then it may be necessary to change to a cocktail that can accept higher strength ionic samples. Color quench problems can be reduced by either diluting the sample with water (if practicable), or by using a cocktail which is more resistant to color quenching e.g., any of the PerkinElmer Ultima Gold and OptiPhase HiSafe cocktails. Of the classical solvent-based cocktails, Insta-Gel Plus (10 mL) can accommodate greater than 2.5 mL of certain sample types and forms a stable gel (usually at greater than 3 mL sample volume), thus making large sample volumes possible. Pico-Fluor 40 (10 mL) can accommodate greater than 2.5 mL of certain samples and remains in a single liquid phase. Cocktails suitable for these samples are shown in Table 8.3.

C. Medium Ionic Strength Buffers

Aqueous buffer concentrations encountered in this group range from 0.1 M up to 0.5 M and typical buffers are 0.1–0.5 M PBS, 0.15–0.5 M NaCl, 0.25 M ammonium acetate, etc. The cocktails suitable for these sample types are shown in Table 8.4.

TABLE 8.4 Cocktail Selection for Medium Ionic Strength Samples (Based on the use of 10 mL of Cocktail)

	Safer cocktail	Classical cocktail
Low sample volume (0–2.5 mL)	Ultima Gold, XR, AB, LLT, and other equivalent cocktails.	Insta-Gel Plus (Insta-Gel XF), Pico-Fluor 40, Hionic-Fluor, and other equivalent cocktails.
High sample volume (> 2.5 mL)	Ultima Gold XR and other equivalent cocktails.	Pico-Fluor 40, Hionic-Fluor and other equivalent cocktails.

TABLE 8.5 Sample Capacity of Selected Cocktails for Various Ionic Strength Buffers (Sample Capacities are for 10 mL Cocktail at 20°C).

Ionic strength	Ultima Gold XR	Hionic-Fluor	Pico-Fluor 40	Ultima Gold	Ultima Gold MV	Opti-Fluor	Pico-Fluor 15
0.5 M NaCl	9.0 mL	1.4 mL	3.0 mL	1.5 mL	1.25 mL	1.1 mL	1.2 mL
0.75 M NaCl	6.5 mL	2.25 mL	2.75 mL	0.75 mL	0.75 mL	0.75 mL	0.5 mL
1.0 M NaCl	5.5 mL	8.5 mL	2.3 mL	0.5 mL	0.5 mL	0.5 mL	0.25 mL

The cocktails are designed to overcome any phase instability problems and therefore cocktail selection is limited by volume and concentration factors. The Ultima Gold and OptiPhase HiSafe families will, in general, give higher quench resistance than the classical cocktails, i.e., higher efficiency at the same sample load. Ultima Gold and OptiPhase HiSafe 2 works well with low sample volumes of aqueous buffers up to 0.5 M. Although Insta-Gel Plus will accept small volumes of certain sample types, Pico-Fluor 40 is the recommended classical cocktail for these samples.

D. High Ionic Strength Buffers

With high ionic strength buffers, the choice of cocktails includes Ultima Gold XR, Optiphase Hi-Safe 3, LumaSafe Plus, Hionic-Fluor, or Pico-Fluor 40. Certain high ionic strength samples can be accommodated in other cocktails, however the capacity is usually very low (less than 0.5 mL). The only other method of overcoming the problem of low sample acceptance of high ionic strength samples is to dilute the buffer sample with water and convert it into a medium strength buffer which simplifies cocktail selection. The sample capacity of selected cocktails for increasing ionic strength solutions is shown in Table 8.5.

E. Acids

This group includes commonly encountered mineral acids such as hydrochloric acid, nitric acid, sulfuric acid, perchloric acid, orthophosphoric acid, and hydrofluoric acid as well as some aqueous miscible organic acids

TABLE 8.6 Cocktail Selection for Acids (Based on the Use of 10 mL of Cocktail)

Acid	Concentration	Safer cocktail	Classical cocktail
Mineral acids	0–2 M	Ultima Gold AB, LLT, and other equivalent cocktails.	Insta-Gel Plus (Insta-Gel XF*), Pico-Fluor 40, Hionic-Fluor, and other equivalent cocktails.
	>2 M	Ultima Gold AB, LLT and other equivalent cocktails.	
Trichloroacetic acid	0–20%	Ultima Gold LLT and other equivalent cocktails.	Hionic-Fluor and other equivalent cocktails.

such as acetic acid, formic acid, and trichloroacetic acid (TCA). Acids are commonly used as extractants (Thomson and Burns, 1996a) pH modifiers, and solubilizers (Thomson, 1996b). There are a number of potential problems associated with this particular sample group and these include quenching, reaction with cocktail components, and chemiluminescence. Strong mineral acids can also cause marked quenching effects due primarily to interaction with the scintillators. This can be overcome by using a cocktail that is known to be compatible with mineral acids or, preferably, by diluting the acid with water prior to the addition of the cocktail. Certain strong mineral acids will react with cocktail components causing both color development and changes in surfactant characteristics. For example, adding even small amounts of concentrated sulfuric acid to a cocktail will result in almost immediate color formation and eventual sulphonation of the surfactants (emulsifiers). This alteration to the surfactants will result in a change or loss of emulsifying properties and lead to phase instability. The color formation is due to sulphonation of minor impurities in the solvent and, in addition to color, significant amounts of chemiluminescence may be produced. Another example involves adding small amounts of concentrated nitric acid which results in the formation of a yellow/brown color, due to the dissociation of nitric acid and release of NO_2. This problem can be overcome by diluting the acid with water prior to adding it to the cocktail. With some cocktails, the addition of TCA can produce chemiluminescence. Although a rare occurrence, acid-induced chemiluminescence can be avoided by using a cocktail that is resistant to it such as Ultima Gold LLT. An overview of suitable cocktails for acids is shown in Table 8.6.

F. Alkalis

This group of samples includes bases such as sodium hydroxide, potassium hydroxide, and ammonium hydroxide. Alkaline samples are produced from applications involving pH modification, cell lysis, CO_2 trapping (Thomson and Burns, 1994) and solubilization. The major problem normally encountered is chemiluminescence and in general the amount of chemiluminescence is influenced by both the volume and concentration of alkali added. The standard method of avoiding this problem is to use a cocktail that is known to be resistant to chemiluminescence. Other methods of overcoming the problem include diluting the base with water to reduce the effect, allowing the chemiluminescence to decay in the dark before counting, and neutralizing the base with acid. Prolonged storage of cocktails with alkalis present is not recommended due to the potential for color formation. Where possible, counting should be performed within one or two days.

G. Other Types

This final group covers other aqueous samples/mixtures that are occasionally used in LSC methods and assays. These aqueous mixtures are usually fairly specific for certain types of assays, e.g., sucrose gradients in DNA and RNA separation, urea as a denaturing and reducing buffer, and imidazole as a biological buffer. Ultima Gold and OptiPhase HiSafe 2 are the cocktails of choice for these sample types. A slightly expanded compilation of one manufacturer's cocktails for different sample types is shown in Table 8.7 (Safer Cocktails), Table 8.8 (Classical Cocktails), and Table 8.9 (Safer Cocktails for Acids). Other available cocktails are suitable for most of these sample types and the correct selection can be made using the cocktail equivalents table (Table 8.2). This may prove useful in providing a basic guide to cocktail selection. To further help with cocktail selection, sample capacities are presented to help with both the selection and suitability of a cocktail.

Appendix: B. Selection and Suitability of a Cocktail Based on Ionic Strength

1. Determine the approximate ionic strength using the sample molarity:

$$\text{Mixture Molarity} = \frac{[(v_a \times m_a) + (v_b \times m_b) + \cdots]}{(v_a + v_b + \cdots)}$$

Where:

$$v_a = \text{Volume of solution A}$$
$$m_a = \text{Molarity of solution A}$$
$$v_b = \text{Volume of solution B}$$
$$m_b = \text{Molarity of solution B}$$

Example: 10 mL of 0.2 M NaCl (solution A) added to 2 mL of 0.5 M KH_2PO_4 (solution B)

$$\text{Mixture Molarity} = \frac{[(10 \times 0.2) + (2 \times 0.5)]}{(10 + 2)}$$
$$= 0.25 \text{ M (a medium molarity sample)}$$

TABLE 8.7 Performance of Safer Cocktails with Various Aqueous Buffers at 20°C (Based on the Use of 10 mL of Cocktail). Dashes Indicate No or Very Limited Sample Capacity

Sample	Ultima Gold (mL)	Ultima Gold XR (mL)	Ultima Gold AB (mL)	Ultima Gold LLT (mL)	Ultima Gold MV (mL)	Opti-Fluor (mL)	Poly-Fluor (mL)
Deionised water	3.2	10.0	10.0	10.0	2.0	1.8	3.2
0.01 M PBS (pH 7.4)	6.5	10.0	8.5	8.0	4.0	3.0	3.0
0.1 M PBS (pH 7.4)	4.0	8.5	< 0.2	< 0.2	3.0	2.0	0.6–1.2
0.5 M PBS (pH 7.4)	0.5	1.25	–	–	< 0.5	–	–
0.05 M Tris-HCl (pH 7.4)	4.5	10.0	10.0	9.0	2.75	2.7	3.2
0.15 M Sodium Chloride	6.5	10.0	7.5	6.5	5.0	3.5	3.1
0.5 M Sodium Chloride	1.5	9.0	8.0	6.0	1.25	1.1	2.2
1.0 M Sodium Chloride	0.5	5.5	4.5	3.5	0.5	0.5	1.4
0.04 M NaH_2PO_4 (pH 7.4)	8.0	10.0	0.75–8.0	0.7–7.0	2.25	4.25	0.6–2.5
0.1 M NaH_2PO_4 (pH 4.9)	10.0	10.0	8.0	8.0	7.0	7.5	2.0
0.2 M NaH_2PO_4 (pH 4.9)	3.5	10.0	1.0–6.5	1.0–5.0	2.75	3.0	2.0
0.25 M Amm. Acetate	3.25	8.0	5.0	5.5	3.0	1.5	2.0
0.1 M Amm. Sulfate	3.25	10.0	1.0–7.0	1.0–5.5	2.25	3.0	2.0
0.1 M Sodium Sulfate	4.25	10.0	–	–	3.25	4.0	1.75
0.1 M Hydrochloric Acid	6.5	7.0	10.0	10.0	4.5	4.0	2.7
10% TCA	3.0	7.0	4.5	4.0	1.5	2.5	2.3
20% TCA	2.0	5.0	3.0	3.0	0.5	2.0	1.5
0.1 M Sodium Hydroxide	2.5	10.0	10.0	7.5	1.5	5.0	3.0
1.0 M Sodium Hydroxide	3.0	1.0	–	–	1.75	0.25	0.75
0.1 M Imidazole (pH 7.4)	10.0	10.0	10.0	10.0	2.0	4.5	2.5
8 M Urea	1.0	2.5	2.0	3.5	0.5	0.5	1.0

TABLE 8.8 Performance of Classical Cocktails with Various Aqueous Buffers at 20°C (Based on the Use of 10 mL of Cocktail). Dashes Indicate No or Very Limited Sample Capacity

Sample	Insta-Gel XF/Plus (mL)	Pico-Fluor 15 (mL)	Pico-Fluor 40 (mL)	Hionic-Fluor (mL)
Deionised water	0–1.7 2.9–10.0	1.6	2.3	1.2
0.01 M PBS (pH 7.4)	0.2–1.6 3.1–10.0	2.9	10.0	1.4
0.1 M PBS (pH 7.4)	1.0–2.0	< 0.25	6.4	1.6
0.5 M PBS (pH 7.4)	–	–	2.0	7.0
0.05 M Tris-HCl (pH 7.4)	0–1.8 3.0–10.0	2.0	3.0	3.0
0.15 M Sodium Chloride	0–1.8 4.9–10.0	4.0	10.0	1.1
0.5 M Sodium Chloride	0–2.1 3.0–10.0	1.2	3.0	1.4
1.0 M Sodium Chloride	0.4–7.0	< 0.3	2.3	8.5
0.04 M NaH_2PO_4 (pH 7.4)	0.3–1.9 3.0–10.0	2.0	4.0	1.75
0.1 M NaH_2PO_4 (pH 4.9)	0.6–2.0	6.0	10.0	1.75
0.2 M NaH_2PO_4 (pH 4.9)	1.0–2.0 4.0–0.0	2.0	7.1	1.75
0.25 M Amm. Acetate	0–1.75 3.5–10.0	1.5	5.0	1.75
0.1 M Amm. Sulfate	0.5–1.75 3.5–10.0	2.5	8.5	1.75
0.1 M Sodium Sulfate	–	2.5	8.5	1.75
0.1 M Hydrochloric Acid	0–1.5 3.0–10.0	1.8	7.25	1.3
10% TCA	0–2.1 (no gel phase)	1.5	4.0	1.5
20% TCA	0–3.75 (no gel phase)	1.25	3.25	4.5
0.1 M Sodium Hydroxide	0–2.0 3.0–10.0	3.3	2.5	1.2
1.0 M Sodium Hydroxide		0.5	4.5	1.2
0.1 M Imidazole (pH 7.4)	0–1.75 3.0–10.0	3.0	5.5	1.75
8 M Urea	0–1.5*	1.0*	2.5	1.0

*Clears only after extended agitation.

2. Select appropriate cocktail based on ionic strength.
3. Test for sample/cocktail compatibility.
    3.1 Dispense 10.0 mL cocktail into a 20 mL glass vial. The use of a glass vial allows a clear view of the mixture.
    3.2 Add the desired sample volume, cap, and shake thoroughly.

TABLE 8.9 Recommended Safer Cocktails for Mineral Acids (Based on the Use of 10 mL of Cocktail). Dashes Indicate No or Very Limited Sample Capacity

Sample	Ultima Gold (mL)	Ultima Gold XR (mL)	Ultima Gold AB (mL)	Ultima Gold LLT (mL)	Opti-Fluor (mL)	Poly-Fluor (mL)
0.1 M HCl	6.5	7.0	10.0	10.0	4.0	2.7
1.0 M HCl	0.5	2.5	5.5	5.0	0.5	3.0
2.0 M HCl	–	1.0	2.25	3.0	–	4.5
5.0 M HCl	–	< 0.5	2.0	1.5	–	0.5
1.0 M HNO₃	–	2.5	3.25	3.5	0.75	3.5
2.0 M HNO₃	0.5	2.0	2.25	2.5	0.75	3.5
3.0 M HNO₃	–	1.0	2.0	2.25	0.5	1.0
1.0 M H₂SO₄	–	0.25	6.5	7.0	–	2.0
2.0 M H₂SO₄	–	–	4.0	4.0	–	2.75
1.0 M HClO₄	2.0	2.0	2.25	2.25	1.5	1.0
2.0 M HClO₄	1.5	1.5	2.0	2.5	1.0	0.75
1.0 M H₃PO₄	–	1.5	0.5–10.0	0.5–10.0	0.5–1.5	3.0
2.0 M H₃PO₄	–	0.5	0.5–4.0	0.5–6.0	0.5–1.0	3.0

3.3 If the mixture is clear, proceed with the stability test.

3.4 If the mixture is cloudy or hazy, try increasing the cocktail volume and/or decreasing the sample volume. If the mixture does not clear, select a cocktail that can accept higher ionic strength samples such as Ultima Gold XR, OptiPhase HiSafe 3 or an equivalent cocktail.

3.5 If the mixture separates into two distinct phases (like oil and water), or is milky, select a cocktail which can accept higher ionic strength samples.

4. Test for stability: Use a mixture that has passed the sample/cocktail compatibility test. Allow it to stand at the LSC counting temperature for a minimum of two hours or the proposed count time for the sample, if this is greater than two hours. If the mixture remains stable, successful counting will be possible.

5. Count.

Note: Once the proper proportions and stability of the sample/cocktail mixture have been established, plastic vials can be considered for routine counting.

IV. SOLUBILIZATION

In its simplest terms, solubilization is the action of certain chemical reagents on the chemical bonds of a macromolecular structure (such as animal or plant tissue) that effects a structural breakdown (or digestion) into smaller, simpler subunits which can then be directly dissolved in a liquid scintillation cocktail (Kobayashi and Mondsley, 1974). The tissue sample may be whole, homogenized, macerated, or in some other state of subdivision prior to solubilization. When the digested samples are added to an appropriate liquid scintillation cocktail they should yield clear, colorless, homogeneous liquids exhibiting a minimum of quench, a minimum of chemiluminescence, and a maximum of counting stability. The chemical reagents used should be capable of rapid and complete digestion with respect to both small and large sample sizes, and should not require any complex care or methodology. Also, the combination of reagents and the method of digestion should allow accurate determination of the isotopic content with a minimum of systematic error. Solubilizers are predominantly used for the traditional animal metabolism studies, and more recently have been increasingly used in cell and tissue culture applications. Another area of growing interest is the fate of radionuclides in the environment; in this field of interest, solubilizers have been found to be an invaluable tool in sample preparation. Fundamentally there are still only three major classifications of solubilizing reagents and these are:

1. Alkaline systems (e.g., Soluene-350 and Solvable)
2. Acidic systems (e.g., Perchloric Acid)
3. Other systems (e.g., Sodium Hypochlorite)

A. Systems

The mode of action of "alkaline systems" is solubilization by hydrolysis, and a wide range of samples including animal tissues, blood, urine, bone tissue, muscle, amino acids, nucleic acids, and proteins can be digested with these reagents. With "acidic systems" the sample is oxidized to soluble products by the action of certain strong acids, usually oxidizing acids. Samples such as cartilage, bone, collagen fibers, and dried and hard plant samples can be digested by these reagents. Occasionally mixed acid reagents, and acids with an added oxidizing agent, are preferred due to their increased oxidative power. Under "other systems" a number of different reagents can be considered, however the most useful reagent is sodium hypochlorite whose mode of action is by the process of oxidative bleaching. This is particularly useful when dealing with plant samples, especially those containing chlorophyll, where the sodium hypochlorite effectively prevents color quench in subsequent liquid scintillation counting by bleaching out all of the color present. It is not possible to cover the use of every solubilizer; the intention, therefore, is to focus on the most commonly used solubilizers and their usefulness for LSC applications. This section will identify those sample types that are routinely encountered in solubilization work and will offer helpful hints on sample preparation as well as recommending suitable reagents.

B. Sample Preparation Methods

The following sample preparation techniques, using the reagents detailed in Table 8.10, were carried out using High Performance Glass Vials. All ^{3}H

TABLE 8.10 **Characteristics of Solubilizers**

Reagent	Type	Concentration	Flash point	Density (g/mL)	Warning
Soluene-350 and equivalents	Alkaline*	~0.5 M in toluene	5°C	0.88	Corrosive, flammable
Hyamine hydroxide	Alkaline*	1.0 M in toluene	18°C	0.93	Corrosive, flammable
Solvable	Alkaline	0.4 M in water		1.02	Corrosive
Perchloric acid	Acidic	70%		1.70	Corrosive
Nitric acid	Acidic	68–70%		1.42	Corrosive
Sodium hypochlorite	Other	5–7% available chlorine		1.16	Corrosive
Hydrogen peroxide	Other	30% (100 volumes)		1.11	Corrosive

*Quaternary Ammonium Hydroxide type.

TABLE 8.11 **Reagents for Solubilization and LSC Counting of Muscle**

Sample size	Solubilizer (1.0 mL used)	Digestion time at 50–60°C (h)	Sample appearance	LSC cocktail	^{3}H Counting efficiency (%)
50–200 mg	Soluene-350	1½–4	Clear	Hionic-Fluor	41–33
50–200 mg	Soluene-350	1½–4	Clear	Ultima Gold	49–37
50–150 mg	Soluene-350	1½–4	Clear	Pico-Fluor	42–36
50–200 mg	Solvable	2–3½	Clear	Hionic-Fluor	42–33
50–200 mg	Solvable	2–3½	Clear	Ultima Gold	48–41
50–150 mg	Solvable	2–3½	Clear	Pico-Fluor	39–34

counting efficiencies presented were determined using a Tri-Carb®3100 with 67% absolute ^{3}H efficiency (sealed argon purged standard) operating at 19°C.

1. Whole Tissue

The method of solubilizing whole tissue is relatively straightforward and, apart from color formation with certain tissue types, no major problems should be encountered during sample preparation and counting. Although this section only mentions Soluene-350 and Solvable, it is also possible in certain cases to use Hyamine Hydroxide.

2. Muscle (50–200 mg)

The method for processing muscle samples is shown below and typical results are illustrated in Table 8.11. The choice of LSC cocktail influences the maximum sample size that can be processed.

Procedure:

1. Place selected sample size in a 20 mL glass scintillation vial.
2. Add an appropriate volume of solubilizer (1–2 mL depending on sample size).

TABLE 8.12   Reagents for Solubilization and LSC Counting of Liver

Sample size	Solubilizer (1.0 mL used)	Digestion time at 50–60°C (h)	Sample appearance*	LSC cocktail	^{3}H Counting efficiency (%)
100 mg	Soluene-350	4	Clear, yellow	Hionic-Fluor	15
50 mg	Solvable	1	Clear, pale yellow	Ultima Gold	47
100 mg	Solvable	1½	Clear, pale yellow	Hionic-Fluor	23

*appearance after decolorization with hydrogen peroxide.

3. Heat in an oven or water bath at 50–60°C for the specified time with occasional swirling.
4. Cool to room temperature and add 10 mL of a selected cocktail.
5. Temperature and light adapt for at least one hour before counting.

3. Liver

The method for processing liver samples (50–100 mg) is shown below and typical results obtained are illustrated in Table 8.12. As before, the choice of LSC cocktail influences the maximum sample size that can be processed.
Procedure:

1. Place selected sample size in a 20 mL glass scintillation vial.
2. Add 1–2 mL of solubilizer.
3. Heat in an oven or water bath at 50–60°C for the specified time with occasional swirling.
4. Cool to room temperature.
5. Add 0.2 mL of 30% hydrogen peroxide in two aliquots of 0.1 mL, with swirling between additions. Allow any reaction to subside between additions of the hydrogen peroxide.
6. Heat again at 50–60°C for 30 minutes to complete decolorization.
7. Add 10 mL of a selected cocktail and temperature and light adapt for at least one hour before counting.

Notes: Solubilization of liver always results in highly colored samples due to the presence of bilirubin. The above work was therefore restricted to a viable sample size (which should ideally not exceed 75 mg). In our experience, Solvable has proved to be better than Soluene-350 for this particular sample type, mainly due to more rapid solubilization.

4. Kidney, Heart, Sinew, Brains, and Stomach Tissue

The method for processing the above five sample types is shown below and typical results are illustrated in Table 8.13. As previously stated, the choice of LSC cocktail influences the maximum sample size which can be processed.
Procedure:

1. Place selected sample size in a 20 mL glass scintillation vial.
2. Add 1–2 mL of solubilizer.

TABLE 8.I3 Reagents for solubilization and LSC of Various Tissues

Sample size	Solubilizer (1.0 mL used)	Digestion time at 50–60°C (h)	Sample appearance*	LSC cocktail	³H Counting efficiency (%)
Kidney					
50–100 mg	Soluene-350	1½–5	Clear, pale yellow	Hionic-Fluor	41–34
50–100 mg	Solvable	1–2	Clear, pale yellow	Hionic-Fluor	40–38
Heart					
50–100 mg	Soluene-350	2–3	Clear, pale yellow	Hionic-Fluor	40–38
50–100 mg	Solvable	1–3	Clear, pale yellow	Hionic-Fluor	40–38
Sinew					
50–100 mg	Soluene-350	1–4	Clear, pale yellow	Hionic-Fluor	44–38
50–100 mg	Solvable	1–2	Clear, pale yellow	Hionic-Fluor	42–39
Brain					
50–100 mg	Soluene-350	1½–2	Clear, pale yellow	Hionic-Fluor	43–41
50–100 mg	Solvable	1–2	Clear, pale yellow	Hionic-Fluor	42–40
Stomach					
50–100 mg	Soluene-350	1½–3	Clear, pale yellow	Hionic-Fluor	41–39

Note: It is not possible to digest stomach tissue with Solvable.
*Appearance after decolorization with hydrogen peroxide.

3. Heat in an oven or water bath at 50–60°C for the specified time with occasional swirling.
4. Cool to room temperature.
5. Add 0.2 mL of 30% hydrogen peroxide in two aliquots of 0.1 mL with swirling between additions. Allow any reaction to subside between additions of hydrogen peroxide.
6. Heat again at 50–60°C for 30 minutes to complete decolorization.
7. Add 10 mL of selected cocktail, temperature and light adapt for one hour before counting.

5. Feces

The digestion of feces (Morrison and Franklin, 1978) strongly depends on the type of animal. It is possible to use both Soluene-350 and Solvable; however, there can be problems with residual color and incomplete digestion due to the presence of cellulose type material present in feces from species such as rabbit. As an alternative, the use of a sodium hypochlorite solution is recommended. Sodium hypochlorite resolved a problem for one researcher (unpublished work) who was attempting to digest guinea pig feces. Sodium hypochlorite substantially digested this sample rapidly and isotope recoveries of greater than 98% for ³H were achieved. This recovery level was confirmed by combustion in a sample oxidizer (PerkinElmer Tri-Carb Sample Oxidizer, Model 307). The solubilization method used for processing this feces sample follows.

Procedure:

1. Weigh 50 to 150 mg of feces into a 20 mL glass scintillation vial.
2. Add 0.5 mL of sodium hypochlorite solution and cap tightly.
3. Heat in an oven or water bath at 50–55°C for about 30 to 60 minute with occasional swirling.
4. Cool to room temperature.
5. Remove the cap and blow out any remaining chlorine using a gentle stream of air or nitrogen.
6. Add 15 mL of Hionic-Fluor and shake to form a clear mixture.
7. Temperature and light adapt for one hour before counting.

Note: After digestion, a small amount of white residual matter may remain, however this should not affect the recovery.

6. Blood

The successful preparation of blood samples (see Moore, 1981) for LSC counting can often be technically difficult, and successful digestion can be largely dependent on the practical experience of the researcher. The source of blood and the correct choice of solubilizer also influence the results of digestion. Consequently, methods are given for both Soluene-350 and Solvable, and the final choice of method rests with the individual researcher. Some typical results, obtained in an independent laboratory, are shown in Table 8.14.

7. Soluene-350 Method

Procedure:

1. Add a maximum of 0.4 mL of blood to a glass scintillation vial.
2. Add, while swirling gently, 1.0 mL of a mixture of Soluene-350 and isopropyl alcohol (1 : 1 or 1 : 2 ratio). Ethanol may be substituted for the isopropyl alcohol if desired.
3. Incubate at 60°C for two hours. The sample at this stage will be reddish-brown.
4. Cool to room temperature.
5. Add 0.2 mL to 0.5 mL of 30% hydrogen peroxide dropwise or in small aliquots. Foaming will occur after each addition, therefore, gentle agitation is necessary. Keep swirling the mixture until all

TABLE 8.14 Reagents for Solubilization and LSC of Blood

Sample size (mL)	Solubilizer (1.0 mL used)	Sample appearance*	LSC cocktail	^{3}H Counting efficiency
0.2–0.4 mL	Soluene-350	Clear, pale yellow	Hionic-Fluor	29–19%
0.25–0.5 mL	Solvable	Clear, pale yellow	Hionic-Fluor	37–27%
0.5 mL	Solvable	Clear, pale yellow	Pico-Fluor 40	~29%
0.2 mL	Solvable	Clear, pale yellow	Ultima Gold	~44%

*Appearance after decolorization with hydrogen peroxide.

foaming subsides and then continue swirling until all of the hydrogen peroxide has been added.

6. Allow to stand for 15–30 min at room temperature to complete the reaction.

7. Cap the vial tightly and place in an oven or water bath at 60°C for 30 min. The sample at this stage should now have changed to pale yellow.

8. Cool to room temperature and add 15 mL of Hionic-Fluor.

9. Temperature and light adapt for one hour before counting.

8. Solvable Method

Procedure:

1. Add a maximum of 0.5 mL blood to a glass scintillation vial.

2. Add 1.0 mL Solvable.

3. Incubate the sample at 55–60°C for one hour. Sample at this stage will be brown/green in appearance.

4. Add 0.1 mL of 0.1 M EDTA-di-sodium salt solution which helps reduce foaming when the subsequent hydrogen peroxide is added.

5. Add 0.3 mL to 0.5 mL of 30% hydrogen peroxide in 0.1 mL aliquots. Gently agitate between additions to allow reaction foaming to subside.

6. Allow to stand for 15–30 min at room temperature to complete the reaction.

7. Cap the vial tightly and place in an oven or water bath at 55–60°C for one hour. The color will change from brown/green to pale yellow.

8. Cool to room temperature and add 15 mL of Pico-Fluor 40.

9. Temperature and light adapt for one hour before counting.

9. Plant Material

There are two main problems associated with digestion of plant material: the presence of pigments (especially chlorophyll) and the difficulty of digesting cellulose. Some of the colored pigments can be bleached with hydrogen peroxide, but not all, and therefore many samples will remain highly colored. The primary problem with cellulose is that it is not soluble in the alkaline solubilizers, and, in general, some form of skeletal cellulose material remains after attempted solubilization. To overcome these two problems a number of different reagent systems have been devised with the overall result that plant material can be processed. Since there is such a wide variety of plant sample types, this section will be confined to describing the solubilization methods together with their associated advantages and drawbacks. Soluene-350 can be used to solubilize various plant materials, but in general the sample size must be kept small (< 50 mg). With such small samples it is often possible to achieve limited decolorization with hydrogen peroxide; however color quenching remains a problem and the cellulose is not dissolved. This classifies the use of Soluene-350 for the digestion of plant material in the "of limited use" category. Solvable is also not ideally suited to solubilize plant materials and suffers from the same drawbacks associated with Soluene-350. Perchloric acid/nitric acid solution and perchloric acid/hydrogen peroxide solution reagent systems are suitable for digesting samples by the method of "wet oxidation" and are particularly useful for solubilizing samples such as hard and dried plant material. They have also proved useful for the digestion of cartilage, bone, collagen fibers and even some highly colored samples such as blood and liver. The general methods for these latter two reagents are as follows

a. Perchloric Acid/Nitric Acid (Wahid et al., 1985)

Procedure:

1. Prepare the solubilizing reagent by adding one volume of 70% perchloric acid to one volume of 70% nitric acid.
2. Where possible, the sample should be oven dried and then finely cut.
3. Place prepared sample (up to 200 mg) in a glass scintillation vial fitted with a poly-cone lined urea screw cap.
4. Add approximately 0.6 mL of the prepared $HClO_4/HNO_3$ reagent (1 : 1).
5. Digest the sample in the closed vial at 50–70°C for one hour or until an almost colorless solution is obtained.
6. Cool the vial to room temperature and add 15 mL of Hionic-Fluor.
7. Temperature and light adapt for one hour before counting.

b. Perchloric Acid/Hydrogen Peroxide (Sun et al., 1988; Mahin and Lofberg, 1966; Recalcati et al., 1982; Fuchs and De Vries, 1972)

Procedure:

1. Where possible, the sample should be oven dried and then finely cut.
2. Place prepared sample (up to 200 mg) in a glass scintillation vial fitted with a poly-cone lined urea screw cap.
3. Add 0.2 mL of 70% perchloric acid and swirl gently to completely wet the sample.
4. Add 0.4 mL of 30% hydrogen peroxide and again swirl gently. (This sequence of addition prevents frothing.)
5. Digest the sample in a closed vial at 50–70°C for one hour or until an almost colorless solution is obtained.
6. Cool the vial to room temperature and add 15 mL Hionic-Fluor.
7. Temperature and light adapt for one hour before counting.

Notes (for both methods): With both these acidic reagent systems, if the isotope label is ^{14}C, there is a potential for loss of the label as radiocarbon dioxide ($^{14}CO_2$) during solubilization. Tritium (^{3}H) losses are virtually prevented due to the formation of ^3H_2O which condenses inside the vial. Ensure that a poly-cone insert is used in the cap in place of a foil lined insert as these aggressive reagents will oxidize the foil (aluminum) insert and may contaminate the digest. Do not heat these mixtures above 90°C; perchloric acid may decompose violently above this temperature. Due to the aggressive nature of these mixtures the use of gloves is recommended during all handling steps. The perchloric acid/nitric acid method has also been successfully used for the determination of ^{45}Ca and ^{35}S in cartilage and bone (unpublished work).

c. Sodium Hypochlorite

It is also possible to use sodium hypochlorite, as reported by Fuchs and De Vries (1972), Porter (1980), and Smith and Lang (1987). Although sodium hypochlorite does not completely dissolve cellulose, it is capable of decolorizing not only chlorophyll but also almost all other pigments found in plant materials. Therefore, providing that the radioisotope is not assimilated within the cellulose structure, this reagent should be considered for preparing plant materials for LSC. The advantages of using this reagent for plant solubilization are that it is simple, rapid, and does not result in loss of ^{14}C as radiocarbon dioxide gas. In practice, sodium hypochlorite penetrates the plant material and rapidly decolorizes the pigments, leaving behind a white skeleton of cellulose material. The general method is described next.

Procedure:

1. Place the sample (up to 200 mg) in a glass scintillation vial.
2. Add 1.0 mL of sodium hypochlorite solution.
3. Swirl gently until all of the sample has been completely wetted.
4. Cap tightly and place in an oven or water bath at 50–60°C for approximately one to two hours.
5. Completeness of digestion is usually indicated by removal of pigmentation and/or the appearance of white skeleton of cellulose.
6. Cool the vial to room temperature and carefully vent each vial under a fume hood (decolorization is by action of released chlorine and some residual chlorine remains).
7. Blow out any remaining chlorine with a gentle stream of air or nitrogen.
8. Add 15 mL Hionic-Fluor.
9. Temperature and light adapt for one hour before counting.

Note: Sodium hypochlorite is more commonly known as bleach and should have greater than 5% available chlorine if it is to be an effective solubilizer.

10. Electrophoresis Gels

Polyacrylamide gel electophoresis (PAGE) is a technique commonly used for the separation and identification of biological species. Electrophoresis can be best described as the movement of small ions and macromolecules in solution under the influence of an electric field. The rate of migration depends on the size and shape of the molecule, the charge carried, the applied current, and the resistance of the medium. Electrophoresis is carried out in gels cast either in tubes or as slabs. A number of gel materials have been used successfully, including agar, agarose, and polyacrylamide. Agar and agarose gels are prepared by heating the granular material in the appropriate electrolyte buffer, casting the gels and allowing them to set on cooling. Polyacrylamide gels are made from acrylamide and N,N'-methylene bisacrylamide (bisacrylamide) mixtures dissolved in electrolyte and polymerised by the addition of chemical catalysts. Crosslinking agents other than bisacrylamide have been used and include DATDA (diallyltartardiamide),

BAC (N,N'-bisacrylylcystamine) and ethylene diacrylate (Choules and Zimm, 1965). The location of the various compounds within the gel is determined by staining and by the presence of radioactivity in discrete zones. Sample preparation for analysis by LSC involves either elution of the sample from the gel or complete dissolution of the gel.

a. Elution

The complete solution of the gel is not required for satisfactory counting. This is indicated by the observation that when acrylamide gels are treated with a solubilizer, such as Soluene-350 or its equivalent, the gel swells rapidly and the entrapped macromolecules diffuse out into the solubilizer. Addition of an appropriate cocktail results in a suitable counting medium. During this diffusion process, it is presumed that the macromolecules (proteins) are partially hydrolysed by the strong organic base (Terman, 1970; Basch, 1968). After 10 to 20 hours of digestion with solubilizer at ~45°C of 1 mm gel slices, the undissolved gel can be removed after being counted and no perceptible loss of either ^{3}H or ^{14}C counts, or changes in the isotope ratios of the samples, are observed (Terman, 1970). Another method of preparing the gel slices for counting is to add the 1–2 mm gel slice to 5 mL water and macerate the slice with the aid of a stirring rod. This mixture is allowed to stand at 50°C for about two hours, cooled, and then 10 mL of a suitable cocktail, such as Insta-Gel Plus or equivalent, is added.

b. Dissolution

The alternative to elution is the complete dissolution of gels and this can be accomplished using a variety of reagents and these include hydrogen peroxide (Diener and Paetkau, 1972), periodic acid (Spath and Koblet, 1979), 2-mercaptoethanol (Hansen et al., 1980), piperidine (containing EDTA), alkalis, and ammonia. The selection of a suitable reagent depends upon the gel, in that the choice of crosslinker dictates the dissolving reagent required. Acrylamide gels crosslinked with DATDA are dissolved in 0.5 mL of 2% periodic acid in two hours at room temperature (Anderson and McClure, 1973), and the digest can be counted in 10 mL of a suitable cocktail such as Insta-Gel Plus, Ultima Gold, OptiPhase HiSafe 2 or equivalents. Acrylamide gels crosslinked with BAC are dissolved in 2-mercaptoethanol (Hansen et al., (1980). Polyacrylamide gels that are soluble in alkalis can be constructed using ethylene diacrylate as the crosslinker instead of bisacrylamide. Polyacrylamide gels crosslinked with bisacrylamide can be dissolved by adding the 1–2 mm gel slice to 0.5 mL of 30% (100 vol) hydrogen peroxide and heating at 50°C until the gel dissolves. Some authors use a one hour digestion at 55°C (Diener and Paetkau, 1972), overnight at 40°C (Benjamin, 1971) and overnight at 60°C (Dion and Moore, 1971). The hydrogen peroxide also decolorizes the stained gels. When using hydrogen peroxide cocktail selection is extremely important due to the potential for chemiluminescence. When peroxides are present in an alkaline medium, the chances of chemiluminescence are high. Consequently, after dissolving the gel slice in hydrogen peroxide it is advisable to add a small amount of 1–2 M HCl to keep the mixture neutral or slightly acidic. In addition, sometimes traces of copper ions have been added to aid the decomposition of the excess hydrogen peroxide and the presence of these ions intensifies the chemiluminescence in an alkaline medium. Cocktails suitable for this type of sample include Hionic-Fluor and Aquasol-2 (both acidified with 0.5 mL 2 M HCl per 10 mL cocktail). Certain researchers (Goodman and Matzura, 1971) have criticized the use of hydrogen peroxide and heat on the grounds that during digestion radioactive carbon dioxide and water may be lost. They proposed the digestion of 1 mm gel slices using 0.25 mL of a mixture containing 1 part concentrated ammonium hydroxide and 99 parts 30% hydrogen peroxide at room temperature for four to eight hours. This method is reported to be free from the danger of loss of radioactivity because of the low temperature digestion. Agarose gels can be rapidly and easily dissolved by treatment with sodium hypochlorite solution for about one hour at ~45°C. In this instance cocktail selection is very important as any residual hypochlorite solution has a tendency to decompose under slightly acidic conditions producing chlorine, and this can lead to chemiluminescence. This chemiluminescence only becomes apparent after standing for >24 hours and therefore cocktail selection is important. At present the recommended cocktail for this application is Hionic-Fluor. All the above methods are suitable for the isotopes normally encountered in gel electrophoresis, and in one modification ^{32}P can be detected and counted in wet gel slices, either alone or covered in buffer, by Cerenkov counting. As an alternative to elution or dissolution, the gel slices can be combusted using a commercially available oxidizer system. Although only applicable to ^{3}H and ^{14}C, combustion offers a rapid and accurate means of sample preparation free from the chemiluminescence problems associated with most acrylamide gel procedures.

A variety of LSC cocktails, of both high flash point and of classical solvent types, are suitable for use with the various solubilizing reagents. The information presented in this section is condensed into a quick reference guide (Table 8.15); and this may prove particularly useful in selecting the most appropriate solubilizer and cocktail for a specific sample type.

V. COMBUSTION

Combustion (sample oxidation) is the complete combustion of a sample using the "open/closed flame combustion technique." This technique is principally used for ^{3}H and ^{14}C. During combustion the organic portion of the sample is completely converted to water and carbon dioxide, while the inorganic portion remains in the particle trap. The respective reactions are:

$$4H + O_2 \rightarrow 2H_2O$$
$$C + O_2 \rightarrow CO_2$$

Since the products of combustion are water and carbon dioxide, problems such as color quenching, self-absorption and chemiluminescence are eliminated. The two systems operating within a combustion process are

TABLE 8.15   **Solubilization Selection Guide**

Sample type	Solubilizer	Maximum sample size	Suitable LSC cocktails
Muscle	Soluene-350	150 mg 200 mg	Pico-Fluor 40, Ultima Gold or Hionic-Fluor
	Solvable	150 mg 200 mg	Pico-Fluor 40, Ultima Gold or Hionic-Fluor
Liver	Soluene-350	100 mg	Hionic-Fluor
	Solvable	50 mg 100 mg	Ultima Gold Hionic-Fluor
Kidney	Soluene-350	100 mg	Hionic-Fluor
	Solvable	100 mg	Hionic-Fluor
Sinew	Soluene-350	150 mg	Hionic-Fluor
	Solvable	150 mg	Hionic-Fluor
Brain	Soluene-350	150 mg	Hionic-Fluor
	Solvable	150 mg	Hionic-Fluor
Stomach	Soluene-350	100 mg	Hionic-Fluor
Feces	Soluene-350	150 mg	Hionic-Fluor
	Solvable	20 mg	Hionic-Fluor
	Sodium hypochlorite	20 mg	Hionic-Fluor
Blood	Soluene-350	0.4 mL	Hionic-Fluor
	Solvable	0.5 mL 0.2 mL	Hionic-Fluor or Pico-Fluor 40 Ultima Gold
Plant material	Soluene-350	< 50 mg	Hionic-Fluor
	Solvable	< 50 mg	Hionic-Fluor
	$HClO_4/H_2O_2$	200 mg	Hionic-Fluor
	$HClO_4/HNO_3$	200 mg	Hionic-Fluor
	Sodium hypochlorite	200 mg	Hionic-Fluor
Gels			
bisacrylamide	H_2O_2	1–2 mm slices	Hionic-Fluor + HCl Aquasol-2 + HCl
	Soluene-350	1–2 mm slices	Hionic-Fluor
	Solvable	1–2 mm slices	Hionic-Fluor or Ultima Gold
	Water	1–2 mm slices	Insta-Gel XF
DATDA	periodic acid	1–2 mm slices	Hionic-Fluor or Ultima Gold

(1) water production and incorporation into a cocktail and (2) carbon dioxide production, absorption, and incorporation into a cocktail. On the ^{3}H side the process is straightforward in that the water formed is incorporated into an emulsifying cocktail such as Monophase S. On the ^{14}C side the basic chemical reaction involves the production of carbon dioxide by the combustion (oxidation) of organic carbon, followed by absorption by an amine. The absorption reaction scheme is that 2 moles of amine react with 1 mole of carbon dioxide to form a carbamate. This carbamate is then added to a special cocktail (e.g., Permafluor E+) for counting. Since the counting

quench of the ^3H is only water and the quench for ^{14}C is carbamate/Carbo-Sorb E, the sample size processed can be as much as 2 g for ^3H and as much as 1.5 g for ^{14}C labeled materials. For low level ^3H counting, multiple combustions are possible with collection of the combusted material (water) in the same counting vial, hence, better counting statistics and shorter counting time. There are two principal oxidizer systems in use today and these are the PerkinElmer Sample Oxidizer and the "Harvey Oxidizer" (also known as R. J. Harvey Biological Oxidizer). Both are able to both combust and separate ^3H and ^{14}C quantitatively but the respective mechanisms by which this is achieved are slightly different. Differences in combustion chamber, plumbing, heating techniques, timing cycles, and reagents account for the main operational differences. Consultation with each manufacturer's product brochures show how these individual instruments operate. A complete description of the PerkinElmer system, together with details on alternative methods of combustion are given in Liquid Scintillation Analysis—Science and Technology (Kessler, 1989). Combustion can be found in many areas including biochemistry, metabolic studies, pharmacokinetics, and agrochemical studies. Some typical applications include the determination of ^3H and ^{14}C in waste oil and the determination of ^3H and ^{14}C in soils. Combustion is the only technique that, prior to counting, can physically separate ^3H and ^{14}C, thereby eliminating many of the errors associated with the counting of dual labeled samples.

VI. COMPARISON OF SAMPLE OXIDATION AND SOLUBILIZATION TECHNIQUES

The primary objective of all sample preparation methods is to obtain a stable homogeneous solution suitable for analysis by liquid scintillation counting (LSC). There are no absolutes in sample preparation; whichever method produces a sample that lends itself to accurate and reproducible analysis is acceptable. However, there will be occasions when more than one method will be both suitable and available and the selection of either method will depend on other factors. It is precisely this situation that occurs when considering solubilization and sample combustion for sample preparation. Both techniques are routinely used to process samples that are not directly soluble in LSC cocktails. There are numerous sample types that fall into this category and typical examples include tissue, muscle, kidney, liver, feces, blood, plant material, etc. Many of these samples are encountered in ADME (absorption, distribution, metabolism, and excretion) studies in which the biological behavior and potential toxicological effects of a test substance are investigated.

A. Solubilization

Solubilization is the action of certain chemical reagents on organic materials (such as animal or plant tissue) that effects a structural breakdown (or digestion) into a liquid form that can then be directly dissolved in a liquid scintillation cocktail. Typical solubilizers include organic and inorganic alkalis which act by the process of alkaline hydrolysis, and certain mineral acids which effect solubilization by acidic oxidation. The solubilization process usually involves heating the sample/solubilizer at elevated temperature (40–65°C) for periods ranging from <1 to 24 h, until a homogeneous mixture is formed. Certain samples that remain colored after solubilization are optionally treated with hydrogen peroxide, and following this a recommended LSC cocktail is added and the sample is ready for analysis by LSC.

B. What is Sample Combustion?

The principle of sample oxidation is that the sample is combusted in an oxygen rich atmosphere and any hydrogen present is oxidized to water while any carbon is oxidized to carbon dioxide. If tritium is present then the combustion product will be ^3H_2O and, if ^{14}C is present, then the combustion product will be $^{14}CO_2$. In the PerkinElmer Model 307 Sample Oxidizer the water is condensed in a cooled coil and then washed into a vial where it is mixed with an appropriate LSC cocktail. The CO_2 is trapped by vapor-phase reaction with an amine and the resulting product is mixed with an appropriate LSC cocktail. At the end of the combustion cycle two separate samples (a tritium sample and a ^{14}C sample) are trapped at ambient temperature, thus minimizing cross contamination.

C. Advantages and Disadvantages

The oxidation process described above can be compared with the steps in a typical solubilization method in Fig. 8.1. As can be seen from Fig. 8.1 there are more steps involved in solubilization as compared to combustion and while this may seem less attractive there are other important factors to consider before deciding which methodology is best.

I. Solubilization Methods and Suitability

The decision to use solubilization as the preferred method of sample preparation depends primarily on the nature of the sample and the number of samples that need to be processed. When the number of samples is low then solubilization is usually the method of choice. Solubilization is suitable for many organic sample types but certain of these are prone to problems such as color formation, limited sample size, and time to complete solubilization. These include biological samples such as whole blood, plasma, serum, liver, kidney, fatty tissue, and most plant materials. Many other biological sample types including, muscle, whole tissue, brain, stomach, intestines, nerve cells, cornea, and cartilage can be easily and rapidly processed. There is no major cost involved in setting up to do solubilization. The cost of 1.0 L solubilizer reagent and 10.0 L appropriate LSC cocktail is relatively small and this is sufficient for 1,000 analyses, assuming 1.0 mL solubilizer used with 10.0 mL cocktail. The method is relatively straightforward and is described earlier. To process a sample simply add the solubilizer and

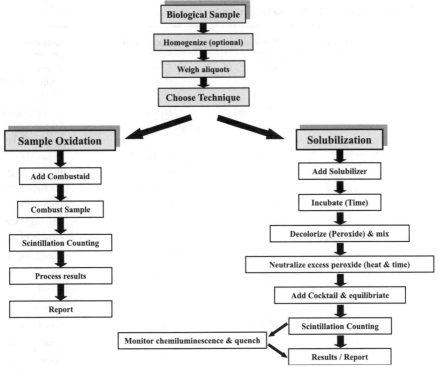

FIGURE 8.1   Sample oxidation and solubilization techniques compared.

heat at 50 to 60°C until the sample is dissolved. After solubilization the sample may be colored and this color can usually be removed or reduced by treatment with hydrogen peroxide. The final step is to add the recommended LSC cocktail and the sample is ready for counting. Using this method many samples can be processed simultaneously and then counted sequentially.

a. The typical advantages of solubilization are:

- Capital outlay is low.
- Homogeneous samples are produced.
- Many samples can be processed simultaneously.
- Color quench is corrected using a quench curve.
- It is suitable for a diversity of isotopes.

b. The disadvantages of solubilization are:

- Sample sizes are generally ≤ 200 mg.
- Time to complete solubilization can vary from a few hours to many hours.
- Certain samples require modified techniques or longer solubilization times.
- With acidic solubilizers loss of radioactivity by volatilization may occur.

TABLE 8.16   Sample Types that can be Processed by Solubilization

Sample type	Solubilizer	Max. sample size	Suitable LSC cocktails
Muscle	Soluene®-350	150 mg 200 mg	Pico-Fluor™ 40 Ultima Gold™ or Hionic-Fluor™
	SOLVABLE™	150 mg 200 mg	Pico-Fluor 40 Ultima Gold or Hionic-Fluor
Liver	Soluene-350	100 mg	Hionic-Fluor
	SOLVABLE	50 mg 100 mg	Ultima Gold Hionic-Fluor
Kidney	Soluene-350	100 mg	Hionic-Fluor
	SOLVABLE	100 mg	Hionic-Fluor
Heart	Soluene-350	100 mg	Hionic-Fluor
	SOLVABLE	150 mg	Hionic-Fluor
Sinew	Soluene-350	150 mg	Hionic-Fluor
	SOLVABLE	150 mg	Hionic-Fluor
Brains	Soluene-350	150 mg	Hionic-Fluor
	SOLVABLE	150 mg	Hionic-Fluor
Stomach	Soluene-350	100 mg	Hionic-Fluor
Feces	Hypochlorite	150 mg	Hionic-Fluor
	Soluene-350	20 mg	Hionic-Fluor
	SOLVABLE	20 mg	Hionic-Fluor
Blood	Soluene-350	0.4 mL	Hionic-Fluor
	SOLVABLE	0.5 mL 0.2 mL	Hionic-Fluor or Pico-Fluor 40 Ultima Gold
Plant Material	Soluene-350	< 50 mg	Hionic-Fluor
	SOLVABLE	< 50 mg	Hionic-Fluor
	$HClO_4/H_2O_2$	200 mg	Hionic-Fluor
	$HClO_4/HNO_3$	200 mg	Hionic-Fluor
	Hypochlorite	200 mg	Hionic-Fluor

- Excess peroxide must be destroyed after decolorization.
- Chemiluminescence may be present.
- If ^3H and ^{14}C are both present in the sample then dual label DPM calculation is necessary.

Solubilization has been in use for many years and experienced researchers are typically able to optimize the methodologies through attention to detail and judicious selection of reagents and cocktails. Examples of sample types, reagents and cocktails are shown in Table 8.16.

2. Sample Combustion Methods and Suitability

Since sample combustion is suitable for any organic and even some inorganic samples, the selection of this method is usually governed by the number of samples that need to be processed. When the sample load exceeds 50 per day then sample oxidation becomes the method of choice for many sample types. In manual mode the PerkinElmer Model 307 oxidizer can process 100 samples daily, while the robotic version can process 240 samples in 8 h. High temperature flame combustion at 1300°C enables wet, dry or freeze-dried samples up to 1.5 g to be processed. For those samples containing dual label $^3H/^{14}C$ the combustion cycle produces single label ^3H and ^{14}C samples in separate vials with no cross contamination.

a. Advantages of Sample Combustion

- Sample processing time is rapid.
- Robotic sample processing is possible.
- Sample can be wet, dry, or freeze-dried.
- Any sample containing H and/or C can be combusted.
- It is ideally suited for both single and dual label ^3H and ^{14}C.
- Sample sizes up to 1.5 g are possible.
- Radioactive recovery is excellent (>97%).
- Memory effect is <0.08%.
- There is no loss of radioactivity by volatilization.
- There is no chemiluminescence interference.
- There is no color quench interference.

b. Disadvantages of Sample Combustion

- Initial capital investment.
- It is only suitable for ^3H and ^{14}C.
- Need a gas supply (oxygen and nitrogen).
- Must be operated in a fume hood.
- Reagents are corrosive and flammable.

The flame combustion technology has proven to be a simple and reliable means of sample preparation and can process a diverse array of samples with a high degree of precision and accuracy. The technique requires a minimal amount of time and sample handling, and eliminates any potential interference from color quenching or chemiluminescence. In addition, with the Perkin Elmer System 387 robotic option the entire procedure can be automated to process up to 80 samples per run, without supervision. The diversity of samples that can be processed using flame combustion is shown in Table 8.17.

Both solubilization and flame combustion are viable sample preparation techniques for a diverse array of animal and plant tissues. Flame combustion and solubilization procedures each have specific advantages and some disadvantages. Solubilization is more suited to those situations where the sample numbers are low. There can be problems such as color formation, chemiluminescence, and sample size limitations when solubilizing with certain sample types, but by careful attention to detail and experience these can be overcome. Flame combustion, operated in either manual or robotic format provides a very powerful tool to rapidly process many different sample types with a high degree of precision and accuracy. The technique requires a minimal amount of time and sample handling and eliminates color quench

TABLE 8.17   Sample Types that can be Processed by Flame Combustion

Liver	Whole blood	Bone	Gels
Spleen	Lung	Egg shell	Plastics
Skin	Heart	Plant tissue	Filters
Plasma	Fat	Bacteria	Crude oil
Muscle	Intestines	Insects	TLC's
Kidney	Hair	Glands	Toluene
Brain	Adipose	Water	Synthetic fibers
Feces	Bladder	Urine	Soil

TABLE 8.18   Trapping Capacity of Suitable Reagents for Carbon Dioxide

	$mM\ CO_2$ per mL	mL Required for 1 mM CO_2	mL Required for 5 mM CO_2	mL Required for 10 mM CO_2	Flash point
Sodium hydroxide 0.1 M	0.05	20.0	–	–	–
Sodium hydroxide 1.0 M	0.50	2.0	10.0	–	–
Hyamine hydroxide 1.0 M (in methanol)	0.50	2.0	10.0	–	11°C
Ethanolamine	8.10	0.12	0.62	1.23	93°C
Carbo-Sorb E	4.80	0.21	1.04	2.08	27°C

and chemiluminescence interferences. The only limitation is the initial capital investment and the restriction to ^{3}H and ^{14}C. The PerkinElmer Model 307 manual oxidizer is well suited for sample loads of up to 50 per day, but where high sample throughput is critical the model 387 robotic option, with its capability of processing 240 samples per day, is the ideal solution.

VII. CARBON DIOXIDE TRAPPING AND COUNTING

Certain assays result in the generation of discrete gaseous $^{14}CO_2$ samples (not originating from combustion) and these require a modified approach for the trapping of the gas and subsequent liquid scintillation counting (LSC) (Qureshi *et al.*, 1985; Schadewaldt *et al.*, 1983; Pfeiffer *et al.*, 1981; Riffat *et al.*, 1985). $^{14}CO_2$ gas samples originate from a variety of sources, including, $^{14}CO_2$ in expired breath (Bird, 1997); expired by plants; expulsion from blood (Kaczmar and Manet, 1987); and release in enzymatic studies (Sissons, 1976). There are a number of potentially useful reagents available for trapping carbon dioxide; some of these are shown in Table 8.18.

A. Sodium Hydroxide

Sodium hydroxide absorbs/traps CO_2 by a reaction that produces a sodium carbonate solution. Potassium hydroxide performs in a similar way by forming potassium carbonate. Some of the recommended LSC cocktails for use with this mixture include:

- Emulsifier Safe, which accepts up to 2 mL of 0.1 M sodium hydroxide/ CO_2 in 10 mL cocktail;
- Opti-Fluor, which accepts up to 5 mL of 0.1 M sodium hydroxide/CO_2 in 10 mL cocktail;
- Hionic-Fluor, which accepts up to 5 mL of 1.0 M sodium hydroxide fully saturated with CO_2 in 10 mL cocktail;
- Ultima-Flo AF which accepts 10 mL of 0.5 M NaOH/CO_2 or 5 mL of 1.0 M NaOH/CO_2 in 10 mL cocktail.

Notes: Hionic-Fluor is suitable for use with hydroxide/carbonate solutions due to its sample capacity for concentrated solutions and alkaline pH. Ideally the final pH should be above 9 to avoid liberation of trapped CO_2. Cocktails containing mixed surfactant systems, such as Ultima Gold or Ultima Gold XR can be used, however counting should be performed the same day as these cocktails have the potential for slow release of CO_2 on prolonged storage (characterized by dropping CPM levels). Cocktails containing free-acid phosphate esters such as AquaSafe should not be used as their acid nature will cause rapid loss of ^{14}C as $^{14}CO_2$.

B. Hyamine Hydroxide

Chemically, hyamine hydroxide performs similarly to sodium and potassium hydroxide in that it forms hyamine carbonate on reaction with CO_2 (Bird, 1997). Some of the recommended LSC cocktails include: Insta-Fluor Plus, which will accept up to 7.5 mL of hyamine hydroxide saturated with carbon dioxide in 10 mL cocktail and Emulsifier Safe which will accept up to 3 mL of hyamine hydroxide saturated with carbon dioxide in 10 mL cocktail, providing a safer system due to the high flash-point of this LSC cocktail.

Note: Foaming of hyamine hydroxide used to absorb carbon dioxide expelled in rat breath has been reported. This can be overcome by the addition of one drop of silicone antifoam per 10 mL of hyamine hydroxide. Both of the above cocktails are resistant to any potential chemiluminescence possible with this system.

C. Ethanolamine

Ethanolamine reacts with CO_2 in a rather different way than previously discussed reagents in that a carbamate is formed as opposed to a carbonate. The main difference between these two species is that a carbamate is more stable under slightly acidic conditions whereas a carbonate reacts rapidly with acids to release carbon dioxide. Ethanolamine/CO_2 is notoriously difficult to incorporate into LSC cocktails and thus the recommended LSC solutions may seem a little unusual. The main system known which is capable of accepting this unusual reagent is shown in Table 8.19. The recommended LSC cocktail requires the use of a co-solvent, e.g., methyl cellosolve to

TABLE 8.19   Suitable LSC Solutions and Capacity for Ethanolamine/CO_2.

Cocktail	Ethanolamine (mL)	Methyl cellosolve (mL)	mM CO_2 trapping capacity (mM)
Hionic-Fluor (10.0 mL)	1.0 mL	4.0 mL	8.1 mM
Hionic-Fluor (10.0 mL)	2.0 mL	6.5 mL	16.2 mM
Hionic-Fluor (10.0 mL)	3.0 mL	8.5 mL	24.3 mM

facilitate the take up of the ethanolamine/CO_2. There may be systems based on other equivalent cocktails but at present they are unreported.

D. Carbo-Sorb E

Carbo-Sorb E will react with CO_2 to form a carbamate as well, and was developed to work in the PerkinElmer sample oxidizer model 307. It is however, possible to use this reagent as a carbon dioxide trapping agent outside of the model 307. As with the model 307, the recommended cocktail is Permafluor E+, and the following conditions are recommended. For ratios of Carbo-Sorb E to Permafluor E+ from 1:10 up to 1:1, maximum saturation of carbon dioxide is possible with no phase separation of the resulting carbamate. Should any minor haziness develop then either add 2–4 mL methanol to the mixture until a clear solution forms, or add extra Carbo-Sorb E to the mixture until a clear solution forms. This latter method simply dilutes the absorption capacity below the critical 90% level. The use of Carbo-Sorb E is not recommended in enzymatic, plant or human studies due to the corrosive nature of the volatile amine present.

This information is condensed into a quick reference table (Table 8.20), and it may prove particularly useful when the total trapping capacity per standard 20 mL LSC vial is required.

VIII. BIOLOGICAL SAMPLES

Typical biological samples commonly encountered include blood, plasma, serum, urine, feces, homogenates, bacteria, and water-soluble protein. These sample types are generally processed and counted by either direct counting, solubilization, or combustion. Direct counting is the addition of raw sample to a cocktail and is suitable for urine, serum, plasma, and water soluble protein. Solubilization, which is the dissolution or digestion of a sample, is suitable for processing blood, plasma, serum, feces, homogenates, and bacteria. Combustion involves the oxidation/burning of a sample to convert the organic carbon and hydrogen to $^{14}CO_2$ and $^{3}H_2O$ respectively, and is suitable for all of the above sample types without exception.

A. Urine

In the direct counting of urine the two main obstacles are color quench and separation/ precipitation of proteins. The color quench problem can be

TABLE 8.20 Reference Table for CO_2 Trapping and LSC Counting

CO_2 absorber	mm CO_2 per mL	mL Reqd for 1 mM CO_2	mL Reqd for 5 mM CO_2	mL Reqd for 10 mM CO_2	LSC cocktail	Cocktail volume (mL)	mL of Absorber	Max CO_2 capacity (mM)
NaOH (0.1 M)	0.05	20.0	—	—	Emulsifier safe Opti-Fluor	15.0 14.0	3.0 7.0	0.15 0.35
NaOH (0.5 M)	0.25	4.0	—	—	Ultima-Flo AF	10.0	10.0	2.50
NaOH (1.0 M)	0.50 0.50	2.0 2.0	10.0 10.0	—	Hionic-Fluor Ultima-Flo AF	14.0 14.0	7.0 7.0	3.50 3.50
Hyamine hydroxide (in methanol)	0.50 0.50	2.0 2.0	10.0 10.0	—	Emulsifier Safe Insta-Fluor	15.0 12.0	4.5 9.0	2.25 4.50
Ethanolamine	8.10	0.12	0.62	1.23	Hionic-Fluor / Methyl cellosolve	10.0 8.5	3.0	24.3
Carbo-Sorb E	4.80	0.21	1.04	2.08	Permafluor E+	10.0	10.0	48.0

resolved either by using a cocktail which has good color quench resistance properties (e.g., Ultima Gold LLT or AB, or equivalent cocktail), or by reducing the sample size to obtain an acceptable balance between quench levels and counting efficiency, especially important when measuring ^{3}H in urine. ^{14}C is much more resistant to color quenching and routine counting is usually possible. Precipitation of proteins is almost impossible to prevent but the correct cocktail selection can reduce this problem to an acceptable level. It is advisable to arrange to count the samples within 24 hours if possible and carry out the counting at or slightly below 20°C. Urine from different animal species may require modified treatments or even different cocktails. Particularly difficult urine samples are from small animals such as dog and rat. It may be necessary to reduce the sample volume considerably or even change to the use of a solubilizer with a classical cocktail (e.g., Soluene-350 with Hionic-Fluor).

B. Plasma and Serum

When counting plasma or serum there are potential problems with sample volume, color quench, and separation–precipitation of proteins. Plasma/serum samples are particularly difficult due mainly to the large amounts of proteinaceous material present. Sample size should be kept below 1.0 mL in 10 mL cocktail. Recommended cocktails are Ultima Gold (sample volume up to 750 μL) and Pico-Fluor 30 (sample volume up to 1.0 mL). The color quench problem can be resolved either by using a cocktail which has good color quench resistance properties (e.g., Ultima Gold or equivalent cocktail), or by keeping the sample size below the limits indicated above to obtain an acceptable balance between quench levels and counting efficiency, especially important when measuring ^{3}H in urine. ^{14}C is much more resistant to color quenching and routine counting is usually possible. Precipitation of proteins is almost impossible to prevent but the correct cocktail selection can reduce this problem to an acceptable level. With certain cocktails it is sometimes of benefit to add about 10% ethanol or isopropyl alcohol. The addition of alcohol helps prevent the precipitation of the proteinaceous material. It is advisable to arrange to count the samples within 24 hours if possible and carry out the counting at or slightly below 20°C. Again, plasma and serum from different animal species may require modified treatments or even different cocktails. Particularly difficult plasma/serum samples are from dog and rat. It may be necessary to reduce sample volume considerably or even change to the use of a solubilizer with a classical solvent-based cocktail (e.g., Soluene-350 and Hionic-Fluor). When preparing samples the use of glass vials is recommended so that the contents can be visually checked for stability. If a stable mixture results then a change to plastic vials is possible. Alternatively the use of combustion (sample oxidation) is a more than suitable alternative.

C. Homogenates

The only problem with direct counting of homogenates is sample–cocktail compatibility. This can be resolved by either using a cocktail which is compatible with the sample (e.g., Ultima Gold, Insta-Gel Plus or Hionic-Fluor), reducing the sample size (recommended sample volume is 0–2 mL), or increasing the cocktail volume to obtain a clear mixture. Usually the homogenate is prepared in either water or 7:3 ethanol:water, and the counting is carried out at or slightly below 20°C. With some homogenates it may be necessary to reduce sample volume considerably or even change to the use of a solubilizer with a classical cocktail (e.g., Soluene-350 and Hionic-Fluor). Alternatively the use of combustion (sample oxidation) is a more than acceptable alternative.

D. Solubilization

Solubilization sample preparation methods for blood, plasma-serum, red blood cells, water-soluble proteins, homogenates, tissue samples, bacteria and cells, organs (heart, liver, kidney, etc.), and feces has already been covered.

E. Combustion

Combustion (sample oxidation) is applicable for all biological sample types. Combustion of a biological sample depends principally on its physical form and can be divided into solid samples and liquid samples where the latter includes homogenates. The size of solid sample that can be processed can be as much as 2 g for ^{3}H and as much as 1.5 g for ^{14}C labeled materials. The two main problems with solid samples are either poor or incomplete combustion, or too rapid combustion. Poor combustion can be resolved by either increasing the combustion time or adding PerkinElmer Combustaid to assist the combustion process. Too rapid combustion can be resolved by adding cellulose powder that helps moderate the burn. With liquid samples that are generally restricted to a maximum sample size of 1.0 mL, poor or incomplete combustion is the only problem. This can be rectified by increasing the combustion time or by adding Combustaid to assist the combustion process. For low level tritium determinations, multiple combustions are possible with collection of the combustion product (water) in the same counting vial, thus improving detection efficiency.

IX. FILTER AND MEMBRANE COUNTING

Filter counting, or solid support counting as it is sometimes known, is probably best described as heterogeneous counting. The main difference between homogeneous counting and heterogeneous counting is that homogeneous counting relies on 4π geometry while 2π geometry applies to heterogeneous counting. An explanation of the terms 4π and 2π geometry is needed to appreciate the differences between the two counting techniques. In homogeneous counting, the sample is completely "dissolved" in the liquid scintillation cocktail; therefore the photons of light (scintillations), which are emitted as the end result of the energy transfer process within the liquid scintillation cocktail, are free to radiate in any direction. In geometric terms

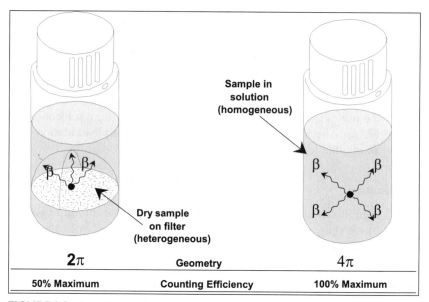

FIGURE 8.2   Sample counting geometries encountered in liquid scintillation analysis.

this freedom of radiation is described by a sphere or globe whose surface area is $4\pi r^2$. In heterogeneous counting, this freedom is restricted by the presence of the filter or membrane which absorbs both the kinetic energy of β-particles and the photons of light passing in one plane; therefore the emitted light can only occupy the surface area of a hemisphere which is $2\pi r^2$. Hence the derivation of the expressions 4π and 2π counting geometries and these are illustrated in Figure 8.2:

In essence filter counting can be a relatively simple technique where the sample is isolated or collected on a filter and usually dried. This filter is placed in a scintillation vial; a small volume of an appropriate LSC cocktail such as Ultima Gold F is added and, after ensuring that the filter is completely wet, the vial is counted.

A. Elution Situations

The difficulty in counting on filters or other solid supports is that when the sample is immersed in a cocktail, four situations may develop:

1. The sample may remain bound to the filter or solid support—*no elution* situation.
2. The sample may be partially eluted by the cocktail—*partial elution* situation.
3. The sample may have a certain solubility in the cocktail—*equilibrium* situation.
4. The sample may be completely dissolved in the cocktail—*complete elution* situation.

Of these, the one to avoid is partial elution as the soluble fraction is counted with 4π geometry, whereas the insoluble (filter bound) portion is counted with 2π geometry. This will make measurement under these circumstances irreproducible. However, a partial elution situation may go to equilibrium with time and therefore should not be discounted out of hand. Repeat counting of the sample, over several hours, will determine if an equilibrium situation has been reached and this is characterized by a constant count rate being obtained over time. If the sample is insoluble (no elution), the efficiency and reproducibility of counting will depend on the magnitude of the β-energy, the nature of the filter or solid support, its orientation in the vial, and the size of the sample molecule. If the sample is completely dissolved or eluted into the cocktail (complete elution), counting considerations will be similar to those of solubilized samples, where a true homogeneous state is obtained. Other factors which affect counting are the presence and composition of the sample precipitate and the amount of sample that becomes soluble in the cocktail (Bransome and Grower, 1970).

B. Sample Collection and Filters

Self-absorption affects the efficiency of counting on solid supports with low energy ^{3}H samples being more susceptible than higher energy isotope samples such as ^{14}C. The type and amount of sample, thickness, absorption level, and the material of construction of the solid support also influence the self-absorption effect. The order of counting efficiency of solid supports is: glass fiber > cellulose acetate > standard cellulose > chromatographic paper (Wang and Jones, 1959; Gill, 1967). This order of efficiency will vary depending upon the size of the molecules. Smaller molecules can readily diffuse into amorphous regions of the cellulose fibers while the larger molecules may remain on the surface. Microscopically, glass fiber filters appear as an impermeable virtual network of threads whereas the paper filters appear as capillary tubes. For glass fiber filters, the efficiency can be markedly different for sample material trapped on the surface as opposed to that embedded in the pores. This is particularly true for low energy beta-particles from ^{3}H. In some cases, reproducible counting efficiencies can be obtained by addition of a carrier of known weight (many times more than the sample), which subsequently induces the same amount of self-absorption for each sample (Wang and Willis, 1965). The carrier must be added before the filtration step, and time must be allowed for complete mixing with the real sample. When using chromatographic paper to isolate or collect samples, one should remember that some grades contain a UV enhancer and this can be eluted into the cocktail producing unwanted chemiluminescence. Such spurious counts can lead to an overestimation of the activity of the sample. Providing all of these factors and effects are taken into consideration during sample preparation, successful and reproducible counting can be accomplished using this technique.

C. Filter and Membrane Types

In practice there are a number of filter types which can be used to isolate or collect various sample types for LSC. The choice of filter type will depend upon both the nature and particle size of the sample, however glass fiber filters are recommended if at all practical. Other filter types which have been used include cellulose nitrate, cellulose acetate, mixed cellulose esters, polyvinyl chloride (PVC), polyacrylonitrile, normal paper, polycarbonate, Teflon, nylon and polythene terephthalate (PET). The categories of sample types which can be analyzed by this technique include: precipitates of macromolecules (such as nucleic acids and proteins) aquatic and terrestrial ecosystem samples (such as algae and phytoplankton) and other deposits (such as airborne particulate matter).

D. Sample Preparation Methods

The different elution situations influence both the choice of the sample preparation method and the recommended liquid scintillation cocktail.

I. No Elution

This situation is highly desirable since sample preparation for counting by LSC is both simple and rapid. Sample quench is constant and simple CPM (counts per minute) mode on the LSC is preferred as external standard quench correction cannot be employed. With constant quench and therefore constant efficiency, the CPM results obtained are as accurate as DPM (disintegrations per minute) results obtained through normal 4π homogeneous counting. After collection of the sample on the filter, the filter is dried and placed in the vial. Approximately 2–3 mL of cocktail is added (ensuring that the filter is completely wet) and counting can be carried out immediately. For best counting performance using this method, it is recommended that the filter is completely dried prior to the addition of cocktail. Additionally a knowledge of the solubility characteristics of the sample will aid in the selection of the most appropriate cocktail. In general, the most applicable cocktail for dried filter counting is an organic cocktail such as Ultima Gold F or BetaPlate Scint, which provide the highest counting efficiency. Occasionally it is not practical to completely dry the filters; and in these cases a cocktail such as Ultima Gold MV should be used with the slightly damp filters for highest counting performance. The type of samples routinely counted using this method include precipitates from DNA and RNA studies, phytoplankton from sea water, algae from aquatic environments, as well as samples from enzyme activity assays, cell proliferation, and receptor binding assays.

Note: A simple method to confirm that a no elution situation exists is to decant the cocktail into another vial and recount the cocktail—absence of activity confirms that no elution has occurred and that the sample is completely bound to the filter. If necessary, accurate quantitation of the total isotope activity (i.e., DPM) can be carried out by removing the filter and using either solubilization or combustion techniques.

2. Partial Elution

As previously mentioned this situation is the least desirable due to the presence of both 2π and 4π geometry within the counting mixture. Any results from this situation will be inaccurate and cannot be reproduced. It is possible however, using one or a combination of the following methods, to convert the system from partial to an equilibrium or complete elution situation:

1. After sample preparation, shake the contents for a fixed time period and recount. Repeat this procedure until constant CPM's are obtained, (i.e., equilibrium situation).
2. Change to a cocktail in which the sample has either zero or complete solubility.
3. Extract the sample from the filter with a suitable solvent prior to adding the appropriate cocktail.

3. Complete Elution

The goal with complete elution is to convert from 2π to 4π geometry. Two slightly different sample preparation methods are employed in that either the entire filter is dissolved in an appropriate cocktail or the sample is extracted or eluted from the filter prior to the addition of cocktail. In the first instance the cocktail of choice is Filter-Count and filter types which can be dissolved by this cocktail include cellulose nitrate, mixed cellulose esters and polyvinyl chloride (PVC). Filter-Count will not dissolve cellulose acetate, glass fiber, normal paper, PTFE, nylon, or phosphocellulose filters. With cellulose acetate and glass fiber filters a transparent appearance results, while the others remain relatively unaltered. Filter-Count will not give color formation with any filter whether soluble or insoluble. The use of Filter-Count is extremely simple in that sample preparation involves adding cocktail (Filter-Count) to the filter, allowing it to dissolve (with optional heating); and then counting. Cellulose acetate is not dissolved by Soluene-350 or by Filter-Count. This filter type can be dissolved by strong hydrochloric acid, and after dissolution and dilution with water, the LSC cocktail Ultima Gold AB is recommended for trouble free counting. It is important to note that with the exception of cellulose acetate, normal paper and PET filters, virtually all other filters produce color when used with Soluene-350. In this variant of the technique, dissolving the filter overcomes the self-absorption problems, saves on drying time (accepts wet or dry filters) and provides reproducible results. In the second case, as previously described, the sample is extracted or eluted from the filter with a suitable solvent and then counted using the appropriate cocktail.

Note: It is possible to adapt this technique for alpha/beta counting of airborne particulates. Providing the correct filter type is used, the sample filter can be dissolved in Filter-Count; then Ultima Gold AB can be added (ratio of 2:1 Filter-Count:Ultima Gold AB). The benefits of such a method include the removal of self-absorption problems (especially important for alphas) and significant time saved on sample preparation (ashing and acid extraction steps are eliminated).

The introduction of the PerkinElmer TopCount® microplate scintillation and luminescence counters together with the development of various filter plates offers the ability to count labeled samples in filter plates (24 or 96 samples per plate), minimizing sample preparation steps and increasing sample throughput. A number of publications on the applicability of this assay method are available (PerkinElmer, TopCount Topics 11, 12 and 18).

A quick reference table (Table 8.21), can be used as a guide for selecting the correct LSC cocktail type for a particular filter or membrane type.

TABLE 8.21 Cocktail Selection Guide for Filter Counting

Filter type		Filter-Count	Ultima Gold F	Ultima Gold MV	Soluene-350/ Hionic-Fluor	Filter-Count/ Ultima Gold AB
Glass Fiber	Dry	✓	✓	✓		✓
	Wet	✓		✓		✓
	Dissolved					
Cellulose Nitrate	Dry		✓	✓		
	Wet			✓		
	Dissolved	✓				✓
Cellulose Acetate	Dry	✓	✓	✓		✓
	Wet	✓		✓		✓
	Dissolved					
Mixed Cellulose Esters	Dry		✓	✓		
	Wet			✓		
	Dissolved	✓				✓
PVC	Dry		✓	✓		
	Wet			✓		
	Dissolved	✓				✓
Polyacrylonitrile	Dry	✓	✓	✓		✓
	Wet	✓		✓		✓
	Dissolved					
Polycarbonate	Dry	✓	✓	✓		✓
	Wet	✓		✓		✓
	Dissolved					
Teflon	Dry	✓	✓	✓		✓
	Wet	✓		✓		✓
	Dissolved					
Nylon	Dry	✓	✓	✓		✓
	Wet	✓		✓		✓
	Dissolved					
PET	Dry	✓	✓	✓	✓	✓
	Wet	✓		✓	✓	✓
	Dissolved					
Normal paper	Dry	✓	✓	✓	✓	✓
	Wet	✓		✓	✓	✓
	Dissolved					

X. SAMPLE STABILITY TROUBLESHOOTING

Sample stability troubleshooting is the identification and resolution of problems which can occur as a result of incorrect sample preparation. These problems become apparent during counting and are sometimes not visible during the sample preparation process. These problems can be found wherever samples are mixed with cocktail, irrespective of whether the sample is solid, dissolved in an organic solvent or dissolved in an aqueous system.

Fundamentally this problem becomes apparent when one or more of the following LSC situations occur:

1. Unstable count rate (decreasing count rate).
2. Unstable count rate (increasing count rate).
3. Reduced counting efficiency (reduced more than expected).

All three above problems can usually be resolved by minor changes to sample preparation procedures, and/or by careful selection of the correct cocktail.

A. Decreasing Count Rate

Potential reasons for a decreasing count rate are two phase separation of the sample–cocktail mixture (sometimes described as "milky" appearance), precipitation of the radiolabeled sample, and chemiluminescence. When a two phase situation occurs and the activity is more soluble in the aqueous phase than the organic phase, then as the two phases separate the activity migrates with the aqueous phase and results in lower CPM's. This can be remedied by increasing the cocktail volume or changing to a cocktail more suited to the sample type. In addition, check that the cocktail is suitable for use at the operating temperature (both sample preparation and counting temperature). Precipitation of the radiolabeled sample can be overcome by either diluting the sample with water or a suitable cosolvent, or by changing to a more suitable cocktail. A chemiluminescence problem is usually characterized by higher than expected CPM's that decay with time. Chemiluminescence can be overcome by either changing to a chemiluminescence resistant cocktail such as Hionic-Fluor, Ultima Gold or equivalent cocktail, or by waiting for the chemiluminescence to decay. Additional suggestions include insuring that the correct cocktail type is in use (i.e., do not use a lipophilic cocktail with aqueous samples) using a glass vial to check that the sample/cocktail mixture is homogeneous and checking that the sample and cocktail) have been thoroughly mixed (shaken) and formed a stable microemulsion. Finally, consult the cocktail phase diagram to verify that the selected cocktail is suitable for use with the sample type.

B. Increasing Count Rate

Potential reasons for an increasing count rate are two-phase separation of the sample–cocktail mixture (sometimes described as "milky" appearance), inadequate mixing of sample and cocktail, incomplete elution of the sample from a solid support, and incomplete or slow solubilization of the sample in the cocktail. When a two phase situation occurs and the activity is more soluble in the organic phase than the aqueous phase, then as the two phases separate the activity remains in the organic phase and results in increasing CPM's due to the reduction in quench in the organic phase. This can be remedied by increasing the cocktail volume or changing to a cocktail more suited to the sample type. Inadequate mixing of the sample and cocktail, which in essence is the same as the preceding situation can be easily remedied by insuring that the mixture is thoroughly shaken. If a sample is incompletely or only slowly eluted from a solid support, increasing CPM will probably result. This can be corrected by modifying the elution conditions to get complete elution. Reagents such as water, alkali-solubilizer, acid, or other solvent may be useful in this instance. In certain cases, provided the appropriate filter material is in use, it may be possible to use a cocktail such as Filter-Count to dissolve the filter and thus separate the sample from the solid support. Where the sample is not completely soluble in the cocktail then dissolve the sample in a solvent that is more compatible with the cocktail, or add a suitable cosolvent.

C. Reduced Counting Efficiency

Potential reasons for reduced counting efficiency are either high color quench or high chemical quench in the samples. If the color quench is a result of solubilization, then bleach the sample with hydrogen peroxide, prior to adding cocktail. If the sample is naturally colored (e.g., some metallic salts), dilute with water or use a color quench-resistant cocktail (e.g., Ultima Gold family cocktail or equivalent cocktail). Usually chemical quench results from the use of an inappropriate solvent which is highly quenching and therefore the remedy may simply be to change to a less chemical quenching alternative (e.g., dichloromethane is less quenching than chloroform). The sequence of chemical quench strength is shown in Table 8.22. Other remedies include

TABLE 8.22   Strength of Chemical Quenchers

Solvent	Quench strength
Nitro groups (nitromethane)	Strongest Quencher
Sulfides (diethyl sulfide)	↓
Halides (chloroform)	↓
Amines (2-methoxyethylamine)	↓
Ketones (acetone)	↓
Aldehydes (acetaldehyde)	↓
Organic acids (acetic acid)	↓
Esters (ethyl acetate)	↓
(Water)	↓
Alcohols (ethanol)	↓
Ethers (diethyl ether)	↓
Other hydrocarbons (hexane)	Mildest Quencher

either increasing cocktail volume or decreasing sample volume. Overall if the sample preparation method cannot be changed, it may be necessary to increase the count time to obtain better statistical results.

XI. SWIPE ASSAYS

All laboratories handling radionuclides are required to conduct radiation safety surveys to maintain their license. These surveys include those necessary to evaluate external exposure to personnel, surface contamination levels; and concentrations of airborne radioactive material in the facility and in effluents from the facility. In Klein *et al.* (1992) the wisdom of relying on wipe tests for detecting removable radioactivity due to low-energy beta emitters from surfaces has been questioned. Their opinion is that "Without considering the kind and amount of radioactivity present, wipe testing is scientifically misguided and wasteful of resources and personnel...." They conclude by saying well-reviewed operating procedures and good working practice are the best protection in the work place. It is true that wipe testing may not be very efficient but it does give some measure of what is removable. For weak beta emitters, especially tritium, there is little choice but to use a wipe test and a liquid scintillation counter to detect its presence. For better or for worse, wipe testing remains an important requirement for all licensees.

A. Wipe Media and Cocktails

Various wipe media have been evaluated for their ability to gather and release radiolabeled compounds into counting solutions, and include filter paper, glass fiber filter, cotton swab, and plastic squares. Virtually any emulsifying cocktail is suitable for use in this particular area; and the only major prerequisites are that the cocktail has a high tritium counting efficiency and is able to accommodate small volumes of aqueous and alcohol–aqueous samples in a clear microemulsion. Paper filter circles are not soluble in any of the counting solutions used. Glass fiber filters, in general, became translucent when placed into the counting solution and the faint outline of the glass fiber filter can be seen on the bottom of the vial. The plastic swabs show different solubility patterns depending on the LSC cocktail used. In some cases, the blue plastic stem can dissolve into a series of blue globs that adhere to the sides and bottom of the vial. When shaken, these blobs remained immiscible. Other cocktails can extract the blue color of plastic tube. The cotton swab can appear dissolved as a clear layer of fluid on top. When shaken, the sample can turn cloudy, again depending upon the cocktail used. The thin plastic squares dissolve in most LSC cocktails but the speed of dissolution depends upon the type of cocktail: fast in classical and slow in safer. The plastic wipes dissolve rapidly in classical cocktails and appear as droplets either floating on top or laying on bottom of the vial. When shaken, the samples become cloudy. In all cases, when the samples containing the swabs or plastic squares became cloudy on shaking, the count rates did not change when recounted (remain within ±3% of the last count rate).

For quantitation of wiped radioactive contaminants by liquid scintillation counting, it is essential that all the radioactive material be in solution. The contaminants are generally classified as being either water- or organic-soluble material. Organic-soluble material will normally dissolve in the organic solvents contained in the LSC cocktail. To accommodate water-soluble material, an emulsifying cocktail is used with added water. Without water, water-soluble material such as leucine cannot be emulsified and will remain undissolved on the solid support resulting in poor recoveries.

B. Regulatory Considerations

Any laboratory using radioactivity would be classed as a restricted area. Therefore, the action level for any α–β contamination would be the presence of radioactivity exceeding 20,000 DPM/100 cm^2. Assuming that the wiping procedure picked up at least 1 percent of the surface contamination (a very conservative estimate), there would be a minimum of 200 DPM deposited on the wipe. For tritium 100% recovery means that 200 DPM would be recovered. With a counting efficiency of 50%, this translates into 100 net CPM. This amount of activity, being greater than five times background (about 20 CPM), can be detected easily. Even if this level of activity were quenched 50%, the activity of 50 net CPM is still easily detected. In a real situation dealing with removable contamination, it would be reasonable to expect that an amount greater than 1% would be picked up by wiping. With this assumption, it would appear that wiping is not an unreasonable method for detecting and following the removal of radioactive contamination from a surface in any biomedical laboratory.

C. Practical Considerations

For most isotopes it is important to have the sample completely dissolved in the cocktail, especially with weak beta emitters such as ^{3}H. Wipe medium selection is largely up to the individual, but the consensus of opinion is that glass fiber filter is the best overall performer. However, paper is probably the most popular medium because of its availability, ease of use and low cost. In one institution, waste computer paper is cut to size and used as wipes. Cotton swabs are also easy to use but recoveries are relatively low for both paper and swabs compared to glass fiber and plastic squares. The plastic squares are principally intended for psuedocumene- and xylene-based cocktails in which they are readily soluble. Safer cocktails based on linear alkyl benzene, phenylxylylethane or diisopropylnaphthalene solvents slowly dissolve the plastic wipes but form an immiscible second phase. Surprisingly, it appears that the radioactivity is released into the cocktail during this process, based upon reported recoveries. Klein *et al.* (1992), found that wetting the wipe medium greatly increased the amount of radioactive contamination removed. The problem of wetting wipes is that any solvent will increase the possibility of spreading the contamination. The solvent can cause contamination to further penetrate a surface as well as increase the probability of transfer to the hands. If paper is used, wet paper will lose its strength and tend to shred when moderate pressure, which is recommended, is applied during wiping. The contaminated surface should be exposed to solvents only during the decontamination procedure. Furthermore, it would be an added burden to control the amount of water added to each wipe before wiping a surface.

D. General Procedure for Wipe Testing

Most regulatory guides do not contain any detail on how to conduct these surveys, and the protocols are left to the licensee. However, guidelines do suggest an area of $100\,cm^2$ be monitored and the action level of contamination for weak beta emitters to be 2,000 DPM/100 cm^2 in unrestricted areas and 20,000 DPM/100 cm^2 in restricted areas. In an earlier publication, U.S.A.E.C. Regulatory Guide 1.86, the action level for removable β–γ contamination was set at 1000 DPM/100 cm^2. Before doing any wipes, it is advisable to prepare a map of the area to be monitored. The locations to be wiped should be circled and numbered to identify the wipes after they are completed. Use the same dry wipe medium for all wipes to maintain uniform monitoring conditions. Wipe an area about 100 cm^2 (approximately 4×4 inches) using a moderate amount of pressure. The wipe can be numbered or it can be placed directly into a numbered counting vial. After the wipe is placed into a vial, counting solution containing ~2% water is added. The vial is capped and vortexed a few seconds to promote elution from the wipe into the counting solution. The vial is assayed by liquid scintillation counting. One of the most useful consequences of the presence of multichannel analyzers in liquid scintillation counters is that the spectrum of any sample being counted can be visualized on-line. An inspection of the spectrum can be very helpful in identifying the nature of the sample being counted. For example, if the contamination is due to a single radionuclide, then the visualization of the expected spectrum would confirm this fact. If the sample contained two radionuclides, then the spectrum would also reflect this fact. The ability to identify the components of a mixed sample depends on the relative content of each component. The spectrum can be visualized in either the log or linear form. In general, the log mode is more helpful in detecting the presence of more than one radionuclide in the sample. Finally, any laboratory handling radioactivity should be staffed with properly trained personnel. This assumption is generally true when laboratories first begin this type of work. However, experience has shown that as time goes by, people tend to forget or neglect the safeguards designed into approved protocols. This is why wipe tests are necessary. Its purpose is to continually ensure safe working conditions within laboratories handling radionuclides. If contamination is detected, viewing the sample spectrum should be helpful in identifying the radionuclides present. Although the wipe test is qualitative at best, the detection of radioactivity significantly over background levels should be cause for further investigation.

XII. PREPARATION AND USE OF QUENCH CURVES IN LIQUID SCINTILLATION COUNTING

Basically, the liquid scintillation process is the conversion of the energy of a radioactive decay event into photons of light. Photomultiplier tubes (PMTs) are used to detect and convert the photons into electrical pulses. Both the sample and the scintillator are dissolved in an aromatic solvent, which allows energy to be transferred. Any factor, which reduces the efficiency of the energy transfer or causes the absorption of photons (light), results in quenching in the sample. There are two main types of quench: chemical quench (Gibson, 1980; Birks, 1971) and color quench (Gibson, 1980; Ross, 1965; Ten Haaf, 1975).

A. Chemical Quench

Chemical quench occurs during the transfer of energy from the solvent to the scintillator. Any chemical species that is electronegative (electron capturing) will affect the energy transfer process by capturing or stealing the pi electrons associated with the aromatic solvent and thus reduce the availability of pi electrons necessary for efficient energy transfer.

B. Color Quench

Color quench is an attenuation of the photons of light. The photons produced are absorbed or scattered by the color in the solution, resulting in reduced light output available for measurement by the PMTs. The steps in the energy transfer process affected by chemical and color quenching are indicated in Fig. 8.3.

The collective effect of quench is a reduction in the number of photons produced and therefore detected CPM (counts per minute). Counting efficiency is affected by the degree of quenching in the sample. To determine absolute sample activity in DPM (disintegrations per minute or absolute activity), it is necessary to measure the level of quench of the samples first, then make the corrections for the measured reduction in counting efficiencies.

C. Measurement of Quench

It is possible to measure quench accurately via high-resolution spectral analysis. Quenching manifests itself by a shifting of the energy spectrum toward lower energy channels in the multichannel analyzer (MCA). On Perkin-Elmer's Tri-Carb series LSA's, there are two methods of spectral analysis for measuring quench.

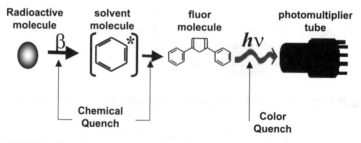

FIGURE 8.3 Quenching in the energy transfer process.

The first method is the Spectral Index of the Sample (SIS) which uses the sample isotope spectrum to monitor the quench of the solution. The SIS value decreases as quench increases, reflecting the shift of the spectrum to lower energy. The second method used to measure quench is the transformed Spectral Index of the External Standard (t-SIE), which is calculated from the Compton spectrum induced in the scintillation cocktail by an external ^{133}Ba gamma source. The source is positioned under the sample vial, causing a Compton spectrum to be produced in the cocktail solution. From a mathematical transformation of this spectrum, the t-SIE value is determined, and t-SIE is a relative value, on a scale from 0 (most quenched) to 1000 (unquenched). The calculated t-SIE value is adjusted to 1000 when the instrument is calibrated. Like SIS, t-SIE decreases as quench increases. Both SIS and t-SIE are used as quench indicating parameters (QIPs). t-SIE is independent of the sample isotope and of the activity in the vial, and has a large dynamic range. This makes it a very reproducible means of tracking the quench of the cocktail. SIS uses the sample isotope spectrum to track quench; it is most accurate with high-count rate samples. The range of SIS values reflects the energy range of the isotope. Both can be used as QIPs to create quench curves, although use of the external standard is preferred for samples containing low activity and is required for multilabeled samples.

D. Quench Curve

A quench standard curve is prepared with a series of standards in which the absolute radioactivity (DPM) per vial is constant and the amount of quench increases from vial to vial. A quench curve uses the relationship between counting efficiency and QIP to correct the measured CPM to DPM. When a quench curve is made, the DPM value in each standard is known. Each standard is counted and the CPM is measured. The counting efficiency is calculated using the following relationship:

$$\frac{CPM \times 100}{DPM} = \% \text{ Counting Efficiency}$$

At the same time, the QIP is measured for each standard. A correlation is made using the QIP on one axis (X) and the % efficiency on the other axis (Y). A curve is fitted to the standard points. Figure 8.4 shows the quench curves for ^{3}H and ^{14}C using SIS as the QIP, and Figure 8.5 shows the quench curves for the same isotopes using t-SIE as the QIP. Once the quench curve is stored in the instrument computer, it can be used for automatic DPM calculations. When unknowns are counted, the sample CPM and the QIP are measured. Using the QIP, the counting efficiency is determined from the quench curve. Sample DPM are then calculated by applying the appropriate efficiency to the CPM of the sample.

$$DPM = \frac{CPM}{\text{Efficiency (expressed as a decimal)}}$$

Sample Spectrum SIS Quench Curves

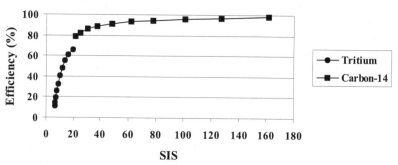

FIGURE 8.4   Quench curves for ^{3}H and ^{14}C using **SIS** as the **QIP.**

External Standard (t-SIE) Quench Curves

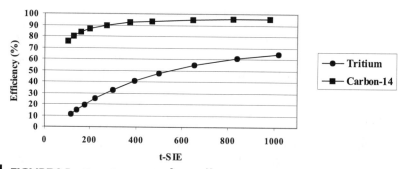

FIGURE 8.5   Quench curves for ^{3}H and ^{14}C using **t-SIE** as the **QIP.**

The standards and unknowns must be counted with the same energy regions. PerkinElmer LSCs with spectra based libraries (2500 series, 2700 series, and the new 2900 and 3100 series) store the curve in a 0-E_{max} window and allow the curve to be recalculated for the windows used in the protocol. For other LSCs (1600,1900, 2100, 2200, and 2300) the windows used to acquire the quench curve must be used in the actual DPM determination.

I. Preparation of Quench Curves

a. Method I

As mentioned, a quench standard curve is prepared with a series of standards in which the absolute radioactivity (DPM) per vial is constant and the amount of quench increases from vial to vial. The quench is increased from vial to vial by the addition of a quenching agent. A quenching agent is any chemical or color material added to the vial which causes a shift in the standard spectrum to a lower energy and a subsequent decrease in the counting efficiency of the radioactive standard. Usually a series of 6 to 10 quench standards are prepared per radionuclide. This series is sometimes referred to as the quench set.

PerkinElmer provides factory stored quench curves in the instrument for ^{3}H and ^{14}C. On occasion, it is necessary for the investigator to prepare a quench curve, for isotopes other than ^3H or ^{14}C, for example, ^{35}S. There are some basic considerations to note before preparing a quench curve:

- It is necessary to obtain a calibrated source of radioactivity to use as the source of the activity (DPM). It is essential that a known amount of activity be added per vial. Also, the standard material must be compatible with the cocktail chosen.
- A suitable quenching agent must be chosen. It is desirable to closely approximate the chemical environment in the samples. If samples contain water with various other constituents, add the same material in increasing amounts to the standards. Additional quenching agents that are most often used and available in the laboratory are carbon tetrachloride (CCl_4), acetone (CH_3CH_3CO), chloroform ($CHCl_3$), and nitromethane (CH_3NO_2).

To prepare the quench standards, perform the following steps:

1. Make a batch of radioactive solution in the chosen cocktail such that the desired DPM are transferred to each individual vial when dispensing the cocktail. Prepare the standards with a sufficient level of activity, typically 50,000–200,000 DPM per vial, in order to be able to count the standards with good statistics in a short time. If ten standards are to be made with 10 mL of cocktail per vial, then 100+ mL of radioactive cocktail solution are required. If 15 mL of cocktail is to be used then 150+ mL of radioactive cocktail is required. Note: If the unknowns to be counted contain two radio-isotopes (both ^3H and ^{14}C for example), then individual standard curves must be prepared for each isotope.
2. Count the individual standards for at least five minutes to check for constant activity (CPM). Any sample that deviates more than 2% from the mean should be discarded.
3. Add incremental amounts of the quenching agent to vials 2...n (*quenching agent is not added to vial 1*) to obtain the desired quench range. It may be necessary to predetermine the amounts to add per vial by testing various volumes of quenching agent added to the cocktail only (no radioactivity), and monitor the amount of cocktail quench using tSIE. Otherwise add the suggested amounts of nitromethane based on the information given in Table 8.23.
4. Count the complete set under the conditions described in the instrument operation manual for storing a quench curve. Practically we suggest that the standards are counted to a pre-selected level of statistical accuracy (generally 0.5%2s), and this is usually achieved within 5 minutes per sample with the sealed standards which we provide.
5. Once the quench curve(s) are counted and stored, count unknown samples using the stored quench curve(s) to determine the DPM value for each sample.

TABLE 8.23   Volume of Nitromethane Needed for Quench Curve

Quench level	Toluene standards (15 mL) (in μ L)	Ultima Gold standards (15 mL) (in μ L)
A (1)	0	0
B (2)	1	5
C (3)	5	10
D (4)	11	15
E (5)	17	26
F (6)	25	45
G (7)	35	70
H (8)	45	110
I (9)	55	150
J (10)	66	230

b. Method 2

Basically the preparation of a quench curve with any LSC cocktail is relatively straightforward and the following procedure is given as a guideline. Many researchers use their own methods and equipment and the procedure is therefore open to modification.

1. Dispense 10.0 mL or 15.0 mL of LSC cocktail into ten high performance glass vials.
2. Add activity to each vial (200,000 DPM for ^{3}H or 100,000 DPM for ^{14}C)
3. Count all ten vials to ensure that the same amount of activity is in each vial. A count time of about 5 minutes per vial will be sufficient. Any sample that deviates more than 2% from the mean should be discarded.
4. Number the vials 1–10 or A to J and add the suggested amounts of nitromethane based on the information given in Table 8.23.
5. Count the complete set under the conditions described in the instrument operation manual for storing a quench curve. PerkinElmer recommends counting each standard for 30 minutes or until a pre-selected level of statistical accuracy (using %2S terminator, generally 0.5%) is reached.

Notes:

1. For dispensing the activity use a glass barreled microliter syringe fitted with a Chaney adapter. Such an adapter ensures reproducible dispensing of activity.
2. After preparation the standards should be stored in the dark, preferably at 5°C to 10°C for best stability.

TABLE 8.24 Techniques for Reducing Color in Certain Samples

Nature of sample	Suggested remedy
Color from sample solubilization	Treat with hydrogen peroxide to bleach out the color
Plant material	Consider sample oxidation
Inorganic matrix	Change to alternate colorless anion

2. Notes on Using the Quench Curves

1. t-SIE is independent of the sample isotope and of the activity in the vial, and has a large dynamic range. This makes it a very reproducible means of tracking quench in the cocktail.
2. SIS should only be used when there is at least 500 CPM activity in the sample. Remember that SIS uses the sample isotope spectrum to track quench; it is most accurate with high-count rate samples. For an accurate SIS a good sample spectrum needs to be acquired.
3. SIS should not be used for low activity samples since an accurate sample spectrum cannot be acquired.
4. Most customers prefer to purchase quench standards. For cocktails based on toluene, xylene, pseudocumene or LAB (linear alkyl benzene) as the solvent, toluene quench standards should be used. For cocktails based on DIN (di-isopropylnaphthalene) or PXE (phenylxylylethane) as the solvent, Ultima Gold quench standards should be used. If the wrong quench standard is used there can be an error in DPM. This error is most pronounced with low energy isotopes such as tritium (see Tables 8.25–8.28).
5. Be sure that your prepared quench curve covers a wide tSIE range (i.e. 800–300) in order to provide accurate DPM results.

3. Color Quench

When a small amount of color is present in a sample there is virtually no difference between chemical and color quenching and the standard chemical quench curves are suitable. This applies to samples where the tSIE is in the range 100 to 400. However if a significant amount of color is present in the sample (tSIE is <100) it may be necessary to consider preparing a color quench curve. When preparing a color quench curve the selection of a suitable color quench agent is important. Aqueous soluble food dyes are usually a good choice as they provide both a stable color and a wide range of colors. The color of the sample must match the color used in the quench curve. Do not use a pH indicator since some cocktails contain acidic components and these will alter the color. Preparing a color quench curve is very similar to the method used for a chemical quench curve and the only difference is the quench agent. If considerable color is present in the sample it may be wise to modify or change the sample preparation method to either remove or reduce the level of color. Some techniques that have proved useful are shown in Table 8.24. Finally the easiest way to reduce color quench is to either decrease the sample size or increase the cocktail volume, or both.

4. Quench Curve Errors

The errors that can be present when the wrong quench curve is used with an LSC cocktail are shown in Tables 8.25–8.28.

Table 8.25 shows that only Ultima Gold cocktails should be used for tritium DPM measurements with the Ultima Gold tritium quench curve.

Table 8.26 shows that the Ultima Gold cocktails should not be used for tritium DPM measurements with the Toluene tritium quench curve.

Table 8.27 shows that only Ultima Gold cocktails should be used for carbon-14 DPM measurements with the Ultima Gold carbon-14 quench curve.

Table 8.28 shows that the Ultima Gold cocktails should not be used for carbon-14 DPM measurements with the Toluene carbon-14 quench curve.

5. Using a Quench Curve

Once the quench curve is stored in the instrument computer, it can be used for automatic DPM calculations. When unknowns are counted, the

TABLE 8.25 DPM Errors for Various Cocktails vs. Ultima Gold Quench Curve (Tritium)

	Ultima Gold	Toluene	Opti-Fluor	Insta-Gel Plus	Pico-Fluor 15
No quench	− 0.12%	− 1.04%	+ 6.00%	+ 2.70%	+ 4.89%
Low quench	− 0.46%	+ 4.24%	+ 7.06%	+ 5.14%	+ 6.45%
Medium quench	+ 0.04%	+ 5.87%	+ 8.43%	+ 5.82%	+ 6.91%
High quench	− 0.14%	+ 10.10%	+ 14.41%	+ 10.02%	+ 11.89%
Highest quench	− 0.20%	+ 13.42%	+ 18.01%	+ 13.36%	+ 13.43%

TABLE 8.26 DPM Errors for Various Cocktails vs. Toluene Quench Curve (Tritium)

	Ultima Gold	Toluene	Opti-Fluor	Insta-Gel Plus	Pico-Fluor 15
No quench	− 4.10%	− 0.49%	− 0.57%	− 1.58%	− 0.32%
Low quench	− 5.33%	− 0.27%	− 0.13%	− 1.22%	+ 0.23%
Medium quench	− 6.51%	+ 0.01%	+ 0.45%	− 0.19%	+ 0.60%
High quench	− 10.39%	− 0.01%	+ 1.21%	− 0.79%	− 0.49%
Highest quench	− 16.16%	− 0.70%	+ 0.11%	+ 0.56%	− 0.21%

TABLE 8.27 DPM Errors for Various Cocktails vs. Ultima Gold Quench Curve (Carbon-14)

	Ultima Gold	Toluene	Opti-Fluor	Insta-Gel Plus	Pico-Fluor 15
No quench	+ 0.06%	− 1.37%	+ 2.51%	+ 2.25%	+ 1.96%
Low quench	+ 0.03%	+ 0.12%	+ 2.04%	+ 0.78%	+ 0.80%
Medium quench	− 0.13%	+ 0.84%	+ 1.72%	+ 1.15%	+ 1.20%
High quench	+ 0.02%	+ 1.30%	+ 2.51%	+ 3.11%	+ 1.71%
Highest quench	− 0.63%	+ 4.52%	+ 3.81%	+ 3.59%	+ 2.77%

TABLE 8.28 **DPM Errors for Various Cocktails vs. Toluene Quench Curve (Carbon-14)**

	Ultima Gold	Toluene	Opti-Fluor	Insta-Gel Plus	Pico-Fluor 15
No quench	+0.45%	+0.42%	+1.71%	+1.90%	+1.96%
Low quench	+0.27%	−0.03%	+0.89%	+1.20%	+0.43%
Medium quench	−0.54%	−0.37%	+1.28%	+0.31%	+0.83%
High quench	−0.81%	+0.01%	+1.86%	+0.77%	+0.81%
Highest quench	−6.51%	+0.33%	+1.04%	+0.49%	+0.88%

External Standard (t-SIE) Quench Curve

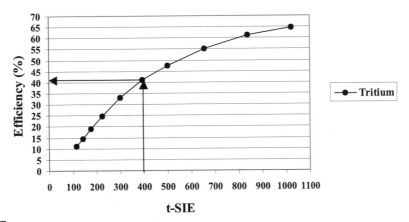

FIGURE 8.6 **Tritium Quench Curve using t-SIE as QIP.**

sample CPM and the QIP are measured. Using the QIP (SIS or t-SIE) the counting efficiency is determined from the appropriate quench curve.

For example:

1. A Tritium sample is analyzed and has

$$\text{Count rate} = 10,000\,\text{CPM}$$

$$\text{t-SIE} = 400$$

From the quench curve shown in Fig. 8.6 the instrument uses the t-SIE of 400 to determine that the counting efficiency is 42%. Since we now know the CPM and the counting efficiency it is possible to calculate the DPM i.e.

$$\begin{aligned} \text{DPM} &= \frac{\text{DPM}}{\text{Efficiency (expressed as a decimal)}} \\ &= \frac{10,000}{0.42} \\ &= 23,809\,\text{DPM} \end{aligned}$$

TABLE 8.29 Recommended Quench Curves for Various Cocktails

Cocktail	Recommended quench curve (^{3}H and ^{14}C)
Ultima Gold ™	Ultima Gold
Ultima Gold AB	Ultima Gold
Ultima Gold LLT	Ultima Gold
Ultima Gold MV	Ultima Gold
Ultima Gold XR	Ultima Gold
OptiPhase HiSafe 2	Ultima Gold
OptiPhase HiSafe 3	Ultima Gold
OptiPhase TriSafe	Ultima Gold
OptiPhase SuperMix	Ultima Gold
Ultima Gold F	Ultima Gold
OptiScint HiSafe	Ultima Gold
BetaPlate Scint	Ultima Gold
StarScint	Ultima Gold
Opti-Fluor®/Opti-Fluor O	Toluene
Emulsifier-Safe™	Toluene
Insta-Gel® Plus	Toluene
Pico-Fluor™ 15	Toluene
Pico-Fluor 40	Toluene
Insta-Fluor™	Toluene
Hionic-Fluor™	Toluene
Filter-Count™	Toluene
Carbo-Sorb® E/Permafluor® E+	Toluene
Monophase® S	Toluene
Formula 989	Toluene
Aquasol™/Aquasol II	Toluene
Aquassure®/Biofluor®	Toluene
Atomlight®	Toluene
Econofluor®-2	Toluene

A compilation of recommended quench curves for different LSC cocktails are shown in Table 8.29, and this will provide a basic guide to correct quench curve selection. For accurate DPM recovery, it is imperative that the quench curve selected is appropriate for the LSC cocktail being used. PerkinElmer manufactures quench curves using either a safer, high flash point (Ultima Gold) cocktail or a classical solvent (toluene) that are suitable for use with all Perkin-Elmer's LSC cocktails. Whenever there is doubt about the appropriateness of a quench curve researchers should always prepare their own using the same chemistry as that found in the sample.

REFERENCES

Anderson, L. E. and McClure, W. O. (1973). An improved scintillation cocktail of high solubilizing power. *Anal. Biochem.* **51**, 173–179.

Basch, R.S. (1968). An improved method for counting tritium and carbon-14 in acrylamide gels. *Anal. Biochem.* **26**, 184–188.

Benjamin, W. B. (1971). Selective in vitro methylation of rat chromatin associated histine following partial hepatectamy. *Nature (London)*, **234**, 18.

Bird, N. (1997). *Packard Counter Intelligence Applications*, Detection of helicobacter pylori infection using a 14C-urea breath test. CIA-001, PerkinElmer Life and Analytical Sciences, Boston.

Birks, J. B. (1971). Liquid scintillator solvents. *In* "Organic Scintillators and Liquid Scintillation Counting" (C. T. Peng, D. L. Horrocks and E. L. Alpen, Eds.), pp. 3–23. Academic Press, New York.

Bransome, E. D. Jr. and Grower, M. F. (1970). Liquid scintillation counting of ^{3}H and ^{14}C on solid supports: A warning. *Anal. Biochem.* **38**, 401–408.

Choules, G. L. and Zimm, B. H. (1965). An acrylamide gel soluble in scintillation fluid. *Anal. Biochem.* **13**, 336–344.

Diener, E. and Paetkau, V. H. (1972). Antigen recognition: Early surface-receptor phenomena induced by binding of a tritium labeled antigen. *Proc. Nat. Acad. Sci. U.S.* **69**, 2364–2368.

Dion, A. S. and Moore, D. H. (1972). Gel electrophoresis of reverse transcriptase of musine mammary tumor virions. *Nature (London)* **240**, 17.

Freiberg, S. (1990). Scintillation medium and method. International Patent WO/91/11735.

Fuchs, A. and De Vries, F.W. (1972). A comparison of methods for the preparation of ^{14}C-labeled plant tissues for liquid scintillation counting. *Int .J. Appl. Radiat. Isot.* **23**, 361–369.

Gibson, J. A. B. (1980). Modern techniques for measuring the quench correction in a liquid scintillation counter. *In* "Liquid Scintillation Counting, Recent Applications and Developments" (C. T. Peng, D. L. Horrocks and E. L. Alpen, Eds.), pp. 153–172. Academic Press, New York.

Gill, D. M. (1967). Licintillation counting of tritiated compounds supported by solid filters. *Int. J. Appl. Radiat. Isot.* **18**, 393 –398.

Goodman, D. and Matzura, H. (1971). An improved method of counting radioactive acrylamide gels. *Anal. Biochem.* **42**, 481–486.

Hansen, J. N., Pheiffer, P. H., and Boehnert, J. A. (1980). Chemical and electrophorestic properties of solubilizable disulfide gels. *Anal. Biochem.* **105**, 192–201.

Kaczmar, B. U. and Manet, R. (1987). Accurate determination of ^{14}CO$_2$ by expulsion from blood. *Appl. Radiat. Isot.* **38**(7), 577–578.

Kessler, M. J., Ed. (1989). "Liquid Scintillation Analysis — Science and Technology," pp. 7–1 to 7–9, Packard publication, PerkinElmer Life and Analytical Sciences, Boston.

Klein, R. C., Linins, I., and Gershey, E. L. (1992). Detecting removable surface contamination. *Health Phys.* **62**, 186–189.

Kobayashi, Y. and Maudsley, D. V. (1974a). *In* "Biological Applications of Liquid Scintillation Counting," pp. 153–167. Academic Press, New York.

Kobayashi, Y. and Maudsley, D. V. (1974b). *In* "Biological Applications of Liquid Scintillation Counting," pp. 58–68. Academic Press, New York.

Mahin, D. T. and Lofberg, R. T. (1966). A simplified method of sample preparation for determination of tritium, or sulfur-35 in blood or tissue by liquid scintillation counting. *Anal. Biochem.* **16**, 500–509.

Mirsky, J. (1990). Liquid scintillation medium with a 1,2-dicumylethane solvent. U.S. Patent 5,135,679.

Moore, P. A. (1981). Preparation of whole blood for liquid scintillation counting. *Clin. Chem.* **27**(4), 609–611.

Morrison, B. J. and Franklin, R. A. (1978). A rapid, hygienic method for the preparation of fecal samples for liquid scintillation counting. *Anal. Biochem.* **85**, 79–85.

PerkinElmer, TopCount Topics #11, Direct Counting of Millipore® MultiScreen® Filtration Plates, PerkinElmer Life and Analytical Sciences, Boston.

PerkinElmer, TopCount Topics #12, Biological Applications of Microplate Scintillation Counting. PerkinElmer Life and Analytical Sciences, Boston.

PerkinElmer, TopCount Topics #18, Counting Radioisotopic and Luminescent Labels on Filters and Membranes with the FlexiFilter® Plate. PerkinElmer Life and Analytical Sciences, Boston.

Patterson, M. S. and Greene, R. C. (1965). Measurement of low energy beta-emitters in aqueous solution counting. *Anal. Chem.* **37**, 854–857.

Peng, C. T. (1983). Sample preparation in liquid scintillation counting. *In* "Advances in Scintillation Counting," (S. A. McQuarrie, C. Ediss, and L. I. Wiebe, Eds.), pp. 279–306. University of Alberta, Alberta, Canada.

Pfeiffer K., Rank, D. and Tschurlovits, M. (1981). A method of counting ^{14}C as $CaCO_3$ in a liquid scintillator with improved precision. *Int. J. of App. Radiat. Isot.* **32**, 665–667.

Porter, N. G. (1980). A method for bleaching plant tissues prior to scintillation counting. *Lab. Pract.* **29**(1), 28–29.

Qureshi, R. M., Fritz, P., and Dsimmie, R. J. (1985). The use of CO_2 absorbers for the determination of specific ^{14}C activities. *Int. J. Appl. Radiat. Isot.* **36**(2), 165–170.

Recalcati, L. M., Basso, B., Albergoni, F. G., and Radice, M. (1982). On the determination of ^{14}C-labeled photosynthesis products by liquid scintillation counting. *Plant Sci. Lett.* **27**, 21–27.

Riffat M., Qureshi, R. M., and Fritz, P. (1985). ^{14}C dating of hydrological samples using simple procedures. *Int. J. Appl. Radiat. Isot.* **36**(10), 825.

Ross, H. H. (1965). Color quench correction in liquid scintillator systems. *Analytical Chemistry*, **37**(4), 621–623.

Schadewaldt, P., Förster, M. E. C., Münch, U., and Staib, W. (1983). A device for the liberation and determination of $^{14}CO_2$. *Anal. Biochem.* **132**, 400–404.

Sissons, C. H. (1976). Improved technique for accurate and convenient assay of biological reactions liberating $^{14}CO_2$. *Anal. Biochem.* **70**, 454–462.

Smith, I. K. and Lang, A. L. (1987). Decolorization and solubilization of plant tissue prior to determination of ^{3}H, ^{14}C and ^{35}S by liquid scintillation counting. *Anal. Biochem.* **164**, 531–536.

Späth, P. J. and Koblet, H. (1979). Properties of SDS-polyacrylamide gels highly cross-linked with N,N'-diallyltartardiamide and the rapid isolation of macromolecules from the gel matrix. *Anal. Biochem.* **93**, 275–285.

Sun, D., Wimmers, L. E., and Turgeon, R. (1988). Scintillation counting of ^{14}C-labeled soluble and insoluble compounds in plant tissue. *Anal. Biochem.* **169**, 424–428.

Ten Haaf, F. E. L. (1975). Color quenching in liquid scintillation counters. *In* "Liquid Scintillation Counting", (M. A. Crook, and P. Johnson, Eds.), Vol. 3, pp. 41–43. Heyden & Son, Ltd, London.

Terman, S. (1970). Relative effect of transcription-level and transcription-level control of protein synthesis during early development of the sea urchin. *Proc. Nat. Acad. Sci. U.S.* **65**, 985–992.

ter Wiel, J. and Hegge, Th.C. J. M. (1983). Mixture for use in the LSC analysis technique. U.S. patent 4,624,799.

ter Wiel, J. and Hegge, Th.C. J. M. (1991). Advances in scintillation cocktails. *In* "Liquid Scintillation Counting and Organic Scintillators" (H. H. Ross, J. E. Noakes, and J. D. Spalding, Eds.), pp. 51–67. Lewis Publishers, Chelsea, Michigan.

Thomson, J. (1985). Scintillation counting medium and counting method, US Patent 4,657,696, April 14th 1987.

Thomson, J. (1991). Di-isopropylnaphthalene-A new solvent for liquid scintillation counting. *In* "Liquid Scintillation Counting and Organic Scintillators" (H. H. Ross, J. E. Noakes, and J. D. Spalding, Eds.), pp. 19–34. Lewis Publishers, Chelsea, Michigan.

Thomson, J. and Burns, D. A. (1994). Packard LSC Counting Solutions, "Radio-Carbon Dioxide ($^{14}CO_2$) Trapping and Counting", CS-001. PerkinElmer Life and Analytical Sciences, Boston.

Thomson, J. and Burns, D. A. (1996a). Packard LSC Counting Solutions, "Environmental Sample Preparation for LSC," CS-004. PerkinElmer Life and Analytical Sciences, Boston.

Thomson, J. and Burns, D. A. (1996b). Packard LSC Counting Solutions, "LSC Sample Preparation by Solubilization," CS-003. PerkinElmer Life and Analytical Sciences, Boston.

Wahid, P. A., Kamalam, N. V., and Jayasree Sankar (1985). Determination of ^{32}P in wet-digested leaves by Cerenkov counting. *Int . J. Appl. Radiat. Isot.* **36**(4), 323–324.

Wang, C. H. and Jones, D. E. (1959). Liquid scintillation counting of paper chromatograms. *Biochem. Biophys. Res. Commun.* **1**, 203–205.

Wang, C. H. and Willis, D. L. (1965). "Radiotracer Methodology in Biological Science," pp. 202–206. Prentice-Hall, Englewood Cliffs, New Jersey.

Wunderley, S.W. (1986). Composition for liquid scintillation counting. U.S. Patent 4,867,905.

9

CHERENKOV COUNTING

MICHAEL F. L'ANNUNZIATA
The Montague Group, P.O. Box 5033, Oceanside, CA 92052-5033, USA

I. INTRODUCTION
II. THEORY
III. QUENCHING AND QUENCH CORRECTION
    A. Internal Standardization
    B. Sample Channels Ratio
    C. Sample Spectrum Quench Indicating Parameters
    D. External Standard Quench Correction
IV. CHERENKOV COUNTING PARAMETERS
    A. Sample Volume
    B. Counting Vials
    C. Wavelength Shifters
    D. Refractive Index
    E. Sample Physical State
V. CHERENKOV COUNTING IN THE DRY STATE
VI. RADIONUCLIDE ANALYSIS WITH SILICA AEROGELS
VII. CHERENKOV COUNTING IN MICROPLATE FORMAT
    A. Sample-to-Sample Crosstalk
    B. Sample Volume Effects
    C. Quench Correction
VIII. MULTIPLE RADIONUCLIDE ANALYSIS
    A. Sequential Cherenkov and Liquid Scintillation Analysis
    B. Cherenkov Analysis with Wavelength Shifters
IX. RADIONUCLIDE STANDARDIZATION
X. GAMMA-RAY DETECTION
XI. PARTICLE IDENTIFICATION
    A. Threshold Cherenkov Counters
    B. Ring Imaging Cherenkov (RICH) Counters
    C. Time-of-Propagation (TOP) Cherenkov Counters
XII. APPLICATIONS IN RADIONUCLIDE ANALYSIS
    A. Phosphorus-32
    B. Strontium-89 and Strontium-90 (Yttrium-90)
    C. Strontium-90(Yttrium-90) Exclusive of Strontium-89
    D. Yttrium-90
    E. Other Applications
XIII. ADVANTAGES AND DISADVANTAGES
XIV. RECOMMENDATIONS
    REFERENCES

I. INTRODUCTION

Photons of light are produced when a charged particle travels through a transparent medium at a speed greater than the speed of light in that medium. The medium may be any liquid or solid provided it is transparent. These photons are referred to as Cherenkov radiation in honor of the Russian physicist P. A. Čerenkov for his basic research on this phenomenon (see Čerenkov, 1934). In the scientific literature, three variations of the contemporary spelling of Cherenkov radiation can be found, namely, (i) Čerenkov after the Russian, (ii) Cherenkov, which provides the English pronunciation from the Russian Č, and (iii) Cerenkov. For consistency throughout this text, the author will adopt the spelling Cherenkov, which conforms with that used by *Chemical Abstracts*.

Cherenkov radiation was first observed by Marie Curie in 1910 as reported by E. Curie (1941). The radiation was researched by Mallet (1929), whose studies were not as extensive as those of Cherenkov (1934, 1937) after whose work the radiation is now known. Frank and Tamm (1937), who shared the Nobel Prize in physics with Cherenkov (see Frank, 1960), are responsible for much of the theoretical work that went into the understanding of Cherenkov light. Comprehensive treatments on the theory and properties of Cherenkov radiation are available from Jelley (1958, 1962) and Ritson (1961).

Cherenkov radiation consists of a continuous spectrum of wavelengths extending from the ultraviolet region into the visible part of the spectrum peaking at about 420 nm (see Kulcsar *et al.*, 1982; Claus *et al.*, 1987). Only a negligible amount of photon emissions is found in the infrared or microwave regions.

Cherenkov photon emission is the result of local polarization along the path of travel of the charged particle with the emission of electromagnetic radiation when the polarized molecules return to their original states (see Gruhn and Ogle, 1980). This has been described by Marshall (1952) as the electromagnetic "shock" wave that is analogous to the acoustical shock wave or sonic boom created by supersonic aircraft. The Cherenkov effect is depicted clearly by Burden and Hieftje (1998), as illustrated in Fig. 9.1, where the charged particle (e.g., electron or β particle) distorts the electron clouds of atoms in close proximity to the high-speed particle traversing a transparent medium. The Cherenkov radiation is propagated as a conical wave front, that is, the radiation is emitted as a cone in the direction of particle travel.

Cherenkov radiation, when produced at significant levels, can be employed for the efficient measurement of radioactivity. This was first demonstrated by Belcher (1953), who used a liquid-nitrogen cooled single photomultiplier to measure Cherenkov radiation intensity in terms of count rates per mCi of various radionuclides in aqueous solution. However, it was not until dual photomultiplier liquid scintillation analyzers, namely the Packard Tri-Carb 314 liquid scintillation analyzer,[1] became available commercially and in widespread use in the early 1960s did practical research

[1]Modern Tri-Carb liquid scintillation counters are now produced by PerkinElmer Life and Analytical Sciences, Boston

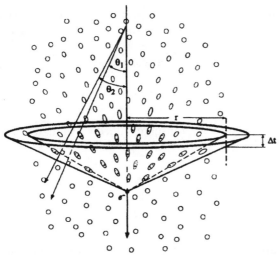

FIGURE 9.1 Depiction of the production of Cherenkov radiation in a dispersive medium and the resulting wave front expansion. The wave front spreading lengthens the excitation pulse on a time scale that is small in comparison to the fluorescence decay. Δt is the duration of the light pulse along a line parallel to the axis of the particle at a distance r from the axis. (From Burden and Hieftje, 1998, reprinted with permission Copyright 1998 American Chemical Society.)

into Cherenkov counting for the measurement of radioactivity begin. Although the production of Cherenkov radiation does not involve the scintillation phenomenon, a conventional liquid scintillation counter can detect and count Cherenkov photons emitted from a given sample in a standard counting vial. Certain radionuclides may be counted in water or other suitable transparent medium without the use of scintillation fluor or any other chemical reagents. They may even be counted in the dry state in air, albeit at a diminished counting efficiency, where the glass or plastic vial wall containing the sample acts as the transparent medium.

Cherenkov counting in an aqueous medium (i.e., without scintillation cocktail) leaves the sample in an uncontaminated state that is often suitable for any other chemical tests deemed necessary in a given research study. This technique is popular for the radioassay of relatively high-energy β particle-emitting nuclides. The threshold β particle energy for the production of Cherenkov radiation is discussed subsequently in this chapter. The ease of sample preparation, low cost (no need for fluors and fluor solvents), and the automatic counting techniques available with a liquid scintillation analyzer are among the principal reasons for the widespread use of this counting method.

II. THEORY

The threshold condition for the production of Cherenkov radiation in a transparent medium is given by

$$\beta n = 1 \tag{9.1}$$

where β is the relative phase velocity of the particle, that is, the velocity of the particle divided by the speed of light in a vacuum, and n is the refractive index of the medium (i.e., the ratio of the velocity of light in a vacuum to its velocity in the medium). Only charged particles that possess

$$\beta \geq 1/n \tag{9.2}$$

produce Cherenkov photons in transparent media (Sundaresan, 2001). The value of β of the charged particle is dependent on its kinetic energy, as reported by Jelley (1962):

$$E = m_0 c^2 \left[\frac{1}{(1 - \beta^2)^{1/2}} - 1 \right] \tag{9.3}$$

where m_0 is the particle rest mass in grams, c is the velocity of light (equal to 2.99×10^{10} cm s^{-1}), and E is the kinetic energy in ergs (where 1.602×10^{-12} ergs $= 1$ eV). The value of the rest energy for an electron is $m_0 c^2 = (9.11 \times 10^{-28}$ g$)(2.99 \times 10^{10}$ cm s$^{-1})^2 = 8.14 \times 10^{-7}$ ergs. The energy in ergs may be converted to electron volts as follows: $(8.14 \times 10^{-7}$ ergs$)/(1.602 \times 10^{-12}$ ergs eV$^{-1}) = 511$ keV.

Substituting the value of 511 keV for $m_0 c^2$ in Eq. 9.3 and solving for β gives

$$\beta = \left[1 - \left(\frac{1}{E/511 + 1} \right)^2 \right]^{1/2}. \tag{9.4}$$

Thus, the value of β, where electrons or beta particles are concerned, is dependent upon the electron energy, E, in keV according to Eq. 9.4. From Eqs. 9.2 and 9.4 the minimum or threshold energy that β particles must possess for the production of Cherenkov radiation as a function of the index of refraction of the medium is derived as

$$E_{th} = 511 \, \text{keV} \left[\left(1 - \frac{1}{n^2} \right)^{-1/2} - 1 \right] \tag{9.5}$$

When water is the transparent medium, where $n = 1.332$, the threshold energy for the production of Cherenkov photons is calculated according to Eq. 9.5 to be 263 keV (or 0.263 MeV). Thus, only β particles or electrons that possess energy in excess of 263 keV produce Cherenkov photons in water. The threshold energy for Cherenkov production will vary according to the index of refraction of the medium, and will be lower for media of higher index of refraction. The practical applications of the effect of refractive index on the threshold energy are discussed in Section IV.D of this chapter.

An important property of Cherenkov radiation which should be considered is the directional emission of the Cherenkov photons. When

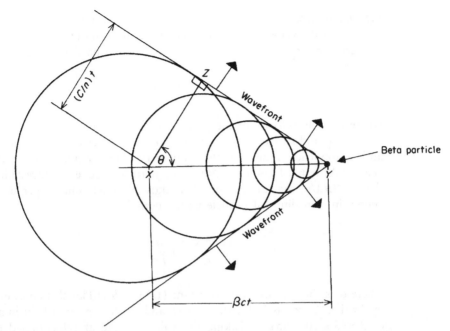

FIGURE 9.2 Huygens' construction of a conical Cherenkov wave front. The position of the charged particle at time t = 0 is x; the distance of travel of the particle from x to y in time t is βct; the distance of travel of the Cherenkov photon from x to z is (c/n)t. The angle of emission of the Cherenkov photons is represented by θ. (From L'Annunziata, 1987.)

$\beta > 1/n$, Cherenkov photons are emitted as a cone at an angle, θ, to the direction of the beta particle as depicted in Fig. 9.2 and described by Marshall (1952). The angle, θ, is defined by the equation

$$\cos \theta = \frac{(c/n)t}{\beta ct} = \frac{1}{\beta n}, \tag{9.6}$$

where β and n are as defined for Eq. 9.1 and 9.4. When $\beta = 1/n$, the threshold condition for the production of Cherenkov photons (Eq. 9.1), the Cherenkov radiation is emitted in the exact forward direction of the charged particle, that is, $\cos \theta = 1$ and $\theta = 0$.

As the beta-particle energy, E, of Eq. 9.4 increases, the value of β approaches unity and $\cos \theta$ approaches 0.7508. Consequently, the maximum angle of emission of a Cherenkov photon is 41.3° in water. Cherenkov photons are emitted at a more acute angle for electrons just above the threshold energy and cannot exceed an emission angle of 41.3°, even for the most energetic electrons in water.

In the fields of particle and nuclear physics the Cherenkov effect has been exploited over the last decade for the identification of particles based on their velocity, which is discussed briefly further on in this chapter and reviewed in detail by Va'vra (2000), Križan (2001), and Joram (2002). The term $(c/n)t$ of

Fig. 9.2 represents the distance of travel of the wave front measured by the product of the velocity of the particle in the medium (c/n) and time (t), while the term βct of Fig. 9.2 measures the distance of particle travel according to

$$\beta ct = (v/c)ct = vt \tag{9.7}$$

where v is the velocity of the particle in the medium, c the speed of light in a vacuum, and t is time. Also of relevance in studies of particle identification and radionuclide analysis is the number of photons created per unit photon energy. As noted by Križan (2001) for particles above the threshold $\beta = 1/n$, that is $\beta > 1/n$, the number of Cherenkov photons emitted per unit photon energy in a medium of particle path length L is

$$\frac{dN}{dE} = L \frac{\alpha}{\hbar c} \sin^2 \theta \tag{9.8}$$

where α is the fine structure constant (1/137), $\hbar$ is Planck's constant divided by 2π (i.e., $h/2\pi = 6.582 \times 10^{-16}$ eV s), and c is the speed of light in a vacuum or 2.998×10^{10} cm s^{-1}. Taking the values of the aforementioned constants the term $\alpha/\hbar c$ of Eq. 9.8 becomes 370 eV^{-1}cm^{-1}. From Eq. 9.8 Križan (2001) calculates, that in 1 cm of water, a particle track where $\beta = 1$ (most energetic particle possible) emits $N = 320$ photons in the spectral range of visible light ($\Delta E \approx 2$ eV), and with an average detection efficiency of $\varepsilon = 0.1$ over the spectral interval, only $N = 32$ photons would be measured. The detection efficiency for Cherenkov photons should not be confused with the term counting efficiency of each particle that interacts with the medium, which is employed in radionuclide analysis, as it will be seen in this chapter that Cherenkov counting efficiencies can reach 85% or 0.85 (see Table 9.4). It is important to note that the amount of energy loss by Cherenkov radiation does not add a significant amount to the total energy loss in the medium suffered by the charged particle (Grichine, 2001; Sundaresan, 2001).

The number of photons emitted per path length of electron travel in the wavelength interval between λ_1 and λ_2 is calculated by Sowerby (1971) according to

$$\frac{dN}{dl} = 2\pi \alpha z^2 \left(\frac{1}{\lambda_2} - \frac{1}{\lambda_1} \right) \sin^2 \theta \tag{9.9}$$

where α is the fine structure constant ($e^2/\hbar c = 1/137$), z is the particle charge ($z = 1$ for electrons or beta particles), and the refractive index of the medium and particle velocity appear in the $\sin^2 \theta$ term. Over the visible range of wavelengths from $\lambda_1 = 400$ to $\lambda_2 = 700$ nm, Sundaresan (2001) estimated the number of photons N per path length L to be

$$N/L = 490 \sin^2 \theta \text{ cm}^{-1} \tag{9.10}$$

The number of Cherenkov photons emitted per electron in the spectral region of 3000–6000 Å was calculated by Sowerby (1971) assuming a linear decrease in electron energy to zero according to the range of electrons in the medium.

The Cherenkov effect is the result of a physical disturbance caused by the high-energy charged particle along its path of travel resulting in a directional anisotropic emission of light. Therefore, there is no chemical fluorescence nor the relatively long excitation decay times associated with fluorescence. Burden and Hieftje (1998) calculated the width of a typical Cherenkov-generated pulse or the duration of the photon flash, Δt illustrated in Fig. 9.1, which is a function of the spread of the Cherenkov wave front and the position of observation with respect to the particle trajectory. They approximated the duration of a light pulse observed at a distance r parallel to the particle path, as illustrated in Fig. 9.1, according to

$$\Delta t = \frac{r}{\beta c} \left(\sqrt{\beta^2 \, n^2(\lambda_2) - 1} - \sqrt{\beta^2 \, n^2(\lambda_1) - 1} \right) \tag{9.11}$$

where λ_2 and λ_1 define the range of Cherenkov photon wavelengths. For example, taking a 1-MeV electron traversing water and detecting the light pulse at a distance $r = 1$ cm parallel to the particle path of travel between the wavelengths of 300 and 350 nm Burden and Hieftje (1998) calculated a light pulse duration of 326 fs. The Cherenkov light duration is approximately a million-fold shorter that the nanosecond decay times of liquid scintillation fluorescence photons described in Chapter 5.

The directional emission of Cherenkov photons is a disadvantage when conventional liquid scintillation spectrometers are used for counting. The photocathodes of most liquid scintillation counters consist of two photomultiplier tubes (PMTs) positioned at 180° relative to each other (see Fig. 5.3 of Chapter 5). This is not an optimum arrangement for the detection of Cherenkov photons; however, reflector material on the surface of the counting chamber walls facilitates the detection of Cherenkov photons, which otherwise would not reach the PMTs in coincidence. Thus, when conventional liquid scintillation counters are employed for Cherenkov counting, the counting efficiencies are inferior to the theoretical maximum efficiencies. For optimum counter response Ross (1969) suggested that four photocathodes be operated in coincident pairs, each pair located at 180° to the other.

Extensive treatments of the origin and interpretation of Cherenkov radiation are given by Marshall (1952), Jelley (1962), Gruhn and Ogle (1980), and Kulcsar et al. (1982). Practical reviews on the application of Cherenkov counting to the measurement of radionuclides are available from L'Annunziata (1979, 1984a, 1987, 1997, and 1998), Takiue et al. (1993, 1996), and Al-Masri (1996), and a comprehensive theoretical and practical treatment of Cherenkov radiation and its application to radionuclide measurement is available in a book by Grau Carles and Grau Malonda (1996).

III. QUENCHING AND QUENCH CORRECTION

Chemical quenching in Cherenkov counting is non-existent because Cherenkov photons arise from a physical disturbance of the molecules of the medium in contrast to chemical fluorescence, which occurs in the liquid scintillation phenomenon. However, color quenching can be very significant in Cherenkov counting, greater than that which may occur in standard liquid scintillation counting techniques, as reported by Elrick and Parker (1968), because Cherenkov photons are weaker and cover a wider spectrum of wavelengths than those produced in scintillation. The counting efficiency in Cherenkov radioassay is determined by one of the quench correction techniques subsequently discussed.

A. Internal Standardization

This method is described in Chapter 5, Section V.A. It may be highly accurate, but it is time consuming. An added disadvantage of this quench correction technique in Cherenkov counting is possible contamination of the sample by the addition of the internal standard, that is, addition of the standard could render the sample unsuitable for subsequent chemical analysis. One of the great advantages of Cherenkov counting is the radioactivity analysis of the sample in its original state (e.g., aqueous solution) without the need to add any reagents.

B. Sample Channels Ratio

This method is described briefly in Chapter 5, Section V.B and in greater detail in previous texts by the author (see L'Annunziata, 1979, 1984a, and 1987). The sample channels ratio (SCR) technique was a most popular method of monitoring for quench with the older-generation liquid scintillation counters equipped with a single channel analyzer for data storage and pulse height gain control. Contemporary liquid scintillation analyzers are equipped with a multichannel analyzer (MCA), which permit the facile measurement of a sample spectrum quench indicating parameter such as SIS or SQP(I) described in Chapter 5. Nevertheless, the SCR technique is accurate and is still used occasionally by some researchers. Some examples of its applications to the Cherenkov counting analysis of particular radionuclides will be cited here.

The sample channels ratio technique has been applied to color quench correction in the Cherenkov counting of several radionuclides. Stubbs and Jackson (1967) prepared a series of color-quenched standards in liquid scintillation counting vials each containing ^{32}P of the same activity (1.85 kBq) and increasing amounts of Scarlet R or Naphthol Yellow S dye as color quencher. The quantity of dye per vial ranged from 0.01 to 3.00 mg. Channel A of the liquid scintillation counter was set successively to give channel widths from between 50–100 and 50–600, where a wide-open channel of 50–1000 was chosen for Channel B. For each width selected for Channel A, the channels ratio of A to B was plotted against detection

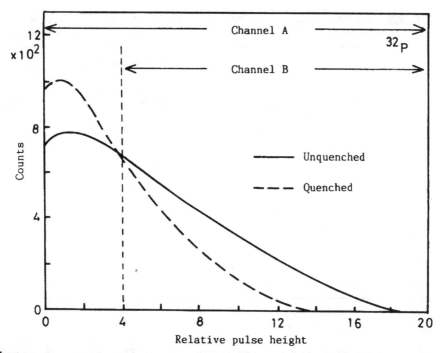

FIGURE 9.3 ^{32}P Cherenkov pulse height distribution and channel settings for the sample channels ratio (SCR) method. The pulse height distribution shifts toward low pulse height with color quenching (From Fujii and Takiue, 1988a, reprinted with permission from Elsevier Science.)

(counting) efficiency. After testing various widths for Channel A, the optimum width was selected as that which gave the most linear plot over the greatest range of detection efficiency. The plot of counting efficiency versus sample channels ratio for a selected channel width of channels A and B is used as the quench correction curve. The counting efficiency of any unknown sample is then obtained from its particular SCR value and the quench correction curve.

A good example of the SCR technique applied to color quench correction in the Cherenkov counting of ^{32}P and ^{36}Cl can be taken from the work of Fujii and Takiue (1988a). The pulse height discriminators of Channel A were selected to encompass all pulse events, and the discriminator settings for Channel B were selected to encompass only a part of the pulse height spectrum for the least quenched sample as illustrated in Fig. 9.3. When quench occurs, the ratio of counts in Channel B (Fig. 9.3) to counts in Channel A (B : A) will change as pulse heights shift from right to left (higher to lower magnitude) along the pulse height spectrum. Therefore, counting a series of quenched standards containing variable color quench levels similar to those described in Table 9.1 will produce a sample channels ratio quench correction curve of percent counting efficiency versus SCR. An example of such a curve is illustrated in Fig. 9.4 as reported by Fujii and Takiue (1988a) from counting channels illustrated in Fig. 9.3. In practice, the SCR value of an unknown sample will give the percent counting

TABLE 9.1 Constituents of ^{32}P Nonquenched (Vial 1) and Quenched (Vials 2 to 10) Standards in Water for the Preparation of a Cherenkov Counting Quench Correction Curve[a]

Vial number	Vial constituents (mL)		
	H$_2$O (mL)	Quenching agent (0.01% CrO$_3$)	Standard [^{32}P] phosphate (mL)
1	14.0	0.0	1.0
2	13.5	0.5	1.0
3	13.0	1.0	1.0
4	12.0	2.0	1.0
5	11.0	3.0	1.0
6	10.0	4.0	1.0
7	9.0	5.0	1.0
8	8.0	6.0	1.0
9	6.0	8.0 .	1.0
10	4.0	10.0	1.0

[a]From L'Annunziata (1987).

efficiency from the quench correction curve. The activity of the sample in DPM is then obtained by dividing the count rate from Channel A (the wide-open counting region) by the decimal value of the percent counting efficiency.

The sample channels ratio technique has been employed successfully for counting efficiency determinations in the Cherenkov counting of ^{42}K in biological samples by Moir (1971) and of ^{34m}Cl, ^{36}Cl, and ^{38}Cl as well as ^{32}P by Wiebe et al. (1980), who used a 1% solution of potassium dichromate in water as the color quenching agent for the preparation of standard quench correction curves. The sample channels ratio technique was also used for the determinations of ^{36}Cl and ^{32}P in tissue solubilizer solutions (see Bem et al., 1983). The author has employed the above quench correction technique successfully for the assay of ^{32}P in water using 0.01% CrO$_3$ as a quenching agent (see L'Annunziata, 1984a, 1987). Increasing amounts of CrO$_3$ solution, which is yellow in color, are added to a series of liquid scintillation vials, each containing the same activity of ^{32}P standard (e.g., 3.7 kBq mL^{-1}) as described in Table 9.1. Water is added to bring the quenched standards to a uniform and optimum volume. Also 0.01% aqueous K$_2$Cr$_2$O$_7$, which is yellow in color, serves as a good quenching agent for the preparation of quenched standards (L'Annunziata, 1987).

Although a yellow color is used in the above procedure, Stubbs and Jackson (1967) and Bem et al. (1980) report identical SCR quench correction curves when either a yellow or a red dye is employed as the quenching agent. The sample channels ratio technique for quench correction has been employed successfully by Frič and Palovčikova (1975) and Karamanos et al. (1975) for the assay of ^{32}P and ^{210}Pb, Carmon and Dyer (1987) for the Cherenkov assay of ^{106}Ru, Takiue et al. (1993, 1996) for the Cherenkov measurements of ^{36}Cl, ^{32}P, and ^{90}Sr(^{90}Y), and Vaca et al. (1998) and

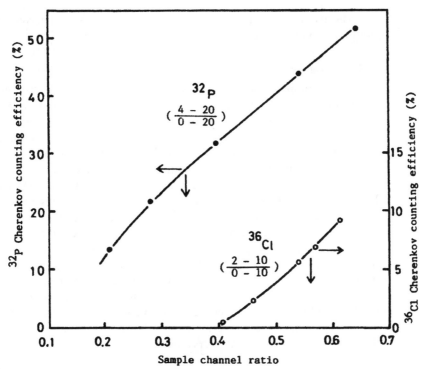

FIGURE 9.4 Color quench correction curves for the sample channels ratio method in the Cherenkov measurement of ^{32}P and ^{36}Cl. (From Fujii and Takiue, 1988a, reprinted with permission from Elsevier Science.)

Tarancon *et al.* (2002) for the Cherenkov efficiency calibration of ^{90}Y colored samples.

C. Sample Spectrum Quench Indicating Parameters

Modern liquid scintillation analyzers are equipped with a multichannel analyzer, which permits the measurement of a sample spectrum quench-indicating parameter (QIP) such as SIS or SQP(I) described in Chapter 5. The use of a sample spectrum quench-indicating parameter to correct for color quench in Cherenkov counting offers several advantages over the previously described SCR quench correction technique, among which are the following: (1) the time consuming procedure of testing several channel region settings to produce the most linear SCR quench correction curve is avoided, (2) possible sources of error arising from differing errors associated with determining the different count rates in two counting regions required to produce a sample channels ratio are avoided, and (3) computer automation in the measurement of the sample spectrum and the QIP associated with the sample spectrum results in ease and preparation and error reduction in generating the plot of the quench correction curve and in the automated reading of the quench correction curves to determine counting efficiencies of unknown samples.

The use of a sample spectrum QIP requires first a determination of an optimum counting region, which is described subsequently.

I. Counting Region

The counting region must first be defined by observing the pulse height spectrum of the Cherenkov photons with the computer monitor display from the multichannel analyzer of the liquid scintillation analyzer. A colorless sample of the radionuclide of interest is placed into a plastic liquid scintillation vial and the sample pulse height spectrum is observed as illustrated in Fig. 9.5. Plastic counting vials are recommended for highest Cherenkov photon detection efficiency and lower backgrounds (see Section IV.B of this chapter). A colorless sample is selected, as this would represent the least quenched sample achievable. The pulse heights originating from Cherenkov radiation in water do not generally exceed the equivalent of ~50 keV for pulse height spectra calibrated on an energy scale. This may differ for other instruments that do not use such a scale; however, the limits of the pulse height spectra can be determined easily on the computer monitor output from the MCA. For this particular type of instrument with pulse heights calibrated in keV energy equivalents, the counting region can be defined by setting the lower level (LL) discriminators and upper level (UL) discriminators at 0–30 keV, respectively.

2. Quench Correction

Although there is no chemical quench in Cherenkov counting, color quench correction is necessary if samples are colored even to a slight extent. In many applications of biology, biochemistry, environmental monitoring,

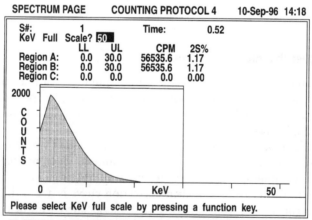

FIGURE 9.5 Pulse height spectrum produced by Cherenkov photons from a sample of ^{32}P in water as seen on the computer monitor of a PerkinElmer Tri-Carb 2300TR liquid scintillation analyzer. A counting region of 0–30 keV was selected to encompass all pulse heights originating from the ^{32}P radionuclides in the sample. (From L'Annunziata, 1997, reprinted with permission from PerkinElmer Life and Analytical Sciences.)

and agricultural research, among other fields, the radionuclide of interest [e.g., ^{32}P, ^{89}Sr, ^{90}Sr(^{90}Y)] must be analyzed in animal or plant tissue. The tissue may be ashed in a muffle furnace to destroy the organic matter and to put the radionuclide in soluble form. Then the ash may be dissolved in dilute acid. This may leave the sample colorless or at times with only a slight yellow color. Other applications may find the radionuclide already in solution, and it may be desirable not to alter the sample chemically so that the sample may be counted directly. Also, this would permit other studies to be carried out on unaltered samples. If samples are colored to any extent and decolorization is not performed, a color quench correction curve will be needed.

A color quench correction curve is prepared from a series of color-quenched standards of the nuclide of interest. The quenched standards are prepared by adding varying amounts of aqueous coloring agent, such as 0.01% CrO_3 or 0.01% $K_2Cr_2O_7$, to a series of scintillation counting vials as illustrated in Table 9.1. An exact known activity of the radionuclide of interest is added to each vial and the sample volumes are brought up the optimum volume or the volume to be used for the experimental samples. It is important to keep in mind that the volume of the quenched standards, the type of counting vials used (e.g., plastic or glass), and the counting region selected for the quench correction curve should be the same as those used for the analysis of unknown samples.

The quench standards are counted in the optimum region, such as the region of 0–30 keV found in Section III.C.1, and the count rates (CPM) and quench indicating parameter of each of the color-quenched standards are recorded. From the count rates the counting efficiency of each color-quenched standard is calculated from the known disintegration rates of each standard as follows

$$E_{std} = \frac{CPM_{std}}{DPM_{std}} \tag{9.12}$$

where E_{std} is the counting efficiency of the quenched standard, CPM_{std} is the count rate of the standard obtained in the previously defined counting region, and DPM_{std} is the known disintegration rate of the quenched standard. Each calculated value of E_{std} is plotted against the pulse height spectral QIP (e.g., SIS, or SQP(I)) as illustrated in Fig. 9.6. The quench correction curve illustrated in Fig. 9.6 is a plot of the counting efficiency versus the Spectral Index of the Sample (SIS) for the radionuclide ^{86}Rb. It was prepared from seven quenched standards in seven counting vials containing a known and exact activity of ^{86}Rb and increasing concentrations of CrO_3 over the range of 0–0.05% as reported by L'Annunziata and coworkers (see Noor et al., 1996). This CrO_3 concentration range fits within the range of CrO_3 concentrations of the standards described in Table 9.1.

The SIS is derived from the mean pulse height of the spectrum. The magnitude of the SIS provides a measure of the photon intensities escaping the counting vials. This serves as an excellent QIP for Cherenkov counting. Via multichannel analysis the liquid scintillation analyzer measures the SIS

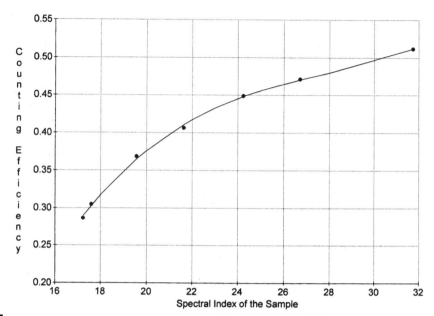

FIGURE 9.6   Color quench correction curve for the Cherenkov counting efficiency determination of ^{86}Rb in water using the Spectral Index of the Sample (SIS) as the quench-indicating parameter. Polyethylene plastic vials were used and the volumes of standards were 15 mL. [From L'Annunziata and coworkers (see Noor et al., 1996), reprinted with permission from Elsevier Science.]

for each quench standard and calculates the counting efficiency of each standard according to Eq. 9.12. The quench correction curve is stored in the memory of the LSA computer, and it is used for the automatic determination of counting efficiencies of all unknown experimental samples from the value of the SIS of each sample. Therefore, in practice, when an experimental sample of unknown activity is counted, its count rate (CPM) is measured in the specified counting region and the SIS of the sample is determined by the LSA. The instrument computer program will take the value of the SIS and extract the sample counting efficiency from the stored color quench correction curve of counting efficiency versus SIS. The instrument will then calculate and print out the activity of the experimental sample according to the equation

$$DPM_S = \frac{CPM_S}{E} \tag{9.13}$$

where DPM_S is the activity in disintegrations per minute of the radionuclide sample, CPM_S is the Cherenkov count rate of the radionuclide sample in the counting region selected (e.g., 0–30 keV), and E is the Cherenkov counting efficiency of the radionuclide sample obtained from the quench correction curve.

Quantitative activity recoveries are obtained when using SIS as the quench indicating parameter as tested by the author for ^{32}P sample activities above approximately 1700 DPM providing Cherenkov count rates of approximately 850 CPM (see L'Annunziata, 1997). However, when count rates are below 1000 CPM, error can occur in the measurement of the SIS, because at low activity levels a long counting time is required to measure the sample spectrum quench indicating parameter. Samples should be counted for sufficient time to reach a % 2 sigma (% 2S) of 0.5% standard deviation of the count rate in order to get a good measurement of the sample spectrum upon which the value of the QIP depends. Consequently, when count rates are <1000 CPM, conventional liquid scintillation analysis using tSIE as a quench indicating parameter may be preferred over Cherenkov counting.

D. External Standard Quench Correction

Ross (1980), Smith (1981), and Takiue *et al.* (1996) reported that the use of external standards is not applicable generally to quench correction in Cherenkov counting, because the external standards commonly supplied with liquid scintillation counters (e.g., ^{133}Ba, ^{137}Cs, and ^{152}Eu) produce low-energy Compton electrons, which are too far below the threshold energy for the production of Cherenkov radiation. However, commercial liquid scintillation analyzers of LKB Rack Beta Instruments are equipped with a ^{226}Ra external standard, which emits considerable quantities of high-energy gamma radiation to produce appreciable numbers of Compton electrons with energies above the Cherenkov threshold of 263 keV in water. A quench-indicating parameter based on the ^{226}Ra external standard, known as the G-number (G#), was patented by Grau Carles and Grau Malonda (1992) to permit the liquid scintillation quench correction of samples under extreme levels of quench where Cherenkov counting in liquid scintillation analysis would predominate. The writer envisages that the G# could be applied also to color quench correction when the Cherenkov effect is the only source of photons. The G# is described in Section V.C.6 of Chapter 5 and by Grau Carles *et al.* (1993a).

IV. CHERENKOV COUNTING PARAMETERS

The detection efficiencies in Cherenkov counting may be optimized by controlling several counting parameters in addition to the optimum counting region settings previously described. These parameters, which are easily controlled, are discussed subsequently.

A. Sample Volume

The detection efficiency of Cherenkov photons can be dependent significantly on the sample volume. The optimum volume may differ from one instrument to another and even from one nuclide to another measured with the same instrument as well as the type of vial used (e.g., glass or plastic). For example, Clausen (1968) reported an optimum volume of 10–11 mL for the Cherenkov counting of ^{32}P with the liquid scintillation counter used. This volume dependence was also tested by Wiebe *et al.* (1980) for the Cherenkov counting of ^{34m}Cl, ^{36}Cl, ^{38}Cl, and ^{32}P and, for the same instrument, the volume dependence was greatest for ^{38}Cl. The volume in this case also affects the pulse-height spectral shape. This effect was explained by Wiebe *et al.* (1980) on the basis of the extended range (up to 2.5 cm in water) of the high-energy beta particles emitted from ^{38}Cl ($E_{max} = 4.91$ MeV), which would, for the larger volumes (>4 mL), give rise to more pulses at greater pulse heights. The extended range in water provided by the larger sample volumes also results in an increase in the number of photons produced owing to the longer particle path length of travel. This consequently results in an increase in the probability of producing a detectable pulse. Several other works may be cited where volume effects on the Cherenkov counting efficiency of various nuclides were studied including the following: ^{214}Bi and ^{214}Pb by Al-Masri and Blackburn (1995a); ^{119m}Cd by Ramesh and Subramanian (1997); ^{32}P by L'Annunziata (1997), BenZikri (2000), and L'Annunziata and Passo (2002); ^{86}Rb by L'Annunziata and coworkers (see Noor *et al.*, 1996); and ^{90}Y by Coursey *et al.* (1993) and L'Annunziata and Passo (2002).

Volume effects are clearly discernible between different vial types, such as glass and plastic vials, as illustrated in Fig. 9.7. The figure also illustrates that the effect of volume on counting efficiency is more significant in glass then plastic vials. This was also observed in studies by Vaca *et al.* (1998) and L'Annunziata and Passo (2002) comparing plastic and glass vials in the Cherenkov counting of ^{90}Y. The effects of vial type are discussed in the next section of this chapter. It is important therefore that each instrument, vial type, and radionuclide to be analyzed by Cherenkov counting be tested for sample volume effects. These tests can be performed as follows:

1. Place a small volume (e.g., 1.0 mL) of a known activity (DPM) of the aqueous solution of the radionuclide of interest into a 20-mL capacity counting vial.
2. Determine the count rate (CPM) of the sample in the appropriate counting region (e.g., 0–30 keV), and calculate the percent counting efficiency (%E) according to the equation

$$\%E = \frac{CPM}{DPM}(100) \qquad (9.14)$$

3. Add an additional 1.0 mL of pure water to the sample in the counting vial to increase the total volume to 2.0 mL.
4. Recount the sample in the same counting region and calculate again the new counting efficiency according to Eq. 9.12.
5. Repeat steps 3 and 4 until a total sample volume of 20 mL is reached.
6. The %E versus sample volume is then plotted.

As illustrated in Fig. 9.7 sample volume can have a significant effect on sample count rate and detection efficiency. The effect of sample volume on counting efficiency is greater in glass counting vials, higher counting efficiencies are observed with plastic vials, and only very small changes in counting efficiency (~1%) occurred in polyethylene plastic vials when

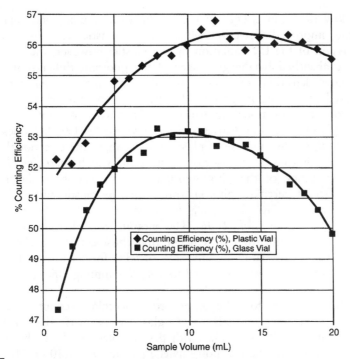

FIGURE 9.7   Effect of sample volume on the Cherenkov counting efficiency of ^{32}P in water with polyethylene plastic and glass counting vials measured with a PerkinElmer Tri-Carb 2300TR liquid scintillation analyzer. (From L'Annunziata, 1997, reprinted with permission from PrekinElmer Life and Analytical Sciences.)

sample volumes varied from 9 to 20 mL for the particular vials, radionuclide, and instrument used for these tests. The effect of sample volume on detection efficiency is due to the directional characteristics of Cherenkov radiation. Because Cherenkov photons are emitted at specific angles to the direction of travel of the β particle, changes in volume can affect the angles of deflection by the reflector material in the surrounding counting chamber walls *vis-à-vis* the face of the photomultiplier tubes. However, plastic counting vials disperse the radiation, reducing its directional properties, whereby the effect of sample volume on detection efficiency is further reduced. The advantages of plastic counting vials in Cherenkov counting are explained further in Section IV.B of this chapter.

Obviously, if reproducible counting efficiencies are to be maintained, it is necessary that the optimum counting volume be determined experimentally for the particular instrument, vial type, and nuclide employed and that this volume be used for every sample.

B. Counting Vials

It is well known that plastic counting vials produce higher Cherenkov counting efficiencies than glass vials. Because a significant fraction of

Cherenkov radiation is emitted in the ultraviolet region, one might assume that the improved detection efficiency in plastic vials is due to the transmission of the UV radiation by the plastic; however, the UV radiation could be absorbed by the glass face of the photomultiplier tube. Increased Cherenkov detection efficiency provided by plastic counting vials is considered to be due in part to the dispersive or scattering effects of the plastic on the directional (anisotropic) Cherenkov photons according to Kellogg (1983) and Grau Carles and Grau Malonda (1996). As described by Takiue et al. (1996), the use of plastic vials changes the directional nature of Cherenkov photons to an isotropic emission, which increases the photon capture efficiency of the photomultiplier tubes. The pulse height spectra resulting from Cherenkov photons transmitted by plastic and glass counting vials containing ^{86}Rb in water were compared by L'Annunziata and coworkers (see Noor et al., 1996). The photons emitted from the plastic vials were of higher number and pulse height than those emitted by glass vials. The optimum Cherenkov counting efficiency of 53% for ^{86}Rb was obtained with polyethylene plastic vials compared to a 43% counting efficiency when glass vials were used.

Another factor for the improved Cherenkov counting efficiencies obtained with plastic vials over glass vials was demonstrated by L'Annunziata and Passo (2002) with dry samples of ^{90}Y to be due to the higher index of refraction of polyethylene plastics ($n = 1.50-1.54$) compared to that of borosilicate glass ($n = 1.468-1.487$). The higher index of refraction of the plastic yields a lower beta particle-threshold energy for the production of Cherenkov photons when beta particles pass through the walls of the counting vial. The threshold energies for the production of Cherenkov photons in polyethylene plastic and borosilicate glass are calculated according to Eq. 9.5 to be 167 and 183 keV, respectively.

Ross (1980) and Kellogg (1983) reported that polyethylene vials give significantly higher counting efficiencies and lower backgrounds than glass vials. Furthermore, polystyrene vials can offer an improved counting efficiency, which may be due to a weak scintillation effect in the polystyrene. In addition to improved counting efficiency, plastic counting vials produce lower backgrounds than glass vials as demonstrated in the Cherenkov counting of ^{90}Y by Passo and Cook (1994), Vaca et al. (1998), and L'Annunziata and Passo (2002), Cherenkov counting for ^{32}P by L'Annunziata (1997), and Cherenkov counting of ^{188}Re by Kushita and Du (1998). Therefore, plastic counting vials provide higher Cherenkov counting figures of merit. The term figure of merit (FOM) is used to optimize sample counting conditions. It is calculated as

$$\text{FOM} = \frac{E^2}{B}, \tag{9.15}$$

where E is the counting efficiency expressed as a percent and B is the background count rate in counts per minute (CPM). By increasing counting efficiency and/or reducing background, we improve the FOM and the

TABLE 9.2 ^{32}P Cherenkov Counting Performance[a]

Vial type	Count mode	Counting efficiency (%)	Background[b] (CPM)	Figure of merit (E^2/B)
Plastic[c]	NCM[d]	56.7	16.6	193.7
Glass	NCM	50.2	29.1	86.6

[a]From L'Annunziata (1997) reprinted with permission from PerkinElmer Life and Analytical Sciences.

[b]Background measurements were made with triplicate samples of pure water in the counting region of 0–30 keV (LL-UL) and counted for 200 minutes each with a %2 sigma (%2S) standard deviation of 3.0% with a PerkinElmer Tri-Carb 2300TR liquid scintillation analyzer.

[c]Polyethylene plastic.

[d]NCM represents normal count mode where high-sensitivity count mode for low-level counting is not activated.

TABLE 9.3 ^{90}Sr(^{90}Y) Cherenkov Counting Performance in a Wide-Open Counting Region of 0–30 keV and in an Optimized Counting Region Providing Optimized FOMs[a]

Vial type	Count mode	% Efficiency	Background (CPM)	Figure-of-merit (E^2/B)
0–30 keV Window				
Plastic	NCM	71.7	15.8	325
Plastic	LLCM	67.2	11.1	407
Glass	NCM	62.7	24.0	164
Glass	LLCM	53.8	8.0	362
Optimized Window				
Plastic	NCM	68.8	13.1	361
Plastic	LLCM	65.6	9.7	444
Glass	NCM	61.5	21.2	178
Glass	LLCM	49.8	6.6	376

[a]From Passo and Cook (1994) reprinted with permission from PerkinElmer Life and Analytical Sciences.

sensitivity of the instrument to measure the radioactivity of the sample. The counting efficiencies, backgrounds, and calculated FOMs for polyethylene plastic and low-potassium glass vials in the Cherenkov counting of ^{32}P in water are listed in Table 9.2, and similar data for the Cherenkov counting of ^{90}Sr(^{90}Y) in water are provided in Table 9.3. The FOM of the Cherenkov counting of ^{32}P is more than doubled on shifting from glass to polyethylene plastic counting vials, and the FOM for the Cherenkov counting of ^{90}Sr(^{90}Y) is improved by almost threefold when shifting glass vials in normal count mode (NCM) to polyethylene plastic vials in low-level count mode (LLCM). This is the case for either a wide-open counting region of 0–30 keV according to data provided in Table 9.3 or an optimized counting region (i.e., a narrower region providing optimized FOMs), as demonstrated by data in Table 9.3. The low-level count mode listed in Table 9.3 refers to a patented pulse discrimination (TR-LSC) method of reducing background described in Chapters 5 and 6.

In the thorough studies made on the accuracy and reliability of Cherenkov counting, Takiue *et al.* (1993, 1996) demonstrate a further advantage of plastic over glass counting vials. They used methyl red and bromocresol green as color quenching agents to study the accuracy of quench correction curves based on the SCR or SIS as quench-indicating parameter for the Cherenkov counting of $^{90}Sr(^{90}Y)$, ^{32}P, and ^{36}Cl in glass and plastic vials for both turbid and clear samples. Turbid samples were prepared by adding small amounts of milk to the aqueous counting medium. The quench correction curves for turbid and clear solutions of the radionuclides were different in glass vials, where the turbid solutions provided higher counting efficiencies as illustrated in Fig. 9.8. This is probably a result of the dispersion of the directional Cherenkov radiation by the turbidity. However, Takiue *et al.* (1996) demonstrate that plastic vials give identical quench correction curves for either turbid or clear solutions. All values of percent counting efficiency and quench-indicating parameter fall directly on the same quench correction curve for both turbid and clear solution when plastic counting vials are used, as illustrated in Fig. 9.8.

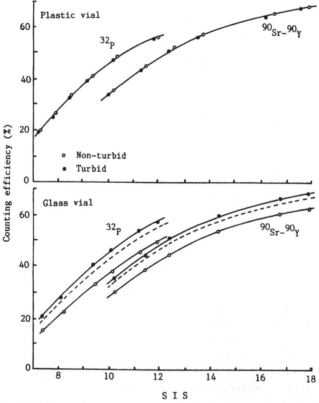

FIGURE 9.8 Color quench correction curves for the Cherenkov measurement of ^{32}P and $^{90}Sr(^{90}Y)$ for clear and turbid samples in plastic and glass vials using SIS as the quench-indicating parameter. (From Takiue et al., 1996, reprinted with permission of Elsevier Science.)

Cherenkov counting with aqueous samples in plastic vials offers an added advantage because samples may be stored in these vials indefinitely. Relatively short-lived nuclides (e.g., ^{32}P, $t_{1/2} = 14.3$ days) can be stored as aqueous solutions in the plastic vials until they decay to background levels, permitting sample disposal as simple water (see Kellogg, 1983).

From the data and examples described above, it is clear that polyethylene plastic vials give the highest Cherenkov counting efficiencies and most accurate results when compared with glass counting vials. Glass vials can give higher counting efficiencies if wavelength shifters are used, as discussed in the following.

C. Wavelength Shifters

The ultraviolet component of Cherenkov radiation, which may pass plastic vials, can be absorbed partially by the glass of the photomultiplier tube. Certain compounds, such as the sodium and potassium salts of 7-amino-1,3-naphthalenedisulphonic acid (ANDA) in $100 \, \text{mg} \, \text{L}^{-1}$ concentration, are successfully employed as wavelength shifters by absorbing the ultraviolet photons and reemitting them isotropically in the visible region, as demonstrated by Parker and Elrick (1966, 1970) and Elrick and Parker (1968). Thus, the directional character of Cherenkov photons and the absorption of their ultraviolet wavelengths are reduced by the emission of the radiation in all directions at wavelengths more efficiently detected by photomultiplier tubes.

Lauchli (1969, 1971) reported that weak beta particle-emitters such as ^{36}Cl require 5 mM ANDA, and strong beta particle-emitters such as ^{86}Rb and ^{40}K require 2.5 mM ANDA for optimum counting efficiencies. ANDA was demonstrated to be useful for the routine counting of ^{36}Cl, ^{40}K, and ^{86}Rb in plant materials, providing counting efficiencies of 13, 79, and 61%, respectively, for samples containing 1 g of digested plant tissue, and 23, 93, and 70%, respectively, for samples not containing digested plant tissue. Only a small increase in Cherenkov counting efficiency of ^{188}Re from 53 to 55% in glass vials with the addition of $100 \, \text{mg} \, \text{L}^{-1}$ of ANDA was demonstrated by Kushita and Du (1998); however, the counting efficiency slightly dropped from 58 to 54% in plastic vials with the addition of ANDA wavelength shifter. This effect is unusual and unexplained. The writer can only note that ^{188}Re emits higher-energy beta particles [i.e., three beta transitions of 2120 keV (79% intensity), 1970 keV (20% intensity), and ~ 1900 keV (1% intensity)] than all other radionuclides studied with wavelength shifters, and possibly a greater fraction of the Cherenkov photons are created in the visible rather than the ultraviolet.

In most cases significant increases in Cherenkov counting efficiencies are achievable with wavelength shifters, which could be advantageous if the sample is not needed for further studies or chemical analysis, for which the wavelength shifter may act as a contaminant. ANDA may also be employed as a wavelength shifter in the Cherenkov counting of ^{59}Fe as reported by Kannan (1975).

Frič and Finocchiaro (1975) found anthranilic acid and quinine to be effective as wavelength shifters for the Cherenkov counting of ^{32}P. Concentrations of 10^{-3} M anthranilic acid and 2×10^{-3} M quinine increased the counting efficiency of ^{32}P from 47.2 to 58.4% and from 47.2 to 54%, respectively. Bezaguet et al. (1979) reported an increase in the light output by a factor of about 3.8 when ANDA is employed as a wavelength shifter at a concentration of $10 \, \text{mg} \, \text{L}^{-1}$.

Many chemical compounds have been tested for their wavelength shifting properties. Some are sensitive to pH, salt concentration, and storage in the sample solution for periods of several hours, a few days, or weeks. van Ginkel (1980) tested the stability and wavelength-shifting properties of several organic compounds, and some of their properties are listed in Table 9.4. The pH sensitivity of wavelength shifters has been studied in detail

TABLE 9.4 **Effect of Different Wavelength Shifters, Dissolved in Distilled Water, on the Counting of Cherenkov Radiation from ^{32}P, using Polyethylene Counting Vials, Vial Constituents of 10 mL, an Amplifier gain of 30%, and Discriminator Settings of 20–1000**[a]

Compound	Concentration used (g L^{-1})	Counting efficiency (%)
Water in glass vial		51.0
Water in quartz vial		53.0
Water in polyethylene vial		55.5
β-Methylumbelliferone[a]	0.2	67.0
Esculin[b]	0.4	73.0
Quinine sulfate[c]	1.0	68.4
Quinine[c]	1.0	66.2
	10.0	76.0
	25.0	77.0
Thymine	0.5	65.9
Alloxazin	0.1	55.1
Pyrimidine-HCl	5.0	56.4
Quinolinic acid	0.5	53.3
Sodium dihydrobenzoate	1.0	66.3
Pentobarbitol	1.0	57.3
Phenobarbitol	1.0	57.0
Lumichrome	0.001	57.2
Sodium salicylate[d]	0.5	66.8
	1.0[e]	68.0
	100[e]	76.0
	1000[e]	85.0

[a]From van Ginkel (1980) reprinted with permission from Elsevier Science.
[b]Stable in the pH range 3–11.
[c]Can only be used at pH 0–3.
[d]Highly sensitive to pH change and salt concentration.
[e]Refractive index is increased.

by Kellogg (1983), and it was demonstrated that the use of wavelength shifters is not appropriate when the sample pH can vary to a significant degree. Sample pH can vary the fluorescent properties of the wavelength-shifting molecule and consequently cause the counting efficiency to vary.

Also Ross (1971), Paredes *et al.* (1980), and Bem *et al.* (1983) indicated that care should be exercised when wavelength shifters are used because some shifters show chemical decomposition over several hours. This can lead to color quenching and a reduction in counting efficiency.

PPO has excellent wavelength shifting properties in Cherenkov counting. However, it cannot be incorporated directly into water. Takiue *et al.* (1984) devised a method for introducing PPO into water in the form of a micelle by first preparing a PPO–ethanol solution. A 2-mL sample of PPO–ethanol is added to 13 mL of aqueous solution of ^{32}P for Cherenkov counting. The counting efficiency for ^{32}P was 68.4% (optimum), which was 1.6 times as high as that measured with conventional Cherenkov counting (42.3%). The optimum concentration of PPO in the counting vial was reported to be $0.02\,g\,L^{-1}$. Takiue *et al.* (1996) increased the Cherenkov counting efficiency of ^{32}P in water from 50 to 65% using 4-methylumbelliferone as a wavelength shifter while demonstrating that turbid and clear samples of ^{32}P in glass scintillation vials give the same quench correction curve when wavelength shifter is used.

Although the use of wavelength shifters can significantly increase the counting efficiency, it must be kept in mind that the detection process is no longer a purely physical one, and thus, chemical quenching becomes possible.

A further disadvantage is that the counting sample may be rendered useless for subsequent chemical tests because the wavelength shifter may act as a contaminant. Some aromatic wavelength shifters and solvents of high refractive index act as scintillants and, for this reason, Cherenkov counting with these compounds as solvents is at times referred to as Cherenkov–scintillation counting. An example is methyl salicylate, which has a high refractive index ($n = 1.5369$); according to Eq. 9.5, the calculated beta particle energy threshold for Cherenkov production is reduced to 162 keV.

Wiebe *et al.* (1978) reported counting efficiencies of 50.3 and 14.4% for ^{18}F and ^{14}C in methyl salicylate in the coincidence counting mode. Higher counting efficiencies were obtained in the singles counting mode. This mode of counting is discussed later in this chapter. The sample channels ratio method was employed to determine counting efficiencies, and the standard curves (%E vs. SCR) were prepared using chlorobenzene or dimethyl sulphoxide as a quenching agent. In the case of ^{18}F measurement, the methyl salicylate acts as an efficient Cherenkov–scintillation medium, because both Cherenkov and scintillation phenomena occur. Positrons are emitted from ^{18}F ($E_{max} = 636$ keV), well over the beta particle-threshold energy for Cherenkov production. However, the beta particles emitted from ^{14}C ($E_{max} = 155$ keV) do not reach the threshold energy for Cherenkov production in methyl salicylate, and consequently this solvent only serves as an inefficient scintillation flour for ^{14}C.

To keep the advantages of using a wavelength shifter while still counting under pure Cherenkov conditions free of chemical quenching, Ross (1976) and Takiue and Ishikawa (1978) fabricated a counting vial containing two chambers, which separate the sample from the wavelength-shifting solution (see also L'Annunziata, 1984a, 1987). The vial consists of an outer chamber used to contain the wavelength-shifting solution, and an inner chamber of approximately 10-mL volume contains the radionuclide and solvent (e.g., water). The wavelength-shifting solution is sealed permanently in the outer chamber by an all-glass seal. The entire inner vial is made of quartz so that the ultraviolet radiation can pass the inner chamber wall before reaching the solution of wavelength shifter. The vials are reusable, and to prevent contamination the sample can be contained in a polyethylene bag placed within the inner chamber. Ross (1976) reported an increase in counting efficiency of ^{89}Sr by a factor of 1.88 when dimethyl-POPOP is used in the wavelength-shifting chamber as compared to pure water. Greater improvements in counting efficiency would be expected for lower energy beta particle-emitters such as ^{36}Cl. The main disadvantage of this technique is that the volume of sample that may be conveniently counted is reduced by a factor of two. This method of Cherenkov counting has not had many recent applications.

Another approach to the application of wavelength shifters external to the sample solution is the application of a thin film or coat of wavelength shifter onto the faces of the photomultiplier tubes devised by Grande and Moss (1983). Pure quartz and plastic vials are transparent to the ultraviolet fraction of Cherenkov photons. However, this ultraviolet radiation can be absorbed in part by the glass faces of the photomultiplier tubes. This absorption is obviated by the application of a coating of wavelength shifter to the glass photomultiplier tube face. The coating designed by Grande and Moss (1983) consists of 5.7% p-terphenyl and 3.1% bis-MSB in a polyethyl methacrylate matrix. The transparent coating has an average quantum efficiency of 91% in the wavelength range of 200–400 nm.

D. Refractive Index

From Eq. 9.5 it is evident that the energy threshold for the production of Cherenkov radiation by electrons (beta particles) is a function of the refractive index. An increase in the refractive index of the medium should lower the energy threshold and increase the detection efficiency for a beta particle-emitting radionuclide. The detection efficiency should increase to some extent for all beta-emitting nuclides with an E_{max} above the threshold energy in pure water, because the beta particle-energies of all beta particle-emitting nuclides constitute a broad spectrum between zero and E_{max}.

Ross (1969) found that the Cherenkov effect and detection efficiencies are significantly increased only for radionuclides that have a low beta particle E_{max}, as illustrated in Table 9.5. This may be explained by the fact that a smaller fraction of the total beta particle-energy spectra of high-energy beta particle-emitting nuclides is below 263 keV. However, the use of solvents of high refractive index as Cherenkov counting media for low-energy beta particle-emitters such as ^{99}Tc may not always be a practical approach, since very high counting efficiencies are obtained by liquid scintillation techniques.

TABLE 9.5 Cherenkov Response as a Function of Solvent Refractive Index[a]

Nuclide	E_{max} (MeV)	Detection efficiency (%)			
		$n = 1.3220$[b]	$n = 1.4026$[c]	$n = 1.4353$[d]	$n = 1.4644$[e]
^{99}Tc	0.295	0.01	–	–	1.02
^{204}Tl	0.765	16.1	19.6	21.4	23.2
^{32}P	1.710	50.2	50.9	51.3	51.6

[a]From Ross (1969), reprinted with permission Copyright American Chemical Society.
[b]Water.
[c]53 wt% glycerol.
[d]75 wt% glycerol.
[e]95 wt% glycerol.

Wiebe and Ediss (1976) reported very high Cherenkov counting efficiencies for ^{36}Cl using a high-refractive index methyl salicylate ($n = 1.5369$) as the counting medium. Counting efficiencies of 82.4 and 91.6% were reported, depending on the liquid scintillation spectrometer employed. The high counting efficiencies were reported to be due not only to the increased refractive index (thus lowered Cherenkov threshold energy) but also to the wavelength-shifting and scintillation properties of this solvent, as discussed in Section IV.C of this chapter.

The high refractive indexes of concentrated solutions of sodium salicylate in water (100–1000 g L^{-1}) provided significant increases in counting efficiency of ^{32}P (see Table 9.4). However, these solutions may not render themselves as practical for the routine measurement of radionuclides because of the high concentrations of salicylate required.

E. Sample Physical State

The production and detection of Cherenkov radiation is not restricted to liquid samples. Rather, Cherenkov radiation may be produced with dry and solid samples in plastic or glass counting vials. This is discussed in the following Section V of this chapter.

V. CHERENKOV COUNTING IN THE DRY STATE

The Cherenkov counting of radionuclides in the dry state can under certain circumstances be more advantageous such as in the analysis of biochemicals on filter material, electrophoresis gels, thin-layer chromatograms, and paper swipes taken for radioactivity monitoring. In some circumstances, particularly with high-energy beta emitters, such as ^{90}Y, the Cherenkov counting in the dry state can yield figures of merit comparable to Cherenkov counting in aqueous solution.

The first report of the Cherenkov counting of a radionuclide in the dry state was provided by Hülsen and Prenzel (1968), who analyzed the amounts of ^{32}P tracer in green algae by counting the Cherenkov photons. The dry algae was collected on glass fiber filter and deposited into glass or plastic counting vials. They reported a counting efficiency of 13% for ^{32}P in the dry state in glass vials and a twofold increase in counting efficiency when plastic vials were used. More modern instrumentation yield higher counting efficiencies. Berger (1984) reported a 25% counting efficiency of Cherenkov radiation from ^{32}P in the dry state compared to a 56 per cent counting efficiency with the nuclide in the aqueous state. The method of Berger (1984) involved measuring ^{32}P on dry glass filters placed flat at the bottom of air-filled glass scintillation vials. He demonstrated the analysis of dry samples of ^{32}P in recombinant DNA procedures where enzyme-catalyzed reactions were carried out in volumes of $10\,\mu$L or less. In this case the ^{32}P samples were deposited onto glass fiber filters, dried under heat lamps, and then inserted in glass counting vials yielding a counting efficiency of $25\pm1\%$ using a conventional LSA. The Cherenkov radiation is produced in the glass wall of the scintillation vial. It can also be produced in polyethylene vials with a slightly higher counting efficiency. The air in the vial makes no detectable contribution to the production of Cherenkov radiation.

Bunnenberg *et al.* (1987) describes the Cherenkov counting of radionuclides on aerosol filters in the dry state. The filters of 50-mm diameter are sealed in polyethylene foil to avoid contamination and loss of activity. The filters are rolled and inserted into the borehole of a Plexiglas solid with the shape of a normal 20-mL scintillation vial. The Cherenkov photon emissions are measured with a conventional liquid scintillation analyzer. Cherenkov counting efficiencies of 44.2 and 65.8% were reported for the measurement ^{89}Sr and ^{90}Sr(^{90}Y) in the Plexiglas vials.

Counting efficiencies of 10 and 14% are reported for the Cherenkov counting of ^{32}P in the dry state in the sample wells of a 24-well and 96-well OptiPlates, which are white microplates used with the PerkinElmer TopCount microplate scintillation and luminescence counter (see Anonymous, 1996). Counting efficiencies are increased with the addition of water to the sample in the microplate wells; however, the possibility of analyzing ^{32}P samples in the dry state without any additive, even water, is a good option to have available. More information on counting samples in a microplate format is provided in Section XI of Chapter 5, Section V.B of Chapter 11 and in Section V of this chapter.

There are numerous examples of the Cherenkov counting of ^{32}P in the dry state and a few will be cited including the following: lyophilized 5'-nucleotide monophosphates (Lehmann and Bass, 1999), precipitated covalent histone–DNA complexes (Angelov *et al.*, 2000), dry samples on Whatman phosphocellulose paper Luciani *et al.*, 2000), dry polyacrylamide gels (Bélanger *et al.*, 2002; Gittens *et al.*, 2002), and dry P81 paper associated with protein kinase activity studies (Moore *et al.*, 2002).

A linear logarithmic relationship between Cherenkov counting efficiencies and average energies of the emitted beta particles and internal conversion electrons was demonstrated by Morita-Murase *et al.* (2000). For this study they used standards of ^{32}P, ^{36}Cl, ^{60}Co, and ^{137}Cs in the dry state, and the counting efficiencies reported for the particular dry sample geometries were $38.8\pm0.1\%$, $5.2\pm0.03\%$, $0.69\pm0.02\%$, and $4.98\pm0.13\%$, respectively. Based on these

TABLE 9.6 Cherenkov Counting of ^{90}Sr(^{90}Y) in the Dry State and in Water[a]

Medium	Volume[b] (mL)	^{90}Y Counting efficiency (%)		Background (CPM)		FOM[c]	
		Glass[d]	Plastic[e]	Glass[d]	Plastic[e]	Glass[d]	Plastic[e]
Air[f]	20	51.3	66.9	22.2	20.4	119	219
Water	20	66.2	73.4	21.4	16.1	205	334

[a]From L'Annunziata and Passo (2002) reprinted with permission from Elsevier Science.
[b]Volume of the medium (air volume for dry samples or water volume).
[c]Figure of merit calculated as the percent counting efficiency squared divided by the background count rate (CPM), i.e. E^2/B.
[d]Borosilicate glass vials.
[e]Polyethylene plastic vials (1-mm thickness).
[f]The medium for air-dried samples. Samples are air-dried at the bottom of the counting vials.

results and the correlation that exists between Cherenkov counting efficiency and logarithm of the γ-ray energy, they estimated that the Cherenkov counting efficiencies of ^{214}Pb and ^{214}Bi due to γ-rays would be 0.003 ± 0.0006 and $0.09 \pm 0.003\%$, respectively.

Much attention is currently focused on the applications of ^{90}Y in the form of radiopharmaceuticals for medical research and cancer treatment (e.g., Campbell et al., 2000, 2001; Pandeya et al., 2001; Abbas Rizvia et al., 2002; Chimura et al., 2002) as well as the concern for ^{90}Sr(^{90}Y) in the environment (Scarpita et al., 1999). Cherenkov counting has proven to be among the most facile and inexpensive methods of analysis for ^{90}Y in aqueous solutions at relatively high detection efficiencies of 39–68% (Coursey et al., 1993). The Cherenkov counting of ^{90}Y in the dry state was proposed as a practical alternative and tested by L'Annunziata and Passo (2002). The detection efficiency for ^{90}Y in the dry state in polyethylene plastic vials was slightly higher that that obtained for ^{90}Y in 20-mL of water in glass vials (Table 9.6). Also, the Cherenkov counting backgrounds are lower in the plastic vials than those obtained with glass counting vials. The counting efficiencies of ^{90}Y in Table 9.6 were measured with standards of ^{90}Sr(^{90}Y) in secular equilibrium, and the ^{90}Sr contribution to the Cherenkov counting efficiencies can be ignored since the maximum Cherenkov counting efficiency for ^{90}Sr does not exceed 1% as reported by Rucker (1991), Chang et al. (1996), and Cook et al. (1998).

Higher Cherenkov counting efficiencies are obtained with plastic counting vials over glass vials as explained in Section IV.B of this chapter, and this remains regardless of whether or not the samples are counted in the dry state or in solution. The higher index of refraction of polyethylene plastic ($n = 1.50$–1.54) over borosilicate glass ($n = 1.468$–1.487) yields a lower beta particle-threshold energy (167 keV) for the production of Cherenkov photons in plastic as compared to the energy threshold in glass (183 keV). See Section IV.B. In the dry state, air is the only medium in addition to the vial walls for the production of Cherenkov photons. Air has a low index of refraction ($n = 1.00027712$) at the Sodium D line at STP (Lide, 2001). According to Eq. 9.5 electrons would have to exceed the threshold energy of

2.12×10^4 keV or 0.0212 GeV for the production of Cherenkov photons in air. Consequently, no Cherenkov photons are produced by beta particles in air, and only the counting vial walls are the source of Cherenkov photons from sample beta particle-emissions when the samples are in the dry state. Gases as media for the production of Cherenkov photons are limited to the measurement of high-energy particles in the MeV and GeV regions. For example, Iodice *et al.* (1998) calculated the threshold energy for the production of Cherenkov photons by electrons or positrons in CO_2 at STP ($n = 1.00041$) to be 0.017 GeV.

When counting radionuclide samples in the dry state it is expected that increasing the wall thickness of the plastic would increase the Cherenkov counting efficiency (L'Annunziata and Passo, 2002). Increasing the path length of travel of the beta particles in the plastic should increase the number of Cherenkov photons according to Eq. 9.8 or, according to Sundaresan (2001), over the visible range of Cherenkov wavelengths (400–700 nm) the number of Cherenkov photons produced per path length of travel can be aproximated according to Eq. 9.10 described previously in this chapter. The Cherenkov counting efficiencies for ^{90}Y in the dry state in polyethylene plastic vials listed in Table 9.6 were produced in vials of 1-mm thickness. Thicker-walled (2 mm) plastic counting vials are available commercially (MaxiVial, PerkinElmer Life and Analytical Sciences, Boston, MA). The counting efficiencies listed in Table 9.6 for air-dried samples leave the samples at the bottom of the counting vial. Because of the counting geometry effects described previously in Section IV.A for samples of low volume *vis-à-vis* the photomultiplier tubes centers, increased Cherenkov counting efficiencies are expected if the dried samples (e.g. dried aqueous samples, filter or swipe material) are elevated on a 1-cm thick polyethylene plastic disk of a suitable diameter placed on the bottom of the counting vial. The elevated sample should reduce the geometry effect as well as increase the amount of plastic medium in the vial to enhance the production of Cherenkov photons.

Caution is recommended when counting samples in the dry state because of self-absorption that will occur to varying degrees depending on the sample thickness and density. Self-absorption, that is, the absorption of beta particle radiation by the sample, can reduce significantly the detection efficiency. Fortunately, the lower-energy beta-particles are the first to be absorbed by the sample depending on sample thickness and density. If the beta particle-threshold energy for Cherenkov photon production in polyethylene plastic is 167 keV, the sample thickness and density should not be capable of absorbing beta particles of energy in excess of 167 keV or even reduce significantly the energy of beta particles escaping from the dry sample to energies below the threshold energy of 167 keV. Several empirical formulae, such as those provided in Eqs. 1.26–1.28 of Chapter 1, are available for calculating the average range for electrons or beta particles in matter. These may be applied to calculate the average range of beta particles from samples in the dry state. If the sample thickness is such that self-absorption may occur at any appreciable extent or be suspect, an internal radionuclide standard could be used to determine the degree to which self-absorption occurs. Such an internal standard is first applied to a dissolved sample, which is then dried before counting. The effect of the amount of dry residue on the Cherenkov counting efficiency of the standard can be measured by varying the amounts of residue that would occur after sample drying. If, on the other hand, Cherenkov counting is carried out on dry filter material placed within counting vials, the effect of the filter material on self-absorption can be tested by adding radionuclide standard solution to the filter material. The filter material is then dried and counted. Cherenkov detection efficiencies for standards counted with and without the presence of filter material will provide a quantitative measure of self-absorption.

VI. RADIONUCLIDE ANALYSIS WITH SILICA AEROGELS

One of the principal advantages of analyzing radionuclides by counting Cherenkov photons is that no interference will arise from radionuclides in the sample with decay emissions below the threshold energy ($E_{th} = 263$ keV in water) for the production of Cherenkov photons, such as ^{3}H ($E_{max} = 18.6$ keV), ^{14}C ($E_{max} = 156$ keV), and ^{35}S ($E_{max} = 167$ keV). However interfering radionuclides, such as ^{60}Co ($E_{max} = 315$ keV), ^{89}Sr ($E_{max} = 1490$ keV), and ^{137}Cs (two beta branches: $E_{max} = 514$ keV at 94% intensity and $E_{max} = 1180$ keV at 6% intensity), which have beta-spectral maximal energies above the Cherenkov threshold energy are present in samples sometimes encountered in the environment particularly where fission-product contamination is of concern. For the Cherenkov analysis of ^{90}Sr(^{90}Y) in the environment, Brajnik et al. (1994, 1995) and Pestotnik et al. (2002) demonstrated the discrimination against the interfering radionuclides by the use of low-refractive index silica aerogel counting medium.

Silica aerogel is a highly porous and transparent solid material. It has an index of refraction that varies according to its density, which can be controlled during its manufacture over the range of $n = 1.01$–1.06. Higher indexes of refraction up to 1.20 are feasible. Some reviews on the production, properties, and applications of silica aerogels are given by Cantin et al. (1974), Brinker and Sherer (1999), Gougas et al. (1999), and Ishino et al. (2001). An excellent history of silica aerogel discovery and production together with aerogel characteristics and applications are related by Fricke and Tillotson (1997). The first aerogels were produced by Kistler (1931), who reasoned that gels consisted of a suspension of independent solid and liquid phases whereby, if the liquid phase were removed from the gel quickly in a non-disturbing manner, the solid phase would remain behind as a porous material with the same shape as the original gel suspension. Modern silica aerogels were later produced by Nicolaon and Teichner (1968), who devised the use of organosilanes for the preparation of aerogels, which has developed into the modern sol–gel method.

The production of silica aerogels is described briefly by Cantin et al. (1974), Fricke and Tillotson (1997), and Ishino et al. (2001). Detailed procedures for small-scale production in the laboratory are reported by Adachi et al. (1995) and in a patent held by Matsushita Electric Works, Ltd. (1992). In brief silica aerogels are produced firstly with the formation of silica alcogel $\{SiO_2\}_n$ by the polycondensation of orthosilicic acid, $nSi(OH)_4$, formed via the hydrolysis of tetramethoxysilane (TMOS), $Si(OCH_3)_4$, in methanol solution in the presence of ammonia as catalyst according to the following:

$$nSi(OCH_3)_4 + 4nH_2O \xrightarrow{NH_3} nSi(OH)_4 + 4nCH_3OH \qquad (9.16)$$

$$nSi(OH))_4 \rightarrow \{SiO_2\}_n + 2nH_2O \qquad (9.17)$$

The concentrations of ammonia and solvent (e.g. ethanol) used in the polymerization are varied to adjust to the desired density and refractive index of the final product. For example, silica aerogel refractive indices are reported to range from 1.0275–1.0324 for aerogel densities that vary from 0.0985–0.1140 $g\,cm^{-3}$ (Ishino *et al.*, 2001). The final product of silica aerogel is obtained by drying at a supercritical point to remove the ethanol solvent and transform the alcogel into silica aerogel with creating cracks in the porous substance or volume shrinkage. Aerogels exhibit a crushing strength of about 10 $kg\,cm^{-2}$ and are prepared in various sizes to accommodate particular instrumental configurations (e.g., 10–30 cm in diameter and 2–4 cm thickness) for radionuclide analysis or studies of relativistic particles in high energy physics. The development of aerogels detectors with light guides containing wavelength shifter between the aerogel and phototubes to increase light collection has been studied by Barnykov *et al.* (1998). Also silica aerogels have been prepared doped with fluors to increase the efficiency of detection of Cherenkov photons (Bockhorst *et al.*, 1995). Silica aerogels are available commercially from several suppliers, among which are the following: Matsushita Electric Works, Ltd., Osaka, Japan; Airglass AB, Staffanstorp, Sweden; NanoPore Inc., Albuquerque, NM, USA; Cabot Corp., Boston, MA, USA or Cabot GmbH, Frankfurt/Hoechst, Germany; and Boreskov Institute of Catalysis, Novosibirsk, Russia.

Brajnik *et al.* (1994, 1995) and Pestotnik *et al.* (2002) selected a silica aerogel with an index of refraction of $n = 1.055$, which would yield a threshold energy of 1092 keV (1.09 MeV) according to Eq. 9.5 to create a detector system for the analysis of $^{90}Sr(^{90}Y)$ in the environment via the exclusive Cherenkov counting of the daughter ^{90}Y ($E_{max} = 2280$ keV). Other beta-emitting radionuclides such as ^{137}Cs, ^{60}Co, and ^{89}Sr, which are normally encountered in the environment would not cause any significant interference, as their beta-energy spectral end points (E_{max}) are well under the aerogel Cherenkov threshold energy with the exception of ^{89}Sr. In the case of ^{89}Sr, only a small fraction of its beta particles extend beyond 1092 keV, and covering the aerogel detector entrance and sides with aluminum foil further discriminated against any interference of this nature while, at the same time, improve the light collection efficiency. They placed a multiwire proportional chamber (MWPC) between the radionuclide source and silica aerogel detector. The beta source is placed directly on the MWPC, and in coincidence mode the background count rates were reported to be as low as 2 counts per hour or 0.033 CPM yielding detection limits as low as 0.1 Bq for $^{90}Sr(^{90}Y)$. The activity of ^{90}Sr is easily determined from the ^{90}Y activity, as both radionuclides would have the same activity if in secular equilibrium at the

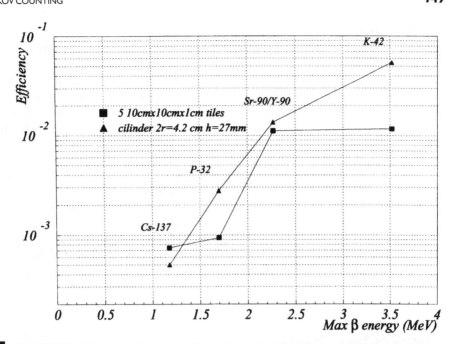

FIGURE 9.9 Efficiencies (count rate divided by activity) for different point sources as a function of β-spectrum end-point energy for two different radiator cases: five 1-cm thick 10×10 cm² tiles (squares) and 27-mm thick cylindrical aerogel with aluminized side walls (triangles). (From Pestotnik *et al.*, 2002, © 2002 IEEE.)

time of analysis, or by calculation, from the time interval that the ^{90}Y was separated from its ^{90}Sr parent. Detection efficiencies of two silica aerogel detector geometries are illustrated in Fig. 9.9 as a function of radionuclide beta particle energy maxima. The low backgrounds achieved demonstrate the potential of this silica aerogel Cherenkov detector design as a radiation monitor for ^{90}Sr(^{90}Y) in the environment.

VII. CHERENKOV COUNTING IN MICROPLATE FORMAT

The measurement of radionuclides in samples in a microplate format is described in detail in Section XI of Chapter 5 (Liquid Scintillation Analysis: Principles and Practice) and Section V.B of Chapter 11 (Solid Scintillation Analysis). These sections describe in detail the design and operation of multiple detector systems for the simultaneous analysis of up to 12 samples in a 24-, 96-, or 384-sample well microplate. In addition to the multiple simultaneous analyses of samples by liquid and solid scintillation as described in Chapters 5 and 11, the microplate scintillation and luminescence counter can be used to analyze Cherenkov radiation from high-energy β particle-emitting radionuclides such as ^{32}P.

The sample wells of microplates have a considerably smaller sample capacity than liquid scintillation counting vials. Consequently, the Cherenkov counting efficiencies and sample count rates achievable are lower with samples in the microplate format. Nevertheless, the high sample throughput available with microplate counters with a 12-phototube assembly allowing the simultaneous counting of up to 12 samples (see Chapters 5 and 11) make Cherenkov counting of samples in the microplate format an attractive option. Because of the reduced Cherenkov detection efficiencies in microplate sample wells, it is recommended that Cherenkov counting with multiple-detector microplate counters be limited to radionuclides with β particle energies of $E_{max} > 1 \, MeV$.

The radionuclide ^{32}P with an $E_{max} = 1710 \, keV$ provides suitable Cherenkov counting efficiencies in the limited sample volumes of 1.5 mL and 350 μL for the 24-well and 96-well microplates, respectively. Many applications in molecular biology and biochemistry require ^{32}P applications, and the high sample throughput and inexpensive sample analysis possible with Cherenkov counting make Cherenkov counting in a microplate format very attractive. Some detector and sample properties should be considered to optimize Cherenkov counting of samples in microplates, and these are described subsequently.

A. Sample-to-Sample Crosstalk

In the case of the microplate counter design, where up to 12 photomultiplier tubes are automatically situated above or on top of as many as 12 sample wells in close proximity there exists the possibility of sample-to-sample crosstalk. This can occur when the β particle energy is of sufficient magnitude to be able to travel from one sample well to an adjacent well and cause interference by creating Cherenkov photons in the well of a neighboring sample.

Sample-to-sample crosstalk can be determined by adding a radioactivity spike (e.g., ^{32}P in water) to one sample well and only pure water to the surrounding eight sample wells of the microplate. The percent crosstalk can then be calculated by determining the count rates above background in the wells containing only water as a percentage of the radioactivity detected by the PMT above the sample well containing the ^{32}P spike.

The percent crosstalk and percent detection efficiencies in the Cherenkov counting of ^{32}P were tested with the PerkinElmer TopCount microplate counter, which utilizes up to 12 adjacent PMTs above 12 adjacent microplate sample wells. These are listed in Table 9.7 together with background measurements for samples in 24- and 96-well OptiPlates as well as the 96-well UniFilter and 96-well FlexiFilter microplates. The filter microplates are described in Sections XI of Chapter 5 and V.B of Chapter 11. The percent crosstalk values for Cherenkov counting listed in Table 9.7 are low in all cases and generally negligible for the 200 μL and 1.0 mL samples in the 96- and 24-well OptiPlates. The detection efficiencies of 25% are about one-half the efficiencies attainable with Cherenkov counting using the conventional liquid scintillation analyzer. However, the high sample throughput for unattended counting offered by microplate counting is the trade-off, which might be preferred. For example, up to 40 microplates can be stacked in a TopCount microplate counter allowing for the unattended analysis of up to 3840 samples when a 96-well microplate is used (i.e., 96-wells × 40 plates).

B. Sample Volume Effects

The effects of sample volume on the Cherenkov counting efficiency of ^{32}P in 96- and 24-well microplates are provided in Table 9.8. The volume effects are significant for the relatively small sample volumes tested in the 96-well microplates, which can have a total capacity of 350 μL per sample well. For the larger sample sizes of 500–1500 μL, there is no significant effect of sample volume of Cherenkov detection efficiency.

Apparently the larger volumes of water in the 24-well microplate provide a greater probability for a larger number of high-energy β particles to have a longer path of travel in the aqueous medium needed for the production of

TABLE 9.7 Cherenkov Counting Performance in 96- and 24-Well OptiPlate Microplates and 96-Well UniFilter and FlexiFilter Microplates with a PerkinElmer TopCount Scintillation and Luminescence Counter[a]

Microplate type	Crosstalk (%)	Counting efficiency (%)	Background (CPM)
96-well OptiPlate (200 μL sample)	0.4	25	20
24-well OptiPlate 1.0 mL sample)	0.3	25	93
UniFilter-96 GF/B (wet, 20 μL)	1.0	24	38
FlexiFilter nylon (wet, 20 μL)	2.0	25	38

[a]From Anonymous (1996) reprinted with permission from PerkinElmer Life and Analytical Sciences.

TABLE 9.8 Cherenkov Counting Performance Versus Volume of Aqueous Sample in 96-Well and 24-Well OptiPlate Microplates with a PerkinElmer TopCount Scintillation and Luminescence Counter[a]

96-well OptiPlate		24-well OptiPlate	
Sample Size (μL)	Counting efficiency (%)	Sample size (μL)	Counting efficiency (%)
Dry	10	Dry	14
10	12	500	22
25	14	1000	22
50	15	1500	23
100	18	–	—
300	26	–	—

[a]From Anonymous (1996) reprinted with permission from PerkinElmer Life and Analytical Sciences.

Cherenkov photons. Samples of very small aqueous volume will have less medium within which to travel and produce Cherenkov photons. A significant number of beta particles near the sample surface can escape into the air or microplate wall without producing any significant Cherenkov effect. The larger sample sizes of the 24-well microplate provide, for the most part, higher counting efficiencies than smaller samples of the 96-well plates.

Cherenkov counting of ^{32}P in samples in the dry state in 96- and 24-well OptiPlates are also possible, albeit at a lower detection efficiency (see Table 9.8).

C. Quench Correction

As previously discussed there is no chemical quench in Cherenkov counting. However, color in the sample will absorb Cherenkov photons, reduce pulse heights, and consequently reduce counting efficiency. It is, therefore, important to determine the degree of quench in each sample well of the microplate sample holder and the counting efficiency of each sample. From the counting efficiency, the sample activities in disintegrations per minute can be determined.

Because up to 12 samples can be counted simultaneously in a microplate scintillation counter and high sample throughput is generally a requirement with this type of counter and sample format, the most automated method of measuring sample quench and counting efficiency is to utilize a sample spectrum quench-indicating parameter (QIP), such as SIS, as previously described in Section III.C.2 of this chapter.

The TopCount microplate scintillation counter utilizes the transformed Spectral Index of the Sample (tSIS) as a sample spectrum QIP similar to the Spectral Index of the Sample (SIS) used with the conventional liquid scintillation analyzers. Other commercial microplate scintillation counters will utilize another sample spectrum QIP such as SQP(E) described in Chapter 5. The tSIS measures the magnitude of the sample pulse height spectrum. Color quench in Cherenkov counting reduces Cherenkov photon intensities emitted by a sample and consequently reduces the sample pulse height spectrum according to the degree of quench. The tSIS is a unitless number that reflects the magnitude of the pulse height spectrum; that is, as the quench level of a sample increases, the magnitude of the sample pulse height spectrum diminishes together with the value of the tSIS. The scintillation and luminescence counter can measure automatically the value of tSIS for each sample directly from the Cherenkov pulse height spectrum of the sample. It is necessary, therefore, only to establish a quench correction curve of percent counting efficiency versus tSIS from a set of color quench standards of ^{32}P (or other high-energy beta emitter) in water. From such a curve stored in the memory of the instrument computer, the microplate counter can determine automatically the Cherenkov counting efficiency of each sample in each well from the tSIS value of each sample.

A series of quench standards and a color quench correction curve obtained with the quenched standards can be prepared according to the procedure described by Anonymous (1993). It is adaptable to any microplate scintillation counter, and presented as follows modified by the author:

1. Calculate the total volume of water needed to fill the desired number of sample wells with radionuclide standard (e.g., ^{32}P), taking into account the number of standards, number of replicates, sample volume, and well size, adding 100% to the total. For example, eight samples in triplicate at 250 μL per sample well plus 100% equals 12 mL total volume. The sample size should be selected according to well size and size of the unknown experimental samples to be measured after the quench correction curve is prepared.

2. A bulk aqueous solution of ^{32}P standard will be prepared taking into account that we need a total volume of at least 12 mL according to the preceding calculation. Therefore, add sufficient activity of ^{32}P from a radioactive stock solution to make a solution to contain approximately 100,000 DPM of ^{32}P per 250 μL for the 12 mL needed. This would require, for example, the following calculated activity of ^{32}P: (100,000 DPM per 250 μL)(12 mL) $= 4.8 \times 10^6$ DPM $= 2.16$ μCi. Mix the solution thoroughly.

3. Aliquot equal volumes of the ^{32}P solution into eight liquid scintillation vials, labeled "Standard 1" through "Standard 8", which is the minimum number of standards recommended. A maximum of 20 standards is possible. Verify the activity in each vial by Cherenkov counting in a conventional liquid scintillation analyzer, collecting at least 160,000 counts (0.5% 2σ value) per vial, and discarding any vials that do not fall within 2% of the average.

4. Color quenching solutions will be made next. Prepare a stock solution of a color quench compound. Yellow food dye concentrate may be used directly as a stock solution, or titan yellow can be used at 5 mg/mL or Sudan red at 1 mg/mL.

5. Dispense 1 mL of water into eight liquid scintillation vials. Make eight color quenching solutions by adding the following amounts of the stock solution of color quench compound to each vial:

Color quenching solution number	Stock solution of color quench compound (μL)
1	0
2	0.5
3	1.1
4	2.3
5	4.5
6	9.0
7	16.0
8	25.0

6. To each of the standard vials 1 through 8 prepared in step 3, add 20 μL per mL of the corresponding color quenching solution.

7. Aliquot solutions of the color-quenched standards into the desired microplate wells. The accuracy of this operation should be better than ±2%. It is recommended that the microplate used for the quench standards be the same type as the microplate used for the assay of experimental samples. Because the location of each standard will later be defined using the "Plate Mapping" feature in the counter, the standards can be placed anywhere on the microplate. A maximum of ten replicates is allowed. Heat seal the microplate and mix thoroughly.

8. Assay an identical aliquot of the unquenched Standard 1 solution for DPM in a standard liquid scintillation analyzer after adding 10 mL of fluor cocktail to determine the DPM of the standard set. The accuracy of this measurement will directly affect the accuracy of the quench curve. It is recommended that this measurement be done in triplicate and at least 160,000 total counts be collected. Check for chemiluminescence after the addition of fluor cocktail, to ensure this is not occurring in the sample, when the DPM measurements are made.

9. Set up and count the standards by Cherenkov counting with the microplate scintillation analyzer (e.g., TopCount). Enter the DPM value of the ^{32}P standard, which was determined in Step 8, and define the plate map in accordance with the layout of standards. If tSIS, or any other sample spectrum quench-indicating parameter, is used as the quench indicating parameter, it is recommended that the samples be counted long enough to collect at least 10,000 total gross counts per sample to ensure maximum accuracy.

The microplate analyzer automatically calculates the counting efficiencies of each standard using the DPM value determined in Step 8 above utilizing Eq. 9.12, which was described previously. The analyzer will also determine automatically the values of tSIS from the pulse height spectra of each color-quenched standard in each sample well, and the quench correction curve of percent counting efficiency versus QIP (e.g., tSIS) will be plotted automatically and stored in the memory of the analyzer computer. An example of a typical color quench correction curve is illustrated in Fig. 9.10. The DPM of ^{32}P of subsequent unknown experimental samples can be determined automatically by the analyzer by measuring the count rate (CPM) and the tSIS of each sample. From the tSIS value of a given sample and the color quench correction curve, the instrument will determine the counting efficiency of the sample. It will use the decimal value of the percent counting efficiency and the count rate of the sample to calculate the disintegration rate (DPM) of the sample according to Eq. 9.11 described previously in this chapter.

VIII. MULTIPLE RADIONUCLIDE ANALYSIS

A. Sequential Cherenkov and Liquid Scintillation Analysis

The conventional measurement of dual β particle-emitting radionuclides in the same sample by liquid scintillation analysis requires three or four quench

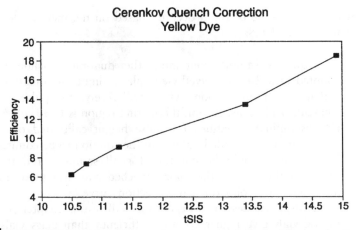

FIGURE 9.10 Example of a typical color quench correction curve for the Cherenkov counting of ^{32}P with a TopCount microplate scintillation and fluorescence counter. (From Anonymous, 1996, reprinted with permission from PerkinElmer Life and Analytical Sciences.)

correction curves for the determination of the same number of counting efficiencies depending on the method used, for example, exclusion and inclusion dual-region analysis, as described in Section VIII of Chapter 5. The methods can be considered tedious sometimes for the inexperienced in these techniques, particularly when preset optimum counting regions have not been established and set by the instrument manufacturer. The multiple variable counting efficiencies in these double-radionuclide liquid scintillation analysis methods can also be a source of error. With these points in mind, Fujii and Takiue (1988a,b) developed a simplified method of analyzing mixtures of dual radionuclides, which has a single requirement, namely that one of the radionuclides of the dual mixture must emit β particles with an $E_{max} > 263$ keV in sufficient number to be measurable by Cherenkov counting. The method of Fujii and Takiue (1988a,b) involves sequential Cherenkov and LSA counting and it was developed for both dual β emitters and dual α-β emitters as mixtures. The method is described subsequently.

Examples of nuclide pairs that meet the above β particle energy criterion are ^{32}P–^{14}C, ^{32}P–^{35}S, ^{32}P–^{45}Ca, ^{32}P–^{33}P, ^{36}Cl–^{35}S and ^{36}Cl–^{45}Ca, among others. Some of these nuclide pairs have proved to be useful tools in the elucidation of reaction mechanisms and pathways in studies of biosynthesis as reviewed by L'Annunziata (1984b). Another example of a popular nuclide pair is purified ^{89}Sr–^{90}Sr (after removal of ^{90}Y daughter) from liquid waste (Walter *et al.*, 1993; Poletico *et al.*, 1994; Lee *et al.*, 2002). The method involves first counting the Cherenkov photons of an aqueous solution of a nuclide pair to determine the activity or DPM of the high-energy radionuclide of the pair (e.g., ^{32}P or ^{36}Cl of the above examples). Scintillation cocktail is then added to the aqueous solution of the nuclide pair, and the total DPM of the sample is determined by the efficiency tracing (ET-DPM) method. The activity of the low-energy β emitter of the dual-radionuclide mixture is determined by difference between the total DPM of the sample and the DPM of the higher energy β emitter measured by Cherenkov counting. The following is an outline of the procedure based on the method developed by Fujii and Takiue (1988a):

1. An aqueous sample containing the dual-radionuclide mixture is counted in a 20-mL polyethylene plastic liquid scintillation counting vial in a counting region, which will receive the pulse events from Cherenkov photons. A typical counting region is 0–30 keV for LL and UL discriminator settings for pulse height calibrated on an energy scale. The recommended volume of the solution to be counted is 5 mL. The volume should be constant for all samples, and this volume should be the same as the color quenched standards used to determine the Cherenkov color quench correction curve.

   Note 1: As described in Section IV.B of this chapter polyethylene plastic vials give higher counting efficiency than glass vials, and the effect of sample volume on Cherenkov counting efficiency is reduced with the polyethylene plastic vials. Therefore, plastic vials are recommended over glass vials.

   Note 2: The sample volume of 5 mL is selected because the Cherenkov counting efficiency is reduced at volumes of less than 5 mL (see Fig. 9.7), and room must remain in the vial to add scintillation fluor cocktail at a later step in the procedure. It is important to keep the aqueous sample at a constant volume (e.g., 5 mL), because sample volume will affect Cherenkov counting efficiency. If the sample volume is less than 5 mL add water to the counting vial to bring the volume up to 5 mL before analysis by Cherenkov counting.

2. Use a color quench correction curve to determine the Cherenkov counting efficiency of the sample and to convert the count rate (CPM) to disintegration rate (DPM) of the higher energy β emitter of the dual-nuclide mixture. See Sections III.B, III.C, and VII of this chapter on methods of color quench correction.

   Note: Chemical quench does not occur in Cherenkov counting. Even if samples are colorless, it is convenient to use a color quench correction curve for Cherenkov counting, because most modern liquid scintillation analyzers are programmed to provide automatic DPM calculations of experimental samples via a quench correction curve. Also, the use of an established color quench correction curve in Cherenkov counting will eliminate a possible source of error in counting efficiency determinations, as the author has experienced that even a very slight color, barely visible to the "naked eye" can produce significant quench in Cherenkov counting.

3. Add 10 mL of a liquid scintillation fluor cocktail to the aqueous sample in the scintillation vial after Step 2 is complete.

   Note: A fluor cocktail should be selected that will mix homogeneously with the 5 mL of aqueous sample. See Chapter 8 for information on the selection of fluor cocktails. Some examples of suitable fluor cocktails are Ultima Gold XR and Insta-Gel, which have high water holding capacities; the latter fluor cocktail forms a homogeneous gel at high water loads.

4. Determine the total DPM of the sample using the Efficiency Tracing with ^{14}C DPM method (ET-DPM). This method is simple and it is described in detail in Section V.F of Chapter 5. The ET-DPM method is a standard program or option on many liquid scintillation analyzers of PerkinElmer Life and Analytical Sciences. The method will give the total DPM of the sample automatically without any need for a quench correction curve. The total DPM of the sample, therefore, represents the sum of the activities of the two radionuclides in the mixture.
5. Subtract the activity (DPM) of the higher energy radionuclide determined by Cherenkov counting of the sample in Step 2 from the total sample DPM determined in Step 4 to yield the activity in DPM of the lower energy β emitter of the dual radionuclide mixture.

This method can also be applied to the analysis of dual mixtures of alpha- and beta-emitting radionuclides as demonstrated by Fujii and Takiue (1988b). Alpha emitters will not give rise to Cherenkov photons, because the threshold energy for the production of Cherenkov radiation by α particles in water medium is 1000 MeV or 10^6 keV as explained by Fujii and Takiue (1988b). Many radionuclides that decay by the emission of α particles also emit γ-rays. However, γ-radiation from α-emitting radionuclides does not interfere with this method, because γ-radiation must possess at least 0.43 MeV to produce 0.263-MeV Compton electrons of energy sufficient to produce Cherenkov photons, and α-emitters scarcely emit γ-radiation in excess of 0.43 MeV. Consequently, the above method of analysis of dual-radionuclide mixtures is applicable to α- and β-emitting radionuclide pairs when the β emitter of the pair has β particle-emissions of sufficient number in excess of 263 keV energy to produce measurable Cherenkov radiation. The method was demonstrated by Fujii and Takiue (1988b) with the ^{32}P–^{241}Am radionuclide pair. Quantitative activity determinations of the two radionuclides are obtained according to the procedure just described. When the mixture is first counted in water, the Cherenkov radiation is measured from the ^{32}P to provide the DPM of that radionuclide only. Adding liquid scintillation fluor cocktail to the aqueous mixture and subsequent analysis by the ET-DPM method provides the total DPM of the ^{32}P + ^{241}Am mixture. Subtraction of the ^{32}P DPM obtained by Cherenkov counting from the total DPM of the mixture provides the activity of the ^{241}Am. The theory and practice of ET-DPM for β emitters as single nuclides or mixtures in described in detail in Section V of Chapter 5. The application of the method to mixtures of α- and β- emitters was demonstrated by Fujii and Takiue (1988b), and an example of efficiency tracing curves for a ^{32}P–^{241}Am mixture is given in Fig. 9.11. As illustrated in the figure, the curves are extrapolated to 100% counting efficiency where CPM = DPM irrespective of quench level. See Section V.F of Chapter 5 for detailed information on this technique. The method should be applicable to most dual mixtures of α and β emitters when the β emitters can be measured by Cherenkov counting.

The sequential Cherenkov and LSA method was successfully employed by Morel and Fardeau (1989) for the analysis of ^{32}P–^{33}P in studies of phosphate fertilizer use efficiency. L'Annunziata and coworkers (see Noor *et al.*, 1996)

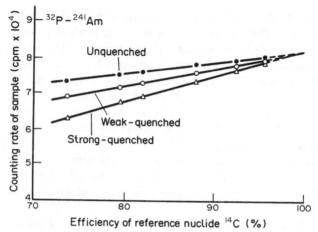

FIGURE 9.11 Efficiency tracing curves of a ^{32}P–^{241}Am mixture at three levels of quench. Each extrapolated value up to 100% counting efficiency provides the total activity of the sample to be measured. (From Fujii and Takiue, 1988b, reprinted with permission from Elsevier Science.)

TABLE 9.9 Percent Recoveries of Calculated Activities of Five Composite Mixtures or ^{86}Rb–^{35}S–^{33}P Determined by Efficiency Tracing DPM (ET-DPM) Technique and the Component Activities of ^{86}Rb Determined by Cherenkov Counting and Combined ^{35}S+^{33}P Obtained by Difference[a]

Sample DPM[b] ^{86}Rb : ^{35}S : ^{33}P	Total DPM (actual)	Total DPM (ET)	Total DPM recovery (%)	^{86}Rb DPM (Cherenkov)	^{86}Rb DPM recovery (%)	^{35}S–^{33}P DPM (difference)	^{35}S+^{33}P DPM recovery (%)
4326 : 7294 : 7194	18814	18671	99.2	4387	101.4	14284	98.6
2146 : 3620 : 3424	9190	9185	99.9	2199	101.2	6986	99.2
1042 : 1794 : 1550	4386	4408	100.5	1070	102.7	3338	99.8
3113 : 5510 : 4620	13243	13237	100.0	3171	101.9	10666	99.4
432.3 : 742.6 : 646.1	1821	1819	99.9	445.5	103.1	1373.5	98.9

[a]From L'Annunziata and coworkers (Noor *et al.* 1996) reprinted with permission from Elsevier Science.
[b]In hundreds.

have applied the sequential Cherenkov counting and LSA analysis by ET-DPM method to analyze the triple-radionuclide mixture of ^{86}Rb–^{35}S–^{33}P. The total DPM of the mixture was determined by the ET-DPM method and the activity of ^{86}Rb of the triple-radionuclide mixture was determined by Cherenkov counting in water with 100% recoveries for several activity proportions of triple-radionuclide mixture as illustrated in Table 9.9. The difference of the two measurements provided the activities of the ^{35}S+^{33}P mixture. The activities of the ^{35}S and ^{33}P are determined by liquid scintillation analysis of the dual label after the ^{86}Rb decays to a negligible level according to its half-life ($t_{1/2} = 18.8$ days).

B. Cherenkov Analysis with Wavelength Shifters

Dual-radionuclide analysis by Cherenkov counting by the same conventional methods (e.g., dual-region counting) used in liquid scintillation analysis is not practical, because of the high degree of overlap of the pulse height spectra of Cherenkov photons of β-emitting radionuclides. To overcome this difficulty and still make dual-radionuclide analysis possible by Cherenkov counting Fujii and Takiue (1988c) developed a method that utilizes only one counting region while measuring the changes in the pulse height distributions and counting efficiencies of the two radionuclides in the mixture before and after the addition of wavelength shifter.

As described in Section IV.C of this chapter, the addition of wavelength shifter to a sample containing β emitters, which produce Cherenkov radiation in a medium such as water, changes both the emission wavelength and the directional anisotropic nature of the Cherenkov photons into isotropic emissions. Hence, these effects of wavelength shifters improve the photon capture potential of the Cherenkov radiation by the photomultiplier tubes, which increases the Cherenkov counting efficiencies of the two radionuclides in a mixture. In general, the effect of wavelength shifter on counting efficiency is greater for lower energy β emitters (e.g., ^{36}Cl) than for higher energy β emitters (e.g., ^{32}P). The effects of wavelength shifter on the Cherenkov pulse height spectra produced by ^{32}P and ^{36}Cl are illustrated in Fig. 9.12. If the effects of wavelength shifter on the counting efficiencies are significantly different, the method of Fujii and Takiue (1988c) may be applied to the DPM measurements of dual radionuclides in a mixture by Cherenkov

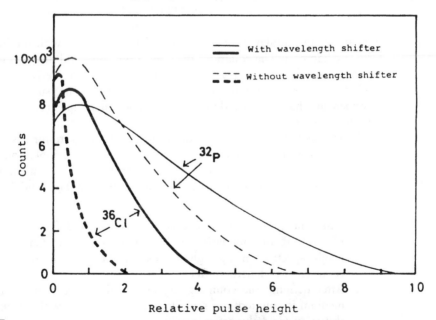

FIGURE 9.12 Pulse height distributions in ^{32}P and ^{36}Cl Cherenkov measurements without and with wavelength shifter (PPO: 0.03 g L^{-1}). The samples have the same activity. (From Fujii and Takiue, 1988c, reprinted with permission from Elsevier Science.)

counting. The method is based on the following simple conditions and equations:

A sample containing a mixture of two β-emitting radionuclides, which produce Cherenkov radiation in a non-scintillating medium (e.g., water), will give different count rates n_A and n_B when counted without and with wavelength shifter, respectively according to the equations

$$n_A = D_1 \, e_{1A} + D_2 \, e_{2A}, \tag{9.18}$$

and

$$n_B = D_1 \, e_{1B} + D_2 \, e_{2B}, \tag{9.19}$$

where D_1 and D_2 are the activities (e.g., DPM) of the two radionuclides, and e_{1A}, e_{1B}, e_{2A}, and e_{2B} are the counting efficiencies of the nuclides in the counting conditions without and with the wavelength shifter, respectively.

Equations 9.18 and 9.19 are solved simultaneously for the two unknown values D_1 and D_2 to give

$$D_1 = \frac{n_A \, e_{2B} - n_B \, e_{2A}}{e_{1A} \, e_{2B} - e_{1B} \, e_{2A}}, \tag{9.20}$$

$$D_2 = \frac{n_A \, e_{1B} - n_B \, e_{1A}}{e_{1B} \, e_{2A} - e_{1A} \, e_{2B}}. \tag{9.21}$$

We can see the similarities of the above Eqs. 9.18–9.21 with the Eqs. 5.34–5.37 of Chapter 5. The equations defining count rates and disintegration rates for radionuclides producing Cherenkov radiation are derived from data produced in one counting region only, whereas, the similar equations described in Chapter 5 are derived from data originating from two different counting regions.

Equations 9.20 and 9.21 are solved for the disintegration rates of the two radionuclides by determining the count rates from the sample and counting efficiencies of the radionuclides without and with the introduction of wavelength shifter. An outline of the procedure used by Fujii and Takiue is as follows:

1. Only one counting region is used. This can be defined by setting the LL and UL discriminators to encompass the entire Cherenkov pulse height spectrum in its highest pulse height distribution possible, that is, when wavelength shifter is present.
2. After defining the counting region, 10 mL of the sample in a water medium is counted in a scintillation counting vial without wavelength shifter to yield the count rate n_A.
3. Wavelength shifter is introduced into the sample by adding 2 mL of one of the following solutions: $0.03 \, \mathrm{g \, L^{-1}}$ PPO in ethanol or $0.2 \, \mathrm{g \, L^{-1}}$

of 4-methyl-umbelliferone. The sample is counted again to yield the count rate n_B.

4. Counting efficiencies are determined by taking two other 10-mL aliquots of the unknown sample, adding 0.1 mL of reference standards of the two radionuclides of the mixture, and counting before and after the addition of wavelength shifter. The internal standardization technique is described in Section V.A of Chapter 5

Fujii and Takiue (1988c) demonstrate quantitative recoveries with this technique for the $^{32}P-^{36}Cl$ and $^{86}Rb-^{36}Cl$ radionuclide mixtures.

Other radionuclide mixtures could be analyzed by this method provided that the ratios e_{1A}/e_{1B} and e_{2A}/e_{2B} of increase of the Cherenkov counting efficiencies are different from each other, because Eqs. 9.20 and 9.21 cannot be solved when e_{1A}/e_{1B} is equal to e_{2A}/e_{2B}. According to Fujii and Takiue (1988c) the rate of increase of Cherenkov counting efficiency for lower energy β-emitters is larger than that for higher energy β-emitters (Ross, 1971; Takiue $et\ al.$, 1984). For example, in the case of the ^{36}Cl and ^{32}P, Fujii and Takiue (1988c) report that the rate of increase of Cherenkov counting efficiency (27%) for ^{36}Cl with PPO wavelength shifter is 2.9 times as high as the counting efficiency (9.5%) without wavelength shifter. This was larger than the rate of increase of 1.2 for the ^{32}P Cherenkov counting efficiencies before and after the addition of wavelength shifter.

IX. RADIONUCLIDE STANDARDIZATION

The concepts of radionuclide standardization, that is, the determination of absolute activity of samples, upon which the activity of other samples are traceable, were described in detail in Section IX of Chapter 5. Radionuclide standardization is not used generally to determine sample activities in the routine measurement of radionuclides in applied research or radionuclides in the environment; rather, standardization is a technique required to determine the absolute activity of a sample often to less than 1% discrepancy upon which the activities of other standards may be traced.

Liquid scintillation analysis (LSA) is often used for the standardization of radionuclides, because of the higher detection efficiencies achieved in LSA compared to Cherenkov counting. However, under certain circumstances, as explained by Grau Carles and Grau Malonda (1995) and Grau Malonda and Grau Carles (1998a,b), Cherenkov counting can be advantageous in the calibration of certain radionuclides where the detection of low-energy electrons by LSA can complicate the elaboration of the standardization model. In the case of Cherenkov counting, the Cherenkov energy threshold (263 keV in pure water) serves as a discriminator for low-energy electrons (i.e., $E < 263$ keV in pure water).

The computation of the Cherenkov counting efficiency according to Grau Carles and Grau Malonda (1995) should correct for previously unexplained discrepancies in the predicted and experimental values. The differences in experimental counting efficiencies between ^{36}Cl and ^{204}Tl serve as an example of such a discrepancy, where the counting efficiency of ^{36}Cl ($E_{\text{max}} = 710$ keV) is 70% greater than the counting efficiency of ^{204}Tl ($E_{\text{max}} = 763$ keV). Although the maximum energy of the β emissions of ^{36}Cl is less than that of ^{204}Tl, the counting efficiency of ^{36}Cl is extraordinarily much greater than that of ^{204}Tl.

New concepts are introduced by Grau Carles and Grau Malonda (1995) to explain unusual Cherenkov counting efficiencies such as those presented in the above example of ^{36}Cl and ^{204}Tl. These concepts are (1) intrinsic Cherenkov counting efficiency, which is the ratio between counted pulses and emitted particles over the Cherenkov energy threshold, and (2) Cherenkov yield, which is the ratio between emitted β particles over the Cherenkov energy threshold and the total number of emitted β particles. These terms to be further defined subsequently, were used by Grau Carles and Grau Malonda (1995) to explain the unusual behavior of ^{36}Cl and ^{204}Tl by showing that the intrinsic Cherenkov counting efficiencies of the two radionuclides were similar (10%), whereas the Cherenkov yields of the two differ significantly, namely 65% for ^{36}Cl and 40% for ^{204}Tl.

The method requires prior calibration with two standards, namely ^{36}Cl and ^{32}P, upon which the counting efficiency of any other radionuclide with β emissions greater than the Cherenkov energy threshold may be determined. Grau Carles and Grau Malonda (1995) use the new concepts of intrinsic Cherenkov counting efficiency and Cherenkov yield to define the Cherenkov counting efficiency ε_c according to the equation

$$\varepsilon_c = r_k \varepsilon_k \tag{9.22}$$

where ε_k is the intrinsic counting efficiency (i.e., the ratio of the counted pulses and emitted β particles over the Cherenkov threshold energy), and r_k is the Cherenkov yield (i.e., the ratio of the emitted β particles over the Cherenkov energy threshold to the total number of emitted β particles). The Cherenkov yield is obtained from the β particle distributions as follows:

$$r_k = \frac{\int_{W_k}^{W_m} N(W)\, dW}{\int_1^{W_m} N(W)\, dW} \tag{9.23}$$

where W_m and W_k are the maximum β particle energy and the Cherenkov threshold energy, respectively, and $N(W)$ is the β particle distribution where

$$W = \frac{E}{511} + 1 \tag{9.24}$$

and E is the β particle energy in keV. The intrinsic Cherenkov efficiency involves counting the sample and relating the number of counted pulses to the detection probability $f(W)$ according to the equation

$$\varepsilon_k = \frac{\int_{W_k}^{W_m} N(W)f(W)\,dW}{\int_{W_k}^{W_m} N(W)dW}. \tag{9.25}$$

Equation 9.25 can be used to calculate the counting efficiency of any β-emitter [with known β particle distribution $N(W)$] once the function $f(W)$ is known.

According to the solution of Grau Carles and Grau Malonda (1995), the function $f(W)$ is zero for energies below W_k and unity for electrons (β particles) that have sufficient energy for total detection, that is, the detection probability is nearly 1 for electrons of $E > 1$ MeV or 1000 keV in water. In the regions intermediate to these limits, Grau Carles and Grau Malonda explain that the detection probability increases exponentially according to the following:

$$\begin{aligned} f(W) &= a(W - W_k)^n, & W_k \leq W < W_u, \\ f(W) &= 1, & W \geq W_u, \end{aligned} \tag{9.26}$$

where W_u is the minimum energy that corresponds to the total detection and a and n are parameters defined by least squares fitting using the radionuclides ^{36}Cl and ^{32}P as standards. When all of the β particles are partially detected $W_m < W_u$, the minimum condition is defined as

$$\min \sum_{i=1}^{\nu} \left[\varepsilon_{ki} - a \int_{W_k}^{W_m} N_i(W)(W - W_k)^n \, dW \right]^2 \tag{9.27}$$

where ν is the total number of radionuclides involved in the fitting. When $W_m > W_u$ the minimum condition is defined as

$$\min \sum_{i=1}^{\nu} \left[\varepsilon_{ki} - a \int_{W_k}^{W_u} N_i(W)(W - W_k)^n \, dW - \int_{W_u}^{W_m} N_i(W)dW \right]^2. \tag{9.28}$$

Both of the above conditions are used to characterize the parameter n and obtain the values of a and W_u by least-squares fitting.

For example, using ^{36}Cl and ^{32}P as standards, Grau Carles and Grau Malonda applied Eqs. 9.22–9.28 to obtain the detection probability function $f(W)$ for the particular liquid scintillation spectrometer they used as follows:

$$f(W) = 0.5424\,(W - 1.5004)^{1.60},$$
$$1.5005 \leq W < 2.9662$$
$$f(W) = 1, \quad W \geq 2.9662$$

The value of 1.5004 for W_k was calculated in light of the fact that the ^{36}Cl and ^{32}P standards were counted in 15 mL of 1 M HCl instead of pure water.

The index of refraction of 1 M HCl reduces the Cherenkov threshold energy from 263 to 255.7 keV. Therefore, W_k is calculated as

$$W_k = \frac{255.7\,\text{keV}}{511\,\text{keV}} + 1 = 1.5004$$

The importance of this technique is that the above detection probability function, once determined, can then be used to calculate the counting efficiency of any other radionuclide that emits β particles or γ-rays (that produce Compton electrons) over the Cherenkov threshold energy of 255.7 keV in 15 mL of 1 M HCl with the particular liquid scintillation spectrometer used. If the solution type, solution volume, vial type, or instrument is changed, the detection probability function is refitted.

The method provides good agreement between experimental and computed counting efficiencies as illustrated in Table 9.10. The counting efficiencies of ^{36}Cl and ^{32}P of Table 9.10 show no discrepancy between experimental and calculated values, because these two nuclides are used as the standards. The nuclides ^{60}Co and ^{137}Cs deserve attention, because both are β and γ emitters.

In the case of ^{60}Co the β particle emissions make a negligible contribution to the production of Cherenkov photons; however, its two γ-ray emissions of 1.33 and 1.17 MeV produce Compton electrons of sufficient energy to yield Cherenkov radiation. Therefore, the Cherenkov counting efficiency for ^{60}Co is calculated according to

$$\varepsilon = \varepsilon(\gamma_1) + \varepsilon(\gamma_2) - \varepsilon(\gamma_1)\varepsilon(\gamma_2) \tag{9.29}$$

where $\varepsilon(\gamma_1)$ and $\varepsilon(\gamma_2)$ are the counting efficiencies for the Compton electrons produced by each of the two γ-rays, and the product of the two efficiencies considers the simultaneous Compton interaction of the two γ-rays of ^{60}Co.

The application of this method to the standardization of ^{137}Cs is more complicated, because this nuclide decays by two β branches over the Cherenkov energy threshold. It reaches secular equilibrium with its daughter nuclide ^{137m}Ba, which undergoes a 662-keV γ-ray decay transition that results in 89% abundance of 662-keV γ-ray emissions and a 11% abundance of internal conversion electrons, a fraction of which are above the Cherenkov threshold energy.

The above method was successfully used by Navarro et al. (1997) to determine the counting efficiency and activity of ^{234}Th by Cherenkov counting to within 1.5% uncertainty. The computer program, CHEREN, used for the calculation of the Cherenkov counting efficiency is described by Grau Carles and Grau Malonda (1998).

In an effort to yet reduce further the amount of uncertainty in radionuclide standardization by Cherenkov counting Grau Malonda and Grau Carles (1998a,b) introduced a new model based on two parameters that depend on the measurement conditions and on the equipment, in addition to parameters based on the nature of the radionuclide decay emissions. The anisotropic character of Cherenkov radiation reduces the experimental counting efficiency and complicates its theoretical computation. According to the new model, the directional character of Cherenkov radiation and the amount of energy that Cherenkov light must transfer to create one photoelectron at the photocathode are defined as two new free parameters (see Section IX, Chapter 5 for the definition of free parameter). The current model offers the advantage of calculating the Cherenkov counting efficiency regardless of the refractive index or acid concentration of the medium.

As explained by Grau Malonda and Grau Carles (1998a,b) the calculation of the counting efficiency of an electron of energy E requires the evaluation of the total number of emitted photons k emitted in the wavelength interval (λ_1, λ_2) per unit path length of electron travel according to the Frank and Tamm theory (Tamm, 1939; Jelley, 1958) described by the equation

$$\frac{dk}{dx} = 2\pi\alpha\left(\frac{1}{\lambda_1} - \frac{1}{\lambda_2}\right)\left(1 - \frac{1}{n_2\,\beta_2}\right) \qquad (9.30)$$

where α is the fine structure constant, n and β are the index of refraction of the medium and electron relative phase velocity previously described in Eq. 9.1, and the wavelength interval includes those Cherenkov radiation frequencies to which the photocathode of the photomultiplier tube is sensitive. Because Cherenkov light is emitted as a cone, a given Cherenkov event is mainly oriented towards one of the two photomultipliers of a conventional LSA (e.g., photopmultiplier A), while the other photomultipler (B) receives light from the same Cherenkov event that is reflected or diffused. Grau Malonda and Grau Carles (1998a,b) define the anisotropy coefficient α as the rate of Cherenkov photons directed towards photomultiplier A, and β as the rate of photons directed towards photomultiplier B where

$$\alpha + \beta = 1 \qquad (9.31)$$

TABLE 9.10 Experimental and Computed Cherenkov Counting Efficiencies[a]

Nuclide	Efficiency exp.	Efficiency comp.	Discrepancy (%)
^{36}Cl	0.0666	0.0666	0.0
^{204}Tl	0.0402	0.0404	0.6
^{89}Sr	0.374	0.377	0.9
^{32}P	0.468	0.468	0.0
^{90}Sr+^{90}Y	0.620	0.631	1.8
^{60}Co	0.0561	0.0556	0.7
^{137}Cs	0.0493	0.0475	3.7

[a]From Grau Carles and Grau Malonda (1995) reprinted with permission from Elsevier Science.

When $\alpha = 1$ and $\beta = 0$, the anisotorpic properties of the LSA is displayed at a maximum when only photomultiplier A detects the Cherenkov and no event is recorded by the LSA in coincidence counting. The minimum value for α is 0.5, corresponding to the maximum value for β. The number of Cherenkov photons traveling directly towards photomultipliers A and B are defined as $k\alpha$ and $k(1-\alpha)$, respectively. If s and w represent the transmission probabilities and photocathode quantum efficiencies, respectively, the number of photons generated at the photocathodes of photomultipliers A and B are written as $ksw\alpha$ and $ksw(1-\alpha)$, respectively. With these variables defined and taking q as the product of the probabilities s and w, Grau Malonda and Grau Carles (1998a,b) derived the Cherenkov counting efficiency ε_B of pure β emitters as

$$\varepsilon_B = \int_{E_k}^{E_0} N(E)\{1 - \exp[-qk(E)\alpha]\}\{1 - \exp[-qk(E)(1-\alpha)]\} \, dE \qquad (9.32)$$

where $N(E)$ is the β particle energy distribution, E_0 and E_k are the maximum β particle and Cherenkov threshold energies, respectively, and the function $N(E)$ was normalized according to the following:

$$\int_0^{E_0} N(E) \, dE = 1 \qquad (9.33)$$

For radionuclides that decay by internal conversion the counting efficiency calculation defined by Eq. 9.32 was modified by Grau Malonda and Grau Carles (1998a,b) to account for nine possible electron conversion possibilities. Also, for radionuclides that decay by pure gamma-ray emission the counting efficiency calculation was expressed to account for Compton electrons with energies above the Cherenkov threshold of the counting medium, as well as to include the gamma-photon escape probability. When both gamma-ray emission and internal conversion processes occur in radionuclide decay the counting efficiency calculation was expressed as a sum according to both decay probabilities.

The robustness of the calculations for Cherenkov counting efficiency was demonstrated by Grau Malonda and Grau Carles (1998a,b) for several radionuclides in a variety of media and in glass or plastic vials with very close discrepancies between experimental and computed counted efficiencies. The following serve as examples of computed counting efficiencies expressed as a decimal (i.e., $100\% = 1.0$) with the discrepancy between the computed and experimental counting efficiencies expressed as a percent: [36]Cl, 0.0913 (−0.88%); [204]Tl, 0.0526 (−1.94%); ^{89}Sr, 0.428 (0.92%); ^{32}P, 0.522 (−1.36%); ^{90}Sr, 0.673 (1.17%), [210]Bi, 0.1310 (−0.61%); [234m]Pa, 0.542 (0.73%); [137]Cs + [137m]Ba, 0.0722 (0.41%); and [40]K, 0.394 (0.76%). The close agreement between the experimental and computed theoretical Cherenkov counting efficiencies was further demonstrated by Grau Malonda and Grau Carles (2002).

X. GAMMA-RAY DETECTION

Gamma radiation can produce Cherenkov photons indirectly through gamma-ray photon–electron interactions as the gamma radiation travels through a transparent medium. The number of photons emitted by a Cherenkov detector is generally only approximately 1% of the number emitted by a good scintillator for the same gamma-ray energy loss (Sowerby, 1971). Although the Cherenkov detection efficiencies of gamma radiation are low, unique applications of the Cherenkov effect for the analysis of gamma radiation exist, and the effect plays an important role as a source of background in various methods of radioactivity analysis. One should always be aware of the potential for gamma radiation to produce Cherenkov photons.

The transfer of gamma-ray photon energy to an atomic electron via a Compton interaction produces a Compton electron with energy, E_e, within the range between zero and a maximum defined by

$$0 < E_e \leq E_\gamma - \frac{E_\gamma}{1 + 2E_\gamma/0.511} \tag{9.34}$$

where E_γ is the gamma-ray photon energy in MeV and the term $E_\gamma - (E_\gamma/(1 + 2E_\gamma/0.511))$ defines the Compton-electron energy at $180°$ Compton scatter according to equations previously defined in Chapter 1. To produce Cherenkov photons the Compton electron must possess energy in excess of the threshold energy, E_{th}, defined by Eq. 9.5 previously in this chapter. For example, the threshold energy for electrons in water ($n = 1.332$) according to Eq. 9.5 is calculated to be 263 keV. A Compton electron must possess, therefore, energy in excess of 263 keV to produce Cherenkov photons in water. In this case, however, the gamma-ray photon must possess an energy in excess of 422 keV calculated according to the inverse of Eq. 1.109 or

$$E_\gamma = E_e + E_\gamma' + \phi \tag{9.35}$$

where E_e is the Compton electron energy, E_γ' is the energy of the Compton-scattered photon, and ϕ is the electron binding energy. The electron binding energy is negligible and can be ignored. Thus, Eq. 9.35 can becomes

$$E_\gamma = E_e + \frac{E_\gamma}{1 + 2E_\gamma/0.511} \tag{9.36}$$

For example, if we take E_e to be 0.263 MeV, the threshold electron energy for Cherenkov production in water, and E_γ' as the scattered-photon energy at $180°$ Compton scatter, Eq. 9.36 becomes

$$E_\gamma = 0.263\,\text{MeV} + \frac{E_\gamma}{1 + 2E_\gamma/0.511} \tag{9.37}$$

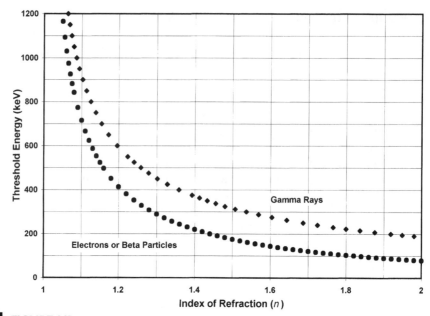

FIGURE 9.13   Threshold energy for Cherenkov radiation as a function of index of refraction of the detector medium for gamma rays and electrons or beta particles. The threshold energies for electrons or beta particles are calculated according to Eq. 9.5, and the gamma-ray threshold energies are calculated according to Eq. 9.36 as the gamma rays that yield electrons of the threshold energy via 180° Compton scatter.

where $E_\gamma = 0.422$ MeV is the threshold gamma-ray energy for the production of Cherenkov photons in water. Threshold energies will vary according to the index of refraction of the medium, and these are provided graphically in Fig. 9.13 for gamma radiation and electrons or beta particles.

Although Cherenkov detection efficiencies for gamma radiation are low, the phenomenon is applied to create threshold detectors. A variety of media, which vary significantly in refractive index, can be selected to discriminate against gamma radiation of specific energy. For example, silica aerogels of low refractive index ($n = 1.026$) can be used to discriminate against gamma rays of relatively high energy (2.0 MeV) while a transparent medium of high refractive index such as flint glass ($n = 1.72$) can serve to discriminate against relatively low-energy gamma radiation (0.25 MeV). Figure 9.13 illustrates the potential for gamma-ray energy discrimination according to refractive index of the detector medium.

Another application of gamma-ray detection is the Cherenkov verification technique used in nuclear safeguards to verify the authenticity of irradiated nuclear fuel, which is one of the important tasks performed by the International Atomic Energy Agency (IAEA). The IAEA nuclear safeguards program audits the national declarations of fuel inventories to insure that no illicit diversion of nuclear material has occurred. High levels of gamma radiation are emitted by fission products in irradiated nuclear fuel. The irradiated fuel stored under water will produce Cherenkov light as a result of Compton scattering in the water surrounding the fuel. A Cherenkov Viewing

Device containing a UV-transmitting lens coupled to a UV-sensitive charge-coupled device (CCD) and image monitor enables the real-time imaging of the UV light portion of the Cherenkov radiation in the presence of normal room lighting (Attas *et al.*, 1990, 1992, 1997; Attas and Abushady, 1997; Kuribara, 1994; Kuribara and Nemeto, 1994, Lindsey *et al.*, 1999). The presence of fission products and the nature of their distribution, as indicated by the Cherenkov glow, is used as evidence of fuel verification.

XI. PARTICLE IDENTIFICATION

Cherenkov counters are applied in particle physics research for the determination of particle mass (m), velocity (β), and particle identification (PID). Cherenkov detectors of various designs are applied to the discrimination and identification of high-energy particles, among which are threshold Cherenkov counters, ring-imaging Cherenkov (RICH) detectors, as well as time of flight (TOF) and time of propagation (TOP) measurements.

A. Threshold Cherenkov Counters

Threshold Cherenkov counters consist of Cherenkov detectors of differing refractive index employed to discriminate particles of different mass based on the differing Cherenkov threshold energies that the particles have in the detectors. For example, if we consider a beam of two types of particles of different mass (m), such as pions ($\pi^{\pm}$, $m = 139.6 \, \text{MeV}/c^2$) and kaons $K^{\pm}$, $m = 493.7 \, \text{MeV}/c^2$), a Cherenkov detector may be selected of a given refractive index (n) such that the particle of higher mass does not produce Cherenkov radiation. This would be the case if the threshold condition for the production of Cherenkov radiation is not met by the particle of higher mass, that is, $\beta < 1/n$. Sundaresan (2001) describes another example of the application of two Cherenkov detectors of different refractive index, namely silica aerogel ($n = 1.01 - 1.03$) and pentane ($n = 1.357$) whereby particles of lower mass, such as 10 GeV kaons ($m = 493.7 \, \text{MeV}/c^2$) produce Cherenkov photons in the two Cherenkov detector media, whereas particles of higher mass such as protons ($m = 939.3 \, \text{MeV}/c^2$) produce Cherenkov photons only in the silica aerogel. The difference in count rates from Cherenkov photons produced in the two detectors yield the relative numbers of the heavier and lighter particles. Adachi *et al.* (1995) describe a threshold Cherenkov counter for the identification of $\pi^{\pm}$ and $K^{\pm}$ in a particle beam with momentum in the region of 1.0–2.5 GeV/c. Silica aerogel with refractive index of 1.0127 with 14-cm thickness was used as the detector and photomultiplier tubes for the measurement of Cherenkov photons. The threshold momentum for the detection of $\pi^{\pm}$ and $K^{\pm}$ was determined to be 0.863 GeV/c and 3.05 GeV/c, respectively. A Cherenkov detector arrangement reported by Perrino *et al.* (2001) provides an example of two Cherenkov detectors operated in tandem together with time-of-flight (TOF) measurements to discriminate between pions (π^{+}), positrons (e^{+}), and protons (p^{+}). The experimental set-up consisted of CO_2 gas detectors ($n = 1.00041$) providing excellent detection for positrons, and a silica aerogel detector with a refractive index $n = 1.025$ providing (π, p) discrimination in the 1–4 GeV/c range with Cherenkov thresholds of 0.62 and 4.2 GeV/c for pions and protons, respectively. Pions and protons at 1 and 2 GeV/c are below the Cherenkov threshold in the CO_2 gas. Complementary data provided by time-of-flight measurements between two BC408 scintillation detectors separated at a 23 m distance along the particle beam enabled the tagging of protons. TOF measurements are determined by signals between detectors permitting the determination of the speed of a particle, and with its total energy signal, the mass of the ion can be identified (Lilley, 2001).

B. Ring Imaging Cherenkov (RICH) Counters

The Ring Imaging Cherenkov (RICH) detector is designed principally for particle identification, as it can provide information on the velocity, β, and the charge of the particle, z, and complementary information provided by rigidity measurements using a magnetic tracker can provide the identity of the particle according to its mass (Pinto da Cunha, 2000). The detector is designed to accept particles that originate from any 4π direction. Several detector geometries and designs are reviewed by Glässel (1999), and the classical RICH geometry is illustrated in Fig. 9.14. The distance (d = 2f) from the source of the charged particles or interaction vertex (Fig. 9.14) defines the radius of a spherical mirror, R_s. The Cherenkov photon detector has a concentric spherical surface of smaller radius. The space between the outer surface of the photon detector and inner spherical mirror is filled with a transparent medium of a given refractive index [e.g. gas, C_4F_{10} ($n = 1.0015$), liquid, C_6F_{14} ($n = 1.276$), crystalline NaF ($n = 1.33$), or silica aerogel

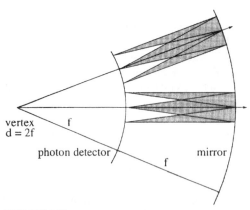

FIGURE 9.14 Classical RICH detector geometry. A spherical mirror surrounds a spherical photon detector. The two arrows illustrate two charged-particle trajectories. The shaded areas surrounding the particle trajectories illustrate the emitted conical Cherenkov light and cone image (ring) reflected onto the detector surface. The radius (r) of the light cone (not indicated in the figure) is that distance on the detector surface from the line of particle trajectory to the focal point (ring) of the reflected light. (From Glässel, 1999, reprinted with permission from Elsevier Science.)

$(n = 1.01 - 1.02)$ to serve as the Cherenkov radiator (Pinto de Cunha *et al.*, 2000). The Cherenkov radiator is chosen according to the mass and momenta of the particles to be identified, as the emission of Cherenkov radiation at an angle θ must satisfy the threshold condition $\beta > 1/n$. At the moment the particle penetrates the Cherenkov radiator, the Cherenkov radiation is emitted at an angle θ according to the particle velocity, β, and the refractive index (n) of the Cherenkov radiator as defined by Eq. 9.6, that is, $\cos\theta = 1/\beta n$. The radiator dimensions used by Pinto da Cunha *et al.* (2000) were 2 cm thickness and 50 cm radius. The Cherenkov photons are reflected off the inner surface of the outer spherical mirror as a ring of light onto the conical detector surface. The ring has a radius, r, which is measured to determine the particle velocity. An imaging detector is used to provide an image of the ring of Cherenkov light and its coordinates relative to the vertex. According to the derivations of Sundaresan (2001), the focal length, f, of the mirror is defined as $f = R_s/2$; and if $r = f\theta$, we can write $r = R_s\theta/2$ and

$$\cos\theta = \cos(r/f) = \cos(2r/R_s) \tag{9.38}$$

When the threshold condition for the emission of Cherenkov photons is met, that is, $\beta n > 1$, the Cherenkov photons are emitted at an angle θ to the particle trajectory according to Eq. 9.6, namely, $\cos\theta = 1/\beta n$ and

$$\beta n = \frac{1}{\cos(2r/R_s)} \tag{9.39}$$

and

$$\beta = \frac{1}{n\cos(2r/R_s)} \tag{9.40}$$

Thus, the particle velocity, β, in units of speed of light can be obtained from the radius r of the Cherenkov ring image and its coordinates from which the emission angle θ can be derived. The particle charge can be derived from the Cherenkov photon intensity.

C. Time-of-Propagation (TOP) Cherenkov Counters

A relatively new concept in the application of Cherenkov detectors for particle identification is via the measurement of the time-of-propagation (TOP) and horizontal emission angle, Φ, of Cherenkov photons described by Akatsu *et al.* (2000) and Ohshima (2000). The basic structure of the Time-of-Propagation Cherenkov counter is illustrated in Fig. 9.15. The TOP detector consists of a quartz Cherenkov radiator bar (20 mm-thick, 60 mm-wide, 3150 mm-long). Two mirrors are located at both ends for focusing the Cherenkov photons. The mirror is flat at the backward end to reflect the Cherenkov light towards the forward end where butterfly-shaped mirrors are

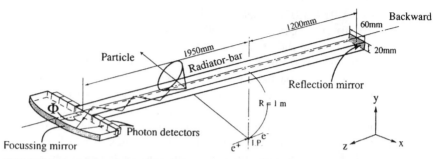

FIGURE 9.15 Structure of the TOP Cherenkov counter. Basic parameters are indicated in the figure. The bar (Cherenkov radiator) and mirrors are made of synthetic optical quartz ($n = 1.47$ at $\lambda = 390$ nm) is configured z-asymmetric to the interaction point (IP) of an asymmetric collider. The counters are placed 1 m radially distant from the interaction point to form a cylindrical structure. (From Ohshima, 2000, reprinted with permission from Elsevier Science.)

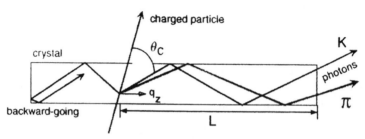

FIGURE 9.16 Side view of propagating photons. TOP is inversely proportional to z-component q_z of the light velocity. $\mathbf{TOP} = (L \times n(\lambda))/cq_z = 4.90\,\mathbf{ns} \times L\,m/q_z$. θ_c is the Cherenkov angle and L is the particle injection position from the bar-end in meters. At the opposite end, a mirror reflects the backward-going photons. (From Ohshima, 2000, reprinted with permission from Elsevier Science.)

located. The Cherenkov photons are focused horizontally onto the photon detector plane, and the time-of-propagation and angle Φ are measured by position-sensitive multi-channel phototubes. The method is based on the following principles: (1) the Cherenkov photon emission angle (θ_c) illustrated in Fig. 9.16 is a function of the particle velocity (β) according to the relation $\cos\theta_c = 1/\beta n$ where n is the refractive index of the Cherenkov radiator, (2) the TOP of photons in a light guide with internal-total-reflection characteristics can be calculated as a function of the photon emission angle, and (3) a correlation between TOP and photon emission angle would provide information on particle identification. Notice from the illustration provided in Fig. 9.16, the Cherenkov photon emission angle created by the pion (π) is less acute than that created by the kaon (K) and the TOP of the Cherenkov photons derived from the pion is shorter than that derived by the kaon. TOP differences of 100 ps or more are found for normal incident 4 GeV/c K and π at 2 m long propagation (Ohshima, 2000). A historical review and a thorough treatment of the theory of RICH counters are available from Sequinot and Ypsilantis (1994) and Ypsilantis and Sequinot (1994).

XII. APPLICATIONS IN RADIONUCLIDE ANALYSIS

The application of Cherenkov counting to the activity analysis of radionuclides is popular in those cases where the Cherenkov counting efficiency of the radionuclide of interest is adequate to meet particular detection limits required. Table 9.11 provides approximate Cherenkov counting efficiencies of radionuclides measured and listed according to the E_{max} of the β particles emitted by each radionuclide. The E_{max} is listed, as the Cherenkov detection efficiency of radionuclides is a function of the threshold energy (E_{th}) and the refractive index (n) of the medium calculated according to Eq. 9.5. For example, if water is the medium ($n = 1.332$), the Cherenkov counting efficiency would be a function of the number of β particles of $E > 263$ keV relative to the total number of β particles emitted by the radionuclide. Cherenkov counting is popular, when counting efficiencies are adequate, because of the ease of sample preparation and low expense incurred in the preparation and disposal of samples. Because water is generally the medium of counting, and fluor scintillation cocktail is not used, samples are often left in a state suitable for subsequent tests such as chemical analysis, spectrometric analysis, chromatographic tests, or even chemical compound extraction and isolation. Some references to the application of Cherenkov counting to the analysis of specific radionuclides not already cited in this chapter are provided in the following paragraphs.

A. Phosphorus-32

Cherenkov counting of ^{32}P in aqueous extracts or in the dry state has become popular, particularly because of the easy and inexpensive sample preparation techniques involved as well as the relatively high counting efficiencies obtained (L'Annunziata, 1997; L'Annunziata and Passo, 2002). As chemical quenching does not exist in pure Cherenkov counting, sample preparation techniques may be employed with little concern for the type of reagents used, and sample color may be bleached by chemicals with no quenching effect.

Cherenkov counting of ^{32}P in the dry state applied to research in the biological and physical sciences was reviewed previously in Section V of this chapter. The popularity of Cherenkov counting of ^{32}P in aqueous media is more popular as illustrated by numerous reports including those of Baxter *et al.* (2002), Bem *et al.* (1980), BenZikri (2000), Chow (1980), Fardeau (1984), Fric and Palovcikova (1975), L'Annunziata (1997), L'Annunziata and Passo (2002), Lefebvre and Glass (1981), Lickly *et al.* (1988), Morel and Fardeau, (1989), Østby and Krøkje (2002), Rowinska *et al.* (1975 and 1987), Smith *et al.* (1972), Tuffen *et al.* (2002), Tuininga *et al.* (2002), Uz *et al.* (2002), Warshawsky *et al.* (2002), and Wiebe *et al.* (1971), among others. Glass (1978) described a very simple and rapid sample preparation technique where ^{32}P- or ^{86}Rb-labelled plant material is ashed directly in glass liquid scintillation vials. An aqueous solution of 2.5 mM ANDA wavelength shifter can be added before counting to increase counting efficiencies. Although the addition of wavelength shifter is not necessary. Wahid *et al.* (1985) described

TABLE 9.II **Experimentally Determined Cherenkov Counting Efficiencies**

Nuclide	E_{max} (keV)[a]	Counting efficiency[b] (%)	References
[99]Tc	292 (100%)	1.0[c]	Scarpitta and Fisenne (1996)
[59]Fe	273 (48.5%)	5.8[c]	Scarpitta and Fisenne (1996)
	475 (51.2%)		
[90]Sr	546 (100%)	1.0	Rucker (1991); Chang et al. (1996)
[60]Co	Compton electrons[d]	5.6	Grau Carles and Grau Malonda (1995)
[36]Cl	710 (98%)	6.6	Grau Carles and Grau Malonda (1995)
[204]Tl	763 (98%)	4.0	Grau Carles and Grau Malonda (1995)
[137]Cs	510 (92%)	4.8	Grau Carles and Grau Malonda (1995)
	1170 (8%)		
[198]Au	960 (99%)	5.4	Parker and Elrich (1970)
[47]Ca	660 (83%)	7.5	Parker and Elrich (1970)
	1940 (17%)		
[210]Pb([210]Bi)	1160 (from [210]Bi > 99%)	18	Blais and Marshall (1988)
[115m]Cd	680 (3%)	35	Bem et al. (1978); Ramesh and Subramanian (1997)
	1620 (97%)		
[89]Sr	1490 (100%)	42	Rucker (1991); Chang et al. (1996)
[228]Th	580 (from [212]Pb)[e]	53	Al-Masri and Blackburn (1994)
	1790 (from [208]Tl)[e]		
	2250 (from [212]Bi)[e]		
[86]Rb	680 (8.5%)	53	L'Annunziata and coworkers (see Noor et al., 1996)
	1770 (91.5%)		
[40]K	1310 (89%)	55	Pullen (1986)
[32]P	1710 (100%)	57	Takiue et al. (1993)
[234]Th	(from [234m]Pa)[e]	60	Navarro et al. (1997); Nour et al. (2002)
	2290 (99.8%)		
[238]U	(from [234m]Pa)[e]	62	Blackburn and Al-Masri (1994)
[90]Y	2280 (100%)	72	L'Annunziata and Passo (2002)
[188]Re	2120 (79%)	53	Kushita and Du (1998)
	1970 (20%)		
	<1900 (1%)		
[106]Ru([106]Rh)	2000 (from [106]Rh, 3%)	62	Carmon and Dyer (1987)
	2440 (from [106]Rh, 12%)		

(*continued*)

◼ **TABLE 9.11** Continued

Nuclide	E_{max} (keV)[a]	Counting efficiency[b] (%)	References
	3100 (from ^{106}Rh, 11%)		
	3530 (from ^{106}Rh, 68%)		
^{234m}Pa	2290 (99.8%)	54	Grau Malonda and Grau Carles (1986a,b, 2000)
^{42}K	1970 (18.4%)	75	Buchtela and Tschurlovits (1975)
	3560 (81.3%)		
^{226}Ra/^{222}Rn	1505 (from ^{214}Bi, 16%)[e]	77	Blackburn and Al-Masri (1993)
	1540 (from ^{214}Bi, 16%)[e]		Al-Masri and Blackburn (1995, 1999)
	3270 (from ^{214}Bi, 24%)[e]		
	650 (from ^{214}Pb)[e]		
^{34m}Cl	Several β-decay branches up to a energy maximum of 4,500	57	Wiebe et al. (1980)
^{38}Cl	4913 (57.6%)	66	Wiebe et al. (1980)
	2770 (11.1%)		
	1111 (31.3%)		
^{18}F	635 (97%)	50[f]	Wiebe et al. (1978)
^{214}Bi(^{214}Pb)	(See ^{226}Ra, this table)		

[a]Where more than one β-decay process occurs for a given radionuclide, the % abundance of each decay process is provided in parenthesis.

[b]Counting efficiencies will vary from instrument to instrument and vial type (e.g., glass or plastic). Also, measurements made with older less efficient photomultiplier tubes provided lower counting efficiencies than can be expected from contemporary more efficient phototubes. No attempt is made in this table to categorize the counting efficiencies according to the aforementioned governing factors that will affect slightly the counting efficiency values reported. However, the counting efficiencies listed are intended to provide a general idea of their orders of magnitude relative to the energies of the β-particle emissions. The counting efficiencies are values reported of nuclides in water without wavelength shifter unless otherwise noted.

[c]Measured in 10 mL aqueous 25 mM 7-amino-1,3-naphthalene disulfonic acid (ANSA) wavelength shifter.

[d]The Cherenkov photons from ^{60}Co are due to Compton electrons produced by its 1.33 and 1.17 MeV γ-ray emissions.

[e]Daughter nuclide.

[f]A combination of Cherenkov and scintillation counting in high refractive index methyl salicylate ($n = 1.5369$).

a nitric-perchloric acid wet digestion procedure for the preparation of ^{32}P in plant material for Cherenkov counting. However, the author prefers to ash the plant material at $500°C$ for 1 hour in a muffle furnace, because perchloric acid digestions are hazardous and tedious. The plant ash is dissolved with 2 mL of concentrated HNO_3 and transferred directly into a scintillation counting vial for Cherenkov counting. Subsequent to ashing and Cherenkov counting the analysis of total phosphorus by colorimetry via a phosphomolybdate blue complex is possible (L'Annunziata, unpublished work). Chemical analysis can be conveniently performed on samples before or after Cherenkov counting as no fluors or scintillants are added, which would interfere with chemical analyses. An example may be taken from the work of Palmer (1969) and Schneider (1970). These investigators determined the amounts of total phosphorus in biological samples by means of colorimetry using the phosphomolybdate blue complex method. The intensity of the blue color is proportional to the concentration of phosphorus in the sample according to Beer's law. After total phosphorus analysis, the activity of ^{32}P in the sample is determined by bleaching or decolorizing the blue complex at a high pH followed by Cherenkov counting of the colorless solution. However, the author of this chapter prefers to carry out the Cherenkov counting of the samples before the colorimetric analysis of total phosphorus, because (1) the relative short half-life of ^{32}P ($t_{1/2} = 14.3$ days) often warrants counting the samples without too much delay when sample activities are at their highest, and (2) bleaching of the samples to remove the blue color of the phosphomolybdate complex is not required when Cherenkov counting precedes colorimetry.

B. Strontium-89 and Strontium-90 (Yttrium-90)

The radionuclide ^{90}Sr reaches secular equilibrium with its daughter ^{90}Y. The ^{90}Sr is traditionally assayed in biological and radio-toxicological samples by separation and assay of the short-lived daughter ^{90}Y after its chemical separation from its parent (L'Annunziata, 1971) as described in Chapter 1, Section VI.C of this book. However, the analysis of $^{90}Sr(^{90}Y)$ often requires also the analysis of ^{89}Sr, because the two radioisotopes of strontium are often encountered together as described by L'Annunzata and Fuller (1968).

The first reports on the application of Cherenkov counting to the analysis of ^{89}Sr and $^{90}Sr(^{90}Y)$ were by Buchtela and Tschurlovits (1975) from the Atominstitut in Vienna, Austria and Randolf (1975) from the U.S. Atomic Energy Commission, Health Services Laboratory in Idaho Falls. Both papers were printed in the same journal volume. Since these pioneering papers, Cherenkov counting has become a popular procedure for the analysis of ^{89}Sr and $^{90}Sr(^{90}Y)$ via Cherenkov counting of the ^{90}Y daughter separated from the parent. Contemporary procedures are divided into two groups, those that include the analysis of samples containing mixtures of combined ^{89}Sr–$^{90}Sr(^{90}Y)$ and the more simplified case of samples containing only the parent/daughter radionuclides $^{90}Sr(^{90}Y)$. These two cases are described in this section with examples of contemporary methods.

I. Cherenkov Counting of ^{89}Sr with ^{90}Sr(^{90}Y)

The Cherenkov counting of samples containing ^{89}Sr with ^{90}Sr(^{90}Y) is appealing, because both ^{89}Sr and ^{90}Y produce significant Cherenkov radiation. See Table 9.11 where ^{89}Sr and ^{90}Y are listed with Cherenkov counting efficiencies in water of ~40% and ~70%, respectively. The low Cherenkov counting efficiency of ~1% for ^{90}Sr in water permits the determination of ^{89}Sr in a freshly isolated ^{89}Sr–^{90}Sr mixture, that is, a ^{89}Sr–^{90}Sr mixture from which the ^{90}Y daughter was removed fresh [e.g., via precipitation of yttrium, ion exchange chromatography, or extraction chromatography, (e.g., Eichrom resins), among other methods]. The relative Cherenkov counting efficiencies of the three radionuclides are evidenced by the Cherenkov radiation pulse height spectra produced in a liquid scintillation analyzer by equal activities of ^{89}Sr, ^{90}Sr, and ^{90}Y illustrated in Fig. 9.17. Banavali *et al.* (1992) used carbonate precipitation and extraction chromatography to isolate the radioactive strontium, which is analyzed by Cherenkov counting yielding primarily ^{89}Sr activity. After a few days to permit ingrowth of ^{90}Y the sample is recounted by Cherenkov counting to permit the activity determination of ^{90}Y. The ^{90}Sr in the sample is calculated by the amount of ^{90}Y ingrowth from the time of radiostrontium isolation. See Section VI.C of Chapter 1 for calculations of ingrowth of daughter nuclide with parent after the separation of radionuclides in secular equilibrium. The contribution of ^{90}Sr activity to the Cherenkov count rate of freshly prepared mixtures of ^{89}Sr–^{90}Sr cannot be ignored when the ^{90}Sr activity is higher than that of ^{89}Sr.

Chang *et al.* (1996) report a method utilizing only Cherenkov counting for the analysis of mixtures of ^{89}Sr and ^{90}Sr(^{90}Y). A freshly prepared sample of isolated radiostrontium (i.e., yttrium was removed and discarded) is counted in a 15 mL aqueous solution contained in a plastic counting vial. The

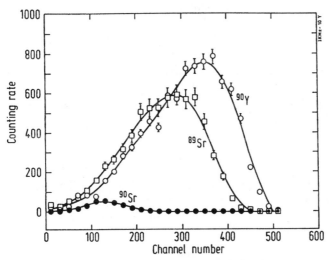

FIGURE 9.17 Pulse height spectra of the Cherenkov radiation produced by the β radiations of ^{90}Sr, ^{89}Sr, and ^{90}Y in aqueous solution at a constant activity level of 250 Bq for all three nuclides. (From Walter *et al.*, 1993.)

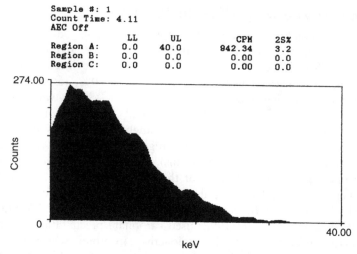

```
Sample #: 1
Count Time: 4.11
AEC Off
                LL       UL          CPM      2S%
Region A:      0.0     40.0       942.34      3.2
Region B:      0.0      0.0         0.00      0.0
Region C:      0.0      0.0         0.00      0.0
```

FIGURE 9.18   Cherenkov pulse height spectrum of ^{90}Sr(^{90}Y) in 10 mL H$_2$O in a plastic counting vial using a Packard 2750TR/LL liquid scintillation analyzer and a counting region of 0–40 keV. The sample contains 1,300 DPM each of ^{90}Sr and ^{90}Y in secular equilibrium. Approximately 1% of the Cherenkov photons originate from ^{90}Sr; 71% arise from ^{90}Y. (From L'Annunziata, 1997, unpublished work.)

sample is counted in a region of 0–50 keV, which is sufficient to encompass all pulse heights produced by Cherenkov radiation when pulse height spectra are plotted on a scale equivalent to keV energy. The channel numbers illustrated in Fig. 9.17 are equivalent to approximately 0–50 keV, when the channel numbers are calibrated against beta-particle energies in keV as illustrated in Fig. 9.18. The purified strontium samples are counted for sufficient time to provide a 2 sigma counting error of <2%, and the count rate, time, and ^{90}Y separation date are recorded. The sample is allowed to sit for 2 to 3 days to permit significant ingrowth of ^{90}Y with ^{90}Sr, and the sample is then recounted in the same vial. The count rate and time of the second counting are recorded. The two count rates are used in Eqs. 9.45 and 9.46 to determine the activities of ^{89}Sr and ^{90}Sr of the mixture. The equations used are found according to the derivations of Chang *et al.* (1996) as follows:

When a sample containing ^{89}Sr and ^{90}Sr is purified by separation from the ^{90}Y daughter, at time t_0, the total activity of the purified strontium is the sum of the two activities or

$$A_{t_0} = A_{^{90}Sr} + A_{^{89}Sr} \qquad (9.41)$$

During ^{90}Y ingrowth, the ^{89}Sr decays slightly ($t_{1/2} = 50.5$ days) while the ^{90}Sr will not undergo any significant decay ($t_{1/2} = 28.8$ years). Therefore, at time t_1, when the first count rate is determined, the total activities in the sample are expressed as

$$
\begin{aligned}
A_{t_1} &= A_{^{90}Sr}[e^{-\lambda_1(t_1-t_0)}] + A_{^{90}Y} + A_{^{89}Sr}[e^{-\lambda_3(t_1-t_0)}] \\
&= A_{^{90}Sr}[e^{-\lambda_1(t_1-t_0)}] + A_{^{90}Sr}[1 - e^{-\lambda_2(t_1-t_0)}] + A_{^{89}Sr}[e^{-\lambda_3(t_1-t_0)}]
\end{aligned}
\qquad (9.42)
$$

where A_{90Sr}, A_{90Y}, A_{89Sr}, λ_1, λ_2, and λ_3 are the activities in DPM and the decay constants for ^{90}Sr, ^{90}Y, and ^{89}Sr, respectively.

At time t_1, the Cherenkov count rate (CPM) C_{t_1} of the sample is expressed as

$$C_{t_1} = E_{90Sr} A_{90Sr}[e^{-\lambda_1(t_1-t_0)}] + E_{90Y} A_{90Sr}[1 - e^{-\lambda_2(t_1-t_0)}]$$
$$+ E_{89Sr} A_{89Sr}[e^{-\lambda_3(t_1-t_0)}] + C_b \tag{9.43}$$

where E_{90Sr}, E_{90Y}, and E_{89Sr}, are the counting efficiencies (CPM/DPM) of each nuclide and C_b is the background count rate.

At time t_2, the Cherenkov count rate C_{t_2} is expressed as

$$C_{t_2} = E_{90Sr} A_{90Sr}[e^{-\lambda_1(t_2-t_0)}] + E_{90Y} A_{90Sr}[1 - e^{-\lambda_2(t_2-t_0)}]$$
$$+ E_{89Sr} A_{89Sr}[e^{-\lambda_3(t_2-t_0)}] + C_b \tag{9.44}$$

Equations 9.43 and 9.44 are solved simultaneously to give the equations

$$A_{89Sr} = \frac{C_{t_1} - C_b - A_{90Sr} E_{90Sr} e^{-\lambda_1 \Delta t_1} - A_{90Sr} E_{90Y} + A_{90Sr} E_{90Y} e^{-\lambda_2 \Delta t_1}}{E_{89Sr} e^{-\lambda_3 \Delta t_1}} \tag{9.45}$$

$$A_{90Sr} = \frac{e^{-\lambda_3 \Delta t_1}(C_{t_2} - C_b) - e^{-\lambda_3 \Delta t_2}(C_{t_1} - C_b)}{\left(\begin{array}{l} E_{90}[e^{-\lambda_1 \Delta t_2} e^{-\lambda_3 \Delta t_1} - e^{-\lambda_1 \Delta t_1} e^{-\lambda_3 \Delta t_2}] + E_{90Y}[e^{-\lambda_3 \Delta t_1}(1 - e^{-\lambda_2 \Delta t_2})) \\ - e^{-\lambda_3 \Delta t_2}(1 - e^{-\lambda_2 \Delta t_1})] \end{array}\right)} \tag{9.46}$$

where $\Delta t_1 = t_1 - t_0$ and $\Delta t_2 = t_2 - t_0$.

When ^{89}Sr is absent from the sample Chang et al. (1996) show that Eq. 9.46 for the activity (DPM) of ^{90}Sr becomes simplified to

$$A_{90Sr} = \frac{C_{t_2} - C_{t_1}}{E_{90Y}[e^{-\lambda_2 \Delta t_1} - e^{-\lambda_2 \Delta t_2}]} \tag{9.47}$$

This method involves only the Cherenkov counting of the sample at two time intervals following the separation of radiostrontium from the radioyttrium daughter. For several standard mixtures of ^{89}Sr–^{90}Sr(^{90}Y) tested by Chang et al. (1996) a mean error of only 2.5% was reported for calculated ^{89}Sr and ^{90}Sr mixtures.

Chu et al. (1998) demonstrated the analysis of ^{89}Sr–^{90}Sr(^{90}Y) by separating the strontium from yttrium on a column of crown-ether Eichrom Sr resin (Horwitz et al., 1992 and Vajda et al., 1992). The ^{90}Y and ^{89}Sr+ ^{90}Sr fractions are separated and eluted from a 8-mm dia. × 40-mm long column of Eichrom Sr resin, and the two fractions are analyzed by Cherenkov counting. The Cherenkov detection efficiency for ^{90}Sr is considered negligible and neglected. The ^{90}Sr activity is derived from the calculated activity of ^{90}Y, as the ^{90}Sr and ^{90}Y were in secular equilibrium,

that is their activities were equal prior to chemical separation. A similar
analytical approach was taken by Grahek *et al.* (1999).

2. Sequential Cherenkov Counting and Liquid Scintillation Analysis

Other methods used for the analysis of ^{89}Sr–^{90}Sr(^{90}Y) mixtures involve
the sequential counting of the Cherenkov radiation followed by liquid
scintillation counting, which was first introduced by the pioneering work of
Buchtela and Tschurlovits (1975) and Randolph (1975).

a. Sequential Analysis Without Wavelength Shifter

Rucker (1991) describes a sequential Cherenkov radiation and liquid
scintillation analysis procedure for the analysis of ^{89}Sr–^{90}Sr(^{90}Y). The method
involves purification of the sample, which includes removal of yttrium from
the sample by hydroxide scavenging followed by strontium precipitation as
a carbonate. The strontium carbonate is washed, dried, and weighed to
calculate the percent chemical recovery of standardized strontium carrier. The
precipitate is dissolved in 5 mL of 0.3 M HCl, and the aqueous solution is
analyzed by Cherenkov counting without the use of any wavelength shifter to
improve Cherenkov counting efficiencies, which provides a count rate
primarily from the ^{89}Sr in the sample. After Cherenkov counting, scintillation
fluor cocktail is added to the sample to provide a liquid scintillation count
rate in one wide counting region not confined to spectral resolutions for the
measurement of ^{89}Sr and ^{90}Sr. The two count rate determinations, the first
Cherenkov counting and the second liquid scintillation counting, used
optimized discriminator settings for the counting regions. Count ratios in
each region for Cherenkov and LSC are used to correct for the fraction of the
Cherenkov counts due to ^{90}Sr and the fraction of both counts (Cherenkov
and LSC) due to the early ingrowth of ^{90}Y.

The following equations derived by Rucker are provided to illustrate the
basis of the method that lead to the working equations:

The Cherenkov count rates, which are the first ones determined for a
sample, originate primarily from ^{89}Sr (e.g., Cherenkov counting efficiency
40%), with only a very small contribution from ^{90}Sr (1% counting efficiency)
and the early ingrowth of ^{90}Y described by the equation

$$C_1 = {}^{89}\text{Sr cpm}_c + {}^{90}\text{Sr cpm}_c + {}^{90}\text{Y cpm}_c \tag{9.48}$$

where C_1 is the net (background subtracted) Cherenkov count rate and cpm$_c$
represents the Cherenkov counts per minute obtained from each nuclide. The
liquid scintillation count rate comes primarily from the ^{89}Sr and ^{90}Sr
of the sample with only a small contribution from the early ingrowth of ^{90}Y.
The net count rate from liquid scintillation analysis, C_2, is therefore described
by the following equation:

$$C_2 = {}^{89}\text{Sr cpm}_l + {}^{90}\text{Sr cpm}_l + {}^{90}\text{Y cpm}_l, \tag{9.49}$$

where cpm_l is the liquid scintillation count rate from each nuclide. Rucker (1991) then defines the count rates due to the early ingrowth of ^{90}Y, in Cherenkov counting and LSC, in terms of the ^{90}Sr count rates when the elapsed times are known from the time of chemical separation of ^{90}Y to the time of Cherenkov counting and the time of liquid scintillation counting as follows:

$$C_1 = {}^{89}\text{Sr cpm}_c + ({}^{90}\text{Sr cpm}_c)(A_1) \tag{9.50}$$

and

$$C_2 = {}^{89}\text{Sr cpm}_l + ({}^{90}\text{Sr cpm}_l)(A_2) \tag{9.51}$$

where

$$A_1 = 1 + (D_1)(1 - e^{-\lambda t_c}) \tag{9.52}$$

and

$$A_2 = 1 + (D_2)(1 - e^{-\lambda t_l}) \tag{9.53}$$

where D_1 is the ratio of the Cherenkov counting efficiency of ^{90}Y to the Cherenkov counting efficiency of ^{90}Sr (i.e., $\varepsilon^C_{90Y}/\varepsilon^C_{90Sr}$), t_c is the time from chemical separation of strontium from yttrium to Cherenkov counting, D_2 is the ratio of the liquid scintillation counting efficiency of ^{90}Y to the liquid scintillation counting efficiency of ^{90}Sr (i.e., $\varepsilon^l_{90Y}/\varepsilon^l_{90Sr}$), t_l is the time from chemical separation of strontium from yttrium to liquid scintillation counting, and λ is the ^{90}Y decay constant.

The Cherenkov count rates and liquid scintillation count rates of each nuclide are demonstrated by Rucker to be related to their ratio of counting efficiencies by the two counting methods as follows:

$$^{89}\text{Sr cpm}_c = ({}^{89}\text{Sr cpm}_l)(B_1) \tag{9.54}$$

and

$$^{90}\text{Sr cpm}_c = ({}^{90}\text{Sr cpm}_l)(B_2) \tag{9.55}$$

where B_1 is the ratio of the ^{89}Sr Cherenkov counting efficiency to ^{89}Sr liquid scintillation counting efficiency (i.e., $\varepsilon^C_{89Sr}/\varepsilon^l_{89Sr}$) and B_2 is the ratio of the ^{90}Sr Cherenkov counting efficiency to ^{90}Sr liquid scintillation counting efficiency (i.e., $\varepsilon^C_{90Sr}/\varepsilon^l_{90Sr}$). Equations 9.54 and 9.55 are then incorporated into Eq. 9.50 to give

$$C_1 = ({}^{89}\text{Sr cpm}_l)(B_1) + ({}^{90}\text{Sr cpm}_l)(B_2)(A_1). \tag{9.56}$$

The ratios of the counting efficiencies are determined experimentally by counting standards of known activity. The use of standard quench correction curves of counting efficiency versus a quench indicating parameter should also be feasible. It is important to keep in mind, however, that an accurate determination of the counting efficiency of ^{90}Sr is best carried out with a freshly prepared sample, which has undergone only negligible ingrowth of daughter ^{90}Y from the time of chemical separation of parent and daughter to the time of determination of the counting efficiency of ^{90}Sr.

Rucker solved simultaneously Eqs. 9.51and 9.56 to give

$$^{90}\text{Sr cpm}_l = \frac{[(B_1)(C_2) - C_1]}{[(B_1)(A_2) - (B_2)(A_1)]} \tag{9.57}$$

and

$$^{89}\text{Sr cpm}_l = C_2 - (A_2)(^{90}\text{Sr cpm}_l). \tag{9.58}$$

The count rates of ^{89}Sr and ^{90}Sr obtained from Eqs. 9.57 and 9.58 can then be converted to activities (DPM) by dividing by their respective liquid scintillation counting efficiencies. Using the activity conversion of Rucker, the sample disintegration rates can be expressed in terms of activity units per sample volume by including the sample volume, strontium recovery rate from the chemical separation, and a conversion factor to activity units as follows:

$$^{90}\text{Sr in pCi L}^{-1} = \frac{(^{90}\text{Sr cpm}_l)}{[(\varepsilon^l_{90_{Sr}})(V)(R)(2.22\,\text{DPM/pCi})]} \tag{9.59}$$

and

$$^{89}\text{Sr in pCi L}^{-1} = \frac{(^{89}\text{Sr cpm}_l)}{[(\varepsilon^l_{89_{Sr}})(V)(R)(2.22\,\text{DPM/pCi})]} \tag{9.60}$$

where $\varepsilon^l_{90_{Sr}}$ and $\varepsilon^l_{89_{Sr}}$ are the liquid scintillation counting efficiencies of ^{90}Sr and ^{89}Sr, respectively, V is the sample volume in liters (L), R is the chemical recovery in decimal percent, and 2.22 DPM/pCi is the conversion factor from sample DPM to pCi.

Quantitative recoveries are reported by Rucker (1991) using various mixtures of ^{89}Sr and ^{90}Sr standards. Obviously, the analysis of ^{90}Sr activity represents also the activity of its daughter nuclide ^{90}Y, which reaches secular equilibrium as ^{90}Sr(^{90}Y) after approximately 18 days (See Fig. 1.35 of Chapter 1).

b. Sequential Analysis with Wavelength Shifter

Walter et al. (1993) describe the activity analysis of mixtures of ^{89}Sr and ^{90}Sr(^{90}Y) using Cherenkov counting with wavelength shifter followed by liquid scintillation analysis. A wavelength shifter was used during Cherenkov counting to increase the Cherenkov counting efficiencies of ^{89}Sr and ^{90}Sr.

As in the previous method, the ^{89}Sr and ^{90}Sr of the sample are purified from ^{90}Y by chemical means. The γ-ray emitters ^{85}Sr and ^{88}Y are used as tracers to measure the chemical yields in the separation procedures, because they do not cause interference in Cherenkov or liquid scintillation counting with the exception of count rate measurements of low activity samples.

Once purified, the radiostrontium samples are counted for Cherenkov radiation in 6 mL of aqueous solution in a counting region that accepts all pulse events arising from Cherenkov photons. The aqueous solution used for Cherenkov counting was doped with a $4.0\,\mathrm{g\,L^{-1}}$ concentration of the sodium salt of 1-naphthylamine-4-sulfonic acid as wavelength shifter. Walter *et al.* (1993) report increases in Cherenkov counting efficiency with the wavelength shifter by a factor of 1.55 for ^{89}Sr and 3.2 for ^{90}Sr compared to detection efficiencies without a shifter. Cherenkov counting efficiencies reported with wavelength shifter are ^{90}Sr 6.3%, ^{89}Sr 64%, ^{90}Y 81%, ^{85}Sr 0.3% and ^{88}Y 0.2%. The latter two nuclides are used as tracers to determine the yields of the chemical separation of Sr and Y, respectively from the sample.

After the Cherenkov counting of the samples, 10 mL of liquid scintillation fluor cocktail are added to the 6 mL of aqueous solution in each sample counting vial. The liquid scintillation assay was carried out without counting region limitations, and quench correction was determined with calibration curves of counting efficiency versus an external standard quench-indicating parameter. However, in the previous step of Cherenkov counting, an external standard QIP cannot be used in most cases as described in Section III.D of this chapter.

After completion of the two count rate determinations on the same sample, Walter *et al.* (1993) used the following equations to convert the count rates to ^{89}Sr and ^{90}Sr disintegration rates:

$$D_{89} = \frac{1}{f_\varepsilon} \left[R_C\, \varepsilon_{90}^L - R_L\, \varepsilon_{90}^C \right] \tag{9.61}$$

$$D_{90} = \frac{1}{f_\varepsilon} \left[R_L\, \varepsilon_{89}^C - R_C\, \varepsilon_{89}^L \right] \tag{9.62}$$

with

$$f_\varepsilon = \varepsilon_{90}^L\, \varepsilon_{89}^C - \varepsilon_{89}^L\, \varepsilon_{90}^C \tag{9.63}$$

where D_{89} and D_{90} are the disintegration rates (DPM) of ^{89}Sr and ^{90}Sr, R_C and R_L are the net count rates (CPM) corrected for background for the Cherenkov counting and liquid scintillation counting of a sample, respectively, ε_{89}^C and ε_{90}^C are the Cherenkov counting efficiencies for ^{89}Sr and ^{90}Sr, respectively, and ε_{89}^L and ε_{90}^L are the liquid scintillation counting efficiencies of ^{89}Sr and ^{90}Sr, respectively.

A sequential Cherenkov counting followed by dual-region liquid scintillation analysis technique described by Walter *et al.* (1993), permits the assay of ^{89}Sr and ^{90}Sr even after significant ingrowth of ^{90}Y up to 6.6 half-lives of ^{90}Y with average deviations of only 6.6% from the true values.

C. Strontium-90(Yttrium-90) Exclusive of Strontium-89

The Cherenkov analysis of ^{90}Sr(^{90}Y) in the absence of ^{89}Sr generally involves the chemical separation of the ^{90}Y daughter from its parent nuclide ^{90}Sr followed by the Cherenkov counting of ^{90}Y in aqueous media (L'Annunziata, 1971). Once the ^{90}Y is separated from ^{90}Sr, it will decay by its half-life of 64.06 h as illustrated in Fig. 1.35 of Chapter 1. The activity of ^{90}Y at a given time of chemical separation from its parent is used to determine the activity of ^{90}Sr. The time of chemical separation and the time of Cherenkov counting of the isolated ^{90}Y are recorded. The ^{90}Y decay that occurs during that period is calculated to arrive at the ^{90}Y activity at the time of chemical separation from its parent ^{90}Sr, when the parent and daughter nuclides were in secular equilibrium, that is, when the activities or DPM values of the two radionuclides were equal. See Section VI.C of Chapter 1 for additional information on secular equilibrium.

Several examples of the application of the above technique to the analysis of ^{90}Sr are available from the literature including works by Bjornstad *et al.* (1992), Dewberry *et al.* (2000), Ghods, *et al.* (1994), Llaurad\u00f3 *et al.* (2001), Martin (1987), Poletiko *et al.* (1994), Rao *et al.* (2000), Shabana *et al.* (1996), Shawky and El-Tahawy (1999), Scarpitta *et al.* (1999), Suomela *et al.* (1993), Torres *et al.* (1999, 2002), and Vaca *et al.* (1999, 2001a,b). Some researchers elect to separate the ^{90}Sr(^{90}Y) by chemical precipitation of ^{90}Y followed by Cherenkov counting of ^{90}Y (Shawky and El-Tahawy, 1999), while others will follow the ingrowth of ^{90}Y by Cherenkov counting after the chemical separation of the parent and daughter nuclides via either strontium or yttrium precipitation (Scarpitta *et al.*, 1999 and Rao *et al.*, 2000). Several researchers have adopted the method of Bjornstad *et al.* (1992) and Suomela *et al.* (1993), which involves the separation of ^{90}Sr and ^{90}Y by chemical extraction of ^{90}Y with 5–10% HDEHP (di-2-etyl-hexyl-phosphoric acid) in toluene (Llaurad\u00f3 *et al.* 2001; Torres *et al.*, 1999, 2002; Vaca *et al.*, 1999, 2001a,b). The ^{90}Y in the organic phase is then back-extracted with 3 M HNO_3, precipitated as $Y(OH)_3$, and then redissolved in aqueous solution for Cherenkov counting of ^{90}Y. A brief description of the method used is provided below. Detailed procedures are available from the original papers.

The Swedish Radiation Protection Institute has issued a detailed procedure for the Cherenkov analysis of ^{90}Sr in food and environmental samples authored by Suomela *et al.* (1993). The procedure entails the following basic steps (1) ashing of the organic samples at 610°C, (2) dissolution of the ash in dilute acid, (3) the addition of yttrium carrier (10 mg of stable Y^{3+}) to facilitate extraction and chemical separation of yttrium, (4) extraction with HDEHP in toluene at low pH to remove ^{90}Y, (The Sr–Y separate at this stage of the procedure. The time of the extraction is noted, because the separated ^{90}Y can be observed to decay at its half-life of 64.06 h.

The Y collects in the organic phase, and the Sr remains in the aqueous phase), (5) back-extraction of the organic phase with $3\,M\,HNO_3$ to transfer the Y into the aqueous phase, (6) precipitation of yttrium from the aqueous phase as yttrium hydroxide at high pH, (7) dissolution of the yttrium hydroxide in $1\,mL$ of conc. HNO_3, (8) dilution of the dissolved yttrium hydroxide with $14\,mL$ of distilled water, (9) transfer of the solution to a 20-mL plastic scintillation counting vial, and finally (10) counting of the ^{90}Y in a counting region set to receive pulse events arising from Cherenkov photons. The ^{90}Sr of the sample is calculated according to the equation

$$A_{^{90}Sr} = \frac{(R - R_0)\,e^{\lambda\Delta t}}{(E)(f)} \tag{9.64}$$

where $A_{^{90}Sr}$ is the activity of ^{90}Sr in bequerels or disintegrations per second (DPS), R is the gross ^{90}Y Cherenkov count rate of the sample in counts per second, R_0 is the background Cherenkov count rate, that is, count rate of a blank counting vial in counts per second, λ is the decay constant of ^{90}Y in units of hours^{-1},

$$\lambda = 0.693/t_{1/2} = 0.693/64.06\,\text{hours} = 0.0108\,\text{hours}^{-1}$$

Δt is the time interval in hours between the chemical separation of ^{90}Y and the counting time, E is the decimal equivalent of the percent Cherenkov counting efficiency of ^{90}Y, and f is the chemical yield of the extracted yttrium.

If samples are color quenched, the Cherenkov counting efficiency of ^{90}Y can be determined with a color quench correction curve of percent counting efficiency versus a sample spectrum quench indicating parameter as described by Poletiko *et al.* (1994) or by the channels ratio method employed by Vaca *et al.* (1998) and Tarancon *et al.* (2002) and also expounded upon in detail in Sections III.B and III.C of this chapter. The Cherenkov counting efficiencies of ^{90}Y may be determined with a standard of $^{90}Sr(^{90}Y)$ in secular equilibrium. As described previously in this section the optimum Cherenkov counting efficiency of ^{90}Y in water is ~70%, while that of ^{90}Sr is only ~1% (see Fig. 9.17). Therefore, the use of standards of $^{90}Sr(^{90}Y)$ in secular equilibrium for the preparation of ^{90}Y Cherenkov color quench correction curves is convenient, as very little contribution (~1%) to the total Cherenkov count rate is made by Cherenkov photons from ^{90}Sr.

The chemical yield of yttrium is included in Eq. 9.64, as the chemical extraction of yttrium from the sample will not be complete. The method used by Suomela *et al.* (1993) is practical, because it involves chemical titration of the sample taken from the counting vial after Cherenkov counting is completed to determine the chemical yield. The aqueous sample of yttrium is transferred from the counting vial to an Erlenmeyer flask and diluted with water to permit titration with xylenolorange indicator with $0.1\,M$ Titriplex III to a yellow color. The mL of titrant required to reach the endpoint are recorded as T_{sample}. The procedure is repeated with a standard by titrating the same volume of yttrium carrier as had been added to the sample during the chemical separation and extraction procedure for yttrium. The mL of titrant required for the standard is recorded as $T_{standard}$. The chemical yield f of Eq. 9.64 is calculated as $f = T_{sample}/T_{standard}$.

D. Yttrium-90

The Cherenkov counting of ^{90}Y in the dry state and in aqueous media was previously described in Section V of this chapter. It is noted here because of the increased attention the nuclide is receiving in terms of its medical applications in the pure form, that is, extracted or "milked" in pure form without contamination from its parent ^{90}Sr. Dietz and Horwitz (1992) underscore the therapeutic applications of ^{90}Y including the radiolabeling of antibodies for tumor therapy and the production of radiolabeled particles for the treatment of liver malignancies, among others. The applications and Cherenkov counting of ^{90}Y is expounded upon in detail in Section V of this chapter.

Although the liquid scintillation counting efficiency of ^{90}Y is 100% with the exception of samples under unusually high quench conditions, Cherenkov counting is often the method of choice for the routine analysis of this nuclide, because of the high Cherenkov counting efficiencies achieved and low cost of analysis. Cherenkov counting efficiencies of ^{90}Y reach up to ~70% in pure water and > 80% with wavelength shifter under optimum conditions (see L'Annunziata and Passo, 2002; Passo and Cook, 1994; Walter et al. (1993).

E. Other Applications

In addition to the previously cited popular applications of Cherenkov counting to the analysis of ^{32}P, ^{89}Sr, ^{90}Sr(^{90}Y) in secular equilibrium and pure ^{90}Y, numerous other, less common applications are reported in the literature. Some of these are cited next.

Bem et al. (1978) described the Cherenkov counting of ^{115m}Cd in blood and animal tissue with counting efficiencies of approximately 35%.

For studies involving the removal of ^{106}Ru from nuclear waste solutions, Carmon and Dyer (1987) report the Cherenkov counting of this radionuclide in aqueous solution in the presence of weak β emitters. The optimum Cherenkov counting efficiency of ~62% was reported. Fujii et al. (2002) describe the design of a large volume (1 L) Cherenkov counting system to monitor high-energy β emitters, namely ^{89}Sr, ^{90}Y, and ^{32}P, in liquid waste detection limits of 0.003–0.007 and 0.001–0.003 Bq mL^{-1} for 5 and 20 min measurements, respectively.

^{228}Th and ^{234}Th can be analyzed via Cherenkov counting of their β-emitting daughters. Al-Masri and Blackburn (1994) and Nour et al. (2002) demonstrate the analysis of these nuclides when in secular equilibrium with their daughters. The ^{234m}Pa daughter nuclide of ^{234}Th yields a Cherenkov counting efficiency of up to 62% and the ^{212}Bi, ^{212}Pb and ^{208}Tl daughter nuclides of ^{228}Th can be detected with a reported Cherenkov counting efficiency of up to 53%. The Cherenkov analysis of ^{234}Th can be used for the determination of uranium when ^{234}Th is in secular equilibrium with its parent ^{238}U (see Blackburn and Al-Masri, 1994); Al-Masri and Blackburn, 1994); Bower *et al.*, 1994).

^{210}Pb has been applied to studies of container adsorption of Pb(II) by Pacer (1990). The analysis of ^{210}Pb by Cherenkov counting is possible only through the detection of the Cherenkov photons produced by the daughter nuclide ^{210}Bi, which reaches secular equilibrium with its parent nuclide. The Cherenkov counting efficiency of ^{210}Pb(^{210}Bi) without wavelength shifter is reported to be approximately 18% (Blais and Marshall, 1988).

Blackburn and Al-Masri (1993) and Al-Masri and Blackburn (1995) demonstrated the analysis of ^{222}Rn and ^{226}Ra via the Cherenkov counting of their β-emitting daughters ^{214}Bi and ^{214}Pb with a Cherenkov counting efficiency of 77%. These techniques were subsequently applied to the analysis of ^{222}Rn in surface waters (Al-Masri and Blackburn, 1999).

The Cherenkov counting efficiency of ^{36}Cl is low. An optimum counting efficiency of only 6.6% can be expected due to the low E_{max} of its β particles (see Table 9.11). Nevertheless, Holst *et al.* (2000) elected to analyze ^{36}Cl tracer nuclide in biological studies by Cherenkov counting because analysis could be carried out without liquid scintillation cocktail, which would render the sample useless for further studies.

Buchtela and Tschurlovits (1975) reported a Cherenkov counting efficiency of ~75% for ^{42}K in water. It has very limited use as a tracer nuclide, because of its relative short half-life of 12.36 hours. The isotope ^{40}K can be analyzed also in water with a Cherenkov detection efficiency of 55% as reported by Pullen (1986), and the Cherenkov counting technique has been applied by Rao *et al.* (2000) to the analysis of ^{40}K in natural water samples. Because of the lack of a suitable radiotracer for K and the close chemistry of the Rb and K elements, it is at times possible to use ^{86}Rb as a tracer for potassium. L'Annunziata and coworkers (Noor *et al.* 1996) and Tuininga *et al.* (2002), report the use of ^{86}Rb as a radiotracer with a Cherenkov counting efficiency of up to 55% in water.

XIII. ADVANTAGES AND DISADVANTAGES

It is obvious that Cherenkov counting cannot compete with liquid scintillation counting as far as highest possible counting efficiency is concerned. In addition, Cherenkov counting cannot be used for very low-energy beta particle-emitters. However, if very high counting efficiencies are not necessary, Cherenkov counting of relatively high-energy beta particle-emitters has the following advantages over liquid scintillation counting:

1. Samples may be counted directly in aqueous solution or in organic solvents.
2. No floors, fluor cocktails, or other compounds need be added, and thus the sample remains unadulterated can be used for subsequent studies or analysis.
3. A larger volume of sample may be counted in a 20-mL counting vial since floor solution is not needed.

4. Chemical quenching in nonexistent. Solutes may change the refractive index of the medium and consequently the beta particle-energy threshold for Cherenkov production, but chemical quench *per se* does not occur.

5. Time and expense are saved in sample preparation.

6. There is no interference from other radionuclides that cannot produce Cherenkov photons (e.g., ^{3}H, ^{14}C, ^{125}I and other nuclides that do not emit beta-particles of energy equal to or in excess of the threshold energy for the production of Cherenkov photons or radionuclides that do not yield significant Compton electrons of sufficient energy to create Cherenkov photons.

7. There are no chemical fluor disposal problems associated with Cherenkov counting.

XIV. RECOMMENDATIONS

The following recommendations can be made to facilitate the Cherenkov analysis of radionuclides and optimize Cherenkov counting performance:

1. Always use plastic vials for Cherenkov counting unless a wavelength shifter is used. Plastic vials provide higher Cherenkov counting efficiencies and lower background than glass vials (i.e., higher figures of merit). In addition, they are less expensive than glass and unbreakable.

2. Unless samples are totally colorless or have a constant quench level, prepare a color quench correction curve based on the color of the experimental samples. A color quench correction curve is preferred at all times, because even slight amounts of color could cause significant quench and loss of counting efficiency. Also, most contemporary liquid scintillation analyzers can use the sample spectrum quench-indicating parameter and a quench correction curve stored in the instrument computer for automatic conversion of Cherenkov count rate to activity, which facilitates automation and high sample throughput.

3. For samples of low Cherenkov count rates (e.g. <1000 CPM), the liquid scintillation analyzer will require long counting times to determine a sample spectrum quench-indication parameter [e.g., SIS or SQP(I)]; error can still occur. In such cases, an internal standard method may be used to determine the Cherenkov counting efficiency.

4. Maintain the samples for counting at a constant volume in counting vials unless it can be verified through testing that there are no significant volume effects on counting efficiency over a certain range of volume changes.

5. All samples should be counted for a period of time sufficient to achieve a % 2 sigma (% 2S) error of 1% or lower to ensure accurate count determinations and good measurements of the sample spectrum quench indicating parameter.

REFERENCES

Abbas Rizvia, S. M., Hennikerb, C. A. J., Goozeea, G., and Allen, B. J. (2002). In vitro testing of the leukaemia monoclonal antibody WM-53 labeled with alpha and beta emitting radioisotopes. *Leukemia Res.* **26**(1), 37–43.

Adachi, I., Sumiyoshi, T., Hayashi, K., Iida, N., Enomoto, R., Tsukada, K., Suda, R., Matsumoto, S., Natori, K., Yokoyama, M., and Yokogawa, H. (1995). Study of a threshold Cherenkov counter based on silica aerogels with low refractive indices. *Nucl. Instrum. Methods Phys. Res., Sect. A* **355**, 390–398.

Al-Masri, M. S. (1996). Cerenkov counting technique. *J. Radioanal. Nucl. Chem., Articles* **207**, 205–213.

Akatsu, M., Aoki, M., Fujimoto, K., Higashino, Y., Hirose, M., Inami, K., Ishikawa, A., Matsumoto, T., Misono, K., Nagai, I., Ohshima, T., Sugi, A., Sugiyama, A., Suzuki, S., Tomoto, M., and Okuno, H. (2000). Time-of-propagation Cherenkov counter for particle identification. *Nucl. Instrum. Methods Phys. Res., Sect. A* **440**, 124–135.

Al-Masri, M. S. and Blackburn, R. (1994). Simultaneous determination of ^{234}Th and ^{228}Th in environmental samples using Cerenkov counting. *Radiochim. Acta* **65**, 133–136.

Al-Masri, M. S. and Blackburn, R. (1995a). Application of Cerenkov radiation for the assay of ^{226}Ra in natural water. *Sci. Total Environ.* **173/174**, 53–59.

Al-Masri, M. S. and Blackburn, R. (1995b). Radiochemical determination of ^{226}Ra using Cerenkov Counting. *J. Radioanal. Nucl. Chem., Articles* **195**, 339–344.

Al-Masri, M. S. and Blackburn, R. (1999). Radon-222 and related activities in surface waters of the English Lake District. *Appl. Radiat. Isot.* **50**, 1137–1143.

Angelov, D., Charra, M., Seve, M., Côte, J., Saadi, K., and Dimitrov, S. (2000). Differential remodeling of the HIV-1 nucleosome upon transcription activators and SWI/SNF complex binding. *J. Molec. Biol.* **302**(2), 315–326.

Anonymous (1993). Quench and quench correction. *TopCount Topics*, TCA-015, pp. 8. PerkinElmer Life and Analytical Sciences, Boston.

Anonymous (1996). Cerenkov counting performance on the TopCount® microplate scintillation and luminescence counter. *TopCount Topics*, TCA-024, pp. 4. PerkinElmer Life and Analytical Sciences, Boston.

Attas, M. and Abushady, I. (1997). "Cerenkov Viewing Device for Spent Fuel Verification at Light Water Reactors." IAEA Inspector Training Manual. © IAEA. Atomic Energy of Canada, Ltd., Pinawa, Manitoba.

Attas, E. M., Chen, J. D., and Young, G. J. (1990). A Cherenkov viewing device for used-fuel verification. *Nucl. Instrum. Methods Phys. Res., Sect. A* **299**, 88–93.

Attas, E. M., Chen, J. D., and Young, G. J. (1992). An ultraviolet imager for nuclear safeguards inspectors. Proceedings of the International Meeting on Electron Tubes and Image Intensifiers, San Jose, CA, Feb. 10-12, 1992. *Proc SPIE* **1655**, 50–57.

Attas, E. M., Burton, G. R., Dennis Chen, J., Young, G. J., Hildingsson, L., and Trepte, O. (1997). A nuclear fuel verification system using digital imaging of Cherenkov light. *Nucl. Instrum. Methods Phys. Res., Sect. A* **384**, 522–530.

Banavali, A. D., Moreno, E. M., and McCurdy (1992). Strontium-89, strontium-90 analysis by Eichrom column chemistry and Cerenkov counting. Abstract in *Proc. 38th Annual Conf. On Bioassay, Analytical and Environmental Radiochem.* Santa Fe, New Mexico.

Barnykov, M. Yu., Buzykaev, A. R., Danilyuk, A. F., Ganzhur, S. F., Goldberg, I. I., Kolachev, G. M., Kononov, S. A., Kravchenko, E. A., Mikerov, V. I., Muraviova, T. M., Minakov, G. D., Onuchin, A. P., Sidorov, A. V., and Tayursky, V. A. (1998). Development of aerogel Cherenkov counters with wavelength shifters and phototubes. *Nucl Instrum. Methods Phys Res., Sect. A* **419**, 584–589.

Baxter, J. W., Pickett, S. T. A., Dighton, J., and Carreiro, M. M. (2002). Nitrogen and phosphorus availability in oak forest stands exposed to contrasting anthropogenic impacts. *Soil Biol. Biochem.* **34**, 623–633.

Bélanger, M., Desjardins, P., Chatauret, N., and Butterworth, R. F. (2002). Loss of expression of glial fibrillary acidic protein in acute hyperammonemia. *Neurochem. Int.* **41**, 155–160.

Belcher, E. H. (1953). The luminescence of irradiated transparent media and the Čerenkov effect. I. The luminescence of aqueous solutions of radioactive isotopes. *Proc. Royal Soc.* **A216**, 90–102.

Bem, E. M., Bem, H., and Reimschussel, W. (1978). The use of Cherenkov radiation in the measurement of cadmium-115m in blood and other tissues. *Bull. Environm. Contam. Toxicol.* **19**, 677–683.

Bem, E. M., Bem, H., and Reimschuessel, W. (1980). Determination of phosphorus-32 and calcium-45 in biological samples by Čerenkov and liquid scintillation counting. *Int. J. Appl. Radiat. Isotopes* **31**, 371–374.

Bem. E. M., Bem, H., and Reimschussel, W. (1983). Direct determination of phosphorus-32 or chlorine-36 in tissue solubilizer solutions by Cherenkov scintillation counting. *J. Radioanal. Chem.*, **79**, 69–76.

BenZikri, A. (2000). Cerenkov counting. *Health Phys.* **79**(5), S70-S71.

Berger, S. L. (1984). The use of Cherenkov radiation for monitoring reactions performed in minute volumes: examples from recombinant DNA technology. *Anal. Biochem.*, **136**, 515–519.

Bezaguet, A., Geles, C., Reucroft, S., Marini, G., and Martellotti, G. (1979). Experience with di-potassium-2-amino-6,8-naphthalene-disulphonic acid as a wavelength shifter in a water Cherenkov detector. *Nucl. Instrum. Meth.* **158**, 303–306.

Bjornstad, H. R., Lien, H. N., Yu-Fu, Y., and Salbu, B. (1992). Determination of ^{90}Sr in environmental and biological materials with combined HDEHP solvent extraction – low liquid scintillation counting technique. *J. Radioanal. Nucl. Chem. Articles* **156**(1), 165–173.

Blackburn, R. and Al-Masri, M. S. (1993). Determination of radon-222 and radium-226 in water samples by Cerenkov counting. *Analyst* **118**, 873-876.

Blackburn, R. and Al-Masri, M. S. (1994). Determination of uranium by liquid scintillation and Cerenkov counting. *Analyst* **119**, 465–468.

Blais, J. S. and Marshall, W. D. (1988). Determination of lead-210 in admixture with bismuth-210 and polonium-210 in quenched samples by liquid scintillation counting. *Anal. Chem.* **60**, 1851–1855.

Bockhorst, M., Heinloth, K., Pajonk, G. M., Begag, R., and Elaloui, E. (1995). Fluorescent dye doped aerogels for the enhancement of Cherenkov light detection. *J. Non–Cryst. Solids* **186**, 388–394.

Bower, K., Angel, A., Gibson, R., Robinson, T., Knobeloch, D., and Smith, B. (1994). Alpha-beta discrimination liquid scintillation counting for uranium and its daughters. *J. Radioanal. Nucl. Chem. Articles* **181**, 97–107.

Brajnik, D., Korpar, S., Medin, G., Starič, M., and Stanovnik, A. (1994). Measurement of ^{90}Sr activity with Cherenkov radiation in a silica aerogel. *Nucl. Instrum. Methods Phys. Res., Sect A.* **353**, 217–221.

Brajnik, D., Medin, G., Stanovnik, A., and Starič, M. (1995). Determination of high energy β-emitters with a Ge spectrometer or Cherenkov radiation in an aerogel. *Sci. Total Environ.* **173/174**, 225–230.

Brinker, C. J. and Sherer, G. W. (1999). "Sol-Gel Science." Academic Press, San Diego, pp. 97.

Buchtela, K. and Tschurlovits, M. (1975). A new method for determination of ^{89}Sr, ^{90}Sr and ^{90}Y in aqueous solution with the aid of liquid scintillation counting. *Int. J. Appl. Radiat. Isot.* **26**, 333–338.

Bunnenberg, C., Kraul, K., and Kühn, W. (1987). Analysis of beta-nuclides by Cherenkov-spectrometry. *Nucl. Instrum. Methods Phys. Res., Sect. A* **255**, 346–350.

Burden, D. L. and Hieftje, G. M. (1998). Cerenkov radiation as a UV and visible light source for time-resolved fluorescence. *Anal. Chem.* **70**(16), 3426–3433.

Campbell, A. M., Bailey, I. H., and Burton, M. A. (2000). Analysis of the distribution of intra-arterial microspheres in human liver following hepatic yttrium-90 microsphere therapy. *Phys. Med. Biol.* **45**(4), 1023–1034.

Campbell, A. M., Bailey, I. H., and Burton, M. A. (2001). Tumour dosimetry in human liver following hepatic yttrium-90 microsphere therapy. *Phys. Med. Biol.* **46**(2), 487–498.

Cantin, M., Casse, M., Koch, L., Jouan, R., Mestreau, P., Roussel, D., Bonnin, F., Moutel, J., and Teichner, S. J. (1974). Silica aerogels used as Cherenkov radiators. *Nucl. Instrum. Methods* **118**, 177–182.

Carmon, B. and Dyer, A. (1987). Čerenkov spectroscopic assay of fission isotopes. II. Cherenkov counting of ^{106}Ru on UV-colour-quenched solutions containing other β-emitters. *J. Radioanal. Nucl. Chem. Articles* **109**, 229–236.

Čerenkov, P. A. (1934). Visible glow of pure liquids under the influence of γ-rays. *Compt. Rend. Acad. Sci. URSS*, **2**, 451–454.

Čerenkov, P. A. (1937). Visible radiation produced by electrons moving in a medium with velocities exceeding that of light. *Phys. Rev.* **15**, 378.

Chang, T.-M., Chen, S.-C., King, J.-Y., and Wang, S.-J. (1996). Rapid and accurate determination of $^{89/90}$Sr in radioactive samples by Cerenkov counting. *J. Radioanal. Nucl. Chem. Articles* **204**, 339–347.

Chimura, A. J., Schmidt, B. D., Corson, D. T., Traviglia, S. L., and Meares, C. F. (2002). Electrophilic chelating agents for irreversible binding of metal chelates to engineered antibodies. *J. Controlled Release* **78**(1–3), 249–258.

Chow, N. P. (1980). *In* "Liquid Scintillation Counting, Recent Applications and Development," Vol. 1, (C. T. Peng, D. L. Horrocks and E. L. Alpen, Eds.) Academic Press, New York and London.

Chu, T.-C., Wang, J.-J., and Lin, Y.-M. (1998). Radiostrontium analytical method using crown-ether compound and Cerenkov counting and its applications in environmental monitoring. *Appl. Radiat. Isot.* **49**(12), 1671–1675.

Claus, R., Seidel, S., Sulak, L., Bionta, R. M., Blewitt, G., *et al.*, numerous authors. (1987). A waveshifter light collector for a water Cherenkov detector. *Nucl. Instrum. Methods Phys. Res., Sect. A* **261**, 540–542.

Clausen, T. (1968). Measurement of phosphorus-32 activity in a liquid scintillation counter without the use of scintillator. *Anal. Biochem.*, **22**, 70–73.

Cook, G. T., Passo Jr., and Carter, B. D. (1998). Environmental liquid scintillation analysis. *In* "Handbook of Radioactivity Analysis" (M. F. L'Annunziata, Ed.), 1st ed., pp. 331-386. Academic Press, San Diego.

Coursey, B. M., Calhoun, J. M., and Cessna, J. T. (1993). Radioassays of yttrium-90 used in nuclear medicine. *Nucl. Med. Biol.* **20**(5), 693–700.

Curie, E. (1941). "Madam Curie." Heinemann, London.

Dewberry, R. A., Leyba, J. D., and Boyce, W. T. (2000). Observation and measurement of ^{79}Se in Savannah River Site high level waste tank fission product waste. *J. Radioanal. Nucl. Chem.* **245**(3), 491–500.

Dietz, M. L. and Horwitz, E.P. (1992). Improved chemistry for the production of yttrium-90 for medical applications. *Appl. Radiat. Isot.* **43**, 1093–1101.

Elrick, R. H. and Parker, R. P. (1968). The use of Cerenkov radiation in the measurement of β-emitting radionuclides. *Int. J. Appl. Radiat. Isotopes*, **19**, 263–271.

Fardeau, J.-C. (1984). *Fert. Boric.*, 86, 23–30.

Frank, I. M. (1960). Optics of light sources moving in refractive media. *Science* **131**, 702–712.

Frank, I. M. and Tamm, I. G. (1937). Coherent visible radiation of fast electrons passing through matter. *Dokl. Akad. Nauk. SSSR*, **14**, 109.

Frič, F. and Finocchiaro, V. (1975). Cherenkov counting efficiency increase of phosphorus-32 using anthranilic acid or quinine as a wavelength shifter. *Radiochem. Radioanal. Lett.* **21**, 205–210.

Frič, F. and Palovčíková, V. (1975). Automatic liquid scintillation counting of ^{32}P in plant extracts by measuring the Cerenkov radiation in aqueous solutions. *Int. J. Appl. Radiat. Isotopes* **26**, 305–311.

Fricke, J. and Tillotson, T. (1997). Aerogels: production, characterization, and applications. *Thin Solid Films* **297**, 212–223.

Fujii, H. and Takiue, M. (1988a). Radioassay of dual-labeled samples by sequential Cherenkov counting and liquid scintillation efficiency tracing technique. *Nucl. Instrum. Methods Phys. Res., Sect. A* **273**, 377–380.

Fujii, H. and Takiue, M. (1988b). Radioassay of alpha- and beta-emitters by sequential Cherenkov and liquid scintillation countings. *Appl. Radiat. Isot.* **39**, 327–330.

Fujii, H. and Takiue, M. (1988c). Radioassay of dual-labeled samples with a Cherenkov counting technique. *Nucl. Instrum. Methods Phys. Res., Sect. A* **265**, 558–560.

Fujii, H., Matsuno, K., and Takiue, M. (2002). Construction of a liquid waste monitor using the Cerenkov technique for nuclear medicine. *In* "Proceedings of the LSC 2001 Conference," Karlsruhe, Germany, May 7–11, 2001. Radiocarbon, Tucson (In Press).

Ghods, A., Hussain, M., and Mirna, A. (1994). Rapid determination of ^{90}Sr in foodstuffs and environmental material. *Radiochim. Acta* **65**, 271–274.

Gittens, J. R., Schuler, M. A., and Strid, Å. (2002). Identification of a novel nuclear factor-binding site in the *Pisum sativum sad* gene promoters. *Biochim. Biophys. Acta* **1574**, 231–244.

Glass, A. D. M. (1978). An improved method of sample preparation for the radioassay of β-particle emitters by Čerenkov counting. *Int. J. Appl. Radiat. Isotopes* **29**, 75–76.

Glässel, P. (1999). The limits of the ring image Cherenkov technique. *Nucl. Instrum. Methods Phys. Res., Sect. A* **433**, 17–23.

Gougas, A. K., Ilie, D., Ilie, S., and Pojidaev, V. (1999). Behavior of hydrophobic aerogel used as a Cherenkov medium. *Nucl. Instrum. Methods Phys. Res., Sect. A* **421**, 249–255.

Grahek, Ž., Zečević, N., and Lulić, S. (1999). Possibility of rapid determination of low-level ^{90}Sr activity by combination of extraction chromatography separation and Cherenkov counting. *Anal Chim. Acta* **399**, 237–247.

Grande, M. and Moss, G. R. (1983). An optimized thin film wavelength shifting coating for Cherenkov detection. *Nucl. Instrum. Meth.* **215**, 539–548.

Grau Carles, A. and Grau Malonda, A. (1992). "Precise System for the Determination of the Quench Parameter of Radioactive Samples in Liquid Phase," Patent P. 9202639, December 29, 1992, Registry of Industrial Property, Madrid, Spain.

Grau Carles, A. and Grau Malonda, A. (1995). Radionuclide standardization by Cherenkov counting. *Appl. Radiat. Isot.* **46**, 799–803.

Grau Carles, A. and Grau Malonda, A. (1996). "Aplicación de la Radiación Cherenkov a la Metrología de Radionucleidos." Editorial CIEMAT, Avda. Complutense, 22, 28040 Madrid, pp. 140.

Grau Carles, A. and Grau Malonda, A. (1998). CHEREN, the Cherenkov counting efficiency. *Comp. Phys. Commun.* **111**, 258–264.

Grau Carles, A., Grau Malonda, A., and Rodríguez Barquero, L. (1993a). Cherenkov radiation effects on counting efficiency in extremely quenched liquid scintillation samples. *Nucl. Instrum. Methods Phys. Res., Sect. A* **334**, 471–476.

Grau Malonda, A. and Grau Carles, A. (1998a). The anisotropy coefficient in Cerenkov counting. *Proc. ICRM 1997*, National Institute of Standards and Technology, Gaithersburg, MD.

Grau Malonda, A. and Grau Carles, A. (1998b). The anisotropy coefficient in Cerenkov counting. *Appl. Radiat. Isot.* **49**(9–11), 1049–1053.

Grau Malonda, A., and Grau Carles, A. (2002). Half-life determination of ^{40}K by LSC. *Appl. Rdaiat. Isot.* **56**, 153–156.

Grichine, V. M. (2001). On irreducible fluctuations of Cherenkov radiation. *Nucl. Instrum. Methods Phys. Res., Sect. A* **463**, 418–423.

Gruhn, C. R. and Ogle, W. (1980). *In* "Liquid Scintillation Counting, Recent Applications and Development" (C.-T. Peng, D. L. Horrocks and E. L. Alpen, Eds.), Vol. 1, pp. 357–374, Academic Press, New York and London.

Holst, P. B., Christophersen, C., and Engvild, K. C. (2000). In vivo incorporation of radioactive ^{36}Cl, a method for monitoring chloro compounds in biological material. *J. Chromatogr. A* **903**, 267–270.

Horwitz, E. P., Chiarizia, R., and Dietz, M. L. (1992). A novel and strontium-selective extraction chromatographic resin. *Solvent Extract. Ion Exchange* **10**(2), 313–336.

Hülsen, W. and Prenzel, U. (1968). "Dry" counting of ^{32}P in liquid scintillation counters. *Anal. Biochem.* **26**, 483-484.

Iodice, M., Cisbani, E., Colilli, S., Crateri, R., Frullani, S., Garibaldi, F., Giuliani, F., Gricia, M., Lucentini, M., Mostarda, A., Pierangeli, L., Santavenere, F., Urciuoli, G. M., De Leo, R., Lagamba, L., Leone, A., Perrino, R., Kerhoas, S., Lugol, I. C., Mazaev, B., Vernin, P., and Zaccarian, A. (1998). The CO_2 gas Cherenkov detectors for the Jefferson Lab Hall-A spectrometers. *Nucl. Instrum. Methods Phys. Res., Sect. A* **411**, 223–237.

Ishino, M., Chiba, J., En'yo, H., Funahashi, H., Ichikawa, A., Ieiri, M., Kanda, H., Masaike, A., Mihara, S., Miyashita, T., Murakami, T., Nakamura, A., Naruki, M., Muto, R., Ozawa, K.,

Sato, H. D., Sekimoto, M., Tabaru, T., Tanaka, K. H., Yoshimura, Y., Yokkaichi, S., Yokoyama, M., and Yokogawa, H. (2001). Mass production of hydrophobic silica aerogel and readout optics of Cherenkov light. *Nucl. Instrum. Methods Phys. Res., Sect. A* **457**, 581–587.

Jelley, J. V. (1958). "Cerenkov Radiation and its Applications." Pergamon Press, New York.

Jelley, J. V. (1962). Cerenkov radiation: its origin, properties and applications. *Contemp. Phys.* **3**, 45–57.

Joram, C. (2002). The evolution of the Cherenkov imaging technique in high energy physics. *Nucl. Phys. B (Proc. Suppl.)* **109B**, 153–161.

Kannan, S. (1975). Measurement of ^{59}Fe in plant materials using Cerenkov radiation. *Int. J. Appl. Radiat. Isotopes* **26**, 557–558.

Karamanos, R. E., Bettany, J. R., and Rennie, D. A. (1975). Lead-210 assay in soil and plant material using Cerenkov radiation. *Can. J. Soil Sci.* **55**, 407–413.

Kellogg, T. F. (1982). Triton-X-100 acts as a scintillant when used as a wavelength shifter in Čerenkov counting systems. *Int. J. Appl. Radiat. Isotopes* **33**, 165–170.

Kellogg, T. F. (1983). The effect of sample composition and vial type on Cerenkov counting in a liquid scintillation counter. *Anal. Biochem.* **134**, 137–143.

Kistler, S. S. (1931). Coherent expanded aerogels and jellies. *Nature* **127**, 741-

Križan, P. (2001). Recent progress in Čerenkov counters. *IEEE Trans. Nucl. Sci.* **48**(4), 941–949.

Kulcsar, F., Teherani, D., and Altmann, H. (1982). Study of the spectrum of Cherenkov light. *J. Radioanal. Chem.*, **68**, 161–168.

Kuribara, M. (1994). Spent fuel burnup estimation by Cerenkov glow intensity measurement. *IEEE Trans. Nucl. Sci.* **41**(5), 1736–1739.

Kuribara, M. and Nemoto, K. (1994). Development of new UV – I.I. Cerenkov viewing device. *IEEE Trans. Nucl. Sci.* **41**(1), 331–335.

Kushita, K. N. and Du, J. (1998). Radioactivity measurement of 188Re by Cherenkov counting. *Appl. Radiat. Isot.* **49**(9–11), 1069–1072.

L'Annunziata, M. F. (1971). Birth of a unique parent-daughter relation: secular equilibrium. An experiment in radioisotope techniques. *J. Chem. Educ.* **48**, 700–703.

L'Annunziata, M. F. (1979). "Radiotracers in Agricultural Chemistry," Academic Press, New York and London.

L'Annunziata, M. F. (1984a). The detection and measurement of radionuclides. *In* "Isotopes and Radiation in Agricultural Sciences" (M. F. L'Annunziata and J. O. Legg, Eds.), Vol. 1, pp. 142–231, Academic Press, New York and London.

L'Annunziata, M. F. (1984b). Agricultural biochemistry: Reaction mechanisms and pathways in biosynthesis. *In* "Isotopes and Radiation in Agricultural Sciences" (M.F. L'Annunziata and J. O. Legg, Eds.), Vol. 2, pp. 106–182, Academic Press, New York and London.

L'Annunziata, M. F. (1987). "Radionuclide Tracers, Their Detection and measurement." pp. 505. Academic Press, New York and London.

L'Annunziata, M. F. (1997). Cerenkov counting of ^{32}P-Instrument performance data. *Counter Intel. Tri-Carb LSC Application Note*, **002**, p. 6. PerkinElmer Life and Analytical Sciences, Boston.

L'Annunziata, M. F. and Fuller, W. H. (1968). The chelation and movement of ^{89}Sr-^{90}Sr(^{90}Y) in a calcareous soil. *Soil Sci.* **105**, 311–319.

L'Annunziata, M. F. and Passo, C. J., Jr. (2002). Cherenkov counting of yttrium-90 in the dry state; correlations with phosphorus-32 Cherenkov counting data. *Appl. Radiat. Isot.* **56**, 907–916.

Läuchli, A. (1969). Radioassay for β-emitters in biological materials using Cerenkov radiation. *Int. J. Appl. Radiat. Isotopes* **20**, 265–270.

Läuchli, A. (1971). Application of Cerenkov counting to ion transport studies in plants. *In* "Organic Scintillators and Liquid Scintillation Counting" (D. L. Horrocks and C.-T. Peng, Eds.), pp. 771–782, Academic Press, New York and London.

Lee, M. h., Chung, K. H., Choi, G. K., and Lee, C. W. (2002). Measurement of ^{90}Sr in aqueous samples using liquid scintillation counting with full spectrum DPM method. *Appl. Radiat. Isot.* **57**, 257–263.

Lefebvre, D. D. and Glass, A. D. M. (1981). A rapid method for the separation of mixed ^{32}P- and ^{86}Rb-labelled plant extracts by anion-exchange chromatography. *Int. J. Appl. Radiat. Isotopes*, **32**, 116–117.

Lehmann, K. A. and Bass, B. L. (1999). The importance of internal loops within RNA substrates of ADAR1. *J. Molec. Biol.* **291**(1), 1–13.

Lickly, T. D., Quinn, T., Blanchard, F. A., and Murphy, P. G. (1988). Determination of migration of phosphorus-based additives from food packaging material into food-simulating solvents by neutron activation/Cerenkov counting. *Appl. Radiat. Isot.* **39**, 465–470.

Lide, D. R. (2001). "CRC Handbook of Chemistry and Physics," 81st ed., pp. 10–220, CRC Press, Boca Raton, FL.

Lilley, J. S. (2001). "Nuclear Physics," pp. 393, John Wiley & Sons, Ltd., New York.

Lindsey, C. S., Lindblad, T., Waldemark, K., and Hildingsson, L. (1999). Nuclear fuel assembly assessment project and image categorization. Proceedings of the Ninth Workshop on Virtual Intelligence, Dynamic Neural Networks. Royal Institute of Technology, KTH, Stockholm, Sweden, June 22–26, 1998. *Proc. SPIE* **3728**, 491–499.

Llauradó, M., Torres, J. M., Tent, J., Sahuquillo, A., Muntao, H., and Rauret, G. (2001). Preparation of a soil reference material for the determination of radionuclides. *Anal. Chim. Acta* **445**, 99–106.

Luciani, M. G., Hutchins, J. R. A., Zheleva, D., and Hupp, T. R. (2000). The C-terminal regulatory domain of p53 contains a functional docking site for cyclin A. *J. Molec. Biol.* **300**(3), 503–518.

Mallet, L. (1929). The ultra-violet radiation of substances subjected to γ-rays. *Compt. Rend. Acad. Sci. Paris*, **188**, 445–447.

Marshall, J. (1952). Particle counting by Cerenkov radiation. *Phys. Rev.* **86**, 685–693.

Martin, J. E. (1987). Measurement of ^{90}Sr in reactor wastes by Cerenkov counting of ^{90}Y. *Appl. Radiat. Isot.* **38**, 953–957.

Matsushita Electric Works, Ltd. (1992). Japanese Patent Early Publication (KOKAI) No. 1992-198238.

Moir, A. T. B. (1971). Channels ratio quench correction using Cherenkov radiation for the assay of ^{42}K in biological samples. *Int. J. Appl. Radiat. Isot.* **22**, 213–216.

Moore, F., Asbóth, G., and López Bernal, A. (2002). Thromboxane receptor signaling in human myometrial cells. *Prostaglandins Lipid Mediators* **67**, 31–47.

Morel, C. and Fardeau, J. C. (1989). The uptake by crops of fresh and residual phosphatic fertilizers by simultaneous measurements with ^{32}P and ^{33}P. *Appl. Radiat. Isot.* **40**, 273-278.

Morita-Murase, Y., Murakami, I., and Homma, Y. (2000). Cerenkov counting efficiencies for β^--emitters in dry state in glass vials. *Chem Lett.* **10**, 1158–1159.

Navarro, N., Grau Carles, A., Alvarez, A., Salvador, S., and Gomez, V. (1997). Standardization of ^{234}Th by Cerenkov counting. *Appl. Radiat. Isot.* **48**(7), 949–952.

Nicolaon, G. A. and Teichner, S. J. (1968). New preparation process for silica xerogels and aerogels, and their textural properties. *Bull. Soc. Chim. France*, **1968**(5), 1900–1906. (U.S. Patent No. 3,672,833 (1972).

Noor, A., Zakir, M., Rasyid, B., Maming, and L'Annunziata, M. F. (1996). Cerenkov and liquid scintillation analysis of the triple label ^{86}Rb-^{35}S-^{33}P. *Appl. Radiat. Isot.* **47**, 659–668.

Nour, S., Burnett, W. C., and Horwitz, E. P. (2002). ^{234}Th analysis of marine sediments via extraction chromatography and liquid scintillation counting. *Appl. Radiat. Isot.* **57**, 235–241.

Ohshima, T. (2000). Test result of time-of-propagation Cherenkov counter. *Nucl. Instrum. Methods Phys. Res., Sect. A* **453**, 331-335.

Østby, L. and Krøkje, Å. (2002). Cytochrome P450 (CYP1A) induction and DNA adducts in a rat hepatoma cell line (Fao), exposed to environmentally relevant concentrations of organic compounds, singly and in combinations. *Environ. Toxicol. Pharmacol.* **12**, 15–26.

Pacer, R. A. (1990). Effect of container adsorption on the ^{210}Pb/^{210}Bi secular equilibrium relationship in spiked aqueous solutions. *J. Radioanal. Nucl. Chem. Articles* **139**, 255–262.

Palmer, F. B. St. C. (1969). Measurement of phosphorus-32 following colorimetric phosphorus determination. *Anal. Biochem.* **31**, 493–501.

Pandeya, U., Mukherjeea, A., Chaudharyb, P. R., Pillaib, M. R. A., and Venkatesh, M. (2001). Preparation and studies with ^{90}Y-labelled particles for use in radiation synovectomy. *Appl. Radiat. Isot.* **55**(4), 471–475.

Paredes, C. H., Landolt, R. R., Kessler, W. V., and Shoemaker, R. L. (1980). A comparison of proportional and Cherenkov counting for sulfur tablet neutron dosimeters. *Health Phys.* **39**, 564–566.

Parker, R. P. and Elrick, R. H. (1966). The assay of β-emitting radioisotopes using Cerenkov counting. *Int. J. Appl. Radiat. Isotopes* **17**, 361–362.

Parker, R. P. and Elrick, R. H. (1970). Cerenkov counting as a means of assaying β-emitting radionuclides. *In* "The Current Status of Liquid Scintillation Counting" (E. D. Bransome, Jr., Ed.), pp. 110–122, Grune and Stratton, New York and London.

Passo Jr., C. J. and Cook, G. T. (1994). "Handbook of Environmental Liquid Scintillation Spectrometry." PerkinElmer Life and Analytical Sciences, Boston.

Perrino, R., Lagamba, L., Cisbani, E., Colilli, S., Crateri, R., LeLeo, R., Frullani, S., Garibaldi, F., Giuliani, F., Gricia, M., Iodice, M., Iommi, R., Leone, A., Lucentini, M., Mostarda, A., Nappi, E., Pierangeli, L., Santavenere, F., and Urciuoli, G. M. (2001). Performances of the aerogel threshlof Cherenkov counter for the Jefferson lab Hall A spectrometers in the 1–4 GeV/*c* momentum range. *Nucl. Instrum. Methods Phys. Res., Sect. A* **457**, 571–580.

Pestotnik, R., Križan, P., Korpar, S., Bračko, M., Starič, M., and Stanovnik, A. (2002). Investigation of ^{90}Sr detection with Cherenkov radiation in silica aerogels. *In* "Proceedings of IEEE Symposium on Nuclear Science and Medical Imaging including Nuclear Power Systems." Nov. 4–9, 2001, San Diego, IEEE, Piscataway, NJ.

Pinto da Cunha, J., Neves, F., and Lopes, M. I. (2000). On the reconstruction of Cherenkov rings from aerogel radiators. *Nucl. Instrum. Methods Phys. Res., Sect. A* **452**, 401–421.

Poletiko, C., Lemercier, R., Leiba, B., and Llug, P. (1994). Measurement of strontium-90 by the Cerenkov counting technique. *J. Radioanal. Nucl. Chem. Articles* **178**, 131–141.

Pullen, B. P. (1986). Cerenkov counting of ^{40}K in KCl using a liquid scintillation spectrometer. *J. Chem. Educ.* **63**, 971.

Ramesh, A. and Subramanian, M. S. (1997). Trace determination of tellurium using liquid scintillation and Cerenkov counting. *Analyst* **122**, 1605–1610.

Randolph, R. (1975). Determination of strontium-90 and strontium-89 by Cerenkov and liquid-scintillation counting. *Int. J. Appl. Radiat. Isot.* **26**, 9–16.

Rao, D. D., Mehendarge, S. T., Chandramouli, S., Hegde, A. G., and Mishra, U. C. (2000a). Application of Cherenkov radiation counting for determination of ^{90}Sr in environmental samples. *J. Environ. Radioact.* **48**, 49–57.

Rao, D. D., Sudheendran, V., Baburajan, A., Chandramouli, S., Hegde, A. G., and Mishra, U. C. (2000b). Measurement of high energy gross beta and ^{40}K by Cerenkov counting in liquid scintillation analyzer. *J. Radioanal. Nucl. Chem.* **243**(3), 651–655.

Ritson, D. M. (1961). "Techniques of High Energy Physics." Interscience. New York.

Ross, H. H. (1969). Measurement of β-emitting nuclides using Cherenkov radiation. *Anal. Chem.* **41**, 1260–1265.

Ross, H. H. (1971). Performance parameters of selected waveshifting compounds for Cerenkov counting. *In* "Organic Scintillators and Liquid Scintillation Counting." (D. L. Horrocks and C.-T. Peng, Eds.), pp. 757–769. Academic Press, New York and London.

Ross, H. H. (1976). Theory and application of Cherenkov counting. *In* "Liquid Scintillation Science and Technology" (A. A. Noujaim, C. Ediss and L. I. Wiebe, Eds.), pp. 79–91 Academic Press, New York and London.

Ross, H. H. (1980). Recent applications of Cherenkov radiation. *In* "Liquid Scintillation Counting, Recent Applications and Development", (C.-T. Peng, D. L. Horrocks and E. L. Alpen, Eds.), Vol. I, pp. 375–385. Academic Press, New York and London.

Rowinska, L., Jaskolska, H., and Radwan, M. (1975). Determination of phosphorus-32 activity using Cherenkov radiation. *Isotopenpraxis*, **11**, 223–225.

Rowinska, L., Walis, L., Panczyk, E., Dalecki, W., and Kusowski, M. (1987). Application of Cherenkov radiation to study some technical processes. *Nukleonika* **32**, 111–130.

Rucker, T. L. (1991). Calculational method for the resolution of ^{90}Sr and ^{89}Sr counts from Cerenkov and liquid scintillation counting. *In* "Liquid Scintillation Counting and Organic Scintillators" (H. Ross, J. E. Noakes, and J. D. Spaulding, Eds.), pp. 529–535. Lewis Publishers, Chelsea, MI.

Scarpitta, S. C., and Fisenne, I. M. (1996). Cerenkov counting as a compliment to liquid scintillation counting. *Appl. Radiat. Isot.* **47**, 795–800.

Scarpita, S., Odin-McCabe, J., Gaschott, R., Meier, A., and Klug, E. (1999). Comparison of four ^{90}Sr groundwater analytical methods. *Health Phys.* **76**(6), 644–656.

Schneider, P. B. (1971). Determination of specific activity of ^{32}P-labeled compounds using Cherenkov counting. *J. Nucl. Med.*, **12**, 14–16.

Sequinot, J. and Ypsilantis, T. (1994). A historical survey of ring imaging Cherenkov counters. *Nucl. Instrum. Methods Phys. Res., Sect. A* **343**, 1–29.

Shabana, E. I., Al-Hussan, K. A., and Al-Jaseem, Q. K. (1996). Adaptation of a radioanalytical method for strontium-90 to different kinds of environmental samples and its application for the investigation of some samples in the Riyadh region. *J. Radioanal. Nucl. Chem. Letters* **212**, 229–240.

Shawky, S. and El-Tahawy, M. (1999). Distribution pattern of ^{90}Sr and ^{137}Cs in the Nile Delta and the adjacent regions after Chernobyl accident. *Appl. Radiat. Isot.* **50**, 435–443.

Sowerby, B. D. (1971). Čerenkov detectors for low-energy gamma rays. *Nucl Instrum. Methods Phys. Res.* **97**, 145–149.

Smith, B. S. W. (1981). The contribution of the counting vial to variable quench in Cerenkov counting. *Int. J. Appl. Radiat. Isot.* **32**, 760–762.

Smith, R. D., Anderson, J. J. B., Ristic, M. M., and Huxsoll, D. L. (1972). The use of Cerenkov radiation in ^{32}P-labeled platelet survival studies. *Int. J. Appl. Radiat. Isot.* **23**, 513–517.

Stubbs, R. D. and Jackson, A. (1967). Channels ratio colour quenching correction in (Cherenkov counting. *Int. J. Appl. Radiat. Isot.* **18**, 857–858.

Sundaresan, M. K. (2001). "Handbook of Particle Physics." CRC Press LLC, Boca Raton, FL, pp. 446.

Suomela, J., Wallberg, L., and Melin, J. (1993). Methods for determination of strontium-90 in food and environmental samples by Cerenkov counting. SSI-Report 93-11, ISSN 0282-4433, Swedish Radiation Protection Institute, 171 16 Stockholm.

Takiue, M. and Ishikawa, H. (1978). Improvement of Cherenkov counting method. *Radioisotopes* **27**, 123–127.

Takiue, M., Fujii, H., and Ishikawa, H. (1984). PPO-Ethanol system as wavelength shifter for the Cherenkov counting technique using a liquid scintillation counter. *Nucl Instrum. Meth. Phys. Res.* **227**, 571–575.

Takiue, M., Fujii, H., and Aburai, T. (1993). Reliability of activity determined by Cerenkov measurements in a liquid scintillation counter. *In* "Liquid Scintillation Spectrometry 1992" (J. E. Noakes, F. Schönhofer and H. A. Polach, Eds.), *Radiocarbon* 1993, pp.69–73.

Takiue, M., Natake, T., Fujii, H., and Aburai, T. (1996). Accuracy of Cerenkov measurements using a liquid scintillation spectrometer. *Appl. Radiat. Isot.* **47**, 123–126.

Tamm, I. (1939). Radiation emitted by uniformly moving electrons. *J. Phys.* **1**, 439–454.

Tarancon, A., García, J. F., and Rauret, G. (2002). Mixed waste reduction in radioactivity determination by using plastic scintillators. *Anal. Chim. Acta* **463**, 125–134.

Torres, J. M., Llauradó, M., and Rauret, G. (1999). Microwave-assisted pre-treatment of environmental samples for the determination of ^{90}Sr. *Anal. Chim. Acta* **379**, 135–142.

Torres, J. M., Tent, J., Llauradó, M., and Rauret, G. (2002). A rapid method for ^{90}Sr determination in the presence of ^{137}Cs in environmental samples. *J. Environ. Radioact.* **59**, 113–125.

Tuininga, A. R., Dighton, J., and Gray, D. M. (2002). Burning, watering, litter quality and time effects on N, P, and K uptake by pitch pine (*Pinus rigida*) seedlings in a greenhouse study. *Soil Biol. Biochem.* **34**, 865–873.

Tuffen, F., Eason, W. R., and Scullion, J. (2002). The effect of earthworms and arbuscular mycorrhizal fungi on growth of and ^{32}P transfer between *Allium porrum* plants. *Soil Biol. Biochem.* **34**, 1027–1036.

Uz, T., Qu, T., Sugaya, K., and Manev, H. (2002). Neuronal expression of arylalkylamine N-acetyltransferase (AANAT) mRNA in the rat brain. *Neurosci. Res.* **42**, 309–316.

Vaca, F., Manjón, G., and García-León, M. (1998). Efficiency calibration of a liquid scintillation counter for ^{90}Y Cherenkov counting. *Nucl. Instrum. Methods Phys. Res., Sect. A* **406**, 267–275.

Vaca, F., Manjón, G., and García-León, M. (1999). ^{90}Sr in an alkaline pulp mill located in the south of Spain. *J. Environ. Radioact.* **46**, 327–344.

Vaca, F., Manjón, G., and García-León, M. (2001). The presence of some artificial and natural radionuclides in a Eucalyptus forest in the south of Spain. *J. Environ. Radioact.* **56**, 309–325.

Vaca, F., Manjón, G., Cuéllar, S., and García-León, M. (2001). Factor of merit and minimum detectable activity for ^{90}Sr determinations by gas-flow proportional counting or Cherenkov counting. *Appl. Radiat. Isot.* **55**, 849–851.

Vajda, N., Ghods, E. A., Cooper, E., and Danesi, P. R. (1992). Determination of Radiostrontium in soil samples using a crown ether. *J. Radioanal. Nucl. Chem. Articles* **162**(2), 307–323.

van Ginkel, G. (1980). An investigation of various wavelength-shifting compounds for improving counting efficiency when ^{32}P-Čerenkov radiation is measured in aqueous samples. *Int. J. Appl. Radiat. Isot.* **31**, 307–312.

van Ginkel, G. (1980). An investigation of various wavelength-shifting compounds for improving counting efficiency when ^{32}P-Čerenkov radiation is measured in aqueous samples. *Int. J. Appl. Radiat. Isot.* **31**, 307–312.

Va'vra, J. (2000). Particle identification methods in high-energy physics. *Nucl. Instrum. Methods Phys. Res., Sect. A* **453**, 262–278.

Wahid, P. A., Kamalam, N. V., and Sankar, S. J. (1985). Determination of phosphorus-32 in wet-digested plant leaves by Cerenkov counting. *Int. J. Appl. Radiat. Isot.* **36**, 323–324.

Walter, K., Trautmann, N., and Herrmann, G. (1993). Simultaneous determination of strontium-90 and strontium-89 by Cerenkov-radiation and liquid-scintillation counting: a re-evaluation for low-level counting. *Radiochim. Acta* **62**, 207–212.

Warshawsky, D., Dowty, H. V., LaDow, K., Succop, P., and Talaska, G. (2002). Reduction of a 7,12-dimethylbenz[a]anthracene DNA adduct in rat mammary tissue in vivo when pretreated with tamoxifen. *Toxicol. Lett.* **132**, 71–79.

Wiebe, L. I. and Ediss, C. (1976). Methyl salicylate as a medium for radioassay of chlorine-36 using a liquid scintillation spectrometer. *In* "Liquid Scintillation Science and Technology" (A. A. Noujaim, C. Ediss and L. I. Wiebe, Eds.), pp. 93–102, Academic Press, New York and London.

Wiebe, L. I., Helus, F., and Maier-Borst, W. (1978). Čerenkov counting and (Cherenkov-scintillation counting with high refractive index organic ligands using a liquid scintillation counter. *Int. J. Appl. Radiat. Isot.* **29**, 391–394.

Wiebe, L. I., McQuarrie, S. A., Ediss, C., Maier–Borst, W., and Helus, F. (1980). Liquid scintillation counting of radionuclides emitting high-energy beta radiation. *J. Radioanal. Chem.* **60**, 385–394.

Wiebe, L. I., Noujaim, A. A., and Ediss, C. (1971). Some aspects of the measurement of ^{32}P Cerenkov radiation in water by a liquid scintillation spectrometer. *Int. J. Appl. Radiat. Isot.* **22**, 463–467.

Ypsilantis, T. and Sequinot, J. (1994). Theory of ring imaging Cherenkov counters. *Nucl. Instrum. Methods Phys. Res., Sect. A* **343**, 30–51.

10

RADIOISOTOPE MASS SPECTROMETRY

GERHARD HUBER, GERD PASSLER, AND KLAUS WENDT
Institut für Physik, Universität Mainz, 55099 Mainz, Germany

JENS VOLKER KRATZ AND NORBERT TRAUTMANN
Institut für Kernchemie, Universität Mainz, 55099 Mainz, Germany

I. INTRODUCTION

Mass spectrometric methods are very sensitive and enable in many cases a multielement determination of trace and ultratrace elements combined with a good isotopic analysis (Bacon *et al.*, 2001). Therefore, these techniques are also applied for the detection of long-lived radionuclides (Becker and Dietze, 2000) mainly in environmental samples (Adriaens *et al.*, 1992; Bailey *et al.*, 1993; Bibler *et al.*, 1998; Eroglu *et al.*, 1998; Edmonds *et al.*, 1998; Becker and Dietze, 1999; Wendt *et al.*, 1999), nuclear materials (Betti, 1997; Chartier *et al.*, 1999), glass and ceramics (Rohr *et al.*, 1994; Fukuda and Sayama, 1997), and in high-purity substances (Beer and Heumann, 1992;

Herzner and Heumann, 1992). Radiometry, used as standard method for the determination of radionuclides, has some disadvantages for ultratrace analysis of long-lived nuclides because the detection limit depends on the half-life and the decay type of the isotope to be measured. Besides that, for beta measurements, as applied for trace analysis of pure β-emitters like ^{90,89}Sr or ^{99g}Tc, careful and time consuming chemical separations are needed to remove other β-emitters while for α-spectroscopy with surface barrier detectors, carrier-free or almost carrier-free samples are a prerequisite for a good energy resolution, and even then, an unambiguous isotopic analysis is very difficult as, e.g., for ^{239}Pu/^{240}Pu due to the very similar α-energies. Here, mass spectrometric techniques, which apply direct atom counting and different experimental set-ups and ionization methods to achieve a good sensitivity and isotopic as well as isobaric selectivity, are superior. The sensitive and fast determination of long-lived radioisotopes is of great interest in many areas such as risk assessment, low-level surveillance of the environment, studies of biological effects, radioactive waste control, management of radioactive waste for disposal, or investigations of the migration behaviour of actinides etc. The most important radioisotope mass spectrometric techniques are thermal ionization mass spectrometry TIMS (Callis and Abernathey, 1991; Platzner, 1997), glow discharge mass spectrometry GDMS (Betti, 1996), secondary ion mass spectrometry SIMS (Adriaens *et al.*, 1992; Betti *et al.*, 1999), inductively coupled plasma mass spectrometry ICP-MS or laser ablation inductively coupled plasma mass spectrometry LA-ICP-MS (Kim *et al.*, 1991; Crain, 1996; Becker and Dietze, 1998; Becker and Dietze, 1999; Becker *et al.*, 2002), resonance ionization mass spectrometry RIMS (Wendt *et al.*, 2000), and accelerator mass spectrometry AMS (Rucklidge, 1995; Fifield *et al.*, 1996). RIMS and AMS are used for sensitive monoelemental ultratrace analysis and precise determination of isotopic ratios whereas the other mass spectrometric methods represent very sensitive multielemental techniques permitting the determination of the concentrations and isotopic abundances of trace and ultratrace elements. Their limits of detection (LOD) are in the concentrations range of ng/g to sub-ng/g for solids and down to sub-pg/L for aqueous solutions. A precision as low as 0.02% relative standard deviation (RSD) for isotope ratio measurements can be obtained (Becker, 2002). With AMS, the precision is almost comparable whereas with RIMS, it is in the low percentage range. The detection limits for RIMS and AMS are 10^{6}–10^{7} (fg) and 10^{4} atoms (atg) per sample, respectively.

The different mass spectrometric methods can be classified according to the evaporation and ionization processes. For the analysis of compact solids and powder samples, various evaporation steps like thermal evaporation, laser ablation, evaporation and atomization in plasmas or electron beam, or ion bombardment are possible and the ionization of the atoms or molecules can be performed by electron impact, multiphoton excitation with laser light, or ionization during the sputtering process. Either simultaneous evaporation and ionization can occur in the ion source (TIMS, SIMS, ICP-MS) or a postionization is used separating in time and space the process of evaporation and atomization from the one of ionization of the atomic species (TIMS,

GDMS, LA-ICP-MS, RIMS, and AMS). The mass separation system to be used is defined by the physical properties of the ions and the formation of disturbing molecular and cluster ions which can be produced in a great surplus. Both static magnetic field or combinations of electric and magnetic sector field systems, e.g., Mattauch–Herzog or Nier–Johnson mass spectrometers, are in operation for ion separation as well as dynamic ion separation systems like quadrupole-, time-of-flight-, ion cyclotron resonance-, and ion trap mass spectrometer. Ion detection can be performed by photo-multiplier, channel plates, channeltrons, Faraday cups or even by ion-sensitive photographic plates. A direct analysis of the sample material without any chemical separation (GDMS, SIMS, LA-ICP-MS) is of advantage because a possible contamination introduced during the chemical treatment (TIMS, RIMS, and AMS) and interferences in aqueous solutions (ICP-MS) cannot be excluded. The quantification in solid state mass spectrometry, however, is more difficult than in chemically treated samples because very often no suitable standard reference materials are available (Becker and Pickhardt, 2000).

In the following, the most frequently used mass spectrometric methods and their applications for the determination of long-lived radionuclides and isotope ratio measurements are described.

II. THERMAL IONIZATION MASS SPECTROMETRY (TIMS)

A. Principle

Thermal ionization mass spectrometry (TIMS) is well suited for the precise measurement of isotopic ratios with a precision better than 0.01% (Platzner, 1997). For TIMS measurements, a small volume ($\sim 10 \mu L$) of an aqueous solution is deposited on a filament. Very often, a two filament arrangement is applied. One of the filaments is used for the evaporation of the analyte by thermal heating and the other for ionization of the evaporated atoms or molecules on the hot filament surface.

The release of atoms or molecules from hot surfaces populates all final states according to statistical thermodynamics. For species with a low first ionization potential I_p, this includes neutral and singly ionized states, according to the Saha–Langmuir equation. The ratio of ions to neutrals in thermal equilibrium is expressed by

$$N_+/N_o = g_+/g_o \exp \left[(W_e - I_p)/kT \right]$$

for positive ions. For negative ions from thermal ionization with a low electron affinity E_a,

$$N_-/N_o = g_-/g_o \exp \left[(E_a - W_e)/kT \right].$$

The statistical weights g_i are given by the multiplicity of the neutral and ionized states. The work function W_e determines the electron emission from the (metal) surface. T is the temperature of the filament and k the Boltzmann constant.

Extraction to Spectrometer

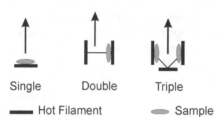

Single   Double   Triple

━ Hot Filament   ⬭ Sample

FIGURE 10.1 Schematic view of thermal ionization sources. Single filaments are used for alkali elements. When the sample, e.g., Ca may evaporate before the required ion current is reached, multiple filaments (2 or 3) are used.

Under clean vacuum conditions, all elements with ionization potentials I_p well below 7 eV may be efficiently ionized at $T = 800–2000°$ C from filaments of the refractory elements Ta, Re, or W. These metals have high electron work functions W_e of 4.30–4.98 eV.

Singly charged positive ions are obtained with high thermal ionization efficiency for alkali elements, alkaline earth elements, lanthanides, and actinides without molecular background from hydrocarbons present in the vacuum system (P-TIMS). For elements with a first ionization potential above 7 eV, special ion emitting reagents like silica gel and H_3PO_4 and/or H_3BO_3 are needed and the formation of MO_x^+ ions is possible.

Negative ions of halogens, S, Se, Te are produced by thermal ionization (N-TIMS) on thoriated W or LaB_6 surfaces with low electron work functions $W_e \sim 2.7$ eV. When volatilizing refractory elements as MO_x molecules, the corresponding negative oxide ions are produced by thermal ionization. The control of N-TIMS may show memory effects and is much more delicate than P-TIMS.

The ion source is built as a single, double, or triple filament (Kawai *et al.*, 2001) as schematically illustrated in Fig. 10.1 using thin ribbons, which are cleaned by extensive heating in high vacuum.

The samples are very carefully prepared using pure reagents in order to avoid isobaric contaminants. These controlled conditions reduce errors due to fractionation of the isotope content during the measurement (Johnson and Beard, 1999). The correction for the isotope fractionation in thermal ionization has been investigated very extensively (Ramakumar and Fiedler, 1999; Habfast, 1998).

B. Applications

TIMS measurements for the precise determination of isotope ratios and absolute contents of elements in the samples imply (single, double, and triple) isotope spiking and isotope dilution techniques (Vanhaecke *et al.*, 1998; Fassett and Paulsen, 1989; Thirlwall, 2000).

Fast magnet switching or multicollector detection (Ramakumar and Fiedler, 1999) is used for [187]Re, [238]U, and [239]Pu isotopes in order to minimize errors due to fluctuations in the ion source operation (House *et al.*,

2002). The analytical topics in nuclear applications are related with precise isotope ratios and with high sensitivity sample analysis. A high dynamic range and a good stability of thermal ionization is needed for the determination of isotope ratios of the elements under investigation and of reference samples like U, Pb, and other elements. The thermal ionization is used in many on-line mass separators (ISOL) facilities for a highly selective and efficient separation of alkali and other elements (Lettry et al., 1996).

I. Isotope Ratios with TIMS

Branching ratios of the ^{40}K decay to Ca and the dating of Madagascar sanidine has been reported. The interference at mass 40 could be solved by double spiking with ^{43}Ca/^{48}Ca (Nägler and Villa, 2000). A procedure for routine high-precision isotope analyses of the K–Ca system was free of Ca fractionations (Fletcher et al., 1997). The various interests in Li-isotope studies including its relevance for nucleosynthetic processes in the solar system motivated the precise measurement of Li in open sea water and several international rock standards (James and Palmer, 2000). The monitoring of Pu accidental intake of workers has been performed by TIMS on a large number of urine samples in Los Alamos (Wagner et al., 2000; Inkret et al., 1998). The IRMM in Geel reports on U abundances in natural U ore samples using synthetic reference isotope mixtures for calibration (Richter et al., 1999). The inter-laboratory and other errors in Pb isotope analyses has been carefully investigated with a combined TIMS and GDMS study using a ^{207}Pb–^{204}Pb double spike (Thirlwall, 2000). A reliable analytical technique for Mo isotope abundances in Mo fission products and also in samples for double-beta decay studies has been tested on molybdenites and high purity metal and compared with IUPAP references (Wieser and de Laeter, 2000). A modified thermal ionization mass spectrometer with deceleration components was tested with ^{234}U/^{238}U standards and on ^{230}Th/^{232}Th samples (Van Calsteren and Schwieters, 1995). The method of total evaporation has been tested with U and Pu samples in a thermal ionization quadrupole mass spectrometer (Fiedler, 1995).

2. High Sensitivity Measurements with TIMS

Some work has been reported on new developments which are partly related to high sensitivity. Liu and co-workers (2000) have developed an oxide species measurement with TIMS using enriched ^{16}O gas inlet for light rare earths. A high efficiency thermal ionization source based on a W crucible has been tested at Los Alamos (Duan et al., 1997). Competing with AMS, an improved value of the half-life of ^{126}Sn has been obtained by thermal mass spectrometry (Oberli et al., 1999). In principle, all TIMS measurements can be tuned to high sensitivity, when isobaric contamination can be excluded, like the Pu intake monitoring mentioned above (Wagner et al., 2000; Inkret et al., 1998). For stable and long-lived isotopes, this is the exception, but on-line TIMS at ISOL facilities can be found for all short-lived alkali elements, e.g., at the forefront of nuclear physics research on the halo nucleus ^{11}Li with a half-life of 8.7 ms (Lettry et al., 1996; Evensen et al., 1996).

III. GLOW DISCHARGE MASS SPECTROMETRY (GDMS)

A. Principle

Glow discharge mass spectrometry (GDMS) is a very powerful and efficient analytical method for direct trace analysis of solids (Harrison, 1988; Betti, 2002). Among others it has found widespread applications for the determination of trace elements in conducting and nonconducting solids and liquids (Winchester *et al.*, 1990; Duckworth *et al.*, 1993; Barshick *et al.*, 1993) as well as for the characterization of nuclear samples (Betti *et al.*, 1996; Betti, 1996). Compared to other mass spectrometric techniques where often time consuming chemical pre-treatment steps are required, the glow discharge-based methods have the advantage of simpler sample preparation procedures due to the fact that the measurements are carried out directly on solid samples. Glow discharge sources have not only been used as ion sources in mass spectrometry but also for optical emission spectrometry. The principle of such a source, operated at a pressure of 0.1–10 torr with argon as plasma gas, is based on Ar^+-ions which are formed in a low-pressure plasma and are accelerated towards a cathode consisting of the sample material. The sample material is sputtered at the surface and the evaporated neutral species are ionized in a glow discharge plasma by Penning and/or electron impact ionization. After acceleration in an electric field, the positive ions are separated in a double-focussing mass spectrometer according to their mass-to-charge and energy-to-charge ratios and detected by a Faraday cup, photomultiplier or a channeltron. A schematic drawing of a glow discharge mass spectrometer is shown in Fig. 10.2. Depending on the solid samples either a direct current (dc) or a radiofrequency powered device is used as ion source. For the handling of nuclear materials, it might be necessary to install the Glow Discharge Mass Spectrometer in a glove box (Betti *et al.*, 1994). For this, the ion source chamber, the sample interlock, and the pumping system are placed in the glove box. Furthermore, it is necessary that all

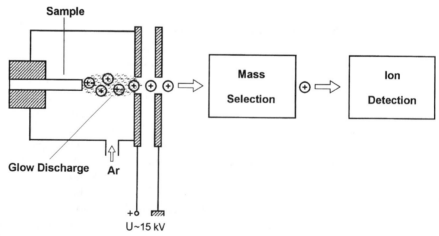

FIGURE 10.2   Schematic picture of a glow discharge mass spectrometer.

supplies to the ion source and the pumping parts are equipped with absolute filters to prevent any external contamination.

A commercially available glow discharge mass spectrometer like the VG-9000 instrument (Winsford, UK) consists of a dc glow discharge ion source coupled to a double-focusing mass spectrometer with reverse (Nier–Johnson) geometry providing a high transmission (> 75%) and a mass resolution $m/\Delta m$ of up to 10,000. With dc-GDMS (direct current-glow discharge mass spectrometry) trace elements in electrically conducting materials with detection limits in the ng/g range with a reproducibility of ~ 10% could be determined (van Straaten et al., 1994; Venzago and Weigert, 1994). Due to charge effects on the sample surface, the analysis of nonconducting materials is more difficult with this technique. Therefore, for dc-GDMS measurements, mixing of nonconducting powdered samples with high-purity metal or graphite powder (de Gendt et al., 1995) or the use of a secondary cathode (Schelles et al., 1996) have been applied. Radiofrequency GDMS allows the analysis of conducting as well as semiconducting and nonconducting samples without restriction, but this method is still under development (Marcus, 1996).

B. Applications

In the fields of nuclear technology and research, the characterization of fuel elements, cladding materials, nuclear waste glasses, and smuggled nuclear samples is important as well as the determination of radioisotopes in the environment requiring the detection of trace, minor, and major elements and of their isotopic composition. Mainly dc-GDMS has been applied for the investigation of trace elements and for bulk and isotopic analysis.

I. Trace and Bulk Analysis of Nuclear Samples

Metallic alloy fuels consisting of UNdZr and UPuZr with an uranium content of 81% and 71%, respectively, were analyzed semiquantitatively by GDMS using the signal intensity of the analyte and taking into account the element sensitivity of uranium. The relative sensitivity factor (RSF) values were obtained from a comparison with other methods like ICP-MS and ICP-AES (inductively coupled plasma-atomic emission spectometry) analysing uranium metal specimens of different origin. The agreement between the concentrations determined by GDMS and other techniques is rather good (Betti, 2002).

Another application of GDMS is the analysis of zirconium alloys (Robinson and Hall, 1987), which are used as cladding materials for nuclear fuel elements. Quantitative data of zircaloy cladding material were obtained by applying RSF matrix specific and also RSF values for a uranium metal sample yielding for both RSF values a good agreement. This is an indication that for metallic samples matrix effects are negligible (Betti et al., 1996).

Nonconductive nuclear samples, especially oxide-based compounds like uranium and plutonium oxide samples have also been investigated by GDMS (Betti et al., 1996). Two versions have been applied, either a secondary cathode placed directly in front of the nonconducting sample surface or the use of powdered samples mixed with a conducting host matrix in an appropriate ratio (Winchester and Marcus, 1988; Tong and Harrison, 1993).

As host matrix, tantalum and titanium are well suited because the formation of UO^+ and PuO^+ species is hindered. Oxygen as a major matrix element causes problems due to its release during the discharge process. In the GD plasma, it influences the signal by quenching metastable atoms and by forming oxide complexes with the analytical species. Tantalum should be used for a secondary cathode due to its getter capability for oxygen. It could be shown that for several elements the RSFs depend on the oxygen content in the sample (Betti *et al.*, 1996), which implies a specific matrix reference sample for the quantitative analysis of oxygen-containing samples. A good agreement for plutonium oxide samples using ICP-MS and GDMS with tantalum as secondary cathode was obtained for a number of radioisotopes (Aldave de las Heras *et al.*, 2000). Nuclear waste glasses could also be analysed with the same GDMS techniques (Betti, 1996).

Materials for nuclear fuel production must be characterized for the isotopic composition of the major elements and the concentration of trace elements. In order to detect contaminations during the fabrication process, the trace elements have to be detected in the starting material as well as in the final pellets or fresh fuels. For this purpose, dc-GDMS and ICP-MS have been used successfully (Aldave de las Heras *et al.*, 2000). The data from the analysis of an uranium oxide reference sample (Morille, CEA, France) with the secondary cathode-GDMS technique show precisions on the order of 10% RSD or better (Betti, 2002). The detection limits for several trace elements are at the low μg/g level.

2. Determination of Radioisotopes in the Environment

The migration of radionuclides in the environment is a major aspect in safety assessment, e.g., of nuclear waste repositories, and very sensitive methods are required for their determination. GDMS has been used for the detection of uranium in soil samples (Duckworth *et al.*, 1993) and of trace radioisotopes in soil, sediment, and vegetation (Betti *et al.*, 1996; Giannarelli *et al.*, 1996). Another application is the determination of ^{237}Np in Irish Sea sediment samples (Aldave de las Heras *et al.*, 2002). Again, the secondary cathode technique was used for such environmental samples. In trace analysis, the conductive host matrix can have an influence on the sensitivity; e.g., tantalum is not suited for isotope measurements on uranium compounds because it forms polyatomic ions which interfere with $^{235}U^+$ and $^{236}U^+$, whereas silver does not produce interferences with the uranium isotopes. This aspect has to be considered in the determination of ^{236}U with detection limits in the pg/g range. This isotope indicates the presence of irradiated uranium in a sample.

3. Determination of Isotopic Compositions

The precise and accurate measurement of isotope ratios is a further application of GDMS. Here, the main interest is the isotopic composition of nuclear materials, especially of uranium and plutonium, which was investigated by means of GDMS (Betti *et al.*, 1996; Rasmussen *et al.*, 1996; Young *et al.*, 1998; Barshick *et al.*, 1998). A comparison with TIMS showed that GDMS is competitive in terms of precision and accuracy for some important elements.

4. Depth Measurements

GDMS has also been used for the study of the mechanism of corrosion of zircaloy cladding of nuclear fuels by measuring the diffusion of impurities in ZrO_2 layers by depth profiling (Actis-Dato et al., 2000). The investigation of the diffusion mechanism of trace elements is a general problem in environmental processes like sorption or desorption of actinides in host rocks. Such studies are important for the conception of a nuclear waste repository and GDMS is a versatile tool for that.

IV. SECONDARY ION MASS SPECTROMETRY (SIMS)

A. Principle

Secondary ion mass spectrometry (SIMS) is mainly applied for mono- and multielement trace analysis on surfaces of solid materials or thin layers, depth profiling measurements, and for the mapping of the elemental and molecular distributions at surfaces with high sensitivity, the so-called imaging technique. SIMS is based on the mass spectrometric analysis of ions, which are generated by the interaction of a primary ion beam in the 1–25 keV range with a solid sample. Depending on the energy of the ions (Ar^+, O_2^+, O_2^-, Cs^+, Ga^+ etc.) and the nature of the sample, the ions penetrate into the solid substance to depths between 1–10 nm and transfer their kinetic energy to the atoms of the solid. Impact cascades cause the sputtering of positively or negatively charged secondary ions and neutral particles from the surface (Benninghoven et al., 1992). The emitted secondary ions which can be atomic, molecular, or cluster ions, are separated in a mass spectrometer according to their mass-to-charge ratio and measured with a detection system consisting of, e.g., an electron multiplier and a Faraday cup to cover a wide dynamic range. For a given element, the yield of secondary ions depends not only on its ionization potential but also on the physico-chemical properties of the specimen such as its chemical composition or thickness. There are two operational regimes in SIMS, namely dynamic SIMS, which uses a high primary ion current density for layer-by-layer erosion and depth profiling studies, and static SIMS which operates at a low primary ion current. Dynamic SIMS is restricted to elemental analysis but allows the three-dimensional characterization of the sample composition by scanning the primary ion beam over the surface and combining this imaging with depth profiling. Imaging can be performed either via the microscope or the microprobe mode. In the microscope mode, the sample is illuminated by a broad beam of primary ions and the produced secondary ions are accelerated by an electrostatic field and focused with an electrostatic lens. Here, the lateral resolution is about 0.4 μm and determined by the ion optics. For microprobe investigations, a narrow illuminating beam is used and an immersion lens acts as the collecting optical system. In this mode, the resolution of $\sim 20\,\mu$m is defined by the beam diameter (Chabala *et al.*, 1995).

Static SIMS (van Vaeck *et al.*, 1999; Adriaens *et al.*, 1999) enables the detection of structural ions from organic compounds with an information depth limited to the uppermost surface layer, i.e., speciation and organic analysis are feasible, but depth profiling is excluded.

The used primary ion guns are based on the electron or plasma ionization of reactive or inert gases (e.g., O_2^+, Ar^+), surface ionization (Cs^+), and liquid metal ion field emission (Ga^+).

The mass spectrometric analysis of SIMS is normally performed by singly focusing mass separators like magnetic field, quadrupole, or time-of-flight mass spectrometers, but high-resolution double-focusing mass separation systems are to be preferred because they are able to resolve the interferences of molecular or cluster ions compared to the atomic ions.

In general, the SIMS instruments are equipped with a secondary electron multiplier such as a discrete dynode or channeltron type device. Furthermore, a channel plate-scintillator-photomultiplier or a Faraday cup can be used for the ion detection. A schematic presentation of a secondary ion mass spectrometer is given in Fig. 10.3.

With a double-focusing mass spectrometer, a resolution of $m/\Delta m = 10,000$ is achievable. A detection limit for SIMS in the ng/g–pg/g range can be reached (Tamborini *et al.*, 1998). In order to quantify the analytical results obtained by SIMS, the ion count rates must be corrected with relative sensitivity coefficients (RSC) which vary, among others, as a function of the primary ion beam, the physical and chemical properties of the

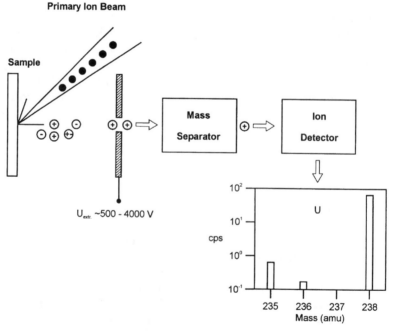

FIGURE 10.3 Schematic presentation of the secondary ion mass spectrometry principle.

element to be investigated, the composition of the matrix, and the charge of the secondary ions. These have to be determined by using standard reference materials.

B. Applications

The application of SIMS for the mass spectrometric analysis of radioisotopes is manifold. Due to the good lateral resolution of the primary ion beam, SIMS is successfully applied for particle identification and for isotope analyses in astrophysics, e.g., of presolar dust grains. The characterization of environmental samples contaminated with radioisotopes, nuclear medical applications using ^{99}Tc, ^{89}Sr, and ^{76}Br as radiotracers (Fourre et al., 1997), surface investigations, depth profile measurements, and material science studies are further examples for the use of SIMS.

I. Particle Analysis

SIMS has been applied for the identification of single particles of uranium and for the determination of their isotopic composition (Tamborini et al., 1998). This is important in the environmental monitoring control for non-proliferation in nuclear safeguards. Such particles can be transported as aerosols and recovered in environmental and swipe samples. The isotopic composition delivers information on the origin of the particles. SIMS was also used for the characterization of plutonium and highly enriched uranium particles with a diameter of $\sim 10\,\mu m$ in nuclear forensic studies (Betti et al., 1999; Tamborini and Betti, 2000). The isotopic composition of the particles could be measured with an accuracy and precision of 0.5%. Furthermore, trace elements could be determined in the particles by SIMS, and their composition enabled to reveal the history of the sample. SIMS was also applied for the identification of plutonium particles and their characterization according to their isotopic ratio, especially with respect to the ratio ^{239}Pu/^{240}Pu. Other examples are the isotopic determination of uranium and plutonium in glass microparticles (Simons, 1986) and in clay microspheres (Stoffels et al., 1994). The isotopic composition of artificially produced monodisperse uranium oxide particles could be verified with SIMS (Erdmann et al., 2000). SIMS applied in combination with electron probe microanalysis and glow discharge mass spectrometry, allowed the characterization of environmental samples contaminated with traces of radioisotopes (Lefèvre et al., 1996). Here, the main emphasis was put on the radionuclides ^{137}Cs, ^{129}I, and ^{238}U, which could be localized in soil and grass samples. The chemical analysis of meteorites and the isotope analyses of presolar dust grains are two examples for the application of SIMS in astrophysics. The obtained results make it possible to detect their origin and their path through the interstellar medium and thus give new insights into the cosmic nucleosynthesis (Diehl and Hillebrandt, 2002). For the isotope analyses of dust grains consisting mainly of SiC, a new generation of instruments, nano SIMS, has been used enabling investigations for ranges of $\sim 100\,nm$. Calcium-aluminum-rich inclusions (CAI) from the Allende Meteorite were investigated by means of SIMS to get information on the ^{11}B/^{10}B and ^{9}Be/^{11}B compositions. The $^{10}B/^{11}B$ values from various spots show ^{10}B excesses which are correlated with the Be/B ratio in a manner indicative of the in situ decay of the radioactive isotope ^{10}Be ($T_{1/2} = 1.6 \times 10^6$ a). ^{10}Be is produced only by nuclear spallation reactions and not by stellar nucleosynthesis, so its existence in early solar system material (CAI) attests intense irradiation processes in the solar nebula (Keegan *et al.*, 2000).

2. Trace Analysis

The application of the SIMS technique for the assessment of trace concentrations of uranium and plutonium in urine samples was explored (Amaral *et al.*, 1997). Concentrations in the order of 10^{-10} g/L of ^{238}U and 10^{-11} g/L ^{239}Pu can be determined with this method. The accurate quantification of a number of elements among them Th and U in complex matrices is another example (Ottolini and Oberti, 2000). Among the physical methods for in vitro analytical imaging in the microscopic range, SIMS plays an important role because it images the isotopes of quite a number of elements with very good detection limits and an excellent resolution (Thellier *et al.*, 2001).

V. INDUCTIVELY COUPLED PLASMA MASS SPECTROMETRY (ICP-MS)

A. Principle and Instrumentation

Inductively coupled plasmas (ICPs) are flamelike electrical discharges that have revolutionized the practice of elemental and isotopic ratio analyses. In particular, argon ICPs have made these plasmas a very effective vaporization-atomization-excitation-ionization source for atomic emission spectrometry (AES), atomic fluorescence spectrometry (AFS), and mass spectrometry (MS). A very comprehensive discussion of ICP-MS can be found in Montaser (1998). The following outline can only be a brief introduction to the principle and the relevant instrumentation. Briefly, the plasma is formed in a stream of argon gas (usually 8–20 L/min) flowing through an assembly of three concentric quartz tubes known as the plasma torch. The most common torch for ICP-MS has an internal diameter of 18 mm. The torch is encircled at the top by an induction coil, also called the load coil, connected to a free-running or crystal-controlled radiofrequency (RF) generator. This induction coil is made from copper and is cooled, either by water or by the argon gas. The generator can operate at frequencies ranging from 4 to 50 MHz. Most ICP-based instruments operate at either 27 or 40 MHz at typical output powers of 1–2 kW. The magnetic field generated by the RF current through the load coil induces a current in the argon gas stream. A plasma is formed almost instantaneously when the Ar gas is seeded with energetic electrons. These electrons are produced by a high-voltage Tesla discharge or a solid-state piezoelectric transducer. A stable, self-sustaining plasma is maintained as long as the magnetic field is sufficiently high and the gas flows in a symmetric pattern. The largest current flow occurs on the periphery of the plasma, which gives the ICP a distinctive annular configuration.

This doughnut-type structure ensures the efficient introduction of sample aerosol into the central channel of the plasma, thus resulting in the efficient desolvation, vaporization, atomization, excitation, and ionization of the sample. The Ar-ICP is capable of ionizing a wide range of elements, particularly metals, and it therefore allows simultaneous multielement determination.

Since its introduction four decades ago (Reed, 1961a, b), the ICP has exhibited a number of special attributes: As a result of the axial channel in the ICP discharge, the sample aerosol passes through the narrow, central core of the discharge without interacting significantly with the surrounding reactive atmosphere. The analyte passway is effectively isolated and confined by the argon plasma gas. This sequestering of the analyte ionic species provides nearly optimum conditions for their sampling from the ICP. The sampling cone orifice is positioned along the axial channel in a region exhibiting maximum analyte signal and minimal background. The Ar-ICP exhibits a high gas temperature (4500–8000 K) and a high electron temperature (8000–10000 K). Such conditions coupled with a relatively long plasma sample interaction time (2–3 ms) lead to nearly complete vaporization–atomization of the sample aerosol and also reduce the chemical and physical interferences in the plasma. In addition, because of the high electron density (1–3×10^{15} cm^{-3}), the extent of ionization-type interferences in Ar-ICP spectrometries is small in comparison with combustion flame spectrometries. The Ar-ICP is a robust electrical flame whose plasma properties and analytical characteristics are not significantly affected when the sample composition is altered. This robustness also offers a low-noise condition in detecting a wide range of elements in diverse materials at major, minor, trace, and ultratrace concentration levels.

With ICP-MS, the tailflame of the plasma is extracted into a differentially pumped two-stage interface (Fig. 10.4) containing both a sampler and a

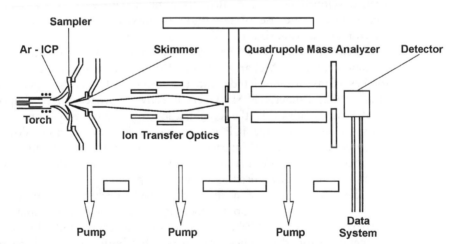

FIGURE 10.4 Principal components of a quadrupole ICP-MS system. Ions are created in the ICP and transferred into the quadrupole mass spectrometer through the MS interface. After separation according to the mass-to-charge ratio, ions are collected at the detector and the ion-beam intensities are reported by the data system.

skimmer first developed by Campargue (Campargue, 1964, 1984) for molecular beam techniques.

The sampler provides the first differential pumping aperture of approximately 1 mm resulting in a pressure of typically 5 mbar. Behind the sampler, a concentric shock wave structure is formed. This structure surrounds the so-called zone of silence and terminates in a perpendicular shock wave front called the Mach disc. The skimmer orifice is located in the zone of silence for optimal extraction of the analyte ions. The velocity of the particles passing through the MS interface is around 2.5×10^5 cm/s resulting in a mass-dependent axial kinetic energy in the range 0.5–10 eV for ions having masses of 6–240 amu. A cylindrically symmetric electrostatic lens system is used to transfer the sampled ions to the mass analyser. Most spectroscopic studies of the ions produced in the ICP discharge have been conducted using quadrupole mass spectrometers. Paul and Raether (1955) proposed quadrupole mass analysers in 1955 and developed their ideas over the next four years into a practical measuring device. Dawson and Whetten (1969) and Dawson (1995) presented major reviews of RF quadrupole mass spectrometers. Currently, the quadrupole mass spectrometer offers two major advantages: ease of use with an ion source near ground potential and low cost. The major drawback is the poor peak shape compared to magnetic double-focusing mass analysers and the intrinsically low mass resolution of about 0.5 amu. This is insufficient to separate analyte ions from interfering ions generated by the ICP ion source producing isobaric or interelement spectral overlaps, arising primarily from molecular species formed by a combination of sample, plasma gas, and matrix or solvent constituents. Examples are the interferences of $^{40}Ar^{16}O$ with ^{56}Fe, or $^{35}Cl^{16}O$ with ^{51}V. These interferences generate signals at the same nominal masses as the analyte and contribute to the analyte signal.

To overcome these problems, different strategies and techniques have been developed which are currently applied with varying levels of success. These include the use of nonargon or mixed plasmas (Montaser, 1998), electrothermal vaporization (Becker and Hirner, 1994) or laser ablation (Richner *et al.*, 1994), matrix separation by chromatography and preconcentrations (McLaren *et al.*, 1993; Ebdon *et al.*, 1993), shielded torches (Gray, 1986; Uchida and Ito, 1994), cold plasma operation (Jiang *et al.*, 1988; Tanner, 1995). A very effective approach for the improvement of ICP-MS analysis is the combination of a high-resolution sector field mass analyser with the ICP ion source. This configuration allows proper interpretation of spectral interferences in elemental analyses and also permits the operation of the plasma in its normal "hot" mode so that levels of oxides are not preferentially raised as in the cold plasma.

The combination of a magnetic mass analyzer with an electrostatic analyzer leads to the double-focusing sector field mass analyzer. The electrostatic sector establishes ion trajectories by energy only. In contrast, the magnetic sector determines trajectories by both mass and energy, i.e., momentum. When these ion-optical elements are combined, ions are sorted by mass only. The two sectors also provide angular focusing; that is, the total system is considered to be double-focusing. Mass resolutions are sufficient to separate the interfering components for most of the well-known "difficult" elements. In general, high-resolution ICP-MS is the preferred method for investigating low levels of components in complicated matrices and for measurements of isotope ratios. In the last few years, instrumental progress for improving figures of merit in isotope ratio measurements in ICP-MS with a single ion detector has been achieved by the introduction of the collision cell interface (Boulyga and Becker, 2000) in order to dissociate disturbing argon-based molecular ions, to reduce the kinetic energy of ions and neutralize the disturbing argon ions of the plasma (Ar^+). The application of the collision cell in ICP-MS results in higher ion transmission, improved sensitivity, and better precision of isotope ratio measurements compared to ICP-MS without the collision cell (Becker, 2002). The most important instrumental improvement for isotope ratio measurements by sector-field ICP-MS was the application of a multiple ion collector device (MC-ICP-MS) in order to obtain better precision of isotope ratios of up to 0.002%, RSD (Becker, 2002).

Other potential mass analysers for ICP-MS include

- ion cyclotron resonance (ICR) mass spectrometers
- ion trap mass spectrometers
- time-of-flight mass spectrometers
- and others.

After the mass analyser, the ions are detected by one or more ion-counting detector(s) or, for higher intensities, by some form of analogue measurement.

B. Sample Introduction

The introduction of the sample into the plasma plays a key role in the production of ions and interfering species. Sample introduction into plasmas has been reviewed in several major publications (e.g., Greenfield and Montaser, 1992; Heitkemper *et al.*, 1992; Gustavsson, 1992; Browner, 1987; McLeod *et al.*, 1992; Montaser, 1998). Generally, the sample can be introduced in gaseous, liquid, and solid form. Here, we restrict ourselves to liquids and solids and those techniques that have been applied in the analysis of long-lived radionuclides.

I. Nebulization

Liquids are dispersed into fine aerosols before being introduced into the ICP. This is achieved with pneumatic nebulizers (PNs) and ultrasonic nebulizers (USNs). Other devices such as the thermospray nebulizer are also used. About two dozen sample introduction devices are commercially available from ICP manufacturers and other vendors. A comprehensive review of PNs is presented by Sharp (1988). Two basic configurations are in use. In the concentric type, the sample solution passes through a capillary surrounded by a high-velocity gas stream parallel to the gas stream axis. The crossflow type has a liquid-carrying capillary set at a right angle to the tube carrying the high-velocity gas stream. In both configurations, a pressure differential created across the sample draws the solution through the capillary. The analyte transport efficiency of commonly used PNs ranges from 1 to 5% in ICP spectrometry. One drawback of pneumatic nebulization is nebulizer clogging. In ultrasonic nebulization, the sample solution is introduced onto the surface of a piezoelectric transducer driven by an ultrasonic generator at a frequency of 200 kHz to 10 MHz. Because the total production rate of droplets generated by the USN is greater than that obtained with PNs, a desolvation device is essential for removing excess solvent to avoid plasma cooling. Analyte transport efficiencies approaching 20% are reached for USNs operating at an uptake rate of 1 mL/min. Higher analyte transport efficiency may be obtained if the spray chamber is heated or if the solution uptake rate is reduced to a few microliters. For example, analyte transport efficiencies close to 100% are achieved with a micro-flow USN at solution uptake rates of 5–20 μL/min (Montaser, 1998). Direct injection nebulizers (DIN) introduce 100% of the aerosol into the plasma, thereby increasing the solvent load. Most recent versions of the DIN are microconcentric pneumatic nebulizers positioned inside the ICP torch. Further developments such as the high efficiency nebulizer (HEN) or the direct injection high efficiency nebulizer (DIHEN) and many others are described in detail (Montaser, 1998).

2. Hyphenated Systems

The combination of liquid chromatography (LC) with ICP-MS provides attractive opportunities for element speciation. In most cases, the interfacing of LC to the ICP is performed by a direct transfer of the eluate from the column to a variety of nebulizing devices. Recent applications of LC-ICP-MS are surveyed in Montaser (1998). Capillary electrophoresis (CE) is another attractive separation technique with excellent speciation capabilities. The first coupling of CE with ICP-MS was reported by Olesik *et al.* (1995) who used a conventional concentric nebulizer and a conical spray chamber as interface to the ICP. The CE connection to ground inside the nebulizer shell was since solved in a number of tricky ways. Figures of merit are detection limits usually in the low ppb range and precision in the range of 1–5% RSD.

3. Laser Ablation

Laser ablation is the most versatile solid sampling technique for ICP spectrometry. Gray (1985) first used laser ablation for ICP-MS. When sufficient energy in the form of a focused laser beam is directed onto the sample, material from the surface is sputtered and vaporized. The plume of vapor and particulate matter is transported in an argon carrier gas to the plasma for atomization and ionization, see Fig. 10.5.

Interest in LA-ICP-MS stems from the ability to sample diverse materials ranging from conducting and nonconducting inorganic and organic compounds as solids or powders. In addition to conventional bulk analysis, the focusing characteristics of lasers permit sampling of small areas, such that localized microanalysis and spatially resolved studies are feasible.

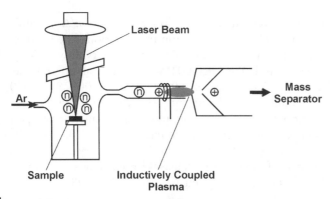

FIGURE 10.5   Principle of LA-ICP-MS. By the interaction of a focused laser beam with a solid surface, sample material mostly in form of neutral particles (n) is ablated and evaporized and the ablated material is transported using argon as the carrier gas in the inductively coupled plasma.

C. Applications to Radionuclides

The application of ion chromatography (IC) for the determination of trace radioisotopes as well as for the separation of fission products and actinides is reviewed by Betti (1997). The coupling of IC and a low-resolution ICP-MS has been found to be effective in eliminating isobaric interferences in the determination of ^{90}Sr, ^{137}Cs, lanthanides and actinides etc. This coupling on-line has been installed inside a glove-box for the determination of the complete inventory of fission products and actinides in spent nuclear fuel and high-level liquid waste samples. With this system, e.g., for the radioactive isotopes of Cs at nominal masses 134, 135, and 137, the isobaric interferences by the natural isotopes of Ba were eliminated giving a detection limit of $16\,\mathrm{pg\,g^{-1}}$ for total Cs with a precision of 2.5% at a concentration level of 100 ppb (Barrero Moreno *et al.*, 1999). The accuracy of the ratio ^{134}Cs/^{137}Cs was found to be 2.5%. The same system was applied successfully to the determination of actinides in aqueous leachate solutions from uranium oxide (Solatie *et al.*, 2000). The results were compared to those of classical radiometric methods and determinations by α-, γ spectrometry, and liquid scintillation counting (LSC). The comparison confirmed that IC-ICP-MS is a powerful method for the detection of long-lived radionuclides. IC-ICP-MS was found to be the most convenient choice being much less time consuming than the radiometric methods as for sample preparation (α spectrometry) as well as for counting time (γ-spectrometry and LSC).

Soil samples from the former Semipalatinsk nuclear test site in Eastern Kasachstan were investigated (Yamamoto *et al.*, 1996) in order to determine the contamination level of plutonium and to determine the local plutonium isotope ratios. After dissolution of the soil samples, the Pu fraction was purified by anion-exchange chromatography, evaporated to dryness and dissolved in 1 M HNO_3 for introduction into the ICP-MS. Isotope ratios were determined with a double-focusing ICP-MS. The sample solution was introduced into the instrument through an ultrasonic nebulizer which allowed the amount of sample reaching the plasma to be significantly increased.

The measurements were at relatively low resolution $m/\Delta m = 400$ by opening the slits in order to obtain a higher ion transmission. The isotope ratios were determined at the level of 0.1 pg/g solution. Very low $^{240}Pu/^{239}Pu$ ratios of 0.036 ± 0.001 and 0.067 ± 0.001 were found at different test sites and $^{242}Pu/^{239}Pu$ ratios of 0.000048 ± 0.000004 and 0.00029 ± 0.00005, respectively. These values are significantly lower than the values commonly accepted for global fallout from the atmospheric nuclear weapons tests. For site 1, the isotopic composition was found to be ^{238}Pu 0.006%, ^{239}Pu 96.5%, ^{240}Pu 3.48%, ^{241}Pu 0.01%, and ^{242}Pu 0.004%, very similar to the isotopic composition of Pu found in the Nagasaki area. For site 2, the composition ^{238}Pu 0.19%, ^{239}Pu 93.4%, ^{240}Pu 6.30%, ^{241}Pu 0.08%, and ^{242}Pu 0.03% was found.

Highly radioactive waste from defense-related activities at the Savannah River Site in South Carolina are to be incorporated into borosilicate glass in the Defense Waste Processing Facility (DWPF) for long-term geological isolation. Processing and repository safety considerations require the determination of 24 radioisotopes that meet the repository criteria. These isotopes include fission products, activation products, actinides, and daughter nuclei that grow in the waste. A routine determination of four isotopes (Kinard *et al.*, 1997), ^{137}Cs, ^{90}Sr, ^{238}Pu, and ^{238}U is performed by ICP-MS in the DWPF operation for process control. This work shows that the concentrations of the other 20 radioisotopes in the final glass product can be predicted from a thorough characterization of the high level waste tanks and a knowledge of the concentrations of the major nonradioactive components in the vitrification process.

The central analytical laboratory (ZCH) in Jülich has been applying mass spectrometry for trace and ultratrace analyses of elements, isotope ratio measurements, and surface analysis for many years. Also in this laboratory, ICP-MS and LA-ICP-MS have increasingly replaced thermal ionization mass spectrometry (TIMS) which has been used as the dominant analytical technique for precise isotope ratio measurements for many decades. An analytical method for the ultratrace and isotopic analysis of uranium in radioactive waste samples using a double-focusing sector field ICP mass spectrometer is described (Kerl *et al.*, 1997). In high purity water, a detection limit for uranium in the lowest fg/mL range has been achieved. Under optimum experimental conditions ($^{235}U/^{238}U = 1$), the precision in $^{235}U/^{238}U$ isotopic ratio determinations was 0.07% RSD. With the isotopic standard U-020 ($^{235}U/^{238}U = 0.0208$) a precision of 0.23% RSD at the 100 pg/mL level using ultrasonic nebulization was achieved. With $^{234}U/^{238}U$ ratios of down to 10^{-5}, the values obtained by double-focusing sector-field ICP-MS and α spectrometry were in agreement (Kerl *et al.*, 1997). The nuclide analysis of an irradiated tantalum target is another example (Becker and Dietze, 1997). This tantalum was used as the target in a spallation neutron source with 800 MeV protons. The determination of the concentration of spallation nuclides in a highly radioactive matrix in the concentration range from $1 \, \mathrm{ng \, g^{-1}}$ to $50 \, \mu\mathrm{g \, g^{-1}}$ was performed to verify the theoretical results for spallation yields in tantalum. For the determination of spallation nuclides, a double-focusing sector-field ICP-MS was used after liquid–liquid extraction of the tantalum matrix in order to remove the high ^{182}Ta activity. The method for the determination of trace impurities after matrix separation was developed using high-purity inactive tantalum. The ICP-MS results were compared to those obtained by neutron activation analysis yielding fair agreement and showing that less elements can be determined with NAA. The main problem in determining long-lived radionuclides in tantalum is possible interferences of radionuclides with stable isotopes at the same mass with a different atomic number. Therefore, in order to separate isobars for rare earth elements, HPLC was coupled with ICP-MS. Day *et al.* (2000) have developed capillary electrophoresis interfaced with double-focusing sector-field ICP-MS for further abundance determination of lanthanides produced via spallation reactions in the irradiated tantalum target. A MicroMist nebulizer with a small volume cyclonic spray chamber was used for ICP-MS sample introduction. The CE-ICP-MS interface featured a self-aspirating electrolyte make-up solution for electrical ground connection and control of nebulizer suction. The CE-ICP-MS features fast run times and small sample sizes ($\sim$35 nL injection volume). Detection limits for the most abundant lanthanide isotopes were 0.72–3.9 ppb, an improvement of as much as one order of magnitude compared to a quadrupole ICP-MS system using a similar experimental arrangement. The nuclide abundances were in agreement with those previously detected by HPLC-ICP-MS. The deviation from theoretically predicted yields was typically 20–30%.

Precise long-term measurements of uranium and thorium isotope ratios were carried out in 1 μg/L solutions using a standard quadrupole inductively coupled plasma mass spectrometer (Becker and Dietze, 1999a) The isotopic ratios of uranium (^{235}U/^{238}U = 1, 0.02, and 0.00725) were determined using a cross-flow nebulizer (CFN, at a solution uptake of 1 mL/min) and a low-flow microconcentric nebulizer (MCN, at a solution uptake of 0.2 mL/min) over 20 h. For 1 μg/L uranium solution, (^{235}U/^{238}U = 1) relative standard deviations of 0.05 and 0.044% using CFN and MCN, respectively, were achieved. Additional short-term isotope ratio measurements using a direct injection high-efficiency nebulizer (DIHEN) of 1 μg/L uranium solution (^{235}U/^{238}U = 1) at a solution uptake of 0.1 mL/min yielded a RSD of 0.06–0.08%. The sensitivity of solution introduction for uranium, thorium, and plutonium (about 150 MHz/ppm) increased significantlly compared to CFN and MCN, and the solution uptake could be reduced to 1 μL/min in DIHEN-ICP-MS. Isotope ratio measurements at an ultralow concentration level (e.g., determination of ^{240}Pu/^{239}Pu isotope ratio in a 10 ng/L Pu waste solution) were carried out for the characterization of radioactive waste and environmental samples.

In another systematic investigation (Becker and Dietze, 1999b), the capability of double-focusing sector-field ICP-MS coupled with a plasma-shielded torch using different nebulizers was investigated. The total amount of analyte for each long-lived radionuclide (^{226}Ra, ^{230}Th, ^{237}Np, ^{238}U, ^{239}Pu, and ^{241}Am; each 1 ng/L in aqueous solution) using different nebulizers varied between 0.4 and 10 pg for the different nebulizers. The application of the shielded torch yielded an increase in sensitivity of up to a factor of 5 compared with the configuration without shielded torch. Sensitivities of about

2000 MHz/ppm were measured with the shielded torch. The detection limits were in the sub-pg/L range and the precision ranged from 1 to 2% RSD for the 1 ng/L concentration level. At low solution uptake rates (down to several μL/min), the uranium solutions were analyzed by DIHEN-ICP-MS using a double-focusing sector-field instrument with higher sensitivity than quadrupole based ICP-MS. The precision of Pu isotopic analysis by DF-ICP-MS with a shielded torch was 0.2, 2, and 14% RSD for 1000, 100, and 10 pg/L.

The figures of merit of ICP-MS with a hexapole collision cell (HEX-ICP-QMS) were studied. A significant improvement in sensitivities and detection limits was achieved for ions affected by interferences with argon ions or argon based molecular ions due to the removal of Ar^+ and ArX^+ ions ($X = H$, C, N, O, Ar) as well as for heavy elements due to better ion transmission through the hexapole ion guide (Boulyga and Becker, 2000a). HEX-ICP-MS was applied in the determination of uranium in soil samples from the Chernobyl area (Boulyga and Becker, 2000b). A limit of $^{236}U/^{238}U$ ratio determination in 10 μg/L uranium solution was 3×10^{-7} corresponding to a detection limit of 3 pg/L. The precision of uranium isotopic ratio measurements in 10 μg/L standard solutions was 0.13% ($^{235}U/^{238}U$) and 0.33% ($^{236}U/^{238}U$).

The capability of LA-ICP-MS for measurements of long-lived radionuclides in non-conducting concrete matrix which is a very common matrix in waste packages, was investigated. The limits of detection of long-lived radionuclides were determined in the low pg/g range in a quadrupole LA-ICP-MS and in double-focusing sector field LA-ICP-MS (Becker and Pickhardt, 2000). Different calibration procedures—the correction of analytical results with experimentally determined relative sensitivity coefficients (RSCs), the use of calibration curves and solution calibration by coupling LA-ICP-MS with an ultrasonic nebulizer—were applied for the determination of long-lived radionuclides, especially for Th and U in different solid samples.

Becker (2002) contains a very recent review of the state-of-the-art and progress in precise and accurate isotope ratio measurements by ICP-MS and LA-ICP-MS including a comprehensive paragraph on the determination of long-lived radionuclides.

Danesi *et al.* (2003) investigated the isotopic composition of uranium and plutonium in selected soil samples collected in Kosovo from locations where depleted uranium (DU) ammunition was expended during the 1999 Balkan conflict. These analyses were conducted using gamma spectrometry (^{235}U, ^{238}U), α spectrometry (^{238}Pu, $^{239+240}Pu$), ICP-MS (^{234}U, ^{235}U, ^{236}U, ^{238}U) and accelerator mass spectrometry (^{236}U). The results indicated that, whenever the uranium concentration exceeded the normal environmental values (2–3 mg/kg), the increase was due to DU contamination. ^{236}U was present in the released DU at a constant ratio of ^{236}U (mg/kg)/^{238}U (mg/kg) $= 2.6 \times 10^{-5}$, indicating that the DU used in the ammunition was from a batch that had been irradiated with neutrons and was then reprocessed. The plutonium concentration in the soil (undisturbed) was about 1 Bq/kg $^{239+240}Pu$ and could be entirely attributed to the fallout of the nuclear weapons tests of the 1960s (no appreciable contribution from DU). Similar conclusions were arrived in Pöllänen *et al.* (2003).

VI. RESONANCE IONIZATION MASS SPECTROMETRY (RIMS)

A. Principle

Generally, the various traditional mass spectrometric methods are limited by isobaric interferences and neighbouring masses, due to their finite abundance sensitivity. This is, because the common types of ion sources feature only limited or even negligible elemental and no isotopic selectivity at all. Using resonant excitation and ionization of the analyte by laser light, Resonance Ionization Mass Spectrometry (RIMS) achieves highest elemental and, if needed, even very high isotopic selectivity in the ionization process. In order to apply this selective laser ionization, the analyte needs to be volatilized in form of neutral atoms.

RIMS offers outstanding properties, which may be summarized as follows:

- Almost complete suppression of atomic and molecular isobaric interferences.
- Very good overall sensitivity in the fg-range ($\sim 10^6$ atoms).
- Feasibility of ultra-high isotope selectivity by taking advantage of the isotope shift in the atomic transitions in addition to the abundance sensitivity of the mass spectrometer.

However, in contrast to other mass spectrometric methods, RIMS is dedicated to single element determination and not suitable for multi-element analysis.

The principle of resonant ionization (RIS) is shown in Fig. 10.6. Starting from the ground state or from a thermally populated low-lying excited state,

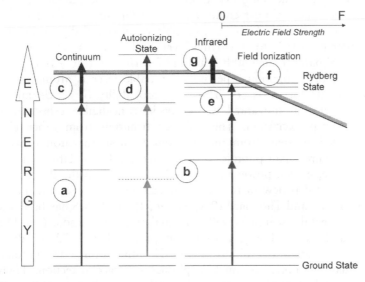

FIGURE 10.6 Principle of two- or three-step resonant excitation and ionization of an atom by laser light. Besides photon absorption, also other ionizing mechanisms may be exploited for the ionization step.

the sample atom is excited to a high-lying excited state by one (a) or two (b) resonant photon absorptions. If two resonant steps are used, either a real or a virtual intermediate state can be involved. Finally, the highly excited atom is ionized by another photon, which either nonresonantly raises the electron energy beyond the ionization limit to the continuum (c), or resonantly populates an autoionizing state (d), i.e., a bound state above the first ionization potential. Such an autoionizing state immediately decays under emission of the electron and formation of an ion. Alternatively, high-lying Rydberg states may be resonantly populated (e) and subsequently ionized by, e.g., application of an electric field (f), far infrared photons (g), by collisions with buffer gas atoms, or by any other state selective ionization method.

Due to the high cross section for the photon–atom interaction, an efficiency of the optical excitation and ionization near 100% can be realized with laser light. Total photon fluxes of $\geq 10^{15}$–10^{18} photons per second from cw-lasers or per pulse ($\sim 10\,\text{ns}$) from pulsed laser systems are easily achievable, while the cross section for optical excitation from the ground state is in the order of $\lambda^2/2\pi$, i.e., $\sim 10^{-10}$–$10^{-9}\,\text{cm}^2$ or $\sim 10^{14}$–10^{15} barn, respectively. The bottleneck of the RIS process is the ionizing step with a cross section of only 10^{-17}–$10^{-19}\,\text{cm}^2$ for the nonresonant process to the continuum. If an autoionizing state is populated, the efficiency is increased by two to three orders of magnitude. Nearly the same holds, if a Rydberg state is involved. Thus, power and bandwidth of the lasers are crucial factors in order to achieve highest possible ionization efficiency. The bandwidth of the lasers compared to the linewidths of the excited transitions determines both the optical selectivity and the laser power needed for the saturation of these transitions.

The most outstanding property of RIMS is its extremely high elemental selectivity, i.e., the suppression of isobars and neighbouring masses from other elements is almost complete. This is due to the fact that in an atom, the density of levels which are accessible via electrical dipole transitions, is of the order of $\sim 1/\text{eV}$ for low-lying levels and $\sim 100/\text{eV}$ around $n = 20$, while the typical natural line width is around $7 \times 10^{-8}\,\text{eV}$ (for $\sim 10\,\text{ns}$ lifetime). The linewidth of a common type, pulsed tunable laser is around $10^{-4}\,\text{eV}$, and even far less for cw-lasers. Thus, the probability of accidentally matching a transition of an unwanted atomic species is negligible, especially for a two- or three-step excitation. The only contribution from other elements or molecules may stem from thermal ionization in the atomic beam source, or from nonresonant photoionization. Particularly, the latter may be of some importance, if high power ultraviolet light is used in the RIS process.

A detailed review of resonance ionization mass spectrometry is given by Payne, Deng, and Thonnard (Payne *et al.*, 1994). RIMS has been widely used in fundamental research as well as in analytical applications (Letokhov, 1987; Hurst and Payne, 1988), since it was proposed in 1972 for the first time (Ambartzumian and Letokhov, 1972). The studies range from atomic physics and quantum optics to nuclear physics or trace detection. Highlights in fundamental research are the measurements on isotope shift and nuclear moments of very short-lived, neutron-deficient gold, platinum, and iridium isotopes produced on-line at ISOLDE/CERN (Wallmeroth *et al.*, 1989;

Krönert *et al.*, 1991; Hilberath *et al.*, 1992; Sauvage *et al.*, 2000) or fission isomers of americium (Lauth *et al.*, 1992; Backe *et al.*, 2000; Backe *et al.*, 2001) in order to study nuclear properties. Also, the determination of the ionization potentials of the actinides up to einsteinium (Köhler *et al.*, 1997; Erdmann *et al.*, 1998; Peterson *et al.*, 1998) with samples of only $\sim 10^{12}$ atoms per element, as well as the first atomic spectroscopy on fermium (Sewtz *et al.*, 2003) show the outstanding possibilities RIMS offers.

One particular field for the analytical application of RIMS is the ultratrace determination of very long-lived radioisotopes (Kluge *et al.*, 1994; Wendt *et al.*, 1995; Wendt *et al.*, 1999; Wendt *et al.*, 2000). Many long-lived α- and β-emitters belong to the most important hazardous radioisotopes. Depending on their half-life, decay type, and individual interferences with other radioisotopes, such as $^{239}Pu/^{240}Pu$, $^{238}Pu/^{241}Am$ or $^{89}Sr/^{90}Sr$, the determination via radiometric techniques suffers from comparably high limits of detection (LOD) and often insufficient selectivity for some of these isotopes. However, for effective surveillance of the environment and a rapid risk assessment in case of a nuclear accident, a fast determination of the most hazardous isotopes at low detection limits, i.e., in the fg range (10^6–10^7 atoms) is indispensable and, in some cases, the isotopic composition provides valuable information on the source of such a contamination. Beyond that, ultratrace determination of long-lived radioisotopes is also applied in biomedical tracer investigations, for radiodating and in cosmochemistry (Willis *et al.*, 1991; Ham and Harrison, 2000; Bushaw *et al.*, 2001; Müller *et al.*, 2001).

B. RIMS Systems and Applications

I. RIMS with Pulsed Lasers

When the demands on isotope selectivity are moderate (i.e., the resolution of a particular mass spectrometer is sufficient), pulsed lasers usually are used for RIMS. In this way, saturation of the optical excitation steps is normally achieved, and the ionization process is easily combined with a mass separation by a time-of-flight mass spectrometer, where the laser pulse delivers the start signal. For good sensitivity, the temporal and spatial overlap of the evaporized sample and the laser beams needs to be considered. If standard low repetition rate pulsed lasers (10–100 Hz) are used, the sample may be volatilized by pulsed laser desorption or ablation, respectively (Krönert *et al.*, 1991). Though this method has been and still is successfully used for fundamental research at the on-line facility ISOLDE/CERN in the COMPLIS-experiment (Sauvage *et al.*, 2000), it usually leads to high pulse-to-pulse fluctuations, as the process depends extremely critical on the pulse-to-pulse stability of power and spatial profile of the heating laser. Thus, it is not very well suited for quantitative ultratrace determination. If the sample is vaporized continuously by resistive heating, the use of high repetition rate pulsed lasers (5–25 kHz) may, if not complete, reduce the duty-cycle losses. Such a system is routinely used for ultratrace determination of actinides, primarily plutonium, in environmental and biological samples (Passler *et al.*, 1997; Grüning *et al.*, 2001).

After chemical separation of the plutonium from the sample, plutonium hydroxide is electrolytically deposited on a tantalum backing. Subsequently, the plutonium hydroxide is covered with a thin layer ($\sim 1\,\mu m$) of titanium by sputtering technique. Thus, when such a so-called sandwich filament is heated to about 1000°C inside the vacuum recipient, the plutonium hydroxide is converted to the oxide, which then is reduced to the metallic state during diffusion through the titanium layer. Reaching the titanium surface, the plutonium is efficiently evaporated. Subsequently, the atoms are ionized near the filament surface by a three-step three-colour laser excitation. The ions are then accelerated in an electric field and mass-selectively detected with a multi-channel plate detector behind a time-of-flight drift tube equipped with an ion reflector to improve the mass resolution. In order to obtain quantitative results, a known amount of a tracer isotope — usually ^{244}Pu in the case of environmental samples — is added to the sample, prior to the chemical treatment. Additionally, in most cases a very small amount of ^{236}Pu is added for monitoring the chemical yield by α-spectroscopy.

As explicated above, a high repetition rate laser system is used for maximum temporal overlap. Until recently, three dye lasers pumped by two powerful copper vapor lasers were in operation. The repetition rate was 6.6 kHz, the pulse length ~ 10–20 ns and the linewidth of the dye lasers varied from 1–8 GHz, depending on the wavelength and the mode of operation. As copper vapor lasers suffer from high maintenance efforts and costs, a new full solid-state laser system has been established, consisting of three Ti : Sapphire-lasers pumped by a high power, high repetition rate Nd : YAG-laser (1–25 kHz) with an average output power of up to 50 W at 532 nm. Though the pulse duration of the pump laser exceeds 400 ns, the Ti : Sapphire lasers provide pulses of ~ 70 ns. The tuning range of the lasers is 725–895 nm, while the linewidth is 2–5 GHz. Each laser delivers up to 3 W output power. The tuning range can be extended by frequency doubling in a nonlinear crystal. This is done for one of the lasers, as the ionization energy of ~ 6 eV for the actinides must be provided by the sum of the corresponding three photons. Figure 10.7 shows the setup as used at the University of Mainz for sensitive and isotope-selective determination of plutonium and other actinides.

Another way to improve the temporal overlap with pulsed lasers is to use a so-called "laser ion source," i.e., to confine the sample atoms inside a small heated chamber, where they are resonantly ionized by the laser beams entering through a small aperture. The ions are extracted by an electric field, e.g., the penetrating part of an external acceleration field, via the same aperture or a second one, while the neutral atoms are "stored" inside the chamber by multiple collisions with the wall for the period of a few laser shots. Combined with a double-focusing mass spectrometer (Mattauch-Herzog-type), such a system has been used for ultratrace analysis of technetium (Passler *et al.*, 1997) with a LOD of 10^6 atoms of ^{99g}Tc, which compares to a β-spectroscopic $LOD_{\beta\text{-spectr.}}$ of 10^{11} atoms in a 24 h measurement. ^{95m}Tc, produced by the ^{93}Nb$(\alpha, 2n)^{95m}$Tc-reaction from a

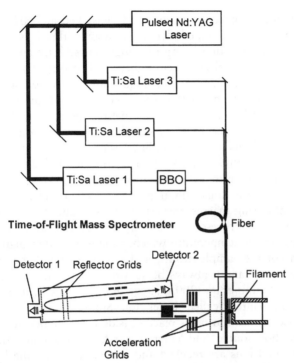

FIGURE 10.7 Sketch of the setup for pulsed laser RIMS as used at the University of Mainz for sensitive and isotope-selective determination of plutonium and other actinides. Frequency doubling of laser 1 is achieved with a BBO crystal.

niobium target, was used as tracer in these measurements. Meanwhile, similar types (hot cavity-type) of laser ion sources are used with great success for selective ionization of short-lived artificially produced radioisotopes at the on-line mass separators ISOLDE/CERN (Lettry *et al.*, 1998), IRIS/Gatchina (Barzakh *et al.*, 1997; Alkhazov *et al.*, 1991), TIARA/Takasaki (Koizumi *et al.*, 1997), and IMP/Lanzhou (Yuan *et al.*, 1994). An ion guide type laser ion source is used at the LISOL on-line isotope separator at Leuven. In this case, the operational principle is based on the resonant laser ionization of the nuclear reaction products that are thermalized and neutralized in a high pressure noble gas, while the ions are flushed out by the gas jet through a nozzle (Kudryavtsev *et al.*, 2002). Both types of laser ion sources have been applied to a large number of elements, including Be, Ti, Mn, Co, Ni, Cu, Zn, Ga, Ru, Rh, Ag, Cd, In, Sn, Ba, Pb, Fr, Ho, Er, Tm, Yb, and Am (Köster, 2002). The on-line work will be extended to many other elements; successful off-line tests have already been performed for about 30 additional elements (Köster *et al.*, 2002).

A more sophisticated type of an ion guide laser ion source — Ion Guide detected RIS (IGRIS) — as described by Backe *et al.* (1997) has enabled the first atomic spectroscopy investigations on fermium (Sewtz *et al.*, 2003) and seems to be a very promising approach for ultratrace analysis, too. At an argon pressure of about 35 mbar, the storage time within the interaction volume illuminated by the laser light is about 40 ms. This value is limited by diffusion and thermal convection in the buffer gas. Yet, at 200 Hz repetition rate of the excimer laser-pumped dye laser system, this is sufficient time for efficient resonant ionization by up to 8 high power laser pulses. The ions are guided to a nozzle by electric fields, then flushed out of the cell by the gas jet. By means of skimmers and a quadrupole ion guide structure, the ions are separated from the buffer gas, mass selected in a quadrupole mass spectrometer, and counted on a channeltron detector with an overall sensitivity of $\sim 10^{-4}$ (counted ions per sample atom in the source).

The ionization efficiencies of the on-line laser ion sources are in the order of a few percent, typically. Compared to a conventional ion source (e.g., plasma ion source) with up to 50% ionization efficiency, the advantage of the laser ion source lies in the drastically improved selectivity, i.e., suppression of isobars and neighbouring masses. Nevertheless, the gain in selectivity is still limited by non-resonant processes – primarily thermal ionization at the cavity walls. Depending on the temperature which, usually, is rather high to reduce the sticking time of the sample atoms, and on the material the chamber is made of, interferences cannot always be suppressed in a hot cavity laser ion source. Therefore, Blaum *et al.* (2003) suggested a novel scheme for a highly selective laser ion source, called Laser Ion Source Trap (LIST). In this approach, the laser ionization takes place outside the chamber, just when the atoms leave the hot source and enter a gas-filled segmented quadrupole structure. All thermal ions are repelled and driven back into the hot source, while the laser ions are trapped in a RF quadrupole trap, where they are accumulated, cooled and bunched for optimum mass resolution and background reduction. These features will also benefit ultratrace determination where very high elemental selectivity and low detection limits are essential.

2. RIMS with Continuous Wave Lasers

In some cases, the demands on isotope selectivity are extremely high, in particular, if one aims to determine a very rare isotope of an omnipresent element with one or more stable isotopes at ultratrace level. Isotope selectivities above 10^9 are not accessible by conventional mass spectrometry (Heumann *et al.*, 1995). By using narrow-bandwidth lasers, i.e., cw-lasers, a remarkable isotope selectivity of several orders of magnitude can be achieved already in the ionization process by taking advantage of the isotope shift in the atomic transitions. This selectivity is multiplied with the abundance sensitivity of the mass spectrometer. Thus, extremely high isotope selectivities up to 10^{15} seem feasible. For ultratrace determination of the long-lived radioisotope ^{41}Ca ($T_{1/2} = 1.04 \times 10^5$ a) in the presence of stable calcium, an overall isotope selectivity $> 10^{12}$ has been demonstrated at an LOD of $<10^6$ atoms ^{41}Ca (Müller *et al.*, 2001). This enables various applications like using ^{41}Ca as a biomedical tracer isotope for in-vivo studies of human calcium kinetics, cosmochemical investigations on the ^{41}Ca content in meteorites and correlated fundamental research on the production cross section in nuclear reactions, and the determination of ^{41}Ca activity in reactor concrete and other materials for an integrated neutron dosimetry. For the mentioned purposes, ^{41}Ca abundances of 10^{-9} to 10^{-12} relative to the predominant stable isotope ^{40}Ca (96.94%), must be determined. The use of ^{41}Ca for radiodating would make a new time window of 10^4 to 10^6 years before present accessible. As the ambient ^{41}Ca concentration is estimated to be only about 10^{-15}, the selectivity still needs to be further improved for this application.

Besides the laser system, the setup for ultratrace determination of ^{41}Ca consists of a vacuum recipient with a graphite furnace for efficient atomization of the sample and a quadrupole mass filter for mass-selective detection of the ions induced by selective laser excitation. The calcium atomic beam is crossed perpendicularly by the laser beams at a distance of about 2 cm from the orifice. For maximum isotope selectivity, a three-step three-colour excitation scheme ($\lambda_1 = 422.8$ nm, $\lambda_2 = 732.8$ nm, and $\lambda_3 = 845.6$ nm) is used to populate the 4s15f $^1F_3$ Rydberg state, which is nonresonantly ionized by the 10.6 μm radiation of a CO_2-laser. All three resonant excitation steps are driven by the narrow-band radiation from extended cavity diode lasers (ECDL). Up to 5 mW of blue laser light for the first excitation step is generated by frequency doubling the 845.6 nm laser output of an ECDL in a $KNbO_3$-crystal, which is placed inside an external enhancement cavity. The optical frequencies of all three ECDLs are computer controlled and stabilized to an accuracy of < 300 kHz and a maximum long-term drift of < 3 MHz/day. In order to compensate for the residual Doppler shift perpendicular to the atomic beam axis, the two red laser beams are fed in counter propagating to the blue laser beam. Using this triple resonant excitation scheme, an optical isotope selectivity $S_{opt}(^{41}Ca/^{40}Ca) > 10^9$ has been shown, which results in an overall selectivity of $S \approx 5 \times 10^{12}$ for neighbouring isotopes (Müller et al., 2001). The stated overall detection efficiency is about 5×10^{-5}.

To determine the ^{41}Ca-content in the presence of stable Ca, the low abundance isotope ^{43}Ca is used as a reference isotope as, like ^{41}Ca, it is an odd-mass isotope with nuclear spin $I = 7/2$, and its isotopic abundance of only 0.134% reduces the required dynamic range. In this way, measurements of the ^{41}Ca-content of concrete samples from the shield of a nuclear research reactor have been performed successfully. The results show a very low relative abundance of the order of a few times 10^{-10} corresponding to contents of only 20–60 mBq/g, which were calibrated by quantifying the total calcium content via x-ray fluorescence spectroscopy (Müller et al., 2000). In the frame of biomedical and nutritional studies in connection with osteoporosis prevention and therapy, urine samples have been measured with equally low abundances. For synthetic calcium test samples, ^{41}Ca ratios down to 3×10^{-13} have already been demonstrated, limited by the overall detection efficiency, so far.

The isotope selectivity that is achievable by the resonant excitation and ionization process depends on the isotope shift of the atomic levels involved. This atomic isotope shift consists of two components with opposite sign, the mass shift and the field shift, where the first one is predominant in the light elements while, for the heavy elements, the latter one plays the major role. For some elements in between, i.e., in the mass region $A \approx 100$, the two effects nearly compensate each other. In strontium, for instance, fast and sensitive determination of the radiotoxic isotopes ^{89}Sr and ^{90}Sr in air samples after a nuclear accident requires an overall selectivity of $S \approx 10^{11}$ (Monz *et al.*, 1993), if $\sim 10^8$ atoms of ^{90}Sr and ^{89}Sr are to be detected among 10^{18} atoms of stable strontium, mainly ^{88}Sr, as assumed to be collected in $1000\,m^3$ of air. Due to the small isotope shifts in Sr, this requirement is hardly met by triple-resonance diode laser based RIMS (Bushaw and Nörtershäuser, 2000).

Therefore, the high optical selectivity of collinear laser spectroscopy has been used for fast strontium determination in environmental samples. In this setup, the sample is ionized in a conventional ion source, accelerated to a comparatively high energy of $\sim 30\,keV$ and mass separated. The enriched ion beam passes a cesium vapour cell, where it is neutralized by resonant charge exchange. Thus, a fast neutral atomic beam is formed, where nearly all atoms are in the metastable 5s4d and 5s5p triplet states, which are predominantly populated in the charge exchange process. Residual ions and collisionally produced Rydberg atoms from the charge exchange process are then removed from the beam by an electrostatic filter deflector, before the fast atoms are optically excited by superposition of a quasi anti-collinear laser beam. UV radiation of 363.8 nm from a stabilized narrow band-width cw argon laser is used to excite the Sr atoms from the 5s4d ^{3}D$_3$ state to the 5s23f ^{3}F$_4$ Rydberg state at an intersection angle of 2° between laser and fast atomic beam. Subsequently, the Rydberg atoms are field-ionized, deflected from the residual beam and counted by a channeltron detector after passing an electrostatic energy filter.

Due to the anticollinear excitation, the transition line is strongly Doppler shifted. Therefore, a fixed frequency laser can be used, while the resonance is tuned via the beam energy. Moreover, at a given beam energy, the various isotopes have different velocities, due to their different masses, i.e., they are separated by a large differential Doppler shift, so to speak an artificial isotope shift, which is about two orders of magnitude larger than the natural isotope shift in strontium for a typical beam energy of 30 keV.

With this approach, an overall efficiency of 1×10^{-5} and a selectivity of $S(^{90}Sr/^{88}Sr) > 10^{10}$ have been achieved for ^{90}Sr measurements in quite a number of various environmental and technical samples (Wendt *et al.*, 1997). The LOD for ^{90}Sr has been determined to be 3×10^6 atoms, corresponding to an activity of 1.5 mBq.

VII. ACCELERATOR MASS SPECTROMETRY (AMS)

A. Principle

Accelerator Mass Spectrometry (AMS) is a modern and sophisticated mass spectrometric technique, specialized to provide highest selectivity with respect to isobaric and isotopic contaminations. This feature is realized by identifying and counting individual atoms with special detection techniques after acceleration to energies in the MeV range. Any interferences from atomic or molecular species are efficiently suppressed in the mass spectrometric measuring process of AMS. Thus, it permits the determination of ultrarare radioisotopes with abundances far below 10^{-9} of a dominant neighbouring isotope and surpasses the usual selectivity limits in isotopic ratio measurements of conventional mass spectrometric techniques by far (Purser *et al.*, 1981). The development of this technique has been boosted primarily in connection with the determination of the cosmogenic radioisotope ^{14}C, which has an unique status offering numerous applications, e.g., for radiodating, atmospheric and oceanographic circulation studies, for the determination of anthropogenic radioactive contamination levels, or in bio-medicine for studies on cancer prevention by using ^{14}C-labelled compounds. Many of these fields of ^{14}C-determination have just become possible by AMS, which has replaced the formerly used technique of low level radiometric counting (Libby, 1955), due to its higher sensitivity, smaller sample size requirement and faster response. Radiocarbon dating, carried out on ^{14}C isotopic levels as low as 10^{-15} is the broadest field of application of AMS with thousands of samples per year, but further applications on other radioisotopes have been discovered, which contributes to the value and acceptance of the technique (Michel, 1999).

Following a suggestion by Oeschger and coworkers (Williams *et al.*, 1969), which was triggered by a first, very early mass spectrometric experiment at an accelerator by Alvarez and Cornog (1939), it was realized in the late 1970s that isobaric contaminations can often be eliminated very efficiently by performing mass spectrometry with high-energy ($\sim$ MeV) ion beams (Muller, 1977; Nelson *et al.*, 1977; Bennett *et al.*, 1977). In this regime, a variety of experimental techniques can be applied and quantitative and highly selective counting of the species of interest with unrivalled background suppression is enabled:

- Already in the ion source of the mass spectrometer, atomic isobars can be expelled by exploiting the stability property of negative ions most often used for AMS in the first acceleration stage of a tandem accelerator. For example, ^{14}N^{-}, the only abundant atomic isobar of ^{14}C^{-}, is not stable, and consequently is not produced nor accelerated.
- Molecular isobars can be efficiently removed by passing the accelerated beam through a stripping process within a thin foil or a gaseous target. Here negative atomic ions are efficiently converted into positive charge states by stripping off some electrons while molecules are caused to disintegrate.
- Finally versatile and discriminatory ion detection techniques, e.g., elemental selective energy-loss measurements in ionization chambers, can be applied on high-energy ions ($\sim$MeV) to identify the species via its mass and proton number and further reduce the influence of isobaric interferences.

Examples of those interferences in the case of the ^{14}C^{+} ion are ^{28}Si^{2+}, ^{14}N^{+} or ^{13}CH^{+}. Other limiting effects include background effects originating from gas kinetics, wall scattering, and charge-exchange collisions. Nevertheless, these contributions can be strongly suppressed by optimizing the experimental conditions. A typical AMS system is based on a tandem accelerator with a terminal voltage of 2.5–10 MV, usually involving a beam line length of several tens of meters. On the side of the ion source, the so called low-energy side, intense beams of negative atomic or molecular ions of the species to be analyzed are produced. This is accomplished usually by sputtering, applying Cs^+ ion bombardment with about 10 keV energy, focused onto the target of the analyte. The latter is introduced in form of about 10 mg of material, fixed within a target wheel or a sample exchange mechanism with more than 60 positions, which enables rapid change between analytical sample, blank, and standard. The importance of proper chemical pretreatment of the sample, including removal of extraneous material, dissolution, and, when appropriate, addition of carrier, separation of the element of interest and conversion into the most suitable form for the sputtering process, must be pointed out (Tuniz *et al.*, 1998). A minor drawback of the AMS technique is the fact that rather large sample sizes around 10 mg are required for the initial ionization process. Negative target ions from the sputtering process are accelerated towards a first low-energy (typically ~ 50 keV) mass separator for a first mass selection. By varying the electric potential applied at the magnet chamber a rapid change from one mass to another is possible, avoiding complications which arise from the scanning of the magnetic field, e.g., hysteresis or slow switching speed. Subsequently, acceleration to the high terminal voltage of several MV takes place within the first half of the tandem accelerator. On the high positive potential of the terminal, the ions undergo a stripping process in a thin foil ($\sim 5\ \mu g/cm^2$) or a gas target and a spectrum of positive charge states is populated. Now positively charged, these ions are accelerated a second time, this time from the terminal voltage down to ground, e.g., in the case of carbon, typically charge states between C^{2+} and C^{7+} are populated, depending on the ion velocity. For the case of the C^{5+} ion, this leads to a total beam energy of 15 MeV on the high energy side, when a typical 2.5 MV tandem accelerator is used. All molecular ions are dissociated during the stripping process and remaining fragments have significantly altered mass to charge ratio and can easily be separated afterwards. For this purpose, the high energy ion beam undergoes another mass selective step in a second sector field magnet as well as a charge-state and energy-filtering within an electrostatic deflector before reaching the detectors. Due to the high energy of the incoming ions, some kind of species identification with respect to charge number Z and mass A is possible by measuring energy and specific energy loss in a sensitive semiconductor telescope or an energy sensitive ionization chamber. A typical layout of an AMS machine, which shows these individual components, is given in Fig. 10.8.

Generally, AMS determines the concentration of the rare radioisotope under study by comparing its count rate to the ion current of an abundant stable isotope of the same element as reference. The measured quantity is therefore, in most cases, an isotope ratio, which might cover many orders of magnitude. In most applications, a high accuracy in the order of $\leq 1\%$ for this value is required. Such precise isotope ratio measurements with unrivalled dynamical range of up to 15 orders of magnitude are accomplished experimentally by accelerating both isotopes alternately in a fast cyclic mode. While the transmitted atoms of the rare ultratrace isotope are counted individually as ion counts in a two-dimensional Z versus A

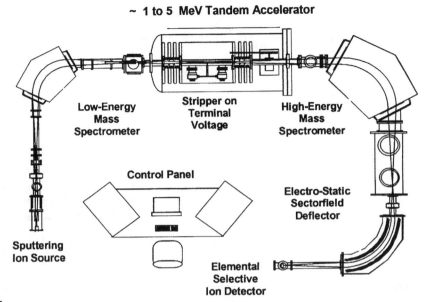

FIGURE 10.8   Schematic of a typical AMS machine.

plane, the reference isotope, which typically is up to a factor of 10^{12} more abundant, is measured as electrical current using a Faraday cup detector for this intense beam. Proper calibration and stability of both detectors during the measurement is mandatory. Hence, each analytical measurement is flanked by measurements on blinds and calibration samples for background surveillance and proper quantification of results. For the heavier isotopes, the energy resolution of semiconductor detectors or ionization chambers is not sufficient for a proper species identification and suppression of isobars and/or neighboring isotopes. A better background suppression is possible by combining the energy determination in the detector with a velocity analysis of the incoming ions by an additional time-of-flight measurement (Coffin and Engelstein, 1984). In a different approach, a gas filled magnet in front of the detector leads to a spatial separation of projectiles with different proton number Z and hence significantly increases the selectivity with respect to those isobars, which have not been filtered out sufficiently during the stripping process (Paul, 1990). For some isotopes, the process of fully stripping the ions after gaining sufficiently high energy is favorable for complete background suppression, but can only be realized at large accelerators. Recently, enormous effort can be observed within the AMS community to reduce size and costs of AMS systems. By incorporating well-matched technical features like gas strippers and state-of-the-art silicon detectors, low-voltage (300 kV–1 MV) and very compact AMS machines have been developed (Synal *et al.*, 2000). By now, these systems have demonstrated their broad range of applicability not only for ^{14}C measurements but also in analyzing other radioisotopes like ^{3}H, ^{26}Al, ^{41}Ca, ^{129}I, and Pu-isotopes with selectivities close to that of a standard large-frame AMS machine (Suter *et al.*, 2000).

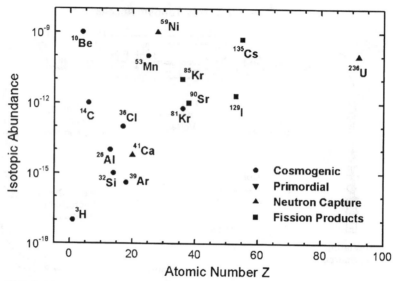

FIGURE 10.9 Natural radioisotopes with abundances below 10^{-9} together with their process of origin.

The technique of AMS has nowadays become a standard method, applied for investigations on a variety of long-lived radioisotopes. Commonly utilized isotopes in routine operation are ^{10}Be, ^{14}C, ^{26}Al, ^{36}Cl, ^{41}Ca, and ^{129}I. Many other isotopes, namely ^{3}H, ^{3}He, ^{7}Be, ^{22,24}Na, ^{32}Si, ^{39}Ar, ^{44}Ti, ^{53}Mn, ^{55,60}Fe, ^{59,63}Ni, ^{79}Se, ^{81}Kr, ^{90}Sr, ^{93}Zr, ^{93}Mo, ^{99}Tc, ^{107}Pd, ^{151}Sm, ^{205}Pb, ^{236}U, ^{237}Np, and $^{238-244}$Pu have also been investigated by AMS during recent years. Most of the radioisotopes mentioned in both categories have natural abundances in the range of 10^{-9} down to 10^{-18} as it is illustrated in Fig. 10.9. Here some isotopes are missing as their natural abundance is either not known precisely or is varying across the earth, due to an inhomogeneous distribution of natural or anthropogenic sources. The isotopes shown in Fig. 10.9 are of primordial origin or produced by cosmic radiation, neutron capture, or as fission product. More details about AMS and a compilation of the different AMS versions for individual radioisotopes can be found in the textbook on AMS (Tuniz *et al.*, 1998) and a number of review articles (Litherland *et al.*, 1987; Elmore and Phillips, 1987; Woelfli, 1987; Vogel *et al.*, 1995; Fifield, 1999; Kutschera *et al.*, 2000). The application of AMS in various fields is increasing rapidly since the late 1970s. As of the year 2000, there were approximately 50 uniquely dedicated AMS facilities in use around the world, while quite some more nuclear physics accelerator facilities have been adapted to perform AMS measurements, often also on a routine basis.

B. Applications

I. Radiodating in Archaeology and other Applications of the Isotope ^{14}C

AMS has become best known due to its application on the isotope ^{14}C, where thousands of samples per year are investigated worldwide. ^{14}C is produced by the neutron component of cosmic rays from ^{14}N in the upper atmosphere with a rate of about 2 atoms/cm^2 s. Through the formation of CO_2 it is very homogeneously distributed throughout the world and is easily assimilated by living organisms. With a half-life of $T_{1/2} = 5730$ a and an average abundance in living organisms of about 10^{-12}, the determination of the remaining ^{14}C-abundance in fossils not only gives access to a dating-range of about 40,000 years before present but, in addition, opens up a broad field of further investigations, e.g., circulation studies in the atmosphere or the oceans, authentication of minerals, plants, and even artificial items like precious pieces of art. Most important is the fact that a well adapted continuous calibration curve of the ^{14}C content in the biosphere, dating back to 7000 B.C. can be established from counting tree rings. Outstanding applications for ^{14}C-radiodating by AMS include the dating of the Ice Man from the Tyrolean Alps (3120–3350 B.C.), the Shroud of Turin (1260–1390 A.D.), as well as numerous approaches of dating ice cores from the artic regions or glaciers.

2. AMS Applications in Geo- and Cosmoscience

Numerous additional studies in the field of geosciences are nowadays carried out with the AMS technique using other radioisotopes than ^{14}C. The cosmogenic production mechanism and the short residence time of ^{10}Be in the atmosphere makes it an ideal cosmic ray and hence sun intensity monitor, which is preserved in arctic ice for hundreds of years and serves as a paleoclimatic recorder. Together with data on a few other long-lived radionuclides, e.g., ^{26}Al and ^{36}Cl both measured by AMS, further information like the strength of the earth's magnetic field, the mode of atmospheric mixing and precipitation patterns are extracted. In this way, e.g., the 11 year solar cycle as well as severe changes during the end of the last glacial period, 40 ka back, have been obtained. From investigations of the same three isotopes, the age of the Arizona Meteor crater has been determined to about 49 ka (Vogel *et al.*, 1995).

The production mechanisms for long-lived radioisotopes are significantly different in space than on the earth surface, due to the different composition of cosmic radiation. Thus, AMS analyses in supposed extraterrestrial material like micro meteorites, which exhibit isotopic anomalies from the well known terrestrial abundances, serve as the most stringent test for its authenticity. In addition, if data of at least some of the different routine AMS isotopes, i.e., ^{10}Be, ^{14}C, ^{26}Al, ^{36}Cl, ^{41}Ca, and ^{129}I, are combined, these measurements provide valuable information, e.g., on the terrestrial age of the material, the exposure age and the geometrical conditions of exposure (Herzog, 1994).

3. Noble Gas Analysis

Due to their chemical inertness, noble gases form an ideal tracer for dating ground- and surface waters, for performing related circulation studies or for the analysis of gaseous inclusions in ice cores. Since noble gases generally do not form stable negative ions, long-lived noble gas isotopes, e.g., ^{39}Ar or ^{81}Kr, cannot be analyzed at a standard AMS facility. As a resource, W. Kutschera and coworkers used an electron cyclotron resonance source to produce positive ions and a high energy cyclotron accelerator to accelerate them. To separate ^{81}Kr from its abundant isobar ^{81}Br, they applied the full-stripping technique, accelerating the ions up to 4 GeV beam energy and removing all electrons of the ions (Collon *et al.*, 1997). Once fully stripped, ^{81}Kr^{+36} can be well separated from the maximum charge stage of ^{81}Br^{+35} with $Z = 35$. With a detection efficiency of $\sim 1 \times 10^{-5}$, a 10% precision measurement required 0.5 cc STP of a krypton sample (Collon *et al.*, 2000). For ground water dating, this amount had been extracted from 16 tons of groundwater. In a closely related development, ^{39}Ar has been detected at an isotopic abundance level of 1×10^{-16} with an efficiency of 1×10^{-3} using a high-energy (~ 200 MeV) linear accelerator (Collon *et al.*, 2003). Since the charge number of ^{39}Ar ($Z = 18$) is lower than that of its abundant isobar, ^{39}K ($Z = 19$), the full stripping technique cannot be used here. Instead, ^{39}Ar has been separated from ^{39}K using the standard technique of a gas-filled magnet, which uses the fact that the energy loss of ^{39}K is significantly different from that of ^{39}Ar.

4. AMS in Life Sciences

Studies using tracer isotopes have become enormously relevant for various kinds of in-vivo studies about influences of nutrition or deficiencies as well as on analyzing actuators and prevention of numerous diseases. In most cases, no suitable enriched stable isotope for the studies is available and radioisotopes must be chosen, and high donation doses would lead to a significant radiation dose for the test person. Furthermore, such high doses could affect the proper physiological response itself. Hence, AMS with its high sensitivity and selectivity has become a valuable system for analyzing bio-medical samples (Jackson *et al.*, 2001). Again ^{14}C is the key isotope, used for labelling numerous bio-medically relevant compounds, while their influence can be studied by AMS measurements on their content in excretions or blood (Vogel, 2000). Further radioisotopes used in this field are ^{10}Be, ^{26}Al, ^{41}Ca, ^{79}Se, and ^{129}I. Aluminum up to recently has been considered an essential trace element, while only in the last decade its toxicity has been recognized. A number of AMS studies with ^{26}Al on gastrointestinal and intravenous absorption have been carried out, which indicated pathways and established exchange rates for Al crossing the blood–brain barrier (Tuniz *et al.*, 1998). ^{41}Ca has become extremely versatile in the analysis of nutritional and other effects on calcium-kinematics and the prevention of diseases like osteoporosis (Freeman *et al.*, 1997). The high demand on measuring capacity for these life science applications from the bio-medical and nutritionists side has recently led to the commissioning of dedicated commercial AMS instruments (Barker and Garner, 1999).

5. AMS Measurements on Long-lived Radionuclides in the Environment

Most of the anthropogenic long-lived radioisotopes in the environment are stemming from the nuclear fuel cycle or nuclear explosions. They are produced either by neutron capture (e.g., actinides) or as fission products (e.g., ^{90}Sr, ^{99}Tc, ^{135}Cs). The determination of low level contaminations of these isotopes is of major importance for nuclear safeguards but often complicated by the necessity to distinguish these contributions from the naturally existing level. A few high transmission, high voltage AMS machines are specialized for measuring these radioisotopes (Hotchkis *et al.*, 2000). Overall efficiencies of up to 10^{-4} enabling measurements with as few as 10^6 atoms of the radioisotope of interest have been demonstrated in favorable cases with typical selectivities above 10^{10}. AMS investigations in this field have been reported for ^{90}Sr (Paul *et al.*, 1997), ^{99}Tc (Fifield *et al.*, 2000; Bergquist *et al.*, 2000), ^{206}Pb (Paul *et al.*, 2000), ^{236}U (Zhao *et al.*, 1994; Zhao *et al.*, 1997; Berkovits *et al.*, 2000) and Pu-isotopes (Fifield *et al.*, 1996; Wallner *et al.*, 2000; McAnnich *et al.*, 2000; Hotchkis *et al.*, 2000). Applications in Pu also include bio-medical studies on the uptake and retention of Pu in living organisms (Fifield *et al.*, 2000).

REFERENCES

Actis-Dato, L. O., Aldave de las Heras, L., Betti, M., Toscano, E. H., Miserque, F., and Gouder, T. (2000). Investigation of mechanisms of corrosion due to diffusion of impurities by direct current glow discharge mass spectrometry depth profiling. *J. Anal. At. Spectrom.* **15**, 1479–1484.

Adriaens, A. G., Fassett, J. D., Kelly, W. R., Simons, D. S., and Adams, F. C. (1992). Determination of uranium and thorium concentrations in soils: comparison of isotope dilution a secondary ion mass spectrometry and isotope dilution-thermal ionization mass spectrometry. *Anal. Chem.* **64**, 2945–2950.

Adriaens, A., van Vaeck, L., and Adams, F. (1999). Static secondary ion mass spectrometry (S-SIMS) Part 2: Material science applications. *Mass Spectrom. Rev.* **18**, 48–81.

Aggarwal, S. K., Kumar, S., Saxena, M. K., Shah, P. M., and Jain, H. C. (1995). Investigations for isobaric interferences of ^{238}Pu at ^{238}U during thermal ionization mass spectrometry of uranium and plutonium from the same filament loading. *Int. J. Mass Spectrom.* **178**, 113–120.

Aldave de las Heras, L., Bocci, F., Betti, M., and Actis Dato, L. O. (2000). Comparison between the use of direct current glow discharge mass spectrometry and inductively coupled plasma quadrupole mass spectrometry for the analysis of trace elements in nuclear samples. *Fresenius J. Anal. Chem.* **368**, 95–102.

Aldave de las Heras, L., Hrnecek, E., Bildstein, O., and Betti, M. (2002). Neptunium determination by dc glow discharge mass spectrometry (dc-GDMS) in Irish Sea Sediment samples. *J. Anal. At. Spectr.* **17**, 1011–1014.

Alkhazov, G. D., Batist, L. K., Bykov, A. A., Vitman, V. D., Letokhov, V. S., Mishin, V. I., Panteleyev, V. N., Sekatsky, S. K., and Fedoseyev, V. N. (1991). Application of a high efficiency selective laser ion source at the IRIS facility. *Nucl. Instrum. Methods Phys. Res., Sect. A* **306**, 400–402.

Alvarez, L. W. and Cornog, R. (1939). ^{3}He in helium. *Phys. Rev.* **56**, 379.

Amaral, A., Galle, P., Cossonnet, C., Franck, D., Pihet, P., Carrier, M., and Stephan, O. (1997). Perspectives of uranium and plutonium analysis in urine samples by secondary ion mass spectrometry. *J. Radioanal. Nucl. Chem.* **226**, 41–45.

Ambartzumian, R. V. and Letokhov, V. S. (1972). Selective two-step (STS) photoionization of atoms and photodissociation of molecules by laser radiation. *Appl. Opt.* **11**, 354–358.

Backe, H., Eberhardt, K., Feldmann, R., Hies, M., Kunz, H., Lauth, W., Martin, R., Schöpe, H., Schwamb, P., Sewtz, M., Thörle, P., Trautmann, N., and Zauner, S. (1997). A compact apparatus for mass selective resonance ionization spectroscopy in a buffer gas cell. *Nucl. Instrum. Methods Phys. Res., Sect. B* **126**, 406–410.

Backe, H., Dretzke, A., Hies, M., Kube, G., Kunz, H., Lauth, W., Sewtz, M., Trautmann, N., Repnow, R., and Maier, H. J. (2000). Isotope shift measurement at ^{244f}Am. *Hyperfine Interact.* **127**, 35–39.

Backe, H., Dretzke, A., Habs, D., Hies, M., Kube, G., Kunz, H., Lauth, W., Maier, H. J., Repnow, R., Sewtz, M., and Trautmann, N. (2001). Stability of superdeformation for americium fission isomers as function of the neutron number. *Nucl. Phys.* **A690**, 215C–218C.

Bacon, J. R., Crain, J. S., Van Vaeck, L., and Williams, J. G. (2001). Atomic spectrometry update. Atomic mass spectrometry. *J. Anal. At. Spectrom.* **16**, 879–915.

Bailey, E. H., Kamp, A. J., and Ragnarsdottir, V. (1993). Determination of uranium and thorium in basalts and uranium in aqueous solution by inductively coupled plasma mass spectrometry. *J. Anal. Atom. Spectrom.* **8**, 551–556.

Barker. J. and Garner, R. C. (1999). Biomedical applications of accelerator mass spectrometry/ isotope measurements at the level of the atom. *Rap. Comm. Mass. Spec.* **13**, 285–293.

Barrero Moreno, J. M., Betti, M., and Nicolaou, G. (1999). Determination of caesium and its isotopic composition in nuclear samples using isotope dilution-ion chromatography-inductively coupled plasma mass spectrometry. *J. Anal. At. Spectrom.* **14**, 875–879.

Barshick, C. M., Duckworth, D. C., and Smith, D. H. (1993). Analysis of solution residues by glow discharge mass spectrometry. *J. Am. Soc. Mass Spectrom.* **4**, 47–53.

Barshick, C. M., Goodner, K. L., Clifford, H. W., and Eyler, J. R. (1998). Application of glow discharge Fourier-transform ion cyclotron resonance mass spectrometry to isotope ratio measurements. *Int. J. Mass Spectrom.* **178**, 73–79.

Barzakh, A. E., Denisov, V. P., Fedorov, D. V., Orlov, S. Y., and Seliverstov, M. D. (1997). A mass-separator laser ion source. *Nucl. Instrum. Methods Phys. Res., Sect. B* **126**, 85–87.

Becker, J. S. (2002). State-of-the-art and progress in precise and accurate isotope ratio measurements by ICP-MS and LA-ICP-MS. *J. Anal. At. Spectrom.* **17**, 1172–1185.

Becker, J. S. and Dietze, H.-J. (1997). Double-focusing sector field inductively coupled plasma mass spectrometry for highly sensitive multi-element and isotopic analysis. *J. Anal. At. Spectrom.* **12**, 881–889.

Becker, J. S. and Dietze, H.-J. (1998). Determination of long-lived radionuclides by double-focusing sector field ICP mass spectrometry. *Adv. Mass Spectrom.* **14**, 681–689.

Becker, J. S. and Dietze, H.-J. (1999a). Precise isotope ratio measurements for uranium, thorium, and plutonium by quadrupole-based inductively coupled plasma mass spectrometry. *Fresenius J. Anal. Chem.* **364**, 482–488.

Becker, J. S. and Dietze, H.-J. (1999b). Application of double-focusing sector field ICP mass spectrometry with shielded torch using different nebulizers for ultratrace and precise isotope analysis of long-lived radionuclides. *J. Anal.At. Spectrom.* **14**, 1493–1500.

Becker, J. S. and Pickhardt, C. (2000). Trace, ultratrace and isotope analysis of long-lived radionuclides by laser ablation inductively coupled plasma mass spectrometry. *Report Jül-* **3821**, 197–227.

Becker, J. S. and Dietze, H.-J. (2000). Mass spectrometry of long-lived radionuclides. In "Encyclopedia of Analytical Chemistry" (R. A. Meyers, Ed.), pp. 12947–1296. John Wiley & Sons, Chicester.

Becker, J. S., Pickhardt, C., and Dietze, H.-J. (2000). Laser ablation inductively coupled plasma mass spectrometry for the trace, ultratrace and isotope analysis of long-lived radionuclides in solid samples. *Int. J. Mass Spectrom.* **202**, 283–297.

Becker, S. and Hirner, A. V. (1994). Coupling of inductively coupled plasma mass spectrometry (ICP-MS) with electrothermal vaporisation (ETV). *Fresenius J. Anal. Chem.* **350**, 260–263.

Beer, B. and Heumann, K. G. (1992). Trace analyses of U, Th and other heavy metals in high purity aluminum with isotope dilution mass spectrometry. *Fresenius J. Anal. Chem.* **343**, 741–745.

Bennett, C. L., Beukens, R. P., Clover, M. R., Gove, H. E., Liebert, R. B., Litherland, A. E., Purser K. H., and Sondheim, W. E. (1977). Radiocarbon dating using electrostatic accelerators: negative ions provide the key. *Science* **198**, 508–510.

Benninghoven, A., Janssen, K. T. F., Tümpner, J., and Werner, H. W. (1992). "Secondary Ion Mass Spectrometry, SIMS VIII." John Wiley & Sons, New York.

Berkovits, D., Feldstein, H., Ghelberg, S., Hershkowitz, A., Navon, E., and Paul, M. (2000). ^{236}U in uranium minerals and standards. Nucl. *Instrum. Methods Phys. Res.*, Sect. B **172**, 372–376.

Betti, M. (1996). Use of a direct current glow discharge mass spectrometer for the chemical characterization of samples of nuclear concern. *J. Anal. Atom. Spectrom.* 11, 855–860.

Betti, M. (1997). Use of ion chromatography (ICP/MS) for the determination of fission products and actinides in nuclear applications. *J. Chromatog.* A789, 369–379.

Betti, M. (2002). Analysis of samples of nuclear concern with glow discharge atomic spectrometry. *In* "Glow Discharge Plasmas in Analytical Chemistry" (R. K. Marcus, J. A. C. Broekaert, Eds.), pp. 273–290. John Wiley & Sons, New York.

Betti, M., Giannarelli, S., Hiernaut, T., Rasmussen, G., and Koch, L. (1996). Detection of trace radioisotopes in soil, sediment and vegetation by glow discharge mass spectrometry. *Fresenius J. Anal. Chem.* 355, 642–646.

Betti, M., Rasmussen, G., and Koch, L. (1996). Isotopic abundance measurements on solid nuclear-type samples by glow discharge mass spectrometry. *Fresenius J. Anal. Chem.* 355, 808–812.

Betti, M., Rasmussen, G., Hiernaut, T., Koch, L., Milton D. M. P., and Hutton, R. C. (1994). Adaption of a glow discharge mass spectrometer in a glove-box for the analysis of nuclear materials. *J. Anal. At. Spectrom.* 9, 385–391.

Betti, M., Tamborini, G., and Koch, L. (1999). Use of secondary ion mass spectrometry in nuclear forensic analysis for the characterization of plutonium and highly enriched uranium particles. *Anal. Chem.* 71, 2616–2622.

Bibler, N. E., Kinard, W. F., Boyce, W. T., and Coleman, C. J. (1998). Determination of long-lived fission products and actinides in Savannah River Site HLW sludge and glass for waste acceptance. *J. Radioanaly. Nucl. Chem.* 234, 159–163.

Blaum, K., Geppert, C., Kluge, H.-J., Mukherjee, M., Schwarz, S., and Wendt, K. (2003). A novel scheme for a highly selective laser ion source. *Nucl. Instrum. Methods Phys. Res., Sect. B*, in press.

Boulyga, S. F. and Becker, J. S. (2000a). Inductively coupled plasma mass spectrometry with hexapole collision cell: figures of merit and applications. *Report Jül-3821*, 13–44.

Boulyga, S. F. and Becker J. S. (2000b). Determination of uranium from nuclear fuel in environmental samples using inductively coupled plasma mass spectrometry. *Report Jül-3821*, 107–133.

Browner, R. F. (1987). Fundamental aspects of aerosol generation and transport. In "Inductively Coupled Plasma Emission Spectroscopy, Part II" (P. W. J. M Boumans, Ed.), Wiley, New York.

Bushaw, B. A. and Nörtershäuser, W. (2000). Resonance ionization spectroscopy of stable strontium isotopes and ^{90}Sr via 5s^2 ^{1}S$_0$ → 5s5p ^{1}P$_1$ → 5s5d ^{1}D$_2$ → 5s11f ^{1}F$_3$ → Sr$^+$. *Spectrochim. Acta* B55, 1679–1692.

Bushaw, B. A., Nörtershauser, W., Müller, P., and Wendt K. (2001). Diode-laser-based resonance ionization mass spectrometry of the long-lived radionuclide ^{41}Ca with $< 10^{-12}$ sensitivity. *J. Radioanal. Nucl. Chem.* 247, 351–356.

Callis, E. L. and Abernathey, R. M. (1991). High-precision isotopic analyses of uranium and plutonium by total sample volatilization and signal integration. *Intern. J. Mass Spectrom. Ion Proc.* 103, 93–105.

Campargue, R. (1964). High intensity molecular beam apparatus. *Rev. Sci. Instrum.* 35, 111–112.

Campargue, R. (1984). Progress in overexpanded supersonic jets and skimmed molecular beams in free-jet zones of silence. *J. Phys. Chem.* 88, 4466–4474.

Chabala, J. M., Soni, K. K., Li, J., Gavrilov, K. L., and Levi-Setti, R. (1995). High-resolution chemical imaging with scanning ion probe SIMS. *Int. J. Mass Spectrom. Ion Processes* 143, 191–212.

Chartier, F., Aubert, M., and Pilier, M. (1999). Determination of Am and Cm in spent nuclear fuels by isotope dilution inductively coupled plasma mass spectrometry and isotope dilution thermal ionization mass spectrometry after separation by high-performance liquid chromatography. *Fresenius J. Anal. Chem.* 364, 320–327.

Coffin, J. P. and Engelstein, P. (1984). Time-of-flight systems for heavy ions. *In* "Treatise on Heavy-Ion Science," (D. A. Bromley, Ed.), Plenum Press, New York.

Collon, P., Ahmad, I., Bichler, M., Broecker, W. S., Caggiano, J., Dewayne Cecil, L., El Masri, Y., Golser, R., Jiang, C. L., Heinz, A., Henderson, D., Kutschera, W., Lehmann, B., Leleux, P.,

Loosli, H. H., Pardo, R. C., Paul, M., Schlosser, P., Scott, R. H., Smethie, Jr., W. M., and Vondrasek, R. (2003). Tracing the ocean with ^{39}Ar. *Nucl. Instrum. Methods Phys. Res., Sect. B*, in press.

Collon, P., Antaya, T., Davids, B., Fauerbach, M., Harkewicz, R., Hellstrom, M., Kutschera, W., Morrissey, D., Pardo, R., Paul, M., Sherrill, B., and Steiner, M. (1997). Measurement of Kr-81 in the atmosphere. *Nucl. Instrum. Methods Phys. Res., Sect. B* **123**, 122–127.

Collon, P., Kutschera, W., Loosli, H. H., Lehmann, B. E., Purtschert, R., Love, A., Sampson, L., Anthony, D., Cole, D., Davids, B., Morrissey, D. J., Sherrill, B. M., Steiner, M., Pardo, R. C., and Paul, M. (2000). Kr-81 in the Great Artesian Basin, Australia: a new method for dating very old groundwater. *Earth. Plan. Sci. Lett.* **182**, 103–113.

Crain, J. S. (1996). Applications of inductively coupled plasma mass spectrometry in environmental radiochemistry. *Atom. Spectrosc. Perspectives* **11**, 30–39.

Danesi, P. R., Bleise, A., Burkart, W., Cabianca, T., Campbell, M. J., Makarewicz, M., Moreno, J., Tuniz, C., and Hotchkis, M. (2003). Isotopic composition and origin of uranium and plutonium in selected soil samples collected in Kosovo. *J. Environ. Radioac.* **64**, 121–131.

Dawson, P. H. and Whetten, N. R. (1969). Mass spectroscopy using rf quadrupole fields. *Adv. Electron. Electron Phys.* **27**, 59–185.

Dawson, P. H., Ed. (1976). "Quadrupole Mass Spectrometry and its Applications." Elsevier, Amsterdam, reissued (1995) by AIP Press, Woodbury, N. Y.

Day, J. A., Caruso, J. A., Becker, J. S., and Dietze, H.-J. (2000). Application of capillary electrophoresis interfaced to double focusing sector field ICP-MS for nuclide abundance determination of lanthanides produced via spallation reactions in an irradiated tantalum target. *J. Anal. At. Spectrom.*, **15**, 1343–1348.

De Gent, St., van Griecken, R., Hang, W., and Harrison, W. W. (1995). Comparison between direct current and radiofrequency glow discharge mass spectrometry for the analysis of oxide-based samples. *J. Anal. At. Spectrom.* **10**, 689–695.

Diehl, R. and Hillebrandt, W. (2002). Astronomy with radioactivity. *Physik Journal* **1**, 47–53.

Duan, Y., Chamberlin, E. P., and Olivares, J. A. (1997). Development of a new high efficiency thermal ionization source for mass spectrometry. *J. Mass Spectrom.* **161**, 27–39.

Duckworth, D. C., Barshick, C. M., and Smith, D. H. (1993). Analysis of soils by glow discharge mass spectrometry. *J. Anal. At. Spectrom.* **8**, 875–879.

Ebdon, L., Fisher, A., Handley, H., and Jones, P. (1993). Determination of trace metals in concentrated brines using inductively coupled plasma mass spectrometry on-line preconcentration and matrix elimination with flow injection. *J. Anal. At. Spectrom.* **8**, 979–981.

Edmonds, H. N., Moran, S. B., Hoff, J. A., Smith, J. N., and Edwards, R. L. (1998). Protactinium-231 and thorium-230 abundances and high scavenging rates in the western arctic ocean. *Science* **280**, 405–407.

Elmore, D. and Phillips, F. M. (1987). Accelerator mass spectrometry for measurement of long-lived radioisotopes. *Science* **236**, 543–550.

Erdmann, N., Betti, M., Stetzer, O., Tamborini, G., Kratz, J. V., Trautmann, N., and van Geel, J. (2000). Production of monodisperse uranium oxide particles and their characterization by scanning electron microscopy and secondary ion mass spectrometry. *Spectrochim. Acta* **B55**, 1565–1575.

Erdmann, N., Nunnemann, M., Eberhardt, K., Herrmann, G., Huber, G., Köhler, S., Kratz, J. V., Passler, G., Peterson, J. R., Trautmann, N., and Waldek, A. (1998). Determination of the first ionization potential of nine actinide elements by resonance ionization mass spectroscopy (RIMS). *J. Alloys Comp.* **271–273**, 837–840.

Eroglu, A. E., McLeod, C. W., Leonard, K. S., and McCubbin, D. (1998). Determination of technetium in sea-water using ion exchange and inductively coupled plasma mass spectrometry with ultrasonic nebulisation. *J. Anal. Atom. Spectrom.* **13**, 875–878.

Evensen, A. H. M., Catherall, R., Jonsson, O. C., Kugler, E., Lettry, J., Ravn, H. L., Drumm, P., Van Duppen, P., Tengblad, O., and Tikhonov, V. (1996). Release and yields from thorium and uranium targets irradiated with a pulsed proton beam. *Nucl. Instr. Methods Phys. Res., Sect. B* **126**, 160–165.

Fassett, J. D. and Paulsen, P. J. (1989). Isotope dilution mass spectrometry for accurate elemental analysis. *Anal. Chem.* **61**, 643A–649A.

Fiedler, R. (1995). Total evaporation measurements: experience with multi-collector instruments and a thermal ionization quadrupole mass spectrometer. *J. Mass Spectrom.* **146**, 91–97.

Fifield, L. K. (1999). Accelerator mass spectrometry and its applications. *Rep. Prog. Phys.* **62**, 1223–1274.

Fifield, L. K. (2000). Advances in accelerator mass spectrometry. *Nucl. Instrum. Methods Phys. Res., Sect. B* **172**, 134–143.

Fifield, L. K., Cresswell, R. G., di Tada, M. L. Ophel, T. R., Day, J. P., Clacher, A. P., King, and S. J., Priest, N. D. (1996). Accelerator mass spectrometry of plutonium isotopes. *Nucl. Instrum. Methods Phys. Res., Sect. B* **117**, 295–303.

Fletcher, I. R., Maggi, A. L., Rosmanand, K. J. R., and McNaughton, N. J. (1997). Isotopic abundances of K and Ca using a wide-dispersion multi-collector mass spectrometer and low fractionation Ionization techniques. *Int. J. Mass Spectrom.* **163**, 1–17.

Fourre, C., Clerc, J., and Fragu, P. (1997). Contribution of mass resolution to secondary ion mass spectrometry microscopy imaging in biological microanalysis. *J. Anal. Atom. Spectrom.* **12**, 1105–1110.

Freeman, S. P. H. T., King, J. C., Vieira, N. E., Woodhouse, L. R., and Yergey, A. L. (1997). Human calcium metabolism including bone resorption measured with a Ca-41 tracer. *Nucl. Instrum. Methods Phys. Res., Sect. B* **123**, 266–270.

Fukuda, M. and Sayama, Y. (1997). Determination of traces of uranium and thorium in (Ba, Sr)TiO₃ ferroelectrics by inductively coupled plasma mass spectrometry. *Fresenius J. Anal. Chem.* **357**, 647–651.

Gray, A. L. (1985). Solid sample introduction by laser ablation for inductively coupled plasma source mass spectrometry. *Analyst* **110**, 551–556.

Gray, A. L. (1986). Influence of load coil geometry on oxides and doubly charged ion response in inductively coupled plasma mass spectrometry. *J. Anal. At. Spectrom.* **1**, 247–249.

Greenfield, S. and Montaser, A. (1992). Common RF generators, torches, and sample introduction systems. *In* "Inductively Coupled Plasmas in Analytical Atomic Spectrometry" (A. Montaser and D. W. Golightly, Eds.), 2nd ed., VCH, New York.

Grüning, C., Huber, G., Kratz, J. V., Passler, G., Trautmann, N., Waldek, A., and Wendt, K. (2001). Determination of trace amounts of plutonium in environmental samples by RIMS using a high repetion rate solid state laser system. *In* "Resonance Ionization Spectroscopy 2000," AIP Conf. Proc. (J. E. Parks and J. P. Young, Eds.), Vol. 584, pp. 255–260. AIP Press, Melville New York.

Gustavsson, A. (1992). Liquid sample introduction into plasmas. *In* "Inductively Coupled Plasmas in Analytical Atomic Spectroscopy" (A. Montaser, D. W. Golightly, Eds.), 2nd. ed. VCH, New York.

Habfast, K. (1998). Fractionation correction and multiple collectors in thermal ionization isotope ratio mass spectrometry. *Int. J. Mass Spectrom.* **176**, 133–148.

Ham, G. J. and Harrison, J. D. (2000). The gastrointestinal absorption and urinary excretion of plutonium in male volunteers. *Radiat. Prot. Dosim.* **87**, 267–272.

Harrison, H. W. (1988). Glow discharge mass spectrometry. *In* "Inorganic Mass Spectrometry, Chemical Analysis" (F. Adams, R. Gijbels, R. van Griecken, Eds.), Vol. 95, pp. 85–123. John Wiley & Sons, New York.

Hedges, R. E. M. and Gowlett, J. A. J. (1986). Radiocarbon dating by accelerator mass spectrometry. *Sci. Ameri.* **254**, 82–89.

Heitkemper, D. T., Wolnik, K. A., Fricke, F. L., and Caruso, J. A. (1992). Injection of gaseous samples into plasmas. *In* "Inductively Coupled Plasmas in Analytical Atomic Spectrometry" (A. Montaser, D. W. Golightly, Eds.), 2nd. ed. VCH, New York.

Herzner, P. and Heumann, K. G. (1992). Ultratrace analysis of U, Th, Ca and selected heavy metals in high-purity refractory metals with isotope dilution mass spectrometry. *Mikrochim. Acta* **106**, 127–135.

Herzog, G. F. (1994). Applications of accelerator mass spectrometry in extraterrestrial materials. *Nucl. Instrum. Methods Phys. Res., Sect. B* **92**, 492–499.

Heumann, K. G., Eisenhut, S., Gallus, S., Hebeda, E. H., Nusko, R., Vengosh, A., and Walczyk, T. (1995). Recent developments in thermal ionization mass spectrometric techniques for isotope analysis – a review. *Analyst* **120**, 1291–1299.

Hilberath, Th., Becker, St., Bollen, G., Kluge, H.-J., Krönert, U., Passler, G., Rikovska, J., Wyss, R., and the ISOLDE Collaboration (1992). Ground-state properties of neutron-deficient platinum isotopes. *Z. Physik* **A342**, 1–15.

Hotchkis, M., Fink, D., Tuniz, C., and Vogt, S. (2000). Accelerator mass spectrometry analyses of environmental radionuclides: sensitivity, precision and standardisation. *Appl. Rad. Isot.* **53**, 31–37.

House, A., Eisenhower, A., Jason, N., Bock, B., Hansen, B. T., and Nägler, Th.F. (2002). Measurement of calcium isotopes ^{44}Ca using a multicollector TIMS technique. *Int. J. Mass Spectrom.* **220**, 385–397.

Huett, T., Ingram, J. C., and Delmore, J. E. (1995). Ion-emitting molten glass–silicagel revisited. *Int. J. Mass Spectrom.* **146**, 5–14.

Hurst, G. S. and Payne, M. G. (1988). "Principles and Applications of Resonance Ionization Spectroscopy." Hilger Publications, Bristol.

Inkret, W. C., Efurd, D. W., Miller, G., Roskop, D. J., and Benjamin, T. M. (1998). Applications of thermal ionization mass spectrometry to the detection of ^{239}Pu and ^{240}Pu intakes. *Int. J. Mass Spectrom.* **178**, 113–120.

Jackson, G. S., Weaver, C., and Elmore, D. (2001). Use of accelerator mass spectrometry for studies in nutrition. *Nutr. Res. Rev.* **14**, 317–334.

James, R. H. and Palmer, M. R. (2000). The Li isotope composition of international rock standards. *Chem. Geol.* **166**, 319–326.

Jiang, S.-J., Houk, R. S., and Stevens, M. A. (1988). Alleviation of overlap interferences for determination of potassium isotope ratios by inductively coupled plasma mass spectrometry. *Anal. Chem.* **60**, 1217–1220.

Johnson, C. M. and Beard, B. L. (1999). Correction of instrumentally produced mass fractionation during isotopic analysis of Fe by thermal mass spectrometry. *Int. J. Mass Spectrom.* **193**, 87–99.

Kawai, Y., Nnomura, M., Murata, H., Susuki, T., and Fujii, Y. (2001). Surface ionization of alkaline earth iodides in double filament system. *Int. J. Mass Spectrom.* **206**, 1–5.

Kerl, W., Becker, J. S., Dietze, H.-J., and Dannecker, W. (1997). Isotopic and ultratrace analysis of uranium by double-focusing sector field ICP mass spectrometry. *Fresenius J. Anal. Chem.* **359**, 407–409.

Kim, C. K., Seki, R., Morita, S., Yamasaki, S., Tsumura, A., Takaku, Y., Igarashi, Y., and Yamamoto, M. (1991). Application of a high resolution inductively coupled plasma mass spectrometer to the measurement of long-lived radionuclides. *J. Anal. Atom. Spectrom.* **6**, 205–209.

Kinard, W. F., Bibler, N. E., Coleman, C. J., and Dewberry, R. A. (1997). Radiochemical analyses for the defense waste processing facility startup at the Savannah River Site. *J. Radioanal. Nucl. Chem.* **219**, 197–201.

Kluge, H.-J., Bushaw, B. A., Passler, G., Wendt, K., and Trautmann, N. (1994). Resonance ionization spectroscopy for trace analysis and fundamental research. *Fresenius J. Anal. Chem.* **350**, 323–329.

Köhler, S., Deißenberger, R., Eberhardt, K., Erdmann, N., Herrmann, G., Huber, G., Kratz, J. V., Nunnemann, M., Passler, G., Rao, P. M., Riegel, J., Trautmann, N., and Wendt, K. (1997). Determination of the first ionization potential of actinide elements by resonance ionization mass spectroscopy. *Spectrochim. Acta* **B52**, 717–726.

Koizumi, M., Osa, A., Sekine, T., Kubota, M. (1997). Development of a laser ion source with pulsed ion extraction. *Nucl. Instrum. Methods Phys. Res., Sect. B* **126**, 100–104.

Köster, U. (2002). Resonance ionization laser ion sources. *Nucl. Phys.* **A701**, 441C–451C.

Krönert, U., Becker, St., Bollen, G., Gerber, M., Hilberath, Th., Kluge, H.-J., Passler, G., and the ISOLDE Collaboration (1991). On-line laser spectroscopy by resonance ionization of laser-desorbed, refractory elements. *Nucl. Instrum. Methods Phys. Res., Sect. A* **300**, 522–537.

Kudryavtsev, Y., Bruyneel, B., Franchoo, S., Huyse, M., Gentens, J., Kruglov, K., Mueller, W. F., Prasad, N. V. S. V., Raabe, R., Reusen, I., Van den Bergh, P., Van Duppen, P.,

Van Roosbroeck, J., Vermeeren, L., and Weissman, L. (2002). The Leuven isotope separator on-line laser ion source. *Nucl. Phys.* **A701**, 465C 469C.

Kutschera, W., Golser, R., Priller A., and Strohmaier, B., Eds. (2000). Accelerator Mass Spectrometry. *Nucl. Instr. Meth. Phys. Res.*, Sect.B **172**, 1–977 and earlier Proceedings of the Series of International Conference on Accelerator Mass Spectrometry, all printed *Nucl. Instrum. Methods Phys. Res., Sect. B.*

Lauth, W., Backe, H., Dahlinger, M., Klaft, I., Schwamb, P., Schwickert, G., Trautmann, N., and Othmer, U. (1992). Resonance ionization spectroscopy in a buffer gas cell with radioactive decay detection, demonstrated using ^{208}Tl. *Phys. Rev. Lett.* **68**, 1675–1678.

Lefèvre, O., Betti, M., Koch, L., and Walker, C. T. (1996). EPMA and mass spectrometry of soil and grass containing radioactivity from the nuclear accident at Chernobyl. *Mikrochim. Acta (Suppl.)* **13**, 399–408.

Letokhov, V. S. (1987). "Laser Photoionization Spectroscopy." Academic Press, Orlando.

Lettry, J., Catherall, R., Focker, G. J., Jokinen, A., Kugler, E., Ravn, H., Drumm, P., Van Duppen, P., Evensen, A. H. M., and Jonsson, O. C. (1996). Pulse shape of the ISOLDE radioactive ion beams. *Nucl. Instr. Methods Phys. Res., Sect. B* **126**, 130–134.

Lettry, J., Catherall, R., Focker, G. J., Jonsson, O. C., Kugler, E., Ravn, H., Tamburella, C., Fedoseyev, V., Mishin, V. I., Huber, G., Sebastian, V., Koizumi, M., and Köster, U. (1998). Recent developments of the ISOLDE laser ion source. *Rev. Sci. Instrum.* **69**, 761–763.

Libby, W. F. (1955). "Radiocarbon Dating." University of Chicago Press, Chicago.

Litherland, A. E., Allen, K. W., and Hall, E. T., Eds. (1987). Proc. Royal Discussion Meeting, *Phil. Trans. Roy. Soc. London* **A323**, 1–173.

Liu, Y., Huang, M., Masuda, A., and Masao, I. (1998). High-precision determination of Os and Re isotope ratios by in situ oxygen isotope rate correction using negative thermal ionization mass spectrometry. *Int. J. Mass Spectrom.* **173**,163–175.

Liu, Y., Masuda, M., and Inoue, M. (2000). Measurement of isotopes of light rare earth elements in the form of oxide ions: a new development in thermal ionization mass spectrometry. *Anal. Chem.* **72**, 3001–3005.

Marcus, R. K. (1996). Radiofrequency powered glow discharges: Opportunities and challenges. *J. Anal. At. Spectrom.* **11**, 821–828.

McAninch, J. E., Hamilton, T. F., Brown, T. A., Jokela, T. A., Knezovich, J. P., Ognibene, T. J., Proctor, I. D., Roberts, M. L., Sideras-Haddad, E., Southon, J. R., and Vogel, J. S. (2000). Plutonium measurements by accelerator mass spectrometry at LLNL. *Nucl. Instrum. Methods Phys. Res., Sect.* **B172**, 711–716.

McKeegan, K. D., Chaussidon, M., and Robert, F. (2000). Incorporation of short-lived ^{10}Be in a calcium-aluminium-rich inclusion from the Allende meteorite. *Science* **289**, 1334–1337.

McLaren, J. W., Lam, J. W. H., Berman, S. S., Akatsuka, K., and Azeredo, M. A. (1993). On-line method for the analysis of sea-water for trace elements by inductively coupled plasma mass spectrometry. *J. Anal. At. Spectrom.* **8**, 279–286.

McLeod, C. W., Routh, M. W., and Tikkanen, M. W. (1992). Introduction of solids into plasmas. *In* "Inductively Coupled Plasmas in Analytical Atomic Spectrometry" (A. Montaser, D. W. Golightly, Eds.), 2nd. ed. VCH, New York.

Michel R. (1999). Long-lived radionuclides as tracers in terrestrial and extraterrestrial matter. *Radiochim. Acta* **87**, 47–73

Montaser, A. (1998). "Inductively coupled plasma mass spectrometry." Wiley-VCH, New York.

Monz, L., Hohmann, R., Kluge, H.-J., Kunze, S., Lantzsch, J., Otten, E. W., Passler, G., Senne, P., Stenner, J., Stratmann, K., Wendt, K., Zimmer, K., Herrmann, G., Trautmann, N., and Walter, K. (1993). Fast, low-level detection of strontium-90 and strontium-89 in environmental samples by collinear resonance ionization spectroscopy. *Spectrochim. Acta* **48B**, 1655–1671.

Müller, P., Blaum, K., Bushaw, B. A., Diel, S., Geppert, Ch., Nähler, A., Nörtershäuser, W., Trautmann, N., and Wendt, K. (2000). Trace detection of ^{41}Ca in nuclear reactor concrete by diode-laser-based resonance ionization mass spectrometry. *Radiochim. Acta* **88**, 487–493.

Müller, P., Bushaw, B. A., Blaum, K., Diel, S., Geppert, C., Nähler, A., Trautmann, N., Nörtershauser, W., and Wendt, K. (2001). ^{41}Ca ultratrace determination with isotopic selectivity $> 10^{12}$ by diode-laser-based RIMS. *Fresenius J. Anal. Chem.* **370**, 508–512.

Muller, R. A. (1977). Radioisotope dating with a cyclotron. *Science* **196**, 489–494.

Nägler, Th.F., and Villa, I. M. (2000). In pursuit of the ^{40}K branching ratios: K-Ca and ^{39}Ar-^{40}Ar dating of gem silicates. *Chem. Geology* **169**, 5–16.

Nelson, D. E., Korteling R. G., and Stott, W. R. (1977). Carbon-14: direct detection at natural concentrations. *Science* **198**, 507–508.

Oberli, F., Gartenmann, P., Meier, M., Kutschera, W., Suter, M., and Winkler, G. (1999). The half-life of ^{126}Sn refined by thermal ionization mass spectrometry measurements. *Int. J. Mass Spectrom.* **184**, 145–152.

Olesik, J. W., Kinzer, J. A., and Olesik, S. V. (1995). Capillary electrophoresis inductively coupled plasma spectrometry for rapid elemental speciation. *Anal. Chem.* **67**, 1–12.

Ottolini, L. and Oberti, R. (2000). Accurate quantification of H, Li, Be, B, F, Ba, REE, Y, Th, and U in complex matrixes: A combined approach based on SIMS and single-cyrstal structure refinement. *Anal. Chem.* **72**, 3731–3738.

Passler, G., Erdmann, N., Hasse, H.-U., Herrmann, G., Huber, G., Köhler, S., Kratz, J. V., Mansel, A., Nunnemann, M., Trautmann, N., and Waldek, A. (1997). Application of laser mass spectrometry for trace analysis of plutonium and technetium. *Kerntechnik* **62**, 85–90.

Paul, M. (1990). Separation of isobars with a gas-filled magnet. *Nucl. Instrum. Methods Phys. Res., Sect. B* **52**, 315–321.

Paul, M., Berkovits, D., Cecil, L. D., Feldstein, H., Hershkovitz, A., Kashiv, Y., and Vogt, S., (1997). Environmental ^{90}Sr measurements. *Nucl. Instrum. Methods Phys. Res., Sect. B* **123**, 394–399.

Paul, M., Berkovits, D., Ahmad, I., Borasi, F., Caggiano, J., Davids, C. N., Greene, J. P., Harss, B., Heinz, A., Henderson, D. J., Henning, W., Jiang, C. L., Pardo, R. C., Rehm, K. E., Rejoub, R., Seweryniak, D., Sonzogni, A., Uusitalo, J., and Vondrasek, R. (2000). AMS of heavy elements with an ECR ion source and the ATLAS linear accelerator. *Nucl. Instrum. Methods Phys. Res., Sect. B* **172**, 688–892.

Paul, W. and Raether, M. (1955). Das Elektrische Massenfilter. *Z. Phys.* **140**, 262–273.

Payne, M. G., Deng, L., and Thonnard, N. (1994). Applications of resonance ionization mass spectrometry. *Rev. Sci. Instrum.* **65**, 2433–2459.

Peterson, J. R., Erdmann, N., Nunnemann, M., Eberhardt, K., Huber, G., Kratz, J. V., Passler, G., Stetzer, O., Thörle, P., Trautmann, N., and Waldek, A. (1998). Determination of the first ionization potential of einsteinium by resonance ionization mass spectroscopy (RIMS). *J. Alloys Comp.* **271–273**, 876–878.

Platzner, I. T. (1997). Modern isotope ratio mass spectrometry. *In* "Chemical Analysis," Vol. 145, pp. 1–530. John Wiley & Sons, Chicester.

Pöllänen, R., Ikäheimonen, T. K., Klemola, S., Vartti, V.-P., Vesterbacka, K., Ristonmaa, S., Honkamaa, T., Sipilä, P., Jokelainen, I., Kosunen, A., Zilliacus, R., Kettunen, M., and Hokkanen, M. (2003). Characterisation of projectiles composed of depleted uranium. *J. Environ. Radioactivity* **64**, 133–142.

Purser, K. H., Williams, P., Litherland, A. E., Stein, J. D., Storms, H. A., Gove, H. E., and Stevens, C. M. (1981). Isotopic ratio measurement at abundance sensitivities greater than $1:10^{15}$: A comparison between mass spectrometry at keV and MeV energies. *Nucl. Instrum. Methods* **186**, 487–498.

Ramakumar, K. L. and Fiedler, R. (1999). Calibration procedures for a multicollector mass spectrometer for cup efficiency, detector amplifier linearity, and isotope fractionation to evaluate the accuracy in the total evaporation method. *Int. J. Mass Spectrom.* **184**, 109–118.

Reed, T. B. (1961a). Growth of refractory crystals using the induction plasma torch. *J. Appl. Phys.* **32**, 2534–2535.

Reed, T. B. (1961b). Induction-coupled plasma torch. *J. Appl. Phys.* **32**, 821–824.

Richner, P., Evans, D., Wahrenberger, C., and Dietrich, V. (1994). Applications of laser ablation and electrothermal vaporization as sample introduction techniques for ICP-MS. *Fresenius J. Anal. Chem.* **350**, 235–241.

Richter, S., Alfonso, A., De Bolle, W., Wellun, R., and Taylor, P. D. P. (1999). Isotopic "fingerprints" of natural uranium ore samples. *Int. J. Mass Spectrom.* **193**, 9–14.

Robinson, K. and Hall, E. F. H. (1987). Glow discharge mass spectrometry for nuclear materials. *J. Metal* **39**, 14–16.

Rohr, U., Meckel, L., and Ortner, H. M. (1994). Ultratrace analysis of uranium and thorium in glass; Part 1: ICP-MS, classical photometry and chelate-GC. *Fresenius J. Anal. Chem.* **348**, 356–363.

Rucklidge, J. (1995). Accelerator mass spectrometry in environmental geoscience. *Analyst.* **120**, 1283–1290.

Sauvage, J., Boos, N., Cabaret, L., Crawford, J. E., Duong, H. T., Genevey, J., Girod, M., Huber, G., Ibrahim, F., Krieg, M., Le Blanc, F., Lee, J. K. P., Libert, J., Lunney, D., Obert, J., Oms, J., Peru, S., Pinard, J., Putaux, J. C., Roussière, B., Sebastian, V., Verney, D., Zemlyanoi, S., Arianer, J., Barre, N., Ducourtieux, M., Forkel-Wirth, D., Le Scornet, G., Lettry, J., Richard-Serre, C., and Veron, C. (2000). COMPLIS experiments: Collaboration for spectroscopy measurements using a Pulsed Laser Ion Source. *Hyperfine Interact.* **129**, 303–317.

Schelles, W., de Gent, St., Maes, K., and van Grieken, R. (1996). The use of a secondary cathode to analyse solid non-conducting samples with direct current glow discharge mass spectrometry: Potential and restrictions. *Fresenius J. Anal. Chem.* **355**, 858–860.

Sewtz, M., Backe, H., Dretzke, A., Kube, G., Lauth, W., Schwamb, P., Eberhardt, K., Grüning, C., Thörle, P., Trautmann, N., Kunz, P., Lassen, J., Passler, G., Dong, C. Z., Fritzsche, S., and Haire, R. G. (2003). First observation of atomic levels for the element fermium (Z = 100). *Phys. Rev. Lett.*, in press.

Sharp, B. L. (1988). Pneumatic nebulizers and spray chambers for inductively coupled plasma spectrometry, A Review: Part 1. Nebulizers. *J. Anal. At. Spectrom.* **3**, 613–652.

Simons, D. S. (1986). Single particle standards for isotopic measurements of uranium by secondary ion mass spectrometry. *J. Trace Microprobe Tech.* **4**, 185–195.

Solatie, D., Carbol, P., Betti, M., Bocci, F., Hiernaut, T., Rondinella, V. V., and Cobos, J. (2000). Ion chromatography inductively coupled plasma mass spectrometry (IC-ICP-MS) and radiometric techniques for the determination of actinides in aqueous leachate solutions from uranium oxide. *Fresenius J. Anal. Chem.*, **368**, 88–94.

Stoffels, J. J., Briant, J. K., and Simons, D. S. (1994). A particulate isotopic standard of uranium and plutonium in an aluminosilicate matrix. *J. Am. Soc. Mass Spectrom.* **5**, 852–858.

Suter, M., Jacob, S. W. A., and Synal, H.-A. (2000). Tandem AMS at sub-MeV energies – status and prospects. *Nucl. Instrum. Methods Phys. Res., Sect. B* **172**, 144–151.

Synal, H.-A., Jacob, S., and Suter, M. (2000). The PSI/ETH small radiocarbon dating system. *Nucl. Instrum. Methods Phys. Res., Sect. B* **172**, 1–7.

Tamborini, G., and Betti, M. (2000). Characterisation of radioactive particles by SIMS. *Mikrochim. Acta* **132**, 411–417.

Tamborini, G., Betti, M., Forcina, V., Hiernaut, T., Giovannone, B., and Koch, L. (1998). Application of secondary ion mass spectrometry to the identification of single particles of uranium and their isotopic measurement. *Spectrochim. Acta* **B53**, 1289–1302.

Tanner, S. D. (1995). Characterization of ionization and matrix suppression in inductively coupled "cold" plasma mass spectrometry. *J. Anal. At. Spectrom.* **10**, 905–921.

Thellier, M., Derue, C., Tafforeau, M., Le Sceller, L., Verdus, M.-C., Massiot, P., and Ripoll, C. (2001). Physical methods for in vitro analytical imaging in the microscopic range in biology, using radioactive or stable isotopes. *J. Trace Microprobe Techn.* **19**, 143–162.

Thirlwall, M. F. (2000). Inter-laboratory and other errors in Pb isotope analyses investigated using a ^{207}Pb–^{204}Pb double spike. *Chem. Geology* **163**, 299–322.

Tong, S. L. and Harrison, W. W. (1993). Glow discharge mass spectrometric analysis of non-conducting materials. *Spectrochim. Acta* **B48**, 1237–1245.

Tuniz, C., Bird, J. R., Fink, D., and Herzog, G. F. (1998). Accelerator mass spectrometry. CRC Press LLC, Boca Raton.

Uchida, H. and Ito, T. (1994). Comparative study of 27.12 and 40.68 MHz inductively coupled argon plasmas for mass spectrometry on the basis of analytical characteristic distributions. *J. Anal. At. Spectrom.* **9**, 1001–1006.

Van Calsteren, P. and Schwieters, J. B. (1995). Performance of a thermal ionization mass spectrometer with a deceleration lens system and post-deceleration detector selection. *J. Mass Spectrom.* **146**, 119–129.

Van Straaten, M., Swenters, K., Gijbels, R., Verlinden, J., and Adriaenssens, E. (1994). Analysis of platinum powder by glow discharge mass spectrometry. *J. Anal. At. Spectrom.* **9**, 1389–1397.

Van Vaeck, L., Adriaens, A., and Gijbels, R. (1999). Static secondary ion mass spectrometry (S-SIMS) Part 1: Methodology and structural interpretation. *Mass Spectrom. Rev.* **18**, 1–47.

Vanhaecke, F., Diemer, J., Heumann, K. G., and Moens, L. (1998). Use of thermal ionization isotope dilution mass spectrometry (TI-IDMS) as an oligo-element method for the determination of photographically relevant trace elements in AgCl emulsions. *Fresenius J. Anal. Chem.* **362**, 553–557.

Venzago, C. and Weigert, M. (1994). Application of the glow discharge mass spectrometry (GDMS) for the multielement trace and ultratrace analysis of sputtering targets. *Fresenius J. Anal. Chem.* **350**, 303–309.

Vogel, J. S. (2000). Accelerator mass spectrometry for human biochemistry: the practice and the potential. *Nucl. Instrum. Methods Phys. Res., Sect. B* **172**, 884–891.

Vogel, J. S., Turteltaub, K. W., and Nelson, D. E. (1995). Accelerator mass spectrometry. *Anal. Chem.* **367A**, 353–359.

Wagner, S., Boone, S., Chamberlin, J. W., Duffy, C. J., Efurd, D. W., Israel, K. M., Koski, N. L., Kottmann, D. L., Lewis, D., Lindahl, P., Roensch, F. R., and Steiner, R. E. (2000). Practical application of thermal ionization mass spectrometry for the determination of Pu for the LANL bioassay program. V. Int. Conf. on Meth. and Appl. of Radioanalyt. Chemistry (MARC V) Kailua-Kona, HI, USA, April 9–14, 2000.

Wallmeroth, K., Bollen, G., Dohn, A., Egelhof, P., Krönert, U., Borge, M. J. G., Campos, J., Rodriguez, Yunta A., Heyde, K., de Coster, C., Wood, J. L., Kluge, H.-J., and the ISOLDE Collaboration (1989). Nuclear shape transition in light gold isotopes. *Nucl. Phys.* **A493**, 224–252.

Wallner, C., Faestermann, T., Gerstmann, U., Hillebrandt, W., Knie, K., Korschinek, G., Lierse, C., Pomar, C., and Rugel, G. (2000). Development of a very sensitive AMS method for the detection of supernova-produced longliving actinide nuclei in terrestrial archives. *Nucl. Instrum. Methods Phys. Res., Sect. B* **172**, 333–337.

Wendt, K., Bhowmick, G. K., Bushaw, B. A., Herrmann, G., Kratz, J. V., Lantzsch, J., Müller, P., Nörtershäuser, W., Otten, E. W., Schwalbach, R., Seibert, U.-A., Trautmann, N., and Waldek, A. (1997). Rapid trace analysis of ^{89,90}Sr in environmental samples by collinear laser resonance ionization mass spectrometry. *Radiochim. Acta* **79**, 183–190.

Wendt, K., Blaum, K., Bushaw, B. A., Grüning, C., Horn, R., Huber, G., Kratz, J. V., Kunz, P., Müller, P., Nörtershäuser, W., Nunnemann, M., Passler, G., Schmitt, A., Trautmann, N., and Waldek, A. (1999). Recent developments in and applications of resonance ionization mass spectrometry. *Fresenius J. Anal. Chem.* **364**, 471–477.

Wendt, K., Passler, G., and Trautmann, N. (1995). Trace detection of radiotoxic isotopes by resonance ionization mass spectrometry. *Phys. Scr.* **T 58**, 104–108.

Wendt, K., Trautmann, N., and Bushaw, B. A. (2000). Resonant laser ionization mass spectrometry: an alternative to AMS? *Nucl. Instrum. Methods Phys. Res., Sect. B* **172**, 162–169.

Wieser, M. E. and De Laeter, J. R. (2000). Thermal ionization mass spectrometry of Mo isotopes. *Int. J. Mass Spectrom.* **197**, 253–261.

Williams, P. M., Oeschger, H., and Kinney, P. (1969). Natural radiocarbon activity of the dissolved organic carbon in the north-east pacific ocean. *Nature* **224**, 256–258.

Willis, R. D., Thonnard, N., Eugster, O., Michel, T., and Lehmann, B. E. (1991). In "Resonance Ionization Spectroscopy 1990," Inst. Phys. Conf. Ser. (J. E. Parks and N. Omenetto, Eds.), Vol. 114, pp. 275–278. IOP Publishing Ltd., Bristol.

Winchester, M. R. and Marcus, R. K. (1988). Glow discharge sputter atomization for atomic absorption analysis of nonconducting powder samples. *Appl. Spectrosc.* **42**, 941–944.

Winchester, M. R., Duckworth, D. C., and Marcus, R. K. (1990). In "Glow Discharge Spectroscopies" (R. K. Marcus, Ed.), Chapter 7. Plenum Press, New York.

Woelfli, W. (1987). Advances in accelerator mass spectrometry. *Nucl. Instrum. Methods Phys. Res., Sect. B* **29**, 1–13.

Yamamoto, M., Tsumura, A., Katayama, Y., and Tsukatani, T. (1996). Plutonium isotopic composition in soil from the former Smipalatinsk nuclear test site. *Radiochim. Acta* **72**, 209–215.

Yuan, P., Zhou, S., and Wei, B. (1994). The preliminary experiments on laser ion source at IMP. *Rev. Sci. Instrum.* **65**, 1275–1277.

Zhao, X. L., Kilius, L. R., Litherland, A. E., and Beasley, T. (1997). AMS measurement of environmental U-236: Preliminary results and perspectives. *Nucl. Instrum. Methods Phys. Res., Sect. B* **126**, 297–300.

Zhao, X. L., Nadeau, M. J., Kilius, L. R., and Litherland, A. E. (1994). The first detection of naturally-occurring ^{236}U with accelerator mass spectrometry. *Nucl. Instrum. Methods Phys. Res., Sect. B* **92**, 249–253.

SOLID SCINTILLATION ANALYSIS

MICHAEL F. L'ANNUNZIATA

The Montague Group, P.O. Box 5033, Oceanside, CA 92052-5033, USA

I. INTRODUCTION

Solid scintillation is a process whereby the absorption of energy from ionizing radiation (e.g., alpha particles, beta particles, heavier charged particles, and x- or gamma-rays) by certain crystalline inorganic or organic materials results in the emission of flashes of visible light from the solid absorbing material. Even neutron radiation, which is nonionizing, can produce the scintillation effect in certain solid scintillators, which may be used specifically for the detection and measurement of neutrons. The effect of neutrons in scintillators is generally the result of ionization and scintillation caused by (n, p) or (n, α) nuclear reactions within the scintillator material, which will be described further on in this chapter.

The phenomenon of solid scintillation was observed in 1903 by Elster and Geitel. They were able to see the individual flashes of light emitted by a ZnS screen placed in front of a sample emitting alpha particles. Also in 1903 Sir William Crookes developed the apparatus for counting the light flashes. It consisted of a brass tube fitted with a lense at one end and a ZnS screen at the other. The position of the radioactive source, radium, from the ZnS screen could be adjusted by means of a thumbscrew. The closer the radium was adjusted to the screen, the more numerous and frequent were the flashes of light. He named the apparatus the Spinthariscope from the Greek "spintharis" meaning spark. The instrument was used by Earnest Rutherford (1919, 1920) and his students Hans Geiger and Ernest Marsden. They were able to visibly count the individual scintillations produced by alpha particles as the particles bombarded the ZnS screen. It was a crucial instrument in their research with alpha particles and the atomic nucleus described in Chapter 1. The first report of scintillation in organic compounds was documented much later in a thesis by Hereforth (1948), who studied under the leadership of H. Kallman. An historical account of this discovery is related by Niese (1999). In her thesis presented on September 13, 1948 at the Technical University Berlin − Charlottenburg, Hereforth reported that certain crystalline organic compounds could convert absorbed energy of alpha- and beta-particles. The discovery of this solid scintillation phenomenon in organic crystals lead to liquid scintillation analysis through the work of Kallman (1950) and Reynolds *et al.* (1950), which is covered in Chapter 5.

The advances made in the solid scintillation analysis of nuclear radiation detection since these early discoveries are astounding. These may be best compared with advances made in development from the abacus to the modern electronic computer.

This chapter focuses mainly on solid scintillation analysis with inorganic crystalline materials. Certain solid inorganic crystalline materials have a high "stopping power" for x- and gamma-radiation, because of the high densities and high atomic numbers (Z) that can be achieved in the preparation of these crystals. Therefore, many inorganic scintillators are used throughout many sectors of basic and applied research and in almost all medical diagnostic imaging instrumentation for the detection and measurement of penetrating x- and gamma-radiation (van Eijk et al., 1994; van Eijk, 2001). However, depending on detector types and properties, solid inorganic scintillators can also be applied to the detection and measurement of particulate radiation, such as alpha particles, beta particles, and even neutrons. The latter are detected indirectly through nuclear reactions in crystal detectors. Special applications and solid scintillation detectors used for the analysis of alpha, beta, and neutron radiation will also be discussed in this chapter. Applications of solid scintillation analysis are broad. We can find applications in the measurement of radioactivity and radionuclides in almost every field of science including physics, chemistry, industry, space exploration, and the clinical and biochemical fields. This chapter will cover the basic principles and current practice of solid scintillation for the analysis of nuclear radiation and radionuclides.

II. PRINCIPLES OF SOLID SCINTILLATION

A. Solid Scintillators and Their Properties

In Chapter 5 we saw that nuclear radiation produces a scintillation when the radiation energy is absorbed by certain organic materials (fluors) such as anthracene, PTP, and PPO dissolved in suitable solvents. The interaction of radiation with the organic compounds occurs at the molecular level; that is, the organic molecule in a ground energy state absorbs radiation energy and is elevated to an excited energy state. On returning to its ground state, the molecule emits absorbed energy as visible light. The detection and analysis of electromagnetic radiation by these organic materials, either solid or liquid, are generally not efficient due to the high penetrating power of electromagnetic radiation with the exception of low-energy gamma or x-radiation. The relatively low density of organic materials (e.g., anthracene, $1.3\,\mathrm{g\,cm^{-3}}$ and PTP, $1.2\,\mathrm{g\,cm^{-3}}$) diminishes the "stopping power" of these scintillators for high-energy gamma radiation.

The higher density (σ) and higher atomic number (Z) of certain inorganic crystalline materials such as NaI ($3.7\,\mathrm{g\,cm^{-3}}$), CsI ($4.5\,\mathrm{g\,cm^{-3}}$), and $Bi_4Ge_3O_{12}$ also known as BGO ($7.1\,\mathrm{g\,cm^{-3}}$) (see Table 11.1) give these solids a greater gamma-ray stopping power, that is, a greater ability to absorb energy of impinging gamma radiation. In addition to the higher density and atomic number offered by these inorganic crystalline materials, there are other basic

TABLE II.I Commercially Available Inorganic Scintillators (See van Eijk, 2001)[a]

	NaI:Tl	CsI:Tl	CsI:Na	BGO[b]	CdWO$_4$	BaF$_2$	CsF	CeF$_3$	CsI	GSO:Ce[b]	YAP:Ce[b]	^{6}Li glass:Ce	^{6}LiF/ZnS:Ag	^{6}LiI:Eu
Emission max., nm, Slow[c]	410	565, 420	420	480	500	310			450	440	350	400	450	470
Fast[c]						220	390	285, 305	320					
Light yield, gammas Photons/MeV, slow	4.1×10^4	6.6×10^4	4.0×10^4	0.9×10^4	2.8×10^4	1.1×10^4				1000	2.1×10^4	~4000	7.5×10^4	1.2×10^4
Fast						1500	2000	~4000	2000	8000				
Light yield, Photons/neutron												~6000	1.6×10^4	5.0×10^4
Decay time, Slow, μs	0.23	0.8–>6	0.63	0.30	~3/~17	0.60	2.9		1.0	0.6			~1	1.4
Fast, ns						0.8		5/31	~6/~28	60	30	75		
Refractive index	1.85	1.80	1.84	2.15	2.2	1.50	1.48	1.62	1.95	1.85	1.93	1.56	2.36	1.96
Density, g/cm^3	3.67	4.51	4.51	7.13	7.9	4.88	4.11	6.16	4.51	6.71	5.5	2.5–2.7	4.1	4.1
ρZ_{eff}^4,[d] 10⁶	24.5	38	38	227	134	38	33	50	38	84	7			31
Radiation length[e], cm	2.6	1.8	1.8	1.1	1.1	2.0	2.0	1.7	1.8	1.4	2.9			
AE/E (662 keV) FWHM,%	5.6	4.3	7.4	9.0	6.8	7.7				7.8	4.4			
Hygroscopicity	Yes	Slightly	Yes	No	No	No	Very	No	Slightly	No	No	No	No	Very
References (footnotes)	f	g	h	i	j	k	l	m	n	o	p	q	r	s

[a]Reprinted with permission from Elsevier Science; [b]BGO = Bi$_4$Ge$_3$O$_{12}$, GSO = Gd$_2$SiO$_5$, YAP = YAlO$_3$; [c]Components of emission; [d]High density ρ and atomic number Z for efficient gamma-ray absorption by the photoelectric effect per cm $\propto \rho Z_{eff}^{3-4}$, where $Z_{eff} = \sqrt[4]{\left(\sum_i W_i Z_i^4\right)}$ and W_i and Z_i are the weight fraction and atomic number of element i of the scintillation detector [see also Knittel (1998)]; [e]Inversly proportional to density and atomic number, radiation length $X_0 \propto \sigma^{-1} Z^{-1}$, expressed in cm; [f]Hofstadter (1948), Holl et al. (1988), Kinloch et al. (1994), Miyajima et al. (1984), Moszynski et al. (1981) and Sakai (1987); [g]Hofstadter (1950), Holl et al. (1988), Sakai (1987), Schotanus et al. (1990), Valentine et al. (1993), and Fiorini and Perotti (1997); [h]Aitken et al. (1967), Brinkman (1965), Holl et al. (1988), Ryskin and Dorenbos (1994) and Sakai (1987); [i]Averkiev et al. (1990), Holl et al. (1988), Moszynski et al. (1981), Nestor and Huang (1975), Okajima et al. (1982), Sakai (1987) and Weber and Monchamp (1973); [j]Holl et al. (1988), Kinloch et al. (1994), Kröger (1948), Sakai (1987), and Melcher et al. (1989); [k]Dorenbos et al. (1993), Ershov et al. (1983), Laval et al. (1983), Sperr (1987), and Zhu et al. (1986); [l]Farukhi (1982), Moszynski et al. (1981), Moszynski et al. (1983), Mullani et al. (1980) and Tailor et al. (1986); [m]Auffray et al. (1996), Anderson (1990), and Moses and Derenzo (1990); [n]Kobayashi et al. (1987), Kubota et al. (1988), Schotanus et al. (1990), Woody et al. (1990, 1991); [o]Ishibashi et al. (1989), Melcher et al. (1990, 1991), Sakai (1987), and Takagi and Fukazawa (1983); [p]Dorenbos and de Haas (1994), Kapusta et al. (1999), ...(1990), Atkinson et al. (1987), Brollo et al. (1990), Crismatec (1991), Nuclear Enterprises Ltd. (1994), and Sakamoto (1990);

requirements for many applications involving gamma-radiation measurements. These requirements are principally a fast response or decay time for light output (10–100 ns) and high light yield ($\geq 20{,}000$ photons per gamma-ray or particle energy absorbed) as described in review papers by van Eijk (1993, 1994, 1997, 2001). However, as explained by van Eijk (1997, 2001), these inorganic scintillators can be categorized into two groups. In the first group are included NaI:Tl, CsI:Tl, CsI:Na, BGO, and $CdWO_4$ with light yields of $\geq 10^4$ photons per MeV and decay times of > 200 ns. The second group includes BaF_2, CsF, CeF_3, and CsI with faster decay times (~ 1–30 ns), but lower light yields (a few thousand photons per MeV). The GSO:Ce scintillator falls in between these two groups with a 60-ns decay time and 8000 photons per MeV light output. The use of a colon in the nomenclature of these detectors separates the molecular formula from the atomic activation sites or luminescence centers of the detector, respectively. Parentheses are also used. For example, NaI:Tl or NaI(Tl) signifies thallium-activated sodium iodide. The mechanisms of activation and luminescence in inorganic scintillators is described further on in this chapter. To improve solid scintillation performance for the measurement of gamma radiation in many applications, researchers are searching actively for a scintillator that meets all of the basic requirements of fast response, high light yield, high density, and high Z.

Table 11.1 provides the properties of several commercially available and commonly used inorganic crystalline scintillators. Many others are available, and some of these scintillators and their specific applications will be described further on in this chapter. A comprehensive review of the properties of numerous inorganic scintillators and the state-of-the-art on the development of new scintillators is provided by Moses (2002), van Eijk (2001), and van Eijk et al. (2001). New scintillators are appearing as a result of the quest for more efficient scintillators with faster response, higher light yields and energy resolution, and low production costs. Examples are $LaCl_3$:Ce and $LaBr_3$:Ce, which are recently discovered scintillators that exhibit energy resolutions of 3.1 and 3.0% FWHM approaching resolutions of direct solid state detectors (Moses, 2002; van Eijk et al., 2001). Another scintillation detector, that has attracted much attention, is lead tungstate because of its high density ($7.9 \, g/cm^3$) and radiation hardiness often required properties in high-energy physics research (Moses, 2002). Another scintillator worthy of mention is bismuth silicate ($Bi_4Si_3O_{12}$, BSO). It is a faster scintillator than bismuth germanate (BGO) by a factor of about 2. However, its light yield is smaller by a factor of one-fifth with respect to BGO, which is not a problem at the high gamma radiation energies often encountered in high-energy physics experiments (Kobayashi et al., 1996; Ishii et al., 2002). Many industrial, research, and medical applications have different demands for detector characteristics and efficiencies. Nevertheless, the basic properties of fast response (10–100 ns), high light yield ($> 25{,}000$ photons/MeV), high density, and high atomic number (Z) are required for many applications. Much research has been focused on the Ce^{3+}-doped crystals of lutetium (Lu) to improve the energy transfer processes for faster response with high light yields as reviewed by van Eijk (1995, 2001 and Rodnyi, 2001). Among these new scintillators are the Ce^{3+}-doped lutetium listed in Table 11.2. Included in Table 11.2 is Ce-activated lutetium aluminum

TABLE 11.2 Scintillation Detectors Prospective for PET Applications (From Korshik and Lecoq, 2001, © 2001 IEEE)

Material	Density (g/cm^3)	Z_{eff} [a]/Ph. Abs. Coef. at 511 keV, cm^{-1}/X_0^c,cm	LY[b] (Ph./MeV)	τ (ns)	λ (nm)	Ref.
Lu_2S_3:Ce	6.2	66.7/0.241/1.25	28000	32	592	g
$LuAlO_3$:Ce[d]	8.34	64.9/0.29/1.1	11400	17	365	h
Lu_2SiO_5:Ce[e]	7.4	66/0.28/1.1	27000	40	420	i
$Lu_2Si_2O_7$:Ce[f]	6.23	64.4/0.21/1.39	30000	30	380	j
LuF_3:Ce	8.3	61.1/0.31/1.0	8000	23 + slow	310	g
$LuBO_3$:Ce	7.4	64.5/0.28/1.32	26000	39	410	g

[a] See footnote d of Table 11.1.
[b] Light Yield (photons/MeV).
[c] See footnore e of Table 11.1.
[d] LuAP:Ce.
[e] LSO:Ce = cerium-activated lutetium oxyorthosilicate.
[f] LPS:Ce = cerium-activated lutetium pyrosilicate.
[g] van Eijk (1997).
[h] Moszyński et al. (1997).
[i] Melcher et al. (1995).
[j] van Eijk (1999).

perovskite (Ce:LuAlO$_3$ or LuAP), which displays a 2.7 times faster decay than Ce-activated lutetium oxyorthosilicate (Ce:Lu$_2$SiO$_5$ or LSO). This crystal scintillator is available commercially and it is an efficient detector in positron emission tomography (PET) and other applications in which stopping power, speed, and energy resolution are crucial (Saoudi et al., 1999; Melcher et al., 2000; Korzhik and Lecoq, 2001). For reviews on research in this field over the past decade, works by Belsky et al. (2001), Dorenbos et al. (1991), Hell et al. (2000), Korzhik and Lecoq (2001), Lempicki et al. (1995, 1996, 1997), Melcher and Schweitzer (1992), Moszynski et al. (1996, 1997), Rodnyi (2001), Ryzhikov et al. (1999), Trower (1994, 1995), van Eijk (1993, 1994, 1997, 2001), Visser et al. (1991), and Zazubovich (2001) may be consulted.

The inorganic materials in the solid crystalline state are usually "doped" with varying concentrations (mol%) of an element impurity such as thallium, cerium, or other elements, which act as centers of activation or luminescence centers in the crystal. For example, a commonly used crystal detector of this type is NaI:Tl, which is sodium iodide activated with approximately 1.3×10^{-3} mole fraction (Mundy and Rothman, 1983) of thallium. It is referred to as a thallium-activated sodium iodide detector. The concentration element impurity, that creates the luminescence centers in the crystal detector, will affect the light output or pulse heights. For example, Hamada et al. (2001) demonstrated that increasing the Tl$^+$ concentration from 10^{-3} to 10^{-2} mol in CsI:Tl detectors increased the pulse heights originating from heavy particle (alpha) excitation compared with those of electron excitation. Thus, in studies of heavy-ion physics CsI:Tl detectors with increased concentrations of luminescence centers may be recommended. Other examples are listed in Tables 11.1 and 11.2, where a colon separates the element acting as the center of activation from the detector formula. The thallium or other such ions serve as activation centers in the inorganic crystals, where deexcitation of absorbed energy from interacting nuclear radiation and subsequent emission of visible light or scintillation occur. Other scintillators do not require doping with element impurities to improve the scintillation efficiency. Some of these scintillators are also described in Table 11.1.

The mechanisms of the conversion of nuclear radiation energy to photons of visible light in such crystal detectors is described subsequently.

B. The Scintillation Process

I. Gamma- and X-Ray Interactions

The scintillation process is explained in detail by Dietrich *et al.* (1973), Heath *et al.* (1979), Kabler (1975), Lempicki (1995), Lempicki *et al.* (1993), and Wojtowicz *et al.* (1994b, 1995, 1996). Recent advances on the mechanisms of scintillation in inorganic crystal detectors are reported by Saoudi *et al.* (1999), Belsky *et al.* (2001), Hamada *et al.* (2001), Rodnyi (2001) and Zazubovich (2001). Lempicki (1995) outlines the scintillation process in three stages: (1) creation of electron–hole pairs by an absorbed gamma- or x-ray photon, (2) energy transfer after electron–hole pair production to a luminescent center, and (3) light emission at a luminescent center. The treatment that follows will refer at times to the interaction of electromagnetic gamma- or x-radiation with the crystal detector. However, the same basic principles apply to any radiation type, that produces ionization and electron–hole pairs in matter.

As described in Chapter 1, the principal mechanisms of the interaction of gamma- or x-ray photons with matter are the Compton effect, photoelectric effect, and pair production. In each of these mechanisms of interaction with scintillating crystals gamma photon energy is absorbed by the crystal. As a result of the initial interaction of each gamma photon in the crystal an electron will be produced, either a Compton electron, photoelectron, or an electron from pair production. The Compton and photoelectrons are the result of ionization providing an electron–hole pair. The Compton electron, photoelectron, and electron from pair production will travel on and produce further ionization and additional electron–hole pairs in the crystal detector until their energy is absorbed and they come to a stop. Electron–hole pair production is, therefore, the first step in the absorption of gamma- or x-radiation in crystal scintillators.

Let us now look at the mechanisms whereby the electron-hole pairs produced in the crystal scintillators can result in the emission of visible light photons from the crystal lattice. To explain this process we should consider first that there are two types of crystal scintillator detectors. Lempicki (1995) classifies the detectors into the following two types:

1. Activated materials. This type of crystal detector contains a trace amount of element impurity such as Tl^+ or Ce^{+3}, which are introduced into the crystal lattice to serve as centers of activation to enhance luminescence. Examples of this type are NaI:Tl and CsI:Tl as

seen in Table 11.1 and LuAlO$_3$:Ce(LuAP) and Lu$_2$SiO$_5$:Ce(LSO) as seen in Table 11.2.

2. Stoichiometric materials. These crystals do not contain any doped elements as centers of activation. Examples are BGO, CdWO$_4$, BaF$_2$, and CsI cited in Table 11.1. In these cases luminescence may be considered a property of the lattice itself.

The crystal detector still in extensive use today since its discovery by Hofstadter in 1948 is thallium-activated sodium iodide [NaI(Tl) or NaI:Tl]. We can confine our explanation of scintillation to this crystal, because of its widespread use and the extensive study devoted to this detector.

In brief, the principal mechanisms of photon emission subsequent to electron–hole production in the NaI(Tl) crystal occur in the following two sequences: (1) electron and hole trapping processes and (2) radiative recombination mechanisms (Lempicki, 1995). The predominate electron–hole trapping processes are explained as follows:

The Tl is normally found in the crystal lattice as Tl$^+$ ions. Electrons, e$^-$, are trapped by Tl$^+$ to form Tl^0 as described by

$$e^- + Tl^+ \rightarrow Tl^0 \qquad (11.1)$$

and holes are trapped by Tl$^+$ to form Tl^{++} according to

$$h^+ + Tl^+ \rightarrow Tl^{+2} \qquad (11.2)$$

The predominate radiative recombination mechanisms then come into play to include the interaction of trapped electrons with free holes to form activated thallium or (Tl$^+$)* followed by the emission of visible light upon thallium deactivation according to

$$h^+ + Tl^0 \rightarrow (Tl^+)^* \rightarrow Tl^+ + h\nu\,(335,\,420\,\text{nm}) \qquad (11.3)$$

and the interaction of trapped holes with free electrons to produce activated thallium or (Tl$^+$)* followed by the emission of visible light upon thallium deactivation as illustrated by

$$e^- + Tl^{+2} \rightarrow (Tl^+)^* \rightarrow Tl^+ + h\nu\,(335,\,420\,\text{nm}) \qquad (11.4)$$

Although of lower density than other available inorganic scintillator crystal detectors, the NaI(Tl) crystal in combination with low-noise photomultipliers remains among the most commonly used today for x- and gamma-ray photon measurements (Hell et $al.$, 2000). This is due partly to the development of polycrystalline NaI(Tl) by Harshaw/Filtrol Partnership (Solon, OH) under the trade name Polyscin, which is prepared by the recrystallization of single-crystal ingots under heat and pressure. Polyscin is more durable than single-crystal material and can be fabricated in a wide range of sizes and shapes to accommodate specific applications without any cost to scintillation performance. Mechanical and thermal shock inflicted on single-crystal NaI(Tl) can cause fractures, which can cleave among many planes and propagate across the entire crystal, resulting in an overall reduction in light output efficiency and pulse height resolution. Fractures in Polyscin, however, usually remain confined to localized areas of the crystal with no overall interference with light collection efficiency and pulse height resolution.

Other important factors that have contributed to the popularity of NaI(Tl) are the relative short decay time of this detector and its high light output or relative scintillation conversion efficiency. The decay time is the time necessary for deactivation. Shorter decay times permit greater resolution of succeeding deactivation events, referred to as resolving time, in the crystal detector and allow higher count rates without coincidence loss. The resolving time of an NaI(Tl) crystal detector is around $0.3\,\mu s$ (Quaranta and Piccini, 1984).

Optical emission in the scintillation process is governed by an exponential decay with decay constants varying with detector type (see Tables 11.1 and 11.2). Heath (1983) explained the effect of variable optical emission decay on the basis of detector afterglow. That is, although most excited states of a scintillator or phosphor may have the same lifetime, phosphors are characterized by the presence of long-lived states referred to as afterglow. Grabmaier (1984) defines afterglow as the emission of light from the scintillator after the cessation of radiation excitation. This is a property of all scintillation crystal phosphors, and NaI(Tl) and bismuth germanate ($Bi_4Ge_3O_{12}$, abbreviated BGO) display this property to a minimal degree. The high density and atomic number of BGO give it a greater stopping power than NaI(Tl) for electromagnetic radiation. This is of particular importance in improving detection efficiencies of high-energy gamma radiation. Therefore, bismuth germanate has become a very popular detector. The properties of BGO detectors and their applications are discussed later in this chapter.

Research during the past decade has produced scintillation crystal detectors with higher density, higher light yields, and faster responses than BGO. Some of these new and fast solid scintillation detectors are the Ce^{3+}-activated crystals found in Table 11.2. Belsky *et al.* (2001), Korzhik and Lecoq (2001), Lempicki *et al.* (1993), Pauwels *et al.* (2000), Saoudi *et al.* (1999), Rodnyl (2001), Wojtowicz *et al.* (1994a, 1995, 1996), and van Loef *et al.* (2001) describe properties and the scintillation process in these materials. Following the production of electron–hole pairs by the impinging gamma radiation, a two-step scintillation process involves first the trapping of holes by Ce^{3+} according to

$$Ce^{3+} + h^+ \rightarrow Ce^{4+} \qquad (11.5)$$

and the interaction of the trapped holes with conduction band electrons yielding excited Ce^{3+} ions as follows

$$Ce^{4+} + e^- \rightarrow (Ce^{3+})^* \qquad (11.6)$$

The excited cerium ion $(Ce^{3+})^*$ falls to a stable state with the emission of a scintillation photon, $h\nu$, according to

$$(Ce^{3+})^* \rightarrow Ce^{3+} + h\nu \qquad (11.7)$$

where h^+ and e^- represent a valence band hole and conduction band electron, respectively. Wojtowicz et al. (1994a,b) demonstrated that the high hole capture cross section of the Ce^{3+} ions and good transport properties of electrons are responsible for the high scintillation light yield of these Ce-doped crystals.

The scintillation processes previously described with Tl^+ and Ce^{3+} activation centers can be similarly described with other ions doped in a crystal lattice to serve as centers of luminescence. Bressi et al. (2001) modeled the scintillation process in the following steps: (1) absorption of radiation, (2) production of an electron–hole pair, (3) thermalization of the electron and hole, (4) energy transfer to luminescence or activation center, and (5) emission. They define the light yield (LY) in units of photons/MeV as

$$LY = 4.35 \times 10^5 \frac{\beta SQ}{E_g[eV]} \frac{photons}{MeV} \qquad (11.8)$$

where β is the conversion efficiency (Step 1), E_g is the band-gap energy (Step 2), that is, the energy difference between the valence and conduction band of the detector, S is the transport or transfer efficiency of the e–h pair/ energy to the luminescence center (Steps 3 and 4), and Q is the luminescence efficiency (Step 5). A similar definition for photon production or light yield are provided by Rodnyi et al. (1995) and van Eijk (2001). The above relationship indicates that the light yield could be controlled by the selection of a scintillation detector with a small band gap. Band-gap energies may vary from < 1 eV to several eV. Values are provided by van Eijk (2001) for several types of detectors.

The relative light output of all scintillation crystal detectors is a function of temperature, and the effect of temperature varies significantly from one detector type to another over the temperature range −100 to +140°C. These temperature effects are generally less significant for NaI(Tl) detectors in normal room measurements of radionuclides, because this detector displays a rather stable light output response at ±20°C at either side of room temperature (Heath et al., 1979; Sabharwal et al., 1982). In space exploration and borehole logging, where more extremes in temperature are experienced, temperature effects on the light output of scintillation crystal detectors are very significant. NaI(Tl) crystals are often the detectors of choice for in situ gamma spectroscopy including borehole logging, because of their high light output (40,000 photons MeV^{-1}), high effective atomic number, and high scintillation efficiency (Asfahani, 1999; Rudin et al., 2001). However, Rudin et al. (2001) demonstrate that $CaF_2(Eu)$ detectors may be a suitable replacement for NaI(Tl) particularly as groundwater probes and liquid waste sites. The $CaF_2(Eu)$ detectors have a lower light yield (24,000 photons MeV^{-1}), but they are nonhygroscopic and more fracture resistant. The low effective atomic number of $CaF_2(Eu)$ renders it less suitable for gamma-ray spectroscopy at energies above a few hundred keV.

2. Neutron Interactions

Due to the lack of charge on the neutron, it will not produce direct ionization and scintillation in a solid scintillation crystal detector. The interaction of neutrons and consequent scintillation effects in crystals are the result of a nuclear reaction between the neutron and a particular nucleus in the crystal. The products of these nuclear reactions, such as protons, deuterons, alpha particles, and electromagnetic radiation, then interact with the crystal to cause ionization, activation, and luminescence in the detector. The solid scintillation detection of neutrons is dealt with in detail in Sections VI, VII, and VIII of this Chapter.

3. Neutrino Interactions

As discussed in Chapter 1 the neutrino is an elusive particle of zero charge and very low mass. Nuclear power reactors are an abundant source of electron anti-neutrinos ($\bar{\nu}_e$) in the low-energy (MeV) range, and these are observed traditionally via their interactions with protons via inverse beta decay or

$$\bar{\nu}_e + p^+ \rightarrow n + \beta^+ \tag{11.9}$$

similar to the reaction illustrated in Eq. 1.24 of Chapter 1. The positron yields a prompt signature with the emission of 511 keV annihilation photons followed by a delayed neutron capture reaction. Liquid scintillator serves as a good detector and source of protons for the above observations. In the search for more efficienct detector types Li *et al.* (2001) carried out preliminary studies on the use of CsI(Tl) crystal detectors for the study of low-energy (MeV) neutrinos. They point out that neutral current excitation on nuclei by neutrinos according to the following:

$$\bar{\nu}_e + (A, Z) \rightarrow \bar{\nu}_e + (A, Z)^* \tag{11.10}$$

has been observed in the case of ^{12}C nuclei (Armbruster *et al.*, 1998), and that crystal scintillation detectors would provide good gamma resolution and capture efficiency to the neutrino interactions. Li *et al.* (2001) note that with CsI(Tl) as the active target, the gamma lines with M1 transitions would be 81 and 160 keV for ^{133}Cs and 58, 202, and 418 keV for ^{127}I. These would be observed in the pulse height spectra from the CsI(Tl) detector. Added advantages to the CsI(Tl) crystals are their relatively low cost, high light yield, high photon absorption (short radiation length), mechanical stability (easy to machine), and weak hygroscopicity. The mechanical stability of CsI(Tl) permits the design of large detectors with masses in the ton to several ton range. Also the weakly hygroscopic nature of the crystal obviates the need to hermetically seal it from atmospheric humidity. A CsI(Tl) detector has been designed for construction near a reactor core for the study of these low-energy neutrino interactions (Wong and Li, 1999). Also Li *et al.* (2001) point out other neutrino interactions with ^{133}Cs and ^{127}I of the CsI(Tl) crystal detector. These are the following: (1) inverse beta decay

$$\bar{\nu}_e + (A, Z) \rightarrow \beta^+ + (A, Z-1)^*, \tag{11.11}$$

which has the distinct signatures of two 511 keV photons from the positron annihilation, the gamma lines characteristic of the daughter nuclei, and the positron itself; and (2) resonant orbital electron capture

$$\bar{\nu}_e + e^- + (A, Z) \rightarrow (A, Z-1)^*, \tag{11.12}$$

which can take place only at the narrow energy range where the neutrino energy is equal to the Q value of the transition. Again, in this case, the characteristic gamma lines of the excited daughter nuclei from the scintillation detector would provide the signature for this interaction.

Bressi *et al.* (2001) and Antonini *et al.* (2001) propose the use of Yb-doped $Y_3Al_5O_{12}$ (YAG:Yb) scintillator for the detection of low-energy neutrinos, which would react with the ^{176}Yb as follows:

$$\nu_e + {}^{176}\text{Yb} \rightarrow {}^{176}\text{Lu}^* + e^- \quad (Q = 301 \text{ keV}) \tag{11.13}$$

$$^{176}\text{Lu}^* \rightarrow {}^{176}\text{Lu} + \gamma \quad (E_\gamma = 72 \text{ keV}) \tag{11.14}$$

They note that the signature for this neutrino event is a prompt electron and delayed gamma signal, and to have good discrimination against background noise, a delayed coincidence within 50 ns is required as well as high light yield. Tests on YAG:Yb crystals with as much as 15% Yb have been tested by Bressi *et al.* (2001) and Antonini *et al.* (2001). They demonstrated a fluorescence time of approximately 10 ns and a light yield of 8000 photons/MeV, which would be suitable for neutrino detection.

4. Heavy Ion Interactions

Charged particles will produce electron–hole pairs and luminescence in solid scintillator crystals according to mechanisms previously described for the cases of x- and gamma-radiation. Heavy ions, such as alpha particles or ions encountered in accelerator beams have shorter ranges of travel in scintillator crystals requiring crystals of small dimensions. The scintillation response of cerium-doped $YAlO_3$ (YAP:Ce) to heavy ions was studied by Klamra *et al.* (2000), and Westman *et al.* (2002) demonstrated the utility of a small (10 mm × 10 mm × 1 mm) YAP:Ce crystal as a heavy-ion detector within an accelerator. Semiconductor detectors can provide excellent resolutions for charged particles; however, they cannot withstand the extreme conditions including ultrahigh vacuum (10^{-12} Torr) requirements on materials used inside the accelerator. YAP:Ce detectors are reported by Moszynski *et al.* (1998) among which are good light yield, short fluorescence decay times, robustness, and chemical inertness (m.p. $= 1875°C$). An example of the resolution achievable for ^{226}Ra and its daughter alpha particles with a YAP:Ce crystal is illustrated in Fig. 11.1. Westman *et al.* (2002) demonstrate the durability and efficiency of YAP:Ce detectors for measuring the accelerator beam widths for a beam of noncooled ^{40}Ar^{13+} ions and a

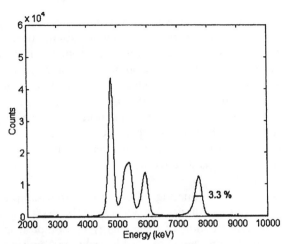

FIGURE II.I Energy spectrum of ^{226}Ra α-particles obtained with a YAP:Ce crystal. A resolution of 3.3% FWHM for the 7.7 MeV α-line is illustrated. (From Westman *et al.*, 2002, reprinted with permission from Elsevier Science.)

cooled beam of $^{19}F^{6+}$ ions. The solid scintillation detector has also been used in studies of dielectric recombination of 1 GeV Pb^{53+} (Lindroth *et al.*, 2001).

C. Conversion of Detector Scintillations to Voltage Pulses

The employment of the solid scintillation phenomenon for the analysis of nuclear radiation requires the conversion of the visible photon emissions of the inorganic scintillator crystal to voltage pulses. The voltage pulses can be measured to determine their magnitude and number. As in the case of liquid scintillation analysis (LSA), a photomultiplier tube (PMT) with a photo-sensitive cathode is used in many instruments to convert a given photon emission to photoelectrons, which are amplified by a series of dynodes to an avalanche of electrons (see Fig. 5.4, Chapter 5). Other types of instrumental components can also do this job, and these are described in Section III.B of this chapter. The magnitude of the resulting pulse height produced by the PMT is proportional to the photon intensity emitted by the scintillation crystal. The photon intensities depend, in turn, on the amount of nuclear radiation energy absorbed by the scintillation crystal. In Table 11.1 from van Eijk (2001), it is reported that NaI(Tl) produces around 40 photons per keV absorbed. Therefore, in summary, we can state that a given pulse height in volts produced by the PMT (within the resolving time of the PMT, e.g., 5 ns) is proportional to the scintillation photon intensity; and the scintillation photon intensity is, in turn, proportional to the energy of the nuclear radiation absorbed by the detector crystal. The following equation can be written to describe this process, which is the basis of the principle behind solid or liquid scintillation analysis of the intensity and energies of nuclear radiation:

Pulse height (volts) ∝ scintillation photon intensity ∝ gamma-ray

energy absorbed   (11.15)

The absorption of a monoenergetic beam of gamma radiation by the crystal detector is governed by the exponential attenuation Eq. 1.121 (Chapter 1), $I = I_0 e^{-\mu x}$, and the attenuation coefficient, μ, is a sum of the independent coefficients due to the Compton, photoelectric, and pair production effects in the crystal, defined by Eq. 1.128 (Chapter 1). The magnitude of the attenuation coefficients for these radiation absorption processes in a given absorber material is a function of the incident gamma-ray energy. Attenuation coefficients due to the Compton, photoelectric, and pair production processes as well as the total attenuation coefficient (sum of these three processes) in NaI as a function of gamma-ray energy are plotted in Fig. 1.30 of Chapter 1.

Because gamma rays, originating either from the sample or from annihilation radiation and x-radiation, produced in the crystal and detector shielding, may impart any fraction of their energy in Compton interactions, a wide range of photon intensities of visible light are emitted from the crystal detector. Consequently, a spectrum of pulse magnitudes in volts is produced as an output from the PMT. Pulses low in magnitude are also produced by thermal noise, that is, thermionic emission of electrons from the cathode of the PMT. All pulses produced may be amplified and then registered as counts over a period of time to provide a count rate and analyzed by a multichannel analyzer (MCA) to determine the spectrum of pulse heights produced and gamma-ray energies absorbed by the crystal detector. The principles of operation and applications of the PMT as well as other photomultipliers and the MCA as components of the solid scintillation analyzer are discussed further on in this chapter and other parts of this book.

III. SOLID SCINTILLATION ANALYZER

From the information provided up to this point of the chapter we might conclude that solid scintillation is used most often for the measurement of electromagnetic x- and gamma-radiation. This is the case, and this section of the chapter will focus mostly on the analysis of this radiation type. Special applications of solid scintillation for the measurement of charged particle radiation and neutrons are covered separately in this chapter.

A modern state-of-the-art solid scintillation analyzer is computer controlled to enable the automatic hands-off analysis of many samples, sample data processing, instrument calibration and control, and programmed instrument performance assessment (IPA). Linked to the computer are the hardware components, including the sample changer, solid scintillation detector, or even several detectors in one instrument to enable the high-throughput simultaneous analysis of samples, and the electronic components required to convert scintillation photons to voltage pulses and pulse height MCAs required to correlate these voltage pulses with radiation energies and to quantify their intensities. The basic components of the solid scintillation analyzer will be discussed subsequently.

A. Scintillation Crystal Detectors

The chemical structure, properties, and performance of a wide range of solid crystal scintillation detectors are provided in Table 11.1 of this chapter. The properties of many other inorganic scintillators are available from van Eijk (2001).

Scintillation crystal detectors may be employed in various shapes or geometries. These are referred to mainly as (1) the planar detector, (2) the well-type detector, and (3) the through-hole detector. The crystalline scintillation detectors are discussed in this section of the chapter. Yet other detector formats exist of the noncrystalline plastic or glass type. There are 24- or 96-well plastic microplate solid scintillators and microsphere plastic or glass scintillators for high-throughput scintillation counting and scintillation proximity assay. Also, there are plastic and glass fiber-optic scintillators. These as well as other scintillator detector types are discussed separately further on in this chapter.

I. Planar Detector

The planar solid scintillation detector usually consists of a solid NaI(Tl) crystal that is cylindrical in shape, as NaI(Tl) is the most commonly employed detector material. The dimensions will vary as described in the following. One flat surface of the cylindrical crystal surface is exposed to the radiation source and the other end of the crystal is connected to a PMT by means of a light pipe and light-transmitting grease. The PMT converts the individual scintillation events into measurable voltage pulses as described later in this section. The crystal surfaces that are not in contact with the PMT are covered with an aluminum coating (about 0.05 cm thick), which blocks out external light and protects the scintillation crystal detector from atmospheric moisture (particularly hygroscopic crystal detectors) and from physical damage. The aluminum may also serve as a shield for alpha and low-energy beta radiation. The inner part of the aluminum is often whitened or covered with a reflector material (e.g., MgO) in order to reflect photons of visible light toward the PMT.

The surface of a crystal of this type can receive, at best, only 50% of the gamma radiation emitted from a given sample, since the crystal can be exposed to a maximum of 180° of a possible 360° of radiation emission. This counting geometry is often referred to as 2π geometry. This is not a very efficient counting geometry, because of the very limited exposure the detector can have to the sample radiation, as this type of detector can "see" only 50% of the sample at best. Detection efficiencies with this counting geometry are relatively low compared with other detector geometries. Consequently, planar detectors are generally not used for the analysis of samples, which are handled and prepared in the laboratory, as more efficient detector geometries are available for sample analysis. The planar detector is used in other applications such as the monitoring of surface radioactivity, area monitors, and space exploration.

The crystal detector is most commonly 75×75 mm in size. Smaller crystals are used, but they provide lower gamma-ray detection efficiencies because the probabilities become greater that a gamma ray may pass the crystal detector without dissipating its energy within. Belle *et al.* (1974) demonstrated that oversized crystals are characterized by an enhanced background without increased sensitivity. They concluded that a 70 × 70 mm NaI(Tl) crystal is best with gamma-ray energies up to 0.4 MeV, whereas a 150 × 150 mm crystal is recommended for gamma-ray energies between 0.4 and 3.0 MeV.

2. Well-Type Detector

The well-type detector consists of a crystal detector with a cavity drilled at the center. For example, a cavity of 15 mm diameter may be drilled into the center of a 75-mm-diameter cylindrical crystal. As described in reference to the planar detector design, the crystal surfaces not in contact with the PMT are covered with a thin aluminum coat for the reasons given in Section III.A.1. The sample to be counted is enclosed within a plastic or glass tube and placed within this cavity, which is referred to as the detector well. This detector geometry was designed in the early years of solid scintillation analysis and used by the author in the early 1960s. Manufacturers of this type of detector utilize a very thin aluminum coat (0.01–0.02 inch) to achieve the highest counting efficiency possible. The detector therefore becomes vulnerable to damage if the sample tube is dropped into the well. Contamination of the well cavity over long-term use is also a concern.

The well-type detector was envisaged with the objective of surrounding the sample with detector as much as possible. Optimum counting geometry is obtained when a sample is only a point source and that source is at the very center and completely surrounded with a radiation detector. Such an arrangement enables the detector to receive all of the possible radiation emissions, which may emanate in all directions (360°) from a given sample. This is referred to as 4π counting geometry. Obviously, such a counting geometry with a well-type crystal detector is not possible. The well-type detector was designed to try to approach, as closely as possible, a 4π counting geometry.

The well-type detector is still used. However, it is not the best detector geometry for the analysis of samples contained in tubes in many circumstances, particularly when relatively high-energy gamma-emitting radionuclides are analyzed. In such circumstances there is a higher degree of variability in detection efficiency as a function of sample size or volume. For example, if the sample is a point source at the bottom of a sample tube and the tube is inserted into a well-type detector, the sample is as far as possible from the well opening. As the sample size or volume is increased, the surface of the sample in the sample tube approaches the well entrance. As the sample volume is increased, the detection efficiency may increase and then diminish as illustrated in Fig. 11.2 for ^{65}Zn, or the detection efficiency may remain stable for small volumes of sample and then drop as the sample volume is increased as in the case of ^{22}Na (Fig. 11.2). The different response of the well-type detector to counting geometry is due to the energies of the gamma-ray emissions. The effect of sample volume on detection efficiency in

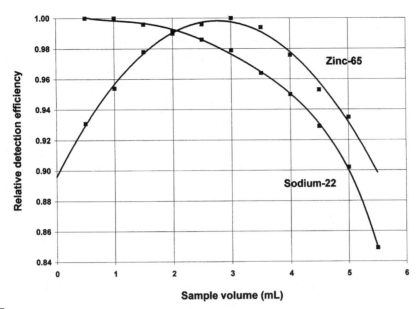

FIGURE II.2 Variation of detection efficiency of a NaI(Tl) well-type detector for the analysis of ^{22}Na and ^{65}Zn as a function of sample volume [From L'Annunziata, unpublished work (1965) for ^{22}Na and Garcia Burciaga (1976) and Garcia Burciaga et al. (1978) for ^{65}Zn].

a well-type detector is more significant for radionuclides that emit gamma radiation of higher energy. In the example provided in Fig. 11.2 we can see that there is a drop of about 15% detection efficiency for ^{22}Na as the sample volume increases from that of a point source (0.0 mL) to 5.5 mL. However, for the measurement of ^{65}Zn there is a smaller but very significant 10% change in detection efficiency over the sample volume range 0.0–5.5 mL. The optimum volumes for measuring the activities in these two radionuclide samples are also very different. If we compare the gamma-ray energies of ^{22}Na and ^{65}Zn, it is possible to explain the different sample geometry effects on detection efficiencies in the well-type detector. The decay schemes of the two radionuclides are illustrated in Chapter 1 (see Eqs. 1.35–1.38). Sodium-22 emits a relatively high-energy gamma of 1.275 MeV with 100% intensity and 0.511-MeV annihilation gamma with 180% intensity (90% of the ^{22}Na radionuclides decay by positron emission). On the other hand ^{65}Zn emits a lower-energy gamma ray (1.115 MeV) with an intensity of 50% and a lower abundance of 0.511-MeV annihilation gamma (3.0% arising from the 1.5% of the ^{65}Zn nuclides decaying by positron emission). Most of the ^{65}Zn nuclides (98.5%) decay by the electron capture process, which give rise to a large number of low-energy x-rays of the copper daughter nuclide. See Appendix A for data on radiation types, energies, and radiation intensities of many of the radionuclides. The effect of sample volume on the detection efficiency of the well-type detector diminishes with gamma-ray energy of the radionuclide. However, even for the low-energy radionuclides ^{57}Co, with gamma energies of 0.122 MeV (87% intensity) and 0.014 MeV (9% intensity) and Fe daughter K x-rays ($\sim$55% intensity), and ^{125}I, with gamma energy of 0.035 MeV (7% intensity) and Te daughter K x-rays (138% intensity), sample volumes should be constant and under 1 mL for the 15-mm detector well dimension in the 75-mm diameter cylindrical crystal cited previously or corrections can be made for changes in detection efficiency as a function of sample volume. Hunter *et al.* (1975) also demonstrated significant decreases in detection efficiency of these radionuclides with the well-type detector with sample volumes over 1 mL. In many cases detector dimensions will vary and the user should plot the detection efficiencies against sample volume as illustrated for the particular radionuclide. Corrections for the effect of sample geometry on detection efficiency can then be made.

3. Through-Hole Detector

The through-hole solid scintillation detector consists of the cylindrical crystal with a hole machined all the way through the center of the crystal. Figure 11.3 illustrates four crystal detectors of the through-hole type with a sample tube in the central cavity of each detector. Such a detector geometry permits the automatic loading of samples from the bottom and automatic instrument adjustment of the sample elevation to position the sample at the very center of the crystal detector. The four samples illustrated in Fig. 11.3 are of different volume but positioned automatically by the instrument to provide optimum detection efficiency and minimal variation of detection efficiency as a function of volume. The benefits of the through-hole detector can be appreciated from the data provided in Fig. 11.4, which illustrates the effect of sample volume on the detection efficiency of ^{59}Fe. In the through-hole detector the ^{59}Fe detection efficiency remains constant over the sample volume range 0.0–1.0 mL and diminishes by about 4% over the range 1.0–2.0 mL. However, with the well-type detector, the detection efficiency

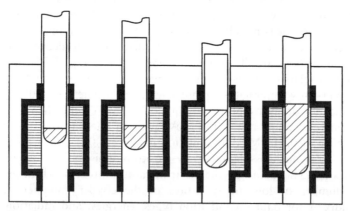

Sample elevation is protocol-adjustable to four positions to provide optimal counting of any sample volume.

FIGURE 11.3 Four through-hole solid scintillation crystal detectors with sample tubes containing different volumes of sample. The samples are illustrated in four positions centered in the detectors automatically by instrument counting protocols. (Courtesy of PerkinElmer Life and Analytical Sciences.)

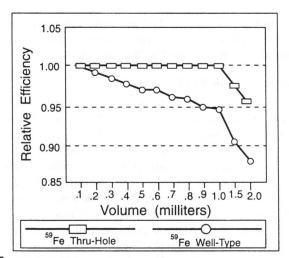

FIGURE II.4 Effects of sample volume on detection efficiencies of ^{59}Fe in through-hole and well-type solid scintillation detectors. (Courtesy of PerkinElmer Life and Analytical Sciences.)

varies for all samples over the sample volume range 0.0–2.0 mL with up to a 14% drop in detection efficiency for a 2.0-mL sample. One can appreciate, therefore, the benefits of the through-hole detector in providing optimum and more constant detection efficiencies.

Automatic gamma analyzers with through-hole detectors are available commercially with up to 10 detectors to provide for the simultaneous analysis of 10 samples in tubes of variable dimension up to a maximum size of 14×100 mm for a 1.5-inch-diameter crystal. Other models of automatic gamma analyzers can accommodate sample tubes or tubes up to 16×100 mm in size for crystal detectors 2 or 3 inches in diameter. The larger crystal detectors provide higher detection efficiencies, particularly desirable for the higher energy gamma-emitting radionuclides. These gamma analyzers have automatic sample changers, which can accommodate up to 1000 samples for hands-off analysis. An array of 10 through-hole detectors of an automatic gamma analyzer and the underlying samples in cassettes of an automatic sample changer are illustrated in Fig. 11.26. The sample tubes are moved automatically in cassettes, loaded from below, and centered automatically in the through-hole detectors.

B. Photomultipliers

A photomultiplier is an electronic device used to measure low levels of light by converting light photons, which strike the light-sensitive surface of the photomultiplier, into a photocurrent (e.g., flow of electrons). The magnitude of the electric pulse produced by the photomultiplier is a function of the intensity of the light that strikes the photosensitive surface of the photomultiplier. The most common type of photomultiplier used with commercial automatic laboratory gamma analysis instrumentation is the dynode photomultiplier, more commonly known as the photomultiplier tube.

Other types of photomultipliers may be used where miniature devices are required and for field applications, space exploration, and so on. The various types of photomultipliers are described in this section.

I. Dynode Photomultiplier or PMT

The dynode photomultiplier or PMT is placed in direct contact with the solid scintillation crystal via a plastic light pipe and clear translucent grease to ensure optimum light transmission from the scintillation crystal to the PMT. The PMT is constructed of an outer glass surface with an inner photocathode (see Fig. 5.4). The photocathode consists of a photosensitive substance that produces photoelectrons when bombarded with photons of visible light. The photoelectrons produced at the photocathode are then accelerated toward a positively charged dynode in the PMT. The dynode is made of a metal plate containing a substance on the surface such as a bialkali compound, which emits secondary electrons upon impact with accelerated electrons. The acceleration of the photoelectrons and the impact of these on the dynode produce multiple secondary electrons. A series of several additional dynodes are contained in the PMT, and each subsequent dynode is given a higher positive potential to further accelerate the secondary electrons and produce a larger number of secondary electrons. The final outcome is an avalanche of electrons at the last dynode. This final avalanche of secondary electrons is collected as a current pulse, which is of a magnitude that can be handled by the electronic circuitry and further amplified, counted, and analyzed for its pulse height.

From the description of the dynode photomultiplier or PMT provided in the previous paragraph, it is easy to see why this electronic component is also referred to as an electron multiplier phototube. Therefore, from the preceding account, we can state that the PMT has a twofold purpose: (1) to convert any given scintillation of visible light emitted from the scintillation detector into a current pulse of secondary electrons and (2) to amplify the current pulse to a magnitude that can be handled by the counting and pulse height analysis circuitry associated with the radiation analyzer. It is important to note at this point that the number of photoelectrons produced at the PMT photocathode and the magnitude of the final current pulse collected after the series of dynode amplifications are a function of the light intensity, which is in turn a function of the radiation energy absorbed by the scintillation detector. As a consequence, the current gain from the PMT will depend on the initial scintillation photon intensity and it will vary within the range 10^6–10^9. This range of current gain provides a source of variable pulse heights, which may be analyzed by other components of the radiation analyzer, such as the pulse height analyzer and MCA discussed later on in this chapter.

2. Microchannel Plate Photomultiplier

Microchannel plate photomultipliers (MCP-PMs) can have advantages over the conventional multidynode PMTs described in the previous section, particularly when the size of the photomultiplier is critical (Fraser, 2002; Hayashi, 1982; Lees and Fraser, 2002; Spaulding and Noakes, 1983). A microchannel plate (MCP) consists of closely packed and fused micro glass capillary tubes with inner walls containing a material that can easily yield secondary electrons. A photocathode is placed at the entrance of the fused capillary tubes and an anode collector at the exit to produce a MCP-PM. Certain microchannel plates are manufactured as radioactivity detectors and imagers (Lees and Fraser, 2002; Lees and Hales, 2001; Lees *et al.*, 1997; Price and Fraser, 2001). These are referred to as MCP detectors, which are manufactured to detect directly nuclear radiation (e.g., beta particles and x-rays) when the entrance to the MCP is made of a nickel-based electrode, which may be coated also with an alkali halide (e.g., CsI) capable of producing electrons directly from the impinging nuclear radiation.

Light photons emitted from a scintillation crystal detector are first absorbed by the photocathode, yielding photoelectrons in the process. The photoelectrons are then accelerated across the ends of the microchannel plate capillary tubes by a voltage potential applied across the channels. As the photoelectrons travel through the microchannels, secondary electrons are produced via impact or collisions with the inner walls of the channels. These secondary electrons, in turn, produce other secondary electrons through further collisions with the capillary walls. The result is a cascade or avalanche of electrons. An illustration of the production of secondary electrons and consequent electron amplification in a microchannel is illustrated in Fig. 11.5. For clarity only a few secondary electrons are illustrated; the avalanche of electrons or output charge cloud produced at the end collector plate are not shown. The avalanche of electrons along the length of the microchannels may yield an overall electron multiplication of 10^4 when operated at near-saturation voltage (Spaulding and Noakes, 1983). Certain MCP designs can provide electron multiplications as high as 10^6 (Lo and Leskovar, 1981; Timothy, 1981). Commercial MCPs have diameters of 25–100 mm, channel pore (channel) sizes of 6–15 μm and are 0.5–1.5 mm thick (Lees and Fraser, 2002; Lees and Hales, 2001; Leutz, 1995; Oba and Rehak, 1981; Spaulding and Noakes, 1983).

Advantages and disadvantages of MCP-PM assemblies over conventional dynode PMTs are reviewed by Spaulding and Noakes (1983). The advantages are (1) improved time resolution with pulse rise times of approximately

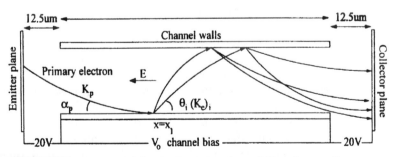

FIGURE 11.5 Signal amplification in a microchannel. K_p and α_p are the energy and grazing angle for the incident primary electron initiating the avalanche. θ_i ($i =$1, 2... in the schematic shown here) and $(K_e)_i$ are the emission angles and energies of the emitted secondaries. (From Price and Fraser, 2001, reprinted with permission from Elsevier Science.)

250 ps and fall times of approximately 780 ps, (2) low background noise; and (3) small size. The principal disadvantages of current state-of-the-art MCPs compared with conventional PMTs are their relatively high cost of fabrication and the low gains or electron magnifications provided. Another disadvantage of MCPs is that moderate magnetic flux densities (≥ 0.5T) prevent electrons from hitting the tube walls and, therefore, stop electron multiplication (Leutz, 1995). The MCP-based intensifiers cannot be used coupled to detectors in high gamma-ray fields (Hoslin et al., 1994).

Multichannel plate detectors were developed originally for their use in photon counting in astronomical x-ray telescopes (Weisskopf et al., 1995) and certain detectors are commercially available (Photonis SAS, France) for use as imagers (e.g., beta autoradiography) for commonly used radioisotopes in the biological sciences. These detectors contain a nickel-based electrode at the face or entrance to the microchannel, which may be coated also with an alkali halide (e.g., CsI) as a photocathode. Beta particles originating from a radionuclide (e.g., ^{3}H, ^{14}C, ^{35}S, ^{32}P) or gamma- or x-radiation from certain radionuclides (e.g., ^{125}I) are capable of producing primary electrons upon colliding with the microchannel photocathode. The primary electrons then collide with the walls of the microchannel resulting in an electron multiplication or avalanche as described in the previous paragraph for the MCP-PM. Microchannel plates of this type may be only 1.5 mm thick with channel diameters ranging from 6 to 12.5 μm. The microchannels are bundled side-by-side to form a relatively large area detector with dimensions of 100×100 mm (Lees and Hales, 2001). This forms a radioactivity imaging device as each microchannel acts as an individual detector. Spatial resolutions of $\sim 70\,\mu$m full width half maximum are reported (Lees et al., 1997) for the large area 100×100 mm detector). The detector has proven to be a useful device for the imaging of radionuclides in the biological sciences (Lees and Fraser, 2002; Lees et al., 1997, 1998, 1999).

3. Semiconductor Photomultipliers

The quest is on to find a good substitute for the relatively large albeit very efficienct photomultipler tube in applications of radioactivity analysis that demand small size and low weight such as in the fields of nuclear medicine and space exploration. In particular, imaging technologies, such as positron emission tomography (PET) and computed tomography, are advancing rapidly placing a high demand for efficient and small photomultiplier detectors that can provide high resolution images. Much research is currently underway in improving the performance of semiconductor photomultipliers (photodiodes) for this purpose. Although photodiodes show no promise as general replacements for the commonly used dynode PMTs in the general laboratory analysis of radionuclides, there are certain applications in which photodiodes are essential or more advantageous than PMTs. Photodiodes with wafer dimensions are used as substitutes for PMTs in high-energy physics experiments where magnetic fields and space limitations preclude the use of PMTs (Grassman et al., 1985a,b; Iwanczyk et al., 1999; Lecomte et al., 1999; Lorenz, 1983; Lorenz et al., 1986; Marler,

2000; Moszyński *et al.*, 2000a, 2001, 2002a; Olschner, 1996; Renker, 2002; Summer, 1988; van Driel and Sens, 1984; Zimmermann *et al.*, 2002). In the field of nuclear medicine imaging, in particular, positron emission tomography (PET) and single-photon emission computed tomography (SPECT) there is increased demand to improve on image resolution from photon emissions, which only miniscule photodiodes can provide. Research in the improvement of photodiode performance in nuclear medicine image of x- and gamma-radiation is currently intense (Allier *et al.*, 1999, 2000a,b, 2002; Aoyama *et al.*, 2002; Balcerzyk *et al.*, 2001; Castoldi *et al.*, 2000; Fiorini *et al.*, 2000, 2001; Kubo *et al.*, 2002; Lecomte *et al.*, 1999; Moszyński *et al.*, 1999, 2000, 2001, 2002a; Patt *et al.*, 2000; Renker, 2002; Rocha and Correia, 2001; Ziegler *et al.*, 2001; Zimmermann *et al.*, 2002). In space exploration the characteristics of low weight, small volume, and low power consumption exclude the use of the bulky PMTs. Compact x- and gamma-ray spectrometers consisting of scintillators resistant to intense cosmic radiation coupled to small wafer photodiodes are in use on planetary probes and orbital vehicles. Research on the discovery of new photodiodes and their improvement for application in astrophysics and planetary exploration is underway (Eisen *et al.*, 2002; Fiorini *et al.*, 2000; Renker, 2002; Zimmermann *et al.*, 2002).

a. p-i-n Photodiodes

The classical PMT can be substituted, at least for certain applications, by the solid-state light amplifier reported by Orphan *et al.* (1978). As described by Heath *et al.* (1979), the light amplifier consists of a conventional photocathode coupled to a silicon diode. The photoelectrons produced at the photocathode of the device are accelerated at a potential of 15 keV and focused on to the reverse-biased p-i-n silicon junction diode. Because about 3.6 eV is required to produce an electron–hole pair in silicon, the light amplifier produces a net single-stage gain of about 4000 (i.e., 15,000 eV/ 3.6 eV). A subsequent slight gain produces an amplification equivalent to that of a PMT.

A photocathode need not be used in conjunction with the p-i-n semiconductor diode because, as explained by Groom (1983, 1984), photons with energies not much in excess of 1 MeV can produce electron–hole (e–h) pairs directly in a semiconductor diode with nearly 100% efficiency. These devices are called p-i-n diode detectors. For example, PIN-diode detector arrays were used by Laitinen *et al.* (2002) directly without scintillator detectors to determine the depth profile of ^{31}Si radiotracer (β^-, $E_{\max} = 1.5$ MeV) in amorphous ceramic.

When a PIN diode is coupled to a scintillation detector they are referred to as PIN photodiodes. The photodiodes (PDs) are exposed directly to the photons of visible light emitted by the scintillation crystal detector. The e–h pairs produced in the depletion layer are collected to provide what is referred to as a photocurrent signal. The principle of the photodiode is described by Lorenz (1983). In brief, the photodiodes used by Lorenz (1983) were made of high-resistivity silicon (> 1 kohm cm) with a very thin p-type diffusion front layer and an e-type diffusion layer on the back of the wafer. A reverse bias voltage of a few tens of volts generates a depletion layer over nearly the entire depth of a 200-μm-thick diode.

A common photodiode similar to that used by Lorenz is the Hamamatsu p-i-n photocell described by Kurahashi (1983), also referred to as the p-i-n photodiode or PIN photodiode. Electron–hole pairs are produced in these diodes when photons are absorbed by the diodes as illustrated in Fig. 11.6. The electron–hole pairs formed in the depletion layer separate, and the electrons and holes migrate to generate a photocurrent. The e–h pairs formed in the p and n regions recombine and produce no charge effects (see Fig. 11.6). Surface reflection of photons, illustrated in Fig. 11.6, accounts for a reduced quantum efficiency (about 70%). The pulse height is dependent on the number of electron–hole pairs produced and, in turn, dependent on the intensity of photons emitted from the solid scintillation detector. See Chapter 4 in this book and also L'Annunziata (1987) for additional reading on the mechanisms of interaction of radiation with semiconductor materials. Groom (1984) studied the use of a silicon photodiode for the detection of light emitted from a bismuth germanate crystal detector and demonstrated that between 1000 and 1400 electron–hole pairs are produced per MeV of energy deposited in the BGO detector. Meisner *et al.* (1995) report the application of two side-mounted PIN photodiodes as photomultipliers for a cylindrical ($15 \times 61\,mm$) CsI(Tl) scintillation detector. The detector was used successfully in a soil probe as a rugged and magnetic field-insensitive means of identifying gamma ray-emitting ^{137}Cs and ^{60}Co contaminants at soil depths of 20 m. Schotanus (1995) of SCIONIX Holland BV demonstrated the energy resolutions achieved with the coupling of a small 1-cm^3 CsI(Tl) crystal to a $10 \times 10\,mm^2$ silicon PIN photodiode. The resolutions reported were 8.7 and 4.7% FWHM for 511- and 1275-keV photopeaks of ^{22}Na, respectively.

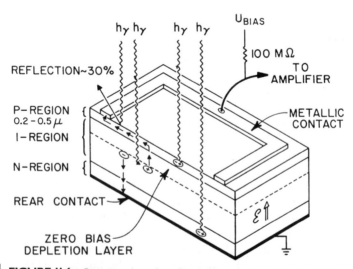

FIGURE 11.6 Cross section of a p-i-n semiconductor photodiode illustrating photon interactions with different components of the diode including the migration of electron–hole pairs formed in the depletion layer of the diode. (From Lorenz, 1983.)

For 662- and 1332-keV photopeaks of ^{137}Cs and ^{60}Co sources, energy resolutions obtained from the coupled miniature crystal and photodiode were 7.4% (662 keV) and 4.4% (1332 keV). Photomultiplier tubes are sensitive to weak magnetic fields in the gauss range and can be shielded to only a few hundred gauss, whereas photodiodes are completely insensitive to magnetic fields. Photomultiplier tubes also occupy much more space than photodiodes, and applications employing many crystal detectors and their individual photomultiplying device are bulky and awkward, particularly when shielding is also required. Photodiode wafers are small and can be placed in direct intimate contact with the crystal detector.

Silicon photodiodes can display high electronic noise due to capacitance and dark leakage current. As explained by Holl *et al.* (1995), the PD noise can be reduced only by introducing long filter time constants unfit for high-count-rate applications. In addition, the PDs are highly sensitive to charged particles that may pass the depletion layer and produce false signals of magnitude similar to that of expected signals from the incident photon radiation. With a PIN photodiode readout coupled to a CsI(Tl) scintillator Mikhailov and Panteleev (2001) observed a long lasting "light noise" originating from the crystal similar to a crystal afterglow. This noise together with photodiode leakage current are contributing factors to the energy resolution of the CsI(Tl)-photodiode gamma detector. Reduced temperature can improve energy resolution. They report a resolution of < 5.5% FWHM for the 1275 keV ^{22}Na line at 5°C.

A prototype 61-pixel hybrid PIN photodiode for use in an imaging scintillation camera for nuclear medicine or high-energy physics research was tested by Datema *et al.* (1999). Photoelectrons are generated at the photocathode from light emissions from a thin (0.5, 1, or 2 mm) CsI(Tl) crystal coupled to the photocathode. The photoelectrons are accelerated toward the silicon PIN-diode via a 12 kV potential difference providing a gain of 3000 through electron-impact ionization in the silicon anode. The photocathode measures only 18 mm in diameter. A 0.5 mm CsI(Tl) crystal could stop 90% of 60 keV photons, and the PIN-photodiode coupled to the crystal could provide a spatial resolution of 5.3 mm FWHM of the light pool on the photocathode.

For a compact planetary rover spectrometer, Eisen *et al.* (2002) describe a small-volume, low-mass gamma-ray spectrometer, which would be useful for planetary vehicles during lander missions. The spectrometer consists of a central 3 cm diameter by 3 cm thick $CdWO_4$ scintillation detector surrounded by a Compton suppressor shield made of 1.5×1.5 segments of $CdWO_4$ detectors coupled to PIN photodiodes. Resolutions of 6.8% FWHM for 662 keV ^{137}Cs gamma ray and 4.7% FWHM for 1.33 MeV ^{60}Co gamma ray was reported. The poorer resolution at 662 keV is due to a large leakage current and capacity of the diode. Also for applications in astrophysics and nuclear physics research, Zimmerman *et al.* (2002) tested a lateral-design silicon-on-insulator (SOI) p-i-n photodiode. The photodiode area was 3.66 mm². For the detections of light from scintillator detectors it exhibited external quantum efficiencies of 66.8, 78.6, and 68.4% for wavelengths of 638, 430, and 400 nm, respectively. For a reverse bias of 100 V the photodiode displayed both a low dark current and low capacitance of 0.66 nA and 0.72 pF/mm², respectively.

PIN photodiodes are used in modern nuclear medicine gamma cameras. For example, Kubo *et al.* (2002) report the use of a scintillator-PIN-photodiode camera to acquire single photon emission computed tomography (SPECT) in medical diagnosis. They report the scintillator-photodiode system as a suitable alternative to the traditional Anger camera system. As a direct radiation sensor, i.e., the scintillator is omitted, the PIN-silicon photodiode was used by Aoyama *et al.* (2002) to measure organ doses in x-ray computed tomography (CT) and other diagnostic radiology measurements.

b. Avalanche Photodiodes

Advances have been made in the development of avalanche photodiodes (APDs) for scintillator readout capable of accepting high-count-rate applications for large multichannel systems used in positron emission tomography and studies of high-energy particle (HEP) physics. Holl *et al.* (1995) underscored the remaining key role of PMTs as classical readouts for scintillation detectors in many fields; however, they review advances made over the past decade in the development of the APD for the photomultiplier readout with fast scintillators.

As explained by Holl *et al.* (1995) and Moszyński *et al.* (2002a), light photons produce electron–hole pairs as they penetrate the silicon of the APD, illustrated in Fig. 11.7. The depth of photon penetration is dependent on the

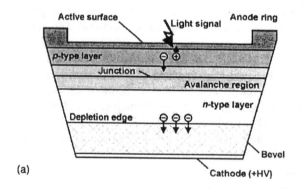

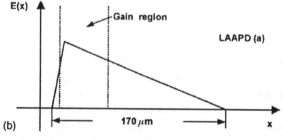

FIGURE 11.7 Schematic cross-section of a large-area avalanche photodiode (a) and electric field profile (b) according to Advanced Photonix, Inc. (From Moszyński *et al.*, 2002, reprinted with permission from Elsevier Science.)

wavelength defined by their attenuation coefficients. The e–h pairs combine instantaneously in the very thin p+ front layer, while electrons in the depleted volume drift toward the n+ layer under the applied electric field potential. When the electrons enter the high-field region of the pn transition, impact ionization occurs. Impact ionization refers to collisions of electrons with other electrons, whereby more e–h pairs are produced and avalanche multiplication of e–h pairs occurs. Only electrons are able to produce the e–h pairs at a certain field strength. Therefore, the initial e–h pairs produced by the incident photon, which penetrate only a shallow depth result in an avalanche of electrons (pulse amplification). A relatively wide depletion layer extends into the n-type region resulting in a low dark current. The broad gain region of the electric field profile (Fig. 11.7) permits operation at high gain with low excess noise.

The advantages of the APD over the conventional PD described in the previous section are much lower noise for fast readout and lower interference (by a factor of 10–100) from charged particles that traverse the APD. Moszyński *et al.* (2002a) outline the advantages and limitations of the APD compared to the conventional PMT. Among the advantages are high quantum efficiency, good spatial uniformity, insensitivity to magnetic fields, compactness, low power consumption, rugged and flexible geometrical configuration. Some limitations to performance are lower gain, higher noise contribution, and susceptibility to radiation damage. For the measurement of gamma radiation the APD is used in conjunction with a solid scintillator whereby the photons from the scintillator are detected by the APD. Energy resolution with the APD coupled to an appropriate solid scintillator can yield energy resolutions superior to those provided by the conventional PMT (Allier *et al.*, 2002; Balcerzyk *et al.*, 2002; Moszyński *et al.*, 2002a). An energy resolution of 3.65% FWHM for the 662 keV gamma ray from ^{137}Cs measured with a small (ø 8 mm × 5 mm) $LaCl_3(Ce^{3+})$ scintillator coupled to an APD serves as an example (Fig. 11.8). The APD can be used directly as a

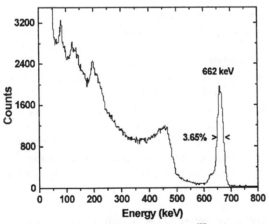

FIGURE 11.8 Pulse height spectrum of ^{137}Cs measured with $LaCl_3(Ce^{3+})$ (ø 8 mm × 5 mm) coupled to an APD no. 70-73-510 from Advanced Photonix, Inc. The energy resolution (FWHM) at 662 keV is 3.65 ± 0.05%. (From Allier *et al.*, 2002, reprinted with permission from Elsevier Science.)

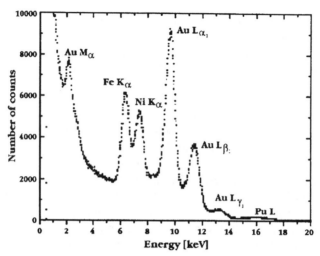

FIGURE II.9 X-ray spectrum from a large-area avalanche photodiode used as a target excited by x-rays from a ^{239}Pu source. (From Moszyński *et al.*, 2002, reprinted with permission from Elsevier Science.)

radiation detector (e.g., low-energy x-rays) without the use of a scintillation detector. For example, Moszyński *et al.* (2002a) demonstrated the excellent energy resolutions and symmetrical peaks obtained with a 10-mm large area avalanche photodiode as the target excited by x-rays from a ^{239}Pu source at a gain illustrated in Fig. 11.9. As apparent from the figure the useful energy range for x-rays has an upper limit of about 10 keV.

A modified silicon avalanche photodiode referred to as the "buried-junction APD," "reverse APD," or "reach-through APD" is described and characterized by McIntyre *et al.* (1996) and Lecomte *et al.* (1999). It was designed for use exclusively with scintillators. As described by Lecomte *et al.* (1999) the device is designed so that the high-field multiplying region is close to the front (entrance window) with the peak field only about 4 μm deep. With such an arrangement only e–h pairs generated within the first few micrometers of the depletion layer will result in electron collection with full multiplication. An e–h pair generated further in depth within the wide drift region behind the multiplying region results in only the hole entering the multiplying region where it undergoes a much reduced gain. Because most solid scintillators produce light at wavelengths of 500 nm or less for which the optical coefficient is > 1/μm, most of the light is absorbed within the first 1–3 μm of the depletion layer generating electrons with full multiplication. The dark current in such a device undergoes only hole multiplication, and consequently its contribution to noise is reduced significantly. These devices, when coupled to scintillation detectors, have potential application as imaging detectors for PET and detectors resistant to high magnetic fields required in high-energy physics research.

Several researchers have tested APDs as scintillator readout devices in lieu of PMTs (Allier *et al.*, 2002; Balcerzyk *et al.*, 2002; Carrier and Lecomte, 1990; Farrel *et al.*, 1990; Fioretto *et al.*, 2000; James *et al.*, 1992; Marler *et al.*, 2000; Moszyński *et al.*, 2002b); and APD designs for practical

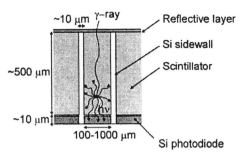

FIGURE II.IO Schematic of the proposed well-type detector (not to scale). (From Allier et al., 2000, reprinted with permission from Elsevier Science.)

application in positron emission tomography and x-ray imaging have been tested by Balcerzyk *et al.* (2001), Holl *et al.* (1995), Marriott *et al.* (1994), Moszyński *et al.* (1999, 2000, 2001), Pichler *et al.* (1999), Renker (2002), and Schmelz *et al.* (1995). A prototype high-resolution positron tomograph equipped with avalanche photodiode arrays and LSO crystal detectors is reported by Ziegler (2001).

To take advantage of micromachining technology and achieve sub-millimeter resolutions required in x-ray and gamma-camera imaging for medical diagnosis, Allier *et al.* (1999, 2000a,b) have devised an array of scintillator crystals encapsulated in well-type silicon sensors as illustrated in Fig. 11.10. An x-ray or gamma-ray that would strike the scintillator crystal would produce light photons that would remain confined by vertical silicon sidewalls. The photons would then be collected onto an avalanche photodiode at the bottom of the well. Image resolutions would be governed by the well dimensions and any degree of Crosstalk between wells that may be caused by x- or gamma-radiation striking more than one well at acute angles. In combination with CsI(Tl) scintillator a light detection efficiency of $\sim 7\%$ yielding a semiconductor conversion of $8\,e^-/keV$. An energy resolution of 19% FWHM at 662 keV gamma-ray peak from ^{137}Cs is achieved with a shaping time of $6\,\mu s$. With similar objectives Rocha and Correia (2001) fabricated a CsI(Tl) scintillator-silicon-well x-ray microdetector to improve medical imaging technology. The detector consists of CsI(Tl) encapsulated in a well that is etched into a silicon substrate (Fig. 11.11a,b). The light produced by an x-ray striking the scintillator produces visible light that is detected by a silicon photodiode at the well bottom. The scintillator wells are only $100\,\mu m$ pixel square size and $515\,\mu m$ deep. The inner walls of the wells are made of aluminum, which serves the dual purpose of confining the scintillation light within a given well while at the same time reflecting light to enhance photon collection by the photodiode.

The APD is finding its way into the development of equipment for the identification of hidden explosives particularly antipersonnel mines. Explosive materials characterized by a high concentration of nitrogen can be identified by the 10.8 MeV γ-ray emitted in the thermal neutron capture reaction on nitrogen nuclei. Fioretto *et al.* (2000) developed a CsI(Tl)-APD sensor for the spectroscopic analysis required for the explosive detector. Scintillation detector sizes varied from $2.5 \times 2.5\,cm$ to $10.2 \times 10.2\,cm$. The scintillation

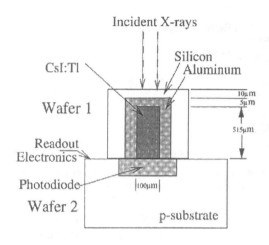

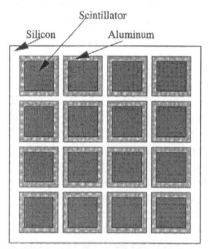

FIGURE 11.11 The scintillator-silicon-well x-ray microdetector illustrated as (a) cross-section of one complete well (not to scale), and (b) a 4 × 4 array of silicon wells filled with scintillators (bottom view of wafer). (From Rocha and Correia, 2001, reprinted with permission from Elsevier Science.)

detectors were tested coupled to a Hamamatsu APD, which varied in size from 1, 3.24, and 7.84 cm^2 and of 200 or 300 μm thickness. The energy resolutions obtained were comparable or better than those of the more bulky and less durable PMT. The CsI(Tl)-APD detector has the following advantages over the CsI(Tl)-PMT sensor: low cost, robustness, low hygroscopicity, and immunity to magnetic fields.

c. Silicon Drift Photodiodes

The application of silicon drift photodiodes in conjunction with solid scintillators for improved x-ray and gamma-ray imaging in computed tomography for medical diagnosis is a new technology receiving much current attention in research development. However, the first silicon drift photodiodes were developed several years ago (Kemmer *et al.*, 1987;

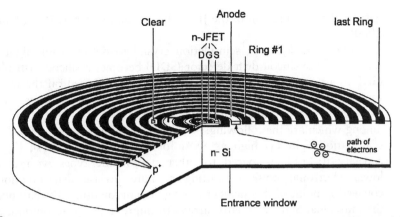

FIGURE II.I2 Schematic cross-section of a cylindrical silicon drift detector with on-chip JFET. (From Fiorini et al., 2000, reprinted with permission from © Elsevier Science.)

Avset et al., 1990) for studies in high-energy physics after the first reports of silicon drift detectors (Gatti and Rehak, 1984; Gatti et al., 1984; Rehak et al., 1985).

The silicon drift photodiode (SDP) is coupled to the solid scintillator to measure the photons of visible light emitted by the scintillator following x- or gamma-ray interaction with the scintillator detector. The principle of operation of the SDP is based on the function of early silicon drift detectors (SDDs) reported by Rehak et al. (1985), and a schematic cross section of a modern silicon drift detector described by Fiorini et al. (2000, 2001) is illustrated in Fig. 11.12. The drift chamber is based on the principle that the semiconductor wafer, with rectifying junctions implanted on both surfaces, becomes fully depleted through a small n^+ collecting anode. Ionizing x- or gamma-radiation passing the entrance window will produce free electrons that remain confined in a buried potential channel by a superimposed electrostatic field parallel to the surface. The field transports the electrons along the buried potential toward the collecting anode (Fig. 11.12). As noted by Rehak et al. (1985) the transit time of the electrons inside the channel measure the distance of the incident radiation from the anode. The SDD therefore, can be used to measure both position and energy of the incident radiation.

To maximize the benefit from the intrinsic low-output capacitance of the SDD, an input n-channel JFET (Junction Field Effect Transistor) is directly integrated on the detector chip close to the collecting anode. The JFET is a voltage controlled resistor where the resistive element is a bar of silicon available from the InterFET Corp., Garland, Texas. For the n-channel JFET the bar is an n-type material sandwiched between two types of p-type material. The two layers of p-type material are electrically connected to comprise a gate (G), while one end of the n-type bar is called the source (S) and the other end the drain (D) as noted in Fig. 11.12. The JFET minimizes the stray capacitance of the connection between detector and front-end electronics. Also the leakage current from the detector and the signal charges accumulated on the anode are discharged continuously by the reverse biased gate-to-channel junction of the JFET (Fiorini and Lechner, 1999; Fiorini *et al.*, 2000).

When coupled to a scintillation crystal to detect visible light from the scintillator the silicon drift detector (SDD) becomes a silicon drift photodiode (SDP). Currently research and development are focused on the modern SDP for x- and gamma-ray computed tomography because of several advantages over the photomultiplier tube (PMT) as well as other photodiode types among which are the following: (1) higher quantum efficiency for detection of scintillation light (as high as 85% for 565 nm with CsI:Tl), (2) very low capacity (0.1 pF) compared to other photodiode types with concomitant lower electronic noise permitting a more precise determination of the conversion point of the incident x- or gamma-radiation, (3) small pixel-size of the individual SDP-scintillator arrays to improve image resolution, and (4) a given resolution obtained with a conventional SPECT photomultiplier system can be obtained with a SPECT silicon drift photodiode-scintillation system with a much lower radiation dose to the human body (Castoldi *et al.*, 2000; Fiorini *et al.*, 1998, 2000, 2001; Olschner, 1996; Patt *et al.*, 2000).

The design of the silicon drift photodiode has taken on different sizes and arrays for x- and gamma-ray imaging among which may be cited the following: photodiode pixels measuring 3 mm × 3 mm coupled to 5-mm thick CsI:Tl scintillator (Olschner, 1996), 3 mm diameter SDP coupled to 10 mm thick GSO scintillator (Fiorini *et al.*, 1997, 1998), 2 mm × 2 mm SDP pixels in 64 element arrays coupled to 4 mm thick crystal scintillator (Patt *et al.*, 2000), and 3 mm^2 hexagonal SDP pixels coupled to 1.4 mm thick CsI(Tl) crystal scintillator (Fiorini *et al.*, 2000, 2001). Typical resolutions reported are 5.8% FWHM for the 662 keV gamma ray of ^{137}Cs with a 4 mm^2 SDP coupled to a 4 mm thick CsI(Tl) scintillator (See Fig. 11.13), and a spatial resolution of 0.6 mm (Fiorini *et al.*, 2001).

In the absence of a scintillator crystal, silicon drift detectors (SDDs) provide excellent energy and position resolutions for x-radiation. At near-room temperature (only slight cooling to −25°C yield energy resolutions close to values of classical Si(Li) detectors at cryogenic temperatures. The SDD therefore has excellent applications for near-room temperature x-ray

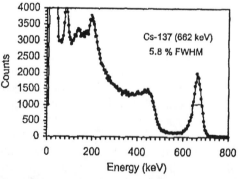

FIGURE 11.13 Cs-137 spectral response for 4 mm^2 SDP pixels elements (of 64 element array) coupled to a 2 × 2 × 4 mm high Cs(Tl) scintillator. (From Patt *et al.*, 2000, © IEEE.)

absorption fine structure applications (EXAFS), in space exploration, and studies of high-energy physics (Hartmann *et al.*, 1997; Iwanczyk *et al.*, 1996, 1999; Piemonte *et al.*, 2002). Typical energy resolutions achieved are 227 eV FWHM for the $Mn_{K\alpha}$ line at room temperature decreasing to 152 keV at $-20°C$ (Hartmann *et al.*, 1997) and 1.4 keV FWHM at the 13.9 keV peak of ^{241}Am (Šonský *et al.*, 2002).

d. HgI₂ Photodiodes

Early work of Iwanczyk *et al.* (1983a,b,c and 1984) has demonstrated that the solid-state photodiode mercuric iodide (HgI_2) can be efficiently used as a photodetector when coupled to a solid scintillation NaI(Tl) or CsI(Tl) crystal detector in lieu of a PMT. Wang *et al.* (1994a,b) describe the construction and performance of the HgI_2 photodetector and provide evidence that these photodetectors have certain advantages over PMTs. The solid-state device operates similarly to the previously described silicon diode photodetector: light photons emitted from the CsI(Tl) scintillation crystal pass through a semitransparent entrance electrode and are then absorbed by the mercuric iodide crystal to produce electron–hole pairs in a very thin layer of the HgI_2 near the surface electrode. With the application of a negative bias to the surface electrode, electron flow within the semiconductor produces a current pulse referred to as the photocurrent. A cross-sectional diagram of the HgI_2 photodetector is illustrated in Fig. 11.14. Its diameter is 0.5 or 1 inch. A quartz window serves as the outer front surface that is placed in contact with the solid scintillation crystal.

The photocurrent response of HgI_2 to light was measured by Iwanczyk *et al.* (1983c) and found to be favorable for the detection of light from scintillation crystals that have emission maxima between 400 and 560 nm. Wang *et al.* (1994b) point out that the CsI(Tl) scintillation crystal detector with an emission spectrum centered at 530 nm provides the best match for the most sensitive wavelength range of the HgI_2 photodetector. A typical gamma-ray spectrum from a 0.5×0.5 in. CsI(Tl) scintillation crystal coupled to a 0.5-inch HgI_2 photodetector provides a FWHM energy resolution of 4.5% for the 662-keV gamma-ray of ^{137}Cs (see Fig. 11.15). When larger 1.5×1.5 in. CsI(Tl) scintillation crystal detectors were coupled to larger 1-inch HgI_2 photodetectors, Wang *et al.* (1994a) found a drop in the HWHM energy resolution to 5.9% for the 662-keV ^{137}Cs gamma-ray. For larger volume CsI(Tl) scintillation detectors (3×4 in.), Wang *et al.* (1994b)

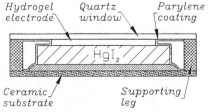

FIGURE II.I4 Cross section of an HgI₂ photodiode employed as a photodetector (photomultiplier) when coupled with a CsI(Tl) scintillation detector. (From Wang *et al.*, 1994a, reprinted with permission from Elsevier Science.)

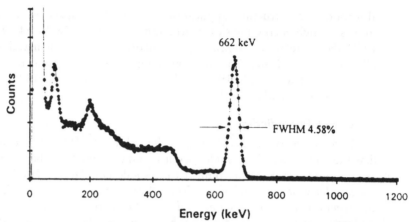

FIGURE 11.15 Spectrum taken with a 0.5-inch HgI_2 photodetector coupled to a 0.5 × 0.5 inch CsI(Tl) scintillator for gamma rays from a ^{137}Cs source. The FWHM energy resolution for 662 keV is 4.58%. (From Wang et al., 1994a, reprinted with permission from Elsevier Science.)

proposed the use of more than one HgI_2 photodetector coupled to the large scintillation crystal.

The HgI_2 photodetector has several advantages over the conventional PMT when applied to solid scintillation counting and spectrometry. These advantages have been summarized by Iwanczyk et al. (1983c, 1984) and Wang et al. (1994a,b). In brief, these advantages are as follows: (1) The quantum efficiencies of currently developed HgI_2 photodetectors for scintillation light from NaI(Tl) and CsI(Tl) are 40 and 57% respectively, compared with the low value of 20% for typical PMTs. (2) The physical mechanisms of operation of HgI_2 photodetectors do not limit their size, whereas there is an irreducible minimum size of PMTs due to the spacing requirements of the PMT dynodes. (3) Solid-state photodetectors offer combined compact construction and low-power operation; however, PMTs are bulky and draw relatively large currents (mA) at high voltages (> 1 kV). (4) The higher quantum efficiency of solid-state photodetectors provides the potential for achieving better energy resolutions than are attainable with PMTs. (5) Greater spatial resolutions in imaging are possible with the use of solid-state photodetectors. (6) Solid-state detectors are more rugged than PMTs and need not be shielded from magnetic fields.

Polycrystalline and amorphous HgI_2, of course, may serve alone as wide bandgap semiconductor detectors for room temperature x-ray imaging and spectroscopy in the fields of high-energy physics, nuclear medicine, and astrophysics (Owens et al., 2002; Schrieber et al., 1997, 1998a,b; Turchetta et al., 1999). For large area imaging applications Turchetta et al. (1999) demonstrated low-cost polycrystalline HgI_2 detectors fabricated by depositing HgI_2 directly on an insulating substrate having electrodes in the form of microstrips forming pixels 1.49 mm wide with 0.1 mm spacing and 300 μm detector thickness. The detectors were tested against $^{90}Sr(^{90}Y)$ beta particles and a high-energy beam of 100 Gev muons yielding good charge collection, radiation hardiness, and good spatial (~75 μm) resolution. The energy resolutions at room temperature of a small $7 \, mm^2$, 0.5 mm thick HgI_2 for x-radiation were reported by Owens *et al.* (2002) to be 600 eV FWHM at 5.9 keV rising to 6 keV FWHM at 100 keV.

It deserves mention to note that a new field is budding in research on scintillation in wide-bandgap semiconductors. Interest in these detectors rests in their potential applications as detectors in high-energy physics research and x-ray and gamma-ray imaging. As noted by Derenzo *et al.* (2002) and Yu and Cardona (2001), several compound semiconductors have fast, near-band-edge emission of potential interest as scintillation detectors. The early work of Lehmann (1965, 1966) illustrated intentional doping to enhance fast, near-band-edge emission in ZnO and CdS semiconductors. Scintillation from semiconductors was also demonstrated by doping them with isovalent impurities (Ryzhikov *et al.*, 2001; Schotanus *et al.*, 1992). Derenzo *et al.* (2002) tested several semiconductor detectors namely, HgI_2, PbI_2, CuI, ZnO:Ga, and CdS:I against 18-keV x-ray excitation over the temperature range of 11–365 K. The observed luminosities at the low temperatures were comparable to that of BGO at 295 K (8200 photons/MeV). Currently these materials are potentially useful scintillators only at low temperature.

C. Pulse Height Discriminators

Subsequent to amplification and pulse shaping, pulse height discriminators come into play. A pulse height discriminator is an electronic gate that permits pulses of a certain magnitude to continue on in the circuit to be registered and counted. In general, there are two discriminators, which are used to identify the pulses, which may continue on to be registered. These are the lower level (LL) and the upper level (UL) discriminators. The LL discriminator can be used to eliminate electronic noise pulses, which will be a function of the photomultiplier cathode to anode voltage setting (Pujol *et al.*, 2000). The settings of both the LL and UL discriminators can be used to define a counting region or counting window. The heights of the LL and UL discriminators may be input manually when setting up a computer-controlled counting protocol. According to the settings of the LL and UL discriminators, the instrument circuitry will allow only pulses of magnitude (volts) higher than the LL discriminator and lower than the UL discriminator to continue on to be registered or counted by the scaler. Although the amplifier output pulse heights are measured in volts, the LL and UL discriminator settings are calibrated in other units, such as, kiloelectron volts (keV), which allows correlation between amplifier output pulse heights from the crystal detector and the actual gamma-ray energies absorbed by the detector.

D. Single-Channel Analyzer

The positive linear correlation between the gamma-ray energy dissipation within the crystal scintillation detector and pulse height amplitude output from the PMT, preamplifier, and linear amplifier are used to measure the gamma-ray energies emitted by a given radionuclide or mixture of radionuclides in a sample. For this purpose a single-channel analyzer (SCA) may be employed.

An SCA contains two electronic pulse height discriminators described in the previous section of this chapter, namely the LL and UL discriminators. The output pulses from the amplifier, that have heights or amplitudes below the LL discriminator setting or above the UL discriminator setting, are rejected. Only pulses that have magnitudes higher than the LL discriminator setting and lower than the UL discriminator setting, referred to as the discriminator window, pass the discriminator circuit and go on to be registered by a scaler circuit. If we expose a solid scintillation detector to a sample containing gamma ray-emitting radionuclides, we may use the LL and UL discriminators to analyze the pulse heights produced and consequently the energies of the gamma rays emitted by the sample. For example, if we begin by arbitrarily setting the LL discriminator at its lowest possible pulse height, namely 0.0 keV, and the UL discriminator at 1.0 keV, the discriminator circuit will allow only pulses of magnitude between 0.0 and 1.0 keV equivalents to go on to be registered and counted by a scaler circuit. A time interval may be selected to give the instrument a specific time to register pulses passing the discriminator circuit between 0.0 and 1.0 keV. The pulses collected by the scaler circuit in the allotted time interval are summed. The total counts collected in the 0.0–1.0 keV counting region (window) are registered; these counts may also be reported as a count rate (e.g., counts per minute, CPM) if the total counts collected are divided by the time interval. The procedure as described will produce one data value, the total number of counts or count rate in the one counting region (channel) defined by the discriminator settings of 0.0–1.0. We may repeat this counting procedure after changing the LL and UL discriminator settings to 1.0 and 2.0 keV, respectively. This will provide a second value of counts or count rate collected of pulses in the energy range 1.0–2.0 keV. The procedure may be continued as described by collecting counts in the regions 2.0–3.0, 3.0–4.0, 4.0–5.0, and so on up to a maximum, for example, 1999–2000 keV. The counts or count rate collected in each channel selected are then plotted against the window height or channel height, which is the LL discriminator setting in keV of each selected channel. This plot will produce a spectrum of gamma-ray energies absorbed by the scintillation crystal detector and consequently a spectrum of the gamma-ray energies emitted by a given sample. Figure 11.16 provides an example of gamma-ray spectra plotted using a single-channel analyzer according to the procedure described. The use of two discriminators in this fashion is referred to as single-channel gamma-ray analysis. The difference between LL and UL in keV is referred to as the window width or channel width, and the LL discriminator setting for a given channel setting is referred to as the window height or channel height. Manual variation of the channel height while maintaining the channel width constant permits scanning of the entire spectrum of pulse-height volts produced by a given gamma radiation source. The procedure is simple but very tedious and time consuming. Most modern solid scintillation analyzers use a multichannel analyzer (MCA), which provides automatically the gamma-ray spectrum of

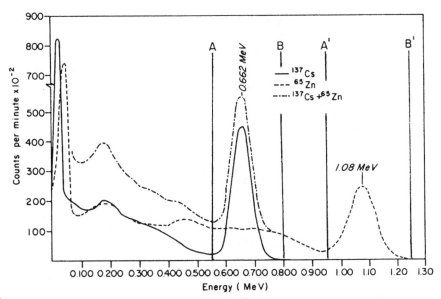

FIGURE 11.16 Gamma-ray spectra of ^{137}Cs and ^{65}Zn as separate radiation sources and as a mixture. The lines **A** and **B** represent lower level (**LL**) and upper level (**UL**) discriminator settings for a counting region which receives all the pulses arising from the photopeak of 0.662-MeV gamma rays from the ^{137}Cs sample in secular equilibrium with its daughter radionuclide ^{137m}Ba. The lines **A′** and **B′** represent LL and UL discriminator settings for another counting region, which receives pulses from the photopeak of the 1.08-MeV gamma rays of ^{65}Zn. (From L'Annunziata, 1965, unpublished work.)

a sample without the need to adjust channel widths or heights. The MCA is described subsequently.

E. Multichannel Analyzer

With the SCA previously discussed, one can count at one time only the pulses received by a given channel (counting window or region) after selection of the channel position (window height) and channel width. Numerous channel positions corresponding to definite pulse height ranges must be tediously chosen in order to construct energy spectra similar to those illustrated in Fig. 11.16. For example, approximately 40 channel heights of 80 keV width were manually selected to produce the gamma-ray spectrum of ^{137}Cs in Fig. 11.16. At each of these channel positions the count rate was determined for a 1-min time interval. A minimum of 40 min was thus necessary to scan the ^{137}Cs gamma-ray spectrum, exclusive of the time involved in changing the channel heights. If a narrower channel width is selected (e.g., 20 keV rather than 80 keV), as many as 160 count rate determinations in the same number of counting channels would be required.

On the other hand, the MCA permits the simultaneous accumulation of pulses in as many as several thousands of energy levels (channels) in a semiconductor core memory. The stored spectra of pulses in the core memory may be viewed simultaneously in real time on the computer monitor of a solid scintillation analyzer and saved on disk or printed whenever desired.

The following is a description of the basic principle of operation of an MCA. Gamma rays from a sample enter a NaI(Tl) crystal or other scintillator detector. As products of photoelectric, Compton scattering interactions among other mechanisms of gamma-ray energy dissipation within the crystal, signal pulses of different magnitude exit the PMT or photodiode and preamplifier. As previously described in this chapter, pulses of different magnitude arise from different energy quanta absorbed by the solid scintillation crystal detector. The pulses pass on to an analog-to-digital converter (ADC), where each pulse amplitude in volts is converted into an equivalent time interval by the production of an equivalent number of pulses from a periodic oscillator. An address register then counts the number of pulses produced by the periodic oscillator, which gives a quantified measure of the amplitude of each input signal. This quantification identifies the channel among many thousands within which the pulse is to be stored in the core memory. In most MCAs, counts in excess of 10^6 can be stored in each channel. The contents of the core memory are viewed on the computer monitor while the sample is being counted and, if desired, saved on disk or printed.

The ADC works on the principle of charging a capacitor to a level proportional to the peak height of a sampled pulse (Fluss *et al.*, 1983). The capacitor is then discharged at a constant rate to a reference zero level and, during the discharge time, a high-frequency clock (10–200 MHz) is run into an address register. This register then contains digital information that is proportional to the pulse height.

Multichannel analyzers are available commercially with thousands of channels. Modern solid scintillation analyzers are equipped with a computer-controlled 4000-channel MCA, which can provide automated sample processing and spectral analysis over the gamma-ray energy range 0–2000 keV. An example of gamma spectra produced by an automatic modern solid scintillation analyzer is illustrated in Fig. 11.17. The simultaneous accumulation and readout of pulses in many thousands of channels corresponding to pulses of defined amplitudes is very timesaving, and the hands-off construction of the gamma spectra is effortless.

F. Other Components

The previous paragraphs in this section have provided details of the physical detector, the scintillation process, and the electronic components needed to convert x- or gamma-radiation from a sample into voltage pulses that may be analyzed in terms of number (counts) and magnitude (pulse height or keV energy) and the data converted into count rate (CPM) or gamma spectra. There are other components of the modern solid scintillation analyzer that are vital to the automatic processing and analysis of samples. These components include (1) computer system control for the hands-off analysis of samples and processing of data; (2) a sample changer for the automatic analysis of 750–1000 samples tubes; (3) automatic and adjustable counting regions for single-, dual-, and multiple-label radionuclide measurements; (4) many (up to 60) different analysis protocols that can be set by users for

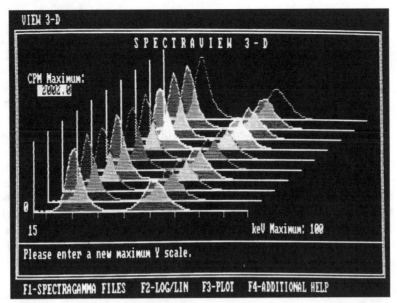

FIGURE 11.17 Multichannel gamma spectra of ^{125}I over the energy range 15–100 keV as seen on the computer monitor of a modern solid scintillation gamma analyzer. A gamma spectrum of a given sample is determined by MCA and plotted and analyzed in real time. (Courtesy of PerkinElmer Life and Analytical Sciences.)

the automatic analysis of samples; (5) automatic protocol-adjustable sample positions in detectors for optimum sample counting geometry (for through-hole detectors, see Section III.A.3 of this chapter); (6) live real-time spectral display with an MCA; (7) computer, hard disk, and printer for the processing, storage, and recording of data; (8) computer-controlled automatic instrument performance assessment (IPA), detector normalization, decay corrections, background reductions, spillover calculations, and radionuclide activity (DPM) calculations; and (9) computer-controlled special application programs for instrumental analysis of samples, such as single- and multiple-site receptor binding assays, radioactive waste tracking, radioimmunoassay data processing, chromium release assays, hepatitis screening, and traditional and radioallergosorbent test (RAST) allergy screening. These components, among others, that are characteristic of automatic gamma analyzer systems are discussed in Section V of this chapter.

IV. CONCEPTS AND PRINCIPLES OF SOLID SCINTILLATION ANALYSIS

A. Gamma-Ray Spectra

The spectrum of pulse heights in volts, being directly proportional to the gamma-ray energies dissipated in the crystal detector, provides a measure of the gamma-ray energy spectrum of the radiation source. Such energy spectra are illustrated in Figs. 11.8, 11.13, 11.15, and 11.16. The gamma-ray spectrum is simply a plot of the relative abundances of the pulses versus the pulse height (voltage) calibrated in gamma ray energies in units of keV or MeV. The portion of the peaks enclosed within the A and B and A′ and B′ discriminators of Fig. 11.16 represent the LL and UL discriminator settings of two counting regions. The two peaks of Fig. 11.16, known as the photopeaks, are the result of the interaction of 0.662 MeV gamma rays of ^{137}Cs or 1.08 MeV gamma rays of [65]Zn with the crystal detector via the photoelectric effect or a combination of Compton effects resulting in the total dissipation of the gamma-ray energy in the crystal.

Gamma-ray spectra are very useful for the identification of nuclides. For example, the photopeak due to the photoelectric effect or a combination of other effects resulting in total photon absorption is usually the highest energy peak. An exception is the "doubles peak," which can occur at a higher energy as it is due to the coincidence detection of two gamma-ray photons (see Section IV.C of this chapter). The photopeak gives a measure of the gamma-ray energy, which is a characteristic of each nuclide, and may be used to identify unknown nuclides.

Gamma-ray spectra may also be characterized by several other peaks such as those illustrated in Fig. 11.18. These are often not well defined because of the Compton smear or background produced by a wide spectrum of Compton electron energies. The Compton edge and backscatter peaks are characteristic of most gamma-ray spectra. Compton electrons of maximum energy produce the Compton edge. A portion of the radiation, which may pass through the detector without interaction, may be scattered at 180° by the detector shielding back into the detector. Absorption of the backscattered radiation by the crystal detector produces the backscatter peak.

Three other peaks illustrated in Fig. 11.18 may also be observed in gamma spectra. These are the annihilation peak, double-escape peak, and single-escape peak. These peaks are the result of pair production in the crystal detector and cannot be observed in the spectra of gamma rays of energy below the theoretical threshold of 1.02 MeV. In practice, they generally are not observed to any appreciable extent and significant numbers of pair-production events are confined to high-energy (> 1.5 MeV) gamma ray–photon interactions. The annihilation peak is located at an energy value of 0.51 MeV and is the result of the absorption of one of the two 0.51-MeV gamma-ray photons following positron annihilation accompanying pair production. When one of the two annihilation gamma rays escapes from the detector without interaction, the single-escape peak is produced at an energy level of $E_\gamma - 0.51$ MeV, where E_γ is the gamma-ray energy characterized by the photopeak. Both of the annihilation gamma rays may escape the detector without interaction and produce the double escape peak at the energy level of $E_\gamma - 1.02$ MeV.

Gamma-ray spectrometry may also be applied to the assay of more than one gamma ray-emitting nuclide in a mixture. If these nuclides emit gamma rays of characteristic energies where photoelectric peaks of the gamma spectra are well separated, they may be assayed separately by setting the counting regions or windows directly over the photopeak of each nuclide. For example, Fig. 11.16 illustrates the combined spectrum of a mixture of [137]Cs

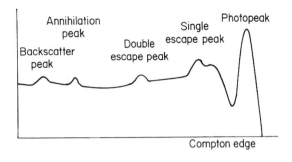

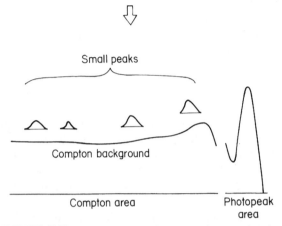

FIGURE 11.18 Segmentation of a spectral curve showing the case of a typical spectrum for gamma rays of about 2 MeV. (From Tominaga, 1983, reprinted with permission from © Elsevier Science.)

and ^{65}Zn and discriminator settings that encompass the photopeak of each nuclide. The ^{65}Zn activity can be determined with LL and UL discriminator settings A′ and B′ providing a 0.950–1.25 MeV counting window, whereas the 0.550–0.800 MeV counting window defined by LL and UL discriminator settings A and B provides pulse contributions from both nuclides. The contribution due to ^{65}Zn in this window, which may be determined with a pure ^{65}Zn standard of known activity, must be subtracted from the total number of counts in this counting region to provide the net counts due to the nuclide ^{137}Cs. The determination of the gamma spectrum of a sample and the analysis of the gamma-emitting nuclides in a given sample are most effectively carried out with an MCA. The use and mode of operation of a MCA are discussed in Section III.E. of this chapter.

B. Counting and Detector Efficiencies

Solid scintillation gamma counting fortunately does not have the disadvantages of quenching normally associated with the liquid scintillation counting technique. This is because the sample is kept physically separate from the scintillator, which prevents any possible physical or chemical interference with the gamma energy transfer and scintillation processes. Nevertheless, the sample composition and the counting geometry, namely the sample's shape and volume as well as its distance from the scintillation detector, can have a marked effect on the count rate or the counting efficiency of a particular sample. Also, in the literature it is possible to encounter several terms related to counting and detection efficiency, which can be confusing. These terms and the factors that affect counting and detection efficiency will be discussed in more detail in this section.

I. Counting Efficiency

The counting efficiency is defined as the ratio of the number of counts for a given period of time (count rate) to the disintegration rate or activity of the source. It is calculated as

$$E = \text{cpm}_{\text{std}}/\text{dpm}_{\text{std}} \tag{11.16}$$

where cpm_{std} is the count rate in counts per minute of a standard source measured by the analyzer and dpm_{std} is the disintegrations per minute or activity of the radiation standard source. The value of the counting efficiency, also expressed as a percentage, is a characteristic of a particular detector, the counting region selected as defined by the LL and UL discriminator settings, the radionuclide, and the counting geometry (Abbas, 2001a,b; Abbas and Basiouni, 1999; Heath, 1964; Moss *et al.*, 1984; Mishra *et al.*, 1965; Selim and Abbas, 2000; Selim *et al.*, 1998; Swailem and Riad, 1983). The counting geometry will include the sample, its composition (e.g., density governing self absorption), geometry (e.g., point source, shape, or volume), and the combined geometry of the sample, sample tube or container, and detector.

The counting efficiency of a single-nuclide sample is easily determined from the count rate of an internal radionuclide standard of known disintegration rate. By internal radionuclide standard is meant a radionuclide of known activity that is counted with counting conditions and geometry similar to those of the sample when it is counted alone without the internal standard. The procedure involves first counting the sample without the standard. This is followed by adding the internal standard to the sample and counting a second time. The counting efficiency of the sample can be calculated as follows:

$$E = \frac{\text{cpm}_{s+i} - \text{cpm}_s}{\text{dpm}_i} \tag{11.17}$$

where cpm_s is the count rate of the sample, cpm_{s+i} is the count rate of the sample after addition of the internal standard, and dpm_i is the disintegration rate or activity of the internal standard. This method of determining counting efficiencies is valid when the following conditions are met: (1) The volume and composition of the internal standard used have no effect on the counting geometry and counting efficiency of the sample; (2) The internal standard is counted with the same instrumental parameters (e.g., LL and UL discriminator settings) as the sample; (3) The standard used is the same nuclide as that of the sample of unknown activity.

Once known, the counting efficiency for a particular sample can be used to convert the count rate of the sample to disintegration rate (activity) according to the equation

$$dpm_s = cpm_s / E \qquad (11.18)$$

For certain radionuclides the disintegration rates of unknown sample activities can be determined directly by the "doubles-peak" method, which does not require any measure of counting efficiency. This method is described in Section IV.C of this chapter.

2. Detector Efficiency

Detector efficiencies are often used to compare and evaluate different types of solid scintillator crystals and counting geometries. Several terms are used to define detector efficiency, and these will be discussed subsequently.

a. Full-Energy Peak Efficiency

This term for detector efficiency may be referred to as the full-energy peak efficiency, photopeak efficiency, or photoelectric efficiency. The full-energy peak efficiency or photopeak efficiency is defined as the ratio of the number of counts in the photopeak, N_p, to the number of gamma rays, N_0, of the photopeak energy emitted by the source. The value of the full-energy peak efficiency is a function of the distance from the radiation source to the scintillator detector and the solid angle subtended by the front surface of the crystal detector (Moss *et al.*, 1984). It can be used to evaluate the properties of the solid crystal detector to fully stop or absorb impinging gamma radiation by the photoelectric effect or a series of Compton interactions resulting in complete absorption of the gamma-ray energy. Evans (1980) compared the full-energy peak efficiencies of BGO and NaI(Tl) detectors of the same dimensions with radionuclides that span a wide range of gamma energies. The full-energy peak efficiencies versus gamma energy for the two crystal detectors are illustrated in Fig. 11.19. The work of Evans (1980) indicated that the photopeak resolution of BGO was only about half as good as that obtained with equivalent NaI(Tl) detectors (see Section IV.F); however, the higher photopeak efficiency of BGO and the near absence of Compton continua (Compton smear) make BGO the detector of choice when a high-resolution detector is not needed.

b. Total or Absolute Efficiency

The total efficiency, which should not be confused with the counting efficiency previously discussed (Eq. 11.16), is defined as the probability that the incident gamma rays will interact with the crystal detector via at least one of the predominate mechanisms of gamma-ray energy absorption, namely the photoelectric effect, Compton effect, and pair production. It is defined as the

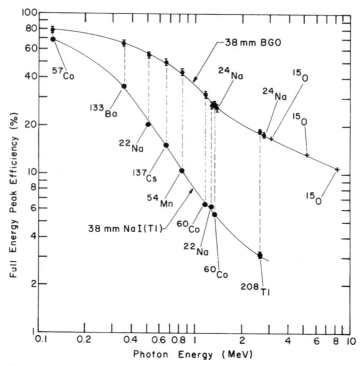

FIGURE 11.19 Full-energy peak efficiencies of 38-mm BGO and NaI(Tl) scintillators as a function of gamma-ray energy. (From Evans, 1980, © IEEE.)

ratio of the number of counts produced by the detector to the number of gamma rays emitted by the source (in all directions). The total efficiency is different from the counting efficiency defined by Eq. 11.16, because the counting efficiency is calculated on the basis of the disintegration rate (dpm) of a source; whereas the absolute efficiency is calculated on the basis of the number of gamma rays emitted per radionuclide disintegration. Because gamma emission is often characteristic of the daughter nuclide in various decay processes, gamma emission may occur via several energy states and may not occur with each disintegration of a given radionuclide (see Chapter 1, Section III). According to the radionuclide, the gamma rays emitted per disintegration may vary from less than one gamma ray to several gamma rays.

The relationship between photopeak efficiency, E_p, and total efficiency, E_t, can be expressed by

$$E_p = E_t P \tag{11.19}$$

where P is the peak-to-total ratio, that is, the ratio of the full-energy peak efficiency to the total efficiency (Abbas, 2001).

The magnitudes of the photoelectric efficiency (or photopeak efficiency) and total detector efficiency diminish as the photon energy increases. The relationship between photon energy and detector efficiency is nearly linear when plotted on a logarithmic scale (Brinkman and Veenboer, 1979; Suzuki,

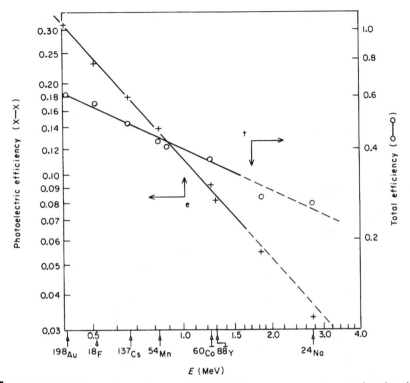

FIGURE 11.20 Photoelectric, e, and total efficiency, t, curves for samples placed on the bottom of the well of a 60 × 60 mm NaI(Tl) crystal. Diameter of well is 17 mm, depth of well is 38 mm. (From Brinkman and Veenboer, 1979, reprinted with permission from © Elsevier Science.)

1982; Sudarshan *et al.*, 1992). As illustrated in Fig. 11.20, the photoelectric and total detection efficiencies decrease as the photon energy increases. Thus, nuclides that emit high-energy gamma radiation (e.g., ^{24}Na, 1.38 and 2.76 MeV) are detected with lower efficiencies than nuclides that emit lower energy gamma radiation (e.g., ^{54}Mn, 0.83 MeV). The inverse relation between gamma energy and detector efficiency can be extrapolated to counting efficiency. Such comparisons can be made, of course, when there is negligible self-absorption among the nuclide samples. The relationship between counting efficiency and photon energy is explained by the fact that most of the photons of low energy (< 0.100 MeV) hitting the detector are absorbed due to high absorption coefficients (Holmberg *et al.*, 1972; Sudarshan *et al.*, 1992). As the photon energy increases, the absorption coefficients decrease or, in other words, the stopping power of the detector crystal decreases.

c. Relative Full-Energy Peak Efficiency

The relative full-energy efficiency, E_{rel}, is the full-energy peak efficiency of a given detector (e.g., germanium semiconductor detector) relative to that of a 3 × 3 inch (76 × 76 mm) NaI(Tl) scintillation detector at a source-to-detector distance of 25 cm using the 1.33-MeV photopeak of ^{60}Co (L'Annunziata, 1987). It is calculated as

$$E_{rel} = A/A_{NaI}$$ (11.20)

where A is the number of counts collected in the 1.33-MeV full-energy peak for the germanium and NaI(Tl) detectors, respectively, over a fixed period of time.

C. Sum-Peak Activity Determinations

The activity or disintegration rate of some radionuclides may be determined directly from the x- or gamma-ray spectra, without the use of standards, if they decay by the emission of two photons in coincidence (i.e., x- or gamma-ray photons). The technique, known as the sum-peak or doubles-peak method, was developed following the pioneering work of Brinkman *et al.* (1963a,b, 1965), Eldridge and Crowther (1964), Sutherland and Buchanan (1967), and Horrocks and Klein (1975). The method has been applied most often to the activity analysis of ^{60}Co and ^{125}I. In the case of ^{60}Co, it decays by β emission with the concomitant emission of two gamma rays from the daughter nuclide ^{60}Ni in coincidence, namely, 1.17 and 1.33 MeV gamma rays of equal intensity. The radionuclide ^{125}I decays exclusively by electron capture followed by a 35-keV gamma-ray emission of the daughter nuclide ^{125}Te. Following electron capture the K x-rays of 27.4 and 31 keV are emitted in coincidence with the 35-keV gamma-ray photons of the ^{125}Te daughter.

Based on the pioneering work of Brinkman *et al.* (1963a,b) the disintegration rate of a nuclide source emitting two gamma rays in coincidence or an x-ray and gamma ray in coincidence, such as the cases for ^{60}Co or ^{125}I, can be determined by the sum-peak equation

$$N_0 = T + \frac{A_1 A_2}{A_{12}}$$ (11.21)

where T is the total count rate of the entire spectrum extrapolated to zero energy (or the area under the total spectrum), A_1, A_2, and A_{12} are the count rates of the photopeaks (or areas under these peaks) of γ_1, γ_2, and the sum-peak, respectively (Kawano and Ebihara, 1990, 1991, 1992; Miyahara *et al.*, 2000; Yuan and Hwang, 2000). The photopeaks of γ_1 and γ_2 are also referred to as the singles peaks, and these may consist of two separate photopeaks of two gamma rays of different energy as for the case of ^{60}Co or, they may consist of a gamma-ray and x-ray peak or two x-ray peaks as in the case of ^{125}I. The sum-peak is also called the doubles-peak as it is the result of coincidence summing of two photons. Equation 11.21 serves as the basis for working equations used in the experimental activity determinations of ^{60}C and ^{125}I, subsequently described. Miyahara *et al.* (2000) applied the

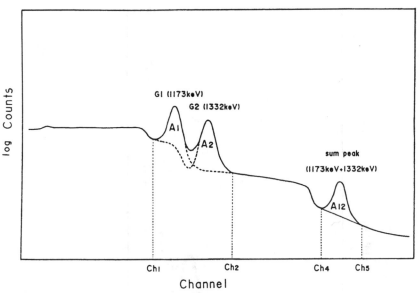

FIGURE II.2I Typical gamma spectrum of ^{60}Co as measured with a NaI(Tl) crystal. The photopeaks are identified as follows: A_l, photopeak of γ_1 (GI); A_2, photopeak of γ_2 (G2); A_{l2}, sum peak. (From Kawano and Ebihara, 1990, 1992; reprinted with permission from Elsevier Science.)

sum-peak method to the activity analysis of ^{60}Co using a high-resolution semiconductor detector yielding well-resolved singles and doubles peaks.

An example of a typical gamma spectrum of ^{60}Co measured with a NaI(Tl) detector is illustrated in Fig. 11.21. Kawano and Ebihara (1990, 1991, 1992) measured the activity N_0 of ^{60}Co according to Eq. 11.21 and the measured gamma spectrum according to the following calculated spectral areas:

$$T = I_T - B_T \tag{11.22}$$

$$A_1 = A_2 = \frac{(I_{1+2} + B_s)}{(2 + \alpha)} \tag{11.23}$$

$$A_{12} = I_{12} - \tfrac{1}{2}(Ch_5 - Ch_4 + 1)(C5 + C4) \tag{11.24}$$

where I_T and B_T are the area and background, respectively, under the entire spectrum; I_{1+2} is the total area between channel 1 (Ch$_1$) and channel 2 (Ch$_2$) and B_S the background for the same area; α is the ratio of the area of the Compton scattering of γ_2 under the photopeak of γ_1 to A_2; I_{12} is the total area between channel 4 (Ch$_4$) and channel 5 (Ch$_5$), and C4 and C5 are the counts in channels Ch$_4$ and Ch$_5$, respectively.

The sum-peak method for the activity analysis and absolute counting (standardization) of ^{125}I has been used by several researchers (De Felice and Myteberi, 1995; Dias and Koskinas, 1995; Martin and Taylor, 1992;

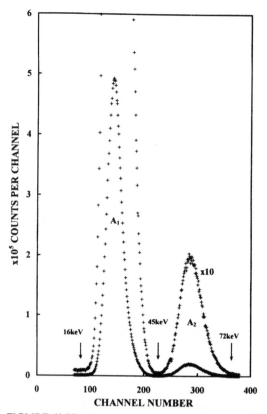

FIGURE 11.22 Typical pulse height spectrum of ^{125}I obtained using a single 3×3 inch NaI(Tl) crystal. (From Dias and Koskinas, 1995; reprinted with permission from Elsevier Science.)

Wang *et al.*, 2001; Yuan and Hwang, 2000). The case of ^{125}I is treated differently than the previous example, as it decays by electron capture with the emission of a 27.4- and 31-keV x-ray in coincidence with the 35-keV gamma-ray photon emitted by the excited state of the ^{125}Te daughter. The solid scintillation gamma-ray spectrum produced by these photons is illustrated in Figs. 11.17 and 11.22. The photopeak occurring at approximately 28.5 keV is due to the unresolved single-photon detection of the x-rays and the 35-keV gamma rays, while the coincidence photopeak (sum peak) occurring at 56.8 keV is the result of coincident summing of two x-rays or an x-ray and a 35-keV gamma ray. The singles peaks for the two x-rays and the one gamma ray remain unresolved under one peak at ~ 28.5 keV. Only with the use of a high-resolution semiconductor detector can the singles peaks for each of the photon emissions be observed (Wang *et al.*, 2001).

For the more complex case of the unresolved singles peaks of ^{125}I, the following equation of Eldridge and Crowrther (1964) is applied:

$$N_0 = \frac{P_1 P_2}{(P_1 + P_2)^2} \frac{(A_1 + 2 A_2)^2}{A_2} \qquad (11.25)$$

where A_1 is the count rate of the unresolved K x- and gamma-ray singles peak area, A_2 is the count rate of the K x- plus gamma-ray sum-peak area, P_1 is the K x-ray emission probability per electron capture decay, i.e., $P_1 = P_K \omega_K = 0.697 \pm 0.015$ where P_K is the K shell electron capture probability and ω_K is the fluorescence yield of K x-ray, P_2 is the sum of the K x-ray emission probability for internal conversion events plus the 35-keV gamma-ray emission probability; i.e., $P_2 = (1 + \alpha_K \omega_K)/(1 + \alpha) = 0.786 \pm 0.040$, where α and α_K are total and K shell conversion coefficients (Dias and Koskinas, 1995; Martin and Taylor, 1992; Wang et al., 2001; Yuan and Hwang, 2000). De Felice and Myteberi (1995) demonstrated the application of the sum-peak method to the absolute activity measurement of ^{125}I with an overall uncertainty of 0.3% (1σ). Also, Dias and Koskinas (1995) and Yuan and Hwang (2000) found excellent agreement between the sum-peak method and other methods of radionuclide standardization of ^{125}I.

Taking advantage of the high resolution and efficiency for x-rays and low-energy gamma rays provided by a well-type HPGe semiconductor detector, Wang et al. (2001) was able to define the count rates of the individual resolved single peaks for two K x-rays and the gamma ray as well as the sum peaks for the ^{125}I spectrum illustrated in Fig. 11.23. The spectrum illustrates the three singles peaks for the two K x-ray emissions ($K_\alpha X$ and $K_\beta X$) from the electron capture decay of ^{125}I as well as the 35-keV gamma emission. The five sum peaks for all probabilities of coincidence summing between the two K x-rays and the one gamma ray are also discernable. From the peak areas and the following equation derived by Wang et al. (2001), the ^{125}I activity was calculated:

$$N_0 = T + \frac{N_\gamma N_{\alpha\beta}}{N_{\beta\gamma}(2N_{\alpha\alpha} + N_{\alpha\beta})}\left[N_\alpha + N_\beta - N_\gamma\left(\frac{N_{\alpha\beta}}{2N_{\beta\gamma}} + \frac{N_6}{N_{\beta\gamma}} - \frac{N_{\alpha\alpha}}{N_{\alpha\beta}}\right)\right] \quad (11.26)$$

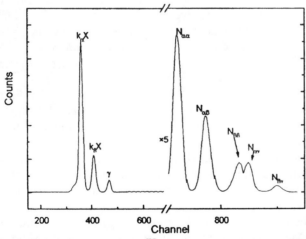

FIGURE 11.23 Spectrum of ^{125}I obtained with a well-type HPGe semiconductor detector. The singles peaks are identified as $K_\alpha X$, $K_\beta X$, and γ, while the doubles or sum peaks are denoted as $N_{\alpha\alpha}$, $N_{\alpha\beta}$, $N_{\beta\beta}$, $N_{\alpha\gamma}$, and $N_{\beta\gamma}$. (From Wang et al., 2001, reprinted with permission from Elsevier Science.)

where N_α, N_β, and N_γ are the singles peaks for the 27.3 keV K_α x-ray, 31.0 K_β x-ray, and 35.49 keV γ ray, respectively, $N_{\alpha\alpha}$, $N_{\alpha\beta}$, $N_{\beta\beta}$, $N_{\alpha\gamma}$, and $N_{\beta\gamma}$ are sum peaks for all possible combinations of coincidence detection of the two x-ray emissions or one of the x-ray emissions with the gamma ray, and N_6 represents the unresolved sum peak of K_β x-rays ($E_{\beta\beta} = 62.0$ keV) and sum peak of K_α x-rays and γ rays ($E_{\alpha\gamma} = 62.8$ keV), which could not be separated and therefore treated as one peak, $N_6 = N_{\beta\beta} + N_{\alpha\gamma}$. With this method Wang *et al.* (2001) determined the ^{125}I activity with 0.6% uncertainty.

In Chapter 1, Section II.C.2, examples were provided of radionuclides that exhibited different electron-capture decay rates as a function of the chemical environment of the radionuclide. Because electron capture involves extranuclear atomic electrons, which can be affected by chemical environments, the decay rates of EC radionuclides can be affected. Singh *et al.*, (2001) utilized this phenomenon and the sum-peak method (gamma–gamma summing) to correlate variations in nuclear precession frequencies with electric field gradient components in different chemical environments. They measured the variations in the ^{75}Se sum-peak intensity at the 400 keV line due to the simultaneous detection of the $121 + 279$ keV gamma rays relative to the singles peak intensities at 121 and 279 keV.

D. Self-Absorption

The counting efficiencies of experimental samples can be affected significantly by self-absorption, particularly when the nuclide tracer emits low-energy x- or gamma-radiation. Self-absorption may be defined as the absorption of radiation by the sample emitting the radiation. This absorbed radiation does not reach the detector and, if appreciable, counting efficiencies can be markedly reduced. Self-absorption is a function of the sample composition and geometry, namely its effective atomic number, density, thickness, and the x-ray or gamma-ray photon energy. Photons of low energy such as the 28- and 32-keV x-rays and 35-keV gamma rays emitted from ^{125}I are much more susceptible to self-absorption than the higher-energy gamma rays of, for example, ^{60}Co (1.17 and 1.33 MeV).

The x- and gamma-rays of ^{125}I can be attenuated significantly before escaping from a typical 1-cm-diameter sample tube (see Malcolme-Lawes and Massey, 1980). The effect of salt concentration and ions of different atomic mass absorption coefficients on the solid scintillation counting efficiency of ^{125}I is illustrated in Fig. 11.24. Malcolme-Lawes and Massey (1980) reported a channels ratio counting technique as a monitor for such effects on the counting efficiency of ^{125}I. Rieppo *et al.* (1975) and Rieppo (1976a,b) have studied extensively the relationship of counting efficiency and sample density (e.g., water, bone, and aluminum) for low-energy photons in the energy range 0.010–0.200 MeV. Self-absorption can be significant in the solid scintillation analysis of weak (low-energy) x- or gamma-ray photons (< 400 keV), and it is a function of the thickness and density of the samples (Elane Streets, 1994), because of the relatively poor penetrating power of low-energy photons.

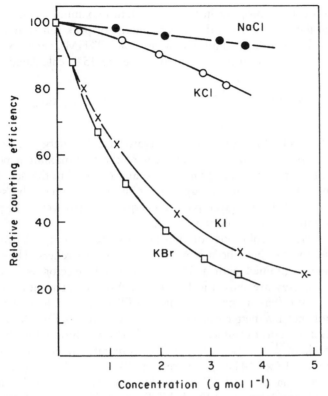

FIGURE II.24 Variation of counting efficiency for ^{125}I (relative to efficiency in water $= 100$) with concentration of materials in aqueous samples. Sample volumes are 5 mL and bottle diameter is 1.3 cm. (From Malcolme-Lawes and Massey, 1980, reprinted with permission from Elsevier Science.)

E. Counting Geometry

The counting geometry is the geometric relationship between the sample and detector, including the sample and detector size and shape, the distance between the sample and detector, sample thickness, and angle of orientation of the detector around the sample (e.g., $180°$ or 2π for a flat sample and planar detector or $\gg 2\pi$ geometry for a well detector or through-hole detector). The effect of sample geometry on the detection and counting efficiencies for various detector types including the planar, well-type, and through-hole detectors is discussed in Section III.A of this chapter.

F. Resolution

The resolution of a solid scintillation detector is a measure of the "sharpness" of the x-ray or gamma-ray energy peaks of the pulse height spectra. The sharper or narrower these peaks, the greater the resolution and the ability of the crystal detector to resolve or distinguish between the photon energy lines of one or more radionuclides in a sample. The resolution is expressed most often as the width of the photon peak at half-maximum peak height.

The width of the photopeak in keV at the half-maximum height is expressed in terms of the percentage of the photon energy or peak center and is referred to as % full width at half-maximum (% FWHM). For example, the resolution of the photopeak illustrated in Fig. 11.15 is calculated as follows:

$$\% \text{ FWHM} = \frac{675.2 \text{ keV} - 644.8 \text{ keV}}{662 \text{ keV}} \times 100 = 4.58\% \qquad (11.27)$$

where 675.2 and 644.8 keV in the numerator are the energy values obtained from the energy axis taken at both sides of the half-maximum peak height (position of the two arrows of Fig. 11.15) and 662 keV in the denominator is the energy value from the energy axis at the peak center. The smaller the value of % FWHM for a given photon energy, the narrower is the photopeak and the greater is the resolution.

The energy resolutions of scintillators are energy dependent and experimental values have been reported by many researchers. The resolutions experimentally attained with a 38×38 mm BGO detector as a function of gamma-ray energy are shown in Fig. 11.25. When comparing resolutions for photopeaks of different energies, a high % FWHM for a low-energy photon peak could mean a better resolution than a lower % FWHM for a high-energy photon. For example, a 17% FWHM resolution for the 514-keV gamma peak of ^{137}Cs (see Fig. 11.25) represents an energy resolution of 87 keV (i.e., 0.17×514 keV), whereas a 10% FWHM resolution for the 1275-keV gamma peak of ^{22}Na (see Fig. 11.25) can be calculated as an energy resolution of only 127.5 keV. Evans (1980) pointed out that the resolution of the BGO detector is only about half as good as that which can be obtained with equivalent NaI(Tl) crystals. However, the higher efficiency

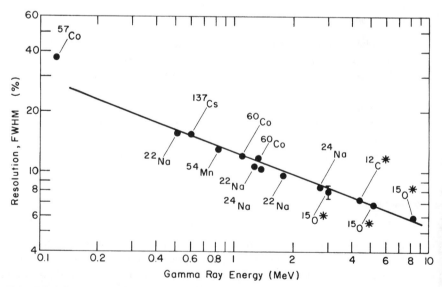

FIGURE 11.25 Resolution of 38-mm BGO scintillator as a function of gamma-ray energy. (From Evans, 1980, © IEEE.)

of BGO and the near absence of Compton continua make this detector the choice when a high-resolution detector is not needed.

A scintillation crystal detector can become damaged when the crystal undergoes fracture via thermal or mechanical shock or when it absorbs atmospheric moisture, which can occur with the hygroscopic NaI(Tl) crystals, as well as other hygroscopic crystal detectors, when the protective cover is damaged. When damage occurs the energy transfer scintillation processes and light transmission of the crystal are impaired and the energy resolution will drop. The photopeaks may become broadened and/or transformed. Modern gamma scintillation analyzers are equipped with an MCA and automatic calibration programs to check the integrity of the photopeak resolution of calibrated NIST-traceable gamma-emitting standards.

G. Background

Background includes counts collected by the scintillation detector produced by radiation events arising from sources outside the sample under analysis. The sources of background radiation are cosmic rays, natural radionuclides in the sample tube and neighboring materials (environment), detector contamination, and radioactivity from samples that may be in close proximity to the sample under analysis. The latter source of background should be considered particularly in automatic gamma-counting systems where multiple detectors arrays and multiple sample-changing cassettes are all in close proximity and when high-energy gamma-emitting radionuclides are present in sample tubes.

Background count rates should always be determined for the particular counting region of interest defined by the lower level (LL) and upper level (UL) discriminator settings. The background for a particular counting protocol may be determined by using a blank sample, that is, a sample tube or microplate well containing the volume of reagents used without any radioactive sample. This is similar to a reagent blank. When background count rates are significantly high relative to the sample count rates, background counts must be subtracted from the sample counts. Under these circumstances, sample counts and background counts should be collected over a sufficient period of time to get a measure of the sample and background count rates with a low %2S (%2 sigma standard deviation).

The measurement of background also serves another purpose, that of testing the stability of the instrument detector systems and associated electronics. This fits within a program of instrument performance assessment, which is dealt with in Chapter 5. The stability of the background count rates serves as an index of the stability of the counting system.

Modern solid scintillation analyzers are equipped with an MCA, which can be programmed to collect the background counts according to pulse height over the entire spectrum of pulse heights for a predetermined time period. The background for any counting region of interest defined by LL and UL discriminator settings is easily obtained from the entire pulse height spectrum. The instrument need only sum the pulse heights in those channels for any region of interest.

V. AUTOMATED SOLID SCINTILLATION ANALYZERS

Automated solid scintillation analyzers are referred to here as high-throughput computer-based multiple-detector instruments capable of analyzing simultaneously as many as 10–12 samples with automated instrument performance assessment, multiple-user programming, "walk-away" sample analysis, and "hands-off" data processing of 750–1000 samples with potential links to computer mainframe and local area network (LAN) assay automation and data management. Two types of automated scintillation analyzers will be described in this section: (1) automated gamma analyzers, which utilize up to 10 NaI(Tl) detectors for the simultaneous measurement of gamma- and x-ray-emitting radionuclides in sample tubes, and (2) microplate scintillation analyzers, which utilize 24-, 96-, or 384-well microplates for the solid or liquid scintillation analysis of beta particle- and atomic electron-emitting radionuclides.

A. Automated Gamma Analysis

An automated gamma analyzer should possess all the components needed to provide a complete unattended analysis of large volumes of samples containing x-ray and gamma-emitting radionuclides. Among the main components of such a system are (1) a large-capacity sample changer capable of holding and moving up to 750–1000 sample tubes to and from detectors, (2) multiple detectors for the simultaneous analysis of up to 10 samples when a high sample throughput is required, (3) automatic computer-controlled system with numerous (up to 60) independent programmable assay protocols, (4) sample number and sample cassette identification via binary code, which can be linked to a local area network data processing and management system through a programmable RS-232 computer interface, (5) automatic isotope decay correction according to date and time, (6) live gamma spectrum displays, (7) automatic detector normalization and detector crosstalk measurement for multiple-detector instruments, (8) correction for radionuclide interference between two counting regions via automatic spillover calculations, (9) automatic background correction in all counting regions, (10) multiple (up to three) simultaneous counting regions, either preset for certain radionuclides, such as ^{125}I, ^{57}Co, ^{60}Co, ^{75}Se, ^{51}Cr, ^{131}I, ^{59}Fe, and ^{22}Na, or defined arbitrarily with LL and UL discriminator settings over a wide energy region (e.g., 15–2000 keV), (11) automatic instrument performance assessment, and (12) special applications software for sample data processing and calculations.

I. Multiple Detector Design

In an early part of this chapter (Section III.A) a description of the various types of detector geometries was given together with a comparison of their performance. The advantages of the through-hole detector geometry were described. This detector geometry is used in one of the leading state-of-the-art automated gamma analyzers. An illustration of an array of 10 detectors of the through-hole type of an automatic high-throughput NaI(Tl) gamma

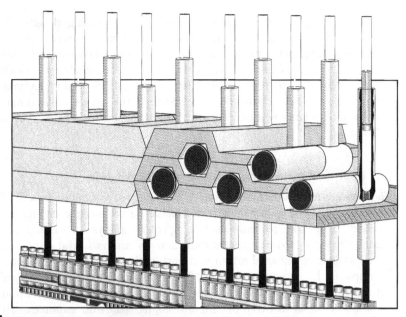

FIGURE II.26 Staggered array of 10 NaI(Tl) scintillation detectors and photomultiplier tubes of a high-throughput automatic gamma analyzer. At the right two exposed photomultiplier tubes and NaI(Tl) detectors are illustrated. A see-through diagram of one of the NaI(Tl) detectors is illustrated at the far right with a sample tube in position for analysis. A lead counterweight and elevator are shown above and below the sample tube. Sample tubes in cassettes directly below the detectors are moved upward by elevators into counting position. (Courtesy of PerkinElmer Life and Analytical Sciences.)

scintillation analyzer is illustrated in Fig. 11.26. The proximity of the detectors to each other and to samples in the sample changer combined with the penetrating power or range of high-energy gamma radiation emitted by some radionuclides could give rise to possible error from crosstalk. We can define crosstalk as the registration of radiation events by a detector when those events originate from samples other than the sample in the detector.

There are two types of detector crosstalk, namely (1) detector-to-detector crosstalk, which occurs when a detector registers one or more events produced by radiation emanating from a sample located in another adjacent detector, and (2) deck-to-detector crosstalk, which occurs when a detector registers events arising from radiation emitted by samples on the sample changer deck. In multidetector automatic gamma analysis instrumentation, the gamma radiation from one or more samples may be of sufficient energy and intensity to reach neighboring detectors and register events in the neighboring detectors.

Detector-to-detector crosstalk can be calculated by determining the count rate (CPM) from a detector containing no sample and dividing that count rate by the disintegration rate (DPM) or count rate of a sample in an adjacent detector. The ratio is expressed as a percentage. Also, deck-to-detector crosstalk could arise from samples on the deck that are in a sample cassette on the sample deck adjacent to and directly below a sample that is being counted. The staggered configuration of the shielding design illustrated in Fig. 11.26 provides 18 mm of lead shielding between detectors. This shielding, together with a lead sample elevator platform and lead overhead sample counterweight, provides 4π (360°) lead shielding, which virtually eliminates detector crosstalk for radionuclides with gamma energies < 500 keV. When crosstalk is suspected, the modern automatic gamma analyzer can employ a mathematical matrix calculation to correct for detector crosstalk.

2. Multiuser Automatic Gamma Activity Analysis

Modern automatic gamma counters are computer based. The entire operation of the gamma analyzer, including the sample changing, counting, single- and multiple-radionuclide spectral analysis, count rate statistical analysis, isotope decay corrections, activity analysis, and data processing as well as instrument calibration and performance assessment, is controlled and completed by computer. It is necessary only to place the sample tubes in the appropriate sample cassettes and to input the basic information that the computer needs to carry out the automatic hands-off gamma analysis. Some of the major features of a typical automatic gamma analyzer will be outlined briefly here.

The modern automatic gamma analyzer has no toggle switches, buttons, or dials for the operator to set up and carry out an instrumental gamma analysis program. The one and only switch on the instrument is the main power switch, which is kept in the "On" position at all times. The only means of controlling the instrument is via the computer keyboard. Up to 60 user-definable analysis protocols are available, and a user will select and name or identify one of the analysis protocols and input the necessary information to the instrument computer to define how the instrument will analyze his or her samples. Therefore, an analysis protocol is simply a customized procedure edited by the user that the instrument will implement to count and analyze a particular set of samples. The user will first identify and edit a protocol. The computer keyboard procedure for editing a counting protocol should be user friendly, because the instrument should provide on the computer screen the available options for the user on what may be input as instructions to the computer. Editing an analysis protocol on the computer screen involves (1) naming the protocol, which helps the user identify that protocol for future applications and also informs others that the protocol belongs to a particular application or user, (2) defining the count conditions parameters, which control the counting of samples, and (3) defining the data reduction parameters, which can include assay-specific data calculations, curve fitting, and data printout.

Counting efficiencies obtained for ^{125}I in 12 × 75 mm polyethylene plastic tubes with the 1.5-, 2-, and 3-inch NaI(Tl) detectors on the PerkinElmer Cobra automatic gamma counting systems are > 72–74%. The backgrounds obtained for the same three detectors for the analysis of ^{125}I are ≤ 30, ≤ 50, and ≤ 55 CPM. The slight increase in background with increase in detector size is expected. Resolutions achievable are ≤ 34% for ^{125}I and the 1.5-inch detector, $\leq 32\%$ for ^{129}I and the 2-inch detector, and $\leq 12\%$ for ^{137}Cs and the 3-inch detector.

3. Multiple Gamma-Emitting Nuclide Analysis

There are two options for measuring multiple gamma-emitting radio-nuclides in mixtures, namely the solid scintillation crystal detector or the germanium semiconductor detector. The solid scintillation detectors provide higher detection efficiencies than the semiconductor detector requiring shorter counting times for lower activity samples, but the semiconductor detector yields much higher energy peak resolutions permitting the more facile quantification of complex mixtures particularly when the photopeaks of different nuclides in the mixture are close (L'Annunziata, 1984, 1987; Roig et al., 1999). The differences between the detection efficiencies and resolutions achieved by the two types of detectors was measured and demonstrated by Roig et al. (1999) and illustrated in Fig. 11.27. They reported a detection efficiency and photopeak resolution of 22% and 60 keV FWHM at 661.7 keV of ^{137}Cs, respectively, for a 3-inch NaI(Tl) detector, whereas the lower 1.7% efficiency but much improved 1.4 keV resolution for the 661.7 keV photopeak of ^{137}Cs was obtained with the Germanium semiconductor detector. The high detection efficiencies of solid scintillation detectors makes these attractive for

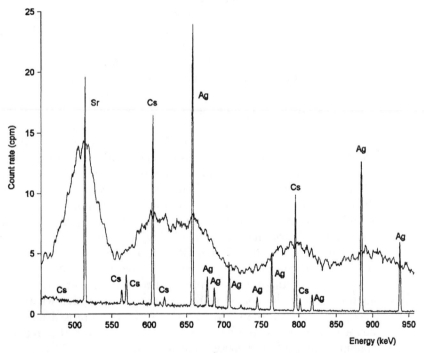

FIGURE 11.27 Spectra of a mixture of ^{85}Sr, ^{134}Cs, and ^{110m}Ag obtained with a 3-inch Packard Cobra NaI:Tl with a counting time of 15 min (upper spectrum), and with a Canberra GR2020 intrinsic germanium detector with a counting time of 170 min (lower spectrum). Radionuclide activities: 76.5 Bq of ^{85}Sr; 78.4 Bq of ^{134}Cs; 139 Bq of ^{110m}Ag. (From Roig et al., 1999, reprinted with permission from Elsevier Science.)

the analysis of single radionuclides or, of multiple radionuclides when the photopeaks of these have energies separated enough to count the areas under the photopeaks of each radionuclide of the mixture or where corrections for any overlap or spillover of activity from interfering nuclides may be calculated.

Simultaneous analysis of more than one gamma-emitting nuclide in the same sample with the NaI(Tl) scintillation analyzer is common. It is necessary only that the photopeaks of the nuclides be separated enough so that counting regions can be defined to encompass the photopeaks of each nuclide. This will maximize the counts collected for each nuclide in each region, although there can be overlap of counts from other radionuclides in each counting region. Let us look at two examples of the simplest case of two nuclides in a mixture and a more complicated case of several nuclides in the same sample.

a. Dual-Nuclide Analysis

The gamma spectrum of two nuclides in the same sample was seen previously in Fig. 11.16 for the case of a mixture of ^{137}Cs and ^{65}Zn. In this case we see two counting regions defined by setting the lower level and upper level energy or pulse height discriminator settings of A and B (550 and 800 keV) for the ^{137}Cs counting region and A' and B' (950 and 1250 keV) for the ^{65}Zn counting region, respectively. The counting regions encompass the photopeaks of each radionuclide; however, it should be noted that the region set for counting the lower energy gamma emitter ^{137}Cs also contains counts arising from Compton smear of the higher energy nuclide ^{65}Zn. This is known as spilldown, because counts from the high-energy gamma emitter "spill down" into the counting region of the lower energy gamma emitter. It is necessary, therefore, to calculate the percentage of the counts in ^{137}Cs region that are due to ^{65}Zn to correct for the spilldown. Often, the photopeaks of both radionuclides of the mixture have counts arising from each nuclide. We would, therefore, also have to correct for spillup, which is the percentage of counts that "spill up" into the counting region of the higher energy gamma emitter from the lower energy gamma emitter of the mixture.

The spilldown and spillup corrections that must be made can be illustrated easily by using the dual label ^{125}I–^{57}Co as an example. This is a dual label used for the common folic acid-vitamin B$_{12}$ assays in serum samples. Two counting regions are fixed, namely region A defined by setting the LL–UL discriminators of 15–75 keV for the ^{125}I counting region and region B defined by the LL and UL discriminator settings of 75–165 keV for the ^{57}Co counting region. The spectra of the two radionuclides overlap. The CPM of ^{125}I in the mixture is obtained from the count rate in region A after subtracting out the counts in that region due to spilldown from the higher energy ^{57}Co. Also, the CPM of ^{57}Co in the mixture is determined from the count rate in region B after subtracting the counts in that region due to the spillup from the lower energy nuclide ^{125}I.

The spilldown and spillup correction factors are determined by counting the pure samples of each nuclide in each counting region. If regions A and B are the counting regions for the lower and higher energy nuclides, respectively, we can determine the spilldown and spillup correction factors as follows. First, a pure sample of the higher energy nuclide ^{57}Co is taken and counted in both regions and the spilldown correction factor calculated as

$$\text{spilldown} = (\text{CPM region A/CPM region B})(100) \qquad (11.28)$$

The spilldown calculation gives the percentage of counts in the counting region of the lower energy nuclide generated by the higher energy radionuclide. Then a pure sample of the lower energy radionuclide is taken and counted in the two counting regions. From the count rates in these regions the spillup correction factor is calculated as

$$\text{spillup} = (\text{CPM region B/CPM region A})(100) \qquad (11.29)$$

The spillup calculation gives the percentage of counts in the counting region of the higher energy nuclide generated by the lower energy nuclide.

Once the percent spilldown and spillup correction factors are determined, the values are input into the counting protocol, and the automatic gamma analyzer will make the calculations required for the corrected and final count values for the two radionuclides in a mixture. The following equations form the basis upon which the count rates of the two nuclides in the mixture are calculated:

$$y_1 = x_1 + s_2\, x_2 \qquad (11.30)$$

$$y_2 = x_2 + s_1\, x_1 \qquad (11.31)$$

where y_1 and y_2 are the total number of counts from a dual-label nuclide mixture in photopeak counting regions 1 and 2 (regions A and B), x_1 and x_2 are the unknown number of counts of nuclides 1 and 2 in their photopeak regions, and s_1 and s_2 are constants, which are the spillover values for nuclides 1 and 2, respectively. Typical values for the spilldown and spillup for the ^{125}I–^{57}Co dual label with the PerkinElmer Cobra automatic gamma analyzer are 3.00% or 0.03 for the spilldown value (s_2) of nuclide 2 (^{57}Co) into the counting region of ^{125}I and 0.80% or 0.008 for the spillup value (s_1) of nuclide 1 (^{125}I) into the counting region of ^{57}Co. Equations 11.30 and 11.31 are solved simultaneously for the two unknown values of x_1 and x_2, which are the corrected count rate values of nuclides 1 and 2 in their respective counting regions.

Some other dual-label radionuclides that are commonly measured by solid scintillation gamma spectrometry are ^{125}I–^{131}I (Nozaki and Saito, 1995), ^{57}Co–^{58}Co used in the Shilling test to diagnose deficiencies in vitamin B$_{12}$ absorption, and ^{134}Cs–^{137}Cs in the environment. The technique of calculating percent spilldown and percent spillup of the overlapping photopeaks is also required in these double-label measurements.

b. More Complex Multiple Nuclide Analysis

The simplest case of the solid scintillation gamma analysis of two nuclides in a mixture was just discussed. Let us now look at the more complex case of a mixture of more than two gamma-emitting nuclides. The solid scintillation analysis of multiple-nuclide mixtures gives rise to a more complex pulse height spectrum where the counting regions, which are selected to surround the photopeaks of each nuclide of the mixture, will register counts from the nuclide of interest as well as counts from other nuclides spilling up and spilling down into the counting regions. An example of such a case is illustrated in Fig. 11.28, which illustrates the gamma spectrum of a mixture of six radionuclides and six counting regions (regions A through F) selected to encompass the photopeaks of each nuclide. In the fashion of Eqs. 11.30 and 11.31, six equations can be written to describe the counts collected

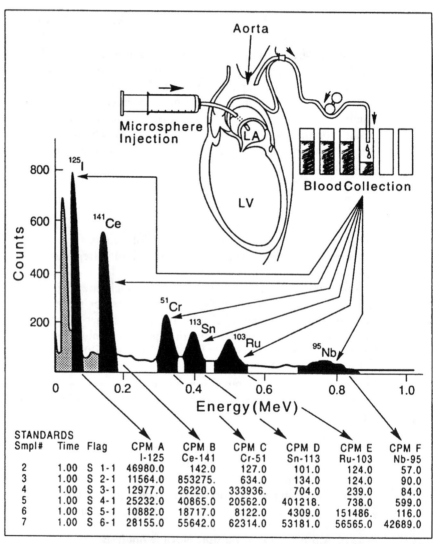

STANDARDS Smpl#	Time	Flag	CPM A I-125	CPM B Ce-141	CPM C Cr-51	CPM D Sn-113	CPM E Ru-103	CPM F Nb-95
2	1.00	S 1-1	46980.0	142.0	127.0	101.0	124.0	57.0
3	1.00	S 2-1	11564.0	853275.	634.0	134.0	124.0	90.0
4	1.00	S 3-1	12977.0	26220.0	333936.	704.0	239.0	84.0
5	1.00	S 4-1	25232.0	40865.0	20562.0	401218.	738.0	599.0
6	1.00	S 5-1	10882.0	18717.0	8122.0	4309.0	151486.0	116.0
7	1.00	S 6-1	28155.0	55642.0	62314.0	53181.0	56565.0	42689.0

FIGURE II.28 Six photopeaks from a composite spectrum of ^{125}I, ^{141}Ce, ^{51}Cr, ^{113}Sn, ^{103}Ru, and ^{95}Nb and count rates (CPM) obtained from six counting regions after spillover corrections were applied using **PCGERDA** software for calculating regional blood flow with radionuclide-labeled microspheres. (Courtesy of PerkinElmer Life and Analytical Sciences.)

in each of the six counting regions illustrated in Fig. 11.28, including the correction factors for the spillup and spilldown of counts from radionuclides in each counting region. The six equations need to be solved simultaneously. To facilitate the analysis of multiple labels of this type, computer software such as PCGERDA can be employed with the automatic gamma analyzer for data processing and printout of the results of the count rates of each nuclide in the mixture. The application of PCGERDA software is described in detail in the literature (PerkinElmer, 1993).

The lack of resolution in solid scintillation detectors may be overcome with the use of multivariate calibration methods (Roig *et al.*, 1999). Multivariate calibration techniques were discussed in Chapter 5 Sections V.F.4 and VIII.E.3. Among these methods, multivariate regression (MLR) and partial least squares regression (PLS) can be used to determine amounts of radionuclides of interest although other radionuclides are present that affect the experimental signal provided the isotopes are present in both the unknown samples and a calibration set (Martin and Naes, 1989; Toribo *et al.*, 1995). The advantage is that these methods have proven successful for the analysis of samples with unknown interferences. The MLR and PLS methods were tested by Roig *et al.* (1999) to determine activities of ^{85}Sr, ^{134}Cs, and ^{110m}Ag in a mixture, which produced solid scintillation spectra with overlapping photopeaks and Compton smear illustrated in Fig. 11.29. They used the multivariate calibration, MATLAB™ (The Mathworks, Inc., Natick, Massachusetts) and the PLS-Toolbox for MATLAB™ written by Wise (1992). They tested the influence of different calibration sets on the prediction of radionuclide activities, and prediction errors to samples without interferences for the aforementioned mixture were demonstrated to be < 4%, which are comparable to errors obtainable with germanium detectors. Other spectrum unfolding methods for the solid scintillation analysis of gamma ray-emitting nuclides are described by Muravsky *et al.* (1998), Nguyen *et al.* (1996), Skipper and Hangartner, (2002), Sükösd *et al.* (1995) and Suzuki *et al.* (1993). For laboratory and field measurements of soil samples Chiozzi *et al.* (2000a,b) used an EG&G Ortec NaI(Tl) detector and multichannel analyzer for the measurement of radionuclides of uranium, thorium, and potassium. Samples of 0.65–0.85 kg were measured in polyethylene Marinelli beakers. The analysis was based on solid scintillation gamma spectra of the environmental samples, which provided adequately separated photopeaks for ^{214}Bi (^{238}U series), ^{208}Tl (^{232}Th series), and ^{40}K. Takiue *et al.* (1999) and Fujii *et al.* (1999, 2000) used the combined spectral information from a NaI(Tl) solid scintillation detector and liquid scintillation analyzer to measure seven beta-, gamma-, and x-ray-emitting nuclides in a mixture. The method is described in detail in Section VIII.E.1 of Chapter 5 and an illustration of the NaI(Tl) pulse height spectrum from the nuclides ^{125}I, ^{51}Cr, and ^{22}Na in the mixture is illustrated in Fig. 5.39 of Chapter 5.

B. Microplate Scintillation Analysis

The automated scintillation analyzers with the highest throughput today are the microplate counters, which can analyze all beta particle- and atomic

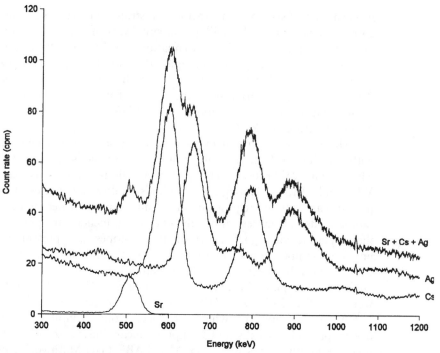

FIGURE 11.29 Smoothed spectra obtained with a NaI(Tl) solid scintillation detector (PerkinElmer Auto-Gamma® 5530) of single solutions of ^{85}Sr, ^{134}Cs, and ^{110m}Ag and a sample with the three radionuclides (upper spectrum). Radionuclide activities (when present): 100 Bq of ^{85}Sr; 1390 Bq of ^{134}Cs; 1220 Bq of ^{110m}Ag. (From Roig et al., 1999, reprinted with permission from Elsevier Science.)

electron-emitting nuclides by liquid or solid scintillation counting as well as carry out the nonradioactive luminescence counting. These counters are referred to as microplate scintillation and luminescence counters. In this chapter, we will limit our treatment of this counter to only the solid scintillation counting of beta- and atomic electron-emitting nuclides.

I. Solid Scintillation Counting in Microplates

One of the most advanced designs of microplate scintillation counters is the TopCount Microplate Scintillation and Luminescence Counter (PerkinElmer Life and Analytical Sciences). Other similar counters are available such as the Wallac MicroBeta radioactivity plate counter (PerkinElmer Life and Analytical Sciences). The TopCount microplate scintillation counter utilizes 2, 6, or 12 photomultiplier tubes for the simultaneous counting of 2, 6, or 12 samples. The samples are counted in 24-, 96-, or 384-well microplates, which are stacked in the counter's microplate changing mechanism. The instrument moves the microplates one at a time under the PMTs for the simultaneous counting of up to 12 samples in separate wells. Figure 11.30 illustrates the location of a PMT relative to one of the microplate sample wells. The highly reflective surface of the microplate well assures reflection of light photons from scintillation events toward the PMT. Although only one PMT per sample is employed, the

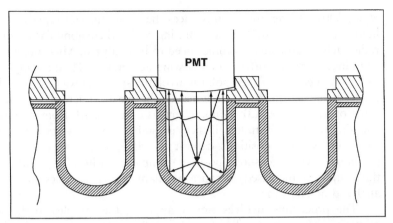

FIGURE II.30 Diagram illustrating the position of a photomultiplier tube (PMT) and sample in a microplate well. In TopCount high-throughput automatic scintillation analysis up to 12 PMTs are positioned simultaneously above 12 sample wells. In scintillation proximity assay (SPA) aqueous liquid medium is in the sample well, which has a white reflective surface. In solid scintillation analysis the sample is dry and adhered to the surface of a LumaPlate plastic scintillator-coated well. (Courtesy of PerkinElmer Life and Analytical Sciences.)

backgrounds are low due to a patented method of time-resolved liquid scintillation counting (TR-LSC), which can distinguish between true scintillation events and background noise (U.S. patent 4,651,006). The theory of TR-LSC for background reduction in TopCount liquid and solid scintillation counting is described by PerkinElmer (1991a). In the solid scintillation analysis of beta particle- and atomic electron-emitting nuclides the samples are analyzed in the dry state in the microplate wells as discussed further on in this section.

The 24-, 96-, and 384-well microplates contain wells oriented in a 6×4, 12×8, and 24×16 fashion along the length and width of each microplate. The maximum usable well volumes are 1500, 350, and 80 μL for the 24-, 96-, and 384-well microplates, respectively. Up to 40 microplates may be stacked in the counter microplate changing mechanism. If we consider the largest 384-well microplate, the stacking of 40 microplates in the counter would allow the unattended counting of 15,360 samples. The counter is computer based and fully automated for complete hands off sample analysis.

Solid scintillation counting in microplates involves depositing a sample into the well of a special plastic microplate (e.g., LumaPlate and Scintiplate available from PerkinElmer Life and Analytical Sciences, Boston, MA, or FlashPlate available from New England Nuclear, NEN, Wilmington, DE). The microplate contains a solid scintillator at the bottom of the well. The material is dried and then counted in the microplate scintillation counter. By this technique the activities (DPM) of small volumes of any nonvolatile beta particle- or atomic electron-emitting nuclide-labeled samples may be determined. The small samples involved are particularly applicable to cytotoxicity, immunoassay, receptor binding, enzyme activity, and other metabolic studies (Boernsen *et al.*, 2000; Earnshaw and Pope, 2001; Turlais *et al.*, 2001). In certain circumstances the technique can replace conventional liquid scintillation analysis in counting vials when nonvolatile forms of the radionuclides and small volumes need to be analyzed. Microplate scintillation counting, therefore, offers the following advantages: (1) the samples are safer to handle, because of their solid nonvolatile state; (2) waste disposal costs are reduced due to the small volumes of the plastic microplates and the solid state of the waste; (3) samples may be recovered from the wells after counting for further studies, if they are stable in the dry state; (4) once in the wells, the sample positions cannot be changed mistakenly by laboratory personnel, which ensures positive sample identification; and (5) sample throughput is increased, because as many as 12 samples in wells can be analyzed at one time.

The procedure involves simply applying the samples to the wells of a plastic-scintillator microplate (e.g., LumaPlate). The plates are dried in a fume hood overnight, or a heat lamp, hair dryer, or centrifugal evaporator may be used to facilitate sample drying. Prior to counting, the microplate is sealed with cover film (e.g., PerkinElmer TopSeal-P) to facilitate handling and to prevent contamination of the samples and counter. A paper by the PerkinElmer (1996) provides detailed information on sample handling and performance for solid scintillation counting with the LumaPlate.

The TopCount solid scintillation counting efficiencies (%) reported for various radionuclide-labeled compounds deposited in separate wells of the 96-well LumaPlate are ^{3}H (49%), ^{14}C (85%), ^{32}P (87%), ^{51}Cr (24%), and ^{125}I (75%) with backgrounds of only 8–9 CPM. The low backgrounds provide high figures of merit (FOM $= E^{2}/B$). The 24-well LumaPlate will provide significantly higher counting efficiencies, for example, ^{3}H (55%), ^{14}C (95%), ^{32}P (93%), ^{51}Cr (48%), and ^{125}I (83%), but with higher backgrounds of about 19–20 CPM.

Because the scintillation phenomenon occurs in the solid scintillation detector (wall of the microplate well) separate from the sample and its radionuclides, there is no chemical quench. However, if the sample residues along the walls of the microplate wells are colored, it is possible to have quenching from the absorption of the scintillation photons from color. The effect of color quenching can be determined with colored agents as described in Section V.B.2.g of this chapter and in a report by the PerkinElmer (1996). A quench correction curve of %E versus a quench-indicating parameter (e.g., transformed spectral index of the sample, tSIS) is prepared with color-quenched radionuclide standards. The quench correction curve is, in turn, used by the instrument to convert count rates (CPM) to activity (DPM) according to Eq. 11.18.

The close proximity of the wells and PMTs could contribute to well-to-well crosstalk, which can be defined as pulse events registered by one PMT arising from nuclear radiation or light produced in a neighboring well. The LumaPlate structural material is opaque to light, which prevents optical well-to-well crosstalk. High-energy beta radiation and low-energy gamma rays could theoretically pass through the walls of one well and produce a scintillation event in the walls of another. Crosstalk expressed as percentage is defined and calculated as described in Section V.A.1. The following reported crosstalk values (%) are negligible and generally can be ignored: ^{3}H (not detectable), ^{14}C (0.005%), ^{32}P (0.3%), ^{51}Cr (<0.02%), and ^{125}I (<0.4%).

2. Scintillation Proximity Assay (SPA)

Scintillation proximity assay (SPA) is a technology for the analysis of binding reactions, commonly studied in the medical and biochemical sciences, which circumvents the need to separate bound from free fractions. Glass or plastic solid scintillator microspheres are used in this assay together with an isotope-labeled ligand. With the use of modern liquid handling equipment and an automatic scintillation counter for samples in a microplate format, hundreds of samples may be prepared and analyzed in a single day, because traditional processes for the separation of bound and free fractions are not required. This section provides a brief treatment of SPA. Additional sources of general information on SPA can be obtained from Udenfriend *et al.* (1987), PerkinElmer (1991b), Neumann and van Cauter (1991), Roessler *et al.* (1993), Jessop (1993, 1996), and Cook (1996). More detailed information and procedures may be obtained from references cited in the following.

a. Basic Principles

Scintillation proximity assay makes use of the weak (low-energy) radiation emitted by certain radionuclides, such as the very low-energy beta-particles emitted by ^{3}H ($E_{\max} = 18.6$ keV) or the weak Auger electrons emitted by ^{125}I ($E_{\max} = 30$ keV). The particulate radiations emitted by these radionuclides have very short ranges of travel in water. For example, the maximum range of ^{3}H beta particles in water is only about 6 μm (see Table 1.11 in Chapter 1), and beta particles on the average (mean) lose all of their energy and come to a full stop before they reach 1.5 μm in water (Amersham, 1996). The Auger electrons of ^{125}I, which are of slightly higher energy, travel only a few micrometers further in water. Therefore, if these radionuclides are in aqueous solution their radiations will not be detectable unless they are in intimate or very close contact with a scintillating compound. SPA, therefore, makes use of yttrium silicate scintillating glass microspheres or polyvinyltoluene (PVT) scintillator microspheres, which emit visible light when the beta particles or Auger electrons from ^{3}H or ^{125}I are able to penetrate the microspheres. Because of the short ranges of travel of these radiations in water, their penetration into microsphere scintillators can happen only when the ^{3}H or ^{125}I nuclides are in intimate or very close contact with the microspheres. If the microsphere scintillators are coated with an antibody or receptor molecule and the ^{3}H or ^{125}I resides as a label on an antigen or ligand molecule, a binding reaction can occur between the antibody and antigen or receptor and ligand molecules as illustrated in Fig. 11.31. The close proximity of the ^{3}H or ^{125}I through the binding reactions enables the beta-particle or Auger-electron radiations to penetrate the microsphere scintillator with concomitant isotropic emission of photons of visible light by the scintillator microspheres. A scintillation counter can register each flash of light emitted by any aqueous medium containing a

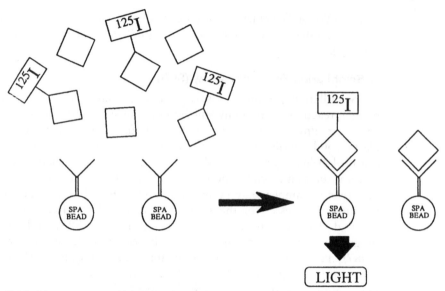

FIGURE 11.31 Illustration of the principle of scintillation proximity assay (SPA). A micro-sphere scintillator (SPA bead) containing an attached antibody or receptor molecule binds with an ^{125}I-labeled antigen or ligand molecule (a ^{3}H, ^{14}C, ^{35}S, or ^{33}P label instead of ^{125}I may also be used). The binding brings the ^{125}I-nuclide label in close proximity to the SPA bead. An Auger electron from the ^{125}I (or beta particle from an alternative radioisotope) on the bound antigen or ligand molecule can penetrate the SPA bead, producing scintillation with the emission of visible light in all directions. In contrast, unbound ^{125}I ligands or bound unlabeled ligands do not produce light emission. Ligands that compete for radionuclide-labeled ligand binding sites reduce the detected light emission. (Adapted from Buchan et al., 1993.)

mixture of (1) microsphere scintillator with surrounding antibody or receptor molecule, (2) free radioisotope-labeled antigen or ligand molecules, and (3) bound radioisotope-labeled antibody–antigen or bound radioisotope-labeled receptor–ligand molecules on scintillator microspheres. The count rate (CPM) of the photon emissions will serve as a measure of the bound radioisotope-labeled fraction, because radiations emitted from labeled molecules not bound close to the microsphere surface dissipate their energy in the aqueous medium and are not detected.

Scintillation proximity assays are not limited to applications with radionuclides emitting very low-energy particulate radiations, such as ^{3}H or ^{125}I. However, the use of radionuclides emitting higher-energy beta-particle radiation, such as ^{14}C ($E_{max} = 155\,keV$), ^{35}S ($E_{max} = 167\,keV$), and ^{33}P ($E_{max} = 249\,keV$) has become popular over the past few years (Delaporte et al., 2001; DeLapp et al., 1999; Evans et al., 2002; Jeffery et al., 2002; Macarrón et al., 2000; McDonald et al., 1999; Park et al., 1998; Sorg et al., 2002). Because of the relative high energies of the beta particles emitted by ^{14}C, ^{35}S, or ^{33}P and the consequent relative long lengths of travel of these particles in water (see ^{14}C in Table 1.11 of Chapter 1), it is necessary to minimize scintillation proximity bead contact with free radionuclide label in the solution medium. This is accomplished by either (1) letting the scintillation proximity beads settle for several hours or overnight, (2) microplate

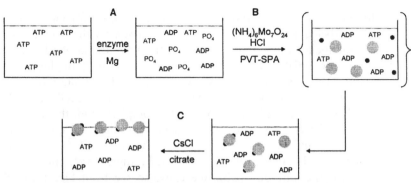

FIGURE 11.32 Schematic representation of the scintillation proximity ATPase assay. (Hydrolysis of ATP to generate radioactive inorganic phosphate [^{33}P]PO$_4^{3-}$. (B) Addition of polyvinyltoluene scintillation proximity beads (PVT-SPA beads) suspended in a solution of ammonium molybdate and HCl. The [^{33}P]phosphomolybdate anion (^{33}PMo$_6$O$_{40}$)$^{3-}$ (small dark circles) forms and then absorbs onto the surface of SPA beads (larger gray circles). (C) Addition of CsCl to float the beads prior to radioactivity analysis. (From Jeffery et al., 2002, reprinted with permission from Elsevier Science.)

centrifugation, or (3) suspension (flotation) of the scintillation proximity beads over 7 M cesium chloride.

The latter suspension technique used by Jeffery *et al.* (2002) is illustrated in Fig. 11.32, which also exemplifies the use of SPA in the study of enzyme activity. In this example ATPase catalyses the hydrolysis of ^{33}P-labeled ATP to generate radioactive inorganic phosphate. Scintillation proximity beads suspended in a solution of ammonium molybdate are added. The ammonium molybdate forms an anion with the ^{33}P-phosphate, and the phosphomolybdate anion then absorbs onto the SPA beads. The addition of 3.5 M CsCl solution causes the SPA beads to float to the surface of the suspension separating the beads from the unbound radioactive ATP. The amount of light emitted by the beads provides a measure of the ^{33}P-phosphate adhered to the SPA beads. The technique was demonstrated as useful for kinetic studies of the enzyme activity.

b. Immunoassay Applications

Let us consider the case of the binding of ^{3}H- or ^{125}I-labeled antigen to an antibody adhered to an SPA bead as illustrated in Fig. 11.31 and the following equation adapted from Watson (1996):

Step 1:
$$Ag^* + Ab-SPA\ bead \longleftrightarrow Ag^*-Ab-SPA\ bead \qquad (11.32)$$
$$+$$
Step 2: $$Ag\ (unlabeled\ antigen)$$
$$\uparrow\downarrow$$
$$Ag-Ab-SPA\ bead\ (unlabeled\ complex)$$

where "Ag*" denotes the radioisotope-labeled antigen, "Ab–SPA bead" represents the antibody adhered to a SPA bead, and "Ag*–Ab–SPA bead" the complex between the radioisotope-labeled antigen and antibody on the SPA bead, which reach equilibrium when mixed together (Step 1).

Immunoassays are quantitative analytical methods for measuring the amount of unlabeled antigen "Ag" in a sample based on the competitive binding of the unlabeled antigen and labeled antigen for the antibody. In other words, if we add unlabeled antigen to Eq. 11.32 (step 2) in the form of either a standard solution or an unknown sample, the amount of labeled antigen bound to the antibody on the SPA bead will be reduced by the formation of unlabeled "Ag–antibody–SPA bead" complex, and the amount of radioactivity detected by the microsphere will diminish corresponding to the amount of unlabeled antigen added to the system. This is also referred to as competitive binding between the labeled antigen (Ag*) and unlabeled antigen (Ag) for the antibody (Ab). We can also see from Fig. 11.31 that given a limited concentration of antibody molecules attached to microspheres, adding unlabeled antigen to the system will reduce the number of labeled antigen molecules associated with the microsphere. Consequently the rate of scintillation photon emission and radioactivity measured by the scintillation detector will be reduced correspondingly.

If we can compare the inhibition produced by unlabeled antigens in an experimental unknown sample to the inhibition produced by known concentrations of unlabeled antigens used as standards, we can determine the concentration of the antigens in our unknown sample. With the preparation of an appropriate standard curve, the count rate of an unknown sample can be used to determine the amount or concentration of unlabeled antigen or ligand in the sample. An example of such a standard curve is illustrated in Fig. 11.33, which was prepared from a commercial scintillation

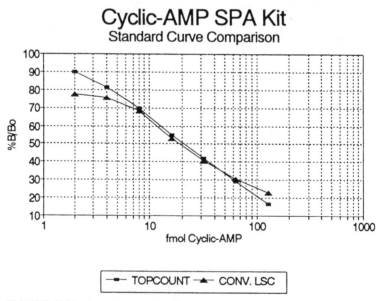

FIGURE 11.33 Standard curves (Scatchard plots) for the scintillation proximity assay (SPA) of cyclic AMP determined with a TopCount microplate scintillation counter and a conventional liquid scintillation counter using a commercially available SPA kit. (From PerkinElmer, 1991b, reprinted with permission from PerkinElmer Life and Analytical Sciences.)

proximity assay kit. The standard curve, also known as a Scatchard plot, is named after George Scatchard, who carried out many of the pioneering studies of macromolecule–ligand interactions. The ordinate gives the ratio of the radioactivity or quench-corrected count rate of the antibody-labeled antigen complex (B) to the radioactivity or quench-corrected count rate (B_0) of the antibody-labeled antigen complex in the absence of any unlabeled antigen. In other words, the nomenclature B/B_0 arises from the definitions of $B =$ the antibody-bound radioactivity and $B_0 =$ the antibody-bound radio-activity at zero dose of unlabeled antigen. The amount (e.g., fmol) or concentration (e.g., nM) of unlabeled antigen is plotted on the abscissa. The ratio B/B_0 may also be expressed as a percentage. It is obvious that, at zero dose of unlabeled antigen, $B = B_0$, and therefore, the ratio is unity or 100% when no unlabeled antigen is added to the medium. The ordinate may also be expressed in other units, which define the distribution of the radioisotope label, such as the ratio of the bound to the free fraction. The curve illustrated in Fig. 11.33 illustrates the similarity in standard curves obtained with two types of detector systems, namely TopCount or MicroBeta microplate scintillation counting and conventional liquid scintillation counting. Both detector systems may be used for SPA; however, the microplate scintillation counter is preferred for high-throughput assays, because the microplate scintillation counter provides the possibility of assaying up to 12 samples simultaneously as discussed further on in this chapter.

The standard curve is prepared from a given quantity of microspheres containing adsorbed antibody ($20-200\,\mu g$), a given activity of ^{3}H- or ^{125}I-labeled antigen, and various amounts of unlabeled antigen. Adequate buffer, which may contain a low concentration (e.g., 0.1%) of bovine serum albumin (BSA), detergent, and 10–12% glycerol are also added in a total volume of about $200\,\mu L$. The BSA serves as a nonspecific protein which, together with the detergent, helps reduce any nonspecific binding of the isotope-labeled antigen with the microspheres. The glycerol prevents settling of the microspheres, keeping them in suspension. The total sample volume is, therefore, very small and generally does not exceed $400\,\mu L$. Detailed procedures and reagents used in SPA and a comparison of SPA with enzyme immunoassay (EIA) and radioimmunoassay (RIA) are given by Horton and Baxendale (1995). Many studies involving scintillation proximity immuno-assays are available in the scientific literature, among which the works of Akisu *et al.* (1998), Amersham (1999, 2000), Frolik *et al.* (1999), Holland *et al.* (1994), Mansfield *et al.* (1996), Patel *et al.* (1996), Poggeler and Heuther (1992); Swinkels *et al.* (1990), Schoenfeld and Luqmani (1995), and de Serres *et al.* (1996), can serve as examples.

The small sample volumes involved in SPA ($< 400\,\mu L$) and the large numbers of samples that must be analyzed by most laboratories make scintillation proximity assays of samples prepared in microplate formats most favorable. Samples may be prepared in sample microplates equipped with 24-, 96-, 384-, and 1536 sample-well formats. The application of SPA to very-high throughput 1536 sample-well format is discussed further on in this chapter. With the exception of the 1536 sample-well format, the plates measure only $12.8 \times 5.6 \times 1.9\,cm$, $L \times W \times H$). Automatic scintillation analyzers are equipped to analyze up to 12 samples in microplates simultaneously. The sample wells are organized in three formats, namely, 24-well plates (6 × 4 wells), 96-well plates (12 × 8 wells), and 384-well plates (24 × 16 wells), which accommodate the simultaneous analysis of up to 6–12 samples, respectively, by an equal number of PMTs in a TopCount microplate scintillation counter (see Section V.B for information on microplate scintillation counting). Papers by PerkinElmer (1991b), Komesli et al., (1999), and Roessler et al. (1993) provide reviews of the microplate scintillation counters available for high-throughput SPAs in general, including receptor binding assays.

c. Receptor Binding Assays

A receptor binding assay is another type of SPA, which provides information on the relative competitiveness of ligands for binding sites on receptor molecules. The technique involves mixing microsphere scintillator beads with solubilized or membrane-bound receptors such as membrane protein. The high affinity of the membrane protein for the microspheres is the basis of this assay. Once the receptor molecules are immobilized on the microspheres, a radioisotope-labeled ligand is added. The isotope-labeled ligand then binds to the receptor molecules on the microsphere beads (Fig. 11.31). This brings the isotope-labeled ligand into close proximity to the microsphere scintillator beads, which will excite the scintillator. The scintillation photons emitted from the microspheres can be assayed on the microplate scintillation counter. If, however, we then add another competing ligand or drug, which is not labeled with radioisotope, it will displace the isotope-labeled ligand on the receptor molecules according to its competitive affinity for the receptor and its concentration. The displacement of isotope-labeled ligand from the receptor into the medium solution and away from the scintillator microspheres will result in a reduction of the measured radioactivity. Different ligands over a range of ligand concentrations can be tested in this fashion against a given receptor. A Scatchard plot can then be drawn for each ligand as illustrated in Fig. 11.34 over the entire range of unlabeled competitive ligand concentrations tested where $B =$ the bound ligand radioactivity and $B_0 =$ the bound ligand radioactivity at zero dose of unlabeled competitive ligand. We can determine from the Scatchard plots which ligands or drugs tested have greater or lesser affinity for the receptor. The SPA Scatchard plots illustrated in Fig. 11.34 from the work of Banks et al., (1995) depict the competitive binding of wild-type interleukin-5 (IL-5) and three IL-5 mutants to IL-5 receptor protein, which contains a specific ligand binding α-chain and a signal-transducing β-chain. Interleukin-5 plays an important role in the pathology of inflammatory lung conditions observed in asthma. It is believed that antagonists of IL-5 binding to its receptor α-chain would have therapeutic value. The work of Banks et al. (1995) illustrated in Fig. 11.34 demonstrates that the IL-5 mutant E55A binds slightly more strongly to the receptor protein than the wild-type IL-5, whereas the IL-5 mutants K82... and H37... bind increasingly much less strongly to the receptor protein. In receptor binding assays the term IC_{50} and its value are often used to compare the competitive

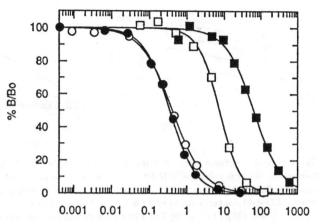

FIGURE II.34 Comparison of interleukin-5 (IL-5) and three mutants of IL-5 in a receptor binding assay. Wild-type IL-5 and three mutated forms of IL-5 were tested for receptor α-chain binding by scintillation proximity assay. The results show competition as percentage maximum binding (%B/B₀) by mutant or wild-type IL-5 for receptor binding over the IL-5 concentration range 0.00I–I000 nM. The competitors shown are (○) wild-type IL-5, (●) E55A mutant, (□) K82AK83AK84A mutant, and (■) H37AK38AH40A mutant. (From Banks et al., 1995, reprinted with permission from Elsevier Science.)

binding of antagonists for the receptor molecule. The IC_{50} is calculated as the concentration required to inhibit 50% of the labeled ligand binding to its receptor. The relative IC_{50} values for the four antagonists illustrated in Fig. 11.34 from left to right are 1.0, 0.5, 18.0, and 40.0, which are calculated by normalizing one antagonist to the value of 1.

Temperature can have a significant effect on SPAs. The work of Blair et al. (1995) demonstrates the significant differences in IC_{50} values obtained over the temperature range 15–27°C. It can be important, therefore, to control assay temperature by using a scintillation counter, which has a temperature control. Numerous examples of the application of SPA to receptor binding studies are available in the current literature, among which the following are cited: the use of ^{3}H-labeled ligands (Gevi and Domenici, 2002; Leesnitzer et al., 2002; Ye et al., 2002), ^{125}I-labeled ligands (Dayton et al., 2000; Holland et al., 1994; Mandine et al., 2001), ^{35}S-labeled ligands (DeLapp et al., 1999), and ^{33}P-labeled ligands (McDonald et al., 1999).

d. Enzyme Assays

Enzyme assays can be included among the list of applications of scintillation microspheres in SPA. Enzyme assays can be of two types: (1) the enzyme activity toward a substrate is determined by the hydrolytic action of the enzyme (e.g., protease or nuclease) on an isotope-labeled substrate attached to the SPA microspheres, resulting in the release of radioactivity from the scintillation microspheres; and (2) the enzyme activity toward a substrate is determined by the polymerizing action of the enzyme (e.g., polymerase or transferase) on a donor isotope-labeled substrate in

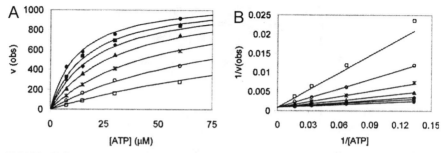

FIGURE 11.35 Observed reaction rates (ATPase activity) of HPV6 El (a El helicase from human papillomavirus) via SPA using [γ-³³P]ATP substrate. (A) The observed rates (in pM inorganic phosphate, Pᵢ, produced per s) for substrate concentrations ranging from 7.5 to 60 μM at inhibitor (ATP-γ-S) concentrations of 0 (●), 3.1 μM (■), 6.3 μM (♦), 12.5 μM (▲), 25 μM (*), 50 μM (○), and 100 μM (□). (B) The same data points plotted in double reciprocal format. (From Jeffery *et al.*, 2002, reprinted with permission from Elsevier Science.)

the free medium, which becomes attached to the SPA microspheres, resulting in an increase in radioactivity detection from the scintillation proximity beads.

An illustrative of a scintillation proximity enzyme assay is that of an ATPase assay reported by Jeffery *et al.*, (2002) illustrated previously in Fig. 11.32. In this case, the enzyme catalyzes a substrate, whereby the product of the reaction (³³P-inorganic phosphate) becomes attached to the SPA bead. As enzyme activity increases an increase in radioactivity is observed from the SPA beads. Consequently enzyme kinetics and the effects of inhibitors on enzyme activity can be easily observed without the need of tedious procedures and separation steps. For example, Fig. 11.35 illustrates the ATPase reaction rates as a function of substrate (ATP) concentration and inhibitor ATP-γ-S concentration as measured with SPA. This serves as only one example of many enzyme activity measurements that can be carried out without the tedious preliminary preparations such as high-performance liquid chromatography (HPLC), electrophoreses, or centrifugation. Numerous enzyme assays using SPA technology can be found in the current literature, among which the following are excellent examples: the use of ³H-labeled substrates (Cheung and Zhang, 2000; He *et al.*, 2002; Nare *et al.*, 1999; Skorey *et al.*, 2001; Yang *et al.*, 2002; Zhang *et al.*, 2002), ³H-, ¹⁴C-, and ³⁵S-labeled substrates (Macarrón *et al.*, 2000), and ³³P-labeled substrate (Evans *et al.*, 2002).

e. SPA in 1536-Well Format

Although fluorescence techniques have grown in importance, current estimates by various surveys of high-throughput screening laboratories indicate that radiometric assays constitute between 20 and 50% of all screens performed (Hertzberg and Pope, 2000). In the quest to further increase sample throughput in scintillation proximity analyses, advances have recently been made with respect to increasing assay miniaturization and reducing read-time. Microplates with 1536 sample wells (Greiner Labortechnik GmbH, Frickenhausen, Germany) are receiving attention for application to

SPA for higher sample throughput (Beveridge *et al.*, 2000; Hertzberg and Pope, 2000; Sorg *et al.*, 2002). The reaction volume of a sample well of the new 1536-well format microplate is only $7 \mu L$ compared to the larger sample-well volume ($50 \mu L$) of the 384-well microplate format. The sample sizes and levels of radioactivity are about 7-fold less with the higher-throughput 1536-well microplate. Because of the lower level of radioactivity, imaging platereaders have emerged such as the ViewLux™ (PerkinElmer Life and Analytical Sciences, Boston, MA), Leedseeker™ (Amersham Pharmacia, Piscataway, NJ) and CLIPR™ (Molecular Devices, Sunnyvale, CA). The platereader uses high sensitivity charge-coupled device (CCD) cameras and lenses capable of providing rapid and quantitative images of the plates. The platereaders provide sensitivity and speed, as they can rapidly quantify the light intensity from each well with low background and low variability (Hertzberg and Pope, 2000). In a thorough comparison of the 384- and 1536-well formats for automated SPA screening for serine kinase inhibitors Sorg *et al.* (2002) demonstrated that a throughput of 106 of the 384-well plates, corresponding to 37,312 compounds and 3392 control panels could be analyzed in one working day. However, for the 1536-well format, thirty 1536-well assay plates corresponding to 42,240 compounds and 7680 controls could be assayed in a working day.

f. Other Assays and SPA Kits

Many new applications of SPA are being found involving the measurement of various biomolecular interactions and compound screening. Also, numerous kits for specific SPAs are available commercially from Amersham Biosciences, Piscataway, NJ, or from Amersham International plc, Little Chalfont, Buckinghamshire, England. These kits include SPA beads or microspheres, buffer, isotope-labeled ligand or antigen, and instructions on use. The various kits, their applications, and the reaction mechanisms involved are described in detail in the Amersham catalog (Amersham Biosciences, 2002). An excellent review of the principles and applications of SPA is available in a paper by Cook (1996). A broad range of specific SPA applications may be found in many published reports, among which the following may be cited for further reference: Baker *et al.* (1996), Chandrakala *et al.* (2001), Delaporte *et al.* (2001), Hancock *et al.* (1995), Horton *et al.* (1995), Lerner and Carter (1995), and Liu *et al.* (2001).

A unique application of SPA has been in the field of health physics. Han *et al.* (1998) demonstrated that membranes containing cerium-activated yttrium silicate can be used for wipe tests for ^{3}H-cortisol. The wipe membranes are assayed with a liquid scintillation analyzer without scintillation cocktail, and their performance was almost as efficient as wipe tests that require scintillation cocktail. The new technique reduces the amount of radioactive waste as well as cost as no scintillation cocktail is required.

g. Color Quench Correction

The main advantages of SPA are (1) the circumvention of separation steps usually linked to immunoassays, receptor binding assays, and enzyme activity assays, among other assays carried out in the medical and biochemical sciences; (2) the reduced costs per analysis arising from a reduction in reagent volumes consumed and the waste accumulated; and (3) the consequent high-throughput analysis possible when SPA is linked to automatic liquid handling and microplate sample counting techniques. The first advantage of the SPA technique just listed can give rise to colored samples, because purification steps in the assay procedure are avoided. Color in the assayed samples can absorb the visible photons of light emitted by the microsphere scintillation beads before this light can reach the photomultipliers. Absorption of the light emission will cause a reduction of the light intensity reaching the PMTs and consequently a reduction in the pulse heights registered by the scintillation counter. A reduction in count rate is generally the result, because many reduced pulse heights fall below the detectable levels set by a lower level pulse height discriminator setting. The phenomenon of reduced pulse heights and concomitant reduction in count rate of samples due to color of the sample is known as color quenching. It is important, therefore, that sample assays be corrected for color quench. We should keep in mind that chemical quench does not occur in SPA, because this is a solid scintillation technique in which the beta particles or Auger electrons from ^{3}H and ^{125}I radionuclides, respectively, or beta particles from ^{14}C, ^{35}S, or ^{33}P, enter into the solid scintillator to cause scintillation where no chemical quenching agents can be found.

The correction of quenched sample count rates (CPM) requires a measurement of the detection efficiency of each sample via a quench correction curve of counting efficiency or relative efficiency versus a suitable quench-indicating parameter (QIP) such as that provided by an external standard (e.g., tSIE). However, fortunately the SPA technique does not require absolute activity measurements and the count rate of the sample (CPM) will suffice provided the sample count rates are corrected for color quench. In reports by Amersham Life Science (1996a,b), Hughes *et al.* (2001) for PerkinElmer Life Sciences, Neumann *et al.* (1994), and PerkinElmer (1997), it is demonstrated that the quench-corrected count rate (QC-CPM) produced by a microplate scintillation counter is simple and provides accurate results for most SPAs. For example, the PerkinElmer TopCount® and MicroBeta® microplate scintillation counters can count simultaneously up to 12 wells of a microplate and at the same time determine the color quench of each sample from the sample pulse height spectrum. The quench-indicating parameter measured is the transformed spectral index of the sample (tSIS) determined by the TopCount® or spectral quench parameter [SQP(I)] or asymmetric quench parameter [AQP(I)] measured with the MicroBeta®. A thorough treatment of quench-indicating parameters and their use are provided in Chapter 5. These QIPs are excellent for monitoring the color quench level of samples, because they are not affected by the color of the sample; that is, all color quench correction curves will be identical regardless of the color of the sample. A quench correction curve can be prepared using standard (e.g., ^{3}H-PVT, ^{125}I-PVT or other isotope) SPA beads according to the procedure described in detail in the references cited earlier in this paragraph. Of course, the standard

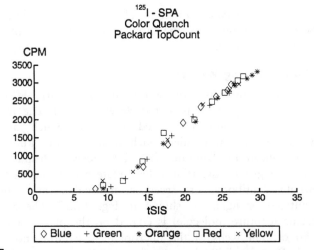

FIGURE 11.36 Count rates versus the quench-indicating parameter tSIS observed with ^{125}I-labeled SPA beads on TopCount microplate scintillation counter with various concentrations of five different colors of dye. (From Neumann et al., 1994, courtesy of PerkinElmer Life and Analytical Sciences.). Similar curves are obtained with PerkinElmer MicroBeta microplate scintillation counter using the quench-indicating parameter [SQP(I)] or [AQP(I)] (See Hughes et al., 2001.)

isotope-labeled beads must contain the same radioisotope used for the scintillation proximity assays. Figure 11.36 illustrates the constant relationship between count rate and tSIS one can expect regardless of sample color. The count rate (CPM) on the ordinate of this curve can be converted to read percent relative efficiency by placing the highest count rate (corresponding to the highest tSIS value) at 100% with all other count rates converted to a percentage relative to the highest count rate. In practice, a researcher will determine the average CPM of an unquenched radionuclide-bound microsphere scintillator sample (colorless sample) and then input this value into a counting protocol of the instrument to prepare a quench correction curve. Several quenched standards are prepared by adding the same amount of radionuclide-bound microspheres to sample wells including increasing amounts of color-quenching agent. The instrument will then count each sample (unquenched and quenched) and plot a quench curve of percent counting efficiency versus the quench-indicating parameter [e.g., tSIS, SQP(I), or AQP(I)]. Once the curve is prepared, it is stored in the memory of the microplate scintillation analyzer protocol so that each experimental sample that is counted can be corrected for color quench according to the quench-indicating parameter (QIP) of the sample. For example, if an experimental sample gave a QIP value corresponding to a relative efficiency of 25% from a standard curve, and the count rate of the sample $(CPM)_s$ was 5250, the instrument would calculate quench corrected count rate (QC-CPM) as

$$QC\text{-}CPM = (CPM)_s/E = 5250/0.25 = 21,000 \qquad (11.33)$$

Hughes *et al.* (2001) and Neumann *et al.* (1994) for PerkinElmer Life and Analytical Sciences demonstrate the application of this color quench correction technique to SPAs.

h. SPA with Scintillating Microplates

Microsphere scintillators are not the only means of carrying out SPA. It is also possible to use specially prepared 96-well polystyrene microplates, which contain plastic scintillator-coated wells to which are bonded the antibody or receptor molecules. Such a microplate, known as the FlashPlate, is produced by DuPont NEN Life Science Products, Boston, Massachusetts. A diagram illustrating a FlashPlate microplate well, a bound capture molecule (antibody or receptor molecule), and the bound radiolabeled antigen or ligand is shown in Fig. 11.37. The antibody or receptor molecule (capture molecule) is coated on the plastic scintillator in each of the 96 or 384 wells of the microplate. Therefore, no scintillator beads are used. Radionuclide-labeled antigen or ligand molecule is added and allowed to incubate with the capture molecule. Only radionuclide label bound to the receptor molecule is in sufficiently close proximity to the plastic scintillator to produce a scintillation effect in the scintillator and the concomitant emission of light. Unbound radionuclide label will be too far away from the capture molecule and scintillator to produce any scintillation effect. The addition of unlabeled antigen or an unlabeled competitive binding ligand will reduce the number of radionuclide-labeled compounds on the capture molecules and thereby reduce the count rates detected by the microplate scintillation analyzer. The polystyrene-based scintillator on a microplate format provides the platform for nonseparation assays using

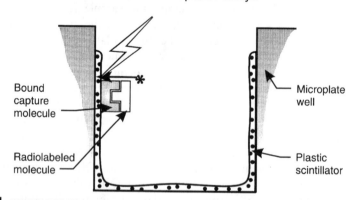

FIGURE 11.37 Diagram illustrating a magnified sample well of a FlashPlate plastic micro-plate with coated plastic scintillator. A magnified capture molecule (antibody or receptor molecule) coated on the plastic scintillator is illustrated bonded to a radioisotope-labeled antigen or ligand molecule. Radioactivity from the radiolabeled molecule can penetrate the plastic scintillator and produce scintillation. (Courtesy of DuPont NEN Life Science Products, Inc.)

low- and moderate-energy beta particle-emitting nuclides, such as ^{3}H ($E_{max} = 18.6$ keV), ^{14}C ($E_{max} = 155$ keV), ^{35}S ($E_{max} = 167$ keV), and the atomic-electron emitter ^{125}I ($E_{max} = 30$ keV).

FlashPlates are available in 96-well microplate format for specific assays, including immunoassays, receptor binding assays (Allan *et al.*, 1999; Bosse *et al.*, 1998; Coupert *et al.*, 1998; Komesli *et al.*, 1999; Watson and Selkirk, 1998), enzyme activity assays (Earnshaw and Pope, 2001; Turlais *et al.*, 2001), among other specific assays (Kariv *et al.*, 1999) used in molecular biology studies. The FlashPlate was first introduced for use in the 96-well format; however, it has been expanded to the 384-well format to permit increased high-throughput screening capabilities (Sissors and Casto, 1998).

VI. DETECTION OF NEUTRONS

The neutron-induced nuclear reactions that occur in inorganic scintillator detectors as well as in scintillators of other types (e.g., plastic or glass media) depend on the energy of the neutron. See Sections VII and VIII for treatment of neutron detection in plastic and glass media. Fast neutrons are detected in plastic scintillators via neutron and proton recoil reactions or via converter material containing ^{232}Th or ^{238}U (Yamane *et al.*, 1998), which undergo fast-neutron induced fission yielding fission products detectable with inorganic scintillators. Thermal neutrons, on the other hand, are detected on the basis of the strong neutron reaction capture cross sections with particular elements and nuclides such as ^{155}Gd, ^{157}Gd, ^{6}Li, ^{10}B, ^{235}U, and ^{239}Pu (Knitel, 1998; van Eijk, 1994; Yamane *et al.*, 1998). See Table 1.6, Chapter 1 for values of thermal neutron capture cross sections. The element Gd and nuclide ^{6}Li are used in some inorganic crystal scintillator detectors for the measurement of neutron radiation, whereas ^{10}B is used in some organic plastic scintillator neutron detectors, discussed further on in this chapter. Some specific examples of thermal neutron interactions with inorganic crystal detectors are described subsequently. A review of advances in neutron detection is provided by Peurrung (2000) and a thorough text on thermal neutron detection in inorganic scintillators and storage phosphors is given by Knitel (1998).

A. Gadolinium Orthosilicate, Gd$_2$SiO$_5$:Ce (GSO:Ce) Scintillator

A relatively new scintillation crystal, which has received attention in this respect, is gadolinium orthosilicate, Gd$_2$SiO$_5$, abbreviated GSO, or as the cerium-activated form (GSO:Ce) discussed previously in this chapter (Table 11.1). The properties of this crystal as a thermal neutron detector have been researched in detail by Reeder (1994a,b). This detector is particularly applicable to thermal neutron detection, because (1) the cross section of Gd for the capture of thermal neutrons is enormous (see Table 1.6 of Chapter 1) and (2) thick GSO crystals are not required as thermal neutrons will penetrate only the first 11 μm from the surface of the GSO crystal, and (3) as noted by Knitel (1998) the thermal neutron capture reactions of gadolinium yield high probabilities of internal conversion. The internal conversion electrons have short ranges in the scintillator yielding fluorescence close to the point of neutron interaction. The gamma radiation produced has a higher probability of escape from the scintillation detector. Knitel (1998) provides an excellent example of the relevance of high probability for internal conversion. He provides the example of ^{113}Cd, which has a very high thermal neutron capture, but a relatively low probability of internal conversion, only 4% that of ^{155}Gd and ^{157}Gd.

As explained by Reeder (1994b), about 80% of the neutron captures in Gd occur in ^{157}Gd and 18% of the neutron captures occur in ^{155}Gd, both natural isotopes of Gd according to the following:

$$^{155}_{64}\text{Gd} + {}^{1}_{0}\text{n} \longrightarrow {}^{156}_{64}\text{Gd}^* \longrightarrow \gamma \text{ rays} + \text{x-rays} + \text{conv. electrons} \qquad (11.34)$$

$$^{157}_{64}\text{Gd} + {}^{1}_{0}\text{n} \longrightarrow {}^{158}_{64}\text{Gd}^* \longrightarrow \gamma \text{ rays} + \text{x-rays} + \text{conv. electrons} \qquad (11.35)$$

The thermal neutron capture cross section of ^{157}Gd is an enormous 2.54×10^5 barns, and that of ^{155}Gd is a very large 6.1×10^4 barns (Table 1.6 of Chapter 1). The neutron capture product nuclides ^{158}Gd and ^{156}Gd are themselves stable nuclides; however, the binding energies of the neutron capture processes leave the nuclides in an excited state. Deexcitation of ^{156}Gd and ^{158}Gd results in a release of a cascade of gamma rays (up to 2 MeV), x-rays, and conversion electrons (Tokanai *et al.*, 2000). The 29-keV conversion electrons, because of their charge, have a range of only about 2 μm in the GSO and these, therefore, do not escape the crystal and produce excitation and luminescence. The 44-keV x-rays produced have a range of about 75 μm in GSO. Reeder (1994b) explains that about half of these escape from the GSO crystal without detection, because the thermal neutron capture occurs near the crystal surface. The pulse height spectra from the crystal luminescence show two peaks at 35 and 80 keV. The lower energy peak corresponds to excitation resulting from absorption of a conversion electron or x-ray as products of the neutron capture processes, and the higher energy peak corresponds to the simultaneous detection of a conversion electron and x-ray. The small size requirement of this detector for thermal neutrons is an advantage in applications in which detector size is a limiting factor. In addition, GSO is reported to have a light output of about 20% of that of NaI(Tl) and a short decay time constant of 60 ns (Table 11.1) suitable for measuring arrival times of neutrons from a pulsed source at high count rates (Ishibashi, *et al.*, 1998; Takani *et al.*, 2000).

For the detection of 6 Å cold neutrons Takanai *et al.* (2000) characterized a 0.5 mm-thick GSO:Ce activated with a Ce concentration on 0.5 mol% coupled to a photomultiplier tube. They obtained a pulse height spectra with clear 31 and 81 keV peaks corresponding to conversion electron or x-ray singles peaks and their coincidence doubles peaks, respectively. A position resolution of 1.3 mm FWHM of a two-dimensional neutron image was measured. Takanai *et al.* (2000) concluded that with a GSO scintillator as thin as 20 μm, more than 90% of 80 keV gamma rays and 44 keV x-rays, induced by neutron absorption by Gd, would escape the GSO scintillator, and the main component of the pulse height spectrum would be the 31 keV x-ray conversion electron peak. This is ideal for the reduction of background from gamma- and x-radiation and, in addition, the spatial resolution would be improved because the range of 31 keV conversion electrons in GSO is only $2\,\mu m$. The neutron position resolution in such a detector system would be limited only by the light imaging system.

B. LiBaF$_3$:Ce Scintillator

The scintillator LiBaF$_3$ doped with small amounts of Ce is a relatively new scintillator. Since it was demonstrated by the research group at the Technical University of Delft (Knitel *et al.*, 1996; Combes *et al.*, 1997) that LiBaF$_3$ scintillator could be used to discriminate against neutron and gamma radiation much attention has been drawn toward the unique characteristics of this detector and its applications (Baldochi *et al.*, 1999; Combes *et al.*, 1998; Gektin *et al.*, 1998; Knitel, 1998; Reeder and Bowyer, 2000a,b,c, 2001a,b, 2002). Cerium doping at a low concentration of about 0.005 mol % plays an essential role with this detector in pulse shape analysis for neutron/gamma discrimination (Reeder and Bowyer, 2002).

The detection of thermal neutrons in scintillators containing Li is based on the high thermal neutron capture cross section of ^{6}Li, equivalent to 940 barns (see Table 11.1, Chapter 1), which favors the reaction

$$^6_3\text{Li} + ^1_0\text{n} \rightarrow {}^3_1\text{H} + {}^4_2\text{He} + 4.8\,\text{MeV} \tag{11.36}$$

The tritium and alpha particle share the 4.8-MeV kinetic energy liberated, which produces the scintillation in the Li-containing scintillators either enriched with the isotope ^{6}Li or in its natural abundance (7.5% ^{6}Li). Combes *et al.* (1998) calculated that a LiBaF$_3$:Ce,Rb (cerium and rubidium co-coped) 96% ^{6}Li-enriched crystal of 0.235 cm thickness would have a thermal neutron absorption probability of 0.96 compared to that of a crystal at ^{6}Li natural abundance, which would display a thermal neutron absorption probability of 0.23.

On the other hand, fast neutron reactions in LiBaF$_3$ scintillator are several, and these are classified into elastic scattering [^{6}Li(n,n)^{6}Li or ^{7}Li(n,n)^{7}Li] or n-capture reactions including ^{6}Li(n,t)^{4}He, ^{6}Li(n,nd)^{4}He, ^{7}Li(n,nt)^{4}He, ^{19}F(n,p)^{19}O, ^{19}F(n,α)^{16}N, and ^{19}F(n,nα)^{15}N. The degree of occurrence of these fast neutron reactions in LiBaF$_3$ scintillator will depend on the cross sections as a function of incident neutron energy and the relative isotopic abundance of ^{6}Li, ^{7}Li, and ^{19}F in the scintillator (Reeder and Bowyer, 2002).

What has drawn much attention to LiBaF$_3$:Ce scintillator is its ability to distinguish heavy charged particles (p, d, t, or α), which are products of neutron reactions from electrons, which are products of gamma-ray interactions. The n/γ discrimination by this scintillator is based on the presence or absence of a fast component in the scintillation light output. Under gamma irradiation $LiBaF_3:Ce$ yields a very fast ($\sim 1\,ns$) light output due to core-valence (CV) transitions or luminescence followed by an additional light output with a short lifetime of about 35 ns arising from Ce^{3+} excitation. Both of these light outputs arising from gamma interactions are considered fast, and both of these fast components are absent under neutron irradiation. Both neutron and gamma irradiation produce light output with a relatively long component with a lifetime of about 2.5 μs due to self-trapped-excitation (STE) luminescence. The n/γ radiation is discriminated, therefore, on the basis of the presence or absence of the short (ns) lifetime components in the scintillation pulse events.

The data acquisition system used by Reeder and Bowyer (2001b, 2002) consisted of a multi-input charge-to-digital converter (QDC). The scintillator is coupled to a photomultiplier tube (PMT), and the signal from the PMT is set to two separate inputs to the QDC. Each input is gated to record the amount of charge in two portions of the pulse. A "short" gate of about 30 ns records the fast components due to the CV and Ce^{3+} fluorescence arising from gamma interactions, and the "total" gate of about 1.6 μs records both the short and long components due to STE excitations as a consequence of both gamma and neutron interactions. A two-dimensional array of the "short" signal versus the "total" signal is illustrated in Fig. 11.38 illustrates the discrimination of n/γ events. Discrimination between fast and thermal neutrons pulse events are also possible (Reeder and Bowyer, 2002).

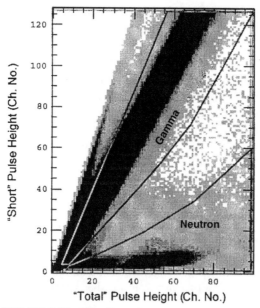

FIGURE 11.38 Two-dimensional array of pulse heights from "short" to pulse heights from "total" gate for ^{252}Cf at 11.55 cm from crystal. Data were obtained with a Cd shield around the scintillator and a 1.27-cm thick Pb absorber between the source and scintillator. Outlined regions define events attributed to neutrons or to gammas. (From Reeder et al., 2002, reprinted with permission from Elsevier Science.)

Reeder and Bowyer (2002) underscore the potential of LiBaF$_3$:Ce scintillator applications in the fields of nuclear arms control, accountability for nuclear safeguards, and smuggling scenarios. They call attention particularly to the following applications: (1) characterization of neutron sources according to fast to thermal neutron ratios often needed for neutron activation analysis and other analytical applications, (2) fast neutron spectroscopy, (3) identification of (α,n) and spontaneous neutron sources, (4) measurement of neutron/gamma fingerprints permitting the identification of various radiation sources, and (5) applications in neutron time-of-flight experiments.

C. Ce^{3+}-Activated Borates

Inorganic scintillators containing boron have potential as thermal neutron detectors due to the high likelihood of their capture by the presence of the natural isotope ^{10}B according to the reaction

$$^{10}_{5}B + ^{1}_{0}n \rightarrow ^{7}_{3}Li + ^{4}_{2}He + 2.8\,MeV \tag{11.37}$$

Boron-10 has a high natural abundance of approximately 20% and a large neutron-capture cross section (Table 11.1, Chapter 1). The reaction leaves alpha particles in the MeV energy range, which will dissipate their energy close to the point of neutron interaction.

Numerous borate scintillators have been tested for their neutron detection properties and these are reviewed thoroughly by Knitel (1998). Among the borates reviewed by Knitel, the metaborate LaB$_3$O$_6$:Ce^{3+} (0.05% Ce) shows promise with a relatively high scintillation light yield of ~8000 photons/MeV. Also the orthoborates YBO$_3$:Ce^{3+}, GdBO$_3$:Ce^{3+}, and ScBO$_3$:Ce^{3+} doped with 1% Ce display a relative high scintillation light yield of ~10,000 photons/MeV. According to the review, the orthoborate with the highest potential as a thermal neutron scintillator is probably YAl$_3$B$_4$O$_{12}$:Ce^{3+}, as it combines a relatively high scintillation light yield of ~10,000 photons/MeV, a short scintillation decay time of 30 ns, a Z$_{eff}$ of 28, and a boron weight fraction of 12%.

The orthoborate scintillator Li$_6$Gd(BO$_3$)$_3$:Ce^{3+} has received recent attention as an efficienct detector for thermal neutrons as it contains lithium (^{6}Li), gadolinium (^{155}Gd and ^{157}Gd), and boron (^{10}B), all of which have large neutron-capture cross sections. The scintillation light yield of a 1% Ce-doped crystal displays a high scintillation light yield of ~14,000 photons/ MeV (Knitel, 1998); however, Chaminade et al. (2001) report that a maximum scintillation efficiency is achieved at a 3–5% Ce concentration emitting light at 390 nm with a pulse height from neutron capture of ^{6}Li that is 6 times greater than that of Li-glass. Its' longer primary scintillation decay time of 200 ns with a smaller component at 700 ns can be a drawback Knitel (1998). An interesting property is that any combination of the neutron-capturing nuclei ^{6}Li, ^{10}B, or ^{157}Gd may be employed in the manufacturing of the crystals. For example, the ^{6}Li (n,alpha) and ^{10}B (n, alpha) reactions yield high-energy alpha particles for easily detected signals in the scintillator, while the ^{157}Gd with its enormous thermal neutron cross section (Table 11.1, Chapter 1) would provide a high neutron capture probability in thin layers of scintillator (Chaminade *et al.*, 2001).

D. Barium Fluoride (BaF$_2$) Detectors

Attention has been drawn to the inorganic scintillator barium fluoride, BaF$_2$, for the detection of high-energy neutrons. The properties of BaF$_2$ are given in Table 11.1, as it is also used for the detection of gamma rays of wide energy range.

Kryger *et al.* (1994) demonstrate the efficiency of BaF$_2$ for the detection of high-energy neutrons (10–150 MeV), while at the same time it is possible to discriminate from simultaneous gamma-ray interactions in the crystal using pulse shape discrimination techniques. Pulse shape n/γ discrimination, as discussed previously for the case of LiBaF$_3$:Ce scintillator, is based on the fact that high-energy neutron reactions in the BaF$_2$ produce the heavier charged particles, such as alpha particles, protons, and deuterons, as products of neutron reactions (see Chapter 1, Section II.G.3.d), in contrast to gamma-ray interactions, which produce only the much lighter electrons. Kryger *et al.* (1994) demonstrate that absolute neutron detection efficiencies are dependent on neutron energy, and efficiencies as high as 0.4 or 40% for neutrons of the highest energy are possible.

E. Other Scintillators

Isotopes such as ^{6}Li in LiI(Eu) and BaLiF$_3$ crystals, in plastic and glass scintillators, and even ^{6}Li mixed with ZnS(Ag) powder scintillator or ZnS(Ag) + ^{6}LiF scintillator plates are used for the detection and imaging of thermal neutrons (Combes *et al.*, 1997; Gorin *et al.*, 2002; Knitel *et al.*, 1995, 1996a,b; Nagarkar *et al.*, 2001; van Eijk, 1997). The scintillation properties of the silicate crystal LiYSiO$_4$:Ce^{3+} was reviewed by Knitel (1998) as a promising fast thermal neutron scintillator with a light yield of 2000 photons/Mev of absorbed alpha particle equivalent to $\sim$10^4 photons per absorbed neutron considering the ^{6}Li reaction (Eq. 11.36). The gamma-ray induced scintillation in this crystal has a short lifetime of 28 ns, and the alpha particle-induced scintillation has both a fast component of several tens of nanosecond followed by a slow component of tens of microseconds (Knitel, 1998). Neutron/gamma-ray discrimination has not been demonstrated yet with this detector. Numerous examples of ^{6}Li-containing scintillators at natural or enriched abundances are cited in various parts of this chapter.

Another isotope, not yet mentioned in this chapter, with a large thermal neutron-capture cross section, is ^{3}He (see Table 11.1, Chapter 1). The case of ^{3}He is a special example, because it can be incorporated only in gaseous form in scintillator detectors (see Section VII.B.2.g of this chapter).

VII. SCINTILLATION IN PLASTIC MEDIA

Plastic scintillators may be considered as solid solutions of one or more organic scintillation fluors (fluorescent dopants) in translucent plastic. The solid plastic medium plays a role similar to that of the liquid solvent previously discussed in liquid scintillation counting. Many types of plastic scintillators are in use today mainly for the measurement of low penetrating radiation, such as particulate radiation, x-rays, and low-energy gamma rays. Plastic scintillators have also become popular for the measurement of neutrons via neutron-induced nuclear reactions in the plastic media. Two main types of plastic scintillator designs exist: (1) the integral plastic scintillators, which are constructed of one whole piece of plastic polymer containing scintillator varying in size from a few millimeters to a meter or more in diameter, and (2) fiber-optic arrays of plastic scintillator consisting of as many as several thousand plastic scintillating fibers each $25-60\,\mu m$ in diameter as individual light pipes bundled together into only a few millimeters diameter. In this section some of these detectors and their applications are discussed.

A. The Scintillation Process in Plastic

Plastic scintillators are classified into binary, ternary, or even higher order scintillator solutions. A binary scintillator consists of two molecular species, namely the solvent X (plastic or organic liquid) and the solute Y (organic fluor or dopant). The solvent (plastic) medium, which is the bulk material, absorbs the major part of the energy dissipated by incident radiation (see Brookes, 1979). This absorbed energy, that is, excitation energy of the solvent molecules, is then transferred to the solute molecules, which is represented as follows:

$$X^* + Y \rightarrow X + Y^* \tag{11.38}$$

where X^* and Y^* represent excited states of solvent X and solute Y, respectively. Decay of Y^* from its excited state to the ground state results in fluorescence or the emission of visible light.

The addition of a wavelength shifter or secondary solute, Z, to the medium constitutes a ternary system. In such a system the energy transfer process is extended to a third molecular species, namely the secondary solute. In other words, the excited primary solute Y^* transfers its energy of excitation both radiatively and nonradiatively to the secondary solute, Z, as follows:

$$Y^* + Z \rightarrow Y + Z^* \tag{11.39}$$

Decay of Z^* occurs with the emission of visible radiation of longer wavelength than that emitted by Y^*. Secondary solutes are used primarily to provide fluorescent radiation with a maximum emission wavelength that may be better matched to the sensitivity of the photomultiplier tube or other photomultiplying device used to convert the visible radiation of scintillation to a current pulse. A unitary scintillator is one that consists of a single pure molecular species such as the organic crystalline phosphor (e.g., anthracene). Unitary scintillators are not discussed here because plastic scintillators consist of only the binary or higher order scintillators.

Polystyrene-based plastic scintillators have become popular, as polystyrene (PS) is easily drawn into fibers used in scintillator fiber-optic arrays treated further on in this chapter. In a study of polystyrene-based scintillators, Bross (1990) reviews the scintillation process and its efficiency. As explained by Bross (1990), solutions of polystyrene with an efficient fluorescent dopant can give a 100% quantum yield, which is defined as the probability that an excited state in the polymer leads to an emission of a photon by the dopant (fluor). The primary dopant referred to as Y in Eqs. 11.38 and 11.39, which couples directly to the primary scintillation of the solvent, is of higher concentration ($\cong 1\%$). The secondary dopant (Z of Eq. 11.39), which is added to the polystyrene polymer in low concentration ($\cong 0.01\%$), shifts the mean fluorescent wavelength of the primary dopant from the deep blue 350–390 nm further into the red.

Bross (1990) underscores the ideal conditions for fluorescence in polystyrene scintillators. These are (1) a wavelength shifter with emission and absorption spectra that do not overlap, such as exhibited by compounds that undergo intramolecular proton transfer (IPT compounds) upon excitation; (2) a primary dopant, with an absorption spectrum that overlaps the fluorescence emission distribution of polystyrene; and (3) a fluorescence spectrum of the primary dopant with emission peaks longer than 500 nm, which obviates the need for a secondary dopant. The primary dopants 3-hydroxyflavone (3HF), 2-(2-hydroxyphenyl)-benzothiazole (HBT), p-ter-phenyl (PTP), and 1-phenyl-3-mesityl-2-pyrazoline (PMP), among others, have been tested as a new generation of favorable dopants in polystyrene (Zorn et al., 1988; Destruel et al., 1989; Bross, 1990). A review of the development, composition, and applications of older generations of plastic scintillators is given in the previous text (L'Annunziata, 1987).

B. Integral Scintillators

Integral plastic scintillators are classified here as those that constitute one entire piece of plastic polymer containing scintillator. These are fabricated in a wide range of sizes and shapes from microspheres 1–10 μm in diameter to large sheets up to over 3 m long and cylinders over 1 m in diameter. Even meltable plastic scintillator is available in the form of a flexible sheet, which is placed on top of 24- or 96-well format microplates for the automatic counting of beta particle- or Auger electron-emitting radionuclides (PerkinElmer Life and Analytical Sciences, Boston, MA) as discussed further on in this section. Almost any shape of plastic scintillator polymer can be fabricated to accommodate the requirements of any particular instrument (Thomson, 1993). This versatility in detector design opens the door to a wide spectrum of applications; and it is, therefore, one of the major advantages of plastic scintillators.

I. Composition

The composition of plastic scintillators have varied considerably over the years as the development of more efficient plastic scintillators has continued. Some of the plastic scintillators that may be encountered in their chronological order of development are as follows:

1. $10\,g\,L^{-1}$ of PTP (solute) in polystyrene or polyvinyltoluene (solvent) as a binary system, with the addition of 0.01–$0.1\,g\,L^{-1}$ POPOP as secondary solute to constitute a ternary system (see Brooks, 1979).

2. 0.5–3.0% concentrations of butyl-BPD or PTP as the primary solute plus 0.02–0.03% POPOP, BBOT, or bis-MSB as the secondary solute (i.e., wavelength shifter) dissolved in cross-linked polystyrene or a copolymer of methyl methacrylate and styrene (see Inagaki and Takashima, 1982).

3. 3% naphthalene plus 1% butyl-PBD as activators (joint primary solutes) plus 0.01–0.001% bis-MSB wavelength shifter in Plexiglas GS 218 (trade product of Rohm GmbH, Darmstadt, FRG) polymerizing base material (see Klawonn et al., 1982).

4. 1% PPO plus 10–12% naphthalene in polymethylmethacrylate (PMMA) or 1% butyl-PBD in polystyrene (see Kononenko et al., 1983).

5. Acrylic scintillator containing 12% naphthalene, 1% PPO, and 0.1% POPOP manufactured by Polytech Inc., Owensville, Missouri (see Fields and Jankowski, 1983).

6. Commercial plastic scintillators such as NE 102A, NE 104, NE 104B, NE 110, NE 114, Pilot U, and Pilot 425 (manufactured by Nuclear Enterprises Ltd. Sighthill, Edinburgh, Scotland) and BC404, BC408, BC412, BC418, BC454, and BCF-10 (manufactured by Bicron Corporation, Newbury, OH).

7. 3-hydroxyflavone (3HF) or PTP in polystyrene or polyvinyltoluene plastic (see Zorn et al., 1988).

8. $0.05\,mol\,L^{-1}$ 1-phenyl-3-mesityl-2-pyrazoline (PMP) in polystyrene or polyvinyltoluene (see Destruel et al., 1989).

9. 2-(2-hydroxyphenyl)-benzothiazole (HBT) in polystyrene (see Bross, 1990).

10. Polystyrene doped with butyl-PBD and POPOP (see Kulkarni et al., 1993; Hudson et al., 1994).

11. Dow 663 polystyrene doped with 1–1.5% PPO and 0.01–0.03% POPOP (see Pla-Dalmau et al., 2001).

In some of the plastic scintillators just described, high concentrations (10–15%) of naphthalene are used. Under such circumstances the naphthalene is added to act as a secondary solvent. The role of the secondary solvent is to facilitate solvent-to-solute energy transfer by acting as an intermediary in the energy transfer from the primary solvent (plastic) to the primary solute (primary fluor). As explained by Brooks (1979), the secondary solvent effect is particularly important for the transfer of energy between π-electron triplet energy states of solvent and solute and for applications such as pulse height discrimination that depend on the associated delayed scintillation component. However, in general, secondary solvents and secondary solutes degrade the timing resolution of a scintillation fluor system.

2. Radiation Detection

Applications of plastic scintillators are found where other detector types (e.g., gas, liquid, semiconductor, and solid crystal scintillators) often cannot be employed. See also Section V.B.2 of this chapter for a treatment of the applications of plastic scintillators in scintillation proximity assay. Some examples of the applications of integral plastic scintillators are the following:

a. Beta Probes and Gauges

Lerch *et al.* (1982) developed a beta probe that utilized a 10-mm-diameter and 3-mm-thick Nuclear Enterprise plastic scintillator detector. This detector was used to measure regional myocardial kinetics of positron-emitting radionuclide tracers *in vivo* without interference from radioactivity outside the region of interest. The positron-emitting radionuclide tracers were administered to dogs as $[^{15}O]H_2O$ or $[^{11}C]$palmitate. The positrons emitted by these radiotracers have a maximum range in body tissue of < 5 mm, and the gamma radiation resulting from positron annihilation is detected by the plastic scintillator with a very low efficiency accounting for $< 5\%$ of the total observed count rate. Consequently, this detector is highly specific to a selected region of interest. The plastic scintillator is optically coupled to a dynode photomultiplier tube provided with a high-voltage power supply. A thin (0.05 mm) aluminum window protects the photomultiplier from extraneous light while not attenuating the positron radiation to an appreciable extent. The pulse events are processed through an amplifier and single-channel analyzer operated in the integral mode (i.e., only one pulse height discriminator). The discriminator level was set at approximately 400 keV to exceed the Compton edge of ^{11}C gamma radiation. Although the gamma detection efficiency was low, this setting provided an even higher sensitivity of the counting system to the positive beta particles. A ratemeter monitors the output pulses from the single-channel analyzer, which are then fed to a computer to record and collect the count rate as a function of time and correct for radionuclide decay.

A plastic scintillator (NE 102) coupled to a photomultiplier tube has been used by Jaklevic *et al.* (1983) as a detector assembly for the precision beta gauge measurement of small mass deposits ($\pm 2\,\mu g\,cm^{-2}$) on thin substrates. A beta particle-thickness gauge consists of a beta particle-source and a detector that can measure the beta particle-intensity after it is transmitted through a sample (see Chapter 1, Section II.D). Low-energy beta particle-emitters such as ^{147}Pm ($E_{max} = 225$ keV) or ^{14}C ($E_{max} = 156$ keV) were used as radiation sources. Jaklevic *et al.* (1983) chose a plastic scintillator detector for this type of application because it had the following advantages over the alternative semiconductor detector: (1) the high count rates used ($> 10^5$ cps) are easily tolerated with plastic scintillators, contrary to semiconductor detectors. (2) Plastic scintillators exhibit negligible noise currents, which allow very low discriminator threshold settings and the detection of low-amplitude pulses.

b. Gas and Liquid Flow Detectors

A plastic scintillator gas chromatography detector (flow cell) was devised by Knickelbein *et al.* (1983) for the measurement of energetic beta-emitting nuclide tracers (e.g., ^{18}F, ^{11}C, and ^{31}Si) as labels on organic compounds separated by gas chromatography. The radionuclide label is detected while traveling the effluent stream of a gas chromatograph through an appropriate chamber (flow cell) manufactured with plastic scintillator. The gas flow channel was fabricated by cementing rods of highly polished NE 102 plastic ($1 \, cm^2 \times 10 \, cm$ length) to NE 102 plates ($7 \times 10 \times 1cm$). The entire flow cell assembly consisted of a 4×4 array of the four rod-plate assemblies. The flow cell is positioned between two photomultiplier tubes to enable coincidence counting and reduced backgrounds. The two walls of the flow cell not exposed to the PM tubes were made of nonscintillating lucite. The detection efficiency of this scintillator device was determined to be 80–85% for ^{18}F positrons. Detection efficiency is reduced by (1) inefficient optical coupling, (2) loss of decay events involving low-energy positrons, and (3) loss of approximately 5% of the positrons deposited in the ends of the flow cell not exposed to the two nonscintillating ends of the cell. Plastic scintillator flow cells of many sizes and shapes for gases and liquids are available commercially (PerkinElmer Life and Analytical Sciences, Boston, MA, and Nuclear Enterprises, Edinburgh). It should be noted that plastic is sensitive to certain organic solvents and oxidizing acids. When these substances must be assayed, glass scintillators must be used. Glass scintillators consist of cerium-activated lithium silicate glass, which is inert to all organic and inorganic reagents except hydrofluoric acid. Automated flow scintillation analyzers are produced by PerkinElmer Life and Analytical Sciences. A thorough treatment of flow scintillation analysis is provided in Chapter 12.

A high pressure BCF-10 plastic scintillation detector was designed by Schell *et al.* (1997, 1999a,b) for the measurement of low levels of radioactive gases in flow systems. Such a detector is useful for the measurement of radioactive gases released into the atmosphere from nuclear power plants, fuel processing facilities, nuclear weapons test sites, and hospitals that discard xenon used in diagnostic medicine.

c. Microsphere Scintillators

Even weak beta-particle emitters such as ^{3}H ($E_{max} = 18.6 \, keV$) and the Auger electron emitter ^{125}I ($E_{max} = 30 \, keV$) can be assayed in solution with plastic scintillator provided the radionuclides are in close proximity to the plastic scintillator or in direct contact with it. The pioneering work of Gruner *et al.* (1982) and Kirk and Gruner (1982) demonstrated that plastic scintillator microspheres 1 to $10 \, \mu m$ in diameter, encapsulated in gel permeable to diffusible substances, could be used in aqueous solutions to monitor concentrations of ^{3}H-labeled solute. Since the maximum range of tritium beta particles is only a few μm in materials with a density equivalent to that of water, a gel layer of only a few μm thickness would suffice to shield the plastic scintillator microspheres from external beta particle-radiation. However, tritium label that is in solution and free to diffuse into the gel could excite the scintillator, while bound or insoluble label was excluded. The light output, therefore, from a medium containing the gel-coated scintillator beads and an aqueous solution of tritium-labeled substrate or molecule would provide a measure of the solution concentration of the tritium label. If the medium also contained a macrophase impermeable to the gel (e.g., microorganisms, cells, vesicles, or macromolecules) that may absorb or bind with the solute, the light output from the scintillator could serve as a measure of the absorption or release of tritium label by the macrophase. Plastic scintillator microspheres were also used by Hart and Greenwald (1979) for the immunoassay of albumen at concentrations <1 ppb. This pioneering work has led to the development of scintillation proximity assay (SPA), which is currently a very popular analytical technique in the biological and biochemical sciences. Scintillation proximity assay is discussed in detail in Section V.B of this chapter.

d. Meltable Wax Scintillators

Although wax scintillators cannot be classified among plastic scintillators, they are included in this section, because meltable wax and meltable plastic techniques are similar and the application of meltable wax has led to the development of the use of meltable plastic in the automatic solid scintillation analysis of multiple samples in microplate formats. Fujii and Takiue (1989) and Fujii and Roessler (1991) reported the use of meltable paraffin scintillator for the scintillation analysis of ^{3}H and ^{14}C on solid support material, such as glass fiber filters and membrane filters.

The procedure entails the application of 0.3 mL of melted paraffin scintillator at 40°C to the radionuclide sample on the solid support. Upon cooling to room temperature, the melted scintillator solidifies, and the sample is placed into a plastic vial and counted with a standard LSA without any liquid fluor cocktail. The meltable scintillator formulation consists of 10 g of PPO, 1.0 g of bis-MSB, 670 mL of paraffin, and 330 mL of p-xylene. Depending on the solid support used, the counting efficiencies for ^{3}H and ^{14}C varied from 0 to 30% and 70 to 87%, respectively. Reproducible results (constant counting efficiencies) are obtained for a given type of solid support. Chemical quenching and color quenching are negligible in most circumstances. This technique has three main advantages: (1) the use of large volumes of liquid scintillation fluor cocktails is eliminated; (2) radioactive waste disposal costs are highly reduced, because of the small volumes of paraffin scintillator used; and (3) sample preparation and measurement are simplified. This work was extended by Takiue et al. (1995) and Fujii et al. (1996) to the continuous counting of samples from effluents dried on a solid support. The counting efficiencies reported for ^{3}H, ^{14}C, ^{32}P, ^{45}Ca, and ^{90}Sr(^{90}Y) by this solid scintillation method were 16, 85, 97, 95, and 95%, respectively.

The application of meltable wax to the analysis of ^{3}H in 96-well microplate formats was tested and reported by Hinrichs and Ueffing (1995).

They used a FlexiScint scintillation wax and a ViewPlate 96-well microplate available from PerkinElmer Life and Analytical Sciences, Boston, MA, for the containment and counting of samples. The ViewPlate measures approximately $12.8 \times 1.9 \times 8.6$ cm in length, height, and width. The 96 wells in the plate are arranged 12 wells along the length and 8 wells along the width of the plate. The volume of each well may be 75 or $350\,\mu$L for shallow- and deep-well plates, respectively. Hinrichs and Ueffing (1995) used the ViewPlate, because it is constructed with a clear bottom to enable the microscopic visualization of cell growth in each well. [^{3}H]thymidine incorporated into the DNA of living cells was analyzed in each well by treating with 10% trichloroacetic acid (TCA) to fix the cells and precipitate DNA followed by washing with water. To the precipitated cellular material in each well was pipetted $25\,\mu$L of FlexiScint scintillation wax after melting at 90°C in a water bath. The wells were then counted in a microplate scintillation counter (PerkinElmer TopCount) capable of automatically counting up to 12 wells of the microplate at one time. For more information on microplate counting for automatic scintillation analysis, see Section V.B of this chapter. The use of solid scintillation wax provided reproducible results comparable to those obtained with LSA. This microplate analysis technique with solid scintillator material provides several advantages over LSA: (1) fewer pipetting steps with radioactive material are required, (2) smaller volumes of reagents are used with spill-free handling of the scintillation-radionuclide mixtures, and (3) the smaller quantities of solid radioactive waste are safer to handle and the waste disposal costs are lower. The FlexiScint meltable scintillator is now manufactured as a meltable plastic described subsequently.

e. Meltable Plastic Scintillator

FlexiScint is now available as a meltable plastic scintillator sheet, which facilitates the addition of an equal amount of solid scintillator to each well of a 24- or 96-well microplate. Radionuclide samples are first harvested onto a suitable filter medium, such as glass fiber filter, membrane filter, nylon, nitrocellulose, or ion exchange paper. Harvesting is best done with a harvester that can handle 24 or 96 samples simultaneously in the microplate format, such as the FilterMate harvester (PerkinElmer Life and Analytical Sciences). The harvested samples are dried, and a precut sheet of FlexiScint plastic scintillator is placed on top of the filter. The plastic scintillator on top of the filter is melted at 60°C for 10 min. The filter with melted scintillator is then placed into a reusable FlexiFilter assembly, which consists of a filter tray, collimator, and a carrier designed according to a 24- or 96-well microplate format. The FlexiFilter assembly is then placed into a microplate scintillation counter capable of counting as many as 12 filtered samples simultaneously. This technique can be used with many filter or membrane applications, such as cell proliferation, receptor binding, dot blots, reverse transcriptase, and kinase activity studies. This solid scintillation counting technique has been demonstrated to provide results comparable to those of liquid scintillation analysis in microplate format for the measurement of ^{3}H and ^{125}I used in receptor binding assays.

Another meltable thermoplastic scintillator is Meltilex produced by
PerkinElmer Life and Analytical Sciences. It can be applied to a wide range
of assays such as the receptor binding assays described previously or the
counting of fine powders (e.g., TLC scrapings, ashes, and precipitates) on
solid support in a counting vial with a conventional LSA. The plastic can be
molded and cut to size according to the needs of a particular application. A
review of Meltilex applications is given by Suontausta et al. (1993) and
Potter (1993).

f. X- and Gamma-Radiation Detectors

Plastic scintillators are not generally employed for the detection of
gamma rays because of their low efficiencies of energy conversion and light
yield, that is, low energy resolution, compared with the conventional NaI(Tl)
crystal detector. However, plastic scintillators are the most appropriate
detectors for gamma rays in experiments in which high counting rates
($> 10^6$ cps), good timing properties, and large surface areas are needed (Caria
et al., 1981). As explained by Brooks (1979), the energy resolution attainable
for a particular scintillator-photomultiplier combination depends on the
following: 1. The matching of the scintillator emission spectrum with the
photomultiplier photocathode response or sensitivity. 2. The efficiency of
light transmission to the photocathode of the photomultiplier. 3. The
absolute efficiency of the scintillator. These factors have a combined effect
described in terms of the practical efficiency, ε, which is an expression of the
number of photoelectrons (photocathode electrons) produced per keV of
incident nuclear energy deposited in the scintillator crystal. Caria et al. (1981)
determined that 3.5 ± 0.25 photocathode electrons are produced on the
average with the 5.9-keV x-rays from a ^{55}Fe source with a detection efficiency
of 82%, and that 5.0 ± 0.25 photocathode electrons are produced with the
8.1-keV x-rays from a ^{65}Zn source with a detection efficiency of 94%. They
put the threshold of approximately 3–4 keV for the detection of gamma
rays with plastic scintillators and with efficiencies $> 50\%$. Typical energy
resolutions of gamma-ray spectra obtained with plastic scintillator (NE 102A)
are 15 and 23% HWHM for ^{40}K and ^{137}Cs photopeaks, respectively
(Skoldborn et al., 1972).

As already noted the detection and measurement of electromagnetic
radiation in plastic scintillators are generally limited to x-rays and low-energy
gamma radiation. For example, plastic scintillators have a unique application
in the measurement of the relatively low-energy gamma radiation from fissile
^{239}Pu for the analysis and monitoring of special nuclear materials (SNMs) for
nuclear safeguards (Fehlau, 1994; Gupta et al., 1995). In the monitoring of
weapons-grade plutonium in low-level process-generated waste packages
there is interference from gamma radiation of the ever-present ^{241}Am. Gupta
et al. (1995) report the design of a lead-shielded Chamber Gram Estimator
(CGE) for the determination of plutonium mass in waste packages. The
instrument shielding protects the operator from gamma exposure as well as
reducing the background to the counting chamber. It has a sample chamber
with dimensions of $15 \times 15 \times 18$ inches (height, width, and depth) to
accommodate waste packages. Because of the size of the chamber, plastic

scintillator has the advantage of being easily manufactured to fit the chamber dimensions. Four plastic scintillator detectors were used, each measuring $10 \times 15 \times 1.5$ inches (width, length, and thickness), positioned at the top, bottom, and left and right sides of the chamber. The plutonium-containing waste sample is, therefore, placed in the space between the four detectors. A 1/8-inch-thick copper sheet is installed inside the CGE chamber to minimize interference from the 60-keV gamma rays from ^{241}Am, as the copper shield will have less effect on the stronger gamma radiation emitted from the ^{239}Pu (379 and 129 keV). Using calibration standards of PuO_2, a standard curve of instrument response of net count rate (cps-background) versus grams of plutonium is plotted.

g. Neutron Detectors

Pure Plastic Scintillators. Plastic scintillators are used to detect fast neutrons through collisions of the neutrons with protons in the scintillator. Plastic scintillators are appropriate because of the relatively high hydrocarbon polymer proton (hydrogen) content of plastics. Inorganic scintillator detectors described earlier in this chapter contain no protons (hydrogen atoms) in their molecular structures. (See also Sections VI and VIII of this chapter.)

The collisions of fast neutrons with protons in plastic scintillators result in recoil of the proton and transfer of energy to the proton (see Fig. 1.11, Chapter 1) with subsequent conversion of the proton recoil energy to light in the crystal scintillator (Fehlau, 1994). The light is then subsequently converted to an electric pulse by an attached photomultiplier tube as illustrated in Fig. 11.39. A large-area plastic scintillation detector for high-energy fast neutrons (10–170 MeV) was designed by Karsch *et al.* (2001). The detector consists of 2-meter long blocks of Bicron BC 408 or Nuclear Enterprises NE 102A plastic scintillator with cross sections of $10 \, cm^2$. A total of 12 plastic blocks were bundled to provide a detector with a total area of $2.4 \, m^2$. Photomultiplier tubes were coupled with Lucite light guides to both ends of each of the 2 m long plastic scintillator blocks as illustrated in Fig. 11.40. The detector bars are covered with a common 4 mm thick veto paddle made of NE102A or BC408 plastic scintillator to permit discrimination of charged particles from neutral ones. According to Karsch

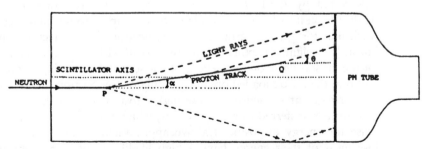

FIGURE II.39 Schematic of a neutron path in a plastic scintillation neutron detector illustrating a recoil proton track and the emitted scintillation light within the scintillator. θ represents the light emission angle and α the angle of the track. **(From De *et al.*, 1993, $^{©}$ IEEE.)**

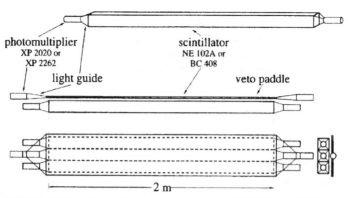

FIGURE II.40 Schematic layout of a module of COSYnus (COSY neutron spectrometer). Three bars as the one shown in the upper part of the figure and a common veto paddle from one module. (From Karsch et al., 2001, reprinted with permission from Elsevier Science.)

et al. (2001) time-of-flight (t_{of}) measurements relative to a reference signal were determined from the sum of the two time signals from the ends of a plastic bar detector according to the equation

$$t_{of} = \frac{t_L + t_R}{2} + t_0 \qquad (11.40)$$

where t_L and t_R are the time signals from the left and right ends of a scintillator bar, and t_0 is a calibration constant. The position (x) of neutron interaction and scintillation is given by the difference of the two time signals

$$x = c_{eff}\frac{t_L - t_R}{2} + x_0 \qquad (11.41)$$

where c_{eff} is the effective speed of light in the plastic scintillator and x_0 is a calibration constant.

Boron-loaded Plastic Scintillators. If fast neutrons are slowed by collisions in a moderating material, they may be rendered undetectable in a plastic scintillator. In such cases, plastic scintillators loaded with ^{10}B, which has a high thermal neutron-capture cross section, are used to detect thermal neutrons. For example, the boron-loaded plastic (organic) scintillator BC454 was tested by Byrd et al. (1992) as a fast neutron detector to analyze warheads on missiles as a possible method to apply to the counting of nuclear warheads for nuclear arms treaty verification. The detector contains 5% boron by weight. As explained by Byrd et al. (1992), a fast neutron incident on the crystal detector loses energy by a series of scatterings from the H, C, and B in the boron-loaded organic scintillator (see Section II.G.3 of Chapter 1 for a treatment of neutron scattering). Most of the incident energy of the neutron is transferred to recoil protons, which produce a detectable first light pulse for energy dispositions above approximately 0.5 MeV. Neutrons that lose most of their energy have a high likelihood of being captured by the ^{10}B(n,α)^{7}Li reaction described previously in this chapter (Eq. 11.37)

The energy dissipated in the organic scintillator produces a second scintillation event and second current pulse in the attached photomultiplier tube. After a 350-ns delay upon registration of a first pulse from proton recoil energy loss in the detector, a coincidence gate opens to accept second pulses occurring within $25.6\,\mu s$ arising from energy dissipation via the $^{10}B(n,\alpha)^{7}Li$ reaction. The two events are stored to memory for off-line spectral analysis. The detector responds to neutrons in the fission energy range 0.5–15 MeV. Neutron detection efficiencies of 5% with a ^{252}Cf fission source are reported (Byrd *et al.*, 1992).

Normand *et al.* (2002) report the low-cost synthesis of a polystyrene scintillator containing 5% by weight of boron and 1.5% by weight of PTP as the primary fluor and 0.01% POPOP as the secondary fluor. They note that boron-loaded plastic scintillators can be advantageous for neutron counting particularly in nuclear waste management. The advantages would be an increase in neutron counting efficiency and a decrease in neutron mean life inside the detector.

n/γ Discrimination with Plastic Scintillator-Lucite Detector. Watanabe *et al.* (2002) report a unique application of plastic scintillator sandwiched together with pure lucite plastic to separate high-energy neutron and gamma radiation. The detector consists of 10 layers of plastic scintillator plates alternately sandwiched between 10 Lucite plastic plates. The Lucite plastic plates are pure, that is, they do not contain scintillator and consequently act as Cherenkov radiators. The plates measure $100 \times 103 \times 3.7\,mm$ (width $\times$ length $\times$ thickness). When sandwiched together and coupled to two photomultipliers at two ends they form a n/γ discriminator, referred to as a scintillator-Lucite sandwich detector (SLSD). The principle behind the SLSD is that high-energy gamma radiation will form e^{+} and e^{-} in the detector, which will result in the emission of scintillation light in the plastic scintillator sheets and Cherenkov photons in the pure Lucite sheets; while the neutron radiation will not produce any light in the pure Lucite, but will produce scintillation in the plastic scintillator through proton and carbon recoil from n–p and n–C collisions. Effective n/γ separation is demonstrated in the energy region of 20 MeV to 12 GeV with a neutron detection efficiency of 7–10% and a γ-misclassification probability of $< 10^{-2}$.

^{3}**He in Plastic.** A unique approach to the detection of thermal neutrons in plastic scintillators is reported by Knoll *et al.* (1987, 1988). They incorporated ^{3}He gas into a plastic scintillator matrix using ^{3}He-filled spherical glass shells 50–200 μm in diameter and 0.5–3.0 μm in wall thickness. The glass spheres were prepared with ^{3}He gas pressures up to 200 atm and dispersed into the plastic scintillator matrix. Incident neutrons react with the ^{3}He according to the equation

$$^{3}_{2}He + ^{1}_{0}n \rightarrow ^{1}_{1}H + ^{3}_{1}H + 765\,keV \qquad (11.42)$$

The energy liberated in the reaction is shared by the proton (574 keV) and the triton (191 keV). These two charged particles dissipate some or all of their energy when traveling from their point of origin in the high-pressure gas through the wall of the glass sphere. Only a fraction of the energy of the reaction will therefore remain when these particles can enter the plastic scintillator. Knoll *et al.* (1987) concluded that the triton loses all of its energy within the confines of the ^{3}He gas and sphere, and only the proton has sufficient energy to escape the sphere and produce a scintillation event and output signal by an attached photomultiplier and be counted. The neutron counting efficiency for this type of detector with ^{3}He spheres at a 50% packing fraction is reported to be equivalent to that of a pure gas ^{3}He detector operated at 100 atm pressure.

C. Scintillating Fiber Detectors (SFDs)

Fiber-optic scintillator arrays are popular as detectors of nuclear radiation in several fields including medical imaging, large-area surface detectors, neutron imaging, and particle physics. These detectors are referred to as scintillating fiber detectors (SFDs). A review of scintillating fiber detectors was provided by Leutz (1995), who limits his review to plastic scintillating fibers as detectors, because plastic fibers are reported to provide better photon yields, longer light attenuation, shorter decay times, and longer nuclear reaction and radiation lengths than glass fiber. Fluor-doped plastic scintillating fibers are available from several manufacturers including Nuclear Enterprises, Edinburgh, Scotland; Bicron Corp., Newbury, Ohio; Kuraray, Tokyo, Japan; and BetaScint Inc., Kennewick, Washington.

I. Basic Principles

As reported by Leutz (1995), the plastic fibers are manufactured with a scintillating plastic core several micrometers up to a few millimeters in diameter, and the cores are surrounded by a cladding layer ($\leq 5\,\mu$m) of lower refractive index than the scintillating core (Fig. 11.41). The scintillating core is usually made of polystyrene (PS) plastic polymer and the cladding of polymethyl methacrylate (PMMA). When the cladding is of lower refractive index than the scintillating core (i.e., $n_g < n_c$) the fiber traps a fraction t of the scintillating light as illustrated in Fig. 11.41 according to the equation

$$t = 1 - (n_2 / n_1) \qquad (11.43)$$

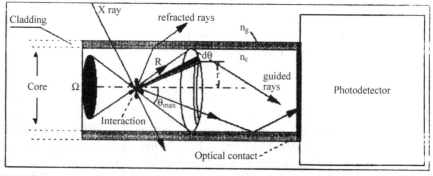

FIGURE 11.41 Schematic diagram of the optical propagation in the plastic scintillating fiber. (From Ikhlef *et al.*, 2000, reprinted with permission from Elsevier Science.)

where n_1 and n_2 are the index of refraction of the scintillating core and that of the cladding material, respectively. Any untrapped light is absorbed by an extramural absorber (EMA), if the fiber is coated with a black layer. As explained by Leutz (1995), even in the absence of an EMA the untrapped light escapes at an angle, which prevents it from being trapped again by a neighboring fiber, and it becomes lost after about 50 mm of fiber length. Therefore, the untrapped light will not contribute to appreciable crosstalk between neighboring scintillating fibers.

The fiber diameter and type of scintillator used in the plastic core material govern the spatial resolution, because proper selection of scintillating fluor in the core minimizes crosstalk between neighboring fibers. Leutz (1995) points out that the ionizing nuclear radiation will first cause excitation of the polystyrene. Since the polystyrene fluorescence yield is poor, an aromatic scintillator must be included in polystyrene core polymer. Both primary and secondary scintillators have been used, such as p-terphenyl (PTP) and POPOP, respectively; however, POPOP has an overlapping absorption and emission spectrum. A single scintillator concentration of > 0.015 molar fraction of 1-phenyl-3-mesityl-2-pyrazoline (PMP) as a one-component fluor in the polystyrene scintillating plastic core is considered optimum as it has a large Stokes shift, that is, practically no overlap of its absorption and emission spectra. Leutz (1995) indicates that crosstalk between scintillating fibers is avoided by using only one fluor component, which displays a large Stokes shift in the polystyrene core, such as PMP (Güsten and Mirsky, 1991). This is explained by the fact that light emitted by the first fluor (dopant) can escape from fibers of small diameter, excite the wavelength shifter (second fluor or dopant) in neighboring fibers, and cause undesirable crosstalk when the absorption and emission spectra of the two dopants overlap. The difference in crosstalk between fiber bundles possessing different fluor components can be appreciated Fig. 11.42, which illustrates two fiber bundles of 30-μm diameter. The fibers illustrated were excited at one end with a laser at 265 nm wavelength, and the emitted light was guided through a 150-mm bundle length and photographed with a charge-coupled device (CCD), which converts the optical image into electronic signals. The bundle in the left photograph of Fig. 11.42 illustrates crosstalk over six fiber layers, and that on the right shows no crosstalk. From the spatial resolution illustrated in Fig. 11.42 we can appreciate the potential of plastic scintillating fiber-optic arrays as imaging detectors.

2. Tomographic Imaging Detectors

The high spatial resolution achievable with plastic scintillating fiber-optic arrays has opened the door to the application of this type of detector to medical imaging. Chaney et al. (1992), Del Guerra et al. (2001), Kulkarni et al. (1993), Nelson et al. (1993), and Worstell et al. (1998) have demonstrated the application of plastic scintillating fibers to single-photon imaging and positron emission tomography (PET).

In brief, PET is a form of tomographic imaging that can provide highly resolved 3-dimensional images of body organs and a display of the dynamics of radionuclide-labeled compound metabolism in organs. A positron-emitting

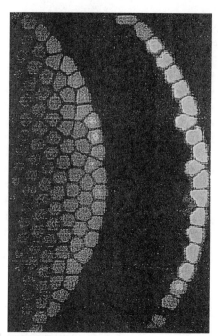

FIGURE II.42 Crosstalk of scintillating fibers (left) in a fiber bundle doped with *p*-terphe-nyl and POPOP (two components). No crosstalk (right) in a fiber bundle with PMP doping (one component), otherwise same excitation and light detection as for the picture at the left. The higher light output of the monolayer at the right also indicates that no scintillating light crosses to the neighboring fiber layer. (From Leutz, 1995, reprinted with permission from Elsevier Science.)

nuclide in the chemical form of a radiopharmaceutical, for example, is first administered to the body through intravenous injection. The positrons emitted are annihilated a few millimeters from their originating atomic nucleus with the concomitant emission of annihilation radiation, namely two 0.511-MeV gamma rays emitted in opposite directions (180° apart). Multiple radiation detectors are mounted in a circle around the body part where the labeled organ or radionuclide localization is expected. Two of the many detectors surrounding the body are activated when the two photons, originating from one positron-electron annihilation, reach detectors simultaneously 180° apart. This coincidence detection of annihilation photons accurately determines the line segment in which the radionuclide resided. Thousands of line segments are analyzed by a computer to reconstruct the distribution of the decayed radionuclides, producing a tomographic image in a cross-sectional slice of the organ where the radiopharmaceutical or radionuclides had concentrated.

For PET measurements, Kulkarni *et al.* (1993) used two cylindrical detectors of 2.8 cm radius with 1.0-mm diameter BCF-10 plastic scintillating fibers (2400 fibers bundled together per detector). The detectors were placed in optical contact with position-sensitive photomultipliers to enable the production of a photon image of the scintillating fibers. PET images from the positron emitter ^{64}Cu administered to a test rat were obtained with a spatial resolution of $\approx 2\,mm$ FWHM and an efficiency of 2.3% for the detection of the two photons in coincidence. Other tests by Nelson *et al.* (1993) for a similar PET system gave an FWHM resolution of 2.5 mm for a 0.5-mm-diameter ^{22}Na positron-emitting source in air and water. The sensitivity of the detectors was 19 cps/μCi.

Del Guerra *et al.* (2001) designed a 900 finger YAP:Ce crystal matrix consisting of a 30×30 array of single finger crystals measuring $2 \times 2 \times 30\,mm^3$. The light produced by each interaction event is transported inside one of the YAP:Ce crystal fingers toward 30 single-cladding wavelength-shifting fibers measuring $2 \times 2\,mm^2$ in diameter oriented in the X and Y directions to collect the light escaping from each crystal matrix row and column. This detector arrangement was tested with a ^{22}Na source producing 511 keV gamma rays from positron annihilation where well-resolved images from each fiber provided 2-dimensional images acquired by Hamamatsu PMTs, which also record radiation intensities.

3. Two-Dimensional Imaging

Two-dimensional images with high resolution are possible using plastic scintillating fiber arrays oriented orthogonally to each other whereby scintillation pulse events are sent along the scintillating fibers in the X- and Y-directions. A device was designed by Morimoto *et al.* (2000) for the high-resolution imaging of two-dimensional gel electrophoresis needed for the analysis of images of ^{32}P in the study of molecular biology including restricted landmark genomic scanning. The detector is designed with four sets of scintillating fiber layers, each layer consisting of 256 fibers of 0.8 mm^2 cross section made of Bicron BCF-12 plastic scintillator. Two fiber layers provide X-directional position and the other two layers give Y-directional position of radiation events. The overall area of the detector is $35 \times 43\,cm^2$. A radiation event such as a beta particle penetrates both fiber planes and produces visible photon emission in an X- and Y-fiber. The light in each fiber travels to both ends of the fiber where it is detected by Hamamatsu multi-anode photomultiplier tubes (MAPMT). The signals from the 256 scintillating fibers in each direction are decoded by two sets of 16-channel MAPMTs. A typical electrophoresis gel with genome spots labeled with ^{32}P takes about 50 h to provide images applicable to quantitative analysis.

A detector that provides 2-dimensional images of cold neutron beams required in high-energy physics was designed by Gorin *et al.* (2002) which contains a Bicron BC-704 plastic detector consisting of $^6LiF/ZnS(Ag)$ scintillator plate. The scintillator plate is sandwiched between two wavelength-shifting fiber arrays optically glued to the scintillator plate and aligned in orthogonal directions to each other as illustrated in Fig. 11.43. The x- and y-coordinate of a neutron hit in the scintillator plate is provided by the light that propagates along the fibers producing signals from one and the other fiber array, which are analyzed by a Hamamatsu MAPMT. The detection efficiency for 10 Å neutrons was reported to be 55%. The resolution obtained is 1.0 mm and 1.1 mm FWHM in the X- and Y-direction, respectively. With smaller diameter fibers (0.5 mm $\times$ 0.5 mm cross section)

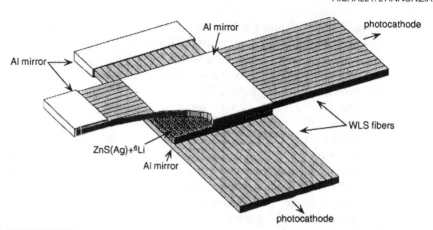

FIGURE 11.43   Schematic view of an imaging device composed of ZnS(Ag) + ^{6}LiF scintillator plate optically glued onto the wavelength-shifting fiber arrays. The scintillator plate has a size of 20 mm × 20 mm × 0.5 mm. The wavelength-shifting fiber has a cross section of 1 mm × 1 mm. (From Gorin et al., 2002, reprinted with permission from Elsevier Science.)

resolutions could be reduced to 0.5 mm FWHM and detection efficiency increased.

4. Neutron and Proton Tracking Detectors

A particle tracking detector that utilizes plastic scintillating fiber is reported by Ryan et al. (1999). The detector measures energy and direction of neutrons by detecting double neutron-proton scatters and recording images of the proton recoil ionization tracks. The tracking detector employs a closely packed bundle of 10 cm long and 250 μm square Bicron BCF-99-55 organic scintillating plastic fibers. The fibers are arranged in a block of stacked planes with the fibers in each plane orthogonal to those in the planes above and below. The alternating orientation of the scintillating fibers permits the recording and tracking of ionizing particles in three dimensions in the fiber block. The fiber scintillation tracks are detected and imaged by photomultipliers and image intensifier/CCD camera optics. The tracking of the recoil protons within the detector provides the energy and direction of incident nonrelativistic neutrons. The Bragg peak, resulting from greater ionization at the end of the proton track, provides information on proton track direction. A second proton scatter by a neutron provides the spatial information necessary to determine the incident neutron energy and direction. Fig. 11.44 shows an example of a CCD image of a double scatter event displaying two recoil proton tracks from a single 65-MeV neutron incident from the top of the figure. The greater ionization at the end of each track (Bragg effect) is seen in the figure measured as increased light intensity (darker image).

5. Avalanche Photodiodes for Scintillating Fiber Readout

Avalanche photodiodes (APDs) for scintillator readout are discussed in detail in Section III.B of this chapter. The application of the APD for the

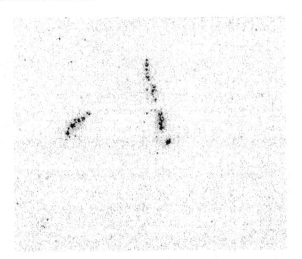

FIGURE II.44 Raw CCD image of double neutron-proton scatter from ~ 65 MeV neutron incident from above. (From Ryan *et al.*, 1999, reprinted with permission from Elsevier Science.)

readout of scintillating fiber arrays has been investigated by Bähr *et al.* (2000) and Okusawa *et al.* (2000, 2001). The advantages that APDs can offer over the conventional photomultiplier tubes are the high quantum efficiency of semiconductors, smaller dimensions, lower operating voltage, and lower sensitivity to magnetic fields often encountered in high-energy physics research. The quantum efficiencies of photomultiplier tube photocathodes are ≤ 25% which limits the detection efficiency of PMT per scintillating fiber. Bähr *et al.* (2000) compared several APDs to conventional photomultiplier tubes with bialkali photocathodes from room temperature down to −150°C and found improved efficiencies of the APDs for low light signals from blue and green scintillating fibers of 0.5 mm diameter. Okusawa *et al.* (2001) tested APD readouts for 0.5 mm diameter polystyrene-based scintillating fiber of 3 m in length containing 1.0% PTP as the primary fluor and 1500 ppm 3HF as the secondary fluor. They measured detection efficiencies of 100% from a triggering β particle from ^{90}Sr(^{90}Y) source when the APD was operated at −40°C or below. At room temperature the detection efficiency is ~50%.

6. Multilayer Scintillator Fiber Radioactivity Monitor

Plastic scintillating fiber technology has developed into large-area (up to 1800 cm^2) monitors for radioactivity contamination of soil surfaces. Research work by Schilk *et al.* (1993, 1994, 1995a,b) and Abel *et al.* (1995) has demonstrated the practical application of a multilayer plastic scintillating fiber detector capable of discriminating between beta-emitting radionuclides in the soil surface environment according to the radionuclide beta particle-emission energies and consequent depths of penetration within the plastic fiber detector. The field radiation monitors are available from BetaScint, Kennewick, Washington.

7. Directional Neutron Scintillating Fiber Detector

The unique characteristics of plastic scintillating fiber have led to the development of a directional fast neutron detector. The work reported by Hoslin *et al.* (1994) demonstrated a scintillating fiber detector (SFD) for fast neutrons that can discriminate against neutrons entering at angles nonparallel to the fiber axis. The detector consists of a fiber bundle constructed of polystyrene plastic scintillating fibers with acrylic cladding. Each fiber is 10 cm in length and 0.3 or 0.5 mm in diameter. The fiber bundle dimensions were 2.5×2.5 cm for a small SFD and 10×10 cm for a larger SFD.

The basic principle underlying this technique is that the angular efficiency of detection may be used to discriminate against background neutrons entering nonparallel to the fiber axis. When a neutron enters the detector along the axis of the fiber and scatters on collision with a proton, the recoil proton tends to stay within and deposit all of its energy in only one or two fibers leaving a bright spot of light at the fiber end, which is coupled to light-sensing devices (gamma ray-insensitive electro-optic intensifiers) coupled to a CCD. If, however, a neutron enters from the side of the bundle and scatters on collision with a proton, the recoil proton tends to travel across several fibers, leaving a track of scintillation excitation events across the fibers, whereby the light created in these fibers is greatly reduced. Therefore, an energy threshold can be set so that pulse heights produced from light intensity of a certain magnitude will discriminate between neutrons incident along the axis of the SFD or at an angle to the axis.

The directionality of this detector makes it insensitive to neutron backgrounds (i.e., neutrons not hitting the face of the fiber bundle parallel to its axis) and insensitive to gamma-radiation backgrounds. The detection efficiency for 2–3 MeV neutrons is reported to be approximately 20%. If high-gamma-radiation backgrounds are not a problem, a microchannel plate (MCP)-based photomultiplier may be used and the detector could be applied to the directional measurement of lower energy neutrons down to 0.5 MeV.

VIII. SCINTILLATING GLASS FIBER NEUTRON DETECTORS

Many new applications are being found for lithium-loaded glass scintillating fiber detectors in the measurement of neutron radiation, because of the high thermal neutron-capture cross section of ^{6}Li (see Table 1.6 of Chapter 1). This section will give a brief description of the principles of the methodology and some recent advances. More information can be obtained from the references cited.

A. Basic Principles

Either cerium-activated or terbium-activated lithium glass scintillators are used for the detection of low-energy neutrons. The mechanism is based on the high neutron-capture cross section of the stable isotope ^{6}Li for thermal neutrons according to the ^{6}Li(n,α)^{3}H reaction illustrated by Eq. 11.36 earlier in this chapter. Capture of thermal neutrons by ^{6}Li produces an alpha particle and triton with 4.8 MeV energy liberated as kinetic energy of the reaction products. The alpha particle possesses 2.1 MeV and the triton carries away 2.7 MeV energy. The charged particles travel and dissipate their energy in the activated lithium glass scintillator, and the energy of the particles is converted into flashes of light or scintillations. Light photons are emitted isotropically (in all directions), such as a light flash for each neutron-capture reaction. A photomultiplier coupled to the glass scintillator converts the individual scintillations into electrical pulses. The pulses are amplified, separated into various pulse heights with a multichannel analyzer, and counted within predefined counting regions according to those defined by pulse height discriminators as described previously in this chapter. For additional information on the principles of this solid scintillation analysis technique for neutrons see Abel *et al.* (1994), Chiles *et al.* (1990), Dalton (1987), Peurrung (2000), Seymour *et al.* (2001), and Zanella *et al.* (1995).

B. Detector Characteristics and Properties

The isotopes ^{6}Li and ^{7}Li constitute natural lithium in isotopic abundances of 7.4 and 92.6%, respectively. The lithium content and isotopic abundance of ^{6}Li (natural abundance or enriched) will affect the thermal neutron absorption efficiency in the lithium-loaded glass scintillating detectors. Detectors of this type, therefore, are available at various concentrations of lithium and different percent abundances of the ^{6}Li isotope. Also, the detectors may be fabricated as integral (one piece) glass scintillators or as optical fibers, which are becoming very popular for applications cited further on in this chapter. In addition, current ^{6}Li-loaded glass scintillating detectors are of two types, cerium activated and terbium activated. The types of glass scintillating detectors and some of their properties are outlined next.

Both cerium-activated and terbium-activated ^{6}Li-loaded glass scintillator detectors are prepared as optical fibers. The structure and function of glass fiber detectors are similar to those of the plastic fiber detectors described in Section VII.C of this chapter. The glass fiber consists of a central scintillating core surrounded by cladding material such as a hard silicone polymer. Based on the differences in refractive indexes of the glass scintillating core and the cladding, a certain percentage of the scintillation light (photons) produced from excitation in the glass will travel along the direction of the fiber and be detected by a photomultiplier device. The light in these fibers will travel along the bend of the fiber even for very extreme fiber contortions. For example, Abel *et al.* (1994) reported that light in glass scintillating fibers can travel around 3.8-cm-radius and bends without losing more than 5% of its intensity.

The diameter of the lithium silicate glass scintillating fiber can vary according to application. However, the fiber diameter may be selected in view of the ranges of the alpha particle and triton reaction products of the thermal neutron reaction with ^{6}Li (Eq. 11.36) in the glass scintillating core, among other considerations. For example, Abel *et al.* (1994) explained that the range of the 2.7-MeV triton is 40 μm in lithium silicate glass, whereas the 2.1-MeV alpha particle travels only $7\,\mu$m. Also, the triton is more effective in producing light in the scintillating glass, because the dense ionizing track left by the alpha particles undergoes a great deal of ion pair recombination. Therefore, according to recommendations of Abel *et al.* (1994), it is essential to select a fiber diameter that would be large enough to stop most of the tritons completely. The diameter of the fibers should not be excessively large, because the gamma sensitivity increases according to the diameter of the fiber. Monte Carlo calculations of Abel *et al.* (1994) showed that 66% of the neutron-capture events would produce tritons, which are completely stopped in a fiber $115\,\mu$m in diameter.

Cerium-activated glass scintillating detectors from Nuclear Enterprises, Sighthill, Edinburgh, contain about $4\,$wt% Ce_2O_3, and terbium-activated glass scintillating detectors developed by Zanella *et al.* (1995) were doped with 5 or $10\,$wt% Tb_2O_3. Both cerium-doped and terbium-doped ^{6}Li glass scintillators have been used; however, they differ in neutron detection properties (Spector *et al.*, 1993a,b; Zanella *et al.*, 1995). Cerium-doped scintillating ^{6}Li glass fibers are characterized by a fast pulse decay ($\sim$50 ns) suitable for high-count-rate applications. Precaution must be taken in their preparation to avoid or diminish formation of Ce^{4+} in the fiber drawing process, because of the high light attenuation caused by Ce^{4+}. Cerium doping in silicate glass matrices is affected when Ce^{3+} is oxidized to Ce^{4+}, which can occur during the fiber drawing process at high temperatures. The emission band of Ce^{3+} is overlapped by the absorption band of Ce^{4+}, which quenches light output and, if present in appreciable amounts, could limit the useful length of the fiber detector. Terbium-doped glass scintillators do not demonstrate any self-quenching; however, terbium-doped glass has a slower pulse decay ($\sim$1 ms), which limits their use to low-count-rate applications.

Scintillating glass fiber detectors are available from several sources, among which are: 1. Collimated Holes, Inc., Campbell, CA; 2. Nuclear Enterprises, Bankhead May, Sighthill, Edinburgh, Scotland; 3. NucSafe LLC, Clinton, TN; 4. Oxford Instruments, Inc. Oak Ridge, TN; and 5. SES Technology, Sandbank, Argyl, Scotland. Abel *et al.* (1994) and Zanella *et al.* (1995) provide some directions on the laboratory and workshop preparation of cerium-doped and terbium-doped scintillating glass fibers, respectively.

C. Applications

I. Neutron Spectrometry in n/γ and n/p Fields

Chiles *et al.* (1990) have developed a combined neutron and gamma-ray spectrometer using a cerium-activated ^{6}Li glass scintillator coupled to a Bicron BC501 liquid scintillator, which can discriminate between thermal neutrons, high-energy neutrons, and gamma radiation. The ^{6}Li glass scintillator is sensitive to the thermal neutrons and emits light with a time constant of $\sim$60 ns, while light emitted in the liquid scintillator by proton recoil from fast neutrons is emitted with a decay constant of $\sim$30 ns, and the time constant for scintillations occurring from gamma-scattered Compton electrons is only ~ 3.7 ns. The three different light decay constants make possible the electronic separation of pulses arising from the three different radiations. This type of detector was fabricated to facilitate neutron monitoring and surveillance in nuclear facilities and weapons storage, where a possibility of fission excursion exists. Spector *et al.* (1993a,b) and Jensen *et al.* (1993) report the use of Li glass scintillating fiber bundles as detectors for fast neutron spectrometry.

Astronauts living and working in orbital space are exposed to various types of radiation including neutrons, protons, gamma radiation, heavier charged particles, etc. A unique neutron spectrometer capable of measuring the neutron spectrum in mixed neutron-gamma and neutron-proton radiation fields was developed by Taniguchi *et al.* (2001). The spectrometer utilizes a pair of ^{6}Li and ^{7}Li glass scintillators in a spherical polyethylene multimoderator spectrometer similar to the Bonner sphere. A schematic view of the spectrometer is provided in Fig. 11.45. The spectrometer is designed to discriminate neutron from other particles, such as protons, which are dominate components of space radiation, by subtracting the light outputs of ^{7}Li glass scintillator from light produced by ^{6}Li glass scintillator. NE912 and NE913 glass scintillator detectors are employed, which measure 2.54 cm long × 2.54 cm diameter. The NE912 glass scintillator is doped with 7.7 wt% of 95% ^{6}Li-enriched lithium, whereas the NE913 glass scintillator contains 8.3 wt% of 99.99 ^{7}Li-enriched lithium. The ^{6}Li has a high sensitivity to thermal neutrons through the ^{6}Li(n,α)^{3}H reaction illustrated by Eq. 11.36 discussed earlier in this chapter, and the ^{7}Li has a low sensitivity to neutrons. The scintillators are coupled to photomultipliers and the output pulse events are amplified and analyzed by a multichannel pulse height analyzer.

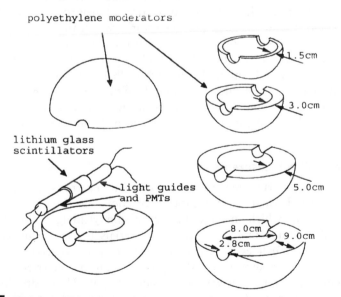

FIGURE 11.45 Schematic view of the multi-moderator spectrometer with a pair of ^{6}Li and ^{7}Li glass scintillators. (From Taniguchi *et al.*, 2001, reprinted with permission from Elsevier Science.)

Taniguchi *et al.* (2001) calculated the response functions to neutrons for the various polyethylene moderator thicknesses and neutron energies (0.25–22 MeV). The detector demonstrated clear separation of neutron from gamma-ray events, when the neutron to gamma-ray flux was >0.3, and neutron from proton events could be clearly discriminated.

2. Neutron-Beam Imaging

Applications of ^{6}Li-glass scintillating fiber detectors to the two-dimensional imaging of neutron beams and sources are reported by Kanyo *et al.* (1992), Schäfer *et al.* (1995), and Ottonello *et al.* (1995). Spatial resolutions for two-dimensional area imaging detectors are 1.0×1.0 mm for a 100-mm-diameter detector and 8.0×8.0 mm for a large 600×600 mm detector; and the neutron absorption efficiencies for the Ce-activated ^{6}Li detectors of 1 mm thickness were 65% for 1 Å (0.1 nm) and 85% for 2 Å (0.2 nm) neutrons (Schäfer *et al.*, 1995). These imaging neutron detectors are used for neutron beam positioning and spatial distribution diagnostics (Ottonello *et al.*, 1995) and neutron diffraction experiments with biological crystals, reflection experiments for the investigation of a near-surface layer of 100 Å depth, and ultracold neutron microscopy (Kanyo *et al.*, 1992).

3. Monitors for Illicit Nuclear Material Trafficking

Other applications of Ce^{3+}-activated ^{6}Li-scintillating glass fiber neutron detectors reported by Abel *et al.* (1994, 1995b), Bliss and Craig (1995), Bliss *et al.* (1995a,b, 1996), and Seymour *et al.* (2000, 2001) are (1) portal, freight and vehicle monitors produced at the Pacific Northwest National Laboratory, Richland, Washington, and NucSafe LLC, Clinton, TN, capable of detecting small quantities of weapons grade plutonium including a 5-m^2 sensor constructed with 250 km of fiber detector; (2) neutron sensors based on flexible ribbons made of glass fiber detector, which can be wrapped around waste barrels or process piping; (3) a neutron sensor that can be mounted inside a small travel case where the scintillating ribbons are wrapped around a 7.6-cm closed-cell foam mold; (4) high efficiency neutron detectors for environmental survey of nuclear storage facilities; (5) wearable neutron and gamma-ray sensors; and (6) neutron dosimeters for boron-capture neutron therapy capable of operation in the presence of large x-ray and gamma-ray fluxes.

4. Neutron Flux Measurements

Possible converter materials for thermal neutrons are ^{6}Li, which undergoes the thermal neutron-capture reactions of ^{6}Li(n,α)^{3}H and fissile ^{235}U and ^{239}Pu, which undergo nuclear fission yielding fission fragments. On the other hand common converter materials for fast neutrons are the fissionable radionuclides ^{232}Th and ^{238}U, which undergo fission upon reaction with fast neutrons (>1 MeV). With neutron converter materials in mind a new type of detector was developed at Nagoya University (Mori *et al.*, 1994), which consists of an optical fiber with its tip covered with a mixture of a neutron converter and ZnS(Ag) scintillator. Along these lines Yamane *et al.* (1998, 1999) designed and tested small detectors for thermal and fast neutron flux measurements in a research reactor. They chose ^{6}Li as converter for the analysis of thermal neutron flux and ^{232}Th as converter for fast neutron flux measurement. The scintillator selected was ZnS(Ag), which was mixed with a converter and glued with epoxy to the tip of glass optical fiber (2 mm diameter quartz). Glass optical fiber was selected, because it is more resistant to radiation damage than plastic fiber. A 0.5 mm thick aluminum cap covers the neutron detector tip, which both serves to protect the scintillator and converter material and reflect light toward the optical fiber. A schematic view of the detector and measurement layout is provided in Fig. 11.46. The electric signals from the photomultiplier tube (PMT) are sent through a preamplifier and amplifier (AMP), and multichannel analyzer

(A)

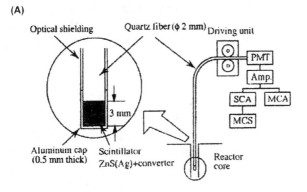

(B)

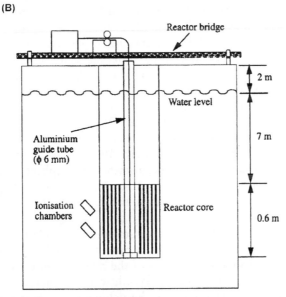

FIGURE 11.46 Schematic view of (A) the quartz fiber tip and the measurement arrangement, and (B) layout of the measurement of neutron flux in a research reactor. (From Yamane et al., 1999, reprinted with permission from Elsevier Science.)

(MCA) for pulse height measurements. A single channel analyzer (SCA) and multichannel scaler were also applied to the measurements.

The advantages of this type of detector for neutron flux measurements demonstrated by Yamane *et al.* (1999) are the following: 1. The detector is slim (2 mm diameter) and capable of fitting into narrow measurement positions such as between reactor fuel plates. 2. Due to the small size of the detector, it has reduced spatial dependence of the neutron flux providing good flux measurement resolution. 3. Small changes or fluctuations in the neutron flux at certain known positions in the reactor core, such as dips caused by spacers can be measured. 4. The neutron current or gradient of the flux can be measured. 4. A neutron flux mapping can be made over large distances of the reactor core in a short period of time. This is a major advantage as conventional measurements of the axial flux profile of a reactor core using cobalt wire activation measurements may take several hours, while the use of glass fiber detector would take only a few minutes for the same measurement.

IX. BONNER-SPHERE NEUTRON SPECTROMETRY

The Bonner-sphere neutron spectrometer was first designed by Bramblett *et al.*, (1960) with a ^{6}LiI(Eu) scintillator mounted on a polystyrene light pipe as coupling to a photomultiplier tube (PMT). It has developed into a popular instrument to measure neutron fields in various environments including space orbital vehicles, cosmic ray interactions in the atmosphere, plutonium facilities, fusion test reactors, etc. The first Bonner-sphere spectrometer consisted of a small 4 mm diameter × 4 mm thick ^{6}LiI(Eu) scintillation crystal containing 96.1% ^{6}Li and some modern Bonner-sphere spectrometers still utilize the ^{6}LiI(Eu) detector (Hajek *et al.*, 2000; Haney *et al.*, 1999; Varela *et al.*, 1999). The spectrometer is calibrated with various polyethylene spherical moderators of differing diameters to provide an isotropic response to neutrons. The moderator spheres are manufactured to dimensions measured to the exact inch or half-inch. The units used to define Bonner-sphere moderator dimensions since the onset was in inches and the dimension in inches in this particular case has remained popular, although centimeters are used by some just as well. A schematic of the apparatus is illustrated in Fig. 11.47. Approximately 10–12 polyethylene moderator spheres are used to calibrate the spectrometer. The sphere dimensions are precisely defined and may range from 2 to 18 in. in diameter. Each sphere, of course, has a hole to accommodate the detector at the center of the sphere.

The size and shape of the ^{6}LiI(Eu) crystal and its high ^{6}Li enrichment was chosen to provide good gamma-ray discrimination and high efficiency for thermal neutrons. Approximately 80% of the thermal neutrons are absorbed in the surface 1.0 mm of the ^{6}LiI(Eu) crystal, while gamma- and fast-neutron interactions would be proportional to the volume of the detector. The small 4 × 4 mm detector provides a large surface to volume ratio to reduce detector response to fast neutrons and gamma rays. While the ^{6}LiI(Eu) crystal is still

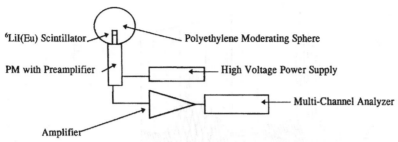

FIGURE 11.47 Schematic of the Bonner neutron spectrometer. (From Haney *et al.*, 1999, reprinted with permission from Elsevier Science.)

employed as the detector, the SP9 spherical ^{3}He proportional counter of 32 mm diameter operated at 2 atm pressure (Centronic Ltd., UK) has become a popular replacement. The ^{3}He proportional counter is sensitive to thermal neutrons ($\sigma = 5{,}330$ barns) reacting according to the ^{3}He(n,p)^{3}H reaction previously described (Eq. 11.42). The detector has superior neutron fluence response and gamma-ray discrimination.

The spherical counter is calibrated as a function of sphere diameter and neutron energy over a broad range of neutron energies (e.g. 10^{-7} to >1000 MeV). The response of small moderator spheres will show an increase, peak out, and then decrease rapidly as the neutron energy is increased. This is what would be expected, because when the moderator is small low-energy neutrons will have a good probability of reaching the detector; but as the neutron energy increases small moderators become increasingly less efficient in thermalizing neutrons. The response of larger spheres is low for the low-energy neutrons, then increases according to neutron energy, reaches a maximum, and finally decreases at very high neutron energies. The low response of large-sphere moderators to low energy neutrons is due to the capture of the thermalized neutrons by hydrogen of the polyethylene moderator. The path-length of travel of neutrons after thermalization is longer in the larger spheres, which increases their probability of capture before reaching the ^{6}LiI(Eu) or ^{3}He detector. The high-energy neutrons, therefore, have increased chances of becoming thermalized in the larger polyethylene spheres and yet reach the detector. At very high neutron energies (>10 MeV) probabilities for thermalization in the larger polyethylene spheres diminishes and the response drops.

A key to the Bonner-sphere spectrometer is its calibration in terms of its response as a function of neutron energy. The response functions for different sphere sizes over a wide range of neutron energies are calculated. The plot of Fig. 11.48 illustrates that the peak of the response function varies according to neutron energy and sphere size, and the response peak moves to higher energies as the sphere size increases. In a review by Thomas and Alevra (2002) they describe the sphere readings according to the following:

$$M_i = \int R_i(E)\phi(E)\,\mathrm{d}E \tag{11.44}$$

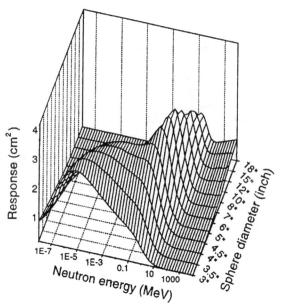

FIGURE 11.48 Response functions for a Bonner sphere set based on a SP9 ^{3}He counter. (From Thomas and Alevra, 2002, © Crown Copyright 2002, reproduced by permission of the Cintroller of HMSO.)

where M_i is the reading for sphere i, $R_i(E)$ is the response function of sphere i exposed to a neutron field with spectral fluence $\phi(E)$. Good numerical methods are available that provide approximations to $R_i(E)$ supported with measurements form monoenergetic neutron sources. From measurements taken by the Bonner-sphere set in an unknown neutron field the extraction of $\phi(E)$ from the above calibration data permits the unfolding of the energy spectrum. As elucidated by Thomas and Alevra (2002) neutron spectra is represented as an array where ϕ_j is the neutron fluence in group j extending over the energy range of E_j to E_{j+1} and a measured reading is given by

$$M_i = \sum_{j=1}^{n} R_{ij}\,\phi_j \tag{11.45}$$

where R_{ij} represents the response function $R_i(E)$ averaged over group j. Equation 11.45 provides a set of m linear equations, one for each sphere; and if $m \geq n$, the equations can be solved by least squares (Zaidins *et al.*, 1978) to provide values of the spectral fluence ϕ_j. However, since the number of spheres is usually small ($m \approx 10$), the solution provides a poor representation of the neutron spectrum (Thomas and Alevra, 2002), and additional a priori information combined with spectral unfolding techniques are used. Many spectral unfolding codes have been developed, and the methods are reviewed by Matzke (2002).

X. LUCAS CELL

The Lucas cell is a solid scintillation chamber designed for the specific purpose of measuring the alpha particle emissions of environmental levels of ^{222}Rn in air. It was first designed by Lucas (1957), and the basic components of the cell have not changed since. The cell consists of a vessel of ~ 100 mL volume or larger shaped like a bell-jar and made of Kovar metal with a narrow brass collar at the top as the only orifice and a clear silica window at the flat bottom (face). The inner wall of the cell is coated with a thin layer (~ 20–50 mg cm^{-2}) of ZnS(Ag) scintillator. The inner surface of the flat window at the face of the Lucas cell is not coated with scintillator; it is covered with a transparent coat of tin oxide to maintain electrical conductivity and prevent any accumulation of radon or its daughters on the window. Environmental ^{222}Rn is first trapped from the air by activated charcoal. The radon is then liberated by heating and transferred with a carrier gas into the Lucas cell. To measure radon activity, the clear window of the Lucas cell is placed in contact with the face of a conventional photomultiplier tube. The alpha particles of radon and its daughters in the Lucas cell interact with the scintillator; and the scintillation light photons from these interactions are detected by the PMT and converted into electrical pulses.

Semkow *et al.* (1994) have determined the detection efficiency of the Lucas cell for ^{222}Rn and its 3 alpha particle-emitting daughters, ^{218}Po, ^{214}Po, and ^{210}Po. The detection efficiency varies between 0.76 and 0.83 counts per alpha particle depending on the alpha particle-energy, which differs for each radionuclide. The Lucas cell can also be used for the measurement of ^{226}Ra in water via the measurement of its emanating ^{222}Rn daughter after secular equilibrium is reached. The detailed procedure is provided by Peters *et al.* (1993). In brief, the principle of the method entails the concentration and separation of the radium-226 in water by coprecipitation with barium sulfate. The concentrated sample is redissolved and sealed in a purged flask (bubbler). After 21 days to allow for ingrowth of the radon-222 daughter, the gaseous ^{222}Rn is transferred to a Lucas cell using helium as a carrier gas and then counted as described. The detection limit is reported to be about 0.05 pCi per sample when counted for 1000 min, depending on the individual background of the Lucas cell used.

Although the detection limits of the Lucas cell for ^{222}Rn are good, the procedure for sample preparation is tedious and time consuming. Modern liquid scintillation methods are available that entail the facile trapping and analysis of ^{222}Rn from air or water (Passo and Floeckher, 1991; Passo and Cook, 1994). For the trapping of ^{222}Rn from air, a liquid scintillation counting vial (Pico-Rad) is used, which contains a canister of activated carbon at its orifice. After 48 h of exposure of the activated carbon to the room air, a xylene-based liquid scintillator (e.g., Insta-Fluor) is placed at the bottom of the vial. Desorption of the ^{222}Rn from the charcoal to the scintillation fluor takes place via the vapor phase within 3 h. The ^{222}Rn in water is also easily determined by LSA after trapping in an organic fluor (e.g., Insta-Fluor or Opti-Fluor O) by simply adding 10 mL of freshly collected water by syringe to the organic fluor. The ^{222}Rn fully partitions into the organic fluor phase in about 3 h, after which conventional LSA may be used. The liquid scintillation techniques provide advantages over the Lucas cell, which are improved sensitivity, minimal sample preparation time, small sample sizes, and higher throughput with automatic counting. The minimal detection limit for ^{222}Rn in air with the Pico-Rad LSC vial is $\sim 2\,\mathrm{mBq}\,\mathrm{L}^{-1}$ as reported by Morishima *et al.* (1992), and the detection limit of ^{222}Rn in water by LSC is approximately $0.37\,\mathrm{Bq}\,\mathrm{L}^{-1}$ (Spaulding and Noakes, 1993).

To improve the detection efficiency for a large volume Lucas cell and reduce background counts Sakamoto and Takahara (2001) designed a scintillation cell with an inside diameter of 7.5 cm and a length of 30 cm constructed of a vinyl chloride tube with acrylic resin windows at each end. Condensing lenses were also placed at each end of the tube to aid in the detection of scintillation light by photomultiplier tubes placed at each end. The inside wall of the tube was coated with ZnS(Ag) scintillator adhered onto aluminum sheet. The two photomultiplier tubes were provided for coincidence counting to reduce background. Pulse height analysis provides a spectrum of the alpha peaks, whereby proper selection of a counting window with discriminator settings would further facilitate background reduction. The average counting efficiency for radon and its progenies was reported to be 0.55 with a background count rate of 0.015 cps. The lower limit of detection (LLD) for a 2-hour counting time was $< 10\,\mathrm{Bq}\,\mathrm{m}^{-3}$.

XI. RADIONUCLIDE STANDARDIZATION

A. $4\,\pi\beta–\gamma$ Coincidence Counting

$4\,\pi\beta–\gamma$ coincidence counting is a direct method of activity determination, as it is independent of any quench indicating parameters and no reference standards are required for counting efficiency determinations. An additional advantage of this method is that the activity of a nuclide can be determined from experimental counting data only without the need for detector efficiencies.

The method is reviewed in detail in the NCRP Report No. 64 (1985) and by Mann *et al.* (1988). As may be construed from the name applied to this method, it involves the coincidence detection of two types of radiations from a given radionuclide. The methods can be applied, therefore, to radionuclides that emit in coincidence more than one distinguishable type of radiation. Two types of detectors are required in coincidence to distinguish exclusively two types of radiation emissions, such as a liquid scintillation detector (or gas proportional counter) to measure a beta emission and a solid scintillation (or semiconductor detector) capable of measuring a gamma emission. Subsequent to pulse height discrimination from each detector, three counting channels are used for data collection. These are a β channel and γ channel, which collect pulses from the beta particle-detector and gamma-ray detector, respectively, and a coincidence channel to register the two emission types in coincidence. The nomenclature used for this technique may identify the type of beta- and gamma-detectors used in coincidence. For example, the term $4\pi\beta(PC)-\gamma(NaI)$ signifies a proportional counter and NaI solid scintillation detector used for the detection of a β and γ ray in coincidence, respectively. Whereas, the term, $4\pi\beta(LS)-\gamma(Ge)$ signifies liquid scintillation and germanium semiconductor detectors are used for the coincidence detection of a β and γ ray, respectively.

Using the notation of NCRP Report No. 64 (1985) for the simple case of a two-stage $\beta-\gamma$-decay scheme from a radionuclide source, the count rates in the three channels are described as

$$N_\beta = N_0\,\varepsilon_\beta\,, \quad N_\gamma = N_0\,\varepsilon_\gamma \text{ and } N_c = N_0\,\varepsilon_\beta\,\varepsilon_\gamma\,, \qquad (11.46)$$

where N_β, N_γ, and N_c are the count rates in the beta, gamma, and coincidence channels, respectively, while ε_β and ε_γ are the respective counting efficiencies in the beta and gamma detectors, and N_0 is the activity of the radionuclide source. From these equations the activity of the radionuclide source can be determined from the relationship

$$N_0 = \frac{N_\beta\,N_\gamma}{N_c}\,. \qquad (11.47)$$

Therefore, the disintegration rate of the sample can be determined from the count rates of the three channels without any detector counting efficiency determinations.

To exploit this theory a beta-efficiency extrapolation technique is widely used for the standardization of both simple and complex $\beta-\gamma$ emitters. The extrapolation method has developed into a very accurate standardization method following the work of several researchers including Houtermans and Miguel (1962), Baerg (1973), Grigorescu (1963, 1973), Funck and Nylandstedt Larsen (1983), Simpson and Meyer (1988, 1992a, 1994a), Yan (1996), Attie *et al.* (1998), Park *et al.* (1998), Grigorescu *et al.* (1998), and Razdolescu *et al.* (2000). The meaning of the term beta-efficiency extrapolation is derived from the technique, which involves successively altering the detection efficiency of the beta particle detector by attenuating the beta emissions of the radionuclide source with absorber material or by altering the beta-energy threshold of the beta particle detector and plotting the consequent linear relationship of the count rate ratio $N_\beta N_\gamma/N_c$ of Eq. 11.47 versus a count rate ratio such as N_γ/N_c-1, N_γ/N_c, or $(1-N_c/N_\gamma)(1/N_c/N_\gamma)$. The linear plot is extrapolated to $N_\gamma/N_c=1$, which is an extrapolation to 100% beta efficiency, where $N_c/N_\gamma=\varepsilon_\beta$ according to the aforementioned Eqs. 11.46, and the count rate ratio $N_\beta N_\gamma/N_c$ would be equal to the radionuclide activity. The following beta-efficiency extrapolation procedure of Grigorescu *et al.* (1998) illustrates the principles involved:

Grigorescu *et al.* (1998) described the $4\pi\beta(PC)-\gamma(NaI)$ standardization of ^{110m}Ag and ^{75}Se. The procedure has been utilized also by Razdolescu *et al.* (2000) for the standardization of ^{169}Yb, and applied also to the $4\pi\beta(LS)-\gamma(NaI)$ standardization of ^{110m}Ag by García-Toraño *et al.* (2000).

The equations that define the count rates for the three channels according to the definitions provided in Eqs. 11.46 are

$$N_\beta = N_0 \sum_r a_r(\varepsilon_{\beta r} + (1 + \varepsilon_{\beta r})\varepsilon_{\beta \gamma r}), \qquad (11.48)$$

$$N_\gamma = N_0 \sum_r a_r \varepsilon_{\gamma r}, \qquad (11.49)$$

and

$$N_c = N_0 \sum_r a_r(\varepsilon_{\beta r}\varepsilon_{\gamma r} + (1 - \varepsilon_{\beta r})\varepsilon_{cr}), \qquad (11.50)$$

where as described for Eqs. 11.46, N_β, N_γ, and N_c are the count rates in the beta, gamma, and coincidence channels, respectively, N_0 is the radionuclide activity, a_r are the branching ratios, $\varepsilon_{\beta r}$ and $\varepsilon_{\beta \gamma r}$ are the beta particle detector efficiencies (proportional or liquid scintillation counter) for the beta spectrum and gamma transitions, including conversion electrons, respectively, and $\varepsilon_{\gamma r}$ is the efficiency of the gamma detector for the gamma-rays associated with the rth branch, and ε_{cr} is the probability of detecting a coincidence event when the beta particle is not detected (e.g., Compton scattering or γ–γ coincidence events). From Eqs. 11.48–11.50 the following is derived by Grigorescu *et al.* (1998):

$$\frac{N_\beta N_\gamma}{N_c} = N_0\left[1 + (1 - K)\left(\frac{N_\gamma}{N_c} - 1\right)\right] \qquad (11.51)$$

where

$$K = \frac{\sum_r a_r(1 - \varepsilon_{\beta r})(1 - \varepsilon_{\beta \gamma r})}{\sum_r a_r \varepsilon_{\gamma r}(1 - \varepsilon_{\beta r})(1 - \varepsilon_{cr}/\varepsilon_{\gamma r})} \sum_r a_r \varepsilon_{\gamma r} \qquad (11.52)$$

From Eq. 11.51 we see that, although the activity, N_0, remains an unknown constant, the ratio of the count rates $N_\beta N_\gamma/N_c$ is a function of $(N_\gamma/N_c - 1)$. If we could consecutively vary the beta particle detection efficiencies by means of absorber material of varying thickness between the source and beta particle detector or by varying the beta threshold of the detector, we would concomitantly reduce the value of $N_\beta N_\gamma/N_c$ as a function of $(N_\gamma/N_c - 1)$. Following the addition of absorbers of increasing thickness (e.g., gold-coated films) onto the source, the count rates in each channel corrected for dead time, background, and resolving time, are recorded, and the ratio $N_\beta N_\gamma/N_c$ is plotted against $(N_\gamma/N_c - 1)$ as illustrated in Fig. 11.49. The linear plot is extrapolated to the value of $N_\gamma/N_c = 1$ or $(N_\gamma/N_c - 1) = 0$ (See Fig. 11.49). From Eq. 11.51 it is seen that when $(N_\gamma/N_c - 1) = 0$, the count rate ratio $N_\beta N_\gamma/N_c = N_0$, the sample activity. The overall uncertainties reported in the use of this technique for radionuclide standardizations are 1% for [110m]Ag

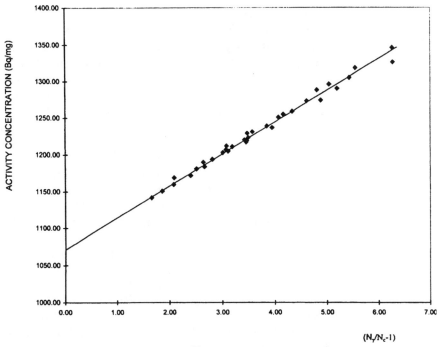

FIGURE II.49 Extrapolation set for ^{75}Se, threshold variation. (From Grigorescu et al., 1998, reprinted with permission from Elsevier Science.)

and ^{75}Se reported by Grigorescu et al. (1998), 0.49% for ^{110m}Ag reported by García-Toraño et al. (2000), and 0.9% for ^{169}Yb reported by Razdolescu et al. (2000).

In a $4\pi(x, e)$–γ coincidence method for the standardization of ^{201}Tl, Simpson and Meyer (1992a, 1994a) used liquid scintillation to detect the x-rays and Auger electrons and a NaI(Tl) detector to count coincident gamma rays. They demonstrated that the ratio of count rates in the three channels was proportional to the source activity as follows:

$$\frac{N_x N_\gamma}{N_c} = N_0 \left[m\frac{N_\gamma}{N_c} + n \right] \tag{11.53}$$

where N_x, N_γ, and N_c are the liquid scintillation count rate, gamma count rate, and coincidence count rates, respectively, N_0 is the source activity and m and n are a combination of decay scheme parameters, such that $m + n = 1$. They altered the liquid scintillation detection efficiencies by varying the discriminator level settings to discriminate against conversion electrons and K Auger electrons, and plotted the linear relationship of the count-rate ratio $N_x N_\gamma/N_c$ against N_γ/N_c. The linear plot was extrapolated to $N_\gamma/N_c = 1$, where, according to Eq. 11.53, the count rate ratio $N_x N_\gamma/N_c = N_0$, the sample activity. They report a total uncertainty of $\pm 0.52\%$ for the ^{201}Tl activity with this technique. Signal processing electronics described by

Simpson and Meyer (1988) allow for the setting of multiple thresholds in the liquid scintillation pulse height spectrum to permit the collection of 15 data points of varying efficiency to be collected simultaneously from the source. A computer program described by Simpson and van Oordt (1997) operating in the Windows® environment is used to control the counting system and data collection. Other reports of $4\pi\beta-\gamma$ coincidence standardization involving similar beta efficiency extrapolations were carried out for the standardization of the following sources with the uncertainties of the determination in parenthesis: ^{177}Lu ($\pm 0.22\%$) and ^{188}Re ($\pm 0.3\%$) reported by Schötzig et al. (2001), ^{133}Ba ($\pm 0.31\%$), ^{109}Cd ($\pm 0.68\%$), ^{125}I ($\pm 0.52\%$), ^{75}Se ($\pm 0.28\%$), and ^{204}Tl ($\pm 0.59\%$) reported by Simpson (2002), and ^{134}Cs ($\pm 0.38\%$) reported by García-Toraño et al. (2002).

In a unique approach, Simpson and Meyer (1996) standardized a ^{204}Tl source via $4\pi(X,e)-X_K$ coincidence system. They used a NaI(Tl) detector for the detection of the x-rays from the source and a liquid scintillation counter to detect the L Auger electrons and L x-rays associated with the escaping K x-rays. They demonstrated that the ratio of count rates in the three counting channels $N_{4\pi}N_X/N_c$ was a linear function of $(1-\varepsilon_{KE})/\varepsilon_{KE}$, where $N_{4\pi}$ and N_X are the count rate in the liquid scintillation and solid scintillation (x-ray) channels, respectively, and ε_{KE} is equivalent to N_c/N_X. Extrapolation of the linear plot to $N_c/N_X = 1$, yielded the source activity. The uncertainty for the ^{204}Tl activity reported was $\pm 0.59\%$.

Schötzig et al. (1999) standardized sources of ^{153}Sm with an overall uncertainty of 0.1% using the $4\pi\beta-\gamma$ coincidence method. The activity of the source was determined by extrapolating the ratio of the count rates in the three channels, namely, $N_\beta N_\gamma/N_c$ against $(1-\varepsilon_\beta)/\varepsilon_\beta$ where ε_β is the beta particle detection efficiency equivalent to the count rate ratio N_c/N_γ, as can be seen from Eqs. 11.46. Extrapolation of the linear plot to $N_c/N_\gamma = 1$ or $\varepsilon_\beta = 1$ provided the source activity Similar studies were carried out for the standardization of the following sources with the uncertainties of the determination in parenthesis: ^{67}Ga ($\pm 1.1\%$) reported by Attie et al. (1998), ^{192}Ir ($\pm 0.7\%$) reported by Hino and Ohgaki (1998), and ^{54}Mn ($\pm 0.19\%$) and ^{85}Sr ($\pm 0.23\%$) reported by Park et al. (1998).

B. Windowless 4π–CsI(Tl) Sandwich Spectrometry

Absolute activity measurements of certain radionuclides are possible with a 4π counting geometry with a solid scintillation detector. Altzitzoglou et al. (2002) and Hult et al. (2000) used a 4π–CsI(Tl) sandwich spectrometer to determine the activity of ^{89}Sr and ^{204}Tl with uncertainties of 0.4 and 1.1%, respectively. They used the sandwich spectrometer described by Denecke (1987, 1994). The windowless 4π–CsI(Tl) sandwich spectrometer consists of two CsI(Tl) crystals measuring 50 mm in diameter and 25 mm in height, which are mounted in a source-interlock chamber filled with dry hydrogen. The radioactive source is sandwiched between the bare front faces of the crystals. Hemispherical cavities of 10 mm diameter at the center of the crystals prevent contact with the radioactive source. The gap between the two crystals is < 4 μm when the crystals are in the closed position providing a >99.98% 4π geometry around the source. The detectors are 100% sensitive to charged particles with energies >10 keV, and the photon detection efficiency is >99% for photons in the energy range of 10–220 keV. Photon detection efficiencies decrease at energies >220 keV dropping to 45% at 1 MeV (Altzitzoglou *et al.*, 2002). In the case of the beta particle-emitter ^{89}Sr ($E_{max} = 1.49$ MeV) no emission escapes correction, and only a very small correction for the cut-off of low-energy events is necessary. The activity analysis is considered as an accurate direct measurement, as no reference to a standard is needed. The activity of the sample is obtained by dividing the measured integral count rate by the efficiency of the instrument for a specific nuclide provided by calculation and correction for dead time and background.

A NIST 4π NaI(Tl) system is described and used by Zimmermann *et al.* (1998, 2001, 2002) for the standardization of the following radionuclides with the % uncertainties of the determination provided in parenthesis: ^{117m}Sn (2.43%), ^{177}Lu (1.95%), and ^{188}W/^{188}Re (2.5%). The NIST 4π NaI(Tl) system consists of two 6 cm × 20 cm NaI(Tl) well crystals mounted face-to-face on slip rods. When the crystal assembly is raised out of its shields the crystals separate permitting placement of the source in the center well. When closed the crystals surround the source providing a 4π counting geometry.

XII. PHOSWICH DETECTORS

A phoswich detector consists of two or more scintillation detectors optically coupled as a phosphor sandwich, from which the scintillation light output is viewed by a single photomultiplier tube (PMT). This unique detector arrangement is designated as a PHOSWICH, which is the acronym for PHOSphor sandWICH. Much current research is directed to the development of phoswich detectors as practical instruments for the simultaneous measurement and discrimination of alpha, beta, gamma, or neutron radiation.

The principle behind phoswich detectors is the different properties of interaction that certain scintillators display with different types of radiation. In particular, different scintillators will differ in their propensity to interact or absorb radiation depending on the type and even the energy of the radiation as well as display differing light output decay times with concomitant different pulse shape events. Consequently placing multiple scintillators of different kinds on top of each other as a sandwich detector and coupling the combined scintillators to one PMT will constitute a detector capable of measuring simultaneously several types of radiation.

The selection of the scintillators, their dimensions, and their arrangement as a sandwich will depend on the types of radiation to be analyzed. For example, a phoswich detector designed to measure alpha/beta/gamma or neutron radiation will consist of scintillators sandwiched in a fashion so that the alpha particles are first absorbed by a thin scintillator sensitive to alpha particles, followed by a second scintillator that may be thicker but capable of absorbing the more penetrating beta particles, and finally another scintillator sensitive to the more penetrating gamma or neutron radiation. Also, the scintillators that constitute the sandwich are carefully selected taking into account their differing scintillation light decay times, which can further facilitate pulse shape analysis for radiation discrimination.

A. Simultaneous Counting of α-, β-, and γ-Rays or α-, $\beta(\gamma)$-Rays, and Neutrons

There is a need for detectors that can measure simultaneously radiations of several types particularly in the different processes and nuclear safety management required in the nuclear fuel cycle. Much research and development had gone initially into the development of phoswich detectors for the simultaneous measurement and discrimination of α-, β-, and γ-rays (Usuda, 1992; Usuda and Abe, 1994; Usuda et al., 1994a,b). An example of such a phoswich detector is the coupled three-detector array of ZnS(Ag), NE102A, and BGO scintillators designed by Usuda et al. (1994a) for the simultaneous counting of α-, β-, and γ-rays. The phoswich consisted of a thin $(10\,\mathrm{mg\,cm}^{-2})$ ZnS(Ag) scintillator sandwiched with a thicker (5 mm) NE102 plastic scintillator, which was sandwiched with a third 5mm thick BGO scintillator. The α particles penetrate only the ZnS(Ag) scintillator, while the β particles and γ-rays continue on to interact with the NE102A plastic scintillator, where β and soft γ interactions occur, and finally the hard γ-rays continue on to interact with the BGO scintillator. This phoswich detector was tested for the discrimination of radiation types from a mixture of several sources such as ^{244}Cm for α particles, ^{90}Sr(^{90}Y) for β particles, ^{241}Am for soft γ-rays as well as α particles, and ^{137}Cs and ^{60}Co for β and γ counting. Usuda et al. (1994) report that ZnS(Ag) is insensitive to β- and γ-rays with a pulse rise time the slowest among the scintillators, and the NE102A plastic scintillator has the highest sensitivity to β particles (low-Z scintillator), the fastest pulse rise time among the scintillators used, and a relatively narrow peak width (FWHM). The third scintillator, BGO, has a rise time intermediate between those of ZnS(Ag) and NE102A and a high sensitivity to γ-rays (high Z and density). The BGO may be replaced with an NaI(Tl) scintillator for the third phoswich for radiation monitoring purposes, as the latter provides lower backgrounds in this triple-phoswich detector.

Usuda (1995) noted that, in the radiation monitoring of high burn-up spent fuels, significant neutron emission must be considered in addition to other forms of radiation. He devised a modification of the triple-phoswich detector previously described to include a ^{6}Li glass scintillator, which has a high reaction cross section for thermal neutrons. A triple-phoswich detector assembly consisting of ZnS(Ag), NE102A plastic, and ^{6}Li glass scintillators permits the simultaneous counting of α-, $\beta(\gamma)$-rays and thermal neutrons. The ^{6}Li glass scintillator used was cerium-activated 7.5% lithium silicate glass containing 95.6% enriched ^{6}Li (NS8 Nikon scintillator). The detector was further developed by Usuda et al. (1997, 1998a,b), and Yasuda et al. (2001) with the careful selection of scintillator detectors according to their pulse decay rise times, light outputs, and emission wavelengths with the employment of optical filters to further enhance the discrimination of pulse events from each scintillator according to pulse shape analysis.

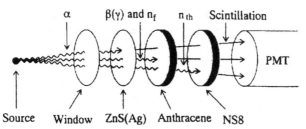

FIGURE 11.50 Arrangement of a ZnS(Ag)/anthracene/NS8 phoswich for simultaneous counting of α-, β(γ)-rays and neutrons. (From Usuda et al., 1997, reprinted with permission from Elsevier Science.)

Their phoswich detectors were developed to measure simultaneously α-, $\beta(\gamma)$-rays, and fast neutrons (n_f) and thermal neutrons (n_{th}). A typical arrangement of a phoswich detector with these capabilities is illustrated in Fig. 11.50. Alpha particles are fully absorbed by the thin ($10\,mg\,cm^{-2}$) ZnS(Ag) scintillator. Beta particles of low energy ($< 20\,keV$) interact slightly with the ZnS(Ag), but the higher-energy beta particles continue on to be fully absorbed by the 10 mm thick anthracene scintillator. The fast neutrons interact with the anthracene via proton recoil while the thermal neutrons undergo neutron capture in the 5 mm thick NS8 ^{6}Li glass scintillator enriched in ^{6}Li to 95.6%. The phoswiches are protected from ambient light with Al-coated Mylar film ($0.25\,mg\,cm^{-2}$). The phoswiches can also be placed into radioactive solutions if protected with Au-coated Mylar film ($0.9\,mg\,cm^{-2}$). The output signals from the PMT have specific rise times and pulse heights that are analyzed by pulse shape discrimination. The rise-time distributions of α-, $\beta(\gamma)$-rays, n_f, and n_{th} from ^{244}Cm, ^{137}Cs, ^{252}Cf, and ^{252}Cf $+^{244}$Cm reported by Usuda et al.(1997, 1998) with the ZnS(Ag)/anthracene/NS8 phoswich is illustrated in Fig. 11.51. The abbreviation FOM in the figure refers to crystal figure-of-merit. The FOMs are separated into different categories, to reflect the ability of the scintillator to discriminate pulse events from different radiation sources. For example, Usuda et al. (1997) classified FOMs into $\beta(\gamma)/\alpha$, $\beta(\gamma)/n_f$, n_f/n_{th}, and n_f/α. In a more recent study Yasuda et al. (2001) used a more simple dual detector phoswich of ZnS(Ag)/^{6}Li glass or ZnS(Ag)/ anthracene and a three-dimensional analysis of pulse event number, height, and rise time to fully separate β/γ, n_f, and α events.

Researchers at the University of Missouri-Columbia designed a triple crystal phoswich detector for the simultaneous analysis of alpha, beta, and gamma radiation (Childress and Miller, 2002; White and Miller, 1999) and they have evaluated each crystal for mischaracterizations of beta and gamma radiation. Their detector design consisted of a sandwich of the following three detectors provided in the order of radiation penetration: (1) a $10\,mg\,cm^{-2}$ thick (0.002445 cm) ZnS:Ag for the detection of alpha particles, (2) a 0.254 cm thick CaF$_2$:Eu crystal for the detection of beta particles and some low-energy gamma radiation, and (3) a 2.54 cm thick NaI:Tl crystal for the measurement of gamma radiation. They used MCNP, which is a Monte Carlo simulation program (Briesmeister, 2000) capable of simulating electron, photon, and neutron interactions in detectors of simple

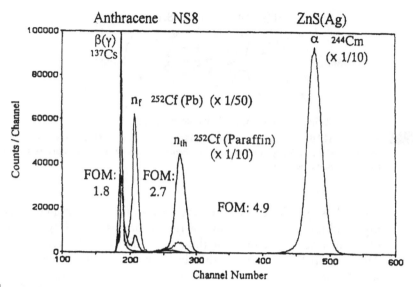

FIGURE 11.51 Pulse shape spectra of α, β particles (including γ rays), thermal and fission neutrons observed with ZnS(Ag)/anthracene/NS8 phoswich. (From Usuda et al., 1998, reprinted with permission from Elsevier Science.)

geometry including spheres, cylinders, planes, cones, and ellipsoids. The application of Monte Carlo computer simulations to estimate interactions in detectors has been demonstrated to yield low errors ($< 5\%$) when compared to collected data (Bronson and Wang, 1996; Kamboj and Kahn, 1996; Rodenas et al., 2000). Analysis by Childress and Miller (2002) show that the most probable beta particle energy loss in the first scintillator ZnS:Ag is around 20 keV due to the short path length that electrons can traverse in ZnS:Ag. The Monte Carlo code showed that the ZnS:Ag attenuation of gamma rays was restricted to gamma photons below 50 keV. The CaF$_2$:Eu preferentially interacts with beta particles. The Monte Carlo simulations found that the phoswich design has minimum inherent energy limits of 250 keV E_{max} for beta particles and 50 keV for gamma rays. The 2.54 cm thick NaI:Tl crystal yielded intrinsic gamma efficiency ranges from a maximum of 80% for 100 keV to 26% for 2 MeV photons. Mischaracterization of gamma events in the CaF$_2$:Eu crystal can be calculated and corrected.

B. Remote Glass–Fiber Coupled Phoswiches

Additional studies by Yasuda et al. (2000a) found that a YAP scintillator (YAlO$_3$:Ce) has a fast rise time and sharp peak (small FWHM) of only 10 mg cm^{-2} thickness in a phoswich detector coupled to a 5 mm thick YAG (Y$_3$Al$_5$O$_{12}$:Ce) scintillator, which was sufficient to accept high count rates and clearly distinguish alpha particles from beta and gamma rays. A new type of phoswich detector was developed by coupling a ZnS(Ag)/NE102A phoswich to the photomultiplier tube via a quartz optical fiber. A NE172 wavelength shifter was positioned between the phoswich and the optical fiber for optimum scintillation light transmission via the optical fiber and readout by the PMT. The light from the phoswich detector is transmitted through the optical fiber for remote pulse shape analysis and discrimination of alpha and beta(gamma) rays (Yasuda *et al.*, 2000b). This phoswich detector can be used to measure radioactivity in liquids as a dip-type inline monitor or to measure radioactivity in narrow or isolated spaces such as glove boxes or hot cells.

C. Low-Level Counters

Phoswich detectors are also employed in low-level counting of radionuclides in the environment in the presence of interfering γ radiation. For example, Wang *et al.*, (1994) utilized a double $CaF_2(Eu)$–$NaI(Tl)$ phoswich central (β, γ) detector in an anticoincidence low-background detector assembly for the measurement of β emitting $^{89}Sr + {}^{90}Sr({}^{90}Y)$ activities in waste effluent of nuclear power plants in the presence of γ-emitting radionuclides. The window of the $CaF_2(Eu)$ crystal used in the phoswich assembly was coated with aluminized Mylar; and the crystal was very thin (0.0635 mm) compared with the coupled $NaI(Tl)$ detector (6.36 cm thick). This allowed the low-atomic-number $CaF_2(Eu)$ to perform as an effective β detector with negligible interaction of γ radiation. A separate detector guard consisting of a surrounding $NaI(Tl)$ crystal operated in anticoincidence mode to reduce backgrounds in a high-background laboratory from 5880 ± 229 cps without shielding to 1.0 ± 0.1 cpm with shielding and anticoincidence counting. An MDA for ^{90}Sr of 20 pCi L^{-1} is reported.

D. Simultaneous Counting of n/γ/p Fields

Neutron spectrometers that can adequately discriminate between neutrons and charged particles and measure equivalent doses to the human body are needed in space exploration such as on board the Space Shuttle and inside large human spacecraft such as the International Space Station. In these environments high-energy charged particles in space produce high-energy neutrons by their interaction with the spacecraft structural materials. With this objective in mind Takada *et al.* (2002) developed a phoswich detector for neutron spectrometry in a mixed field of neutrons and charged particles. They developed a phoswich detector by coupling two organic scintillators of different light-output decay times, which could measure high-energy neutrons up to 130 MeV.

The phoswich neutron detector consists of a 133 mm diameter by 133 mm long NE213 organic liquid scintillator surrounded by a 15 mm thick NE115 plastic scintillator. A glass cell encapsulates the liquid scintillator to protect the optically coupled NE115 plastic scintillator. The phoswich is coupled to a single photomultiplier tube via a light guide. The NE213 scintillator was selected by Takada *et al.* (2002) for the measurement of high-energy neutrons because of its ability to discriminate between gamma rays and neutrons. The light-output decay times of this scintillator are 3.7 ns for gamma-ray induced scintillation and 30 ns for neutron-induced scintillation.

(A) Gamma Ray Enters The Detector.   Signal From The Detector

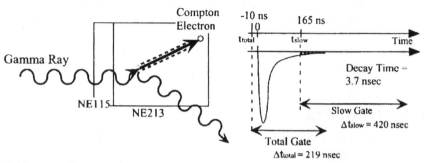

(B) Neutron Enters The Detector.

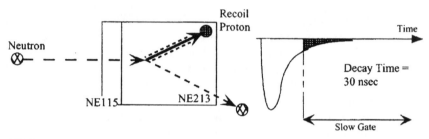

(C) Proton Enters The Detector.

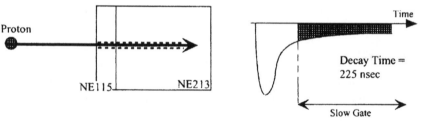

FIGURE 11.52 Schematic models of signals produced by the NE115/NE213 phoswich detector. (A), (B) and (C) are the cases of a gamma ray, neutron, or proton entering into the detector, respectively. The top-right sketch is the time relation between the pulse from the detector and the gate inputs to two ADCs. Two light outputs (total and slow components) are obtained by integrating the gate signal with the time intervals. (From Takada et al., 2002, reprinted with permission from Elsevier Science.)

The surrounding NE115 scintillator provides for charged-particle detection, because it has a much longer 225 ns decay time.

The light decay time constants between the two scintillators employed in this phoswich detector was used to discriminate pulses of the three different particle species (n/γ/p). Fig. 11.52 illustrates how the light output signals are produced in the scintillators via interactions of the three particle species, and how pulse shape discriminations of the light output are used to characterize the signals from the detector. In brief, the Compton electron scattered by a gamma ray dissipates its energy only in the NE213 scintillator, and its signal displays a fast component of 3.7 ns (Fig. 11.52(A)). The neutron is detected via proton recoil, illustrated in Fig. 11.52(B), that occurs only in the NE213 scintillator has a slower decay time of 30 ns compared to the gamma-ray. The proton dissipates its energy in both scintillators (NE115 + NE213), and its signal is much longer as it becomes the sum of the fast and slow components with a decay time of 225 ns for the slow component.

The three particles were detected by Takada *et al.* (2002) using pulse shape discrimination based on the standard Computer Automated Measurement and Control (CAMAC) charge-integration ADCs (Lecroy 2249w, Lecroy Corp., Chestnut Ridge, NY). Analog-to-digital converters (ADCs) can be of two kinds, namely, an ADC that measures either charge or voltage and produces a digital number proportional to the magnitude of the input signal. Takada *et al.* (2002) are developing energy-response functions of the phoswich detector to photons, neutrons, and protons. From the response functions, the energy spectra of the gamma, neutrons, and protons could be obtained through spectrum unfolding. Research in this direction is reported already by Takada *et al.* (2001a,b).

REFERENCES

Abbas, M. I. (2001a). Analytical formulae for well-type NaI(Tl) and HPGe detectors efficiency computation. *Appl. Radiat. Isot.* 55, 245–252.

Abbas, M. I. (2001b). A direct mathematical method to calculate the efficiencies of a parallelepiped detector for an arbitrarily positioned point source. *Radiat. Phys. Chem.* 60, 3–9.

Abbas, M. I., and Basiouni, M. M. (1999). Direct mathematical calculation of the photopeak efficiency for gamma rays in cylindrical NaI(Tl) detectors. *Am. Inst. Phys.* **CP450**, 268–272.

Abel, K. H., Arthur, R. J., Bliss, M., Brite, D. W., Brodzinski, R. L., and Craig, R. A. *et al.* (1994). Scintillating-glass-fiber neutron sensors. *Nucl. Instrum. Methods Phys. Res. Sect. A* **353**, 114–117.

Abel, K. H., Schilk, A.J., Brown, D. P., Knopf, M. A., Thompson, R. C., and Perkins,R. W. (1995a). Characterization and calibration of a large area beta scintillation detector for determination of Sr-90. *J. Radioanal. Nucl. Chem.* **193**, 99–106.

Abel, K. H., Arthur, R. J., Bliss, M., Brite, D. W., Brodzinski, R. L., and Craig, R. A. *et al.* (1995b). Performance and applications of scintillating-glass-fiber neutron sensors. *In* "Scintillating Fiber Detectors: Proceedings of the SCIFI93 Workshop" (A. D. Bross, R. C. Ruchti, and M. R. Wayne, Eds.), pp. 463–472. World Scientific, River Edge, NJ.

Aitken, D. W., Beron, B. L., Yenicay, G., and Zulliger, H. R. (1967). The fluorescent response of NaI(Tl), CsI(Tl), CsI(Na) and CaF$_2$(Eu) to X-rays and low-energy gamma rays. *IEEE Trans. Nucl. Sci.* **14**, 468–477.

Akisu, M., Kultursay, N. *et al.* (1998). Platelet-activating factor levels in term and preterm human milk. *Biol. Neonate* **74**, 289–293.

Allen, G. F., Hutchins, A., and Clancy, J. (1999). An ultrahigh-throughput screening assay for estrogen receptor ligands. *Anal. Biochem.* **275**, 243–247.

Allier, C. P., Hollander, R. W., Sarro, P. M., and van Eijk, C. W. E. (1999). Thin photodiodes for a scintillator-silicon well detector. *IEEE Trans. Nucl. Sci.* **46**(6), 1948–1951.

Allier, C. P., Hollander, R. W., Sarro, P. M., de Boer, M., and van Eijk, C. W. E. (2000a). Scintillation light read-out by low-gain thin avalanche photodiodes in silicon wells. *IEEE Trans. Nucl. Sci.* **47**(4), 1303–1306.

Allier, C. P., Hollander, R. W., Sarro, P. M., and van Eijk, C. W. E. (2000b). Scintillation light read-out by thin photodiodes in silicon wells. *Nucl. Instrum. Methods Phys. Res., Sect. A* **442**, 255–258.

Allier, C. P., van Loef, E. V. D., Dorenbos, P., Hollander, R. W., van Eijk, C. W. E., Krämer, K. W., and Güdel, H. U. (2002). Readout of a LaCl$_3$(Ce^{3+}) scintillation crystal with a large area avalanche photodiode. *Nucl. Instrum. Methods Phys. Res., Sect. A* **485**, 547–550.

Altzitzoglou, T., Denecke, B., Johansson, L., and Sibbens, G. (2002). Standardization of ^{89}Sr using three different methods. *Appl. Radiat. Isot.* **56**, 447–452.

Amersham Biosciences Catalog (2002). "Biodirectory 2002." pp. 10–41. Amersham Biosciences Corp., Piscataway, NJ.

Amersham Life Sciences (1996a). Evaluation of the rigid Wallac 96-well microtitre plates for use in scintillation proximity assays. *Proximity News*, No. 1, p. 4. Amersham International plc, Buckinghamshire, UK.

Amersham Life Sciences (1996b). Colour quench correction in scintillation proximity assays. *Proximity News*, No. 6, p. 4. Amersham International plc, Buckinghamshire, UK.

Amersham Pharmacia Biotech. (1999). The use of secondary antibody coated PVT SPA beads to measure insulin in serum. Amersham Pharmacia Biotech, Piscataway, NJ, *Proximity News*, No. 6, p. 4.

Amersham Pharmacia Biotech. (2000). Miniaturization of the camp SPA direct screening assay to 384-well format. Amersham Pharmacia Biotech, Piscataway, NJ, *Proximity News*, No. 57, p. 4.

Anderson, D. F. (1990). Cerium fluoride: a scintillator for high rate applications. *Nucl. Instrum. Methods Phys. Res. Sect. A* **287**, 606–612.

Antonini, P., Bressi, G., Carugno, G., and Iannuzzi, D. (2001). Scintillation properties of YAG:Yb crystals. *Nucl. Instrum. Methods Phys. Res., Sect. A* **460**, 469–471.

Aoyama, T., Koyama, S., and Kawaura, C. (2002). An in-phantom dosimetry system using pin silicon photodiode radiation sensors for measuring organ doses in x-ray CT and other diagnostic radiology. *Med. Phys.* **29**(7), 1504–1510.

Armbruster, B., Blair, I., Bodmann, B. A., Booth, N. E., Drexlin, G. *et al.* (1998). Measurement of the weak neutral current excitation ^{12}C $(\nu_\mu \, \nu'_\mu)$ $^{12}C^*(1^+, 1; 15.1 \, \text{MeV})$ at $E_{\nu_\mu} = 29.8 \, \text{MeV}$ *Phys. Lett. B* **423**, 15–20.

Asfahani, J. (1999). Optimization of spectrometric gamma-gamma probe configuration using very low radioactivity sources for lead and zinc grade determination in borehole logging. *Appl. Radiat. Isot.* **51**, 449–459.

Atkinson, M., Fent, J., Fisher, C., Freund, P., Hughes, P., Kirby, J., Osthoff, A., and Pretzl, K. (1987). Initial tests of a high resolution scintillating fibre. *Nucl. Instrum. Methods Phys. Res., Sect. A* **254**, 500–514.

Attie, M. R. P., Koskinas, M. F., Dias, M. S., and Fonseca, K. A. (1998). Absolute disintegration rate measurements of Ga-67. *Appl. Radiat. Isot.* **49**(9–11), 1175–1177.

Auffray, E., Baccaro, S., Beckers, T., Benhammou, A. N., Belsky, A. N., Borgia, B., Boutet, D., Chipaux, R., Daninei, I., and de Notaristefani, F. *et al.* (1996). Extensive studies of CeF$_3$ crystals, a good candidate for electromagnetic calorimetry at future accelerators. *Nucl. Instrum. Methods Phys. Res., Sect. A* **383**, 367–390.

Averkiev, V. V., Lyapidevskii, V. K., and Salakhutdinov, G. Kh. (1990). Spectrometric characteristics of bismuth germanate detectors in the X-ray and gamma-quanta energy range of 4.5–662 keV. *Prib. Tekhn. Eksp.* **4**, 80–82.

Avset, B. S., Ellison, J. A., Evensen, L., Hall, G., Hansen, T.-E., Roe, S., and Wheadon, R. (1990). Silicon drift photodiodes. *Nucl. Instrum. Methods Phys. Res., Sect. A* **288**, 131–136.

Bähr, J., Bärwolff, H., Kantserov, V., Kell, G., and Nahnhauer, R. (2000). Investigation of silicon avalanche photodiodes for use in scintillating fiber trackers. *Nucl. Instrum. Methods Phys. Res., Sect. A* **442**, 203–208.

Baerg, A. P. (1973). The efficiency extrapolation method in coincidence counting. *Nucl. Instrum. Methods* **112**, 143–150.

Baker, C. A., Poorman, R. A., Kézdy, F. J., Staples, D. J., Smith, C. W., and Elhammer, Å. P. (1996). A scintillation proximity assay for UDP-GalNAc:polypeptide, N-acetylgalactose-aminyltransferase. *Anal. Biochem.* **239**, 20–24.

Balcerzyk, M., Moszyński, M., Kapusta, M., and Szawlowski, M. (2001). Timing properties of LuAP:Ce studied with large-area avalanche photodiodes. *IEEE Trans. Nucl. Sci.* **48**(6), 2344–2347.

Balcerzyk, M., Klamra, W., Moszyński, M., Kapusta, M., and Szawlowski, M. (2002). Energy resolution and light yield non-proportionality of ZnSe:Te scintillator studied by large area avalanche photodiodes and photomultipliers. *Nucl. Instrum. Methods Phys. Res., Sect. A* **482**, 720–727.

Baldochi, S. L., Shimamura, K., Nakano, K., and Mujilatu, N. (1999). Growth and optical characteristics of Ce-doped and Ce:Na-codoped BaLiF₃ single crystals. *J. Cryst. Growth* **200**, 521–526.

Banks, M., Graber, P., Proudfoot, A. E. I., Arod, C. Y., Allet, B., Bernard, A. R., Sebille, E., McKinnon, M., Wells, T. N. C., and Solari, R. (1995). Soluble interleukin-5 receptor α—chain binding assays: use of screening and analysis of interleukin-5 mutants. *Anal. Biochem.* **230**, 321–328.

Belle, Y. S., Zlobin, L. I., Moroz, G. I., and Yushkevich, G. F. (1974). Selection of an optimum detector size for radioisotope analysis. *Spectrom. Meth. Anal. Radioakt. Zagryaz. Pochv. Aerozol., Dokl. Vses. Soveshch.* 11–14.

Belsky, A. N., Auffray, E., Lecoq, P., Dujardin, C., Garnier, N., Canibano, H., Pedrini, C., and Petrosyan, A. G. (2001). Progress in the development of LuAlO₃-based scintillators. *IEEE Trans. Nucl. Sci.* **48**(4), 1095–1100.

Beveridge, M., Park, Y.-W., Hermes, J. et al. (2000). Detection of p56[lck] kinase activity using scintillation proximity assay in 384-well format and imaging proximity assay in 384- and 1536-well format. *J. Biomol. Screen.* **4**, 205–211,

Blair, J., Smith, J. T., Wood, E., Picardo, M., Skinner, R., and Cook, N. (1995). The effect of temperature on the GAP.SH2/phosphotyrosine peptide interaction; analysis using scintillation proximity assay. (Poster). *2nd European Conference On High Throughput Screening*, Budapest, May. (Copy available from Amersham International plc., Whitechurch, Cardiff CF4 7YT, UK.)

Bliss, M., and Craig, R. A. (1995). A variety of neutron sensors based on scintillating glass waveguides. *In* "Pacific Northwest Fiber Optic Sensor Workshop" (E. Udd, Ed.), pp. 152–158.

Bliss, M., Brodzinski, R. L., Craig, R. A., Geelhood, B. D., Knopf, M. A., Miley, H. S., Perkins, R. W., Reeder, P. L., Sunberg, D. S., Warner, R. A., and Wogman, N. A. (1995a). Glass-fiber-based neutron detectors for high- and low-flux environments. *In* "Photoelectronic Detectors, Cameras, and Systems" (C. B. Johnson, E. J. Fenyves, Eds.), pp. 108–117. Proceedings SPIE.

Bliss, M., Craig, R. A., Reeder, P. L., and Sunberg, D. S. (1995b). Development of a real-time dosimeter for therapeutic neutron radiation. *IEEE Trans. Nucl. Sci.* **42**, 639–643.

Bliss, M., Craig, R. A., and Sunberg, D. S. (1996). A wearable neutron detector. *In* "Proceedings of the 1996 Institute of Nuclear Materials Management".

Boernsen, K. O., Floeckher, J. M., and Bruin, G. J. M. (2000). Use of a microplate scintillation counter as a radioactivity detector for miniaturized separation techniques in drug metabolism. *Anal. Chem.* **72**, 3956–3959.

Bosse, R., Garlick, R., Brown, B., and Menard, L. (1998). Development of non-seperation binding and functional assays for G protein-coupled receptors for high throughput screening: pharmacological characterization of the immobilized CCRS receptor on FlashPlate. *J. Biomol. Screen.* **3**, 285–292.

Bramblett, R. L., Ewing, R. I., and Bonner, T. W. (1960). A new type of neutron spectrometer. *Nucl. Instrum. Methods* **9**, 1–12.

Bressi, G., Carugno, G., Conti, E., Del Noce, C., and Iannuzzi, D. (2001). New prospects in scintillation crystals. *Nucl. Instrum. Methods Phys. Res., Sect. A* **461**, 361–364.

Briesmeister, J. (2000). MCNP—a general Monte Carlo N-particle transport code. Los Alamos National Laboratory Manual, Los Alamos, NM.

Briesmeister, J. (2002). MCNP—a general Monte Carlo N-particle transport code. Version 4C. Los Alamos National Laboratory Publication LA-13709-M, Los Alamos, NM.

Brinckmann, P. (1965). CsI(Na) scintillation crystals. *Phys. Lett.* **15**, 305.

Brinkman, G. A., and Veenboer, J. Th. (1979). Calibration of well-type thallium-activated sodium iodide detectors. *Int. J. Appl. Radiat. Isot.* **30**, 171–176.

Brinkman, G. A., Alten Jr., A. H. W., and Veenboer, J. Th. (1963a). Absolute standardization with a NaI(Tl) crystal. I. *Int. J. Appl. Radiat. Isot.* **14**, 153–157.

Brinkman, G. A., Alten Jr., A. H. W., and Veenboer, J. Th. (1963b). Absolute standardization with a NaI(Tl) crystal. II. *Int. J. Appl. Radiat. Isot.* **14**, 433–437.

Brinkman, G. A., Alten Jr., A. H. W., and Veenboer, J. Th. (1965). Absolute standardization with a NaI(Tl) crystal. IV. *Int. J. Appl. Radiat. Isot.* **16**, 15–18.

Brollo, S., Zanella, G., and Zannoni, R. (1990). Light yield in cerium scintillating glasses under x-ray excitation. *Nucl. Instrum. Methods Phys. Res., Sect. A* **293**, 601–605.

Bronson, F. L., and Wang, L. (1996). *Mater. Manage.* **25**, 154.

Brookes, F. D. (1979). Development of organic scintillators. *Nucl. Instrum. Methods* **162**, 477–505.

Bross, A. D. (1990). Development of intrinsic IPT scintillator. *Nucl. Instrum. Methods Phys. Res. Sect. A* **295**, 315–322.

Buchan, K. W., Sumner, M. J., and Watts, I. S. (1993). Human placental membranes contain predominantly ET_B receptors. *J. Cardiovasc. Pharmacol.* **22**(Suppl. 8), S136-S139.

Byrd, R. C., Auchampaugh, G. F., Moss, C. E., and Feldman, W. C. (1992). Warhead counting using neutron scintillators: detector development, testing and demonstration. *IEEE Trans. Nucl. Sci.* **39**, 1051–1055.

Caria, M., Masciotta, P., and Serci, S. (1981). On the detection of low energy γ-rays with plastic scintillators. *Nucl. Instrum. Methods Phys. Res.* **188**, 473–474.

Carrier, C., and Lecomte, R. (1990). Recent results in scintillation detection with silicon avalanche photo diodes. *IEEE Trans. Nucl. Sci.* **37**, 209–214.

Castoldi, A., Chen, W., Gatti, E., Holl, P., and Rehak, P. (2000). Fast silicon drift diodes free from bias connections on the light entering side. *Nucl. Instrum. Methods Phys. Res., Sect. A* **439**, 483–496.

Chaminade, J. P., Viraphong, O., Guillen, F., Fouassier, C., and Czirr, B. (2001). Crystal growth and optical properties of new neutron detectors $Ce^{3+}:Li_6R(BO_3)_3$ (R=Gd,Y). *IEEE Trans. Nucl. Sci.* **48**(4), 1158–1161.

Chandrakala, B., Elias, B. C., Mehra, U., Umapathy, N. S., Dwarakanath, P., Balganesh, T. S., and DeSousa, S. M. (2001). Novel scintillation proximity assay for measuring membrane-associated steps of peptidoglycan biosynthesis in *Escherichia coli. Antimicrob. Agents Chemother.* **45**(3), 768–775.

Chaney, R. C., Fenyves, E. J., Anderson, J. A., Antich, P. P., and Atac, M. (1992). Testing of the spatial resolution and efficiency of scintillating fiber PET modules. *IEEE Trans. Nucl. Sci.* **39**, 1472–1474.

Cheung, A., and Zhang, J. (2000). A scintillation proximity assay for poly(ADP-ribose) polymerase. *Anal. Biochem.* **282**, 24–28.

Childress, N. L., and Miller, W. H. (2002). MCNP analysis and optimization of a triple crystal phoswich detector. *Nucl. Instrum. Methods Phys. Res., Sect. A* **490**, 263–270.

Chiles, M. M., Bauer, M. L., and McElhaney, S. A. (1990). Multi-energy neutron detector for counting thermal neutrons, high-energy neutrons, and gamma photons separately. *IEEE Trans. Nucl. Sci.* **37**, 1348–1350.

Chiozzi, P., De Felice, P. Fazio, A., Pasquale, V., and Verdoya, M. (2000a). Laboratory application of NaI(Tl) γ-ray spectrometry to studies of natural radioactivity in geophysics. *Appl. Radiat. Isot.* **53**, 127–132.

Chiozzi, P., Pasquale, V., Verdoya, M., and De Felice, P. (2000b). Practical applicability of field γ-ray scintillation spectrometry in geophysical surveys. *Appl. Radiat. Isot.* **53**, 215–220.

Combes, C. M., Dorenbos, P., van Eijk, C. W. E., Gesland, J. Y., and Rodnyi, P.A. (1997). Optical and scintillation properties of $LiBaF_3$: Ce crystals. *J. Lumin.* **72–74**, 753–755.

Combes, C. M., Dorenbos, P., Hollander, R. W., and van Eijk, C. W. E. (1998). A thermal-neutron scintillator with n/γ discrimination $LiBaF_3$:Ce,Rb. *Nucl. Instrum. Methods Phys. Res., Sect. A* **416**, 364–370.

Cook, N. D. (1996). Scintillation proximity assay: a versatile high throughput screening technology. *Drug Discov. Today* **1**, 287–294.

Coupet, J., Dou, Y., and Kowal, D. (1998). A fast and easy FlashPlate assay for high throughput screening of cloned G protein-coupled receptors by [^{35}S]-GTPγS binding. PerkinElmer Life Sciences, FlashPlate File No. 3, p. 6. Boston, MA.

Crismatec Co. (1991). Scintillation Detectors Catalog. Crismatic Co., France.

Crookes, W. (1903a). The emanation of radium. *Proc. Royal Soc.* (London) **A71**, 405–408.

Crookes, W. (1903b). Certain properties of the emanation of Radium. *Chemical News* **87**, 241.

Dalton, A. W. (1987). Light conversion efficiency of small lithium scintillators for electrons, protons, deuterons and alpha particles. *Nucl. Instrum. Methods Phys. Res., Sect. A* **254**, 361–366.

Datema, C. P., Meng, L. J., and Ramsden, D. (1999). Results obtained using a 61-pixel hybrid photodiode scintillation camera. *Nucl. Instrum. Methods Phys. Res., Sect. A* **422**, 656–660.

Dayton, B. D., Chiou, W. J., Opgenorth, T. J., and Wu-Wong, J. R. (2000). Direct determination of endothelin receptor antagonist levels in plasma using a scintillation proximity assay. *Life Sci.* **66**(10), 937–945.

De, A., Dasgupta, S. S., and Sen, D. (1993). Time dispersion in large plastic scintillation neutron detectors. *IEEE Trans. Nucl. Sci.* **40**, 1329–1332.

De Felice, P. and Myteberi, Xh. (1995). Standardization of ^{125}I by the sum-peak method and the results of a bilateral comparison between ENEA (Italy) and INP (Albania). *J. Radioanal. Nucl. Chem. Lett.* **200**, 109–118.

Delaporte, E., Slaughter, D. E., Egan, M. A., Gatto, G. J., Santos, A., Shelley, J., Price, E., Howells, L., Dean, D. C., and Rodrigues, A. D. (2001). The potential for CYP2D6 inhibition screening using a novel scintillation proximity assay-based approach. *J. Biomol. Screening* **6**(4), 225–231.

DeLapp, N. W., McKinzie, J. H., Sawyer, B. D., Vandergriff, A., Falcone, J., McClure, D., and Felder, C. C. (1999). Determination of [^{35}S]guanosine-5'-O-(3-thio)triphosphate binding mediated by cholinergic muscarinic receptors in membranes from Chinese hamster ovary cells and rat striatum using an anti-G protein scintillation proximity assay. *J. Pharmacol. Exp. Ther.* **289**(2), 946–955.

Del Guerra, A., Damiani, C., Zavattini, G., Di Domenico, G., Belcari, N., and Vaiano, A. (2001). X-Y position readout of a YAP:Ce crystal matrix using wavelength-shifting fibers. *IEEE Trans. Nucl. Sci.* **48**(4), 1108–1113.

Denecke, B. (1987). Measurement of the 59.5 keV gamma-ray emission probability in the decay of ^{241}Am with a 4π-CsI(Tl)-sandwich spectrometer. *Int. J. Appl. Radiat. Isot.* **38**(10), 823–830.

Denecke, B. (1994). Absolute activity measurements with the windowless 4π-CsI(Tl)-sandwich spectrometer. *Nucl. Instrum. Methods Phys. Res., Sect. A* **339**, 92–98.

Derenzo, S. E., Weber, M. J., and Klintenberg, M. K. (2002). Temperature dependence of the fast, near-band-edge scintillation from CuI, HgI$_2$, PbI$_2$, ZnO:Ga and CdS:In. *Nucl. Instrum. Methods Phys. Res., Sect. A* **486**, 214–219.

De Serres, M., McNulty, M. J., Christensen, L., Zon, G., and Findlay, J. W. A. (1996). Development of a novel scintillation proximity competitive hybridization assay for the determination of phosphorothioate antisense oligonucleotide plasma concentrations in a toxicokinetic study. *Anal. Biochem.* **233**, 228–233.

Destruel, P., Taufer, M., D'Ambrosio, C., Da Via, C., Fabre, J. P., Kirkby, J., and Leutz, H. (1989). A new plastic scintillator with large Stokes shift. *Nucl. Instrum. Methods Phys. Res., Sect. A* **276**, 69–77.

Dias, M. S., and Koskinas, M. F. (1995). Accidental summing correction in ^{125}I activity determination by the sum-peak method. *Appl. Radiat. Isot.* **46**, 945–948.

Dietrich, H. B., Purdy, A. E., Murray, R. B., and Williams, R. T. (1973). Kinetics of self-trapped holes in alkali halide crystals. Experiments in thallium-activated sodium iodide and thallium-activated potassium iodide. *Phys. Rev. B* **8**, 5894–5901.

Dorenbos, P., and de Haas, J. T. M. (1994). Stratech Report 94–03, Radiation Technology Group, Delft University of Technology, Delf, The Netherlands.

Dorenbos, P., Visser, R., van Eijk, C. W. E., Hollander, R. W., and den Hartog, H. W. (1991). X-ray and gamma ray luminescence of Ce^{+3} doped BaF$_2$ crystals. *Nucl. Instrum. Methods Phys. Res., Sect. A* **310**, 236–239.

Dorenbos, P., de Haas, J. T. M., Visser, R., van Eijk, C. W. E., and Hollander, R. W. (1993). X-ray and gamma-ray luminescence of Ce^{+3} doped BaF$_2$ crystals. *IEEE Trans. Nucl. Sci.* **40**, 424–430.

Earnshaw, D. L., and Pope, A. J. (2001). FlashPlate scintillation proximity assays for characterization and screening of DNA polymerase, primase, and helicase activities. *J. Biomol. Screening* **6**(1), 39–46.

Eisen, Y., Evans, L. G., Starr, R., and Trombka, J. I. (2002). CdWO$_4$ scintillator as a compact gamma ray spectrometer for planetary lander missions. *Nucl. Instrum. Methods Phys. Res., Sect. A* **490**, 505–517.

Elane Streets, W. (1994). Development of self-absorption coefficients for the determination of gamma-emitting radionuclides in environental and mixed waste samples. *Nucl. Instrum. Methods Phys. Res., Sect. A* **353**, 702–705.

Eldridge, J. S., and Crowther, P. (1964). Absolute determination of ^{125}I in clinical applications. *Nucleonics* **22**(6), 56–59.

Elster, J., and Geitel, H. (1903). Uber die durch radioactive emanation erregte scintillierende Phosphoreszenz der Sidot-Blende. *Phys. Z.* **4**, 439–440.

Entine, G. (1990). Large area silicon avalanche photodiodes for scintillation detectors. *Nucl. Instrum. Methods Phys. Res., Sect. A* **288**, 137–139.

Ershov, N. N., Zakharov, N. G., and Rodnyi, P. A. (1983). Spectral and kinetic study of the characteristics of intrinsic luminescence of fluorite-type crystals. *Opt. Spektrosk. (USSR)* **53**(1), 51.

Evans, A. E., Jr. (1980). Gamma-ray response of a 38-mm bismuth germanate scintillator. *IEEE Trans. Nucl. Sci.* **NS-27**, 172–175.

Evans, D. B., Rank, K. B., and Sharma, S. K. (2002). A scintillation proximity assay for studying inhibitors of human tau protein kinase II/cdk5 using a 96-well format. *J. Biochem. Biophys. Methods* **50**, 151–161.

Farrel, R., Olschner, F., Frederick, E., McConchie, L., Vanderpuye, K., Squillante, M. R., and Entine, G. (1990). Large area silicon avalanche photodiodes for scintillation detectors. *Nucl. Instrum. Methods Phys. Res., Sect. A* **288**, 137–139.

Farukhi, M. R. (1982). Recent developments in scintillator detectors for X-ray and positron CT applications. *IEEE Trans. Nucl. Sci.* **29**, 1237–1249.

Fehlau, P. E. (1994). Integrated neutron/gamma-ray portal monitors for nuclear safeguards. *IEEE Trans. Nucl. Sci.* **41**, 922–926.

Fields, T., and Jankowski, D. (1983). Simple scheme for light collection from large-area plastic scintillators. *Nucl. Instrum. Methods* **215**, 131–133.

Fioretto, E., Innocenti, F., Viesti, G., Cinausero, M., Zuin, L., Fabris, D., Lunardon, M., Nebbia, G., and Prete, G. (2000). CsI(Tl)-photodiode detectors for γ-ray spectroscopy. *Nucl. Instrum. Methods Phys. Res., Sect. A* **442**, 412–416.

Fiorini, C., and Lechner, P. (1999). Continuous charge restoration in semiconductor detectors by means of the gate-to-drain current of the integrated front-end JFET. *IEEE Trans. Nucl. Sci.* **46**(3), 761–764.

Fiorini, C., and Perotti, F. (1997). Scintillation detection using a silicon drift chamber with on-chip electronics. *Nucl. Instrum. Methods Phys. Res., Sect. A* **401**, 104–112.

Fiorini, C., Longini, A., Perotti, F., Labanti, C., Lechner, P., and Strüder, L. (1997). Gamma ray spectroscopy with CsI(Tl) scintillator coupled to silicon drift chamber. *IEEE Trans. Nucl. Sci.* **44**(6), 2553–2560.

Fiorini, C., Longini, A., Perotti, F., Labanti, C., Rossi, E., Lechner, P., and Strüder, L. (2001). Gamma-ray imaging detectors based on SDDs coupled to scintillators. *Nucl. Instrum. Methods Phys. Res., Sect. A* **461**, 565–567.

Fiorini, C., Longini, A., and Perotti, F. (2000). New detectors for γ-ray spectroscopy and imaging, based on scintillators coupled to silicon drift detectors. *Nucl. Instrum. Methods Phys. Res., Sect. A* **454**, 241–246.

Fiorini, C., Perotti, F., and Labanti, C. (1998). Performances of a silicon drift chamber as fast scintillator photodetector for gamma-ray spectroscopy. *IEEE Trans. Nucl. Sci.* **45**(3), 483–486.

Fluss, M. J., Mundy, J. N., and Rothman, S. J. (1983). Radiotracer techniques 4. Detection and assay. 4.3. Pulse counting systems. *Methods Exp. Phys.* **21**, 55–59.

Fraser, G. W. (2002). The ion detection efficiency of microchannel plates. *Int. J. Mass Spectrom.* **215**, 13–30.

Frolik, C. A., Black, E. C., Chandrasekhar, S., and Adrian, M. D. (1998). Development of a scintillation proximity assay for high-throughput measurement of intact parathyroid hormone. *Anal. Biochem.* **265**, 216–224.

Fujii, H., and Takiue, M. (1989). Paraffin scintillator for radioassay of solid support samples. *Appl. Radiat. Isot.* **40**, 495–499.

Fujii, H., and Roessler, N. (1991). Solidifying scintillator for solid support samples. *In* "Liquid Scintillation Counting and Organic Scintillators" (H. Ross, J. E. Noakes, and J. D. Spaulding, Eds.), pp. 69–81. Lewis Publishers, Chelsea, MI.

Fujii, H., Takiue, M., and Aburai, T. (1996). Development of a paraffin scintillator-solid support system for liquid radiochromatography. *In* "Liquid Scintillation Spectrometry 1994" (G. T. Cook, D. D. Harkness, A. B. MacKenzie, B. F. Miller, and E. M. Scott, Eds.), pp. 237–244. Radiocarbon, Tucson, AZ.

Fujii, H., Matsuno, K., and Takiue, M. (1999). Construction of analytical beta ray monitor for liquid waste. *Radioisotopes* **48**, 465–471.

Fujii, H., Matsuno, K., and Takiue, M. (2000). Hybrid radioassay of multiple radionuclide mixtures in waste solutions by using liquid and NaI(Tl) scintillation monitors. *Health Phys.* **79**(3), 294–298.

Funck, E., and Nylandstedt Larsen, A. (1983). The influence from low energy x-rays and Auger electrons on $4\pi\beta$-γ coincidence measurements of electron-capture-decaying nuclides. *Int. J. Appl. Radiat. Isot.* **34**, 565–569.

Garcia Burciaga, G. (1976). The use of ^{65}Zn in the study of the organic compounds of zinc in chicken manure and its transformations upon incorporation into a sandy soil. M.Sc. Thesis, National School of Agriculture, Chapingo, Mexico.

Garcia Burciaga, G., L'Annunziata, M. F., Ortega, M. L., and Alvarado, R. (1978). Chemistry of ^{65}Zn-labelled chicken manure in the soil. Proceedings IAEA International Symposium, Dec. 11–15, Colombo, Sri Lanka, IAEA-SM-235/43, pp. 393–406.

García-Toraño, E., Roteta, M., and Rodríguez Barquero, L. (2000). Standardization of ^{110m}Ag by liquid scintillation and $4\pi\beta$-γ coincidence counting. *Appl. Radiat. Isot.* **52**, 637–641.

García-Toraño, E., Rodríguez Barquero, L., and Roteta, M. (2002). Standardization of ^{134}Cs by three methods. *Appl. Radiat. Isot.* **56**, 211–214.

Gatti, E., and Rehak, P. (1984). Semiconductor drift chamber – an application of a novel charge transport scheme. *Nucl. Instrum. Methods Phys. Res.* **225**, 608–614.

Gatti, E., Rehak, P., and Walton, J. T. (1984). Silicon drift chambers – first results and optimum processing of signals. *Nucl. Instrum. Methods Phys. Res.* **226**, 129–141.

Gektin, C. M., Shiran, N., Voloshinovski, A., Voronova, V., and Zimmerer, G. (1998). Scintillation in LiBaF$_3$(Ce) crystals. *IEEE Trans. Nucl. Sci.* **45**, 505–507.

Gevi, M., and Domenici, E. (2002). A scintillation proximity assay amenable for screening and characterization of DNA gyrase B subunit inhibitors. *Anal. Biochem.* **300**, 34–39.

Gorin, A., Kuroda, K., Manuilov, I., Morimoto, K., Oku, T., Ryazantsev, A., Shimizu, H. M., Suzuki, J., and Tokanai, F. (2002). Development of scintillation imaging device for cold neutrons. *Nucl. Instrum. Methods Phys. Res., Sect. A* **479**, 456–460.

Grabmaier, B. C. (1984). Crystal scintillators. *IEEE Trans. Nucl. Sci.* **NS-31**, 372–376.

Grassmann, H., Moser, H. G., Dietl, H., Eigen, G., Fonseca, V., Lorenz, E., and Mageras, G. (1985a). Improvements in photodiode readout for small CsI(Tl) crystals. *Nucl. Instrum. Methods Phys. Res., Sect. A* **234**, 122–124.

Grassmann, H., Lorenz, E., Moser, H. G., and Vogel, H. (1985b). Results from a CsI(Tl) test calorimeter with photodiode readout between 1 and 20 GeV. *Nucl. Instrum. Methods Phys. Res., Sect. A* **235**, 319–325.

Grigorescu, L. (1963). Nesure absolue de l'activité des radionucléides par la méthode des coincidences beta-gamma. Corrections des coincidences instrumentals et de temps morts. MESUCORA Congress, Paris.

Grigorescu, L. (1973). Accuracy of coincidence measurements. *Nucl. Instrum. Methods* **112**, 151–155.

Grigorescu, E. L., Sahagia, M., Razdolescu, A., Luca, A., and Radwan, R. M. (1998). *Appl. Radiat. Isot.* **49**(9–11), 1165–1170.

Groom, D. E. (1983). *In* "Proceedings of the International Workshop on Bismuth Germanate." (C. Newman Holmes, Ed.), pp. 256–265. Princeton University, November 1982, CONF-821160, DE83 011369.

Groom, D. E. (1984). Silicon photodiode detection of bismuth germanate scintillation light. *Nucl. Instrum. Methods Phys. Res.* **219**, 141–148.

Gruner, S. M., Kirk, G. L., Patel, L., and Kaback, H. R. (1982). A method for rapid, continuous monitoring of solute uptake and binding. *Biochemistry* **21**, 3239–3243.

Gupta, V. P., Dulco, G. B., and Balmer, D. K. (1995). A simple system for estimating gram quantities of plutonium in small waste packages using plastic scintillator detectors. *Radiat. Prot. Man.* **12**, 47–59.

Güsten, H., and Mirsky, J. (1991). PMP, a novel scintillation solute with a large Stokes's shift. *In* "Liquid Scintillation Counting and Organic Scintillators" (H. Ross, J. E. Noakes, and J. D. Spaulding, Eds.), pp. 1–7. Lewis Publishers, Chelsea, MI.

Hajek, M., Berger, T., and Schöner, W. (2000). Comparison of measurements with active and passive Bonner sphere spectrometers. *Trans. Amer. Nucl. Soc. TANSAO* **83**, 263–265.

Hamada, M. M., Costa, F. E., Pereira, M. C. C., and Kubota, S. (2001). Dependence of scintillation characteristics in the CsI(Tl) crystal on Tl^+ concentrations under electron and alpha particle excitations. *IEEE Trans. Nucl. Sci.* **48**(4), 1148–1153.

Han, M.-J., Bummer, P. M., and Jay, M. (1998). Solid scintillation proximity membranes. II. Use in wipe test assays for radioactive contamination. *J. Membr. Sci.* **148**, 223–232.

Hancock, A. A., Vodenlich, A. D., Maldonado, C., and Janis, R. (1995). α_2-adrenergic agonist-induced inhibition of cyclic AMP formation in transfected cell lines using a microtiter-based scintillation proximity assay. *J. Recept. Signal Transduct. Res.* **15**, 557–579.

Haney, J. H., Barnhart, T. E., and Zaidins, C. S. (1999). Extraction of neutral spectral information from Bonner-sphere data. *Nucl. Instrum. Methods Phys. Res., Sect. A* **431**, 551–555.

Hart, H. E., and Greenwald, E. B. (1979). Scintillation proximity assay (SPA): a new method of immunoassay. Direct and inhibition mode detection with human albumin and rabbit antihuman albumin. *Mol. Immunol.* **16**, 265–267.

Hartmann, R., Strüder, L., Kemmer, J., Lechner, P., Fries, O., Lorenz, E., and Mirzoyan, R. (1997). Ultrathin entrance windows for silicon drift detectors. *Nucl. Instrum. Methods Phys. Res., Sect. A* **387**, 250–254.

Hayashi, T. (1982). Recent developments in photomultipliers for nuclear radiation detectors. *Nucl. Instrum. Methods* **196**, 181–186.

He, X., Mueller, J. P., and Reynolds, K. A. (2000). Development of a scintillation proximity assay for β-ketoacyl-acyl carrier protein synthase III. *Anal. Biochem.* **282**, 107–114.

Heath, R. L. (1964). "Scintillation Spectrometry, Gamma-Ray Spectrum Catalogue." 2nd Edition, USAEC Report Number IDO-16880-1.

Heath, R. L. (1983). *In* "Advances in Scintillation Counting" (S.A. McQuarrier, C. Ediss, and L. I. Wiebe, Eds.), pp. 156–175. University of Alberta, Edmonton.

Heath, R. L., Hofstadter, R., and Hughes, E. B. (1979). Inorganic scintillators, a review of techniques and applications. *Nucl. Instrum. Methods* **162**, 431–456.

Hell, E., Knüpfer, W., and Mattern, D. (2000). The evolution of scintillating medical detectors. *Nucl. Instrum. Methods Phys. Res., Sect. A* **454**, 40–48.

Hereforth, L. (1948). "Die Fluoreszensanregung organisher Substanzen mit Alphateilchen, schnellen Elektronen und Gammastrahlen." Thesis presented Sept. 13, 1948, Technical University, Berlin-Charlottenburg.

Hertzberg, R. P., and Pope, A. J. (2000). High-throughput screening: new technology for the 21st century. *Curr. Opin. Chem. Biol.* **4**, 445–451.

Highes, K. T., Ireson, J. C., Jones, N. R. A., and Kivalä, P. (2001). Colour quench correction in scintillation proximity assays using ParaLux Count Mode. Application Notes, pp. 12. PerkinElmer Life and Analytical Sciences, Boston.

Hino, Y, and Ohgaki, H. (1998). Absolute measurement of ^{192}Ir. *Appl. Radiat. Isot.* **49**(9–11), 1179–1183.

Hinrichs, R., and Ueffing, M. (1995). Rapid microplate mitogenic assay using a meltable scintillation wax. *Biotechniques* **19**, 234–238.

Hofstadter, R. (1948). Alkali halide scintillation counters. *Phys. Rev.* **74**, 100–101.

Hofstadter, R. (1950). Properties of scintillation materials. *Nucleonics* **6**(5), 70–72.

Holl, I., Lorenz, E., and Mageras, G. (1988). A measurement of the light yield of common inorganic scintillators. *IEEE Trans. Nucl. Sci.* **35**, 105–109.

Holl, I. Lorenz, E., Natkaniez, D., Renker, D., Schmelz, C., and Schwartz, B. (1995). Some studies of avalanche photodiode readout of fast scintillators. (1994). *IEEE Trans. Nucl. Sci.* **42**, 351–356.

Holland, J. D., Singh, P., Brennand, J. C., and Garman, A. J. (1994). A nonseparation microplate receptor binding assay. *Anal. Biochem.* **222**, 516–518.

Holmberg, P., Rieppo, R., and Passi, P. (1972). Calculated efficiency values for well-type thallium-doped sodium iodide detectors. *Int. J. Appl. Radiat. Isot.* **23**, 115–120.

Horrocks, D. L., and Klein, P. R. (1975). Theoretical considerations for standardization of ^{125}I by the coincidence method. *Nucl. Instrum. Methods* **124**, 585–589.

Horton, J. K., and Baxendale, P. M. (1995). Mass measurements of cyclic AMP formation by radioimmunoassay, enzyme immunoassay, and scintillation proximity assay. *In* "Methods in Molecular Biology," Vol. 41, "Signal Transduction Protocols" (D. A. Kendall, and S. J. Hill, Eds.), pp. 91–105. Humana Press, Totowa, NJ.

Horton, J. K., Smith, L., Ali, A., Baxendale, P. M., Neumann, K., and Kolb, A. (1995). High throughput screening for cAMP formation by scintillation proximity radioimmunoassay. *TopCount Topics*, TCA-021, PerkinElmer Life and Analytical Sciences, Boston.

Hoslin, D., Armstrong, A. W., Hagan, W., Shreve, D., and Smith, S. (1994). A directional fast neutron detector using scintillating fibers and an intensified CCD camera system. *Nucl. Instrum. Methods Phys. Res., Sect. A* **353**, 118–122.

Houtermans, H., and Miguel, M. (1962). 4π-β-γ coincidence counting for the calibration of nuclides with complex decay schemes. *Int. J. Appl. Radiat. Isot.* **13**, 137–142.

Hudson, C., Chaney, R. C., Fenyves, E. J., Hammack, H., and Antich, P. P. (1994). Measurement of the energy resolution of a scintillating fiber detector. *SPIE* **2281**, 65–70.

Hughes, K. T., Ireson, J. C., Jones, N. R. A., and Kivelä, P. (2001). Colour quench correction in scintillation proximity assays using ParaLux Count Mode. Application Notes. pp.12. PerkinElmer Life and Analytical Sciences, Boston.

Hult, M., Altzitzoglou, T., Denecke, B., Persson, L., Sibbens, G., and Reher, D. F. G. (2000). Standardization of ^{204}Tl at IRMM. *Appl. Radiat. Isot.* **52**, 493–498.

Hunter, D., Dratz, A. F., Rohrer, R. H., and Coberly, J. C. (1975). Potential errors in radioassay of ^{125}I. *J. Nucl. Med.* **16**, 952–954.

Ikhlef, A., Skowronek, M., and Beddar, A. S. (2000). X-ray imaging and detection using plastic scintillating fibers. *Nucl. Instrum. Methods Phys. Res., Sect. A* **442**, 428–432.

Inagaki, T., and Takashima, R. (1982). New types of plastic scintillators. *Nucl. Instrum. Methods* **201**, 511–517.

Ishibashi, H., Kurashige, K., Kurata, Y., Susa, K., Kobayashi, M., Tanaka, M., Hara, K., and Ishii, M. (1998). Scintillation performance of large Ce-doped Gd_2SiO_5 (GSO) single crystal. *IEEE Trans. Nucl. Sci.* **45**, 518–521.

Ishibashi, H., Shimizu, K., Susa, K., and Kubota, S. (1989). Cerium doped GSO scintillator and its application to position sensitive detectors. *IEEE Trans. Nucl. Sci.* **36**, 170–172.

Ishii, M., Harada, K., Hirose, Y., Senguttuvan, N., Kobayashi, M., Yamaga, I., Ueno, H., Miwa, K., Shiji, F., Yiting, F., Nikl, M., and Feng, X. Q. (2002). Development of BSO ($Bi_4Si_3O_{12}$) crystal for radiation detector. *Opt. Materials* **19**, 201–212.

Iwanczyk, J. S., Barton, J. B., Dabrowski, A. J., Kusmiss, J. W., and Szymczyk, W. M. (1983a). *In* "Proceedings of the International Workshop on Bismuth Germanate" (C. Newman Holmes, Ed.), pp. 285–289. November 1982, Princeton University, CONF. 821160, DE83 011369.

Iwanczyk, J. S., Barton, J. B., Dabrowski, A. J., Kusmiss, J. W., and Szymczyk, W. M. (1983b). A novel radiation detector consisting of an HgI_2 photodetector coupled to a scintillator. *IEEE Trans. Nucl. Sci.* **NS-30**, 363–367.

Iwanczyk, J. S., Barton, J. B., Dabrowski, A. J., Kusmiss, J. W., Szymczyk, W. M., Huth, G. C., Markakis, J., Schnepple, W. F., and Lynn, R. (1983c). Scintillation spectrometry with HgI_2 as the photodetector. *Nucl. Instrum. Methods* **213**, 123–126.

Iwanczyk, J. S., Dabrowski, A. J., Markakis, J. M., Ortale, C., and Schnepple, W. F. (1984). Large area mercuric iodide photodetectors. *IEEE Trans. Nucl. Sci.* **NS-31**, 336–339.

Iwanczyk, J. S., Patt, B. E., Segal, J., Plummer, J., Vilkelis, G., Hedman, B., Hodgson, K. O., Cox, A. D., Rehn, L., and Metz, J. (1996). Simulation and modeling of a new silicon x-ray drift detector design for synchrotron radiation applications. *Nucl. Instrum. Methods Phys. Res., Sect. A* **380**, 288–294.

Iwanczyk, J. S., Patt, B. E., Tull, C. R., Segal, J. D., Kenney, C. J., Bradley, J., Hedman, B., and Hodgson, K. O. (1999). Large area silicon drift detectors for x-rays – new results. *IEEE Trans. Nucl. Sci.* **46**(3), 284–288.

Jaklevic, J. M., Madden, N. W., and Wiegand, C. E. (1983). A precision beta gauge using a plastic scintillator and photomultiplier detector. *Nucl. Instrum. Methods* 214, 517–518.

James, K. M., Masterson, M. J., and Farrell, R. (1992). Performance evaluation of new large-area avalanche photodiodes for scintillation spectroscopy. *Nucl. Instrum. Methods Phys. Res., Sect. A* 313, 196–202.

Jeffery, J. A., Sharom, J. R., Fazekas, M., Rudd, P., Welchner, E., Thauvette, L., and White, P. W. (2002). An ATPase assay using scintillation proximity beads for high-throughput screening or kinetic analysis. *Anal. Biochem.* 304, 55–62.

Jensen, G. L., Wang, J. C., and Czirr, J. B. (1993). High-efficiency fast neutron detectors. *Nucl. Instrum. Methods Phys. Res., Sect. A* 333, 474–483.

Jessop, R. A. (1993). Recent advances in scintillation proximity assay (SPA) technology. *In* "Liquid Scintillation Spectrometry 1992" (J. E. Noakes, F. Schönhofer, and H. A. Polach, Eds.), pp. 285–289. Radiocarbon, Tucson, AZ.

Jessop, R. A. (1996). New applications of scintillation proximity assay. *In* "Liquid Scintillation Spectrometry 1994" (G. T. Cook, D. D. Harkness, A. B. MacKenzie, B. F. Miller, and E. M. Scott, Eds.), pp. 175–182. Radiocarbon, Tucson, AZ.

Kalber, M. N. (1975). *In* "Proceedings of the NATO Advanced Study Institute on Radiation Damage in Material," Corsica, 1973. Hordhof International Publishing.

Kallman, H. (1950). Scintillation counting with solutions. *Phys. Rev.* 78(5), 621–622.

Kamboj, S., and Kahn, B. (1996). Evaluation of Monte Carlo simulation of photon counting efficiency for germanium detectors. *Health Phys.* 70, 512–529.

Kanyo, M., Reinartz, R., Schelten, J., and Müller, K. D. (1992). Two-dimensional neutron scintillation detector with optimal gamma discrimination. *Nucl. Instrum. Methods Phys. Res., Sect. A* 320, 562–568.

Kapusta, M., Balcerzyk, M., Moszyński, M., and Pawelke, J. (1999). A high-energy resolution observed from a YAP:Ce scintillator. *Nucl. Instrum. Methods Phys. Res., Sect. A* 421, 610–613.

Kariv, I., Stevens, M. E., Behrens, D. L., and Oldenburg, K. R. (1999). High throughput quantitation of cAMP production mediated by activation of seven transmembrane domain receptors. *J. Biomol. Screen.* 4, 27–32.

Karsch, L., Böhm, A., Brinkmann, K.-Th., Demirörs, L., and Pfuff, M. (2001). Design and test of a large-area scintillation detector for fast neutrons. *Nucl. Instrum. Methods Phys. Res., Sect. A* 460, 362–367.

Kawano, T., and Ebihara, H. (1990). Determination of disintegration rates of a ^{60}Co point source and volume sources by the sum-peak method. *Appl. Radiat. Isot.* 41, 163–167.

Kawano, T., and Ebihara, H. (1991). Measurement of disintegration rates of small ^{60}Co sources in lead containers by the sum-peak method. *Appl. Radiat. Isot.* 42, 1165–1168.

Kawano, T., and Ebihara, H. (1992). Error estimation for the sum-peak method by use of two ^{60}Co point sources in place of extended sources. *Appl. Radiat. Isot.* 43, 705–711.

Kemmer, J., Lutz, G., Belau, E., Prechtel, U., and Welser, W. (1987). Low capacity drift diode. *Nucl. Instrum. Methods Phys. Res., Sect. A* 253, 378–381.

Kierstead, J. A., Stoll, S. P., and Woody, C. L. (1994). *Mater. Res. Soc. Symp. Proc.* 348, 469.

Kinloch, D. R., Novak, W., Raby, P., and Toepke, I. (1994). IEEE Conference Record Nuclear Science Symposium and Medical Imaging Conference, Oct. 31–Nov. 6, 1993, San Francisco, p. 661.

Kirk, G., and Gruner, S. (1982). Encapsulated scintillators monitor ^{3}H-solute concentrations. *IEEE Trans. Nucl. Sci.* NS-29, 769–772.

Klamra, W., Kerek, A., Moszyński, M., Norlin, L.-O., Novák, D., and Possnert, G. (2000). Response of BaF$_2$ and YAP:Ce to heavy ions. *Nucl. Instrum. Methods Phys. Res., Sect. A* 444, 626–630.

Klawonn, F., Kleinknecht, K., and Pollmann, D. (1982). A new type of acrylic scintillator. *Nucl. Instrum. Methods* 195, 483–489.

Knickelbein, M. B., Root, J. W., and Hurlbut, C. R. (1983). A plastic scintillation counter for applications in radio-gas-chromatography. *Nucl. Instrum. Methods* 216, 121–125.

Knitel, M. J. (1998). "New Inorganic Scintillators and Storage Phosphors for Detection of Thermal Neutrons". Delft University Press, Delft, The Netherlands, pp 180.

Knitel, M. J., Dorenbos, P., Combes, C. M., Andriessen, J., and van Eijk, C. W. E. (1996b). Luminescence and storage properties of LiYSiO$_4$: Ce. *J. Lumin.* 69, 325–334.

Knitel, M. J., Dorenbos, P., de Haas, J. T. M., and van Eijk, C. W. E. (1995). LiBaF$_3$, as a thermal neutron scintillator. *Radiat. Measure.* **24**, 361–363.

Knitel, M. J., Dorenbos, P., de Haas, J. T. M., and van Eijk, C. W. E. (1996a). LiBaF$_3$, a thermal neutron scintillator with optimal n-γ discrimination. *Nucl. Instrum. Methods Phys. Res., Sect. A* **374**, 197–201.

Knoll, G. F., Henderson, T. M., and Felmlee, W. J. (1987). A novel ^{3}He scintillation detector. *IEEE Trans. Nucl. Sci.* **NS-34**, 470–474.

Knoll, G. F., Knoll, T. F., and Henderson, T. M. (1988). Light collection in scintillation detector composites for neutron detection. *IEEE Trans. Nucl. Sci.* **35**, 872–875.

Kobayashi, H. (1987). Kyoto University preprint, KUNS-900, November 1987.

Kobayashi, M., Ishii, M., Harada, K., and Yamaga, I. (1996). Bismuth silicate Bi$_4$Si$_3$O$_{12}$, a faster scintillator than bismuth germanate Bi$_4$Ge$_3$O$_{12}$. *Nucl. Instrum. Methods Phys. Res., Sect. A* **372**, 45–50.

Komesli, S., Page, M., and Dutartre, P. (1999). Use of Flashplate technology *in vitro* measurement of ^{125}I-labeled TGF-β1 binding on chimeric extracellular domain of type II transforming growth factor receptor. FlashPlate File #14, pp. 6. PerkinElmer Life and Analytical Sciences, Boston.

Kononenko, W., Selove, W., and Theodosoiu, G. E. (1983). Scintillator-waveshifter light measurements. *Nucl. Instrum. Methods* **206**, 91–97.

Korzhik, M., and Lecoq, P. (2001). Search of new scintillation materials for nuclear medicine applications. *IEEE Trans. Nucl. Sci.* **48**(3), 628–631.

Kroger, F. A. (1948). "Some Aspects of the Luminescence of Solids." Elsevier, Amsterdam.

Kryger, R. A., Azhari, A., Ramakrishnan, E., Thoennessen, M., and Yokoyama, S. (1994). Efficiency of a BaF$_2$ scintillator detector for 15–150 MeV neutrons. *Nucl. Instrum. Methods Phys. Res. Sect. A* **346**, 544–547.

Kubo, N., Mabuchi, M., Katoh, C., Arai, H., Morita, K., Tsukamoto, E., Morita, Y., and Tamaki, N. (2002). Validation of left ventricular function from gated single photon computed emission tomography by using a scintillator-photodiode camera: a dynamic myocardial phantom study. *Nucl. Med. Commun.* **23**(7), 639–643.

Kubota, S., Sakuragi, S., Hashimoto, S., and Ruangen, I. (1988). A new scintillation material: pure CsI with 10 ns decay time. *Nucl. Instrum. Methods Phys. Res., Sect. A* **268**, 275–277.

Kulkarni, P. V., Anderson, J. A., Antich, P. P., Prior, J. O., Zhang, Y., Fernando, J., Constantinescu, A., Goomer, N. C., Parkey, R. W., Fenyves, E., Chaney, R. C., Srivastava, S. C., and Mausner, L. F. (1993). New approaches in medical imaging using plastic scintillating detectors. *Nucl. Instrum. Methods Phys. Res., Sect. B* **79**, 921–925.

Kurahashi, H. (1983). In "Proceedings of the International Workshop on Bismuth Germanate," Princeton University, November 1982 (C. Newman Holmes, Ed.), pp. 290–293. CONF-821160, DE83 011369.

Laitinen, P., Tiourine, G., Touboltsev, V., and Räisänen, J. (2002). Detection system for depth profiling of radiotracers. *Nucl. Instrum. Methods Phys. Res., Sect. B* **190**, 183–185.

L'Annunziata, M. F. (1984). The detection and measurement of radionuclides. In "Isotopes and Radiation in Agricultural Sciences" (M. F. L'Annunziata and J. O. Legg, Eds.), Vol. 1, pp. 141–231.

L'Annunziata, M. F. (1987). "Radionuclide Tracers: Their Detection and Measurement." Academic Press, London.

Laval, M., Moszyński, M., Allemand, R., Cormoreche, E., Guinet, P., Odru, R., and Vacher, I. (1983). Barium fluoride—inorganic scintillator for subnanosecond timing. *Nucl. Instrum. Methods* **206**, 169–176.

Lecomte, R., Pepin, C., Rouleau, D., Dautet, H., McIntyre, R. J., McSween, D., and Webb, P. (1999). Radiation detection measurements with a new "Buried Junction" silicon avalanche photodiode. *Nucl. Instrum. Methods Phys. Res., Sect. A* **423**, 92–102.

Lees, J. E., and Fraser, G. W. (2002). Efficiency enhancements for MCP-based beta autoradiography imaging. *Nucl. Interim. Methods Phys. Res., Sect. A* **477**, 239–243.

Lees, J. E., and Hales, J. M. (2001). Imaging and quantitative analysis of tritium-labelled cells in lymphocyte proliferation assays using microchannel plate detectors originally developed for x-ray astronomy. *J. Immunol. Methods* **247**, 95–102.

Lees, J. E., and Richards, P. G. (1999). Rapid, high-sensitivity imaging of radiolabeled gels with microchannel plate detectors. *Electrophoresis* **20**, 2139–2143.

Lees, J. E., Fraser, G. W., and Carthew, P. (1998). Microchannel plate detectors for ^{14}C autoradiography. *IEEE Trans. Nucl. Sci.* **45**, 1288–1292.

Lees, J. E., Fraser, G. W., and Dinsdale, D. (1997). Direct beta autoradiography using microchannel plate (MCP) detectors. *Nucl. Instrum. Methods Phys. Res., Sect. A* **392**, 349–353.

Lees, J. E., Pearson, J. F., Fraser, G. W., Hales, J. M., and Richards, P. G. (1999). An MCP-based system for beta autoradiography. *IEEE Trans. Nucl. Sci.* **46**, 636–638.

Leesnitzer, L. M., Parks, D. J., Bledsoe, R. K., Cobb, J. E., Collins, J. L., Consler, T. G., Davis, R. G., Hull-Rude, E. A., Lenhard, J. M., Patel, L., Plunket, K. D., Shenk, J. L., Stimmel, J. B., Therapontos, C., Willson, T. M., and Blanchard, S. G. (2002). Functional consequences of cysteine modification in the lingand binding sites of peroxisome proliferator activated receptors by GW9662. *Biochem.* **41**, 6640–6650.

Lehmann, W. (1965). Optical absorption edges of zinc oxide and cadmium sulfide. *J. Electrochem. Soc.* **112**(11), 1150–1151.

Lehmann, W. (1966). Edge emission of n-type conducting ZnO and CdS. *Solid-State Electron.* **9**(11–12), 1107–1110.

Lempicki, A. (1995). The physics of inorganic scintillators. *J. Appl. Spectrosc.* **62**, 209–231.

Lempicki, A., Berman, E., Wojtowicz, A. J., Balcerzyk, M., and Boatner, L. A. (1993). Cerium-doped orthophosphates: new promising scintillators. *IEEE Trans. Nucl. Sci.* **40**, 384–387.

Lempicki, A., Brecher, C., Wisniewski, D., Zych, E., and Wojtowicz, A. J. (1996). Lutetium aluminate: spectroscopic and scintillation properties. *IEEE Trans. Nucl. Sci.* **43**, 1316–1320.

Lempicki, A., Randles, M. H., Wisniewski, D., Balcerzyk, M., Brecher, C., and Wojtowicz, A. J. (1995). $LuAlO_3$:Ce and other aluminate scintillators. *IEEE Trans. Nucl. Sci.* **42**, 280–284.

Lempicki, A., Wojtowicz, A. J., and Brecher, C. (1997). Inorganic scintillators. *In* "Wide-Gap Luminescent Materials: Theory and Applications" (S. R. Rotman, Ed.), pp. 235–301. Kluwer Academic Publishers, Norwell, MA.

Lerch, R. A., Ambos, H. D., Bergmann, S. R., Sobel, B. E., and Ter-Pogossian, M. M. (1982). Kinetics of positron emitters *in vivo* characterized with a beta probe. *Am. J. Physiol.*, **242**, H62-H67.

Lerner, C. G., and Carter, D. C. (1995). PCR colony screening using the scintillation proximity assay to detect inserts in cloning vectors. *Biotechniques* **19**, 914–917.

Leutz, H. (1995). Scintillating fibers. *Nucl. Instrum. Methods Phys. Res. Sect. A* **364**, 422–448.

Litton Airtron SYNOPTICS, Technical Brief. (1997). Ce:LuAP scintillator crystals. Litton Airtron SYNOPTICS, Synthetic Crystals and Optical Products, Charlotte, NC.

Li, H. B., Liu, Y., Chang, C. C., Chang, C. Y., Chao, J. H., Chen, C. P. and Chen, T. Y. *et al.* (2001). A CsI(Tl) scintillating crystal detector for the studies of low-energy neutrino interactions. *Nucl. Instrum. Methods Phys. Res., Sect. A* **459**, 93–107.

Lindroth, E., Danared, H., Glans, P., Pešić, Z., Tokmen, M., Vikor, G., and Schuch, R. (2001). QED effects in Cu-like Pb recombination resonances near threshold. *Phys. Rev. Lett.* **86**, 5027–5030.

Liu, J., Feldman, P. A., Lippy, J. S., Bobkova, E., Kurilla, M. G., and Chung, T. D. Y. (2001). A scintillation proximity assay for RNA detection. *Anal. Biochem.* **289**, 239–245.

Lo, C. C., and Leskovar, B. (1981). Performance studies of high gain photomultipliers having Z-configuration of microchannel plates. *IEEE Trans. Nucl. Sci.* **NS-28**, 698–704.

Lorenz, E. (1983). *In* "Proceedings of the International Workshop on Bismuth Germanate." Princeton University, November 1982 (C. Newman Holmes, Ed.), pp. 229–255. CONF-821160, DE83 011369.

Lorenz, E., Mageras, G., and Vogel, H. (1986). Test of barium fluoride calorimeter with photodiode readout between 2 and 40 GeV incident energy. *Nucl. Instrum. Methods Phys. Res., Sect. A* **249**, 235–240.

Lucas, H. F. (1957). Improved low-level alpha-scintillation counter for radon. *Rev. Sci. Instrum.* **28**, 680–683.

Macarrón, R., Mensah, L., Cid, C., Carranza, C., Benson, N., Pope, A. J., Díez, E. (2000). A homogeneous method to measure aminoacyl-tRNA synthetase aminoacylation activity using scintillation proximity assay technology. *Anal. Biochem.* **284**, 183–190.

Malcolme-Mawes, D. J., and Massey, S. (1980). The variation of γ-counting efficiency for iodine-125 with sample composition. *Int. J. Appl. Radiat. Isot.* **31**, 155–158.

Mandine, E., Gofflo, D., Jean-Baptiste, V., Sarubbi, E., Touyer, G., Deprez, P., and Lesuisse, D. (2001). Src homology-2 domain binding assays by scintillation proximity and surface plasmon resonance. *J. Mol. Recogn.* **14**, 254–260.

Mann, W. B., Rytz, A., Spernol, A., and McLaughlin, W. L. (1988). "Radioactivity Measurements, Principles and Practice". Pergamon Press, Oxford.

Mansfield, R. K., Bhattacharyya, D., Hartman, N. G., and Jay, M. (1996). Scintillation proximity radioimmunoassay with microporous membranes. *Appl. Radiat. Isot.* **47**, 323–328.

Marler, J., McCauley, T., Reucroft, S., Swain, J., Budil, D., and Kolaczkowski, S. (2000). Studies of avalanche photodiode performance in a high magnetic field. *Nucl. Instrum. Methods Phys. Res., Sect. A* **449**, 311–313.

Marriott, C. J., Cadorette, J. E., Lecomte, R., Scasner, V., Rousseau, J., and van Lier, J. E. (1994). High resolution PET imaging and quantitation of pharmaceutical biodistributions in a small animal using avalanche photodiode detectors. *J. Nucl. Med.* **35**, 1390–1396.

Martens, H., and Naes, T. (1989). "Multivariate Calibration." Wiley, New York.

Martin, R. H., and Taylor, J. G. V. (1992). The standardization of ^{125}I: a comparison of three methods. *Nucl. Instrum. Methods Phys. Res., Sect. A* **312**, 64–

Matzke, M. (2002). Propagation of uncertainties in unfolding procedures. *Nucl. Instrum. Methods Phys. Res., Sect. A* **476**, 230–241.

McDonald, O. B., Chen, W. J., Ellis, B., Hoffman, C., Overton, L., Rink, M., Smith, A., Marshall, C. J., and Wood, E. R. (1999). A scintillation proximity assay for the Raf/MEK/ERK kinase cascade: high-throughput screening and identification of selective enzyme inhibitors. *Anal. Biochem.* **268**, 318–329.

McIntyre, R. J., Webb, P. P., and Dautet, H. (1996). A short-wavelength selective reach-through avalanche photodiode. *IEEE Trans. Nucl. Sci.* **43**(3), 1341–1346.

Meisner, J. E., Nicaise, W. F., and Stromswold, D. C. (1995). CsI(Tl) with photodiodes for identifying subsurface radionuclide contamination. *IEEE Trans. Nucl. Sci.* **42**, 288–291.

Melcher, C. L., and Schweitzer, J. S (1992). Cerium-doped lutetium oxyorthosilicate: a fast, efficient new scintillator. *IEEE Trans. Nucl. Sci.* **39**, 502–505.

Melcher, C. L., Manente, R. A., and Schweitzer, J. S. (1989). Applicability of barium fluoride and cadmium tungstate scintillators. *IEEE Trans Nucl. Sci.* **36**, 1188–1192.

Melcher, C. L., Schweitzer, J. S., Utsa, T., and Akiyama, S. (1990). Scintillation properties of GSO. *IEEE Trans. Nucl. Sci.* **37**, 161–164.

Melcher, C. L., Schweitzer, J. S., Manente, R. A., and Peterson, C. A. (1991). Applicability of GSO scintillators for well logging. *IEEE Trans. Nucl. Sci.* **38**, 506–509.

Melcher, C. L. *et al.* (1995). "Proceedings of the 3rd International Conference on inorganic Scintillator Applications." (P. Dorenbos and C. W. E. van Eijk, Eds.), pp. 309–316. Delft University Press, Delft.

Melcher, C. L., Schmand, M., Eriksson, M., Eriksson, L., Casey, M., Nutt, R., Lefaucheur, J. L., and Chai, B. (2000). Scintillation properties of LSO:Ce boules. *IEEE Trans. Nucl. Sci.* **47**(3), 965–968.

Mikhailov, M. A., and Panteleev, L. (2001). Noise contributions to energy resolutions in CsI(Tl) scintillation detector with PIN photodiode readout. *Nucl. Instrum. Mehtods Phys. Res., Sect. A* **463**, 288–292.

Miyahara, H., Narita, N., Tomita, K., Katoh, Y., Mori, C., Momose, T., and Shinohara, K. (2000). Activity measurement by using γ-ray sum peak method considering intake radioactivity and its application. *Radioisotopes* **49**, 253–259.

Miyajima, M., Sasaki, S., and Shibamura, E. (1984). Number of photoelectrons from a photomultiplier cathode coupled with a NaI(Tl) scintillator. *Nucl. Instrum. Methods* **224**, 331–334.

Mishra, U. C., Brar, S. S., and Gustafson, P. F. (1965). Calculation of the gamma-ray detection efficiency of annular and well-type sodium iodide crystals. *Int. J. Appl. Radiat. Isot.* **16**, 697–704.

Mori, C., Osada, T., Yanagida, K., Aoyama, T., Uritani, A., Miyahara, H., Yamane, Y., Kobayashi, K., Ichihara, C., and Shiroya, S. (1994). Simple and quick measurement of neutron flux distribution by using an optical fiber with scintillator. *J. Nucl. Sci. Technol.* **31**, 248–249.

Morimoto, K., Tokanai, F., Tanihata, I., and Hayashizaki, Y. (2000). Development of scintillating fiber imager. *IEEE Trans. Nucl. Sci.* **47**(6), 2034–2038.

Morishima, H., Koga, T., Kawai, H., Kondo, S., Mifune, M., Konishi, Y., and Shirai, C. (1992). Radon measurement using a liquid scintillation spectrometer. *Proceedings 3rd Low Level Counting Conference Liquid Scintillation Analysis*, Tokyo, pp. 179–184.

Moses, W. W., and Derenzo, S. E. (1990). The scintillation properties of cerium-doped lanthanum fluoride. *Nucl. Instrum. Methods Phys. Res., Sect. A* **299**, 51–56.

Moses, W. W. (2002). Current trends in scintillator detectors and materials. *Nucl. Instrum. Methods Phys. Res., Sect. A* **487**, 123–128.

Moss, C. E., Tissinger, E. W., and Hamm, M. E. (1984). Efficiency of 7.62 cm bismuth germanate scintillators. *Nucl. Instrum. Methods Phys. Res.* **221**, 378–384.

Moszyński, M., Allemand, R., Laval, M., Odru, R., and Vacher, J. (1983). Recent progress in fast timing with CsF scintillators in application to time-of-flight positron tomography in medicine. *Nucl. Instrum. Methods* **205**, 239–249.

Moszyński, M., Czarnacki, W., Kapusta, M., Szawlowski, M., Klamra, W., and Schotanus, P. (2002b). Energy resolution and light yield non-proportionality of pure NaI scintillator studied with large area avalanche photodiodes at liquid nitrogen temperatures. *Nucl. Instrum. Methods Phys. Res., Sect. A* **486**, 13–17.

Moszyński, M., Gresset, C., Vacher, J., and Odru, R. (1981a). Properties of CsF, a fast inorganic scintillator in energy and time spectroscopy. *Nucl. Instrum. Methods* **179**, 271–276.

Moszyński, M., Gresset, C., Vacher, J., and Odru, R. (1981b). Timing properties of BGO scintillator. *Nucl. Instrum. Methods* **188**, 403–409.

Moszyński, M., Kapusta, M., Balcerzyk, M., Szawlowski, M., and Wolski, D. (2000). Large area avalanche photodiodes in x-ray and scintillation detection. *Nucl. Instrum. Methods Phys. Res., Sect. A* **442**, 230–237.

Moszyński, M., Kapusta, M., Balcerzyk, M., Szawlowski, M., Wolski, D., Węgrzecka, I., and Węgrzecki, M. (2001). Comparative study of avalanche photodiodes with different structures in scintillation detection. *IEEE Trans. Nucl. Sci.* **48**(4), 1205–1210.

Moszyński, M., Kapusta, M., Wolski, D., Klamra, W., and Cederwall, B. (1998). Properties of the YAP:Ce scintillator. *Nucl. Instrum Methods Phys. Res., Sect. A* **404**, 157–165.

Moszyński, M., Kapusta, M., Zalipska, J., Balcerzyk, M., Wolski, D., Szawlowski, M., and Klamra, E. (1999). Low energy γ-ray scintillation detection with large area avalanche photodiodes. *IEEE Trans. Nucl. Sci.* **46**(4), 880–885.

Moszyński, M., Szawlowski, M., Kapusta, M., and Balcerzyk, M. (2002a). Large area avalanche photodiodes in scintillation and x-ray detection. *Nucl. Instrum. Methods Phys. Res., Sect. A* **485**, 504–521.

Moszyński, M., Wolski, D., Ludziejewski, T., Kapusta, M., Lempicki, A., Brecher, C., Wisniewski, D., and Wojtowicz, A. J. (1997). Properties of the new LuAP:Ce scintillator. *Nucl. Instrum. Methods Phys. Res., Sect. A* **385**, 123–131.

Moszyński, M., Wolski, D., Ludziejewski, T., Lempicki, A., Brecher, C., Wisniewski, D., and Wojtowicz, A. J. (1996). LuAP, a new fast scintillator. Proceedings International Conference on Scintillators and Their Applications, SCINT95, pp. 348–351. Delft University Press, Delft, The Netherlands.

Mullani, N. A., Ficke, D. C., and Ter-Pogossian, M. M. (1980). Cesium fluoride: a new detector for positron emission tomography. *IEEE Trans. Nucl. Sci.* **27**, 572–575.

Mundy, J. N. and Rothman, S. J. (1983). Radiotracks techniques. 4. Detection and assay. 4.2. Radiation detectors for pulse counting. *Methods Exp. Phys.* **21**, 50–54.

Muravsky, V. A., Tolstov, S. A., and Kholmetskii, A. L. (1998). Comparison of the least squares and the maximum likelihood estimators for gamma-spectrometry. *Nucl. Instrum. Methods Phys. Res., Sect. B* **145**, 573–577.

Murray, R. B. (1958). Use of ^{6}LiI(Eu) as a scintillation detector amd spectrometer for fast neutrons. *Nucl. Instrum. Methods* **2**, 237–248.

Nagarkar, V. V., Tipnis, S. V., Gaysinskiy, V., Klugerman, Y., Squillante, M. R., and Entine, G. (2001). Structured LiI scintillator for thermal neutron imaging. *IEEE Trans. Nucl. Sci.* **48**(6), 2330–2334.

Nare, B., Allocco, J. J., Kuningas, R., Galuska, S., Myers, R. W., Bednarek, M. A., and Schmatz, D. M. (1999). Development of a scintillation proximity assay for histone deacetylase using a biotinylated peptide derived from histone-H4. *Anal. Biochem.* **267**, 390–396.

NCRP (1985). "A Handbook of Radioactivity Measurement Procedures". Report No. 64. National Council on Radiation Protection and Measurements., Washington, DC, 20014.

Nelson, G. S., Chaney, R. C., Fenyves, E. J., Hammack, H., Hudson, C., Antich, P. P., and Anderson, J. A. (1993). Testing of a scintillation fiber PET system. *SPIE* **2007**, 132–136.

Nestor, O. H., and Huang, C. Y. (1975). Bismuth germanate: a high-Z gamma-ray and charged particle detector. *IEEE Trans. Nucl. Sci.* **22**, 68–71.

Neumann, K. E., and van Cauter, S. (1991). Scintillation proximity assay: instrumentation requirements and performance. *In* "Liquid Scintillation Counting and Organic Scintillators" (H. Ross, J. E. Noakes, and J. D. Spaulding, Eds.), pp. 365–374. Lewis Publishers, Chelsea, MI.

Neumann, K., Kolb, A., Englert, D., Hughes, K. T., Jessop, R. A., and Harris, A. (1994). Quench correction in scintillation proximity assays. *Top Count Topics*, **TCA-019**, PerkinElmer Life and Analytical Sciences, Boston.

Nguyen, H. V., Campbell, J. M., Couchell, G. P., Li, S., Pullen, D. J., Schier, W. A., Seabury, E. H., and Tipnis, S. V. (1996). Programs in C for parameterizing measured 5″ × 5″ NaI gamma response functions and unfolding of continuous gamma spectra. *Comput. Phys. Commun.* **93**, 303–321.

Niese, S. (1999). The discovery of organic solid and liquid scintillators by H. Kallman and L. Hereforth 50 years ago. *J. Radioanal. Nucl. Chem.* **241**(3), 499–501.

Normand, S., Mouanda, B., Haan, S., and Louvel, M. (2002). Study of a new boron loaded plastic scintillator. *IEEE Trans. Nucl. Sci.* **49**(2), 577–582.

Nozaki, T., and Saito, J. (1995). Isotopic double tracers for the measurement of the biological half-life of an element or compound, exemplified by ^{125}I-^{131}I taken up in seaweed. *Appl. Radiat. Isot.* **46**, 1299–1305.

Nuclear Enterprises Ltd. (1994). "Scintillators for the Physical Sciences." Catalogue, Sighthill, Edinburgh, Scotland.

Oba, K., and Rehak, P. (1981). Studies of high-gain micro-channel plate photomultipliers. *IEEE Trans. Nucl. Sci.* **NS-28**, 683–688.

Okajima, K., Takami, K., Ueda, K., and Kawaguchi, F. (1982). Characteristics of gamma-ray detector using a bismuth germanate scintillator. *Rev. Sci. Instrum.* **53**, l285–1286.

Okusawa, T., Sasayama, Y., Yamasaki, M., and Yoshida, T. (2000). Readout of a scintillating-fiber array by avalanche photodiodes. *Nucl. Instrum. Methods Phys. Res., Sect. A* **440**, 348–354.

Okusawa, T., Sasayama, Y., Yamasaki, M., and Yoshida, T. (2001). Readout of a 3 m long scintillating fiber by an avalanche photodiode. *Nucl. Instrum. Methods Phys. Res., Sect. A* **459**, 440–447.

Olschner, F. (1996). Silicon drift photodiode array detectors. *IEEE Trans. Nucl. Sci.* **43**(3), 1407–1410.

Orphan, V., Polichar, R., and Ginaven, R. (1978). A solid-state photomultiplier tube for improved gamma counting techniques. *Trans. Am. Nucl. Soc.* **28**, 119–121.

Ottonello, P., Palestini, V., Rottigni, G. A., Zanella, G. and Zannoni, R. (1995). Slow neutron beam diagnostics with a scintillating fiber detector. *Nucl. Instrum. Methods Phys. Res., Sect. A* **366**, 248–253.

Owens, A., Bavdaz, M., Brammertz, G., Krumrey, M., Martin, D., Peacock, A., and Tröger, L. (2002). The hard x-ray response of HgI$_2$. *Nucl. Instrum. Methods Phys. Res., Sect. A* **479**, 535–547.

Park, T. S., Hwang, H. Y., and Lee, J. M. (1998). An improved coincidence counting technique for standardization of radionuclides. *Appl. Radiat. Isot.* **49**(9–11), 1147–1151.

Park, Y.-W., Garyantes, T., Cummings, R. T., and Carter-Allen, K. (1998). Optimization of ^{33}P scintillation proximity assays using cesium chloride bead suspension. *TopCount Topics*, **TCA-030**, PerkinElmer Life and Analytical Sciences, Boston.

Passo, C. J., Jr., and Cook, G. T. (1994). "Handbook of Environmental Liquid Scintillation Spectrometry," pp. 8–1 to 8–15. PerkinElmer Life and Analytical Sciences, Boston.

Passo, C. J., Jr., and Floeckher, J. M. (1991). The LSC approach to radon counting in air and water. *In* "Liquid Scintillation Counting and Organic Scintillators" (H. Ross, J. E. Noakes, and J. D. Spaulding, Eds.), pp. 375–384. Lewis Publishers, Chelsea, MI.

Patel, S., Harris, A., O'Beirne, G., Cook, N. D., and Taylor, C. W. (1996). Kinetic analysis of inositol triphosphate binding to pure inositol triphosphate receptors using scintillation proximity assay. *Biochem. Biophys. Res. Commun.* **221**, 821–825.

Patt, B. E., Iwanczyk, J. S., Tull, C. R., Segal, J. D., MacDonald, L. R., Tornai, M. P., Kenney, C. J., and Hoffman, E. J. (2000). Fast-timing silicon photodetectors. *IEEE Trans. Nucl. Sci.* **47**(3), 957–964.

Pauwels, D., Le Masson, N., Viana, B., Kahn-Harari, A., van Loef, E. V. D., Dorenbos, P., and van Eijk, C. W. E. (2000). A novel inorganic scintillator. *IEEE Trans. Nucl. Sci.* **47**(6), 1787–1790.

PerkinElmer (1991a). Theory of TopCount operation. *TopCount Topics*, **TCA-003**, PerkinElmer Life and Analytical Sciences, Boston.

PerkinElmer (1991b). Scintillation proximity assay on the TopCount microplate scintillation counter. *TopCount Topics*, **TCA-004**, PerkinElmer Life and Analytical Sciences, Boston.

PerkinElmer (1993). PCGERDA: a new software package for calculating regional blood flow using the radiolabeled microsphere technique. *Gamma-Notes*, **GN007**, PerkinElmer Life and Analytical Sciences, Boston.

PerkinElmer (1996). Solid scintillation counting. *TopCount Topics*, **TCA-002**, PerkinElmer Life and Analytical Sciences, Boston.

PerkinElmer (1997). High efficiency count mode for SPA and Cytostar-T™ assays. *TopCount Topics*, **TCA-029**, PerkinElmer Life and Analytical Sciences, Boston.

Peters, R. J., Bates, B. B., and Knab, D. (1993). Radon-226 in water-radon emanation. *In* "Health and Environmental Chemistry: Analytical Techniques, Data Management, and Quality Assurance", Vol. II (L. S. Tamura, Ed.), pp. ER180–1–ER180–18. Los Alamos National Laboratory Manual, LA-10300-M.

Peurrung, A. J. (2000). Recent developments in neutron detection. *Nucl. Instrum. Methods Phys. Res., Sect. A* **443**, 400–415.

Pichler, B. J., Böning, G., Rafecas, M., Schlosshauer, M., Lorenz, E., and Ziegler, S. I. (1999). LGSO scintillation crystals coupled to new large area APDs compared to LSO and BGO. *IEEE Trans. Nucl. Sci.* **46**(3), 289–291.

Piemonte, C., Bonvicini,, V., Rashevsky, A., Vacchi, A., and Wheadon, R. (2002). Electric performance of the ALICE silicon drift detector irradiated with 1 GeV electrons. *Nucl. Instrum. Methods Phys. Res., Sect. A* **485**, 133–139.

Pla-Dalmau, A., Bross, A. D., and Mellot, K. L. (2001). Low-cost extruded plastic scintillator. *Nucl. Instrum. Methods Phys. Res., Sect. A* **466**, 482–491

Poggeler, B., and Heuther, G. (1992). Versatile one tube scintillation proximity homogeneous radioimmunoassay of melatonin. *Clin. Chem.* **38**(2), 314–315.

Potter, C. G. (1993). Filterscint: a high efficiency scintillation technique for ^{3}H-labeled harvested samples. *In* "Liquid Scintillation Spectrometry 1992" (J. E. Noakes, F. Schönhofer, and H. A. Polach, Eds.), pp. 313–318. Radiocarbon, Tucson, AZ.

Pujol, Ll., Suarez-Navarro, J. A., and Montero, M. (2000). A method for the selection of the optimum counting conditions in a ZnS(Ag) scintillation detector. *Appl. Radiat. Isot.* **52**, 891–897.

Price, G. J., and Fraser, G. W. (2001). Calculation of the output charge cloud from a microchannel plate. *Nucl. Instrum. Methods Phys. Res., Sect. A* **474**, 188–196.

Quaranta, H. O., and Piccini, J. L. (1984). Use if the random sum peak in the determination of the resolving time of thallium-doped sodium iodide scintillation detectors. *Int. J. Appl. Radiat. Isot.* **35**, 410–411.

Razdolescu, A. C., Grigorescu, M., Sahagia, M., Luca, A., and Ivan, C. (2000). Standardization of [169]Yb by the $4\pi\beta$-γ method. *Appl. Radiat. Isot.* **52**, 505–507.

Reeder, P. L. (1994a). Neutron detection using GSO scintillator. *Nucl. Instrum. Methods Phys. Res. Sect. A* **340**, 371–378.

Reeder, P. L. (1994b). Thin GSO scintillator for neutron detection. *Nucl. Instrum. Methods Phys. Res. Sect. A* **353**, 134–136.

Reeder, P. L., and Bowyer, S. M. (2000a). Neutron/gamma discrimination in LiBaF$_3$ scintillator. Proc. Conf. Methods and Applications of Radioanalytical Chemistry – MARC V, Kailua-Kona, HI, 9–14 April 2000.

Reeder, P. L., and Bowyer, S. M. (2000b). Neutron and simultaneous gamma detection with LiBaF$_3$ scintillator. Proc. Conf. Applications of Accelerators for Research and Industry, CAARI 2000, Denton, TX, 1–4 November 2000.

Reeder, P. L., and Bowyer, S. M. (2000c). Fast neutron and alpha detection using LiBaF$_3$ scintillator. Proc. Conf. IEEE/NSS-MIC 2000, Lyon, France, 15–20 October 2000.

Reeder, P. L., and Bowyer, S. M. (2001a). Neutron/gamma discrimination in LiBaF$_3$ scintillator. *J. Radioanal. Nucl. Chem.* **248**, 707–711.

Reeder, P. L., and Bowyer, S. M. (2001b). Fast neutron and alpha detection using LiBaF$_3$ scintillator. *IEEE Trans. Nucl. Sci.* **48**(3), 351–355.

Reeder, P. L., and Bowyer, S. M. (2002). Calibration of LiBaF$_3$:Ce scintillator for fission spectrum neutrons. *Nucl. Instrum. Methods Phys. Res., Sect. A* **484**, 469–485.

Rehak, P., Gatti, E., Longini, A., Kemmer, J., Holl, P., Klanner, R., Lutz, G., and Wylie, A. (1985). Semiconductor drift chambers for position and energy measurements. *Nucl. Instrum. Methods Phys. Res., Sect. A* **235**, 224–234.

Renker, D. (2002). Properties of avalanche photodiodes for applications in high energy physics, astrophysics and medical imaging. *Nucl. Instrum. Methods Phys. Res., Sect. A* **486**, 164–169.

Reynolds, G. T., Harrison, F. B., and Salvini, B. (1950). Liquid scintillation counters. *Phys. Rev.* **78**(4), 488.

Rhodes, N. J., and Johnson, M. W. (1996). *In* "Proceedings of the SCINT95 Conference" (P. Dorenbos, Ed.), p. 73. Delft University Press, Delft, The Netherlands.

Rieppo, R. (1976a). Calculated 10–150 keV γ-ray efficiency values of sodium iodide detectors for a 4-pi-detector geometry. Intercomparison with the respective values of well- and face-type sodium iodide detectors. *Int. J. Appl. Radiat. Isot.* **27**, 457–459.

Rieppo, R. (1976b). Efficiency values of sodium iodide well-type detectors in the γ-region 10–150 keV for point and needle-shaped source geometries. *Int. J. Appl. Radiat. Isot.* **27**, 453–456.

Rieppo, R., Blomster, K., and Holmberg, P. (1975). Calculated 10–150 keV gamma-ray efficiency values for well-type sodium iodide detectors. *Int. J. Appl. Radiat. Isot.* **26**, 558–561.

Rocha, J. G., and Correia, J. H. (2001). A high-performance scintillator-silicon-well x-ray microdetector based on DRIE techniques. *Sens. Actuators A* **92**, 203–207.

Ródenas, J., Martinavarro, A., and Rius, V. (2000). Validation of the MCNP code for the simulation of Ge-detector calibration. *Nucl. Instrum. Methods Phys. Res., Sect. A* **450**, 88–97.

Rodnyi, P. A. (2001). Progress in fast scintillators. *Radiat. Meas.* **33**, 605–614.

Rodnyi, P. A., Dorenbos, P., and van Eijk, C. W. E. (1995). Energy loss in inorganic scintillators. *Phys. Status Solidi B* **187**, 15–29.

Roessler, N., Englert, D., and Neumann, K. (1993). New instruments for high throughput receptor binding assays. *J. Recept. Res.* **13**, 135–145.

Roig, M., de Juan, A., García, J. F., Toribo, M., Vidal, M., and Rauret, G. (1999). Determination of a mixture of gamma-emitting radionuclides using solid scintillation detectors and multivariate calibration. *Anal. Chim. Acta* **379**, 121–133.

Rudin, M. J., Richardson, W. M., Dumont, P. G., and Johnson, W. H. (2001). In situ measurement of transuranics using a calcium fluoride scintillation detection system. *J. Radioanal. Nucl. Chem.* **248**(2), 445–448.

Rutherford, E. (1919). Collision of α-particles with light atoms. *Nature* (London) **103**, 415–418.

Rutherford, E. (1920). Nuclear constitution of atoms. *Proc. Royal Soc.* (London) **97A**, 374–401.

Ryan, J. M., Castaneda, C. M., Holslin, D., Macri, J. R., McConnell, M. L., Romero, J. L., and Wunderer, C. B. (1999). A scintillating plastic fiber tracking detector for neutron and proton imaging and spectroscopy. *Nucl. Instrum. Methods, Phys. Res., Sect. A* **442**, 49–53.

Ryskin, N. N., and Dorenbos, P. (1994). Unpublished results.

Ryzhikov, V., Chernikov, V., Gal'chinetskii, L., Galkin, S., Lisetskaya, E., Opolonin, A., and Volkov, V. (1999). The use of semiconductor scintillation crystals A^{II}B^{VI} in radiation instruments. *J. Cryst. Growth* **197**, 655–658.

Ryzhikov, V., Starzhinsky, M., Gal'chinetskii, L., Gashin, P., Kozin, D., and Danshin, E. (2001). New semiconductor scintillators in ZnSe(Te,O) and integrated radiation detectors based thereon. *IEEE Trans. Nucl. Sci.* **48**, 356–359.

Sabharwal, S. C., Ghosh, B., Phiske, M. R., and Navalkar, M. P. (1982). High temperature performance characteristics of a NaI(Tl) detector. *Nucl. Instrum. Methods* **195**, 613–616.

Sakai, E. (1987). Recent measurements on scintillator-photodetector systems. *IEEE Trans. Nucl. Sci.* **34**, 418–422.

Sakamoto, S. (1990). A Li-glass scintillation detector for thermal-neutron TOF measurements. *Nucl. Instrum. Methods Phys. Res., Sect. A* **299**, 182–186.

Sakamoto, S., and Takakura, H. (2001). Detection efficiency improvement of a large volume scintillator. *J. Radioanal. Nucl. Chem.* **248**(2), 345–351.

Saoudi, A., Pepin, C., Pépin, C., Houde, D., and Lecomte, R. (1999). Scintillation light emission studies of LSO scintillators. *IEEE Trans. Nucl. Sci.* **46**(6), 1925–1928.

Schäfer, W., Jansen, E., Will, G., Szepesvary, A., Reinartz, R., and Müller, K. D. (1995). Update on the Juelich linear and area neutron scintillation detectors. *Physics B* **213** and **214**, 972–974.

Schell, W. R., Tobin, M. J., Marsan, D. J., Schell, C. W., Vives-Batlle, J., and Yoon, S. R. (1997). Measurement of fission product gases in the atmosphere. *Nucl. Instrum. Methods Phys. Res., Sect. A* **385**, 277–284.

Schell, W. R., Vives-Batlle, J., Yoon, S. R., and Tobin, M. J. (1999a). High-pressure swing system for measurements of radioactive fission gases in air samples. *Nucl. Instrum. Methods Phys. Res., Sect. A* **420**, 416–428.

Schell, W. R., Vives-Batlle, J., Yoon, S. R., and Tobin, M. J. (1999b). High-pressure plastic scintillation detector for measuring radiogenic gases in flow systems. *Nucl. Instrum. Methods Phys. Res., Sect. A* **421**, 591–600.

Schenck, J. (1953). Activation of lithium iodide by europium. *Nature* **171**, 518–519.

Schieber, M., Zuck, A., Braiman, M., Melekhov, J., Nissenbaum, J., Turchetta, R., Dulinski, W., Husson, D., and Riester, J. L. (1998b). Radiation-hard polycrystalline mercuric iodide semiconductor particle counters. *Nucl. Instrum. Methods Phys. Res., Sect. A* **410**, 107–110.

Schieber, M., Zuck, A., Braiman, M., Nissenbaum, J., Turchetta, R., Dulinski, W., Husson, D., and Riester, J. L. (1997). Novel mercuric iodide polycrystalline nuclear particle counters. *IEEE Trans. Nucl. Sci.* **44**(6), 2571–2575.

Schieber, M., Zuck, A., Braiman, M., Nissenbaum, J., Zuck, A., Turchetta, R., Dulinski, W., Husson, D., and Riester, J. L. (1998a). Evaluation of mercuric iodide ceramic semiconductor detectors. *Nucl. Phys. B (Proc. Suppl.)* **61**, 321–329.

Schilk, A. J., Abel, K. H., and Perkins, R. W. (1995b). Characterization of uranium contamination in surface soils. *J. Environ. Radiat.* **26**, 147–156.

Schilk, A. J., Abel, K. H., Brown, D. P., Thompson, R. C., Knopf, M. A., and Hubbard, C. W. (1995a). Sensitive, high-energy beta scintillation sensor for real-time, in situ characterization of uranium-238 and strontium-90. *J. Radioanal. Nucl. Chem. Articles* **193**, 107–111.

Schilk, A. J., Knopf, M. A., Thompson, R. C., Hubbard, C. W., Abel, K. H., Edwards, D. R., and Abraham, J. R. (1994). Real-time, in situ detection of ^{90}Sr and ^{238}U in soils via scintillating fiber-sensor technology. *Nucl. Instrum. Methods Phy. Res. Sect. A* **353**, 477–481.

Schilk, A. J., Perkins, R. W., Abel, K. H., and Brodzinski, R. L. (1993). Surface and subsurface characterization of uranium contamination at the Fernald environmental management site. Battelle Pacific Northwest Laboratory, PNL-8617/UC-606.

Schmelz, C., Bradbury, S. M., Holl, I., Lorenz, E., Renker, D., and Ziegler, S. (1995). Feasibility study of an APD readout for a high resolution PET with nsec time resolution. *Nucl. Sci. Symp. Med. Imaging Conf.* IEEE Conf. Rec. 1994, **3**, 1160–1164.

Schoenfeld, A., and Luqmani, Y. A. (1995). Semiquantification of polymerase chain reaction using a bead scintillation proximity assay and comparison with the Southern blot method. *Anal. Biochem.* **228**, 164–167.

Schotanus, P. (1995). Miniature radiation detection instruments. *Radiat. Measure.* **24**, 331–335.

Schotanus, P. (1996). "SCIONIX Scintillation Detectors." SCIONIX Holland B.V., Bunnik, The Netherlands. pp. 48.

Schotanus, P., Dorenbos, P., and Ryzhikov, V. D. (1992). Detection of CdS(Te) and ZnSe(Te) scintillation light with silicon photodiodes. *IEEE Trans. Nucl. Sci.* **39**, 546–550.

Schotanus, P., Kamermans, R., and Dorenbos, P. (1990). Scintillation characteristics of pure Tl-doped CsI crystals. *IEEE Trans. Nucl. Sci.* **37**, 177–182.

Schötzig, U., Schönfeld, E., Günther, E., Klein, R., and Schrader, H. (1999). Standardization and decay data of ^{153}Sm. *Appl. Radiat. Isot.* **51**, 169–175.

Schötzig, U., Schrader, H., Schönfeld, E., Günther, E., and Klein, R. (2001). Standardization and decay data of ^{177}Lu and ^{188}Re. *Appl. Radiat. Isot.* **55**, 89–96.

Selim, Y. S., and Abbas, M. I. (2000). Analytical calculations of gamma scintillator efficiencies. II: total efficiency for wide co-axial disk sources. *Radiat. Phys. Chem.* **58**, 15–19.

Selim, Y. S., Abbas, M. I., and Fawzy, M. A. (1998). Analytical calculations of the efficiencies of gamma scintillators. I: total efficiency for co-axial disk sources. *Radiat. Phys. Chem.* **53**, 589–592.

Semkow, T. M., Parekh, P. P., Schwenker, C. D., Dansereau, R., and Webber, J. S. (1994). Efficiency of the Lucas scintillation cell. *Nucl. Instrum. Methods Phys. Res. Sect. A* **353**, 515–518.

Seymour, R., Richardson, B., Morichi, M., Bliss, M., Craig, R. A., and Sunberg, D. S. (2000). Scintillating-glass-fiber neutron sensors, their application and performance for plutonium detection and monitoring. *J. Radioanal. Nucl. Chem.* **243**(2), 387–388.

Seymour, R., Hull, C. D., Crawford, T., Coyne, B., Bliss, M., and Craig, R. A. (2001). Portal, freight and vehicle monitor performance using scintillating glass fiber detectors for the detection of plutonium in the Illicit Trafficking Radiation Assessment Program. *J. Radioanal. Nucl. Chem.* **248**(3), 699–705.

Simpson, B. R. S. (2002). Radioactivity standardization in South Africa. *Appl. Radiat. Isot.* **56**, 301–305.

Simpson, B. R. S., and Meyer, B. R. (1988). A multiple-channel 2- and 3-fold coincidence counting system for radioactivity standardization. *Nucl. Instrum. Methods Phys. Res., Sect. A* **263**, 436–440.

Simpson, B. R. S., and Meyer, B. R. (1992). The standardization of ^{201}Tl by liquid scintillation coincidence counting. National Accelerator Centre Report. NAC/92–01, PO Box 72, Faure, 7131 South Africa.

Simpson, B. R. S., and Meyer, B. R. (1994). Standardization and half-life of ^{201}Tl by the $4\pi(xe)-\gamma$ coincidence method with liquid scintillation counting in the 4π-channel. *Appl. Radiat. Isot.* **45**, 669–673.

Simpson, B. R. S., and Meyer, B. R. (1996). Activity measurement of ^{204}Tl by direct liquid scintillation measurements. *Nucl. Instrum. Methods Phys. Res., Sect. A* **369**, 340–343.

Simpson, B. R. S., and van Oordt, E. J. (1997). Data acquisition program for the NAC radioactivity standards laboratory. NAC Report NAC/97–03, National Accelerator Centre, PO Box 72, Faure, 7131 South Africa.

Singh, K., Kawaldeep, and Sahota, H. S. (2001). Study of nuclear quadrupole interactions in different environments of decaying atoms of ^{75}Se by sum peak method. *Appl. Radiat. Isot.* **54**, 261–267.

Sissors, D. L., and Casto, S. (1998). Converting 96-well assays to 384-well FlashPlate assays. PerkinElmer Life Sciences, FlashPlate File No. 11, pp. 6, Boston, MA.

Skipper, J. A., and Hangartner, T. N. (2002). Deblurring of x-ray spectra acquired with a NaI-photomultiplier detector by constrained least-squares deconvolution. *Med. Phys.* **29**(5), 787–796.

Skoldborn, H., Arvidsson, B., and Andersson, M. (1972). *Acta Radiol. Suppl.* **319**, 233–241.

Skorey, K. I., Kennedy, B. P., Friesen, R. W., and Ramachandran, C. (2001). Development of a robust scintillation proximity assay for protein tyrosine phosphatase 1B using the catalytically inactive (C215S) mutant. *Anal. Biochem.* **291**, 269–278.

Sorg, G., Schubert, H.-D., Büttner, F. H., and Heilker, R. (2002). Automated high throughput screening for serine kinase inhibitors using a LEADSeeker™ scintillation proximity assay in the 1536-well format. *J. Biomol. Screening* **7**(1), 11–19.

Sonský, J., Hollander, R. W., Sarro, P. M., and van Eijk, C. W. E. (2002). X-ray spectroscopy with a multi-anode sawtooth silicon drift detector: the diffusion process. *Nucl. Instrum. Methods Phys. Res., Sect. A* **477**, 93–98.

Spaulding, J. D., and Noakes, J. E. (1983). In "Advances in Scintillation Counting" (S. A. McQuarrie, C. Ediss, and L. I. Wiebe, Eds.), pp. 112–122. University of Alberta Press, Edmonton.

Spaulding, J. D., and Noakes, J. E. (1993). Determination of ^{222}Rn in drinking water using an alpha/beta liquid scintillation counter. *In* "Advances in Scintillation Spectrometry 1992" (J. E. Noakes, F. Schoenhofer, and H. Polach, Eds.) pp. 373–381. Radiocarbon, Tucson, AZ.

Spector, G. B., McCollum, T., and Spowart, A. R. (1993a). Improved terbium-doped, lithium-loaded glass scintillator fibers. *Nucl. Instrum. Methods Phys. Res., Sect. A* **329**, 223–226.

Spector, G. B., McCollum, T., and Spowart, A. R. (1993b). Advances in terbium-doped lithium-loaded scintillator glass development. *Nucl. Instrum. Methods Phys. Res., Sect. A* **326**, 526–530.

Sperr, P. (1987). Timing measurements with barium fluoride scintillators. *Nucl. Instrum. Methods Phys. Res., Sect. A* **254**, 635–636.

Sudarshan, M., Joseph, J., and Singh, R. (1992). Full energy peak efficiency of NaI(Tl) gamma detectors and its analytical and semi-empirical representations. *J. Phys. D: Appl. Phys.* **25**, 1561–1567.

Sükösd, Cs., Galster, W., Licot, I., and Simonart, M. P. (1995). Spectrum unfolding in high energy gamma-ray detection with scintillation detectors. *Nucl. Instrum. Methods Phys. Res., Sect. A* **355**, 552–558.

Summer, R. (1988). The L3 BGO electromagnetic calorimeter. *Nucl. Instrum. Methods Phys. Res., Sect. A* **265**, 252–257.

Sun, R.-K. S. (1987). Photo-energy calibration of ^{6}LiI(Eu) crystals in mixed radiation fields using ^{24}Na. *Health Phys.* **53**(2), 191–196.

Suontausta, J., Oikari, T., and Webb, S. (1993). A meltable thermoplastic scintillator. *In* "Liquid Scintillation Spectrometry 1992" (J. E. Noakes, F. Schönhofer, and H. A. Polach, Eds.), pp. 173–178. Radiocarbon, Tucson, AZ.

Sutherland, L. G., and Buchanan, J. D. (1967). Error in the absolute determination of disintegration rates of extended sources by coincidence counting with a single detector—Application to I-125 and Co-60. *Int. J. Appl. Radiat. Isot.* **18**, 786–787.

Suzuki, S. (1982). Detection efficiency of NaI(Tl) crystals and loss of position resolution caused by photon interactions in the crystals in γ-cameras. *Int. J. Appl. Radiat. Isot.* **33**, 411–414.

Suzuki, T., Goda, K., and Suzuki, N. (1993). Analysis of intensity of two closed gamma ray spectra with NaI scintillator. *J. Nucl. Sci. Technol.* **30**, 1071–1074.

Swailem, F. M., and Riad, M. M. (1983). The response of the 7F8(A) NaI(Tl) well crystal to γ-radiation. *Int. J. Appl. Radiat. Isot.* **34**, 1469–1472.

Swinkels, L. M. J. W., Ross, H. A., and Benraad, Th. J. (1990). Scintillation proximity assay determination of steroid hormones. *Eur. J. Nucl. Med.* **16**(7), 553.

Tailor, R. C., Nestor, O. H., and Utts, B. K. (1986). Investigation of cerium-doped barium fluoride. *IEEE Trans. Nucl. Sci.* **33**, 243–246.

Takada, M., Taniguchi, S., Nakamura, T., Nakao, N., Uwamino, Y., Shibata, T., and Fujitaka, K. (2001a). Neutron spectrometry in a mixed field of neutrons and protons with a phoswich neutron detector. Part I: response functions for photons and neutrons of the phoswich neutron detector. *Nucl. Instrum. Methods Phys. Res., Sect. A* **465**, 498–511.

Takada, M., Taniguchi, S., Nakamura, T., and Fujitaka, K. (2001b). Neutron spectrometry in a mixed field of neutrons and protons with a phoswich neutron detector. Part II: application of the phoswich neutron detector to neutron spectrum measurements. *Nucl. Instrum. Methods Phys. Res., Sect. A* **465**, 512–524.

Takada, M., Taniguchi, S., Nakamura, T., Nakao, N., Uwamino, Y., Shibata, T., and Fujitaka, K. (2002). Characteristics of a phoswich detector to measure the neutron spectrum in a mixed field of neutrons and charged particles. *Nucl. Instrum. Methods Phys. Res., Sect. A* **476**, 332–336.

Takagi, K., and Fukazawa, T. (1983). Cerium-activated gadolinium orthosilicate (Gd_2SiO_5) single crystal scintillator. *Appl. Phys. Lett.* **42**, 43–45.

Takiue, M., Fujii, H., Aburai, T., and Yanokura, M. (1995). A continuous scintillation counter using a paraffin scintillator and a solid support. *Appl. Radiat. Isot.* **46**, 191–198.

Takiue, M., Natake, T., and Fujii, H. (1999). A hybrid radioassay technique for multiple beta-emitter mixtures using liquid and NaI(Tl) scintillation spectrometers. *Appl. Radiat. Isot.* **51**, 429–434.

Taniguchi, S., Takada, M., and Nakamura, T. (2001). Development of multi-moderator neutron spectrometer using a pair of ^{6}Li and ^{7}Li glass scintillators. *Nucl. Instrum. Methods Phys. Res., Sect. A* **460**, 368–373.

Thomas, D. J., and Alevra, A. V. (2002). Bonner sphere spectrometers – a critical review. *Nucl. Instrum. Methods Phys. Res., Sect. A* **476**, 12–20.

Thomson, J. (1993). Plastic scintillators: some novel applications. *In* "Liquid Scintillation Spectrometry 1992" (J. E. Noakes, F. Schönhofer, and H. A. Polach, Eds.), pp. 179–184, Radiocarbon, Tucson, AZ.

Timothy, J. G. (1981). Curved-channel microchannel array plates. *Rev. Sci. Instrum.* **52**, 1131–1142.

Tokanai, F., Morimoto, K., and Oku, T., Ino, T., and Suzuki, J. *et al.* (2000). Cold neutron imaging with a GSO scintillator. *Nucl. Instrum. Methods Phys. Res., Sect. A* **452**, 266–272.

Tominaga, S. (1983). Generation of precise scintillator response curves by an interpolation technique. *Nucl. Instrum. Methods* **215**, 231–233.

Toribo, M., García, J. F., Izquierdo-Ridorosa, A., Tauler, R., and Rauret, G. (1995). Simultaneous determination of plutonium alpha emitters by liquid scintillation counting using multivariate calibration. *Anal. Chim. Acta* **310**, 297–305.

Trower, W. P. (1994). Cerium-doped yttrium aluminum perovskite (YAP): properties of commercial crystals. *In* "Scintillators and Phosphor Materials" (M. J. Weber, P. Lecoq, R. C. Ruchti, C. Woody, W. M. Yen, and R-y. Zhu, Eds.), pp. 131–136. Material Research Society Symposium Proceedings, Pittsburgh.

Trower, W. P. (1995). New wine for old bottles: sodium iodide detector applications better accomplished with YAlO$_3$:Ce scintillators. *Appl. Radiat. Isot.* **46**, 517–518.

Turchetta, R., Dulinski, W., Husson, D., Riester, J. L., Schieber, M., Zuck, A., Melekhov, L., Saado, Y., Hermon, H., and Nissenbaum, J. (1999). Imaging with polycrystalline mercuric iodide detectors using VLSI readout. *Nucl. Instrum. Methods Phys. Res., Sect. A* **428**, 88–94.

Turlais, F., Hardcastle, A., Rowlands, M., Newbatt, Y., Bannister, A., Kouzarides, T., Workman, P., and Aherne, G. W. (2001). High-throughput screening for identification of small molecule inhibitors of histone acetyltransferases using scintillation microplates (FlashPlate). *Anal. Biochem.* **298**, 62–68.

Udenfriend, S., Gerber, L., and Nelson, N. (1987). Scintillation proximity assay: a sensitive and continuous isotopic method for monitoring ligand/receptor and antigen/antibody interactions. *Anal. Biochem.* **161**, 494–500.

Usuda, S. (1992). Development of ZnS(Ag)/NE102A and ZnS(Ag)/stilbene phoswich detectors for simultaneous α and $\beta(\gamma)$ counting. *J. Nucl. Sci. Technol.* **29**(9), 927–929.

Usuda, S. (1995). Simultaneous counting of α, $\beta(\gamma)$-rays and thermal neutrons with phoswich detectors consisting of ZnS(Ag). ^{6}Li-glass and /or NE102A scintillators. *Nucl. Instrum. Methods Phys. Res., Sect. A* **356**, 334–338.

Usuda, S., and Abe, H. (1994). Phoswich detectors for flow monitoring of actinide solutions with simultaneous α and β γ counting. *J. Nucl. Sci. Technol.* **31**(1), 73–79.

Usuda, S., Abe, H., and Mihara, A. (1994a). Phoswich detectors combining doubly or triply ZnS(Ag), NE102A, BGO and/or NaI(Tl) scintillators for simultaneous counting of α, β and γ rays. *Nucl. Instrum. Methods Phys. Res., Sect. A* **340**, 540–545.

Usuda, S., Abe, H., and Mihara, A. (1994b). Simultaneous counting of α, β and γ rays with phoswich detectors. *J. Alloys Compd.* **213/214**, 437–439.

Usuda, S., Sakurai, S., and Yasuda, K. (1997). Phoswich detectors for simultaneous counting of α-, $\beta(\gamma)$-rays and neutrons. *Nucl. Instrum. Methods Phys. Res., Sect. A* **388**, 193–198.

Usuda, S., Yasuda, K., and Sakurai, S. (1998a). Development of phoswich detectors for simultaneous counting of alpha particles and other radiations (emitted from actinides). *Appl. Radiat. Isot.* **49**, 1131–1134.

Usuda, S., Yasuda, K., and Sakurai, S. (1998b). Simultaneous counting of radiation emitted from actinides with improved phoswich detectors by applying an optical filter. *J. Alloys Compd.* **271–273**, 58–61.

Valentine, J. D., Moses, W. W., Derenzo, S.E., Wehe, D. K., and Knoll, G.F. (1993). Temperature dependence of CsI(Tl) gamma-ray excited scintillation. *Nucl. Instrum. Methods Phys. Res., Sect. A* **325**, l47–157.

Van Driel, M. A., and Sens, J. C. (1984). Linearity and resolution of photodiodes. *IEEE Trans. Nucl. Sci.* **NS-31**, 83–86.

van Eijk, C. W. E. (1993). Research and development of scintillation crystals and glasses. *In* "Heavy Scintillators for Scientific and Industrial Applications," Proceedings of the "Crystal 2000" International Workshop, Sept. 22–26, 1992, Chamonix, France (F. De Notaristefani, P. Lecoq and M. Schneegans, Eds.), Editions Frontieres, Gif-sur-Yvette Cedex, France.

van Eijk, C. W. E. (1994). Inorganic scintillator requirements. *Proceedings International Symposium PHYSCI 94*, St. Petersburg, Sept. 30-Oct 1, TUD-SCIP-94–11, pp. 1–12.

van Eijk, C. W. E. (1995). Fast lanthanide doped inorganic scintillators. *Proceedings Tenth Feofilov Symposium*, St. Petersburg, July 3–7.

van Eijk, C. W. E. (1997a). Development of inorganic scintillators. *Nucl. Instrum. Methods Phys. Res., Sect. A.* **392**, 285–290.

van Eijk, C. W. E. (1997b). *In* "Proceedings of the 4th International Conference on Inorganic Scintillator Applications." (Z. W. Win, P. J. Li, X. Q. Feng, and Z. L. Xue, Eds.), pp. 5–12. Shanghai, China, 1997.

van Eijk, C. W. E. (1999). *In* "Proceedings 8th International Symposium of the Electrochemical Society." (C. Ronda, L. Shea, and A. Srivastava, Eds.), Vol. 99, p. 40.

van Eijk, C. W. E. (2001). Inorganic-scintillator development. *Nucl. Instrum. Methods Phys. Res., Sect. A* **460**, 1–14.

van Eijk, C. W. E., Andriessen, J., Dorenbos, P., and Visser, R. (1994). Ce^{3+} doped inorganic scintillators. *Nucl. Instrum. Methods Phys. Res., Sect. A* **348**, 546–550.

van Eijk, C. W. E., Dorenbos, P., van Loef, E. V. D., Krämer, F., and Güdel, H. U. (2001). Energy resolution of some new inorganic-scintillator gamma-ray detectors. *Radiat. Meas.* **33**, 521–525.

van Loef, E. V. D. (1999). Stratech Report IRI-ISO-990033, Radiation Technology Group, Delft University of Technology, Delft, The Netherlands.

van Loef, E. V. D., Dorenbos, P., van Eijk, C. W. E., Krämer, K., and Güdel, H. U. (2001). High-energy-resolution scintillator: Ce^{3+} activated $LaBr_3$. *Appl. Phys. Lett.* **79**(10), 1573–1575.

Varela, A., Policroniades, R., Jiménez, F., and Calvillo, J. (1999). The use of a Bonner sphere spectrometer for determining the spatial distribution of neutron fields. *Nucl. Instrum. Methods Phys. Res., Sect. A* **428**, 439–445.

Visser, R., Dorenbos, P., van Eijk, C. W. E., Hollander, R. W., and Schotanus, P. (1991). Scintillation properties of Ce^{+3} doped BaF_2 crystals. *IEEE Trans. Nucl. Sci.* **38**, 178–183.

Wang, C.-F., Lee, J.-H., and Chiou, H.-J. (1994). Rapid determination of Sr-89/Sr-90 in radwaste by low-level background beta counting system. *Appl. Radiat. Isot.* **45**, 251–256.

Wang, Y. J., Patt, B. E., and Iwanczyk, J. S. (1994a). The use of HgI_2 photodetectors combined with scintillators for gamma-ray spectroscopy. *Nucl. Instrum. Methods Phys. Res., Sect. A* **353**, 50–54.

Wang, Y. J., Iwanczyk, J. S., and Patt, B. E. (1994b). New concepts for scintillator/HgI_2 gamma ray spectroscopy. *IEEE Trans. Nucl. Sci.* **41**, 910–914.

Wang, Z., Zhang, X., Chang, Y., and Liu, D. (2001). The determination of ^{125}I activity using sum-peak method with a well-type HPGe-detector-based spectrometer. *Nucl. Instrum. Methods Phys. Res., Sect. A* **459**, 475–481.

Watanabe, H., Abe, K., Harada, E., Inoue, S., and Inagaki, T. *et al.* (2002). Scintillator-Lucite sandwich detector for n/γ separation in the GeV energy region. *Nucl. Instrum. Methods Phys. Res., Sect. A* **484**, 118–128.

Watson, J. (1996). *In vitro* measurement of the second messenger cAMP: RIA vs FlashPlate®. FlashNews No. FN001, NEN Life Sciences Products, Boston, MA.

Watson, J., and Selkirk, J. V. (1998). Use of FlashPlate technology for *in vitro* measurement of [^{35}S]-GTPγS binding in CHO cells expressing the human 5-HT$_{1B}$ receptor. PerkinElmer Life Sciences, FlashPlate File No. 2, pp. 7, Boston, MA.

Weber, M. J., and Monchamp R. R. (1973). Luminescence of $Bi_4Ge_3O_{12}$: spectral and decay properties. *J. Appl. Phys.* **44**, 5495–5499.

Weisskopf, M. C., Odell, S. L., Elsner, R. F., and van Speybroeck, L. P. (1995). Advanced x-ray astrophysical observatory AXAF – an overview. *Proc. SPIE – Int. Soc. Opt. Eng.* **2515**, 312.

Westman, S., Kerek, A., Klamra, W., Norlin, L.-O., and Novák, D. (2002). Heavy ion detection at extreme high vacuum by means of a YAP:Ce scintillator. *Nucl. Instrum. Methods Phys. Res., Sect. A* **481**, 655–660.

White, T. L., and Miller, W. H. (1999). A triple-crystal phoswich detector with digital pulse shape discrimination for alpha/beta/gamma spectroscopy. *Nucl. Instrum. Methods Phys. Res., Sect. A* **422**, 144–147.

Wise, B. M. (1992). PLS-Toolbox for use with MATLAB®. Center for Process Analytical Chemistry (CPAC), University of Washington, Seattle.

Wojtowicz, A. J., Lempicki, A., Wisniewski, D., and Boatner, L. A. (1994a). *Mater. Res. Soc. Symp. Proc. 1994*, Conf. Record 348 (Scintillators and Phosphor Materials), pp. 123–129.

Wojtowicz, A. J., Lempicki, A., Wisniewski, D., Balcerzyk, M., and Brecher, C. (1995). The role of charge transfer states in Ln^{+3}-activated scintillators. IEEE Conference Record, Nuclear Science Symposium and Medical Imaging Conference Oct. 30–Nov 5, 1994 (R. C. Trendler, Ed.), Vol. 1, pp. 134–138.

Wojtowicz, A. J., Lempicki, A., Wisniewski, D., Balcerzyk, M., and Brecher, C. (1996). The carrier capture and recombination processes in Ln^{+3}-activated scintillators. *IEEE Trans. Nucl. Sci.* **43**, 2168–2173.

Wojtowicz, A. J., Wisniewski, D., Lempicki, A., and Boatner, L. A. (1994b). Scintillation mechanisms in rare earth orthophosphates. Conference Record EURODIM94.

Wong, H. T., and Li, J. (1999). A pilot experiment with reactor neutrinos in Taiwan. *Nucl. Phys. B (Proc. Suppl.)* **77**, 177–181.

Woody, C. L., Kierstead, J. A., Levy, P. W., and Stoll, S. (1991). Radiation damage in undoped CsI and CsI(Tl). Conference Record of the IEEE Nuclear Science Symposium Nov. 2–9, Santa Fe, pp. 1516–1523.

Woody, C. L., Levy, P. W., Kierstead, J. A., Skwarnicki, T., Sobolewski, Z., Goldberg, M., Horwitz N., Souder, P., and Anderson, D. F. (1990). Readout techniques and radiation damage of undoped cesium iodide. *IEEE Trans. Nucl. Sci.* **37**, 492–499.

Worstell, W., Johnson, O., Kudrolli, H., and Zavarzin, V. (1998). First results with high-resolution PET detector modules using wavelength-shifting fibers. *IEEE Trans. Nucl. Sci.* **45**(6), 2993–2999.

Wunderly, S. W. (1989). Solid scintillation counting: a new technique for measuring radiolabeled compounds. *Appl. Radiat. Isot.* **40**, 569–573.

Wunderly, S. W. (1993). Simultaneous measurement of alpha and beta emissions on Ready Cap. In "Liquid Scintillation Spectrometry 1992" (J. E. Noakes, F. Schönhofer, and H. A. Polach, Eds.), pp. 217–223. Radiocarbon, Tucson, AZ.

Yamane, Y., Lindén, P., Karlsson, J. K.-H., and Pázsit, I. (1998). Measurement of 14.1 MeV neutrons with a Th-scintillator optical fiber detector. *Nucl. Instrum. Methods Phys. Res., Sect. A* **416**, 371–380.

Yamane, Y., Uritani, A., Misawa, T., Karlsson, J. K.-H., and Pázsit, I. (1999). Measurement of the thermal and fast neutron flux in a research reactor with a Li and Th loaded optical fiber detector. *Nucl. Instrum. Methods Phys. Res., Sect. A* **432**, 403–409.

Yan, C. G. (1996). Improvement of accuracy of efficiency extrapolation method in 4π β-γ coincidence counting *Nucl. Instrum. Methods Phys Res., Sect. A* **369**, 383–387.

Yang, D. (1993a). Alpha liquid scintillation analysis: some recent developments and applications. In "Liquid Scintillation Spectrometry 1992" (J. E. Noakes, F. Schönhofer, and H. A. Polach, Eds.), pp. 201–201. Radiocarbon, Tucson, AZ.

Yang, D. (1993b). Alpha-counting with a solid scintillator. *J. Radioanal. Nucl. Chem. Letters* **175**, 393–400.

Yang, D. (1993c). Alpha-particle counting with solid scintillator cap. *Radioact. Radiochem.* **4**, 8–13.

Yang, F., Dicker, I. B., Kurilla, M. G., and Pompliano, D. L. (2002). PolC-type polymerase III of *Streptococcus pyogenes* and its use in screening for chemical inhibitors. *Anal. Biochem.* **304**, 110–116.

Yasuda, K., Usuda, S., and Gunji, H. (2000a). Properties of a YAP powder scintillator as alpha-ray detector. *Appl. Radiat. Isot.* **52**, 365–368.

Yasuda, K., Usuda, S., and Gunji, H. (2000b). Development of scintillation-light-transmission type phoswich detector for simultaneous alpha- and beta (gamma)-ray counting. *IEEE Trans. Nucl. Sci.* **47**(4), 1337–1340.

Yasuda, K., Usuda, S., and Gunji, H. (2001). Simultaneous alpha, beta/gamma, and neutron counting with phoswich detectors by using a dual-parameter technique. *IEEE Trans. Nucl. Sci.* **48**(4), 1162–1164.

Ye, L., Surugiu, I., and Haupt, K. (2002). Scintillation proximity assay using molecularly imprinted microspheres. *Anal. Chem.* **74**, 959–964.

Yu, P. Y., and Cardona, M. (2001). "Fundamentals of Semiconductors." 3rd Edition, Springer, Berlin.

Yuan, M.-C., and Hwang, W.-S. (2000). The absolute counting of ^{125}I. *Appl. Radiat. Isot.* **52**, 523–526.

Zaidins, C. S., Martin, J. B., and Edwards, F. M. (1978). A least-squares technique for extracting neutron spectra from Bonner sphere data. *Med. Phys.* **5**(1), 42–47.

Zanella, G., Zannoni, R., Dall'Igna, R., Polato, P., and Bettinelli, M. (1995). Development of a terbium-lithium glass for slow neutron detection. *Nucl. Instrum. Methods Phys. Res., Sect. A* **359**, 547–550.

Zazubovich, S. (2001). Physics of halide scintillators. *Radiat. Meas.* **33**, 699–704.

Zhang, Y., Yang, F., Kao, Y.-C., Kurilla, M. G., Pompliano, D. L., and Dicker, I. B. (2002). Homogenous assays for *Escherichia coli* DnaB-stimulated DnaG Primase and DnaB Helicase and their use in screening for chemical inhibitors. *Anal. Biochem.* **304**, 174–179.

Zhu, Y. C., Li, J. G., Shao, Y.Y., Sun, H.S., Dong, B. Z., Zhou, G. P., Zheng, Zp., Cui, F. Z., and Yu, C. J. (1986). New results on the scintillation properties of BaF_2. *Nucl. Instrum. Methods Phys. Res. Sect. A* **244**, 577–578.

Ziegler, S. I., Pichler, B. J., Boening, G., Rafecas, M., Pimpl, W., Lorenz, E., Schmitz, N., and Schwaiger, M. (2001). A prototype high-resolution animal positron tomograph with avalanche photodiode arrays and LSO crystals. *Eur. J. Nucl. Med.* **28**(2), 136–143.

Zimmerman, B. E., Cessna, J. T., and Schima, F. J. (1998). The standardization of the potential bone palliation radiopharmaceutical ^{117m}Sn(+4)DTPA. *Appl. Radiat. Isot.* **49**(4), 317–328.

Zimmerman, B. E., Cessna, J. T., and Unterweger, M. P. (2002). The standardization of ^{188}W/^{188}Re by $4\pi\beta$ liquid scintillation spectrometry with the CIEMAT/NIST ^{3}H-standard efficiency tracing method. *Appl. Radiat. Isot.* **56**, 315–320.

Zimmerman, B. E., Unterweger, M. P., and Brodack, J. W. (2001). The standardization of ^{177}Lu by $4\pi\beta$ liquid scintillation spectrometry with ^{3}H-standard efficiency tracing. *Appl. Radiat. Isot.* **54**, 623–631.

Zimmermann, H., Müller, B., Hammer, A., Herzog, K., and Seegebrecht, P. (2002). Large-area lateral P-I-N photodiode on SOI. *IEEE Trans. Nucl. Sci.* **49**(2), 334–336.

Zorn, C., Bowen, M., Majewski, S., Walker, J., Wojcik, R., Hurlbut, C., and Moser, W. (1988). Pilot study of new radiation-resistant plastic scintillators doped with 3-hydroxyflavone. *Nucl. Instrum. Methods Phys. Res., Sect. A* **273**, 108–116.

12

■ FLOW SCINTILLATION ANALYSIS

MICHAEL F. L'ANNUNZIATA

The Montague Group, P.O. Box 5033 Oceanside, CA 92052–5033, USA

I. INTRODUCTION

Flow scintillation analysis (FSA) is the application of scintillation detection methods for the quantitative analysis of radioactivity in a flowing system. The technology is applied most commonly to the measurement of radionuclide activities in high performance liquid chromatography (HPLC) effluent streams referred to as radio-HPLC. The applications of HPLC have become widespread in many fields of science including agricultural and food chemistry, biochemistry, molecular biology, immunology, and associated fields such as medical and drug research, environmental monitoring, and radioactive waste monitoring, among others. These fields often require the use of HPLC to separate molecular compounds and, when radionuclides are used to study the metabolic rates of organic compounds or ligands, radio-HPLC is often the method of choice to separate and quantify the activity levels of radionuclides associated with these compounds.

The alternative to FSA is the collection of the HPLC effluent in fractions, which is a common technique known as fraction collection, followed by the liquid or solid scintillation analysis of each fraction for radionuclide activity. The measurement of radionuclide activity in HPLC fractions by conventional liquid scintillation analysis (LSA) requires the preparation of a separate scintillation vial or scintillation microplate well for each fraction, which can number in the hundreds for a given HPLC run. This is tedious, time-consuming, and expensive in terms of fluor cocktail consumption, laboratory staff time, and waste disposal costs to say the least. Also the analysis of fractions cannot provide an online real-time radionuclide activity measurement of the HPLC effluent. Only post-run analysis are possible with collected fractions; and the success or failure of a given HPLC run can only be known after the run is completed and samples are analyzed at a high cost of fluor cocktail and work time. Only FSA can provide real-time analysis during the HPLC run and with minimal fluor cocktail consumption. A thorough study made by Rapkin (1993) compared the sensitivity and accuracy of low-level analysis (e.g., lower limits of detection of 20.8 CPM for conditions of 10 CPM background) for continuous flow scintillation analysis and liquid scintillation analysis of fractions. Statistical considerations suggested that fraction collection followed by LSA offered the possibility of greater sensitivity and accuracy of radioactivity measurements in HPLC eluates, although with considerable inconvenience. However, Rapkin (1993) demonstrated that, even for low-activity analysis, when the randomness of fraction collection is accounted for, the assumed advantages of fraction collection disappear and the trend to replace fraction collection with the continuous real-time radioactivity measurement of HPLC eluates is justified.

The first liquid radiochromatography flow cells were developed following the observation of Steinberg (1958, 1960) that the fluorescence of solid anthracene crystals was useful for the detection of the beta radiations of ^{3}H and ^{14}C when suspended in aqueous solutions containing these radionuclides. Simultaneous independent studies by Rapkin and Packard (1960) and Schram and Lombaert (1960, 1961) demonstrated the feasibility of counting ^{3}H and ^{14}C in flowing aqueous streams when these streams were passed over anthracene crystals within a cell placed between two photomultiplier detectors. Anthracene is no longer used as a scintillator for radioactivity measurements in flowing systems, as many developments have occurred in this field since these pioneering works were reported. For early reviews on the development of FSA see previous works by the author (L'Annunziata, 1979) and Parvez et al. (1988).

This chapter will focus on the state-of-the-art and current techniques of FSA used to measure radioactivity in HPLC effluents; however, other applications will be cited such as the measurement of radioactivity in streams associated with the nuclear power industry and the application of FSA to the measurement of radionuclides in the environment including the on-line discrimination of mixtures of α and β emitters.

II. BASICS OF FLOW SCINTILLATION ANALYSIS INSTRUMENTATION

There are three basic detector types applied to flow scintillation analysis. These are (1) the liquid scintillator (homogeneous) flow cell, (2) the solid scintillator (heterogeneous) flow cell, and (3) the gamma-cell. The latter gamma-cell is manufactured according to three detector types to improve detection efficiencies for high-energy-, low-energy-, and annihilation-gamma radiation in HPLC effluents. The various scintillation flow cells currently used are listed in Table 12.1 together with their characteristics, applications, and advantages and disadvantages. Scintillation flow cells are interchangeable within a given scintillation analyzer designed specifically for this purpose. It is important to make the proper selection of the cell, which will provide the optimum detection efficiencies witsh lowest background as well as meet the needs of any particular FSA. A description of the scintillation flow cells will be provided in this section together with an account of the advantages and disadvantages of each.

A. HPLC and Scintillation Analyzer

The principal components of a typical flow scintillation analyzer (FSA) is illustrated as a block diagram in Fig. 12.1. The connection of the effluent stream from a HPLC system is illustrated at the upper left-hand corner of the figure. The scintillation flow cell is placed between the two photomultiplier tubes (PMTs), and it is interchangeable as several types are available. The scintillation flow cells differ according to the modes of radiation detection, efficiency for radiation types and radiation energy, and particular experimental requirements including narrowbore and microbore radio-HPLC systems. These are described in the next section. The flow cell is placed between the two PMTs, so that the two opposite side windows of the cell are in direct contact with the opposite faces of the photomultiplier tubes. The orientation of the flow cell vis-à-vis the PMTs is illustrated in Fig. 12.2.

If a liquid scintillator or homogeneous flow cell is used, liquid fluor cocktail must be uniformly mixed with the HPLC effluent stream. Special low-viscosity nongelling fluor cocktails are recommended, which are described further on in this chapter. The liquid scintillator is added to the HPLC effluent stream by a variable volume scintillator pump (LS Pump) and a static fluid mixer (Mixing Tee) before reaching the scintillation flow cell. Mixing the effluent stream with fluor cocktail renders the separated chemical components useless for any subsequent chemical or biological tests. If further studies are required on the chemical compounds separated by the HPLC such as mass spectrometry (MS) or nuclear magnetic resonance (NMR)

███ **TABLE 12.1** **Types of Scintillation Flow Cells Used for the Measurement of Radioactivity in HPLC Effluents**

I. Liquid scintillator (homogeneous) cell

  A. Characteristics

    Coiled Teflon tubing, 35–5000 μL volume

  B. Radionuclides analyzed and %E (in parenthesis)

    All β emitters, e.g., ^{3}H (20–60%), ^{14}C (70–95%), ^{35}S and ^{33}P (70–95%), ^{32}P (85–95%)

    Low-energy γ emitters, e.g., ^{125}I (60–90%)

    All α emitters (> 95%)[a]

  C. Advantages

    1. High %E

    2. No adsorption of radionuclide on Teflon tube walls

  D. Disadvantages

    1. HPLC eluent is mixed with cocktail fluor and thus separated compounds cannot be further studied or isolated for structural analysis.

    2. An electronic stream splitter is required to obtain part of the eluent to enable other chemical and biological studies on separated compounds;

II. Solid scintillator (heterogeneous) cell

  A. Characteristics

    Fine solid scintillator particles packed within coiled tubing, 150–420 μL volume, among which are:

    1. Monocrystalline SolarScint[b]

    2. Yttrium Glass (YG)

    3. Polycrystalline cerium-activated yttrium silicate [YSi(Ce)]

    4. High-pressure (high flow rate) cells containing one of the following scintillators:

      a. Europium-activated calcium fluoride [CaF$_2$(Eu)]

      b. Cerium-activated lithium glass [lithium glass(Ce)]

      c. Cerium-activated yttrium silicate [YSi(Ce)]

      d. Scintillating plastic beads (plastic scintillator)

  B. Radionuclides analyzed (cell detector types and %E in brackets)

    All β emitters

      ^{3}H [SolarScint[b], 3.0%; YG, 1.5%; YSi(Ce), 2.8%]

      ^{14}C [SolarScint[b], 70%; YG, 63–86%; YSi(Ce), 68%]

      High-pressure, high-rate cells:

      ^{3}H [CaF$_2$(Eu), > 5%; Lithium Glass(Ce), > 1%; YSi(Ce), > 2%; Plastic, > 2.5%]

      ^{14}C [CaF$_2$(Eu), > 85%; Lithium Glass(Ce), > 45%; YSi(Ce), > 65%; Plastic, > 15%]

      All α emitters (50–60%E)

  C. Advantages

    1. Good %E except for ^{3}H.

    2. Sample in effluent is not destroyed (no cocktail fluor is used).

    3. No chemical quench effects associated with fluor cocktail occur.

    4. Costs associated with the use and disposal of fluor cocktail are avoided.

    5. High salt, buffer, or pH gradients do not effect %E.

    6. High-pressure cells permit up to 3000 psi cell pressure and high flow rates.

(*continued*)

███ **TABLE 12.1** **Continued**

    D. Disadvantages

        1. Sample may bind reversibly or irreversibly onto scintillator yielding peak broadening and high backgrounds.

        2. Low %E for ^{3}H.

III. High-energy gamma cell

    A. Characteristics

        Coiled Teflon tubing with 6-mm thick BGO windows; 35–550 μL volume

        (e.g., Gamma-B^b flow cell) See Fig. 12.4.

    B. Radionuclides analyzed and %E (in parenthesis)^c

        High-energy γ emitters (> 70 keV), e.g., ^{67}Ga (44%), ^{131}I (58%), ^{111}In (25%), ^{99m}Tc (63%), ^{201}Tl (35%), ^{54}Mn (19%), ^{60}Co (20%), ^{133}Ba (28%), ^{85}Sr (19%), ^{51}Cr (3%).

    C. Advantages

        1. No liquid scintillation fluor cocktail required

        2. No solid scintillator required in cell tubing

        3. High signal-to-noise ratio

        4. Good spectral resolutions for high-energy gamma (> 300 keV)

    D. Disadvantages

        1. Used exclusively for γ emitters or high-energy β emitters (e.g., ^{32}P, ^{90}Y)

IV. Low-energy gamma cell

    A. Characteristics

        Coiled Teflon tubing with 4-mm thick CaF$_2$(Eu) windows; 10–650 μL volume; (e.g., Gamma-C^b flow cell), See Fig. 12.4.

    B. Radionuclides analyzed and %E (in parenthesis)^c

        Low-energy γ emitters (1–70 keV), e.g., ^{55}Fe (14%), ^{125}I (90%), ^{109}Cd (34%), ^{201}Tl (28%), ^{99m}Tc (14%), ^{67}Ga (15%)

        High-energy β emitters, e.g. ^{32}P ($\sim 60\%E$)^b, ^{90}Y (60%E)

    C. Advantages

        1. No liquid scintillation fluor cocktail required

        2. No solid scintillator required in cell tubing

        3. High counting efficiencies (%E) for low-energy γ emitters (< 70 keV)

    D. Disadvantages

        Used exclusively for γ emitters or high-energy β emitters (e.g., ^{32}P ($\sim$60%E), ^{90}Y (60%E)

V. PET Cell

    A. Characteristics

        Coiled Teflon tubing with 6-mm thick BGO windows opaque on one side; see Fig. 12.7;

    B. Radionuclides analyzed and %E (in parenthesis)

        Positron (β^+-emitters such as those used in positron emission tomography (PET), e.g., ^{18}F (25%), ^{11}C (43%), ^{13}N, ^{15}O, ^{82}Rb

    C. Advantages

        1. Low backgrounds approximately 1/10 of backgrounds from Gamma-B^b flow cell with only 1/3 drop in %E

        2. No liquid scintillation fluor cocktail required

        3. No solid scintillator required in cell tubing

(*continued*)

TABLE 12.1   Continued

D. Disadvantages

   1. Used exclusively for positron β^+ emitters

VI. Narrow-bore cell

  A. Characteristics

    1. Designed for applications using 2 mm ID narrow-bore HPLC columns

    2. Cell sizes range from 15 to 250 μL

    3. Effluent (sample) flow rates range from 133 to 1 mL/min

    4. For homogeneous cells liquid scintillation cocktail flow rates (3 : 1 cocktail : sample ratio) range from 400 μL to 3 mL/min

    5. Heterogeneous cells use yttrium silicate, lithium glass, calcium fluoride, plastic scintillator and SolarScint formats

  B. Radionuclides analyzed and %E (in parenthesis)

    1. All radionuclides emitting radiations that produce significant scintillation with fluor cocktail in homogeneous cells or scintillation in solid scintillator heterogenous cell including α-, β^--, β^+-, x-rays, weak-γ-, and Auger electrons, etc.

    2. ^{3}H (35–42%), ^{14}C (80–90%) in homogeneous cells

  C. Advantages

    1. Improved peak resolutions without peak tailing

    2. Applied to simultaneous HPLC/FSA/mass spectrometry structure analysis

    3. Reduced cocktail consumption and waste disposal costs compared to conventional FSA cells

VII. Micro-bore cell

  A. Characteristics

    1. Designed for applications using 1 mm ID narrow-bore HPLC columns

    2. Cell sizes range from 3 to 100 μL

    3. Effluent (sample) flow rates range from 25 to 150 μL/min

    4. For homogeneous cells liquid scintillation cocktail flow rates (3 : 1 cocktail : sample ratio) range from 75 to 450 μL/min

    5. Heterogeneous cells use yttrium silicate, lithium glass, calcium fluoride, plastic scintillator and SolarScint formats

  B. Radionuclides analysed and %E (in parenthesis)

    1. All radionuclides emitting radiations that produce significant scintillation with fluor cocktail in homogeneous cells or scintillation in solid scintillator heterogenous cells including α-, β^--, β^+-, x-rays, weak-γ-, and Auger electrons, etc.

    2. ^{3}H (25–35%), ^{14}C (75–80%) in homogeneous cells

  C. Advantages

    1. Up to 10-fold improved peak resolutions without peak tailing compared to conventional FSA cells

    2. Applied to simultaneous HPLC/FSA/mass spectrometry structure analysis

    3. Reduced cocktail consumption and waste disposal costs compared to conventional FSA cells

[a]Personal communication with C. J. Passo, Jr., PerkinElmer Life and Analytical Sciences.
[b]Trademark of PerkinElmer Life and Analytical Sciences, Boston.
[c]Counting efficiency values as %E of gamma-emitting radionuclides are from Anonymous (1995).

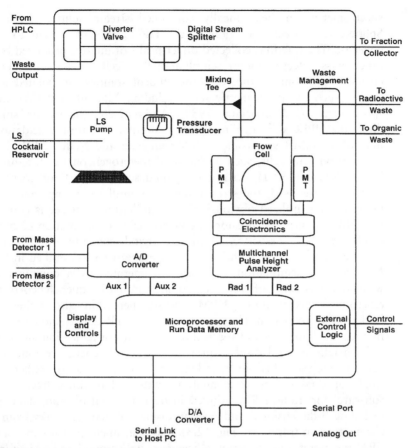

FIGURE 12.1 Block diagram of a state-of-the-art fully configured flow scintillation analysis system. (Courtesy of PerkinElmer Life and Analytical Sciences.)

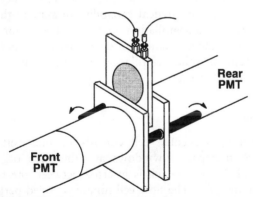

FIGURE 12.2 Drawing illustrating the orientation of a flow cell vis-à-vis the two photomultiplier tubes. The flow cell is illustrated extended over its normal location and in the process of being either installed or removed from the center of the two PMTs. Thumbscrews illustrated as solid black with arrows are used to facilitate the interchange of flow cells to accommodate different radionuclide measurements and methods of detection. (Courtesy of PerkinElmer Life and Analytical Sciences.)

spectrometry, an electronically controlled stream splitter (Digital Stream Splitter) can direct a portion of the HPLC effluent stream to a fraction collector. The fractions collected under peaks of interest produced by a UV or other mass detector and/or scintillation flow cell can be combined and the chemical compounds studied in the effluent solution or isolated for further studies such as spectrometric tests including MS, infrared (IR), and NMR spectrometry (L'Annunziata, 1984). Alternatively, the stream splitter may direct the HPLC effluent directly for molecular structural elucidation via on-line MS or NMR spectrometry (L'Annunziata and Nellis, 2001a,b). Certain scintillation flow cells, such as the solid (heterogeneous) cell, gamma cell or PET cell (Table 12.1), do not use liquid scintillation fluor cocktail. When these cells are used, the stream splitter, scintillator pump, and static fluid mixer are not utilized obviously. The scintillation analyzer is equipped with two PMTs, high voltage supply, coincidence and summation circuitry, pulse height analyzer, associated analog-to-digital converter, and multichannel analyzer similar to the components of a modern LSA described in Chapter 5. Modern flow scintillation analyzers are operated by a computer equipped with multitasking software including automatic quench correction and efficiency determination, DPM measurements in the effluent stream, background reduction electronics, such as TR-LSC (Anonymous, 1996), multichannel analysis and spectral display, self normalization and calibration, pulse height spectral display, preset and variable energy regions in keV for activity analysis, dual independent counting regions for either automatic single or dual radionuclide analysis with update times from 1 to 120 s, software for radio-HPLC direct instrument control and data reduction including 3-dimensional and overlay display of activity peaks from different chromatogram traces (see Fig. 12.3), and instrument performance assessment (IPA) for monitoring detection efficiencies, background and Chi-square values with ^{3}H and ^{14}C standards. Detailed descriptions and the practical applications of most of these features are given in Chapter 5 "Liquid Scintillation Analysis." Actually there is great similarity between the electronic components and radioanalytical capability of state-of-the-art flow scintillation analyzers and liquid scintillation analyzers. The major difference between the two analyzers is the mechanism of detection, which in the case of FSA may be either liquid or solid scintillation and the radionuclides are analyzed in a dynamic liquid stream.

B. Liquid (Homogeneous) Flow Cells

The liquid scintillator (homogeneous) cell consists of fine Teflon tubing coiled flat between two transparent windows at either side of the coiled tubing as seen in Fig. 12.4, and the cell is inserted between the two PMTs of the flow scintillation analyzer. The flow cell tubing is coiled perpendicular to the planar faces of two PMTs. Because the liquid flow cell consists only of tubing, the detection of radionuclides requires a prior mixing of the entire or a fractional part of the radioactive HPLC effluent with liquid scintillation fluor cocktail while the effluent stream is in motion and prior to the arrival of the effluent–fluor cocktail mixture at the orifice of a flow cell.

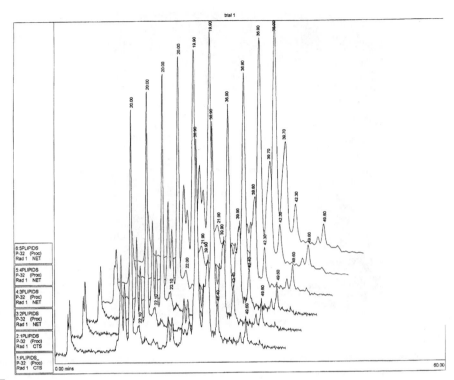

FIGURE 12.3 3D-analysis overlay of six flow scintillation analysis traces of HPLC separations of ^{32}P-labeled lipids (radioactivity vs. retention time in minutes) with PerkinElmer FLO-ONE for Windows. Up to 32 traces can be overlaid to compare differences or offset into a three-dimensional display as illustrated. From the display it is possible to analyze changes in data over a series of runs, and reports of results can be produced. (Courtesy of PerkinElmer Life and Analytical Sciences.)

This has traditionally been referred to as the homogeneous method, because the effluent stream and fluor cocktail are homogeneously mixed, and the flow cell is referred to as a liquid cell.

This cell type provides the highest counting efficiencies for low-energy β emitters such as ^{3}H over the range of 20–60%, and intermediate-energy β emitters such as ^{14}C and ^{35}S with counting efficiencies in the range of 70–95%. The high detection efficiencies are expected with the liquid (homogeneous) flow cells, because a liquid scintillation detection method is used where the radionuclides are in solution with fluor cocktail. Special fluor cocktails should be used that do not gel when mixing with HPLC effluents and suppress chemiluminescence, which can occur following the immediate mixing of cocktail with HPLC effluents.

The wide range of chemicals used as eluates in HPLC and the changing chemical characteristics of HPLC effluents when gradient elution is carried out and when particular compounds are eluted from the column will cause variable quench. As the percent counting efficiency of any radionuclide is a function of quench a gradient quench correction curve is needed to convert on-line count rates to disintegration rates.

FIGURE 12.4 Photograph of a scintillation flow cell. The finger-tight connectors for the HPLC effluent inlet and outlet tubing are at the very top of the flat flow cell. The Teflon tubing is wound flat between two transparent windows with outer stainless steel casing for the liquid homogeneous or solid heterogeneous cells. In the case of the gamma cells, the Teflon tubing is wound flat between solid scintillator windows consisting of 6-mm-thick BGO or 4-mm-thick $CaF_2(Eu)$. (Courtesy of PerkinElmer Life and Analytical Sciences.)

A disadvantage of the liquid (homogeneous) flow cell is that the separated compounds in the HPLC effluents are rendered useless for further chemical or biological tests due to their mixing with fluor cocktail. However, an effluent stream splitter can be used to divert a fraction of the HPLC effluent to the on-line analysis of FSA peaks via MS or NMR spectrometry or on to a fraction collector to permit further studies on the HPLC separated components. As in liquid scintillation counting radionuclides that emit low-energy gamma rays, Auger electrons, and internal conversion electrons can be measured also in the liquid flow cell with high counting efficiency. In the case of ^{125}I, counting efficiencies in the range of 60–90% are possible. Alpha-emitting radionuclides are detected with high counting efficiencies (>95%) in the liquid flow cell.

C. Solid (Heterogeneous) Flow Cells

The solid scintillator (heterogeneous) flow cell is manufactured with fine beads or particles of an insoluble solid scintillator placed within the Teflon tubing of the flow cell. The tubing is coiled flat between two transparent windows and placed between the faces of two PMTs. The HPLC radioactive effluent stream will flow through the cell and make intimate contact with the solid scintillator beads. This latter approach is referred to as the heterogeneous method, because the effluent stream and scintillator do not mix, and the flow cell is called a solid cell. The photomultiplier tubes will detect and measure the scintillation light photons emitted from the tubing, and the radioactivity determined during the sample residence time in the flow cell according to conventional liquid scintillation technology described in Chapter 5 and further on in this chapter.

Various types of solid scintillators are used to make up the heterogeneous cell including yttrium glass, polycrystalline cerium-activated yttrium silicate [YSi(Ce)], europium-activated calcium fluoride [CaF_2(Eu)], cerium-activated lithium glass, and plastic scintillator, among others. These will vary in their detection efficiency for various β and α emitters. The counting efficiencies are low for weak β emitters like ^{3}H ($1.5-5\%E$) and good for the intermediate-energy β emitters such as ^{14}C ($45-85\%E$). High-energy β emitters are detected as expected with yet higher counting efficiency.

A major advantage of the heterogeneous flow cell is that the sample in the HPLC effluent is not destroyed, because no flour cocktail is used. The entire effluent stream can be analyzed by on-line nuclear magnetic resonance or mass spectrometry, or it may be collected by a fraction collector and further chemical or biological tests can be carried out on the separated fractions. Also, chemical quench, which is a problem with the liquid homogeneous flow cells, does not occur with the solid (heterogeneous) cells. Therefore, high salt, buffer solutions or pH gradients used in HPLC eluates will not affect counting efficiency. The only problem is that, due to the high surface area and structure of the solid scintillator packed in these flow cells, compounds undergoing separation in the HPLC often bind reversibly or irreversibly onto the scintillator. This can result in high backgrounds and peak broadening. Irreversible binding of sample onto the scintillator may require replacement of the flow cell. A relatively new solid heterogeneous flow cell containing monocrystaline SolarScint (trademark of PerkinElmer Life and Analytical Sciences, Boston, MA) undergoes minimal sample binding. The inositol phosphates are one example of compounds that undergo adsorption onto yttrium silicate (YSi) scintillator resulting in peak broadening. SolarScint flow cells operated in the pH range of 3–8 show minimal effects of compound binding resulting in optimal peak resolutions. This is evidenced by Fig. 12.5, which illustrates the high resolutions of the HPLC-separated ^{3}H-labeled inositol phosphates detected with a 210-μL SolarScint heterogeneous flow cell.

Certain designs of heterogeneous flow cells are manufactured with thick glass tubing to withstand high pressures of up to 3000 psi (200 bar) with fast flow rates.

D. Gamma and PET Flow Cells

The gamma- and PET flow cells are made to provide highest detection efficiencies for gamma radiation. The cell types can be classified into those designed to provide optimum detection efficiencies for (1) high-energy γ emitters, (2) low-energy γ emitters and (3) positron (β^+) emitters.

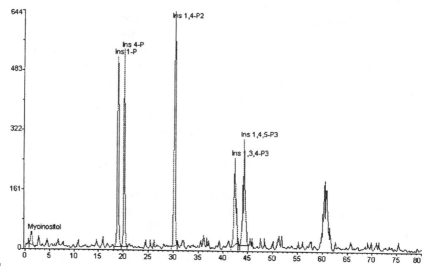

FIGURE 12.5 Activity peaks of ³H-labeled inositol mono-, di-, and triphosphates separated by HPLC and measured with a 210-μL solid (heterogeneous) flow cell containing SolarScint solid scintillator, from the work of Dr. K. E. Nye at The Medical College, St. Bartholomew's Hospital, University of London, UK. (Courtesy of PerkinElmer Life and Analytical Sciences.)

The positron (β^+) emitters, which are principally those radionuclides used in positron emission tomography (PET), produce gamma radiation via annihilation (see Chapter 1). Table 12.1 provides the main characteristics and detection properties of these scintillation flow cells.

I. High-Energy Gamma Cell

The high-energy gamma flow cell will detect all γ emitters and high-energy β emitters; however, the high-energy gamma cell is designed specifically to provide optimum detection efficiencies for γ-emitting radionuclides of energy > 70 keV. This type of flow cell (e.g., Gamma-B flow cell, PerkinElmer Life and Analytical Sciences) consists of Teflon tubing coiled flat between two 6-mm-thick bismuth germanate (BGO) solid scintillator windows as illustrated in Fig. 12.6. The scintillator BGO has a very high density ($7.13\,\mathrm{g\,cm^{-3}}$), high atomic number (high Z), which provides this scintillator with a high "stopping power" for high-energy gamma radiation. At the same time BGO has a high light output and short scintillation decay time, among other favorable properties for high-energy gamma-ray detection. The properties of BGO are described in detail in Chapter 11.

Figure 12.6 illustrates two annihilation gamma-rays traveling at 180° angles to each other from the sample in the flow cell tubing to the scintillator crystal windows at either side of the flow cell tubing. If we ignore annihilation radiation, which is produced exclusively by positron emitters, and consider the more common radionuclide gamma decay, we should keep in mind that gamma-rays are emitted from nuclei as single events and they are monoenergetic. A gamma-ray emitted from a radionuclide in the flow cell tubing would interact, therefore, with only one of the crystal scintillator

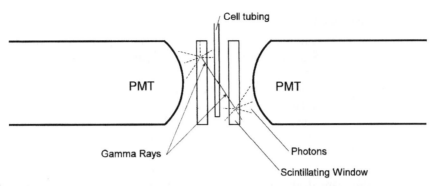

FIGURE 12.6 Basic diagram of a gamma flow cell and photomultiplier tube configurations used for the measurement of all gamma-emitting radionuclides. Annihilation gamma rays are illustrated interacting with solid scintillation detector windows, resulting in light photon emission in each of the scintillating windows. The scintillator windows may be manufactured from a 6-mm-thick BGO crystal for high-energy gamma-ray (> 70 keV) detection or 4-mm-thick $CaF_2(Eu)$ for low energy (< 70 keV) gamma-ray measurements. (From Anonymous, 1995, vreprinted with permission from PerkinElmer Life and Analytical Sciences.)

windows. The scintillation light produced in one of the flow cell windows would be "seen" or detected simultaneously by the two photomultiplier tubes and consequently registered as a voltage pulse in coincidence according to the principles of coincidence scintillation counting (see Chapter 5).

2. Low-Energy Gamma Cell

The low-energy gamma flow cell will detect all γ emitters and high-energy β emitters; however, the cell is designed specifically to provide optimum detection efficiencies for gamma-ray energies < 70 keV. This type of flow cell (e.g. Gamma-C flow cell, PerkinElmer Life and Analytical Sciences) consists of Teflon tubing coiled flat between two 4-mm-thick $CaF_2(Eu)$ solid scintillator windows as illustrated in Fig. 12.6. The $CaF_2(Eu)$ scintillator has a much lower density ($3.19\,\mathrm{g\,cm^{-3}}$) and lower atomic number than the BGO scintillator employed in the previously described high-energy gamma-cells. Therefore, the low-energy scintillation flow cell has a lower gamma-ray stopping power than the previously described high-energy gamma flow cell. The high gamma stopping power of BGO is not needed for gamma rays of energy below 70 keV, and $CaF_2(Eu)$ scintillator has a three-fold higher scintillation conversion efficiency than BGO (L'Annunziata, 1987). The mechanisms of coincidence scintillation detection described in the previous section, however, are the same for both the $CaF_2(Eu)$ and BGO detector windows.

The major advantages in utilizing the gamma cells for the measurement of γ or high-energy β emitters are that no liquid or solid scintillator is required in the cell tubing. The solid scintillator exists outside the flow cell tubing, and therefore, no chemical or color quench is possible. After flow scintillation analysis, the entire HPLC effluent can be diverted to a mass spectrometer or nuclear magnetic resonance spectrometer for identification of HPLC peaks or collected by a fraction collector and further chemical and biological tests performed on the effluent fractions.

3. PET Cell

The PET scintillation flow cell is a patented design (PerkinElmer Life and Analytical Sciences, Boston), which will detect all positron emitters; and it is designed to be used exclusively for the measurement of positron emitters at highly reduced background count rates. The positron-emitters are the short-lived radionuclides used in positron emission tomography (PET) in the field of nuclear medicine, which is described in a previous text by the author (L'Annunziata, 1987). Among the positron-emitters the most commonly used for PET are ^{11}C, ^{13}N, ^{15}O, ^{18}F, and ^{82}Rb.

The emission of positrons during radionuclide decay is always accompanied by the 511 keV gamma-rays due to the annihilation of a positron and an electron, when the positron comes to rest in matter and eventually makes contact with an electron, its antiparticle. Actually two 511 keV gamma-rays are produced simultaneously for each positron emitted by a radionuclide, and the two gamma rays are emitted in opposite directions or 180° to each other. A detailed treatment of positron–electron annihilation and the gamma radiation produced by this phenomenon is given in Chapter 1. An illustration of two gamma-ray photons produced by positron annihilation is provided in Fig. 12.7. The two gamma-rays are illustrated as traveling simultaneously in opposite directions from the flow cell. When annihilation radiation is produced in the flow cell, the two gamma-ray photons will produce scintillation light in the two BGO crystal windows of the flow cell. The BGO windows have an opaque coating on the inner sides facing the flow cell tubing. With such a detector design the light produced in one BGO window can be "seen" or detected by only one photomultiplier tube, that PMT facing the uncovered side of the BGO window. However, for any given annihilation gamma rays originating from the sample in the flow cell the two light flashes produced in the two BGO windows will be detected simultaneously by the two photomultiplier tubes within the coincidence time of the instrument, and the two scintillation events

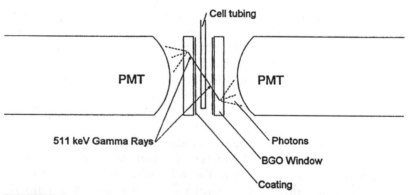

FIGURE 12.7 Basic diagram of a patented PET flow cell and photomultiplier tube configurations used exclusively for the measurement of positron emitters. Annihilation gamma rays are illustrated interacting with the bismuth germanate (BGO) scintillator windows; however, the opaque coatings on the inner sides of the BGO detectors optically isolate the PMTs. (Courtesy of PerkinElmer Life and Analytical Sciences.)

(one in each window) will be registered and counted as a single coincident event. This detector design is intended to reduce background counts. Background events in the two BGO windows will not occur generally at the same time within the coincidence time of the scintillation analyzer; and these events will be rejected by the coincidence circuitry.

A similar flow cell detector for the measurement of positron emitters eluted from HPLC utilized for the analysis of PET radiopharmaceuticals was designed by Takei *et al.* (2001). Flow cells of various volumes containing coiled Teflon tubing are accommodated between two BGO crystal detectors as illustrated in Fig. 12.8. The BGO detectors are housed in aluminum casings, whereby only gamma–photon pairs originating from positron annihilation are detected in coincidence. A 15 mm distance between the two aluminum housings for the BGO detectors allows for the easy insertion of flow cells of different volumes. Two counting window widths for the measurement of ^{11}C, namely, 400–600 keV and 90–680 keV yielded detection efficiencies of $32 \pm 1\%$ and $43 \pm 1\%$, respectively (Takei *et al.*, 2001). The narrower counting region yielded a five-fold higher FOM (i.e., E^2/B) with a background count rate of 1.7 ± 1 cpm and detection limit of 0.3 Bq.

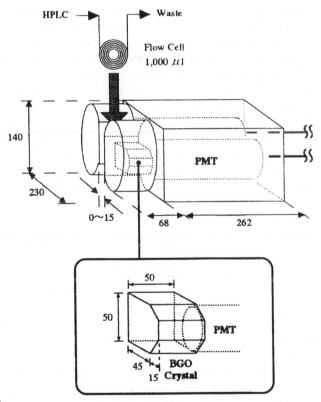

FIGURE 12.8 Block diagram of the sensitive positron detector. A pair of the BGO housings coupled to photomultipliers fixed in an aluminum frame. Dimensions are given in mm. (From Takei *et al.*, 2001, reprinted with permission from Elsevier Science.)

Background count rates in the PET scintillation flow cells are reduced to 1/10 the backgrounds encountered in the previously described high-energy gamma-cells, which also employ BGO crystal windows. The counting efficiencies of the PET scintillation flow cells are reduced by only a third of those obtained by the high-energy gamma cells. Therefore, the PET flow cells provide considerably higher figures of merit (E^2/B) and lower detection limits for the measurement of positron emitters.

E. Narrow-Bore and Micro-Bore Flow Cells

Micro-volume flow scintillation analyzer cells are available and these differ from the conventional FSA cells mainly in their dimensions of inner-diameter (ID) and cell volume. The narrow-bore and micro-bore flow cells are made to accommodate narrow-bore (~ 2 mm ID) and micro-bore (~ 1 mm ID) HPLC systems that yield improved HPLC peak resolutions over the standard FSA systems. The standard or conventional FSA systems use larger cell volumes that accommodate wider HPLC columns (~ 5 mm ID). The major properties that distinguish narrow-bore and micro-bore FSA systems from the standard FSA systems are provided in Table 12.2. The high resolutions provided by the micro-bore HPLC-FSA systems provide improved molecular species separations and identification via direct coupling to the electrospray ionization-mass spectrometer. To demonstrate the potential of micro-bore flow cells Schultz and Alexander (1998) tested a 1.2 μL volume on-column flow cell for use with 250-μm-internal-diameter HPLC columns capable of detecting < 200 DPM for a 1-min-wide chromatogram peak with a PerkinElmer Radiomatic TR150 flow scintillation analyzer (see Fig. 12.9). They demonstrated that a solid scintillation cell packed with silanated lithium glass with a mean particle size of 9.6 ± 2.6 μm yielded a 24% increase in peak area compared to a cell packed with a mean particle size of 21.0 ± 2.5 μm. Also, they demonstrate that a 508-μm-ID cell yielded a 15% increase in peak area compared to a 762 μm-ID cell. As illustrated in Fig. 12.9 the micro-bore FSA cell is equipped with a 30 μm-ID tube for linkage

TABLE 12.2 Characteristics of Standard (Conventional), Narrow-bore, and Micro-bore FSA Systems[a]

Characteristics	Standard FSA	Narrow-bore FSA	Micro-bore FSA
Cocktail pump	Digital	Digital	Precision syringe
HPLC columns	5 mm ID	2 mm ID	1 mm ID
Sample flow rate–FSA	333 μL–2 mL/min	133 μL–1 mL/min	25 μL–150 μL/min
LS cocktail flow rate (at 3:1 cocktail:sample ratio)	1 mL–6 mL/min	400 μL–3 mL/min	75 μL–450 μL/min
Mixing Tee	> 250 μL	10 μL	10 μL
Cell size	35 μL–5 mL	15 μL–250 μL	3 μL–100 μL

[a]Courtesy of PerkinElmer Life and Analytical Sciences.

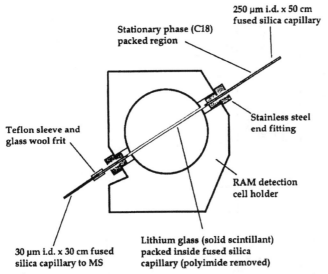

250 μm i.d. x 50 cm
fused silica capillary

Stationary phase (C18)
packed region

Stainless steel
end fitting

Teflon sleeve and
glass wool frit

RAM detection
cell holder

30 μm i.d. x 30 cm fused
silica capillary to MS

Lithium glass (solid scintillant)
packed inside fused silica
capillary (polyimide removed)

FIGURE 12.9 An on-column radioactivity monitor flow cell for use with 250-μm-i.d. microcolumns. The flow cell volume is 1.2 μL. The detection window is 5 cm in length. (From Schultz and Alexander IV, 1998, reprinted with permission of John Wiley and Sons, Inc., Copyright © 1998.)

directly to the inlet of a electrospray ionization chamber of a mass spectrometer.

F. Criteria for Flow Cell Selection

When making a decision on which type of scintillation flow cell would be best for a particular application, several factors should be considered among which are (1) the detection efficiency of the flow cell for the radionuclide of interest, (2) the costs involved in using scintillation fluor cocktail with a liquid (homogeneous) flow cell, (3) the chemical composition of the HPLC eluent used, (4) whether further analysis of the HPLC effluent is required (i.e., must the compounds separated and detected in the HPLC effluent be analyzed by on-line mass spectrometry or NMR spectrometry, or must the components be isolated and other chemical and biological tests be performed on the separated components?), (5) the level of radioactivity of the sample components, that is, the sensitivity required, and (6) the flow cell detector background when the minimal detectable activity is an important factor. Data listed in Tables 12.1 and 12.3 are helpful in making a decision on which type of scintillation flow cell to use for a particular application. Some examples will be cited to illustrate the use of the data in these tables.

The highest counting efficiencies for all β or α emitters are obtained with the liquid scintillator (homogeneous) flow cell. However, a special low-viscosity nongelling liquid scintillation fluor cocktail must be mixed with the HPLC eluent stream with a controlled pump and static fluid mixer. The separated compounds or molecular entities in the HPLC effluent are thereby rendered useless for any other chemical or biological tests that may be

TABLE 12.3 Performance Characteristics of High-Energy (BGO) and Low-Energy [CaF$_2$(Eu)] Bremsstrahlung and Gamma Scintillation Flow Cell Detectors for the Detection of Radionuclides Listed in Order of Increasing Photon Radiation Emission Energies[a]

Radio-nuclide	Radiation energy (keV)[b]	High-energy Gamma-B, BGO				Low-energy Gamma-C, CaF$_2$(Eu)			
		Region (keV)	%E	Background	FOM (E^2/B)	Region (keV)	%E	Background (CPM)	FOM (E^2/B)
^{55}Fe	Bremsstrahlung up to 23.2 keV	0–35	0.2	5	0.008	0–10	14.2	40	5.0
^{125}I	27.4-keV (30%) and 31-keV (>100%) x-rays, 35-keV gamma (7.0%)	0–120	27.1	48	15.3	10–85	81.1	108	60.8
^{109}Cd	22-keV (>100%) and 70-keV (95%) x-rays, 88-keV gamma (3.8%)	5–60	5.1	14	1.85	5–60	34	109	10.6
^{99m}Tc	2-keV x-rays (100%) 140-keV gamma (100%)	40–300	63.2	198	20.2	40–300	14.3	158	1.3
^{201}Tl	x-rays up to 84 keV (195%) 135-keV gamma (2.6%) 167-keV gamma (10%)	30–400	34	275	4.2	30–400	27.7	196	3.9
^{51}Cr	5-keV x-rays (22%) 320-keV gamma (10%)	200–500	3	233	0.039	200–500	0.1	72	0.00014
^{67}Ga	9-keV x-rays (56.4%) 84-keV x-rays (29%) 93-keV gamma (38%) 185-keV gamma (21%) 300-keV gamma (17%)	35–600	44.2	394	5.0	35–600	14.8	217	1.0

Isotope	Radiations							
^{111}In	23-keV x-rays (67%) 26-keV x-rays (14%) 171-keV gamma (91%) 245-keV gamma(94%)	100–650	24.1	385	1.5	100–650	6.2	143	0.3
^{133}Ba	31-keV x-rays (63%) 35-keV x-rays (23%) 81-keV gamma (34%) 303-keV gamma (18%) 356-keV gamma (62%) 384-keV gamma (9%)	240–700	27.5	289	2.6	240–700	2.1	78	0.05
^{131}I	284-keV gamma (6%) 364-keV gamma (81%) 637-keV gamma (7%)	220–950	57.9	379	8.8	220–950	20.2	99	4.1
^{85}Sr	13-keV x-rays (50%) 514-keV gamma (100%)	350–850	19.3	241	1.5	350–850	0.6	57	0.006
^{54}Mn	5-keV x-rays (22%) 835-keV gamma (100%)	600–1200	19.2	153	2.4	600–1200	1.9	32	0.11
^{60}Co	1330-keV gamma (100%) 1170-keV gamma (100%)	900–1700	19.9	99	4.0	900–1700	4.6	16	1.3

[a] Gamma-B and Gamma-C are flow cell trademarks of the PerkinElmer Life and Analytical Sciences. %E and background data are from Anonymous (1995) with permission of PerkinElmer Life and Analytical Sciences.

[b] The most abundant radiations are given, and in parenthesis the approximate intensities or relative abundances of the radiations are provided as a percentage, that is, radiation emissions per 100 disintegrations.

required. A stream splitter could be employed to separate part of the HPLC effluent for on-line mass spectrometry, on-line NMR spectrometry, or for fraction collecting prior to the mixing with the scintillation fluor cocktail. Stream splitting permits only a portion and more often small part of the separated components of the sample to be collected or analyzed on-line by the mass- and NMR spectrometers. In the case of ^{3}H measurements when the separated ^{3}H-labeled components are of low activity (close to background) there is no alternative but to use the liquid homogeneous flow cell.

If sample activities are well above background, ^{3}H can be detected with a solid (heterogeneous) cell containing the solid scintillator monocrystalline SolarScint (PerkinElmer Life and Analytical Sciences, Boston) with a 3% detection efficiency. See Fig. 12.5 for an example of the detection of ^{3}H-labeled inositol phosphates by a heterogeneous flow cell containing SolarScint. The advantages in this case are no fluor cocktail is used, no chemical quench problem exists, and the entire effluent stream can be collected for on-line mass- or NMR spectrometry or other subsequent chemical or biological tests or even isolated from solution following fraction collection. All other beta emitters (e.g., ^{14}C, ^{35}S, ^{33}P, ^{32}P) or alpha emitters are detected with relatively high counting efficiencies with the heterogeneous flow cell, including those containing SolarScint scintillator. The only disadvantage of the heterogeneous cell is the reversible and sometimes irreversible adsorption of the sample molecules (e.g., peptides, lipids, proteins) onto the solid scintillator that cause peak broadening and increased background from one HPLC run to another. SolarScint solid scintillator, however, exhibits minimal adsorption of molecular components from the sample as illustrated in Fig. 12.5, and this solid scintillator in a flow cell should be tried before discarding the possibility of using the heterogeneous flow cell for radio-HPLC.

The solid scintillator (heterogeneous) flow cell is particularly applicable to radioactivity measurements when very high pressure (up to 3000 psi) and high flow rates are required. Table 12.1 lists some solid scintillator (heterogeneous) cells used for high pressure/high flow rate conditions.

The gamma cells are classified into two types: (1) the high-energy gamma cell, which yields optimum detection efficiencies for radionuclide x-ray or gamma emissions > 70 keV energy, and (2) the low-energy gamma cell, which provides higher counting efficiencies for radionuclides with x-ray or gamma emissions of energy < 70 keV. Table 12.3 provides the counting efficiencies (%E), background count rates, and calculated figures-of-merit (FOM) for a number of gamma-emitting radionuclides with both the high-energy Gamma-B and low-energy Gamma-C scintillation flow cells, which are trademarks of the PerkinElmer Life and Analytical Sciences. The counting regions, from which the counting efficiencies and backgrounds were determined, are defined by lower level (LL) and upper level (UL) discriminator settings of the multichannel analyzer with pulse height equivalents in keV. The most abundant x-ray and gamma emissions of each radionuclide are also listed in Table 12.3 to help the reader interpret the efficiencies provided by the two types of gamma cells. From Table 12.3 it can be seen that ^{55}Fe, ^{125}I, and ^{109}Cd are more efficiently measured with the low-level Gamma-C flow cell manufactured with $CaF_2(Eu)$ scintillator windows, which is evidenced by the superior $\%E$ and FOM values. The remaining nuclides listed in Table 12.3 in order of increasing gamma-ray energy are detected more efficiently with the high-energy Gamma-B flow cell, which is manufactured with the higher density and thicker BGO scintillator windows.

The reader should notice that varying the counting region LL and UL discriminator settings will govern the counting efficiency and background for any particular radionuclide to be measured and flow cell used. By adjusting the region settings an optimum figure-of-merit (FOM) can be found, which is calculated as

$$FOM = (E)^2/B \qquad (12.1)$$

where E is the percent counting efficiency and B is the background count rate. An optimum FOM can provide higher sensitivity for radionuclide detection by reducing the minimal detectable activity (MDA) as described further on in this chapter. In Table 12.3 counting region settings can be compared to the x-ray and gamma emissions of each radionuclide. Limiting the counting region settings will limit the radiation energies that can be detected by the flow cells; however the limited region settings will also reduce background count rates. The FOM, therefore, is a good means of calculating the optimum tradeoff of highest counting efficiency and lowest background. An interesting example may be taken from the radionuclide ^{51}Cr listed in Table 12.3. Its 320-keV gamma-ray emission has only 10% intensity or relative abundance, which is defined as the number of emissions per 100 radionuclide disintegrations. In the case of ^{51}Cr, there are only 10 gamma-rays emitted per 100 ^{51}Cr radionuclide disintegrations. Therefore, with the selected counting region of 200–500 keV (Table 12.3) a maximum counting efficiency of only 10% can be achieved, if all gamma rays could be detected without loss. The reported counting efficiency is 3% for the high-energy Gamma-B flow cell. A significantly higher counting efficiency for ^{51}Cr could be obtained by increasing the counting region to 0–500 keV to include detection of the 5-keV x-rays; however, this would result in a much higher background count rate.

The high-energy and low-energy gamma cells have the advantage that the HPLC effluent is not mixed with any fluor cocktail as the x- or gamma-radiation emissions are detected outside the Teflon tubing of the flow cell by the external BGO or $CaF_2(Eu)$ cell windows. Also, there is no adsorption of radioactive sample as occurs in the heterogeneous flow cell types. After measurement of radioactivity in the separated components of the HPLC effluent, the entire effluent stream can be analyzed on-line by mass- or NMR spectrometry or collected in fractions with the traditional fraction collector to permit subsequent chemical and biological studies on fractions of interest.

The reader should also note from Table 12.1 that high-energy beta-particle emitters, such as ^{32}P and ^{90}Y can be detected by the gamma-cells at a significantly high counting efficiency ($\sim 60\%$). Therefore, if samples are not near background and the minimum detectable activity is not of concern, the gamma cell could be the flow cell of choice for these high-energy beta-particle emitters, because no fluor cocktail is required and there is no adsorption of sample in the gamma-cell. Even for the intermediate-energy β emitters, such as ^{14}C, ^{35}S, ^{33}P, where counting efficiencies with the low-energy Gamma-C flow cell may be very low ($\sim 0.1\%$), the low-energy gamma cell may be the cell of choice when radioactivity levels are very high, such as in the μCi or mCi levels, commonly encountered in laboratories involved in the preparation or synthesis of radioisotope-labeled compounds or sources. The low counting efficiencies of the gamma-cell for the intermediate-energy β emitter is an advantage in this case, because sample activities are high, count rates are reduced to useable levels due to the low counting efficiency, and the labeled compounds or sources can be fully recovered with a fraction collector.

III. PRINCIPLES OF FLOW SCINTILLATION COUNTING

A. Count Rates

As noted in Chapter 7 of this book, the sample counting time is an important factor in the measurement of sample radioactivity. Longer counting times provide statistically more accurate measurements of the true count, that is, the standard deviations of the count determinations are reduced as counting time is increased. In flow counting, however, the sample counting time is a function of the flow rate and flow-cell volume, as the sample consists of a stream which passes through a tube (flow cell) situated perpendicular to the faces of the photon detectors (PMTs) as illustrated in Fig. 12.2. The photon detectors, therefore, will see the radioactive sample only for the time that the sample resides in the flow cell. This period of time is referred to as the residence time (T_R), which is calculated as

$$T_R = \frac{V}{F} \tag{12.2}$$

where V is the cell volume in mL and F is the sample flow rate in mL min^{-1}. When a liquid (homogeneous) flow cell is used, the flow rate (F) is a function of the sum of the HPLC and cocktail flow rates. Therefore, the calculation of the flow rate for a homogeneous cell must take into account the mix ratio of the HPLC mobile phase to cocktail. For example, for a 3:1 cocktail/mobile phase ratio, the flow rate is calculated as

$$F = \text{HPLC flow rate} + \text{Cocktail flow rate} \tag{12.3}$$

or

$$F = 1.0 \text{ mL min}^{-1} + 3.0 \text{ mL min}^{-1} = 4.0 \text{ mL min}^{-1}$$

Because the sample flows through the cell at a given rate, sample will enter and leave the cell simultaneously. Therefore, the flow scintillation analyzer will measure the flowing sample in segments according to the sample residence time. The instrument will automatically calculate the count rates according to the calculations illustrated subsequently. The following example illustrates the calculation of residence time and count rate:

Example 12.1 If we have a flow cell of 600 μL volume and a flow rate of 4.0 mL min^{-1} and the counts collected (observed counts) were 1200, we can calculate the residence time and count rate as follows

$$T_R = \frac{0.6\,\text{mL}}{4.0\,\text{mL min}^{-1}} = 0.15\,\text{min}$$

The count rate (CPM) can be calculated by dividing the observed counts by the residence time, as in this example,

$$\text{Count rate} = \frac{\text{Counts}}{T_R}$$

$$\text{Count rate} = \frac{1200\,\text{counts}}{0.15\,\text{min}} = 8000\,\text{CPM} \qquad (12.4)$$

or more easily the count rate is calculated as described previously by the author (L'Annunziata, 1979) by multiplying the counts collected by the inverse of the flow equation ratio (Eq. 12.2) or

$$\text{Count rate} = c\frac{F}{V} \qquad (12.5)$$

where c are the total observed counts. From Eq. 12.5 and the above example, we can calculate the count rate as

$$\text{Count rate} = 1200\,\text{counts} \cdot \frac{4.0\,\text{mL min}^{-1}}{0.6\,\text{mL}} = 8000\,\text{CPM}$$

The observed sample counts and count rate calculations on a flowing sample are repeated or "updated" by the instrument during fixed or variable update periods, which can be set by the operator. The update times can be adjusted to any value over the range of 1–120 seconds in modern flow scintillation analyzers. Both observed sample counts and background counts are, therefore, calculated per update time and net count rates calculated as described in the subsequent section.

B. Background and Net Count Rate

The computer programs of the flow scintillation analyzer are designed to subtract random background events from sample radionuclide decay events to provide a net sample count rate according to the equation

$$\text{Net CPM} = \frac{\text{observed counts} - \text{background counts}}{T_R} \tag{12.6}$$

The background counts are subtracted from the gross sample counts before dividing by the residence time. Otherwise, as explained by Kessler (1986) the background counts are amplified by the residence time of the sample in the flow cell. Reich *et al.* (1988) also explain that the true background of the system is the result of external cosmic radiation and electronic noise, which are independent of the flow. Therefore, the "static" (nonflow) background must be eliminated before the flow equation is applied. If not, the background would be overstated by a multiple of the flow rate.

In the calculation of the net count rate according to Eq. 12.6 both the observed counts and the background counts are per update time. If the update time selected is 6 seconds, then the counts accumulated during a 6-second update time are inputted into the equation. Also, the background counts subtracted in the above equation, whether determined by the system or entered into the software by the operator, is divided from a background CPM into the equivalent of an update-time worth of background counts before being entered automatically into the equation. Hence, the net count rate is calculated each update time by subtraction of the background before dividing by the residence time.

Example 12.2 If a radionuclide standard is injected into a flow cell detector via the inlet line of the liquid scintillator pump, the cell volume is $400\,\mu L$, the flow rate is $4.0\,\text{mL min}^{-1}$, an update time of 6 seconds is selected, a background count rate of 25 counts per minute is entered into the software program, and the observed counts in a given update time is 400, then the Net CPM is calculated according to Eq. 12.6 as

$$\text{Net CPM} = \frac{400\,\text{counts} - 2.5\,\text{counts}}{0.4\text{mL}/4.0\,\text{mL min}^{-1}}$$

$$= \frac{397.5\,\text{counts}}{0.1\,\text{min}}$$

$$= 3975\,\text{CPM}$$

Note that both the observed sample counts and background counts in the above calculation are per update time. In the above example the background represents less than 1% of the net count rate. However, when either the update counts get lower, background increases, or residence time is reduced, the background becomes more significant and an accurate measurement and proper subtraction of background may be necessary.

The background is determined obviously without radioactive sample; however, the homogeneous flow cell must contain the fluor cocktail and HPLC eluent used for a particular HPLC run, and the heterogeneous cell must contain the HPLC eluent when background is determined. If a homogeneous cell is used, the background is determined by filling the cell with the same ratio of cocktail to HPLC mobile phase as will be used during the HPLC runs. A run in counts is then carried out for 10 minutes or more with a minute scaler time for the summary. The statistical method used to calculate the background is as follows

$$\text{BKG} = \left[\frac{B_S}{N} + 2\frac{\sqrt{B_S}}{\sqrt{N}} \right] (N) \qquad (12.7)$$

where BKG is the background subtracted in the calculation of net count rate, B_S is the measured background expressed in counts in one scaler minute, and N is the number of samplings per minute. In the previous example, where the update time was 6 seconds, the calculation is carried out every 6 seconds or 0.1 minute. Therefore, N in this case would be equal to 10 samplings. The division and multiplication by the number of samplings (N) is required, because the calculation is based on an update time, while the background is given in counts each minute.

C. Counting Efficiency and Disintegration Rates

The counting efficiency (E) as defined in Chapter 5 is the ratio of the sample count rate as measured by the instrument and the actual disintegration rate or activity of the measured sample. The ratio, when multiplied by 100 is expressed as the percent counting efficiency or

$$\%E = \frac{\text{CPM}_S}{\text{DPM}_S}(100) \qquad (12.8)$$

where CPM_S and DPM_S are the measured sample count rate and the actual sample disintegration rate, respectively. The counting efficiency is a function of the radionuclide, the type of liquid or solid flow cell detector used, counting region selected, and the level of quench in the sample, which is governed, in turn, by the chemistry of the HPLC eluent and scintillator. To express the flow scintillation analysis results in DPM it is necessary, therefore, to determine the counting efficiency for each new application or HPLC run.

The counting efficiency can be determined with a radionuclide standard in the flow cell while under a static (isocratic) or gradient (dynamic) mode.

When quenching is constant in a HPLC run, that is, the HPLC sample components and eluent have no changing effect on the counting efficiency, a static efficiency correction can be used. Under this type of correction the counting efficiency is constant throughout the entire HPLC run.

When the HPLC sample components and eluent affect the counting efficiency, it is necessary to carry out a gradient counting efficiency run to determine the counting efficiencies at different points in time during the HPLC run as the sample components elute from the HPLC column.

I. Static Efficiency Runs

There are two ways of performing the static efficiency runs. When the sample components have absolutely no quenching effect on the counting efficiency of the scintillation system, it is possible to determine the counting efficiency *independent of the HPLC system*. When the sample components do have a quenching effect on the counting efficiency, but the effect is constant throughout the length of the HPLC run, it is necessary to determine the counting efficiency *dependent on the HPLC system*. The latter method requires carrying out an HPLC run with the same eluent that is used during a normal sample run, but without radioactive sample. The two methods are subsequently described. The detailed procedures are available from Anonymous (1997).

a. Independent of the HPLC System

This method requires spiking a volume of the flow scintillation cocktail with a radionuclide standard and filling the scintillation flow cell with the standard in cocktail. Solid (heterogeneous) flow cells and gamma cells obviously do not use scintillation cocktail, and counting efficiency determinations with these types of cells require only an eluent solution of the radioactivity standard. An outline of the procedure used for liquid (homogeneous) flow cells is as follows:

(1) The normal background of the system is determined first. (2) A known activity (DPM) of the radionuclide of interest is added to an accurately measured volume of the flow scintillation cocktail. A minimum volume of 25 mL is used for each radionuclide to be measured. To ensure sufficiently high count rates and good counting statistics, the activity of the standards used can be estimated by taking into account the flow cell volume and estimated counting efficiency of the flow cell. For example, the final activity concentrations of standard should be at least approximately 10,000 DPM/mL for ^{14}C and approximately 25,000 DPM/mL for ^{3}H. (3) The counting parameters for the particular radionuclide and efficiency run parameters are set in the flow scintillation analyzer including the DPM in the flow cell, which is calculated as

$$\text{DPM in cell} = (\text{DPM/mL})(\text{flow cell volume in mL}) \qquad (12.9)$$

(4) The inlet line for the liquid scintillation (LS) pump is placed into the container holding the solution of radionuclide standard in flow scintillation cocktail. A separate inlet line and line filter are used to avoid possible contamination of the liquid scintillation cocktail source used during normal runs. (5) The LS pump is kept running for at least 5 minutes to assure the complete filling of the flow cell. (6) After filling of the cell, the LS pump is stopped and the Efficiency Run program of the computer-controlled flow scintillation analyzer is initiated. The counting efficiency is calculated according to the basic Eq. 12.8 as

$$\%E = \frac{\text{net CPM in cell}}{\text{DPM in cell}}(100) \qquad (12.10)$$

(7) The efficiency run is saved and flow scintillation cocktail containing no radionuclide is pumped through the cell until background levels are reached. (8) When dual radionuclide analysis is required (e.g., ^{3}H–^{14}C), the procedure is repeated with the second radionuclide with the counting efficiency run set to a second counting region.

b. Dependent on the HPLC System

When the sample components have a quenching effect on the counting efficiency, but the effect is constant, it is necessary to spike a volume of the flow scintillation cocktail with a known activity of radionuclide standard. The flow cell is then filled with a mixture of the spiked liquid scintillation cocktail and HPLC eluent. This method and the previously described procedure are static methods, and the procedures are quite similar with the exception that step (6) in the previous procedure will, in this case, include running both the LS pump and the HPLC to fill the flow cell with a mixture of spiked scintillation cocktail and HPLC eluent. Both pumps are turned off when the flow cell is filled and the Efficiency Run program is initiated as described in the previous static procedure.

^{3}H and ^{14}C are low- and intermediate-energy β-emitting radionuclides, respectively, and the liquid scintillation yields for these radionuclides are easily quenched. Phosphorus-32 is not easily quenched, as described in Chapter 5 of this book. One can expect, therefore, that HPLC eluent will have an effect on the counting efficiency of ^{3}H and ^{14}C in most flow scintillation systems. The preceding counting efficiency determination procedure, which includes HPLC eluent, is recommended when low- to intermediate-level radionuclides are used and the HPLC sample components will have a constant quench effect on the counting efficiency.

2. Gradient Efficiency Run

When the HPLC eluent consists of a gradient mixture, quench will vary during the HPLC run and the counting efficiencies will, therefore, not be constant. In this case a gradient efficiency run is required. The setup procedure for the gradient efficiency run is similar to the previously described efficiency run (dependent on the HPLC system) with the exception that the efficiency run is a dynamic one. The dynamic run requires that both the LS pump and the HPLC gradient run simultaneously during the efficiency run. The computer program linked to the flow scintillation analyzer (e.g., FLO-ONE for Windows, PerkinElmer Life and Analytical Sciences, Boston) will construct a gradient efficiency correction curve or table, which can be used to correct for counting efficiency changes during a specific gradient HPLC run.

When the counting efficiency of the detection system is known the scintillation analyzer can convert the count rates of the eluting components in the HPLC effluent by use of the following converted form of the basic Eq. 12.8:

$$DPM = \frac{net\,CPM}{\%E/100} \qquad (12.11)$$

where net CPM is that defined by Eq. 12.6, whereby the above equation can be written in detail as follows:

$$\text{DPM} = \frac{(\text{observed counts} - \text{background counts})/T_R}{\%E/100} \tag{12.12}$$

Eq. 12.12 can also be expressed as

$$\text{DPM} = \frac{\text{net CPM}}{(T_R)(E)} \tag{12.13}$$

where net CPM is the background-subtracted count rate and E is the decimal equivalent of the percent counting efficiency. For the column chromatographic separation of radionuclides Grate et al. (1996) and Grate and Egorov (1998) take Eq. 12.13 further to include the efficiency of recovery (E_{rec}), which is the ratio of the activity of a particular radionuclide recovered from a separation column (e.g., HPLC column) compared to the activity of that nuclide loaded on the column. They quantify radionuclides eluted from separation columns according to flow scintillation peak areas described by the equation

$$\text{DPM} = \frac{C_n}{(E_d)(E_{\text{rec}})(T_R)} \tag{12.14}$$

where DPM is the radionuclide activity, C_n is the background-corrected peak area or, in other words, the net counts, that is, the observed counts minus the background counts determined in a blank run, E_d is the decimal equivalent of the % detection efficiency, and E_{rec} and T_R are the radionuclide efficiency of recovery and sample residence time in the flow cell, described previously. Radionuclide recovery efficiencies can sometimes be measured during a run with the use of a tracer nuclide, such as the gamma-emitting nuclide ^{85}Sr as a tracer for the beta-emitting nuclide ^{90}Sr. If a known activity of ^{85}Sr is added to a sample, separated on a column, and collected and analyzed for its gamma activity, the recovery efficiency for radiostrontium can be calculated.

D. Minimal Detectable Activity

In many forms of chromatography peaks that are sharp, detectable above the baseline and well defined or proportioned are easier to detect than a broad peak or one that displays tailing and lack of form. In radio-HPLC the minimal detectable activity is a calculated activity of a peak from a flow detector expressed in disintegration per minute (DPM) or disintegrations per second (Bq) based on the relative peak height and the presumption that the limit of detectability is twice the count rate of the background (Reich et al., 1988 and Anonymous, 1990), which is calculated as

$$\text{MDA} = (B)(W)/(T_R)(E) \tag{12.15}$$

where B is the background count rate, W is the width of the peak in minutes, T_R is the residence time defined by Eq. 12.2 and E is the decimal equivalent of the % counting efficiency or $\%E/100$.

As explained by Reich *et al.* (1988) the MDA of a flow detector is not directly related to the total amount of radioactivity in any given peak, but rather to the specific activity of any flow segment residing in the detector at any given time. Therefore, flow rate, cell volume, and peak width in addition to the obvious background and detection efficiency all play a key role in defining the minimal detectable activity. Borai and Mady (2002) measured the minimum detectable activity or lower limit of detection (LLD) in Bq units for ^{238}U and ^{232}Th separated on ion exchange chromatograph columns. They used the expression of Currie (1968) where

$$LLD = K\left[2.71 + 4.65(B)^{1/2}\right] \qquad (12.16)$$

where K is a proportionality constant relating the flow scintillation detector response to the activity, and B is the number of background counts for a given counting period. The value of K has units of reciprocal time and is calculated according to the following:

$$K = W/(T_R)(E)(V) \qquad (12.17)$$

where W is the peak width in units of volume, T_R is the sample residence time in the flow cell, E is the detector counting efficiency, and V is the flow cell detector volume (Reeve, 1977). Using a 0.4 mL heterogenous flow cell consisting of Ce-activated glass scintillator in a IN/US flow scintillation analyzer (IN/US Systems, Tampa, FL) and a Dionex 2000 ion chromatographic system Borai and Mady (2002) measured LLDs of 3.0 ± 0.1 Bq for ^{238}U and 6.0 ± 0.1 Bq for ^{232}Th.

The variables of flow rate, cell volume, and peak width should now be considered as factors governing the optimization of flow scintillation analysis in a radio-HPLC system. The following section will describe these variables in terms of sensitivity, speed of analysis, and resolution.

E. Sensitivity, Flow Rate, and Resolution

The sensitivity of a flow detector in a given radio-HPLC run is another term reflecting the minimal detectable activity, that is, the sensitivity is increased when the MDA is decreased. The sensitivity can be increased by (1) reducing the flow rate, which will increase the residence time (T_R), or (2) increasing the size of the flow cell, which increases the detection efficiency. Reducing the flow rate would increase the HPLC run time, but it could as well diminish the resolution of the HPLC-separated sample components, because flow rate is a key factor in the chromatographic separation of sample components. On the other hand, increasing the size of the flow cell would also reduce resolution. The term resolution refers to the ability to distinguish between activity peaks that are in close proximity to each other. Increasing the flow cell size will lower the resolution. When compounds or molecules are difficult to separate, it is best to use a flow cell size that will give the best resolution although at a diminished sensitivity. On the other hand, when radioactivity levels are low and activity peaks are well separated, it is recommended to sacrifice resolution for increased sensitivity. Therefore, the selection of cell size will depend on the characteristics of any particular HPLC separation run.

As a general rule, the best flow cell volume for a given HPLC application would be a cell with a volume of one-half to one-fourth the volume of the smallest peak of interest. As described by Anonymous (1990) the following equations can be used to estimate the optimum flow cell size for two basic cell types:

For solid (heterogeneous) cells or gamma cells:

$$V_C = K(V_P) \tag{12.18}$$

where V_C is the cell volume, K is a constant between 1/4 and 1/2 where smaller values yield higher resolution and larger values higher sensitivity, and V_P is the volume in mL of the smallest peak of interest. The value of V_P is calculated as

$$V_P = (W)(F) \tag{12.19}$$

where W is the peak width in minutes and F is the flow rate in mL min^{-1}.

For the liquid (homogeneous) cells:

$$V_C = K(V_P + V_S) \tag{12.20}$$

where V_S is the volume of liquid scintillator for the same duration as the peak.

As demonstrated by Kessler (1986) the resolution of HPLC activity peaks from flow detectors are dependent not only on the flow rate and flow cell size as described but also on the update time, which is the time interval in seconds over which the detector pulses are summed. Kessler demonstrated that, maintaining the flow rate and flow cell size constant, two very close and overlying peaks can be separated into two clearly defined peaks by reducing the update time. This may be intuitively obvious as the activity peaks are plotted on a time scale (activity vs. time in minutes). Reducing the pulse summation update time would provide activity changes over shorter time intervals. In summary, we can conclude that the higher sensitivity (lower minimal detectable activity) of a flow detector is highly dependent on the residence time of the sample in the flow detector, cell volume, and update time; however, a tradeoff must be made where sensitivity is sacrificed for resolution by controlling flow rate, cell size, and update time.

F. Precision

As described in Chapters 1 and 7 radioactivity decay is a random event, that is, it cannot be predicted when a given radionuclide would decay. However, we could predict that one-half of a radionuclide sample would decay in one half-life. In view of the random character of radioactivity decay, we can say that the precision of a count rate determination is a function of the total number of counts collected and the counting time, as the count number will be greater for longer counting times.

The precision of a given count determination is expressed in terms of its standard deviation, which is calculated as

$$SD = \sqrt{\text{total counts}} \qquad\qquad (12.21)$$

In the case of flow scintillation analysis, we can look at some specific examples.

Example 12.3 If a flow rate of $4.0\,\text{mL min}^{-1}$ and a flow cell detector of $250\,\mu\text{L}$ were used for a particular HPLC application, we could calculate that the radioactivity from the sample in the flow cell would be observed for 3.75 seconds determined as follows:

$$T_R = V/F = 0.25\,\text{mL}/4.0\,\text{mL min}^{-1} = 0.0625\,\text{min} = 3.75\,\text{seconds}.$$

If the count rate for the 3.75-second period was 20,000 CPM, the standard deviation according to Eq. 12.21 would be calculated as

$$SD = \sqrt{(20{,}000\,\text{cpm})(0.0625\,\text{min})} = \sqrt{1250\,\text{counts}} = 35.4$$

or the counts collected can be expressed as 1250 ± 35.4 counts at one standard deviation and the percent standard deviation (%SD) would be

$$\frac{35.4}{1250}(100) = 2.8\%$$

and the results expressed as $1250\,\text{counts} \pm 2.8\%$.

If we take the same flow rate and flow cell volume but detect a lower count rate of 2000 CPM, the standard deviation could be calculated as

$$SD = \sqrt{(2{,}000\,\text{cpm})(0.0625\,\text{min})} = \sqrt{125} = 11.2$$

and the percent standard deviation (%SD) would be calculated as

$$\frac{11.2}{125}(100) = 8.9\%$$

The percent standard deviation calculation can be simplified to the following

$$\%SD = \frac{100}{\sqrt{\text{total counts}}} \tag{12.22}$$

Example 12.4 If we take the above Example 12.3 and increase the cell volume from 250 to 500 μL and keep the flow rate (4.0 mL min^{-1}) and the two observed count rates (20,000 and 2000 CPM) the same, we can calculate the new percent standard deviations (%SD) as follows to see how increasing cell volume will improve sensitivity and precision, albeit at an expected loss of resolution:

The residence time or the duration that the radioactivity in the cell would be observed can be calculated as

$$T_R = V/F = 0.50 \, \text{mL}/4.0 \, \text{mL min}^{-1} = 0.125 \, \text{min} = 7.5 \, \text{seconds}$$

The percent standard deviations for the example taken for a count rate of 20,000 CPM is calculated as

$$\%SD = \frac{100}{\sqrt{(20,000 \, \text{cpm})(0.125 \, \text{min})}} = \frac{100}{\sqrt{2500 \, \text{counts}}} = 2.0\%.$$

and that for the count rate of 2000 CPM is

$$\%SD = \frac{100}{\sqrt{(2000 \, \text{cpm})(0.125 \, \text{min})}} = \frac{100}{\sqrt{250 \, \text{counts}}} = 6.3\%.$$

Examples 12.3 and 12.4 illustrate that increasing the flow cell volume from 250 to 500 μL increased the residence time and consequently, the precision of the measurement was improved by reducing the %SD of the observed counts. However, larger flow cell volumes will reduce activity peak resolutions as described previously.

Equation 12.22 can be used to calculate the percent standard deviation (%SD) for any count rate (CPM) or total counts collected, that is, (CPM $\cdot$ T_R) of any flow scintillation detector. The residence time (T_R) in minutes required to achieve a desired %SD can be calculated by manipulation of Eq. 12.22 as follows

$$\sqrt{\text{total counts}} = \frac{100}{\%SD} \tag{12.23}$$

or

$$\text{total counts} = \left(\frac{100}{\%SD}\right)^2 \tag{12.24}$$

or

$$(\text{CPM})(T_R) = \left(\frac{100}{\%\text{SD}}\right)^2 \qquad (12.25)$$

and

$$T_R = \frac{1}{\text{CPM}}\left(\frac{100}{\%\text{SD}}\right)^2 \qquad (12.26)$$

which is similar to the equation described by Kessler (1986).

G. Detection Optimization

Major advances in real-time flow scintillation analysis have been made in detection optimization, which has been achieved by including technology currently available to modern liquid scintillation analyzers (Anonymous, 1996; Wasyl and Nellis, 1996). This technology includes (1) multichannel analysis for counting region optimization, (2) time-resolved liquid scintillation counting (TR-LSC) for background reduction, (3) detection and correction of chemiluminescence, which can be a significant interference when fluor cocktail and HPLC eluate are mixed, and (4) operation software, such as FLO-ONE for Windows (PerkinElmer Life and Analytical Sciences, Boston) and Win-Flow for Windows (IN/US Systems, Inc., Tampa, FL), which facilitates flow scintillation analysis setup, optimization, control, performance assessment, and result reporting.

I. Multichannel Analysis

The multichannel analyzer (MCA) in liquid scintillation analysis is described in detail in Chapter 5 on liquid scintillation analysis (LSA). The application of the MCA in flow scintillation analysis as in LSA will sort all signals according to pulse height into individual channels calibrated in keV. The PerkinElmer Radiomatic flow scintillation analyzers are equipped with a 1024-channel MCA and FLO-ONE software, which permit a visual observation of the sample and background pulse height spectra. As in liquid scintillation analysis, visual observation of the pulse height spectra from the MCA facilitates counting region optimization for a flow sample at a given quench level. Region optimization is achieved by setting the lower level discriminator (LLD) and upper level discriminator (ULD) to provide the highest figure of merit calculated as E^2/B, where E is the percent counting efficiency and B is the background count rate. The term figure of merit is analogous to the term signal-to-noise (S/N) ratio, as the sample activity peak is the signal of interest and the background radioactivity is equivalent to the noise we must reduce.

The pulse height spectral display offered by the MCA will also permit visualization of the sample pulse height spectrum originating from more than one radionuclide. If two radionuclides are present in the sample the pulse height spectrum of the MCA can be used to select the LLD and ULD settings for two counting regions. This will enable the activity analysis of two radionuclides in the same sample as described in Section VI.B of this chapter and in more detail in Chapter 5.

2. Chemiluminescence Detection and Correction

The chemiluminescence pulse height spectrum, can also be observed via the MCA. Chemiluminescence can occur only when the liquid homogeneous flow cell is used, and it would occur when HPLC eluate is mixed with scintillation fluor cocktail prior to continuing on to the flow cell. Chemiluminescence is treated as an interference like background; however, unlike background, chemiluminescence can be eliminated altogether from the sample count rate. The occurrence of chemiluminescence can be tested easily by mixing nonradioactive sample with HPLC eluate through a flow cell and observation of the MCA pulse height spectral output. The chemiluminescence pulses are found in the region of 0–6 keV. For high-energy β emitters (e.g., ^{32}P), this portion of the pulse height spectrum can be excluded from the sample counting region if chemiluminescence is of concern. When analyzing low-energy β emitters, such as tritium the chemiluminescence detection and correction, available in the operation setup of the flow scintillation analyzer software (e.g. FLO-ONE for Windows), can be enabled.

Chemiluminescence detection and correction is described in Section VII of Chapter 5. Also see Section IV of this chapter, which describes certain flow scintillation cocktails. Scntillation cocktails are available, which can suppress the occurrence of chemiluminescence.

3. Time-Resolved Liquid Scintillation Counting (TR-LSC)

Time-Resolved Liquid Scintillation Counting (TR-LSC) provides yet a further means of optimizing detection in the flow scintillation analyzer by reducing background. TR-LSC is a patented method of reducing backgrounds by discriminating against sample and background pulses by means of counting the number of afterpulses that occur following an initial pulse event. Afterpulses are more numerous in nonquenchable events, which are pulse events origination from outside the scintillation solution, such as cosmic radiation, that might strike the flow cell or face of one of the photomultiplier tubes. Quenchable events are pulse events originating in the scintillation fluor cocktail of the flow cell, and these have few if any distinguishable afterpulses. By counting afterpulses, the instrument circuitry and software discriminate between quenchable pulses originating from the scintillation solution in the flow cell and nonquenchable pulses originating from outside the flow cell. The instrument circuitry can, therefore, reject pulses that do not originate from within the flow cell as background radiation. A guard scintillator also surrounds the sample chamber but does not come in contact with the low-energy β emissions origination from within the flow cell. External radiation of cosmic origin or external radiation in the laboratory environment can strike or pass through the scintillator guard. The pulse events occurring from external radiation interactions with the guard detector are nonquenchable pulses, and these have numerous afterpulses compared with sample events in the flow cell. The combination of the scintillator detector guard surrounding the flow cell and TR-LSC results in the virtual elimination of nonquenchable background with a dramatic reduction in background count rate of up to 75%. Typical TR-LSC backgrounds are 2–3 CPM for ^{3}H and 4–5 CPM for ^{14}C with region optimization for these radionuclides.

The combined effect of region optimization and TR-LSC can be illustrated in Figs. 12.10 and 12.11. Figure 12.10 illustrates a HPLC run of ^{14}C-labeled drug metabolites. The counting region (0–156 keV) is not optimized, and TR-LSC is not enabled (background = 30 CPM). Only three activity peaks are discernible. When the counting region is optimized to the highest FOM (E^2/B) by adjusting the LLD and ULD settings to 4–100 keV respectively, and TR-LSC is enabled, the background is reduced to 2 CPM and as many as nine activity peaks can be identified as illustrated in Fig. 12.11. The concept and practice of region optimization is discussed in Chapter 5.

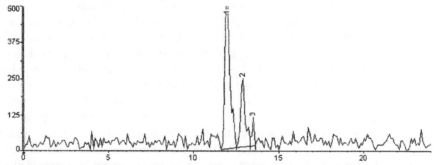

FIGURE 12.10 Flow scintillation analysis trace of ^{14}C metabolites separated via HPLC. A wide-open counting region of 0–156 keV is used without TR-LSC background rejection. [Courtesy of Paul Riska, Boehringer Ingelheim Pharmaceuticals, Inc., Ridgefield, CT (From Anonymous, 1996, reprinted with permission from PerkinElmer Life and Analytical Sciences).]

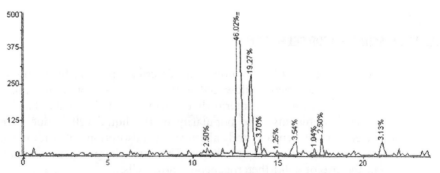

FIGURE 12.11 Flow scintillation analysis trace of ^{14}C metabolites separated via HPLC. An optimized counting region of 4–100 keV is used with TR-LSC background rejection. [Courtesy of Paul Riska, Boehringer Ingelheim Pharmaceuticals, Inc., Ridgefield, CT (From Anonymous, 1996, reprinted with permission from PerkinElmer Life and Analytical Sciences).]

H. Instrument Performance Assessment (IPA)

Commercial flow scintillation analyzers (e.g., PerkinElmer Radiomatic) are equipped with software and radionuclide standards that will assist the operator in setting up the instrument performance assessment. The performance of the flow scintillation analyzer can be tested on a routine basis to provide assurance that the instrument is operating within acceptable parameters and to have a standing record of the instrument performance over a period of time. To maintain good laboratory practice (GLP) and, at times, to satisfy regulatory agencies, it is necessary to have records of the instrument performance on a routine basis, for example, daily, weekly, or monthly, whichever may satisfy our needs as evidence that deviations in instrument performance do not have any effect on the analytical results. The IPA is carried out with ^{3}H and ^{14}C and background standards. The radionuclide standards are NIST traceable. The standards are in sealed vials, which are mounted in a vial holder that duplicates the mounting characteristics of a flow cell.

Eight parameters can be assessed using ^{3}H and ^{14}C standards. These are (1) ^{3}H efficiency, (2) ^{3}H background, (3) ^{3}H figure of merit, (4) ^{3}H chi-square, (5) ^{14}C efficiency, (6) ^{14}C background, (7) ^{14}C figure of merit, and (8) ^{14}C chi-square. The parameters are discussed in more detail in Section XVII.B of Chapter 5. The results are stored and printed in tabular and/or graphic form with a time and date stamp. Data points for each of the above parameters can be stored in memory as a binary file (unchangeable by the user) and plotted. The data can be displayed, stored in computer memory and printed whenever needed. In any given graph for each of the eight parameters listed above, an average line is calculated as illustrated in Fig. 5.65a of Chapter 5. The one, two, and three sigma values are provided to help the user evaluate any trends or outlying points. If a value of any of the parameters fails the limits test set for that parameter, a warning message is received on the computer monitor with suggestions that the user can follow to assess the problem further and to take preventive action if any parameters are obtained too frequently outside of the recommended limits.

IV. FLOW SCINTILLATOR SELECTION

Among the scintillation flow cells described previously in this chapter the liquid (homogeneous) flow cell is the type most commonly utilized for the detection of low- and intermediate-energy β emitters (e.g., ^{3}H, ^{14}C, ^{35}S, ^{33}P) in HPLC effluents. The popularity of the liquid cell is due to the highest detection efficiencies and absence of adsorption of radionuclide-labeled compounds onto solid scintillator. The latter characteristic is a major concern in the use of solid (heterogeneous) flow cells.

Homogeneous flow cell counting requires special cocktails. The development and characteristics of modern flow scintillation cocktails are described in reviews by Thomson (1994, 1997) and in Chapter 8. Because the flour cocktail must be added mechanically and mixed readily with the HPLC effluent, it must possess certain physical and performance characteristics. These characteristics outlined by Thomson (1997) are the following:

- Low viscosity
- Rapid and easy mixing with the HPLC eluate
- High sample acceptance capacity
- Compatibility with complex samples and HPLC gradients
- No gel formation
- Good counting performance
- Low background contribution
- Chemiluminescence resistance
- Safe to handle
- High flash point
- Biodegradable

Obviously, not all of these characteristics can always be achieved to the optimum; however, all of these characteristics are achieved to a certain degree in most circumstances by modern flow scintillation cocktails.

The older generation or classical flow scintillation cocktails, still used by many researchers, have relatively low flash points and are less safe to handle. Modern research for improved flow cocktails has provided now safer high-flash-point cocktails, which are biodegradable and capable of mixing readily with a wide range of sample types and gradients used in HPLC. A major advance was the patented development of a flow scintillation cocktail with the chemical components capable of removing luminescence while simultaneously minimizing background without sacrificing counting efficiency (Hegge and ter Viel, 1986; Thomson, 1993).

Some of the classical flow cocktails such as the Flo-Scint I, II, III, and IV are based on pseudocumene or trimethylbenzene solvents with flash points of about 48°C (118°F). The newer Ultima-Flo cocktails (PerkinElmer Life and Analytical Sciences, Boston) are readily biodegradable and have a high flash point of 120°C (248°F). Examples of these new generation flow cocktails described by Thomson (1997) are the following:

- Ultima-Flo M for multiple dilute sample types and for micro-bore HPLC-FSA systems.
- Ultima-Flo AF specifically formulated for ammonium formate samples and gradients.
- Ultima-Flo AP for ammonium phosphate samples and gradients and an almost all-purpose FSA cocktail accepting a wide variety of sample types with high sample loading capacity.

Ultima-Flo M flow cocktail will accept multiple dilute sample types including methanol/water and acetonitrile/water gradients common in Reverse Phase HPLC applications. The sample-holding capacities of Ultima-Flo M for a wide range of sample types are listed in Table 12.4. Counting efficiencies for ^3H will range from 33% to 47% when mixed with 50% methanol, 50% acetonitrile, pure methanol and pure acetonitrile in cocktail/sample ratios of $2:1-5:1$ v/v (Thomson, 1997).

TABLE 12.4 Typical Sample Load Capacities for Ultima-Flo M at 20°C[a]

Sample	Maximum sample uptake (%)	Optimal mixing ratio cocktail : sample
Deionized water	50.0	1 : 1
Methanol/water (50 : 50)	31.0	3 : 1
Methanol	50.0	1 : 1
Acetonitrile/water	41.2	2 : 1
Acetonitrile	50.0	1 : 1
0.2 M sodium chloride	41.2	2 : 1
0.15 M sodium chloride	41.2	2 : 1
0.05 M sodium chloride	50.0	1 : 1
0.1 M PBS buffer (pH 7.4)	33.3	2 : 1
0.01 M PBS buffer (pH 7.4)	41.2	2 : 1
0.01 M PBS/plasma (10%)	45.9	2 : 1
1.0 M sodium hydroxide	21.6	4 : 1
0.5 M sodium hydroxide	35.5	2 : 1
0.1 M sodium hydroxide	50.0	1 : 1
0.2 M HEPES (pH 7.2)	50.0	1 : 1
0.1 M HEPES (pH 7.2)	50.0	1 : 1
50 mM Tris-HCl	50.0	1 : 1
0.05 M Na_2HPO_4	50.0	1 : 1
0.02 M ammonium formate	50.0	1 : 1

[a]From Thomson (1997), reprinted with permission from PerkinElmer Life and Analytical Sciences.

Ultima-Flo AF is used for the flow scintillation analysis of ammonium formate samples and gradients. Gradients of up to > 2.0 M ammonium formate (pH 3.8 with formic acid) can be mixed readily with Ultima-Flo AF in cocktail : sample ratios of up to 1 : 1 v/v. Thomson (1997) reports ^{3}H counting efficiencies in the range of 28–40% when mixed with up to 2.0 M ammonium formate gradients in a range of cocktail/sample mixtures of 2 : 1 to 5 : 1 v/v.

Ultima-Flo AP is formulated to mix readily with 0–2.0 M ammonium phosphate gradients (pH 3.8 with orthophosphoric acid). This flow scintillation cocktail can accept other gradients of up to 1.0 M phosphate-buffered saline (PBS) and 1.0 M NaOH among sample types listed in Table 12.5. When viscosity is a problem, Ultima-Flo AP is generally recommended (Thomson, 1997). Also, Ultima-Flo AP is considered to be an all-purpose flow scintillation cocktail, because it can accept a very wide range of samples and gradients. Users interested in making the change from the classical flow cocktails to the new generation of safer biodegradable flow cocktails may use the information provided in the previous paragraphs to select a new compatible cocktail for a particular HPLC eluent. Also Table 12.6 may serve as a quick guide for making a cocktail replacement.

TABLE 12.5 Typical Sample Load Capacities for Ultima-Flo AP at 20°C[a]

Sample	Maximum sample uptake (%)	Optimal mixing ratio cocktail : sample
Deionized water	50.0	1 : 1
Methanol/water (50 : 50)	28.6	3 : 1
Methanol	50.0	1 : 1
Acetonitrile/water	31.0	3 : 1
Acetonitrile	50.0	1 : 1
1.0 M sodium hydroxide	23.1	4 : 1
0.5 M sodium hydroxide	28.6	3 : 1
0.1 M sodium hydroxide	50.0	1 : 1
1.0 M PBS	31.0	3 : 1
0.5 M PBS	37.5	2 : 1
0.1 M PBS	50.0	1 : 1
0.01 M PBS	33.3	2 : 1
0.2 M sodium chloride	33.3	2 : 1
0.05 M Na_2HPO_4	50.0	1 : 1
0.01 M PO_4/methanol (50 : 50)	28.6	3 : 1
0.01 M PO_4/acetonitrile (50 : 50)	33.3	2 : 1
0.01 M PBS/methanol (50 : 50)	28.6	3 : 1
0.01 M PBS/acetonitrile (50 : 50)	28.6	3 : 1
5.0 M guanidine	19.4	4 : 1
2.0 M guanidine	31.0	3 : 1

[a]From Thomson (1997), reprinted with permission from PerkinElmer Life and Analytical Sciences.

TABLE 12.6 Replacement table to facilitate the Changeover from the Classical Flow Cocktails to the New Generation of Safer, High-Flash-Point and Biodegradable Flow Cocktails[a]

Classical flow cocktail		New, safer flow cocktail
Pico-Agua Pico-Fluor 30 Flo-Scint A Flo-Scint I, II, III, and IV	can be replaced by →	Ultima-Flo M
Flo-Scint IV	can be replaced by →	Ultima-Flo AF for ammonium formate gradients and imidazole
Flo-Scint IV	can be replaced by →	Ultima-Flo AP for ammonium phospate, PBS, and sodium hydroxide gradients

[a]Courtesy of PerkinElmer Life and Analytical Sciences.

V. STOPPED-FLOW DETECTION

In Section III.E of this chapter the relationship of flow detector sensitivity or efficiency and flow rate were discussed. In summary, it was concluded that higher sensitivity (lower minimum detectable activity) of a flow detector could be achieved by increasing the residence time of a sample in the flow detector (T_R) and the cell volume. Increasing the cell volume would, however, reduce HPLC peak resolutions. During the normal continuous flow of an HPLC run the residence time of a sample (radioactivity peak) in the detector is generally short (< 5 min). If the sample activity is low (near background), it may be difficult or impossible to distinguish the sample radioactivity from background with the short sample residence times of continuous-flow runs. This is obvious because longer counting times are required to reduce counting error (Eq. 12.22) and lower limits of detection (see Eq. 5.107 and Fig. 5.67 of Chapter 5). If we stop the flow of an HPLC run when the radionuclide is detected by the flow detector at its highest count rate (HPLC peak maximum), the sample residence time can be increased indefinitely at its highest count rate. This is referred to as stopped-flow detection, whereby longer counting times can be selected to achieve acceptable counting statistics and lower limits of detection.

The theory and practice of stopped-flow detection are described by Grate et al. (1996), Grate and Egorov (1998), and Egorov et al. (1998). Under the assumption of no secondary mixing or phase separation in the flow cell under stopped-flow conditions, Egorov et al. (1998) note that the fraction of the sample zone present in the detector flow cell at the peak maximum, D_m, is obtained from the transient continuous-flow peak signal using the equation

$$D_m = \frac{(C_{max})(T_R)}{(T_i)(C_n)} \tag{12.27}$$

where C_{max} is the number of net counts at the peak maximum, T_R is the sample residence time as previously defined (Eq. 12.2), T_i is the pulse summation update time, and C_n is the net peak area counts. If the flow is stopped at the peak maximum, the background-subtracted (net) count rate, C_{cpm}, of the radionuclide is related to the sample activity A_{dpm} according to the following equation of Egorov et al. (1998):

$$C_{cpm} = (D_m)(E_d)(E_{rec})(A_{dpm}) \tag{12.28}$$

where E_d and E_{rec} are the detection efficiency and sample recovery efficiency, respectively, described previously (Eq. 12.14).

Stopped-flow detection was tested by Grate and Egorov (1998) and Egorov et al. (1998) for the analysis of ^{99}Tc in a 2.5 mL flow cell. Using a ^{99}Tc standard in the continuous-flow detection mode establishes the time position of the peak maximum as illustrated in Fig. 12.12 (curve A). At the time position of the peak maximum the instrument is programmed to direct

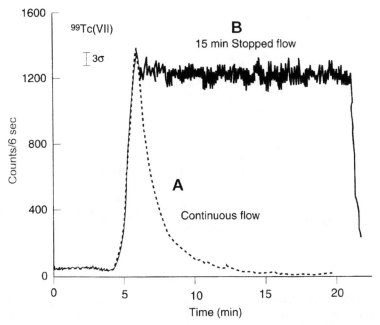

FIGURE 12.12 Comparison of continuous-flow (Curve A) and stopped-flow (Curve B) detection of ^{99}Tc. The error bar indicates 3 standard deviations of the peak maximum counts. (From Grate and Egorov, 1998 and Egorov et al., 1998, reprinted with permission from Elsevier Science.)

the eluent diverter valve to waste and simultaneously stop the cocktail pump initiating a stopped-flow mode. The detector traces (B) in Fig. 12.12 illustrate a duplicate 15-min stopped flow analysis demonstrating reproducibility of the stopped-flow detection. Stopped flow analysis of ^{99}Tc at three activity levels is illustrated in Fig. 12.13. The lowest activity of the standard (268 DPM) (trace 2, Fig. 12.13) would not be statistically distinguishable from the background (trace 1, Fig. 12.13) in a continuous-flow measurement. However, a 15-min stopped-flow measurement provided sufficient signal accumulation (counts) to yield a reliable analysis of the ^{99}Tc activity (268 DPM) with 8% (3σ) counting error (Egorov et al., 1998).

VI. APPLICATIONS

Flow scintillation analysis is applied to the activity analysis of alpha-, beta-, and gamma-emitting radionuclides in research and the applied sciences. The greatest interest in FSA has been traditionally and remains in the research sciences of biochemistry and molecular biology when used in conjunction with HPLC such as in the field of drug metabolism and disposition. Applications have been directed to the use of FSA linked to high performance ionic chromatography (HPIC) in the separation and measurement of radionuclides in the environment. Flow scintillation analysis applied to the on-line measurement of radioactivity in the environment of facilities related

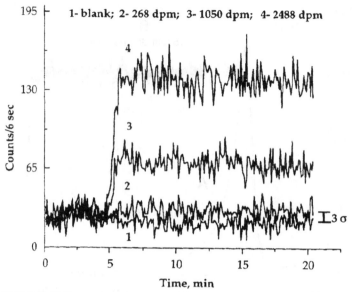

FIGURE 12.13 Selected detector traces from the analysis of ^{99}Tc(VII) standards using 15-min stopped-flow detection. The activities of the standards are listed. The error bar corresponds to 3 standard deviations of the background count rate. (From Egorov et al., 1998, reprinted with permission from Elsevier Science.)

to the nuclear power industry is but one example. Some typical examples of a broad spectrum of FSA applications will be provided subsequently.

A. Single Radionuclide Analysis

Most applications of FSA are in the biological sciences. These studies generally involve FSA for the measurement of radionuclide-labeled bioorganic molecules separated by HPLC, which is one of the most popular chromatographic methods used for molecular separations and characterization. The method provides the means for further analysis such as mass-, infrared-, nuclear magnetic resonance spectroscopy and x-ray diffraction among other tests used for the molecular structural determination of natural products, synthetic compounds, and products of catabolic and anabolic reactions, and so forth. The current trend is the FSA of radioisotope-labeled bioorganic molecules separated by HPLC with on-line mass spectral (MS) and nuclear magnetic resonance (NMR) spectral analysis. Examples will be provided in this section of the chapter.

Many applications of FSA to studies of anabolic and catabolic reactions in the biological sciences are found in the journal literature, and these are too numerous to cite. Among the many examples available from the literature we can cite applications of FSA in the following fields of study: alkaloid metabolism (Mannens et al., 2002); amino acid, protein, and peptide chemistry (Ahmed et al., 1998; Boogaard et al., 1996; Smith and Lutz, 1996); arachidonic acid metabolism (Hankin et al., 1998; Pageaux et al., 1996;

Paulson *et al.*, 2000; Tamby *et al.*, 1996; Zeldin *et al.*, 1996); biosynthesis (Bai and Esko, 1996; Black *et al.*, 1999; Glasgow *et al.*, 1996; Grard *et al.*, 1996); carcinogens and mutagens (Chen *et al.*, 2001b; Gautier *et al.*, 2001; Pritchett *et al.*, 2002; Upadhyaya *et al.*, 2002); drug metabolism and pharmaceutical analysis (Andersson *et al.*, 1998; Chen *et al.*, 2001a; Dockens *et al.*, 2000; Halpin *et al.*, 2002; He *et al.*, 2000; Maggs *et al.*, 2000; Maurizis, *et al.*, 1998; Patrick *et al.*, 2002; Ramu *et al.*, 2000; Riska *et al.*, 1999; Scarfe *et al.*, 2000; Sekiya *et al.*, 2000; Singh *et al.*, 2001; Slatter *et al.*, 2000; Smith *et al.*, 2002; Sohlenius-Sternbeck *et al.*, 2000; Sweeny *et al.*, 2000; Vickers *et al.*, 1998, 2001; Wynalda *et al.*, 2000; Yuan *et al.*, 2002; Zalko *et al.*, 1998); energetics, CO_2, and ion metabolism (Nguyen *et al.*, 1999; Zhang and Hu, 1995; Zhang *et al.*, 1995); enzymology (Kumar *et al.*, 1999; Laethem *et al.*, 1996; Shet *et al.*, 1996; Van Kuilenburg *et al.*, 1999; Zheng *et al.*, 2002); herbicide metabolism (Moghaddam *et al.*, 2001); hormones, catecholamines, and neurotransmitters (Schwahn *et al.*, 2000; Shirley *et al.*, 1996); microbiology (Bezalel *et al.*, 1996; Bogan and Lamar, 1996); natural product characterization and metabolism (Addas *et al.*, 1998; Boulton *et al.*, 1999; Chen *et al.*, 2001b; Ghosheh *et al.*, 2001; Hansen *et al.*, 1999); nucleic acids (Pluim *et el.*, 1999; Xie and Plunkett, 1996); oligosaccharides and glucoproteins (Bai and Esko, 1996; Cacan *et al.*, 1996; Grard *et al.*, 1996); phospholipids (Falasca *et al.*, 1995; Ribbes *et al.*, 1996); prostagladins and leukotrienes (Capdevila *et al.*, 1995; Cortese *et al.*, 1995; Dargel, 1995); steroids (Carsol *et al.*, 1996; Carruba *et al.*, 1996; Dumas *et al.*, 1996; Mensah-Nyagan *et al.*, 1996; Niiyama *et al.*, 2001; Yeoh *et al.*, 1996); sugars, lipoproteins (Alary *et al.*, 1995; Folcik *et al.*, 1995); toxin metabolism (Chen *et al.*, 2001b; Noort *et al.*, 1999; Wormhoudt *et al.*, 1998); and vitamin metabolism (Chen and Gudas, 1996; Kuo *et al.*, 1995; Li *et al.*, 1995).

B. Dual Radionuclide Analysis

Modern flow scintillation analyzers are equipped with a multichannel analyzer, computer display of the sample pulse height spectrum, and pulse height discriminators, which can be set to define two counting regions. The proper setting of these counting regions will permit the activity analysis of two different radionuclides in the sample provided their beta-energy maxima (E_{max}) are significantly different as described in detail in Chapter 5. It is necessary only that the counting efficiencies of the two radionuclides (lower energy and higher energy emitter) in the two counting regions be determined as described in Section III.C of this chapter. The flow scintillation analyzer will automatically determine the activities of the two radionuclides in the flow cell. An example of simultaneous ³H and ³²P traces together with the UV absorption trace of the same HPLC run can be seen in Fig. 12.14.

Some examples of the flow scintillation analysis of dual radionuclide mixtures in HPLC effluents are the following: ³H–¹⁴C (Dayhuff *et al.*, 1986; Kusche and Lindahl, 1990; Sabourin *et al.*, 1988; Seidegård *et al.*, 1990; Shirley and Murphy, 1990; Wells and Digenis (1988); ³H–³³P (Morgan *et al.*, 1987); ³H–³²P (Balla *et al.*, 1987; Guillemette *et al.*, 1989; Nolan and

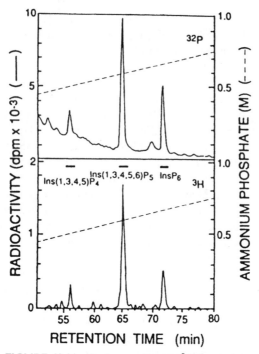

FIGURE 12.14 Elution profile of [³H]inositol tetraphosphate, [³H]inositol pentaphosphate, and [³H]inositol hexaphosphate labeled with ³²P and produced in intact adrenal chromaffin cells measured with a PerkinElmer Radiomatic flow scintillation analyzer. The major peaks for the ³H–³²P double-labeled inositol phosphates are eluted at identical positions (—). Upper panel, ³²P radioactivity; lower panel, ³H radioactivity. (From Sasakawa *et al.*, 1990, reprinted with permission from Elsevier Science.)

Lapetina, 1991; Rubiera *et al.*, 1990; Sasakawa *et al.*, 1990); and ³H–³⁵S (Hughes *et al.*, 1992; Lyon *et al.*, 1994; Mays *et al.*, 1987); and ⁸⁹Sr–⁹⁰Sr (Desmartin *et al.*, 1997).

C. Alpha/Beta Discrimination

The possibility of analyzing α and $\beta(\gamma)$ emitting radionuclides in the same sample and reporting these as gross α and gross β activities has widespread applications in the environmental measurement of radioactivity. See Chapters 5 and 6 for additional information on this subject. Because the radionuclides are often extracted in mixtures, HPIC linked to flow scintillation analysis with alpha/beta discrimination offers great promise for the facile environmental analysis of mixtures of α- and β-emitting radionuclides. Separation procedures for fission products utilizing HPLC and HPIC have been developed (Bradbury *et al.*, 1990; Reboul and Fjeld, 1994). The separation procedure for six common actinides by HPIC are reported by Reboul and Fjeld (1995). Figure 12.15 from the work of Reboul and Fjeld (1995) provides an excellent illustration of actinide separation by HPIC and on-line flow cell measurements of the separated nuclides. Such excellent (clean) radioisotope separations obviously preclude

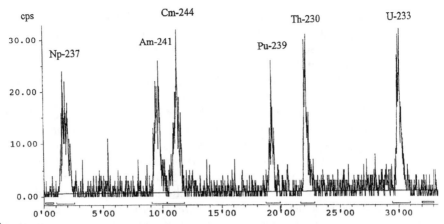

FIGURE 12.15 **A typical chromatogram of select actinides using cationic elution. (From Reboul and Fjeld, 1995, reprinted with permission © Waverly, Williams and Williams.)**

the need for gross α and gross β activity analysis by pulse shape analysis. However, when radionuclides are not clearly separated analysis of gross alpha and gross beta activity with flow cells is possible.

As described in detail in Chapters 5 and 6, it is possible to use pulse decay analysis (PDA) or pulse shape analysis (PSA) to discriminate between alpha- and beta- or gamma-decay events in the same sample. Events originating from alpha-particle interactions with scintillation cocktail have 35–40 ns longer decay lifetimes than events originating from beta-particle or gamma interactions. This is a result of the longer deexcitation and light emission processes in scintillation fluors after α particle interactions. In pulse shape analysis (PSA) or pulse shape discrimination (PSD), the area of the tail of a pulse is compared to the total pulse area, which provides a method of assigning a pulse to that of an α pulse (long pulse) or β–γ pulse (short pulse). Such pulse discrimination for α–β analysis requires two multichannel analyzers (MCAs), where pulses of longer decay times originating from α events are registered in the α-MCA and those of shorter decay times are registered in the β-MCA as originating from β or $\beta(\gamma)$ events.

Usuda and Abe (1992) and Usuda *et al.*, (1992) tested several solid scintillators including CsI(Tl), NaI(Tl), CaF$_2$(Eu), BaF$_2$, BGO, NE102A and stilbene for the pulse shape discrimination of events arising from α and $\beta(\gamma)$ interactions. Planar solid scintillators were coupled to a single photomultiplier tube. They used a flow cell consisting of a spiral channel (not tubing), because tubing will absorb all of the α emissions. Solutions of the $\alpha + \beta(\gamma)$ sources flow along the spiral channel in contact with a Au-coated Mylar film, which protects the solid scintillator. The surface of the planar solid scintillator is also protected with Au-coated Mylar film. The α count rate obviously is a function of the thickness of the Au-Mylar protective layers, and the system suffers from low α detection efficiency. Their work demonstrated the potential of solid scintillators for the simultaneous analysis of $\alpha + \beta(\gamma)$ sources using PSD, and CsI(Tl) crystal scintillator was found to provide the best discrimination between α and $\beta(\gamma)$ rays. The scintillator

CsI(Tl), however, is water soluble as well as hygroscopic. Therefore, it cannot be applied as a scintillator in a solid (heterogeneous) flow cell where the radioactive solution would flow through cell tubing containing a fine powder of solid scintillator.

DeVol and Fjeld (1995) tested three solid heterogeneous flow cells consisting of 63–90 μm particles of CaF$_2$(Eu), glass scintillator (GS-20) and BaF$_2$ packed into translucent Teflon tubing as potential flow cells, which can be coupled to HPIC systems for on-line α–β discrimination for the routine on-line analysis of alpha-emitters. The contribution from beta-emitters in the chromatograph effluents can be removed by time discriminator settings of a pulse shape analyzer (see Chapters 5 and 6). By selecting the optimum time discriminator setting the $\beta(\gamma)$ contribution to the background count rate was reduced by a factor of 4.1 with a BaF$_2$ heterogeneous flow cell at the expense of only a slight < 9% loss in α detection efficiency. An alpha detection efficiency of ~ 50% for ^{233}U is reported by DeVol and Fjeld (1995) providing minimal detectable activities of 0.6 Bq for flow cells containing CaF$_2$(Eu) and BaF$_2$ and 1.1 Bq for the cell containing GS-20. Like most heterogeneous flow cells the lack of inertness and intrinsic contamination of the scintillator remains a problem.

Lochny *et al.* (1998) and Wenzel *et al.* (1999) devised a plastic scintillator flow cell for the on-line measurement of alpha- and beta-emitting radionuclide mixtures in high-level waste. The detector consisted of the plastic scintillator Meltilex (Wallac, Finland) and transparent perfluoroalkoxy (PFA) tubing (Dupont, USA). Meltilex is a meltable plastic containing the fluors PPO (diphenyloxazole) and MSB (methylstyrylbenzene). The inner walls of the flow cell tubing was coated with Meltilex permitting detection of both alpha and beta particles. With the use of appropriate energy windows and varying the flow cell geometry the detector is able to suppress beta particles to a great extent, and the detector capable of monitoring the decontamination of radioactive waste.

A series of research studies by Hastie *et al.* (1999), DeVol *et al.* (1999), and Tan *et al.* (2000) at Clemson University's Department of Environmental Engineering and Science Department has culminated in a very efficient flow cell detector for alpha/beta pulse shape discrimination. The detector reported by Tan *et al.* (2000) consisted of a flow cell containing granular 63–93 μm Parylene C polymer coated CsI:Tl scintillator situated between two photomultiplier tubes in coincidence detection mode. The scintillation pulses were analyzed by pulse shape discrimination using the charge integration technique (see Sections XIII of Chapter 5 and VI.B of Chapter 11 for detailed treatments of pulse shape discrimination). Pulse shape discrimination is achieved by comparing the fast decay component (pulse integrated over a short period of time, I_f) to the total integrated current from the pulse, I_{tot}, that is, I_f/I_{tot}, as described in Section VI.B of Chapter 11. The excellent pulse shape discrimination of alpha and beta events achieved is illustrated in Fig. 12.16. Tan *et al.* (2000) report a spillover of 2.1% for α events to β and 1.4% spillover of β events to α at a pulse shape discriminator setting $I_f/I_{tot} = 72$ (Fig. 12.16) for a mixture of ^{233}U and ^{90}Sr(^{90}Y) yielding alpha and beta detection efficiencies and backgrounds of 31.3% and 26.4%, and 0.84 cps and 0.31 cps, respectively. The figure of

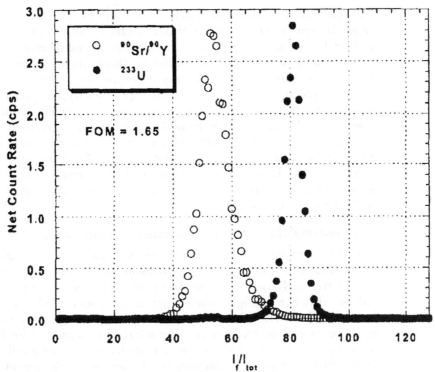

FIGURE 12.16 Pulse shape spectrum of CsI:Tl with aqueous solutions of ^{90}Sr/^{90}Y and ^{233}U. (From Tan *et al.*, 2000 © IEEE.)

merit (FOM) reported in Fig. 12.16 represents the ratio of the difference between alpha and beta peaking times to the sum of FWHMs for the alpha and beta pulse shape spectrum peaks.

A unique application of alpha/beta pulse shape discrimination reported by Wierczinski *et al.* (2001) was applied to the first on-line separation and detection of a subsecond α-decaying nuclide. Subsecond ^{224}Pa ($t_{1/2}=0.85$ s) was produced by the ^{209}Bi(^{18}O, 3n)^{224}Pa reaction at the 88-inch cyclotron at the Lawrence Berkeley National Laboratory. After production the nuclide was transported via a gas-jet system to a fast centrifuge system followed by on-line extraction with trioctylamine/scintillation solutions and on-line flow scintillation counting with pulse shape discrimination to minimize beta events.

D. On-Line FSA and Mass Spectrometry (MS)[1]

In the biological sciences, mass spectrometry is one of the most popular methods employed for the determination of the molecular weights and structures of metabolites. The popularity of mass spectrometry is due to

[1]Taken in part from L'Annunziata, M. F. and Nellis, S. W. (2001). Metabolism studies with on-line HPLC and mass spectrometry (MS) interfaced with the flow scintillation analyzer (FSA). *FSA Application Note FSA-005.* PerkinElmer Life and Analytical Sciences, Boston. (Reprinted with permission from PerkinElmer.)

the high sensitivity of the analytical method and the possibility of obtaining molecular weights and structure of metabolites directly from HPLC effluents without further sample treatment. The traditional and more time-consuming methods of structure analysis, including chromatography of metabolites, followed by chromatogram fraction collection, and purification prior to submission of isolated metabolites to mass spectrometry, have been applied for many years (L'Annunziata, 1970, 1984; L'Annunziata and Fuller, 1971a). Advanced techniques include the on-line mass spectral structural analysis of metabolites directly off the HPLC column after detection of peaks of interest. This section describes the state of the art of on-line mass spectrometry of radioisotope-labeled metabolites following HPLC separation and FSA detection of chromatogram peaks of interest, also referred to as the hyphenated radio-HPLC-FSA-MS analysis.

I. Radio-HPLC-FSA-MS Instrumentation and Interfacing

When the metabolism of a radioisotope-labeled compound is studied and the metabolites are separated by HPLC, flow scintillation analysis provides for the quantitative analysis of metabolites in terms of percentage of total recovered radioactivity. For example, when a parent compound labeled with a radio-isotope, such as ^{3}H, ^{14}C, ^{32}P, ^{33}P is administered with a known radioactivity to a test animal or medium and the metabolites separated by HPLC, the percentage of the total radioactivity administered is automatically measured by the FSA prior to mass spectrometry. Consequently, the use of FSA prior to mass spectrometry provides advantages over the UV detector, which include (1) irrefutable evidence that a certain HPLC peak is one of interest, (2) the measurement of radioactivity from the isotope label is performed by the FSA without a miss, unless the isotope label is near or essentially at background levels, (3) the FSA reports the radioactivity of the HPLC-separated parent compound and metabolite fractions in quantitative units of disintegrations per minute (DPM) providing valuable data for the quantitative percentages of total radioactivity administered to a test organism, and (4) the FSA can store quantitative data on metabolites over a series of HPLC runs carried out over a time span to determine the time course of a metabolism study (see Fig. 12.3).

The FSA provides the real-time radioactivity levels of metabolites as these are eluted from the HPLC column, and the radioactivity peaks from the FSA can provide the signal to initiate mass spectrometric analysis. The FSA is connected directly to the MS if using a heterogeneous (solid) flow cell (see Fig. 12.9). If a homogeneous (liquid) flow cell is used, the flow is split to both the FSA and MS. The homogeneous flow cell arrangement requires HPLC eluate splitting, because scintillation cocktail is mixed with eluate for radioisotope analysis. Stream splitting is often set to provide most of the stream to the FSA and a small portion to the mass spectrometer, because of the high sensitivity of mass spectrometers (pg/μL) that utilize electrospray ionization techniques for sample introduction. For example, Ramu et al. (2000) split the HPLC eluate at 1 mL/min in the ratio of one to nine to provide ca. 100 μL/min into the mass spectrometer and ca. 900 μL/min into the PerkinElmer Radiomatic 150TR FSA. Similarly Andersson et al. (1998) used a stream splitter that diverted ca. 80% of the HPLC eluate to a UV monitor and PerkinElmer Radiomatic A525 FSA and ca. 20% of the eluate to the mass spectrometer. The homogeneous flow cell setup provides higher detection efficiencies (up to 45% for ^{3}H and 88% for ^{14}C) depending on the quench level of HPLC solvents. Stream splitting of HPLC eluate to the flow scintillation analyzer and mass spectrometer is a common practice (see also Singh *et al.*, 2001; Vickers *et al.*, 2001; Yuan *et al.*, 2002) with most of the eluate going to the FSA and only a small fraction to the mass apectrometer.

Interfaces with the HPLC effluent and mass spectrometer must liberate the biochemical or bioorganic molecular species of interest (e.g., metabolite) from the aqueous solvent molecules and ionize the molecular species prior to mass spectrometric separation of the molecular ions and molecular ion fragments. This is performed most commonly by spray ionization (SI) techniques, which involve a combination of processes including spraying the HPLC effluent from a fine capillary into minute droplets, pneumatic heating with a drying gas, applied electric potential, and in some cases chemical ionization. The most common MS interfaces used in conjunction with HPLC are electrospray ionization (ESI) and atmospheric pressure chemical ionization (APCI). In the ESI method a nebulizer gas and electric field is introduced at the interface to produce charged droplets of the HPLC effluent. The combination of electric field energy and pneumatic heating via a warm concurrent dry gas stream cause the charged droplets of the HPLC effluent to subdivide and yield eventually single ionized molecules. Figure 12.17 illustrates the interface between the effluent of a micro-bore HPLC chromatograph and FSA effluent to an electrospray ionization source of the mass spectrometer. APCI is a chemical ionization technique, which employs a mechanism similar to ESI. The differences exist in the establishment of a plasma of the nebulized HPLC effluent by a DC discharge at atmospheric pressure. Reactions between ions and molecules occur in the plasma to produce molecular ion species of the biochemical or bioorganic compounds

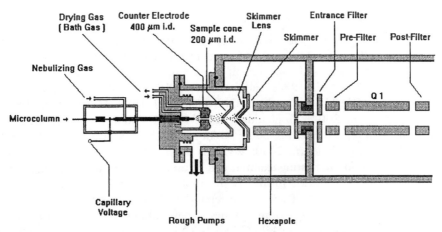

FIGURE 12.17 Schematic of a modified electrospray ion source for Quattro I mass spectrometer. The counterelectrode is designed with a 400-μm-i.d. hole for applications with flow rates of 1–20 μL/min. (From Schultz and Alexander, IV, 1998, reprinted with permission of John Wiley and Sons, Inc., Copyright © 1998.)

present, which are then introduced via vacuum into the mass spectrometer. Spray ionization techniques yield high ionization efficiencies and consequently mass spectral detection limits as low as the pg/μL level. Also, spray ionization techniques are soft, which yield relatively minor molecular ion fragmentation providing accurate molecular weight determinations in the mass range up to 10^5–10^6 daltons, the sensitivity depending on the specific type of mass spectrometer used. The theory and principles of electrospray ionization are treated in detail by Kebarle and Tang (1993).

The molecular ions and molecular ion fragments produced via the spray ionization interface are separated in the mass spectrometer according to their mass to charge ratios (m/z) using electric and/or magnetic fields. Several types of mass spectrometers are available for use on-line with HPLC including time of flight analyzers, quadrupole ion filters, and quadrupole ion trap instruments. The characteristics of these instruments are described in detail by Lambert *et al.* (1998). A popular mass spectrometer used on-line with HPLC in metabolism studies is the triple quadrupole mass spectrometer, which is a tandem mass spectrometer (Andersson *et al.*, 1998; Boulton *et al.*, 1999; Halpin *et al.*, 2002; Maggs *et al.*, 2000; Mannens *et al.*, 2002; Moghaddam *et al.*, 2001; Noort *et al.*, 1999; Patrick *et al.*, 2002; Riska *et al.*, 1999; Wormhoudt *et al.*, 1998). Tandem mass spectrometry is often abbreviated as MS/MS, because it consists of dual mass analyzers coupled in a tandem instrument useful in selectively separating the parent molecular ions from the product ion fragments. Dual mass analysis with the tandem analyzer is accomplished by first selecting the parent molecular ions after initial ionization with a sector magnet, and the molecular ions are further dissociated via collision with a gas such as He, Ne, N_2, or Ar referred to as collision induced dissociation (CID). The ion fragmentation products of CID are analyzed subsequently in the second mass analyzer according to their m/z and abundance. The tandem MS is highly sensitive and operates up to mass to charge ratios of 4000 described in detail by de Hoffman (1996).

2. Representative Data

During the past five years, numerous research papers employing on-line HPLC-FSA-MS have appeared in the scientific journals. Only a few will be cited here and some examples of representative data described in this section (Adas *et al.*, 1998; Andersson *et al.* 1998; Boulton *et al.*, 1999; Chen *et al.*, 2001b; Halpin *et al.*, 2002; Kumar *et al.* 1999; Maggs *et al.*, 2000; Mannens *et al.*, 2002; Moghaddam, *et al.*, 2001; Noort *et al.*, 1999; Patrick *et al.*, 2002; Ramu *et al.*, 2000; Riska *et al.*, 1999; Scarfe *et al.*, 2000; Sekiya *et al.*, 2000; He *et al.*, 2000; Wynalda *et al.*, 2000).

An interesting example can be taken from the work of Maggs *et al.* (2000), who studied the rat biliary metabolites of β-artemether (AM), an antimalarial endoperoxide. They administered [^{14}C]AM i.v. to rats (10 μCi/kg) and collected bile hourly up to five hours. They submitted the bile to HPLC-FSA-MS including tandem HPLC-FSA-MS/MS. The HPLC eluate at 0.9 mL/min was split between the FSA and the LC-MS interface by taking only ca. 40 μL/min to the mass spectrometer. The LC eluate directed to the FSA was mixed with PerkinElmer UltimaFlo AP scintillation cocktail (1 mL/min). Electrospray ionization mass spectra were acquired with a tandem quadrupole mass spectrometer. The presence of ammonium acetate in the LC buffer produced the MS molecular ion (M) cationized with ammonium ion, i.e. $(M + NH_4)^+$. In tandem MS/MS, the collision induced dissociation (CID) of the ammonium adducts of major metabolites was achieved with argon gas at a collision energy of 20 eV.

HPLC-FSA-MS of the [^{14}C]AM and bile metabolites provided quantitative analysis of the metabolites as well as evidence for structural confirmations. The HPLC radiochromatogram and radiometric quantification data of the HPLC peaks provided by the FSA are illustrated in Figure 12.18 and Table 12.7. The quantitative analytical data of metabolites, provided by FSA,

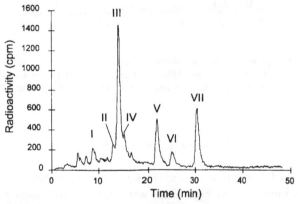

FIGURE 12.18 HPLC radiochromatogram of the biliary metabolites (pooled 0- to 3-h collections) of [^{14}C]AM (35 μmol/kg, i.v.) in male rats. (From Maggs et al., 2000, reprinted with permission of The American Society for Pharmacology and Experimental Therapeutics.)

TABLE 12.7 Biliary Metabolites of [^{14}C]AM in Male Rats. Bile (0- to 3-h Pooled Collections) from Anesthized and Cannulated Male Rats Administered [^{14}C]β-Artemether (AM) 35 μmol/kg i.v. Analyzed by Reverse Phase HPLC with Radiomatic Quantification. Data are Means ± S.D. (n = 6)[a]

Metabolite[b]	% Chromatographed radioactivity
I (dihydroxy AM.glucuronide)	6.0 ± 2.1
II (hydroxy AM.gluuronide)	3.1 ± 0.9
III (9α-hydroxy AM.glucuronide)	33.4 ± 6.8
IV (hydroxy AM.glucuronide)	4.4 ± 1.7
V (hydroxy AM.glucuronide)	21.4 ± 3.0
VI (hydroxy AM.glucuronide)	3.0 ± 1.1
VII (dihydro AM)	22.5 ± 4.4

[a]From Maggs et al. (2000) reprinted with permission from The American Society for Pharmacology and Experimental Therapeutics.
[b]Proposed identity of (glucuronide) metabolite. Roman numerals refer to peaks in the radiochromatogram (Fig. 12.18).

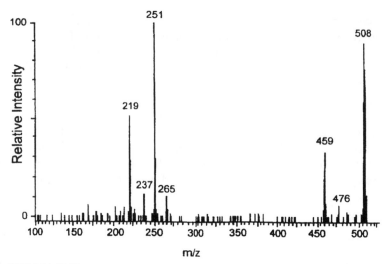

FIGURE 12.19 Daughter spectra obtained by CID of $[M+NH_4]^+$ for compound **V** (hydroxyAMglucuronide; parent ion, *m/z* 508) in rat bile. (From Maggs *et al.*, 2000, reprinted with permission of **The American Society for Pharmacology and Experimental Therapeutics.**)

is a major advantage of FSA over UV detectors for the identification of HPLC peaks of interest.

The proposed identities of the metabolites quantified in the radio-chromatogram illustrated in Fig. 12.18 were derived from data provided by the electrospray mass spectra of $[^{14}C]AM$ metabolites and daughter ion spectra of the metabolites created by CID with tandem LC-MS/MS. An example of the daughter spectra of one of the major metabolites is illustrated in Fig. 12.19. The daughter ions at *m/z* 476, 459, 265, 251, 237, and 219 were in agreement with the fragmentation pathway of compound V, a hydroxyAMglucuronide.

E. On-Line FSA and Nuclear Magnetic Resonance (NMR) Spectroscopy[2]

As described previously in this chapter flow scintillation analysis is commonly used to quantify the radioisotope label on organic compounds such as biochemicals, drugs, and metabolites separated from complex mixtures by HPLC. The subsequent task of determining the molecular structure of the separated substances can be formidable. Traditional methods of structure determination involve collecting the radio-HPLC separated fractions that correspond to activity peaks measured by the flow scintillation analyzer. The collected fractions are then isolated, further purified, and then submitted to spectroscopic methods of analysis such as mass spectrometry (MS) and

[2]Taken in part from L'Annunziata, M. F. and Nellis, S. W. (2001). Flow scintillation analyzer (FSA) interfaced with the HPLC and nuclear magnetic resonance (NMR) spectrometer. A state-of-the-art application of the Packard Radiomatic FSA. *FSA Application Note FSA-004.* PerkinElmer Life and Analytical Sciences, Boston. (Repritned with permission from PerkinElmer.)

nuclear magnetic resonance (NMR) spectroscopy. Both MS and NMR methods provide complementary information that can be used to derive a molecular structure. Mass spectrometry, described in Section VI.D of this chapter, can provide the molecular weight, molecular formula, and structure from ion fragmentation patterns, while NMR can provide additional important structural information including the spatial orientation of atoms in the molecular structure. A good example of this can be taken from early work of the author concerning the mass and NMR spectra of the inositol diastereomers, all of which produce virtually identical electron impact mass spectra, but different NMR spectra (L'Annunziata, 1970, 1984; L'Annunziata and Fuller, 1971a,b, 1976). A new and increasingly popular approach to drug metabolism and natural product studies involves the on-line measurement of the NMR spectra of compounds directly off the HPLC column obviating the need for compound isolation. This section will describe new developments in linking the flow scintillation analyzer from the HPLC to the NMR spectrometer to provide on-line (in-situ) molecular structure analysis of radioisotope labeled compounds.

1. Principle of NMR Spectroscopy

NMR spectroscopy has been used to derive the molecular structure of organic compounds from the magnetic properties of the atomic nuclei (e.g., ^1H and ^{13}C) and the surrounding molecular electrons since the first commercial NMR spectrometer appeared in 1960. Nuclei of certain atoms of odd mass such as ^1H and ^{13}C, or even mass and odd charge have a net charge and a spin. The spinning charge of the nucleus creates a magnetic dipole (μ). If one places the spinning proton nuclei, which are a component of most organic compounds, in a magnetic field (H), the axis of the magnetic dipoles of these nuclei will precess at an angle (θ) with respect to the magnetic field axis as illustrated in Fig. 12.20. The precession of the nuclei with respect to the applied magnetic field axis occurs somewhat like the way a spinning

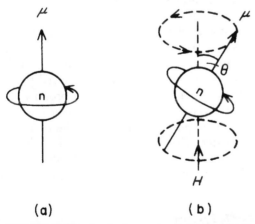

(a)　　　　　　**(b)**

FIGURE 12.20 A spinning proton in (a) the absence and (b) the presence of an externally applied magnetic field, *H*. (From L'Annunziata, 1984, reprinted with permission of Academic Press, Inc., San Diego.)

top precesses under the force of the Earth's gravitational field. The angular velocity of this precession is a function of the strength of the applied magnetic field (H) and the effects of *shielding* caused by spinning electrons in the environment of the proton nuclei. While under the forces of a stable magnetic field (H) of the NMR spectrometer, the proton nuclei are irradiated with radio frequency energy tunable over a narrow range. When the variable frequency is attuned to the precessional angular velocity of a given proton nucleus of a molecule, the two frequencies are in *resonance*. The applied energy at this resonance frequency is absorbed by the proton nucleus, and the nucleus is caused to flip or become aligned against the applied magnetic field (H). The energy absorbed by the proton nucleus that causes it to reach the higher energy spin state (i.e., flip) is the energy measured by the NMR spectrometer.

Fortunately in NMR spectroscopy the resonance absorption by proton nuclei is complicated by the shielding effect of electron clouds of varying densities in the environment of organic molecules. The electron cloud surrounding a nucleus also has charge, spin, and therefore produce their own characteristic magnetic field, which apposes or shields the externally applied field. The degree of *shielding* is a function of the electron cloud density, which will differ from nucleus to nucleus in the organic molecule, because of the differing electronegativities of neighboring atoms. Therefore, protons in a molecule will absorb different resonance frequencies depending on their location in the molecule. This effect is referred to as the *chemical shift*. A proton, which is highly shielded, absorbs at a lower resonance frequency than a proton with reduced shielding. The presence of atoms of differing degrees of electronegativity (electron-withdrawing ability) in molecules as well as the differing three dimensional orientation of atoms within molecules will cause a wide spectrum of shielding effects on neighboring protons. This gives rise to a wide spectrum of resonance absorption frequencies for protons depending on the structural group to which the protons are attached, their neighboring atoms, and their spatial orientation in the molecule. Therefore, the differing resonance absorption frequencies or chemical shifts of protons in NMR spectroscopy provide an absorption spectrum, which serves as a means for identifying chemical groups and their spatial positions in organic molecules.

The chemical shift of a particular proton nucleus in a molecule is recorded with respect to the chemical shift of the protons on the reference molecule, tetramethylsilane or $(CH_3)_4Si$ most often referred to as TMS. The difference in chemical shifts of a proton or group of protons in a molecule with respect to that of TMS is recorded and calculated in units of Hz, whereas the magnitude of the applied frequency is in the order of magnitude of MHz, a million fold greater. The difference in chemical shift of a proton nucleus with respect to that of TMS in Hz is divided by the applied frequency in MHz to record chemical shifts in convenient units of parts per million (ppm).

2. Radio-HPLC-FSA-NMR System

Because time is money, the trend is to analyze samples as fast as possible with as much automation that current technology will permit. This has led to recent advances in metabolism studies where the molecular structure of isotope labeled metabolites must be determined. A major and relatively recent advance in this field has been the direct linking of the NMR spectrometer to the high performance liquid chromatograph (HPLC). Several papers on this technology serve as excellent examples (Bailey *et al.*, 2000; Hansen *et al.*, 1999; Scarfe *et al.*, 2000; Shockcor *et al.*, 1996; Singh *et al.*, 2001; Smith *et al.*, 1999).

When the metabolism of a radioisotope labeled compound is studied and the metabolites are separated by HPLC, flow scintillation analysis provides for the quantitative analysis of metabolites in terms of percentage of total recovered radioactivity (see Table 12.7). For example, when a parent compound labeled with a radioisotope, such as ^3H, ^{14}C, ^{32}P, ^{33}P, is administered with a known radioactivity to a test animal and the metabolites separated by HPLC, the percentage of the total radioactivity administered is automatically measured by the FSA prior to NMR spectroscopy. This was illustrated previously in Table 12.7 and will be demonstrated also later in this section with data taken from the work of Sweeny *et al.* (2000). Consequently, the use of FSA (flow scintillation analysis) prior to NMR spectroscopy provides advantages over the UV detector, outlined previously in Section VI.D.1 of this chapter.

Radioisotope tracers are commonly used in metabolic studies, and there remains the need to quantify the isotope label on the metabolites eluted from the HPLC prior to their molecular structure analysis by NMR spectroscopy. The FSA provides the real time radioactivity levels of metabolites as these are eluted from the HPLC column, and the radioactivity peaks from the FSA can provide the signal to initiate NMR spectroscopic analysis. This will allow the researcher using HPLC-FSA-NMR to accurately stop the flow and capture the HPLC peak of interest in the NMR flow probe for molecular structure analysis. The FSA is connected between the UV and NMR if using a heterogeneous (solid) flow cell. If a homogeneous (liquid) flow cell is used, the flow is split to both the FSA and NMR. The heterogeneous flow cell uses a solid scintillant detector of radioisotope label (e.g., ^3H, ^{14}C) providing full recoveries of the HPLC eluate for subsequent NMR analysis.

A popular heterogeneous flow cell for the FSA utilizes SolarScint (trademark of PerkinElmer Life and Analytical Sciences, Boston), which is a solid scintillator that undergoes minimal compound binding in most cases, providing optimal peak resolutions, high detection efficiencies for ^{14}C (70%), and full sample recovery for NMR spectroscopy (i.e., no effluent splitting). The homogeneous flow cell arrangement requires HPLC effluent splitting, because scintillation cocktail is mixed with effluent for radioisotope analysis. The latter homogeneous flow cell setup is most appropriate for ^3H analysis with detection efficiencies of up to 45% depending on the quench level of HPLC solvents.

A specially designed flow probe is inserted into the NMR sample chamber. The probe is constructed to permit the sample to flow into the NMR spectrometer and the resonance spectra obtained while either flowing through, or more commonly stopped and analyzed for a required period of time. The probe placed in the bore of the magnet holds the sample with a commonly employed cell volume of 120 μL. It contains the antennae for sample radio frequency energy irradiation and the receipt of the weak radio frequency resonance signal. A stop flow mode is commonly employed for measurement of the NMR spectra, because of the low sample concentrations in the HPLC peaks. Suitable NMR spectra are obtained with samples as small as $1\,\mu g$ (or even sub-microgram) depending on sample molecular weight and analysis time (Beery, 2000 and Silva-Elipe, 2000). Sample analysis times can vary from 1–2 hours to 1–2 days. Stop flow NMR measurements of single peaks of the liquid chromatogram are governed by the signal from the UV absorbance detector or the signal from the FSA radioactivity detector. A signal from the radioactivity detector also confirms a metabolite of an isotope labeled parent compound and quantifies the isotope label in that metabolite, while peaks observed from the UV detector, that do not coincide with radioisotope peaks can be ignored. The FSA detector, therefore, can be used to not only trigger stop flow for NMR analysis and save valuable experimental time by permitting the researcher to ignore unlabelled UV peaks, but also to provide valuable quantitative data for metabolic studies.

3. Radio-HPLC-FSA-NMR Representative Data

The application of the FSA in HPLC-NMR setups for the chromatographic purification, radioactivity label analysis, and molecular structure analysis of isotope labeled metabolites can be found in several fairly recent reports in the scientific journals. Only a few will be cited here (Bailey *et al.*, 2000; Dockens, *et al.*, 2000; Hansen *et al.*, 1999; Kumar *et al.*, 1999; Maurizis *et al.*, 1998; Paulson *et al.*, 2000; Scarfe *et al.*, 2000; Singh *et al.*, 2001; Sweeny *et al.*, 2000; Vickers *et al.*, 1998). Some researchers will use the stop flow method described above, where the signal of a liquid chromatogram peak from the FSA radioactivity detector or UV detector will trigger the stop flow needed for in-situ NMR spectroscopy in the HPLC eluate. Others will utilize the same FSA or UV signal to collect the entire peak in a suitable vial and then submit the sample to further purification prior to NMR analysis in a suitable solvent. Representative examples of the application of HPLC FSA and the subsequent NMR spectroscopic results obtained will be cited subsequently.

In a study on the metabolism of the prodrug oseltarmivir of the influenza neuramidase inhibitor GS-4071 Sweeny *et al.* (2000) administered oral doses of $[^{14}C]$oseltamivir to rats. Metabolites in rat urine, plasma, liver, and lung were separated by HPLC and on-line radioactivity of metabolite liquid chromatogram peaks were determined with a PerkinElmer Radiomatic FSA with PerkinElmer FLO-ONE for Windows. The elution times of the metabolites were determined with the PerkinElmer Radiomatic FSA and via UV absorbance. A representative radiochromatogram printout from the FSA obtained from the urine fraction is illustrated in Fig. 12.21.

The PerkinElmer FLO-ONE for Windows used with the FSA in this work is a comprehensive radio-HPLC workstation software package developed to exploit the graphical user interface and multitasking capabilities offered by Windows. The liquid chromatogram peak labeled GS4104 is that of the radioisotope labeled parent compound $[^{14}C]$oseltamivir. The peak labeled

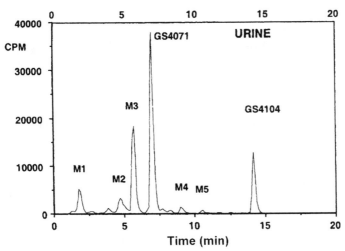

FIGURE 12.21 Representative radiochromatogram of rat urine after a single oral dose of [¹⁴C]oseltamivir. (From Sweeny et al., 2000, reprinted with permission of The American Society for Pharmacology and Experimental Therapeutics.)

GS4071 is the influenza inhibitor and peaks labeled M1 to M5 are metabolites, some of which are illustrated in the metabolic sequence in Fig. 12.22. The PerkinElmer FLO-ONE software provided quantitative analysis of the metabolites as these were eluted from the HPLC column. For example, the percentages of total recovered ^{14}C radioactivity as oseltamivir and metabolites in rat urine were, according to the FSA trace of Fig. 12.21, GS4104 (15.50%), prodrug GS4071 (47.00%), M1 (7.47%), M2 (6.02%), M3 (22.10%), M4 (1.34%), and M5 (0.66%).

The molecular structure of metabolites were derived from evidence provided by mass and NMR spectral data. Among the metabolites purified by radio-HPLC the metabolite peak labeled M3 in the PerkinElmer Radiomatic FSA printout serves as an excellent example of the structural derivation from NMR data alone. By comparing the ^{1}H-NMR resonance assignments (i.e., chemical shifts in ppm) for the oseltamivir parent compound to the assignments for the M3 metabolite, the structure for the M3 peak from the FSA could easily be deduced as that (R)-ω-carboxylic acid oseltamivir illustrated in Fig. 12.22. The ^{1}H-NMR resonance assignments taken from the NMR spectra for oseltamivir and the metabolite M3, for example, are provided in Table 12.8. This work of Sweeny et al. (2000) clearly demonstrated the power of the use of FSA to monitor the presence and quantitative data for the amounts of radioisotope labeled metabolites in HPLC effluent prior to NMR molecular structure analysis.

In the previously described example the proton resonances (chemical shifts in ppm) and the proton coupling constants (J) provided by the NMR spectra of a metabolite were interpreted to derive at the molecular structure. The following example taken from the work of Vickers et al. (1998) demonstrates also the use of HPLC with FSA and NMR. However, in this work the NMR spectrum of the metabolite was confirmed in addition to other structural evidence by comparing its spectrum to that of a reference

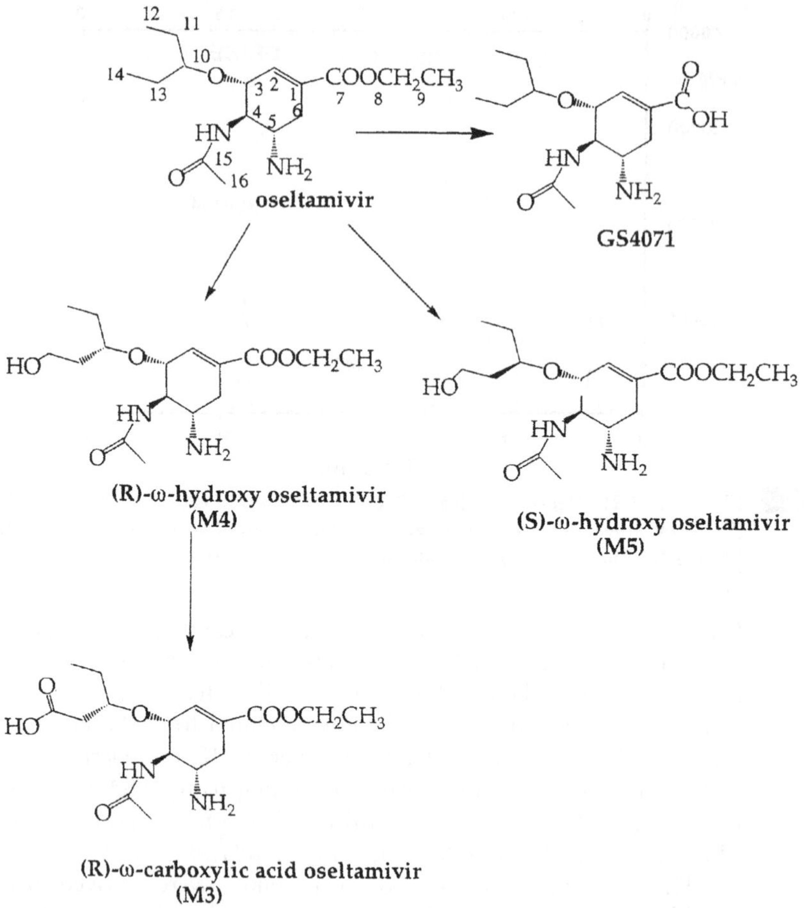

FIGURE 12.22 Scheme of oseltamivir metabolism in the rat determined with ^{14}C labeling, FSA, mass spectrometry, and NMR spectroscopy. (From Sweeny et al., 2000, reprinted with permission of The American Society for Pharmacology and Experimental Therapeutics.)

compound as a "fingerprint" for structural confirmation. In this work the disposition and metabolism of ^{14}C-labeled MK-499 was studied in rats and dogs. This drug is an antiarrhythmic agent for treatment of malignant ventricular tachyarrhythmias. Metabolites in urinary and biliary samples were separated by HPLC and analyzed with a PerkinElmer Radiomatic FLO-ONE flow scintillation analyzer. Vickers et al. (1998) used mass and NMR spectral data to determine the molecular structure of several ^{14}C-labeled metabolites. For brevity the NMR spectrum of one metabolite is illustrated here (Fig. 12.23) together with the spectrum of a compound of known structure as a reference.

Proton NMR resonances of the metabolite can be assigned as illustrated in the lower spectrum of Fig. 12.23, which identifies the resonance chemical shifts (δ) for the protons on the aromatic rings. Also the resonances of the uncoupled protons of the CH_3 group occur as a singlet at the chemical shift (δ) of approximately 2.8 ppm. The remaining CH_2 protons of the molecule

TABLE 12.8 ¹H-NMR Assignments (ppm), Multiplicities, and Coupling Constants (Hz) for Oseltamivir and M3 in Deuterium Oxide; (s) Singlet; (d) Doublet; (t) triplet; (q) quadruplet; (m) multiplet[a]

Assignment	Oseltamivir		M3	
	Peak (ppm)	J (Hz)	Peak (ppm)	J (Hz)
CH at C-2	6.87 (s)		6.87 (s)	
CH at C-3	4.36 (d)	8.9	4.36 (d)	8.6
CH₂ at C-8	4.27 (q)	7.3	4.27 (q)	7.3
CH at C-4	4.06 (q)	11.6	4.06 (t)	10.0
CH at C-10	3.60 (m)		4.00 (m)	
CH at C-5	3.59 (m)		3.60 (m)	
CHₐ at C-6	3.00 (q)		3.00 (m)	
CH_b at C-6	2.53 (m)		2.56 (m)	2
CHₐ at C-13	1.59 (m)		2.53	
CH_b at C-13	1.57 (m)		2.40	
CH₃ at C-16	2.10 (s)		2.09 (s)	
CH₂ at C-11	1.57 (m)		1.60 (m)	
CH₃ at C-9	1.31 (t)	7.3	1.31 (t)	7.3
CH₃ at C-12	0.90 (t)	7.3	.86 (t)	7.3
CH₃ at C-14	0.87 (t)			

[a]From Sweeny *et al.*, 2000. Reprinted with permission of The American Society for Pharmacology and Experimental Therapeutics.

provide resonances at chemical shifts at approximately 3.3 and 4.8 ppm. The most valuable evidence provided by the NMR of the metabolite is that it demonstrates to be an identical "fingerprint" of the reference compound (upper spectrum). The derived molecular structure can thus be conclusive.

F. On-line Radio-HPLC-FSA-MS-NMR

Several researchers are already using on-line HPLC-UV-FSA-NMR-MS instrumentation. For on-line spectroscopic analysis of natural products, Bailey *et al.* (2000), Hansen *et al.* (1999), Scarfe *et al.* (2000), and Shockcor *et al.* (1996) split the HPLC effluent in the proportions of 95% to the NMR spectrometer and the remaining 5% of the effluent to the mass spectrometer in light of the relative sensitivities of the two spectrometers. The NMR spectra were obtained using the stop flow method with resonance signal acquisitions varying from several minutes to hours with a 500.13 MHz Bruker DRX-500 NMR spectrometer. The various acquisition times were dependent on compound concentrations off the HPLC column. Smith *et al.* (1999) also report the use of a splitter of HPLC effluent to the MS and NMR

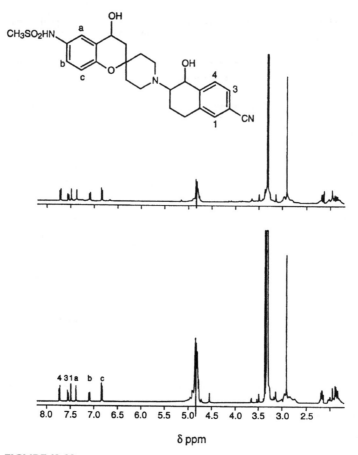

FIGURE 12.23 NMR spectra of reference compound (above) and metabolite of the drug MK-499. (From Vickers *et al.*, 1998, reprinted with permission of The American Society for Pharmacology and Experimental Therapeutics.)

spectrometers. In the near future we can expect to see a growing number of scientific reports with the hyphenated analytical methods of HPLC-UV-FSA-NMR-MS, as illustrated in Fig. 12.24, for the on-line separation, radioisotope label analysis, and molecular structural elucidation of complex mixtures. As reported by Hansen *et al.* (1999) these techniques will cut the time needed to carry out such complex studies to short durations from one day to a few weeks compared to the span of months required when traditional techniques of compound isolation, purification, and subsequent spectroscopy are undertaken.

G. On-Line Nuclear Waste Analysis

I. ^{3}H Effluent Water Monitors

A characteristic of heavy water reactor operation is the production of tritium in the moderator. Consequently as reported by Sigg *et al.* (1994) and

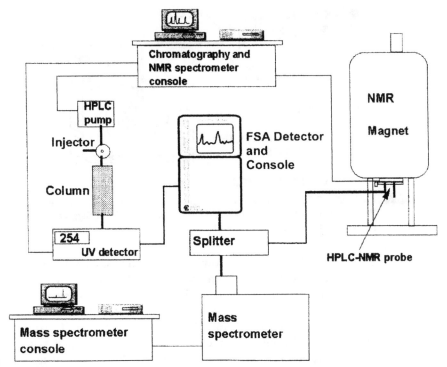

FIGURE 12.24 Instrumental setup of the HPLC-UV-FSA-NMR-MS apparatus. [Modified from Hansen *et al.*, 1999, reprinted with permission; Copyright (1999) American Chemical Society.]

Hofstetter (1995) the potential release of tritium into the environment through aqueous effluents is of concern at CANDU reactors and at other tritium handling facilities. Developments in flow scintillation analysis have provided a rapid on-line detection system, which can monitor continuously tritium levels in effluents permitting corrective action to mitigate unwanted tritium release into the environment.

The first flow cells used at the Savannah River Site (SRS), Aiken, South Carolina were of the heterogeneous type containing solid scintillator beads, such as yttrium silicate, reported by Hofstetter (1991). Continuous monitoring of tritium levels in the aqueous effluents of selected SRS heavy water purification facilities makes use of a heterogeneous flow cell containing plastic scintillator coupled with coincidence electronics providing a sensitivity of $\sim 25\,\text{kBq/L}$ (670 pCi/mL). The system requires a water purification pretreatment process, which includes microfiltration, ultraviolet sterilization, activated charcoal and ion exchange resin columns, and a phase separator before continuation of the aqueous stream to the flow cell at a rate of $3\,\text{mL/min}$. The solid scintillators provide limited detection sensitivity with a counting efficiency of only 0.18% for ^3H, but adequate for the needs of the facilities where they are employed (Hofstetter, 1993, 1995).

Tritium effluent water monitors reported by Sigg *et al.* (1994) and those utilized at CANDU systems (Cutler *et al.*, 1993) make use of a homogeneous

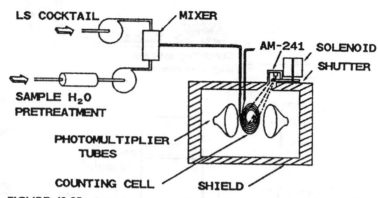

FIGURE 12.25 Overview of a tritium effluent water monitor utilizing a modified PerkinElmer Radiomatic flow scintillation analyzer. (From Sigg et al., 1994, reprinted with permission from Elsevier Science.)

2.5 mL liquid scintillation flow cell containing coiled Teflon tubing installed in a modified PerkinElmer Radiomatic A250 flow scintillation analyzer. The effluent monitoring method requires water pretreatment, and the liquid scintillation (homogeneous) cell system provides tritium quantification as well as rapid detection at several pCi/mL. Sensitivities of 16 and 1 pCi/mL are reported for 5-minute and daily averages of counting data, respectively, which is reported to provide approximately 200 times greater sensitivity than similar methods utilizing the solid scintillator heterogeneous flow cells. A simple diagram of the liquid scintillation tritium effluent water monitor is provided in Fig. 12.25. As described by Sigg et al. (1995) a centrifugal pump moves environmental water to be monitored at about 10 L/min from the sampling location to the instrument position where it passes across the surface of a cross-flow filter. Sample water at only 2 mL/min is drawn through the filter for further pretreatment by flash distillation. A 0.25-mL static mixer blends 0.15 mL/min of liquid scintillation cocktail with the sample water before passing through the 2.5 mL flow cell. It is reported that, at the lowest flow rates, full response to changes in environmental tritium is possible in 30 min, and liquid scintillation cocktail consumption is less than 7 L a month. Flow rates are increased by as much as sixfold if increased response time is needed. Ultima Gold XR (PerkinElmer Life and Analytical Sciences, Boston) liquid scintillation cocktail is used for low flow rates. Sigg et al. (1995) report that other fluor cocktails with lower viscosity are available when higher flow rates are required. Low viscosity fluor cocktails were described previously in this chapter, and they are listed in Tables 12.6. A tritium counting efficiency of 36% is reported with the liquid scintillation homogeneous flow cell system, which is many orders of magnitude more sensitive than the plastic solid scintillator heterogeneous flow cell. As noted earlier the plastic solid scintillator flow cell could provide a tritium counting efficiency of only 0.18%. The flow scintillation analyzer was equipped with an external [241]Am standard with a 3-mm thick tungsten shutter assembly to expose the flow cell to the 59.5 keV gamma-rays for the production of a quench-indicating parameter.

2. ^{89}Sr and ^{90}Sr(^{90}Y) Analysis

Since the 1960s nuclear atmospheric weapons tests, there has been great concern over the levels of ^{90}Sr in the environment as a serious threat to internal contamination in the human body. This is due mainly to the relatively long half-life of ^{90}Sr ($t_{1/2} = 28.8$ years) and the similarities in the chemistry of calcium and strontium whereby ^{90}Sr will accumulate in the body by residing with a long biological half-life in the human skeleton along with calcium (L'Annunziata and Fuller, 1968). A paper by Desmartin *et al.* (1997) underscores the remaining concern for ^{90}Sr residues in the environment, as it is produced in a relatively high yield (5.8% from ^{235}U) during the fission of heavy elements. The classical method of analysis of ^{90}Sr involves precipitation of its daughter ^{90}Y as a carbonate or oxalate followed by counting of the precipitated ^{90}Y (L'Annunziata, 1971; Piltingsrud and Stencel, 1972).

Current techniques in the analysis of ^{90}Sr in the environment involves the combination of HPIC and FSA also referred to as On-Line Liquid Scintillation Counting (OLLSC) when homogeneous flow cell counting is involved [Knöchel and coworkers (see Alfaro *et al.*, 1995); Desmartin *et al.* 1997]. Desmartin *et al.* (1997) describes the coupling of HPIC to separate ^{90}Sr from natural or power plant reactor water and on-line flow scintillation analysis before the daughter radionuclide ^{90}Y has time to grow to any significant level following ^{90}Sr separation. They used a high performance ionic chromatographic Dionex 2010i system (Dionex Corporation, Sunnyvale, CA) coupled to a PerkinElmer Radiomatic flow scintillation analyzer with FLO-ONE operation software (PerkinElmer Life and Analytical Sciences, Boston) equipped with a 2.5 mL homogeneous flow cell operated with an eluent flow rate of 1 mL/min and a scintillator flow rate of 2.5 mL/min.

Desmartin *et al.* (1997) set the pulse height discriminators to provide two counting regions of the flow scintillation analyzer, which is equipped with a multichannel analyzer, to measure the ^{89}Sr and ^{90}Sr simultaneously. Synthetic reactor water containing various concentrations of ^{134}Cs$^+$, ^{90}Sr^{2+}(^{90}Y^{3+}), ^{60}Co^{2+}, ^{133}Ba^{2+}, and ^{135}I$^-$ were easily separated by HPIC to isolate radiostrontium. Trivalent ions such as ^{90}Y^{3+} are stopped at the top of the HPIC column. Therefore, the ^{90}Sr(^{90}Y) parent daughter radionuclides are separated during the chromatography run. The closest chromatograph peaks to ^{90}Sr^{2+} are those of ^{134}Cs$^+$ and ^{60}Co^{2+}, which are well separated as illustrated in Fig. 12.26. The reported minimal detectable activities (MDAs), calculated according to Eq. 12.15 in Section III.D of this chapter, are 480 DPM for ^{89}Sr and 60 DPM for ^{90}Sr.

Grate and Egorov (1998a) and Grate *et al.* (1996, 1999) have automated the analysis of ^{90}Sr in nuclear waste based on sequential injection analysis, which rapidly separates ^{90}Sr from ^{90}Y, ^{137}Cs, among other radionuclides. They used a sorbent extraction microcolumn containing Sr-Spec resin (EiChrom Industries, Darien, IL) that selectively binds ^{90}Sr as a crown ether complex under acidic conditions. The ^{90}Sr is analyzed on-line with a flow scintillation analyzer, which can be operated in the continuous- or stopped flow modes described previously in Section V of this chapter. More information concerning the automation techniques are described in Chapter 14.

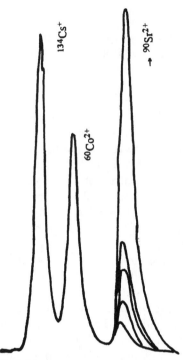

FIGURE 12.26 Activity peaks from on-line flow scintillation analysis of HPIC separated nuclides in synthetic reactor water containing increasing amounts of $^{90}Sr^{2+}$ measured with a PerkinElmer Radiomatic flow scintillation analyzer. The retention times of the $^{134}Cs^{+}$, $^{60}Co^{2+}$ and $^{90}Sr^{2+}$ are 3.5, 6.3 and 10.0 min. respectively. (From Desmartin et al., 1997 with kind permission from Klewer Academic Publishers.)

Several extractive scintillation sensor materials were tested by Devol et al. (2001) for either off-line or on-line monitoring of radiostrontium. On-line separation techniques offer the advantage for more automated and rapid analysis cited in the previous paragraph and discussed in detail in Chapter 14. The most effective extractive sensor found by Devol et al. (2001) consisted of a bed of strontium-specific resin marketed under the trademark Sr-Spec Resin (100 to 150 μm from EiChrom Industries) and BC-400 scintillating beads (100 to 200 μm from Bicron, Inc.) combined together 1:2 by weight as a mixed-bed. The Sr-Spec Resin consists of an inert, porous organic polymer support impregnated with a solution of *bis*-4,4′(5′)-*tert*-butyl-cyclohexano-18-crown-6 (DtBuCH18C6) in 1-octanol. It displays a high selective affinity to strontium at low pH while excluding alkali and alkaline earths. For on-line tests Devol et al. (2001) packed < 0.5 g of the Sr Resin/BC-400 mixture into 3 mm OD × 1.6 mm ID × 140 mm long polytetrafluoroethylene tubing yielding an approximate pore volume (PV) of 200 to 400 μL, which was 60–70% of the total void volume. The flexible tubing was coiled in an approximate diameter of 2.54 cm and placed between two PMT tubes of the flow scintillation analysis detector (β-Tam Model 1, IN/US Systems, Inc.). The column was loaded with 1–5 PVs of sample in 4 M HNO$_3$

and the column washed with additional ~ 25 PVs of 4 M HNO_3 and the sorbed ions were eluted with 50 PVs of distilled-deionized water. The detection efficiency reported was 76% for ^{89}Sr ($E_{max} = 1.49$ MeV). The detection efficiency for ^{90}Sr, which was not tested, should be slightly lower since the beta-particle emissions from ^{90}Sr are of lower energy ($E_{max} = 0.546$ MeV).

3. Other Radionuclides

a. Automated On-Line Sorbent Column Extraction Separations

Several automated on-line analytical procedures with flow scintillation analysis have been developed including the following: (1) flow injection and sequential injection analysis combined with on-column redox reactions for the separation of americium, curium, and plutonium from other actinides (Grate and Egorov, 1998b; Egorov *et al.*, 1998b; Grate *et al.*, 1999) employing TRU-resin from EiChrom Industries, Inc., Darien, IL; (2) sequential injection renewable column separation of ^{90}Sr, ^{241}Am, and ^{99}Tc in aged nuclear waste using EiChrom TRU- and TEVA-resins (Egorov *et al.*, 1999); and (3) sequential injection analysis of ^{99}Tc separated from stable and radioactive interferences on a TEVA-resin column (Egorov *et al.*, 1998a). The automated techniques are described in Chapter 14.

b. On-Line Capillary Electrophoresis Analysis

The development of on-line radioactivity detection in conjunction with capillary electrophoresis (CE) has seen limited development, and these works are reviewed by Klunder *et al.* (1997). Most applications have been directed to the analysis of radiopharmaceuticals and radioisotope-labeled biochemicals including ^{32}P (Gordon *et al.*, 1993; Pentoney *et al.*, 1989, 1990), ^{99m}Tc (Altria *et al.*, 1990; Poirier *et al.*, 1995), and the PET radionuclides ^{11}C and ^{18}F (Westerberg *et al.*, 1993). Success in metal ion analysis by capillary electrophoresis sparked the work of Klunder *et al.* (1997) to develop rapid high-resolution separations of fission products by capillary electrophoresis with on-line radioactivity analysis. The reduced sample volumes involved with CE techniques offer the added advantage of minimizing worker exposure when analyzing high-level radioactive waste.

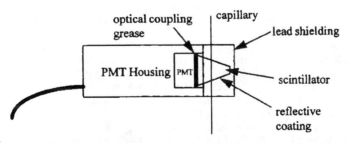

FIGURE 12.27 Diagram of the on-line capillary electrophoresis detector. Scintillator is dye-doped polyvinyltoluene (Bicron BC-400). (From Klunder *et al.*, 1997; Copyright © 1997 American Chemical Society.)

For the capillary electrophoresis analysis Klunder *et al.* (1997) designed an on-line scintillation detector using Bicron BC-400 plastic scintillator, which consists of scintillator-doped polyvinyltoluene (PVT), which yields a light output of 40–50% relative to NaI(Tl). The plastic detector was machined into cones of 9 mm dia., tapering to 3 mm, with a total length of 10 mm as illustrated in Fig. 12.27. The CE capillary passes through a 400 μm hole drilled through the middle of the cone providing 4π counting geometry for a 6 mm length of the capillary. Klunder *et al.* (1997) demonstrated excellent on-line radionuclide peak resolutions including the separation of ^{137}Cs(^{137m}Ba) parent daughter nuclides. On-line capillary electrophoresis detection efficiencies are reported to be $\sim 60\%$ for ^{152}Eu and $\sim 80\%$ for ^{137}Cs.

REFERENCES

Abbruscato, T. J., Williams, S. A., Misicka, A., Lipkowski, A. W., Hruby, V. J., and Davis, T. P. (1996). Blood-to-central nervous system entry and stability of biphalin, a unique double-enkephalin analog, and its halogenated derivatives. *J. Pharmacol. Exp. Ther.* **276**, 1049–1057.

Adas, F., Berthou, F., Picart, D., Lozac'h, P., Beaugé, F., and Amet, Y. (1998). Involvement of cytochrome P450 2E1 in the (ω-1)-hydroxylation of oleic acid in human and rat liver microsomes. *J. Lipid Res.* **39**, 1210–1219.

Ahmed, L. S., Moorehead, H., Leitch, C. A., and Liechty, E. A. (1998). Determination of the specific activity of sheep plasma amino acids using high-performance liquid chromatography: comparison study between liquid scintillation counter and on-line flow-through detector. *J. Chromatogr. B* **710**, 27–35.

Alary, J., Bravais, F., Cravedi, J.-P., Debrauwer, L., Rao, D., and Bories, G. (1995). Mercapturic acid conjugates as urinary end metabolites of the lipid peroxidation product 4-hydroxy-2-nonenal in the rat. *Chem. Res. Toxicol.* **8**, 34–39.

Alfaro, J., Apfel, T., Diercks, H., Knöchel, A., Sen Gupta, R., and Tödter, K. (1995). Trace analysis of the nuclides ^{90}Sr and ^{89}Sr in environmental samples III: Development of a fast analytical method. *Angew. Chem. Int. Ed. Eng.* **34**, 186–191.

Altria, K. D., Simpson, C. F., Bharij, A. K., and Theobald, A. E. (1990). A gamma-ray detector for capillary zone electrophoresis and its use in the analysis of some radiopharmaceuticals. *Electrophoresis* **11**, 732–734.

Andersson, S. H. G., Lindgren, A., and Postlind, H. (1998). Biotransformation of tolterodine, a new muscarinic receptor antagonist, in mice, rats, and dogs. *Drug Metab. Dispos.* **26**(6), 528–535.

Anonymous (1990). An introduction to flow radiochromatography. *Application Note* **9008**, pp. 8. PerkinElmer Life and Analytical Sciences, Boston.

Anonymous (1995). Detection of gamma emitting radionuclides using FSA, Gamma-B and Gamma-C flow cells. *FSA Application Note.* **FSA-001**, pp. 8. PerkinElmer Life and Analytical Sciences, Boston.

Anonymous (1996). New technology improves radio-HPLC detection. *FSA Application Note.* **FSA-002**, pp. 3. PerkinElmer Life and Analytical Sciences, Boston

Anonymous (1997). "Radiomatic® 500TR Series Flow Scintillation Analyzers, Reference Manual." pp. 454–456. PerkinElmer Life and Analytical Sciences, Boston.

Bai, X. and Esko, J. D. (1996). An animal cell mutant defective in heparan sulfate hexuronic acid 2-O-sulfation. *J. Biol. Chem.* **271**, 17711–17717.

Bailey, N. J. C., Cooper, P., Hadfield, S. T., Lenz, E. M., Lindon, J. C., Nicholson, J. K., Stanley, P. D., Wilson, I. D., Wright, B., and Taylor, S. D. (2000). Application of directly coupled HPLC-NMR-MS/MS to the identification of metabolites of 5-trifluoromethylpyridone (2-hydroxy-5-trifluoromethylpyridine) in hydroponically grown plants. *J. Agric. Food Chem.* **48**(1), 42–46.

Balla, T., Guillemette, G., Baukal, A. J., and Catt, K. J. (1987). Metabolism of inositol 1,3,4-triphosphate to a new tetrakisphosphate isomer in angiotensin-stimulated adrenal glomerulosa cells. *J. Biol. Chem.* **262**, 9952–9955.

Baranowska-Kortylewicz, J., Helseth, L. D., Lai, J., Schneiderman, M. H., Schneiderman, G. S., and Dalrymple, G. V. (1994). Radiolabeling kit/generator for 5-radiohalogenated urinides. *J. Labelled Comp. Radiopharm.* **34**, 513–521.

Beery, J. W. (2000). BASF Corporation. Agricultural Research Center. Research Triangle Park, NC 27709 (Personal Communication, August 21, 2000).

Bezalel, L., Hadar, Y., Fu, P. P., Freeman, J. P., and Cerniglia, C. E. (1996). Metabolism of phenanthrene by the white rot fungus *Pleurotus ostreautus*. *App. Environ. Microbiol.* **62**, 2547–2553.

Black, A. E., Hayes, R. N., Roth, B. D., Woo, P., and Woolf, T. F. (1999). Metabolism and excretion of atorvastatin in rats and dogs. *Drug. Metab. Dispos.* **27**(8), 916–923.

Bogan, B. W. and Lamar, R. T. (1996). Polycyclic aromatic hydrocarbon – degrading capabilities of *Phanerochaete laevis* HHB-1625 and its extracellular ligninolytic enzymes. *Appl. Environ. Microbiol.* **62**, 1597–1603.

Boogaard, P. J., Summer, S. C.-J., and Bond, J. A. (1996). Glutathione conjugation of 1,2,:3,4-diepoxybutane in human liver and rat and mouse liver and lung *in vitro*. *Toxicol. Appl. Pharmacol.* **136**, 307–316.

Borai, E. H. and Mady, A. S. (2002). Separation and quantification of ^{238}U, ^{232}Th and rare earths in monazite samples by ion chromatography coupled with on-line flow scintillation detector. *Appl. Radiat. Isot.* **57**, 463–469.

Boulton, D. W., Walle, U. K., and Walle, T. (1999). Fate of the flavonoid quercetin in human cell lines: chemical instability and metabolism. *J. Pharm. Pharmacol.* **51**, 353–359.

Bradbury, D., Elder, G. R., and Dunn, M. J. (1990). Rapid analysis of non-gamma radionuclides using the ANABET system. Proceedings of the Waste Management Conference, Tucson, Arizona, pp. 327–329.

Cacan, R., Dengremont, C., Labiau, O., Kmiecik, D., Mir, A.-M., and Verbert, A. (1996). Occurrence of a cytosolic neutral chitobiase activity involved in oligomanniside degradation: a study with Madin-Darby bovine kidney (MDBK) cells. *Biochem. J.* **313**, 597–602.

Capdevila, J. H., Morrow, J. D., Belosludtsev, Y. Y., Beauchamp, D. R., DuBois, R. N., and Falck, J. R. (1995). The catalytic outcomes of the constitutive and the mitogen inducible isoforms of prostaglandin H_2 synthase are markedly affected by glutathione and glutathione peroxidase(s). *Biochem.* **34**, 3325–3337.

Carruba, G., Granata, O. M., Farruggio, R., Cannella, S., Lo Bue, A., Leake, R. E., Pavone-Macaluso, M., and Castagnetta, L. A. M. (1996). Steroid-growth factor interaction in prostate cancer. 2. Effects of transforming growth factors on androgen metabolism of prostate cancer cells. *Steroids* **61**, 41–46.

Carsol, J.-L., Adanski, J., Guirou, O., Gerard, H., Martin, P.-M., and de Launoit, Y. (1996). 17β-hydroxysteroid dehydrogenase mRNA abundance in human meningioma tumors. *Neuroendocrinology* **64**, 70–78.

Chen, A. C. and Gudas, L. J. (1996). An analysis of retinoic acid-induced gene expression and metabolism in AB1 embryonic stem cells. *J. Biol. Chem.* **271**, 14971–14980.

Chen, S. C., Zhou, L., Ding, X., and Mirvish, S. S. (2001a). Depentylation of the rat esophageal carcinogen, methyl-*N*-pentylnitrosamine, by microsomes from various human and rat tissues and by cytochrome P450 2A3. *Drug Metab. Dispos.* **29**(9), 1221–1228.

Chen, L.-J., Lebetkin, E. H., and Burka, L. T. (2001b). Metabolism of (*R*)-(+)-pulegone in F344 rats. *Drug Metab. Dispos.* **29**(12), 1567–1577.

Cortese, J. F., Spannhake, E. W., Eisinger, W., Potter, J. J., and Yang, V. W. (1995). The 5-lipoxygenase pathway in cultured human intestinal epithelial cells. *Prostaglandins* **49**, 155–166.

Currie, L. A. (1968). Limits for qualitative detection and quantitative determination. *Anal. Chem.* **40**(3), 586–593.

Cutler, J. M., Mina, N., Swami, L., and Ely, F. A. (1993). Proceedings ANS Topical Meeting on Environmental Transport and Dosimetry, Sep. 1–3, 1993, Charleston, SC, p. 95.

Dargel, R. (1995). Metabolism of leukotrienes in impaired hepatocytes from rats with thioacetamide-induced liver cirrhosis. *Prostaglandins, Leukotrienes Essent. Fatty Acids* **53**, 309–314.

Dayhuff, T. J., Atkins, J. F., and Gesteland, R. F. (1986). Characterization of ribosomal frameshift events by protein sequence analysis. *J. Biol. Chem.* **261**, 7491–7500.

De Hoffmann, E. (1996). Tandem mass spectrometry: a primer. *J. Mass Spectrom.* **31**(2), 129–137.

Desmartin, P., Kopajtic, Z., and Haerdi, W. (1997). Radiostrontium-90 (^{90}Sr) ultra-traces measurements by coupling ionic chromatography (HPIC) and on line liquid scintillation counting (OLLSC). *Environ. Monitor. Assess.* **44**, 413–423.

DeVol, T. A. and Fjeld, R. A. (1995). Development of an on-line scintillation flow-cell detection system with pulse shape discrimination for quantification of actinides. *IEEE Trans. Nucl. Sci.* **42**, 959–963.

DeVol, T. A., Keillor, M. E., and Burggraf, L. W. (1995). Influence of radionuclide adsorption on detection efficiency and energy resolution for flow-cell radiation detectors. IEEE Nucl. Sci. Symp. And Medical Imaging Conf., Vol. 2, Oct 21–28 1995, San Francisco, pp. 795–799.

DeVol, T. A., Chotoo, S. B., and Fjeld, R. A. (1999). Evaluation of granulated BGO, GSO:Ce, YAG:Ce, CaF$_2$:Eu and ZnS:Ag for alpha/beta pulse shape discrimination in a flow-cell radiation detector. *Nucl. Instrum. Methods Phys. Res., Sect. A* **425**, 228–231.

DeVol, T. A., Duffey, J. M., and Paulenova, A. (2001). Combined extraction and scintillation detection for off-line and on-line monitoring of strontium in aqueous solutions. *J. Radioanal. Nucl. Chem.* **249**(2), 295–301.

Dockens, R. C., Santone, K. S., Mitroka, J. G., Morrison, R. A., Jemal, M., Greene, D. S., and Barbhaiya, R. H. (2000). Disposition of radiolabeled ifetroban in rats, dogs, monkeys, and humans. *Drug. Metab. Dispos.* **28**(8), 973–980.

Dumas, B., Cauet, G., Lacour, T., Degryse, E., Laruelle, L., Ledoux, C., Spagnoli, R., and Achstetter, T. (1996). 11β-hydroxylase activity in recombinant yeast mitochondria *In vivo* conversion of 11-deoxycortisol to hydrocortisone. *Eur. J. Biochem.* **238**, 495–504.

Egorov, O., O'Hara, M. J., Ruzicka, J., and Grate, J. W. (1998a). Sequential injection separation system with stopped-flow radiometric detection for automated analysis of ^{99}Tc in nuclear waste. *Anal. Chem.* **70**(5), 977–984.

Egorov, O., Grate, J. W., and Ruzicka, J. (1998b). Automation of radiochemical analysis by flow injection techniques: Am-Pu separation using TRU-resin sorbent extraction column. *J. Radioanal. Nucl. Chem.* **234**(1–2), 231–235.

Egorov, O., O'Hara, M. J., Grate, J. W., and Ruzicka, J. (1999). Sequential injection renewable separation column instrument for automated sorbent extraction separations of radionuclides. *Anal. Chem.* **71**, 345–352.

Falasca, M., Silletta, M. G., Carvelli, A., Di Francesco, A. L., Fisco, A., Ramakrishna, V., and Corda, D. (1995). Signalling pathways involved in the mitogenic action of lysophosphatidylinositol. *Oncogene* **10**, 2113–2124.

Folcik, V. A., Nivar-Aristy, R. A., Krajewski, L. P., and Cathcart, M. K. (1995). Lipoxygenase contributes to the oxidation of lipids in human atherosclerotic plaques. *J. Clin. Invest.* **96**, 504–510.

Gautier, J.-C., Richoz, J., Welti, D. H., Markovic, J., Gremaud, E., Guengerich, F. P., and Turesky, R. J. (2001). Metabolism of ochratoxin A: absence of formation of genotoxic derivatives by human and rat enzymes. *Chem. Res. Toxicol.* **14**, 34–45.

Ghosheh, O., Vashishtha, S. C., and Hawes, E. M. (2001). Formation of the quarternary ammonium-linked glucuronide of nicotine in human liver microsomes: identification and stereoselectivity in the kinetics. *Drug Metab. Dispos.* **29**(12), 1525–1528.

Glasgow, W. C., Hill, E. M., McGown, S. R., Tomer, K. B., and Eling, T.E. (1996). Regulation of 13(S)-hydroxyoctadecadienoic acid biosynthesis in Syrian hamster embryo fibroblasts by the epidermal growth factor receptor tyrosine kinase. *Mol. Pharmacol.* **49**, 1042–1048.

Gordon, J. S., Vasile, S., Hazlett, T. and Squillante, M. (1993). High sensitivity radiation detector for capillary electrophoresis. *IEEE Trans. Nucl. Sci.* **40**, 1162–1164.

Grard, T., Herman, V., Saint-Pol, A., Kmiecik, D., Labiau, O., Mir, A.-M., Alonso, C., Verbert, A., Cacan, R., and Michalski, J.-C. (1996). Oligomannosides or oligosaccharide-lipids as potential substrates for rat liver cytosolic *alpha*-D-mannosidase. *Biochem. J.* **316**, 787–792.

Grate, J. W. and Egorov, O. (1998a). Automating analytical separations in radiochemistry. *Anal. Chem. News & Features, December,* **70**, 779A-788A.

Grate, J. W. and Egorov, O. (1998b). Investigation and optimization of on-column redox reactions in the sorbent extraction separation of americium and plutonium using flow injection analysis. *Anal. Chem.* **70**, 3920–3929.

Grate, J. W., Strebin, R. S., Janata, J., Egorov, O., and Ruzicka, J. (1996). Automated analysis of radionuclides in nuclear waste: rapid determination of ^{90}Sr by sequential injection analysis. *Anal. Chem.* **68**, 333–340.

Grate, J. W., Fadeff, S. K., and Egorov, O. (1999). Separation-optimized sequential injection method for rapid automated analytical separation of ^{90}Sr in nuclear waste. *Analyst* **124**(2), 203–210.

Grate, J. W., Egorov, O., and Fiskum, S. K. (1999). Automated extraction chromatographic separations of actinides using separation-optimized sequential injection techniques. *Analyst* **124**, 1143–1150.

Guillemette, G., Balla, T., Baukal, A. J., and Catt, K. J. (1989). Metabolism of inositol 1,4,5-triphosphate to higher inositol phosphates in bovine adrenal cytosol. *Am. J. Hyperten.* **2**, 387–394.

Halpin, R. A., Porras, A. G., Geer, L. A., Davis, M. R., Cui, D., Doss, G. A., Woolf, E., Musson, D., Matthews, C., Mazenko, R., Schwartz, J. I., Lasseter, K. C., Vyas, K. P., and Baillie, T. A. (2002). The disposition and metabolism of rofecoxib, a potent and selective cyclooxygenase-2 inhibitor, in human subjects. *Drug Metab. Dispos.* **30**(6), 684–693.

Hankin, J. A., Clay, C. E., and Murphy, R. C. (1998). The effects of ethanol and acetaldehyde on the metabolism of prostaglandin E_2 and leukotriene B_4 in isolated rat hepatocytes. *J. Pharmacol. Exp. Ther.* **285**(1), 155–161.

Hansen, S. H., Jensen, A. G., Cornett, C., Bjørnsdottir, I., Taylor, S., Wright, B., and Wilson, I. D. (1999). High-performance liquid chromatography on-line coupled to high-field NMR and mass spectrometry for structure elucidation of constituents of *Hypericum perforatum* L. **71**, 5235–5241.

Hastie, K. H., DeVol, T. A., and Fjeld, R. A. (1999). Development of a homogeneous pulse shape discriminating flow-cell radiation detection system. *Nucl. Instrum. Methods Phys. Res., Sect. A* **422**, 133–138.

He, M. M., Abraham, T. L., Lindsay, T. J., Chay, S. H., Czeskis, B. A., and Shipley, L. A. (2000). Metabolism and disposition of moxonidine in Fischer 344 rats. *Drug Metab. Dispos.* **28**(4), 446–459.

Hegge, Th. C. J. M. and ter Wiel, J. (1986). Mixture for use in the LSC analysis technique. U.S. patent no. 4,624,799. PerkinElmer Life and Analytical Sciences, Boston.

Hofstetter, K. J. (1991). Development of aqueous tritium effluent monitor. *In* "Liquid Scintillation Counting and Organic Scintillators" (Harley Ross, John E. Noakes and Jim D. Spaulding, Eds.), pp. 421–433. Lewis Publishers, Chelsea, Michigan.

Hofstetter, K. J. (1995). Continuous aqueous tritium monitoring. *Fusion Technol.* **25**, 1527–1531.

Hofstetter, K. J. and Eakle, R. F. (1993). Proceedings ANS Topical Meeting on Environmental Transport and Dosimetry, Sep. 1–3, 1993, Charleston, SC, p. 98.

Hughes, M. F., Smith, B. J., and Eling, T. E. (1992). The oxidation of 4-aminobiphenyl by horseradish peroxidase. *Chem. Res. Toxicol.* **5**, 340–345.

Kebarle, P. and Tang, L. (1993). From ions in solution to ions in the gas phase. The mechanism of electrospray mass spectrometry. *Anal. Chem.* **65**(22), 972A–986A.

Kessler, M. J. (1986). Quantitation of radiolabeled molecules separated by high pressure liquid chromatography. *In* "Analytical and Chromatographic Techniques in Radiopharmaceutical Chemistry" (Donald M. Wieland, Michael C. Tobes, and Thomas J. Mangner, Eds.), pp. 149–170, Springer-Verlag, New York.

Klunder, G. L., Andrews, J. E. Jr., Grant, P. M., Andersen, B. D., and Russo, R. E. (1997). Analysis of fission products using capillary electrophoresis with on-line radioactivity detection. *Anal. Chem.* **69**(15), 2988–2993.

Kuo, B.-S., Mathews, B. S., Stuhler, J. D., Carrel, B. K., Ho, J., and Rose, J. Q. (1995). Disposition and cardioselectivity of MDL 74,405, a vitamin E-like free radical scavenger, in rats and dogs after intravenous infusion. *Drug Metab. Dispos.* **23**, 757–764.

Kumar, G. N., Jayanti, V., Lee, R. D., Whittern, D. N., Uchic, J., Thomas, S., Johnson, P., Grabowski, B., Sham, H., Betebenner, D., Kempf, D. J., and Denissen, J. F. (1999). In vitro metabolism of the HIV-1 protease inhibitor ABT-378: species comparison and metabolite identification. *Drug. Metab. Dispos.* **27**(1), 86–91.

Kusche, M. and Lindahl, U. (1990). Biosynthesis of heparin. *J. Biol. Chem.* **265**, 15403–15409.

Laethem, R. M., Balazy, M., and Koop, D. R. (1996). Epoxidation of C_{18} unsaturated fatty acids by cytochromes P4502C2 and P4502CAA. *Drug. Metab. Dispos.* **24**, 664–668.

Lambert, J. B., Shurvell, H. F., Lightner, D. A., and Cooks, R. G. (1998). "Organic Structural Spectroscopy." Prentice Hall, NJ.

L'Annunziata, M. F. (1970). "Soil-Plant Relationships and Spectroscopic Properties of Inositol Stereoisomers: The Identification of D-Chiro- and Muco-inositol in a Desert Soil-Plant System," Ph.D. Dissertation, University of Arizona, Tucson.

L'Annunziata, M. F. (1971). Birth of a unique parent-daughter relation: secular equilibrium. *J. Chem. Educ.* **48**, 700–703.

L'Annunziata, M. F. (1979). "Radiotracers in Agricultural Chemistry." Academic Press, New York and London, pp. 378–397.

L'Annunziata, M. F. (1984). Reaction mechanisms and pathways in biosynthesis. *In* "Isotopes and Radiation in Agricultural Sciences," Vol. 2, pp. 105–182. Academic Press, London.

L'Annunziata, M. F. (1987). "Radionuclide Tracers, Their Detection and Measurement." Academic Press, New York and London.

L'Annunziata, M. F. and Fuller, W. H. (1968). The chelation and movement of ^{89}Sr-^{90}Sr(^{90}Y) in a calcareous soil. *Soil Sci.* **105**, 311–319.

L'Annunziata, M. F. and Fuller, W. H. (1971a). Soil and plant relationships of inositol phosphate stereoisomers; the identification of D-Chiro- and muco-inositol phosphates in a desert soil and plant system. *Soil Sci. Soc. Amer. Proc.* **35**(4), 587–595.

L'Annunziata, M. F. and Fuller, W. H. (1971b). Nuclear magnetic resonance spectra of acetate derivatives of soil and plant inositol phosphates. *Soil Sci. Soc. Amer. Proc.* **35**(4), 655–658.

L'Annunziata, M. F. and Fuller, W. H. (1976). Evaluation of the mass spectral analysis of soil inositol, inositol phosphates, and related compounds. *Soil Sci. Soc. Amer. J.* **40**(5), 672–678.

L'Annunziata, M. F. and Nellis, S. W. (2001a). Flow scintillation analyzer (FSA) interfaced with the HPLC and nuclear magnetic resonance (NMR) spectrometer. A state-of-the-art application of the Packard radiomatic FSA. *FSA Application Note.* **FSA-004**, pp. 7. PerkinElmer Life and Analytical Sciences, Boston.

L'Annunziata, M. F. and Nellis, S. W. (2001b). Metabolism studies with on-line HPLC and mass spectrometry (MS) interfaced with the flow scintillation analyzer (FSA). *FSA Application Note.* **FSA-005**, pp. 7. PerkinElmer Life and Analytical Sciences, Boston.

Li, X.-Y., Astrom, A., Duell, E. A., Qin, L., Griffiths, C. E. M., and Voorhees, J. J. (1995). Retinoic acid antagonizes basal as well as coal tar and glucocortecoid-induced cytochrome P4501A1 expression in human skin. *Carcinogenesis* **16**, 519–524.

Lochny, M., Ullrich, W., and Wenzel, U. (1998). Simple on-line monitoring of α- and β-emitters by solid scintillation counting. *J. Alloys Compd.* **271–273**, 31–37.

Lyon, M., Deakin, J. A., and Gallagher, J. T. (1994). Liver heparan sulfate structure. *J. Biol. Chem.* **269**, 11208–11215.

Maggs, J. L., Bishop, L. P. D., Edwards, G., O'Neill, P. M., Ward, S. A., Winstanley, P. A., and Park, B. K. (2000). Biliary metabolites of β-artemether in rats: biotransformations of a antimalarial endoperoxide. *Drug Metab. Dispos.* **28**(2), 209–217.

Mannens, G. S. J., Snel, C. A. W., Hendrickx, J., Verhaeghe, T., Le Jeune, L., Bode, W., Van Beijsterveldt, L., Lavrijsen, K., Leempoels, J., Van Osselaer, N., Van Peer, A., and Meuldermans, W. (2002). The metabolism and excretion of galantamine in rats, dogs, and humans. *Drug Metab. Dispos.* **30**(5), 553–563.

Maurizis, J.-C., Rapp, M., Azim, E. M., Gaudreault, R. C., Veyre, A., and Madelmont, J.-C. (1998). Disposition and metabolism of a novel antineoplastic agent, 4-*tert*-butyl-[3-(chloroethyl)ureido]benzene, in mice. *Drug Metab. Dispos.* **26**(2), 146–151.

Mays, D. C., Hecht, S. G., Unger, S. E., Pacula, C. M., Climie, J. M., Sharp, D. E., and Gerber, N. (1987). Disposition of 8-methoxypsoralen in the rat. *Drug Metab. Dispos.* **15**, 318–328.

Mensah-Nyagan, A. G., Do-Rego, J.-L., Feuilloley, M., Marcual, A., Lange, C., Pelletier, G., and Vaudry, H. (1996). *In vivo* and *in vitro* evidence for the biosynthesis of testosterone in the telencephalon of the female frog. *J. Neurochem.* **67**, 413–422.

Moghaddam, M. F., Brown, A., Budevska, B. O., Lam, Z., Payne, W. G. (2001). Biotransformation, excretion kinetics, and tissue distribution of an N-pyrrolo[1,2-C]imidazolylphenyl sulfonamide herbicide in rats. *Drug Metab. Dispos.* **29**(8), 1162–1170.

Morgan, R. O., Chang, J. P., and Catt, K. J. (1987). Novel aspects of gonadotropin-releasing hormone action on inositol polyphosphate metabolism in cultured pituitary gonadotrophs. *J. Biol. Chem.* **262**, 1166–1171.

Niiyama, S., Happle, R., and Hoffmann, R. (2001). The feasibility of quantitative analysis of androgen metabolism by use of single dermal papillae from human hair follicles. *Exp. Dermatol.* **10**, 124–127.

Nguyen, C., Todorovic, C., Robin, C., Christophe, A., and Guckert, A. (1999). Continuous monitoring of rhizosphere respiration after labeling of plant shoots with $^{14}CO_2$. *Plant Soil* **212**(2), 191–201.

Nolan, R. D. and Lapetina, E. G. (1991). The production of phosphatidylinositol triphosphate is stimulated by thrombin in human platelets. *Biochem. Biophys. Res. Commun.* **174**, 524–528.

Noort, D., Hulst, A. G., de Jong, L. P. A., and Benschop, H. P. (1999). Alkylation of human serum albumin by sulfur mustard in vitro and in vivo: mass spectrometric analysis of a cysteine adduct as a sensitive biomarker of exposure. *Chem. Res. Toxicol.* **12**, 715–721.

Pageaux, J.-F., Bechoua, S., Bonnot, G., Fayard, J.-M., Cohen, H., Lagarde, M., and Laugier, C. (1996). Biogenesis and metabolic fate of docosahexaenoic and arachidonic acids in rat uterine stromal cells in culture. *Arch. Biochem. Biophys.* **327**, 142–150.

Parvez, H., Reich, A., Lucas-Reich, S., and Parvez, S. (1988). Flow through radioactivity detection in HPLC. *In* "Progress in HPLC," Vol. 3, VSP, Utrecht, The Netherlands.

Patrick, J. E., Kosoglou, T., Stauber, K. L., Alton, K. B., Maxwell, S. E., Zhu, Y., Statkevich, P., Iannucci, R., Chowdhury, S., Affrime, M., and Cayen, M. N. (2002). Disposition of the selective cholesterol absorption inhibitor ezetimibe in health male subjects. *Drug Metab. Dispos.* **30**(4), 430–437.

Paulson, S. K., Hribar, J. D., Liu, N. W. K., Hajdu, E., Bible, R. H. Jr., Piergies, A., and Karim, A. (2000). Metabolism and excretion of [^{14}C]celecoxib in healthy male volunteers. *Drug Metab. Dispos.* **28**(3), 308–314.

Pentoney, S. L. Jr., Quint, J. F., and Zare, R. N. (1989). On-line radioisotope detection for capillary electrophoresis. *Anal. Chem.* **61**, 1642–1647.

Pentoney, S. L. Jr., Zare, R. N., and Quint, J. F. (1990). *In* "Analytical Biochemistry: Capillary Electrophoresis and Chromatography" (C. Horvath, J. G. Nikelly, Eds.), ACS Symposium Series 434, pp. 60–89. American Chemical Society, Washington, DC

Piltingsrud, H. V. and Stencel, J. R. (1972). Determination of yttrium-90 and strontium-90 in samples by use of liquid scintillation beta spectroscopy. *Health Phys.* **23**, 121–122.

Pluim, D., Maliepaard, M., van Waardenburg, R. C. A. M., Beijnen, J. H., and Schellens, J. H. M. (1999). *Anal. Biochem.* **275**(1), 30–38.

Poirier, M. J., Glajch, J. L., and Barry, E. F. (1995). "Wintergreen Conference on Chromatography," 17th International Symposium on Capillary Chromatography and Electrophoresis, May 7–11, 1995, pp. 58–59, Wintergreen, VA.

Pritchett, J. J., Kuester, R. K., and Sipes, I. G. (2002). Metabolism of bisphenol A in primary cultured hepatocytes from mice, rats, and humans. *Drug Metab. Dispos.* **30**(11), 1180–1185.

Rajagopal, S., Venkatachalam, T. K., Conway, T., and Diksic, M. (1992). Synthesis of ^{14}C-labelled α-methyl tyrosine. *Appl. Radiat. Isot.* **43**, 979–987.

Ramu, K., Lam, G. N., and Hughes, H. (2000). In vivo metabolism and mass balance of 4-[4-fluorophenoxy]benzaldehyde semicarbazone in rats. *Drug. Metab. Dispos.* **28**(10), 1153–1161.

Rapkin, E. (1993). β-particle detection in HPLC by flow-through monitoring vs. liquid scintillation counting. *J. Liq. Chromatogr.* **16**, 1769–1781.

Rapkin, E. and Packard, L. E. (1960). *In* "University of New Mexico Conference on Organic Scintillator Detectors," Albuquerque, New Mexico.

Reboul, S. H. (1994). Qualitative and quantitative analysis of alpha- and beta-emitting radionuclides by ion chromatography and on-line scintillation counting. *Diss. Abstr. Int. B* **55**, 895–896.

Reboul, S. H. and Fjeld, R. A. (1994). A rapid method for determination of beta-emitting radionuclides. *Radioact. Radiochem.* **5**(3), 42–49.

Reboul, S. H. and Fjeld, R. A. (1995). Potential effects of surface water components in actinide determinations conducted by ion chromatography. *Health Phys.* **68**, 585–589.

Reeve, D. R. and Crozier, A. (1977). Radioactivity monitor for high-performance liquid chromatography. *J. Chromatogr.* **137**, 271–282.

Reich, A. R., Lucas-Reich, S., and Parvez, H. (1988). Radioactive flow detectors: history and theory. pp. 1–10, *In* "Progress in HPLC," Vol. 3. VSP, Utrecht, The Netherlands.

Ribbes, G., Cane, A., Planat, V., Breton, M., Chap, H., Bereziat, G., Record, M., and Colard, O. (1996). Transacylase-mediated alkylacyl-GPC synthesis and its hydrolysis by phospholipase D occur in separate cell compartments in the human neutrophil. *J. Cell. Biochem.* **62**, 56–68.

Riska, P., Lamson, M., Macgregor, T., Sabo, J., Hattox, S., Pav, J., and Keirns, J. (1999). Disposition and biotransformation of the antiretroviral drug nevirapine in humans. *Drug Metab. Dispos.* **27**(8), 895–901.

Rubiera, C., Lazo, P. S., and Shears, S. B. (1990). Polarized subcellular distribution of the 1-, 4- and 5-phosphatase activities that metabolize inositol 1,4,5-triphosphate in intestinal epithelial cells. *Biochem. J.* **269**, 353–358.

Sabourin, P. J., Bechtold, W. E., and Henderson, R. F. (1988). A high pressure liquid chromatographic method for the separation and quantitation of water-soluble radiolabeled benzene metabolites. *Anal. Biochem.* **170**, 316–327.

Sasakawa, N., Nakaki, T., and Kato, R. (1990). Rapid increase in inositol pentakisphosphate accumulation by nicotine in cultured adrenal chromaffin cells. *FEBS Lett.* **261**, 378–380.

Scarfe, G. B., Lindon, J. C., Nicholson, J. K., Martin, P., Wright, B., Taylor, S., Lenz, E., and Wilson, I. D. (2000). Investigation of the metabolism of $^{14}C/^{13}C$-practolol in rat using directly coupled radio-HPLC-NMR-MS. *Xenobiotica* **30**(7), 717–729.

Schockcor, J. P., Unger, S. E., Wilson, I. D., Foxall, P. J. D., Nicholson, J. K., and Lindon, J. C. (1996). Combined HPLC, NMR spectroscopy, and ion-trap mass spectrometry with application to the detection and characterization of xenobiotic and endogenous metabolites in human urine. *Anal. Chem.* **68**(24), 4431–4435.

Schram, E. and Lombaert, R. (1960). Continuous estimation of carbon-14 in chromatographic effluents by means of anthracene powders. *Archs. Int. Physiol. Biochim.* **68**, 845–846.

Schram, E. and Lombaert, R. (1961). Microvalve and connector for automatic column chromatography. *Anal. Chem.* **33**, 1134–1135.

Schultz, G. A. and Alexander, J. N., IV. (1998). Incorporation of a RAM cell in a microcolumn for on-column detection of radiolabeled molecules for LC-RAM-ESI-MS. *J. Microcolumn Separations* **10**(5), 431–437.

Schwahn, M., Schupke, H., Gasparic, A., Krone, D., Peter, G., Hempel, R., Kronbach, T., Locher, M., Jahn, W., and Engel, J. (2000). Disposition and metabolism of cetrorelix, a potent luteinizing hormone-releasing hormone antagonist, in rats and dogs. *Drug Metab. Dispos.* **28**(1), 10–20.

Seidegård, J. , Grönquist, L., and Ginnarsson, P. O. (1990). Metabolism of a novel nitrosurea, tauromustine, in the rat. *Biochem. Pharmacol.* **39**, 1431–1436.

Sekiya, K., Tezuka, Y., Tanaka, K., Prasain, J. K., Namba, T., Katayama, K., Koizumi, T., Maeda, M., Kondo, T., and Kadota, S. (2000). Distribution, metabolism and excretion of butylidenephthalide of *Ligustici chuanxiong* rhizoma in hairless mouse after dermal application. *J. Ethnopharmacol.* **71**, 401–409.

Shet, M. S., Fisher, C. W., Holmans, P. L., and Estabrook, R. W. (1996). The omega-hydroxylation of lauric acid: oxidation of 12-hydroxylauric acid to dodecanedioic acid by a purified recombinant fusion protein containing P450 4A1 and NADPH-P450 reductase. *Arch. Biochem. Biophys.* **330**, 199–208.

Shirley, M. A. and Murphy, R. C. (1990). Metabolism of leokotriene B_4 in isolated rat hepatocytes. *J. Biol. Chem.* **265**, 16288–16295.

Shirley, M. A., Bennani, Y. L., Boehm, M. F., Breau, A. P., Pathirana, C., and Ulm, E. (1996). Oxidative and reductive metabolism of 9-cis-retinoic acid in the rat: identification of 13,14-dihydro-9-cis-retinoic acid and its taurine conjugate. *Drug Metab. Dispos.* **24**, 232–237.

Sigg, R. A., McCarty, J. E., Livingston, R. R., and Sanders, M. A. (1994). Real-time aqueous tritium monitor using liquid scintillation counting. *Nucl. Instrum. Methods Phys. Res., Sect. A* **353**, 494–498.

Silva-Elipe, M. V. (2000). Merck & Co., Inc., Drug Metabolism, Rahway, NJ 07065 (Personal Communication, August 22, 2000).

Singh, R., Chen, I.-W., Jin, L., Silva, M. V., Arison, B. H., Lin, J. H., and Wong, B. K. (2001). Pharmacokinetics and metabolism of a *Ras* farnesyl transferase inhibitor in rats and dogs: in vitro-in vivo correlation. *Drug. Metab. Dispos.* 29(12), 1578–1587.

Slatter, J. G., Schaaf, L. J., Sams, J. P., Feenstra, K. L., Johnson, M. G. *et al.* (2000). Pharmacokinetics, metabolism, and excretion of irinotecan (CPT-11) following I.V. infusion of [^{14}C]CPT-11 in cancer patients. *Drug Metab. Dispos.* 28(4), 423–433.

Smith, K. D. and Lutz, C. T. (1996). Peptide-dependent expression of HLA-B7 on antigen processing-deficient T2 cells. *J. Immunol.* 156, 3755–3764.

Smith, R. M., Chienthavorn, O., Wilson, I. D., Wright, B., and Taylor, S. D. (1999). Superheated heavy water as the effluent for HPLC-NMR and HPLC-NMR-MS of model drugs. *Anal. Chem.* 71(20), 4493–4497.

Smith, C. J., Wilson, I. D., Abou-Shakra, F., Payne, R., Grisedale, H., Long, A., Roberts, D., and Malone, M. (2002). Analysis of a [^{14}C]-labelled platinum anticancer compound in dosing formulations and urine using a combination of HPLC-ICPMS and flow scintillation counting. *Chromatographia Suppl.* 55, S-151-S-155.

Sohlenius-Sternbeck, A.-K., von Euler Chelpin, H., Orzechowski, A., and Halldin, M. M. (2000). Metabolism of sameridine to monocarboxylated products by hepatocytes isolated from the male rat. *Drug Metab. Dispos.* 28(6), 695–700.

Steinberg, D. (1958). Proceedings of the Symposium on the Advances in Tracer Applications of Tritium, New York.

Steinberg, D. (1960). A new approach to radioassay of aqueous solutions in the liquid scintillation spectrometer. *Anal. Biochem.* 1, 23–29.

Sweeny, D. J., Lynch, G., Bidgood, A. M., Lew, W., Wang, K.-Y., and Cundy, K. C. (2000). Metabolism of the influenza neuraminidase inhibitor prodrug oseltamivir in the rat. *Drug. Metab. Dispos.* 28(7), 737–741.

Takatsuka, J., Takahashi, N., and De Luca, L. M. (1996). Retinoic acid metabolism and inhibition of cell proliferation: an unexpected liaison. *Cancer Res.* 56, 675–678.

Takei, M., Kida, T., and Suzuki, K. (2001). Sensitive measurement of positron emitters eluted from HPLC. *Appl. Radiat. Isot.* 55, 229–234.

Tamby, J. P., Reinaud, P., and Charpigny, G. (1996). Preferential esterification of arachidonic acid into ethanolamine phospholipids in epithelial cells from ovine endometrium. *J. Reprod. Fertil.* 107, 23–30.

Tan, H., DeVol, T. A., and Fjeld, R. A. (2000). Digital alpha/beta pulse shape discrimination of CsI:Tl for on-line measurement of aqueous radioactivity. *IEEE Trans. Nucl. Sci.* 47(4), 1516–1521.

Thomson, J. (1993). Scintillation Counting Medium and Process, European Patent Application No. 93.200718.0, March 1993.

Thomson, J. (1994). The advent of safer flow scintillation cocktails. *In* "Liquid Scintillation Spectrometry 1994," (Gordon Cook, Douglas D. Harkness, Angus B. Mackenzie, Brian E. Miller and E. Marian Scott, Eds.), pp. 257–260. Radiocarbon, Tucson, Arizona, USA.

Thomson, J. (1997). Safer flow cocktails. Counting Solutions, CS-006, pp. 6. PerkinElmer Life and Analytical Sciences, Boston.

Upadhyaya, P., Zimmerman, C. L., and Hecht, S. S. (2002). Metabolism and pharmacokinetics of N'-nitrosonornicotine in the patas monkey. *Drug Metab. Dispos.* 30(10), 1115–1122.

Usuda, S. and Abe, H. (1992). Flow monitor for actinide solutions by simultaneous α and $\beta(\gamma)$ counting using a CsI(Tl) scintillator. *Nucl. Instrum. Methods Phys. Res., Sect. A* 321, 242–246.

Usuda, S., Mihara, A., and Abe, H. (1992). Rise time spectra of α and β γ rays from solid and solution sources with several solid scintillators. *Nucl. Instrum. Methods Phys. Res., Sect. A* 321, 247–253.

Van Kuilenburg, A. B. P., Van Lenthe, H., and Van Gennip, A. H. (1999). Radiochemical assay for determination of dihydropyrimidinase activity using reversed-phase high-performance liquid chromatography. *J. Chromatogr. B* 729, 307–314.

Vickers, A. E. M., Zollinger, M., Dannecker, R., Tynes, R., Heitz, F., and Fischer, V. (2001). In vitro metabolism of tegaserod in human liver and intestine: assessment of drug interactions. *Drug. Metab. Dispos.* 29(10), 1269–1276.

Vickers, S., Duncan, C. A., Slaughter, D. E., Arison, B. H., Greber, T., Olah, T. V., and Vyas, K. P. (1998). Metabolism of MK-499, a class III antiarrhythmic agent, in rats and dogs. *Drug. Metab. Dispos.* **26**(5), 388–395.

Wasyl, M. S. and Nellis, S. W. (1996). Optimizing performance in radioactive HPLC detection using scintillation counting technology. *In* "Liquid Scintillation Spectrometry 1994" (G. T. Cook, D. D. Harkness, A. B. MacKenzie, B. F. Miller and E. M. Scott, Eds.), pp. 357–360. Radiocarbon, The University of Arizona, Tucson.

Wells, D. A. and Digenis, G. A. (1988). Disposition and metabolism of double-labeled [^{3}H and ^{14}C] N-methyl-2-pyrrolidinone in the rat. *Drug Metab. Dispos.* **16**, 243–249.

Wenzel, U., Ullrich, W., and Lochny, M. (1999). The WUW ML bundle detector. A flow through detector for α-emitters. *Nucl. Instrum. Methods Phys. Res., Sect. A* **421**, 567–575.

Westerberg, G., Lundqvist, H., Kilar, F., and Langstrom, B. (1993). β^+-selective radiodetector for capillary electrophoresis. *J. Chromatogr.* **645**(2), 319–325.

Wierczinski, B., Gregorich, K. E., Kadkhodayan, B., Lee, D. M., Beauvais, L. G., Hendricks, M. B., Kacher, C. D., Lane, M. R., Keeney-Shaughnessy, A. A., Stoyer, N. J., Strellis, D. A., Sylwester, E. R., Wilk, P. A., Hoffman, D. C., Malmbeck, R., Skarnemark, G., Alstad, J., Omtvedt, J. P., Eberhardt, K., Mendel, M., Nähler, A., and Trautmann, N. (2001). First chemical on-line separation and detection of a subsecond α-decaying nuclide, ^{224}Pa. *J. Radioanal. Nucl. Chem.* **247**(1), 57–60.

Wormhoudt, L. W., Hissink, A. M., Commandeur, J. N. M., van Bladeren, P. J., and Vermeulen, N. P. E. (1998). Disposition of 1,2-[^{14}C]dibromoethane in male wistar rats. *Drug Metab. Dispos.* **26**(5), 437–447.

Wouters-Ballman, P., Donnay, I., Devleeschouwer, N., and Verstegen, J. (1995). Iodination of mouse EGF with chloramine T at 4°C: characterization of the iodinated peptide and comparison with other labelling methods. *J. Receptor Res.* **15**, 737–746.

Wu, W.-N. and Mutter, M. S. (1995). Biotransformation of linogliride, a hypoglycemic agent in laboratory animals and humans. *J. Pharmac. Biomed. Anal.* **13**, 857–867.

Wynalda, M. A., Hauer, M. J., and Wienkers, L. C. (2000). Oxidation of the novel oxazolidinone antibiotic linezolid in human liver microsomes. *Drug Metab. Dispos.* **28**(9) 1014–1017.

Xie, K. C. A. and Plunkett, W. (1966). Deoxynucleotide pool depletion and sustained inhibition of ribonucleotide reductase and DNA synthesis after treatment of human lymphoblastoid cells with 2-chloro-9-(2-deoxy-2-fluoro-β-D-arabinofuranosyl)adenine. *Cancer Res.* **56**, 3030–3037.

Yuan, J. J., Yang, D.-C., Zhang, J. Y., Bible, R., Jr., Karim, A., and Findlay, J. W. A. (2002). Disposition of a specific cyclooxygenase-2 inhibitor, valdecoxib, in human. *Drug Metab. Dispos.* **30**(9), 1013–1021.

Yeoh, C.-G., Schreck, C. B., Fitzpatrick, M. S., and Feist, G. W. (1996). *In vivo* steroid metabolism in embryonic and newly hatched steelhead trout (*Oncorhynchus mykiss*). *Gen. Comp. Endocrinol.* **1996**, 197–209.

Zalko, D., Perdu-Durand, E., Debrauwer, L., Bec-Ferte, M.-P., and Tulliez, J. (1998). Comparative metabolism of clenbuterol by rat and bovine liver microsomes and slices. *Drug. Metab. Dispos.* **26**(1), 28–35.

Zeldin, D. C., Moomaw, C. R., Jesse, N., Tomer, K. B., Beetham, F., Hammock, B. D., and Wu, S. (1996). Biochemical characterization of the human liver cytochrome P450 arachidonic acid epoxygenase pathway. *Arch. Biochem. Biophys.* **330**, 87–96.

Zhang, L. and Hu, X.-Q. (1995). Serotonin stimulates rapid increase of inositol 1,4,5-triphosphate in ovine uterine artery: correlation with contractile state. *J. Pharmacol. Exp. Ther.* **275**, 576–583.

Zhang, L., Pearce, W. J., and Longo, L. D. (1995). Noradrenaline-mediated contractions of ovine uterine artery: role of inositol 1,4,5-triphosphate. *Eur. J. Pharmacol.* **289**, 375–382.

Zheng, Z., Fang, J.-L., and Lazarus, P. (2002). Glucuronidation: an important mechanism for detoxification of benzo[a]pyrene metabolites in aerodigestive tract tissues. *Drug. Metab. Dispos.* **30**(4), 397–403.

13
RADIONUCLIDE IMAGING

LORAINE V. UPHAM
Myriad Proteomics, Salt Lake City, Utah 84116

DAVID F. ENGLERT
BioConsulting, West Hartford, Connecticut 06107

I. INTRODUCTION

Radionuclide analysis of samples separated in two dimensions can be done by traditional counting methods such as oxidation or solubilization and liquid scintillation counting. The drawback of these traditional methods is that the sample components which were previously separated spatially by thin layer chromatography, animal tissue distribution, or electrophoresis are then rehomogenized before counting (Rogers, 1969). Measuring radionuclides by an imaging method maintains the spatial location in the X–Y plane in addition to providing a measure of the intensity of the activity.

Receptors and RNA transcripts can be measured *in situ* to determine not only their quantity but also their biological location. Organic compounds can be measured in their original position along a gradient of a thin layer chromatography plate, without loss of material that can occur while scraping. Quantity and relative molecular weights of DNA and RNA can be measured directly in a polyacrylamide gel or blot, without destruction of the sample. Although several radionuclide imaging instruments exist for the three-dimensional detection of positron emitters within live organisms, this chapter will be limited in scope to the one- and two-dimensional analysis of radionuclides within flat samples. Readers should refer to Phelps *et al.* (1986) for a comprehensive review of positron emission tomography and Autoradiography.

The following is a review of three methods of radionuclide imaging which are most valuable for the detection of ionizing radiation within flat samples: film autoradiography, storage phosphor screen imaging, and electronic autoradiography using the Microchannel Array Detector (MICAD). A fourth method that utilizes a charge-coupled device (CCD) camera for quantitative image analysis has been shown to be most useful for measuring radionuclides in microplates for high throughput screening. Each method will be discussed in terms of the technology, performance, quantification, and advantages and disadvantages. Techniques of optimizing results will be presented, including several applications of each radionuclide imaging method.

II. FILM AUTORADIOGRAPHY

Film autoradiography is a method of detection of beta particles that is based on the conversion of silver ions to reduced silver atoms within a film emulsion. The latent image is revealed by subsequent development of the film resulting in the reduction of all of the silver atoms of an entire silver halide crystal grain to metallic silver, which produces an autoradiographic image of the radioactivity on the film. Only a single hit from a beta particle or gamma ray is sufficient to convert a grain to a developable state, so the local blackening of film can be directly proportional to the amount of radiation that hits the film (Pelc, 1972). However, until the mid-1970s, film auto-radiography was still considered only qualitative in nature. With the advent of electronic digitizing equipment, such as CCD cameras, light and laser densitometers, and flatbed scanners, it became possible to convert the qualitative film image to a digital image based on optical density of the film (Cross, 1974). The following is an analysis of the use of film for quantitative radionuclide imaging.

A. Micro–Macro Autoradiography

The distinction of "micro" and "macro" autoradiography is based on the resolution requirements of the sample. "Microautoradiography" may be needed if the radionuclides that require visualization need to be localized at cellular or subcellular levels within cells or tissue sections immobilized on microscope slides. In this case, the use of film is limited by the lack of proximity of the silver emulsion layer to the radionuclides of the sample. Two methods are commonly used for microautoradiography. One method is direct emulsion dipping of slides containing covalently bound radioligands or ligands irreversibly bound to receptors. This method requires paraformalde-hyde fixation of the sample, coating with an emulsion, and development of the actual sample such that the silver grains are in direct contact with bound radiolabeled receptor. Quantification is by visual counting of the blackened silver grains. Another method of microautoradiography is the "coverslip technique" in which emulsion-coated coverslips are tightly apposed to sections to generate autoradiograms. This method is ideal for ligands that are not irreversibly or covalently bound such that they would lose their ligand–receptor integrity by the emulsion dipping method. The coverslip emulsion method involves coating a detergent-washed coverslip in the emulsion, drying and attaching to the radiolabeled tissue sections. After exposure, coverslips are then detached, developed, and examined microscopically in dark field to count silver grains depicting the position of the radiolabeled receptors. Detailed protocols for analysis and quantification of radionuclides by microautoradiography can be found in Sharif and Eglen (1993). The remainder of this section will be devoted to the discussion of radionuclide analysis using film for macroautoradiography, the detection and quantification of radionuclides localized in larger anatomical structures (visible with the naked eye), and samples that are separated sufficiently for traditional autoradiography.

B. Performance of Film Autoradiography Methods

The discussion of the performance of film autoradiography for the detection and quantification of radionuclides will be in practical terms that can be compared with newer methods of quantitative imaging. For details regarding history, chemistry, and physics of film autoradiography see Rogers (1969), Gahan (1972), and L'Annunziata (1987).

I. Sensitivity

Direct autoradiography with film is inherently limited in sensitivity by the inefficient transfer of emission energy of radionuclides to the film. Although film has been reported to detect as little as $0.02\,\text{DPM/mm}^2$, the exposure time to achieve this minimum level can take months, depending on the isotope. For practical purposes of this chapter, discussion of sensitivity with respect to radionuclide imaging will be in terms of the speed at which comparable levels of activity can be detected and the efficiency of radioisotopic detection, instead of minimum detectable levels of activity.

Since the early 1980s, film types and compositions have proliferated. The largest manufacturers of film for radionuclide analysis are Kodak (Rochester, NY), Fuji (Tokyo, Japan), Dupont (Wilmington, DE), Ilford (Essex, England), and Agfa Gevaert (Brussels, Belgium). Laskey and Mills (1977) described the most sensitive film for radioisotope detection as Kodak Xomat R, which was able to achieve $5–6\,\text{DPM/mm}^2$ in 24 h. This film was replaced by Kodak Xomat RP, Fuji RX, Dupont Cronex, and Agfa Gevaert Curix RP1, which offered only 50–75% of the efficiency, but the advantages of a longer shelf life, smaller grain size, and higher image quality. In 1980s Kodak developed X-Omat AR film, which is coated with a thick layer of granular silver halide emulsion on two sides of a support. The most recent advances in film include the Kodak BioMax MR and BioMax MS films, which are made with Kodak's "Tabular" or "T" Grain emulsion technology. BioMax MR film is coated only on one side of a clear support, which cuts down on the background build-up that can accumulate with double emulsion films. The BioMax MR film demonstrates a two-fold improvement in sensitivity compared to X-Omat AR (Steinfeld *et al.*, 1994), and provides the best resolution. The BioMax MS film is coated on both sides to provide 4–8 fold better sensitivity when used in conjunction with intensifying screens. Refer to Section II.C.1.a. for information about optimizing performance with film using intensifying screens.

The most difficult challenge for tritium detection by any method remains the high probability of self-absorption of the sample due to the low energy of the predominant beta emission. For better sensitivity to ^3H, special films such as Amersham Ultrofilm have been developed that lack the antiscratch layer applied to the surface of most other films, making them more sensitive to the low energy emissions of tritium.

Film autoradiography is best suited in terms of sensitivity for the detection of isotopes such as ^{14}C and ^{35}S, which emit beta particles with energies of 156 and 167 keV, respectively. Although detection of these isotopes can be limited by the amount of self-absorption of the tissues or sample matrix, the particles emitted will have a cumulative effect on the silver ions of the film. Film is limited in sensitivity to higher energy beta particles of ^{32}P and X-rays, such as those emitted from ^{125}I, ^{131}I, ^{51}Cr, and ^{75}Se, because they pass right through the film whereby only a small proportion are detected by the film.

2. Resolution

In photography, resolution is defined in terms of the distance that must separate two objects before they can be distinguished as separate objects. In film autoradiography, several factors govern the resolution that can be achieved for radionuclide analysis. Factors associated with the source of the activity that affect resolution include:

1. Choice of isotope. Lower energy isotopes that emit particles with shorter path lengths provide better resolution than higher energy isotopes that travel further in the emulsion.
2. Distance between the source and the film emulsion. Increased distance between sample and emulsion decreases resolution significantly.
3. The thickness of the sample source. Samples that are thicker have some particles that are at a greater distance from the emulsion and exhibit less resolution than thinner samples.

Factors of the film that affect resolution include:

1. Thickness of the emulsion. Thicker emulsions improve sensitivity, yet decrease resolution of the film.

2. Size of the silver halide crystals. Smaller crystals result in more precise localization of the ionizing radiation.
3. Length of exposure. Longer exposures result in double hits to crystals and the probability of a hit from an ionizing particle further from the grain is increased, decreasing the resolution.
4. Sensitivity of the emulsion. Less sensitive emulsions require longer exposures and therefore also decrease resolution.

The most significant of these factors is the distance between the sample and the film. The sample should be as close as possible to the film to achieve the highest resolution. The second and third most important factors are the energy of the isotope, which should be as low as possible, and the thickness of the sample, which should be as thin as possible.

Resolution can be defined as the distance from a point source at which the grain density falls to one half of that directly over the source (Rogers, 1969). Although resolution varies significantly between experiments, for a sample that is $1\,\mu$m thick, the best resolutions that one could expect based on isotope and emulsions and light microscopy compared to film are listed in Table 13.1.

For more practical comparisons of film autoradiography to filmless autoradiography methods described later in this chapter, the resolution, R, is described as the distance that would be required between two lines such that the valley between the resulting two peaks is less than or equal to half the peak maximum. The resolution criteria for this is based on the contrast transfer function (CTF), which describes the wave pattern generated by a series of parallel lines and spaces of equal width.

$$\mathrm{CTF} = \frac{\text{average maximum} - \text{average minimum}}{\text{average maximum} + \text{average minimum}} \qquad (13.1)$$

In order to achieve this criterion the CTF must be at least 33% (Hecht and Zajac, 1974). This is a more practical specification for resolution than the point spread function above. Using this criteria, resolution values from Kodak BioMax MR are $300\,\mu$m with ^{35}S and $350\,\mu$m with ^{32}P. These values

TABLE 13.1 Resolution of Film and Emulsions. The potential resolution of film and emulsion autoradiography methods are compared. Values are based on 0.1 μm distance of sample to emulsion, 3 μm distance of sample to film, and thinnest possible emulsions [data from Rogers (1969)].

Isotope	Emulsions (μm)	Film (μm)
^{3}H/^{125}I	0.5–1.0	2.8–5.7
^{35}S/^{14}C	2–5	11–28
^{32}P	5–10	28–56

correspond to 3.3 and 2.9 line pairs per mm, respectively, with a CTF = 33% (Steinfeld *et al.*, 1994).

3. Linear Dynamic Range

A film autoradiogram is a representation of the activity of a sample that can be quantified by measuring the optical density of the film as light passes through it. For any given exposure time there exists a threshold level of activity required to cause a blackening of the film and concomitant measure of optical density. Also for any given exposure, there is a direct relationship, for a limited range of activity, between the activity of a sample and the optical density of the film. The linear dynamic range for any film is between 1.5 and 3 logs of activity. The level of activity at which the film is overexposed or completely black is the upper limit of detection for any given exposure and represents saturation of the silver grains of the film in that area. Fluorography can be used to improve the sensitivity and linear dynamic range of film. Refer to Section II.C.1. regarding techniques for optimization of film. Figure 13.1 is a graphical representation of the limited linear dynamic range of film (Laskey and Mills, 1975).

C. Quantification Methods

A number of systems are commercially available for quantification of the optical density of film. These include light densitometers, the more expensive laser densitometers, very inexpensive flat bed scanners, and a number of video and CCD camera systems. Each of these methods is suitable for measuring the optical density of film that involves transmission of white light through the film and a method of light capture and digitization of data to form a quantitative digital image for software analysis (Orr, 1993).

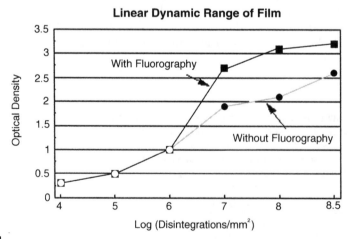

FIGURE 13.1 Graphical representation of the linear dynamic range of film with and without fluorography. Fluorography improves the linear dynamic range slightly [data from Lasky and Mills (1975)].

In order to quantify radionuclide activity of a film autoradiogram, standards must be incorporated into each exposure. The optical density of the film is a function of not only the activity of the sample, but also the exposure time, development procedure, and type of film. The use of standards can eliminate variability associated with these parameters, which exist as a function of this method of detection.

The following steps outline the procedure for quantification of radionuclides using film autoradiography:

1. Calibrate scanner or camera system to recognize total darkness and full brightness (background of the image) such that every pixel of the autoradiogram can be assigned a gray level using a reference table stored in computer memory.

2. Using standards of known activity, calibrate the system with the image of the radiolabeled standards. Quantify the activity of known standards, and enter the activity per mm^2 into the software of the system or a common spreadsheet.

3. Generate a graph of the optical density of the radioactivity of standards versus known standard values, or perform a linear regression to get the best-fit line and a correlation coefficient. If the coefficient of correlation is not >0.95, repeat the calibration to get a more appropriate linear relationship. One possible problem could be under- or overexposure of film, which will exhibit a nonlinear relationship due to the nonlinearity of the response of the film. A nonlinear sigmoidal curve fit is often needed when a large linear dynamic range of activity is included in the standards.

4. Analyze areas of the image by integrating areas of interest and assigning a value for the optical density.

5. Using instrument software or a spreadsheet, apply the relationship between activities of the standards and the optical density readings of samples to obtain quantitative values for areas of interest (Sharif and Eglen, 1993).

Each instrument has various methods of optimizing results; however, the main source of error and area for potential enhancement of results lies with the film. Therefore, the discussion of techniques of optimization of quantitative analysis of radionuclides with film autoradiography centers around methods of improving the capture of the image on film.

I. Techniques for Optimization

a. Intensifying screens

Intensifying screens are thin sheets of inorganic material that can be placed behind film in the exposure cassette to amplify the signal of a radiolabeled sample. Activity of the sample that passes through the film hits the intensifying screen, causing the screen to emit multiple photons of light, which then return through the film. Use of intensifying screens can enhance the detection of low levels of activity by increasing the sensitivity and enhancing the linear dynamic range similar to that which can be achieved with fluorography (see Fig. 13.1). Although the advantages of higher sensitivity and extended linear dynamic range are real, the trade-off is the error in quantification associated with reciprocity-law failure. Films that are exposed directly with no intensifying screens follow the law of reciprocity. The law states that the density of the image formed depends on the absorbed energy of the exposure, which is equal to the product of the exposure intensity and the exposure time. Films exposed with intensifying screens have a density response which varies with exposure time and intensity. This error can be corrected for by the graphical method described by Fujita (1982).

The latest development in the area of intensifying screens has been introduced by Eastman Kodak Company, New Haven, Connecticut, as the BioMax Transcreen system. The BioMax Transcreen HE is suitable for improving the efficiency of detection of isotopes with higher energy beta emissions such as ^{32}P and gamma emissions such as ^{125}I. The BioMax Transcreen LE is more appropriate for improving efficiency of detection of lower energy isotopes such as ^{14}C, ^{35}S, ^{45}Ca, ^{33}P, and even ^{3}H. These durable, hydrophobic, nonporous screens eliminate the need for a thick coating, which would block low energy isotope emissions, making it universally useful for low energy and high energy isotopes. With the Transcreen/film configuration, although about a three-fold loss of linear dynamic range can be anticipated, a three- to ten-fold gain in speed is appreciated compared to direct exposure to film (Vizard et al., 1996).

b. Fluorography

The detection of lower energy isotopes such as ^{3}H and ^{14}C may be enhanced by the use of an organic scintillator such as 2,5-diphenyloxazole (PPO), which converts the energy of a beta particle to visible light (Randerath, 1970; L'Annunziata, 1987). Sensitivity may be increased by a factor of 10–100 for ^{3}H and 5–10 for ^{14}C, depending on the type of film and temperature. In addition, exaggerated resolution results from the suppressed background with respect to the enhanced signal according to Laskey and Mills (1975). Although the advantage of higher sensitivity and exaggerated resolution with fluorography is real, the quantification by film optical density suffers from introduced errors similar to errors introduced by the use of intensifying screens. However, Laskey and Mills (1977) report a method of "preflashing" film and exposing at −70°C, which increases the background "fog" but re-establishes the linear relationship that exists between the activity of the sample and the optical density of the film after exposure, development, and fixation.

2. Advantages of Film Autoradiography

The film autoradiography method of imaging radionuclides still provides the best resolution possible for accurate localization of radiolabeled material, although digitization of the images degrades the resolution to some degree. This advantage is most noticeable with the use of low energy isotopes such as ^{3}H. The resolution possible with isotopes such as ^{32}P is limited by the high energy of the beta emission; therefore, the resolution advantage of the use of film over other methods is reduced.

Film provides a permanent, unalterable record of the sample for incorporation into laboratory notebooks. Unlike other imaging methods, involving capture in digital form and printing through the use of computers, film is a tangible direct representation of the sample. Until recently, most journals still required the original film for publication of an image.

Finally, although film prices vary greatly depending on sensitivity, packaging, size, and volume, and increase with the use of fluorography and intensifying screens, film autoradiography remains the lowest cost alternative for radionuclide imaging. Despite the cost of developing chemicals and the overhead of maintenance of a darkroom, the cost of using film autoradiography requires less initial investment than other methods. Note that this does not include the cost of the increased time to obtain quantitative data, which is difficult, at best, to measure. Also, the instrumentation for digitizing and quantifying by densitometry may be lower cost than other instrumentation for quantification of radionuclides.

3. Disadvantages of Film Autoradiography

Although the minimum detectable levels of activity are low for film autoradiography, the time necessary to achieve these levels are almost prohibitive. Film has a great sensitivity to photon emissions but lacks the efficiency of detection of ionizing radiation. Conversion of ionizing radiation to blackened silver halide crystals can take days, weeks and months of exposure time.

Another limitation of film autoradiography is the linear dynamic range with which radionuclides may be quantified. The response curve of film has a low end threshold at which it exhibits no response to low levels of activity and a saturation point at which additional beta and gamma emissions have no additional affect on the optical density of the blackened film. The linear dynamic range for which radionuclides can be quantified with film autoradiography is therefore limited to 1.5–3 orders of magnitude. This makes determination of the exposure time for any given sample difficult. One may be able to estimate the amount of activity in a sample, but exposure time necessary to obtain a linear representation of all parts of the sample, is difficult to estimate. Some samples may even require two exposures to obtain a linear representation by film autoradiography (Englert et al., 1993). The only way to optimize the exposure time is by trial and error, which is time consuming and can require numerous exposures of the same sample. Standards must be used to calculate what concentrations of activity are in the linear range of the film for any given exposure.

Film autoradiography provides only an analog representation of the activity of any given sample for qualitative analysis. Although the human eye is excellent at visualizing patterns and making a qualitative assessment of the distribution of the radioisotope within a sample using film, the quantification by visual inspection is inaccurate. In order to quantify the activity using film autoradiography, the film must be scanned with a densitometer, or a flat bed scanner, or digitized with a CCD camera. These instruments capture the data in the form of pixels based on the optical density of the film. Film exposure time, developing time, multiple exposures to obtain linear representations of the activity, scanning, and digitizing add up to hours and days of analysis to obtain quantitative data using film autoradiography.

Finally, reports of respiratory and skin effects in radiographers who process x-ray films have been documented. It is suggested that the use of gluteraldehyde as a hardening agent in the developer and synergistic affects with other laboratory chemicals could be the cause of some asthmatic symptoms. Other problems may be associated with poor dark room ventilation and lack of appropriate safe handling techniques (Hewitt, 1993). In addition, regulations for environmental effluent standards have been in question due to the fact that they are based on all forms of silver, the most toxic of which is the free ionic form of silver (Dufficy *et al.*, 1993).

III. STORAGE PHOSPHOR SCREEN IMAGING

Storage phosphor screen imaging was first commercialized by Fuji Photo Film Company as a method of providing long linear dynamic range images for medical x-ray imaging (Sonoda *et al.*, 1983). In the late 1980s and early 1990s, storage phosphor technology was adopted by those laboratories who could afford the technology as the method of choice for imaging radionuclides. Storage phosphor screen imaging began to replace film for a number of widely used applications in molecular biology, pharmacology, and receptor autoradiography.

A. Storage Phosphor Technology

The most critical components of the technology of storage phosphor screen imaging include the phosphor screen chemistry, a scanning mechanism, and light collection optics. Phosphor screens, also referred to as "imaging plates," are used to trap the energy of the radioisotope emissions. Phosphor screens are loaded into the storage phosphor screen scanner to be scanned with laser light to release the latent image. Fuji, Molecular Dynamics (now Amersham Biosciences) and Packard Instrument Company (now Perkin Elmer Life and Analytical Sciences) have developed scanning systems, each with their own scanning mechanism and light collection optics to capture the image. The following section is a discussion of the concept of storage phosphor screen imaging for localization and quantification of radionuclides.

I. Phosphor Screen Chemistry

Radiolabeled samples are exposed to phosphor screens, which store energy in the photostimulable crystals ($BaFBr : Eu^{2+}$) by the mechanism shown in Fig. 13.2. The energy of the radioisotope ionizes Eu^{2+} to Eu^{3+}, liberating electrons to the conduction band of the phosphor crystals. The electrons are then trapped in bromine vacancies, which are introduced during the manufacturing process, and form temporary "F centers". Exposure to a

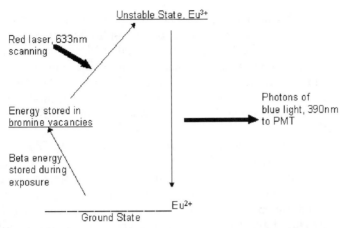

Unstable State, Eu³⁺

Red laser, 633nm
scanning

Energy stored in
bromine vacancies

Photons of
blue light, 390nm
to PMT

Beta energy
stored during
exposure

Eu²⁺
Ground State

FIGURE I3.2 **Schematic representation of the storage phosphor process.**

stimulating laser light of approximately 633 nm releases the trapped electrons from the bromine vacancies back to the conduction band of the crystals, converting Eu^{3+} back to Eu^{2+}, which releases photons at about 390 nm (Hamaoka, 1990). The light emitted from the storage phosphor screen is detected with a conventional high-quantum-efficiency photomultiplier tube (PMT) as described by Amemiya and Miyahara (1988). Another formulation of phosphor screen has been developed that uses different chemistry optimized for detection of luminescence. Details of this screen formulation can be found in Nguyen and Heffelfinger (1995).

2. Scanning Mechanisms and Light Collection Optics

Several systems have been developed and used to scan and create the quantitative images of the radiolabeled samples. Using the phosphor screen chemistry just described, Fuji first developed storage phosphor screen imaging systems for medical x-ray imaging in 1981 to adapt to the need for automation in the hospital environment (Miyahara, 1989). One mechanism for scanning used in the Fuji BAS3000 is described as a modified drum scanner which was originally a conventional densitometer. One drum is used to scan the phosphor screen and the other is used to imprint the image onto normal photographic film. The rotation speed of the drum can be set manually or by the computer as the screen is scanned with a Helium–Neon (He–Ne) laser of about 633 nm (Amemiya *et al.*, 1988). The light is collected by using a "total reflecting glass assembly" and two PMTs with different sensitivities to cover a dynamic range of four orders of magnitude (Amemiya, 1995). Later systems, such as the Fuji BAS1800II and the BAS2500 for life science research, were designed so that the imaging plate is moved on a conveyer belt mechanism. In this case a stationary He–Ne laser is directed by a galvanometer controlled mirror to sweep in the X direction across the screen. The light is collected by a proprietary "light collection guide" that moves along the Y direction and focuses the light into a single PMT (Moron *et al.*, 1995).

Molecular Dynamics licensed the storage phosphor screen technology from Fuji in 1989 and introduced the PhosphorImager products for the life science market. The PhosphorImager 400, 425 and the smaller format, SI models utilize a mechanical design similar to the Fuji BAS2000 in that the phosphor screen is kept in a flat plane while the He–Ne laser beam sweeps across the screen as it is reflected by a galvanometer-controlled mirror. Light is collected by a fiber optic bundle and channeled to a single PMT.

One artifact which results from the BAS2000 design and the PhosphorImager designs is documented as "laser bleed" or "flare". This artifact is described in Molecular Dynamics Technical Note #55 as the result of "stray laser light hitting high intensity signals on the storage phosphor screen around the pixel being excited" (Pickett, 1992). Later instrument designs introduced by Amersham BioSciences and Fuji use a point source light collection device that eliminates this error. In addition, both technologies have been incorporated into more versatile instruments, such as the STORM, Typhoon (Amersham BioSciences, Sunnyvale, CA), and Fuji FLA8000 (Fuji, Tokyo, Japan) that are capable of nonisotopic imaging as well as the storage phosphor screen imaging for radioisotopic detection.

In 1996, Packard BioSciences introduced the Cyclone Storage Phosphor System which is designed with a helical scanning mechanism reminiscent of the high performance Fuji BAS3000. In the Cyclone, flexible storage phosphor screens are loaded onto a cylindrical carousel that spins at 360 revolutions per minute. The screen is scanned with a solid state laser diode (633 nm wavelength) that is reflected by a dichroic mirror and focused to a beam of less than 50 microns in diameter by a lens. The light released from the storage phosphor screen (390 nm) is collected through the same lens and imaged on an aperture, confocal with the laser spot, in front of a single high-quantum-efficiency PMT (Cantu *et al.*, 1997).

B. Comparison of Storage Phosphor Systems

Aside from the obvious differences in physical design and size, it is useful to compare the currently available storage phosphor systems in terms of sensitivity, resolution and linear dynamic range.

I. Sensitivity

Sensitivity can be described as the minimum detectable levels of activity but, as mentioned in Section II.B.1., it is useful to discuss sensitivity with respect to radionuclide imaging in terms of the speed at which comparable levels of activity can be detected, and the efficiency of radioisotopic detection, instead of minimum detectable levels of activity. Specifications for minimum detectable levels of activity for storage phosphor systems have been reported as low as $0.5\,DPM/mm^2/h$ for ^{35}S and $0.1\,DPM/mm^2/h$ for ^{32}P (Johnston *et al.*, 1990).

Although Pickett defines the minimum detectable signal as five times the average background (Pickett, 1992), a more appropriate measure of detection threshold is signal-to-noise ratio. In liquid scintillation counting, the measure of signal (S) to noise (N) ratio is calculated as:

$$S/N \; ratio = E^2/B \qquad (13.2)$$

where E is efficiency of detection and B is the background counts. This formula is appropriate for nuclear counting because the fluctuation in the background on a liquid scintillation counter is proportional to the square root of the background itself as a result of the statistics that govern "nuclear event" counting (L'Annunziata, 1987). In storage phosphor screen imaging, there is no relationship between the background fluctuation and the background itself. Therefore, the fluctuation in the background must be calculated separately in order to measure the signal-to-noise ratio on a storage phosphor system. Storage phosphor systems have background values arbitrarily set in the electronics such that the values are never below zero. Thus, the fluctuation in the noise can be measured and the sensitivity of any given storage phosphor screen imaging system can be determined.

In order to compare storage phosphor systems with each other and other methods of detection the following protocol is used to measure signal-to-noise ratio. Screens are erased using an appropriate light box and exposed for 1 h to a commercially available standard of ^{14}C-labeled material (American Radiolabeled Chemicals, St. Louis, Missouri). Screens are scanned at 300 dpi resolution and quantified by integrating areas of 14 mm^2. Twenty background regions surrounding the standards are quantified also by integrating a 14 mm^2 area. Signal-to-noise is calculated using Eq. 13.3 below where $B1$–$B20$ represent 20 background regions:

$$S/N \; ratio = \frac{Response}{St. \; Dev \; (B1 \ldots B20)} \qquad (13.3)$$

Although the absolute background levels for the various storage phosphor systems vary significantly for the same exposure times, the signal-to-noise ratios and therefore the minimum levels of detection are quite similar.

2. Resolution

As with film autoradiography, CTF or modulation transfer function as described in Johnson et al. (1990) is a good measure of the ability to separate closely spaced lines on a radiolabeled ink source. The same factors associated with samples that affect the resolution of film autoradiography, as described in Section II.B.2, also affect resolution of a storage phosphor system. However, two other factors that also affect the resolution that can be achieved with storage phosphor screen imaging are the characteristics of the storage phosphor system and the characteristics of the phosphor screens used to capture the images. Johnson et al. (1990) also describe a third factor, which is the process of autoradiography, but we consider this a property dependent on the isotope and will discuss it in terms of the variation in resolution between isotopes.

It is impossible to compare the resolution of each instrument independent of the variations between types of phosphor screens because of the mechanical differences between systems. Fuji and the Cyclone systems use similar screen formulations, but the Amersham Bioscience systems use phosphor screens manufactured by Kodak and require that they are mounted to a machined flat support for the scanning mechanism. Resolution comparison between selected instruments in terms of ^{14}C resolution is worth noting, keeping in mind that characteristics of the screens, as well as the characteristics of the imaging systems, affect the resolution performance. The specification for Amersham systems are 1.8–2.1 line pairs/mm with a CTF of 33%. Fuji BAS5000 system specification for resolution is listed as < 3.1 line pairs/mm with a CTF of 10%. The Cyclone Storage Phosphor System is capable of separating 2.5 line pairs/mm, with a CTF of 33%.

The following study was performed to determine the resolution of the Cyclone Storage Phosphor system with four different screen types. These results illustrate both the affects of the storage phosphor screen formulations and the affects of various isotopes used for the autoradiographic process.

Four screen types were used in the study. Section III.C.1., which describes techniques for optimizing storage phosphor screen imaging, includes detailed descriptions of the four screen types used in this study. Since this study, Fuji introduced the MultiSensitive (MS) screen type. The MS screen has been shown to provide better sensitivity and resolution than the original MultiPurpose (MP) screen and has since commercially replaced it (Perkin Elmer Life and Analytical Sciences, Boston, MA).

The most significant features of the screens for this study are that the Super Sensitive (ST), MP (and new MS), and Super Resolution (SR) are coated with a protective layer which makes them incapable of detecting the low energy beta emission of ^3H. The TR screens are uncoated and can therefore detect all isotopes; however, they are considered impractical for the detection of higher energy isotopes due to the fact that they are less durable in nature. TR screens cannot be cleaned and are considered disposable upon contamination. However, Liberatore et al. (1999) describes a method for fixation of tissue sections that can minimize the potential for contamination and extend the life of TR screens. All four screen types were exposed for 1 h to a series of ^{14}C ink lines. TR screens were exposed for 17.75 h to a series of ^3H ink lines. All screens were scanned at 600 dots per inch scanning resolution. The sources contained lines nominally at 1.2, 1.5, 2.0, and 3.0 line pairs/mm. Rectangle lanes of 48 pixels (2.0 mm) wide were placed across the lines for creation of profiles and integration of peaks of activity. The average CTF of the profiles was calculated for each screen and source. Table 13.2 shows the CTF values for each screen at 2.5 line pairs/mm.

The SR screen has the highest CTF for separation of ^{14}C line pairs, as expected, and the ST, which is thicker and optimized for sensitivity, exhibits the lowest CTF. This result is consistent with literature regarding the effects of the thickness of film emulsions (Rogers, 1969). The TR screen exhibits slightly higher resolution of ^{14}C, but most importantly, can detect and does

TABLE 13.2 Comparison of CTF Values of Four Storage Phosphor Screens. The average CTF values represent the profiles of two sources containing lines at a frequency of 2.5 linepairs/mm by storage phosphor screen imaging method

	ST	MP	SR	TR
^{14}C source	17%	25%	33%	38%[a]
^{3}H source	N/A	N/A	N/A	69%

[a]TR screens are uncoated and therefore considered not appropriate for detection of any isotope with higher energy than ^{3}H since other more durable alternatives exist, such as the ST, MP (and now MS), and SR screens.

provide significantly higher resolution for ^{3}H lines. The fact that the same screen and the same lines exhibit higher resolution with ^{3}H indicates that the limiting factor of the resolution of the Cyclone Storage Phosphor System is not the system, but the autoradiographic process, which is a function of the path length of the beta emissions of the isotope being measured. Both the MP and the ST screens have been replaced commercially by the MS screen produced by Fuji. The MS screen exhibits comparable sensitivity of the ST screen, durability of the MP screen and resolution slightly improved, although not as high as the SR screen (unpublished data).

3. Linear Dynamic Range

The linear dynamic range of storage phosphor screen imaging is significantly larger than that of film autoradiography. Several authors have graphically displayed this comparison (Johnston *et al.*, 1990). Typical storage phosphor systems can provide linear data of four to five orders of magnitude. On the Fuji BAS2000 system, Moron *et al.* (1995) noted a proportional relationship between photostimulable light (PSL) units and DPM of activity between 10^1 and 10^5 for a 3 h exposure DPM. Amemiya and Miyahara (1988) illustrate the linearity of the Fuji Imaging Plate method compared to liquid scintillation counting results. Amersham BioSciences specifies the linear dynamic range of 10^4 orders of magnitude as supported by Johnston *et al.* (1990). The Cyclone exhibits linearity of five orders of magnitude as illustrated in Fig. 13.3.

For practical biological applications, typical samples do not require more than three orders of magnitude of linear dynamic range for accurate measurements. However, the significance of having a longer linear dynamic range than that which can be provided by film should not be understated. Due to the nature of film and storage phosphor screen methods, researchers are required to estimate the appropriate exposure time. An exposure time which is too short can result in nonlinear results for lower activity areas of a sample. An exposure time which is too long can result in overexposure or saturation of the phosphor screen readout. A longer linear dynamic range provides less chance of error in estimating exposure time. For those samples that do include very low and very high areas of activity, two different film exposures of different lengths of time may be required to capture the activity

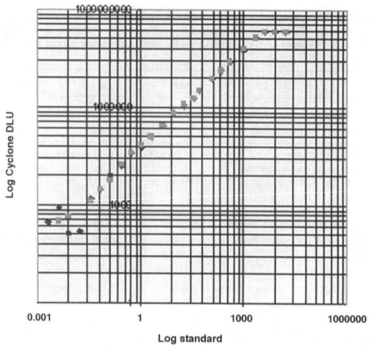

FIGURE 13.3 Linearity of Cyclone Storage Phosphor System using two ^{14}C standards. Cyclone exhibits a linear dynamic range of 5 orders of magnitude.

in the linear range. Phosphor screens are capable of capturing very low activity and very high activity in one exposure that is significantly shorter than typical film exposures for the same sample.

C. Quantification Methods

The use of a storage phosphor system instead of film provides a result much faster than film, but the most significant advantage is the quantitative nature of the image files. Note that the accuracy of the data from a storage phosphor system depends to some extent on user technique, because the phosphor screen is separate from the instrument and cannot be calibrated.

I. Techniques for Optimization

As phosphor screens are scanned, some, but not all, of the data is erased as the phosphor crystals are returned to ground state. If a screen were immediately scanned again, some residual data will create another, more faint image. In order to clear the latent image that is left after scanning, screens should be flooded with white light for 30 s to 5 min, depending on the screen type. Some intense activity samples may leave a residual "ghost image" which can be erased by flooding with bright visible light for 24–48 h (Reichert *et al.*, 1992). A common fluorescent light box can be used for "erasing" screens. In addition, since phosphor screens are sensitive to cosmic radiation, they accumulate background while they are stored in the packaging. Even a new storage phosphor screen should be erased immediately before exposure to achieve the lowest level of background.

Most laboratories have bright fluorescent lighting in the room, which is also efficient at erasing screens. Approximately 85% of the signal accumulated during a 17-h phosphor screen exposure can be erased in less than 1 min of exposure to ambient fluorescent laboratory lights. Therefore, it is recommended that the overhead lights in the lab are turned off while loading a phosphor screen into a scanning instrument after the sample has been exposed.

Screens should be cleaned routinely with a mild, nonabrasive detergent such as Kodak Intensifying Screen Cleaner (Kodak, Rochester, New York) in order to avoid artifacts in images due to buffer or residual salts left from previous samples. Samples that are stained with a dye such as ethidium bromide should never be in direct contact with a storage phosphor screen because they cannot be cleaned off and will always result in a positive signal when the screen is scanned.

Another phenomenon of phosphor screens that could affect accuracy of quantitation is documented as "signal fade." Signal fade is the loss of stored signal that occurs gradually after the sample is removed from the screen. Amemiya and Miyahara (1988) reported that at 20°C, 46% of the stored energy is faded after 2 months. Kodak reports in a technical note that more than 50% of the stored energy available 2 min after exposure remains up to 24 h (Eastman Kodak, 1993). The graph in Fig. 13.4 illustrates the extent to which signal fades after exposure is complete and before scanning.

Wet samples should never be brought in contact with storage phosphor screens. Screens should be protected with plastic film whenever possible, although emissions from weak beta sources will be attenuated.

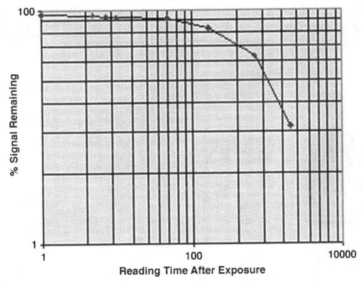

FIGURE 13.4 Signal fade which is exhibited by all storage phosphor screens as a function of time scanned after exposure. Screens which are scanned 1 h after exposure is completed will be stable for many hours.

Signal fade is uniform across a given screen, so the effect is negligible on quantification of samples as compared within one image. When comparing data from one screen exposure to another, one can minimize the effect on the accuracy of data by waiting 1h after exposure is complete before scanning to avoid the exponential decrease in signal that occurs during the first hour at 22–25°C.

In order to improve signal-to-noise ratio and therefore sensitivity, exposure cassettes can be surrounded by lead shielding to eliminate the added background noise, which can be a result of cosmic radiation. It is recommended that low-activity samples that require days of exposure should be shielded with lead for best results (Hamaoka, 1990).

Finally, it is important to choose the appropriate phosphor screen type to optimize quantitative and qualitative results with a storage phosphor system. The following types of screens are available from both Perkin Elmer and Fuji for use with their respective scanning systems. Exposing samples to a screen that is optimized for the specific application will provide the best results from a storage phosphor system.

MultiPurpose (MP) screens, now replaced by MultiSensitive, MS screens (Perkin Elmer Life and Analytical Sciences, Boston, MA and Fuji, Japan) are designed to resist moisture. These screens are ideal for ^{32}P-labeled northern blots, Southern blots, dot and slot blots, TLC plates, preparative gels and differential display gels.

Super Resolution (SR) screens (Perkin Elmer Life and Analytical Sciences and Fuji) are formulated with synthetic phosphor crystals with a highly regular shape so that they exhibit better separation of lines. SR screens resolve ^{14}C-labeled ink line pairs with a CTF which is about 20% higher than an MP screen as can be seen visually in Fig. 13.5. The graph provided in Fig. 13.6 illustrates the difference between MP and SR screens. SR screens are recommended for separation of single base pairs in DNA sequencing gels or imaging the distribution of activity in tissue sections. MS Screens exhibit resolution slightly better than the MP screen, but not as good as the SR screen (unpublished data).

MP   SR

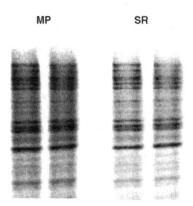

FIGURE 13.5 Qualitative representation of a sample imaged for the same amount of time with an MP screen as compared to an SR screen. The difference in resolution can be seen visually.

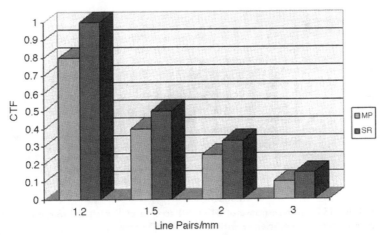

FIGURE 13.6 Graphical representation of the comparison of the CTF with which SR screens can separate line pairs as compared to MP screens.

TABLE 13.3 Comparison of Net Response of Three Phosphor Screens. The Net Response to ^{32}P Dots are Compared with Three Screen Types. Values are Normalized to Maximum Response. ST Screens Exhibit the Highest Net Response

Screen	ST	MP	SR
Net Response	1.00	0.47	0.22
Net Response/Bkgd	1.00	0.79	0.45

Super Sensitive, ST, screens (Perkin Elmer Life and Analytical Sciences and Fuji) were formulated with a thicker coating of phosphor crystals and therefore exhibit the highest net response to ^{125}I-labeled microscales and ^{32}P-labeled dots on a filter. Net response/background measurement is shown in Table 13.3. Figure 13.7 is a graph of the net response of ST screens versus. SR screens to ^{125}I microscales. ST screens are best for samples with low amounts of a high energy isotope such as a low activity ^{125}I-labeled western blot or low activity ^{32}P-labeled northern blot (Veal and Englert, 1997). MS screens exhibit similar sensitivity to the ST screen formulation, and better resolution (unpublished data).

Tritium Sensitive (TR) screens (Perkin Elmer Life and Analytical Sciences and Fuji) are also formulated with the highest grade of phosphor crystals but they are uncoated so that the low energy of ^{3}H emissions can penetrate to the phosphor layer. These screens are best suited for receptor autoradiography in brain tissue (Lidow and Solodkin, 1997). An excellent method for fixation of brain tissue in radioligand binding assays for minimization of contamination of TR screens has been described by Liberatore *et al.* (1999). Moisture accumulation upon repeated use of TR screens can increase the background. Baking at 60°C with a beaker of anhydrous calcium sulfate was shown to reduce the background and help restore the sensitivity of the screen.

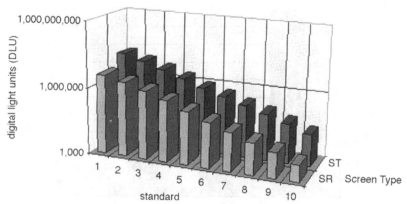

FIGURE 13.7 Net response of ST vs. SR screens to ^{125}I labeled microscales. The signal is accumulated more efficiently with the thicker ST screens.

In addition, silica particle contamination from TLC plate applications can be removed by rinsing with methanol and allowing to air dry (unpublished data).

It should be specifically stated that unlike film, intensifying screens, exposing at low temperature and fluorography do not improve imaging with storage phosphor screens.

2. Advantages of Storage Phosphor Screen Imaging

Storage phosphor screen imaging provides a number of advantages over film autoradiography for quantification of radionuclides. Sensitivity, speed, linear dynamic range, and digital quantification are the most significant.

Johnston *et al.* (1990) found that using the midpoint of the response curve of film, phosphor screens are about 250 times more sensitive than X-ray film for ^{32}P autoradiography and 15 times more sensitive than autoradiography using an intensifying screen. For ^{14}C and ^{35}S autoradiography, storage phosphor screens are 60–100 times more sensitive than direct film autoradiography and 20–30 times more sensitive than film with fluorography. In practice, a sample that would require a 30-day exposure to film can be imaged and quantified using storage phosphor technology in 3 days (Johnston *et al.*, 1990).

Linear dynamic range of storage phosphor screens provides a distinct advantage over film, especially for samples that may require two separate film exposures of different lengths of time to get data in a linear representation. Storage phosphor screens exhibit linearity for 4 to 5 logs of activity (see Fig. 13.3 in III.B.3.).

Storage phosphor screens also provide the practical advantage of reusability, and therefore chemicals and hazardous waste disposal are not required. Although Kodak provides documentation for disposal and treatment of photographic effluent (Eastman Kodak, 1989), storage phosphor screens have an indefinite lifetime and can last for years if handled properly.

Some advantages over direct nuclear imaging methods (see Section IV) make storage phosphor screen imaging a good choice for multiuser environments. First, the fact that the screens are exposed in cassettes independent of the instrument makes storage phosphor technology a good choice for low-activity samples that require days of exposure time. A group of users may own as many screens as necessary, make simultaneous exposures, and use the instrument for only a short scanning time. Second, for high resolution applications, the Super Resolution screen alternative makes higher quality images that will provide the ability to extract more detail from a given image. Finally, the TR screen provides an alternative for ultrahigh-resolution applications that could previously be performed only with film.

3. Disadvantages of Storage Phosphor Screen Imaging

Disadvantages of storage phosphor screen technology compared with film autoradiography are that with storage phosphor methods, the only permanent record of the sample is an electronic file and possibly a printout of that file. Film offers a direct record of the sample for archives. In addition, as a function of the reusability of the screens, artifacts, and "ghost" images can sometimes be seen in future images due to either incomplete erasure or damage from mishandling of screens. Each film is fresh out of the packaging and is not affected by previous images.

Other disadvantages are relative to other radionuclide analysis methods such as direct nuclear imaging described in Section IV. One such disadvantage is the fact that for quantification, the linear dynamic range is more limited on both the low and top end than the direct beta imaging method. As described in Moron *et al.* (1995), when measuring the linearity of a phosphor screen light units, after background subtraction, when compared to ^{14}C standards of known activity, some points were nonlinearly related. In particular, the lowest point on the graph which was the shortest exposure (3 h) of the lowest concentration of activity (50 DPM) and the highest point on the graph, which was 1.05×10^5 DPM at the same exposure time, were nonlinear. The fact that it is possible to get data that are nonlinear by being either below the threshold of detection or beyond the saturation level of the digital image file means that the exposure time is still a somewhat critical factor. Although it is easier to obtain data in the linear range with a phosphor screen than with film because of the significantly longer linear dynamic range, it is still possible to misjudge the exposure time. The exposure time is independent of the instrument and is based on user judgment.

Other challenges to obtaining accurate data with storage phosphor screens are due to the fact that the screens exhibit signal fade, cannot be calibrated individually, require handling in subdued lighting, and require diligence on the part of a user to clean screens, erase background before exposure, and choose the correct exposure time. All of these actions are independent of the scanning system and can profoundly affect the results which can be expected for quantitative radionuclide analysis.

D. Applications of Storage Phosphor Screen Imaging

The following section includes some images and results from autoradiography by the storage phosphor screen method. The most common applications for which researchers use storage phosphor technology are whole body autoradiography, receptor autoradiography, high resolution gel analysis, DNA sequencing, western blotting, and DNA microarray image analysis. Typical applications are those which require high resolution performance and relative quantification of activity within the image file.

I. Whole Body Autoradiography

The fate of potential therapeutic agents in the study of pharmacology and toxicology is often discovered by determining the distribution of radiolabeled compounds in test animals. Animals are sacrificed at specific times after administration of a labeled substance to follow the time course of absorption, distribution, metabolism, and elimination (ADME) functions. One method is to perform necroscopy, oxidation or solubilization followed by liquid scintillation analysis. An alternative method is to section the animals with a cryomicrotome, followed by autoradiography of sections taken at various levels within the animal and quantification of the tissues within the sections (Ullberg, 1977). By utilizing an imaging method for this application, localization is much more precise and distribution in heterogeneous tissues such as kidney and liver can be more revealing than the homogeneous method of oxidization and liquid scintillation counting of whole organs.

Figure 13.8 is a typical whole body section as imaged on the Cyclone Storage Phosphor System. Standards (not shown) are prepared and quantified by liquid scintillation analysis to estimate the DPM/µg of material. Measurement of the digital light units per area $(DLU)/mm^2$ integrated within regions drawn in areas of interest in the sample are compared to the DLU/mm^2 measured using regions drawn within the standards. The amount of radiolabeled material which is distributed within each tissue can be determined on the basis of the known standards.

This ^{14}C-labeled whole body rat tissue section was exposed for 16 h using an SR screen (Perkin Elmer Life and Analytical Sciences, Boston, MA.) to obtain a linear representation of the activity within the tissue sections. The method of film autoradiography requires a 7–10 day exposure for an image, which must then be quantified by measuring optical densities in the area of interest relative to the optical density of the standards. For ^{14}C, the

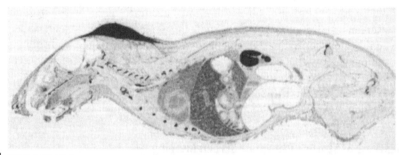

FIGURE 13.8 Autoradiography of ^{14}C labeled drug as distributed in a rat whole body section. This image was exposed for 16 h to an SR screen as compared to 7–10 day exposure to film.

resolution is quite similar to that of film and the quantification is more accurate using storage phosphor screen imaging because of the longer linear dynamic range (Mori and Hamaoka, 1994).

2. Receptor Autoradiography

Storage phosphor technology is most useful for development of receptor autoradiography assays. Typically, ligands that are specific for a brain receptor of interest are labeled with ^{3}H to get the highest possible resolution and therefore the most accurate localization of the receptors. Brain tissue is sectioned and hybridization is performed after the sections are mounted on microscope slides. In order to maximize the level of specific binding by the ex vivo receptor autoradiography method, it is essential to control the concentrations of radiolabeled ligands and blocking agents, and the timing of preincubation, incubation, and washing steps. The use of storage phosphor screen imaging speeds up the development of these assays because the exposure times are so much less using TR screens. Figure 13.9 is an example of results of a receptor autoradiography study as imaged with the Cyclone Storage Phosphor System.

Tissue sections were prepared by methods described by Bigham and Lidow (1995). In previous studies, the exposure times to ^{3}H-sensitive

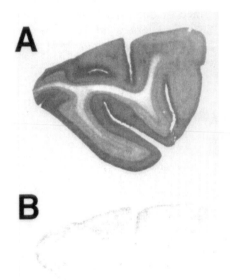

FIGURE 13.9 Total (A) and nonspecific (B) binding of α_1-adrenergic radioligand [^{3}H] prazosin (New England Nuclear, Co., Boston, MA) in rhesus monkey cerebellum. Exposure to TR storage phosphor screen was 18 h in a lead box as compared to 3.5 months exposure to film.

Ultrafilm (Amersham Corp., Arlington, Heights, Illinois) required 3.5 months of exposure (Lidow and Rakic, 1995; Bigham and Lidow, 1995). The images in Fig. 13.9 represent the total and nonspecific binding of α_1-adrenergic radioligand [^{3}H] prazosin (Perkin Elmer, Boston, MA) in rhesus monkey cerebellum. The samples were exposed to a TR screen for only 18 h. Exposures were performed in standard film cassettes, which were enclosed in a 1/4 in. thick lead box to shield from cosmic radiation as described in Section III.C.1. These results suggest that the Cyclone could also be used for the development of assays such as *in situ* hybridization of tissue sections with ribo/oligonucleo-probes for the visualization of mRNA (Lidow and Solodkin, 1997).

3. High Resolution Protein Gels

Molecular biology samples often involve separation of biomolecules by agarose gel or polyacrylamide gel electrophoresis. Polyacrylamide gels can separate DNA fragments which differ in length by only a single base pair. Storage phosphor technology is widely used for autoradiography of high-resolution DNA sequencing or protein gel electrophoresis. Figure 13.10 is a portion of an ^{35}S-labeled sequencing gel that was used to determine the sequence of nucleotides across a junction of cloned fragments. Cyclone storage phosphor screens are available in a 43-cm length so that the entire length of a sequencing gel may be read in one scan. Although most high throughput sequencing is performed by automated DNA sequencers, such as those available from Applied BioSystems (Sunnyvale, California), the Cyclone is appropriate for labs with a variety of molecular biology applications, including radiolabeled DNA sequencing gels.

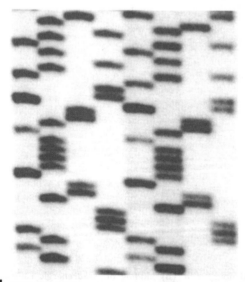

FIGURE 13.10 Image of a portion of a common ^{35}S-labeled sequencing gel which is used to determine the sequence of nucleotides across a junction of cloned fragments. Two hour exposure to SR screen was sufficient to read the entire length of the gel.

Other polyacrylamide gel samples require quantification as well as a high resolution image. Figure 13.11 is an example of a polyacrylamide gel used to separate ^{35}S-labeled proteins. Transfected cell cultures were subjected to various treatments designed to alter expression of a particular protein of interest. In this study, Lanes 1, 2, and 3 contained different concentrations of proteins isolated from a culture that is a negative control for protein expression. Lanes 4, 5, and 6 contained three different concentrations of extracts from cell cultures treated with an agent to stimulate the inducible protein of interest, labeled P1 in the image. HKG represents a "housekeeping" gene used to control for the varying amounts of protein loaded in each lane. The sample was exposed to an SR phosphor screen and quantified using profiles and peak integration (Fig. 13.12) from Veal and Tian (1997a).

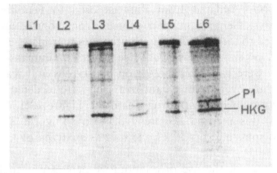

FIGURE 13.11   Image of a polyacrylamide gel used to separate ^{35}S-labeled proteins created with a 1 h exposure with an SR phosphor screen. Lanes 1, 2, and 3 contained different concentrations of proteins isolated form a culture that is a negative control for protein expression. Lanes 4, 5, and 6 are three different concentrations of extracts from cell cultures treated with an agent to stimulate the inducible protein of interest, labeled PI in the image. HKG represents a "housekeeping gene" used to control for the varying amounts of protein loaded in each lane.

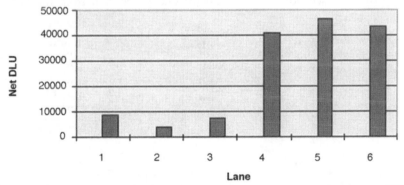

FIGURE 13.12   Graphical representation of induced protein expression as measured by electronic autoradiography. Digital light units (DLU) are reported net of background and normalized based on quantification of the HKG band in each lane.

4. DNA Microarray Applications

The emergence of whole genome sequence data has brought about gene array technology for differential gene expression, mutation screening, sequence analysis, and drug target identification. The commercial availability of mouse, human, and rat genome sequences preprinted on nylon membranes provide a convenient way to conduct gene array assays using radiolabeled sequences and a storage phosphor system. Storage phosphor imaging technology provides the long linear dynamic range and accuracy required for detection of subtle changes in a large range of gene expression levels that can occur within a given experiment. The following are two examples of the use of storage phosphor screen imaging for radiolabeled gene array samples.

Gene arrays can be used to analyze the effects of drug treatments at the molecular level. Atlas Rat Toxicology II arrays (Clontech, Palo Alto, CA) are filters containing rat liver total RNA. Filters containing 465 unique cDNA fragments in duplicate were hybridized with radiolabeled cDNA reverse transcribed from RNAs isolated from Rats exposed to Fenofibrate drug treatment for 10 days. Gene expression profiles from control and treated animals were analyzed to look for clues to the changes that may be a result of drug treatment and potentially cause adverse effects in humans (Jiao and Zhao, 2002). Filters were exposed 18–24 h on SR screens and scanned with the Cyclone. Images are overlaid in QuantArray software to determine which genes are up or down regulated with drug treatment. Figure 13.13a, b are the images obtained by this method. Figures 13.14a,b show the scatter plot display of quantified spots and results of one spot as analyzed by QuantArray (Upham and Fox, 2001).

Another application of the use of quantitative gene array analysis is in research on effects of the environment on human gene expression. For example, it is well documented that exposure to sun causes or results in an increase in actinic keratosis and eventually squamous cell carcinoma (Hodges and Smoller, 2002). Researchers at University of New Mexico collect punch biopsies from patients diagnosed with squamous cell carcinoma (SCC). Four samples are collected from each patient including tissue from (a) the SSC, (b) an actinic keratosis, a precursor lesion of SCC, (c) adjacent sun exposed normal skin, and (d) unexposed skin from the buttocks. Total

FIGURE 13.13 Cyclone images of rat liver total RNA hybridized with control (a) and Fenofibrate treated (b) rat liver total RNA reverse transcribed into cDNA and radiolabeled with ^{33}P-dATP.

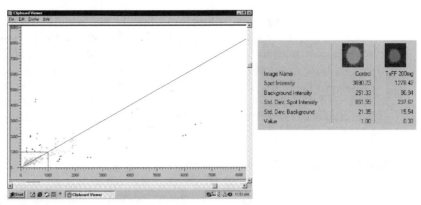

Image Name	Control	7xFF 200mg
Spot Intensity	3890.23	1278.42
Background Intensity	251.33	96.94
Std. Dev. Spot Intensity	851.55	237.67
Std. Dev. Background	21.35	15.54
Value	1.00	0.33

FIGURE 13.14 (a) Scatterplot display of comparison of control and treated filters based on data from Cyclone; (b) Representation of specific spots as selected from Scatterplot.

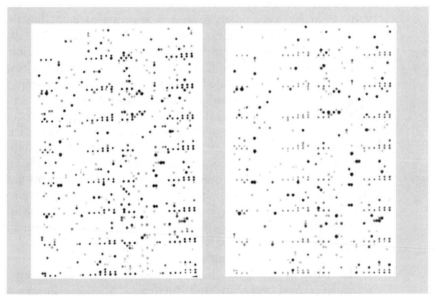

FIGURE 13.15 Gene expression arrays probed with mRNA isolated from normal and tumor tissue. (Courtesy of Dr. Bryan E. Alexander, University of New Mexico.)

RNA is isolated from each sample, reversed transcribed into cDNA and labeled with α-^{33}P-dATP, and hybridized to an ID1001 DermArray Filter containing 5000 human genes from Invitrogen/Research Genetics (Carlsbad, CA). After washing, membranes are exposed to SR screens for 24 hours, to bring out low expressors, and scanned on the Cyclone system. The intensities of the spots correspond to the relative abundance of various transcripts at the time that the RNA was harvested. By comparing multiple filters, differences in gene expression profiles between each of the states, such as tumor versus adjacent normal skin and unexposed versus exposed skin can be observed. Figure 13.15 are quantitative images of these high density gene array filters.

Additional references in literature describe the use of storage phosphor screen imaging for receptor binding assays (Chan *et al.*, 1991); Southern and northern blot analysis (Muller and Gebel, 1994; Robben *et al.*, 2002) western blot analysis (Taylor *et al.*, 1992; Shelton *et al.*, 1994); Gel shift assays (Zinck *et al.*, 1993; Olivas and Maher, 1995); BioChip Imaging (Schena, 2000) and Microarray assays (Popovici *et al.*, 2000), other related isotopes (Gonzalez *et al.*, 2002), and double label autoradiography (Pickett, *et al.*, 1992).

IV. ELECTRONIC AUTORADIOGRAPHY

Electronic autoradiography is a unique method of quantitative radionuclide imaging based on technology of the multiwire gas proportional counter. In the early 1970s, Georges Charpak described his work on multiwire chambers for the detection of high-speed particles in high-energy particle physics applications, for which he later won the Nobel Prize (Charpak and Sauli, 1978). Charpak went on to develop a system that combines the multiwire technology with a CCD that is described in Section V.B.1. In 1985, Bateman *et al.* (1995) reported their development of an electronic, digital beta autoradiography system that utilized a multistep avalanche–multiwire proportional counter that exhibited resolution unsuitable for most molecular biology applications. Sullivan *et al.* (1987) described their use of a multiwire chamber adapted for biological applications; however, the system is no longer available. Likewise, Ambis Systems (now Scanalytics, Fairfax, VA) commercialized the Ambis 100 for TLC analysis and the Ambis 4000 for higher resolution applications. E. G. Berthold and G. Berthold introduced a direct beta scanner that utilizes a multiwire chamber and gas ionization detection in the late 1980s; however, the resolution is marginal for most molecular biology applications. Jeavons *et al.* (1983) described the use of the multiwire chamber together with a high-density avalanche chamber for positron emission tomography. By 1992, Jeavons patented this method and an apparatus for quantitative autoradiography that is commercialized as the MICAD used in the InstantImager Electronic Autoradiography System (Perkin Elmer Life and Analytical Sciences, Boston, MA). This was the first multiwire chamber system to achieve the resolution, accuracy, and ease of use necessary for large-scale use in molecular biology laboratories. The following section is devoted to the InstantImager electronic autoradiography method of imaging and quantifying of radionuclides for biological applications. However, each of the multiwire instruments described, including the InstantImager, has limited availability due to obsolescence of parts.

A. Technology

The technology of the InstantImager Electronic Autoradiography System provides a unique alternative for autoradiography unlike film or storage phosphor screen scanning. It is a true nuclear imaging system in that it includes a nuclear detector for direct quantification of radionuclides in two-dimensional samples. The resultant image is composed of nuclear count data, which has distinct advantages which will be discussed later in this chapter. In this section, we will describe the technology behind this direct beta detection system.

1. The MICAD Detector

The MICAD is composed of two sections: the microchannel array plate and a multiwire chamber (Fig. 13.16). The design was derived from technology initially developed for positron cameras (Jeavons *et al.*, 1983). The purpose of the microchannel array is to localize the particle as precisely as possible as it leaves the sample. The microchannel plate is a structure about 3 mm in thickness and 20 cm × 24 cm in area. It is eight layers of laminated conductive (brass) and nonconductive (fiberglass) material that are drilled with over 210,000 holes in a honey-combed pattern across the entire surface area. The holes are each 0.4 mm in diameter and serve as individual detection elements, referred to as microchannels. A voltage step gradient is imposed on the successive conductive layers to create an electric field of approximately 600 V/mm in the microchannels (Englert *et al.*, 1995). Below the microchannel array is a "window" of aluminized Mylar and a stainless steel mesh that serves to protect the detector from contamination by radioactivity and particulates and to seal the entire detector chamber from air and moisture. Above the microchannel array plate is the multiwire chamber, which consists of (1) an anode plane of 200 gold wires approximately 20 μm in thickness and (2) two cathode planes (*X* and *Y*) formed by metallic cathode tracks, one of which is in the top of the microchannel plate and the other above the anode wires (Jeavons, 1990).

The purpose of the multiwire chamber is to detect each event and convert it to an electronic pulse. The entire MICAD detector is filled with a gas

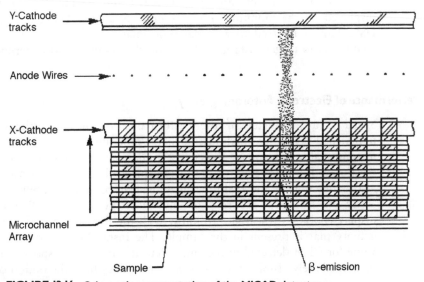

FIGURE 13.16 Schematic representation of the MICAD detector.

mixture of 96.5% Argon (99.99% pure), 2.5% CO_2 (99% pure), and 1.0% isobutane (99% pure) at a flow rate of $25\,cc^3$/min.

2. Digital Signal Processing

Beta particles emitted from samples pass through the protective window, enter the nearest microchannel, and ionize the argon of the gas mixture. The resulting electrons are accelerated by the high electric field in the microchannel to further ionize the gas, producing a cloud of electrons. In this way, the microchannels serve as both collimators and preamplifiers. The cloud of electrons migrates up an electric field gradient into the multiwire chamber. Although any given radioactive event is channeled through one and only one microchannel, the resulting discharge in the multiwire chamber can span up to 40 microchannels in both the X and Y directions. The pattern of this discharge is recorded by measuring the intensity of the discharge every four microchannels; and the data are passed on through a fiber-optic link to the digital signal processor (DSP) located in an external computer. The DSP then performs three operations. First, the DSP validates the data by screening out events that are too large or too small to have resulted from a radioactive event (such as detector background or cosmic events). Second, the DSP detects and removes events that were actually two separate events occurring simultaneously. Third, the DSP determines the exact microchannel in which the event was located by centroiding the intensity values recorded in both the X and Y directions by the following method:

$$\text{Centroid} = \frac{\sum (\text{location})(\text{intensity level})}{\sum (\text{intensity level})} \tag{13.4}$$

The corresponding memory location is then incremented to build the image in real time, as the events are being counted and accumulated. The image is displayed on the monitor with 15-s updates as it is being measured in the DSP. This is referred to as "real time image display and analysis" and it enables users to get instant feedback on the results of an experiment.

B. Performance of Electronic Autoradiography

Speed is a parameter of the performance of the InstantImager that is unmatched by other methods, such as film and storage phosphor screen imaging, although other direct beta imaging methods are similarly fast (Section V.B.). The speed of the InstantImager is due to its sensitivity and is also a function of the real-time imaging feature and the ability to quantify easily the statistical accuracy of any given measurement. Real-time image display enables a researcher to know with the first screen update the level of activity that is present in the sample. The InstantImager software reports a value for CPM detected in the entire detector area. Some spots or bands may be immediately visible on the screen. A user does have the option of stopping the acquisition of the sample when the image is sufficient for a qualitative inspection or of quantifying while counting to get a quantitative measure of the result, even before the image is complete. The effect of the real-time imaging feature is to speed the process of autoradiography with the InstantImager. In the most dramatic example, when an experiment fails, a researcher can know immediately and start the experiment over without waiting for a negative result on film (Englert *et al.*, 1993). Statistical accuracy will be discussed in Section III.C.1. under techniques for optimization of quantification.

1. Sensitivity of Electronic Autoradiography

As with any nuclear detection instrument, the sensitivity improves with the square root of the counting time. The sensitivity is a function of the low background, the efficiency of detection of a given isotope, and the signal-to-noise ratio. In order to determine the sensitivity of the InstantImager, dilution series of ^{14}C activity were spotted in 12-mm-diameter spots on a plastic-backed silica plate and measured for 30 min and for 16 h and 20 min. The average background counts of 32 background regions in both acquisitions was <0.015 CPM/mm^2, which is typical of the InstantImager detector. This amounts to approximately 600 CPM for the entire detector area. In the 30-min acquisition, the 0.75 DPM/mm^2 spot can be seen and quantified with a statistical accuracy of less than 5% error. In the 16 h and 20 min acquisition, <0.2 DPM/mm^2 can be detected with a statistical accuracy of less than 5% error. Thus longer count times result in lower detectable levels of activity (Englert *et al.*, 1993). The background on the InstantImager, unlike that of a phosphor screen or film autoradiography, increases in a linear manner with the acquisition time. Phosphor screen imaging and film autoradiography with densitometry necessarily have two components to their background measurements: background that increases linearly with the acquisition time and baseline background that occurs as a function of the scanner. The latter is constant and independent of screen exposure time. Therefore, although you can expose a phosphor screen or a film for a long time and measure very low levels of activity eventually, the InstantImager obtains a faster result for any given sample. The lowest detectable level of activity is limited by the level of noise that is present in the detector. Typical count times on the InstantImager do not exceed 24 h. Table 13.4 lists the efficiency of detection of various isotopes, where

$$\text{Efficiency } E = \frac{\text{instrument counts/mm}^2}{\text{total disintegrations/mm}^2} \qquad (13.5)$$

For practical purposes, in order to compare the sensitivity of film, storage phosphor screen technology, and electronic autoradiography, it is useful to speak of sensitivity in terms of the time it takes to detect a sample of similar activity. Typically, the acquisition time on the InstantImager is approximately 50–100 times faster than the corresponding film exposure and a factor of 5–10 times faster than storage phosphor screen imaging, depending on the

TABLE 13.4 **Efficiency of Detection of MICAD Detector and Storage Phosphor Screen Methods. The Efficiencies of Detection of Various Isotopes Using the MICAD Detector of the InstantImager and Storage Phosphor Screen Imaging Methods are Listed**

	[14]C	[35]S	[32]P	[125]I
InstantImager[a]	1.5%	2.0%	5%	0.03%
Storage Phosphor Screens[b]	N/A	0.2%	4.0%	>0.05%

[a]InstantImager Specification Sheet, Packard Instrument Company, Meriden, Connecticut (now Perkin Elmer Life and Analytical Sciences, Boston, MA).
[b]Pickett *et al.*, 1992.

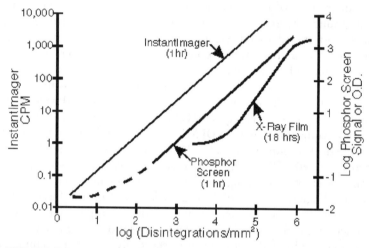

FIGURE 13.17 Comparison of film, storage phosphor screen imaging and electronic autoradiography with the InstantImager with respect to time of exposure or data acquisition and linear dynamic range.

level of activity and the isotopic label of the sample. The graph in Fig. 13.17 illustrates the comparison of sensitivity and linear dynamic range of X-ray film, phosphor screen imaging, and electronic autoradiography in terms of speed to obtain a result.

2. Linear Dynamic Range

The InstantImager does not "store" the activity of the sample; it counts the events. Therefore, the limiting factor of the InstantImager is the speed of the electronics and the number of events that can be recorded in the 16-bit image file. The speed of the electronics and digital signal processing can keep pace with samples of 800,000 CPM in the entire detector area. Typical biological samples do not approach this level of activity but may include some areas of very high activity and some areas of very low activity which require hours of acquisition time. Thus, the limiting factor is the number of events or counts that can be accumulated in any given pixel, which is 65,536 or 2^{16} counts. Often the area of a sample with high activity is not the area of interest, such as in an *in vitro* transcription assay, which may have an area of unincorporated labeled nucleotides (Batt, 1995; Amicone *et al.*, 1997). For these cases, the InstantImager was designed to save automatically an image file of the last update for which no pixel has saturated. The instrument continues to count and accumulate data linearly in the areas that have not exceeded 65,536 in any pixel. This feature extends the linear dynamic range of the InstantImager beyond 10^5 up to 10^6 or even 10^7 depending on the area of the sample being counted. This is the largest linear dynamic range reported for any autoradiography method. Figure 13.17 shows a comparison of linear dynamic ranges of the three methods.

3. Resolution

The MICAD detector design is the limiting factor for the resolution of electronic autoradiography. The microchannels are 0.4 mm in diameter and the centroiding of the digital signal processing can assign events accurately to the appropriate microchannel, but not to finer resolution than the microchannel. Although the microchannels are arranged such that the center-to-center spacing in the X direction is 0.25 mm, the center-to-center spacing in the Y direction is 0.5 mm. The MICAD detector is capable of separating no more than 1 line pair/mm with a CTF of 33%. The pixels of the InstantImager files are 0.5 mm in diameter. In "high resolution mode" the pixels are divided by four and the counts are redistributed by a bilinear interpolation method, which improves the display of the images. The raw data are conserved in the image file and the actual resolution performance in terms of ability to separate line pairs/mm is not improved by this technique.

C. Quantification Methods

Sample preparation for autoradiography with the InstantImager is similar to that which is required for good film autoradiography results. Optimizing gel electrophoresis parameters to get the best separation of bands is one of the best means of optimizing imaging results. Good microtome technique and tissue section preparation methods are also critical for best imaging results (Ullberg, 1977). However, due to the unique nature of the MICAD detector, some precautions should be taken to protect the instrument and obtain the best results with certain sample types.

1. Techniques for Optimization

a. Calibration

The gas mixture is critical to the consistent operation of the InstantImager. Typical A1 size gas tanks will last 3 to 4 months at the appropriate flow settings; however, every bottle is slightly different in composition despite certification from the supplier. The InstantImager has a calibration plate and automatic procedure that should be run on a weekly basis in order to ensure that results are consistent from week to week despite variations in temperature and pressure. The calibration procedure takes approximately 10–15 min and automatically sets voltages to a level based on the transformed spectral index of the sample (tSIS) as described by Kessler (1989) and in Chapter 9 of this book. The calibration source is a ^{14}C-labeled ink drawn paper sealed in plastic. Upon changing gas bottles, the system should be perfused with the new mixture for 48–72 h before calibration. Exact procedures are documented in the InstantImager reference manual. The MICAD detector must be kept on gas at all times in order to avoid diffusion of room air and moisture into the detector, which could be damaged eventually by corrosion. Details of long term storage of a MICAD detector on less expensive, dry CO_2 is also described in the reference manual.

b. Sample Presentation

Since the detector is flexible in nature due to the large number of microchannels per mm^2, it is not perfectly flat. The platform, which is driven by pneumatic pressure, is machined perfectly flat. Therefore, it is recommended that the sample be placed on one of two flexible surfaces provided with the instrument to optimize the sample-to-detector interface. Samples smaller than 20 cm × 20 cm should be placed on the "airbag" sample support to provide even pressure across the detector area. Samples larger than 20 cm × 20 cm should be placed on a foam sample support pad also provided with the instrument. This provides flexibility in the sample to conform to the detector in a most efficient way. In addition, the sample can be placed in the corners or edges of the detector area since the detector is attached in the corners and less flexible there.

Drying down with electrophoresis film (Sigma Chemicals, St. Louis, Missouri) can help to keep dried-down gels flat. However, some gels absorb moisture over time and become curled. Dried-down gels that are curled in the corners or at the edges pose a particular challenge for the InstantImager and should be treated differently. The airbag and the foam sample support pads are so flexible that they enable a curled, dried-down gel to push away from the detector. Distance from the detector causes extremely poor resolution due to the isotropic nature of radioisotope emissions. With film autoradiography and phosphor screen imaging, samples can be pressed tightly against the film or screen without possibility of damage. These samples should be taped down in the corners to the machined flat sample platform, and "matted" with filters of similar thickness to the gel being imaged. This will enable the platform to press against the sample while providing an even distribution of pressure against the detector.

Fluorography or intensifying screens will not improve results from the InstantImager. Recall that the detector is a nuclear detector and recognizes ionizing radiation rather than light. Calibration and proper sample presentation are critical to getting accurate results using the InstantImager Electronic Autoradiography Method.

2. Advantages of Electronic Autoradiography

Electronic autoradiography with the MICAD detector has a number of distinct advantages over film and storage phosphor screen autoradiography. This is evidenced by the fact that approximately 65% of the first users of electronic autoradiography already had access to film autoradiography and storage phosphor screen autoradiography methods in their research labs.

One of the most striking advantages of the InstantImager Electronic Autoradiography System is that the image is displayed on the screen as the data are being accumulated. This feature makes the InstantImager most useful for optimizing experimental conditions. For example, a user may place a northern or western blot into the InstantImager while it is still moist in order to see if washing is sufficient. The sample may then be removed and washed further if the background is still too high. The time saving can be enormous if the experiment failed or the result is negative. Instead of waiting for hours or days to see the result, the InstantImager can produce a first update in 15 s and provide an indication of the results. In addition, samples can be analyzed while the counting is progressing so that when the statistical accuracy of the count is sufficient to provide confidence that the result is significant over background activity, the acquisition may be stopped. Real-time image display and analysis eliminates the need to guess the exposure time of autoradiography. Acquisition times can be determined objectively by the use of electronic autoradiography. See the Southern blot application in Section III.D.5. for an example of this advantage.

Another important advantage of the use of electronic autoradiography for imaging is the accuracy with which quantification can be obtained with direct nuclear detection. Because there are no intermediate capture devices like film emulsion or phosphor screens, the error associated with variations in these devices is also eliminated. Results have been correlated with the "gold standard" of liquid scintillation counting with 99.5% accuracy to prove the technology (Englert *et al.*, 1995). Figure 13.18 is a

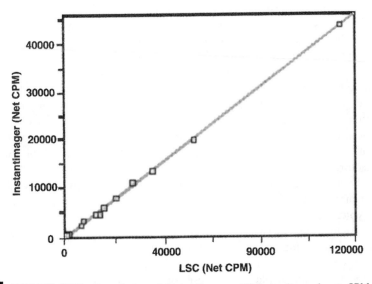

FIGURE 13.18 Correlation of InstantImager CPM results and net CPM results from a Tri-Carb 2300TR (Perkin Elmer Life and Analytical Sciences, Boston, MA). The correlation coefficient, $R^2 = 0.995$ illustrating the accuracy of the direct nuclear imaging method of electronic autoradiography.

graphical representation of the results of experiments to test the relationship between direct nuclear quantification with the MICAD and liquid scintillation analysis.

The system can be calibrated routinely, and variables associated with counting by the MICAD can be controlled. For example, the uniformity of the detector is better than $\pm 5\%$ across the entire detector area and the reproducibility as determined by chi-square statistics is greater than 95%. The benefit of this accuracy is that samples measured on different days can be compared between images without regard for acquisition time and without the use of standards on every image. Unlike the situation in film and phosphor screen methods, the linearity of each image is independent of the counting time as described in Section III.B.2.

Finally, the speed of the InstantImager for those samples that require less than 24 h to obtain a sufficient image is significantly faster than traditional methods. Often results can be obtained with a great degree of statistical accuracy in less than 30 min (Batt, 1995). This provides the ability to move on to the next step in an experiment that depends on the results of an image.

3. Disadvantages of Electronic Autoradiography

One disadvantage is that the sample is placed directly in the instrument. In a multiuser situation in which there are a number of samples that require lengthy exposure times, the InstantImager can be in constant use. Although for any given sample the results will be obtained faster than other methods, it is possible to make only one acquisition at a time. The overall imaging throughput of the laboratory may be less if there are many samples which require long acquisition times. However, if samples are active enough to require less than 24 h for a film exposure, imaging times with the InstantImager are short, and this disadvantage is insignificant.

Samples which require resolution of greater than 1 line pair/mm, such as whole body autoradiography, receptor autoradiography, or *in situ* hybridization in brain tissue sections, and samples with closely spaced bands or spots, are difficult or impossible to measure with spatial accuracy with the InstantImager. The CPM results and counts are accurate but the structures and spatial separation of samples may be insufficient for these applications. This limitation is due to the physical design of the MICAD detector.

In some countries, the availability and cost of the argon gas mixture may be a disadvantage, but in general it can be provided by any local gas supplier of typical laboratory gases.

Finally, the statistical nature of the CPM results of the image file make the qualitative images from the InstantImager less appealing than those from other detection methods such as storage phosphor screen imaging. For publication and display purposes, some features such as "smoothing" and "enhancing" display by interpolation techniques have been implemented, but the actual dot pitch of the images are limited to $250\,\mu$m pixels. Some journals require a higher degree of pixel resolution, which can be achieved only by importing into specialized image presentation software such as Adobe Photoshop or Photo Finish.

D. Applications of Electronic Autoradiography

This section describes selected applications for which the InstantImager is most commonly used. These include metabolism studies, postlabeling of DNA adducts, gel shift assays, enzyme kinetic studies, northern blot analysis, and Southern blot analysis. Typical applications for which the InstantImager is used are those that require accurate quantification and comparisons between images measured on different occasions.

1. Metabolism Studies

Thin layer chromatography (TLC) is commonly used to separate molecules based on differences in affinity toward a particular solvent. The following two examples illustrate the use of InstantImager for quantification of metabolites separated by TLC.

During the approval process of a new pesticide, herbicide, fertilizer, or other agricultural product, metabolism studies on all possible breakdown products must be conducted. Typical samples are ^{14}C extracts of metabolites separated by TLC, as shown in Fig. 13.19. Samples are run in one dimension

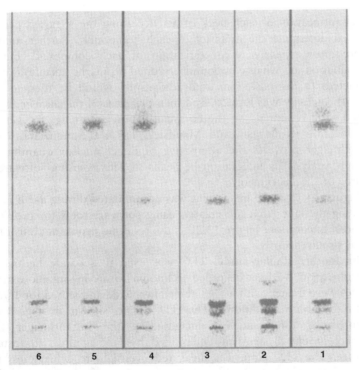

FIGURE 13.19 TLC separation of ^{14}C metabolites, imaged for 2 h using the InstantImager. Lanes which are created in software are used to create profiles shown in Fig. 13.20.

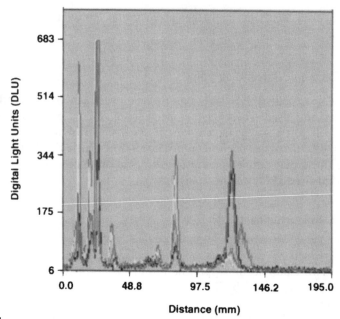

FIGURE 13.20 Profiles which are the sum of the pixels across lanes created with InstantQuant 2.04 which is the InstantImager software. Integration of the peaks can be performed automatically or manually to get CPM results.

and quantification of each peak of activity along the Y-axis is performed in order to determine the quantity of each compound. Further analysis may require more extensive characterization of each compound, but first the determination of what compounds resulted from the metabolism must be completed. In the past, film autoradiography would be used to determine where the activity was located, and then the areas of the silica resin would be excised and counted by liquid scintillation counting to get accurate quantification (Kobayashi and Maudsley, 1974; L'Annunziata, 1979). The InstantImager provides the alternative of direct nuclear quantification and one-step analysis by integration of peaks of activity using software analysis features of the instrument.

Figure 13.19 is the image that was accumulated during a 2-h acquisition, including the lanes that are created using software tools for profile analysis and peak integration. Figure 13.20 is the resulting profile analysis of all lanes shown simultaneously.

In another similar study, TLC was used to separate herbicide parent molecules from various degraded unknown forms in organic extracts. Soil samples were treated with a ^{14}C-labeled herbicide and stored for 1 year under strict laboratory conditions. The TLC plate shown in Fig. 13.21 was measured by direct nuclear imaging on the InstantImager Electronic Autoradiography System and quantified by lane analysis in InstantQuant software. The TLC plate was measured overnight to achieve less than 5% statistical error.

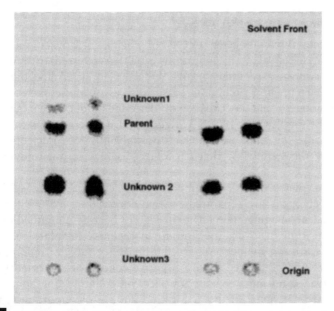

FIGURE 13.21 TLC used to separate herbicide parent molecules from various degraded unknown forms in organic extracts. The image was acquired in 4 h to achieve a high degree of statistical accuracy.

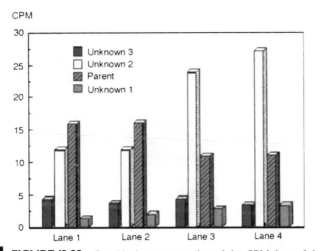

FIGURE 13.22 Graphical representation of the CPM data of the TLC samples in Fig. 13.21.

Figure 13.22 shows a plot of the CPM data obtained from the InstantImager. The graph illustrates an increase in the ratio of degraded forms to parent molecules in Lanes 3 and 4 compared with Lanes 1 and 2. With the CPM results it is also possible to compare exact quantities of each unknown with respect to other samples from different images as well (Veal and Tian, 1997b). For details of sample preparation and methodology of metabolism studies using TLC analysis, refer to Touchstone and Sherma (1985).

2. Postlabeling DNA Adduct Assays

The potential carcinogenic effects of chemicals can be measured by quantification of covalent modifications of DNA referred to adducts, that result from exposure to these agents. In order to determine the potential effects of environmental pollutants or substances of abuse such as cigarette smoke, methods should be capable of detecting adducts that are a small fraction of the total sample of DNA. Reddy and Randerath (1986) describe a method based on the radioactive postlabeling of nucleotides from enzymatically digested DNA that can detect as few as one adduct in 10^{10} nucleotides.

After enzymatic digestion, samples from cells that were subjected to potential carcinogens are separated by two-dimensional TLC and measured by direct nuclear imaging. The image in Fig. 13.23 was quantified during the 25-min acquisition by using the "group of regions" template to draw elliptical regions around the activity. On the image, counts are reported as tags to the ellipses and compared with a negative control of untreated samples.

The most significant benefits of electronic autoradiography for this application are the speed of the acquisition and, most importantly, the long linear dynamic range. The same sample would require two separate film exposures and scraping into liquid scintillation vials for Cherenkov counting (Veal, 1997).

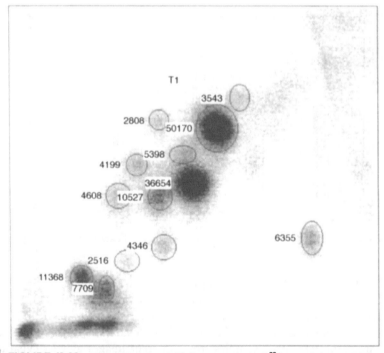

FIGURE 13.23 Two-dimensional TLC separation of ^{32}P post-labeled DNA adducts. The image was acquired in 25 min and analyzed with a "group of regions" template in InstantQuant software. Activities of the regions quantified are shown in the figure as "counts" accumulated during the acquisition time. The range of activity is 300–6000 DPM.

3. Gel Mobility Shift Assays

Gel electrophoresis can be used to visualize and quantify mobility shifts in nucleic acids that can result from the interaction of a particular protein and DNA sequence. Typical gel shift assays are performed using a standard binding mixture that includes a ^{32}P-labeled probe (Chesters *et al.*, 1995).

Figure 13.24 is an example of a gel shift assay quantified using electronic autoradiography. The ^{32}P-labeled probe and probe-protein complexes were separated by polyacrylamide gel electrophoresis and imaged for 5 min using the InstantImager. This type of assay can be used to test a large group of compounds for effects on this binding interaction. Quantification of percent shift can be compared gel to gel throughout the screening process.

Another use of the gel shift assay is to test the inhibitory effect of an enzyme on a particular protein DNA complex. In the example below, LLC-PK1 cells were treated with 1 μM 11-deoxy-16, 16-dimethyl PGE$_2$ to induce activator protein-1 (AP-1):DNA binding activity to a ^{32}P-labeled 12-O-tetradecanoyl phorbol-13-acetate responsive element (TRE). Inhibitory effects of protein kinase C (PKC) inhibitor H-89 on binding activity in nuclear extracts were analyzed using mobility shift assays (Weber *et al.*, 1997). Figure 13.25 shows the image obtained in 22 min using electronic autoradiography. The graph of the results in Fig. 13.26 illustrates the induction effect of ddm PGE$_2$ on AP-1 binding to the TRE element and the

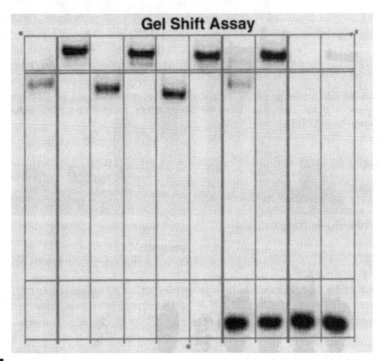

FIGURE 13.24   Gel mobility shift assay used to test the inhibitory effect of an enzyme on a particular protein DNA complex. The ^{32}P-labeled probe and probe-protein complexes were separated by polyacrylamide gel electrophoresis and imaged for 5 min using the InstantImager. Lane analysis and areas of integration are shown.

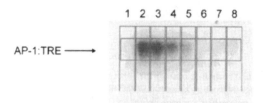

FIGURE 13.25 Image of gel mobility shift assay was acquired in 22 min using electronic autoradiography.

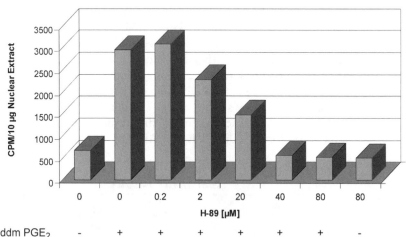

FIGURE 13.26 Graph of the induction effect of ddm PGE$_2$ on AP-1 binding to the TRE element and the dose-dependent inhibitory effect on H-89 on PGE$_2$ induction. [From Weber et al. (1997).]

dose-dependent inhibitory effect on H-89 on PGE$_2$ induction. One other example of the use of the InstantImager for analysis of gel mobility shift assays is described in Coccia et al. (1995).

4. Northern Blot Analysis

Detection and quantification of mRNA as a method of determining expression levels of proteins of interest or the effects of agents that interfere or induce transcription and protein expression levels are often performed by northern blot analysis. Shi et al. (1997) describe their use of northern blot analysis to quantify the effect of N-acetylcysteine (NAC) and dimethyl sulfoxide (DMSO) on macrophage inflammatory protein-1α (MIP-1α) mRNA expression levels in response to lipopolysaccharide (LPS) in a rat alveolar macrophage cell line.

For antioxidant treatment, cells were incubated with 1, 10, or 20 mM NAC or 1% DMSO for 1 h, followed by the addition of LPS for 4 h. The upper panel in Fig. 13.27 is the membrane hybridized with radiolabeled

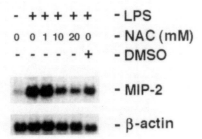

FIGURE 13.27 Northern blot analysis to quantify the effect of *N*-acetylcystein (NAC) and dimethyl sulfoxide (DMSO) on macrophage inflamatory protein-2 (MIP-2) mRNA expression levels in response to lipopolysaccharide (LPS) in a rat alveolar macrophage cell line. The upper panel is the membrane hybridized with radiolabeled MIP-2 cDNA and the lower panel is the same membrane hybridized with mouse *β*-actin cDNA. [From Shi *et al.* (1999).]

MIP-1α cDNA and the lower panel is ethidium bromide-stained 28S ribosomal RNA, indicating the RNA loading per lane. As quantified by the InstantImager, both NAC at 10 and 20 mM and DMSO at 1% significantly reduced the induction of MIP-1α mRNA by LPS. Pretreatment with 1 mM NAC did not influence the LPS induction of MIP-1α mRNA expression. Details of northern blot analysis methods are given in Shi *et al.* (1999). One other example of the use of the InstantImager for northern blot analysis is described in Hesketh *et al.* (1994).

5. Southern Blot Analysis

Southern blot analysis is a common way to detect single-copy sequences of human DNA using a filter transfer of fractionated DNA to hybridize with a ^{32}P-labeled probe. Tyler-Smith and Southern (1993) describe their method of analyzing Taq. I digests of human male and female DNA using a 1.4-kb probe. The results were compared between a film which required 16 h of exposure time and four image acquisitions with the InstantImager. In 3 min, the image could be seen on the screen (Fig. 13.28) and by 1 h (Fig. 13.29), no further data acquisition was necessary (Tyler-Smith and Southern, 1993). A 10-min acquisition time (not shown here) was enough to reduce the statistical counting error of this experiment to less than 10%. A visual comparison of the 3-min acquisition of Figure 13.28 and the 1-h acquisition of Fig. 13.29 illustrates the improvement in image display that occurs with the accumulation of data.

Since the radioactive decay is a random process, the accuracy of the count improves with the square root of the counting time. Figure 13.30 is a graphical illustration of the relationship between acquisition time, net counts, and the two sigma counting error. This is an advantage of the direct nuclear detection method of the InstantImager, which was discussed previously in Section IV.C.2.

Additional references in the literature describe applications of electronic autoradiography using the MICAD detector of the InstantImager for nuclear run-on experiments (Qui *et al.*, 1996; Tian *et al.*, 1997); for radiopharmacy

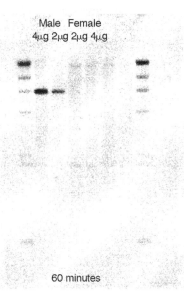

FIGURE 13.28   Three minute acquisition of ^{32}P-labeled Southern blot of Taq. I digests of human chromosomal DNA. [From Tyler-Smith and Southern (1993).]

FIGURE 13.29   One hour acquisition of same Southern blot shown in Fig. 13.28. The visual aspects of the image are improve as more data are accumulated. [From Tyler-Smith and Southern (1993).]

and nuclear medicine (Decristoforo *et al.*, 1997), ^{35}S-labeled protein gel electrophoresis (Buccione *et al.*, 1996), autophosphorylation studies (Treharne *et al.*, 1994; Upton *et al.*, 1995; Zubiaur *et al.*, 1995), enzyme kinetic assays (Michels and Pyle, 1995), and immunochemical blotting (Hansen, 2001).

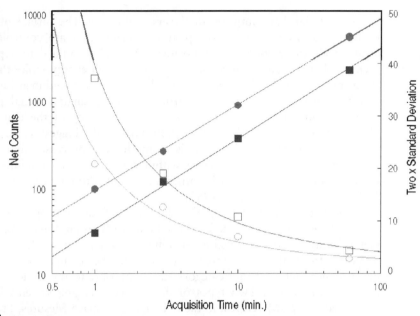

FIGURE 13.30 The relationship between acquisition time, net counts and the two sigma counting error. The accuracy of the measurement improves with the square root of the counting time [From Tyler-Smith and Southern (1993.)

V. CCD CAMERA IMAGING

Charged-coupled device cameras have been applied to astronomy, physics and engineering problems over the last two decades. They are uniquely suited to low light, high resolution applications and have been incorporated into technologies such as facsimile machines, photocopy machines, home video cameras, bar code readers, and other light sensitive detection applications (Weisner, 1995; Slaughter, 2000). For life science imaging, CCD cameras provide a new level of resolution and sensitivity for direct detection of beta events in two-dimensions. The first life science imaging CCD camera systems were developed for gel documentation of samples stained ethidium bromide and illuminated with a UV lightbox, or just colorimetric stain and white light illumination. Detecting radioisotopes with a CCD camera is a much bigger challenge, because the beta energy must be first converted to photons in order to be detected. This section describes technology of the CCD detector, and three unique methods of applying this technology to localization and quantification of beta particles in biological samples.

A. CCD Technology

A CCD camera is a solid state electrical device that is capable of converting light input into electronic signal. The term "charged-coupled" refers to the coupling of electrical potentials that exist within the chemical structure of the silicon material that comprises the layers of the chip. The surface of the CCD is broken down into pixels, or picture elements that are controlled by an array of electrodes or gates. A positive potential is applied to a portion of each pixel such that when photons of high enough energy strike the surface, electrons are effectively captured and binned in a two-dimensional array, which is then detected as an electronic pulse for digital signal processing (Kristian and Blouke, 1982). The quantum efficiency of the conversion of photons to electrons is about 0.70–0.80. One of the biggest sources of error in this type of photon detection is from thermal processes that mask the subtle changes in electron charge within the matrix. To minimize thermal background effects, some CCD cameras are cooled to temperatures as low as 150K (Kristian and Blouke, 1982).

The bigger challenge for the detection of radioisotopic emissions is getting the energy of the beta converted efficiently to photons. Several unique methods have been developed for making this conversion, each of which is so different from the other that they cannot be compared in terms of performance, but are better categorized by appropriate applications for which they are best suited. Two types of digital beta imaging systems have been developed, the β IMAGER and the μ Imager (Biospace Mesures, France) that incorporate a CCD in their detection method. Another type of CCD area imaging system, two of which are the LEADseeker (Amersham BioSciences, Piscataway, NJ) and the ViewLux (Perkin Elmer, Boston, MA), has been developed specifically for the detection of radioisotopes in microplate applications for high-throughput screening assays.

B. CCD Digital Beta Imaging Systems

Two CCD based direct beta imaging systems have been developed independently using two distinct beta energy to photon conversion techniques. Both systems combine precise localization with resolution high enough for tissue section applications with speed and accuracy of direct beta counting (Crumeyrolle-Arias *et al.*, 1996).

I. The β IMAGER

The β IMAGER was developed in collaboration with George Charpak, who was awarded the Nobel Prize in Physics in 1992 for his work on the multiwire chamber, described in Section IV. Whereas the InstantImager Electronic Autoradiography system uses the MICAD detector to localize the beta event from the sample, the β IMAGER uses a scintillating gas mixture of Argon and triethylamine (TEA) and a parallel plate avalanche chamber across which 3000–4000 volts is applied. Radiolabeled samples are placed on the sample holder and automatically positioned so that the gas chamber is closed. Each beta particle leaving the sample ionizes the gas along its trajectory producing about 100 electron/ion pairs per centimeter. These electrons are then multiplied in the parallel plate chamber creating an avalanche that produces UV light at a wavelength of about 280 nm. The photons form a spot that is indicative of the origin of the beta event and its energy of emission. Each spot of light is focused with a UV lens into an image intensifier. The image intensifier converts the photons to electrons using a photocathode, that amplifies the flux in a high vacuum tube with DC current applied. The high energy electrons are converted back into photons by a phosphor screen at the other end of the tube, so that the CCD camera can convert them back to an electronic count in each pixel. Digital signal processing is similar to that described previously in Section IV.A.2, so the image is displayed in real time. In addition, the β IMAGER performs another calculation to discriminate the energy of the original beta emission, which provides the capability of dual-labeled experiments.

a. Performance

The β IMAGER has a detection area similar to the InstantImager and many storage phosphor systems at 20 cm × 25 cm in the full field of view. The zoom option provides higher spatial resolution, but concomitantly decreases the area for samples. Table 13.5 lists the sample area and the resolution capabilities of different isotopes. Smallest pixel size is 10 μm, much smaller than the typical sample feature.

Table 13.5 shows the resolution capability of β IMAGER for various isotopes and two zoom positions (based on full width at half maximum peak discrimination). Two other zoom positions exist in between the highest and lowest zoom, but data for resolution is not available. The resolution for these intermediate zoom positions will be between those listed in the table.

The detection threshold for tritium for this system is particularly interesting. Tritium applications, typically quantified using film (see Section II.1.) can take weeks and months of exposure. The detection threshold for this system for tritium is about 0.007 CPM/mm^2. Detection thresholds for other isotopes are 0.01 CPM/mm^2 for ^{14}C, ^{35}S, and ^{33}P, and 0.1 CPM/mm^2 for ^{32}P with detection efficiency between 50 and 100% of the particles emerging from the sample, depending on the isotope. The linear dynamic range of detection is over 4 orders of magnitude, again not limited by the capture of beta events because they are counted instead of being captured. Therefore the linearity limits have more to do with the electronics of the components of the system rather with a physical feature. Figure 13.31 Graph shows the linearity and detection limit for ^{3}H that can be achieved with the β IMAGER (Dominik *et al.*, 1989).

TABLE 13.5 Resolution capabilities of the β IMAGER for various isotopes at two zoom positions.

Isotope	Full Field of View 20 cm × 25 cm area (μm)	Maximum Zoom 25 mm × 33 mm area (μm)
^{3}H	200	60
^{14}C, ^{35}S, ^{33}P	350	120
^{32}P	500	150

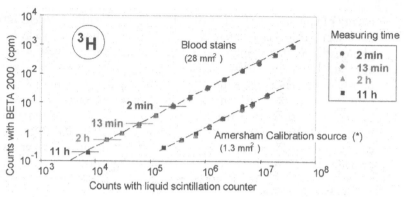

FIGURE 13.31   Linearity of the β IMAGER.

b. Quantitation Methods

Tritium labeled samples are loaded directly into the sample chamber with no cover. Samples labeled with other more energetic isotopes are covered with a Mylar film to protect the detector from contamination. About 10^7 photons are produced by the scintillating gas in the chamber from an average 6 KeV beta particle from a ^{3}H labeled sample. The time required for data accumulation is typically about 10 to 50 times faster than phosphor screen imaging and up to 500 times faster than film. In particular for ^{3}H, sample exposure can be done in less than 8 h as opposed to months on film. Samples are prepared in the same manner as they would for film autoradiography or storage phosphor screen imaging. They should be flat and dry as possible for best results. It is recommended that the Argon gas be C grade or better, with purity of greater than, or equal to 99.998%. The sample platform is 200 mm × 250 mm and can hold 2–4 rat sections at once, or up to 15 slides. For maintenance, the sample platform should be wiped periodically with alcohol. No calibration is required.

c. Advantages

The advantages of the β IMAGER are similar to those of the InstantImager, with the added advantage of availability. Whereas other direct beta counters have parts that have become obsolete, the β IMAGER continues to be manufactured and sold in Europe through Biospace Mesures of France, and in the US through IN/US, Tampa, Florida.

One important advantage of the β IMAGER is the speed at which a sample can be quantified relative to traditional film methods, particularly for tritium applications. Real time imaging enables one to see the image as it is being acquired so that it can be stopped early if the experiment failed, or when the data is sufficient for accurate measurement and localization of activity. The consumable cost is limited to the detection gas, so the cost is less compared to other methods, such as storage phosphor imaging of tritium labeled samples for which screens can be used only once (see Section III.C.1).

The other important advantage of the β IMAGER is the reproducibility and the accuracy of the measurement. The lack of intermediate steps in the process of quantitation that is a function of direct nuclear detection provides fewer sources of error and subjectivity compared to film and phosphor screen technologies in particular. The added feature of dual label measurement by discrimination of beta particle energy in the labels of the sample makes this a more versatile system, particularly for some receptor autoradiography applications.

d. Disadvantages

Disadvantages of this type of system are physical in nature. The ability to make one image at a time is not convenient for sharing of a system among many users with low activity samples. If there are multiple samples in an experiment, they can each be measured with accuracy and compared directly, but they must be measured sequentially. This limits the throughput capabilities of this type of system as compared to storage phosphor imaging or film, by which samples can be measured simultaneously with multiple screens and sample holders. In addition, although no film or phosphor screens are used, the Argon gas laced with TEA is a consumable item that needs to be monitored. Finally, since the sample does go directly into the instrument, the potential for contamination of the system exists and it must be installed in an area approved for the presence of radiolabeled materials.

e. Applications

Although the β IMAGER is capable of a wide range of applications that include detection of radiolabeled material distributed on a flat, dry sample matrix, the high resolution and ^{3}H detection capabilities make receptor autoradiography the predominant application for which the system is used. Blots, gels and TLC plates are more commonly measured on a storage phosphor system or electronic autoradiography system. However, these detection methods have not been able to replace film autoradiography for tritium detection. Although the resolution for ^{3}H using film is still beyond compare, the β IMAGER is so much faster and more quantitative that it useful for some receptor autoradiography applications where exact localization is not the main objective. For measuring penetration of a drug on known receptors, distribution of a labeled drug in whole body autoradiography and other receptor assays, the β IMAGER provides fast, quantitative results.

Receptor Autoradiography: After subcutaneous injection of tritiated known NK3 receptor antagonists, gerbil brain sections were prepared on glass slides for *ex vivo* autoradiography. The quantitation was performed on the β IMAGER to determine the extent to which the compounds penetrated the central nervous system and could be considered potential drug candidates for treatment of psychiatric disorders in which NK3 receptors are involved. Slides were placed in the β IMAGER and imaged in real time as each beta emission was detected. The radioligands binding signal is expressed as counts per minute per square millimeter and the images are shown in Fig. 13.32.

Zoom

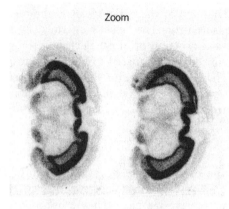

■■■ **FIGURE 13.32** *Ex vivo* measurement of penetration of tritiated compounds in gerbil brain tissue sections. (Courtesy of Dr. X. Langlois Janssen Pharmaceutical, Beerse, Belgium.)

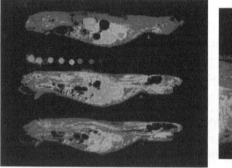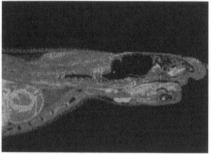

■■■ **FIGURE 13.33** Images of tritium labeled drug distribution in rats, full field of view and with zoom display, after 8 h acquisition with β **IMAGER**. (Courtesy of Dr. A. Mollet, Pfizer, UK.)

It was confirmed that the dose occupying the receptors as determined by this fast and easy method compared positively with *in vivo* behavioral data, providing confidence in the use of this method as a screening tool in drug discovery (Langlois *et al.*, 2001).

Whole Body Autoradiography: In drug discovery, ADME studies to show distribution of radiolabeled compounds in animals is conducted as a part of the regulatory process. Tritium labeled material is injected and animals are sacrificed over a time-course and quantified with respect to localization in tissues in each section and compared with prepared standards. Figure 13.33 shows the distribution in a rat section image acquired for 8 h and displayed with the full field of view and a 75 × 100 mm zoom, for precise localization. Registration of the autoradiographic image with and optical image, such as one acquired by scanning the section on an optical scanner, is also available for better tissue identification and contouring.

2. The μ Imager

The μ Imager is based on the concept of contact imaging, or placing the intensified camera detector in direct contact with the sample. The sample is covered with a thin Mylar sheet coated with a $3\,\mu m$ thick coating of P47, a cerium-doped yttrium silicate solid scintillator. The intensified CCD camera of the μ Imager is put in direct contact with the scintillant-coated sheet covering the radiolabeled sample. By this method, the μ Imager can achieve resolution of about $15\,\mu m$ for ^3H, and $25\,\mu m$ for ^{14}C, ^{35}S, and ^{33}P in the $25\,mm \times 32\,mm$ field of view. The tradeoff for this high level of resolution is a smaller field of view or sample area, which is the size of a conventional microscope slide. The photons emitted from the other side of the scintillator sheet enter directly into a double microchannel plate (MCP) image intensifier that is coupled to the CCD Camera with a fiberoptic taper. Microchannel plates are thin plates of conductive glass containing many holes of about $10\,\mu m$ in diameter. Secondary electron emissions occur successively leading to multiplication of the signal and higher luminous gain (Charon *et al.*, 1998). This image intensifier is used to achieve the highest possible signal and best sensitivity for this highly specialized CCD application. Eliminating the air space between the sample and detection device is the key to achieving this high level of spatial resolution with this direct beta imaging system. Images are displayed in real-time as the sample is counted by a similar process as the β Imager in that the spots of light indicate both origin and energy of the original beta event. In this way, the μ Imager can also discriminate between different isotopes of a dual-labeled sample. In addition, the μ Imager is capable of taking an optical image of the same sample to provide an overlay of the exact tissue morphology to accompany the image of the distribution of radiolabeled material in the sample (Charon *et al.*, 1998).

a. Performance

As a direct beta counter, we can describe the sensitivity in terms of detection efficiencies, and the time to acquire an accurate representation of a radiolabeled sample. The μ Imager is capable of detecting most isotopes with about 60 to 100% efficiency. Background is inherently low as a function (1) the thinness of the scintillator and (2) the ability to discriminate background pulses by energy analysis algorithms in the software. Detection threshold is about $0.4\,CPM/mm^2$ for ^3H and about $0.04\,CPM/mm^2$ from ^{14}C, ^{35}S, and ^{33}P. This is not quite as high in sensitivity as the β IMAGER detection method, but achieves results on the order of 10–20 times faster than storage phosphor imaging and about 100–300 times faster than film (Rapkin, 2001). Counting linearity is in such a small area that the accuracy is $\pm 2\%$ over nearly 5 orders of magnitude. The most significant performance feature of the μ Imager is the resolution capability. The main applications for this system are those that are small and require precise localization of activity within tissue sections or on glass slides small enough to be measured in one image acquisition. As mentioned previously, the FWHM of the smallest feature detected by the system is $15\,\mu m$ for ^3H and $25\,\mu m$ for ^{14}C, ^{35}S, and

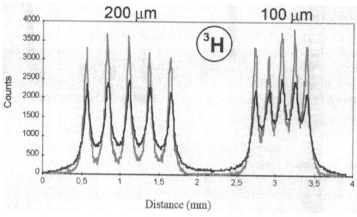

FIGURE 13.34 Resolution capability of the μ Imager.

[33]P (Laniece *et al.*, 1998). Figure 13.34 illustrates the high resolution detection of the μ Imager.

b. Quantitation Methods

Samples covered with a scintillator sheet are placed onto the sample platform and automatically loaded to ensure close contact with the detector. Samples may be labeled with one isotope, or dual labeled with two isotopes of different energy spectra. The sample holder is designed specifically for glass slides, so appropriate applications would be small tissue sections immobilized on glass slides that are used in receptor ligand binding or *in situ* hybridization, or DNA microarrays printed on glass slides. One option includes a four slide automatic slide feeder for multiple sample experiments.

One way to optimize results is to use the optical image morphology to define the areas of interest rather than the radioisotopic image. Whereas some tissues may look homogeneous, but are not relative to the labeled receptor or mRNA of interest, others may look like they are different in nature, and do not show a significant difference in distribution of radiolabeled material. Still other samples contain very small structure that can be more easily defined and separated from the others visually when both the radioisotopic and optical image are considered. This is a key benefit of using this type of system that is capable of both types of detection. In addition, optical images from a microscope system can be overlaid in the software, providing a structural component to act as a template for localization of radiolabeled features.

c. Advantages

The most significant advantage of this system compared to other direct beta detection alternatives is the high resolution at which a sample can be measured. Compared to film and storage phosphor systems, the μ Imager provides the answers and an image significantly faster and with a high degree of quantitative accuracy. In addition, no gas is required and the scintillator sheets are relatively inexpensive. For detection of high resolution samples in tissue sections, this is a unique and highly accurate method of providing quantitative images in real-time.

The sample loader provides some walk away throughput that can be appreciated when multiple sections are created for a given experiment, such as sequential brain tissue sections or multiple DNA microarray samples. The ability to perform discrimination between two labels is also a unique feature that should not be understated. The fact that one image can contain the data of two labels means the imaging time (and error) is half that of a system, such as a storage phosphor system, that would require two exposures to separate screens.

d. Disadvantages

The main disadvantage is the size of the viewing area and the requirement that all samples are mounted on glass slides. The sample viewing area is only specified at 24 mm × 32 mm. Only one sample can be measured at a time and measuring times can be 10–20 h. In addition, for multi-user environments with multiple radioimaging applications, this is a very limited system and lacks versatility for imaging a variety of applications.

e. Applications

The μ Imager is best suited for high resolution, small sample area applications that require precise quantification of activity. As a direct contact imager, it is particularly well-suited for high resolution distribution of activity in tissues and has been adopted for this purpose. In addition, the ability to discriminate between two isotopes provides the ability to perform dual label experiments as described in the following sections.

Dual Label Microarrays Analysis: The major challenges of using microarray technology are mainly associated with sensitivity. Detecting rare mRNA species, comparing mRNA species using small amounts of tissue material, and detection of subtle changes in expression require high sensitivity and accuracy for which a direct beta imaging system with high resolution, such as the μ Imager. The discrimination of isotopes can be exploited to provide more information for each microarray experiment by using two different isotopes to label two mRNA species isolated from separate tissue sources. Figure 13.35 is an example of a high density mRNA array imaged with the μ Imager. Two measurements are reported based on the intensity of each isotope within each spot to show differences in gene expression associated with each tissue. As little as 100 ng of whole-brain polyA RNA was used for probe synthesis, without any amplification, which corresponds to approximately 5 mg of starting neural tissue (Salin *et al.*, 2002).

Dual Label in situ Hybridization: Distribution of mRNA in tissue sections is a way to show localization of areas of unique gene expression referred to as *in situ* hybridization. Using two probes labeled with different isotopes, separate images can be displayed which shows distribution of activity of each isotope in the same tissues using the μ Imager. In the example shown in Fig. 13.36, two probes, one labeled with ^{3}H and one with ^{35}S are hybridized to rat brain hippocampus. The middle image shows the combined activity, the top shows the ^{3}H label in the image and the bottom shows the ^{35}S label in the image. The ability to discriminate the distribution of each

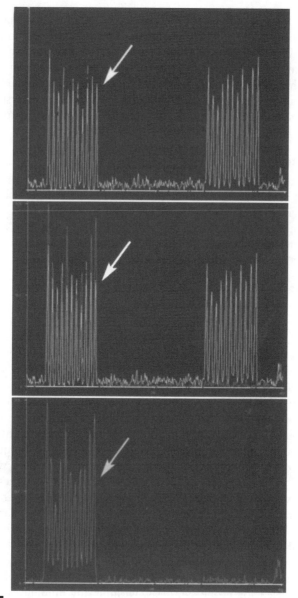

FIGURE 13.35 Dual label Micro array assay using ^{33}P-dATP and ^{3}H-dCTP labeled mRNAs. The isotopes are discriminated by spot size and quantified separately.

isotope makes it possible to show specific areas of expression (Salin *et al.*, 2002).

Superimposing Image of Radiolabeled Section and Optical Image: One challenge with localization of radiolabeled material is to determine the actual position of the activity with respect to the morphology of the organs of interest. Optically, the morphology can be distinguished, but in the radiolabeled image, the tissues can only be distinguished by the levels of

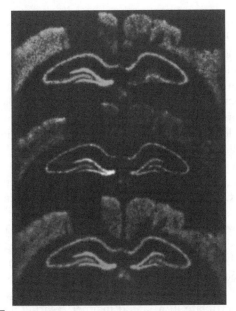

FIGURE 13.36 Dual label in situ hybridization using ^{3}H- and ^{35}S-labeled probes. (Courtesy of Prof. Mallet and Dr. Dumas, LGN, Paris, France.)

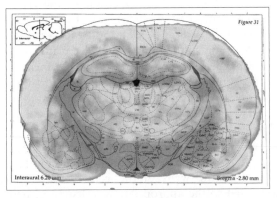

FIGURE 13.37 Example of radiolabeled image superimposed on optical image using μ Imager. (Courtesy of Dr. P Hantraye and Dr. L. Besret, SHFJ, Orsay, France.)

activity which form the image. By using the optical image, or an optical representation of the tissue being imaged, the localization of the activity can be more precisely determined. Figure 13.37 is an overlay of two images, one created by detection of radioactivity distributed in the sample, and the other is a drawing of the expected morphology of the brain tissue (Salin *et al.*, 2000).

3. HTS Imaging Systems: Leadseeker and Viewlux

In response to the ever increasing need to screen microplate based samples for drug discovery, area imaging systems have been developed that are capable of detecting radiolabeled Scintillation Proximity Assays (SPA) (Amersham BioSciences, Piscataway, NJ) or radiolabeled samples in scintillant coated plates known as FlashPlates (Perkin Elmer, Boston, Mass). In addition to radioisotopic detection, these systems are designed for multiple detection methods including fluorescence, luminescence and absorbance. However, detection of radiolabeled material in microplates is one of the most challenging applications because the light output is particularly low and the geometry of detection is difficult due to the three dimensional nature of microplate based samples.

In terms of detection of beta particles, three components of these microplate imaging systems are important: the CCD camera, the lens and the scintillation material. The CCD camera used in these sensitive cameras is a thinned, back illuminated specialized CCD that provides higher quantum efficiencies. Cooled with a recirculated refridgerant, allows the CCD to run at temperatures as low as possible with liquid nitrogen, but without refilling and maintenance (Ramm, 1999). For radioisotopic detection, the light must be gathered efficiently and accurately with respect to the positional coordinates of the sample. Using a typical lens, to view flat samples is problematic in that it views the wells of the microplate in a lateral manner creating a geometric distortion known as "parallax error". The LEADseeker and the Viewlux incorporate a large telecentric lens that views the samples directly and minimizes this effect. In addition, the light collection for this type of system is subject to vignetting, which means the efficiency of detection is as much as 10% higher in the center of the sample or plate. Software correction is used to create a "flat field" of detection so that all samples in a plate are counted with the same efficiency, independent of their position in the camera field of view.

The third important component of CCD camera systems for detection of radioisotopes in microplates is the method for conversion of the beta emission to light. Two methods are available for high-throughput applications, including red-shifted SPA beads from Amersham and the red-shifted scintillant coated FlashPlates from Perkin Elmer. Radiolabeled samples in microplates must be incorporated into an SPA assay using the specific red-shifted beads or bound to the wells of the FlashPlates so that the beta emissions can be converted to the appropriate wavelength of light for which the camera systems have been optimized. The red-shifted proximity beads specialized for use with the LEADseeker camera are polystyrene or yttrium oxide doped with Europium as a scintillant so that they emit a sharp peak a 615 nm, and optimal wavelength detection with the CCD of the LEADseeker system. FlashPlates are polystyrene scintillant coated plates that emit in the 600 nm wavelength region also, and are distributed by Perkin Elmer for use in the ViewLux microplate imager (Perkin Elmer Life and Analytical Sciences, Boston, MA). Like SPA beads, FlashPlates can be coated with high affinity material for binding with appropriate radiolabeled partners to create binding assays for high-throughput screening.

a. Performance

Performance of the area imaging systems for radiolabeled assays depends heavily on the format of the assay, but has been compared with PMT detection with the TopCount Microplate Counter (Perkin Elmer Life and Analytical Sciences, Boston, Mass.) (Graves *et al.*, 2001).

To define minimum detectable limits with this system, the figure of merit used is three times the standard deviation of the background signal. Cooling the CCD chip to temperatures as low as $-100°C$ improves the sensitivity by decreasing the thermal noise. Specifications for sensitivity as published by the manufacturer indicate that as little as 215 and 54 DPM of [^{3}H]biotin coupled to yttrium oxide beads can be detected in 384- and 1536-well plates, respectively. This corresponds to 0.625 fmol biotin, with a specific activity of 40 Ci/mmol, in 1536-well plates and 2.5 fmol of biotin (same specific activity) in 384-well plates. Note that the smaller sample format of the 1536-well plate requires a smaller area of integration. This results in better signal to noise ratios, as defined by the ratio of the signal to the standard deviation of the background (Jessop *et al.*, 2000).

The dynamic range of detection of the chip is limited only by the range of grey levels that can be displayed in the image, similar to the storage phosphor and direct beta imaging technology. The limiting factor to the dynamic range of the assays typically measured on these HTS microplate imaging systems is related much more to assay characteristics, such as background of the assay and displacement curve characteristics. Resolution is not an issue for these applications since the wells are quite large relative to the more challenging applications of tissue sections.

The most important benefit of using a CCD based imaging system for detection of radiolabeled high throughput assays is the speed factor. As compared to more common microplate scintillation counters, speed of the CCD system is excellent since the detection is with a uniform array of light sensitive pixels that can image the entire plate at once, rather than each sample discretely with individually matched PMTs. Where as each sample is typically measured one minute/well in a PMT based system, up to 12 samples at a time with 12 detectors, the CCD system can image the entire plate in one minute and integrate the optical density in the area of the sample. In addition, the uniformity of detection is more consistent, although comparisons require that two different types of beads are used, each optimized for the particular detection system. The comparison of CV's between methods shows better accuracy with the CCD camera system, even at low activity levels (Graves *et al.*, 2001).

b. Quantitation Methods

Each assay requires that the radiolabeled material be bound in some manner to either scintillating beads or the scintillant coated plate for conversion of the beta event to a wavelength of light that the CCD camera can detect and measure. CCD camera systems should be run at temperatures as low as $-100°C$ to minimize dark noise or thermal noise that can increase backgrounds of detection. Samples should be placed in white plates and a smaller sample format, such as 1536-well plates, provides the best geometry of detection. Smaller areas to integrate results in better signal to noise ratios in general. Samples that exhibit visible color should be avoided as they may interfere with the wavelength of detection and appear as false positives.

Similar to other counting methods, longer count times provide a larger signal accumulation over the area of integration.

c. Advantages

Speed is the biggest advantage of using the CCD imaging system for measuring radioisotopes in microplates as compared to a PMT based system for the same assay. Since all samples in the plate are counted simultaneously, the count time per well can be the same or greater as compared to a PMT based system, while the throughput is greatly improved for the whole plate. Based on the same count times, the throughput may be as great as 8 times faster using the CCD imaging system as compared to a 12-detector microplate PMT-based detection system.

Another documented advantage of the LEADseeker system includes the possibility of obtaining fewer false positives with the screening of colored compounds. Cook *et al.* (2002) reported 5.6 fold fewer false positives for a group of selected compounds using the LEADseeker as compared to the TopCount system.

Measuring the whole plate simultaneously using the uniform detection area of the CCD camera provides better accuracy compared to using multiple PMTs. Although the lens creates the vignetting effect, making the middle of the detector more sensitive than the edges, software correction can overcome this and create a more uniform counting area.

These systems are also capable of detecting nonisotopic methods such as luminescent and fluorometric assays, in addition to absorbance and optical density measurements. Therefore, although they are not particularly versatile for detection of radioisotopes in formats other than microplates, they are more versatile in terms of detection methods than other systems, such as the direct beta imagers and storage phosphor systems.

d. Disadvantages

Lack of versatility of detection of radioisotopes in formats other than microplate assays is the most significant disadvantage of this system. Assays must be created using expensive SPA beads or scintillant coated FlashPlates, both produced by single suppliers. This makes the imaging systems linked to the production and availability of the associated beads and plates. Although both beads and plates are produced to be as universally applicable as possible in that they are available pre-conjugated with high affinity binding pairs such as streptavidin and biotin, they are still limited in their range of applications. Finally, software correction to overcome vignetting and achieve uniformity of efficiency presupposes that no wells are so dim that they approach the lower limits of the detector capability. If some samples are so low that they approach the low end of the detection limits, the uniformity may be adversely affected.

e. Applications

Applications for the HTS optimized imaging systems include many nonisotopic assay methods and some not in the microplate format. However, for detection of radioisotopes using the these systems, the red-shifted SPA beads or Europium doped scintillant of the FlashPlate technology is required. Therefore, radioisotopic applications on these systems are almost exclusively microplate assays. The following are examples of assays developed for detection with the LEADseeker system.

Detection of p56^{lck} Kinase Activity: A number of studies have shown that p56^{lck} kinase is important for T-cell activation, and inhibition of this enzyme has potential for providing immunosuppressant effects with few side effects (reference from merck poster). Researchers at Merck in collaboration with Amersham Biosciences performed a study to show that detection of kinase activity and also inhibition with known inhibitors to give response curves for comparing different well formats and detection devices. The enzyme catalyzes the transfer of a γ-phosphate from [γ-^{33}P]-ATP to a biotinylated peptide substrate, which binds to the streptavidin red-shifted SPA beads. Known inhibitors, PP2 (Hanke *et al.*, 1996) and nocodazole (Huby *et al.*, 1998) were added to final concentrations of 50 μM to 0.5 nM in 1% DMSO. Using the LEADseeker system, an entire 1536-well white plate was imaged in 5 min in a fraction of the volume used by the larger formats. Less enzyme and substrate were necessary to show good activity and inhibition (Beveridge *et al.*, 1999). Figure 13.38 shows the results of this high throughput enzyme application.

Measuring Cytochrome P450 Drug Interactions: A competitive binding assay was developed in which residual binding of radioactive ligand is determined by incubating a test compound with the recombinant CYP co-expressed with NADPH-cytochrome P450 and membrane binding SPA beads (Lee *et al.*, 1995). CYP3A4 interactions were measured by competitive inhibition with [^{3}H]ethynylestradiol (Hopkins *et al.*, 2000). Figure 13.39 shows IC$_{50}$ curves generated from a 5 min image acquisition on the LEADseeker system.

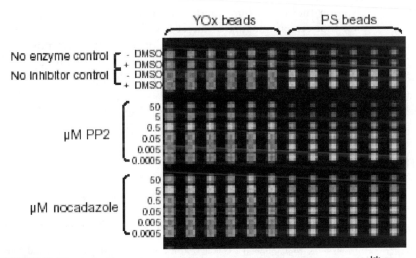

FIGURE 13.38 A 5-minute image of PP2 and nocodazole inhibition of p56^{lck} kinase in 1536-well plates (Beveridge *et al.*, 1999).

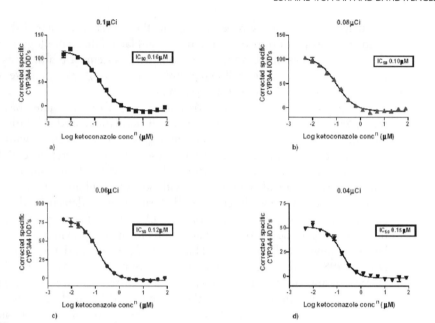

FIGURE 13.39 IC_{50} curves generated using LEADseeker SPA imaging method (Hopkins, 2000).

VI. FUTURE OF RADIONUCLIDE IMAGING

Quantitative imaging of radionuclides has reached a new plateau of performance. Storage phosphor screen imaging has developed to the extent that artifacts of the method are eliminated and the limiting factors of performance are a function of the autoradiographic process. The advantages of storage phosphor technology over traditional film autoradiography make it likely that any life science laboratory utilizing radioisotopes will eventually abandon the use of film for the faster, more quantitative method of storage phosphor screen imaging or direct beta imaging. The new generation of CCD camera based systems has been developed for the detection of radioisotopes in small tissue sections, microarrays and microplate applications. These represent a quantum leap in the area of high resolution radionuclide imaging and the extent to which they are adopted depends on the demand for such expensive, high performance imaging as the existence of film as a standard, low cost alternative remains a viable alternative.

REFERENCES

Amemiya, Y. (1995). Imaging plates for use with synchrotron radiation. *J. Synchrotron Radiation* **2**, 13–21.

Amemiya, Y. and Miyahara, J. (1988). Imaging plate illuminates many fields. *Nature* **336**, 89–90.

Amemiya, Y., Matsushita, T., Nakagawa, A., Satow, Y., Miyahara, Y., and Chikawa, J. (1988). Design and performance of an imaging plate system for X-ray diffraction study. *Nuclear Instruments and Methods in Physics Research* **A266**, 645–653.

Amicone, L., Spagnoli, F. M., Spath, G., Giordano, S., Tommasini, C., Bernardini, S., De Luca, V., Rocca, Weiss, M. C., Comoglio, P. M., and Tripodi, M. (1997). Transgenic expression in the liver of truncated met blocks apoptosis and permits immortilization of hepatocytes. *The Embo Journal* 16 (in press).

Bateman, J. E., Connolly, J. F., and Stephenson, R. (1985). High speed quantitative digital beta autoradiography using a multistep avalanche detector and an Apple II microcomputer. *Nuclear Instruments and Methods in Physics Research* A241, 275–289.

Batt D. (1995). Direct analysis of in vitro transcription products from supercoiled templates containing ribozymes. *Biotechniques* 20, 752–753.

Beveridge, M., Park, Y-W., Hermes, J., Brophy, G., and Santos, A. (1999). Detection of p56[lck] kinase activity using SPA in 384-well format and imaging proximity assay in 384 and 1536-well format. Poster Presentation at 4th Annual International Exposition and Symposium on Drug Discovery Technology, Boston, Mass.

Bigham, M. H. and Lidow, M. S. (1995). Adrenergic and serotenergic receptors in aged monkey neocortex. *Neurobiol. Aging* 16, 91–104.

Buccione, R., Bannykh, S., Santone, I., Baldassarre, M., Facchiano, F., Bozzi, Y., Di Tullio, G., Mironov, A., Luini, A., and De Matteis, M. A. (1996). Regulation of constitutive exocytic transport by membrane receptors. *Journal of Biological Chemistry*, 271(7), 3523–3533.

Cantu, G., Rimsa, J., Gelder, E., Steinberg, D., and Hueton, I. (1997). U.S. Patent V5, 635, 728.

Chan, A. M., Rubin, J. S., Bottaro, D. P., Hirschfield, D. W., Chedid, M., and Aaronson, S. A. (1991). Identification of a competitive HGF antagonist encoded by an alternative transcript. *Science* 29, 254 (5036), 1382–1385.

Charon, Y., Laniece, P., Mastrippolito, R., Siebert, R., Tricoire, H., and Valentin, L. (1998). Quantitative nuclear imaging in biology. *Ann. Phys. Fr.* 22, 707–770.

Charpak, G. and Sauli, F. (1978). The multistep avalanche chamber: a new high-rate, high-accuracy gaseous detector. *Phys. Lett. B.* 78, 523–528.

Chesters, J. K., Boyne, R., Petrie, L., and Lipson, K. E. (1995). Role of the promoter in the sensitivity of human thymidine kinase to lack of Zn2+. *Biochemistry Journal* 308, 659–664.

Coccia, E. M., Marziali, G., Stellacci, E., Perrotti, E., Ilari, R., Orsatti, R., and Battistini, A. (1995). *Virology* 211, 113–122.

Cook, L., Pak, R., Graves, R., Cook, J., Manly, S., and Padmanabha, R. (2002). Benefits from using imaging based LEADseeker and red-shifted emission for radiometric proximity assay screening. *Amersham BioSciences Application Note*.

Cross, S. A. M. (1974). Alteration of ^{14}C-histamine uptake in the presence of H_2-receptor antagonists. *Proc. Aust. Physiol. Pharmacol. Soc.* 5(1), 58.

Crumeyrolle-Arias, M., Jafarian-tehrani, M., Cardona, A., Edelman, L., Roux, P., Laniece, P., Charon, Y., and Haour, F. (1996). Radioimagers as an alternative to film autoradiography for in situ quantitative analysis of 125I-ligand receptor binding and pharmacological studies. *Histochemical Journal* 28, 801–809.

Decristoforo, C., Zaknun, J., Kohler, B., Oberladstaetter, M., and Riccabona, G. (1997). The use of electronic autoradiography in radiopharmacy. *Nuclear Medicine and Biology* 24, 361–365.

Dominik, W., Zaganidis, N., Astier, P., Charpak, G., Santiard, J. C., and Sauli, F. (1989). A gaseous detector for high-accuracy autoradiography of radioactive compounds with optical readout of avalanche positions. *Nucl. Phys. A.* (in process).

Dufficy, T. J., Cappel, R., and Summers, S. M. (1993). Silver discharge regulations questioned. *Water Environmental Technology* 5(4), 52–56.

Eastman Kodak Co., Health Sciences Division, Phillips Lighting Co., and Molecular Dynamics (1993). Storage phosphor screens. *Technical Note #53*.

Eastman Kodak Company (1989). Disposal and treatment of photographic effluent in support of clean water. *Environment* J-55.

Englert, D., Landais, D., and Woisetschlaeger, M. (1993). Assay of Chloramphenicol Acetyl Transferase (CAT) activity. *InstantImager Application Note #002*, Packard Instrument Company, Meriden, CT.

Englert, D., Roessler, N., Jeavons, A., and Fairless, S. (1995). Microchannel array detector for quantitative electronic radioautography. *Cellular and Molecular Biology* 41(1), 57–64.

Fujita, H. (1982). Reciprocity-law failure in medical screen-film systems and its effects on patient exposure and image quality. *In* "Optics in Biomedical Sciences" (G. von Bally and P. Greguss, Eds.), Vol. 31, pp. 78–81. Springer-Verlag, Berlin.

Gahan, P. B. (1972). "Autoradiography for Biologists." Academic Press, London and New York.

Gonzalez, A. L., Li, H., Mitch, M., Tolk, N., and Duggan, D. M. (2002). Energy response of an imaging plate exposed to standard beta sources. *Appl. Radiat. Isot.* **57**(6), 875–882.

Graves, R., Masino, J., Cook, J., Adams, M., and Howells, L. (2001). An evaluation of instrument performance using LEADseeker homogenous imaging system as a radiometric assay detector. *Amersham BioSciences Application Note.*

Hamaoka, T. (1990). Autoradiography of new era replacing traditional X-ray film. *Cell Technology* **9**, 456–462.

Hanke, J. H., Gardner, J. P, Dow, R. L., Changelian, P. S., Brissette, W. H., Weringer, E. J., Pollak, B. A., and Connelly, P. A. (1996). *J. Biol. Chem.* **271**, 696–701.

Hansen, O. (2001). The alpha1 isoform of Na$^+$, K$^+$-ATPase in rat soleus and extensor digitorum longus. *Acta. Physiol. Scand.* **173**(3), 335–41.

Hecht, E. and Zajac, A. (1974). "Optics," pp. 309–311. Addison-Wesley, Reading.

Hesketh, J., Campbell, G., Piechaczyk, M., and Blanchard, J. (1994). Targeting of *c-myc* and β-globin coding sequences to cytoskeletal-bound polysomes by c-*myc* 3' untranslated region. *Biochemistry Journal* **298**, 143–148.

Hewitt, P. J. (1993). Occupational health problems in processing of X-ray photographic films. *Annals of Occupational Hygiene* **37**(3), 287–295.

Hodges, A. and Smoller, B. R. (2002). Immunohistochemical comparison of p16 expression in actinic keratoses and squamous cell carcinomas of the skin. *Modern Pathology* **15**(11), 1121.

Hopkins, A., Price-Jones, M. J., Hughes, K. T., Eddershaw, P. J., Hood, S. R., Woodrooffe, A. J. M., and Tarbit, M. H. (2000). A LEADseeker imaging assay for measuring cytochrome P450 drug interactions. Amersham BioSciences poster presented at 6th Annual Conference of the Society for Biomolecular Screening, Vancouver, Brittish Columbia, Canada.

Huby, R. D., Weiss, A., and Ley, S. C. (1998). Nocodazole inhibits signal transduction by the T cell antigen receptor. *Journal of Biological Chemistry* **273**(20), 12024–31.

Jeavons, A. P. (1990). Method and apparatus for quantitative autoradiography analysis. U.S. Patent 1992, #5, 138, 168

Jeavons, A. P., Hood, K., Herlin, G., Parkman, C., Townsend, D., Magnanini, R., Frey, P., and Donath, A. (1983). The high-density avalanche chamber for positron emission tomography. *IEEE Transact. Nucl. Sci.* **30**, 640–645.

Jessop, R., Waythe, R., Howells, K., Rghei, N., Izzo, J., Price-Jones, M., and Bell, P. (2000). Increasing the sensitivity of the LEADseeker homogeneous imaging system. *Amersham BioSciences Application Note.*

Jiao, H. and Zhou, B. (2002). Cytotoxic effect of perioxisome proliferators fenofibrate on human HepG2 hepatoma cell line and relevant mechanisms. *Toxicol. Appl. Pharmacol.* **185**(3), 172–179.

Johnston, R. F., Pickett, S. C., and Barker, D. L. (1990). Autoradiography using storage phosphor technology. *Electrophoresis* **11**, 355–360.

Kessler, Michael J. (1989). "Liquid Scintillation Analysis Science and Technology." Packard Instrument Company, Meriden, CT.

Kobayashi, Y. and Maudsley, D. V. (1974). "Biological Applications of Liquid Scintillation Counting." Academic Press, New York, San Francisco and London.

Kristian, J. and Blouke, M. (1982). Charged-coupled devices in astronomy. *Scientific American* **247**(4), 66–74.

Laniece, P., Charon, Y., Cardona, A., Pinot, L., Maitrejean, S., Mastrippolito, R., Sandkamp, B., and Valentin, L. (1998). *Journal of Neuroscience Methods* **86**, 1–5.

Laniece, P., Charon, Y., and Dumas, S. (1994). HRRI: a high resolution radioimager for fast, direct quantification in *in situ* hybridization experiments. *BioTechniques* **17**, 338–45.

Laskey, R. A. and Mills, A. D. (1975). Quantitative film detection of ^{3}H and ^{14}C in polyacrylamide gels by fluorography. *Eur. J. Biochem.* **56**, 335–341.

Laskey, R. A. and Mills, A. D. (1977). Enhanced autoradiographic detection of ^{32}P and ^{125}I using intensifying screens and hypersensitized film. *FEBS Letters* **82**, 314–???

L'Annunziata, M. F. (1979). "Radiotracers in Agricultural Chemistry." Academic Press, New York and London.

L'Annunziata, M. F. (1987). "Radionuclide Tracers, Their Detection and Measurement." Academic Press, London and New York.

Langlois, X., Te Riele, P., Wintmolders, C., Leysen, J. E., and Jurzak, M. (2001). Use of β-imager for rapid ex vivo autoradiography exemplified with central nervous system penetrating neurokinin 3 antagonists. *Journal of Pharmacology and Experimental Therapeutics* **299**, 712–717.

Lee, C. A., Kadwell, S. H., Kost, T. A., and Serabjit-Singh, C. J. (1995). CYP3A4 expressed by insect cells infected with a recombinant baculovirus containing both CYP3A4 and human NADPH-cytochrome P450 reductase is catalytically similar to human liver microsomal CYP3A4. *Arch. Biochem. Biophys.* **319**(1), 157–67.

Liberatore, G. T., Wong, J. Y.F., Krenus, D., Jeffreys, B. J., Porritt, M. J., and Howells, D. W. (1999). Tissue fixation prevents contamination of tritium-sensitive storage phosphor imaging plates, *Biotechniques* **26**, 432–434.

Lidow, M. and Solodkin, A. (1997). Use of cyclone storage phosphor system for rapid development of a receptor autoradiographic assay. *Biomedical Products* **24**.

Lidow, M. S. and Rakic, P. (1995). Neurotransmitter receptors in the proliferative zones of the developing primate occipital lobe. *J. Comp. Neurol.* **360**, 392–402.

Michels, W. J., Jr. and Pyle, A. M. (1995). Conversion of a group II intron into a new multiple-turnover ribozyme that selectively cleaves oligonucleotides: elucidation of reaction mechanism and structure/function relationships. *Biochemistry* **34**, 2965–2977.

Miyahara, J. (1989). The imaging plate: a new radiation image sensor. *Chemistry Today* **223**, 29–36.

Mori, K. and Hamaoka, T. (1994). IP Autoradiography Systems (BAS). Protein, Nucleic Acid, and Enzyme **39**(11), 1–13.

Moron, N., Hayama, E., and Shigematsu, A. (1995). Studies on the quantitative autoradiography: radioluminography for quantitative autoradiography of ^{14}C. *Biol. Pharm. Bull.* **18**(I), 89–93.

Muller, T. and Gebel, S. (1994). Heme oxygenase expression in Swiss 3T3 cells following exposure to aqueous cigarette smoke fractions. *Carcinogenesis* **15**(1), 67–72.

Nguyen, Q. and Heffelfinger, D. M. (1995). Imaging and quantitation of chemiluminescence using photoexcitable storage phosphor screen. *Analytical Biochemistry* **226**, 59–67.

Olivas, W. M. and Maher, L. J., III (1995). Competitive triplex/quadruplex equilibria involving guanine-rich oligonucleotides. *Biochemistry* **34**(1), 278–284.

Orr, T. (1993). Instruments for analysis of gel imagery offer wide range of choices. *Genetic Engineering News* pp. 8, 9.

Pelc, S. R. (1972). Theory of autoradiography. In "Autoradiography for biologists." (P. B. Gahan, Eds.), pp. 1–17. Academic Press, London and New York.

Phelps, M. E., Mazziotta, J. C., and Schelbert, H. R. (1986). Positron emission tomography and autoradiography. "Principles and Applications for the Brain and Heart." Raven Press, New York.

Pickett, S. C. (1992). Evaluation of storage phosphor imaging systems. *Molecular Dynamics Technical Note* #55, Sunnyvale, CA.

Pickett, S. C., Barker, D. L., and Johnston, R. F. (1992). Quantitative double-label autoradiography using storage phosphor imaging. *International Labmate*.

Popovici, R. M., Kao, L. C., and Giudice, L. C. (2000). Discovery of new inducible genes in *in vitro* decidualized human endometrial stromal cells using microarray technology. *Endocrinology* **141**(9), 3510–3513.

Qiu, X., Forman, H. J., Schonthal, A. H., and Cadenas, E. (1996). Induction of p21 mediated by reactive oxygen species formed during the metabolism of aziridinylbenzoquinones by HCT116 cells. *Journal of Biological Chemistry* **271**, 31915–31921.

Ramm, P. (1999). Imaging systems in assay screening. *Drug Discovery Today* **4**(9), 401–410.

Randerath, K. (1970). An evaluation of film detection methods for weak β-emitters, particularly tritium. *Analytical Biochemistry* **34**, 188–205.

Rapkin, E. (2001). Real-time radioimaging of biological tissue sections. *American Biotechnology Lab* (reprint).

Reddy, M. V. and Randerath, K. (1986). Nuclease P1-mediated enhancement of sensitivity of ^{32}P-postlabeling test for structurally diverse DNA adducts. *Carcinogenesis* 7, 1543–1551.

Reichert W. L., Stein, J. E., French, B., Goodwin, P., and Varanasi, U. (1992). Storage phosphor imaging technique for detection and quantitation of DNA adducts measured by the 32P-postlabeling assay. *Carcinogenesis* 13(8), 1475–1479.

Robben, J. H., Van Garderen, E., Mol, J. A., Wolfswinkel, J., and Rijnberk, A. (2002). Locally produced growth hormone in canine insulinomas. *Mol. Cel. Endocrinol.* 197(1–2), 187–195.

Rogers, A. W. (1969). "Techniques of autoradiography." Elsevier Publishing Company, New York.

Salin, H., Maitrejean, S., Mallet, J., and Dumas, S. (2000). Sensitive and quantitative co-detection of two mRNA species by double radioactive in situ hybridization. *J. Histochem. Cytochem.* 48(12), 1587–92.

Salin, H., Vujasinovic, A., Mazurie, A., Maitrejean, S., Menini, C., Mallet, J., and Dumas, S. (2002). A novel sensitive microarray approach for differential screening using probes labeled with two different radioelements. *Nucleic Acids Research* 30(4), 17–24.

Schena, M. (2000). "Microarray biochip technology." Eaton Publishing, Natick, Mass.

Sharif, N. A. and Eglen, R. M. (1993). Quantitative autoradiography: a tool to visualize and quantify receptors, enzymes, transporters, and second messenger systems. *In* "Molecular imaging in neuroscience: a practical approach" (N. A. Sharif, Ed.), pp. 71–138. Oxford University Press, IRL.

Shelton, L. S., Albright, A. G., Ruyechan, W. T., and Jenkins, F. J. (1994). Retention of the herpes simplex virus type 1 (HSV-1) UL37 protein on single-stranded DNA columns requires the HSV-1 ICP8 protein. *Journal of Virololgy* 68(1), 521–525.

Shi, M. M., Chong, I. W., Godleski, J. J., and Paulauskis, J. D. (1999). Regulation of macrophage inflammatory protein-2 gene expression by oxidative stress in rat alveolar macrophages. *Journal of Pathology Immunology* 97(2), 309–15.

Sonoda, M., Takano, M., Miyahara, J., and Kato, H. (1983). Computed radiography utilizing scanning laser stimulated luminescence. *Radiology* 148, 833–838.

Steinfeld, R., McLaughlin, W., Vizard, D., and Bundy, D. (1994). Advances in autoradiography: a new film brings quality improvements. *In* "The Biotechnology Report." Campden Publishing Ltd., London.

Sullivan, D. E., Auron, P. E., Quigley, G. J., Watkins, P., Stanchfield, J. E., and Bolon, C. (1987). The nucleic acid blot analyzer. I: high speed imaging and quantitation of ^{32}P-labeled blots. *Biotechniques* 5, 672–678.

Taylor, T. C., Kahn, R. A., and Melancon, P. (1992). Two distinct members of the ADP-ribosylation factor family of GTP-binding proteins regulate cell-free intra-Golgi transport. *Cell* 70(1), 69–79.

Tian, L., Shi, M. M., and Forman, H. J. (1997). Increased transcription of the regulatory subunit of γ-glutamylcysteine synthetase in rat lung epithelial L2 cells exposed to oxidative stress of glutathione depletion. *Archives of Biochemistry and Biophysics* 342, 126–133.

Touchstone, J. C. and Sherma, J. (1985). "Techniques and applications of thin layer chromatography." John Wiley and Sons, New York.

Treharne, K. J., Marshall, L. J., and Mehta, A. (1994). A novel chloride-dependent GTP-utilizing protein kinase in plasma membranes from human respiratory epithelium. *Journal of the American Physiological Society* L592.

Tyler-Smith, C. and Southern, E. M. (1993). Rapid detection and quantitation of single copy sequences in human DNA. *InstantImager Application Note*, 5. Packard Instrument Company, Meriden, CT.

Ullberg, S. (1977). The technique of whole-body autoradiography. Cryosectioning of large specimens. *Science Tools*, 2–29. Special Issue.

Upham, L. V. and Fox, T. (2001). Quantitative analysis of DNA arrays, *Cyclone Application Note*, 1. Packard BioScience, Meriden, CT.

Upton, T., Wiltshire, S., Francesconi, S., and Eisenberg, S. (1995). ABF1 Ser-720 is a predominant phosphorylation site for casein kinase II of *Saccharomyces cerevisiae*. *Journal of Biological Chemistry*, 270(27), 16153–16159.

Veal, L. (1997). Rapid detection and quantitation of covalent DNA adducts by ^{32}P-postlabeling and 2D thin layer chromatography. *Biomedical Products* January, 10.

Veal, L. and Englert, D. (1997). Two high performance storage phosphor screens. *BioMedical Products* **22**, 36.

Veal, L. and Tian, L. (1997a). Two methods of quantitative image analysis. *American Biotechnology Lab*, **July** 8.

Veal, L. and Tian, L. (1997b). Direct nuclear imaging of gel shift and TLC assays. *American Biotechnology Laboratory*, Sept. 1997.

Vizard, D., Attwood, J., and McLaughlin, W. (1996). A universal intensifying screen system for enhanced detection of low- and high-energy isotopes. *American Biotechnology Lab* **15**, 18.

Weber, T. J., Monks, T. J., and Lau S. S. (1997). Prostaglandin E2 (PGE2)-mediated cytoprotection in a renal tubular epthelial cell line: Evidence for phamacologically distinct receptor. *American Journal of Physiology* (in press).

Weisner, D. E. (1992). Charged-coupled devices. "McGraw-Hill Encyclopedia of Science and Technology," 7th ed. McGraw-Hill, Inc. New York, NY.

Zinck, R., Hipskind, R. A., Pingoud, V., and Nordheim, A. (1993). c-fos transcriptional activation and repression correlate temporally with the phosphorylation status of TCF. *EMBO Journal* **12**(6), 2377–2387.

Zubiaur, M., Sancho, J., Ter horst, C., and Faller, D. V. (1995). A small GTP-binding protein, rho, associates with the platelet-derived growth factor type-β receptor upon ligand binding. *Journal of Biological Chemistry* **270**(29), 17221–17228.

14
AUTOMATED RADIOCHEMICAL SEPARATION, ANALYSIS, AND SENSING

JAY W. GRATE AND OLEG B. EGOROV
Pacific Northwest National Laboratory, Richland, Washington 99352

I. INTRODUCTION

Automation in radiochemical analysis offers many significant advantages, including reduced worker exposure to radioactivity, increased reliability, improved safety, and more consistent analytical protocols. Automation is

1129

particularly important in addressing the requirements for carrying out radiochemical separations prior to detecting or quantifying the radionuclides of interest. These aspects of radiochemical analysis can be particularly time consuming and costly. Classical sample preparation methods in radio-chemistry entail tedious manual separations such as precipitations and extractions. Modern radiochemical separations carried out on solid phase separation materials have considerable advantages, and are suitable for automation.

The objectives of automated radiochemical methods are to be able to efficiently obtain radioanalytical results using modern separations, methodology, and technology. Moreover, in many applications, it is desirable to be able to perform measurements at-site or *in situ*. Therefore instruments and sensors are required that perform all the functions previously carried out as sequential manual steps in the laboratory, yet achieve this rapidly and efficiently in an automated instrument or sensor.

Applications for automated radiochemical analysis include more efficient laboratory analyses of nuclear waste (Grate and Egorov, 1998b; Murray, 1994), near real time analysis for nuclear waste process monitoring, detection of radionuclides in the environment, and long-term monitoring of radionuclide contaminants in ground or surface waters.

II. RADIOCHEMICAL SEPARATIONS

A. Separation Requirements

The requirements for radiochemical separations and other sample preparation steps in radiochemical analysis are determined largely by the detection method, which defines the acceptable characteristics of the sample presented to the detector.

There are four primary means of detection used in radiochemical analysis: radiometric detection of gamma rays, beta particles, or alpha particles, and mass spectrometry. Nondestructive analysis is possible primarily for those radionuclides that can be analyzed directly by high resolution gamma-spectroscopy. In almost all other cases, separations are required to separate the radionuclides of interest from other matrix components or other radionuclides that interfere with the detection method. Beta particles are emitted with a range of energies and individual radionuclide species cannot be readily discerned using scintillation detection techniques. For example, radiometric detection of ^{99}Tc in nuclear waste matrixes requires complete separation from other interfering radioactive species. Although alpha particles are emitted with characteristic energies, a number of important radionuclides have unresolvable alpha-energies, including ^{241}Am/^{238}Pu; and ^{237}Np/^{234}U. In addition, matrix effects on the energy resolution of alpha spectroscopy necessitate extensive sample purification and preparation.

An alternative detection technique that is becoming increasingly important in modern radiochemical analysis is inductively coupled plasma mass spectrometry, ICP-MS. The trade-offs between radiation counting and atom counting have been described previously (Ross *et al.* 1993; Smith *et al.* 1992) and in Chapter 10. In general, short-lived fission products are best detected with radiation detection, whereas long-lived (low specific activity) radionuclides can be detected with more sensitivity using ICP-MS.

ICP-MS is also a selective detector since particular mass-to-charge ratios can be monitored. Nevertheless, separations are desirable prior to introducing a sample into the ICP-MS so that the sample is always introduced into the system in a consistent matrix (Colodner *et al.*, 1994; Denoyer, 1992; Thompson and Houk, 1986). Matrix components can affect quantification. Moreover, detection by mass spectrometry is subject to interferences involving isobars and molecular ions, and tailing of peaks when one isotope is at much higher concentration than an isotope with an adjacent mass number (Crain and Alvarado, 1994; Garcia Alonso, 1995; Garcia Alonso *et al.*, 1993; Garcia Alonso *et al.*, 1995; Smith *et al.*, 1995; Tan and Horlick, 1986). Examples include ^{99}Tc/^{99}Ru, ^{151}Sm/, ^{91}Zr^{16}O/^{107}Pd ^{238}UH/^{239}Pu, ^{238}U/^{238}Pu, ^{241}Pu/^{241}Am (Crain and Alvarado, 1994; Garcia Alonso *et al.*, 1995; Smith *et al.*, 1995). In addition, the possibility of interference from a large excess of ^{238}U in the measurement of ^{237}Np has been noted.

In summary, chemical separations are carried out in radiochemical analysis to separate radionuclides with interfering radiation energy spectra for radiometric detection or to remove species leading to isobaric, molecular, or spectral interferences in mass spectrometric detection. In many cases it is also necessary to enable preconcentration and/or separation of radionuclides of interest from stable matrix components that affect subsequent analysis (Kim *et al.* 2000). Whenever chemical separations are required, there may also be a requirement for additional sample preparation steps to control the speciation of the radionuclides prior to separation.

B. Radiochemical Separation Approaches

Separations in conventional radiochemical analysis have been carried out by a variety of classical and chromatographic methods, including precipitation, liquid–liquid extraction, and ion exchange. Often sequential combinations of these methods are used and, in some cases, individual steps must be repeated. Classical radiochemical separation methods are clearly not suitable for use in automated radiochemical analysis and have many disadvantages even for laboratory analyses.

Two column separation formats are generally considered for automated separations in radiochemical analysis. The first involves high performance chromatographic techniques. Reboul *et al.* and Desmartin and coworkers described the use of high performance liquid chromatography coupled with on-line flow-through scintillation detection for intermittent sampling and quantification of actinides and fission products in water (Desmartin *et al.*, 1997; Reboul and Fjeld, 1994, 1995). Smith *et al.* described the use of high performance ion chromatography with on-line radiometric detection for qualitative analysis of Hanford waste samples (Smith *et al.*, 1995). Long separation times (40 minutes to several hours), matrix effects on the

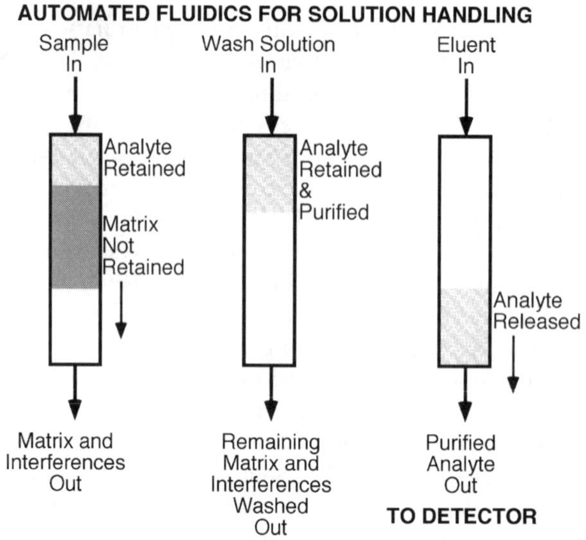

AUTOMATED FLUIDICS FOR SOLUTION HANDLING

FIGURE 14.1 Schematic diagram of the separation approach where the analyte is selectively retained on the solid phase while the matrix and interfering species are removed in the sample load and wash steps. Then an eluent is added that abruptly changes the retention of the analyte, releasing it in purified form for detection. Automated fluidic systems deliver sample, wash solutions and eluents as defined in software.

reproducibility and reliability of the separation procedure, limited capacity of the separation material, and speciation problems are significant issues with these chromatographic approaches that limit their applicability for rapid automated radiochemical analysis on difficult sample matrices.

The second approach uses selective separation chemistries in extraction chromatographic or solid phase extraction formats in order to rapidly and selectively isolate species of interest from stable matrix and radiological interferences. (Cortina and Warshawsky, 1997; Dietz and Horwitz, 1993; Grate and Egorov, 1998b; Izatt et al., 1996). The species of interest are strongly and preferentially retained by the column material under the solution conditions selected for the separation. Following a wash step (or sequence of wash steps) to remove unretained or slightly retained sample components, the species of interest are abruptly released by creating a large drop in the capacity factors via changes in the mobile phase composition or even by doing reaction chemistry on the retained species. Thus, this separation format relies on selective uptake and release properties rather than on high chromatographic efficiency of the separation column. This overall approach is shown in Fig. 14.1.

In general, radiochemical separations using selective sorbent materials can be accomplished rapidly using small columns operated under low pressures. Because of the selective analyte capture during the sample loading step analyte preconcentration can be achieved to improve analysis sensitivity. Compared to the high performance chromatographic format, selective chemical separations are more tolerant of complex sample matrices, and the columns have greater capture capacity (Cortina and Warshawsky, 1997). And unlike conventional chromatography, this approach can use reaction chemistry in addition to simple eluent delivery. Based on these considerations, selective chemistries on solid phases are particularly well suited for automation of radiochemical analysis.

C. Modern Radiochemical Separation Materials

Recently a number of new solid phase materials have been developed to simplify analytical separations of heavy metals and radionuclides, using the separation approach shown in Fig. 14.1. These materials are based on either extraction chromatographic or solid phase extraction materials, and they have made a significant contribution to the development of automated radiochemical analyzers and radionuclide sensors.

Materials using selective or semiselective extractants impregnated on porous polymer and silica gel supports have been developed by Horwitz and coworkers at Argonne National Laboratory and commercialized by Eichrom Technologies, Inc. (Darien, Illinois) (Dietz and Horwitz, 1993; Horwitz et al., 1992a, 1993, 1995; Maxwell, 1997). These materials have also been called solvent-impregnated resins, and in many cases they use chemistry from well-established liquid–liquid extraction separations. Extractant impregnated beads are packed in a column and used to separate species from aqueous solutions by extraction chromatography (Cortina and Warshawsky, 1997). The uptake properties and chemical selectivities of these materials are well characterized in the literature.

A variety of solid phase extraction materials dubbed "AnaLig" have been developed using "molecular recognition" ligands on solid supports and commercialized by IBC Advanced Technologies (American Fork, Utah) (Izatt, 1997; Izatt et al., 1994, 1995, 1996). These ligands are covalently bound to various polymeric or silica-gel supports. The materials have been developed and investigated mainly for wastewater or effluent cleanup processes, and have not been extensively used or characterized for analytical applications. Fundamental metal ion uptake studies of these are not available in the published literature. Nevertheless, recently AnaLig materials selective for Sr, Ra, and Tc were incorporated in Empore™ Rad Disks marketed by 3M for use in analysis of water samples (Fiskum et al., 2000; Schonhofer and Wallner, 2001).

III. AUTOMATION OF RADIOCHEMICAL ANALYSIS USING SEQUENTIAL INJECTION FLUIDICS

A. Sequential Injection Fluidics

The requirements for fluid handling in automation include a means to select samples, reagents, and eluents; bring them into the fluidic system, and deliver them to flow cells, separation columns, and detectors with precise control over volumes, flow rates, and timing. Sequential injection techniques from the field of flow injection (FI) analysis provide a versatile fluid handling approach for meeting these requirements (Fang, 1993; Ruzicka, 1994; Ruzicka and Hansen, 1988). Unlike conventional FI methodology or chromatography,

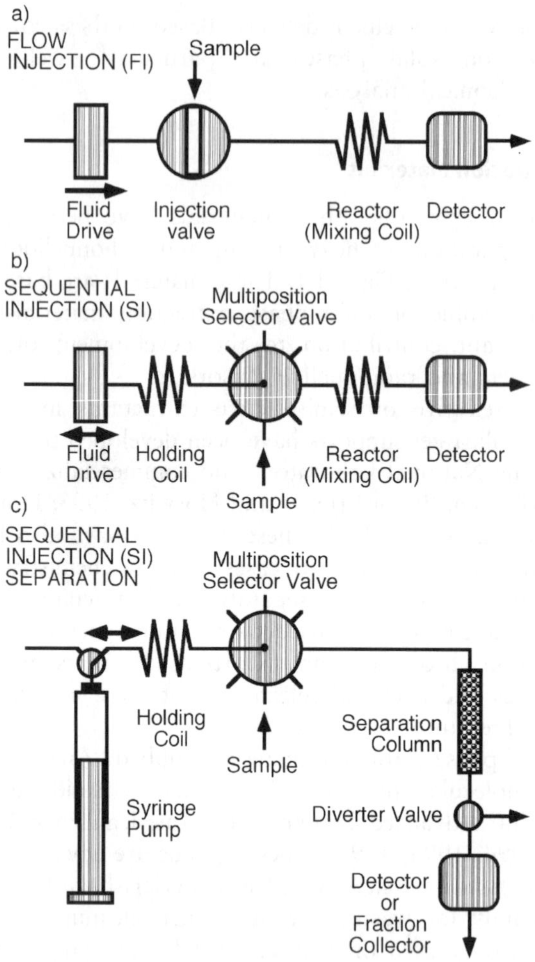

FIGURE 14.2 Schematic diagrams comparing (a) flow injection and (b) sequential injection fluid handling methods. The key components of a sequential injection separation system are shown in c. (Adapted with permission from Anal. Chem., 70 (1998) 779A–788A. Copyright 1998 Am. Chem. Soc.)

which are based on a continuous forward flow paradigm, SI involves fluid flow in both directions and programmed flow reversals (Christian, 1994; Ivaska and Ruzicka, 1993; Ruzicka, 1994; Ruzicka and Marshall, 1990). The pump(s) and valve(s) of a SI system are fully controlled by a computer for automation. FI and SI methods are compared in Fig. 14.2.

The preferred fluid drive for SI is a modern digital syringe pump. A multiposition selection valve is used to select samples, reagents, or eluents that are pulled into a holding coil located between this valve and the pump. Precise control of the flow rate and the timing determine the volume of the solution pulled into the system. After changing the position of the selection valve, the solution or solutions in the holding coil are pushed out of the holding coil (away from the pump) and toward the reactor, separator, and/or detector downstream. Solutions can be pulled into the holding coil and delivered to the flow system downstream one at a time, or multiple solutions can be pulled into the holding coil in sequence and "stacked" there prior to pushing them forward into the flow system. Sequential injection systems are mechanically simple yet quite versatile. Reagents, eluents, samples, and standards can all be nested around the multiposition valve and sequentially selected for delivery into the system. The volume of each solution introduced into the holding coil can be varied simply by timing and flow rate under computer control, rather than by physically reconfiguring the system (e.g., by installing a larger injection loop as in conventional flow injection or chromatography). The pump does not contact either reagent solutions or samples, and it can be located remotely from the remainder of the system.

Typical colorimetric sequential injection analysis (SIA) procedures stack zones of sample and reagent(s) in a holding coil with a narrow internal bore. When the stacked zones are propelled toward the detector mixing occurs by dispersion, yielding a detectable product. Solution volumes are generally microliters, and 500 microns is a typical holding coil inside diameter.

B. Sequential Injection Separations

Use of SI instrumentation for automating radiochemical analysis (Egorov *et al.*, 1998a,b, 1999a,b, 2001b; Grate and Egorov, 1998a,b; Grate *et al.*, 1999a,b, 1996) typically involves use of a separation column, as shown in the apparatus in Fig. 14.2c. A suitable detector or fraction collector follows the column. A diverter valve is useful for directing sample matrix and wash solutions to waste without passing them through the detector; radionuclides released from the separation column can then be directed to the detector.

When using SI instrumentation and methodology to automate column-based separations, the approach must be modified to accommodate larger solution volumes; milliliter quantities of multiple solutions may be necessary to wash separation columns and elute species of interest. Pulling these solutions into the holding coil at typical SIA flow rates is excessively time consuming. However, increasing the flow rate can cause outgasing of solutions in the holding coil due to reduced pressure as the pump pulls the solution in. This is unacceptable, but it can be resolved by using a larger bore holding coil, e.g. 1.6 mm i.d. However, a larger holding core bore leads to increased dispersion between the solution being pulled in and the carrier fluid already in the coil.

To prevent dispersion and mixing of these solutions, an air segment can be pulled into the coil before the reagent solution. ("Reagent" refers generically to reagents and eluents.) Then the reagent solution is delivered to the column. This approach is shown in Fig. 14.3. The air segment is expelled to waste, a new air segment is pulled in, and a new reagent solution is pulled into the holding coil, ready to be delivered to the column. In this way, several different solutions of varying volumes and composition can be delivered to the separation column. The holding coil functions as a zero dispersion

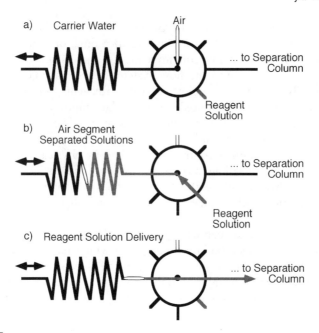

FIGURE I4.3 Schematic diagram for the use of an air segment in the holding coil for separation-optimized sequential injection fluid handling. The air segment is pulled into the holding coil prior to the reagent solution. This keeps the reagent solution separated from the carrier fluid (typically water) in the holding coil and syringe pump. After injecting a fixed amount of the reagent solution past the multiposition selection valve, the remainder of the reagent solution and the air segment are discarded to waste.

volumetric extension of the syringe pump. The use of a large bore holding coil and air segment in this way for sequential injection separations has been referred to as separation-optimized sequential injection (Egorov *et al.*, 1998b; Grate and Egorov, 1998b; Grate *et al.*, 1999a,b).

C. Alternative Fluid Delivery Systems

The choice of the basic fluid handling approach is not necessarily limited to the SI paradigm using a multiposition valve and holding coil. For example, dedicated syringe pumps can be used for sample handling and reagent delivery in an approach typically used in robotic workstations. In addition, modern digital syringe pumps with zero dead volume syringes and multiposition selector valves have become available from Kloehn Ltd. (Las Vegas, NV). In this design, a single syringe pump can be used for sequential delivery of multiple solutions with minimal cross contamination. Thus the syringe pump can serve as both the fluid drive and the holding coil. Finally, in some instrument designs and procedures the combination of digital syringe pumps and positive displacement HPLC pumps can be used to automate solution handling tasks required in the automated sample preparation and separation procedure.

D. Column Configurations

Columns containing the solid phase separation medium can be included in the system directly downstream from the multiposition selection valve, as shown in Fig. 14.2c. However, other configurations are advantageous in some analysis protocols. For example, the column can be configured with a four port two position valve as shown in Fig. 14.4a so that the analyte captured on the "top" of the column is eluted in the opposite direction so that it need not travel through the remainder of the column material.

Another configuration that is useful in some cases is to add a sample injection valve after the multiposition valve and immediately before the separation column, as shown in Fig. 14.4b (Egorov *et al.*, 2001b). In this way, hot radioactive samples are injected directly upstream from the column; the multiposition valve and the holding coil of the sequential injection system are never exposed to the sample. This type of sample injection is similar to conventional FI and the sample volume is determined by hardware, not by

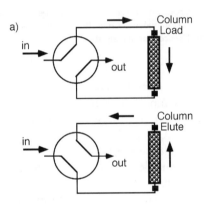

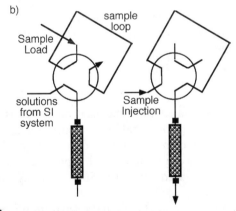

FIGURE 14.4 (a) Schematic illustration of a column coupled to a two position valve so that sample can be loaded in one flow direction and analytes can be eluted in the opposite direction; (b) Schematic illustration of a sample injection valve and loop directly upstream from the separation column.

FIGURE 14.5 Schematic diagram of a column switching system, where the columns are placed between two multiposition selection valves. Only three columns are shown for simplicity, although the valves shown will acccomodate eight.

time and flow rate parameters under software control as in SI systems that use the holding coil as part of the sample injection method.

Yet another useful configuration is to set up column switching capabilities (Grate *et al.*, 1999b). A partial column switching system is shown schematically in Fig. 14.5. Multiple fresh columns can be set up prior to beginning a set of samples. Each sample can be directed to a fresh column for rapid throughput without column wash steps between samples. This approach has been used to set up a rapid automated separation workstation for processing of ^{90}Sr-containing samples along with quality control samples including standards, spiked samples, and blanks (Grate *et al.*, 1999b).

Hence, the basic sequential injection system shown in Fig. 14.2c is often modified to design systems that are optimized for particular analytical approaches under development.

E. Renewable Separation Columns

The separation column for automated radiochemical anlayzers is shown in Fig. 14.2c as a fixed column. The practical life of such fixed columns is determined by the sample type and the separation chemistry involved. With suitable column cleanup steps between samples, such separation columns may be used tens or possibly even hundreds of times. Such columns can then be manually replaced as necessary.

An alternative approach is to automatically pack separation material into the column body on-line, and release it automatically after use. This approach results in "renewable separation columns," and SI systems using them have been referred to as SI-RSC systems. Originally, renewable surface techniques were developed by Ruzicka and coworkers to trap chemically selective particles and beads for observation by microscopic or optical techniques while perfusing with sample (Egorov and Ruzicka, 1995; Mayer and Ruzicka, 1996; Pollema and Ruzicka, 1994; Ruzicka, 1994; Ruzicka and Ivaska, 1997; Ruzicka et al., 1993; Ruzicka and Scampavia, 1999; Willumsen et al., 1997). Changes in the bed properties (e.g., absorbance, reflectance, fluorescence, etc.) induced upon the introduction of the sample solution provided the basis for an analytical measurement. In this approach, the bead or particles serves as sensing surfaces for chemical or biological detection. After use, the beads were released and a new microbed was packed. This process renews the interactive surface for the next measurement.

When the flow cell that captures the beads is not associated with a detector, it serves primarily as a separation medium. This approach can be scaled to handle bed volumes from one microliter to a millliliter or more. Renewable surface methods were introduced to the field of radiochemistry using extraction chromatographic radiochemical separation materials in SI-RSC systems (Egorov et al., 1999a). Slurries of the extraction chromatographic material were captured in a flow cell (column) of well-defined volume, and this packed bed was used to carry out separations. The column bed material was released from the flow cell and fresh suspensions of beads were introduced and captured for subsequent samples. The SI-RSC method has been demonstrated for ^{90}Sr analysis using Sr-resin, ^{99}Tc analysis using TEVA-resin, and ^{241}Am analysis using TRU-resin.

Several approaches exist for capturing and releasing particles in RSC flow cells. The original flow cells for renewable surface techniques were "jet ring cells" where beads were trapped in the end of a tube in contact with a planar surface. The roughened end of the tube allowed fluid to pass but retained beads (Ruzicka, 1994; Ruzicka et al., 1993; Ruzicka and Scampavia, 1999). Beads were removed from the cell either by reversing the fluid flow or lifting the tube away from the surface. Flow cells were also designed to trap beads with a solid rod intersecting the cylindrical channel of a machined flow cell at right angles (Dockendorff et al., 1998; Holman et al., 1997). This rod stopped beads but a leaky tolerance allowed fluid flow around it. Beads were released by an axial motion of the rod that withdraws the rod end from the flow channel. By contrast, a rotating rod design has been developed where the end of the rod is angled; in one rotation position the angled end blocks beads from exiting the outlet channel while allowing fluid flow around a leaky tolerance (Bruckner-Lea et al., 2000). A 180 degree rotation opens outlet channel for bead release. This design in shown in Fig. 14.6a. Frit-based renewable separtion flow cells have also been developed (Egorov et al., 1999a). In these systems the bead containing fluid is directed through a frit to trap beads or to an open channel to discard beads. This approach has been applied to radiochemical separations. The valve for determining which fluidic path is open can be downstream, or it can be integral to the design as shown in Fig. 14.6b.

Thus, there are many ways to implement renewable surface columns and these can be used for either the automated radiochemical analysis methods or the radionuclide sensors to be described in the next section.

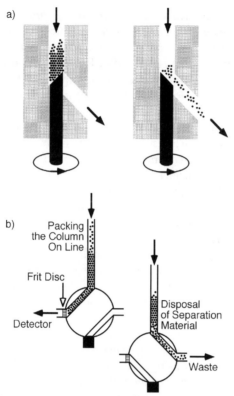

FIGURE 14.6   **Two approaches for renewable separation columns: (a) the rotating rod flow cell; (b) a frit restricted method incorporated into a two position valve.** *Part b adapted with permission from Anal. Chem., 71 (1999) 345–352. Copyright 1999 Am. Chem. Soc.*

F. Detection

Detection of the separated radionuclides can be done in a variety of ways. Flow-through liquid scintillation detectors combine the output of the SI separation system with liquid scintillation cocktail and deliver it into the flow cell of the detector. This flow cell typically consists of a length of tubing in a flat spiral which is placed between two photomultiplier tubes. A schematic diagram of the flow-through liquid scintillation detection approach is shown in Fig. 14.7. Many of the detector traces in this chapter were obtained by this method.

Flow-through scintillation detectors also exist that use particles of a solid scintillator packed into the flow cell. This avoids the use of the liquid scintillation cocktail and the associated waste. The fully automated process analyzer for total Tc in nuclear waste, to be described below, used this type of scintillation detection.

Mass spectrometry represents a detection method that is advantageous for radionuclides with low specific activity or where mixtures of radionuclides are eluted. Mass spectrometry can provide separate signals for each isotope according to their mass-to-charge ratios. ICP-MS has been used as the on-line detector for actinide separations and will be described below.

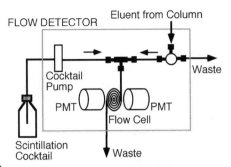

FLOW DETECTOR

Eluent from Column

FIGURE 14.7   Schematic diagram of a flow-through liquid scintillation detection system.

Finally, automation of radiochemical separations represents a significant saving in cost and labor even when the output of the instrument is delivered to a fraction collector for separate off-line counting (Grate *et al.*, 1999a,b). This approach is useful for separating actinides in preparation for alpha spectrometry. In addition, some labs may already be set up to handle radiochemical separations and counting operations separately. Automated radiochemical separations with fraction collection can provide efficiencies for the separation lab without altering the overall work flow from separations to counting.

IV. SELECTED RADIOCHEMICAL ANALYSIS EXAMPLES

A. Strontium-90

Strontium-90 is a high abundance fission product in nuclear waste that decays by beta emission to ^{90}Y, which is also a beta emitter. The half-lives of these isotopes are 29 years and 64 hours respectively. The determination of ^{90}Sr is important in the characterization of stored nuclear wastes from nuclear weapons production and in process streams associated with spent fuel reprocessing. Analytical determination of ^{90}Sr, a pure beta emitter, requires that it be separated from inactive matrix constituents and a variety of interfering radionuclides prior to quantification by counting methods.

Detector traces from the automated SI separation of ^{90}Sr, using flow-through liquid scintillation counting, are shown in Fig. 14.8. This analytical separation was first automated using conventional SI methods with stacked zones, and subsequently with separation-optimized SI (Grate *et al.*, 1999b, 1996). The sample in 8 M nitric acid (HNO$_3$) solution is loaded on an extraction chromatographic material called Sr-resin (Eichrom Technologies, Inc., Darien, Illinois). This resin contains a crown ether in 1-octanol immobilized on a porous polymer support; the crown ether stationary phase retains Sr from 8 M HNO$_3$ with a capacity factor of ~100 (Horwitz *et al.*, 1991, 1992a,b). Other radionuclides (including the ^{90}Y daughter) and sample matrix components are unretained and are removed by washing the resin with additional 8 M nitric acid. Then the purified ^{90}Sr is released from the

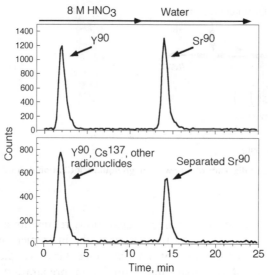

FIGURE I4.8 Detector traces for the separation of a ^{90}Sr and ^{90}Y containing standard in the upper plot, and separation of ^{90}Sr from other radioisotopes in a nuclear waste sample. (Reproduced with permission from Anal. Chem., 68 (1996) 333–340. Copyright I996 Am. Chem. Soc.)

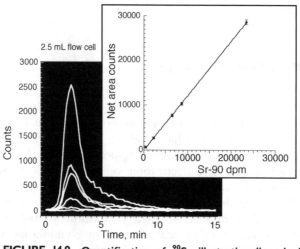

FIGURE I4.9 Quantification of ^{90}Sr illustrating linearly increasing peak areas with increasing standard activity. (Adapted with permission from figures in Anal. Chem., 68 (1996) 333–340. Copyright 1996 Am. Chem. Soc).

column by eluting with water or weak acid (e.g., 0.05 M nitric acid), where the capacity factor is less than one.

The peak area of the eluted ^{90}Sr is used for quantification. Figure 14.9 shows detector traces for different activity standards and the linear calibration curve obtained from peak areas. The theory of flow-through scintillation detection has been described previously in Chapter 12. In evaluating the analytical quantification of radionuclides a number of figure of merits are relevant. The separation or recovery efficiency indicates the fraction of the analyte delivered to the column that is recovered after the separation. The detection efficiency indicates the number of counts observed compared to the number of radioactive decays from the sample in the detector flow cell. These two efficiencies are given as E_{rec} and E_d, respectively, in Chapter 12 and used in Eq. 12.14. The total effective efficiency is the product of the recovery efficiency and the detection efficiency.

The automated separation of ^{90}Sr represents a simple load, wash, and elute procedure for the separation of a species that is present at high activities in nuclear waste samples. The procedure can either be implemented with on-line detection, as just described, or by delivering the eluted Sr to a fraction collector. Using the separation optimized SI method, it has been demonstrated that complete separation and analysis can be carried out in just over 20 minutes.

For high throughput applications, especially in analytical laboratories that customarily count purified samples separately from the sample preparation, fraction collection is advantageous. Using column switching techniques so that each column need not be washed between samples, it is possible to process eight samples in just over an hour (8 minutes per sample) (Grate *et al.*, 1999b). These samples can all be set up in advance and separated under computer control without further operator attention. Elimination of column clean up steps and/or blank runs between sample runs minimizes the amount of time required to complete the separation of the sample set. Then the columns can either be washed for reuse or replaced with fresh columns.

B. Technetium-99

Technetium-99 is a long-lived radioactive isotope that is present in defense related nuclear wastes, stored spent nuclear fuel, and in radioactive waste and process streams associated with spent fuel reprocessing. Due to the high abundance of ^{99}Tc in these wastes, its long radioactive half-life, and the high mobility of technetium in the environment, ^{99}Tc analysis is important throughout nuclear waste characterization and stabilization activities. ^{99}Tc is a pure beta-emitter ($\beta_{max} = 294$ keV) with a half-life of 2.13×10^5 years and specific activity of 629 Bq/μg, decaying to stable ^{99}Ru (Browne and Firestone 1986). Direct nondestructive analysis of ^{99}Tc by gamma-spectroscopy is not possible. Analytical methods using radioactivity detection (beta-counting) require separation of ^{99}Tc from inactive matrix constituents and various interfering radionuclides.

Automated SI separation of ^{99}Tc from a nuclear waste sample is shown in Fig. 14.10 (Egorov *et al.*, 1998b). The separation is carried out using TEVA-resin (Eichrom Technologies, Inc., Darien, Illinois), a porous polymer impregnated with a liquid quaternary amine-based liquid anion exchanger called Aliquat 336 (Horwitz *et al.*, 1995). The sample is loaded on the column and washed with 0.1 M HNO_3, conditions under which Tc is strongly retained on the column and other potentially interfering species are largely unretained. (Inclusion of hydrofluoric acid in the wash solution assures that tetravalent Pu is not retained on the column.) The retained Tc is

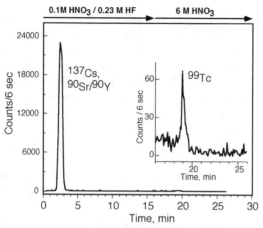

FIGURE 14.10 Detector traces for the separation of ^{99}Tc(VII) from other fission products in the sample. The signal for the Tc is evident in the vertically expanded trace in the inset. Reproduced with permission from Anal. Chem., 70 (1998) 977–984. Copyright 1998 Am. Chem. Soc.)

released from the column using strong nitric acid. The 6 M nitric acid concentration in Fig. 14.10 was selected as a trade-off between higher acid concentrations that provide narrower elution peaks at shorter retention times, and lower acid concentrations that cause less scintillation quenching than strong acid.

The Tc isotope is present at low activities in nuclear waste compared to other fission products such as ^{90}Sr, its daughter ^{90}Y, and ^{137}Cs. The Tc peak is not evident in the full scale detector trace, which is dominated by the abundant fission products. However, it can be easily seen in the vertically expanded inset. Tc activity can be quantified from the scintillation counts, and this is most effectively carried out in a stopped flow mode (Egorov *et al.*, 1998b). The principle of stopped flow detection was described previously in Chapter 12 (see Eq. 12.28) and illustrated in Fig. 12.12, where continuous flow and 15 minute stopped flow detection were compared using a high activity (2.502×10^4 dpm) ^{99}Tc(VII) standard. In stopped flow detection, this time position of the peak maximum in a continuous flow run can be used to stop the flow; then counting can be carried out for whatever time period is necessary to achieve the desired precision. In the experiment shown in Fig. 12.12, 89% of the eluted technetium zone resided in the 2.5 mL detector flow cell during the stopped flow interval. The duplicate detector traces for each type of experiment in Fig. 12.12 demonstrated reproducibility; the high precision of the fluidics and reproducibility of the separation process result in reliable capture of the eluted zone within the detector flow cell sample.

The stopped flow approach facilitates much improved counting statistics and its advantage is most apparent when counting low activity standards or samples where the count rate is statistically indistinguishable from the background. For example, 268 dpm ^{99}Tc(VII) standards gave a stopped flow count *rate* that was statistically indistinguishable from the background. As shown in Fig. 12.13, signal accumulation for a 15 minute stopped-flow interval allows reliable quantification of a 268 dpm ^{99}Tc(VII) standard (with 8% (3σ) counting error) that could not be detected in a continuous flow mode.

Using a diverter valve to isolate the sample in the detector during stopped flow allows the remainder of the SI separation system to be used to process the next sample simultaneously. Thus, stopped flow does not necessarily lengthen the total analysis time (Egorov *et al.*, 1998b).

C. Actinides

Determination of actinides in a variety of sample matrixes is of great importance in the characterization and processing of nuclear wastes, the nuclear industry, and in the remediation of radiologically contaminated sites. Both of the conventional detection methods for actinides, alpha spectoscopy and mass spectrometry, require separation of the actinides from the sample matrix and from one another [as discussed in Separation Requirements (Section II.A)] (Egorov *et al.*, 2001b; Grate and Egorov, 1998a; Grate *et al.*, 1999a).

A number of extraction chromatographic resins can be used to implement a variety of actinide separations. Selected automated radiochemical separations using TRU-resin (Eichrom Technologies, Inc., Darien, Illinois), which is a solvent-impregnated resin loaded with complexant carbamoyl-methylenephosphine oxide (CMPO) in tributylphosphate (Horwitz *et al.*, 1990, 1993) are illustrated in Fig. 14.11. The organic stationary phase in this resin strongly binds trivalent, tetravalent, and hexavalent actinides from nitric acid solutions, with retention increasing with aqueous phase acidity. Tetravalent actinides are more strongly retained than trivalent actinides; and trivalent actinides are not retained in hydrochloric acid solutions. This extraction chromatographic material can be used for simple separations of groups of actinides, or more complicated separations where individual actinides or small groups of actinides are sequentially eluted. In the latter case, several actinides are retained on the column during the sample load and wash steps. Then a number of eluent solutions are delivered in sequence, each releasing a selected actinide or group of actinides. Actinides Am and Pu can be separated from each other and from other actinides if the speciation of Pu, i.e., its oxidation or valence state, is controlled throughout the multistep elution process.

The plot in Fig. 14.11a shows the purification of the actinides as a group. Fission products are unretained or slightly retained and appear during the column wash with 2 M nitric acid (Grate and Egorov, 1998b; Grate *et al.*, 1999a). Ammonium hydrogen oxalate (bioxalate) solution releases the actinides from the column.

The second plot, Fig. 14.11b, shows the release of captured actinides in groups of the same oxidation (or valence) state. First the trivalent actinides are selectively released with 4 M hydrochloric acid. Then an oxalic acid solution is used to selectively release tetravalent actinides. Finally, an ammonium bioxalate solution releases the remaining hexavalent actinides (Grate and Egorov, 1998b; Grate *et al.*, 1999a). This type of separation

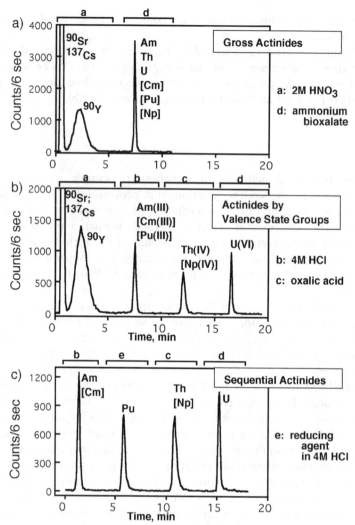

FIGURE 14.11   Three actinides separations performed using automated sequential injection separations with TRU-resin extraction chromatographic material: (a) Separation of the
actinides as a single group (gross actinides) from the fission products and other species in the
sample matrix; (b) Separation of the actinides from the fission products and other species in
the sample matrix, with selective release from the column material in valence state groups;
(c) Sequential elution of the actinides with separation of Pu from Am and the other trivalent
f-element species using on-column redox chemistry. The separation of the actinides from the
fission products and sample matrix is omitted from this plot. (Adapted with permission from
Analyst., 124 (1999) 1143–1150. Copyright 1999 Royal Society of Chemistry.)

requires that the actinides be in well-defined valence states, so samples are
pretreated to adjust the speciation of, for example, Pu.

The final plot, Fig. 14.11c, illustrates a sequential actinide separation.
This separation is based on the valence state separation in the second figure,
but it uses additional steps to separate Pu from Am and the other trivalent
f-elements. This is achieved by converting Pu to the tetravalent state prior to
eluting the trivalent species. Then the tetravalent Pu that remains captured on

the column is reduced to trivalent Pu and released. The remaining tetravalents and hexavalent species are then eluted as before (Grate and Egorov, 1998b; Grate *et al.*, 1999a).

These actinide separations demonstrate that sequential injection separations can be extended from simpler "load, wash, and elute" separations to multicomponent separations. In addition, the separation of Pu from Am demonstrates the use of on-column redox chemistries for selective elution of Pu, where the column serves as both a reactor and separator.

The detector traces in the figures were obtained by flow-through liquid scintillation counting. For trace detection applications, the eluted species from the automated radiochemical separations have been delivered to a fraction collector and quantified with alpha spectrometry off-line (Grate *et al.*, 1999a).

Automated separation methodologies using TRU-resin have been expanded to trace actinide analysis using ICP-MS as the on-line detector. As noted above, radiochemical separations are required for a number of reasons for ICP-MS detection, and this detection method provides separate detection traces for isotopes with different mass-to-charge ratios. The combination of automated separation with ICP-MS is shown schematically in Fig. 14.12. This diagram shows how the main fluid handling system can be maintained in a "cold" zone while the sample injection valve, separation column, and ICP-MS instrument can be placed in a hot zone. Detector traces for the separation of several actindes are shown in Fig. 14.13 (Egorov *et al.*, 2001b). This separation procedure addresses a number of interferences in the actinide analysis noted previously. This separation methodology was used to enable isotope specific determination of actinide species in dissolved vitrified and tank waste sample matrixes.

The use of ICP-MS detection in conjunction with continuous-forward-flow FI to automate extraction chromatographic radionuclide separations was

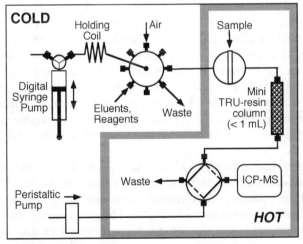

FIGURE 14.12 Schematic diagram for a sequential injection separation system with ICP-MS detection.

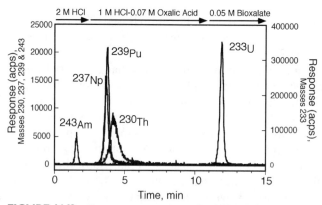

FIGURE 14.13 Detector traces from the ICP-MS for analysis of several actinides, where acps is the area counts per second. (Adapted with permission from Anal. Chem., 70 (1998) 779A–788A. Copyright 1998 Am. Chem. Soc.)

first described in papers by Hollenbach and Aldstadt. Hollenbach *et al.* described the automation of extraction chromatographic methods to separate and preconcentrate Tc, Th, and U from soil samples (Hollenbach *et al.*, 1994). Aldstadt *et al.*, described the use of FI and extraction chromatography to analyze U in environmental samples (Aldstadt *et al.*, 1996). In both cases, the use of an on-line separation system resulted in improved ICP-MS detection limits relative to direct sample introduction.

D. Renewable Separation Column Applications

The renewable separation column approach offers a number of advantages in separation methods where the life of the separation column is limited or tedious regeneration steps are necessary prior to column reuse. These have been demonstrated for ^{90}Sr analysis using Sr-resin, ^{99}Tc separation using TEVA-resin, and Am separation using TRU-resin (Egorov *et al.*, 1999a).

For example, the recovery of the retained ^{90}Sr from the column upon release with water or dilute acid is typically 95%, with 3–4% of the original sample eluting in a subsequent blank run if no additional column wash steps are used. This carryover is associated with the Sr-resin extraction chromatographic material. Instead of washing the column after each analysis, the separation material can be automatically renewed. This eliminates the observed carryover, and it is then reasonable to analyze blanks or low activity standards after high activity samples (Egorov *et al.*, 1999a).

When analyzing Am using TRU resin to separate it from fission products and other actinides, there are several actinides retained on the column after Am elution. If the column is to be reused, these must all be removed from the column (e.g., using ammonium bioxalate). However, using the renewable separation column technique, the column bed can simply be disposed of after Am elution and replaced with new column material for the next sample (Egorov *et al.*, 1999a).

In the ^{99}Tc analysis, the renewable column technique offers an alternative analytical approach. After capturing Tc on TEVA-resin and washing away interferences, one manual method for Tc analysis involves placement of the Tc-containing resin in a vial and adding cocktail for liquid scintillation counting. Thus the Tc-loaded resin is counted directly. The SI-RSC method can perform the separation and deliver the resin material to a liquid scintillation vial automatically (Egorov et al., 1999a).

Finally, the SI-RSC technique enables a new type of automated radiochemical analysis instrument. It enables an "open architecture" where one instrument design can be used for various radiochemical separations simply by using the software to choose which separation material is loaded into the column, and which solutions are selected for the washing and elution steps. A proof of principle demonstration for this approach has been described (Egorov et al., 1999a). Thus, this instrument could perform a series of separations of successive aliquots of one sample, or different analyses on each sample, as required.

V. AUTOMATION USING ROBOTICS

Laboratory robotics represents an important aspect of automation in the analytical laboratory. Modern robotic instrumentation can be used to automate a range of routine analytical operations that may be encountered in wet radiochemical analyses. Some typical examples include sample weighing, dissolution, liquid–liquid and solid phase extraction, serial dilutions, filtration, and sample delivery to instrumentation (Hurst, 1995). Of specific importance is the capability to automate initial sample preparation steps (e.g. dissolution or leaching of the solid samples) that cannot be automated using the fluid handling techniques described in this chapter. Buegelsdijk et al. described a fully automated system for dissolution of Pu metal based on the Zymate II (Zymark Corporation) laboratory robot (Beugelsdijk and Hollen, 1998). The robotic system was set up to perform dissolution of 2–5 g samples of Pu metal. The operational steps include verification of the sample identity through readout of the bar-code label, weighing the sample on the analytical balance, transferring the sample to the dissolution vessel, and sample dissolution. After dissolution, the individual sample aliquots were prepared, weighed, labeled and sorted for subsequent analysis.

Zahradnik and Swietly (1996) described Zymark robotic system for preparation and chemical treatment of diluted spent nuclear fuel samples. The robotic system executed initial sample preparation steps such as batching, weighing, aliquoting and spiking with tracers. Following automated preparation of the sample batch, the robot executes separation steps designed to separate U and Pu from the highly radioactive fission product matrix. Initial chemical treatment steps included drying and redissoltuion of the sample in order to remove volatile fission products such as ^{106}Ru. Following Pu valence state adjustment, an extraction chromatographic separation using immobilized tri-n-octylphosphine oxide (TOPO) was carried out to separate fission products and to collect U and Pu fractions for subsequent analysis using alpha spectroscopy and isotope dilution mass spectroscopy.

In summary, laboratory robotics represent an attractive approach towards automation of sample preparation and treatment steps in radiochemical analysis. Despite substantial promise, the use of robotics in radiochemical analysis has a limited number of applications primarily in the analysis of spent nuclear fuel. Reduction in robot costs may result in more widespread use of robotics in routine radioanalytical practice. In addition, the combination of robotic sample preparation and automated radionuclide analyzer techniques described above can yield workstations capable of fully autonomous radiochemical measurements.

VI. AN AUTOMATED RADIONUCLIDE ANALYZER FOR NUCLEAR WASTE PROCESS STREAMS

The original work on SI methodology for radiochemical analysis was directed to laboratory analysis of beta and alpha-emitters in processed nuclear waste samples. A more challenging objective is to perform rapid radiochemical measurements of nuclear process streams on-line or at-line. This task is especially challenging for the determination of nongamma emitting radionuclides which typically cannot be measured using remote nondestructive methods. Examples of process analysis needs include the determination of specific radionuclides before and after separation processes that remove the radionuclides from the waste streams. Such analyzers can assess the effectiveness of the radionuclide removal process, and can be used to indicate if the output stream will qualify for subsequent conversion to stable waste forms.

In the United States chemical processing of low activity waste (LAW) prior to vitrification at the Hanford site requires removal of ^{90}Sr, transuranics, ^{137}Cs, and ^{99}Tc. The process streams are caustic brine matrixes with complex and varying chemical and radiological composition. For example, the total base content, the concentration of organics, and the complexant concentrations all depend on the source of the feed, as do the aluminum, nitrate, nitrite, dichromate, and radionuclide contents.

An on-line/at-line analytical method is desirable to detect ^{99}Tc in the effluent from the technetium removal process. In waste matrixes with high organic content, a substantial amount of the total ^{99}Tc is present in a reduced, non-pertechnetate form of unknown chemical composition. Thus, in the case of ^{99}Tc, varying sources of feed material result in varying speciation of Tc. The current preferred method of Tc measurement is off-line analysis of a liquid sample using ICP-MS or classical wet radiochemical analysis, which requires taking a process sample and sending it to a centralized laboratory. The overall process of sampling and laboratory analysis is generally too time consuming to provide adequate feedback for continuous process control.

The analytical separation chemistries for ^{99}Tc, such as those described above, are effective only for Tc in the pertechnetate form. Therefore, in order to enable total ^{99}Tc analysis via radiochemical measurement, oxidation

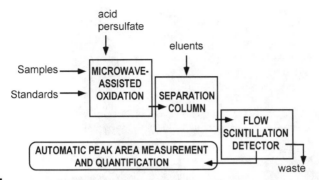

FIGURE 14.14 Schematic diagram illustrating sample processing, separation, and detection for an on-line analyzer for the continuous monitoring of the total Tc content of nuclear waste process streams.

chemistries and procedures are required to convert all Tc species to pertechnetate. Thus, radiochemical analysis of ^{99}Tc entails sample oxidation, Tc separation (e.g. using ion exchange or solvent extraction) and beta scintillation or gas proportional counting.

A fluidic radiochemical analyzer system has been developed to perform all these functions in a fully automated fashion for the rapid analysis of total ^{99}Tc in aged LAW streams (Egorov *et al.*, 2001a, 2002a,b; Egorov and Grate, 2002; Grate and Egorov, 2001a,b). The automated radiochemical analyzer executes fluid handling steps required to perform acidification of the caustic sample, microwave-assisted sample oxidation using peroxidisulfate oxidant, separation of ^{99}Tc(VII) from radioactive interferences using an anion exchange column, and delivery of the separated pertechnetate to a flow-through scintillation detector. A schematic diagram of the automated Tc measurement process is shown in Fig. 14.14. A photograph of the prototypical analyzer unit is shown in Fig. 14.15. The instrument design incorporates advanced digital fluid handling techniques using several zero dead volume syringe pumps, multiple valves for sample and reagent delivery, a two position valve for flow reversal through the column (see Fig. 14.4a), and a diverter valve before the detector.

The sample treatment protocol begins with the sample acidification followed by a first digestion (heating) step. This initial treatment ensures removal of the nitrites, which are expected to interfere with the subsequent oxidation. In addition, initial heating promotes rapid dissolution of the $Al(OH)_3$ precipitate which forms during acidification of the caustic LAW matrix. The nitric acid concentration and volume are selected to ensure complete dissolution of the $Al(OH)_3$ species upon heating, while maintaining relatively high pertechnetate uptake values on the anion exchange sorbent material during sample loading. The initial sample acidification procedure is followed by a second digestion treatment using sodium peroxidisulfate oxidizing reagent. This oxidative treatment of the acidified sample ensures conversion of the reduced Tc species to pertechnetate.

Pertechnetate separation is accomplished using a macroreticular strongly basic anion exchange resin. Compared to a similar extraction chromatographic

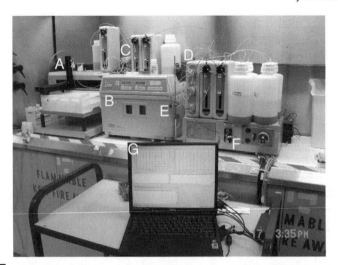

FIGURE 14.15 Photograph of the fully automated total ^{99}Tc analyzer instrument: **(A)** is a robotic autosampler; **(B)** microwave digestion unit; **(C)** fluid handling components for sample injection, automated standard addition, sample acidification/digestion; **(D)** separation fluidics including syringe pumps, flow reversal and diversion valves; **(E)** separation column; **(F)** flow scintillation detector; **(G)** control computer with automation software.

chemistry using TEVA-resin (described above), the solid phase anion exchange material offers much longer column life. However, the pertechnetate elution kinetics of the extraction chromatographic material are superior to those of the solid phase anion exchange material. Nevertheless, rapid pertechnetate elution from the anion exchange material is possible when the eluent flow direction through the column is reversed. The separation selectivity using anion exchange is adequate for the analysis of aged LAW sample matrixes and provides reliable separation of pertechnetate from the major radioactive constituents (^{90}Sr/^{90}Y, ^{137}Cs,) and the minor constituents (e.g. isotopes of Sn, Sb, and Ru). However, a combination of column washes using dilute nitric acid, nitric-oxalic acid, sodium hydroxide, and moderately concentrated nitric acid and is necessary for reliable separation of pertechnetate from anionic species such as Sn, Sb, and Ru.

The purified pertechnetate eluted from the column was detected and quantified with a flow-through scintillation detector using a lithium glass solid scintillator. This scintillator material exhibited excellent stability in strong nitric acid solutions used for pertechnetate elution. The disadvantages of the liquid scintillation detection in this application include increased waste generation associated with the cocktail and sporadic chemoluminescence signals in the presence of oxidants (e.g. Cr(VI)). No chemoluminescence was evident when lithium glass solid scintillator cells were used for on-line quantification of ^{99}Tc in matrixes with high concentrations of Cr(VI) species.

In order to obtain reliable ^{99}Tc quantification in varying sample matrixes, an automated standard addition technique was implemented at a part of the analytical protocol. The standard addition approach is based on the introduction of a ^{99}Tc standard solution during the sample acidification step. The ^{99}Tc standard is in a nitric acid solution with the same acid

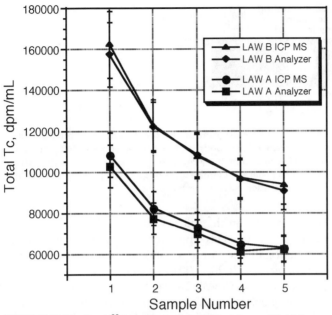

FIGURE 14.16 Total ^{99}Tc analysis in **LAW** matrixes with high organic content comparing ICP-MS and automated radiochemical analyzer instrument results. Samples contain up to 60% of Tc in non-pertechnetate species. Error bar is ±10%.

concentration used for sample acidification. The total effective efficiency (product of the recovery efficiency and the detection efficiency) is calculated based on the difference in analytical response obtained for the analysis of the spiked and unspiked samples. This approach provides a reliable method for remote, matrix matched Tc monitor instrument calibration.

The monitor instrument was fully automated for continuous operation with the control software performing asynchronous device control and detector data acquisition/processing including peak search and integration as well as the instrument calibration via standard addition. Total analysis time was 12.5 minutes per sample and the total analysis time for the standard addition measurement was 22 minutes (including analysis of both sample and spiked sample. In tests against various LAW samples from the US DOE Hanford site, including those with high organic content, the total Tc was accurately quantified as verified by independent sample analysis using ICP-MS. Figure 14.16 compares results for total ^{99}Tc analysis of high organic content LAW samples using ICP-MS and automated radiochemical Tc analyzer.

VII. RADIONUCLIDE SENSORS FOR WATER MONITORING

A. Preconcentrating Minicolumn Sensors

Thus far, the radiochemical separations described entail on-line or off-line detection of species after they have left the separation column. If the

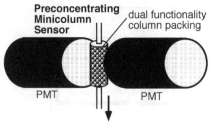

FIGURE 14.17 Schematic diagram of the preconcentrating minicolumn sensor concept where the transparent column body is placed between two PMTs for light collection. The packing material in the column body contains both selective separation chemistry and scintillating materials to generate a luminescent output. The packing material may be delivered using renewable separation column methods.

radionuclides captured on the column could be detected in place, the column would become a sensor. This approach has been successfully demonstrated using dual functionality separation columns as preconcentrating minicolumn sensors (DeVol *et al.*, 2000, 2001a,b; Egorov *et al.*, 1999b; Egorov and Grate, 2002; Roane and DeVol, 2002). These columns perform both selective capture of radionuclides and transduction of the presence of captured radionuclides into a measurable signal. The latter function is achieved by incorporating scintillating materials into the column. Light output is measured by photomultiplier tubes (PMTs) placed around the transparent dual functionality column. The preconcentrating minicolumn sensor concept is shown in Fig. 14.17.

Two approaches have been developed to achieve dual functionality in the columns. In one approach, the solid phase separation material is prepared to be scintillating by combining scintillating fluors with the separation chemistry in individual particles. These particles then represent "selective scintillating microspheres," or SSMs. The alternative approach is to intimately mix scintillating particles with chemically selective particles to obtain both functionalities in the same column. If so desired, renewable separation column techniques can be used to deliver or renew the sensing material or mixture.

This sensing concept is designed to meet the challenging requirements for the sensing of alpha- and beta-emitting radionuclides in water. These radionuclides emit particles with short penetration ranges in condensed media and must be detected to ultratrace analytical detection limits. The close proximity of selective chemistries and scintillating materials in particles of the order of 10–250 microns in diameter enables radiometric detection of alpha or beta emissions from the captured radionuclides. Because of the relatively long range of the beta particles, this approach appears to be particularly suitable for detection of beta-emitters such as ^{99}Tc and ^{90}Sr. The ranges in water for beta particles emitted by ^{90}Y ($E_{max} = 2282\,keV$), ^{90}Sr ($E_{max} = 546\,keV$), and ^{99}Tc ($E_{max} = 294\,keV$) are 1.1 cm, 1.8 mm, and 750 μm respectively.

In this sensing approach, the packed column format provides for efficient fluidic processing of the sample for preconcentration. The selectivity is determined mainly by the separation chemistry used to preferentially capture and preconcentrate the analyte of interest, which localizes the radionuclides in the detection volume of a scintillating detection apparatus. For the analysis of important radionuclide contaminants such as ^{99}Tc radiometric detection offers much lower detection limits than those possible by chemical sensing approaches. Thus, this configuration meets all the functional requirements for effective radionuclide sensors.

B. Sensors for ^{99}Tc(VII)

A radionuclide sensor for ^{99}Tc sensing in groundwater has been reported using SSMs as the column packing material (DeVol et al., 2001b; Egorov et al., 1999b). The SSMs were prepared by impregnating porous acrylic ester beads with a mixture of Aliquat 336 (a liquid anion exchanger based on long chain quaternary ammonium ions) and a pair of scintillating fluors. The pertechnetate uptake characteristics of this material were characterized and found to be as expected for pertechnetate uptake by Aliquat 336, i.e., high uptake at low-acid neutral pH range. The instrumental pulse height spectra of ^{99}Tc obtained using the SSM indicated a detection efficiency of 56%, which is sufficiently high for practical analytical applications. Similar uptake and luminosity (absolute detection efficiency 30–50%) characteristics were observed for Aliquat-336 based sensor materials prepared by diffusing scintillator fluors into the bead matrix or by incorporating scintillator fluors into the bead during the polymerization process.

Sensing properties were characterized with both ^{99}Tc(VII) standards (see Fig. 14.18a) and spiked Hanford site groundwater samples (Egorov et al., 1999b). Injection of an aliquot of ^{99}Tc standard in dilute acid results in analyte capture and measurable scintillation light output. In this example the analyte capture was quantitative and the signal persists as the sensor column is washed with dilute acid. The absolute detection efficiency for the ^{99}Tc detection was 45%. Integration of the light output provides a quantitative measure of the ^{99}Tc in the sample. Because the sensor material exhibits high binding affinity towards Tc(VII), large sample volumes can be preconcentrated using a small sensor column.

On the other hand injection of a radioactive species that does not exhibit affinity to the sensor material results in only a transient peak signal as shown for ^{137}Cs in Fig. 14.18a. Interfering species are promptly removed from the system using a small volume of wash solution. In this manner, the sensing method is selective towards the target analyte.

Analytes can be quantified from either the slope of the uptake signal or from the steady state signal after washing to remove interferences. Detector traces for several standards are shown in Fig. 14.18b (Egorov et al., 1999b). This type of sensor has been successfully demonstrated using actual contaminated groundwater samples from the Hanford site and the standard addition method for quantification. Analytical results were in close agreement with the results of standard radiochemical measurements. Detection limits were below the regulatory drinking water standard.

The preconcentrating minicolumn sensor concept has also been demonstrated using mixed beds to obtain the required dual functionality

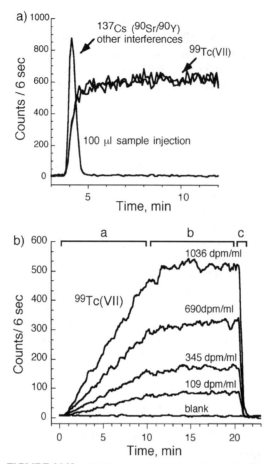

FIGURE 14.18 (a) Sensor response to injected aliquots of $^{99}Tc(VII)$ analyte and potentially interfering species (^{137}Cs) unretained by the sensor material. Following the injection the sensor bed is washed with 10 mL of 0.02 M nitric acid; (b) Detector traces for the capture and quantification of $^{99}Tc(VII)$ from 10 mL standard solution volumes, followed by column wash with 10 mL of 0.02 M nitric acid, and finally ejection of the column bed material. (Reproduced with permission from Anal. Chem., 71 (1999) 5420–5429. Copyright 1999 Am. Chem. Soc.)

(Egorov and Grate, 2002). Nonporous plastic scintillator beads (BC400, particle size 100–250 μm)) were mixed with the strongly basic anion exchange material AGMP1 (particle size 20–50 μm) in a 20:1 weight ratio. The anion exchange beads provide for selective pertechnetate uptake with a very high uptake affinity. As a result, the sensor could be prepared with a total bead volume of only 50 μL. The absolute detection efficiency of the composite sensor column was 34% and analyte recovery (loading efficiency) was 97%. The $^{99}Tc(VII)$ selective composite bed sensor can be regenerated using small volume of 2 M nitric acid solution, resulting in rapid elution of the retained analyte without the loss of the scintillation properties.

This $^{99}Tc(VII)$ sensing method was tested using unacidified Hanford groundwater samples (Egorov and Grate, 2002) Approximately 60 mL of the groundwater sample can be preconcentrated without analyte breakthrough using this sensor, and these experiments showed that radionuclide sensing could be successful on chemically untreated groundwater samples.

Using a mixture of separate solid phase extraction beads and nonporous plastic scintillator beads is advantageous in two ways. The nonporous scintillating beads were found to have much higher chemical stability in sensing and regeneration solutions than extraction chromatographic materials impregnated with scintillating fluors. In addition, this approach facilitates the use of solid phase extraction materials for radionuclide uptake that are not readily impregnated with scintillating fluors.

As originally conceived, these sensors were based on the concept of quantitative capture of the radionuclides of interest on the preconcentrating minicolumn, with either fluidic renewal via reagents or automated renewal of the solid phase sensing material using renewable separation column techniques. However, a new approach has recently been developed based on the concept of equilibrating the preconcentrating minicolumn with ambient radionuclide concentrations (Egorov et al., 2002b,c). In this scenario, the sample is delivered to the sensor column until complete breakthrough has occurred and the entire column reaches dynamic equilibrium with the sample solution. Under these breakthrough conditions, no further analyte preconcentration occurs on the column and analyte concentration in the sample solution before and after the sensor column are equal. Chromatographic theory indicates that under these dynamic equilibrium conditions, analyte concentration on the sensor column is proportional to the analyte concentration in the sample used for equilibration. Under these conditions the sensor column does not require regeneration or renewal, since the signal will go down on equilibration with a new sample of lower concentration. The equilibration approach forms the basis for the reagentless radionuclide selective sensing which is particularly well suited for the development of sensor probes for *in situ* or long term monitoring.

Analyte activity, A, on the sensor after equilibration can be calculated using the following formula:

$$A = DV_sC_l$$

where D is the volume/volume distribution coefficient of the sorbent material, V_s is volume of the stationary sorbent phase in the sensor column, and C_l is analyte activity in the sample used for sensor equilibration

This equilibration approach has been demonstrated successfully for sensing ^{99}Tc(VII), both in the laboratory and on ground water samples from the Hanford site (Egorov et al., 2002c). In addition, the sensor concept has been successfully engineered into a sensor probe in a configuration that fits into well bore holes for down-hole monitoring, using an anticoincidence shield around the sensing flow cell to reduce background counts. Images of the engineered sensor probe are shown in Fig. 14.19. The sensor flow cell with two PMTs is shown in 14.19a, while 14.19b shows the sensor in the anticoincidence shield in the probe housing, and 14.19c shows the complete probe.

(a)

(b)

(c)

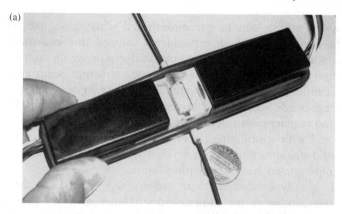

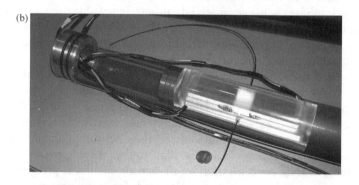

FIGURE 14.19 Images of the sensor probe for Tc monitoring using the reagentless equilibrium sensing approach: (a) the sensor flow cell with two PMTs; (b) the sensor in the anticoincidence shield in the probe housing; (c) the complete probe prior to assembling the outer housing.

Although the above examples all focus on ^{99}Tc(VII) sensing, these analytical concepts have also been used to develop sensors for ^{90}Sr and actinides (DeVol *et al.*, 2000, 2001a,b; Egorov *et al.*, 1999b; Egorov and Grate, 2002; Roane and DeVol, 2002). In the case of ^{90}Sr, it is possible to dispense with the incorporation of scintillating material in the column and instead detect the Cerenkov radiation from the ^{90}Y daughter product.

Radionuclide sensors are particularly important for monitoring contaminated ground and surface waters, although they may also be useful in monitoring nuclear processes. Monitoring ground water is important for detecting expansion of contaminated plumes and assessing remediation approaches such as barriers. The most significant radionuclide analytes are alpha- and beta-emitting radionuclides that cannot be detected by gamma spectroscopy and which are mobile in the environment. Sensors for these species must succeed in obtaining accurate analytical results for species that are normally analyzed by multistep laboratory sample preparation, separation, and detection processes, while minimizing or eliminating reagents.

VIII. MEDICAL ISOTOPE GENERATION

Automated radiochemical separations are not just applicable to the analysis of nuclear waste or contamination. These techniques can also be applied to the isolation and purification of radioisotopes for medical applications, i.e. medical isotope "generation." For example, short-lived radioisotopes are useful in medical imaging and cancer treatment. These short-lived daughter isotopes are generated by the radioactive decay of longer-lived parent isotopes.

In a typical radioisotope generator system, the parent isotope is permanently sorbed on the sorbent column while the daughter product is periodically eluted. One well known example of such generator system is $^{99m}Tc/^{99}Mo$ generator in which alumina column is used to store parent ^{99}Mo isotope and ^{99m}Tc daughter is obtained by eluting the generator with saline (Lambrecht et al., 1997). One disadvantage of this "fixed column" generator approach is that the generator column material may suffer substantial radiolytic damage over the life of the generator.

Recently, short-lived alpha-emitters are becoming increasingly important in radioimmunotherapy of cancer (Geerlings et al., 1993; McDevitt, 1996 p. 369; McDevitt et al., 1998). One of the most promising alpha emitters for the development of viable clinical treatments is ^{213}Bi (half-life 45 minutes). It can be conveniently obtained from the decay of ^{225}Ac (half-life 10 days) Generator systems that isolate ^{213}Bi from a solution containing the parent isotope (^{225}Ac) may be advantageous (Bray and Deschane, 1998; Bray et al., 1998, 2000; Wilbur et al., 1998) in order to avoid the radiolytic damage to "fixed column" generator materials noted above. This approach, however, entails handling of highly radioactive solutions and automation is essential to reduce personnel exposure and maintain containment. Automated radio-chemical separation systems can rapidly isolate the desired daughter isotope from the parent isotope while simultaneously recovering the valuable parent isotope solution for subsequent product ingrowth. Automated medical isotope generators with these characteristics have been demonstrated using SI fluidics and anion exchange chemistry to isolate short-lived alpha emitter ^{213}Bi from the solution of ^{225}Ac in hydrochloric acid (Bray et al., 2000). The automated generator was successfully demonstrated at clinically useful levels (20 mCi) and the separated Bi product was used in the evaluation of the labeling efficiency of proteins.

Automation provides the usual benefits in this application including increased safety, reduced operator exposure, and consistent performance of the procedure. Automated generators will also be useful when the short half-life of the desired isotope precludes shipping isolated isotopes from a central laboratory, i.e., when isotope separations are required at-site. Development of contained automated separation systems will be critical for such applications.

IX. DISCUSSION

The techniques and examples above demonstrate that modern fluidic techniques and instrumentation can be used to develop automated radiochemical separation workstations. In many applications, these can be mechanically simple and key parameters can be controlled from software. If desired, many of the fluidic components and solutions can be located remotely from the radioactive samples and other hot sample processing zones.

There are many issues to address in developing automated radiochemical separation that perform reliably time after time in unattended operation. These are associated primarily with the separation and analytical chemistry aspects of the process. The relevant issues include the selectivity of the separation, decontamination factors, matrix effects, and recoveries from the separation column. In addition, flow rate effects, column lifetimes, carryover from one sample to another, and sample throughput must be considered. Nevertheless, successful approaches for addressing these issues have been developed.

Radiochemical analysis is required not only for processing nuclear waste samples in the laboratory, but also for at-site or *in situ* applications. Monitors for nuclear waste processing operations represent an at-site application where continuous unattended monitoring is required to assure effective process radiochemical separations that produce waste streams that qualify for conversion to stable waste forms. Radionuclide sensors for water monitoring and long-term stewardship represent an application where at-site or *in situ* measurements will be most effective. Automated radiochemical analyzers and sensors have been developed that demonstrate that radiochemical analysis beyond the analytical laboratory is both possible and practical.

ACKNOWLEDGMENTS

The authors would like to acknowledge Matthew O'Hara for his continuing contributions to our automated radiochemistry program, and to Professors Jaromir Ruzicka and Jiri Janata for their support in initiating this program. The authors are very grateful for funding from the U.S. Department of Energy Environmental Management Science Program. The Pacific Northwest National Laboratory is a multiprogram national laboratory operated for the U.S. Department of Energy by Battelle Memorial Institute.

REFERENCES

Aldstadt, J. H., Kuo, J. M., Smith, L. L., and Erickson, M. D. (1996). Determination of uranium by flow injection inductively coupled plasma mass spectrometry. *Anal. Chim. Acta*, **319**, 135–143.
Beugelsdijk, T. J. and Hollen, R. M. (1998). Robotics and automation in radiochemical analysis. *In* "Handbook of Radioactivity Analysis." (M. F. L'Annunziata, Ed.), pp. 693–718. Academic Press, San Diego, CA.

Bray, L. A. and Deschane, J. R. (1998). Separation of bismuth-213 radionuclide from solution. U.S. Patent 5,749,042.

Bray, L. A., Tingey, J. M., DesChane, J. R., Egorov, O. B., and Pacific Northwest National Laboratory, R. W. A. U. S. A. (1998). Development of a unique bismuth (Bi-213) automated generator for use in cancer therapy. Book of Abstracts, 216th ACS National Meeting, Boston, August 23–27, I&EC-049.

Bray, L. A., Tingey, J. M., DesChane, J. R., Egorov, O. B., Tenforde, T. S., Wilbur, S. D., Hamlin, D. K., and Pathare, P. M. (2000). Development of a unique bismuth(Bi-213) automated generator for use in cancer therapy. I&EC Research, 39(9), 3189–3194.

Browne, E. and Firestone, R. B. (1986). "Table of Radioactive Isotopes." John Wiley & Sons, New York.

Bruckner-Lea, C. J., Stottlemyre, M. S., Holman, D. A., Grate, J. W., Brockman, F. J., and Chandler, D. P. (2000). Rotating rod renewable microcolumns for automated, solid-phase DNA hybridization studies. Anal. Chem., 72(17), 4135–4141.

Christian, G. D. (1994). Sequential injection analysis for electrochemical measurements and process analysis. Analyst, 119, 2309–2314.

Colodner, D., Salters, V., and Duckworth, D. C. (1994). Ion sources for analysis of inorganic solids and liquids by MS. Anal. Chem., 66, 1079A–1089A.

Cortina, J. L. and Warshawsky, A. (1997). Developments in solid–liquid extraction by solvent-impregnated resins. Ion Exch. Solvent Extr., 13, 195–293.

Crain, J. S. and Alvarado, J. (1994). Hydride interfernce on the determination of minor actinide isotopes by inductively coupled plasma mass spectrometry. J. Anal. At. Spectrom., 9, 1223–1227.

Denoyer, E. R. (1992). Expanding ICP-MS capabilities using flow injection. Am. Lab., 24, 74–82.

Desmartin, P., Kopajtic, Z., and Haerdi, W. (1997). Radiostrontium-90 (Sr-90) ultra-trace measurements by coupled ion chromatography (HPIC) and on-line liquid scintillation measurements. Enironmental Monitoring and Assessment, 44(1–3), 413–423.

DeVol, T. A., Duffey, J. M., and Paulenova, A. (2001a). Combined extraction chromatography and scintillation detection for off-line and on-line monitoring of strontium in aqueous solutions. J. Radioanal. Nucl. Chem., 249(2), 295–301.

DeVol, T. A., Egorov, O. B., Roane, J. E., Paulenova, A., and Grate, J. W. (2001b). Extractive scintillating resin for 99Tc quantification in aqueous solutions. J. Radioanal. Nucl. Chem., 249(1), 181–189.

DeVol, T. A., Roane, J. E., Williamson, J. M., Duffey, J. M., and Harvey, J. T. (2000). Development of scintillating extraction media for separation and measurement of charged-particle-emitting radionuclides in aqueous solutions. Radioact. Radiochem., 11(1), 34–46.

Dietz, M. L. and Horwitz, E. P. (1993). Novel chromatographic materials based on nuclear waste processing chemistry. LC-GC, 11, 424–426, 428, 430, 434, 436.

Dockendorff, B., Holman, D. A., Christian, G. D., and Ruzicka, J. (1998). Automated solid phase extraction of theophylline by sequential injection on renewable column. Anal. Commun., 35(11), 357–359.

Egorov, O., DeVol, T., and Grate, J. (2001a). Advances in automated radioanalytical chemistry: from groundwater monitoring to nuclear waste analysis. Abstracts of Papers, 222nd ACS National Meeting, Chicago, IL, United States, August 26–30, 2001, NUCL-183.

Egorov, O., Grate, J. W., and Ruzicka, J. (1998a). Automation of radiochemical analysis by flow injection techniques: Am-Pu separation using TRU-resin(sorbent extraction column. J. Radioanal. Nucl. Chem., 234, 231–235.

Egorov, O., O'Hara, M., and Grate, J. (2002a). Automated radiochemical analysis of total Tc-99 in nuclear waste processing streams. Abstracts of Papers, 223rd ACS National Meeting, Orlando, FL, United States, April 7–11, 2002, NUCL-006.

Egorov, O., O'Hara, M., and Grate, J. W. (2002b). Automation of the radiochemical analysis: from groundwater monitoring to nuclear waste analysis. Abstracts of Papers, 223rd ACS National Meeting, Orlando, FL, United States, April 7–11, 2002, NUCL-062.

Egorov, O., O'Hara, M. J., and Grate, J. W. "Radionuclide selective sensors for water monitoring: 99Tc(VII) detection in Hanford groundwater." Spectrum 2002:, Reno, NV, 928–931.

Egorov, O., O'Hara, M. J., Grate, J. W., and Ruzicka, J. (1999a). Sequential injection renewable separation column instrument for automated sorbent extraction separations of radionuclides. *Anal. Chem.*, **71**, 345–352.

Egorov, O. and Ruzicka, J. (1995). Flow injection renewable fiber optic sensor system. Principle and validation on spectrophotometry of chromium(VI). *Analyst*, **120**, 1959–1962.

Egorov, O. B., Fiskum, S. K., O'Hara, M. J., and Grate, J. W. (1999b). Radionuclide sensors based on chemically selective scintillating microspheres: renewable column sensor for analysis of 99Tc in water. *Anal. Chem.* **71**(23), 5420–5429.

Egorov, O. B. and Grate, J. W. (2002). unpublished results.

Egorov, O. B., O'Hara, M. J., Farmer, O. T., III, and Grate, J. W. (2001b). Extraction chromatographic separations and analysis of actinides using sequential injection techniques with on-line inductively coupled plasma mass spectrometry (ICP MS) detection. *Analyst*, **126**(9), 1594–1601.

Egorov, O. B., O'Hara, M. J., Ruzicka, J., and Grate, J. W. (1998b). Sequential injection system with stopped flow radiometric detection for automated analysis of 99Tc in nuclear waste. *Anal. Chem.*, **70**, 977–984.

Fang, Z. (1993). "Flow Injection Separation and Preconcentration." VCH, Weinheim.

Fiskum, S. K., Riley, R. G., and Thompson, C. J. (2000). Preconcentration and analysis of strontium-90 and technetium-99 from Hanford groundwater using solid phase extraction. *Journal of Radioanalytical and Nuclear Chemistry*, **245**(2), 261–272.

Garcia Alonso, J. I. (1995). Determination of fission products and actinides by inductively coupled plasma mass spectrometry using isotope dilution analysis: a study of random and systematic errors. *Anal. Chim. Acta*, **312**, 57–78.

Garcia Alonso, J. I., Babelot, J.-F., Glatz, J.-P., Cromboon, O., and Koch, L. (1993). Applications of glove-box ICP-MS for the analysis of nuclear materials. *Radiochim. Acta*, **62**, 71–79.

Garcia Alonso, J. I., Sena, R., Arbore, P., Betti, M., and Koch, L. (1995). Determination of fission products and actinides in spent nuclear fuels by isotope dilution Ion chromatography inductively coupled plasma mass spectrometry. *J. Anal. At. Spectrom.*, **10**, 381–393.

Geerlings, M. W., Kaspersen, F. M., Apostolidis, C., and Van Der Hout, R. (1993). The feasibility of 225Ac as a source of a-particles in radioimmunotherapy. *Nucl. Med. Comm.*, **14**, 121.

Grate, J. W. and Egorov, O. (1998a). Investigation and optimization of on-column redox reactions in the sorbent extraction separation of americium and plutonium using flow injection analysis. *Anal. Chem.*, **70**, 3920–3929.

Grate, J. W. and Egorov, O. B. (1998b). Automating analytical separations in radiochemistry. *Anal. Chem.*, **70**, 779A–788A.

Grate, J. W., and Egorov, O. B. (2001a). Advances in radioanalytical chemistry using automated sequential injection analysis. Abstracts of Papers, 222nd ACS National Meeting, Chicago, IL, United States, August 26–30, 2001, NUCL-059.

Grate, J. W., and Egorov, O. B. (2001b). Automated radiochemical separation, analysis, and sensing. Abstracts of Papers, 222nd ACS National Meeting, Chicago, IL, United States, August 26–30, 2001, IEC-010.

Grate, J. W., Egorov, O. B., and Fiskum, S. K. (1999a). Automated extraction chromatographic separations of actinides using separation-optimized sequential injection techniques. *Analyst*, **124**(8), 1143–1150.

Grate, J. W., Fadeff, S. K., and Egorov, O. (1999b). Separation-optimized sequential injection method for rapid automated separation and determination of 90Sr in nuclear waste. *Analyst*, **124**, 203–210.

Grate, J. W., Strebin, R. S., Janata, J., Egorov, O., and Ruzicka, J. (1996). Automated analysis of radionuclides in nuclear waste: rapid determination of Sr-90 by sequential injection analysis. *Anal. Chem.*, **68**, 333–340.

Hollenbach, M., Grohs, J., Mamich, S., Kroft, M., and Denoyer, E. R. (1994). Determination of technetium-99, thorium-230 and uranium-234 in soils by inductively coupled plasma mass spectrometry using flow injection preconcentration. *J. Anal. Atom. Spectrom.*, **9**, 927–933.

Holman, D. A., Christian, G. D., and Ruzicka, J. (1997). Titration without mixing or dilution: sequential injection of chemical sensing membranes. *Anal. Chem.*, **69**, 1763–1765.

Horwitz, E. P., Chiarizia, R., and Dietz, M. L. (1992a). A novel strontium-selective extraction chromatographic resin. *Solvent Extr. Ion Exch.*, **10**, 313–336.

Horwitz, E. P., Chiarizia, R., Dietz, M. L., Diamond, H., and Nelson, D. M. (1993). Separation and preconcentration of actinides from acidic media by extraction chromatography. *Anal. Chim. Acta*, **281**, 361–372.

Horwitz, E. P., Dietz, M. L., and Chiarizia, R. (1992b). The application of novel extraction chromatographic materials to the characterization of radioactive waste solutions. *J. Radioanal. Nucl. Chem., Articles*, **161**, 575–583.

Horwitz, E. P., Dietz, M. L., Chiarizia, R., Diamond, H., Maxwell, S. L., and Nelson, M. R. (1995). Separation and preconcentration of actinides by extraction chromatography using a supported liquid anion exchanger: application to the characterization of high-level nuclear waste solutions. *Anal. Chim. Acta*, **310**, 63–78.

Horwitz, E. P., Dietz, M. L., Diamond, H., LaRosa, J. J., and Fairman, W. D. (1990). Concentration and separation of actinides from urine using a supported bifunctional organophosphorus extractant. *Anal. Chim. Acta*, **238**, 263–271.

Horwitz, E. P., Dietz, M. L., and Fisher, D. E. (1991). Separation and preconcentration of strontium from biological, envirnomental, and nuclear waste samples by extraction chromatography using a crown ether. *Anal. Chem.*, **63**, 522–525.

Hurst, W. J. (1995). Robotics in the Laboratory. pp. 91–107. *In* "Automation in the Laboratory." (H. W. J., Ed.), VCH Publishers, New York.

Ivaska, A. and Ruzicka, J. (1993). From flow injection to sequential injection: Comparison of methodologies and selection of liquid drives. *Analyst*, **118**, 885–889.

Izatt, R. M. (1997). Review of selective ion separations at BYU using liquid membrane and solid phase extraction procedures. *J. Incl. Phen. Mol. Rec. Chem.*, **29**, 197–220.

Izatt, R. M., Bradshaw, J. S., and Bruening, R. L. (1995). Accomplishment of difficult chemical separations using solid phase extraction. *Pure & Appl. Chem.*, **68**, 1237–1241.

Izatt, R. M., Bradshaw, J. S., and Bruening, R. L. (1996). Accomplishement of difficult chemical separations using solid phase extraction. *Pure & Appl. Chem.*, **68**, 1237–1241.

Izatt, R. M., Bradshaw, J. S., Bruening, R. L., and Bruening, M. L. (1994). Solid phase extraction of ions of analytical interest using molecular recognition technology. *Am. Lab*, **26**, 28C.

Kim, G., Burnett, W., and Horwitz, E. P. (2000). Efficient preconcentration and separation of actinide elements from large soil and sediment samples. *Anal. Chem.*, **72**(20), 4882–4887.

Lambrecht, R. M., Tomiyoshi, K., and Sekine, T. (1997). Radionuclide Generators. *Radiochim. Acta*, **77**, 103–123.

Maxwell, S. L. (1997). Rapid actinide separation methods. *Radioact. Radiochem.*, **8**, 36–44.

Mayer, M. and Ruzicka, J. (1996). Flow injection Based Renewable Electrochemical Sensor System. *Anal. Chem.*, **68**, 3808–3814.

McDevitt, M. R., Sgouros, G., Finn, R. D., Humm, J. L., Jurcic, J. G., Larson, S. M., Scheinberg, D. A., and Memorial Sloan-Kettering Cancer Center, N.Y. N.Y. U.S.A. (1998). Radioimmunotherapy with alpha-emitting nuclides. *Eur. J. Nucl. Med.*, **25**(9), 1341–1351.

Murray, R. L. (1994). "Understanding Radioactive Waste." Battelle Press, Richland, Washington.

Pollema, C. H., and Ruzicka, J. (1994). Flow injection renewable surface immunoassay: a new approach to immunoanalysis with fluorescence detection. *Anal. Chem.*, **66**, 1825–1831.

Reboul, S. H. and Fjeld, R. A. (1994). A rapid method for determination of beta-emitting radionulides in aqueous samples. *Radioact. Radiochem.*, **5**, 42–49.

Reboul, S. H. and Fjeld, R. A. (1995). Potential effects of surface water components on actinide determinations conducted by ion chromatography. *Health. Phys.*, **68**(4), 584–589.

Roane, J. E. and DeVol, T. A. (2002). Simultaneous separation and detection of actinides in acidic solutions using an extractive scintillating resin. *Anal. Chem.*, **74**(21), 5629–5634.

Ross, R. R., Noyce, J. R., and Lardy, M. M. (1993). Inductively coupled plasma-mass spectrometry: an emerging method for analysis of long-lived radionuclides. *Radioact. Radiochem.*, **4**, 24–37.

Ruzicka, J. (1994). Discovering flow injection; journey from sample to live cell and from solution to suspension. *Analyst*, **119**, 1925–1934.

Ruzicka, J. and Hansen, E. H. (1988). "Flow Injection Analysis." Wiley-Interscience, New York.

Ruzicka, J. and Ivaska, A. (1997). Bioligand interaction assay by flow injection absorptiometry. *Anal. Chem.*, **69**, 5024–5030.

Ruzicka, J. and Marshall, B. D. (1990). Sequential injection: a new concept for chemical sensors, process analysis and laboratory assays. *Anal. Chim. Acta*, **237**, 329.

Ruzicka, J., Pollema, C. H., and Scudder, K. M. (1993). Jet ring cell: a tool for flow injection spectroscopy and microscopy on a renewable solid support. *Anal. Chem.*, **65**, 3566–3570.

Ruzicka, J. and Scampavia, L. (1999). From flow injection to bead injection. *Anal. Chem.*, **71**, 257A–263A.

Schonhofer, F. and Wallner, G. (2001). Very rapid determination of 226Ra, 228Ra and 210Pb by selective adsorption and liquid scintillation spectrometry. *Radioactivity & Radiochemistry*, **12**(2), 33–38.

Smith, M. R., Farmer, O. T., Reeves, J. H., and Koppenaal, D. W. (1995). Radionuclide detection by ion-chromatography and on-line ICP/MS and beta detection: fission product rare earth element measurements. *J. Radianal. Nucl. Chem.*, **194**, 7–13.

Smith, M. R., Wyse, E. J., and Koppenaal, D. W. (1992). Radiounuclide detection by inductively coupled plasma mass spectrometry: a comparison of atomic and radiation detection methods. *J. Radioanal. Nucl. Chem.*, **160**, 341–354.

Tan, S. H. and Horlick, G. (1986). Background spectral features in inductively coupled plasma/mass spectrometry. *App. Spectroscopy*, **40**(445–460), 445–460.

Thompson, J. J. and Houk, R. S. (1986). Inductively coupled plasma mass spectrometric detection for multielement flow injection analysis and elemental speciation by reversed phase liquid chromatography. *Anal. Chem.*, **58**, 2541–2548.

Wilbur, D. S., Hamlin, D. K., Pathare, P. M., Bray, L. A., Tingey, J. M., Egorov, O. B., Brechbiel, M. W., and Sandmaier, B. M. (1998). Studies of labeling proteins with the alpha emitting radionuclide Bi-213. *J. Nucl. Med.*, **39**, 91.

Willumsen, B., Christian, G., and Ruzicka, J. (1997). Kinetic studies of competetive immunoassay. *Anal. Chem.*, **68**, 3482–3489.

Zahradnik, P. and Swietly, H. (1996). The robotized chemical treatment of diluted spent fuel samples prior to isotope dilution analysis. *J. Radioanal. Nucl. Chem.*, **204**(1), 145–157.

15
RADIATION DOSIMETRY

DAVID A. SCHAUER
Department of Radiology and Radiological Sciences,
Uniformed Services University of the Health Sciences, Bethesda, MD

ALLEN BRODSKY
Science Applications International Corporation, McLean, VA

JOSEPH A. SAYEG
Department of Radiation Medicine (Emeritus), University of Kentucky Medical Center,
Lexington, KY

Handbook of Radioactivity Analysis, Second Edition
Published by Elsevier Science B.V.

I. INTRODUCTION

Most of the material in this book is devoted to detecting and quantifying radioactivity. This chapter will focus solely on radiation dosimetry, or the measurement of energy deposited when radiation emissions interact with matter. Radiation dosimetry is the science of measuring or calculating absorbed dose (or dose rate) in matter exposed to ionizing radiation. It is a broad discipline covering measurements and calculations of internal and external irradiation. The scope of this chapter will be limited to the fundamentals, measurements, and applications of external dosimetry. The reader is referred to numerous internal dosimetry publications (ICRP 30, 1979; ICRP 61, 1991; MIRD, 1988).

The aim of this chapter is to provide scientists who perform quantitative measurements and calculations of radioactive materials in various media, or those performing research to improve such measurements and calculations, with fundamental and useful information to assess radiation dose. Therefore, radiation dosimetry in this chapter will refer to the methods of determining amounts of energy deposited per unit mass of the material of interest, averaged over a small volume of interest. In addition, if the distribution of dose may be considered uniform for the practical purposes of assessing radiation effects in inert or biological materials, an average dose over a larger volume may be determined. Thus, dose in this sense is either a density of energy deposition, or an average of this deposition over a material or tissue of interest.

Radiation interactions, which are covered in Chapter 1, are essential to understanding radiation detection and measurement. These interactions create ion pairs that are a fundamental source of radiation effects. These ion pairs leave specific "signatures" in tissues and "tissue-like" substitutes that are studied using various methods.

The measurements section, which follows introductory material on quantities and units and cavity theory fundamentals, has been broadly divided into *physical* (tissue substitutes) and *biological* (tissues) dosimetry. External radiation dosimetry usually requires the use of tissue substitutes and conversion factors. However, actual tissues can also be used to quantify dose. This chapter, which will explore both types of external dosimetry, is not intended to be a comprehensive treatment of all radiation dosimetry methods. The focus will be on the most widely used radiation dosimetry techniques. Physical dosimetry includes ionization, photodosimetry, thermoluminescence (TL), optically stimulated luminescence (OSL), calorimetry, and electron paramagnetic resonance (EPR) spectroscopy. Biological dosimetry or *biodosimetry* is a rapidly expanding field. The majority of this topic will be focused on EPR spectroscopy of teeth/bones. A brief introduction to cytogenetics is also included.

Radiation dosimetry is an integral part of industrial and medical uses of ionizing radiation. The applications section will include an overview of personnel (occupational and retrospective), clinical (therapeutic and diagnostic), and materials processing dosimetry. Applicable standards and performance testing programs will also be discussed. The chapter will conclude with a brief section on radiation dosimetry issues and opportunities for future developments. Beta and neutron dose measurement material is included in the Appendix.

II. QUANTITIES AND UNITS

A. Basic

The International Commission on Radiation Units and Measurements (ICRU) periodically updates its recommendations on quantities and units for use in radiation dosimetry. In the 1970s, the ICRU introduced what is denoted as the SI (Systeme Internationale) and recommended that it replace the previous "conventional" system in regard to the special units used in radiation protection and measurement since the early 1900s. The most recent set of recommendations, ICRU Report 60 (ICRU 1998), has many detailed definitions applicable to radiation dosimetry. Some of the most widely used radiation dosimetry quantities and units will be summarized here.

Quantities are generally called physical quantities when used for the quantitative description of physical phenomena or objects, as stated by the ICRU. A *unit* is a reference sample of a quantity with which the amounts of other quantities of the same kind are compared. The ICRU asserts that every quantity can be expressed as the product of a numerical value and a unit. When the unit in which a quantity is expressed is changed, the quantity stays the same but its numerical value changes accordingly. The following ICRU definitions of quantities and units are those most basic to radiation dosimetry. The following are presented symbolically as in the ICRU report.

Particle Number, N—the number of particles that are emitted, transferred, or received. Unit: 1

Radiant Energy, R—the energy (excluding rest energy) of the particles that are emitted, transferred, or received. Unit: J

Flux, dN/dt—the quotient of dN by dt, where dN is the increment of the particle number in the time interval dt. Unit: s^{-1}

Energy Flux, dR/dt—the quotient dR by dt, where R is the increment of radiant energy in the time interval dt. Unit: W

Fluence, Φ—the quotient dN/da, where dN is the number of particles incident on a sphere of cross-sectional area da. Unit: m^{-2}

Energy Fluence, Ψ—the quotient dR/da, where dR is the radiant energy incident on a sphere of cross-sectional area da. Unit: $J\ m^{-2}$

Fluence Rate, $\emptyset$—the quotient $d\Phi/dt$, where $d\Phi$ is the increment of the fluence in the time interval dt. Unit: $m^{-2}\ s^{-1}$

Energy Fluence Rate, $d\Psi/dt$—the quotient where Ψ is the above quotient dR/da in time interval dt. Unit: $W\ m^{-2}$

Mass Attenuation Coefficient, μ/ρ—for uncharged particles (or photons), is the quotient of dN/N by $\rho\ dl$, where dN/N is the fraction of particles that experience interactions in traversing a distance dl in a material of density ρ. Unit: $m^2\ kg^{-1}$

Mass Energy Transfer Coefficient, μ_{tr}/ρ—for uncharged particles, the quotient dR_{tr}/R by $\rho\,dl$, where dR_{tr}/R is the fraction of incident radiant energy that is transferred to kinetic energy of charged particles by interactions, in traversing a distance dl in a material of density ρ. Unit: $m^2\,kg^{-1}$

Mass Energy Absorption Coefficient, μ_{en}/ρ—for uncharged particles, the quotient dR_{en}/R by $\rho\,dl$, where dR_{en}/R is the fraction of incident radiant energy that is *absorbed* in traversing a distance dl in a material of density ρ. Unit: $m^2\,kg^{-1}$

Note: This definition, although discussed in ICRU 60, is not specifically emphasized there as a basic quantity; it is indicated there as depending on stopping powers of the charged particles for individual elements of a mixture. However, the mass energy absorption coefficient has been measured accurately and published extensively for use in radiation dose determinations in the materials and tissues of most interest. Thus, it is a most important quantity in radiation dosimetry and will be used in the following sections of this chapter.

Mass Stopping Power, S/ρ—for charged particles, is the quotient of dE by $\rho\,dl$, where dE is the energy lost by a charged particle in traversing a distance dl in a material of density ρ. Unit: $J\,m^2\,kg^{-1}$

Note: Mass stopping power is another very important quantity in dosimetry practice, and has been extensively tabulated. In the use of small cavity chambers, discussed in this chapter, the ratio of this quantity between the gas in which ions are collected and the wall of the small cavity provides the ratio of the absorbed doses near the interface. The mass stopping power discussed in this chapter is the collision mass stopping power to differentiate it from the total stopping power, which includes losses due to radiative processes.

The ICRU divided section 4 of Report 60 into two categories (energy conversion and energy deposition) in defining the physical quantities associated with radiation dosimetry. Energy conversion and deposition include the quantities exposure/kerma and absorbed dose, respectively.

After Roentgen's discovery of x-rays and in the early 1900s, x-ray machines were used to diagnose illness and to treat cancers, but there was no accurate way to measure exposures so that physicians could systematically compare results. Dental film packets were used with paper clips, and a visible shadow of a clip on the photographic image was sometimes used as an indication of too much exposure to the physician. The appearance of skin "erythema" (which is a reddening like sunburn) gave the physician treating cancer a signal that he should begin to limit or defer further cancer treatments.

At the same time, physicists were working internationally to define units and methods for measuring x-rays for medical purposes. They had already found that X radiation would produce ionization, and that a voltage between two plates would collect electric charge from a volume of air exposed to x-rays. Physicists knew well how to measure charge accurately. The matter was resolved when, in 1928 at an International Congress of Radiology, exposure was defined.

The ICRU Report 60 definition of exposure, X is the quotient of dQ by dm, where dQ is the absolute value of the total charge of ions of one sign produced in air when all the electrons and positrons liberated or created by photons in air of mass d*m* are completely stopped in air. Ionization from Auger electrons is included in d*Q*. However, ionization due to photons arising from radiative processes is not included. The conventional unit for exposure is the roentgen (R) and the SI unit is $C\,kg^{-1}$ (coulomb per kilogram). It is useful to remember that 1 R equals $2.58 \times 10^{-4}\,C\,kg^{-1}$.

The second important energy conversion quantity is *k*inetic *e*nergy *r*eleased per unit *ma*ss, or *kerma*. Kerma characterizes photon or neutron beams in terms of energy transfer to material. Kerma, K is the quotient of dE_{tr} by d*m*, where dE_{tr} is the sum of the initial kinetic energies of all the charged particles liberated by uncharged particles in a mass d*m* of material. The widest application of kerma involves photon beams and air. This is referred to as air kerma, which is equal to exposure in $C\,kg^{-1}$ multiplied by W/e, (the mean energy per unit charge expended in air by electrons $(J\,C^{-1})$) divided by (1-g), where g is the mean fraction of secondary electron energy that is lost due to radiative processes. The conventional and SI units for kerma are $ergs\,g^{-1}$ and $J\,kg^{-1}$, respectively. The conventional and SI special names for the units of kerma are the rad and gray. One gray equals 100 rads.

The most important energy deposition quantity is absorbed dose. A given exposure to x-rays was easily seen to deliver varying amounts of energy to tissue as the x-ray beam or field decreased in its path through the body, and as it encountered bone as well as soft tissues. Physicians and researchers studying the effects of radiation on different body tissues and organs, as well as cell populations, needed a definition of radiation dose in terms of density of energy deposition near small regions of interest. Thus, definitions of absorbed dose and associated units were developed that expressed dose in terms of a "point" measurement, or estimation, of the density of energy absorbed near any point in tissue or other material. Early definitions of the "rep" (*r*oentgen *e*quivalent *p*hysical) were used in the 1940s and early 1950s, which were defined to be about $95\,ergs\,g^{-1}$, since an exposure of 1 R in air would deliver a dose of about this energy density to a small volume of tissue placed in air at the same point. However, this definition, which varied somewhat with the particular measurement technique, has been replaced with the "rad" (*r*adiation *a*bsorbed *d*ose). Absorbed dose, *D* is the quotient of d*ε* by d*m*, where d*ε* is the mean energy imparted to matter of mass d*m*. The conventional and SI units and special names for absorbed dose are the same as for kerma.

Absorbed dose is the most widely used quantity in nonstochastic processes such as clinical, materials processing and accident level dosimetry when the dose is fairly uniformly distributed over the tissue or organ. It is also the fundamental dosimetric quantity in personnel and environmental radiological protection. However, since stochastic effects depend on the absorbed dose as well as the type and energy of radiation, additional quantities and units were defined.

B. Applied

ICRP Publication 60 (1990) defined personnel protection quantities and units. ICRU Report 51 (1993) defined operational quantities and units.

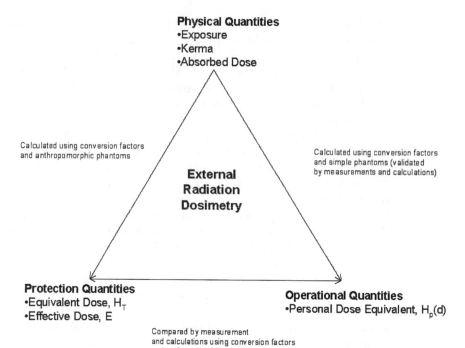

FIGURE 15.1 Relationship among physical, protection, and operational external radiation dosimetry quantities.

ICRU Report 57 (1998) and ICRP Publication 74 (1996) linked the previously described physical quantities with the most widely used protection (equivalent, H_T and effective dose, E) and operational [personal dose equivalent, $H_p(d)$] external radiation dosimetry quantities. Each of these operational and protection quantities has units of $J\,kg^{-1}$ and the special names are the rem (conventional) or sievert (SI). Figure 15.1 presents the relationship among physical, protection and operational external radiation dosimetry quantities.

The protection quantity equivalent dose, denoted H_T, equals the absorbed dose averaged over a tissue or organ and weighted for the radiation quality of interest. Radiation weighting factors (w_R) range from 1 to 20. Specific values, which can be found in Table 1 of ICRP 60 (1990), are representative of the relative biological effectiveness (RBE) of that radiation in inducing stochastic effects at low doses.

Stochastic effects also depend on the organ or tissue irradiated. Therefore, H_T is further weighted by a tissue or organ specific weighting factor, or w_T. This factor represents the contribution of that organ or tissue to the total detriment resulting from uniform whole-body irradiation [see Table 2 of ICRP 60 (1990)]. This doubly weighted absorbed dose is referred to as effective dose, E.

The operational quantity personal dose equivalent [$H_p(d)$] is defined as the dose equivalent in soft tissue at an appropriate depth, d, below a specified point on the body. It has the same units and special names as the previously defined protection quantities. This is the quantity that is actually measured for external radiation protection purposes. It is important to note that this quantity is a conservative approximation of the protection quantity effective dose.

III. FUNDAMENTALS

Evaluation of radiation dose requires measurement with a sensitive radiation detector within the medium of interest. The detector most commonly used for such a measurement is an ionization chamber containing solid walls surrounding a gas cavity, which is constructed according to cavity theory. There is no requirement in cavity theory that the cavity must be restricted to a gas as the cavity may also be a liquid or solid and may be more or less dense than the surrounding wall medium. However, it is more difficult to meet the cavity requirements with liquid or solid cavities.

Many references have documented the extensive research carried out on cavity theory. Excellent reviews have been published by Burlin (1968) and Attix (1986).

W. H. Bragg (1910, 1912) and L. H. Gray (1936) independently applied cavity theory in an attempt to relate the cavity dose to the medium dose surrounding the cavity. Although Bragg was the first to investigate the problem Gray was not aware of Bragg's earlier work and apologized to Bragg in his 1936 publication for not citing his earlier work. Both Bragg and Gray reached similar conclusions in identifying the medium surrounding the cavity as a cavity ionization chamber (or vessel) with the absorption characteristics in different terms. Gray's treatment indicated the specific mathematical formulation relating the cavity dose to any wall dose in terms of modern day nomenclature with more detailed experimental results supporting the theory. The investigations of these two early pioneers were honored by considering their collective work as the Bragg–Gray theory, commonly referred to as the Bragg–Gray Cavity principle.

A. Theoretical Basis of Cavity Theory

Early cavity theory is an idealized model based upon the dose received by a thin medium due to incident monoenergetic charged particles. The assumptions of the theory can be summarized as follows (to be discussed more in detail in a later section):

1. The mass stopping power remains practically constant and characteristic of the initial energy of the charged particle fluence incident upon the medium.
2. The effects of scattering, delta ray production and bremsstrahlung were not considered.
3. The thickness of the cavity is assumed to be small compared to the range of the charged particles emerging from the wall (this defines a thin medium).

4. The presence of the cavity does not change the charged particle fluence passing through it.

5. No charged particles are absorbed or created in the cavity and the dose absorbed by the cavity is entirely due to the charged particles emerging from the wall.

Consider a flow of monoenergetic electrons across an interface (or boundary) between the gas and wall material of a cavity chamber. Although electrons produced in the chamber walls by photon interactions will be scattered in all directions, imagine for the moment that they travel perpendicular to a $1\,\text{cm}^2$ area of the interface, which is nearly a flat plane. Then, the ratio of energy loss, per $\text{g}\,\text{cm}^{-2}$ of path traveled by the electrons between the wall and the gas will be seen to be equal to the ratio of the absorbed doses between regions of the wall and gas close to the interface.

Consider a thin layer of wall of thickness dx near the interface, and a thin layer of air of the same thickness dx on the other side of the interface. Imagine a fluence, ϕ electrons cm^{-2}, through the $1\,\text{cm}^2$ interface.

The absorbed dose in the air portion dx close to the interface would be:

$$D_{\text{air}} = \phi \text{ electrons cm}^{-2} \times S_a \text{ MeV cm}^2\,\text{g}^{-1} \text{ per electron,} \qquad (15.1)$$

in units of $\text{MeV}\,\text{g}^{-1}$, where S_a represents the mass stopping power, $(dE/\rho\,dx)_a$, for these electrons in air. This equation cancels the area units, and gives dose units without consideration of the actual thickness dx. This is because the mass stopping power represents a rate of energy loss along the particle path as it passes through material. The electrons in the atoms do not care from which direction the incoming fast electron is coming; they are spherically symmetric and present the same cross-section to fast electrons from any direction. The fast electron interacts only according to the relative density of the electrons (and hence the relative mass density for low-Z materials) in the material being traversed. The value of the mass stopping power here assumes that the electron energies have not changed substantially in a very small thickness dx.

The absorbed dose in the wall material close to the interface would be:

$$D_{\text{wall}} = \phi S_w \qquad (15.2)$$

Dividing Eq. 15.2 by Eq. 15.1, the ϕ's cancel, and we have

$$D_w/D_a = S_w/S_a = S_a^w, \qquad (15.3)$$

where S_a^w, the ratio of the mass stopping powers, is called the *relative mass stopping power, wall to air*.

Now, with the eyes closed and a little thought, imagine that the $1\,\text{cm}^2$ interface is wrapped around in the shape of a cylinder, somewhat the size of some small cavity chambers. Also, imagine that the electrons enter the interface from all directions. Since the specific ionization of each electron along its path contributes to the mass stopping power in each material, and since electronic collisions depend upon electron densities in the two materials, Eq. 15.3 will hold true for this small cavity as long as the fluence is about the same at all points on the gas–wall interface. Since Eq. 15.3 is generally true for any gas–wall combination, the *Bragg–Gray principle* and *conditions under which it holds* are:

$$D_{\text{wall}} = D_{\text{gas}} S_{\text{g}}^{\text{w}}, \tag{15.4}$$

or as sometimes written

$$D_{\text{wall}} = JW S_{\text{g}}^{\text{w}}, \tag{15.5}$$

where J, charge produced per unit mass of gas (C kg^{-1}); W, energy absorbed per unit charge (J C^{-1}); $S_{\text{g}}^{\text{w}} =$ relative mass stopping power, wall to gas, averaged over the charged particle energy spectrum.

Practical conditions under which the Bragg–Gray principle holds:

1. The gas or air-filled cavity must be small enough in size so that practically all electrons penetrating the wall–gas interface and crossing the cavity have been produced in the wall. This produces a definable characteristic spectrum of electrons, originating from photon–electron interactions in the solid wall that determine the appropriate relative mass stopping power. It means that the cavity must be small compared to the ranges of the electrons in the gas.
2. Negligible absorption of the photon radiation should occur in the gas. This will also generally be true if the cavity dimensions are small compared to the electron ranges.

The above conditions ensure a stable characterizable electron fluence at the interfaces, so that Eq. 15.5 holds true. However, a third condition is necessary to provide a stable instrument that responds consistently over a useful range of photon energies:

3. The wall thickness must be great enough (about equal to the maximum electron range) so that all electrons that can enter the cavity are characteristic of those produced in the wall (and not characteristic of any materials outside the wall). However, the wall thickness should not be so great that it appreciably attenuates the radiation to be measured at the point of interest.

    This condition is called charged particle equilibrium (CPE). For example, very thin-walled chambers designed for use in the x-ray region below 100 keV must have an "equilibrium cap" of thickness of a few millimeters placed over the cavity wall to provide CPE when measuring ^{60}Co fields in the 1.17 and 1.33 MeV photon energy range.

The same relationships would apply to any gas in the cavity, with appropriate relative mass stopping power. They also apply to secondary charged particles other than electrons, produced by radiation other than photons. Some chambers have tissue equivalent gases, and tissue equivalent walls, so that the dose to a small volume of tissue can be measured directly, e.g., for proton recoils ejected by incoming neutrons. The appropriate stopping powers for protons of average energy would be used in this case.

The electron fluence in the case of incoming photons would not be monoenergetic, but would be a spectrum dependent upon the energies of the incoming photons, and the size and design of the chamber. Thus, the relative mass stopping powers need to be those averaged over the proper spectrum. Conveniently, most of the relative mass stopping powers of interest are within 20 percent of unity.

The small cavity chamber can be placed in air to estimate the dose to a small volume of tissue that might be placed at the same point in air, or it can be placed within the tissues of an experimental animal, or within a patient's body cavity to check the value of a delivered dose.

Whether the cavity chamber is exposed in air or within a tissue, the dose to the chamber wall is in both cases obtained from the charge produced in the cavity gas from Eq. 15.5, and then the dose to the tissue is obtained from D_{wall} by multiplying it by the ratio of the (mass-energy absorption coefficient)$_{tissue}$ divided by the (mass-energy absorption coefficient)$_{wall}$:

$$D_{tissue} = D_{wall} \times (\mu_{en}/\rho)_{tissue}/(\mu_{en}/\rho)_{wall}. \qquad (15.6)$$

Note: D_{wall} is *not* multiplied by another ratio of electron mass stopping powers here, but by the ratio of mass-energy absorption coefficients that adjusts for the probability of converting photon energy fluence to absorbed recoil electron energy per gram, in tissue versus in wall material. If the wall material is made of tissue equivalent material, then the value of D_w would be the final dose of interest, since by its composition it would interact with photons and produce the electrons that ionize the gas in the manner desired for the measurement. Caution must be taken in understanding this concept, which is emphasized below:

The dose to the wall is obtained by multiplying the dose to the gas by the ratio of *mass stopping powers (averaged over the electron spectrum)*; while the dose to tissue in which the cavity chamber is imbedded is obtained by multiplying the dose to the chamber wall by the ratio of *mass-energy absorption coefficients (averaged over the incoming photon spectrum)*.

Figure 15.2 is a summary of cavity chamber dosimetry. In this figure, it is assumed that there is a central wire in the cylinder (not shown) that is positively charged with respect to the wall, with a voltage on the ionization plateau of the chamber so that all charge produced within the cavity is collected. Given that fact, or the use of a predetermined "effective volume" of gas in the chamber, the factors shown in the figure are used as follows to calculate dose to the wall and dose to tissue: Dose to the gas (in grays) $= J\,W$, where J is charge collected in coulombs per kilogram of air (or gas), $W = 34$ joules/coulomb, effectively a constant versus electron energy. Dose to the wall near the interface $= (S_w/S_a)_{avg} \times$ (dose to the gas), where the mass stopping power ratio is averaged over the spectrum of electron energies produced by the incoming photons and slowing down in the wall; note that for low-Z wall materials, these average ratios do not differ greatly from 1.

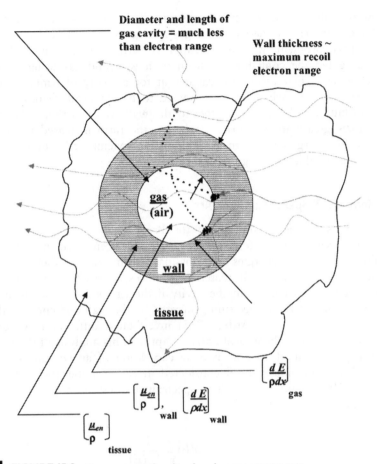

Diameter and length of
gas cavity = much less
than electron range

Wall thickness ~
maximum recoil
electron range

gas
(air)

wall

tissue

$\left[\dfrac{dE}{\rho dx}\right]_{\text{gas}}$

$\left[\dfrac{\mu_{en}}{\rho}\right]_{\text{wall}}$, $\left[\dfrac{dE}{\rho dx}\right]_{\text{wall}}$

$\left[\dfrac{\mu_{en}}{\rho}\right]_{\text{tissue}}$

FIGURE 15.2 Parameters of cavity chamber measurements.

Then, if this (small) chamber is imbedded in tissue, the dose to nearby tissue is equal to $(\mu_{en}/\rho)_{\text{tissue}}/(\mu_{en}/\rho)_{\text{w}} \times$ (dose to wall). The same calculations would be performed if the cavity chamber (with equilibrium wall) were placed in air at a point where a patient's (or worker's) target organ would later be placed. But then, the tissue dose calculated would be the dose to an imaginary "small volume of tissue in air," which for typical gamma-ray energies would be the dose to a small finger extended to the point at arm's length. Then, if the patient were later placed with the target organ at the point where the chamber had been, the target organ dose would be obtained from the dose to the small volume of tissue in air multiplied by the tissue–air ratio (TAR) at the depth of the target organ and beam cross-section at that depth.

A list of mass stopping powers and mass energy absorption coefficients for various elements as a function of photon energy can be found in ICRU Reports (ICRU 37, 1984; ICRU 44, 1989; ICRU 46, 1992).

It is important to note the energy dependence requirement for the cavity. If a detector is constructed for high-energy photons it may not be suitable for low energy photons as the ranges of the secondary electrons may become comparable to the cavity dimensions. This violates the requirements of Bragg–Gray cavity theory where gas interactions could become appreciable.

This problem was investigated by Ma and Nahum (1991). They investigated the contributions from both wall and gas using Monte Carlo techniques. Their results indicated that for a cavity of 6 mm thickness and 6 mm diameter, which was considered to be a typical chamber in use, the contributions of the gas relative to the wall was less than 3% for photon energies greater than 600 keV, whereas the ratio increased to greater than 10% for energies less than 100 keV. For a spacing of 0.5 mm the ratio was still appreciable at energies 50–60 keV.

B. Contributions of L. V. Spencer and F. H. Attix

The physical theory underlying the Bragg–Gray Cavity principle may be briefly explained as follows: Electrons are generated by photons with different starting energies and dissipate their energy solely by means of inelastic collisions. Each material produces a specific spectrum, which is assumed not to change in the cavity if the cavity size is small. In order to understand the basic spectrum generated by the wall and entering the cavity, Spencer (Spencer and Attix, 1955) investigated different types of spectra, which had a firm theoretical basis, to apply in his analysis of the Bragg–Gray principle. He used an approximate expression of an equilibrium spectrum, which had previously been considered in earlier publications (Spencer and Fano, 1954; Fano, 1953). The spectrum expressed in a simplified mathematical form by Attix (1986) is:

$$\phi(E) = \frac{N}{S_w(E)} \tag{15.7}$$

where $\phi(E)$ is the differential energy spectrum of the electrons, N is the number of electrons emitted per unit mass of wall, and $S_w(E)$ is the mass stopping power of the wall. With the above spectrum, including the assumptions of the Bragg–Gray principle previously discussed and the concept of charged particle equilibrium, which was not considered in previous derivations, Spencer was able to reproduce the Bragg–Gray equation.

An investigation of the cavity ionization in air as a function of the atomic number of the wall (Attix *et al.*, 1958) provided valuable information on the validity of the Bragg–Gray theory. Their experimental results indicated that the Bragg–Gray theory did not accurately predict the ionization in air. The ionization per unit mass of air in a specially designed ionization chamber with equilibrium walls of graphite, aluminum, copper, tin, and lead indicated an increased ionization density with decreased cavity spacing, a variation not considered by the theory. In addition, the measured ionization densities were greater than that predicted by the theory, and this effect increased with the atomic number of the wall. Gamma-ray sources of ^{198}Au, ^{137}Cs, and ^{60}Co exhibited similar results. These discrepancies could not be explained by the theory, and Spencer pursued an investigation to consider the production and effect of fast electrons created by "head-on" collisions (called delta rays), which was not specifically considered in the original theory. In the explanations that follow, primary electrons are considered to be the electrons created by photon interactions, whereas the secondary electrons are generated by primary electron interactions.

The Spencer theory (Spencer and Attix, 1955; Attix, 1986), commonly called the Spencer–Attix theory by most authors, considered the delta ray production of a uniformly distributed source of monoenergetic electrons of energy E_0 emitting N particles per gram throughout the medium. Spencer employed the same assumptions as in his previous analysis of the Bragg–Gray principle. In addition, the cavity was characterized by a parameter Δ, which was assumed to be the mean energy of the electrons just large enough to cross the cavity. Thus, Δ defined the cavity size. The equilibrium spectrum of the electrons generated was divided into two components:

1. A fast component whose energies are greater than or equal to Δ.
2. A slow component whose energy is less than Δ.

The Bragg–Gray theory assumes that all the secondary electrons transfer their energy on the spot where they are created (energy transfer is considered to be "dissipative") whereas in the Spencer model it is only for the slow component of the spectrum. For energy transfers greater than Δ (fast component) the energy transfers are not considered "dissipative" and expend their energy outside the cavity. Also, in the Bragg–Gray theory where the effect of delta rays is not considered, the "secondaries" are included in the spectrum and the energy given to the fast component must be subtracted from the stopping power. The restricted stopping power, therefore, must be used that includes only those energy losses due to "secondaries" not exceeding Δ. Hence, this model includes the increased number of charged particles in the equilibrium spectrum due to delta rays as well as taking into consideration the energy lost to delta rays and specifies a variation of cavity size with the energy of the secondary electrons.

The modified electron equilibrium spectrum, as evaluated by Spencer, is given by:

$$\phi(E) = \frac{N}{S_{\mathrm{w}}(E)} \cdot R(E_0, E) \tag{15.8}$$

where $R(E_0, E)$ is the ratio of the differential spectrum, including delta rays, to that of the primary electrons only. The computed results of $R(E_0, E)$ indicated that delta ray production increased the differential electron spectrum markedly for energies that are a small fraction of the initial electron energy E_0, thus producing a major effect on the final spectrum.

Spencer's computed results are summarized in Table 15.1 (Attix, 1986) for different values of the electron energy and cavity size (Δ) with comparisons to the original Bragg–Gray theory. The results indicated that the original Bragg–Gray theory could be used for low-Z materials without appreciable error. The Spencer theory indicated better agreement with the

TABLE 15.1 Values of D_g/D_w Calculated for Air Cavities from Spencer Cavity Theory versus Bragg–Gray theory. Modified from Introduction to Radiological Physics and Radiation Dosimetry (F. H. Attix, Copyright © 1986 by John Wiley & Sons, Inc. This material is used by permission of John Wiley & Sons, Inc.)

Wall Medium	E_o (keV)	Δ (keV) = Air Range (mm) =	2.2 0.015	5.1 0.051	10.2 0.19	20.4 0.64	40.9 2.2	81.8 7.2	Bragg– Gray
Carbon	1308		1.001	1.002	1.003	1.004	1.004	1.005	1.005
	654		0.990	0.991	0.992	0.992	0.993	0.994	0.994
	327		0.985	0.986	0.987	0.988	0.988	0.989	0.989
Aluminum	1308		1.162	1.151	1.141	1.134	1.128	1.123	1.117
	654		1.169	1.155	1.145	1.137	1.131	1.126	1.125
	327		1.175	1.161	1.151	1.143	1.136	1.130	1.134
Copper	1308		1.456	1.412	1.381	1.359	1.340	1.327	1.312
	654		1.468	1.421	1.388	1.363	1.345	1.329	1.327
	327		1.485	1.436	1.400	1.375	1.354	1.337	1.353
Tin	1308		1.786	1.694	1.634	1.592	1.559	1.535	1.508
	654		1.822	1.723	1.659	1.613	1.580	1.551	1.547
	327		1.861	1.756	1.687	1.640	1.602	1.571	1.595
Lead	1308		–	2.054	1.940	1.865	1.811	1.770	1.730
	654		–	2.104	1.985	1.904	1.848	1.801	1.796
	327		–	2.161	2.030	1.946	1.881	1.832	1.876

These data replace those given in Table II of Spencer and Attix (1955), which did not take account of the polarization effect.

experimental studies of Attix and his coworkers (1958), in that, it not only considered delta ray production but also provided a variation of absorbed dose with cavity size. Borg *et al.* (2000) using established Monte Carlo techniques to evaluate the response of air cavities in a graphite medium concluded that their studies were in excellent agreement with the Spencer model (with $\Delta = 10$ keV). They also indicated that the Spencer model could be used as low as 200 keV with little error, despite the conclusion that photon interactions occur in the cavity (Ma and Nahum, 1991).

C. Burlin Cavity Theory

To bridge the region where one could apply the Bragg–Gray or Spencer Cavity theories, where the effect of the wall is the primary consideration, and that region where the effect of the wall is decreasing in importance and becomes negligible with respect to the cavity contribution, Burlin (1966) proposed a theory with the following assumptions:

1. Both media, wall and gas, are uniform in composition.

2. A uniform photon radiation field traverses the wall and gas, and photon attenuation is neglected.
3. Charged particle equilibrium is produced in the wall.
4. The wall electron spectrum is attenuated exponentially in traversing the cavity without a change in the spectral distribution (changes were assumed to be small) and the same attenuation coefficient is used to evaluate the build-up of the gas contribution.

The mathematical equation that expresses Burlin's theory can be considered in terms of three regions from the wall-cavity interface.

1. The first region represents the contribution of only the Spencer or Bragg–Gray theories (which is a limited region).
2. The second region includes the transition from region 1 to region 3.
3. The third region represents the contribution of only the cavity where the primary photon field is not attenuated by the wall or the cavity.

A factor d was introduced to effect the transition from region 1 to region 3. Burlin defined d as the average path length of electrons crossing the cavity, which is also the average ratio of the wall contribution in the cavity divided by the initial wall contribution at the wall–cavity interface. The Burlin theory equation (Attix 1986), where the Bragg–Gray theory would be applicable without appreciable error, can be written as

$$\frac{\overline{D_g}}{D_w} = d\overline{S_w^g} + (1-d)\overline{\left[\frac{(\mu_{en}/\rho)_g}{[\mu_{en}/\rho]_w}\right]} \tag{15.9}$$

where d approaches 1 for small cavities and 0 for large cavities, thus satisfying the limitations for both regions. The factor $(1-d)$ represents the fractional contribution of the gas cavity where the effect of the wall is decreasing. Where the Z value of the wall requires the use of the Spencer theory the reader is referred to the explicit formulation given by Burlin. The factor d is given by:

$$d = \overline{\left[\frac{(\phi(E)_w)_g}{\phi(E)_w}\right]} = \frac{\int_0^G e^{-\beta x}\,dx}{\int_0^G dx} = \frac{1 - e^{-\beta G}}{\beta G} \tag{15.10}$$

and

$$(1-d) = \frac{\beta G + e^{-\beta G} - 1}{\beta G} \tag{15.11}$$

The quantities used in the above evaluations are defined as follows:

$\overline{D_g}$ = Average absorbed dose in the gas cavity.
$D_w = K_w$ = Kerma in wall (absorbed dose in wall under charged particle equilibrium conditions).

$\phi(E)_w =$ Electron fluence spectrum emerging from wall at the wall-cavity interface.

$(\phi(E)_w)_g =$ Electron fluence spectrum from wall in gas cavity due to exponential attenuation,

$(\phi(E)_w)_g = \phi(E)_w e^{-\beta x}$ (x path length in cavity)

$G =$ Mean path length of electrons crossing cavity which is taken as four times the cavity volume divided by the surface area of the cavity

$\beta =$ Exponential attenuation factor, fractional decrease of wall contribution per unit length dx of cavity.

$\overline{S}_w^g =$ Average value of relative mass stopping power of gas to wall.

$(\mu_{en}/\rho)_w =$ Mass-energy absorption coefficient of the wall.

$(\mu_{en}/\rho)_g =$ Mass-energy absorption coefficient of the gas.

Initially, Burlin evaluated β from the empirical equation of Loevinger (1956) for beta rays in air (ignoring a term concerned with forbidden spectra)

$$\beta = \frac{16\rho}{(E_{max} - 0.036)^{1.4}} \ \text{cm}^2 \, \text{g}^{-1} \tag{15.12}$$

where

$E_{max} =$ maximum value of the electron energy generated by the photons in MeV.

$\rho =$ density of air in the cavity in g cm^{-3}

Later he used an exponential relation (Burlin and Snelling, 1969), which was expressed in more explicit terms by Attix (1986) as:

$$e^{-\beta t_{max}} = A(\text{constant}) \quad \text{with } A = 0.01 \tag{15.13}$$

where t_{max} is the greatest distance where only a specified number, A, of the electrons are able to reach (see Table 8.5 in Attix, 1986). Janssens and his coworkers (1974) found that a constant value $A = 0.04$ improved the correlation with experimental results.

The Burlin theory contains no restrictions on the size of the wall or cavity, which can be a gas, liquid, or solid, as long as the photon attenuation is negligible. When a detector has cavity dimensions much larger than the range of the electrons emerging from the wall–cavity boundary the theory reduces to the mathematical formulation that the absorbed dose in any medium is related to the detector wall dose by the ratio of the mass–energy absorption coefficients of medium to wall. However, when the dimensions of the cavity are in the intermediate range between the above limit and a size much smaller than the range of the electrons, the contribution of the Bragg–Gray or Spencer theories must be taken into consideration where the empirical factor d becomes important. Janssens and his coworkers (1974) and Attix (1986) in their early documentation of cavity theory indicated that experimental investigations supported the validity of the Burlin theory and that it was found useful in estimating the absorbed dose over a wide range of sizes. However, the more recent studies of Ma and Nahum (1991) using their Monte Carlo techniques indicated that the contribution of direct photon interactions in the cavity is less than that predicted by the Burlin theory, thus overestimating the departure from the Bragg–Gray theory. Other investigators have provided modifications to the assumptions of Burlin to obtain better agreement between theory, computer stimulation and experiment. These modifications have been discussed by Mobit *et al.* (1997), Attix (1986) and Horowitz (1984) and are beyond the scope of this chapter.

D. Fano Theorem

As previously discussed, the validity of the Bragg–Gray principle is dependent upon the requirement that the electron spectrum from the wall be unchanged by the presence of the cavity and therefore the cavity must be smaller than the maximum ranges of the electrons emerging from the wall. This condition was too restrictive in the construction of ionization chambers. It was assumed by many investigators that this size restriction could be avoided by making the wall and cavity of the same atomic composition regardless of the density difference. This practice was not fully justified, and Fano (1954) presented a proof of this assumption. His theorem is as follows:

> *In a medium of given composition exposed to a uniform flux of primary radiation (such as x-rays or neutrons) the flux of secondary radiation is also uniform and independent of the density of the medium as well as density variations from point to point.*

His proof did not take into consideration the polarization effect at high photon energies (Attix, 1986) and therefore, is applicable only to energies less than 1 MeV. This omission could cause some degree of uncertainty at greater energies.

IV. MEASUREMENTS (PHYSICAL DOSIMETRY)

Ionizing radiation measurements with physical dosimeters entail the use of tissue substitutes to determine dose. Some desired characteristics include tissue equivalence, high radiation sensitivity, minimal environmental degradation, long signal stability, and ease/automation of processing. This section will focus on the most widely used physical dosimeters. Dosimeter characteristics and performance will also be addressed.

A. Ionization Chambers

I. Free-air Chambers

It is useless to define a quantity if it cannot be measured. L. S. Taylor, who participated in the 1928 meetings, built the first free-air chamber in the

United States to measure exposure in units of roentgens. This chamber was built at the National Bureau of Standards (now named National Institute of Standards and Technology (NIST)), and is maintained in the NIST museum in Gaithersburg, Maryland. Other versions of this chamber have been built in the United States and other countries, and the free-air chamber is still used at NIST to calibrate secondary instruments in units of roentgens, as well as in the newer SI units.

Although the average radiation user will not need to use free-air chambers, a schematic view of such a chamber is presented in Fig. 15.3 for its value in helping to understand and remember the concepts of radiation exposure.

The chamber in Fig. 15.3 is designed to collect the positive and negative charges produced in the central box of air defined by the dark rectangle in the center of the figure (which represents a cross-section of an almost cylindrical volume, or a truncated cone, defined by the aperture of the diaphragm and its distance from this volume). In order that all of the positive and negative charges produced (ions and electrons, respectively) are consistently collected, and not recombined, at any intensity of x-rays to be measured, a high voltage (more than 1000 volts) is placed across the central electrodes. This voltage source produces an electric field of more than 100 V/cm, and is held very constant by a system of high voltage resistors. The same voltage is placed on "guard" electrodes—to the right and left of the central electrodes—to ensure that the electric field remains constant and vertical so that all charges are drawn vertically straight to the collecting central electrodes.

Electrons are drawn down through the electrometer toward ground, while the positive ions are drawn upward toward the negative terminal of the voltage

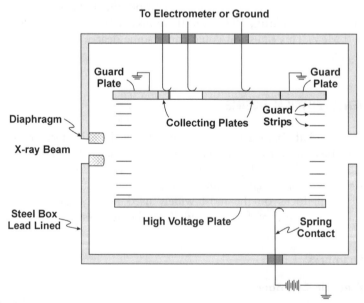

FIGURE 15.3 Sectional view of the Ritz (20- to 100-kV) free-air chamber. From **NBS Special Publication 250.16 (1988).**

source. The electrometer measures the charge collected accurately. The charge per unit volume in the central box is corrected to the charge that would be produced, using conversions to standard temperature and pressure (STP) conditions (0°C and 760 mmHg) in that same volume. This charge per unit mass in the central box is extrapolated, with some small corrections, to the charge per unit mass of air at the center of the hole in the lead beam-defining diaphragm at the left of the chamber. This is the point where the chamber is measuring the roentgen intensity, by the clever design of the device. Since the cross-sectional area of the box perpendicular to the beam increases as the square of the distance, the first thought is that the density of ionization is smaller in the central box than at the diaphragm, and thus the correction to ionization density back at the diaphragm might be large. However, the larger central volume is also collecting charge produced by a beam that is proportionally wider, and is not collecting charge cm^{-3}, but is collecting the total charge produced in a defined volume designed to represent charge density at the diaphragm.

The design and construction of the standard air chamber obviously requires extreme care and many physical considerations. A few more features of the chamber are described here to help establish the important concepts of radiation exposure. The central box must be far enough from the lead diaphragm, so that ionizing electrons produced by the x-radiation that reach the collecting box are only those characteristic of production in air, not lead, for the x-ray energies in the range for which the chamber is designed. Also, electrons produced in, but leaving, the central volume before losing all of their energy must be compensated by electrons entering the volume as produced in air preceding the volume. The collecting electrodes must be far enough apart vertically so that electrons scattered near 90 degrees still lose their energy in air, and produce typical amounts of air ionization that are collected by the electrodes. An exit hole must be provided that is somewhat larger than the beam in order that a negligible amount of electrons, or photons, will be back-scattered to produce irradiation from the rear.

Of course, as higher and higher x-ray energies have been produced and used, larger standard air chambers have been built to meet the above size conditions. Thus, radiation intensities in units of roentgens can be measured as a primary standard only up to photon energies of about 3 MeV—at which energy the size of the standard air chamber must be almost as large as a room. However, x- and gamma energies of interest in radiation protection in the majority of radiation protection applications are below 3 MeV. Also, even above 3 MeV, small thimble chambers can often be used to measure tissue doses, using exposure calibrations in roentgens together with certain available correction factors.

2. Portable R, Thimble, and Cavity Chambers

Since the standard air chamber is a large, sensitive, instrument—not easily portable to the research laboratory, hospital, or industrial laboratory—there was a need for a small instrument to measure exposure, and to interpret doses, at various locations and various points in the diagnostic or treatment regions of patients, or the work areas of employees. Thus, a series of secondary standard instruments were developed; they were called "R chambers" when designed to measure Roentgens, or more generally "thimble chambers" since they are of the size of a thimble or end of finger, or "cavity" chambers since they contained a small gas or air cavity at the center for the collection of charge produced in a defined volume of the gas or air. When the chambers were designed with air in the cavity, and walls made of low Z materials that were "air equivalent" (i.e., had mass stopping powers for electrons that were close to those for air), and had certain other conditions to provide stable measurements of the roentgen, they were called "R chambers." These small chambers could then be compared with ("calibrated against") measurements made with the standard air chamber at the same position in a standard x-ray beam.

While the previous section summarized fundamental principles of cavity theory to provide a better understanding of the principles of cavity ionization measurements, a more simplified presentation is offered for the readers of this text. Here, a few percent accuracy is adequate for dose measurements associated with photons or electrons in the 0.06–3 MeV range.

The measurement of radiation exposures and doses at various points in air and in tissue requires the use of ion collection chambers that are limited in size and are portable. The standard air chamber, although extremely precise, can only be used in a dedicated standards laboratory for calibrating other chambers. Small cavity chambers have been used, with various gases and wall materials, to measure radiation exposure and dose, often within 1 percent accuracy. Their design and use requires the understanding of the basic principles for converting the dose in the cavity gas to the dose to the wall material of the chamber. Since radiation therapy physicists must be able to measure doses in tissue phantoms to within 3%, and within 1% where possible, elaborate equations have been derived incorporating many small corrections to convert charge collected in a cavity chamber to dose, in radiation therapy. The principles described here will include only a very simple presentation of the fundamental Bragg–Gray principle, but will suffice for radiation protection or other research purposes within the energy range of radiations to which personnel or the public might be exposed. The methods and data sources described here are still capable of accuracies within a few percent—well within the accuracies of radiation dose measurements as recommended by the NCRP and ICRP. A brief review of more recent cavity chamber protocols developed for cancer therapy will be presented, with recent references, in the applications section. However, scientists entering the fields of radiological and radiation therapy physics will need to familiarize themselves with the details of the more complex protocols developed in the last three decades to provide precise dosimetry for the therapy accelerators in the 6 MV and greater energy range.

a. Practical Dosimetry with Ionization Chambers

Use of the f-factor   The assignment of dose at a point in a medium from a measurement of the ionizing photon exposure using an ionization chamber involves a well-known conversion factor that includes the ratio of the average mass energy-absorption coefficients (μ_{en}/ρ) for the medium of interest and air. Under conditions of electron equilibrium, the dose in the medium (D_{med}) is:

$$D_{\text{med}} = M \cdot N_x \cdot \overline{f}_{med} \qquad (15.14)$$

where M is the ionization chamber reading (coulombs), N_x is the product of the exposure calibration factor (roentgens per coulomb) and any other chamber-specific correction or perturbation factors and

$$\overline{f}_{\text{med}} = \left(\frac{\overline{W}}{e}\right) \cdot \frac{\int [\mu_{\text{en}}/\rho]_{\text{med}} E\phi(E)\,dE}{\int [\mu_{\text{en}}/\rho]_{\text{air}} E\phi(E)\,dE}, \qquad (15.15)$$

where $\overline{W}/e$ is the mean energy expended in air per ion pair formed. In the integrals of Eq. 15.15 μ_{en}/ρ is the mass-energy absorption coefficient, and $\phi(E)$ is the differential fluence spectrum as a function of photon energy E, at the point of interest. A detailed review of this subject can be found in Schauer *et al.* (1993a, b). These include data for numerous tissues and tissue substitutes.

Use of Tissue–Air Ratios (TARs)   If the cavity chamber is placed within tissue at depth d, in a beam of radius, r_d, during the measurement, then tissue attenuation and backscattering are all taken into account. Eq. 15.6 in general will then give the dose to tissue near the point where the chamber is placed, or if the chamber were small and then removed for the same exposure again at another time Eq. 15.15 would also provide a similar measurement of tissue dose at depth d for an air–wall (R) chamber. However, if an exposure measurement were made with the chamber *not* within the phantom or body, but in air at the same point (phantom removed), then some factor would be needed to obtain the dose at depth d. The most useful factor for this purpose is the *tissue–air ratio (TAR)*, which has been tabulated for a number of photon spectra. The TAR is defined as:

$$\text{TAR} = \frac{\text{dose with chamber in phantom at depth } d, \text{ for field size } r_d}{\text{dose with chamber in air at same point}} \qquad (15.16)$$

In some tables, the TAR might be symbolized as TAR (d, r_d, Q), where the symbols in parentheses indicate that the TAR value pertains to a point at depth d in the body, for a field size of radius r_d at that depth, and for a radiation of "quality" Q. Q refers to the type of photon energy spectrum. TAR tables use field size at depth, and are more convenient for radiation protection purposes than depth-dose tables, which use field size at the surface, but are for fixed source-to-skin distances. TAR tables are independent of source-to-skin distances, so are more useful for varying geometries of exposure, particularly as obtained in radiation protection situations.

When the field cross- section is not circular, an equivalent area can be used with the TAR tables for a square or rectangular field, if the field is not too narrow in one dimension. Otherwise, the tissue–air ratio can be easily measured as follows:

1. Move a water phantom in place with enough water to intercept the total beam, and so that the cavity chamber will be at depth d in the water. Adjust the beam size as desired at depth d, and have ideally about the same amount of water behind the chamber that would simulate the thickness of the individual whose internal dose distribution is being determined. The thickness of an average "standard" or "reference" man is often taken to be 30 cm. Use a container for the water that is made of plastic, so that a small thickness has an attenuation similar to that of water. Water can then be used as the surrogate for soft tissue. Appropriate water phantoms are commercially available.
2. Use a cavity chamber with an equilibrium cap that is covered to protect it against absorbing moisture, but not so sealed hermetically that it does not equilibrate with the environmental pressure and temperature. Place the cavity chamber at the center of the field and at depth d in the water phantom.
3. Obtain a measurement of charge or chamber reading, for a given amount of exposure time sufficient to reach a stable reading. Ensure that the voltage on the chamber is sufficient to achieve complete collection of charge; the chamber should be well onto the ionization collection plateau for the intensity of beam and time of exposure.
4. Remove water phantom, ensure that the chamber still has its equilibrium cap, and take another measurement at the same position in air for the same beam size and distance from the source.
5. Compute the ratio of the phantom measurement taken as described in paragraph 3 to that of the air measurement described in paragraph 4. This is then the tissue–air ratio, TAR (d, r_d, Q).

This is the procedure by which TARs have been obtained. With care, such measurements can be made within a few tenths of a percent, for well-defined beams and phantoms. These measurements can be made with either air-equivalent cavity chambers, R chambers, tissue-equivalent wall chambers, or small chambers made of any light plastic or graphite. Since a ratio is being measured, the mass-energy absorption coefficients that would be used to convert wall dose to tissue dose, in Eqs. 15.6 or 15.15, would cancel out.

In order to obtain an estimate of the maximum dose in tissue for a beam entering the body, the TAR should be measured or available at the depth $d_{\max}$ of maximum value, where the secondary electrons (photoelectric, Compton and pair production) created as the photon beam enters the body build up to "charged particle equilibrium (CPE)." Since the beam size of a TAR determination is always at the depth of measurement, then the TAR at $d_{\max}$ is the same as the *backscatter factor (BSF)* obtained at depth $d_{\max}$ in depth-dose tables. The backscatter factor may be defined as:

$$\text{BSF} = \frac{\text{dose at depth of maximum buildup in field size } A}{\text{dose to a small volume of tissue in air}} \qquad (15.17)$$

The BSF depends on the photon energy spectrum, as well as the field size. For example, the TAR at d_{max} for a $100 \, cm^2$ field is 1.035 for ^{60}Co, and 1.37 for an x-ray spectrum having a first half-value-layer of 1 mm Cu. Thus, although the maximum dose within the body for a monodirectional beam of ^{60}Co gammas would be only 3.5% greater than the measured dose to a small volume of tissue in air, due to the mainly forward scattering of high energy gamma photons, the dose near the surface of the body due to backscattering of x-ray fields can be as much as 40% greater than the dose measured by an instrument in air. It is difficult to measure the BSF for x-ray spectra below a few hundred keV, since the depth of maximum build up is almost at the surface, and there is no room to place an ordinary sized cylindrical chamber in a horizontal beam close enough to the wall surface of the phantom. However, the BSF in such cases can be measured with a thin end-window chamber, placed at the upper surface of a water phantom, with the beam directed downward into the water.

B. Photodosimetry

The physicochemical process of ionizing radiation interacting with silver halide crystals suspended as an emulsion in gelatin was one of the most thoroughly studied fields of dosimetry. Use of these films of emulsion to measure x- or gamma-radiation will be referred to here as "film" dosimetry. The two types of emulsions are typically referred to as x-ray and nuclear (Table 15.2). The emulsion typically is coated as a thin layer on one or both sides of a supporting film of celluloid or glass.

Although x-ray films are still used to check beams in gamma-ray therapy, and dental-type packets of these films are used in some institutions for radiation monitoring (film dosimetry) most of the basic research leading to the establishment of this method was carried out in the first half of the 20th century in research laboratories of the film manufacturers. An extensive treatment of the physical chemistry of light photography and film dosimetry was published by Mees (1954). One of the earliest explanations of the process of latent image formation was published by Gurney and Mott (1938). The birth of personnel monitoring in radiology departments began with the recommendation by Pfahler in 1922 that x-ray and radium workers carry a dental film in their

TABLE 15.2 Photographic Emulsions Used in Radiation Dosimetry (Attix, 1986)

	X-ray	Nuclear
Silver halide content	31–40% By weight	71–80% By weight
Grain size	1–2 μm Diameter	0.3 μm Diameter
Emulsion layer	10–25 μm Thick with density of 2 g/cm^3	1–600 μm Thick with density of 3.3 g/cm^3
Gel layer	0.5 μm Protective layer	0.5 μm Protective layer
Radiation type	Photon, beta	Neutron

pockets for two-week intervals, and further innovations by Edith Quimby in 1926 who used filters in a film badge for energy interpretation, and by Robert S. Landauer, Sr., who initiated the incorporation of easily obtained dental packets into the badge (Brodsky *et al.*, 1995).

Basic research on film dosimetry essentially ended in the latter part of the 20th century, as mass production of dental films became perfected and other solid state dosimeters were under development. However, a number of applied research papers were published that examined environmental effects of the use of films over varied periods of time, and to compare their precision and sensitivity with other types of dosimeters (Corney, 1959; Becker, 1966; Brodsky *et al.*, 1965).

A typical film used for personnel dosimetry is composed of an emulsion of silver bromide (AgBr) grains, 0.3–2 microns in diameter, suspended in gelatin (a proteinaceous substance) and spread on a thin backing of celluloid. The emulsion layer itself is 10 to 20 microns thick before development. The emulsion and its celluloid (cellulose acetate) backing are, in a dental-type packet, wrapped in an opaque paper wrapping that has a thickness of about $30\,mg\,cm^{-2}$, which filters out low-energy x-rays and beta rays. In special circumstances, this can limit the usefulness of film badges for measuring doses to the superficial layers of skin.

The passage through the emulsion of charged particles (i.e., electrons recoiled or emitted by absorption of photon energy, or protons recoiled by neutrons) produces what are termed "latent images." According to Gurney and Mott (1938), a physical process similar to that produced by charged particles in thermoluminescent dosimeters first occurs. The passage of a charged particle through a silver bromide grain (crystal) causes the ejection of atomic electrons from silver or bromide atoms; these electrons then migrate to a conduction band. An electron in the conduction band then migrates to a lattice defect, which has been designed into the crystal by incorporation of a specific concentration and type of impurity into the crystal. The electron is then "trapped" in the defect, releasing a companion "free" Ag+ ion to exist, which then migrates to an electron center in the crystal and deposits as a small clump of silver metal imbedded in the silver bromide matrix. This silver clump is called the "latent image." Only by further chemical action will this image be amplified to cause the entire silver bromide grain to become a grain of metallic silver.

A brief summary of the post irradiation processing follows:

Development: 3–7 minutes at about 20°C (controlled) in developer solution.

Stop bath: Rinse for about 10 seconds to 1 minute in an acetic acid bath to stop reduction of further grains at a given time.

Fixer solution: Place into the fixer bath for 10 minutes, to dissolve and remove silver bromide not activated by the radiation.

Wash and harden: Wash the film in hardener solution, sometimes for 30 minutes to achieve archival quality.

Densitometer: A device that directs a collimated light beam through a film and into a photoelectric cell that converts the remaining unattenuated light into an electric current. Usually, the current is read by a meter that converts it into a logarithmic signal, so that the degree of darkening can be measured over a wide range. The following equation defines the photographic film densiy, D:

$$\text{Density, } D = \log_{10}(I_o/I), \tag{15.18}$$

where I_o is the light beam intensity (photocurrent) with no absorber in place, and, I is the light intensity with the absorber (photographic film of interest) in place.

Note: In this formula, the intensity of the *unattenuated* light beam is placed in the numerator.

Net Density: The density of a film exposed to ionizing radiation is usually compared either to a film that is freshly prepared and not exposed to additional radiation, or (which is of more interest) to a film from the same manufactured batch that has been exposed only to appropriate "background" radiation for the same amount of time. The latter comparison is of interest if one wants to determine the exposure of a worker to occupationally-obtained radiation exposure, compared to the natural background to which he would ordinarily be exposed without his current employment.

In this comparison, the following relationship is obtained for the net density, D_n, from elementary algebra:

$$\begin{aligned} D_n = D_e - D_b &= \log_{10}(I_o/I_e) - \log_{10}(I_o/I_b) \\ &= \log_{10}[(I_o/I_e)(I_b/I_o)] = \log_{10}(I_b/I_e) \end{aligned} \tag{15.19}$$

where D_e and D_b are the radiation-exposed film and background film densities, respectively, and I_e and I_b are the light intensities transmitted through the radiation-exposed and background films, respectively.

A detailed description of developing and fixing of the latent image and subsequent densitometry can be found in Mees (1954).

C. Thermoluminescence (TL)

Daniels *et al.* (1953) were the first to propose the use of thermoluminescence in radiation dosimetry. The *Solid State Dosimetry* textbook provides an excellent review of the definition, history, and theory of thermoluminescence (Becker, 1973).

Thermoluminescent dosimeters (TLDs) are made of solid-state crystalline dielectric materials containing impurities or activators. The activators provide two kinds of centers (traps and luminescent sites). Radiation interactions in the TLD cause electrons to be raised from the valence to the conduction band. From the conduction band they find their way into an electron trap

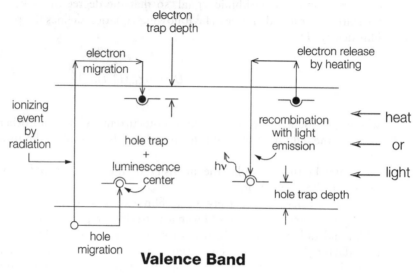

FIGURE 15.4 Energy-level diagram of the thermoluminescence and optically stimulated luminescence processes: (I) ionization by radiation, and trapping of electrons and holes; (II) use of heat or light to release electrons, allowing luminescence production. (Modified from *Introduction to Radiological Physics and Radiation Dosimetry*, F. H. Attix, Copyright © 1986 by John Wiley & Sons, Inc. This material is used by permission of John Wiley & Sons, Inc.)

and an accompanying hole occupies its associated trap. The stability of these traps depends on the material and the trap depth. Most traps are relatively stable at room temperature. As the temperature of the material is increased above room temperature these traps can be dissipated in a controlled manner. The release of these traps is followed by recombination at a luminescent center resulting in light emission. This process, which is referred to as *thermoluminescence*, is summarized in Fig. 15.4. Examples of heat (thermal) sources that give rise to the light output (luminescence) are hot gas, air, lasers, and contact hot planchets. Figure 15.5 is an example of the light output as a function of heating time (or temperature) for two TL materials processed with different heat sources. These plots are referred to as "glow curves."

TL materials have been used in personnel, environmental, and clinical dosimetry for many years. The widely held belief that TLDs would eventually make film dosimetry obsolete has not been realized. Two key issues to consider when selecting a radiation dosimeter are:

1. What is the application? (occupational, medical or environmental measurements)
2. What is/are the anticipated radiation field(s)?
3. What are the anticipated dose levels?

The answers to these questions lead to secondary issues regarding sensitivity, tissue-equivalence (variation in response with photon energy), dose-response relationship, material thickness, and possible holder and filter configurations.

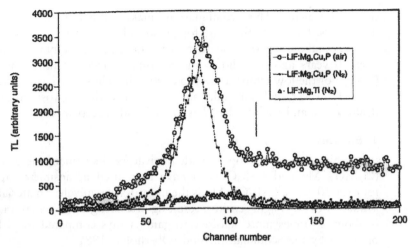

FIGURE 15.5   Glow curves of LiF:Mg, Cu, P and LiF:Mg, Tl using hot air and hot nitrogen heating. (Provided by J. R. Cassata, Naval Dosimetry Center.)

McKeever *et al.* (1995) presented an excellent review of the past and future of TLD materials research. Four categories (fluorides, sulphates, borates, and oxides) are summarized here.

I. Fluorides

Lithium fluoride is widely available in the forms LiF:Mg,Ti and LiF:Mg,Cu,P. The latter phosphor has superior characteristics compared to its long serving counterpart, which has been the most widely used TL personnel dosimeter for the past 30 years. As noted by Moscovitch (1999) the advantages of LiF:Mg,Cu,P over LiF:Mg,Ti, are:

1. higher sensitivity (>10×).
2. less photon energy-dependent response (more tissue equivalent).
3. less fade.
4. linear dose response.

Both phosphors are cost effective and reliable and they will continue to be used for many dosimetry applications.

Natural CaF_2 (fluorite) is used in some limited applications. Synthetic forms of calcium fluoride (CaF_2:Mn, CaF_2:Dy, CaF_2:Tm) are routinely used for radiation dosimetry. For example, the greater sensitivity of calcium fluoride compared to lithium fluoride has made it attractive for environmental dosimetry. However, this greater sensitivity, which equates to less tissue-equivalence has limited its use for personnel dosimetry. The US Navy represents the biggest user of calcium fluoride in the form of CaF_2:Mn to monitor personnel photon doses from nuclear propulsion operations. This system is being replaced by the previously mentioned LiF:Mg,Cu,P.

2. Sulphates

Like synthetic forms of calcium fluoride, calcium sulphates are highly sensitive (30 times greater than LiF:Mg,Ti) and consequently they are not tissue-equivalent. This combination makes them useful environmental dosimeters, but limits their use for personnel dosimetry.

Watanabe prepared the first synthetic calcium sulphate ($CaSO_4:Mn$) over 50 years ago (Oberhofer and Sharmann, 1981). Due to the instability (fading) of the low temperature peak ($90°$) this phosphor has not been widely used. However, calcium sulphate activated with dysprosium or thullium ($CaSO_4:Dy$ and $CaSO_4:Tm$) has gained wide acceptance.

3. Borates

Lithium borates doped with activators like manganese or copper ($Li_2B_4O_7:Mn$ and $Li_2B_4O_7:Cu$) have been used in dosimetry applications for over 20 years. $Li_2B_4O_7:Mn$ is the most tissue-equivalent material presently available, but it is light sensitive resulting in a background signal. $Li_2B_4O_7:Cu$ is about 20 times more sensitive to gamma rays compared to $Li_2B_4O_7:Mn$ prepared by the Schulman method (Oberhofer, 1981).

Magnesium borates ($MgB_4O_7:Dy$ and $MgB_4O_7:Tm$) were introduced over 20 years ago (Oberhofer and Scharmann, 1981). In terms of tissue-equivalence they are comparable to $LiF:Mg, Ti$ and they are approximately seven times more sensitive.

4. Oxides

Oxides of aluminum, beryllium, and magnesium are popular radiation dosimeters. Natural minerals contain aluminum oxide or it can be produced as $Al_2O_3:C$ or $Al_2O_3:Mg, Y$. Beryllium oxide (BeO) is a tissue equivalent competitor to $LiF:Mg, Ti$ with comparable sensitivity. It appears to be more widely used as a thermally stimulated exoelectron emission (TSEE) dosimeter. Magnesium oxide (MgO) is an extensively studied radiation dosimeter that has not found wide acceptance.

Anion deficient $Al_2O_3:C$ is the most widely used oxide dosimeter. It has been used as both a TL and optically stimulated luminescence (OSL) dosimeter. Details of its characteristics and applications as an OSL dosimeter will be addressed in the following section.

D. Optically Stimulated Luminescence (OSL)

The use of OSL for radiation dosimetry was first proposed in 1956 (Akselrod *et al.*, 1998). The mechanism of OSL involves illumination of an irradiated crystal with a specific wavelength of light to initiate the movement of charge trap sites to luminescent centers. Total luminescence, which depends on the amount of stimulation imparted to the crystal, is proportional to dose. The OSL mechanism is similar to TL as depicted in Fig. 15.4. The basic difference is the stimulant, light versus heat.

McKeever (2001) presented an excellent review of the various light stimulation modes. They include continuous wave (CWOSL), linear modulation (LM-OSL) and pulsed (POSL). In brief, CWOSL involves simultaneous sample illumination with constant intensity light and monitoring of the stimulated luminescence emission. Resolution between stimulation and emission light is accomplished by appropriate filter and

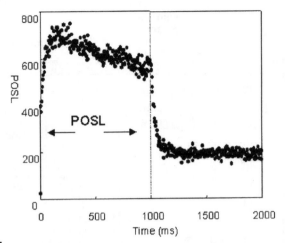

FIGURE 15.6 POSL figure showing the signal from irradiated Al_2O_3, during a 1-second exposure to the laser. The pulse frequency was 4 kHz and the pulse width was 300 ns. The laser used was the 2nd-harmonic (532 nm) from a Nd:YAG laser. After the 1-second stimulation period, the OSL signal decays with a lifetime corresponding to the lifetime of the luminescence centers (35 ms), which are F-centers in this material. (Provided by S. W. S. McKeever, Oklahoma State University.)

wavelength discrimination. Linear increase in the stimulation light intensity is used in LM-OSL resulting in a peak. This peak is the result of a linear increase in OSL output followed by a nonlinear return to zero. In POSL a pulsed stimulation source is used and discrimination between excitation and emission light is accomplished by time resolution. Figure 15.6 is a plot of POSL versus time for irradiated Al_2O_3, during a 1-second exposure to the laser. Each datum point corresponds to the measured output between each laser pulse.

Early applications included archaeological and geological dating, or retrospective dosimetry. Technical weaknesses including phosphor sensitivity and fading precluded the application of OSL to personnel dosimetry. However, recent advances have resolved these issues and OSL is now a widely used personnel dosimeter. Landauer (Glenwood, IL), the largest processor of personnel dosimetry in the world, has replaced its film service with an OSL device (Yoder, personal communication).

Luminescence lifetime is the critical characteristic in the selection of OSL material and readout timing parameters (Akselrod *et al.*, 1998). Studies of Al_2O_3:C concluded that it has a high optical sensitivity, long-lived luminescence lifetime ($\tau = 35$ ms at room temperature), and it can be optimized to a particular application during crystal growth (McKeever, 2001).

E. Calorimetry

Calorimetry provides a direct measurement of absorbed dose. In contrast to the previous methods, which are relative in their response, calorimetry involves the measurement of the temperature rise in tissue-equivalent material. The temperature increase during irradiation is measured and the

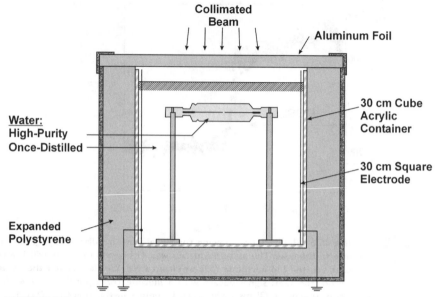

FIGURE 15.7 Major features of the **NIST** sealed water calorimeter. High-purity water is sealed within a thin-wall cylindrical glass container. Entire assembly is immersed in a 30 cm cube acrylic container filled with once-distilled water. (Adapted from Domen (1994).)

energy deposited is equal to heat capacity multiplied by the increase. Heat capacity can be determined in two ways:

1. Measured *in situ* by electrical heating;
2. Use of a known (or previously measured) value.

Water, which has scattering and absorption properties similar to tissue, is the standard reference material for radiation therapy dosimetry. The American Association of Physicists in Medicine (AAPM) TG-51 protocol (AAPM, 1999) describes a methodology for using ionization chambers to determine dose to water. However, a water calorimeter provides the most direct measure of absorbed dose (see Fig. 15.7). Issues regarding heat defect have been addressed by Domen (1994).

F. Electron Paramagnetic Resonance (EPR) Spectroscopy of Alanine

EPR spectroscopy of alanine has replaced Fricke chemical dosimetry in many high dose applications. Dosimeters are typically read on an X-band EPR spectrometer to determine the alanine-derived radical concentration (see Figure 15.8). Details of EPR spectroscopy will be presented in section V.A).

Briefly, the spectrometer should be capable of the following settings:

- Microwave frequency 9–10 GHz with automatic field frequency locking.
- Corresponding magnetic field to set a g-factor of 2.0 (at 9.8 GHz, this equals 350 mT) with a field scan range of 20 mT about the center field.

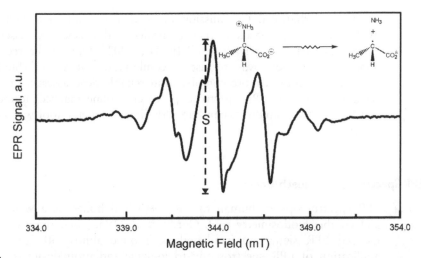

FIGURE 15.8 EPR spectrum of irradiated alanine. The amplitude, s is used for absorbed dose evaluations. (Provided by M. F. Desrosiers, National Institute of Standards and Technology.)

- Radiofrequency modulation amplitude 0.1–1 mT.
- Microwave power 0.1–10 mW (leveled).
- Variable sweep time, time constant, and receiver gain dependent on absorbed dose.
- Sensitivity of the spectrometer should be at least 2×10^{11} spins mT^{-1}.
- Cavity should have a sample access diameter of at least 1 mm greater than the diameter of the dosimeter being analyzed.

Alanine is an amino acid that can be used with an EPR spectrometer to perform accurate and precise dosimetry. The radiation dosimeter is prepared using α-alanine, $CH_3\text{-}CH(NH_2)\text{-}COOH$, in the form of polycrystalline powder. The most commonly used form is L-alanine, however, both stereoisomers are useful for absorbed dose measurements (ASTM, 1999).

V. MEASUREMENTS (BIOLOGICAL DOSIMETRY)

Ionizing radiation measurements with biological dosimeters entail the use of tissues to determine dose. A recent symposium entitled, "21st Century Biodosimetry: Quantifying the Past and Predicting the Future" discussed many of the current biodosimetry tools (NCRP, 2001). One of the specific aims of the symposium was to provide information on the most up-to-date approaches of biomarkers for estimating radiation dose. Chromosomal aberrations have been considered the benchmark for retrospective dose assessment. These include dicentrics and deletions measured by micronuclei. However, the utility of the dicentric assay is limited by the long-term stability of the aberrations and the labor-intensive nature of the assessment.

The International Commission on Radiation Units and Measurements (ICRU) recently released a report entitled, "Retrospective Assessment of Exposures to Ionizing Radiation" (ICRU, 2002). One of the recommendations of the report was to use a combination of several biodosimetric methods to obtain a more complete and reliable dose assessment. The most versatile method is EPR spectroscopy of crystalline matrices found in teeth and bones. This rapidly expanding radiation dosimetry tool will be covered in detail.

A. EPR Spectroscopy of Teeth/Bones

EPR spectroscopy of human tissues (teeth and bones) is a well-established and reliable biodosimetry tool. Gordy *et al.* (1955) first reported radiation-induced EPR signals in irradiated skull bone almost 50 years ago. The application of EPR spectroscopy to ionizing radiation dosimetry was later proposed by Brady *et al.* (1968). Since that time EPR biodosimetry has been applied to accident and epidemiologic dose reconstruction, radiation therapy, food irradiation, quality assurance programs, and archaeological dating. Materials that have been studied include bone, tooth enamel, dentin, alanine and quartz.

This dosimetry method is based on the fact that ionizing radiation interacts with mineralized tissues to produce dose-dependent concentrations of long-lived paramagnetic centers. As a result, the tissue is the dosimeter, and the calibration can be regarded as absolute depending on the tissue of interest. Brady *et al.* (1968) suggested using EPR dosimetry and the additive re-irradiation method to obtain dose estimates from accidental overexposures. EPR biodosimetry of irradiated mineralized tissue was proposed and validated by Desrosiers *et al.* (1991a, 1993) as a quantitative method to measure the absorbed dose from bone-seeking radiopharmaceuticals. Desrosiers (1991b) and Schauer *et al.* (1993c) applied this method to the dosimetry of accidental radiation overexposures in San Salvador (^{60}Co) and Gaithersburg, MD (3 MeV electrons), respectively.

I. EPR Fundamentals

EPR is a nondestructive method applied to materials containing unpaired electrons (i.e., produced by the absorption of ionizing radiation). When paramagnetic materials are placed in a strong magnetic field the absorption of applied microwave energy causes electron spin-flip transitions. The intensity of these transitions is proportional to the number of unpaired spins in the material, which is proportional to the absorbed dose (see Fig. 15.9). In addition, by varying the magnetic field, radical centers with different structures and environments are spectroscopically resolvable. The relationship between microwave frequency and the magnetic field is given by:

$$h\nu = g\beta H_r \tag{15.20}$$

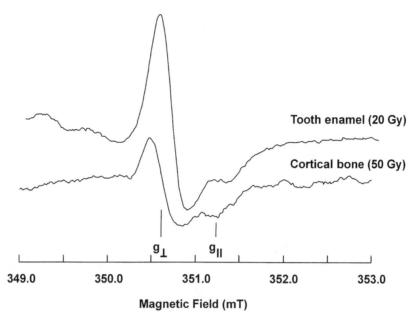

FIGURE 15.9 First derivative of the absorption curve (arbitrary units) with respect to the applied magnetic field (mT) for a human femur (50 Gy) and bovine tooth enamel (20 Gy) irradiated with ^{60}Co gamma rays. The signal of interest, $g_\perp$ (2.0018) is derived from the hydroxyapatite in bone or teeth.

where: h = Planck's constant; ν = microwave frequency; g = spectroscopic splitting factor (typically 2.0); β = Bohr magneton; H_r = magnetic field.

The method of retrospective EPR dosimetry using calcified tissues (bone, enamel, dentin) is based on the measurement of radiation-induced radicals in hydroxyapatite [$Ca_{10}(PO_4)_6(OH)_2$]. During the mineralization process of biological hydroxyapatites, carbonate ions are incorporated into the crystalline lattice substituting for both phosphate and hydroxyl ions. Upon absorption of ionizing energy by the hydroxyapatite crystal, the carbonate ions capture free electrons in the crystal matrix to form free-radical centers (Callens *et al.*, 1987). The dose-dependent formation of carbonate radical centers can be quantified through the use of EPR.

Hydroxyapatite constitutes 95–97% of tooth enamel, 70–75% of dentin, and 60–70% of bones. The predominance of hydroxyapatite along with its high degree of crystallinity makes tooth enamel the most suitable material for retrospective biodosimetry. Human tooth enamel is a calcified tissue with several special features. Acellular in its adult state, tooth enamel is composed of hydroxyapatite crystallites, which can be up to several hundred nanometers in length. The concentration of radiation-induced radicals, and hence the intensity of the EPR signal, increases proportionally with the absorbed dose from about 100 mGy to above 10 kGy. There are no known dose rate effects. The carbonate radical center is extraordinarily stable with a calculated lifetime at 25°C of 10^7 years (Hennig *et al.*, 1981). Free-radical centers in tooth enamel are produced by a wide variety of ionizing radiations, including x-rays, gammas, betas, alphas, protons, and heavy ions (for example, Schauer *et al.*, 1993d).

2. EPR Dosimetry Essentials

The process of EPR dose reconstruction consists of several important steps:

- Sample collection
- Sample preparation
- EPR measurements
- Dose reconstruction
- Interpretation of results

These steps, which are only relevant for tooth enamel, are shown in greater detail in Fig. 15.10 (Desrosiers and Schauer, 2001). It should be noted that although the EPR properties of bone and dentin are very similar to those of enamel, they differ in the procedure of sample preparation (Romanyukha and Regulla, 1996; Weiser *et al.*, 1994).

B. Cytogenetic Techniques

The Proceedings of the International Conference on Low-Level Radiation Injury and Medical Countermeasures included a section on low-level radiation exposure assessment using biodosimetry (AFRRI, 2002). Table 15.3 is a comparison of operational parameters for some conventional and candidate radiation biomarkers. The International Atomic Energy Agency has concluded that the "gold standard" for blood-based biodosimetry is the chromosome aberration based bioassay (IAEA, 1986). This opinion was also the conclusion of the AFRRI Conference attendees (AFRRI, 2002). Chromosome aberrations can occur following exposure to ionizing radiation. In brief, the chromosomes are broken, and then following some period of time, they are able to rejoin.

However, the study of metaphase chromosome aberrations, or micronuclei in peripheral lymphocytes is limited by its relatively short "half life." These aberrations (dicentrics and rings) could disappear only months following irradiation (Turai, 2000). Another promising technique, fluorescence in situ hybridization (FISH) is able to detect stable chromosome aberrations (translocations) in human lymphocytes for many years following exposure to ionizing radiation.

VI. APPLICATIONS

A. Personnel Dosimetry

The primary objective of personnel dosimetry, which can be broadly divided into whole-body and extremity applications, is to accurately and precisely measure occupational and accidental doses. A brief history of the

DOSE RECONSTRUCTION
BY THE EPR METHOD

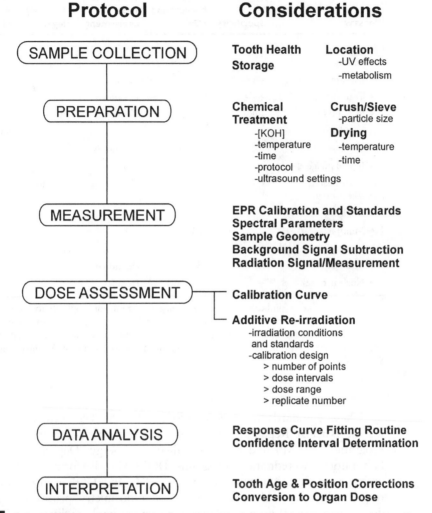

FIGURE 15.10 Schematic of the EPR protocol for retrospective dose assessment. Protocol steps and associated considerations are cited.

development of whole-body personnel dosimetry standards and performance testing is summarized in Table 15.4.

The current whole-body dosimetry testing standard (ANSI, 2001) provides a procedure for testing the performance of dosimetry systems (i.e., hardware, software, and processor supplying services, or in some cases the user of the services). A similar ANSI standard has been published for testing processors of extremity dosimetry (ANSI, 1995).

TABLE 15.3 Comparison of Operational Parameters for Some Conventional and Candidate Radiation Biomarkers

Bioassay	Specificity	Twice background, cGy	Postexposure window for 1-Gy assessment	Medical/ legal	Advanced platforms (clinical/field)
Chromosome aberrations					+++/−−
• Dicentric and ring	+++	2	Months	+++	
• Translocation	+++	6–8	Decades	++	
• PCC-FISH	+/R	10–12	Unknown	R	
DNA adducts					+++/−−
• Single-strand breaks	−−	20–50	<1.5 hours	−−	
• Base damage	−−	200–300	ND	R	
Gene expression/ encoded proteins	−−/R		ND	Unknown	+++/+
• Single		2			
• Multiple		ND			
DNA mutation				R	+++/−
• MtDNA[4977]	R	25	Unknown		
• Nuclear genes	+	Unknown	Decades		

Note: + and − indicate positive and negative rank of applicability (the greater the number of symbols, the greater degree of applicability to a maximum of three symbols) R indicates the need for additional research studies; ND indicates that the required studies have not been done. Reproduced with permission of the Association of Military Surgeons and W. F. Blakely.

These standards have been adopted by the US accrediting programs for the Nuclear Regulatory Commission (National Voluntary Accreditation Program—NVLAP) and the Department of Energy (Department of Energy Laboratory Accreditation Program—DOELAP). Performance testing programs were fully implemented by the NRC and DOE in 1984 and 1987, respectively (see Table 15.4). They test a dosimetry processor's ability to determine personal dose equivalent for occupational conditions and absorbed dose for accident conditions. Blind tests are conducted under controlled conditions that include irradiation with photons, electrons (beta particles), neutrons, and selected mixtures of these radiations.

Another important area of personnel dosimetry is retrospective dose reconstruction. The National Research Council published a textbook addressing many of the relevant aspects including biologic dosimetry and biologic markers (NRC, 1995). Numerous articles have been published detailing the use of EPR biodosimetry for occupational and nonoccupational dose reconstructions (Romanyukha *et al.*, 2001, 2002). Table 15.5 summarizes the population studies that utilized EPR dose reconstruction in tooth enamel.

TABLE 15.4 **Development of Whole-body Personnel Dosimetry Standards and Performance Testing**

Date	Milestones
1955	Film badge intercomparisons by NBS under AEC contract.
1965	R. Gordon published a film badge intercomparison in Radiology.
1973	CRCPD task force met with State and Federal representatives to implement its recommendations for establishing a dosimetry testing program.
1975	ANSI N13.11 committee formed.
1976	Battelle tested dosimetry processor performance against four standards that had been developed. NRC, ERDA and BRH held a public meeting to discuss personnel dosimetry problems. NRC contracted with the University of Michigan for pilot tests.
1978	ANSI N13.11 released for trial use and comment.
1980	NRC published ANPRM on dosimetry testing.
1982	NRC asked NIST to establish a NVLAP accreditation for dosimetry processors.
	University of Michigan completed third round of pilot tests:
	– 1st round (52% fail rate)
	– 2nd round (38% fail rate)
	– 3rd round (25% fail rate)
1983	ANSI N13.11 (1st version) published.
1984	NRC published NPRM on dosimetry testing.
	NVLAP dosimetry testing implemented.
1986	DOE published standard for dosimetry testing (DOE/EH-0027)
1987	NRC published final rule requiring licensees to use NVLAP accredited dosimetry processors.
	DOE Order 5480.15 established DOELAP.
	ANSI N13.11 revision committee formed.
	Several States require NVLAP accreditation.
1993	ANSI N13.11 (2nd version) published.
1995	NVLAP Bulletin issued modifying dose equivalent conversion factors.
1996	ANSI N13.11 revision committee formed.
2001	ANSI/HPS N13.11 (3rd version) published.
2002	NVLAP and NRC announce change of proficiency testing standard in the Federal Register.

Terms
ANPRM: Advance Notice of Proposed Rule Making
ANSI: American National Standards Institute
BRH: Bureau of Radiological Health
CRCPD: Conference of Radiation Control Program Directors
DOE: Department of Energy
ERDA: Energy and Research Development Agency
HPS: Health Physics Society
NIST: National Institute of Standards and Technology
NRC: Nuclear Regulatory Agency
NVLAP: National Voluntary Laboratory Accreditation Program

TABLE 15.5   Population Studies with EPR Dose Reconstruction in Tooth Enamel

Description of group	Year of exposure	Number of doses	Values of reconstructed doses, Gy	Reference
Survivors of A-bombing of Hiroshima, Japan	1945	10, 100	0.3–4.0	Ikeya, 1993; Nakamura *et al.*, 1998
Mayak nuclear workers, Russia	1941–1961	~100	0.2–6.0	Romanyukha *et al.*, 1994, 2000
Techa riverside population, Russia	1941–1958	~100	0.1–10	Romanyukha *et al.*, 2001
Eye-witnesses of Totskoye nuclear test, Russia	1954	10	0.1–0.4	Romanyukha *et al.*, 1999
Chernobyl clean up workers, Ukraine, Russia	1986	660	0–2.0	Chumak *et al.*, 1998
Population of the areas contaminated by Chernobyl fallouts, Russia	1986	2500	0–0.3	Stepanenko *et al.*, 1998
Background population, Russia	–	136	0–0.1	Ivannikov *et al.*, 1997
Population of the areas nearby Semipalatinsk Nuclear Test Site, Kazakhstan	1941–1962	26	0–3.0	Ivannikov *et al.*, 2002; Romanyukha *et al.*, 2002

B. Clinical Dosimetry

The primary objective of clinical dosimetry is to accurately and precisely measure therapeutic (AAPM, 1999 and IAEA, 2000) and to a lesser degree, diagnostic doses (Bushberg *et al.*, 2002) including radiography, fluoroscopy, and computed tomography.

The American Association of Physicists in Medicine (AAPM) TG-51 protocol for clinical reference dosimetry of high-energy photon and electron beams uses ionization chambers as the basis for measurements, but absorbed dose to water calibration factors are also required (AAPM, 1999).

Measurement techniques for clinical diagnostic uses of ionizing radiation are reviewed in Bushberg's textbook (Bushberg *et al.*, 2002). The major variables that determine patient doses include imaging modality, technique factors and in the case of fluoroscopy, beam time. Patient size is another important dose variable. For example, pediatric and small adult patients routinely have computed tomography scans done with the same machine settings as average size adults. This results in unnecessary patient dose that some have argued is in the range of observable excess cancer mortality (Brenner *et al.*, 2001). Recommended diagnostic reference levels for medical imaging modalities have been published by the ICRP (2001).

C. Materials Processing

Industrial applications of ionizing radiation are rapidly expanding. Examples include medical sterilization, food irradiation and most recently, mail irradiation. EPR spectroscopy of alanine is one of the most widely used systems to ensure quality dosimetry in these applications. The standard for practice in this area is ASTM 51607 (2002). Materials processing dosimetry systems must be able to operate in the following conditions:

- absorbed dose range between 1 and 10^5 Gy.
- absorbed dose rate up to 10^2 Gy s^{-1} for continuous radiation fields and 5×10^7 Gy s^{-1} for pulsed radiation fields.
- radiation energy for photons and electrons between 0.1 and 28 MeV.
- irradiation temperature between -60 and $+90°$C.

VII. ISSUES AND OPPORTUNITIES FOR FUTURE DEVELOPMENTS

The use of ionizing radiation in medical and industrial applications continues to increase. Approximately 320 million diagnostic medical and dental x-ray procedures are performed each year in the US. The annual number of CT exams performed increased from approximately 3 million in 1980 to 58 million in 2000. Approximately 750,000 fluoroscopically guided interventional procedures are performed annually in the US. Over 100,000 personnel were monitored for occupational exposure to ionizing radiation at 271 NRC licensed sites during 2000. From 1941–2001 there have been 249 radiation accidents in the US involving 1357 people. These facts highlight the need to study and quantify radiation dose. Improved dosimetry techniques equate to a better understanding of the delayed (stochastic) and acute (deterministic) effects of ionizing radiation.

Most radiation dosimetry applications employ passive devices that are processed following exposure to ionizing radiation, or retrospectively. This equates to days or months for most routine personnel dosimetry applications and many years for some biodosimetry measurements. Issues of signal stability, or fading and background interference are areas for potential improvement.

The invasive nature of biodosimetry tools requires new and innovative thinking. Recent advances in the areas of *in vivo* OSL and EPR measurements of tooth enamel have been published (Godfrey-Smith and Pass, 1997 and Miyake *et al.*, 2000).

VIII. APPENDIX

A. Measurement of Beta Doses to Tissue

Beta-ray exposures and doses to the tissues of hands and fingers of workers are the more frequent dose assessments in university-medical complexes. They can best be estimated routinely by use of special "finger badge" personnel monitoring devices supplied commercially, and distributed as needed by the Radiation Safety Office.

However, when an exposure has been received in the absence of a badge, or an exposure estimate is needed from a particular source for the planning of shielding or other protective measures, beta dose estimates can be made using end-window Geiger tube detectors. The end window should have a thickness of 30 mg/cm^2 or less. The assumption may be made that each beta particle entering a Geiger tube end window causes a count. The counts per minute, divided by the end window area, is an estimate of the beta fluence rate. This rate times the linear stopping power, converted to appropriate energy units, provides an estimate of beta dose rate. For beta radiation encountered in the laboratory that is energetic enough to penetrate to the average depth of sensitive basal cells, a rule of thumb may be used that linear stopping power $\approx 2\,\text{MeV/cm}$.

Then, the beta dose rate may be estimated approximately from the count-rate of the end window Geiger counter as follows:

$$\text{Dose-rate beta (rad/min)} = \frac{C \times (2\,\text{MeV/cm} \times 1.6 \times 10^{-6}\,\text{erg/MeV})}{A\,\text{cm}^2 \times 1\,\text{g/cm}^3\,\text{tissue} \times 100\,\text{erg/g-rad}}$$

$$= 3.2 \times 10^{-8}\,C/A, \tag{15.21}$$

or

$$\text{Dose-rate (rad/hour)} = 60 \times 3.2 \times 10^{-8}\,C/A$$

$$\approx 2 \times 10^{-6}\,C/A, \tag{15.22}$$

or perhaps most useful,

$$\text{Dose-rate (millirad/hour)} \approx 0.002\,C/A \tag{15.23}$$

where

$C =$ counts/minute of end-window Geiger counter, and $A =$ area of end window, in cm^2.

If we assume that a particular Geiger counter has a "dead time" between counts of about 200 μs, and we want to use it only up to about 1,000 counts/s to avoid dead time corrections, then the maximum count-rate range would be 60,000 counts/min. If we assume a cross-sectional area, A, of 10 cm^2, then by Eq. 15.23, we would be limited to measuring $0.002 \times 60,000/10 = 12$ mrad/hour or less, which would often suffice for laboratory surveying and protection design. Higher dose-rate measurements can be made with end window ion chambers for surveying purposes, calibrating the survey instrument with beta sources of known surface dose rate, as calibrated by laboratories traceable to NIST.

For accurate beta dose measurement, small thimble chambers, with both walls and gas, small in dimension compared to average electron ranges, can be used together with the Bragg–Gray principle described in the previous sections. For beta dose measurements, however, condition 3 is not limiting, since the beta rays already are the primary ionizing radiation, and CPE is already established. Here, one wants to be sure that the chamber itself is small and thin enough (compared to the average beta-ray range) so that it does not appreciably absorb or perturb the beta radiation field to be measured.

B. Neutron Dose Measurements

The need for neutron dose measurements in the university or research center environment, except in the vicinity of high-energy machines (>10 MV x-rays) will be infrequent, and can best be made by a dedicated neutron survey meter calibrated to read directly in mrem/hour for a definite neutron spectrum, with corrections for other neutron spectra as necessary. Sometimes, a neutron survey meter, e.g., with a stilbene crystal, can be calibrated directly against the known-intensity radioisotope neutron source used in the laboratory, and then used to survey particular procedures or operations when this source is unshielded. If a neutron detector is constructed with a hydrogen radiator inside a proportional counter, or a BF_3 "long counter" surrounded by a cylinder of paraffin is used to estimate neutron fluence rates, then the neutron fluence rate can be converted to mrem/hour using conversion factors in 10 CFR Part 20, or in ICRU reports. The right-hand column in the table in Part 20 gives the conversion factors in neutrons/cm^2 per rem. A value of 24×10^6 neutrons/cm^2-rem can be seen to be close to the value needed to cover the 0.5–10 MeV neutron range most likely to require monitoring, either from radioisotope sources, or from accelerator sources of neutrons outside the shielded accelerator room. Industrial laboratories or nuclear power plants will maintain appropriate instrumentation for neutron surveys according to license conditions or technical specifications.

REFERENCES

Akselrod, M. S., Lucas, A. C., Polf, J. C., and McKeever, S. W. S. (1998). Optically stimulated luminescence of Al_2O_3. *Radiat. Meas.* **29**, 391–399.

American Association of Physicists in Medicine, Almond, P. R., Biggs, P. J., Coursey, B. M., Hanson, W. F., Saiful Huq, M., Nath, R., and Rogers D. W. O. (1999). *Med. Phys.* **26**, 1841–1870.

American National Standards Institute N13.11 (2001). Personnel Dosimetry Performance: Criteria for Testing.

American National Standards Institute N13.32 (1995). Extremity Dosimetry Performance: Criteria for Testing.

American Society of Testing and Materials (2002). Standard Practice for Use of the Alanine-EPR Dosimetry System (ISP/ASTM 51601-2002E).

Armed Forces Radiobiology Research Institute (2002). *Proceedings of the International Conference on Low-Level Radiation Injury and Medical Countermeasures Military Medicine* **167**, 21–24.

Attix, F. H. (1986). "Introduction to Radiological Physics and Radiation Dosimetry," (Chapter 10, Cavity theory) pp. 231–263, and (Chapter 14, Integrating Dosimeters) p. 412. Wiley, New York.

Attix, F. H., De La Vergne, L., and Ritz, V. H. (1958). Cavity ionization as a function of wall materials, *J. Res. NBS*, **60**, 231–243.

Becker, K. (1966). "Photographic-Film Dosimetry." Focal Press, London.

Becker, K. (1973). "Solid State Dosimetry." CRC Press, Cleveland.

Borg, J., Kawrakow, I. Rogers, D. W. O., and Seuntjens, J. P. (2000). Monte Carlo study of correction factors for Spencer-Attix Cavity theory at photon energies at or above 100 keV. *Med. Phys.* **27**, 1801–1813.

Brady, J., Aarestad, N., and Swartz, H. (1968). *In vivo* dosimetry by electron spin resonance spectroscopy. *Health Phys.* **15**, 41–47.

Bragg, W. H. (1910). Consequences of the corpuscular hypothesis of the gamma and x-rays and the ranges of beta rays. *Philos. Mag.* (London) **20**, 381–416.

Bragg, W. H. (1912). "Studies in Radioactivity." McMillan and Co., Ltd., London.

Brenner, D. J., Elliston, C. D., Hall, E. J., and Berdon W. E. (2001). Estimated risks of radiation-induced fatal cancer from pediatric CT. *Am. J. Roentgen.* **176**, 281–296.

Brodsky, A., Spritzer, A., Feagin, F. E., Bradley, F. J., Karches, G. J., and Mandelberg, H. I. (1965). Accuracy and sensitivity of film measurements of gamma radiation, Part IV, Intrinsic and extrinsic sources of error. *Health Phys.* **11**, 1071–1082.

Brodsky, A., Kathren, R. L., and Willis, C. A. (1995). History of the medical uses of radiation: regulatory and voluntary standards of protection. *Health Phys.* **69**, 781–823.

Burlin, T. E (1966). A general theory of cavity ionization. *Brit. J. Radiol.* **39**, 721–734.

Burlin, T. E. (1968). Cavity Chamber theory (Chapter 8), *In* "Radiation Dosimetry, Vol. 1 Fundamentals" 2nd Ed. (F. H. Attix and C. W. Roesch, Eds.), pp. 331–392. Academic Press, New York.

Burlin, T. E. and Snelling, R. J. (1969). The Application of General Cavity Ionization theory to the Dosimetry of Electron Fields. *Proceedings Second Symposium on Microdosimetry, European, EUR 4452 d-f-e*, 451–473.

Bushberg, J. T., Seibert, J. A., Leidholdt, E. M., and Boone J. M. (2002). "The Essential Physics of Medical Imaging," 2nd ed. Lippincott, Williams and Wilkins, Philadelphia.

Callens, F., Verbeeck, R., Matthys, P., Martens, L., and Boesman, E. (1987). The contribution of $CO_3^{(3-)}$ and CO_2^- to the ESR spectrum near $g = 2$ of powdered human tooth enamel. *Calcif. Tissue Int.* **41**, 121–129.

Chumak, V., Likhtarev, I., Sholom, S., Meckbach, R., and Krjuchkov, V. (1998). Chernobyl experience in field of retrospective dosimetry: Reconstruction of doses to the population and liquidators involved in the accident. *Radiat. Prot. Dosim.* **77**, 91–95.

Corney, G. M. (1959). Photographic Monitoring of Radiation. *In* "Radiation Hygiene Handbook" (H. Blatz, Ed.). McGraw-Hill, New York.

Daniels, F., Boyd, C. A., and Saunders D. F. (1953). Thermoluminescence as a research tool. *Science* **117**, 341–349.

Desrosiers, M. F., Coursey, B., Avila, M., and Parks N. (1991a). Radiopharmaceutical dose assessment. *Nature (London)* **349**, 281–288.

Desrosiers, M. F. (1991b). In vivo assessment of radiation exposure. *Health Phys.* **61**, 851–861.

Desrosiers, M. F., Avila, M., Schauer, D., Coursey, B., and Parks, N. (1993). Experimental validation of radiopharmaceutical absorbed dose to mineralized bone tissue. *Appl. Radiat. Isot.* **44**, 451–463.

Desrosiers, M. F. and Schauer, D. A. (2001). Electron paramagnetic resonance (EPR) biodosimetry. *Nucl. Instrum. Methods Phys. Res., Sect. B* **184**, 211–228.

Domen, S. R. (1994). A sealed water calorimeter for measuring absorbed dose. *J. Res. Nat. Inst. Standards Technol.* **99**, 121–141.

Fano, U. (1953). Degradation and range straggling of high energy radiation. *Phys. Rev.* **92**, 321–349.

Fano, U. (1954). Note on the Bragg–Gray cavity principle for measuring energy dissipation. *Radiat. Res.* **1**, 231–240.

Godfrey-Smith, D. I. and Pass, B. (1997). A new method of retrospective radiation dosimetry: Optically stimulated luminescence in dental enamel. *Health Phys.* **72**, 741–749.

Gordy, W., Ard, W., and Shields, H. (1955). Microwave spectroscopy of biological substances. Paramagnetic resonance in X-irradiated amino acids and proteins. *Proc. Nat. Acad. Sci. U.S.A.* **41**, 981–996.

Gray L. H. (1936). An ionization method for the absolute measurement of gamma ray energy. *Proc. Roy. Soc.* A156, **571–596**.

Gurney, R. W. and Mott, N. F. (1938). The theory of the photolysis of silver bromide and the photographic latent image. *Proc. Roy. Soc.* A164, 151.

Hennig, G., Herr, W., Weber, E., and Xirotiris, N. (1981). ESR dating of the fossil hominid cranium from Petralona cave, Greece. *Nature (London)* **292**, 531–536.

Horowitz, Y. S. (1984). Photon general cavity theory. *Radiat. Prot. Dosim.* **9**, 5–18.

Ikeya, M. (1993). "New Application of Electron Spin Resonance—Dating, Dosimetry and Microscopy." Word Scientific, Singapore.

International Atomic Energy Agency (1986). "Biological Dosimetry: Chromosome Aberrations Analysis for Dose Assessment." Vienna.

International Atomic Energy Agency (2000). "Absorbed Dose Determination in External Beam Radiotherapy." Vienna.

International Commission on Radiation Units and Measurements Report 37 (1984). "Stopping Powers for Electrons and Positrons."

International Commission on Radiation Units and Measurements Report 44 (1989). "Tissue Substitutes in Radiation Dosimetry and Measurement."

International Commission on Radiation Units and Measurements Report 46 (1992). "Photon, Electron, Proton and Neutron Interaction Data for Body Tissues."

International Commission on Radiation Units and Measurements Report 51 (1993). "Quantities and Units in Radiation Protection Dosimetry."

International Commission on Radiation Units and Measurements Report 57 (1998). "Conversion Coefficients for use in Radiological Protection Against External Radiation."

International Commission on Radiation Units and Measurements Report 60 (1998). "Fundamental Quantities and Units for Ionizing Radiation."

International Commission on Radiation Units and Measurements Report 68 (2002). "Retrospective Assessment of Exposures to Ionising Radiation."

International Commission on Radiological Protection Publication 30 (1979). "Limits for Intakes of Radionuclides by Workers."

International Commission on Radiological Protection Publication 60 (1990). "Recommendations of the ICRP."

International Commission on Radiological Protection Publication 61 (1991). "Annual Limits on Intake of Radionuclides by Workers Based on the 1990 Recommendations."

International Commission on Radiological Protection Publication 74 (1996). "Conversion Coefficients for use in Radiological Protection Against External Radiation."

International Commission on Radiological Protection Committee 3 (2001). Draft – "Diagnostic Reference Levels in Medical Imaging."

Ivannikov, A., Skvortzov, V. G., Stepanenko, V. F., Tikunov, D. D., Fedosov, I. M., Romanyukha, A. A. and Wieser, A. (1997). Wide-scale EPR retrospective dosimetry: Results and problems. *Radiat. Prot. Dosim.* **71**, 171–80;.

Ivannikov, A., Zhumadilov, Zh., Gusev, B. I., Miyazawa, Ch., Liao, L., Skvortsov, V. G., Stepanenko, V. F., Takada, J., and Hoshi, M. (2002). Individual dose reconstruction among residents living in the vicinity of the Semipalatinsk nuclear test site using EPR spectroscopy of tooth enamel. *Health Phys.* **83**, 181–196.

Janssens, A., Egermont, G., Jacobs, R., and Thielens, G. (1974). Spectrum perturbation and energy deposition models for stopping power ratio calculations in general cavity theory. *Phys. Med. Biol.* **19**, 611–630.

Loevinger, R. (1956). The dosimetry of beta sources in tissue. The point source function. *Radiol.* **66**, 51–62.

Ma, C. and Nahum, A. E. (1991). Bragg–Gray theory and ion chamber dosimetry for photon beams. *Phys. Med. Biol.* **36**, 411–428.

McKeever, S. W. S., Moscovitch, M., and Townsend, P. D. (1995). "Thermoluminescence Dosimetry Materials: Properties and Uses." Nuclear Technology Publishing, England.

McKeever, S. W. S. (2001). Optically stimulated luminescence dosimetry. *Nucl. Instrum. Methods Phys. Res., Sect B* **184**, 211–228.

Medical Internal Radiation Dose Primer (1988). The Society of Nuclear Medicine, Inc., New York.

Mees, C. E. K. (1954). "The Theory of the Photographic Process." Macmillan Co., New York.

Miyake, M., Liu, K., Walczak, T., and Swartz, H. (2000). In vivo EPR dosimetry of accidental exposures to radiation: experimental results indicating the feasibility of practical use in human subjects. *Appl. Radiat. Isot.* **52**, 1031–1036.

Mobit, P. N., Nahum, A. E., and Mayles, P. (1997). An EGS4 Monte Carlo examination of general cavity theory. *Phys. Med. Biol.* **42**, 1319–1334.

Moscovitch, M. (1999). Personnel dosimetry using LiF:Mg,Cu,P. *Radiat. Prot. Dosim.* **85**, 41–56.

Nakamura, N., Miyazawa, C., Akiyama, M., Sawada, S., and Awa, A. A. (1998). A close correlation between electron spin resonance (ESR) dosimetry from tooth enamel and cytogenetic dosimetry from lymphocytes of Hiroshima atomic-bomb survivors. *Int. J. Radiat. Biol.* **73**, 611–627.

National Council on Radiation Protection and Measurement (2001). 21st century biodosimetry: quantifying the past and predicting the future. *Radiat. Prot. Dosim.* **97**, 1–80.

National Research Council (1995). "Radiation Dose Reconstruction for Epidemiologic Uses." National Academy Press, Washington, DC.

NBS Special Publication 251–16 (1988). "Calibration of X-ray and Gamma-Ray Measuring Instruments." Lamperti, P., Loftus, T. P. and Loevinger R. US Department of Commerce, Washington, DC.

Oberhofer, M. and Scharmann A. (1981). "Applied Thermoluminescence Dosimetry." Adam Hilger Ltd, Bristol.

Romanyukha, A. A., Regulla, D., Vasilenko, E., and Wieser, A. (1994). South Ural nuclear workers—comparison of individual doses from retrospective EPR dosimetry and operational personal monitoring. *Appl. Radiat. Isot.* **45**, 1191–1199.

Romanyukha, A. A. and Regulla, D. (1996). Aspects of retrospective ESR dosimetry. *Appl. Radiat. Isot.* **47**, 1291–1297.

Romanyukha, A. A., Ignatiev, E. A., Ivanov, D. V., and Vasilyev, A. G. (1999). The distance effect on the individual exposures evaluated from the Soviet nuclear bomb test at Totskoye test site in 1954. *Radiat. Prot. Dosim.* **86**, 51–58.

Romanyukha, A. A., Ignatiev, E. A., Vasilenko, E. K., Drozhko, E. G., Wieser, A., Jacob, P., Keriim-Markus, I. B., Kleschenko, E. D., Nakamura, N., and Miyazawa, C. (2000). EPR dose reconstruction for Russian nuclear workers. *Health Phys.* **78**, 11–20.

Romanyukha A. A., Seltzer S., Desrosiers M., Ignatiev E. A., Ivanov D. V., Bayankin S., Degteva M. O., Eichmiller F. C., Wieser A., and Jacob P. (2001) Correction factors in EPR dose reconstruction for the residents of the Middle and Lower Techa riverside. *Health Phys.* **81**, 551–566.

Romanyukha A., Desrosiers M., Sleptchonok O, Land C, Luckyanov N., and Gusev B. I. (2002). EPR dose reconstruction of two Kazakh villages near the Semipalatinsk nuclear test site. *Appl. Magnetic Resonance* **22**, 341–356.

Schauer, D., Seltzer, S., and Links, J. (1993a). Exposure-to-absorbed dose conversion for human adult cortical bone. *Appl. Radiat. Isot.* **44**, 481–489.

Schauer, D. and Links, J. (1993b). Newly computed f-factors for use in radiation dosimetry. *Med. Phys.* **20**, 1371–1373.

Schauer, D. A., Coursey, B., Dick, C., McLaughlin, W., Puhl, J., Desrosiers, M. F., and Jacobson A. (1993c). A radiation accident at an industrial accelerator facility. *Health Phys.* **65**, 131–140.

Schauer, D., Desrosiers, M., Le, F., Seltzer, S., and Links, J. (1993d). Dosimetry of cortical bone and tooth enamel irradiated with x and gamma rays: study of energy dependence. *Radiat. Res.* **138**, 1–8.

Spencer L. V. and Fano U. (1954), Energy spectrum resulting from electron slowing down. *Phys. Rev.* **93**, 1171–1181.

Spencer L. V. and Attix, F. H. (1955). A theory of cavity ionization. *Radiat. Res.* **3**, 231–254.

Stepanenko, V., Skvortsov, V., Tsyb, A., Ivannikov, A., Kondrashov, A., Tikunov, D., Iaskova, E., Shakhtarin, V., Petin, D., Parshkov, E., Chernichenko, L., Snykov, V., Orlov, M., Gavrihn, Yu., Khrousch, V., and Shinkarev, S. (1998). Thyroid and whole-body dose reconstruction in Russia following the Chernobyl accident: Review of progress and results. *Radiat. Prot. Dosim.* **77**, 101–106.

Turai, I. (2000). The IAEA's coordinated research project on biodosimetry. *Appl. Radiat. Isot.* **52**, 1111–1116.

Wieser, A., Haskell, E., Kenner, G., and Bruenger, F. (1994). EPR dosimetry of bone gains accuracy by isolation of calcified tissue. *Appl. Radiat. Isot.* **45**, 521–526.

Yoder, C. (2002). personal communication.

21CFR1020.30 (2002). Diagnostic x-ray systems and their major components.

APPENDIX A:
TABLE OF RADIOACTIVE ISOTOPES

I. INTRODUCTION

Data on the half-life, modes of decay, types of radiation emitted, radiation energies, and intensities of the predominate radiation emissions of radioactive isotopes are given in the following table. The isotopes are listed in order of increasing atomic number. Not all of the isotopes are listed here. The table includes commonly measured radionuclides, fission products, fission radio-activation products, radionuclides used in medical therapy and diagnosis, daughter radionuclides, and specific radionuclides referred to in this book. Decay products of the radionuclides are also provided.

Information available on radionuclide decay modes and radiations is encyclopedic and complex. The most detailed sources of information are obtained from 80 volumes of the journal *Nuclear Data Sheets* published by Academic Press and edited by Martin and Tuli (1997). The data from *Nuclear Data Sheets* are also available online through the National Nuclear Data Center, Brookhaven National Laboratory, Upton, NY 11973. Other detailed and voluminous sources of nuclear data are obtained from Michael Lederer and Shirley (1978), Browne *et al.* (1986), and Firestone *et al.* (1996). Very limited information of practical importance to scientists who need to measure radionuclide activity is provided in the table for ready access. An explanation of the information in this table and examples of practical applications are provided in the following.

Handbook of Radioactivity Analysis, Second Edition
Copyright © 2003 Elsevier Science (USA). All rights reserved.

A. Column I—Nuclide

The radionuclides are listed here in order of increasing atomic number, Z, and under headings by element. When more than one radioisotope is given under the same element, these are listed in order of increasing mass number, A.

B. Column 2—Half-Life

The values of half-life given here are in units of seconds (s), minutes (min), hours (h), days (d), and years (y). The values are those of the total half-life, which would be their decay if measured over a period of time.

C. Column 3—Decay Mode

The modes of decay are designated in this column as follows: α represents decay by alpha-particle emission; β^- signifies decay by negative beta-particle (negatron) emission; β^+ indicates decay by positive beta-particle (positron) emission; EC is electron capture; IT is isomeric transition; $\beta^- n$ is delayed-neutron emission in which β^- emitters decay to unstable nuclides, which undergo instantaneous neutron emission; and SF refers to spontaneous fission. When more than one mode of decay occurs, the percentages of occurrence are given in parentheses beside the decay mode, and the total percentage should sum to 100. No percentage is given when only one mode of decay is cited, as it is understood that this is the only significant decay mode. Some radioisotopes are listed as decaying by $\beta^+ +$ EC without any percentage of occurrence given. In these cases the β^+ decay mode is observed, and the later EC mode is inferred theoretically.

D. Column 4—Radiation Characteristics

The radiations listed are as follows: α is alpha-particle emission, β^- is negative beta-particle or negatron radiation, β^+ is positive beta-particle or positron emission, γ is gamma radiation, $\gamma^\pm$ is annihilation radiation, and n is neutron emission. Internal conversion electron and Auger electron emissions are grouped into the category of atomic electron emissions denoted by the symbol e^-, where the average energies of the electron emissions in MeV are provided in braces {}. Atomic electron emissions are listed only when the average energy and intensity are significant in radionuclide detection and measurement. The intensities of atomic electron emissions are provided in percent enclosed in parentheses beside the average electron energy. X-rays are also given when emitted with significant intensity. Since x-rays are characteristic of the daughter nuclides, these are expressed with the symbol of the daughter nuclide as well as the atomic shell (K or L) of the daughter from which the x-rays originate.

The energies in MeV and intensities (percent) of the major radiations emitted are listed in this column. The energy values (MeV) are given without units and the known percent intensities of the emissions are enclosed in parentheses. The energies of beta particles are those of E_{max}. A percent intensity signifies the percentage of nuclides that would emit a given radiation. For example, the values γ 0.847 (100%) in the table for $^{56}_{27}$Co signifies that ^{56}Co nuclides emit a gamma ray of 0.847 MeV, and one gamma-ray photon of 0.847 MeV energy would be emitted in each ^{56}Co decay. Radiation intensities less than 100% would indicate the number of radionuclides out of 100 that would emit a given radiation type and/or radiation energy. Radiation intensities greater than 100% can also occur. For example, $\gamma^{\pm}$ 0.511 (182%) in the table under radiation characteristics of ^{22}Na signifies that annihilation radiation is produced (the energy of which is invariably 0.511 MeV), and the intensity of this emission is 182%. Thus, for every 100 nuclides of ^{22}Na that decay, there is an expectancy of 182 photons of gamma rays as annihilation radiation. It should be noted here that the percent intensity (abundance) of annihilation radiation throughout this table is expressed as twice the value of the percent intensity of β^{+} radiation, because each positron annihilation is accompanied by the emission of two gamma-ray photons of 0.511 MeV. Note that in the table alongside ^{22}Na, the percent intensity of β^{+} radiation is 90%; that is, for every 100 nuclides of ^{22}Na that decay, 90 are expected to decay by positron emission.

When the absolute intensities of radiation emissions are of low magnitude, the relative intensities may be given in parentheses with the character † to denote a radiation intensity relative to other values listed in the same column. For example, $^{100}_{45}$Rh decay is described in the table as undergoing two decay modes, namely, EC (95.1%) and β^{+} (4.9%). Statistically, only 4.9 of 100 atoms of $^{100}_{45}$Rh decay by β^{+} emission. The intensities of the β^{+} emissions are low and, in this example, the energy maxima of the emissions are followed by the values of their relative intensities in parentheses.

E. Column 5—Decay Product

The daughter nuclide is listed in this column. More than one daughter is possible when more than one decay mode can occur. Under such circumstances the decay modes are listed with the corresponding symbol or abbreviation in parentheses alongside or under the decay product.

F. Some Applications of Radiation Type, Energy, and Intensity Data

A description of the radiation types and their energies and intensities may be used to judge which instrumental detection method could be employed to obtain an optimum detection efficiency for a particular radionuclide. As an example, from the table the nuclide $^{56}_{27}$Co may be cited. Both β^{+} (two decay transitions with $E_{max} = 1.46$ and 0.443 MeV) and γ emissions (several) are listed. However, the relatively low intensity of the β^{+} emissions (19%) indicates that it may be less appropriate to assay for ^{56}Co by gas ionization or liquid scintillation counting of the positron radiation. Rather, solid scintillation counting of the more intense gamma rays and high-energy positrons may be preferable to obtain the highest counting efficiency possible. In contrast, another example can be taken from the table whereby we can conclude from the information provided that the liquid scintillation counting efficiency of $^{129}_{53}$I should be appreciably greater than the solid scintillation counting efficiency due to the greater intensity of beta radiation (100%) and atomic electrons (71%) than gamma rays (9%) emitted during ^{129}I decay.

The beta-particle energies and intensities may also be used to judge whether Cherenkov counting could serve as an appropriate means of radionuclide assay. For example, the table indicates that two beta-decay transitions can occur in the decay of $^{115m}_{48}$Cd. The principal transition occurs with the emission of beta particles of $E_{max} = 1.62$ MeV at a 97% intensity. The other transition ($E_{max} = 0.68$ MeV) is less significant because it has a lower frequency of occurrence (3% intensity). In view of the threshold energy of 0.263 MeV for the production of Cherenkov photons by beta particles in water (see Chapter 9), and taking into account that beta radiation is emitted with a continuous spectrum of energies between zero and E_{max}, it can be estimated that an appreciable number of the beta particles emitted from ^{115m}Cd possess an energy greater than 0.263 MeV. Cherenkov counting of ^{115m}Cd should then be an appropriate technique for the assay of this nuclide, keeping in mind that liquid scintillation counting of the beta particles emitted from ^{115m}Cd would be a more efficient although more expensive counting technique.

In contrast to the preceding case, the energies and intensities of the beta particles emitted from $^{125}_{51}$Sb, listed in the table, serve as an example of radiation data which indicate that only a relatively small percentage of the beta particles emitted by this nuclide have energies greater than 0.263 MeV. Consequently, Cherenkov counting would not be an efficient method for the assay of ^{125}Sb; instead, liquid scintillation counting should be employed.

The nuclide $^{95}_{41}$Nb is another illustrative example that may be taken from the table. In this case, both beta particles and gamma rays are emitted with equal intensity, indicating that detection methods suitable for either the beta particles or gamma rays may be appropriate. However, liquid scintillation detection of the beta radiation would provide higher counting efficiencies and lower backgrounds than solid scintillation detection of the gamma radiation.

In the electron capture (EC) decay process, also known as K-capture, no particle emission results. However, this mode of decay is often accompanied by the emission of gamma radiation, and the transitions produced in the electron orbital energy levels result in the emission of energy as x-rays. In such cases (e.g., $^{49}_{23}$V, $^{53}_{25}$Mn, and $^{71}_{32}$Ge in the table) solid scintillation counting of the x-radiation may be the chosen assay method, although liquid scintillation analysis of atomic electron and x-ray-emitting nuclides is also possible (see Chapter 5).

Table of Radioactive Isotopes

Nuclide $_Z^A X$	Half-life	Decay mode	Radiation characteristics: Energies (MeV) [intensities (%)]		Decay product $_Z^A Y$
Hydrogen					
$_1^3 H$	12.35 y	β^-	β^-	0.0186 (100%)	$_2^3 He$
Beryllium					
$_4^7 Be$	53.3 d	EC	γ	0.478 (10.4%)	$_3^7 Li$
$_4^{10} Be$	1.6×10^6 y	β^-	β^-	0.555 (100%)	$_5^{10} B$
Carbon					
$_6^{11} C$	20.38 min	β^+ (>99%)	β^+	0.961 (>99%)	$_5^{11} B$
		EC (0.2%)	$\gamma^\pm$	0.511 (200%)	
$_6^{14} C$	5730 y	β^-	β^-	0.155 (100%)	$_7^{14} N$
Nitrogen					
$_7^{13} N$	9.96 min	β^+	β^+	1.190 (100%)	$_6^{13} C$
			$\gamma^\pm$	0.511 (200%)	
$_7^{17} N$	4.14 s	$\beta^- n$ (95.1%)	β^-	3.7 (100%)	$_8^{17} O$ (β^-)
		β^- (4.9%)	n	0.383 (34.8%), 0.884 (0.6%), 1.17 (52.7%), 1.70 (7%)	$_8^{16} O$ ($\beta^- n$)
			γ	0.871 (3%), 2.18 (0.3%)	
Oxygen					
$_8^{15} O$	2.03 min	β^+ (>99%)	β^+	1.723 (100%)	$_7^{15} N$
		EC (0.1%)	$\gamma^\pm$	0.511 (200%)	
Fluorine					
$_9^{18} F$	109.7 min	β^+ (97%)	β^+	0.635 (97%)	$_8^{18} O$
		EC (3%)			
Sodium					
$_{11}^{22} Na$	2.60 y	β^+ (90%)	β^+	1.830 (0.06%), 0.540 (90%)	$_{10}^{22} Ne$
		EC (10%)	γ	1.275 (100%)	
			$\gamma^\pm$	0.511 (180%)	
$_{11}^{24} Na$	15.0 h	β^-	β^-	1.390 (100%)	$_{12}^{24} Mg$
			γ	1.37 (100%), 2.75 (100%), 3.87 (0.06%)	
Magnesium					
$_{12}^{28} Mg$	21 h	β^-	β^-	0.459 (100%)	$_{13}^{28} Al$

(continues)

Table of Radioactive Isotopes—Continued

Nuclide $^A_Z X$	Half-life	Decay mode	Radiation characteristics: Energies (MeV) [intensities (%)]	Decay product $^A_Z Y$
Aluminum				
$^{26}_{13}$Al	7.2×10^5 y	β^+ (82%) EC (18%)	γ 1.350 (60%), 0.947 (30%), 0.400 (30%), 0.31 (96%); β^+ 1.160 (82%); γ 1.809 (100%), 1.130 (4%); $\gamma^{\pm}$ 0.511 (164%)	$^{26}_{12}$Mg
$^{28}_{13}$Al	2.24 min	β^-	β^- 2.865 (100%); γ 1.780 (100%)	$^{28}_{14}$Si
Silicon				
$^{31}_{14}$Si	2.62 h	β^-	β^- 1.486 (100%); γ 1.266 (0.07%)	$^{31}_{15}$P
Phosphorus				
$^{32}_{15}$P	14.28 d	β^-	β^- 1.710 (100%)	$^{32}_{16}$S
$^{33}_{15}$P	25.3 d	β^-	β^- 0.249 (100%)	$^{33}_{16}$S
Sulfur				
$^{35}_{16}$S	87.4 d	β^-	β^- 0.167 (100%)	$^{35}_{17}$Cl
Chlorine				
$^{36}_{17}$Cl	3.0×10^5 y	β^- (98.1%) EC (1.9%)	β^- 0.714 (98.1%); S K x-rays ~0.002 (0.1%)	$^{36}_{18}$Ar (β^-) $^{36}_{16}$S (EC)
$^{38}_{17}$Cl	37.3 min	β^-	β^- 4.913 (57.6%), 2.77 (11.1%), 1.11 (31.3%); γ 1.64 (31%), 2.17 (42%)	$^{38}_{18}$Ar
Argon				
$^{37}_{18}$Ar	35.1 d	EC	Cl K x-rays ~0.002 (8.7%)	$^{37}_{17}$Cl
Potassium				
$^{40}_{19}$K	1.26×10^9 y	β^- (89%) EC (11%)	β^- 1.325 (89%); γ 1.460 (11%); Ar K x-rays ~0.003 (9%)	$^{40}_{20}$Ca (β^-) $^{40}_{18}$Ar (EC)
$^{42}_{19}$K	12.36 h	β^-	β^- 3.56 (81.3%), 1.97 (18.4%); γ 1.525 (18%), 0.312 (0.2%)	$^{42}_{20}$Ca
$^{43}_{19}$K	22.2 h	β^-	β^- 1.81 (1.3%), 1.24 (3.5%), 0.825 (87%), 0.465 (8%); γ 1.01 (2%), 0.619 (81%), 0.594 (13%), 0.373 (85%), 0.220 (3%), 0.197 (18%)	$^{43}_{20}$Ca

Element / Isotope	Half-life	Decay mode	Radiation (MeV) and intensity	Product
Calcium				
$^{41}_{20}$Ca	1.1×10^5 y	EC	K K x-rays $\sim$0.003 (12.5%), K L x-rays (0.02%)	$^{41}_{19}$K
$^{45}_{20}$Ca	165.1 d	β^-	β^- 0.258 (100%)	$^{45}_{21}$Sc
$^{47}_{20}$Ca	4.54 d	β^-	β^- 1.98 (16.1%), 0.684 (83.9%); γ 1.30 (77%), 0.808 (5%), 0.489 (5%)	$^{47}_{21}$Sc
Scandium				
$^{43}_{21}$Sc	3.88 h	β^+ + EC	β^+ 1.20 (70†), 0.82 (15†), 0.39 (3†); γ 0.375 (22%); $\gamma^{\pm}$ 0.511 (176%); Ca K x-rays $\sim$0.004 (2%)	$^{43}_{20}$Ca
$^{44m}_{21}$Sc	2.44 d	IT (98.6%) EC (1.4%)	γ 1.14 (2.7%), 1.02 (1.3%), 0.271 (86%); e$^-$ [0.032] (22%); Sc K x-rays (2%)	$^{44}_{21}$Sc (IT) $^{44}_{20}$Ca (EC)
$^{44}_{21}$Sc	3.92 h	β^+ (95%) EC (5%)	β^+ 1.47 (95%); γ 1.159 (100%); $\gamma^{\pm}$ 0.511 (190%)	$^{44}_{20}$Ca
$^{46}_{21}$Sc	83.8 d	β^-	β^- 0.357 (100%); γ 1.120 (100%), 0.889 (100%)	$^{46}_{22}$Ti
$^{47}_{21}$Sc	3.42 d	β^-	β^- 0.610 (26%), 0.450 (74%); γ 0.159 (73%)	$^{47}_{22}$Ti
Titanium				
$^{44}_{22}$Ti	48.2 y	EC	γ 0.078 (98%), 0.068 (90%); e$^-$ [0.011] (100%); Sc K x-rays $\sim$0.004 (19%)	$^{44}_{21}$Sc
Vanadium				
$^{48}_{23}$V	16.0 d	β^+ (49%) EC (51%)	β^+ 0.698 (49%); γ 2.24 (3%), 1.31 (97%), 0.983 (100%), 0.945 (10%); e$^-$ [0.004] (78%)	$^{48}_{22}$Ti
$^{49}_{23}$V	327 d	EC	Ti K x-rays $\sim$0.004 (19%)	$^{49}_{22}$Ti
Chromium				
$^{51}_{24}$Cr	27.7 d	EC	γ 0.320 (10%); e$^-$ [0.003] (67%); V K x-rays $\sim$0.005 (22%)	$^{51}_{23}$V

(continues)

Table of Radioactive Isotopes—Continued

Nuclide ^A_ZX	Half-life	Decay mode	Radiation characteristics: Energies (MeV) [intensities (%)]	Decay product ^A_ZY
Manganese				
$^{52}_{25}$Mn	5.60 d	EC (72%) β^+ (28%)	β^+ 0.575 (28%); γ 1.43 (100%), 0.935 (84%), 0.774 (82%); $\gamma^\pm$ 0.511 (56%); Cr K x-rays ~0.005 (17%)	$^{52}_{24}$Cr
$^{52m}_{25}$Mn	21.1 min	β^+ + EC IT (2%)	β^+ 2.63; γ 1.43 (100%), 0.377 (2%)	$^{52}_{24}$Cr
$^{53}_{25}$Mn	3.7×10^6 y	EC	e^- [0.003] (65%); Cr K x-rays ~0.005 (25%)	$^{53}_{24}$Cr
$^{54}_{25}$Mn	312.2 d	EC	γ 0.835 (100%); e^- [0.003] (64%); Cr K x-rays ~0.005 (24%)	$^{54}_{24}$Cr
$^{56}_{25}$Mn	2.57 h	β^-	β^- 2.83 (47%), 1.03 (34%), 0.718 (18%), 0.30 (~1%); γ 2.11 (15%), 1.81 (29%), 0.847 (99%)	$^{56}_{26}$Fe
Iron				
$^{52}_{26}$Fe	8.27 h	β^+ (56%) EC (44%)	β^+ 0.804 (56%); γ 0.169 (100%), 0.378 (2%); $\gamma^\pm$ 0.511 (112%); Mn K x-rays ~0.006 (13%)	$^{52m}_{25}$Mn
$^{55}_{26}$Fe	2.68 y	EC	e^- [0.004] (102%); Mn K x-rays ~0.006 (27%)	$^{55}_{25}$Mn
$^{59}_{26}$Fe	44.5 d	β^-	β^- 1.57 (0.30%), 0.475 (51.2%), 0.273 (48.5%); γ 1.29 (43%), 1.09 (57%), 0.192 (3%), 0.143 (1%)	$^{59}_{27}$Co
Cobalt				
$^{56}_{27}$Co	78.7 d	EC (81%) β^+ (19%)	β^+ 1.46 (18%), 0.443 (1%); γ 3.26 (13%), 2.60 (17%), 2.02 (11%), 1.76 (15%), 1.24 (66%), 1.04 (15%), 0.847 (100%); $\gamma^\pm$ 0.511 (38%); e^- [0.004] (111%); Fe K x-rays ~0.006 (24%)	$^{56}_{26}$Fe
$^{57}_{27}$Co	271.6 d	EC	γ 0.136 (11%), 0.122 (87%), 0.014 (9%); Fe K x-rays ~0.006 (55%)	$^{57}_{26}$Fe

Nuclide	Half-life	Decay mode	Radiations (MeV)	Product
$^{58}_{27}$Co	70.78 d	EC (85%) β^+ (15%)	β^+ 0.474 (15%) γ 1.67 (0.6%), 0.865 (1.4%), 0.810 (99%) $\gamma^\pm$ 0.511 (30%) e^- [0.004] (118%) Fe K x-rays ~0.006 (26%)	$^{58}_{26}$Fe
$^{60}_{27}$Co	5.27 y	β^-	β^- 1.49 (0.1%), 0.670 (0.2%), 0.315 (99.7%) γ 1.33 (100%), 1.17 (100%)	$^{60}_{28}$Ni
Nickel				
$^{56}_{28}$Ni	6.1 d	EC	γ 0.158 (100%), 0.269 (34%), 0.480 (32%), 0.750 (48%), 0.812 (75%), 1.56 (13%) e^- [0.007] (140%)	$^{56}_{27}$Co
$^{59}_{28}$Ni	8×10^4 y	EC (>99%) β^+ (trace)	e^- [0.004] (136%) Co K x-rays ~0.004 (33%)	$^{59}_{27}$Co
$^{63}_{28}$Ni	100.1 y	β^-	β^- 0.066 (100%)	$^{63}_{29}$Cu
$^{65}_{28}$Ni	2.52 h	β^-	β^- 2.14 (59%), 1.02 (11%), 0.650 (30%) γ 1.49 (25%), 1.12 (16%), 0.368 (4.5%)	$^{65}_{29}$Cu
Copper				
$^{61}_{29}$Cu	3.41 h	β^+ (62%) EC (38%)	β^+ 1.22 (52%), 1.15 (2%), 0.94 (5%), 0.56 (3%) γ 1.19 (5%), 0.657 (11%), 0.373 (3%), 0.284 (12%), 0.067 (4%) $\gamma^\pm$ 0.511 (124%) e^- [0.002] (53%) Ni K x-rays ~0.007 (14%)	$^{61}_{28}$Ni
$^{64}_{29}$Cu	12.70 h	EC (41%) β^- (40%) β^+ (19%)	β^- 0.573 (40%) β^+ 0.657 (19%) γ 1.34 (0.6%) $\gamma^\pm$ 0.511 (38%) e^- [0.002] (60%) Ni K x-rays ~0.007 (16%)	$^{64}_{28}$Ni (EC, β^+) $^{64}_{30}$Zn (β^-)
$^{67}_{29}$Cu	62.0 h	β^-	β^- 0.577 (20%), 0.484 (35%), 0.395 (45%) γ 0.185 (47%), 0.092 (23%) Zn K x-rays ~0.008 (6%)	$^{67}_{30}$Zn

(continues)

Table of Radioactive Isotopes—Continued

Nuclide $^A_Z X$	Half-life	Decay mode	Radiation characteristics: Energies (MeV) [intensities (%)]	Decay product $^A_Z Y$
Zinc				
$^{65}_{30}$Zn	244.0 d	EC (98.5%) β^+ (1.5%)	β^+ 0.325 (1.5%) γ 1.115 (50%) $\gamma^\pm$ 0.511 (3.0%) e^- [0.004] (98%) Cu K x-rays ~0.008 (38%)	$^{65}_{29}$Cu
$^{69m}_{30}$Zn	13.76 h	IT (>99%) β^- (0.03%)	γ 0.439 (95%) Zn K x-rays ~0.008 (2%)	$^{69}_{30}$Zn
$^{69}_{30}$Zn	55.6 min	β^-	β^- 0.897 (100%)	$^{69}_{31}$Ga
Gallium				
$^{66}_{31}$Ga	9.45 h	β^+ (57%) EC (43%)	β^+ 4.15 (51%), 1.84 (1%), 0.935 (3%), 0.747 (1%), 0.367 (1%) γ 4.30 (5%), 2.75 (25%), 2.18 (5%), 1.04 (37%), 0.828 (5%) $\gamma^\pm$ 0.511 (114%) e^- [0.002] (55%) Zn K x-rays ~0.008 (18%)	$^{66}_{30}$Zn
$^{67}_{31}$Ga	78.26 h	EC	γ 0.388 (7%), 0.296 (22%), 0.184 (24%), 0.093 (40%) e^- [0.033] (200%) Zn K x-rays ~0.008 (55%)	$^{67}_{30}$Zn
$^{68}_{31}$Ga	68.33 min	β^+ (90%) EC (10%)	β^+ 1.90 (90%) γ 1.87 (0.15%), 1.24 (0.14%), 1.078 (3.5%), 0.80 (0.4%) $\gamma^\pm$ 0.511 (180%) Zn K x-rays ~0.008 (5%)	$^{68}_{30}$Zn
$^{72}_{31}$Ga	14.12 h	β^-	β^- 3.16 (8%), 2.53 (9%), 1.51 (10%), 0.959 (31%), 0.637 (42%) γ 2.50 (20%), 2.20 (26%), 1.86 (5%), 1.60 (5%), 1.46 (3.5%), 1.05 (7%), 0.894 (10%), 0.835 (96%), 0.630 (27%), 0.601 (8%)	$^{72}_{32}$Ge
Germanium				
$^{68}_{32}$Ge	275 d	EC	e^- [0.004] (121%) Ga K x-rays ~0.009 (44%)	$^{68}_{31}$Ga
$^{71}_{32}$Ge	11.15 d	EC	e^- [0.005] (122%) Ga K x-rays ~0.009 (44%)	$^{71}_{31}$Ga

Nuclide	Half-life	Decay mode		Radiation: energy in MeV (intensity)	Daughter
$^{77}_{32}$Ge	11.30 h	β^-	β^-	2.20 (42%), 1.38 (35%), 0.710 (23%)	$^{77}_{33}$As
			γ	1.09 (6%), 0.93 (5%), 0.80 (6%), 0.73 (14%), 0.632 (11%), 0.553 (18%), 0.417 (25%), 0.368 (15%), 0.263 (45%), 0.21 (61%)	
			e^-	[0.008] (8%)	
Arsenic					
$^{74}_{33}$As	17.79 d	EC (37%)	β^+	1.53 (4%), 0.941 (27%)	$^{74}_{32}$Ge (EC, β^+)
		β^+ (31%)	β^-	1.35 (17%), 0.717 (15%)	$^{74}_{34}$Se(β^-)
		β^- (32%)	γ	0.635 (14%), 0.596 (61%)	
			$\gamma^{\pm}$	0.511 (62%)	
				Ge K x-rays ~0.010 (17%)	
$^{76}_{33}$As	26.32 h	β^-	β^-	2.97 (50%), 2.41 (31%), 1.78 (7%), 1.18 (3%), 0.540 (3%), 0.320 (3%)	$^{76}_{34}$Se
			γ	1.22 (5%), 0.657 (6%), 0.559 (43%)	
$^{77}_{33}$As	38.8 h	β^-	β^-	0.679 (100%)	$^{77m}_{34}$Se
			γ	0.522 (0.8%), 0.239 (2.5%), 0.086 (0.1%)	
Selenium					
$^{75}_{34}$Se	119 d	EC	γ	0.401 (12%), 0.280 (25%), 0.265 (60%), 0.136 (57%), 0.121 (17%), 0.097 (3.3%), 0.066 (1%)	$^{75}_{33}$As
			e^-	[0.014] (145%)	
				As K x-rays ~0.011 (55%)	
$^{77m}_{34}$Se	17.4 s	IT	γ	0.162 (33%)	$^{77}_{34}$Se
			e^-	[0.072] (99%)	
$^{79}_{34}$Se	6.5×10^4 y	β^-	β^-	0.160 (100%)	$^{79}_{35}$Br
Bromine					
$^{77}_{35}$Br	57.0 h	EC (99.3%)	β^+	0.336 (0.7%)	$^{77m}_{34}$Se
		β^+ (0.7%)	γ	1.00 (1.3%), 0.818 (3%), 0.775 (2%), 0.58 (7%), 0.520 (24%), 0.300 (6%), 0.239 (30%)	
			$\gamma^{\pm}$	0.511 (1.4%)	
			e^-	[0.008] (116%)	
				Se K x-rays ~0.012 (52%)	
$^{82}_{35}$Br	35.3 h	β^-	β^-	0.444 (100%)	$^{82}_{36}$Kr
			γ	1.47 (17%), 1.32 (26%), 1.04 (29%), 0.828 (25%), 0.777 (83%), 0.698 (27%), 0.619 (41%), 0.554 (66%)	

(continues)

Table of Radioactive Isotopes—Continued

Nuclide ^A_ZX	Half-life	Decay mode	Radiation characteristics: Energies (MeV) [intensities (%)]	Decay product ^A_ZY
Krypton				
$^{79}_{36}$Kr	35.0 h	EC (93%) β^+ (7%)	β^+ 0.613 (7%) γ 0.836 (2%), 0.606 (10%), 0.398 (10%), 0.261 (12%) $\gamma^{\pm}$ 0.511 (14%) Br K x-rays ~0.012 (51%)	$^{79}_{35}$Br
$^{83m}_{36}$Kr	1.83 h	IT	γ 0.032 (0.05%), 0.009 (5%) Kr K x-rays ~0.013 (16%)	$^{83}_{36}$Kr
$^{85m}_{36}$Kr	4.48 h	β^- (79%) IT (21%)	β^- 0.840 (79%) γ 0.305 (13%), 0.150 (74%) e^- [0.026] (22%)	$^{85}_{36}$Kr (IT) $^{85}_{37}$Rb (β^-)
$^{85}_{36}$Kr	10.70 y	β^-	β^- 0.672 (99.4%), 0.15 (0.6%), Rb K x-rays ~0.013 (2%), Kr K x-rays ~0.013 (4%) γ 0.514 (0.44%)	$^{85}_{37}$Rb
Rubidium				
$^{83}_{37}$Rb	86.3 d	EC	γ 0.79 (0.9%), 0.53 (93%), 0.009 (6%) e^- [0.037] (379%) Kr K x-rays ~0.013 (60%)	$^{83m}_{36}$Kr
$^{84}_{37}$Rb	32.77 d	EC (75%) β^+ (22%) β^- (3%)	β^+ 1.66 (11%), 0.781 (11%) β^- 0.892 (3%) γ 1.90 (0.8%), 1.01 (0.5%), 0.883 (74%) $\gamma^{\pm}$ 0.511 (44%) e^- [0.004] (74%) Kr K x-rays ~0.013 (39%)	$^{84}_{36}$Kr (EC, β^+) $^{84}_{38}$Sr (β^-)
$^{86}_{37}$Rb	18.8 d	β^-	β^- 1.77 (88%), 0.680 (12%) γ 1.08 (9%)	$^{86}_{38}$Sr
Strontium				
$^{85}_{38}$Sr	64.85 d	EC	γ 0.514 (100%) e^- [0.008] (104%) Rb K x-rays ~0.014 (58%)	$^{85}_{37}$Rb
$^{87m}_{38}$Sr	2.81 h	IT (99.7%) EC (0.3%)	γ 0.388 (80%) Sr K x-rays ~0.014 (10%)	$^{87}_{38}$Sr (IT)
$^{89}_{38}$Sr	50.5 d	β^-	β^- 1.49 (100%)	$^{89m}_{39}$Y
$^{90}_{38}$Sr	28.8 y	β^-	β^- 0.546 (100%)	$^{90}_{39}$Y

Element / Nuclide	Half-life	Decay mode	Radiation	Energy MeV (intensity)	Daughter
Yttrium					
$^{87}_{39}$Y	80.3 h	EC (99.8%) β^+ (0.2%)	β^+ γ e^-	0.451 (0.2%) 0.485 (92%), 0.388 (85%) [0.080] (137%) Sr K x-rays ~0.015 (71%)	$^{87m}_{38}$Sr
$^{88}_{39}$Y	106.6 d	EC (99.8%) β^+ (0.2%)	β^+ γ e^-	0.761 (0.2%) 1.84 (100%), 0.898 (91%) [0.005] (100%) Sr K x-rays ~0.015 (60%)	$^{88}_{38}$Sr
$^{89m}_{39}$Y	16.0 s	IT	γ e^-	0.909 (99%) [0.008] (2%)	$^{89}_{39}$Y
$^{90}_{39}$Y	64.06 h	β^-	β^-	2.28 (100%)	$^{90}_{40}$Zr
$^{91}_{39}$Y	58.51 d	β^-	β^- γ	1.54 (100%) 1.21 (0.3%)	$^{91}_{40}$Zr
Zirconium					
$^{93}_{40}$Zr	1.5×10^6 y	β^-	β^- e^-	0.060 (~95%), 0.034 (~5%) [0.028] (170%) Nb K x-rays ~0.017 (10%)	$^{93}_{41}$Nb
$^{95}_{40}$Zr	64.0 d	β^-	β^- γ	0.885 (2%), 0.396 (55%), 0.360 (43%) 0.756 (49%), 0.724 (49%)	$^{95m}_{41}$Nb
$^{97}_{40}$Zr	16.90 h	β^-	β^- e^-	1.91 (~90%), 0.46 (~10%) [0.015] (4%)	$^{97m}_{41}$Nb
Niobium					
$^{93m}_{41}$Nb	13.7 y	IT	e^-	[0.028] (170%) Nb K x-rays ~0.017 (10.5%)	$^{93}_{41}$Nb
$^{94}_{41}$Nb	2×10^4 y	β^-	β^- γ	0.473 (100%) 0.702 (98%), 0.871 (100%)	$^{94}_{42}$Mo
$^{95}_{41}$Nb	34.97 d	β^-	β^- γ	0.924 (0.1%), 0.160 (99.9%) 0.766 (100%)	$^{95}_{42}$Mo
$^{95m}_{41}$Nb	86.6 h	IT (97.5%) β^- (2.5%)	β^- γ e^-	1.16 (2.5%) 0.234 (25%), 0.204 (2.4%) [0.161] (138%) Nb K x-rays ~0.017 (43%)	$^{95}_{41}$Nb (IT) $^{95}_{42}$Mo (β^-)
$^{97m}_{41}$Nb	1.0 min	IT	γ e^-	0.743 (98%) [0.016] (4%)	$^{97}_{41}$Nb
$^{97}_{41}$Nb	72.1 min	β^-	β^- γ	1.27 (100%) 1.02 (1%), 0.658 (98%)	$^{97}_{42}$Mo

(continues)

Table of Radioactive Isotopes—Continued

Nuclide $^A_Z X$	Half-life	Decay mode	Radiation characteristics: Energies (MeV) [intensities (%)]		Decay product $^A_Z Y$
Molybdenum					
$^{93}_{42}$Mo	3500 y	EC	e^-	{0.032} (254%) Nb K x-rays ~ 0.017 (73%), Nb L x-rays ~ 0.002 (6%)	$^{93}_{41}$Nb
$^{99}_{42}$Mo	66.02 h	β^-	β^-	1.21 (84%), 0.840 (2%), 0.450 (14%)	$^{99m}_{43}$Tc
			γ	0.780 (4%), 0.740 (12%), 0.181 (7%), 0.041 (2%)	
			e^-	{0.018} (32%) Tc K x-rays ~ 0.019 (11%)	
Technetium					
$^{95m}_{43}$Tc	61 d	EC (95.8%)	γ	0.204 (66%), 0.582 (31%), 0.786 (9%), 0.835 (28%)	$^{95}_{42}$Mo (EC)
		IT (3.9%)	β^+	0.710 (0.3%)	$^{95}_{43}$Tc (IT)
		β^+ (0.3%)	e^-	0.0139 MeV (102%) Mo K x-rays ~ 0.018 (66%)	
$^{97}_{43}$Tc	2.6×10^6 y	EC	e^-	{0.005} (91%) Mo K x-rays ~ 0.018 (66%)	$^{97}_{42}$Mo
$^{99m}_{43}$Tc	6.00 h	IT	γ	0.140 (90%)	$^{99}_{43}$Tc
			e^-	{0.014} (21%) Tc K x-rays ~ 0.019 (7%)	
$^{99}_{43}$Tc	2.14×10^5 y	β^-	β^-	0.292 (100%)	$^{99}_{44}$Ru
Ruthenium					
$^{97}_{44}$Ru	2.88 d	EC	γ	0.324 (8%), 0.215 (91%)	$^{97}_{43}$Tc
			e^-	{0.013} (97%) Tc K x-rays ~ 0.019 (70%)	
$^{103}_{44}$Ru	39.35 d	β^-	β^-	0.725 (3.5%), 0.225 (91%), 0.117 (5.3%)	$^{103m}_{45}$Rh
			γ	0.610 (6%), 0.497 (88%)	
			e^-	{0.041} (181%) Rh K x-rays ~ 0.021 (6%)	
$^{106}_{44}$Ru	366.5 d	β^-	β^-	0.039 (100%)	$^{106}_{45}$Rh
Rhodium					
$^{100}_{45}$Rh	21 h	EC (95.1%)	β^+	2.61 (45†), 2.07 (39†), 1.26 (13†), 0.540 (3.5†), 0.150 (0.06†)	$^{100}_{44}$Ru
		β^+ (4.9%)	γ	0.446 (11%), 0.539 (78%), 0.822 (20%), 1.107 (13%), 1.362 (15%), 1.553 (21%), 1.929 (12%), 2.375 (35%)	
			e^-	{0.0077} (84%) Ru K x-rays ~ 0.019 (66%)	

Nuclide	Half-life	Decay mode	Radiation	Energy MeV (intensity)	Daughter
$^{102}_{45}$Rh	2.9 y	EC	γ	1.11 (22%), 1.05 (41%), 0.768 (30%), 0.698 (41%), 0.632 (54%), 0.475 (95%), 0.418 (13%), [0.012] (89%)	$^{102}_{44}$Ru
			e⁻	Ru K x-rays ~0.020 (67%)	
$^{103m}_{45}$Rh	56.12 min	IT	γ	0.040 (0.1%)	$^{103}_{45}$Rh
			e⁻	[0.038] (173%) Rh K x-rays ~0.021 (7%)	
$^{105}_{45}$Rh	35.37 h	β⁻	β⁻	0.560 (70%), 0.247 (30%)	$^{105}_{46}$Pd
			γ	0.319 (19%), 0.306 (5%) Pd K x-rays ~0.022 (0.4%)	
$^{106}_{45}$Rh	29.80 s	β⁻	β⁻	3.53 (68%), 3.1 (11%), 2.44 (12%), 2.0 (3%)	$^{106}_{46}$Pd
			γ	1.13 (0.5%), 1.05 (1.5%), 0.622 (11%), 0.512 (21%)	
Palladium					
$^{100}_{46}$Pd	3.6 d	EC	γ	0.074 (98%), 0.084 (100%), 0.126 (11%)	$^{100}_{45}$Rh
$^{103}_{46}$Pd	16.96 d	EC	γ	0.498 (0.011%), 0.362 (0.02%), 0.297 (0.011%)	$^{103m}_{45}$Rh
			e⁻	[0.043] (258%) Rh K x-rays ~0.021 (77%)	
$^{107}_{46}$Pd	7×10^6 y	β⁻	β⁻	0.040 (100%)	$^{107}_{47}$Ag
$^{109}_{46}$Pd	13.43 h	β⁻	β⁻	1.03 (100%)	$^{109m}_{47}$Ag
			γ	0.088 (3.6%) Ag K x-rays ~0.022 (34%)	
Silver					
$^{105}_{47}$Ag	41.29 d	EC	γ	1.09 (2%), 0.618–0.681 complex (12%), 0.443 (10%), 0.344 (42%), 0.280 (32%), 0.064 (10%)	$^{105}_{46}$Pd
			e⁻	[0.019] (117%) Pd K x-rays ~0.022 (78%)	
$^{109m}_{47}$Ag	40 s	IT	γ	0.088 (3.6%)	$^{109}_{47}$Ag
			e⁻	[0.077] (171%) Ag K x-rays ~0.023 (34%)	
$^{110m}_{47}$Ag	252 d	β⁻ (98.5%) IT (1.5%)	β⁻	1.5 (0.6%), 0.529 (36%), 0.087 (61%)	$^{110}_{47}$Ag (IT) $^{110}_{48}$Cd (β⁻)
			γ	1.51 (11%), 1.38 (21%), 0.937 (32%), 0.885 (71%), 0.818 (8%), 0.764 (23%), 0.706 (19%), 0.68 (16%), 0.658 (96%)	
$^{110}_{47}$Ag	24.42 s	β⁻ (99.7%) EC (0.3%)	β⁻	2.89 (93.5%), 2.22 (6.5%)	$^{110}_{48}$Cd
			γ	0.658 (4.5%)	
$^{111}_{47}$Ag	7.45 d	β⁻	β⁻	1.04 (92%), 0.790 (1.1%), 0.695 (6.0%), 0.425 (0.9%)	$^{111}_{48}$Cd
			γ	0.342 (6%), 0.247 (1%)	

(continues)

Table of Radioactive Isotopes—Continued

Nuclide ^A_ZX	Half-life	Decay mode	Radiation characteristics: Energies (MeV) [intensities (%)]		Decay product ^A_ZY
Cadmium					
$^{109}_{48}$Cd	453 d	EC	γ	0.088 (3.8%)	$^{109m}_{47}$Ag
			e⁻	[0.081] (250%)	
				Ag K x-rays ~0.023 (82%)	
$^{113m}_{48}$Cd	14 y	β⁻ (99.9%)	β⁻	0.580 (99.9%)	$^{113}_{49}$In
		IT (0.1%)	γ	0.264 (0.023%)	
$^{115m}_{48}$Cd	44.8 d	β⁻	β⁻	1.62 (97%), 0.68 (3%)	$^{115}_{49}$In
			γ	1.29 (0.9%), 0.935 (1.9%), 0.485 (0.31%)	
$^{115}_{48}$Cd	53.38 h	β⁻	β⁻	1.11 (58%), 0.58 (42%)	$^{115m}_{49}$In
			γ	0.53 (26%), 0.49 (10%), 0.262 (2%), 0.230 (0.6%)	
			e⁻	[0.177] (107%)	
				In K x-rays ~0.025 (41%)	
Indium					
$^{111}_{49}$In	2.83 d	EC	γ	0.247 (94%), 0.173 (89%)	$^{111}_{48}$Cd
			e⁻	[0.034] (103%)	
				Cd K x-rays ~0.025 (83%)	
$^{113m}_{49}$In	99.47 min	IT	γ	0.393 (64%)	$^{113}_{49}$In
				In K x-rays ~0.026 (16%)	
$^{114m}_{49}$In	49.51 d	IT (96.7%)	γ	0.727 (3.5%), 0.558 (3.5%), 0.192 (17%)	$^{114}_{49}$In (IT)
		EC (3.3%)		In K x-rays ~0.025 (34%), Cd K x-rays ~0.023 (3%)	$^{114}_{48}$Cd (EC)
$^{114}_{49}$In	71.9 s	β⁻ (98%)	β⁻	1.98 (98%)	$^{114}_{50}$Sn (β⁻)
		EC (2%)	γ	1.29 (0.2%)	$^{114}_{48}$Cd (EC)
$^{115m}_{49}$In	4.486 h	IT (95%)	β⁻	0.83 (5%)	$^{115}_{49}$In (IT)
		β⁻ (5%)	γ	0.336 (46%)	$^{115}_{50}$Sn (β⁻)
				In K x-rays ~0.026 (34%)	
$^{115}_{49}$In	5 × 10¹⁴ y	β⁻	β⁻	0.495 (100%)	$^{115}_{50}$Sn
Tin					
$^{113}_{50}$Sn	115.1 d	EC	γ	0.255 (1.8%)	$^{113m}_{49}$In
				In K x-rays ~0.026 (98%)	
$^{119m}_{50}$Sn	~250 d	IT	γ	0.024 (16%)	$^{119}_{50}$Sn
				Sn K x-rays ~0.027 (28%)	

Nuclide	Half-life	Decay mode	Radiation: energy in MeV (intensity)	Product
$^{121m}_{50}$Sn	55 y	IT (78%) β⁻ (22%)	β⁻ 0.354 (22%); γ 0.037 (2%); e⁻ [0.008] (161%); Sb K x-rays ~0.028 (16%)	$^{121}_{51}$Sb (β⁻) $^{121}_{50}$Sn (IT)
$^{121}_{50}$Sn	27.06 h	β⁻	β⁻ 0.383 (100%)	$^{121}_{51}$Sb
$^{123}_{50}$Sn	129 d	β⁻	β⁻ 1.42 (100%)	$^{123}_{51}$Sb
$^{126}_{50}$Sn	~2×10⁵ y	β⁻	β⁻ 0.250 (100%); γ 0.088 (37%), 0.087 (9%), 0.064 (10%), 0.023 (6%); e⁻ [0.055] (280%); Sb K x-rays ~0.028 (29%)	$^{126}_{51}$Sb
Antimony				
$^{122}_{51}$Sb	2.68 d	β⁻ (97%) EC (3%)	β⁻ 1.98 (26%), 1.41 (67%), 0.723 (4%); γ 1.26 (0.7%), 1.14 (0.7%), 0.686 (3.4%), 0.564 (66%); Sn K x-rays ~0.026 (2%), Te K x-rays ~0.027 (0.3%)	$^{122}_{52}$Te (β⁻) $^{122}_{50}$Sn (EC)
$^{124}_{51}$Sb	60.20 d	β⁻	β⁻ 2.32 (21%), 1.60 (7%), 0.966 (9%), 0.61 (49%), 0.24 (14%); γ 1.69 (50%), 1.37 (5%), 1.31 (3%), 0.72 (14%), 0.644 (7%), 0.603 (97%)	$^{124}_{52}$Te
$^{125}_{51}$Sb	2.71 y	β⁻	β⁻ 0.612 (14%), 0.444 (12%), 0.300 (45%), 0.125 (29%); γ 0.634 (11%), 0.599 (24%), 0.463 (10%), 0.427 (31%); Te K x-rays ~0.029 (75%)	$^{125m}_{52}$Te
$^{126m}_{51}$Sb	19 min	β⁻ (86%) IT (14%)	β⁻ 2.50 (12†), 1.87 (88†); γ 0.414 (86%), 0.666 (86%), 0.695 (82%); e⁻ [0.011] (27%); Te K x-rays ~0.029 (1%)	$^{126}_{52}$Te (β⁻) $^{126}_{51}$Sb (IT)
$^{126}_{51}$Sb	12.5 d	β⁻	β⁻ 1.90 (100%); γ 0.415 (83%), 0.666 (100%), 0.694 (100%), 0.697 (30%), 0.720 (54%), 0.857 (18%); e⁻ [0.013] (5%); Te K x-rays ~0.029 (2%)	$^{126}_{52}$Te
Tellurium				
$^{123m}_{52}$Te	120 d	IT	γ 0.159 (84%); e⁻ [0.102] (205%)	$^{123}_{52}$Te
$^{125m}_{52}$Te	58 d	IT	γ 0.110 (0.3%), 0.035 (7%); Te K x-rays (~110%)	$^{125}_{52}$Te

(continues)

1225

Table of Radioactive Isotopes—Continued

Nuclide $^{A}_{Z}X$	Half-life	Decay mode	Radiation characteristics: Energies (MeV) [intensities (%)]	Decay product $^{A}_{Z}Y$
$^{127m}_{52}\text{Te}$	109 d	IT (99.2%) β^- (0.8%)	γ 0.089 (0.08%), 0.059 (0.2%) e^- [0.076] (170%) Te K x-rays ~0.028 (36%)	$^{127}_{52}\text{Te}$
$^{127}_{52}\text{Te}$	9.35 h	β^-	β^- 0.695 (100%) γ 0.417 (1%), 0.360 (0.14%), 0.21 (0.03%), 0.058 (0.01%)	$^{127}_{53}\text{I}$
$^{129m}_{52}\text{Te}$	33.52 d	IT (64%) β^- (36%)	β^- 1.60 (36%) γ 0.696 (6%) e^- [0.060] (112%)	$^{129}_{52}\text{Te (IT)}$ $^{129}_{53}\text{I}\ (\beta^-)$
$^{129}_{52}\text{Te}$	69.5 min	β^-	Te K x-rays ~0.030 (28%) β^- 1.45 (70%), 0.989 (15%), 0.69 (4%), 0.29 (11%) γ 1.08 (1.5%), 0.455 (8%), 0.275 (1.7%), 0.027 (19%) I L x-rays ~0.004 (5%)	$^{129}_{53}\text{I}$
$^{132}_{52}\text{Te}$	78.2 h	β^-	β^- 0.215 (100%) γ 0.230 (90%), 0.053 (17%) e^- [0.043] (168%) I K x-rays ~0.030 (71%)	$^{132}_{53}\text{I}$
Iodine $^{123}_{53}\text{I}$	13.02 h	EC	γ 0.159 (83%) e^- [0.028] (108%) Te K x-rays ~0.029 (86%)	$^{123}_{52}\text{Te}$
$^{124}_{53}\text{I}$	4.15 d	EC (75%) β^+ (25%)	β^+ 2.13 (12.3%), 1.53 (11.5%), 0.808 (1.2%) γ 1.69 (14%), 0.73 (14%), 0.645 (1%), 0.605 (67%) $\gamma^{\pm}$ 0.511 (50%) e^- [0.007] (62%)	$^{124}_{52}\text{Te}$
$^{125}_{53}\text{I}$	60.25 d	EC	Te K x-rays ~0.030 (58%) γ 0.035 (7%) e^- [0.018] (246%) Te K x-rays ~0.030 (138%)	$^{125}_{52}\text{Te}$
$^{126}_{53}\text{I}$	13.02 d	EC (53%) β^- (46%) β^+ (1%)	β^- 1.25 (10%), 0.865 (30%), 0.385 (6%) β^+ 1.11 (0.7%), 0.460 (0.3%) γ 0.667 (33%), 0.386 (34%) e^- [0.006] (45%) Te K x-rays ~0.029 (41%)	$^{126}_{52}\text{Te (EC},\beta^+)$ $^{126}_{54}\text{Xe}\ (\beta^-)$

1226

Nuclide	Half-life	Decay mode		Energy MeV (%)	Product
$^{129}_{53}\text{I}$	1.57×10^7 y	β^-	β^-	0.150 (100%)	$^{129}_{54}\text{Xe}$
			γ	0.040 (9%)	
			e^-	[0.014] (165%) Xe K x-rays ~0.032 (71%)	
$^{130}_{53}\text{I}$	12.36 h	β^-	β^-	1.78 (0.4%), 1.04 (51.6%), 0.62 (48%)	$^{130}_{54}\text{Xe}$
			γ	1.15 (12%), 0.743 (87%), 0.669 (100%), 0.538 (99%), 0.419 (35%)	
			e^-	[0.011] (4%) Xe K x-rays ~0.032 (2%)	
$^{131}_{53}\text{I}$	8.04 d	β^-	β^-	0.806 (1%), 0.607 (86%), 0.336 (13%)	$^{131m}_{54}\text{Xe}$
			γ	0.637 (6.8%), 0.364 (82%), 0.284 (5.4%), 0.080 (2.6%)	
			e^-	[0.010] (12%) Xe K x-rays ~0.032 (5%)	
$^{132}_{53}\text{I}$	2.28 h	β^-	β^-	2.16 (18%), 1.61 (21%), 1.22 (24%), 1.04 (15%), 0.802 (21%)	$^{132}_{54}\text{Xe}$
			γ	1.40 (7%), 0.955 (18%), 0.773 (76%), 0.667 (99%), 0.522 (16%)	
			e^-	[0.008] (2.5%) Xe K x-rays ~0.032 (1%)	
$^{133}_{53}\text{I}$	20.9 h	β^-	β^-	1.4 (~94%), 0.5 (~6%)	$^{133m}_{54}\text{Xe}$
			γ	0.530 (90%) Xe K x-rays ~0.032 (1%)	

Xenon

Nuclide	Half-life	Decay mode		Energy MeV (%)	Product
$^{131m}_{54}\text{Xe}$	11.77 d	IT	γ	0.164 (2%)	$^{131}_{54}\text{Xe}$
			e^-	[0.143] (169%) Xe K x-rays ~0.030 (45%), Xe L x-rays ~0.004 (7%)	
$^{133m}_{54}\text{Xe}$	2.19 d	IT	γ	0.233 (10%)	$^{133}_{54}\text{Xe}$
			e^-	[0.192] (156%) Xe K x-rays ~0.031 (56%), Xe L x-rays ~0.004 (7%)	
$^{133}_{54}\text{Xe}$	5.245 d	β^-	β^-	0.346 (100%)	$^{133}_{55}\text{Cs}$
			γ	0.081 (37%)	
			e^-	[0.036] (114%) Cs K x-rays ~0.033 (49%), Cs L x-rays ~0.004 (5%)	

Cesium

Nuclide	Half-life	Decay mode		Energy MeV (%)	Product
$^{131}_{55}\text{Cs}$	9.69 d	EC	e^-	[0.006] (76%)	$^{131}_{54}\text{Xe}$
$^{132}_{55}\text{Cs}$	6.474 d	EC (96.5%)	β^-	0.668 (2%)	$^{132}_{54}\text{Xe (EC, }\beta^+)$
		β^- (2%)	β^+	0.400 (1.5%)	$^{132}_{56}\text{Ba }(\beta^-)$
		β^+ (1.5%)	γ	1.32 (0.6%), 1.14 (0.5%), 0.668 (99%), 0.465 (2%)	

(continues)

Table of Radioactive Isotopes—Continued

Nuclide $^A_Z X$	Half-life	Decay mode	Radiation characteristics: Energies (MeV) [intensities (%)]	Decay product $^A_Z Y$
$^{134}_{55}$Cs	2.062 y		e⁻ [0.008] (73%) Xe K x-rays ~0.032 (73%), Xe L x-rays ~0.004 (7%)	$^{134}_{56}$Ba
		β⁻	β⁻ 0.658 (70%), 0.415 (3%), 0.089 (27%) γ 0.801 (9%), 0.796 (85%), 0.605 (98%), 0.570 (15%), 0.563 (8%)	
$^{135}_{55}$Cs	3 × 10⁶ y	β⁻	β⁻ 0.205 (100%)	$^{135}_{56}$Ba
$^{137}_{55}$Cs	30.17 y	β⁻	β⁻ 1.18 (6%), 0.514 (94%) γ 0.662 (85%) e⁻ [0.062] (17%) Ba K x-rays ~0.035 (7%)	$^{137m}_{56}$Ba
Barium				
$^{131}_{56}$Ba	12.0 d	EC	γ 0.496 (48%), 0.373 (13%), 0.216 (19%), 0.124 (28%) e⁻ [0.045] (135%) Cs K x-rays ~0.035 (98%)	$^{131}_{55}$Cs
$^{133}_{56}$Ba	10.66 y	EC	γ 0.382 (8%), 0.356 (69%), 0.302 (14%), 0.276 (7%), 0.080 (34%) e⁻ [0.055] (211%) Cs K x-rays ~0.034 (123%)	$^{133}_{55}$Cs
$^{137m}_{56}$Ba	2.551 min	IT	γ 0.662 (89%) e⁻ [0.065] (18%) Ba K x-rays ~0.035 (8%)	$^{137}_{56}$Ba
$^{140}_{56}$Ba	12.79 d	β⁻	β⁻ 1.02 (17%), 1.01 (46%), 0.886 (3%), 0.582 (10%), 0.468 (24%) γ 0.537 (24%), 0.438 (2%), 0.424 (3%), 0.305 (4%), 0.163 (6%), 0.030 (14%) e⁻ [0.035] (208%) La K x-rays ~0.035 (2%)	$^{140}_{57}$La
Lanthanum				
$^{140}_{57}$La	40.27 h	β⁻	β⁻ 2.16 (8%), 1.68 (18%), 1.37 (46%), 1.15 (19%), 0.857 (4%), 0.510 (5%) γ 2.53 (3%), 1.60 (96%), 0.925 (10%), 0.815 (24%), 0.487 (46%), 0.329 (20%) e⁻ [0.009] (5%) Ce K x-rays ~0.038 (2%)	$^{140}_{58}$Ce

Element / Isotope	Half-life	Decay mode	Radiation	Energies MeV (%)	Daughter
Cerium					
$^{139}_{58}$Ce	137.2 d	EC	γ	0.165 (80%)	$^{139}_{57}$La
			e^-	[0.033] (101%)	
				La K x-rays ~ 0.036 (185%)	
$^{141}_{58}$Ce	32.55 d	β^-	β^-	0.582 (30%), 0.444 (70%)	$^{141}_{59}$Pr
			γ	0.145 (48%)	
			e^-	[0.026] (38%)	
				Pr K x-rays ~ 0.039 (17%)	
$^{143}_{58}$Ce	33.0 h	β^-	β^-	1.40 (37%), 1.13 (40%), 0.74 (5%), 0.50 (12%), 0.22 (6%)	$^{143}_{59}$Pr
			γ	0.725 (5%), 0.668 (5%), 0.293 (42%), 0.057 (12%)	
			e^-	[0.030] (139%)	
				Pr K x-rays ~ 0.039 (63%)	
$^{144}_{58}$Ce	284.5 d	β^-	β^-	0.316 (75.7%), 0.238 (4.6%), 0.185 (19.7%)	$^{144}_{59}$Pr
			γ	0.134 (11%), 0.080 (2%)	
			e^-	[0.010] (23%)	
				Pr K x-rays ~ 0.037 (8%)	
Praseodymium					
$^{142}_{59}$Pr	19.2 h	β^-	β^-	2.16 (~93%), 0.586 (~7%)	$^{142}_{60}$Nd
			γ	1.57 (3.7%)	
$^{143}_{59}$Pr	13.59 d	β^-	β^-	0.932 (100%)	$^{143}_{60}$Nd
$^{144m}_{59}$Pr	7.2 min	IT	e^-	[0.046] (160%)	$^{144}_{59}$Pr
				Pr K x-rays ~ 0.038 (30%)	
$^{144}_{59}$Pr	17.30 min	β^-	β^-	3.00 (97.8%), 2.30 (1.2%), 0.807 (1%)	$^{144}_{60}$Nd
			γ	2.19 (0.7%), 1.49 (0.3%), 0.695 (1.5%)	
Neodymium					
$^{147}_{60}$Nd	10.98 d	β^-	β^-	0.806 (<0.5%), 0.810 (83%), 0.490 (0.3%), 0.410 (0.7%), 0.369 (15%), 0.215 (1%)	$^{147}_{61}$Pm
			γ	0.533 (13%), 0.439 (1%), 0.319 (3%), 0.091 (28%)	
$^{149}_{60}$Nd	1.73 h	β^-	β^-	1.56 (6%), 1.43 (38%), 1.13 (26%), 1.03 (30%)	$^{149}_{61}$Pm
			γ	0.654 (9%), 0.541 (10%), 0.424 (9%), 0.327 (5%), 0.270 (10%), 0.211 (27%), 0.156 (6%), 0.114 (18%)	
			e^-	[0.048] (87%)	
				Pm K x-rays ~ 0.042 (33%)	

(continues)

Table of Radioactive Isotopes—Continued

Nuclide $^{A}_{Z}X$	Half-life	Decay mode	Radiation characteristics: Energies (MeV) [intensities (%)]	Decay product $^{A}_{Z}Y$
Promethium				
$^{147}_{61}$Pm	2.623 y	β^-	β^- 0.224 (100%)	$^{147}_{62}$Sm
$^{148}_{61}$Pm	41.8 d	β^- (95.4%)	β^- 0.400 (48%), 0.500 (28%), 0.690 (16%)	$^{148}_{62}$Sm (β^-)
		IT (4.6%)	γ 0.288 (12%), 0.414 (19%), 0.550 (94%), 0.599 (13%), 0.630 (89%), 0.725 (33%), 0.915 (17%), 1.014 (20%)	$^{148}_{61}$Pm (IT)
			e^- [0.021] (21%)	
			Sm K x-rays ~0.043 (6%), Pm K x-rays ~0.043 (3%)	
$^{148}_{61}$Pm	5.4 d	β^-	β^- 1.02 (40%), 1.93 (10%), 2.48 (50%)	$^{148}_{62}$Sm
			γ 0.550 (22%), 0.914 (12%), 1.47 (22%)	
$^{149}_{61}$Pm	53.08 h	β^-	β^- 1.07 (90%), 0.784 (9%), 0.470 (0.4%), 0.190 (0.6%)	$^{149}_{62}$Sm
			γ 0.859 (0.2%), 0.590 (0.1%), 0.286 (2%)	
			Sm K x-rays ~0.044 (0.2%)	
$^{151}_{61}$Pm	28.40 h	β^-	β^- 1.19 (10%), 1.13 (~6%), 1.05 (11%), 0.980 (~3%), 0.835 (43%), 0.728 (~10%), 0.500 (~6%), 0.347 (~11%)	$^{151}_{62}$Sm
			γ 0.72 (6%), 0.66 (3%), 0.45 (5%), 0.340 (21%), 0.275 (6%), 0.24 (5%), 0.17 (18%), 0.10 (7%), 0.07 (5%)	
			e^- [0.024] (69%)	
			Sm K x-rays ~0.043 (29%)	
Samarium				
$^{151}_{62}$Sm	87 y	β^-	β^- 0.076 (99.1%), 0.055 (0.9%)	$^{151}_{63}$Eu
			γ 0.022 (0.03%)	
$^{153}_{62}$Sm	46.8 h	β^-	β^- 0.810 (20%), 0.710 (49%), 0.640 (30%)	$^{153}_{63}$Eu
			γ 0.103 (28%), 0.070 (5%)	
			e^- [0.045] (135%)	
			Eu K x-rays ~0.043 (63%), Eu L x-rays ~0.006 (10%)	
Europium				
$^{152}_{63}$Eu	13.2 y	EC (73%)	β^- 1.49 (10%), 1.07 (1%), 0.690 (13%), 0.360 (2%), 0.185 (1%)	$^{152}_{62}$Sm (EC,β^+)
		β^- (27%)	γ 1.41 (22%), 1.13 (14%), 1.09 (12%), 0.965 (15%), 0.779 (14%), 0.344 (27%), 0.245 (8%), 0.122 (37%)	$^{152}_{64}$Gd (β^-)
		β^+ (0.02%)	e^- [0.045] (106%)	
			Gd K x-rays ~0.045 (0.87%), Sm K x-rays ~0.043 (64%)	

Nuclide	Half-life	Decay		Radiations (MeV, intensity)	Product
$^{154}_{63}\text{Eu}$	8.5 y	β^-	β^-	1.87 (11%), 1.20 (1%), 0.976 (4%), 0.843 (17%), 0.579 (38%), 0.274 (29%)	$^{154}_{64}\text{Gd}$
			γ	1.28 (37%), 1.00 (31%), 0.876 (12%), 0.759 (5%), 0.724 (21%), 0.593 (6%), 0.248 (7%), 0.123 (38%), Gd K x-rays ~0.045 (26%), Gd L x-rays ~0.006 (6%)	
$^{155}_{63}\text{Eu}$	4.96 y	β^-	β^-	0.250 (13%), 0.186 (8%), 0.15 (75%), 0.100 (~4%)	$^{155}_{64}\text{Gd}$
			γ	0.105 (20%), 0.087 (32%)	
			e^-	[0.018] (87%)	
				Gd K x-rays ~0.045 (24%), Gd L x-rays ~0.006 (7%)	
$^{156}_{63}\text{Eu}$	15 d	β^-	β^-	2.46 (32%), 1.215 (11%), 0.495 (34%), 0.300 (23%)	$^{156}_{64}\text{Gd}$
			γ	0.089 (9%), 0.646 (7%), 0.723 (6%), 0.812 (10%), 1.065 (5%), 1.153 (7%), 1.230 (9%), 1.242 (7%)	
			e^-	[0.030] (57%)	
				Gd K x-rays ~0.047 (13%)	

Gadolinium

Nuclide	Half-life	Decay		Radiations (MeV, intensity)	Product
$^{153}_{64}\text{Gd}$	241.6 d	EC	γ	0.103 (20%), 0.099 (30%), 0.070 (2.4%)	$^{153}_{63}\text{Eu}$
			e^-	[0.040] (160%)	
				Eu K x-rays ~0.045 (120%), Eu L x-rays ~0.006 (11%)	
$^{159}_{64}\text{Gd}$	18.56 h	β^-	β^-	0.950 (63%), 0.89 (24%), 0.60 (13%)	$^{159}_{65}\text{Tb}$
			γ	0.363 (11%), 0.058 (3%)	
			e^-	[0.006] (42%)	
				Tb K x-rays ~0.047 (20%), Tb L x-rays ~0.007 (4%)	

Terbium

Nuclide	Half-life	Decay		Radiations (MeV, intensity)	Product
$^{160}_{65}\text{Tb}$	72.1 d	β^-	β^-	1.74 (0.4%), 1.55 (0.1%), 0.874 (27%), 0.791 (7%), 0.575 (46%), 0.553 (4%), 0.481 (10%), 0.441 (5%)	$^{160}_{66}\text{Dy}$
			γ	1.27 (7%), 1.18 (15%), 0.966 (25%), 0.879 (30%), 0.299 (30%), 0.197 (6%), 0.087 (12%)	
			e^-	[0.050] (102%)	
				Dy K x-rays ~0.050 (20%), Dy L x-rays ~0.007 (9%)	
$^{161}_{65}\text{Tb}$	6.90 d	β^-	β^-	0.590 (10%), 0.520 (55%), 0.460 (35%)	$^{161}_{66}\text{Dy}$
			γ	0.075 (10%), 0.057 (5%), 0.049 (19%), 0.026 (21%)	
			e^-	[0.042] (193%)	
				Dy K x-rays ~0.050 (21%), Dy L x-rays ~0.007 (19%)	

(continues)

Table of Radioactive Isotopes—Continued

Nuclide $^A_Z X$	Half-life	Decay mode	Radiation characteristics: Energies (MeV) [intensities (%)]	Decay product $^A_Z Y$
Dysprosium $^{165}_{66}$Dy	2.334 h	β^-	β^- 1.31 (80%), 1.22 (16%), others (4%) γ 0.716 (0.7%), 0.633 (0.7%), 0.361 (1.1%), 0.095 (4%) e^- [0.007] (20%) Ho K x-rays ~ 0.052 (10%), Ho L x-rays ~ 0.007 (2%)	$^{165}_{67}$Ho
Holmium $^{166m}_{67}$Ho	1200 y	β^-	β^- 0.065 (100%) γ 0.830 (11%), 0.810 (60%), 0.711 (58%), 0.532 (12%), 0.412 (12%), 0.280 (30%), 0.184 (75%), 0.081 (12%) e^- [0.114] (188%) Er K x-rays ~ 0.053 (39%), Er L x-rays ~ 0.007 (19%)	$^{166}_{68}$Er
Erbium $^{169}_{68}$Er	9.40 d	β^-	β^- 0.340 (58%), 0.332 (42%) γ 0.008 (0.3%)	$^{169}_{69}$Tm
$^{171}_{68}$Er	7.52 h	β^-	β^- 1.49 (2%), 1.07 (90%), 0.575 (8%) γ 0.308 (63%), 0.296 (28%), 0.124 (9%), 0.112 (21%) e^- [0.058] (209%) Tm K x-rays ~ 0.051 (46%), Tm L x-rays ~ 0.007 (13%)	$^{171}_{69}$Tm
Thulium $^{170}_{69}$Tm	128.6 d	β^- (>99%) EC (0.15%)	β^- 0.968 (76%), 0.884 (24%) γ 0.084 (3.3%) e^- [0.015] (32%) Yb K x-rays ~ 0.057 (4%), Yb L x-rays ~ 0.008 (3%)	$^{170}_{70}$Yb
$^{171}_{69}$Tm	1.92 y	β^-	β^- 0.097 (98%), ~0.03 (2%) γ 0.067 (~0.2%)	$^{171}_{70}$Yb
Ytterbium $^{169}_{70}$Yb	32.02 d	EC	γ 0.308 (10%), 0.198 (35%), 0.177 (22%), 0.131 (11%), 0.110 (18%), 0.063 (45%) e^- [0.112] (307%) Tm K x-rays ~ 0.053 (185%), Tm L x-rays ~ 0.008 (45%)	$^{169}_{69}$Tm
$^{175}_{70}$Yb	4.19 d	β^-	β^- 0.466 (86.5%), 0.353 (2.6%), 0.217 (0.8%), 0.073 (10.1%) γ 0.396 (6%), 0.283 (3.7%), 0.114 (2%) Lu K x-rays ~ 0.055 (3.6%)	$^{175}_{71}$Lu

Element / Nuclide	Decay mode	Half-life	Radiation	Energies MeV (intensities)	Daughter
Lutetium $^{177}_{71}$Lu	β^-	6.71 d	β^-	0.497 (90%), 0.384 (3%), 0.249 (0.3%), 0.175 (6.7%)	$^{177}_{72}$Hf
			γ	0.208 (6%), 0.113 (3%)	
			e^-	{0.015} (24%)	
				Hf K x-rays $\sim$0.058 (6%), Hf L x-rays $\sim$0.008 (3%)	
Hafnium $^{175}_{72}$Hf	EC	70 d	γ	0.433 (1.4%), 0.343 (85%), 0.089 (3.4%)	$^{175}_{71}$Lu
			e^-	[0.044] (83%)	
				Lu K x-rays $\sim$0.058 (93%), Lu L x-rays $\sim$0.008 (23%)	
$^{181}_{72}$Hf	β^-	42.4 d	β^-	0.55 (<0.5%), 0.408 (93%)	$^{181}_{73}$Ta
			γ	0.482 (81%), 0.346 (13%), 0.133 (36%)	
			e^-	[0.068] (92%)	
				Ta K x-rays $\sim$0.060 (27%), Ta L x-rays $\sim$0.009 (12%)	
Tantalum $^{182}_{73}$Ta	β^-	115.0 d	β^-	0.540 (45%), 0.408 (24%), 0.246 (31%)	$^{182}_{74}$W
			γ	1.23 (13%), 1.22 (27%), 1.19 (16%), 1.12 (34%), 0.222 (8%), 0.152 (7%), 0.100 (14%), 0.068 (42%)	
			e^-	[0.081] (161%)	
				W K x-rays $\sim$0.065 (36%), W L x-rays $\sim$0.009 (23%)	
Tungsten $^{181}_{74}$W	EC	120.95 d	γ	0.152 (0.1%), 0.136 (0.1%), 0.006 (1%)	$^{181}_{73}$Ta
			e^-	[0.008] (85%)	
				Ta K x-rays $\sim$0.060 (65%), Ta L x-rays $\sim$0.010 (20%)	
$^{185}_{74}$W	β^-	75.1 d	β^-	0.433 (100%)	$^{185}_{75}$Re
$^{187}_{74}$W	β^-	23.85 h	β^-	1.32 (20%), 0.630 (70%), 0.340 (10%)	$^{187}_{75}$Re
			γ	0.773 (4%), 0.686 (27%), 0.618 (6%), 0.552 (5%), 0.479 (23%), 0.134 (9%), 0.072 (11%)	
			e^-	[0.037] (181%)	
				Re K x-rays $\sim$0.065 (20%), Re L x-rays $\sim$0.009 (23%)	
Rhenium $^{183}_{75}$Re	EC	71 d	γ	0.292 (3.2%), 0.209 (3%), 0.162 (24%), 0.109 (3%), 0.108 (2%), 0.099 (3%), 0.052 (2%), 0.046 (8%)	$^{183}_{74}$W
			e^-	[0.105] (278%)	
				W K x-rays $\sim$0.063 (108%), W L x-rays $\sim$0.009 (55%)	

(continues)

1233

Table of Radioactive Isotopes—Continued

Nuclide $^A_Z X$	Half-life	Decay mode	Radiation characteristics: Energies (MeV) [intensities (%)]	Decay product $^A_Z Y$
$^{186}_{75}$Re	90.64 h	β^- (92.2%) EC (7.8%)	β^- 1.07 (77†), 0.934 (23†) γ 0.768 (0.04%), 0.632 (0.03%), 0.137 (9%), 0.122 (0.7%) e^- [0.014] (22%) W K x-rays ~0.065 (6%), Os K x-rays ~0.065 (3.5%)	$^{186}_{76}$Os (β^-) $^{186}_{74}$W (EC)
$^{187}_{75}$Re	4×10^{10} y	β^-	β^- 0.0026 (100%)	$^{187}_{76}$Os
$^{188}_{75}$Re	16.98 h	β^-	β^- 2.12 (79%), 1.97 (20%), <1.9 (1%) γ 0.932 (0.4%), 0.633 (1%), 0.478 (1%), 0.155 (15%) e^- [0.016] (18%) Os K x-rays ~0.065 (5%), Os L x-rays ~0.011 (3%)	$^{188}_{76}$Os
Osmium				
$^{185}_{76}$Os	93.6 d	EC	γ 0.880 (5%), 0.875 (7%), 0.646 (81%) e^- [0.017] (55%)	$^{185}_{75}$Re
$^{191m}_{76}$Os	13.10 h	IT	γ 0.074 (0.07%) e^- [0.060] (127%) Re K x-rays ~0.065 (69%), Re L x-rays ~0.010 (23%)	$^{191}_{76}$Os
$^{191}_{76}$Os	15.4 d	β^-	β^- 0.143 (100%) γ 0.129 (26%) e^- [0.094] (258%) Os K x-rays ~0.070 (8%), Os L x-rays ~0.010 (21%)	$^{191m}_{77}$Ir
$^{193}_{76}$Os	30.6 h	β^-	Ir K x-rays ~0.069 (56%), Ir L x-rays ~0.010 (38%) β^- 1.13 (42%), 1.06 (21%), 0.993 (10%), 0.85 (10%), 0.67 (11%), 0.51 (6%) γ 0.558 (2.1%), 0.460 (3.9%), 0.387 (1%), 0.322 (1.4%), 0.280 (1%), 0.139 (4%) e^- [0.052] (162%) Ir K x-rays ~0.070 (13%)	$^{193}_{77}$Ir
Iridium				
$^{191m}_{77}$Ir	4.96 s	IT	γ 0.129 (25%) e^- [0.094] (258%) Ir K x-rays ~0.070 (56%), Ir L x-rays ~0.010 (40%)	$^{191}_{77}$Ir
$^{192}_{77}$Ir	74.17 d	β^- (95.4%) EC (4.6%)	β^- 0.672 (46%), 0.536 (41%), 0.240 (8%) γ 0.612 (6%), 0.604 (9%), 0.589 (4%), 0.468 (49%), 0.317 (81%), 0.308 (30%), 0.296 (29%) e^- [0.045] (25%) Pt K x-rays ~0.070 (8%), Os K x-rays ~0.065 (4%)	$^{192}_{78}$Pt (β^-) $^{192}_{76}$Os (EC)

Nuclide	Half-life	Decay mode	Radiation	Energy MeV (intensity)	Daughter
$^{194}_{77}$Ir	19.15 h	β^-	β^-	2.24 (89%), 1.92 (5.1%), 1.63 (1.2%), 0.98 (2%), 0.72, 0.61, 0.44	$^{194}_{78}$Pt
			γ	1.48 (0.2%), 1.16 (0.8%), 0.939 (0.6%), 0.64 (1%), 0.328 (13%), 0.294 (2%)	
Platinum					
$^{193m}_{78}$Pt	4.33 d	IT	γ	0.135 (0.11%)	$^{193}_{78}$Pt
			e^-	[0.130] (235%)	
				Pt K x-rays $\sim$0.070 (14%), Pt L x-rays $\sim$0.010 (23%)	
$^{197}_{78}$Pt	18.3 h	β^-	β^-	0.719 (10.6%), 0.643 (79.6%), 0.451 (9.8%)	$^{197}_{79}$Au
			γ	0.191 (6%), 0.077 (20%)	
			e^-	[0.054] (94%)	
				Au K x-rays $\sim$0.075 (3%), Au L x-rays $\sim$0.010 (19%)	
Gold					
$^{195}_{79}$Au	182.9 d	EC	γ	0.129 (1%), 0.099 (10%), 0.031 (1%)	$^{195}_{78}$Pt
			e^-	[0.045] (188%)	
				Pt K x-rays $\sim$0.070 (100%), Pt L x-rays $\sim$0.011 (52%)	
$^{198}_{79}$Au	2.697 d	β^-	β^-	1.37 (0.03%), 0.961 (98.6%), 0.290 (1.3%)	$^{198}_{80}$Hg
			γ	1.09 (0.2%), 0.676 (1%), 0.412 (95%)	
			e^-	[0.015] (6%)	
				Hg K x-rays $\sim$0.075 (3%)	
$^{199}_{79}$Au	3.148 d	β^-	β^-	0.462 (6%), 0.296 (71.6%), 0.250 (22.4%)	$^{199}_{80}$Hg
			γ	0.208 (8%), 0.158 (37%)	
			e^-	[0.058] (65%)	
				Hg K x-rays $\sim$0.075 (17%), Hg L x-rays $\sim$0.011 (12%)	
Mercury					
$^{197m}_{80}$Hg	23.8 h	IT (93.5%) EC (6.5%)	γ	0.279 (7%), 0.134 (34%)	$^{197}_{80}$Hg (IT) $^{197m}_{79}$Au (EC)
			e^-	[0.214] (230%)	
				Hg L x-rays $\sim$0.010 (38%)	
$^{197}_{80}$Hg	64.14 h	EC	γ	0.191 (1%), 0.077 (18%)	$^{197}_{79}$Au
			e^-	[0.057] (136%)	
				Au K x-rays $\sim$0.075 (6%), Au L x-rays $\sim$0.010 (4%)	
$^{203}_{80}$Hg	46.76 d	β^-	β^-	0.214 (100%)	$^{203}_{81}$Tl
			γ	0.279 (82%)	
			e^-	[0.040] (26%)	
				Tl K x-rays $\sim$0.075 (13%), Tl L x-rays $\sim$0.011 (5%)	

(continues)

Table of Radioactive Isotopes—Continued

Nuclide ^A_ZX	Half-life	Decay mode	Radiation characteristics: Energies (MeV) [intensities (%)]	Decay product ^A_ZY
Thallium				
$^{201}_{81}$Tl	74 h	EC	γ 0.135 (3%), 0.167 (9%) Hg K x-rays ~0.075 (93%), Hg L x-rays ~0.011 (40%) e⁻ {0.048} (187%)	$^{201}_{80}$Hg
$^{202}_{81}$Tl	12.23 d	EC	γ 0.961 (0.1%), 0.522 (0.1%), 0.439 (95%) e⁻ {0.021} (49%) Hg K x-rays ~0.075 (78%), Hg L x-rays ~0.011 (28%)	$^{202}_{80}$Hg
$^{204}_{81}$Tl	3.773 y	β⁻ (97.4%) EC (2.6%)	β⁻ 0.763 (97.4%) Hg K x-rays ~0.075 (1.5%), Hg L x-rays ~0.010 (0.7%)	$^{204}_{82}$Pb (β⁻) $^{204}_{80}$Hg (EC)
$^{206}_{81}$Tl	4.18 min	β⁻	β⁻ 1.53 (100%) Pb K x-rays (0.10%)	$^{206}_{82}$Pb
$^{207}_{81}$Tl	4.77 min	β⁻	β⁻ 1.44 (100%) γ 0.897 (0.24%)	$^{207}_{82}$Pb
$^{208}_{81}$Tl	3.1 min	β⁻	β⁻ 1.80 (48.8%), 1.52 (22.7%), 1.29 (23.9%), 1.04 (4.6%) γ 0.510 (22%), 0.583 (86%), 0.860 (12%), 2.614 (100%) e⁻ {0.038} (14%) Pb K x-rays ~0.075 (7%), Pb L x-rays ~0.011 (2.8%)	$^{208}_{82}$Pb
$^{209}_{81}$Tl	2.2 min	β⁻	β⁻ 1.99 (100%) γ 0.117 (81%), 0.467 (81%), 1.56 (98%) e⁻ {0.028} (39%) Pb K x-rays ~0.075 (21%), Pb L x-rays ~0.011 (8%)	$^{209}_{82}$Pb
$^{210}_{81}$Tl	1.30 min	β⁻ β⁻ n (0.007%)	β⁻ 2.34 (19%), 1.87 (56%), 1.32 (25%) γ 2.43 (9%), 2.36 (8%), 2.27 (3%), 2.09 (5%), 2.01 (7%), 1.41 (5%), 1.31 (21%), 1.21 (17%), 1.11 (7%), 1.07 (12%), 0.908 (3%), 0.860 (7%), 0.798 (99%), 0.298 (79%) e⁻ {0.094} (109%) Pb K x-rays ~0.080 (9%), Pb L x-rays ~0.012 (22%)	$^{210}_{82}$Pb
Lead				
$^{209}_{82}$Pb	3.30 h	β⁻	β⁻ 0.645 (100%)	$^{209}_{83}$Bi
$^{210}_{82}$Pb	22.36 y	β⁻ α (trace)	β⁻ 0.061 (19%), 0.015 (81%) γ 0.047 (4%) e⁻ {0.028} (90%) Bi L x-rays ~0.012 (23%)	$^{210}_{83}$Bi

Nuclide	Half-life	Decay mode	Radiations (MeV, intensity)	Product
$^{211}_{82}\text{Pb}$	36.1 min	β^-	β^- 1.36 (92.4%), 0.951 (1.4%), 0.525 (5.5%), 0.251 (0.7%); γ 0.832 (3.8%), 0.704 (0.5%), 0.427 (1.7%), 0.404 (3.8%); e^- {0.005} (5%)	$^{211}_{83}\text{Bi}$
$^{212}_{82}\text{Pb}$	10.64 h	β^-	Bi K x-rays ~0.077 (0.8%), Bi L x-rays ~0.011 (0.9%); β^- 0.568 (12%), 0.331, [0.101]; γ 0.300 (3.3%), 0.238 (44%); e^- {0.074} (66%)	$^{212}_{83}\text{Bi}$
$^{214}_{82}\text{Pb}$	26.8 min	β^-	Bi K x-rays ~0.080 (36%), Bi L x-rays ~0.012 (14%); β^- 1.03 (6%), 0.59 (~52%), 0.65 (~41%); γ 0.052 (1%), 0.241 (7%), 0.295 (19%), 0.351 (37%); e^- {0.074} (57%)	$^{214}_{83}\text{Bi}$
Bismuth $^{206}_{83}\text{Bi}$	6.243 d	EC	Bi K x-rays ~0.080 (23%), Bi L x-rays ~0.012 (13%); γ 1.72 (36%), 1.60 (8%), 1.10 (13%), 1.02 (8%), 0.895 (19%), 0.880 (72%), 0.803 (99%), 0.538 (34%), 0.516 (46%), 0.497 (18%), 0.398 (10%), 0.343 (26%), 0.184 (16%); e^- {0.133} (116%)	$^{206}_{82}\text{Pb}$
$^{207}_{83}\text{Bi}$	38 y	EC	Pb K x-rays ~0.080 (112%), Pb L x-rays ~0.011 (45%); γ 1.77 (9%), 1.06 (77%), 0.570 (98%); e^- {0.116} (56%)	$^{207}_{82}\text{Pb}$
$^{210m}_{83}\text{Bi}$	3.04×10^6 y	α	Pb K x-rays ~0.080 (78%), Pb L x-rays ~0.011 (34%); α 4.95 (55.0%), 4.91 (39.5%), 4.58 (1.4%), 4.57 (3.9%), 4.42 (0.2%); γ 0.304 (23%), 0.266 (45%); e^- {0.047} (29%)	$^{206}_{81}\text{Tl}$
$^{210}_{83}\text{Bi}$	5.013 d	β^- α (trace)	Tl K x-rays ~0.075 (13%), Tl L x-rays ~0.011 (5%); β^- 1.16 (>99%)	$^{210}_{84}\text{Po}$
$^{211}_{83}\text{Bi}$	2.15 min	α (99.7%) β^- (0.3%)	α 6.62 (84%), 6.28 (16%); γ 0.350 (13%); e^- {0.009} (5%); Tl K x-rays ~0.080 (3%), Tl L x-rays ~0.010 (1%)	$^{207}_{81}\text{Tl}$ (α) $^{211}_{84}\text{Po}(\beta^-)$

(continues)

Table of Radioactive Isotopes—Continued

Nuclide $^A_Z X$	Half-life	Decay mode	Radiation characteristics: Energies (MeV) [intensities (%)]	Decay product $^A_Z Y$
$^{212}_{83}$Bi	1.00 h	β^- (64.1%) α (35.9%)	β^- 2.27 (63†), 1.55 (10†), 0.93 (7.5†), 0.67 (6†), 0.45 (8.5†), 0.085 (5†) α 6.09 (9.6%), 6.05 (25.2%), 5.77 (0.1%), 5.61 (0.4%) γ 1.62 (1.5%), 1.51 (0.3%), 1.08 (0.5%), 0.893 (0.4%), 0.785 (1.1%), 0.727 (6.6%), 0.452 (0.4%), 0.288 (0.3%) Po K x-rays ~0.082 (0.2%), Tl K x-rays ~0.080 (0.3%) Po L x-rays ~0.013 (0.1%), Tl L x-rays ~0.010 (7.5%) e^- [0.010] (32%)	$^{212}_{84}$Po(β^-) $^{208}_{81}$Tl(α)
$^{213}_{83}$Bi	47 min	β^- (97.8%) α (2.2%)	β^- 1.39 (68†), 0.96 (32†) α 5.869 (2%), 5.549 (0.2%) γ 0.293 (0.44%), 0.440 (17%) e^- [0.012] (5%)	$^{213}_{84}$Po (β^-) $^{209}_{81}$Tl (α)
$^{214}_{83}$Bi	19.7 min	β^- (>99%) α (0.02%)	β^- 3.26 (19%), 1.88 (9%), 1.51 (40%), 1.02 (23%), 0.42 (9%) γ 0.609 (46%), 1.120 (15%), 1.238 (6%), 1.764 (16%) e^- [0.021] (4%) Po K x-rays ~0.085 (2%), Po L x-rays ~0.012 (1%)	$^{214}_{84}$Po
Polonium $^{208}_{84}$Po	2.897 y	α	α 5.12 (>99%) γ 0.603 (0.006%), 0.291 (0.003%)	$^{204}_{82}$Pb
$^{210}_{84}$Po	138.38 d	α	α 5.30 (100%) γ 0.803 (0.0012%)	$^{206}_{82}$Pb
$^{211}_{84}$Po	0.516 s	α	α 7.45 (98.9%), 6.89 (0.6%), 6.57 (0.5%) γ 0.897 (0.5%), 0.569 (0.5%)	$^{207}_{82}$Pb
$^{212}_{84}$Po	298×10^{-9} s	α	α 8.78 (100%)	$^{208}_{82}$Pb
$^{213}_{84}$Po	4.2×10^{-6} s	α	α 8.375 (>99%), 7.615 (0.006%)	$^{209}_{82}$Pb
$^{214}_{84}$Po	1.6×10^{-4} s	α	α 7.69 (99.98%), 6.90 (0.01%), 6.61 (0.01%) γ 0.798 (0.01%)	$^{210}_{82}$Pb
$^{215}_{84}$Po	1.78×10^{-3} s	α β^- (0.00023%)	α 7.39 (100%) γ 0.438 (0.04%)	$^{211}_{82}$Pb(α)
$^{216}_{84}$Po	0.156 s	α	α 6.78 (100%)	$^{212}_{82}$Pb
$^{218}_{84}$Po	3.05 min	α	α 6.002 (100%)	$^{214}_{82}$Pb

	Half-life	Decay mode	Radiation (energies in MeV, intensities)	Daughter
Astatine				
$^{215}_{85}$At	0.1×10^{-3} s	α	α 8.02 (99.95%), 7.63 (0.05%)	$^{211}_{83}$Bi
$^{217}_{85}$At	0.0323 s	α (99.98%) β^- (0.02%)	γ 0.404 (0.05%) α 7.07 (99.9%), 6.81 (0.06%), 6.61 (0.01%), 6.48 (0.01%)	$^{213}_{83}$Bi
$^{218}_{85}$At	1.6 s	α (99.9%) β^- (0.1%)	α 6.75 (4%), 6.70 (90%), 6.65 (6%)	$^{214}_{83}$Bi
Radon				
$^{219}_{86}$Rn	3.92 s	α	α 6.81 (81%), 6.55 (11.5%), 6.52 (0.12%), 6.42 (7.5%) γ 0.130 (0.13%), 0.271 (10%), 0.402 (7%) e^- [0.006] (4%) Po K x-rays ~ 0.085 (2%), Po L x-rays ~ 0.012 (1%)	$^{215}_{84}$Po
$^{220}_{86}$Rn	54 s	α	α 6.29 (99.93%), 5.75 (0.07%) γ 0.550 (0.1%)	$^{216}_{84}$Po
$^{222}_{86}$Rn	3.824 d	α	α 5.49 (99.92%), 4.99 (0.08%) γ 0.510 (0.07%)	$^{218}_{84}$Po
Francium				
$^{221}_{87}$Fr	4.8 min	α	α 6.34 (83.4%), 6.24 (1.3%), 6.13 (15%), 5.98 (0.5%) γ 0.099 (0.1%), 0.217 (11%), 0.409 (0.1%) e^- [0.008] (8%) At K x-rays ~ 0.085 (3%), At L x-rays ~ 0.012 (2%)	$^{217}_{85}$At
$^{223}_{87}$Fr	21.8 min	β^- (>99.99%) α (trace)	β^- 1.12 (100%) γ 0.235 (3.7%), 0.204 (1.1%) e^- [0.053] (160%) Ra K x-rays ~ 0.095 (8%), Ra L x-rays ~ 0.014 (40%)	$^{223}_{88}$Ra
Radium				
$^{223}_{88}$Ra	11.4 d	α	α 5.87 (0.9%), 5.86 (0.3%), 5.75 (9.5%), 5.72 (52.5%), 5.61 (24.2%), 5.54 (9.2%), 5.50 (1%), 5.43 (2.3%) γ 0.445 (1.3%), 0.338 (2.8%), 0.324 (3.9%), 0.269 (13.6%), 0.154 (5.6%), 0.122 (1.2%) e^- [0.073] (146%) Rn K x-rays ~ 0.090 (52%), Rn L x-rays ~ 0.013 (23%)	$^{219}_{86}$Rn
$^{224}_{88}$Ra	3.665 d	α	α 5.68 (94%), 5.45 (6%) γ 0.645 (0.01%), 0.292 (0.01%), 0.241 (3.7%) Rn K x-rays ~ 0.085 (0.4%), Rn L x-rays ~ 0.012 (0.4%)	$^{220}_{86}$Rn

(continues)

Table of Radioactive Isotopes—Continued

Nuclide $^A_Z X$	Half-life	Decay mode	Radiation characteristics: Energies (MeV) [intensities (%)]		Decay product $^A_Z Y$
$^{225}_{88}$Ra	14.8 d	β^-	β^-	0.320 (100%)	$^{225}_{89}$Ac
			γ	0.040 (29%)	
			e^-	[0.012] (49%)	
				Ac L x-rays ~ 0.014 (13%)	
$^{226}_{88}$Ra	1599 y	α	α	4.78 (94.5%), 4.60 (5.5%)	$^{222}_{86}$Rn
			γ	0.260 (0.01%), 0.186 (4%)	
			e^-	[0.004] (3%)	
				Rn K x-rays ~ 0.085 (0.6%), Rn L x-rays ~ 0.012 (0.7%)	
$^{228}_{88}$Ra	5.77 y	β^-	β^-	0.039 (60%), 0.015 (40%)	$^{228}_{89}$Ac
Actinium					
$^{225}_{89}$Ac	10.0 d	α	α	5.83 (50.6%), 5.80 (0.3%), 5.79 (26.7%), 5.73 (10.1%), 5.72 (3.4%), 5.68 (1.4%), 5.61 (1.2%), 5.29 (0.2%)	$^{221}_{87}$Fr
			γ	0.099 (2.3%), 0.108 (0.3%), 0.111 (0.3%), 0.150 (0.7%)	
			e^-	[0.026] (80%)	
				Fr L x-rays ~ 0.090 (4%), Fr L x-rays ~ 0.013 (20%)	
$^{227}_{89}$Ac	21.77 y	β^- (98.6%)	β^-	0.046 (98.6%)	$^{227}_{90}$Th
		α (1.4%)	α	4.95 (1.2%), 4.86 (0.1%)	
			γ	0.100 (0.03%), 0.015 (0.03%)	
			e^-	[0.003] (39%)	
$^{228}_{89}$Ac	6.13 h	β^-	β^-	2.18 (10.1%), 1.85 (9.6%), 1.70 (6.7%), 1.11 (53%), 0.64 (7.6%), 0.45 (13%)	$^{228}_{90}$Th
			γ	0.209 (4%), 0.270 (4%), 0.338 (12%), 0.463 (5%), 0.794 (5%), 0.911 (29%), 0.964 (6%), 0.969 (17%)	
			e^-	[0.088] (150%)	
				Th K x-rays ~ 0.100 (12%), Th L x-rays ~ 0.015 (41%)	
Thorium					
$^{227}_{90}$Th	18.7 d	α	α	6.04 (24.5%), 6.01 (2.9%), 5.98 (23.4%), 5.96 (3%), 5.92 (0.8%), 5.87 (2.4%), 5.81 (1.3%), 5.76 (20.3%), 5.714 (4.9%), 5.709 (8.2%), 5.701 (3.6%), 5.69 (1.5%), 5.66 (2.1%)	$^{223}_{88}$Ra
			γ	0.330 (2.7%), 0.300 (2.3%), 0.256 (6.7%), 0.236 (11.2%), 0.050 (8.5%)	

Isotope	Half-life	Decay mode		Radiation energies MeV (intensity)	Product
$^{228}_{90}$Th	1.913 y	α	e^-	[0.054] (172%)	$^{224}_{88}$Ra
				Ra K x-rays ~0.095 (6.5%), Ra L x-rays ~0.013 (44%)	
			α	5.42 (72.7%), 5.34 (26.7%), 5.21 (0.4%), 5.17 (0.2%)	
			γ	0.214 (0.3%), 0.167 (0.1%), 0.132 (0.2%), 0.084 (1.6%)	
$^{229}_{90}$Th	7340 y	α	e^-	[0.020] (45%)	$^{225}_{88}$Ra
				Ra L x-rays ~0.013 (9%)	
			α	5.053 (1.6%), 5.051 (5.2%), 4.978 (3.2%), 4.967 (6.4%), 4.901 (10.2%), 4.845 (56.2%), 4.815 (9.3%), 4.797 (1.3%)	
			γ	0.031 (4%), 0.086 (3%), 0.124 (1%), 0.137 (2%), 0.148 (1%), 0.156 (1%), 0.194 (5%), 0.211 (3%)	
$^{230}_{90}$Th	8×10^4 y	α	α	4.69 (76.3%), 4.62 (23.4%), 4.48 (0.1%)	$^{226}_{88}$Ra
			γ	0.253 (0.02%), 0.184 (0.01%), 0.142 (0.05%), 0.068 (0.4%)	
$^{231}_{90}$Th	25.5 h	β^-	β^-	0.302 (52%), 0.218 (20%), 0.138 (22%), 0.09 (6%)	$^{231}_{91}$Pa
			γ	0.084 (6.6%)	
			e^-	[0.094] (476%)	
				Pa K x-rays ~0.100 (1%)	
				Pa L x-rays ~0.015 (97%)	
$^{232}_{90}$Th	1.4×10^{10} y	α	α	4.02 (77%), 3.96 (23%)	$^{228}_{88}$Ra
			γ	0.059 (0.2%)	
$^{234}_{90}$Th	24.1 d	β^-	β^-	0.199 (72.5%), 0.104 (20.7%), 0.060 (5.4%), 0.022 (1.3%)	$^{234}_{91}$Pa
			γ	0.063 (4%), 0.0924 (3%), 0.0928 (3%),	
			e^-	[0.016] (33%)	
				Pa L x-rays ~0.015 (10%)	
Protactinium					
$^{231}_{91}$Pa	3.28×10^4 y	α	α	5.06 (10%), 5.03 (23%), 5.01 (24%), 4.98 (2.3%), 4.95 (22%), 4.93 (2.8%), 4.85 (1.4%), 4.74 (11%), 4.71 (1.4%), 4.68 (2.1%)	$^{227}_{89}$Ac
			γ	0.330 (1%), 0.29 (6%), 0.027 (6%)	
			e^-	[0.048] (236%)	
				Ac K x-rays ~0.100 (2%), Ac L x-rays ~0.015 (54%)	
$^{233}_{91}$Pa	26.95 d	β^-	β^-	0.568 (5%), 0.257 (58%), 0.145 (37%)	$^{233}_{92}$U
			γ	0.341 (4%), 0.312 (37%), 0.300 (6%)	
			e^-	[0.130] (138%)	
				U K x-rays ~0.100 (35%), U L x-rays ~0.016 (43%)	
$^{234m}_{91}$Pa	1.18 min	β^- (99.87%) IT (0.13%)	β^-	2.29 (~98%)	$^{234}_{92}$U (β^-) $^{234}_{91}$Pa (IT)
			γ	0.766 (0.2%), 1.00 (0.7%)	

(continues)

Table of Radioactive Isotopes—Continued

Nuclide ^A_ZX	Half-life	Decay mode		Radiation characteristics: Energies (MeV) [intensities (%)]	Decay product ^A_ZY
$^{234}_{91}$Pa	6.75 h	β^-	β^-	1.51 (~1%), 1.19 (5%), 0.680 (19%), 0.512 (63%), 0.280 (12%)	$^{234}_{92}$U
			γ	1.08 (1%), 0.90 (70%), 0.70 (24%), 0.56 (15%), 0.36 (13%), 0.22 (14%), 0.126 (26%), 0.100 (50%)	
			e^-	[0.265] (380%)	
				U K x-rays ~ 0.100 (53%), U L x-rays ~ 0.015 (110%)	
Uranium					
$^{232}_{92}$U	71.7 y	α	α	5.32 (68.6%), 5.26 (31.2%), 5.14 (0.2%)	$^{228}_{90}$Th
			γ	0.270 (0.004%), 0.129 (0.08%), 0.058 (0.21%)	
$^{233}_{92}$U	1.59×10^5 y	α	α	4.82 (84.4%), 4.80 (0.3%), 4.78 (13.3%), 4.75 (0.2%), 4.73 (1.6%)	$^{229}_{90}$Th
			γ	0.042 (0.06%)	
			e^-	[0.006] (25%)	
				Th K x-rays ~ 0.100 (0.03%), Th L x-rays ~ 0.015 (48%)	
$^{234}_{92}$U	2.4×10^5 y	α	α	4.77 (72%), 4.72 (28%)	$^{230}_{90}$Th
			γ	0.053 (0.12%), 0.121 (0.04%)	
$^{235}_{92}$U	7.1×10^8 y	α	α	4.598 (4.6%), 4.577 (3.7%), 4.503 (1.2%), 4.438 (0.6%), 4.416 (4%), 4.397 (57%), 4.367 (18%), 4.344 (1.5%), 4.324 (3%), 4.267 (0.6%), 4.217 (5.7%), 4.158 (~0.5%)	$^{231}_{90}$Th
			γ	0.143 (11%), 0.163 (5%), 0.186 (53%), 0.205 (5%)	
			e^-	[0.042] (133%)	
				Th K x-rays ~ 0.100 (12%), Th L x-rays ~ 0.015 (39%)	
$^{236}_{92}$U	2.3×10^7 y	α	α	4.494 (74%), 4.445 (26%), 4.331 (0.26%)	$^{232}_{90}$Th
			γ	0.049 (0.08%), 0.112 (0.02%)	
			e^-	[0.011] (35%)	
				Th L x-rays ~ 0.015 (9%)	
$^{237}_{92}$U	6.75 d	β^-	β^-	0.245 (>80%), ~0.09 (~12%)	$^{237}_{93}$Np
			γ	0.026 (2%), 0.059 (33%), 0.065 (1%), 0.164 (2%), 0.208 (22%), 0.332 (1%)	
			e^-	[0.121] (250%)	
				Np K x-rays ~ 0.110 (55%), Np L x-rays ~ 0.015 (64%)	
$^{238}_{92}$U	4.5×10^9 y	α	α	4.196 (77%), 4.149 (23%)	$^{234}_{90}$Th
			γ	0.050 (0.07%), 0.111 (0.02%)	
			e^-	[0.010] (31%)	
				Th L x-rays ~ 0.015 (8%)	

Nuclide	Half-life	Decay mode		Radiation (MeV, %)	Product
$^{239}_{92}$U	23.5 min	β^-	β^-	1.29 (20%), 1.21 (80%)	$^{239}_{93}$Np
			γ	0.043 (4%), 0.075 (52%)	
Neptunium					
$^{237}_{93}$Np	2.14×10^6 y	α	α	4.87 (2.6%), 4.82 (2.5%), 4.80 (3%), 4.79 (47%), 4.77 (25%), 4.76 (8%), 4.66 (3.3%), 4.64 (6%)	$^{233}_{91}$Pa
			γ	0.145 (1%), 0.086 (14%), 0.030 (14%)	
			e^-	[0.064] (184%)	
				Pa K x-rays $\sim$0.100 (5%), Pa L x-rays $\sim$0.015 (52%)	
Plutonium					
$^{237}_{94}$Pu	45.3 d	EC (>99%)	γ	0.060 (5%)	$^{237}_{93}$Np
		α (0.0033%)	e^-	[0.011] (48%)	
				Np K x-rays $\sim$0.110 (43%), Np L x-rays $\sim$0.015 (47%)	
$^{238}_{94}$Pu	86.4 y	α	α	5.50 (71.1%), 5.45 (28.7%), 5.36 (0.2%)	$^{234}_{92}$U
		SF (1.8×10^{-7}%)	γ	0.043 (0.04%)	
			e^-	[0.010] (38%)	
				U L x-rays $\sim$0.015 (11%)	
$^{239}_{94}$Pu	2.41×10^4 y	α	α	5.16 (73.3%), 5.14 (15.1%), 5.10 (11.5%)	$^{235}_{92}$U
		SF	γ	0.375 (0.001%), 0.129 (0.005%), 0.052 (0.02%)	
$^{240}_{94}$Pu	6570 y	α	α	5.16 (75.5%), 5.12 (24.4%), 5.01 (0.1%)	$^{236}_{92}$U
		SF	γ	0.104 (0.007%), 0.045 (0.04%)	
$^{241}_{94}$Pu	14.4 y	β^- (>99%)	β^-	0.021 (>99%)	$^{241}_{95}$Am
		α (0.002%)			
$^{242}_{94}$Pu	3.8×10^5 y	α	α	4.901 (74%), 4.857 (26%)	$^{238}_{92}$U
		SF (0.0006%)	γ	0.045 (0.04%)	
			e^-	[0.008] (30%)	
				U L x-rays $\sim$0.015 (9%)	
Americium					
$^{241}_{95}$Am	432.0 y	α	α	5.54 (0.3%), 5.48 (85.2%), 5.44 (12.5%), 5.38 (1.6%)	$^{237}_{93}$Np
			γ	0.060 (36%)	
			e^-	[0.030] (115%)	
				Np L x-rays $\sim$0.015 (40%)	
$^{242m}_{95}$Am	150 y	IT (>99%)	α	5.207 (0.4%), 5.141 (0.03%)	$^{238}_{93}$Np (α)
		α (0.5%)	γ	0.049 (0.2%)	$^{242}_{95}$Am (IT)
			e^-	[0.040] (118%)	
				Am L x-rays $\sim$0.016 (26%)	

(continues)

Table of Radioactive Isotopes—Continued

Nuclide ^A_ZX	Half-life	Decay mode	Radiation characteristics: Energies (MeV) [intensities (%)]	Decay product ^A_ZY
$^{242}_{95}$Am	16 h	β^- (82.7%) EC (17.3%)	β^- 0.667 (~33%), 0.625 (~49%) γ 0.042 (0.04%), 0.044 (0.02%) e^- [0.019] (77%) Pu K x-rays ~0.110 (12%), Pu L x-rays ~0.015 (11%) Cm L x-rays ~0.016 (18%)	$^{242}_{96}$Cm(β^-) $^{242}_{94}$Pu(EC)
$^{243}_{95}$Am	8 × 10³ y	α SF (2.2 × 10⁻⁸%)	α 5.276 (87%), 5.234 (11.5%), 5.180 (1.1%) γ 0.043 (5%), 0.075 (60%), 0.087 (0.3%), 0.118 (0.6%)	$^{239}_{93}$Np
Curium $^{242}_{96}$Cm	162.76 d	α SF (6.3 × 10⁻⁶%)	α 6.11 (74.0%), 6.07 (26.0%) γ 0.102 (0.006%), 0.044 (0.041%) e^- [0.009] (34%) Pu L x-rays ~0.016 (10%)	$^{238}_{94}$Pu
$^{243}_{96}$Cm	32 y	α (99.7%) EC (0.3%)	α 6.066 (1.5%), 6.059 (5%), 6.010 (1%), 5.992 (6.5%), 5.876 (0.6%), 5.786 (73.3%), 5.742 (10.6%) γ 0.209 (3%), 0.228 (11%), 0.278 (14%) e^- [0.113] (136%) Pu K x-rays ~0.110 (49%), Pu L x-rays ~0.016 (43%)	$^{239}_{94}$Pu (α)
$^{244}_{96}$Cm	18.09 y	α SF (0.0001%)	α 5.81 (76.7%), 5.76 (23.3%) γ 0.100 (0.002%), 0.043 (0.03%) Pu L x-rays ~0.016 (8%)	$^{240}_{94}$Pu
$^{245}_{96}$Cm	9.3 × 10³ y	α	α 5.53 (0.6%), 5.49 (0.8%), 5.36 (93.2%), 5.30 (5%), 5.23 (0.3%) γ 0.042 (0.4%), 0.133 (3%), 0.175 (10%), 0.190 (0.2%) e^- [0.134] (487%) Pu K x-rays ~0.110 (71%), Pu L x-rays ~0.016 (123%)	$^{241}_{94}$Pu
$^{246}_{96}$Cm	5 × 10³ y	α (99.97%) SF (0.03%)	α 5.386 (79%), 5.343 (21%) γ 0.044 (0.03%) e^- [0.007] (27%) Pu L x-rays ~0.016 (8%)	$^{242}_{94}$Pu

Nuclide	Half-life	Decay mode	Radiation energies (MeV) and intensities	Daughter
$^{247}_{96}\text{Cm}$	1.6×10^7 y	α	5.27 (13.8%), 5.21 (5.7%), 5.15 (1.2%), 4.98 (2%), 4.94 (1.6%), 4.87 (71%), 4.82 (4.7%)	$^{243}_{94}\text{Pu}$
		γ	0.279 (3%), 0.289 (2%), 0.347 (1%), 0.404 (72%), Pu K x-rays ~ 0.110 (5%)	
$^{248}_{96}\text{Cm}$	4.7×10^5 y	α (91.7%)	5.08 (75.1%), 5.03 (16.5%)	$^{244}_{94}\text{Pu}$
		SF (8.3%)		
Berkelium				
$^{249}_{97}\text{Bk}$	314 d	β^- (>99%)	0.125 (>99%)	$^{249}_{98}\text{Cf}$ (β^-)
		α (0.0014%)		
Californium				
$^{250}_{98}\text{Cf}$	13.1 y	α (99.92%)	6.03 (84.7%), 5.99 (15%), 5.89 (0.3%)	$^{246}_{96}\text{Cm}$
		SF (0.08%)	γ 0.043 (0.01%)	
			e^- {0.005} (20%)	
			Cm L x-rays ~ 0.017 (6%)	
$^{251}_{98}\text{Cf}$	~ 890 y	α	6.07 (2.7%), 6.01 (11.6%), 5.94 (0.6%), 5.85 (27%), 5.81 (4.2%), 5.79 (2%), 5.76 (3.8%), 5.73 (1%), 5.67 (35%), 5.65 (3.5%), 5.63 (4.5%), 5.60 (0.2%), 5.57 (1.5%), 5.50 (0.3%)	$^{247}_{96}\text{Cm}$
			γ 0.177 (18%), 0.227 (6%), 0.266 (0.5%), 0.285 (1.4%)	
			e^- [0.187] (200%)	
			Cm K x-rays ~ 0.125 (13%), Cm L x-rays ~ 0.090 (109%)	
$^{252}_{98}\text{Cf}$	2.65 y	α (97%)	6.12 (81.6%), 6.08 (15.2%), 5.98 (0.2%)	$^{248}_{96}\text{Cm}$
		SF (3%)	e^- {0.005} (19%)	
			Cm L x-rays ~ 0.017 (6%)	

References

Browne, E., Firestone, R. B., and Shirley, V. S. (1986). "Table of Radioactive Isotopes." John Wiley & Sons, New York.

Firestone, R. B., Shirley, V. S., Baglin, C. M., Frank Chu, S. Y., and Zipkin, J. (1996). "Table of Isotopes," Vols. I and II, 8th ed. John Wiley & Sons, New York.

Martin, M. J. and Tuli, J. J. (Eds.) (1997). "*Nuclear Data Sheets*," Vols. 1–80, Academic Press, San Diego.

Michael Lederer, C. and Shirley, V. S. (Eds.). Brown, E., Dairiki, J. M., Doebler, R. E., Shihab-Eldin, A. A., Jardine, L. J., Tuli, J. K., and Buyrn, A. B. (1978). "Table of Isotopes," 7th ed. John Wiley & Sons, New York.

APPENDIX B:
PARTICLE RANGE-ENERGY CORRELATIONS

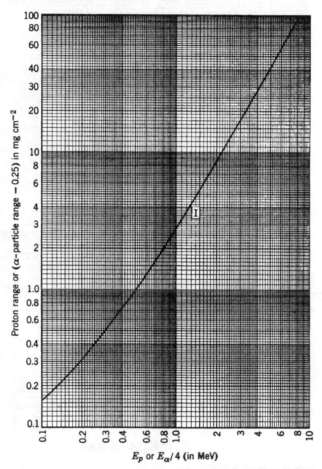

FIGURE B.I Range–energy relations for protons and alpha particles in air. When applying the graph to alpha particles, the alpha-particle energy is divided firstly by 4 and the value of 0.25 mg cm^{-2} added to the range read off the graph. For example, to obtain the range for a 5.5 MeV alpha particle we read $5.5/4 = 1.375 \text{ MeV}$ off the graph and find the range $4.7 \text{ mg cm}^{-2} + 0.25 = 4.95 \text{ mg cm}^{-2}$. The range in cm can be obtained by dividing the above range by the density of air ($\sigma = 1.226 \text{ mg cm}^{-3}$ at **STP**) or $4.95 \text{ mg cm}^{-2}/1.226 \text{ mg cm}^{-3} = 4.0 \text{ cm}$. (From Friedlander et al., 1964, Copyright © John Wiley and Sons, Inc. This material is used by permission of John Wiley & Sons, Inc.)

Handbook of Radioactivity Analysis, Second Edition
Copyright © 2003 Elsevier Science (USA). All rights reserved.

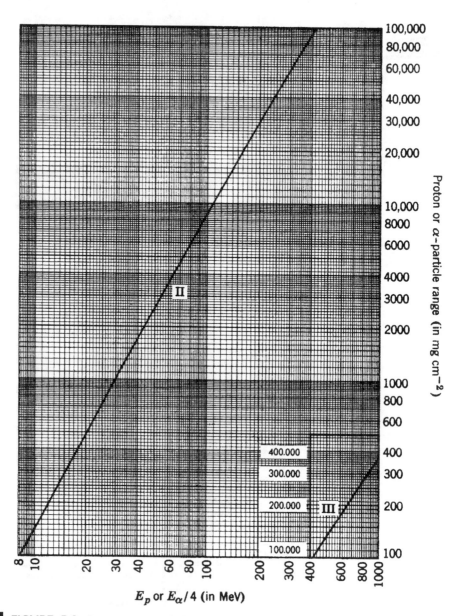

FIGURE B.2 Range-energy relations for protons and alpha particles in air (From Friedlander et al., 1964, Copyright © John Wiley and Sons, Inc. This material is used by permission of John Wiley & Sons, Inc.)

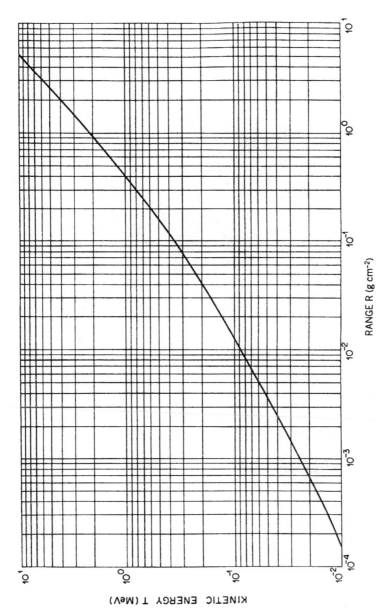

FIGURE B.3 Beta-particle range-energy curve for absorbers of low atomic number. The curve is described by the formulas $R = 0.412T^{1.27-0.0954\ln T}$ for $0.01 \leq T \leq 2.5$ MeV and $R = 0.530T - 0.106$ for $T > 2.5$ MeV where T is the beta-particle energy in MeV. (From US Public Health Service. (1970). Radiological Health Handbook. Publ. No. 2016. Bureau of Radiological Health, Rockville, MD).

INDEX